Advances in FDTD Computational Electrodynamics

Photonics and Nanotechnology

For a listing of recent titles in the
Artech House Antennas and Propagation Series
turn to the back of this book.

Advances in FDTD Computational Electrodynamics

Photonics and Nanotechnology

Allen Taflove
Editor

Ardavan Oskooi and Steven G. Johnson
Coeditors

ARTECH HOUSE
BOSTON | LONDON
artechhouse.com

Library of Congress Cataloging-in-Publication Data
A catalog record for this book is available from the U.S. Library of Congress.

British Library Cataloguing in Publication Data
A catalog record for this book is available from the British Library.

ISBN-13: 978-1-60807-170-8

Cover design by Merle Uuesoo

10 9 8 7 6 5 4 3 2 1

Contents

Preface

Advances in photonics and nanotechnology have the potential to revolutionize humanity's ability to communicate and compute, to understand fundamental life processes, and to diagnose and treat dread diseases such as cancer. To pursue these advances, it is crucial to understand and properly model the interaction of light with nanometer-scale, three-dimensional (3-D) material structures. Important geometric features and material inhomogeneities of such structures can be as small as a few tens of atoms laid side by side. Currently, it is recognized that the most efficient computational modeling of optical interactions with such nanoscale structures is based on the numerical solution of the fundamental Maxwell's equations of classical electrodynamics, supplemented as needed by spatially localized hybrids with (the even more fundamental, but computationally much more intense) quantum electrodynamics.

Aimed at academic and industrial researchers working in all areas of photonics and nanotechnology, this book reviews the current state-of-the-art in formulating and implementing computational models of optical interactions with nanoscale material structures. Maxwell's equations are solved using the finite-difference time-domain (FDTD) technique, investigated over the past 40 years primarily in the context of electrical engineering by one of us (Taflove),[1] and over the past 12 years primarily in the context of physics by two of us (Oskooi and Johnson).[2]

On one level, this book provides an update (for general applications) of the FDTD techniques discussed in the 2005 Taflove-Hagness Artech book, *Computational Electrodynamics: The Finite-Difference Time-Domain Method*, 3rd ed.[3] (It is assumed that the readers of this book are familiar with the fundamentals of FDTD solutions of Maxwell's equations, as documented therein.) On another level, this book provides a wide-ranging review of recent FDTD techniques aimed at solving specific current problems of high interest in photonics and nanotechnology.

Chapters 1 through 7 of this book present important recent advances in FDTD and pseudospectral time-domain (PSTD) algorithms for modeling general electromagnetic wave interactions. Capsule summaries of these chapters follow.

Chapter 1: "Parallel-Processing Three-Dimensional Staggered-Grid Local-Fourier-Basis PSTD Technique," by M. Ding and K. Chen. This chapter discusses a new staggered-grid, local-Fourier-basis PSTD technique for efficient computational solution of the full-vector Maxwell's equations over electrically large, open-region, 3-D domains. The new PSTD formulation scales more efficiently with the size of the computational domain than previous collocated-grid PSTD approaches, and very importantly, *avoids the Gibbs phenomenon artifact*. This allows accurate

[1]See the online article, http://www.nature.com/milestones/milephotons/full/milephotons02.html, in which *Nature Milestones | Photons* cites Prof. Taflove as one of the two principal pioneers of numerical methods for solving Maxwell's equations.

[2]Dr. Oskooi and Prof. Johnson have led the development of a powerful, free, open-source implementation of a suite of FDTD Maxwell's equations solvers at the Massachusetts Institute of Technology (MIT): *Meep* (acronym for MIT Electromagnetic Equation Propagation), available online at http://ab-initio.mit.edu/meep. Meep has been cited in over 600 journal publications, and has been downloaded more than 54,000 times.

[3]As of September 2012, the combined citations of the three editions (1995, 2000, and 2005) of *Computational Electrodynamics: The Finite-Difference Time-Domain Method*, rank this book 7th on the *Google Scholar*® list of the most-cited books in physics, according to the Institute of Optics of the University of Rochester. See: http://www.optics.rochester.edu/news-events/news/google_scholar.html.

PSTD modeling of dielectric structures having high-contrast material interfaces. The complete algorithm is presented, including its implementation for a uniaxial perfectly matched layer (UPML) absorbing boundary condition (ABC).

Chapter 2: "Unconditionally Stable Laguerre Polynomial-Based FDTD Method," by B. Chen, Y. Duan, and H. Chen. This chapter discusses an efficient algorithm for implementing the unconditionally stable 3-D Laguerre polynomial-based FDTD technique with an effective PML ABC. In contrast to the conventional Laguerre-based FDTD method, which requires solving a very large sparse matrix, the new technique requires solving only six tri-diagonal matrices and three explicit equations for a full update cycle. This provides excellent computational accuracy — much better than alternating-direction-implicit FDTD approaches — and can be efficiently parallel-processed on a computing cluster.

Chapter 3: "Exact Total-Field/Scattered-Field Plane-Wave Source Condition," by T. Tan and M. Potter. This chapter discusses an efficient exact FDTD total-field/scattered-field plane-wave source suitable for arbitrary propagation and polarization angles, stability factors, and non-unity aspect ratios. This technique provides *zero leakage* of the incident plane wave into the scattered-field region to machine-precision levels. The incremental computer memory and execution time are essentially negligible compared to the requirements of the primary 3-D grid.

Chapter 4: "Electromagnetic Wave Source Conditions," by A. Oskooi and S. G. Johnson. This chapter provides a tutorial discussion of relationships between current sources and the resulting electromagnetic waves in FDTD simulations. The techniques presented are suitable for a wide range of modeling applications, from deterministic radiation, scattering, and waveguiding problems to nanoscale material structures interacting with thermal and quantum fluctuations. The chapter begins with a discussion of incident fields and equivalent currents, examining the principle of equivalence and the discretization and dispersion of equivalent currents in FDTD models. This is followed by a review of means to separate incident and scattered fields, whether in the context of scatterers, waveguides, or periodic structures. The next major topic is the relationship between current sources and the resulting local density of states. Here, key sub-topics includes the Maxwell eigenproblem and the density of states, radiated power and the harmonic modes, radiated power and the local density of states, computation of the local density of states in FDTD, Van Hove singularities in the local density of states, and resonant cavities and Purcell enhancement. Subsequent major topics include source techniques that enable covering a wide range of frequencies and incident angles in a small number of simulations for waves incident on a periodic surface; sources to efficiently excite eigenmodes in rectangular supercells of periodic systems; moving sources to enable modeling of Cherenkov radiation and Doppler-shifted radiation; and finally thermal sources via a Monte-Carlo/Langevin approach to enable modeling radiative heat transfer between complex-shaped material objects in the near field.

Chapter 5: "Rigorous PML Validation and a Corrected Unsplit PML for Anisotropic Dispersive Media," by A. Oskooi and S. G. Johnson. This chapter discusses a straightforward technique to verify the correctness of any proposed PML formulation, irrespective of its implementation. Several published claims of working PMLs for anisotropic media, periodic media, and oblique waveguides are found to be just instances of adiabatic pseudo-PML absorbers. This chapter also discusses an efficient, corrected, unsplit PML formulation for anisotropic dispersive media, involving a simple refactorization of typical UPML proposals. Appendixes to this chapter provide a tutorial discussion of the complex-coordinate-stretching basis of PML, and the application of coupled-mode theory to analyze and design effective adiabatic pseudo-PML absorbers for FDTD modeling of photonic crystals.

Chapter 6: "Accurate FDTD Simulation of Discontinuous Materials by Subpixel Smoothing," by A. Oskooi and S. G. Johnson. This chapter discusses an efficient local ("subpixel") dielectric smoothing technique for achieving second-order accuracy when modeling non-grid-aligned isotropic and anisotropic dielectric interfaces in a Cartesian FDTD grid. This technique is based on a rigorous perturbation theory (summarized in an Appendix), rather than on an *ad hoc* heuristic. It provides greatly improved accuracy relative to previous approaches without increasing the required computational storage or running time. Subpixel smoothing has an additional benefit: it allows the simulation to respond continuously to changes in the geometry, such as during optimization or parameter studies, rather than changing in discontinuous jumps as dielectric interfaces cross pixel boundaries. Additionally, it yields much smoother convergence of the error with grid resolution, which makes it easier to evaluate the accuracy of a simulation, and enables the possibility of extrapolation to gain another order of accuracy. Unlike methods that require modified field-update equations or larger stencils and complicated position-dependent difference equations for higher-order accuracy, subpixel smoothing uses the standard center-difference expressions, and is easy to implement in FDTD by simply preprocessing the materials.

Chapter 7: "Stochastic FDTD for Analysis of Statistical Variation in Electromagnetic Fields," by S. M. Smith and C. M. Furse. This chapter discusses a new stochastic FDTD (S-FDTD) technique that provides an efficient means to evaluate statistical variations in numerical simulations of electromagnetic wave interactions caused by random variations of the electrical properties of the model. The statistics of these variations are incorporated *directly* into FDTD, which computes an estimate of the resulting mean and variance of the fields at every point in space and time with a *single run*. The field variances computed using only two S-FDTD runs can effectively "bracket" the results using the brute-force Monte Carlo technique, the latter obtained after hundreds or thousands of runs. Hence, the S-FDTD technique offers a potentially huge savings in computation time, and opens up the possibility of assessing statistical parameters for applications in bioelectromagnetics, biophotonics, and geophysics where the material electrical properties have uncertainty or variability.

Chapters 8 through 20 provide a wide-ranging review of recent FDTD techniques aimed at solving specific current problems of high interest in photonics and nanotechnology. In order of presentation, the topics include:

- Plasmonics (emphasizing hybrid models with quantum mechanics), including active plasmonics, nonlocal electrodynamics, and modification of the optical properties of dye molecules closely bound to adjacent metal nanostructures
- Transformation electromagnetics, including non-diagonal anisotropic metamaterial cloaks
- Metamaterials, including periodic sub-wavelength optical structures comprised of non-rectangular-shaped plasmonic components
- Extensive tutorials on computational optical imaging for microscopy and nanoscale lithography
- Biophotonics applications, including imaging/characterization of intracellular structure and sensing of nanoscale intracellular anomalies indicative of early-stage cancer

- Non-paraxial spatial soliton propagation and interactions with nanoscale material structures
- Vacuum quantum phenomena, including blackbody radiation and electromagnetic fluctuations in dissipative open systems, and Casimir forces in arbitrary material geometries
- MIT's flexible, free FDTD software package, *Meep*.

Capsule summaries of these chapters follow.

Chapter 8: "FDTD Modeling of Active Plasmonics," by I. Ahmed, E. H. Khoo, and E. P. Li. This chapter discusses a recent hybrid FDTD/quantum mechanics technique for modeling plasmonic devices having integral semiconductor elements capable of providing gain upon pumping. The new technique integrates a Lorentz-Drude model to simulate the metal components of the device with a multi-level, multi-electron quantum model of the semiconductor component. Two examples of applications are summarized: amplification of a 175-fs optical pulse propagating in a thin, electrically pumped GaAs medium between two gold plates; and the resonance shift and radiation from a GaAs microcavity resonator with embedded gold nanocylinders. An appendix reviews the recent critical-points model for metal optical properties. This model is capable of providing a more accurate treatment of the bulk dielectric dispersion properties of various metals over a wider range of optical wavelengths than previously possible.

Chapter 9: "FDTD Computation of the Nonlocal Optical Properties of Arbitrarily Shaped Nanostructures," by J. M. McMahon, S. K. Gray, and G. C. Schatz. At length scales of less than ~10 nm, quantum-mechanical effects can lead to unusual optical properties for metals relative to predictions based on assuming bulk dielectric values. A full quantum-mechanical treatment of such nanostructures would be best, but is not practical for these structure sizes. However, it is possible to incorporate some quantum effects within classical electrodynamics via the use of a different dielectric model than that for the bulk metal. In this chapter, the quantum effect of primary interest requires a dielectric model wherein the material polarization at a point in space depends not only on the local electric field, but also on the electric field in its neighborhood. This chapter discusses a technique to calculate the optical response of an arbitrarily shaped nanostructure described by such a spatially nonlocal dielectric function. This technique is based on converting the hydrodynamic Drude model into an equation of motion for the conduction electrons, which then serves as a current field in the Maxwell-Ampere law. The latter is incorporated in a self-consistent manner in the FDTD solution of Maxwell's curl equations. Using this hybrid technique, modeling results for one-dimensional (1-D), two-dimensional (2-D), and 3-D gold nanostructures of variable size are presented. These results demonstrate the increasing importance of including nonlocal dielectric phenomena when modeling optical interactions with gold nanostructures as characteristic length scales of interest fall below ~10 nm.

Chapter 10: "Classical Electrodynamics Coupled to Quantum Mechanics for Calculation of Molecular Optical Properties: An RT-TDDFT/FDTD Approach," by H. Chen, J. M. McMahon, M. A. Ratner, and G. C. Schatz. This chapter discusses a new multiscale computational methodology to incorporate the scattered electric field of a plasmonic nanoparticle into a quantum-mechanical optical property calculation for a nearby dye molecule. For a given location of the dye molecule with respect to the nanoparticle, a frequency-dependent scattering response function is first computed using FDTD. Subsequently, the time-dependent scattered electric field at the dye molecule is calculated using this response function through a multidimensional Fourier transform to reflect the effect of polarization of the nanoparticle on the local field at the dye molecule. Finally, a real-time time-dependent density function theory (RT-TDDFT) approach is employed to obtain the desired optical property of the dye molecule in the presence of the nanoparticle's local electric field. Using this technique, enhanced absorption spectra of the N3 dye molecule and enhanced Raman spectra of the pyridine molecule are modeled, assuming proximity to a 20-nm-diameter silver nanosphere. The computed signal amplifications reflect the strong coupling between the wavelike response of the dye molecule's individual electrons and the collective action of the silver nanosphere's dielectric medium. Overall, this hybrid method provides a bridge spanning the gap between quantum mechanics and classical electrodynamics (i.e., FDTD) with respect to both length and time scales.

Chapter 11: "Transformation Electromagnetics Inspired Advances in FDTD Methods," by R. B. Armenta and C. D. Sarris. Transformation electromagnetics exploits the coordinate-invariance property of Maxwell's equations to synthesize the permeability and permittivity tensors of artificial materials to guide electromagnetic waves in specified manners. This chapter shows how employing a coordinate system-independent representation of Maxwell's equations based on the invariance principle provides powerful additional FDTD capabilities, including conformal modeling of curved material boundaries, incorporation of artificial materials providing novel wave-propagation characteristics, and time-dependent discretizations for high-resolution tracking of moving electromagnetic pulses. Examples of these enhanced FDTD capabilities are derived from coordinate transformations of Maxwell's equations involving projections onto the covariant-contravariant vector bases associated with a general curvilinear coordinate system.

Chapter 12: "FDTD Modeling of Non-Diagonal Anisotropic Metamaterial Cloaks," by N. Okada and J. B. Cole. Without proper care, the direct application of FDTD to simulate transformation-based metamaterials having non-diagonal anisotropic constitutive parameters is prone to numerical instabilities. This chapter discusses the basis, formulation, and validation of a technique to solve this instability problem by ensuring that the numerically derived FDTD equations are exactly symmetric. The crucial step in this technique involves finding the eigenvalues and diagonalizing the constitutive tensors. After this diagonalization, any of the previously reported FDTD algorithms for purely diagonal metamaterial cases can be applied. The technique is illustrated with a 2-D FDTD model of a transformation-based elliptical cylindrical cloak comprised of a non-diagonal anisotropic metamaterial. The cloak is found to greatly reduce both the bistatic radar cross-section and the total scattering cross-section of the enclosed elliptical perfect electric conductor (PEC) cylinder at the design wavelength. In fact, as the grid of the FDTD model is progressively refined, scattering by the cloaked PEC cylinder trends rapidly toward zero. However, the bandwidth of the effective scattering reduction is only ~4%; so narrow that it may be described as just a scattering null. This narrow bandwidth appears to limit practical applications of such cloaks.

Chapter 13: "FDTD Modeling of Metamaterial Structures," by Costas D. Sarris. This chapter provides an overview of the application of FDTD to several key classes of problems in metamaterial analysis and design, from two complementary perspectives. First, FDTD analyses of the transient response of several metamaterial structures of interest are presented. These include negative-refractive-index media and the "perfect lens," an artificial transmission line exhibiting a negative group velocity, and a planar anisotropic grid supporting resonance cone phenomena. Second, periodic geometries realizing metamaterial structures are studied. The primary tool used here is the sine-cosine method, coupled with the array-scanning technique. This tool is applied to obtain the dispersion characteristics (and, as needed, the electromagnetic field) associated with planar periodic positive-refractive-index and negative-refractive-index transmission lines, as well as the planar microwave "perfect lens" comprised of sections of 2-D transmission lines exhibiting both positive and negative equivalent refractive indices. The chapter continues with a review of the triangular-mesh FDTD technique for modeling optical metamaterials with plasmonic components, and how this technique could be coupled with the sine-cosine method to analyze periodic plasmonic microstructures requiring much better modeling of slanted and curved metal surfaces than is possible using a Cartesian FDTD grid and simple staircasing. Finally, the periodic triangular-mesh FDTD technique is applied to accurately obtain the dispersion characteristics and electromagnetic modes of a sub-wavelength plasmonic photonic crystal comprised of an array of silver microcylinders.

Chapter 14: "Computational Optical Imaging Using the Finite-Difference Time-Domain Method," by I. R. Capoglu, J. D. Rogers, A. Taflove, and V. Backman. This chapter presents a comprehensive and rigorous tutorial discussion of the theoretical principles that comprise the foundation for emerging electromagnetic-field models of optical imaging systems based on 3-D FDTD solutions of Maxwell's curl equations. These models provide the capability to computationally synthesize images formed by *every* current form of optical microscopy (bright-field, dark-field, phase-contrast, etc.), as well as optical metrology and photolithography. Focusing, variation of the numerical aperture, and so forth can be adjusted simply by varying a few input parameters – literally a *microscope in a computer*. This permits simulations of both existing and proposed novel optical imaging techniques over a 10^7:1 dynamic range of distance scales, i.e., from a few nanometers (the FDTD voxel size within the microstructure of interest) to a few centimeters (the location of the image plane where the amplitude and phase spectra of individual pixels are calculated). This tutorial shows how a general optical imaging system can be segmented into four self-contained sub-components (illumination, scattering, collection and refocusing), and how each of these sub-components is mathematically analyzed. Approximate numerical methods used in the modeling of each sub-component are explained in appropriate detail. Relevant practical applications are cited whenever applicable. Finally, the theoretical and numerical results are illustrated via several implementation examples involving the computational synthesis of microscope images of micro-scale structures. Overall, this chapter constitutes a useful starting point for those interested in modeling optical imaging systems from a rigorous electromagnetic-field point of view. A distinct feature of this approach is the extra attention paid to the issues of discretization and signal processing — a key issue in finite methods such as FDTD, where the electromagnetic field is only computed at a finite set of spatial and temporal points.

Chapter 15: "Computational Lithography Using the Finite-Difference Time-Domain Method," by G. W. Burr and J. T. Azpiroz. This chapter presents a comprehensive and rigorous tutorial discussion of the fundamental physical concepts and FDTD numerical considerations whose understanding is essential to perform electromagnetic-field computations for very large-scale integration (VLSI) optical lithography in the context of semiconductor microchip manufacturing. As the characteristic dimensions of VLSI technology shrink and complexity increases, the usual geometrical approximations of the electromagnetic field interactions underlying optical lithographic technology can become increasingly inaccurate. However, the accurate simulation of both immersion and extreme ultraviolet (EUV) lithographic systems is expected to be an increasingly critical component of semiconductor manufacturing for the foreseeable future. While rigorous computations of the required 3-D electromagnetic fields can be much slower than approximate methods, rapid turnaround is still crucial. The FDTD method offers advantages such as flexibility, speed, accuracy, parallelized computation, and the ability to simulate a wide variety of materials. In fact, FDTD computation of the electromagnetic fields underlying VLSI optical lithography currently offers the best combination of accuracy and turn-around time to understand and model field effects involved with relatively small patterns.

Chapter 16: "FDTD and PSTD Applications in Biophotonics," by I. R. Capoglu, J. D. Rogers, C. M. Ruiz, J. J. Simpson, S. H. Tseng, K. Chen, M. Ding, A. Taflove, and V. Backman. This chapter discusses qualitatively the technical basis and representative applications of FDTD and PSTD computational solutions of Maxwell's curl equations in the area of biophotonics. The FDTD applications highlighted in this chapter reveal its ability to provide ultrahigh-resolution models of optical interactions within individual biological cells, and furthermore to provide the physics kernel of advanced computational microscopy techniques. The PSTD applications highlighted in this chapter indicate its ability to model optical interactions with clusters of many biological cells and even macroscopic sections of biological tissues, especially in regard to developing an improved understanding of the physics of enhanced optical backscattering and turbidity suppression. In all of this, a key goal is to inform readers how FDTD and PSTD can put Maxwell's equations to work in the analysis and design of a wide range of biophotonics technologies. These technologies exhibit promise to advance the basic scientific understanding of cellular-scale processes, and to provide important medical applications (especially in early-stage cancer detection).

Chapter 17: "GVADE FDTD Modeling of Spatial Solitons," by Z. Lubin, J. H. Greene, and A. Taflove. The general vector auxiliary differential equation (GVADE) FDTD method, discussed in this chapter, is a powerful tool for first-principles, full-vector solutions of electromagnetic wave interactions in materials having combined linear and nonlinear dispersions. This technique provides a direct time-domain solution of Maxwell's curl equations without any simplifying paraxial, slowly varying envelope, or scalar approximations. Furthermore, it can be applied to arbitrary inhomogeneous material geometries, and both linear and nonlinear polarizations can be incorporated through the Maxwell-Ampere law. This chapter derives the GVADE FDTD time-stepping algorithm for the electromagnetic field in a realistic 2-D model of fused silica characterized by a three-pole Sellmeier linear dispersion, an instantaneous Kerr nonlinearity, and a dispersive Raman nonlinearity. (Here, the electric field is assumed to have both a longitudinal and a transverse component in the plane of incidence.) Next, the technique is extended to model a plasmonic metal characterized by a linear Drude dispersion. The GVADE FDTD method is then applied to model the propagation of single nonparaxial and overpowered spatial solitons; the interaction of a pair of closely spaced, co-propagating, nonparaxial spatial solitons; spatial soliton scattering by subwavelength air holes; and interactions between nonparaxial spatial solitons and thin gold films. It is concluded that the GVADE FDTD technique will find emerging applications in optical communications and computing involving micro- and nano-scale photonic circuits that require controlling complex electromagnetic wave phenomena in linear and nonlinear materials with important subwavelength features.

Chapter 18: "FDTD Modeling of Blackbody Radiation and Electromagnetic Fluctuations in Dissipative Open Systems," by J. Andreasen. This chapter discusses how the FDTD method can simulate fluctuations of electromagnetic fields in open cavities due to output coupling. The foundation of this discussion is the fluctuation-dissipation theorem, which dictates that cavity field dissipation by leakage is accompanied by thermal noise, simulated here by classical electrodynamics. The absorbing boundary of the FDTD grid is treated as a blackbody that radiates into the grid. Noise sources are synthesized with spectra equivalent to that of blackbody radiation at various temperatures. When an open dielectric cavity is placed in the FDTD grid, the thermal radiation is coupled into the cavity and contributes to the thermal noise for the cavity field. In the Markovian regime, where the cavity photon lifetime is much longer than the coherence time of thermal radiation, the FDTD-calculated amount of thermal noise in a cavity mode agrees with that given by the quantum Langevin equation. This validates the numerical model of thermal noise that originates from cavity openness or output coupling. FDTD simulations also demonstrate that, in the non-Markovian regime, the steady-state number of thermal photons in a cavity mode exceeds that in a vacuum mode. This is attributed to the constructive interference of the thermal field inside the cavity. The advantage of the FDTD numerical model is that the thermal noise is added in the time domain without any prior knowledge of cavity modes. Hence, this technique can be applied to simulate complex open systems whose modes are not known prior to the FDTD calculations. This approach is especially useful for very leaky cavities whose modes overlap strongly in frequency, as the thermal noise related to the cavity leakage is introduced naturally without distinguishing the modes. Therefore, the method discussed here can be applied to a wide range of quantum optics problems.

Chapter 19: "Casimir Forces in Arbitrary Material Geometries," by A. Oskooi and S. G. Johnson. This chapter discusses how FDTD modeling provides a flexible means to compute Casimir forces for essentially arbitrary configurations and compositions of micro- and nanostructures. Unlike other numerical techniques proposed for this application, FDTD is not structure specific, and hence, very general codes such as MIT's freely available *Meep* software (Chapter 20) can be used with no modifications. The chapter begins by establishing the theoretical foundation for the FDTD-Casimir technique. This is followed by a discussion of an efficient implementation employing a rapidly convergent harmonic expansion in the source currents. Then, means to extend the FDTD-Casimir technique to account for nonzero temperatures are presented. The chapter provides five FDTD-Casimir modeling examples, concluding with a fully 3-D simulation where the Casimir force transitions from attractive to repulsive depending on a key separation parameter. Overall, in addition to providing simulations of fundamental physical phenomena, these developments in FDTD-Casimir modeling may permit the design of novel micro- and nano-mechanical systems comprised of complex materials.

Chapter 20: "Meep: A Flexible Free FDTD Software Package," by A. Oskooi and S. G. Johnson. This chapter discusses aspects of the free, open-source implementation of the FDTD algorithm developed at the Massachusetts Institute of Technology (MIT): *Meep* (acronym for MIT Electromagnetic Equation Propagation), available online at http://ab-initio.mit.edu/meep. Meep is a full-featured software package, including, for example, arbitrary anisotropic, nonlinear, and dispersive electric and magnetic media modeling capabilities; a variety of boundary conditions including symmetries and PMLs; distributed-memory parallelism; spatial grids in Cartesian coordinates in one, two, and three dimensions as well as in cylindrical coordinates; and flexible output and field computations. Meep also provides some unusual features: advanced signal processing to analyze resonant modes; subpixel smoothing to accurately model slanted and curved dielectric interfaces in a Cartesian grid; a frequency-domain solver that exploits the time-domain code; complete scriptability; and integrated optimization facilities. This chapter begins with a discussion of the fundamental structural unit of "chunks" that constitute the FDTD grid and enable parallelization. Next, an overview is provided of Meep's core design philosophy of approaching the goal of continuous space-time modeling for inputs and outputs. The discussion continues with an explanation and motivation of Meep's somewhat-unusual design intricacies for nonlinear materials and PMLs; important aspects of Meep's computational methods for flux spectra and resonant modes; a demonstration of the formulation of Meep's frequency-domain solver that requires only minimal modifications to the underlying FDTD algorithm; and how Meep's features are accessible to users via a scripting interface. Overall, a free/open-source, full-featured FDTD package like Meep can play a vital role in enabling new research in electromagnetic phenomena. Not only does it provide a low barrier to entry for standard FDTD simulations, but the simplicity of the FDTD algorithm combined with access to the Meep source code offers an easy route to investigate new physical phenomena coupled with classical electrodynamics.

Acknowledgments

We gratefully acknowledge all of our contributing chapter authors. Their biographical sketches appear in the "About the Authors" section. And of course, we acknowledge our respective family members who exhibited great patience and kept their good spirits during the development of this book.

Allen Taflove, Evanston, Illinois

Ardavan Oskooi, Kyoto, Japan

Steven G. Johnson, Cambridge, Massachusetts

November 2012

Chapter 1

Parallel-Processing Three-Dimensional Staggered-Grid Local-Fourier-Basis PSTD Technique

Ming Ding and Kun Chen

1.1 INTRODUCTION

This chapter discusses an improved pseudospectral time-domain (PSTD) technique suitable for efficient computational solution of the full-vector Maxwell's equations in the case of electrically large three-dimensional (3-D) domains. The improved formulation is based on the scheme of a multi-domain local Fourier transform [1] that combines overlapping domain decomposition within the computation region and carefully constructed mathematical procedures to preserve numerical accuracy. Furthermore, a reformatted derivative operator on the field quantities enables use of the robust staggered Yee space lattice, and eliminates the wraparound effects responsible for the Gibbs phenomenon common in previous collocated-grid PSTD approaches. Actual implementation of the new staggered-grid, local-Fourier-basis PSTD (SL-PSTD) algorithm on a distributed-memory parallel supercomputer is described, and its performance compared with that of conventional PSTD. Finally, the numerical accuracy of the new algorithm is validated using two analytical models.

1.2 MOTIVATION

We seek to obtain the solution of the full-vector Maxwell's equations for electromagnetic waves interacting with essentially continuously inhomogeneous dielectric media. We are especially interested in the case wherein the interaction volume spans many thousands of cubic wavelengths, while the spatial inhomogeneities of the media within this volume have sub-wavelength characteristic distance scales and potentially considerable dielectric contrast. An important current example is the interaction of visible light with biological tissues, which forms the foundation for key biomedical optical techniques. Conventional theoretical and numerical analyses and approximations such as Monte Carlo simulations may not be sufficiently robust to deal with the full fine-grained heterogeneity and morphology of important tissue structures. What is needed is a technique that solves the fundamental Maxwell's equations over an enormous dynamic range of distance scales, from fractions of a wavelength (i.e., structures contained within individual biological cells), to hundreds of wavelengths (i.e., complete tissues comprised of thousands of biological cells).

The finite-difference time-domain (FDTD) method [2] and its close variant, the pseudo-spectral time-domain (PSTD) method [3], discretize Maxwell's equations on a spatial mesh and utilize a leapfrog time-marching scheme to obtain numerical solutions of complex electromagnetic wave interaction problems. To realize an acceptable level of accuracy, FDTD simulations typically require using grids having space cells smaller than ~$\lambda/20$, where λ is the wavelength in the medium. However, for problems where the interacting structure spans hundreds of λ in two dimensions, or many tens of λ in three dimensions, this requirement for a fine spatial resolution poses tremendous demands on computational resources. Furthermore, convergence to the desired final temporal state is very slow due to the small time-step mandated by numerical stability [2].

On the other hand, the PSTD method allows a much larger grid-cell size than does the FDTD method, up to $\lambda/2$ for electromagnetic wave interaction structures having slowly varying properties. In principle, this greatly reduces the PSTD computational burden relative to that of FDTD. However, the *actual* computational burden realized by current PSTD algorithms does not scale down accordingly. This is due to the low efficiency of these algorithms in a distributed-memory supercomputing environment. Several recent illustrative examples of PSTD simulations of large-scale electromagnetic wave interactions with single and multiple dielectric structures include the two-dimensional (2-D) models reported in [4, 5] and the 3-D models reported in [6]. Despite this initial work, PSTD simulations of large-scale problems remain a great challenge.

A key difficulty is that the current formalism of PSTD is only weakly parallelizable because its spatial derivative operator is global. Unlike in FDTD, where the spatial derivative of a field is approximated by a central difference of the field values at adjacent grid cells, PSTD performs the operation in the spectral domain [3, 7, 8]:

$$\frac{\partial \Psi}{\partial x} \equiv \mathcal{F}_x^{-1}\left[jk_x \mathcal{F}_x(\Psi) \right] \tag{1.1}$$

where $\mathcal{F}_x$ and $\mathcal{F}_x^{-1}$ are, respectively, the forward and inverse fast Fourier transforms (FFTs) for the x-variable. (The definitions for the y- and z-variables are analogous.) The FFT is implemented on a global Fourier basis, established on the full data set in the entire computation domain. In a distributed-memory parallel computing environment, the field data along a grid axis are often distributed across many, possibly even all, of the computation nodes, leading to very expensive inter-node communications. The best current PSTD implementations employ a one-dimensional (1-D) domain decomposition and split the entire computation domain along one grid axis (say the x-axis, the longitudinal direction) into a stack of identical slabs [9, 10], as shown in Fig. 1.1.

Here, each slab occupies all the grids in the y-z plane, and the total number of slabs equals the number of nodes m. Then, the derivatives with respect to y and z can be evaluated directly with only local data. However, the calculation of the derivative with respect to x becomes the bottleneck. A massive global data transposition is imposed to shift and realign the data into the same nodes. Non-blocking message-passing interface (MPI) communication is used for the all-to-all exchange in an ordered way to ensure communication balance, following a sequence of all possible node-to-node pairs for MPI mutual calling. Figure 1.1 illustrates only the exchange procedure for node 1. The data in node 1 are sectioned into m blocks. The number of nodes m should be an even number. After the local computation within each node, another massive data transposition is conducted to restore the data points to their original positions.

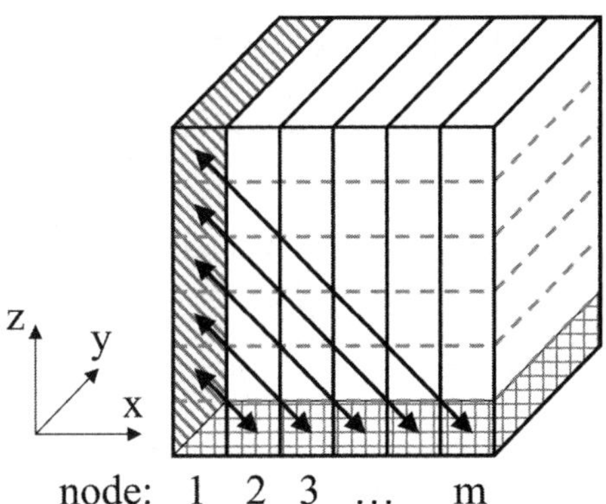

Fig. 1.1 Data transposition in the PSTD algorithm. The 1-D domain decomposition splits the model space into slabs along the x-axis. Each slab occupies the full dimension in the y-z plane. The grid cell number in the x-direction must equal that in the z-direction for the transposition to be executable. Expensive all-node-to-all-node communications are required. A single transposition can only be finished in m–1 separate communications, where m is the number of nodes.

In summary, the major obstacle to the highly efficient parallelization of conventional PSTD algorithms is the requirement for expensive global data exchanges. Basically, all the data points must be exchanged twice per time step, and each transposition contains m–1 separate communications sequentially, accumulating m–1 latencies induced by communication initiations. Furthermore, since each single node must hold all the points on a complete y-z plane, the size of the y-z dimension is restricted by the memory of a single node, and thus the PSTD model cannot grow indefinitely in the transverse dimensions.

1.3 LOCAL FOURIER BASIS AND OVERLAPPING DOMAIN DECOMPOSITION

In the technique discussed in this chapter, we apply a well-established method in computational theory and applications wherein the FFT is supported by a local, rather than a global, Fourier basis, and an associated overlapping domain decomposition scheme is employed [11, 12]. Accordingly, the computational domain can be flexibly divided into an arbitrary number of sub-domains in the x-, y-, and z-directions. In this manner, data exchanges need to occur concurrently only between neighboring nodes, and derivative calculations are performed as independent local tasks executed by all nodes in parallel without degrading accuracy.

Specifically targeting all of the weaknesses of traditional PSTD, we will describe a parallel implementation of a new staggered-grid, local-Fourier-basis PSTD (SL-PSTD). This exploits three elements: (1) a reformatted derivative operator that allows the electric and magnetic fields to be arranged in the staggered-grid Yee lattice; (2) an FFT with a local Fourier basis to synchronize computations on different nodes; and (3) overlapping domain decomposition and an artificial taper function to reduce inter-node data movements, enforce the periodic condition of the FFT on the local data set, and ensure the continuity of derivatives across sub-domain boundaries. Here, data exchanges only occur in the overlapping regions between neighboring nodes, and thus the communication cost is much lower than that for traditional PSTD. Furthermore, the revised derivative operator and the adoption of the Yee lattice naturally handles spatial discontinuities in the model space, such as point sources, without resorting to point-smoothing tricks [13, 14]. The SL-PSTD technique provides a highly efficient and accurate tool for modeling large-scale electromagnetic structures on distributed-memory supercomputers.

1.4 KEY FEATURES OF THE SL-PSTD TECHNIQUE

1.4.1 FFT on a Local Fourier Basis

The FFT in PSTD is a serial process executed by a single computation node. Conventionally it is conceived to be a global data operation. However, the spectral multidomain technique with local Fourier basis breaks the single global computation into multiple independent parallel tasks evenly distributed to all computation nodes [11]. Here, the entire computational domain is first split into multiple nonoverlapping sub-domains. Then, each side of a sub-domain's boundary is translated along the normal to the interface into the neighboring sub-domain by a preset depth. The region enclosed by the new boundaries defines the final sub-domain. Consequently, adjacent sub-domains intersect on their interface by a thickness of twice the preset depth, which sparks the concept of overlapping domain decomposition. From the point of view of each sub-domain, the external half of each intersection is processed differently from the internal half.

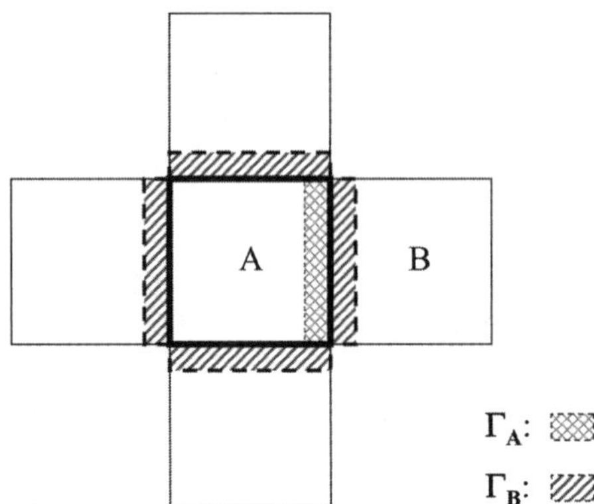

Fig. 1.2 Overlapping domain decomposition in 2-D. The hatched areas Γ_A and Γ_B depict the overlapping of sub-domains A and B. The dashed lines enclose all of sub-domain A, while the thick solid box encloses its nonoverlapping region, and the area between defines its overlapping region.

Figure 1.2 illustrates the definition of nonoverlapping and overlapping regions. Here, sub-domain A and sub-domain B share a common intersection $\Gamma_A + \Gamma_B$. Γ_A and Γ_B belong to the nonoverlapping and overlapping regions of sub-domain A, respectively; and vice versa. Neighbors by corner or edge (in either 2-D or 3-D) do not share an overlapping region if derivatives with respect to only one spatial variable are involved, such as $\partial\psi/\partial x$ and $\partial^2\psi/\partial x^2$, but not $\partial^2\psi/\partial x\partial y$. The local data within each sub-domain consist of two sets: one is comprised of the unmodified intrinsic data in the nonoverlapping region; and the other is comprised of the external data in the overlapping region (first copied from the respective neighbor through inter-node communication and then multiplied by a weighting function). The weighting function artificially tapers both ends of the local data set to zero and thus enforces the periodic condition required for FFT. The smoothness of the taper function ensures that the derivatives in the non-overlapping region are unchanged and remain continuous across sub-domain boundaries. Subsequently, only field values in the nonoverlapping region are updated and the data in the overlapping region are no longer relevant once the derivatives are obtained.

To demonstrate the local FFT algorithm, we consider the following 1-D analytical example function:

$$f(x) = \frac{1}{\sqrt{2\pi\sigma}}\left[\sin x + \sin(1.2x) + 1\right]\exp\left(-x^2/2\sigma^2\right) \tag{1.2}$$

A mesh of N grid points is generated and the whole computational domain is divided into m sub-domains. The overlapping depth is preset to an *even* number n_{ovl}. Hence, the intersection would be $2n_{\text{ovl}}$ wide. Each sub-domain copies n_{ovl} grid values from its left neighbor and another n_{ovl} grid values from its right neighbor. We introduce the following bell-shaped function to serve as the weighting function (noting that other choices of $b(\ell)$ are possible):

$$b(\ell) = \begin{cases} \sin\left(\ell\pi/n_{\text{ovl}}\right) & 0 \le \ell < n_{\text{ovl}}/2 \\ 1 & n_{\text{ovl}}/2 \le \ell \le \ell_2 + n_{\text{ovl}}/2 \\ \sin\left[\left(\ell-\ell_2\right)\pi/n_{\text{ovl}}\right] & \ell_2 + n_{\text{ovl}}/2 < \ell \le \ell_2 + n_{\text{ovl}} \end{cases} \tag{1.3}$$

where ℓ denotes the local grid index, the overlapping depth is preset to n_{ovl}, and the non-overlapping region starts at $\ell_1 = n_{\text{ovl}}$ and ends at ℓ_2 (to be determined later). The local data set is given as

$$\tilde{f}(x_\ell) \equiv f(x_\ell)\, b(\ell) \qquad 0 \le \ell \le \ell_2 + n_{\text{ovl}} \tag{1.4}$$

The top plateau of $b(\ell)$ equalizes $\tilde{f}(x_\ell)$ to $f(x_\ell)$. Note that half of the overlapping grids ($n_{\text{ovl}}/2 \le \ell < n_{\text{ovl}}$ at the left side and $\ell_2+1 \le \ell \le \ell_2 + n_{\text{ovl}}/2$ at the right side) retain the original values from the neighboring sub-domains, providing a mechanism to "stitch" the neighbors together. The left and right edges of $b(\ell)$ gradually taper the remaining overlapping grid values to 0. Therefore, the local data $\tilde{f}(x_\ell)$ satisfy the periodic condition and allow the FFT. The values of $\tilde{f}'(x_\ell)$ within the non-overlapping region ($\ell_1 \le \ell \le \ell_2$) reproduce the value of $f'(x_\ell)$. Effectively, a piecewise concatenation of $\tilde{f}'(x_\ell)$ from all the nonoverlapping regions provides the derivatives in the entire computation region.

As illustrated in Fig. 1.3(a), the x-axis is divided by the vertical dashed lines into multiple sub-domains and one-to-one mapped to all computation nodes. A $b(\ell)$ is aligned with the boundaries of each sub-domain. The curve of $b(\ell)$ in (1.3) has first-order smoothness at the joining point between the rising edges and the top plateau ($\ell = n_{\text{ovl}}/2$ and $\ell_2 + n_{\text{ovl}}/2$), but is not smooth at the taper ends ($\ell = 0$ and $\ell_2 + n_{\text{ovl}}$). A smoother choice of $b(\ell)$ at the taper ends may slightly improve the results. Numerical tests show that the value of n_{ovl} has more influence on the results' accuracy. A sufficiently large n_{ovl} is more important.

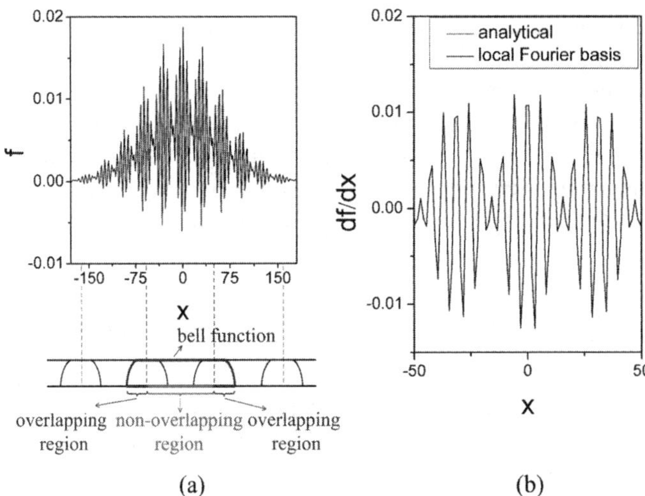

Fig. 1.3 (a) The curve of function $f(x)$ stacked on top of a schematic diagram illustrating overlapping domain decomposition and the bell function. (b) Precise agreement of the first derivative of $f(x)$ given by (1.5) with the analytical results. Only the span $|x| < 50$ is shown for clarity. *Source:* Ding and Chen, *Optics Express*, 2010, pp. 9236–9250, ©2010 The Optical Society of America.

To facilitate FFT calculations and balance the workload on different computation nodes, it is advantageous to keep the length of each local FFT identical. Note that the first (last) sub-domain at the far left (right) has no left (right) neighbor. To compensate for this difference, the nonoverlapping regions of these two sub-domains are wider than those in the middle by n_{ovl} grid cells. After some arithmetic, the width configuration for a decomposition of N grid cells into m nonoverlapping sub-domains should be $n_{\text{ovl}} + (N - 2n_{\text{ovl}})/m$ grid cells for the first and mth sub-domains, and $(N - 2n_{\text{ovl}})/m$ grid cells for all others. Consequently, the length of a local FFT is fixed to a constant $L = (N - 2n_{\text{ovl}})/m + 2n_{\text{ovl}}$ for all sub-domains, and this concludes our last free parameter $\ell_2 = L - n_{\text{ovl}} - 1$.

To validate the above multi-domain FFT procedure, we calculate the first derivative of (1.2) utilizing a staggered-grid formula [15]:

$$\tilde{f}'(x + \Delta/2) \;=\; \mathcal{F}^{-1}\left[jk\, e^{jk\Delta/2} \mathcal{F}\left(\tilde{f}\right) \right](x) \tag{1.5}$$

where Δ denotes the grid size, and $\mathcal{F}$ and $\mathcal{F}^{-1}$ denote, respectively, the forward and inverse FFTs *on the local data* within each sub-domain. Unlike the old operator in (1.1), which places the field function and its derivative on the same grid, the new version in (1.5) calculates the derivative at a location offset by half a grid. Specifically, the $\mathcal{F}(\tilde{f})$ on the right-hand side of (1.5) switches $\tilde{f}(x)$ from the spatial domain to the Fourier domain, and the spatial derivative is identical to the multiplication of factor jk in the Fourier domain, while the extra exponential factor $e^{jk\Delta/2}$ in the Fourier domain is equivalent to shifting the coordinate by $\Delta/2$ in the spatial domain. Thus in (1.5), the value on the right-hand side at coordinate x gives the first derivative at coordinate $x + \Delta/2$.

The derivative of the example function (1.2) can be analytically derived and serve as the gold standard in assessing the accuracy of (1.5) and the validity of the local FFT algorithm.

We choose $\sigma = 20\pi$ to simulate a Gaussian wave packet containing a zero-frequency component plus dozens of oscillations at two frequencies, as in Fig. 1.3(a). Here, we set the grid resolution to $\Delta = 2\pi/5$, or one-fifth the period of one oscillation component. Figure 1.3(b) displays the results calculated using (1.5) with parameters $L = 32$, $m = 50$, $n_{\text{ovl}} = 10$, and the floating-point precision of C (or the equivalent real*4 precision of Fortran). We see that the two curves in this figure are indistinguishable. Moreover, the performance of our staggered-grid local FFT method is directly compared with the collocated-grid global FFT method. The relative deviations of both methods from the analytical standard are at the same order of magnitude $O(10^{-4})$.

1.4.2 Absence of the Gibbs Phenomenon Artifact

With the conventional PSTD technique for Maxwell's equations, spatial derivative operation (1.1) leads to a collocated arrangement of the field vector components (E_x, E_y, E_z, H_x, H_y, H_z) in the space lattice. However, with the SL-PSTD technique, the reformatted derivative operator in (1.5) exhibits the same feature as in classic FDTD wherein the coordinates of the field and its derivative are staggered by one-half grid cell. Therefore, the well-behaved and robust Yee mesh can be directly incorporated into the SL-PSTD algorithm.

A negative consequence of the collocated space lattice and the spatial derivative operator used in the conventional PSTD technique for Maxwell's equations is the susceptibility of that technique to the Gibbs phenomenon artifact caused by a discontinuity of either the electromagnetic field excitation or the geometry or composition of the modeled structure [15]. Concerning the excitation issue, a work-around for point sources was proposed by spreading out such sources to multiple grid cells [13, 14]. In a similar manner, the Gibbs artifact appearing due to an abrupt material change within the PSTD grid, such as a dielectric interface, can be mitigated by spreading out the change in permittivity and/or conductivity to multiple cells on each side of the interface. Of course, this perturbs the geometry of the structure being modeled.

Fortunately, the Gibbs phenomenon artifact is *absent* when the derivative operator in (1.5) and the Yee mesh are implemented in the SL-PSTD algorithm. The reason lies in the behavior of this derivative operator at the Nyquist frequency. Let Δ be the grid-cell size, and correspondingly the maximum frequency of the FFT is the Nyquist frequency, $\pm f_c$, where $f_c = (2\Delta)^{-1}$. A spatial discontinuity in the system generates spectral components at frequencies beyond $\pm f_c$, causing the spectrum, $\mathcal{F}(\psi)$, of the field component, ψ, to no longer vanish at $\pm f_c$. Consequently in the Fourier domain, $jk\mathcal{F}(\psi)$ flips sign when $-f_c \leftrightarrow f_c$ and exhibits a discontinuity at the Nyquist frequency, leaving $\psi' = \mathcal{F}^{-1}[jk\mathcal{F}(\psi)]$ ill-behaved. On the other hand, the extra exponential factor in (1.5) compensates for the sign flip and renders the derivative *invariant* at the Nyquist frequency.

1.5 TIME-STEPPING RELATIONS FOR DIELECTRIC SYSTEMS

To illustrate the methodology and notation of the SL-PSTD algorithm, we first derive its general time-stepping relation for the field component E_x. By Ampere's law:

$$\varepsilon \frac{\partial E_x}{\partial t} + \sigma E_x = -\left(\frac{\partial H_y}{\partial z} - \frac{\partial H_z}{\partial y} \right) \tag{1.6}$$

where ε and σ are the dielectric permittivity and conductivity, respectively.

In the same manner as in the conventional PSTD, the left-hand side of (1.6) is discretized at time step n as:

$$\left.\frac{\partial E_x}{\partial t}\right|_{i,j+1/2,k+1/2}^{n} = \frac{\left.E_x\right|_{i,j+1/2,k+1/2}^{n+1/2} - \left.E_x\right|_{i,j+1/2,k+1/2}^{n-1/2}}{\Delta t} \tag{1.7a}$$

$$\left.E_x\right|_{i,j+1/2,k+1/2}^{n} = \frac{1}{2}\left(\left.E_x\right|_{i,j+1/2,k+1/2}^{n+1/2} + \left.E_x\right|_{i,j+1/2,k+1/2}^{n-1/2}\right) \tag{1.7b}$$

The spatial partial derivatives on the right-hand side of (1.6) are obtained according to (1.5) as:

$$\left.\frac{\partial H_y}{\partial z}\right|_{i,j+1/2,k+1/2}^{n} = \left.\mathcal{F}_z^{-1}\left[jk_z e^{jk_z \Delta z/2}\mathcal{F}_z\left(\tilde{H}_y\right)\right]\right|_{i,j+1/2,k}^{n} \tag{1.8a}$$

$$\left.\frac{\partial H_z}{\partial y}\right|_{i,j+1/2,k+1/2}^{n} = \left.\mathcal{F}_y^{-1}\left[jk_y e^{jk_y \Delta y/2}\mathcal{F}_y\left(\tilde{H}_z\right)\right]\right|_{i,j,k+1/2}^{n} \tag{1.8b}$$

In (1.8), $\mathcal{F}_\xi$ and $\mathcal{F}_\xi^{-1}$ ($\xi = x, y, z$) are, respectively, the one-dimensional forward FFT and one-dimensional inverse FFT with respect to the variable ξ, and $\tilde{H}_\eta$ ($\eta = x, y, z$) are the local magnetic field data defined similarly to that in (1.4). Substituting (1.7) and (1.8) into (1.6) leads to the following time-stepping relation for E_x, assuming that $\sigma = 0$:

$$\left.E_x\right|_{i,j+1/2,k+1/2}^{n+1/2} = \left.E_x\right|_{i,j+1/2,k+1/2}^{n-1/2} + \frac{\Delta t}{\left.\varepsilon_r\right|_{i,j+1/2,k+1/2}\varepsilon_0'}\left\{\begin{array}{l}\left.\mathcal{F}_y^{-1}\left[jk_y e^{jk_y \Delta_y/2}\mathcal{F}_y\left(\tilde{H}_z\right)\right]\right|_{i,j,k+1/2}^{n} \\ -\left.\mathcal{F}_z^{-1}\left[jk_z e^{jk_z \Delta_z/2}\mathcal{F}_z\left(\tilde{H}_y\right)\right]\right|_{i,j+1/2,k}^{n}\end{array}\right\} \tag{1.9}$$

where ε_0' is a scaled value of the vacuum permittivity, as defined in Section 1.6 to mitigate or even eliminate the residual isotropic numerical phase-velocity error. Analogous derivations yield time-stepping relations for the other two E-field components, and for the three H-field components where μ_0' is a scaled value of the vacuum permeability, as defined in Section 1.6:

$$\left.E_y\right|_{i-1/2,j,k+1/2}^{n+1/2} = \left.E_y\right|_{i-1/2,j,k+1/2}^{n-1/2} + \frac{\Delta t}{\left.\varepsilon_r\right|_{i-1/2,j,k+1/2}\varepsilon_0'}\left\{\begin{array}{l}\left.\mathcal{F}_z^{-1}\left[jk_z e^{jk_z \Delta_z/2}\mathcal{F}_z\left(\tilde{H}_x\right)\right]\right|_{i-1/2,j,k}^{n} \\ -\left.\mathcal{F}_x^{-1}\left[jk_x e^{jk_x \Delta_x/2}\mathcal{F}_x\left(\tilde{H}_z\right)\right]\right|_{i-1,j,k+1/2}^{n}\end{array}\right\} \tag{1.10}$$

$$\left.E_z\right|_{i-1/2,j+1/2,k}^{n+1/2} = \left.E_z\right|_{i-1/2,j+1/2,k}^{n-1/2} + \frac{\Delta t}{\left.\varepsilon_r\right|_{i-1/2,j+1/2,k}\varepsilon_0'}\left\{\begin{array}{l}\left.\mathcal{F}_x^{-1}\left[jk_x e^{jk_x \Delta_x/2}\mathcal{F}_x\left(\tilde{H}_y\right)\right]\right|_{i-1,j+1/2,k}^{n} \\ -\left.\mathcal{F}_y^{-1}\left[jk_y e^{jk_y \Delta_y/2}\mathcal{F}_y\left(\tilde{H}_x\right)\right]\right|_{i-1/2,j,k}^{n}\end{array}\right\} \tag{1.11}$$

$$H_x\Big|_{i-1/2,j,k}^{n+1} = H_x\Big|_{i-1/2,j,k}^{n} + \frac{\Delta t}{\mu_r\big|_{i-1/2,j,k}\mu_0'}\left\{\begin{aligned} &\mathcal{F}_z^{-1}\left[jk_z e^{jk_z\Delta_z/2}\mathcal{F}_z\left(\tilde{E}_y\right)\right]\Big|_{i-1/2,j,k-1/2}^{n+1/2} \\ &-\mathcal{F}_y^{-1}\left[jk_y e^{jk_y\Delta_y/2}\mathcal{F}_y\left(\tilde{E}_z\right)\right]\Big|_{i-1/2,j-1/2,k}^{n+1/2}\end{aligned}\right\} \quad (1.12)$$

$$H_y\Big|_{i,j+1/2,k}^{n+1} = H_y\Big|_{i,j+1/2,k}^{n} + \frac{\Delta t}{\mu_r\big|_{i,j+1/2,k}\mu_0'}\left\{\begin{aligned} &\mathcal{F}_x^{-1}\left[jk_x e^{jk_x\Delta_x/2}\mathcal{F}_x\left(\tilde{E}_z\right)\right]\Big|_{i-1/2,j+1/2,k}^{n+1/2} \\ &-\mathcal{F}_z^{-1}\left[jk_z e^{jk_z\Delta_z/2}\mathcal{F}_z\left(\tilde{E}_x\right)\right]\Big|_{i,j+1/2,k-1/2}^{n+1/2}\end{aligned}\right\} \quad (1.13)$$

$$H_z\Big|_{i,j,k+1/2}^{n+1} = H_z\Big|_{i,j,k+1/2}^{n} + \frac{\Delta t}{\mu_r\big|_{i,j,k+1/2}\mu_0'}\left\{\begin{aligned} &\mathcal{F}_y^{-1}\left[jk_y e^{jk_y\Delta_y/2}\mathcal{F}_y\left(\tilde{E}_x\right)\right]\Big|_{i,j-1/2,k+1/2}^{n+1/2} \\ &-\mathcal{F}_x^{-1}\left[jk_x e^{jk_x\Delta_x/2}\mathcal{F}_x\left(\tilde{E}_y\right)\right]\Big|_{i-1/2,j,k+1/2}^{n+1/2}\end{aligned}\right\} \quad (1.14)$$

To implement the pure scattered-field formulation for electric field component $E_\eta\big|_{i,j,k}^{n+1/2}$, we add the following term related to the known incident wave to each bracket on the right-hand side of (1.9) – (1.11):

$$\varepsilon_0'\left(1-\varepsilon_r\big|_{i,j,k}\right)\cdot\left(\frac{\partial E_\eta^{inc}}{\partial t}\right)\Bigg|_{i,j,k}^{n} \quad (1.15a)$$

Similarly, to implement the pure scattered-field formulation for magnetic field component $H_\eta\big|_{i,j,k}^{n+1}$, we add the following term related to the known incident wave to each bracket on the right-hand side of (1.12) – (1.14):

$$\mu_0'\left(1-\mu_r\big|_{i,j,k}\right)\cdot\left(\frac{\partial H_\eta^{inc}}{\partial t}\right)\Bigg|_{i,j,k}^{n+1/2} \quad (1.15b)$$

1.6 ELIMINATION OF NUMERICAL PHASE VELOCITY ERROR FOR A MONOCHROMATIC EXCITATION

According to the Nyquist sampling theorem, the spatial derivatives in PSTD are exact if the meshing density is finer than $\lambda/2$. This leads to a numerical phase velocity $\tilde{v}_p$ within the grid that is independent of the propagation direction, i.e., isotropic. However, since the finite-difference approximation of the time derivative is accurate only to the second order, $\tilde{v}_p$ exhibits a residual numerical error, albeit one that is isotropic within the grid. We now consider a simple means to eliminate this error for a single-frequency (monochromatic) excitation by properly scaling the values of the permittivity and permeability of the dielectric media within the grid.

Consider the PSTD numerical dispersion relation [16] for a wave propagating in 3-D in a material medium having the relative permittivity ε_r and the relative permeability μ_r :

$$(\varepsilon_r\varepsilon_0)\,(\mu_r\mu_0)\left[\frac{2}{\Delta t}\sin\left(\frac{\omega\,\Delta t}{2}\right)\right]^2 = \tilde{k}_x^{\,2} + \tilde{k}_y^{\,2} + \tilde{k}_z^{\,2} = \tilde{k}^2 \quad (1.16)$$

Here, ω is the angular frequency of the wave and $\tilde{k}_\eta$ are the components of the numerical wavevector $\tilde{k}$. To speed up code execution, a large Δt is preferred to decrease the total number of time-steps for the targeted final time. However, this would increase the numerical dispersion error according to (1.16).

Now, assume a monochromatic excitation and define $\alpha = (\omega\Delta t/2) / \sin(\omega\Delta t/2)$. The PSTD numerical dispersion relation of (1.16) with material parameters (ε_r, μ_r) can be written as:

$$\frac{\varepsilon_r \varepsilon_0}{\alpha} \cdot \frac{\mu_r \mu_0}{\alpha} \cdot \omega^2 = \tilde{k}^2 \tag{1.17}$$

Note that α is a number slightly greater than unity for all usable values of the time-step Δt. This implies that $\tilde{v}_p$ is commensurately slightly larger than the correct physical value. We can compensate for this error by using the scaled vacuum permittivity ε_0' and the scaled vacuum permeability μ_0' instead of ε_0 and μ_0, respectively:

$$\varepsilon_0' \equiv \alpha\,\varepsilon_0\,, \qquad \mu_0' \equiv \alpha\,\mu_0 \tag{1.18}$$

Then, (1.17) becomes:

$$\frac{\varepsilon_r (\alpha \varepsilon_0)}{\alpha} \cdot \frac{\mu_r (\alpha \mu_0)}{\alpha} \cdot \omega^2 = \tilde{k}^2 \tag{1.19}$$

Cancellation of the α's yields the correct numerical wavevector value and hence the correct numerical velocity. To this end, we need only to use the scaled values, ε_0' and μ_0' of (1.18), in time-stepping expressions (1.9) – (1.15).

We note that now a relatively large Δt can be chosen to reduce the total number of time-steps without compromising numerical accuracy. However, there remains an upper bound to Δt set by the numerical stability requirement [16]. Furthermore, impulsive excitations have potentially a broad spectrum of angular frequencies, so that at best we could null out the velocity error for the midpoint of this spectrum.

1.7 TIME-STEPPING RELATIONS WITHIN THE PERFECTLY MATCHED LAYER ABSORBING OUTER BOUNDARY

We next consider the formulation of the SL-PSTD time-stepping relations within a standard polynomial-graded uniaxial perfectly matched layer (UPML) absorbing outer boundary. To simply the expressions, we first introduce three parameters:

$$\rho_\xi\big|_\ell = \begin{cases} \rho_\xi^{\max} \cdot \left(\dfrac{n_{\text{PML}} - \ell}{n_{\text{PML}}} \right)^3 & 0 \le \ell < n_{\text{PML}} \\ 0 & n_{\text{PML}} \le \ell < N_\xi - n_{\text{PML}} \\ \rho_\xi^{\max} \cdot \left(\dfrac{\ell - \left(N_\xi - n_{\text{PML}}\right)}{n_{\text{PML}}} \right)^3 & N_\xi - n_{\text{PML}} \le \ell < N_\xi \end{cases} \tag{1.20}$$

where $\xi = x, y, z$ denotes the three coordinates and N_ξ denotes the number of grid cells along the ξ axis. The UPML is assumed to have the same thickness, n_{PML} grid cells, on all six outer-boundary planes. In (1.20), $\rho_\xi^{\max} = 30c/(n_{\text{PML}}\Delta\xi)$ where c is the speed of light in vacuum and $\Delta\xi$ is the grid cell size in the ξ direction.

For each Yee cell inside the PML, we further introduce three time-stepped auxiliary variables (G_x, G_y, G_z) to assist in the calculation of the electric field:

$$G_x\Big|_{i,j+1/2,k+1/2}^{n+1/2} = \left(\frac{2-\rho_y\big|_{j+1/2}\Delta t}{2+\rho_y\big|_{j+1/2}\Delta t}\right) G_x\Big|_{i,j+1/2,k+1/2}^{n-1/2} + \frac{2\Delta t}{\left(2+\rho_y\big|_{j+1/2}\Delta t\right)\varepsilon_0'} \cdot \left\{ \mathcal{F}_y^{-1}\left[jk_y e^{jk_y\Delta_y/2}\mathcal{F}_y\left(\tilde{H}_z\right)\right]\Big|_{i,j,k+1/2}^{n} - \mathcal{F}_z^{-1}\left[jk_z e^{jk_z\Delta_z/2}\mathcal{F}_z\left(\tilde{H}_y\right)\right]\Big|_{i,j+1/2,k}^{n} \right\} \tag{1.21}$$

$$G_y\Big|_{i-1/2,j,k+1/2}^{n+1/2} = \left(\frac{2-\rho_z\big|_{k+1/2}\Delta t}{2+\rho_z\big|_{k+1/2}\Delta t}\right) G_y\Big|_{i-1/2,j,k+1/2}^{n-1/2} + \frac{2\Delta t}{\left(2+\rho_z\big|_{k+1/2}\Delta t\right)\varepsilon_0'} \cdot \left\{ \mathcal{F}_z^{-1}\left[jk_z e^{jk_z\Delta_z/2}\mathcal{F}_z\left(\tilde{H}_x\right)\right]\Big|_{i-1/2,j,k}^{n} - \mathcal{F}_x^{-1}\left[jk_x e^{jk_x\Delta_x/2}\mathcal{F}_x\left(\tilde{H}_z\right)\right]\Big|_{i-1,j,k+1/2}^{n} \right\} \tag{1.22}$$

$$G_z\Big|_{i-1/2,j+1/2,k}^{n+1/2} = \left(\frac{2-\rho_x\big|_{i-1/2}\Delta t}{2+\rho_x\big|_{i-1/2}\Delta t}\right) G_z\Big|_{i-1/2,j+1/2,k}^{n-1/2} + \frac{2\Delta t}{\left(2+\rho_x\big|_{i-1/2}\Delta t\right)\varepsilon_0'} \cdot \left\{ \mathcal{F}_x^{-1}\left[jk_x e^{jk_x\Delta_x/2}\mathcal{F}_x\left(\tilde{H}_y\right)\right]\Big|_{i-1,j+1/2,k}^{n} - \mathcal{F}_y^{-1}\left[jk_y e^{jk_y\Delta_y/2}\mathcal{F}_y\left(\tilde{H}_x\right)\right]\Big|_{i-1/2,j,k}^{n} \right\} \tag{1.23}$$

The resulting time-stepping relations for the electric field are:

$$E_x\Big|_{i,j+1/2,k+1/2}^{n+1/2} = \left(\frac{2-\rho_z\big|_{k+1/2}\Delta t}{2+\rho_z\big|_{k+1/2}\Delta t}\right) E_x\Big|_{i,j+1/2,k+1/2}^{n-1/2} + \left(\frac{2+\rho_x\big|_{i}\Delta t}{2+\rho_z\big|_{k+1/2}\Delta t}\right) G_x\Big|_{i,j+1/2,k+1/2}^{n+1/2} - \left(\frac{2-\rho_x\big|_{i}\Delta t}{2+\rho_z\big|_{k+1/2}\Delta t}\right) G_x\Big|_{i,j+1/2,k+1/2}^{n-1/2} \tag{1.24}$$

$$E_y\Big|_{i-1/2,j,k+1/2}^{n+1/2} = \left(\frac{2-\rho_x\big|_{i-1/2}\Delta t}{2+\rho_x\big|_{i-1/2}\Delta t}\right) E_y\Big|_{i-1/2,j,k+1/2}^{n-1/2} + \left(\frac{2+\rho_y\big|_{j}\Delta t}{2+\rho_x\big|_{i-1/2}\Delta t}\right) G_y\Big|_{i-1/2,j,k+1/2}^{n+1/2} - \left(\frac{2-\rho_y\big|_{j}\Delta t}{2+\rho_x\big|_{i-1/2}\Delta t}\right) G_y\Big|_{i-1/2,j,k+1/2}^{n-1/2} \tag{1.25}$$

$$
\begin{aligned}
E_z\Big|_{i-1/2,j+1/2,k}^{n+1/2} &= \left(\frac{2-\rho_y\big|_{j+1/2}\Delta t}{2+\rho_y\big|_{j+1/2}\Delta t}\right)E_z\Big|_{i-1/2,j+1/2,k}^{n-1/2} + \left(\frac{2+\rho_z\big|_k\Delta t}{2+\rho_y\big|_{j+1/2}\Delta t}\right)G_z\Big|_{i-1/2,j+1/2,k}^{n+1/2} \\
&\quad - \left(\frac{2-\rho_z\big|_k\Delta t}{2+\rho_y\big|_{j+1/2}\Delta t}\right)G_z\Big|_{i-1/2,j+1/2,k}^{n-1/2}
\end{aligned}
\tag{1.26}
$$

Similarly for each Yee cell inside the PML, we introduce three time-stepped auxiliary variables (B_x, B_y, B_z) to assist in the calculation of the magnetic field:

$$
\begin{aligned}
B_x\Big|_{i-1/2,j,k}^{n+1} &= \left(\frac{2-\rho_y\big|_j\Delta t}{2+\rho_y\big|_j\Delta t}\right)B_x\Big|_{i-1/2,j,k}^{n} + \frac{2\Delta t}{\left(2+\rho_y\big|_j\Delta t\right)\mu_0'}\cdot \\
&\left\{\mathcal{F}_z^{-1}\left[jk_z e^{jk_z\Delta_z/2}\mathcal{F}_z\left(\tilde{E}_y\right)\right]\Big|_{i-1/2,j,k-1/2}^{n+1/2} - \mathcal{F}_y^{-1}\left[jk_y e^{jk_y\Delta_y/2}\mathcal{F}_y\left(\tilde{E}_z\right)\right]\Big|_{i-1/2,j-1/2,k}^{n+1/2}\right\}
\end{aligned}
\tag{1.27}
$$

$$
\begin{aligned}
B_y\Big|_{i,j+1/2,k}^{n+1} &= \left(\frac{2-\rho_z\big|_k\Delta t}{2+\rho_z\big|_k\Delta t}\right)B_y\Big|_{i,j+1/2,k}^{n} + \frac{2\Delta t}{\left(2+\rho_z\big|_k\Delta t\right)\mu_0'}\cdot \\
&\left\{\mathcal{F}_x^{-1}\left[jk_x e^{jk_x\Delta_x/2}\mathcal{F}_x\left(\tilde{E}_z\right)\right]\Big|_{i-1/2,j+1/2,k}^{n+1/2} - \mathcal{F}_z^{-1}\left[jk_z e^{jk_z\Delta_z/2}\mathcal{F}_z\left(\tilde{E}_x\right)\right]\Big|_{i,j+1/2,k-1/2}^{n+1/2}\right\}
\end{aligned}
\tag{1.28}
$$

$$
\begin{aligned}
B_z\Big|_{i,j,k+1/2}^{n+1} &= \left(\frac{2-\rho_x\big|_i\Delta t}{2+\rho_x\big|_i\Delta t}\right)B_z\Big|_{i,j,k+1/2}^{n} + \frac{2\Delta t}{\left(2+\rho_x\big|_i\Delta t\right)\mu_0'}\cdot \\
&\left\{\mathcal{F}_y^{-1}\left[jk_y e^{jk_y\Delta_y/2}\mathcal{F}_y\left(\tilde{E}_x\right)\right]\Big|_{i,j-1/2,k+1/2}^{n+1/2} - \mathcal{F}_x^{-1}\left[jk_x e^{jk_x\Delta_x/2}\mathcal{F}_x\left(\tilde{E}_y\right)\right]\Big|_{i-1/2,j,k+1/2}^{n+1/2}\right\}
\end{aligned}
\tag{1.29}
$$

The resulting time-stepping relations for the magnetic field are:

$$
\begin{aligned}
H_x\Big|_{i-1/2,j,k}^{n+1} &= \left(\frac{2-\rho_z\big|_k\Delta t}{2+\rho_z\big|_k\Delta t}\right)H_x\Big|_{i-1/2,j,k}^{n} + \left(\frac{2+\rho_x\big|_{i-1/2}\Delta t}{2+\rho_z\big|_k\Delta t}\right)B_x\Big|_{i-1/2,j,k}^{n+1} \\
&\quad - \left(\frac{2-\rho_x\big|_{i-1/2}\Delta t}{2+\rho_z\big|_k\Delta t}\right)B_x\Big|_{i-1/2,j,k}^{n}
\end{aligned}
\tag{1.30}
$$

$$\begin{aligned} H_y\Big|_{i,j+1/2,k}^{n+1} &= \left(\frac{2-\rho_x\big|_i \Delta t}{2+\rho_x\big|_i \Delta t}\right) H_y\Big|_{i,j+1/2,k}^{n} + \left(\frac{2+\rho_y\big|_{j+1/2} \Delta t}{2+\rho_x\big|_i \Delta t}\right) B_y\Big|_{i,j+1/2,k}^{n+1} \\ &\quad - \left(\frac{2-\rho_y\big|_{j+1/2} \Delta t}{2+\rho_x\big|_i \Delta t}\right) B_y\Big|_{i,j+1/2,k}^{n} \end{aligned} \tag{1.31}$$

$$\begin{aligned} H_z\Big|_{i,j,k+1/2}^{n+1} &= \left(\frac{2-\rho_y\big|_j \Delta t}{2+\rho_y\big|_j \Delta t}\right) H_z\Big|_{i,j,k+1/2}^{n} + \left(\frac{2+\rho_z\big|_{k+1/2} \Delta t}{2+\rho_y\big|_j \Delta t}\right) B_z\Big|_{i,j,k+1/2}^{n+1} \\ &\quad - \left(\frac{2-\rho_z\big|_{k+1/2} \Delta t}{2+\rho_y\big|_j \Delta t}\right) B_z\Big|_{i,j,k+1/2}^{n} \end{aligned} \tag{1.32}$$

1.8 REDUCTION OF THE NUMERICAL ERROR IN THE NEAR-FIELD TO FAR-FIELD TRANSFORMATION

Just as in the FDTD method, the far-field response of an illuminated or radiating structure is obtained through the near-field to far-field transformation in SL-PSTD. The electric and magnetic field vectors in the far-field region are expressed as surface integrals of the equivalent electric and magnetic currents flowing tangentially along a six-sided rectangular virtual surface S that completely encloses the structure of interest. However, the numerical discretization of the required surface integrals must be conducted carefully because of the coarse mesh employed in SL-PSTD. Unlike in FDTD, where the grid cell size is much finer, directly summing the value of the integrand over all of the surface elements of S could accrue large numerical errors. Here, we derive a procedure with spectral accuracy to calculate the surface integral analytically to the maximum extent.

Following the notation of [2], the calculation of the electric and magnetic field vectors in the far-field region ultimately requires the calculation of six surface integrals of the form:

$$\iint_S \breve{J}_\xi e^{jkr'\cos\psi} ds' , \qquad \iint_S \breve{M}_\xi e^{jkr'\cos\psi} ds' \qquad \xi = x, y, z \tag{1.33}$$

In (1.33), $(\breve{J}_x, \breve{J}_y, \breve{J}_z)$ are the equivalent surface electric current vector components, $(\breve{M}_x, \breve{M}_y, \breve{M}_z)$ are the equivalent surface magnetic current vector components, $\vec{r}$ is the position of the observation point at (x, y, z), $\vec{r}'$ is the position of a source point on S at (x', y', z'), and ψ is the angle between $\vec{r}$ and $\vec{r}'$.

In SL-PSTD, the six surface integrals in (1.33) can be evaluated analytically, thereby avoiding the need for discretization. To illustrate this procedure, we consider the integration of $\breve{J}_x$ over the virtual surface $z' = z_0$. In the first step, $\breve{J}_x(x', y', z_0)$ is expressed as the Fourier expansion:

$$\breve{J}_x(x', y', z_0) = \frac{1}{\breve{N}_x \breve{N}_y} \sum_{m,n} \breve{J}_{x,mn}(z_0)\, e^{j(k_{x,m}x' + k_{y,n}y')} \tag{1.34}$$

where $\breve{J}_{x,mn}(z_0)$ is the 2D FFT of $\breve{J}_x(x',y',z_0)$, $\breve{N}_x\breve{N}_y$ is the size of the FFT, and $k_{x,m}$ and $k_{y,n}$ are the corresponding discrete wave vectors. Substituting (1.34) into (1.33), the retarded phase term, which is the main source of error, can now be directly *integrated out*:

$$\begin{aligned}&\int_{x_1}^{x_2}\int_{y_1}^{y_2} \breve{J}_x(x',y',z_0)\, e^{j(k_x x' + k_y y' + k_z z_0)}\, dx'\, dy' \\ &= \frac{1}{\breve{N}_x \breve{N}_y} e^{jk_z z_0} \sum_{m,n} \breve{J}_{x,mn}(z_0) \frac{e^{j(k_{x,m}+k_x)x_2} - e^{j(k_{x,m}+k_x)x_1}}{j(k_{x,m}+k_x)} \cdot \frac{e^{j(k_{y,n}+k_y)y_2} - e^{j(k_{y,n}+k_y)y_1}}{j(k_{y,n}+k_y)}\end{aligned} \quad (1.35)$$

Extensions of this technique to the other five surface integrals about S are straightforward.

1.9 IMPLEMENTATION ON A DISTRIBUTED-MEMORY SUPERCOMPUTING CLUSTER

The SL-PSTD algorithm has been implemented on the Dawning 5000A supercomputer at the Shanghai Supercomputer Center (SSC). The Dawning 5000A is a distributed-memory cluster system interconnected via Infiniband. Each computing unit, a 4-way blade, is comprised of four 2.0-GHz quad-core 64-bit AMD Barcelona CPUs sharing 64 GB of memory on the blade and coordinated by the Linux SMP operation system. The Infiniband interconnect provides a 20-Gbps transfer rate with a measured communication latency of 1.6 microseconds. Message Passing Interface MPI-1 is supported as the package MVAPICH. The Intel C/C++ compiler and the Math Kernel Library (MKL) are installed. In combination, the Linux SMP and the Intel compiler enable OpenMP execution.

Our tasks were allocated to 8 blades consisting of a total of 32 CPUs or 128 cores. The 32 CPUs were configured into a $4 \times 4 \times 2$ topology, corresponding to $4 \times 4 \times 2$ computation nodes. Consequently, this determined the spatial arrangement of the overlapping domain decomposition. The hierarchical structure of the Dawning 5000A leads to a two-level parallelization. On the top level, each CPU served as a single node, and all nodes were synchronized as MPI processes while inter-CPU data exchanges were conducted through MPI communication channels. Within each node, the MPI process spawned 4 OpenMP threads to deploy the loop-level workloads to the 4 cores. Highly efficient FFT procedures from MKL, specifically optimized for the CPU chip at hardware level, were employed. Because the performance of the MKL FFT is sensitive to the length of the data set, the number of grid cells in each sub-domain should be conformed to the characteristic factorization for speedy FFT execution.

The number of sub-domains $m = m_x \times m_y \times m_z$ was known from the above configuration of computation nodes. Additional parameters that needed to be preset were the thickness of the PML, n_{PML}; the overlapping depth, n_{ovl}; and the grid-cell dimensions $(\Delta x, \Delta y, \Delta z)$. The time-step was set as $\Delta t = T/4m_t$ where:

$$m_t = \text{ceil}\left(\frac{c_{\max} T}{8\pi} \sqrt{(\Delta x)^{-2} + (\Delta y)^{-2} + (\Delta z)^{-2}} \right) \quad (1.36)$$

In (1.36), ceil(...) is the ceiling function for real numbers, T is the time period of the electromagnetic wave, and $c_{\max}$ is the maximum speed of light in the entire model space. Because the value of the ceiling function depends on its argument, the evaluation of m_t must follow strictly the formulation of (1.36).

A Yee mesh is created in the model space including the structure of interest, the scattered-field region, and the PML absorbing outer boundary. The size of the scattered-field region is adjusted so that the overall grid dimensions $N_x \times N_y \times N_z$ allow the lengths of the sub-domain data sets to exactly satisfy the requirement of the MKL FFT. We recall from Section 1.4.1 that the first and last sub-domains lack one neighbor, so their nonoverlapping region should be n_{ovl} grid cells wider than in the other sub-domains. This leads to the local data length in each sub-domain:

$$n_{\xi,\text{MKL}} = \frac{N_\xi - 2n_{\text{ovl}}}{m_\xi} + 2n_{\text{ovl}} \qquad \xi = x, y, z \tag{1.37}$$

The choice of N_ξ and thus the size of the scattered-field region should make $n_{\xi,\text{MKL}}$ the right number for speedy execution of the MKL FFT. Now that all model parameters are fixed, the total computation region is divided into $m_x \times m_y \times m_z$ sub-domains following the overlapping domain decomposition procedures, which are one-to-one mapped to $m = m_x \times m_y \times m_z$ computation nodes. The SL-PSTD code was programmed in C and single-precision floating-point arithmetic was utilized.

As in FDTD, SL-PSTD time-stepping involves the alternate updating of electric and magnetic field vector components separated by one-half time-step. Data exchanges between neighboring sub-domains must be completed before the local FFTs can be called. According to Maxwell's curl equations, only the fields tangential to the sub-domain interfaces are subjected to overlapping. A sub-domain can be either completely embedded within or sharing exterior surface(s) with the overall computational domain. An interior sub-domain has six neighbors and requires $4n_{\text{ovl}}$ layers of interprocessor data exchange to update a single field. Surface sub-domains lack some neighbors and hence require fewer layers of data exchanges. Therefore, on a per-time-step and per-node basis, the upper bound of the overall data exchanges would be:

$$\text{Volume of data exchanges} \leq 8n_{\text{ovl}} \left(\frac{N_x N_y}{m_x m_y} + \frac{N_y N_z}{m_y m_z} + \frac{N_z N_x}{m_z m_x} \right) \tag{1.38}$$

In comparison, previous global PSTD techniques implement $12N_y(N_x - N_x/m)\cdot(N_z/m)$ data exchanges for the required double global matrix transposes. Therefore, assuming an $N \times N \times N$ lattice and an $m_x = m_y = m_z = m^{1/3}$ domain decomposition, the SL-PSTD technique gains by a factor of $N/(2n_{\text{ovl}}\, m^{1/3})$ in efficiency in terms of the volume of data exchanges. This is equal to the ratio between the widths of the nonoverlapping and overlapping regions in SL-PSTD. Moreover, because in MPI different nodes can exchange data concurrently, the SL-PSTD algorithm only requires 12 sequences of communications per time-step, whereas for previous global PSTD techniques, the all-to-all communications for the double global matrix transposes require $2(m-1)$ sequences. The extra MPI latency in initiating $2m - 14$ communications per time-step can accumulate to large delays in a slow network environment. The above two analyses indicate that the advantage of using a local Fourier basis in the SL-PSTD technique becomes significant as the size of the model (N) and the scale of the supercomputer (m) grow.

To compare the computational efficacy and scalability of the new SL-PSTD and previous global PSTD (G-PSTD) techniques, we conducted numerical experiments using both methods on two random scattering media of overall dimensions $50 \times 50 \times 50$ μm^3 and $100 \times 100 \times 100$ μm^3. The media under investigation were comprised of mono-dispersed polystyrene spheres in water suspensions. Table 1.1 lists on a per-time-step and per-node basis the average times spent on inter-node data exchange, inner-node computation, and the total over 200 time-steps [1]. (Note that, to facilitate MKL FFT computations, the grid dimensions were slightly different for the G-PSTD and SL-PSTD models, resulting in slightly different grid resolutions.)

TABLE 1.1

Performance Comparison between SL-PSTD and G-PSTD on Two Model Calculations

	Problem Volume and Computational Technique			
	$50 \times 50 \times 50\ \mu m^3$		$100 \times 100 \times 100\ \mu m^3$	
	G-PSTD	SL-PSTD	G-PSTD	SL-PSTD
Grid dimensions	512^3	$580 \times 580 \times 556$	1024^3	$1092 \times 1092 \times 1132$
Data exchange (sec)	1.31	0.38	9.49	1.46
Computation (sec)	0.98	0.91	12.93	7.37
Total (sec)	2.29	1.29	22.42	8.83
Speedup of SL-PSTD vs. G-PSTD	1.77		2.54	

Source: Ding and Chen, *Optics Express*, 2010, pp. 9236–9250, ©2010 The Optical Society of America.

From Table 1.1, we see that the inner-node computation time of SL-PSTD scales approximately linearly with the number of grid cells, increasing by a factor of $7.37 / 0.91 = 8.1$ for a 7.22 rise of the number of grid cells. This compares with a $12.93 / 0.98 = 13.2$ increase in the computation time for G-PSTD. The SL-PSTD scaling here is approximately equal to the volume change of the sub-domains.

We also see that the SL-PSTD data-exchange (communication) time scales approximately linearly with the surface area of the sub-domain interfaces, increasing by a factor of $1.46 / 0.38 = 3.84$ for a $7.22^{2/3} = 3.74$ rise of the surface area. This compares with a $9.49 / 1.31 = 7.24$ increase in the data-exchange time for G-PSTD. The SL-PSTD scaling here is based on the fixed thickness n_{ovl} of the overlapping region (i.e., the third dimension), and thus only the other two dimensions of the interface scale accordingly.

1.10 VALIDATION OF THE SL-PSTD TECHNIQUE

In this section, we verify the accuracy of the SL-PSTD technique using two analytical models: far-field scattering by a plane-wave-illuminated dielectric sphere, and far-field radiation from an electric dipole embedded within a double-layered concentric dielectric sphere. The latter example illustrates the robustness of SL-PSTD relative to the Gibbs phenomenon artifact, as discussed in Section 1.4.2.

1.10.1 Far-Field Scattering by a Plane-Wave-Illuminated Dielectric Sphere

Mie theory is the exact solution of Maxwell's equations for plane-wave scattering by a single sphere. It has been the touchstone for many model simulations. In the Mie model, a plane wave is incident upon a sphere, either lossy or lossless. The scattered electric and magnetic fields in the far field can be exactly derived as series expansions of outgoing spherical waves. Numerical methods to calculate these series to high accuracy have been well developed and tested against various combinations of sphere size, wavelength, and relative refractive index of the sphere to its surrounding medium. The predicted differential cross-section of scattering is the most important physical quality in many applications.

With the intention of setting up a Mie validation of the SL-PSTD technique, we implemented a model of plane-wave scattering by a single 2-μm-diameter polystyrene sphere surrounded by an infinite water region. The vacuum wavelength λ_0 of the incident light was assumed to be 785 nm, and the refractive indices of polystyrene and water were taken as 1.59 and 1.33, respectively. The surface of the sphere was approximated in the Cartesian SL-PSTD grid by simple staircasing using a grid-cell size of 98 nm, corresponding to 5 grid cells per dielectric wavelength λ_d in the polystyrene. A Yee mesh consisting of $108 \times 108 \times 108$ cells was established to enclose the sphere, a portion of its surrounding water medium, and the UPML absorbing outer grid boundary. The mesh was divided into $2 \times 2 \times 2$ sub-domains following the overlapping domain-decomposition procedure.

Figure 1.4 compares the differential scattering cross-section $d\sigma/d\theta$ computed by SL-PSTD with the exact Mie solution [1]. The two curves coincide on the linear scale. Their differences are only visible on the logarithmic scale in the order of 10^{-6} at ~147° scattering angle. We have observed similar levels of accuracy for SL-PSTD vs. Mie in other tests using a variety of combinations of sphere diameters and illumination wavelengths.

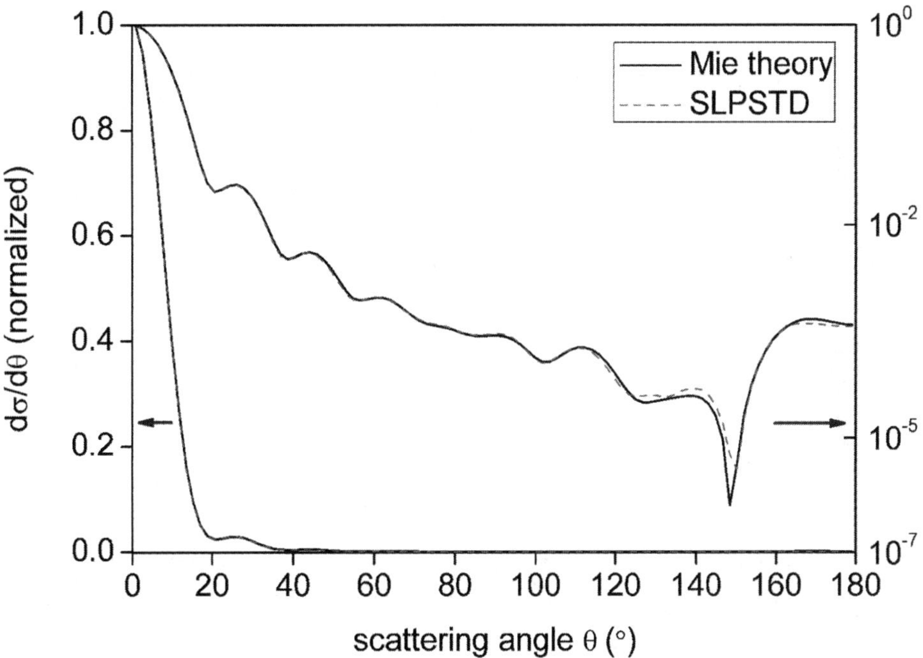

Fig. 1.4 Differential scattering cross-section $d\sigma/d\theta$ as predicted by SL-PSTD vs. the exact Mie solution, shown using a linear scale (left axis) and a logarithmic scale (right axis). *Source:* Ding and Chen, *Optics Express*, 2010, pp. 9236–9250, ©2010 The Optical Society of America.

1.10.2 Far-Field Radiation from an Electric Dipole Embedded within a Double-Layered Concentric Dielectric Sphere

In biological systems, inelastic scattering such as Raman scattering and fluorescence are re-emissions from irradiated bio-molecules. The functional groups responsible for this kind of effect are usually smaller than a few nanometers. In classical modeling, each individual emitter is treated as a point-like electric dipole oscillating at a shifted frequency [17]. The radiation fields of such a dipole embedded within a concentric sphere having an arbitrary number of layers, combinations of radii, and materials can be analytically derived.

Here, we set up a double-layered concentric dielectric sphere with appropriate parameters as a crude model for a biological cell. We assumed that a time-harmonic electric dipole was embedded inside to simulate a single Raman emitter or fluorophore. As shown in the inset of Fig. 1.5, the core layer simulated the cell nucleus of diameter 6 μm and refractive index 1.40, while the outer layer simulated the cell cytoplasm with a diameter of 10 μm and refractive index 1.37. The double-layered sphere itself was assumed to be immersed in water (refractive index 1.33). To show more features of the radiation, the point-like dipole was assumed to be elevated above the center of the core by 1.4 μm along the z-axis and polarized along the z-axis. The oscillation frequency of the dipole was set to 3.82×10^{14} Hz, and the grid-cell size was set to one-fifth the wavelength in the model nucleus. A $172 \times 172 \times 172$-cell grid was established and decomposed into $2 \times 2 \times 2$ overlapping sub-domains.

Figure 1.5 compares the angularly resolved far-field radiation intensities predicted by SL-PSTD and the analytical solution [1]. Both curves coincide with each other on the linear scale.

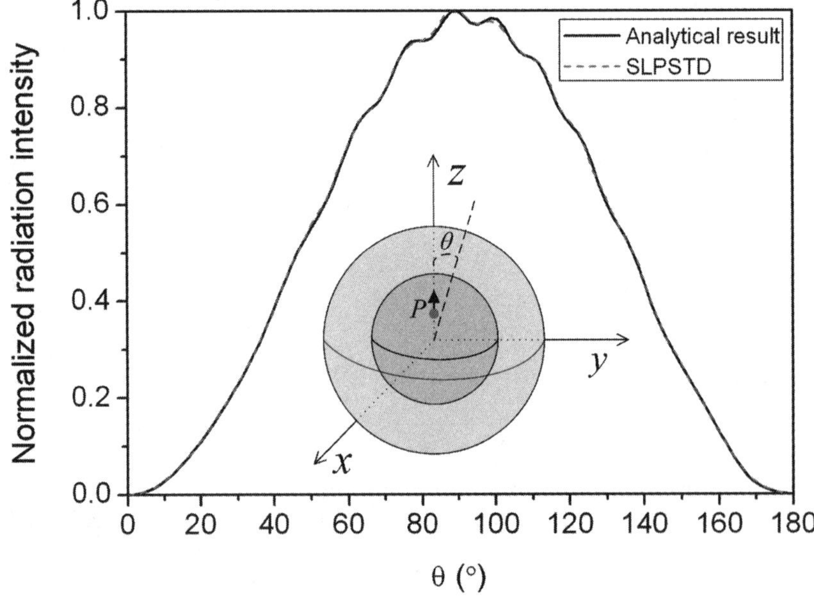

Fig. 1.5 Radiation intensities from an electric dipole embedded in a concentric dielectric sphere, as predicted by SL-PSTD and the analytical solution. The coordinate origin is at the center of the sphere and θ is the inclination angle. *Source:* Ding and Chen, *Optics Express*, 2010, pp. 9236–9250, ©2010 The Optical Society of America.

In Section 1.4.2, we pointed out that the staggered Yee grid provides a natural way to weaken global effects that originate from discontinuities of structures in the modeled system. To illustrate this important feature of the SL-PSTD technique, the concentric dielectric sphere of Fig. 1.5 was also modeled using the standard G-PSTD technique on a collocated grid with global Fourier basis. Note that in both calculations, the dipole was modeled as a true point source occupying only a single grid cell in the 3-D mesh, without resorting to the double-grid or point-smoothing tricks in the literature [13, 14].

Figure 1.6 compares a snapshot of the E_x distribution along the $z = 1.4$ μm plane of Fig. 1.5 (i.e., at the elevated dipole source) computed using both SL-PSTD and G-PSTD [1]. In Fig. 1.6(a), SL-PSTD correctly reproduces the characteristic ring pattern generated by the dipole. On the other hand, in Fig. 1.6(b), the emission field computed by G-PSTD is dominated by spurious artifacts exhibiting a chaotic pattern.

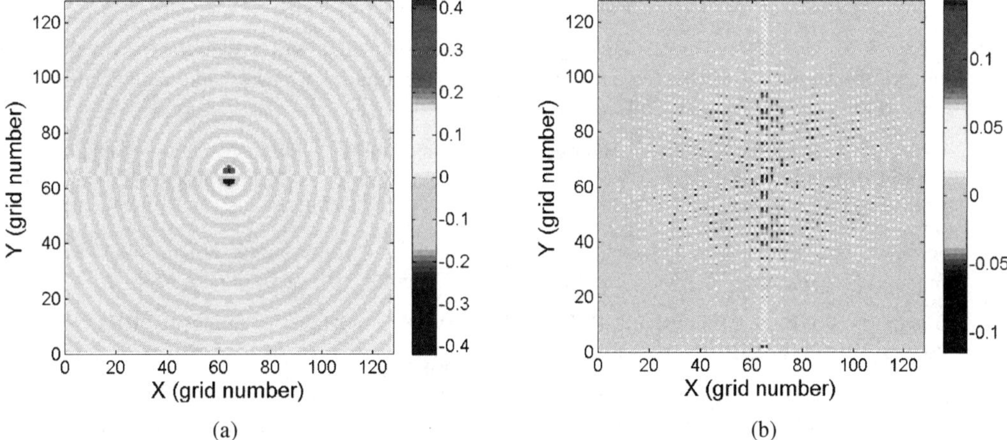

Fig. 1.6 Snapshots of the electric field E_x distribution along the $z = 1.4$ μm plane of Fig. 1.5, i.e., at the elevated dipole source: (a) SL-PSTD solution; (b) G-PSTD solution. *Source:* Ding and Chen, *Optics Express*, 2010, pp. 9236–9250, ©2010 The Optical Society of America.

1.11 SUMMARY

This chapter discussed a new staggered-grid, local-Fourier-basis PSTD (SL-PSTD) technique that is suitable for the efficient computational solution of the full-vector Maxwell's equations over electrically large open-region 3-D domains. The new formulation of PSTD scales more efficiently with the size of the computational domain than previous collocated-grid PSTD approaches, and importantly avoids the Gibbs phenomenon artifact. The complete 3-D SL-PSTD algorithm was presented, including its implementation for a UPML absorbing outer grid boundary. In addition, the implementation of the SL-PSTD algorithm on a distributed-memory parallel supercomputer was described, and its performance evaluated. Finally, the numerical accuracy of the new algorithm was validated by comparison with the results of two analytical models.

REFERENCES

[1] Ding, M., and K. Chen, "Staggered-grid PSTD on local Fourier basis and its applications to surface tissue modeling," *Optics Express*, Vol. 18, 2010, pp. 9236–9250.

[2] Taflove, A., and S. C. Hagness, *Computational Electrodynamics: The Finite-Difference Time-Domain Method*, 3rd ed., Norwood, MA: Artech House, 2005.

[3] Liu, Q. H., "The PSTD algorithm: A time-domain method requiring only two cells per wavelength," *Microwave and Optical Technology Lett.*, Vol. 15, 1997, pp. 158–165.

[4] Tseng, S. H., J. H. Greene, A. Taflove, D. Maitland, V. Backman, and J. T. Walsh, Jr., "Exact solution of Maxwell's equations for optical interactions with a macroscopic random medium," *Optics Lett.*, Vol. 29, 2004, pp. 1393–1395.

[5] Tseng, S. H., Y. L. Kim, A. Taflove, D. Maitland, V. Backman, and J. T. Walsh, Jr., "Simulation of enhanced backscattering of light by numerically solving Maxwell's equations without heuristic approximations," *Optics Express*, Vol. 13, 2005, pp. 3666–3672.

[6] Tseng, S. H., and B. Huang, "Comparing Monte Carlo simulation and pseudospectral time-domain numerical solutions of Maxwell's equations of light scattering by a macroscopic random medium," *Applied Physics Lett.*, Vol. 91, 2007, Article no. 051114.

[7] Witte, D. C., and P. G. Richards, "The pseudospectral method for simulating wave propagation," pp. 1–18 in *Computational Acoustics, Vol. 3*, D. Lee, A. Cakmak, and R. Vichnevetsky, eds., New York: North-Holland, 1990.

[8] Fornberg, B., "Introduction to pseudospectral method via finite differences," Chap. 3 in *A Practical Guide to Pseudospectral Methods*, Cambridge, U.K.: Cambridge University Press, 1996.

[9] Gomez, D. O., P. D. Mininni, and P. Dmitruk, "Parallel simulations in turbulent MHD," *Physica Scripta*, Vol. T116, 2005, pp. 123–127.

[10] Mininni, P. D., D. Rosenberg, R. Reddy, and A. Pouquet, "A hybrid MPI-OpenMP scheme for scalable parallel pseudospectral computations for fluid turbulence," *Parallel Computing*, Vol. 37, 2011, pp. 316–326.

[11] Israeli, M., L. Vozovoi, and A. Averbuch, "Spectral multidomain technique with local Fourier basis," *J. Scientific Computing*, Vol. 8, 1993, pp. 135–149.

[12] Liao, Q. B., and G. A. McMechan, "2-D pseudo-spectral viscoacoustic modeling in a distributed-memory multi-processor computer," *Bull. Seismological Society of America*, Vol. 83, 1993, pp. 1345–1354.

[13] Lee, T. W., and S. C. Hagness, "A compact wave source condition for the pseudospectral time-domain method," *IEEE Antennas and Wireless Propagation Lett.*, Vol. 3, 2004, pp. 253–256.

[14] Liu, Q. H., "Large-scale simulations of electromagnetic and acoustic measurements using the pseudospectral time-domain (PSTD) algorithm," *IEEE Trans. Geoscience and Remote Sensing*, Vol. 37, 1999, pp. 917–926.

[15] Correa, G. J. P., M. Spiegelman, S. Carbotte, and J. C. Mutter, "Centered and staggered Fourier derivatives and Hilbert transforms," *Geophysics*, Vol. 67, 2002, pp. 1558–1563.

[16] Leung, Y. F., and C. H. Chan, "Combining the FDTD and PSTD methods," *Microwave and Optical Technology Lett.*, Vol. 23, 1999, pp. 249–254.

[17] Chew, H., M. Kerker, and P. J. McNulty, "Raman and fluorescent scattering by molecules embedded in concentric spheres," *J. Optical Society of America*, Vol. 66, 1976, pp. 440–444.

Chapter 2

Unconditionally Stable Laguerre Polynomial-Based FDTD Method[1]

Bin Chen, Yantao Duan, and Hailin Chen

2.1 INTRODUCTION

The FDTD method plays an important role in the solution of electromagnetic problems. However, since it is an explicit time-marching technique, its time-step is constrained by the Courant–Friedrich–Levy condition [1]. If the time-step is larger than the Courant limit, the method becomes unstable. Thus, for solving problems with fine details, the time-step must be very small, thereby significantly increasing the computer time. To overcome the Courant limit on the time-step size, an unconditionally stable FDTD method based on the alternating-direction implicit (ADI) technique has been developed [2–4]. The ADI-FDTD method consumes less computer time than does the original FDTD method for solving problems with fine structures. However, it is found that the numerical dispersion error increases markedly for time-steps larger than the Courant stability limit for the maximum-sized grid cells in the model [5]. While the locally-one-dimensional FDTD technique [6, 7] has better computational efficiency than ADI-FDTD, both methods provide comparable error performance.

Reference [8] introduced an alternative, highly accurate, unconditionally stable FDTD scheme that uses weighted Laguerre polynomials as temporal basis functions, and Galerkin's method as the temporal testing procedure to eliminate the time variable. With this "marching-on-in-order" scheme, numerical stability is no longer affected by the time-step. The time-step is used only for calculation of the Laguerre coefficients of the current-source excitation at the pre-processing stage, and for field reconstruction at the post-processing stage. The stability analysis of this order-marching scheme was presented in [9].

Subsequent publications reported incorporation into this scheme of Mur's second-order absorbing boundary condition (ABC) [10] and the perfectly matched layer (PML) ABC [11–13]. Alighanbari *et al.* introduced a time-domain method that combines the scaling function-based multiresolution time-domain technique with the Laguerre polynomial-based time-integration scheme [14]. This approach leads to a reduction in the size of the produced sparse matrix. Shao *et al.* introduced a hybrid time-domain method based on the compact two-dimensional (2-D) FDTD method combined with weighted Laguerre polynomials to analyze the propagation

---

[1]This chapter is adapted from Ref. [18], Y. T. Duan, B. Chen, D. G. Fang, and B. H. Zhou, "Efficient implementation for 3-D Laguerre-based finite-difference time-domain method," *IEEE Trans. Microwave Theory and Techniques*, Vol. 59, 2011, pp. 56–64, ©2011 IEEE.

properties of uniform transmission lines [15]. Subsequently, Chen *et al.* proposed an unconditionally stable Laguerre-based body-of-revolution FDTD method for analyzing structures with circular symmetry [16].

However, these previously reported marching-on-in-order schemes lead to very large sparse matrix equations. Direct solutions of these matrix equations can be challenging, especially for three-dimensional (3-D) models, and are not applicable for many practical problems.

To overcome this difficulty, novel efficient algorithms for implementing 2-D and 3-D unconditionally stable Laguerre-based FDTD techniques were recently reported [17, 18]. This chapter provides the theory and computational simulation results of [18], which advanced and extended the work of [17] to 3-D models and reported the incorporation of the PML ABC. For a full update cycle, the algorithm of [18] solves six tri-diagonal matrices and computes three explicit equations. While preserving accuracy, this leads to greatly reduced computer memory and running-time requirements compared with previous Laguerre-based FDTD implementations.

2.2 FORMULATION OF THE CONVENTIONAL 3-D LAGUERRE-BASED FDTD METHOD

For consistency of notation and completeness, we will first detail the formulation of the conventional 3-D Laguerre-based FDTD method. For simplicity, assuming a linear, isotropic, non-dispersive, and lossless medium, the 3-D differential Maxwell's equations can be written as:

$$\frac{\partial E_x(\boldsymbol{r},t)}{\partial t} = \frac{1}{\varepsilon}\left[D_y H_z(\boldsymbol{r},t) - D_z H_y(\boldsymbol{r},t)\right] - \frac{J_x(\boldsymbol{r},t)}{\varepsilon} \tag{2.1}$$

$$\frac{\partial E_y(\boldsymbol{r},t)}{\partial t} = \frac{1}{\varepsilon}\left[D_z H_x(\boldsymbol{r},t) - D_x H_z(\boldsymbol{r},t)\right] - \frac{J_y(\boldsymbol{r},t)}{\varepsilon} \tag{2.2}$$

$$\frac{\partial E_z(\boldsymbol{r},t)}{\partial t} = \frac{1}{\varepsilon}\left[D_x H_y(\boldsymbol{r},t) - D_y H_x(\boldsymbol{r},t)\right] - \frac{J_z(\boldsymbol{r},t)}{\varepsilon} \tag{2.3}$$

$$\frac{\partial H_x(\boldsymbol{r},t)}{\partial t} = \frac{1}{\mu}\left[D_z E_y(\boldsymbol{r},t) - D_y E_z(\boldsymbol{r},t)\right] \tag{2.4}$$

$$\frac{\partial H_y(\boldsymbol{r},t)}{\partial t} = \frac{1}{\mu}\left[D_x E_z(\boldsymbol{r},t) - D_z E_x(\boldsymbol{r},t)\right] \tag{2.5}$$

$$\frac{\partial H_z(\boldsymbol{r},t)}{\partial t} = \frac{1}{\mu}\left[D_y E_x(\boldsymbol{r},t) - D_x E_y(\boldsymbol{r},t)\right] \tag{2.6}$$

where ε is the electrical permittivity; μ is the magnetic permeability; D_x, D_y, and D_z are the difference operators for the first derivatives along the x-, y-, and z-axes; and $J_x(\boldsymbol{r}, t)$, $J_y(\boldsymbol{r}, t)$, and $J_z(\boldsymbol{r}, t)$ are the excitation sources along the x-, y-, and z-directions, respectively.

The Laguerre polynomials of order p are defined by:

$$L_p(t) = \frac{e^t}{p!}\frac{d^p}{dt^p}(t^p e^{-t}), \qquad p \geq 0;\ t \geq 0 \tag{2.7}$$

Using the orthogonality of Laguerre polynomials with respect to the weighting function e^{-t}, a set of orthogonal basis functions can be constructed, given by:

$$\varphi_p(s,t) = e^{-st/2} L_p(st) \tag{2.8}$$

where $s > 0$ is a time-scale factor. These basis functions are absolutely convergent to zero as $t \to \infty$, and are also orthogonal with respect to st as:

$$\int_0^\infty \varphi_p(st)\,\varphi_q(st)\,d(st) = \begin{cases} 1 & p = q \\ 0 & p \neq q \end{cases} \tag{2.9}$$

Using these basis functions, we can expand the electric and magnetic fields in (2.1) – (2.6) as:

$$E_x(\boldsymbol{r},t) = \sum_{p=0}^{\infty} E_x^p(\boldsymbol{r})\,\varphi_p(st) \tag{2.10a}$$

$$E_y(\boldsymbol{r},t) = \sum_{p=0}^{\infty} E_y^p(\boldsymbol{r})\,\varphi_p(st) \tag{2.10b}$$

$$E_z(\boldsymbol{r},t) = \sum_{p=0}^{\infty} E_z^p(\boldsymbol{r})\,\varphi_p(st) \tag{2.10c}$$

$$H_x(\boldsymbol{r},t) = \sum_{p=0}^{\infty} H_x^p(\boldsymbol{r})\,\varphi_p(st) \tag{2.10d}$$

$$H_y(\boldsymbol{r},t) = \sum_{p=0}^{\infty} H_y^p(\boldsymbol{r})\,\varphi_p(st) \tag{2.10e}$$

$$H_z(\boldsymbol{r},t) = \sum_{p=0}^{\infty} H_z^p(\boldsymbol{r})\,\varphi_p(st) \tag{2.10f}$$

Taking $E_x(\boldsymbol{r}, t)$ as an example, the first derivative of a field component with respect to t is given by [19]:

$$\frac{\partial E_x(\boldsymbol{r},t)}{\partial t} = s\sum_{p=0}^{\infty}\left(0.5E_x^p(\boldsymbol{r}) + \sum_{k=0,\,p>0}^{p-1} E_x^k(\boldsymbol{r})\right)\varphi_p(st) \tag{2.11}$$

Inserting (2.10) into (2.1) – (2.6), respectively, we have:

$$s\sum_{p=0}^{\infty}\left[0.5E_x^p(\boldsymbol{r}) + \sum_{k=0,\,p>0}^{p-1} E_x^k(\boldsymbol{r})\right]\varphi_p(st) = \frac{1}{\varepsilon}\left[D_y\sum_{p=0}^{\infty}H_z^p(\boldsymbol{r})\,\varphi_p(st) - D_z\sum_{p=0}^{\infty}H_y^p(\boldsymbol{r})\,\varphi_p(st)\right] - \frac{J_x(\boldsymbol{r},t)}{\varepsilon} \tag{2.12}$$

$$s\sum_{p=0}^{\infty}\left[0.5E_y^p(\boldsymbol{r}) + \sum_{k=0,\,p>0}^{p-1} E_y^k(\boldsymbol{r})\right]\varphi_p(st) = \frac{1}{\varepsilon}\left[D_z\sum_{p=0}^{\infty}H_x^p(\boldsymbol{r})\,\varphi_p(st) - D_x\sum_{p=0}^{\infty}H_z^p(\boldsymbol{r})\,\varphi_p(st)\right] - \frac{J_y(\boldsymbol{r},t)}{\varepsilon} \tag{2.13}$$

$$s\sum_{p=0}^{\infty}\left[0.5E_z^p(\boldsymbol{r}) + \sum_{k=0,\,p>0}^{p-1} E_y^k(\boldsymbol{r})\right]\varphi_p(st) = \frac{1}{\varepsilon}\left[D_x\sum_{p=0}^{\infty}H_y^p(\boldsymbol{r})\,\varphi_p(st) - D_y\sum_{p=0}^{\infty}H_x^p(\boldsymbol{r})\,\varphi_p(st)\right] - \frac{J_z(\boldsymbol{r},t)}{\varepsilon} \tag{2.14}$$

$$s\sum_{p=0}^{\infty}\left[0.5H_x^p(\boldsymbol{r}) + \sum_{k=0,\,p>0}^{p-1} H_x^k(\boldsymbol{r})\right]\varphi_p(st) = \frac{1}{\mu}\left[D_z\sum_{p=0}^{\infty}E_y^p(\boldsymbol{r})\,\varphi_p(st) - D_y\sum_{p=0}^{\infty}E_z^p(\boldsymbol{r})\,\varphi_p(st)\right] \tag{2.15}$$

$$s\sum_{p=0}^{\infty}\left[0.5H_y^p(\boldsymbol{r}) + \sum_{k=0,\,p>0}^{p-1} H_y^k(\boldsymbol{r})\right]\varphi_p(st) = \frac{1}{\mu}\left[D_x\sum_{p=0}^{\infty}E_z^p(\boldsymbol{r})\,\varphi_p(st) - D_z\sum_{p=0}^{\infty}E_x^p(\boldsymbol{r})\,\varphi_p(st)\right] \tag{2.16}$$

$$s\sum_{p=0}^{\infty}\left[0.5H_z^p(\boldsymbol{r}) + \sum_{k=0,\,p>0}^{p-1} H_z^k(\boldsymbol{r})\right]\varphi_p(st) = \frac{1}{\mu}\left[D_y\sum_{p=0}^{\infty}E_x^p(\boldsymbol{r})\,\varphi_p(st) - D_x\sum_{p=0}^{\infty}E_y^p(\boldsymbol{r})\,\varphi_p(st)\right] \tag{2.17}$$

By using the orthogonality property of the weighted Laguerre functions, we introduce a temporal Galerkin's testing procedure to eliminate the time-dependent terms $\varphi_p(st)$. We multiply both sides of (2.12) – (2.17) by $\varphi_q(st)$ and integrate over $t = [0, +\infty)$. This yields:

$$s\left[0.5E_x^q(\boldsymbol{r}) + \sum_{k=0,q>0}^{q-1} E_x^k(\boldsymbol{r})\right] = \frac{1}{\varepsilon}D_y H_z^q(\boldsymbol{r}) - \frac{1}{\varepsilon}D_z H_y^q(\boldsymbol{r}) - \frac{J_x^q(\boldsymbol{r})}{\varepsilon} \tag{2.18}$$

$$s\left[0.5E_y^q(\boldsymbol{r}) + \sum_{k=0,q>0}^{q-1} E_y^k(\boldsymbol{r})\right] = \frac{1}{\varepsilon}D_z H_x^q(\boldsymbol{r}) - \frac{1}{\varepsilon}D_x H_z^q(\boldsymbol{r}) - \frac{J_y^q(\boldsymbol{r})}{\varepsilon} \tag{2.19}$$

$$s\left[0.5E_z^q(\boldsymbol{r}) + \sum_{k=0,q>0}^{q-1} E_z^k(\boldsymbol{r})\right] = \frac{1}{\varepsilon}D_x H_y^q(\boldsymbol{r}) - \frac{1}{\varepsilon}D_y H_x^q(\boldsymbol{r}) - \frac{J_z^q(\boldsymbol{r})}{\varepsilon} \tag{2.20}$$

$$s\left[0.5H_x^q(\boldsymbol{r}) + \sum_{k=0,q>0}^{q-1} H_x^k(\boldsymbol{r})\right] = \frac{1}{\mu}\left[D_z E_y^q(\boldsymbol{r}) - D_y E_z^q(\boldsymbol{r})\right] \tag{2.21}$$

$$s\left[0.5H_y^q(\boldsymbol{r}) + \sum_{k=0,q>0}^{q-1} H_y^k(\boldsymbol{r})\right] = \frac{1}{\mu}\left[D_x E_z^q(\boldsymbol{r}) - D_z E_x^q(\boldsymbol{r})\right] \tag{2.22}$$

$$s\left[0.5H_z^q(\boldsymbol{r}) + \sum_{k=0,q>0}^{q-1} H_z^k(\boldsymbol{r})\right] = \frac{1}{\mu}\left[D_y E_x^q(\boldsymbol{r}) - D_x E_y^q(\boldsymbol{r})\right] \tag{2.23}$$

where

$$J_x^q(\boldsymbol{r}) = \int_0^{T_f} J_x(\boldsymbol{r},t)\,\varphi_q(st)\,d(st) \tag{2.24}$$

$$J_y^q(\boldsymbol{r}) = \int_0^{T_f} J_y(\boldsymbol{r},t)\,\varphi_q(st)\,d(st) \tag{2.25}$$

$$J_z^q(\boldsymbol{r}) = \int_0^{T_f} J_z(\boldsymbol{r},t)\,\varphi_q(st)\,d(st) \tag{2.26}$$

The time span T_f is chosen in such a way that the waveforms of interest have practically decayed to zero [8].

Now, for convenience, we define a set of auxiliary matrices as:

$$W_E^q = \left[E_x^q(\boldsymbol{r}) \quad E_y^q(\boldsymbol{r}) \quad E_z^q(\boldsymbol{r})\right]^T \tag{2.27}$$

$$W_H^q = \left[H_x^q(\boldsymbol{r}) \quad H_y^q(\boldsymbol{r}) \quad H_z^q(\boldsymbol{r})\right]^T \tag{2.28}$$

$$D_H = \left[D_E\right]^T = \begin{bmatrix} 0 & -D_z & D_y \\ D_z & 0 & -D_x \\ -D_y & D_x & 0 \end{bmatrix} \tag{2.29}$$

$$V_E^{q-1} = \left[-2\sum_{k=0,q>0}^{q-1} E_x^k(\mathbf{r}) \quad -2\sum_{k=0,q>0}^{q-1} E_y^k(\mathbf{r}) \quad -2\sum_{k=0,q>0}^{q-1} E_z^k(\mathbf{r}) \right]^T \tag{2.30}$$

$$V_H^{q-1} = \left[-2\sum_{k=0,q>0}^{q-1} H_x^k(\mathbf{r}) \quad -2\sum_{k=0,q>0}^{q-1} H_y^k(\mathbf{r}) \quad -2\sum_{k=0,q>0}^{q-1} H_z^k(\mathbf{r}) \right]^T \tag{2.31}$$

$$J_E^q = \left[-J_x^q(\mathbf{r}) \quad -J_y^q(\mathbf{r}) \quad -J_z^q(\mathbf{r}) \right]^T \tag{2.32}$$

With some manipulations, (2.18) – (2.20) and (2.21) – (2.23) can be written in the following matrix forms, respectively:

$$W_E^q = aD_H W_H^q + V_E^{q-1} + aJ_E^q \tag{2.33}$$

$$W_H^q = bD_E W_E^q + V_H^{q-1} \tag{2.34}$$

where $a = 2/(s\varepsilon)$ and $b = 2/(s\mu)$.

Finally, inserting (2.34) into (2.33), we obtain the sparse matrix equation for the conventional 3-D Laguerre-based FDTD method:

$$\left(I - abD_H D_E\right)W_E^q = aD_H V_H^{q-1} + V_E^{q-1} + aJ_E^q \tag{2.35}$$

where I is the 3×3 identity matrix. After solving (2.35) for order zero ($q = 0$), one can solve (2.35) for a higher order of q by repeatedly applying back-substitution, ultimately stopping at a q value chosen according to the analytical formulas given in [8]. The expansion coefficients of the magnetic fields can be obtained from (2.21) – (2.23). Then, we can reconstruct the field components in the time domain with (2.10). We again note that the time-step in this Laguerre-based FDTD method is used *only* for calculating the Laguerre coefficients of the current-source excitation in (2.24) – (2.26) at the preprocessing stage, and for field reconstruction at the post-processing stage. The time-step need only be small enough to resolve the temporal variations of the weighted Laguerre polynomials of order q, with stability assured according to [9].

Importantly, we note that $(I - abD_H D_E)$ is, in general, a very large, sparse, irreducible matrix which is independent of the order q. This causes the system of (2.35) to be very expensive to solve. Hence, the conventional 3-D Laguerre-based FDTD method is impractical for many problems of interest. This difficulty motivates the following section, where we present an efficient algorithm for implementing the 3-D Laguerre-based FDTD method.

2.3 FORMULATION OF AN EFFICIENT 3-D LAGUERRE-BASED FDTD METHOD

To solve (2.35) efficiently, we decompose $abD_H D_E$ into the triangular matrices A and B:

$$A = ab\begin{bmatrix} D_{2y} & 0 & 0 \\ -D_x D_y & D_{2z} & 0 \\ -D_x D_z & -D_y D_z & D_{2x} \end{bmatrix} \tag{2.36}$$

$$B = ab\begin{bmatrix} D_{2z} & -D_yD_x & -D_zD_x \\ 0 & D_{2x} & -D_zD_y \\ 0 & 0 & D_{2y} \end{bmatrix} \tag{2.37}$$

where A is a lower triangular matrix and B is an upper triangular matrix. D_{2x}, D_{2y}, and D_{2z} are the difference operators for the second derivatives. Then, (2.35) can be written as:

$$\left(I - A - B\right)W_E^q = a\,D_H V_H^{q-1} + V_E^{q-1} + a\,J_E^q \tag{2.38}$$

Upon adding a perturbation term $AB(W_E^q - V_E^{q-1})$ to (2.38), we obtain the factorized form of (2.38) as:

$$\left(I - A\right)\left(I - B\right)W_E^q = \left(I + AB\right)V_E^{q-1} + aD_H V_H^{q-1} + a\,J_E^q \tag{2.39}$$

Equation (2.39) can be solved into two sub-steps with the following splitting scheme [20–22]:

$$\left(I - A\right)W^* = \left(I + B\right)V_E^{q-1} + a\,D_H V_H^{q-1} + a\,J_E^q \tag{2.40}$$

$$\left(I - B\right)W_E^q = W^* - BV_E^{q-1} \tag{2.41}$$

where $W^* = [E_x^{*q}(\boldsymbol{r}) \;\; E_y^{*q}(\boldsymbol{r}) \;\; E_z^{*q}(\boldsymbol{r})]^T$ is a nonphysical intermediate value. Using the splitting scheme (2.40) and (2.41) to solve (2.36) – (2.38) leads to:

$$\begin{aligned}\left(1 - abD_{2y}\right)E_x^{*q}(\boldsymbol{r}) = {} & 2a\sum_{k=0,q>0}^{q-1} D_zH_y^k(\boldsymbol{r}) - 2a\sum_{k=0,q>0}^{q-1} D_yH_z^k(\boldsymbol{r}) \\ & - 2\sum_{k=0,q>0}^{q-1} E_x^k(\boldsymbol{r}) - aJ_x^q(\boldsymbol{r}) - 2abD_{2z}\sum_{k=0,q>0}^{q-1} E_x^k(\boldsymbol{r}) \\ & + 2abD_yD_x\sum_{k=0,q>0}^{q-1} E_y^k(\boldsymbol{r}) + 2abD_zD_x\sum_{k=0,q>0}^{q-1} E_z^k(\boldsymbol{r})\end{aligned} \tag{2.42a}$$

$$\begin{aligned}\left(1 - abD_{2z}\right)E_y^{*q}(\boldsymbol{r}) = {} & -abD_xD_yE_x^{*q}(\boldsymbol{r}) - 2a\sum_{k=0,q>0}^{q-1} D_zH_x^k(\boldsymbol{r}) \\ & + 2a\sum_{k=0,q>0}^{q-1} D_xH_z^k(\boldsymbol{r}) - 2\sum_{k=0,q>0}^{q-1} E_y^k(\boldsymbol{r}) - aJ_y^q(\boldsymbol{r}) \\ & - 2abD_{2x}\sum_{k=0,q>0}^{q-1} E_y^k(\boldsymbol{r}) + 2abD_zD_y\sum_{k=0,q>0}^{q-1} E_z^k(\boldsymbol{r})\end{aligned} \tag{2.42b}$$

$$\begin{aligned}\left(1 - abD_{2x}\right)E_z^{*q}(\boldsymbol{r}) = {} & -abD_xD_zE_x^{*q}(\boldsymbol{r}) - abD_yD_zE_y^{*q}(\boldsymbol{r}) \\ & + 2a\sum_{k=0,q>0}^{q-1} D_yH_x^k(\boldsymbol{r}) - 2a\sum_{k=0,q>0}^{q-1} D_xH_y^k(\boldsymbol{r}) \\ & - 2\sum_{k=0,q>0}^{q-1} E_z^k(\boldsymbol{r}) - a\,J_z^q(\boldsymbol{r}) - 2abD_{2y}\sum_{k=0,q>0}^{q-1} E_z^k(\boldsymbol{r})\end{aligned} \tag{2.42c}$$

$$\left(1-abD_{2y}\right)E_z^q(\boldsymbol{r}) = E_z^{*q}(\boldsymbol{r}) + 2abD_{2y}\sum_{k=0,q>0}^{q-1} E_z^k(\boldsymbol{r}) \tag{2.42d}$$

$$\begin{aligned}\left(1-abD_{2x}\right)E_y^q(\boldsymbol{r}) &= -abD_zD_yE_z^q(\boldsymbol{r}) + E_y^{*q}(\boldsymbol{r}) \\ &\quad + 2abD_{2x}\sum_{k=0,q>0}^{q-1} E_y^k(\boldsymbol{r}) - 2abD_zD_y\sum_{k=0,q>0}^{q-1} E_z^k(\boldsymbol{r})\end{aligned} \tag{2.42e}$$

$$\begin{aligned}\left(1-abD_{2z}\right)E_x^q(\boldsymbol{r}) &= -abD_yD_xE_y^q(\boldsymbol{r}) \\ &\quad - abD_zD_xE_z^q(\boldsymbol{r}) + E_x^{*q}(\boldsymbol{r}) + 2abD_{2z}\sum_{k=0,q>0}^{q-1} E_x^k(\boldsymbol{r}) \\ &\quad - 2abD_yD_x\sum_{k=0,q>0}^{q-1} E_y^k(\boldsymbol{r}) - 2abD_zD_x\sum_{k=0,q>0}^{q-1} E_z^k(\boldsymbol{r})\end{aligned} \tag{2.42f}$$

Applying spatial central differences to (2.42) with uniform grid cells Δx, Δy, and Δz, we obtain the following discrete space equations for the efficient 3-D Laguerre-based FDTD method:

$$\begin{aligned}-\frac{ab}{\Delta y^2}E_x^{*q}\Big|_{i,j-1,k} + \left(1+\frac{2ab}{\Delta y^2}\right)E_x^{*q}\Big|_{i,j,k} - \frac{ab}{\Delta y^2}E_x^{*q}\Big|_{i,j+1,k} &= \frac{2a}{\Delta z}\sum_{k=0,q>0}^{q-1}\left(H_y^k\Big|_{i,j,k} - H_y^k\Big|_{i,j,k-1}\right) \\ &\quad - 2\sum_{k=0,q>0}^{q-1} E_x^k\Big|_{i,j,k} - \frac{2a}{\Delta y}\sum_{k=0,q>0}^{q-1}\left(H_z^k\Big|_{i,j,k} - H_z^k\Big|_{i,j-1,k}\right) - aJ_x^q\Big|_{i,j,k} \\ &\quad - \frac{2ab}{\Delta z^2}\sum_{k=0,q>0}^{q-1}\left(E_x^k\Big|_{i,j,k+1} + E_x^k\Big|_{i,j,k-1} - 2E_x^k\Big|_{i,j,k}\right) \\ &\quad + \frac{2ab}{\Delta x\Delta y}\sum_{k=0,q>0}^{q-1}\left(E_y^k\Big|_{i+1,j,k} - E_y^k\Big|_{i+1,j-1,k} - E_y^k\Big|_{i,j,k} + E_y^k\Big|_{i,j-1,k}\right) \\ &\quad + \frac{2ab}{\Delta x\Delta z}\sum_{k=0,q>0}^{q-1}\left(E_z^k\Big|_{i+1,j,k} - E_z^k\Big|_{i+1,j,k-1} - E_z^k\Big|_{i,j,k} + E_z^k\Big|_{i,j,k-1}\right)\end{aligned} \tag{2.43a}$$

$$\begin{aligned}-\frac{ab}{\Delta z^2}E_y^{*q}\Big|_{i,j,k-1} + \left(1+\frac{2ab}{\Delta z^2}\right)E_y^{*q}\Big|_{i,j,k} - \frac{ab}{\Delta z^2}E_y^{*q}\Big|_{i,j,k+1} &= \\ &- \frac{ab}{\Delta x\Delta y}\left(E_x^{*q}\Big|_{i,j+1,k} - E_x^{*q}\Big|_{i-1,j+1,k} - E_x^{*q}\Big|_{i,j,k} + E_x^{*q}\Big|_{i-1,j,k}\right) \\ &- \frac{2a}{\Delta z}\sum_{k=0,q>0}^{q-1}\left(H_x^k\Big|_{i,j,k} - H_x^k\Big|_{i,j,k-1}\right) + \frac{2a}{\Delta x}\sum_{k=0,q>0}^{q-1}\left(H_z^k\Big|_{i,j,k} - H_z^k\Big|_{i-1,j,k}\right) \\ &- \frac{2ab}{\Delta x^2}\sum_{k=0,q>0}^{q-1}\left(E_y^k\Big|_{i+1,j,k} + E_y^k\Big|_{i-1,j,k} - 2E_y^k\Big|_{i,j,k}\right) - 2\sum_{k=0,q>0}^{q-1} E_y^k\Big|_{i,j,k} - aJ_y^q\Big|_{i,j,k} \\ &+ \frac{2ab}{\Delta y\Delta z}\sum_{k=0,q>0}^{q-1}\left(E_z^k\Big|_{i,j+1,k} - E_z^k\Big|_{i,j+1,k-1} - E_z^k\Big|_{i,j,k} + E_z^k\Big|_{i,j,k-1}\right)\end{aligned} \tag{2.43b}$$

$$
\begin{aligned}
-\frac{ab}{\Delta x^2} E_z^{*q}\Big|_{i-1,j,k} + \left(1+\frac{2ab}{\Delta x^2}\right) E_z^{*q}\Big|_{i,j,k} - \frac{ab}{\Delta x^2} E_z^{*q}\Big|_{i+1,j,k} &= -aJ_z^q\Big|_{i,j,k} \\
&- \frac{ab}{\Delta x \Delta z}\left(E_x^{*q}\Big|_{i,j,k+1} - E_x^{*q}\Big|_{i-1,j,k+1} - E_x^{*q}\Big|_{i,j,k} + E_x^{*q}\Big|_{i-1,j,k}\right) \\
&- \frac{ab}{\Delta y \Delta z}\left(E_y^{*q}\Big|_{i,j,k+1} - E_y^{*q}\Big|_{i,j-1,k+1} - E_y^{*q}\Big|_{i,j,k} + E_y^{*q}\Big|_{i,j-1,k}\right) \\
&+ \frac{2a}{\Delta y}\sum_{k=0,q>0}^{q-1}\left(H_x^k\Big|_{i,j,k} - H_x^k\Big|_{i,j-1,k}\right) - \frac{2a}{\Delta x}\sum_{k=0,q>0}^{q-1}\left(H_y^k\Big|_{i,j,k} - H_y^k\Big|_{i-1,j,k}\right) \\
&- \frac{2ab}{\Delta y^2}\sum_{k=0,q>0}^{q-1}\left(E_z^k\Big|_{i,j+1,k} + E_z^k\Big|_{i,j-1,k} - 2E_z^k\Big|_{i,j,k}\right) - 2\sum_{k=0,q>0}^{q-1} E_z^k\Big|_{i,j,k}
\end{aligned}
\tag{2.43c}
$$

$$
\begin{aligned}
-\frac{ab}{\Delta y^2} E_z^q\Big|_{i,j-1,k} + \left(1+\frac{2ab}{\Delta y^2}\right) E_z^q\Big|_{i,j,k} - \frac{ab}{\Delta y^2} E_z^q\Big|_{i,j+1,k} &= E_z^{*q}\Big|_{i,j,k} \\
&+ \frac{2ab}{\Delta y^2}\sum_{k=0,q>0}^{q-1}\left(E_z^k\Big|_{i,j+1,k} + E_z^k\Big|_{i,j-1,k} - 2E_z^k\Big|_{i,j,k}\right)
\end{aligned}
\tag{2.43d}
$$

$$
\begin{aligned}
-\frac{ab}{\Delta x^2} E_y^q\Big|_{i-1,j,k} + \left(1+\frac{2ab}{\Delta x^2}\right) E_y^q\Big|_{i,j,k} - \frac{ab}{\Delta x^2} E_y^q\Big|_{i+1,j,k} &= \\
&- \frac{ab}{\Delta y \Delta z}\left(E_z^q\Big|_{i,j+1,k} - E_z^q\Big|_{i,j+1,k-1} - E_z^q\Big|_{i,j,k} + E_z^q\Big|_{i,j,k-1}\right) \\
&+ \frac{2ab}{\Delta x^2}\sum_{k=0,q>0}^{q-1}\left(E_y^k\Big|_{i+1,j,k} + E_y^k\Big|_{i-1,j,k} - 2E_y^k\Big|_{i,j,k}\right) + E_y^{*q}\Big|_{i,j,k} \\
&- \frac{2ab}{\Delta y \Delta z}\sum_{k=0,q>0}^{q-1}\left(E_z^k\Big|_{i,j+1,k} - E_z^k\Big|_{i,j+1,k-1} - E_z^k\Big|_{i,j,k} + E_z^k\Big|_{i,j,k-1}\right)
\end{aligned}
\tag{2.43e}
$$

$$
\begin{aligned}
-\frac{ab}{\Delta z^2} E_x^q\Big|_{i,j,k-1} + \left(1+\frac{2ab}{\Delta z^2}\right) E_x^q\Big|_{i,j,k} - \frac{ab}{\Delta z^2} E_x^q\Big|_{i,j,k+1} &= E_x^{*q}\Big|_{i,j,k} \\
&- \frac{ab}{\Delta x \Delta y}\left(E_y^q\Big|_{i+1,j,k} - E_y^q\Big|_{i+1,j-1,k} - E_y^q\Big|_{i,j,k} + E_y^q\Big|_{i,j-1,k}\right) \\
&- \frac{ab}{\Delta x \Delta z}\left(E_z^q\Big|_{i+1,j,k} - E_z^q\Big|_{i+1,j,k-1} - E_z^q\Big|_{i,j,k} + E_z^q\Big|_{i,j,k-1}\right) \\
&+ \frac{2ab}{\Delta z^2}\sum_{k=0,q>0}^{q-1}\left(E_x^k\Big|_{i,j,k+1} + E_x^k\Big|_{i,j,k-1} - 2E_x^k\Big|_{i,j,k}\right) \\
&- \frac{2ab}{\Delta x \Delta y}\sum_{k=0,q>0}^{q-1}\left(E_y^k\Big|_{i+1,j,k} - E_y^k\Big|_{i+1,j-1,k} - E_y^k\Big|_{i,j,k} + E_y^k\Big|_{i,j-1,k}\right) \\
&- \frac{2ab}{\Delta x \Delta z}\sum_{k=0,q>0}^{q-1}\left(E_z^k\Big|_{i+1,j,k} - E_z^k\Big|_{i+1,j,k-1} - E_z^k\Big|_{i,j,k} + E_z^k\Big|_{i,j,k-1}\right)
\end{aligned}
\tag{2.43f}
$$

Note that exchanging the two matrices A and B leads to slightly different update equations. The systems of (2.43) are six tri-diagonal matrix equations which can be solved efficiently. In an actual simulation, since the right-hand sides of (2.43b) – (2.43f) include unknown values, in order to implement the efficient algorithm, the expansion coefficients of the electric field components must be updated as the following sequence: $E_x^{*q}\big|_{i,j,k}$, $E_y^{*q}\big|_{i,j,k}$, $E_z^{*q}\big|_{i,j,k}$, $E_z^{q}\big|_{i,j,k}$, $E_y^{q}\big|_{i,j,k}$, and $E_x^{q}\big|_{i,j,k}$. The expansion coefficients of the magnetic field components can be calculated explicitly from (2.21) – (2.23) with $E_x^{q}\big|_{i,j,k}$, $E_y^{q}\big|_{i,j,k}$, and $E_z^{q}\big|_{i,j,k}$ already updated. Then, one can reconstruct the field components in the time domain with (2.10).

2.4 PML ABSORBING BOUNDARY CONDITION

In this section, a PML absorbing boundary condition is developed that can maintain the tri-diagonal matrix form of the efficient Laguerre-based FDTD method. In PML regions, Maxwell's equations are expressed as [23]:

$$\frac{\partial E_{xy}}{\partial t} + \frac{\sigma_y}{\varepsilon} E_{xy} = \frac{1}{\varepsilon}\frac{\partial H_z}{\partial y} \tag{2.44a}$$

$$\frac{\partial E_{xz}}{\partial t} + \frac{\sigma_z}{\varepsilon} E_{xz} = -\frac{1}{\varepsilon}\frac{\partial H_y}{\partial z} \tag{2.44b}$$

$$\frac{\partial E_{yz}}{\partial t} + \frac{\sigma_z}{\varepsilon} E_{yz} = \frac{1}{\varepsilon}\frac{\partial H_x}{\partial z} \tag{2.44c}$$

$$\frac{\partial E_{yx}}{\partial t} + \frac{\sigma_x}{\varepsilon} E_{yx} = -\frac{1}{\varepsilon}\frac{\partial H_z}{\partial x} \tag{2.44d}$$

$$\frac{\partial E_{zx}}{\partial t} + \frac{\sigma_x}{\varepsilon} E_{zx} = \frac{1}{\varepsilon}\frac{\partial H_y}{\partial x} \tag{2.44e}$$

$$\frac{\partial E_{zy}}{\partial t} + \frac{\sigma_y}{\varepsilon} E_{zy} = -\frac{1}{\varepsilon}\frac{\partial H_x}{\partial y} \tag{2.44f}$$

$$\frac{\partial H_{xy}}{\partial t} + \frac{\rho_y}{\mu} H_{xy} = -\frac{1}{\mu}\frac{\partial E_z}{\partial y} \tag{2.44g}$$

$$\frac{\partial H_{xz}}{\partial t} + \frac{\rho_z}{\mu} H_{xz} = \frac{1}{\mu}\frac{\partial E_y}{\partial z} \tag{2.44h}$$

$$\frac{\partial H_{yz}}{\partial t} + \frac{\rho_z}{\mu} H_{yz} = -\frac{1}{\mu}\frac{\partial E_x}{\partial z} \tag{2.44i}$$

$$\frac{\partial H_{yx}}{\partial t} + \frac{\rho_x}{\mu} H_{yx} = \frac{1}{\mu}\frac{\partial E_z}{\partial x} \tag{2.44j}$$

$$\frac{\partial H_{zx}}{\partial t} + \frac{\rho_x}{\mu} H_{zx} = -\frac{1}{\mu}\frac{\partial E_y}{\partial x} \tag{2.44k}$$

$$\frac{\partial H_{zy}}{\partial t} + \frac{\rho_y}{\mu} H_{zy} = \frac{1}{\mu}\frac{\partial E_x}{\partial y} \tag{2.44l}$$

where σ_x, σ_y, σ_z and ρ_x, ρ_y, ρ_z are the PML electrical conductivity and magnetic loss, respectively. According to the derivation in the previous section, (2.44) can be written as:

$$\begin{aligned} E_x^q(\boldsymbol{r}) = E_{xy}^q(\boldsymbol{r}) + E_{xz}^q(\boldsymbol{r}) &= -\sigma_z^E\left[aD_z H_y^q(\boldsymbol{r}) + 2\sum_{k=0,q>0}^{q-1} E_{xz}^k(\boldsymbol{r}) \right] \\ &+ \sigma_y^E\left[aD_y H_z^q(\boldsymbol{r}) - 2\sum_{k=0,q>0}^{q-1} E_{xy}^k(\boldsymbol{r}) \right] \end{aligned} \tag{2.45a}$$

$$\begin{aligned} E_y^q(\boldsymbol{r}) = E_{yz}^q(\boldsymbol{r}) + E_{yx}^q(\boldsymbol{r}) &= \sigma_z^E\left[aD_z H_x^q(\boldsymbol{r}) - 2\sum_{k=0,q>0}^{q-1} E_{yz}^k(\boldsymbol{r}) \right] \\ &- \sigma_x^E\left[aD_x H_z^q(\boldsymbol{r}) + 2\sum_{k=0,q>0}^{q-1} E_{yx}^k(\boldsymbol{r}) \right] \end{aligned} \tag{2.45b}$$

$$\begin{aligned} E_z^q(\boldsymbol{r}) = E_{zy}^q(\boldsymbol{r}) + E_{zx}^q(\boldsymbol{r}) &= \sigma_y^E\left[-aD_y H_x^q(\boldsymbol{r}) - 2\sum_{k=0,q>0}^{q-1} E_{zy}^k(\boldsymbol{r}) \right] \\ &+ \sigma_x^E\left[aD_x H_y^q(\boldsymbol{r}) - 2\sum_{k=0,q>0}^{q-1} E_{zx}^k(\boldsymbol{r}) \right] \end{aligned} \tag{2.45c}$$

$$\begin{aligned} H_x^q(\boldsymbol{r}) = H_{xz}^q(\boldsymbol{r}) + H_{xy}^q(\boldsymbol{r}) &= \sigma_z^H\left[bD_z E_y^q(\boldsymbol{r}) - 2\sum_{k=0,q>0}^{q-1} H_{xz}^k(\boldsymbol{r}) \right] \\ &- \sigma_y^H\left[bD_y E_z^q(\boldsymbol{r}) + 2\sum_{k=0,q>0}^{q-1} H_{xy}^k(\boldsymbol{r}) \right] \end{aligned} \tag{2.45d}$$

$$\begin{aligned} H_y^q(\boldsymbol{r}) = H_{yz}^q(\boldsymbol{r}) + H_{yx}^q(\boldsymbol{r}) &= -\sigma_z^H\left[bD_z E_x^q(\boldsymbol{r}) + 2\sum_{k=0,q>0}^{q-1} H_{yz}^k(\boldsymbol{r}) \right] \\ &+ \sigma_x^H\left[bD_x E_z^q(\boldsymbol{r}) - 2\sum_{k=0,q>0}^{q-1} H_{yx}^k(\boldsymbol{r}) \right] \end{aligned} \tag{2.45e}$$

$$
\begin{aligned}
H_z^q(\boldsymbol{r}) = H_{zy}^q(\boldsymbol{r}) + H_{zx}^q(\boldsymbol{r}) &= \sigma_y^H\left[bD_yE_x^q(\boldsymbol{r}) - 2\sum_{k=0,q>0}^{q-1} H_{zy}^k(\boldsymbol{r}) \right] \\
&- \sigma_x^H\left[bD_xE_y^q(\boldsymbol{r}) + 2\sum_{k=0,q>0}^{q-1} H_{zx}^k(\boldsymbol{r}) \right]
\end{aligned}
\tag{2.45f}
$$

where $\sigma_x^E = (1+a\sigma_x)^{-1}$, $\sigma_y^E = (1+a\sigma_y)^{-1}$, $\sigma_z^E = (1+a\sigma_z)^{-1}$, $\sigma_x^H = (1+b\rho_x)^{-1}$, $\sigma_y^H = (1+b\rho_y)^{-1}$, and $\sigma_z^H = (1+b\rho_z)^{-1}$. Writing (2.45) in a matrix form leads to:

$$
W_E^q = a\,\bar{D}_H W_H^q + \bar{V}_E^{q-1} \tag{2.46}
$$

$$
W_H^q = b\,\bar{D}_E W_E^q + \bar{V}_H^{q-1} \tag{2.47}
$$

where

$$
\bar{D}_H = \begin{bmatrix} 0 & -\sigma_z^E D_z & \sigma_y^E D_y \\ \sigma_z^E D_z & 0 & -\sigma_x^E D_x \\ -\sigma_y^E D_y & \sigma_x^E D_x & 0 \end{bmatrix} \tag{2.48}
$$

$$
\bar{D}_E = \begin{bmatrix} 0 & \sigma_z^H D_z & -\sigma_y^H D_y \\ -\sigma_z^H D_z & 0 & \sigma_x^H D_x \\ \sigma_y^H D_y & -\sigma_x^H D_x & 0 \end{bmatrix} \tag{2.49}
$$

$$
\bar{V}_E^{q-1} = \begin{bmatrix} -2\sigma_y^E \sum_{k=0,q>0}^{q-1} E_{xy}^k(\boldsymbol{r}) - 2\sigma_z^E \sum_{k=0,q>0}^{q-1} E_{xz}^k(\boldsymbol{r}) \\ -2\sigma_x^E \sum_{k=0,q>0}^{q-1} E_{yx}^k(\boldsymbol{r}) - 2\sigma_z^E \sum_{k=0,q>0}^{q-1} E_{yz}^k(\boldsymbol{r}) \\ -2\sigma_x^E \sum_{k=0,q>0}^{q-1} E_{zx}^k(\boldsymbol{r}) - 2\sigma_y^E \sum_{k=0,q>0}^{q-1} E_{zy}^k(\boldsymbol{r}) \end{bmatrix} \tag{2.50}
$$

$$
\bar{V}_H^{q-1} = \begin{bmatrix} -2\sigma_y^H \sum_{k=0,q>0}^{q-1} H_{xy}^k(\boldsymbol{r}) - 2\sigma_z^H \sum_{k=0,q>0}^{q-1} H_{xz}^k(\boldsymbol{r}) \\ -2\sigma_x^H \sum_{k=0,q>0}^{q-1} H_{yx}^k(\boldsymbol{r}) - 2\sigma_z^H \sum_{k=0,q>0}^{q-1} H_{yz}^k(\boldsymbol{r}) \\ -2\sigma_x^H \sum_{k=0,q>0}^{q-1} H_{zx}^k(\boldsymbol{r}) - 2\sigma_y^H \sum_{k=0,q>0}^{q-1} H_{zy}^k(\boldsymbol{r}) \end{bmatrix} \tag{2.51}
$$

Inserting (2.47) into (2.46), we obtain:

$$\left(I - ab\bar{D}_H\bar{D}_E\right)W_E^q = a\bar{D}_H\bar{V}_H^{q-1} + \bar{V}_E^{q-1} \tag{2.52}$$

Equation (2.52) is the large sparse matrix equation implementing a PML medium that is generated when using the conventional 3-D Laguerre-based FDTD method. According to the procedure described in the previous section, we decompose $ab\bar{D}_H\bar{D}_E$ into two triangular matrices $\bar{A}$ and $\bar{B}$ and add a perturbation term $\bar{A}\bar{B}(W_E^q - \bar{V}_E^{q-1})$ to (2.52). Then, we can solve (2.52) by factorizing it into two sub-steps using:

$$\left(I - \bar{A}\right)W^* = \left(I + \bar{B}\right)\bar{V}_E^{q-1} + a\bar{D}_H\bar{V}_H^{q-1} \tag{2.53}$$

$$\left(I - \bar{B}\right)\bar{W}_E^q = W^* - \bar{B}\bar{V}_E^{q-1} \tag{2.54}$$

where matrices $\bar{A}$ and $\bar{B}$ are constructed as:

$$\bar{A} = ab\begin{bmatrix} \sigma_y^E D_y \sigma_y^H D_y & 0 & 0 \\ -\sigma_x^E D_x \sigma_y^H D_y & \sigma_z^E D_z \sigma_z^H D_z & 0 \\ -\sigma_x^E D_x \sigma_z^H D_z & -\sigma_y^E D_y \sigma_z^H D_z & \sigma_x^E D_x \sigma_x^H D_x \end{bmatrix} \tag{2.55}$$

$$\bar{B} = ab\begin{bmatrix} \sigma_z^E D_z \sigma_z^H D_z & -\sigma_y^E D_y \sigma_x^H D_x & -\sigma_z^E D_z \sigma_x^H D_x \\ 0 & \sigma_x^E D_x \sigma_x^H D_x & -\sigma_z^E D_z \sigma_y^H D_y \\ 0 & 0 & \sigma_y^E D_y \sigma_y^H D_y \end{bmatrix} \tag{2.56}$$

With some manipulation, we can obtain from (2.53) and (2.54) the discrete space equations of PML for the efficient 3-D Laguerre-based FDTD method.

2.5 NUMERICAL RESULTS

2.5.1 Parallel-Plate Capacitor: Uniform 3-D Grid

To determine the accuracy and computational requirements of the proposed efficient 3-D Laguerre-based FDTD technique, three numerical examples are given. We first apply this technique to model a parallel-plate capacitor in a uniform grid and compare its numerical results with those obtained using both the conventional FDTD method and the ADI-FDTD method.

Figure 2.1 illustrates the configuration of the capacitor, which consists of two parallel 10×10 cm perfect electric conductor (PEC) plates in free space separated by 1 cm [18]. The electromagnetic field near the plate edges exhibits a rapidly varying spatial distribution [24]. Cubic grid cells of size $\Delta = 1$ cm are used uniformly throughout each model.

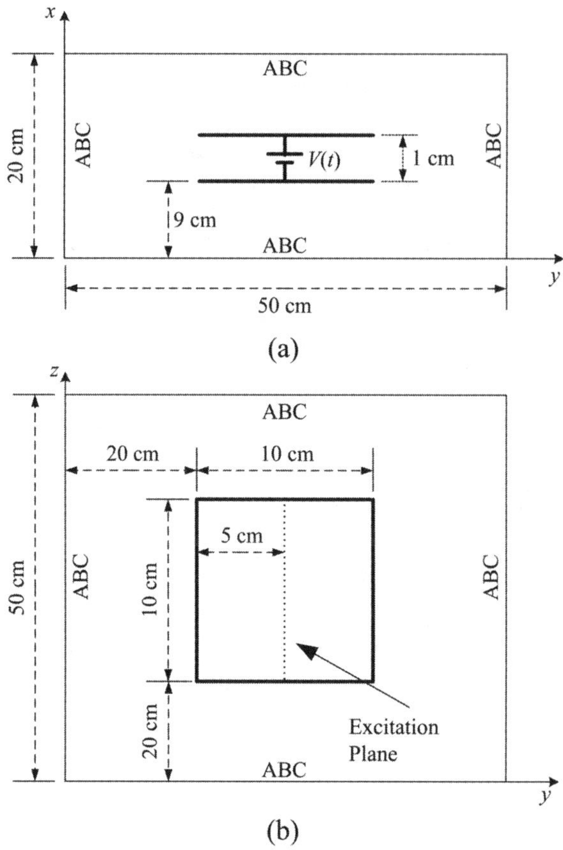

Fig. 2.1 Geometry of the conventional FDTD, ADI-FDTD, and efficient 3-D Laguerre-based FDTD models of the 3-D parallel plate capacitor excited by an ideal voltage source. The computational domain is terminated with the Mur's first-order absorbing boundary condition. (a) Vertical view; (b) horizontal view. *Source:* Duan *et al.*, *IEEE Trans. Microwave Theory and Techniques*, Vol. 59, 2011, pp. 56–64, ©2011 IEEE.

The capacitor is excited with a Gaussian pulse of the form $\exp\{-[(t-T_c)/T_d]^2\}$, where $T_d = 1$ ns and $T_c = 3T_d$. For the conventional FDTD model, we choose as the time-step $\Delta t_{\text{FDTD}} = \Delta/2c = 16.67$ ps, equal to 0.866 times the Courant limit for numerical stability in 3-D. For purposes of comparison, three distinct ADI-FDTD models are run using a time-step Δt_{ADI}, respectively, of Δt_{FDTD}, $4\Delta t_{\text{FDTD}}$, and $8\Delta t_{\text{FDTD}}$. Finally, the efficient 3-D Laguerre-based FDTD technique is exercised for the parameters $q = 52$ and $s = 4.0 \times 10^{10}$ [25, 26]. We note that the upper and lower bounds of q and s can be obtained using an approach proposed in [26]. In addition, as stated earlier, q can be chosen according to the analytical formulas given in [8].

Figure 2.2 shows the normalized frequency-domain E_x field at $f = 0$ Hz along the y-axis at $x = 9$ cm and $z = 25$ cm, calculated by applying a Fourier transformation to the time-domain results [18]. Results obtained using the efficient 3-D Laguerre-based FDTD technique are in excellent agreement with those computed using the conventional FDTD method. However, the ADI-FDTD method provides this level of agreement only for $\Delta t_{\text{ADI}} = \Delta t_{\text{FDTD}}$, and suffers from a loss of accuracy for increasing values of Δt_{ADI}.

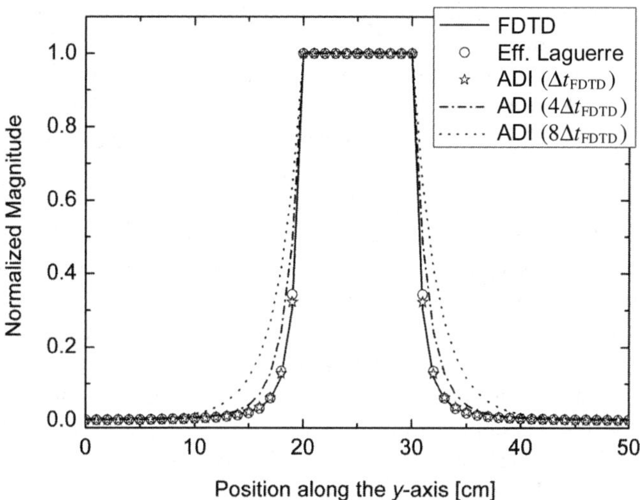

Fig. 2.2 Comparison of the normalized frequency-domain E_x field along the y-axis of the capacitor geometry of Fig. 2.1 at $x = 9$ cm and $z = 25$ cm calculated by the conventional FDTD, ADI-FDTD, and efficient 3-D Laguerre-based FDTD techniques. *Source:* Duan *et al.*, *IEEE Trans. Microwave Theory and Techniques*, Vol. 59, 2011, pp. 56–64, ©2011 IEEE.

2.5.2 Shielded Microstrip Line: Graded Grid in One Direction

We next apply the efficient 3-D Laguerre-based FDTD technique to model a shielded microstrip line and compare its numerical results with those obtained using both the conventional FDTD and ADI-FDTD methods. Figure 2.3 illustrates the modeling geometry [18].

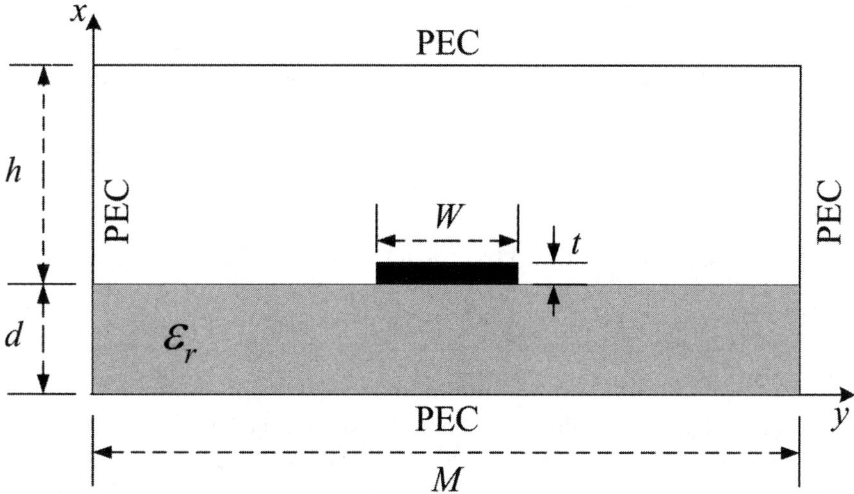

Fig. 2.3 Cross-section of the shielded microstrip line. *Source:* Duan *et al.*, *IEEE Trans. Microwave Theory and Techniques*, Vol. 59, 2011, pp. 56–64, ©2011 IEEE.

The microstrip line shown in Fig. 2.3 has a lossless isotropic dielectric substrate of permittivity $\varepsilon_r = 2.2$ and thickness $d = 0.5$ mm. The signal conductor is a PEC strip of width $W = 0.5$ mm and thickness $t = 0.02$ mm. The overall height (x-dimension) of the rectangular shielding tube is $d + h = 0.5$ mm + 1.0 mm = 1.5 mm, and the width (y-dimension) of the shielding tube is $M = 2.5$ mm. To properly model the thickness of the signal conductor, a 40-cell graded grid is used along the x-direction with $\Delta x_{\text{min}} = 0.02$ mm. In the y- and z-directions, a uniform grid is used with, respectively, 10 cells of $\Delta y = 0.25$ mm and 60 cells of $\Delta z = 0.25$ mm. (Hence, there is a 12.5:1 ratio between maximum and minimum cell dimensions in the grid.) Mur's first-order ABC [10] is applied at the outermost z-planes of the computational domain.

The microstrip line is excited by applying an x-directed electric field at $z = 2.5$ mm having a Gaussian time-waveform of the form $\exp\{-[(t - T_c)/T_d]^2\}$, where $T_d = 30$ ps and $T_c = 3T_d$. For the conventional FDTD model, we choose as the time-step $\Delta t_{\text{FDTD}} = \Delta x_{\text{min}}/2c = 33$ fs, equal to 0.866 times the Courant limit for numerical stability in 3-D. For purposes of comparison, two distinct ADI-FDTD models are run using a time-step Δt_{ADI}, respectively, of $25\Delta t_{\text{FDTD}}$ and $50\Delta t_{\text{FDTD}}$; and the efficient 3-D Laguerre-based FDTD technique is exercised for the parameters $q = 50$ and $s = 1.1 \times 10^{12}$, following the bounds for these parameters proposed in [26]. Laguerre coefficients of the excitation pulse are calculated with a temporal resolution of 0.4 ps.

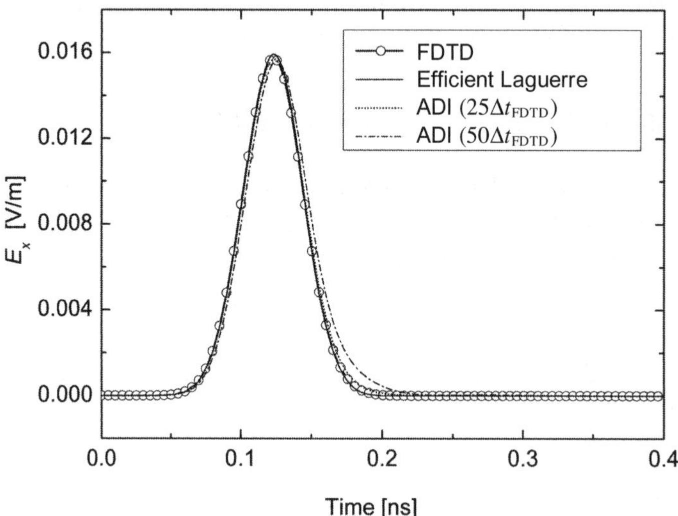

Fig. 2.4 Comparison of the conventional FDTD, ADI-FDTD, and efficient 3-D Laguerre-based FDTD results for the computed E_x field component at observation point ($9\Delta x$, $5\Delta y$, $40\Delta z$) of the shielded microstrip line of Fig. 2.3. *Source:* Duan *et al.*, *IEEE Trans. Microwave Theory and Techniques*, Vol. 59, 2011, pp. 56–64, ©2011 IEEE.

Figure 2.4 compares the conventional FDTD, ADI-FDTD, and efficient 3-D Laguerre-based FDTD results for the computed E_x field component at observation point ($9\Delta x$, $5\Delta y$, $40\Delta z$) up to the final time, $T_f = 0.4$ ns [18]. The agreement between the conventional FDTD and the efficient Laguerre-based FDTD method is excellent. However, the ADI-FDTD results again begin to exhibit a progressive departure from the FDTD benchmark — in this case, just as in Fig. 2.2, when Δt_{ADI} exceeds the Courant limit for the full-size grid cells, here approximately $12.5\,\Delta t_{\text{FDTD}}$.

TABLE 2.1

Computational Requirements for the Shielded Microstrip Line

Scheme	Parameters	No. of Iterations	CPU Time (sec)	Memory (MB)
FDTD	Δt = 0.033 ps	12,121	81.2	1.6
ADI-FDTD	Δt = 0.825 ps	484	18.4	2.5
ADI-FDTD	Δt = 1.65 ps	242	9.1	2.5
Eff. Laguerre	q = 50, s = 1.1×10^{12}	51	3.9	3.7

Source: Duan *et al.*, *IEEE Trans. Microwave Theory and Techniques*, Vol. 59, 2011, pp. 56–64, ©2011 IEEE.

Table 2.1 lists the computer time and memory required for the numerical simulations of the shielded microstrip line of Fig. 2.3 [18]. From this table, we see that the efficient 3-D Laguerre-based FDTD technique runs 2.33 times faster than the ADI-FDTD method which used Δt = 1.65 ps (i.e., 50 × the FDTD time-step). Furthermore, as per Fig. 2.4, the efficient Laguerre-based FDTD technique generates results that are essentially congruent with the FDTD benchmark, rather than exhibiting the deviations seen in the ADI-FDTD modeling results. The only disadvantage for the efficient Laguerre-based technique is that its memory usage is 50% greater than that required by the ADI-FDTD method for this particular model.

It is noteworthy that, when using the conventional Laguerre-based FDTD method to simulate this shielded microstrip line, a sparse 72,000 × 72,000 matrix is set up. Depending on the techniques employed to process this sparse matrix, the computer running time and memory requirements can be significantly greater than the efficient Laguerre-based technique discussed here.

2.5.3 PML Absorbing Boundary Condition Performance

In the third example, our goal is to evaluate the numerical performance of the PML absorbing boundary condition of Section 2.4. To this end, we simulate the radiation of an electric line current source centered within a 30 × 30 × 30 free-space grid comprised of 1-cm cubic unit cells. The line source is x-polarized and has the time dependence of a sinusoidally modulated Gaussian pulse, $\sin[2\pi f_0(t-t_0)]\exp\{-[(t-t_0)/\tau]^2\}$, where $f_0 = 2$ GHz, $\tau = 1/(2f_0)$, and $t_0 = 3\tau$. The computation domain is terminated by either Mur's first-order absorbing boundary condition or a 16-cell-thick PML absorbing boundary, as described in Section 2.4. For this example, we choose $q = 70$ and $s = 1.2\times10^{11}$ as the parameters of the efficient 3-D Laguerre-based FDTD technique, following the bounds for these parameters proposed in [26].

Figure 2.5 shows the relative reflection error, $R_{\text{dB}} = 20\log_{10}\left[\,|E_x^R(t)-E_x^T(t)|/\max|E_x^R(t)|\,\right]$, at observation point (15, 3, 15) in the test grid for the two absorbing boundary conditions [18]. Here, $E_x^T(t)$ is the field computed in the test domain, and $E_x^R(t)$ is the reference field computed using an auxiliary domain that is sufficiently large such that there are no reflections from the outer grid boundary during the observation period. We see that the PML provides a relative reflection error no larger than –60 dB at any time during the observation.

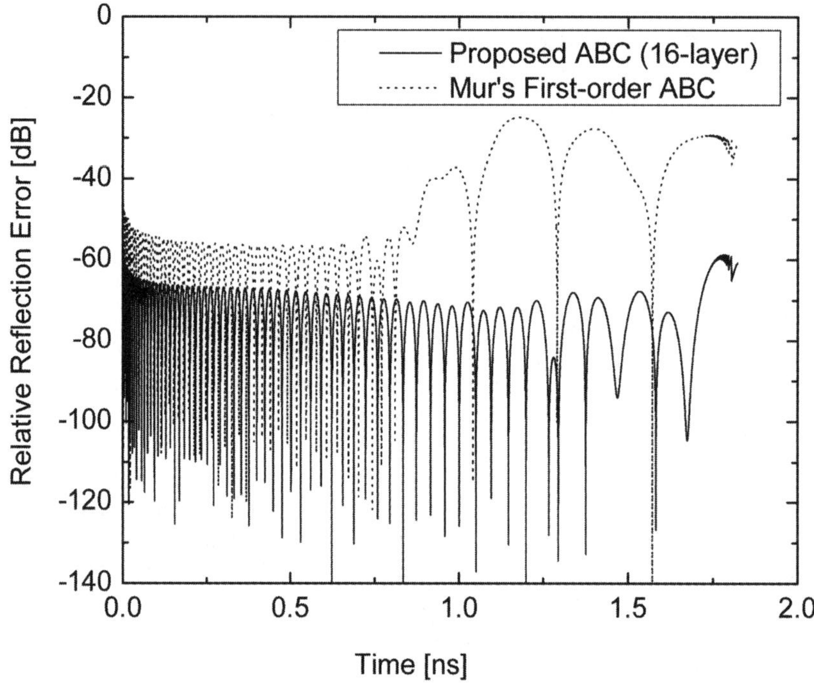

Fig. 2.5 Comparison of the relative reflection errors of the first-order Mur ABC and the PML ABC of Section 2.4 in a $30 \times 30 \times 30$ free-space grid implementation of the efficient Laguerre-based technique. *Source:* Duan *et al.*, *IEEE Trans. Microwave Theory and Techniques*, Vol. 59, 2011, pp. 56–64, ©2011 IEEE.

2.6 SUMMARY AND CONCLUSIONS

This chapter presented an efficient algorithm for implementing the unconditionally stable 3-D Laguerre polynomial-based FDTD technique. In contrast to the conventional Laguerre-based FDTD method, which requires solving a very large sparse matrix, the new approach introduces a perturbation term and a factorization-splitting scheme that allows setting up and solving six tri-diagonal matrices and three explicit equations for a full update cycle.

Numerical examples were presented that indicate the improved computational efficiency of the new Laguerre-based algorithm compared to classical FDTD and ADI-FDTD. Importantly, the new method provides excellent computational accuracy — much better than ADI-FDTD. Specifically, the errors introduced by the perturbation term are significantly less than the second-order truncation error inherent in the ADI-FDTD method. Finally, an effective PML absorbing boundary condition was presented and demonstrated. This PML absorbing boundary retains the computationally efficient tri-diagonal matrix form of the Laguerre-based FDTD interior field calculations.

Although the solutions of the six tri-diagonal matrices for the electric field components are dependent, each tri-diagonal matrix can be efficiently parallel processed on a computing cluster. Future work will involve exploring additional potential optimizations of the new technique and application to solve real problems.

REFERENCES

[1] Taflove, A., and S. C. Hagness, *Computational Electrodynamics: The Finite-Difference Time-Domain Method*, 3rd ed., Norwood, MA: Artech House, 2005.

[2] Namiki, T., "A new FDTD algorithm based on alternating-direction implicit method," *IEEE Trans. Microwave Theory and Techniques*, Vol. 47, 1999, pp. 2003–2007.

[3] Zheng, F., Z. Chen, and J. Zhang, "Toward the development of a three-dimensional unconditionally stable finite-difference time-domain method," *IEEE Trans. Microwave Theory and Techniques*, Vol. 48, 2000, pp. 1550–1558.

[4] Namiki, T., "3-D ADI-FDTD method-unconditionally stable time-domain algorithm for solving full vector Maxwell's equations," *IEEE Trans. Microwave Theory and Techniques*, Vol. 48, 2000, pp. 1743–1748.

[5] Zheng, F., and Z. Chen, "Numerical dispersion analysis of the unconditionally stable 3-D ADI-FDTD method," *IEEE Trans. Microwave Theory and Techniques*, Vol. 49, 2001, pp. 1006–1009.

[6] Shibayama, J., M. Muraki, J. Yamauchi, and H. Nakano, "Efficient implicit FDTD algorithm based on locally one-dimensional scheme," *Electronics Lett.*, Vol. 41, 2005, pp. 1046–1047.

[7] Ahmed, I., E. K. Chua, E. P. Li, and Z. Chen, "Development of the three-dimensional unconditionally stable LOD-FDTD method," *IEEE Trans. Antennas and Propagation*, Vol. 56, 2008, pp. 3596–3600.

[8] Chung, Y. S., T. K. Sarkar, B. H. Jung, and M. Salazar-Palma, "An unconditionally stable scheme for the finite-difference time-domain method," *IEEE Trans. Microwave Theory and Techniques*, Vol. 51, 2003, pp. 697–704.

[9] Chen, Z., and S. Luo, "Generalization of the finite-difference-based time-domain methods using the method of moments," *IEEE Trans. Antennas and Propagation*, Vol. 54, 2006, pp. 2515–2524.

[10] Mur, G., "Absorbing boundary conditions for the finite-difference approximation of the time-domain electromagnetic field equations," *IEEE Trans. Electromagnetic Compatibility*, Vol. EMC-23, 1981, pp. 377–382.

[11] Shao, W., B. Z. Wang, and X. F. Liu, "Second-order absorbing boundary conditions for marching-on-in-order scheme," *IEEE Microwave and Wireless Components Lett.*, Vol. 16, 2006, pp. 308–310.

[12] Ding, P. P., G. F. Wang, H. Lin, and B. Z. Wang, "Unconditionally stable FDTD formulation with UPML-ABC," *IEEE Microwave and Wireless Components Lett.*, Vol. 16, 2006, pp. 161–163.

[13] Yi, Y., B. Chen, H. L. Chen, and D. G. Fang, "TF/SF boundary and PML–ABC for an unconditionally stable FDTD method," *IEEE Microwave and Wireless Components Lett.*, Vol. 17, 2007, pp. 91–93.

[14] Alighanbari, A., and C. D. Sarris, "An unconditionally stable Laguerre-based S-MRTD time-domain scheme," *IEEE Antennas and Wireless Propagation Lett.*, Vol. 5, 2006, pp. 69–72.

[15] Shao, W., B. Z. Wang, X. H. Wang, and X. F. Liu, "Efficient compact 2-D time-domain method with weighted Laguerre polynomials," *IEEE Trans. Electromagnetic Compatibility*, Vol. 48, 2006, pp. 442–448.

[16] Chen, H. L., B. Chen, Y. T. Duan, Y. Yi, and D. G. Fang, "Unconditionally stable Laguerre-based BOR-FDTD scheme for scattering from bodies of revolution," *Microwave and Optical Technology Lett.*, Vol. 49, 2007, pp. 1897–1900.

[17] Duan, Y. T., B. Chen, and Y. Yi, "Efficient implementation for the unconditionally stable 2-D WLP-FDTD method," *IEEE Microwave and Wireless Components Lett.*, Vol. 19, 2009, pp. 677–679.

[18] Duan, Y. T., B. Chen, D. G. Fang, and B. H. Zhou, "Efficient implementation for 3-D Laguerre-based finite-difference time-domain method," *IEEE Trans. Microwave Theory and Techniques*, Vol. 59, 2011, pp. 56–64.

[19] Gradshteyn, I. S., and I. M. Ryzhik, *Table of Integrals, Series, and Products*, New York: Academic, 1980.

[20] Sun, G., and C. W. Trueman, "Unconditionally stable Crank-Nicolson scheme for solving the two-dimensional Maxwell's equations," *Electronics Lett.*, Vol. 39, 2003, pp. 595–597.

[21] Sun, G., and C. W. Trueman, "Approximate Crank-Nicolson schemes for the 2-D finite-difference time-domain method for TEZ waves," *IEEE Trans. Antennas and Propagation*, Vol. 52, 2004, pp. 2963–2972.

[22] Sun, G., and C. W. Trueman, "Efficient implementations of the Crank-Nicolson scheme for the finite-difference time-domain method," *IEEE Trans. Microwave Theory and Techniques*, Vol. 54, 2006, pp. 2275–2284.

[23] Berenger, J. P., "Three-dimensional perfectly matched layer for the absorption of electromagnetic waves," *J. Computational Physics*, Vol. 127, 1996, pp. 363–379.

[24] Garcia, S. G., T. W. Lee, and S. C. Hagness, "On the accuracy of the ADI-FDTD method," *IEEE Antennas and Wireless Propagation Lett.*, Vol. 1, 2002, pp. 31–34.

[25] Yuan, M., J. Koh, T. K. Sarkar, W. Lee, and M. Salazar-Palma, "A comparison of performance of three orthogonal polynomials in extraction of wide-band response using early time and low frequency data," *IEEE Trans. Antennas and Propagation*, Vol. 53, 2005, pp. 785–792.

[26] Yuan, M., A. De, T. K. Sarkar, J. Koh, and B. H. Jung, "Conditions for generation of stable and accurate hybrid TD-FD MoM solutions," *IEEE Trans. Microwave Theory and Techniques*, Vol. 54, 2006, pp. 2552–2563.

Chapter 3

Exact Total-Field/Scattered-Field Plane-Wave Source Condition[1]

Tengmeng Tan and Mike Potter

3.1 INTRODUCTION

The mathematical relations governing the physics of electromagnetic waves are described by Maxwell's equations. Like many other partial differential equations, analytical solutions exist in only a few simple cases. Hence, many electromagnetic wave interaction problems are solved by numerical methods. One of the most important classes of such problems is that of determining the scattered fields from arbitrary objects illuminated by electromagnetic waves in unbounded regions. Such objects can be situated in free space (e.g., airplanes), or embedded within material structures (e.g., land mines).

An efficient technique to compute scattered fields in the context of FDTD modeling is the *total-field/scattered-field* (TF/SF) incident wave source [1], which is employed by almost all current commercial FDTD solvers. Fundamentally, the TF/SF technique is an application of the well-known electromagnetic field equivalence principle [2–4]. By this principle, the original incident wave of infinite extent and arbitrary propagation direction, polarization, and time-waveform is replaced by electric and magnetic current sources appropriately defined on a finite closed surface containing the object of interest. The reformulated problem confines the incident illumination to a compact total-field region, and provides a finite scattered-field region external to the total-field region that is terminated by an absorbing boundary condition (ABC) to simulate the FDTD grid extending to infinity. Scattered fields in the far-field region can be rigorously obtained using a near-to-far-field transform (NFFT) [1].

Figure 3.1 illustrates the application of the equivalence principle to electromagnetic wave scattering by an arbitrary target located in an unbounded region. Here, the incident wave is generated by the electric and magnetic current sources, $\boldsymbol{J}_{\text{source}}$ and $\boldsymbol{M}_{\text{source}}$. In Fig. 3.1(a), the interaction of the incident wave with the target results in the formation of a total electric field $\boldsymbol{E}_{\text{T}}$ and a total magnetic field $\boldsymbol{H}_{\text{T}}$ that fills all of space. We use the word "total" in the sense that, mathematically, each of these fields can be viewed as being the sum of an incident component and a scattered component, i.e., $\boldsymbol{E}_{\text{T}} = \boldsymbol{E}_{\text{inc}} + \boldsymbol{E}_{\text{S}}$ and $\boldsymbol{H}_{\text{T}} = \boldsymbol{H}_{\text{inc}} + \boldsymbol{H}_{\text{S}}$.

---

[1]This chapter is adapted and expanded from Ref. [14], T. Tan and M. Potter, "FDTD discrete planewave (FDTD-DPW) formulation for a perfectly matched source in TFSF simulations," *IEEE Trans. Antennas and Propagation*, Vol. 58, 2010, pp. 2641–2648, ©2010 IEEE.

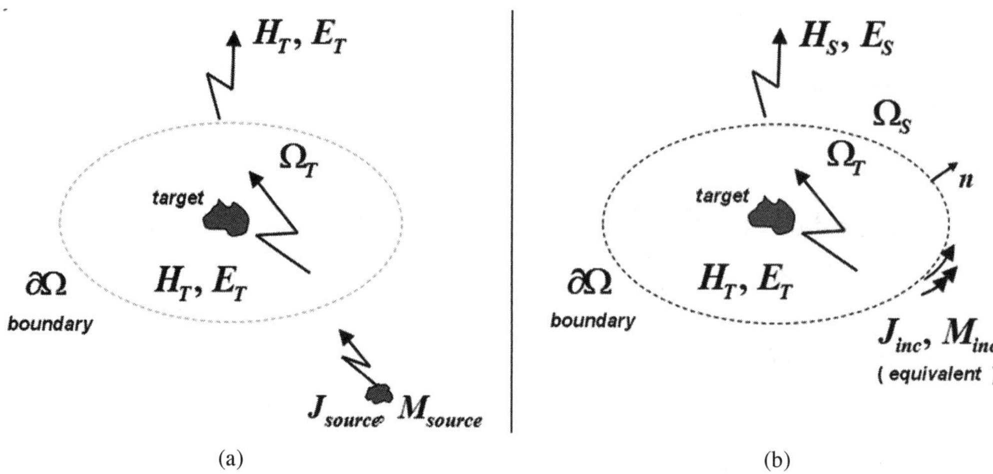

Fig. 3.1 Illustration of the electromagnetic field equivalence principle for electromagnetic wave scattering by a target in an unbounded region: (a) original problem – total fields everywhere in the space; (b) equivalent problem – division of the space into total-field and scattered-field regions.

Using the equivalence principle, the reformulated problem shown in Fig. 3.1(b) separates space into two regions: an interior region where total fields exist, and an exterior region where only scattered fields exist. Here, the original incident field is generated only within the total-field region by equivalent electric and magnetic current sources, $\boldsymbol{J}_{\text{inc}}$ and $\boldsymbol{M}_{\text{inc}}$, existing on the TF/SF boundary, $\partial\Omega$.

As outlined in [1], the TF/SF formulation has the following advantages relative to alternative pure scattered-field wave-source approaches in many FDTD modeling problems:

- *Computational efficiency.* Sources are computed only along the two-dimensional (2-D) surface of the TF/SF interface, $\partial\Omega$, enclosing the target, as opposed to within the entire three-dimensional (3-D) volume of the target.
- *Relatively simple programming of targets.* The required continuity of total tangential $\boldsymbol{E}$ and $\boldsymbol{H}$ across the interface of dissimilar materials within the target is automatically provided by the FDTD algorithm because all material interfaces are located where total fields are time-marched. Furthermore, the TF/SF interface has a fixed shape that is independent of the geometry and material composition of the target.
- *Wide near-field computational dynamic range.* Weak total fields within and near the target are time-marched directly. There is no need to obtain low values of important measurable total fields (e.g., within deep shadow regions or well-shielded internal cavities of the target) via near cancellation of the FDTD computed scattered field by the known incident field. This cancellation process can lead to large errors in the required total-field quantities due to "subtraction noise" wherein small percentage errors in calculating scattered fields can result in large percentage errors in the total fields that are the residue after field cancellation.

While possessing these advantages, the TF/SF formulation also provides the capability to compute the far-field response and to apply a modern perfectly matched layer (PML) ABC. The NFFT virtual surface and the ABC are each located in the well-defined scattered-field region.

3.2 DEVELOPMENT OF THE EXACT TF/SF FORMULATION FOR FDTD

The original TF/SF formulation for FDTD was outlined in [5], and improved on in [6]. Readers are referred to [1] for a thorough treatment of the TF/SF technique and its subsequent improvements. The specific improvement reported in [6] was to incorporate an auxiliary one-dimensional (1-D) FDTD grid acting as an incident field array (IFA) to propagate the plane-wave source concurrently with the primary-grid FDTD simulation. However, because of interpolation errors in using the IFA as a look-up table, and numerical dispersion mismatches between the auxiliary 1-D grid forming the IFA and the primary 2-D or 3-D grid, there existed a nonphysical leakage of the incident wave into the scattered-field region of the primary grid. This leakage was at a level of –30 to –40 dB, depending on the propagation direction of the incident wave in the primary grid. While the subsequent use of signal-processing techniques has yielded reduced leakage, down to a level of approximately –70 dB [7–9], residual leakage at this level can still introduce a noise floor capable of masking the true far-field response of wide-dynamic-range scatterers such as low-observable targets.

An alternative to using an IFA is to construct the incident numerical plane wave directly from the numerical dispersion relation [10, 11]. This technique has become known as the analytic field propagator (AFP) method in the sense that the propagator is known analytically, and interpolation errors are virtually nonexistent. With the AFP method, numerical dispersion and frequency-dependent polarization and nonorthogonality of the field components can be directly accounted. This reduces the incident wave leakage down to approximately –180 dB, at which point it is nonexistent for all practical purposes [11]. However, the AFP method is considerably less efficient than IFA methods because the temporal source function for all points on the TF/SF interface must be generated in a preprocessing step, and then stored for access during the main simulation.

Recently, Tan and Potter have shown in [12–14] that a set of six IFAs can be constructed such that a plane-wave source is created that is exactly matched to the primary FDTD computational grid. In other words, the isolation of the incident and scattered fields is on the level of machine precision (–180 dB for single precision). Through geometric arguments, it can be shown that the propagation direction of a numerical plane wave in a regular FDTD grid can be represented by a countably infinite set of angles [13]. This representation allows a many-to-one correspondence to be made between the fields in the IFAs and the fields in the primary grid, permitting identical numerical dispersion relations with no need for field interpolation. In fact, this mapping can be used to make the AFP method more efficient [15].

In the sections that follow, the basic TF/SF formulation is first described. Then, the methodologies reported in [12–14] are discussed in considerable detail, and illustrative examples of how these lead to an exact plane-wave source condition are presented.

3.3 BASIC TF/SF FORMULATION

For the TF/SF division of space shown in Fig. 3.1(b), Maxwell's equations for the electric field $\boldsymbol{E}$ and magnetic field $\boldsymbol{H}$ propagating in an isotropic nondispersive medium of permittivity ε, permeability μ, electric conductivity σ, and magnetic loss σ^*, can be written as:

$$\varepsilon(\boldsymbol{x})\partial_t \boldsymbol{E}(\boldsymbol{x},t) + \sigma(\boldsymbol{x})\boldsymbol{E}(\boldsymbol{x},t) = \nabla \times \boldsymbol{H}(\boldsymbol{x},t) - \boldsymbol{J}_{\text{inc}}(\boldsymbol{x}_{\partial\Omega},t) \tag{3.1a}$$

$$\mu(\boldsymbol{x})\partial_t \boldsymbol{H}(\boldsymbol{x},t) + \sigma^*(\boldsymbol{x})\boldsymbol{H}(\boldsymbol{x},t) = -\nabla\times\boldsymbol{E}(\boldsymbol{x},t) - \boldsymbol{M}_{\text{inc}}(\boldsymbol{x}_{\partial\Omega},t) \tag{3.1b}$$

where the operator $\partial_t(\cdot)$ is a partial derivative with respect to time t, and $\boldsymbol{x} = [x, y, z] \in \mathbf{R}^3$. Equations (3.1a) and (3.1b) comprise a set of inhomogeneous partial differential equations with forcing functions given by the electric current density $\boldsymbol{J}_{\text{inc}}(\boldsymbol{x}_{\partial\Omega},t)$ and magnetic current density $\boldsymbol{M}_{\text{inc}}(\boldsymbol{x}_{\partial\Omega},t)$. In particular, as shown in Fig. 3.1(b), these sources exist only at the interface $\partial\Omega$ of the total-field region Ω_T and the scattered-field region Ω_S. These current densities are not independent vector fields. They are constructed in such a way that both the electric and magnetic fields interior to the domain Ω_T are identical to those found in the original problem shown in Fig. 3.1(a):

$$\boldsymbol{E}(\boldsymbol{x},t) = \boldsymbol{E}_\text{S}(\boldsymbol{x},t) + \boldsymbol{E}_{\text{inc}}(\boldsymbol{x},t) \quad \text{for } \boldsymbol{x}\in\Omega_\text{T} \tag{3.2a}$$

$$\boldsymbol{H}(\boldsymbol{x},t) = \boldsymbol{H}_\text{S}(\boldsymbol{x},t) + \boldsymbol{H}_{\text{inc}}(\boldsymbol{x},t) \quad \text{for } \boldsymbol{x}\in\Omega_\text{T} \tag{3.2b}$$

where the subscripts S and inc denote, respectively, the scattered and incident waves. Also, the exterior fields in the domain Ω_S are only scattered fields:

$$\boldsymbol{E}(\boldsymbol{x},t) = \boldsymbol{E}_\text{S}(\boldsymbol{x},t) \quad \text{for } \boldsymbol{x}\in\Omega_\text{S} \tag{3.3a}$$

$$\boldsymbol{H}(\boldsymbol{x},t) = \boldsymbol{H}_\text{S}(\boldsymbol{x},t) \quad \text{for } \boldsymbol{x}\in\Omega_\text{S} \tag{3.3b}$$

In other words, both $\boldsymbol{E}(\boldsymbol{x},t)$ and $\boldsymbol{H}(\boldsymbol{x},t)$ are discontinuous for $\boldsymbol{x}\in\partial\Omega$, and these field discontinuities are accounted for by the current densities $\boldsymbol{J}_{\text{inc}}(\boldsymbol{x}_{\partial\Omega},t)$ and $\boldsymbol{M}_{\text{inc}}(\boldsymbol{x}_{\partial\Omega},t)$.

3.4 ELECTRIC AND MAGNETIC CURRENT SOURCES AT THE TF/SF INTERFACE

To facilitate our later discussion of an exact TF/SF formulation for FDTD, we introduce another set of current sources that enables treatment of the field discontinuities for $\boldsymbol{x}\in\partial\Omega$:

$$\boldsymbol{J}_\text{s}(\boldsymbol{x}_{\partial\Omega},t) = \hat{\boldsymbol{n}}\times\boldsymbol{H}_{\text{inc}}(\boldsymbol{x}_{\partial\Omega},t) \equiv \boldsymbol{J}_{\text{inc}}(\boldsymbol{x}_{\partial\Omega},t)\,\partial\Omega_{\hat{n}} \tag{3.4a}$$

$$\boldsymbol{M}_\text{s}(\boldsymbol{x}_{\partial\Omega},t) = -\hat{\boldsymbol{n}}\times\boldsymbol{E}_{\text{inc}}(\boldsymbol{x}_{\partial\Omega},t) \equiv \boldsymbol{M}_{\text{inc}}(\boldsymbol{x}_{\partial\Omega},t)\,\partial\Omega_{\hat{n}} \tag{3.4b}$$

where $\hat{\boldsymbol{n}}$ is the unit normal vector at the interface $\partial\Omega$ pointing from region Ω_T to region Ω_S, and $\partial\Omega_{\hat{n}}$ is the thickness associated with the interface in the direction parallel to $\hat{\boldsymbol{n}}$. These sources have the same units as the corresponding magnetic field and electric field intensities. Note that both $\boldsymbol{J}_\text{s}$ and $\boldsymbol{M}_\text{s}$ assume finite values if the incident wave is not an impulse. This suggests that, for an infinitesimal value of $\partial\Omega_{\hat{n}}$, i.e., an interface that has no thickness, $\boldsymbol{J}_{\text{inc}}$ and $\boldsymbol{M}_{\text{inc}}$ must be an impulse function. However, in the context of FDTD, such impulsiveness is not an issue because $\partial\Omega_{\hat{n}}$ assumes the finite size of the grid discretization. Hence, we can think of $\boldsymbol{J}_{\text{inc}}$ and $\boldsymbol{M}_{\text{inc}}$ as spatially rescaled versions of $\boldsymbol{J}_\text{s}$ and $\boldsymbol{M}_\text{s}$. For example, if $\boldsymbol{J}_{\text{inc}}$ is an electric surface current density, then $\boldsymbol{J}_\text{s}$ becomes its corresponding line current density. The same applies to the $\boldsymbol{M}_{\text{inc}}$ and $\boldsymbol{M}_\text{s}$ pair.

3.5 INCIDENT PLANE-WAVE FIELDS IN A HOMOGENEOUS BACKGROUND MEDIUM

In general, the incident fields in the TF/SF formulation can be any waves that satisfy Maxwell's equations, even in inhomogeneous or random background media. At this point, we assume that the incident fields propagate in a uniform homogeneous medium characterized by the material parameters $\varepsilon(\boldsymbol{x}) = \varepsilon_{\mathrm{B}}$, $\mu(\boldsymbol{x}) = \mu_{\mathrm{B}}$, $\sigma(\boldsymbol{x}) = \sigma_{\mathrm{B}}$, and $\sigma^*(\boldsymbol{x}) = \sigma_{\mathrm{B}}^*$. Maxwell's equations for the incident wave then simplify to:

$$\varepsilon_{\mathrm{B}}\partial_t \boldsymbol{E}_{\mathrm{inc}}(\boldsymbol{x},t) + \sigma_{\mathrm{B}}\boldsymbol{E}_{\mathrm{inc}}(\boldsymbol{x},t) = \nabla\times\boldsymbol{H}_{\mathrm{inc}}(\boldsymbol{x},t) \tag{3.5a}$$

$$\mu_{\mathrm{B}}\partial_t \boldsymbol{H}_{\mathrm{inc}}(\boldsymbol{x},t) + \sigma_{\mathrm{B}}^*\boldsymbol{H}_{\mathrm{inc}}(\boldsymbol{x},t) = -\nabla\times\boldsymbol{E}_{\mathrm{inc}}(\boldsymbol{x},t) \tag{3.5b}$$

Upon solving the system of equations given by (3.5a) and (3.5b), one can immediately evaluate the surface current densities $\boldsymbol{J}_{\mathrm{inc}}(\boldsymbol{x}_{\partial\Omega},t)$ and $\boldsymbol{M}_{\mathrm{inc}}(\boldsymbol{x}_{\partial\Omega},t)$, since these are related to the incident fields through the discontinuity equations (3.4). We note that the equivalence principle, and hence the TF/SF formulation, is applicable for any incident source conditions. However, most commercial FDTD TF/SF implementations include only a plane-wave source. In this chapter, we therefore focus only on a plane-wave implementation.

To establish a foundation for subsequent developments, we now consider the plane wave as a 1-D subspace. Assume that its propagation direction is given by the Cartesian vector $\boldsymbol{p} = [p_x, p_y, p_z]$ with components related to the spherical coordinates θ and φ by [1, 12–14]:

$$p_x = \sin\theta\cos\varphi\,, \quad p_y = \sin\theta\sin\varphi\,, \quad p_z = \cos\theta \tag{3.6}$$

Note that $\boldsymbol{p}$ is effectively a projection vector describing the translation from spherical coordinates to Cartesian coordinates. By definition, a plane wave has field components that have identical amplitude and phase on planar wavefronts that are perpendicular to the direction of propagation. In essence, this means that a plane wave is really a 1-D entity [12–14] that can be fully characterized by using only a projected subspace $r \in \mathbf{R}^1$ such that:

$$\boldsymbol{E}_{\mathrm{inc}}(\boldsymbol{x},t) = \boldsymbol{E}_{\mathrm{inc}}(r,t)\,, \qquad \boldsymbol{H}_{\mathrm{inc}}(\boldsymbol{x},t) = \boldsymbol{H}_{\mathrm{inc}}(r,t) \tag{3.7}$$

Every incident field at a point $\boldsymbol{x} = [x, y, z] \in \mathbf{R}^3$ is related to a point $r \in \mathbf{R}^1$ through a one-to-many mapping determined by projecting the planar wavefronts onto the projection vector:

$$r = \boldsymbol{p}\cdot\boldsymbol{x} = p_x x + p_y y + p_z z \tag{3.8}$$

Expressing (3.5) for the specific case of a plane wave in terms of the subspace r is a straightforward process. One only needs to determine the equivalent curl operator $\nabla\times(\cdot)$ in the r-subspace. This is achieved by using the chain-rule relation:

$$\partial_\xi(\cdot) = \partial_\xi[r]\,\partial_r(\cdot) \tag{3.9}$$

where $\xi \in \{x, y, z\}$. Specifically, the chain-rule substitution is performed using (3.8) as follows:

$$\nabla \times (\cdot) = [\partial_x, \partial_y, \partial_z] \times (\cdot) = [\partial_x r, \partial_y r, \partial_z r] \times \partial_r(\cdot) = \boldsymbol{p} \times \partial_r(\cdot) \tag{3.10}$$

Equations (3.5) for the incident fields, expressed in the r-subspace, now reduce to:

$$\varepsilon_{\mathrm{B}} \partial_t \boldsymbol{E}_{\mathrm{inc}}(r,t) + \sigma_{\mathrm{B}} \boldsymbol{E}_{\mathrm{inc}}(r,t) = \boldsymbol{p} \times \partial_r \boldsymbol{H}_{\mathrm{inc}}(r,t) \tag{3.11a}$$

$$\mu_{\mathrm{B}} \partial_t \boldsymbol{H}_{\mathrm{inc}}(r,t) + \sigma_{\mathrm{B}}^{*} \boldsymbol{H}_{\mathrm{inc}}(r,t) = -\boldsymbol{p} \times \partial_r \boldsymbol{E}_{\mathrm{inc}}(r,t) \tag{3.11b}$$

Note that, because of the cross-product, the vectors $\boldsymbol{p}$, $\boldsymbol{E}_{\mathrm{inc}}$, and $\boldsymbol{H}_{\mathrm{inc}}$ form a mutually orthogonal relationship — in other words, a transverse electromagnetic wave is described.

Projection (3.8) must also be valid for all $\boldsymbol{x} \in \mathbf{R}^3$, and so it must be invariant under translation, resulting in $\boldsymbol{p} \cdot (\boldsymbol{x} + \boldsymbol{d}) = r + \boldsymbol{p} \cdot \boldsymbol{d}$ for any vector $\boldsymbol{d} \in \mathbf{R}^3$. When applied to (3.7), this displacement gives:

$$\boldsymbol{E}_{\mathrm{inc}}(\boldsymbol{x}+\boldsymbol{d}, t) = \boldsymbol{E}_{\mathrm{inc}}(r+\boldsymbol{p}\cdot\boldsymbol{d}, t) \quad \text{and} \quad \boldsymbol{H}_{\mathrm{inc}}(\boldsymbol{x}+\boldsymbol{d}, t) = \boldsymbol{H}_{\mathrm{inc}}(r+\boldsymbol{p}\cdot\boldsymbol{d}, t) \tag{3.12}$$

which will be referred to as translational invariance. Translational invariance will greatly simplify the procedure for constructing a numerical plane wave, particularly when a finite-difference scheme is of interest. Note also that no temporal transformation is necessary because the inner product $r = \boldsymbol{p} \cdot \boldsymbol{x}$ does not depend on t.

3.6 FDTD REALIZATION OF THE BASIC TF/SF FORMULATION

In the FDTD method, Maxwell's equations are discretized in space and time using central differences, with the spatial discretizations in the three Cartesian directions given by Δx, Δy, and Δz, and the temporal discretization given by Δt. To realize the TF/SF formulation in the context of the FDTD method, (3.1) can be expanded into six coupled scalar equations. Using the notation $g(i\Delta x, j\Delta y, k\Delta z, n\Delta t) = g\big|_{i,j,k}^{n}$, (3.1) can be expanded into six scalar equations suitable for FDTD time-stepping. The E_x update is given by:

$$\begin{aligned} E_x\big|_{i,j+1/2,k+1/2}^{n+1/2} &= C_{x,a} E_x\big|_{i,j+1/2,k+1/2}^{n-1/2} + C_{x,y}\left(H_z\big|_{i,j+1,k+1/2}^{n} - H_z\big|_{i,j,k+1/2}^{n}\right) \\ &\quad + C_{x,z}\left(H_y\big|_{i,j+1/2,k}^{n} - H_y\big|_{i,j+1/2,k+1}^{n}\right) - C_{x,b} J_{x,\mathrm{inc}}\big|_{i,j+1/2,k+1/2}^{n} \end{aligned} \tag{3.13}$$

where:

$$C_{x,a} = \frac{2\varepsilon\big|_{i,j+1/2,k+1/2} - \sigma\big|_{i,j+1/2,k+1/2}\Delta t}{2\varepsilon\big|_{i,j+1/2,k+1/2} + \sigma\big|_{i,j+1/2,k+1/2}\Delta t} \tag{3.14a}$$

$$C_{x,b} = \frac{2\Delta t}{2\varepsilon\big|_{i,j+1/2,k+1/2} + \sigma\big|_{i,j+1/2,k+1/2}\Delta t} \tag{3.14b}$$

$$C_{x,y} = C_{x,b} / \Delta y\,, \qquad C_{x,z} = C_{x,b} / \Delta z \tag{3.14c, d}$$

The E_y update is given by:

$$E_y\big|_{i+1/2,j,k+1/2}^{n+1/2} = C_{y,a}E_y\big|_{i+1/2,j,k+1/2}^{n-1/2} + C_{y,z}\left(H_x\big|_{i+1/2,j,k+1}^{n} - H_x\big|_{i+1/2,j,k}^{n}\right) + C_{y,x}\left(H_z\big|_{i,j,k+1/2}^{n} - H_z\big|_{i+1,j,k+1/2}^{n}\right) - C_{y,b}J_{y,\text{inc}}\big|_{i+1/2,j,k+1/2}^{n} \tag{3.15}$$

where:

$$C_{y,a} = \frac{2\varepsilon|_{i+1/2,j,k+1/2} - \sigma|_{i+1/2,j,k+1/2}\Delta t}{2\varepsilon|_{i+1/2,j,k+1/2} + \sigma|_{i+1/2,j,k+1/2}\Delta t} \tag{3.16a}$$

$$C_{y,b} = \frac{2\Delta t}{2\varepsilon|_{i+1/2,j,k+1/2} + \sigma|_{i+1/2,j,k+1/2}\Delta t}, \tag{3.16b}$$

$$C_{y,z} = C_{y,b}/\Delta z, \qquad C_{y,x} = C_{y,b}/\Delta x \tag{3.16c, d}$$

The E_z update is given by:

$$E_z\big|_{i+1/2,j+1/2,k}^{n+1/2} = C_{z,a}E_z\big|_{i+1/2,j+1/2,k}^{n-1/2} + C_{z,x}\left(H_y\big|_{i+1,j+1/2,k}^{n} - H_y\big|_{i,j+1/2,k}^{n}\right) + C_{z,y}\left(H_x\big|_{i+1/2,j,k}^{n} - H_x\big|_{i+1/2,j+1,k}^{n}\right) - C_{z,b}J_{z,\text{inc}}\big|_{i+1/2,j+1/2,k}^{n} \tag{3.17}$$

where:

$$C_{z,a} = \frac{2\varepsilon|_{i+1/2,j+1/2,k} - \sigma|_{i+1/2,j+1/2,k}\Delta t}{2\varepsilon|_{i+1/2,j+1/2,k} + \sigma|_{i+1/2,j+1/2,k}\Delta t} \tag{3.18a}$$

$$C_{z,b} = \frac{2\Delta t}{2\varepsilon|_{i+1/2,j+1/2,k} + \sigma|_{i+1/2,j+1/2,k}\Delta t} \tag{3.18b}$$

$$C_{z,x} = C_{z,b}/\Delta x, \qquad C_{z,y} = C_{z,b}/\Delta y \tag{3.18c, d}$$

Similarly, the H_x update is given by:

$$H_x\big|_{i+1/2,j,k}^{n+1} = D_{x,a}H_x\big|_{i+1/2,j,k}^{n} + D_{x,y}\left(E_z\big|_{i+1/2,j-1/2,k}^{n+1/2} - E_z\big|_{i+1/2,j+1/2,k}^{n+1/2}\right) + D_{x,z}\left(E_y\big|_{i+1/2,j,k+1/2}^{n+1/2} - E_y\big|_{i+1/2,j,k-1/2}^{n+1/2}\right) + D_{x,b}M_{x,\text{inc}}\big|_{i+1/2,j,k}^{n+1/2} \tag{3.19}$$

where:

$$D_{x,a} = \frac{2\mu|_{i+1/2,j,k} - \sigma^*|_{i+1/2,j,k}\Delta t}{2\mu|_{i+1/2,j,k} + \sigma^*|_{i+1/2,j,k}\Delta t} \tag{3.20a}$$

$$D_{x,b} = \frac{2\Delta t}{2\mu|_{i+1/2,j,k} + \sigma^*|_{i+1/2,j,k}\Delta t} \tag{3.20b}$$

$$D_{x,y} = D_{x,b} / \Delta y\,, \qquad D_{x,z} = D_{x,b} / \Delta z \tag{3.20c, d}$$

The H_y update is given by:

$$\begin{aligned} H_y\big|_{i,j+1/2,k}^{n+1} &= D_{y,a} H_y\big|_{i,j+1/2,k}^{n} + D_{y,z}\left(E_x\big|_{i,j+1/2,k-1/2}^{n+1/2} - E_x\big|_{i,j+1/2,k+1/2}^{n+1/2}\right) \\ &+ D_{y,x}\left(E_z\big|_{i+1/2,j+1/2,k}^{n+1/2} - E_z\big|_{i-1/2,j+1/2,k}^{n+1/2}\right) + D_{y,b} M_{y,\text{inc}}\big|_{i,j+1/2,k}^{n+1/2} \end{aligned} \tag{3.21}$$

where:

$$D_{y,a} = \frac{2\mu|_{i,j+1/2,k} - \sigma^*|_{i,j+1/2,k}\Delta t}{2\mu|_{i,j+1/2,k} + \sigma^*|_{i,j+1/2,k}\Delta t} \tag{3.22a}$$

$$D_{y,b} = \frac{2\Delta t}{2\mu|_{i,j+1/2,k} + \sigma^*|_{i,j+1/2,k}\Delta t} \tag{3.22b}$$

$$D_{y,z} = D_{y,b} / \Delta z\,, \qquad D_{y,x} = D_{y,b} / \Delta x \tag{3.22c, d}$$

The H_z update is given by:

$$\begin{aligned} H_z\big|_{i,j,k+1/2}^{n+1} &= D_{z,a} H_z\big|_{i,j,k+1/2}^{n} + D_{z,x}\left(E_y\big|_{i-1/2,j,k+1/2}^{n+1/2} - E_y\big|_{i+1/2,j,k+1/2}^{n+1/2}\right) \\ &+ D_{z,y}\left(E_x\big|_{i,j+1/2,k+1/2}^{n+1/2} - E_x\big|_{i,j-1/2,k+1/2}^{n+1/2}\right) + D_{z,b} M_{z,\text{inc}}\big|_{i,j,k+1/2}^{n+1/2} \end{aligned} \tag{3.23}$$

where:

$$D_{z,a} = \frac{2\mu|_{i,j,k+1/2} - \sigma^*|_{i,j,k+1/2}\Delta t}{2\mu|_{i,j,k+1/2} + \sigma^*|_{i,j,k+1/2}\Delta t} \tag{3.24a}$$

$$D_{z,b} = \frac{2\Delta t}{2\mu|_{i,j,k+1/2} + \sigma^*|_{i,j,k+1/2}\Delta t} \tag{3.24b}$$

$$D_{z,x} = D_{z,b} / \Delta x\,, \qquad D_{z,y} = D_{z,b} / \Delta y \tag{3.24c, d}$$

3.7 ON CONSTRUCTING AN EXACT FDTD TF/SF PLANE-WAVE SOURCE

Some insights can be gleaned from the TF/SF formulation of (3.13) – (3.24) that are valuable for understanding how to construct an exact TF/SF plane-wave source condition. In these E-field and H-field time-stepping expressions, $(J_{x,\text{inc}}\big|^{n}_{i,j+1/2,k+1/2},\ J_{y,\text{inc}}\big|^{n}_{i+1/2,j,k+1/2},\ J_{z,\text{inc}}\big|^{n}_{i+1/2,j+1/2,k})$ and $(M_{x,\text{inc}}\big|^{n+1/2}_{i+1/2,j,k},\ M_{y,\text{inc}}\big|^{n+1/2}_{i,j+1/2,k},\ M_{z,\text{inc}}\big|^{n+1/2}_{i,j,k+1/2})$ are fictitious electric and magnetic current densities that have been mathematically defined to generate a desired incident plane wave that is completely confined within the total-field region, Ω_{T}. In FDTD parlance, these current densities comprise "soft-sources," which necessarily radiate into both Ω_{T} and the scattered-field region, Ω_{S}. It is possible to think of the M sources as radiating a wave into Ω_{S} that counteracts the wave radiated there by the J sources, such that the total wave propagating in Ω_{S} is canceled and the desired incident plane wave propagates only in Ω_{T}. Inexact cancellation would cause a spurious wave to propagate in Ω_{S}. Since the electromagnetic theory forming the foundation of the TF/SF concept does allow exact cancellation, any observed spurious wave in Ω_{S} results from an imperfect algorithmic implementation, i.e., numerical error. This error is often called the *leakage error* due to the appearance of scattered fields in Ω_{S} even in the absence of a scattering structure in Ω_{T}. Note that the incident wave in Ω_{T} would also be corrupted by such an error.

What follows is key to understanding how to completely eliminate the leakage error. Recall that the incident current densities $\boldsymbol{J}_{\text{inc}}$ and $\boldsymbol{M}_{\text{inc}}$ (as well as $\boldsymbol{J}_{\text{s}}$ and $\boldsymbol{M}_{\text{s}}$) are not independent vector fields. They are, in fact, directly related to fields governed by the equivalent Maxwell's equations, which for a plane wave were shown to reduce to the 1-D form (3.11). Again, this is because these current density vectors and the incident fields $\boldsymbol{E}_{\text{inc}}$ and $\boldsymbol{H}_{\text{inc}}$ are related by the discontinuity equations (3.4). Obtaining the wave solution $\boldsymbol{E}_{\text{inc}}$ and $\boldsymbol{H}_{\text{inc}}$ is therefore equivalent to solving for the proper current densities $\boldsymbol{J}_{\text{inc}}$ and $\boldsymbol{M}_{\text{inc}}$ (also $\boldsymbol{J}_{\text{s}}$ and $\boldsymbol{M}_{\text{s}}$).

Hence, constructing a TF/SF formulation designed such that there is no leakage error (i.e., so that the incident wave is perfectly confined within the total-field region) mathematically amounts to ensuring that the same discrete vector calculus operators are used to discretize both (3.1) and (3.5); for the particular case of a plane-wave source, the latter system is (3.11). In this way, the numerical errors for both sets of Maxwell's equations (3.1) and (3.11) are identical, or equivalently, their numerical dispersion relations are identical. We shall call such a construction a *perfectly matched* or an *exact* TF/SF formulation.

3.8 FDTD DISCRETE PLANE-WAVE SOURCE FOR THE EXACT TF/SF FORMULATION

We now turn our attention to the construction of properly discretized FDTD equations such that a plane-wave source in the projected space, $r \in \mathbf{R}^1$, is perfectly equivalent to that which propagates in the full space, $\boldsymbol{x} \in \mathbf{R}^3$. This essentially amounts to implementing the same chain-rule procedure of (3.9) and (3.10), but this time with discretized operators. Specifically, consider the chain-rule applied to a function $g(\boldsymbol{x},t) = g(x,y,z,t) = g(r,t)$, which in the continuous formulation is written as:

$$\partial_\xi g(x,y,z,t) \;=\; \partial_\xi[r]\,\partial_r g(r,t) \tag{3.25}$$

Applying the central-difference scheme to the left-hand side of this equation for the case $\boldsymbol{\xi} = \boldsymbol{x}$ results in:

$$\partial_x g(x,y,z,t) \;\approx\; \frac{g\big|^{n}_{i+1/2,j,k} - g\big|^{n}_{i-1/2,j,k}}{\Delta x} \tag{3.26}$$

Recall that $\boldsymbol{x}$ and r are related by $r = p_x x + p_y y + p_z z$. Fields in the main FDTD grid are discretized such that they only exist at the points $\boldsymbol{x} = [i\Delta x, j\Delta y, k\Delta z]$. Therefore, the corresponding points $r \in \mathbf{R}^1$ are:

$$r_{i,j,k} = p_x i\Delta x + p_y j\Delta y + p_z k\Delta z \tag{3.27}$$

Often, there is a need to refer to fields that lie halfway along the edge of an FDTD voxel. Thus, for notational convenience, a shift in this grid will also be indexed by:

$$r_{i\pm p_x\Delta x/2,\, j\pm p_y\Delta y/2,\, k\pm p_z\Delta z/2} = p_x\Delta x(i\pm 1/2) + p_y\Delta y(j\pm 1/2) + p_z\Delta z(k\pm 1/2) \tag{3.28}$$

For example, a change of Δx must result in a change in this r grid by $\Delta r_x = r_{i+p_x\Delta x,j,k} - r_{i,j,k} = p_x\Delta x$. This means that the same central difference when applied to the right-hand side of the chain-rule $\partial_\xi g(x,y,z,t) = \partial_\xi[r]\,\partial_r g(r,t)$ due to a change in x is evaluated as:

$$\begin{aligned}\partial_x[r]\,\partial_r g(r,t) &\approx \left(\frac{r_{i+p_x\Delta x/2,j,k} - r_{i-p_x\Delta x/2,j,k}}{\Delta x}\right)\cdot\left(\frac{g\big|^n_{r_{i+p_x\Delta x/2,j,k}} - g\big|^n_{r_{i-p_x\Delta x/2,j,k}}}{p_x\Delta x}\right)\\ &\approx \frac{g\big|^n_{r_{i+p_x\Delta x/2,j,k}} - g\big|^n_{r_{i-p_x\Delta x/2,j,k}}}{\Delta x}\end{aligned} \tag{3.29}$$

Equating (3.29) to (3.26) results in $\left(g\big|^n_{i+1/2,j,k} - g\big|^n_{i-1/2,j,k}\right)/\Delta x = \left(g\big|^n_{r_{i+p_x\Delta x/2,j,k}} - g\big|^n_{r_{i-p_x\Delta x/2,j,k}}\right)/\Delta x$, which means that $g\big|^n_{i+1/2,j,k} = g\big|^n_{r_{i+p_x\Delta x/2,j,k}}$ and $g\big|^n_{i-1/2,j,k} = g\big|^n_{r_{i-p_x\Delta x/2,j,k}}$. Effectively, this is a manifestation of the translational invariance in (3.12). The shift relationships for each of the three Cartesian directions are given by:

$$g\big|^n_{i\pm1/2,j,k} = g\big|^n_{r_{i\pm p_x\Delta x/2,j,k}}, \quad g\big|^n_{i,j\pm1/2,k} = g\big|^n_{r_{i,j\pm p_y\Delta y/2,k}}, \quad g\big|^n_{i,j,k\pm1/2} = g\big|^n_{r_{i,j,k\pm p_z\Delta z/2}} \tag{3.30}$$

The FDTD incident plane wave expressed in terms of the discrete $r|_{i,j,k}$ grid is now readily available. This solution can also be formally deduced from (3.11). However, it is much easier to take the E-field updates of (3.13) – (3.18) and the H-field updates of (3.19) – (3.24) directly, but ignoring electric and magnetic current densities and then using (3.30) to translate these into equivalent equations on an $r|_{i,j,k}$-grid. (Here, with no scatterer present, the fields inside the total-field region must be those of the incident wave having no source terms.) This yields the new E-field updates:

$$\begin{aligned}E_{x,\text{inc}}\big|^{n+1/2}_{r_{i,j+p_y\Delta y/2,k+1/2}} &= C_a E_{x,\text{inc}}\big|^{n-1/2}_{r_{i,j+p_y\Delta y/2,k+1/2}} + C_{b,y}\left(H_{z,\text{inc}}\big|^n_{r_{i,j+p_y\Delta y,k+p_z\Delta z/2}} - H_{z,\text{inc}}\big|^n_{r_{i,j,k+p_z\Delta z/2}}\right)\\ &+ C_{b,z}\left(H_{y,\text{inc}}\big|^n_{r_{i,j+p_y\Delta y/2,k}} - H_{y,\text{inc}}\big|^n_{r_{i,j+p_y\Delta y/2,k+p_z\Delta z}}\right)\end{aligned} \tag{3.31}$$

$$\begin{aligned} E_{y,\text{inc}}\Big|_{r_{i+p_x\Delta x/2,\,j,\,k+p_z\Delta z/2}}^{n+1/2} &= C_a E_{y,\text{inc}}\Big|_{r_{i+p_x\Delta x/2,\,j,\,k+p_z\Delta z/2}}^{n-1/2} + C_{b,z}\left(H_{x,\text{inc}}\Big|_{r_{i+p_x\Delta x/2,\,j,\,k+p_z\Delta z}}^{n} - H_{x,\text{inc}}\Big|_{r_{i+p_x\Delta x/2,\,j,\,k}}^{n} \right) \\ &\quad + C_{b,x}\left(H_{z,\text{inc}}\Big|_{r_{i,\,j,\,k+p_z\Delta z/2}}^{n} - H_{z,\text{inc}}\Big|_{r_{i+p_x\Delta x,\,j,\,k+p_z\Delta z/2}}^{n} \right) \end{aligned} \tag{3.32}$$

$$\begin{aligned} E_{z,\text{inc}}\Big|_{r_{i+p_x\Delta x/2,\,j+p_y\Delta y/2,\,k}}^{n+1/2} &= C_a E_{z,\text{inc}}\Big|_{r_{i+p_x\Delta x/2,\,j+p_y\Delta y/2,\,k}}^{n-1/2} + C_{b,x}\left(H_{y,\text{inc}}\Big|_{r_{i+p_x\Delta x,\,j+p_y\Delta y/2,\,k}}^{n} - H_{y,\text{inc}}\Big|_{r_{i,\,j+p_y\Delta y/2,\,k}}^{n} \right) \\ &\quad + C_{b,y}\left(H_{x,\text{inc}}\Big|_{r_{i+p_x\Delta x/2,\,j,\,k}}^{n} - H_{x,\text{inc}}\Big|_{r_{i+p_x\Delta x/2,\,j+p_y\Delta y,\,k}}^{n} \right) \end{aligned} \tag{3.33}$$

where:

$$C_a = \frac{2\varepsilon_B - \sigma_B \Delta t}{2\varepsilon_B + \sigma_B \Delta t} \tag{3.34a}$$

$$C_{b,x} = \frac{2\Delta t / \Delta x}{2\varepsilon_B + \sigma_B \Delta t}, \quad C_{b,y} = \frac{2\Delta t / \Delta y}{2\varepsilon_B + \sigma_B \Delta t}, \quad C_{b,z} = \frac{2\Delta t / \Delta z}{2\varepsilon_B + \sigma_B \Delta t} \tag{3.34b, c, d}$$

Similarly, the new H-field updates are:

$$\begin{aligned} H_{x,\text{inc}}\Big|_{r_{i+p_x\Delta x/2,\,j,\,k}}^{n+1} &= D_a H_{x,\text{inc}}\Big|_{r_{i+p_x\Delta x/2,\,j,\,k}}^{n} + D_{b,y}\left(E_{z,\text{inc}}\Big|_{r_{i+p_x\Delta x/2,\,j-p_y\Delta y/2,\,k}}^{n+1/2} - E_{z,\text{inc}}\Big|_{r_{i+p_x\Delta x/2,\,j+p_y\Delta y/2,\,k}}^{n+1/2} \right) \\ &\quad + D_{b,z}\left(E_{y,\text{inc}}\Big|_{r_{i+p_x\Delta x/2,\,j,\,k+p_z\Delta z/2}}^{n+1/2} - E_{y,\text{inc}}\Big|_{r_{i+p_x\Delta x/2,\,j,\,k-p_z\Delta z/2}}^{n+1/2} \right) \end{aligned} \tag{3.35}$$

$$\begin{aligned} H_{y,\text{inc}}\Big|_{r_{i,\,j+p_y\Delta y/2,\,k}}^{n+1} &= D_a H_{y,\text{inc}}\Big|_{r_{i,\,j+p_y\Delta y/2,\,k}}^{n} + D_{b,z}\left(E_{x,\text{inc}}\Big|_{r_{i,\,j+p_y\Delta y/2,\,k-p_z\Delta z/2}}^{n+1/2} - E_{x,\text{inc}}\Big|_{r_{i,\,j+p_y\Delta y/2,\,k+p_z\Delta z/2}}^{n+1/2} \right) \\ &\quad + D_{b,x}\left(E_{z,\text{inc}}\Big|_{r_{i+p_x\Delta x/2,\,j+p_y\Delta y/2,\,k}}^{n+1/2} - E_{z,\text{inc}}\Big|_{r_{i-p_x\Delta x/2,\,j+p_y\Delta y/2,\,k}}^{n+1/2} \right) \end{aligned} \tag{3.36}$$

$$\begin{aligned} H_{z,\text{inc}}\Big|_{r_{i,\,j,\,k+p_z\Delta z/2}}^{n+1} &= D_a H_{z,\text{inc}}\Big|_{r_{i,\,j,\,k+p_z\Delta z/2}}^{n} + D_{b,x}\left(E_{y,\text{inc}}\Big|_{r_{i-p_x\Delta x/2,\,j,\,k+p_z\Delta z/2}}^{n+1/2} - E_{y,\text{inc}}\Big|_{r_{i+p_x\Delta x/2,\,j,\,k+p_z\Delta z/2}}^{n+1/2} \right) \\ &\quad + D_{b,y}\left(E_{x,\text{inc}}\Big|_{r_{i,\,j+p_y\Delta y/2,\,k+p_z\Delta z/2}}^{n+1/2} - E_{x,\text{inc}}\Big|_{r_{i,\,j-p_y\Delta y/2,\,k+p_z\Delta z/2}}^{n+1/2} \right) \end{aligned} \tag{3.37}$$

where:

$$D_a = \frac{2\mu_B - \sigma_B^* \Delta t}{2\mu_B + \sigma_B^* \Delta t} \tag{3.38a}$$

$$D_{b,x} = \frac{2\Delta t/\Delta x}{2\mu_B + \sigma_B^* \Delta t}, \quad D_{b,y} = \frac{2\Delta t/\Delta y}{2\mu_B + \sigma_B^* \Delta t}, \quad D_{b,z} = \frac{2\Delta t/\Delta z}{2\mu_B + \sigma_B^* \Delta t} \tag{3.38b, c, d}$$

In combination with a source function, the *E*-field updates of (3.31) – (3.34) and the *H*-field updates of (3.35) – (3.38) propagate a 1-D FDTD plane wave that can be coupled into the 3-D FDTD TF/SF equations (3.13) – (3.18) and (3.19) – (3.24) with a perfect match. However, with (3.31) – (3.34) and (3.35) – (3.38) in their current form, these update equations have a computational complexity of $O(N^3)$, because the $r|_{i,j,k}$ grid is still indexed by three integers i, j, and k, and for a 3-D FDTD grid will have $O(N^3)$ discrete points. In the following section, we will show how the complexity of the $r|_{i,j,k}$ grid can be reduced to $O(N)$.

3.9 AN EFFICIENT INTEGER MAPPING

Consider an infinite straight line that is parallel to (and defines) the direction of propagation of the plane-wave source. We introduce a restriction such that this line must pass through at least two field points, $\boldsymbol{x}$, in the FDTD grid defined by $\{\boldsymbol{x} \in [i\Delta x, j\Delta y, k\Delta z]\}$. These two points are separated by integer shifts of the spatial discretization, and do not necessarily have to lie in the region that is simulated. This is equivalent to defining a propagation angle such that:

$$p_x \Delta x = m_x \Delta r, \quad p_y \Delta y = m_y \Delta r, \quad p_z \Delta z = m_z \Delta r \tag{3.39a, b, c}$$

where m_x, m_y, and m_z must be integers, and Δr is a uniform discrete spacing in $r_{i,j,k} \in \mathbf{R}^1$ as yet to be determined. The azimuthal and polar angles that determine this propagation direction can be determined by:

$$\tan\varphi = \frac{p_y}{p_x} = \frac{m_y \Delta x}{m_x \Delta y} \tag{3.40a}$$

$$\tan\theta = \frac{\sqrt{p_x^2 + p_y^2}}{p_z} = \frac{\sqrt{(m_x/\Delta x)^2 + (m_y/\Delta y)^2}}{(m_z/\Delta z)} \tag{3.40b}$$

We call the angles defined by these integers *rational angles* because, for example, (3.40a) is a rational number if the aspect ratio is also a rational number. Note also that these rational angles are unique only up to a scalar factor, i.e., $[m_x, m_y, m_z] = [1, 1, 1]$ and $[m_x, m_y, m_z] = [2, 2, 2]$ propagate at the same angle.

Upon substituting (3.39) into (3.27), we obtain $r_{i,j,k} = (p_x i\Delta x + p_y j\Delta y + p_z k\Delta z) = (m_x i + m_y j + m_z k)\Delta r$. This suggests that there is an $r_{i,j,k}$ grid with uniform grid spacing that can be indexed by:

$$r_{i,j,k} = r_{i_r} = i_r \Delta r \tag{3.41}$$

where

$$i_r = m_x i + m_y j + m_z k \tag{3.42}$$

This new integer-mapped grid reduces the required complexity because (3.42) is now a many-to-one mapping. That is, many points $\{\boldsymbol{x} \in [i\Delta x, j\Delta y, k\Delta z]\}$ in the FDTD grid are mapped to a *single* point on the $r_{i,j,k}$ grid. Specifically, the complexity for this new grid is now only $O[(m_x + m_y + m_z)N]$. As an example, the best-case scenario (ignoring propagation along the grid axes) occurs when the plane wave propagates along the grid diagonal characterized by $[m_x, m_y, m_z] = [1, 1, 1]$. Computing this plane wave requires only $O(3N)$ points compared to $O(N^3)$ points.

Note that in (3.31) – (3.38), all six incident field components must be used to properly propagate the incident field, and each component propagates on a uniform grid with identical spacing. However, in general, due to the coupled nature of (3.31) – (3.38), the discrete field points in the grid for each component do not align with the other components — they are offset by $[\pm m_x/2, \pm m_y/2, \pm m_z/2]$. When relating the shifts on one grid of one field component with another component, these shifts are indexed by:

$$r_{i\pm p_x\Delta x/2,\, j\pm p_y\Delta y/2,\, k\pm p_z\Delta z/2} = i_r\Delta r + (\pm m_x/2 \pm m_y/2 \pm m_z/2)\Delta r \tag{3.43}$$

Using (3.42) together with (3.43), we can now map (3.31) – (3.38) from the (i, j, k) indices to the i_r index. The resulting updating expressions for the plane-wave E-fields are:

$$\begin{aligned} E_{x,\text{inc}}\Big|_{i_r+m_y/2+m_z/2}^{n+1/2} = {} & C_a E_{x,\text{inc}}\Big|_{i_r+m_y/2+m_z/2}^{n-1/2} + C_{b,y}\left(H_{z,\text{inc}}\Big|_{i_r+m_y+m_z/2}^{n} - H_{z,\text{inc}}\Big|_{i_r+m_z/2}^{n}\right) \\ & + C_{b,z}\left(H_{y,\text{inc}}\Big|_{i_r+m_y/2}^{n} - H_{y,\text{inc}}\Big|_{i_r+m_y/2+m_z}^{n}\right) \end{aligned} \tag{3.44}$$

$$\begin{aligned} E_{y,\text{inc}}\Big|_{i_r+m_x/2+m_z/2}^{n+1/2} = {} & C_a E_{y,\text{inc}}\Big|_{i_r+m_x/2+m_z/2}^{n-1/2} + C_{b,z}\left(H_{x,\text{inc}}\Big|_{i_r+m_x/2+m_z}^{n} - H_{x,\text{inc}}\Big|_{i_r+m_x/2}^{n}\right) \\ & + C_{b,x}\left(H_{z,\text{inc}}\Big|_{i_r+m_z/2}^{n} - H_{z,\text{inc}}\Big|_{i_r+m_x+m_z/2}^{n}\right) \end{aligned} \tag{3.45}$$

$$\begin{aligned} E_{z,\text{inc}}\Big|_{i_r+m_x/2+m_y/2}^{n+1/2} = {} & C_a E_{z,\text{inc}}\Big|_{i_r+m_x/2+m_y/2}^{n-1/2} + C_{b,x}\left(H_{y,\text{inc}}\Big|_{i_r+m_x+m_y/2}^{n} - H_{y,\text{inc}}\Big|_{i_r+m_y/2}^{n}\right) \\ & + C_{b,y}\left(H_{x,\text{inc}}\Big|_{i_r+m_x/2}^{n} - H_{x,\text{inc}}\Big|_{i_r+m_x/2+m_y}^{n}\right) \end{aligned} \tag{3.46}$$

The corresponding updating expressions for the plane-wave H-fields are:

$$\begin{aligned} H_{x,\text{inc}}\Big|_{i_r+m_x/2}^{n+1} = {} & D_a H_{x,\text{inc}}\Big|_{i_r+m_x/2}^{n} + D_{b,y}\left(E_{z,\text{inc}}\Big|_{i_r+m_x/2-m_y/2}^{n+1/2} - E_{z,\text{inc}}\Big|_{i_r+m_x/2+m_y/2}^{n+1/2}\right) \\ & + D_{b,z}\left(E_{y,\text{inc}}\Big|_{i_r+m_x/2+m_z/2}^{n+1/2} - E_{y,\text{inc}}\Big|_{i_r+m_x/2-m_z/2}^{n+1/2}\right) \end{aligned} \tag{3.47}$$

$$\begin{aligned} H_{y,\text{inc}}\Big|_{i_r+m_y/2}^{n+1} &= D_a H_{y,\text{inc}}\Big|_{i_r+m_y/2}^{n} + D_{y,z}\left(E_{x,\text{inc}}\Big|_{i_r+m_y/2-m_z/2}^{n+1/2} - E_{x,\text{inc}}\Big|_{i_r+m_y/2+m_z/2}^{n+1/2} \right) \\ &\quad + D_{b,x}\left(E_{z,\text{inc}}\Big|_{i_r+m_x/2+m_y/2}^{n+1/2} - E_{z,\text{inc}}\Big|_{i_r-m_x/2+m_y/2}^{n+1/2} \right) \end{aligned} \tag{3.48}$$

$$\begin{aligned} H_{z,\text{inc}}\Big|_{i_r+m_z/2}^{n+1} &= D_a H_{z,\text{inc}}\Big|_{i_r+m_z/2}^{n} + D_{b,x}\left(E_{y,\text{inc}}\Big|_{i_r-m_x/2+m_z/2}^{n+1/2} - E_{y,\text{inc}}\Big|_{i_r+m_x/2+m_z/2}^{n+1/2} \right) \\ &\quad + D_{b,y}\left(E_{x,\text{inc}}\Big|_{i_r+m_y/2+m_z/2}^{n+1/2} - E_{x,\text{inc}}\Big|_{i_r-m_y/2+m_z/2}^{n+1/2} \right) \end{aligned} \tag{3.49}$$

Equations (3.44) – (3.49) constitute six discrete interrelated time-stepping expressions that propagate fields on six coupled 1-D grids. When supplied with a source function and initial conditions, these equations simulate an electromagnetic wave decomposed into six components and propagating through all space. Figure 3.2 illustrates the layout of the six 1-D grids, as well as the field components that would be used in the E_z update of (3.46). Note that the six grids are physically staggered from each other by half-integer offsets because of the staggered nature of the fields in the main 3-D grid. Furthermore, note that the computational stencil does not use nearest neighbors in the 1-D IFAs. At each time-step, fields are propagated a distance which is an integer multiple of Δr that depends on the field component and on the set of $\{m_x, m_y, m_z\}$.

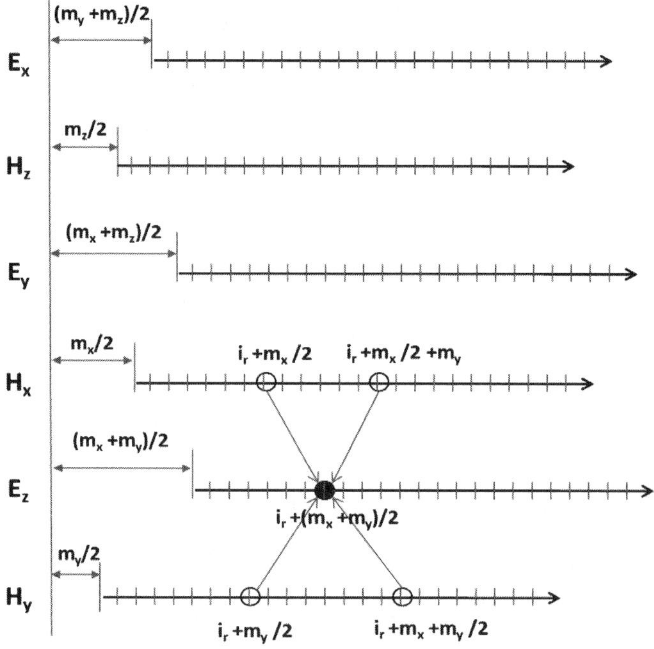

Fig. 3.2 Geometric relationship between the six coupled 1-D grids. The half-integer shifts are a result of the staggered placement of the field components in the main 3-D FDTD grid. Circles denote the field components that are used to update E_z, and their relationship to each other. *Source:* Tan and Potter, *IEEE Trans. Antennas and Propagation*, Vol. 58, 2010, pp. 2641–2648, ©2010 IEEE.

Crucially, the numerical fields in the wave propagated by (3.44) – (3.49), which are essentially arbitrary, have a dispersion relation that is *identical* to that of the main FDTD grid. Hence, when coupled to the main FDTD grid via the TF/SF formulation, there is zero leakage into the scattered-field region of the main grid.

To specifically create a plane-wave source as an incident field, one must define a polarization as well as a propagation direction. A systematic procedure for developing boundary/source conditions such that the desired plane wave is generated is the subject of the next discussion.

3.10 BOUNDARY CONDITIONS AND VECTOR PLANE-WAVE POLARIZATION

The general solution for a linearly polarized vector plane wave in continuous space and time can be compactly written as:

$$\boldsymbol{E}_{\text{inc}}(\boldsymbol{x}, t) = \boldsymbol{P}_e(\psi, \varphi, \theta)\, f\left(t - \sqrt{\mu_{\text{B}}\varepsilon_{\text{B}}}\, \boldsymbol{p}\cdot\boldsymbol{x}\right) \tag{3.50}$$

$$\boldsymbol{H}_{\text{inc}}(\boldsymbol{x}, t) = \boldsymbol{P}_h(\psi, \varphi, \theta)\, f\left(t - \sqrt{\mu_{\text{B}}\varepsilon_{\text{B}}}\, \boldsymbol{p}\cdot\boldsymbol{x}\right) / \eta_{\text{B}} \tag{3.51}$$

where $\boldsymbol{P}_e = [P_{e_x}, P_{e_y}, P_{e_z}]$ and $\boldsymbol{P}_h = [P_{h_x}, P_{h_y}, P_{h_z}]$ account for the Cartesian projection of the polarized incident fields (as shown in Fig. 5.14 of [1]), $f(x, t)$ is a known excitation function, and $\eta_{\text{B}} = \sqrt{\mu_{\text{B}}/\varepsilon_{\text{B}}}$ is the wave impedance of the uniform background medium. The components of the projection vectors are given by:

$$P_{e_x}(\psi, \theta, \varphi) = \cos\psi \sin\varphi - \sin\psi \cos\theta \cos\varphi \tag{3.52a}$$

$$P_{e_y}(\psi, \theta, \varphi) = -\cos\psi \cos\varphi - \sin\psi \cos\theta \sin\varphi \tag{3.52b}$$

$$P_{e_z}(\psi, \theta, \varphi) = \sin\psi \sin\theta \tag{3.52c}$$

$$P_{h_x}(\psi, \theta, \varphi) = \sin\psi \sin\varphi + \cos\psi \cos\theta \cos\varphi \tag{3.53a}$$

$$P_{h_y}(\psi, \theta, \varphi) = -\sin\psi \cos\varphi + \cos\psi \cos\theta \sin\varphi \tag{3.53b}$$

$$P_{h_z}(\psi, \theta, \varphi) = -\cos\psi \sin\theta \tag{3.53c}$$

This formulation makes use of the analytical solution, $f(\boldsymbol{x}, t) = f(t - \sqrt{\mu_{\text{B}}\varepsilon_{\text{B}}}\, \boldsymbol{p}\cdot\boldsymbol{x})$, for the plane wave, where $\sqrt{\mu_{\text{B}}\varepsilon_{\text{B}}}\, \boldsymbol{p}\cdot\boldsymbol{x}$ represents the temporal delay. From the discussion in Section 3.4, mutual orthogonality implies $\eta_{\text{B}}\boldsymbol{H}_{\text{inc}} = \boldsymbol{p}\times\boldsymbol{E}_{\text{inc}}$ and hence $\boldsymbol{P}_h = \boldsymbol{p}\times\boldsymbol{P}_e$, which is valid for any arbitrary source function $f(\boldsymbol{x}, t)$. This means that, for a given propagation direction $\boldsymbol{p}$, knowing the incident vector field $\boldsymbol{E}_{\text{inc}}$ at one location is sufficient to determine the incident vector $\boldsymbol{H}_{\text{inc}}$ uniquely. It then follows that establishing a boundary condition for a plane-wave source requires only one set of these incident vectors. For example, the Cartesian x-components of the electric and magnetic fields are:

$$E_{x,\text{inc}}(\boldsymbol{x}, t) = (\cos\psi \sin\varphi - \sin\psi \cos\theta \cos\varphi)\, f\left(t - \sqrt{\mu_{\text{B}}\varepsilon_{\text{B}}}\, \boldsymbol{p}\cdot\boldsymbol{x}\right) \tag{3.54}$$

$$H_{x,\text{inc}}(\boldsymbol{x}, t) = (\sin\psi \sin\varphi + \cos\psi \cos\theta \cos\varphi)\, f\left(t - \sqrt{\mu_{\text{B}}\varepsilon_{\text{B}}}\, \boldsymbol{p}\cdot\boldsymbol{x}\right) / \eta_{\text{B}} \tag{3.55}$$

The above formulation provides a very simple procedure for establishing the boundary conditions required by the FDTD plane-wave update equations (3.44) – (3.49). Take the particular case of the x-component of the H-field, where the field in the incident wave source grid will be assigned using $\boldsymbol{H}(\boldsymbol{x}, t) = \boldsymbol{H}(r, t)$. It is important to note that, in one time-step, a plane wave in the main FDTD grid propagates at most a distance of $\max\{\Delta\xi\}$ with $\xi \in \{x, y, z\}$. Projecting this into the source grids, and using the rational angle conditions $p_\xi \Delta\xi = m_\xi \Delta r$, the plane wave propagates a distance of $\max\{|m_\xi|\}$ points in the source grids. This means that a set of $\max\{|m_\xi|\}$ points must be sourced in order to properly initiate the plane wave. Using (3.42) and (3.43), the complete set of source conditions at the boundary points in the incident magnetic field grids is given by the following for $0 \le i_r < \max\{|m_\xi|\}$:

$$H_{x,\text{inc}}\Big|_{i_r + m_x/2}^{n+1} = P_{h_x} \cdot f\left[n\Delta t - \sqrt{\mu_{\text{B}}\varepsilon_{\text{B}}}\,(i_r + m_x/2)\Delta r\right] \tag{3.56a}$$

$$H_{y,\text{inc}}\Big|_{i_r + m_y/2}^{n+1} = P_{h_y} \cdot f\left[n\Delta t - \sqrt{\mu_{\text{B}}\varepsilon_{\text{B}}}\,(i_r + m_y/2)\Delta r\right] \tag{3.56b}$$

$$H_{z,\text{inc}}\Big|_{i_r + m_z/2}^{n+1} = P_{h_z} \cdot f\left[n\Delta t - \sqrt{\mu_{\text{B}}\varepsilon_{\text{B}}}\,(i_r + m_z/2)\Delta r\right] \tag{3.56c}$$

Note that this simple boundary/source condition assumes that the wave propagates analytically over this region (i.e., the time delay assumes an analytical phase velocity). Therefore, it is very important that the incident fields coupled into the main grid through the FDTD TF/SF formulation *not* be taken from these boundary points; fields to be coupled have to be from indices $i_r \ge \max\{|m_\xi|\}$. A more elaborate way that allows one to partially account for numerical dispersion in these boundary points can be found in [12].

Lastly, the formulation so far has been constructed for a general 3-D plane-wave problem. To reduce this solution to a 1-D or 2-D problem, one only needs to set the integer propagation values $[m_x, m_y, m_z]$ accordingly. For example, a plane wave propagating along the x-direction is characterized by $[m_x, m_y, m_z] = [1, 0, 0]$.

3.11 REQUIRED CURRENT DENSITIES $\boldsymbol{J}_{\text{inc}}$ AND $\boldsymbol{M}_{\text{inc}}$

We next consider the current densities appearing in (3.13), (3.15), (3.17), (3.19), (3.21), and (3.23) in the context of the exact TF/SF plane-wave source. Mapping the proper current densities $\boldsymbol{J}_{\text{inc}}$ and $\boldsymbol{M}_{\text{inc}}$ from the incident-field grids developed in the previous sections follows directly from the discontinuity equations (3.4). In the FDTD TF/SF implementation, an arbitrary point on the closed virtual surface that defines the interface between the total-field and scattered-field regions has the outward unit normal vector $\hat{\boldsymbol{n}} = [\hat{n}_x, \hat{n}_y, \hat{n}_z]$. For example, expanding (3.4a) results in:

$$J_{x,s} = \hat{n}_y H_{z,\text{inc}} - \hat{n}_z H_{y,\text{inc}} \tag{3.57a}$$

$$J_{y,s} = \hat{n}_z H_{x,\text{inc}} - \hat{n}_x H_{z,\text{inc}} \tag{3.57b}$$

$$J_{z,s} = \hat{n}_x H_{y,\text{inc}} - \hat{n}_y H_{x,\text{inc}} \tag{3.57c}$$

In most Cartesian FDTD grids, the TF/SF interface is rectangular and aligned with major grid axes. In this case, the components of the normal vectors to the TF/SF interface assume values of $\hat{n}_\xi = 0, \pm 1$. For example, $\hat{\boldsymbol{n}} = [0, \pm 1, 0]$ are the outward normal vectors for the two planar sides of the TF/SF interface perpendicular to the y-axis, and $\hat{\boldsymbol{n}} = [0, 0, \pm 1]$ are the outward normal vectors for the two planar sides perpendicular to the z-axis. From (3.57a), these are the only four surfaces where the current density $J_{x,s}$ resides. Hence, the current densities accounting for the H-field discontinuities across the interfaces normal to $\hat{n} = [0, \pm 1, 0]$ reduce to $J_{x,s} = \pm H_{z,\text{inc}}$. According to (3.4a), these values can be rescaled by $\hat{n} = [0, \pm \Delta y, 0]$, resulting in $J_{x,\text{inc}} = \pm H_{z,\text{inc}} / \Delta y$. The current densities responsible for the H-field discontinuities across the interfaces normal to $\hat{n} = [0, 0, \pm 1]$ are $J_{x,s} = \pm H_{y,\text{inc}}$, which after rescaling by $\hat{n} = [0, 0, \pm \Delta z]$ become $J_{x,\text{inc}} = \pm H_{y,\text{inc}} / \Delta z$. The remaining electric and magnetic current densities are found in a similar manner. This rescaling is needed because Maxwell's equations (and their FDTD counterparts) with incident-wave forcing terms are expressed in terms of $\boldsymbol{J}_{\text{inc}}$ and $\boldsymbol{M}_{\text{inc}}$.

Consider a rectangular TF/SF interface having a size indexed from $[I_0, J_0, K_0]$ to $[I_1, J_1, K_1]$. For this interface, the equations shown below provide the full set of required current densities expressed in terms of the electric and magnetic fields of the incident wave, multiplied by the corresponding FDTD coefficients. Note that the integer coordinate transformation follows from the many-to-one integer mapping of (3.42) or $i_r = m_x i + m_y j + m_z k$ [in conjunction with the mapping shift (3.43)]. At each of the six faces of the TF/SF interface, the required surface electric current densities to be used in (3.13), (3.15), and (3.17) are:

y-z faces ($J_0 \le j \le J_1$ and $K_0 \le k \le K_1$)

At $i = I_0$:

$$C_{y,b} J_{y,\text{inc}}\Big|^n_{I_0+1/2,\, j,\, k+1/2} = -C_{y,x} H_{z,\text{inc}}\Big|^n_{I_0,\, j,\, k+1/2} = -C_{y,x} H_{z,\text{inc}}\Big|^n_{i_r+m_z/2} \tag{3.58a}$$

$$C_{z,b} J_{z,\text{inc}}\Big|^n_{I_0+1/2,\, j+1/2,\, k} = C_{z,x} H_{y,\text{inc}}\Big|^n_{I_0,\, j+1/2,\, k} = C_{z,x} H_{y,\text{inc}}\Big|^n_{i_r+m_y/2} \tag{3.58b}$$

At $i = I_1$:

$$C_{y,b} J_{y,\text{inc}}\Big|^n_{I_1+1/2,\, j,\, k+1/2} = C_{y,x} H_{z,\text{inc}}\Big|^n_{I_1+1,\, j,\, k+1/2} = C_{y,x} H_{z,\text{inc}}\Big|^n_{i_r+m_x+m_z/2} \tag{3.58c}$$

$$C_{z,b} J_{z,\text{inc}}\Big|^n_{I_1+1/2,\, j+1/2,\, k} = -C_{z,x} H_{y,\text{inc}}\Big|^n_{I_1+1,\, j+1/2,\, k} = -C_{z,x} H_{y,\text{inc}}\Big|^n_{i_r+m_x+m_y/2} \tag{3.58d}$$

x-z faces ($I_0 \le i \le I_1$ and $K_0 \le k \le K_1$)

At $j = J_0$:

$$C_{x,b} J_{x,\text{inc}}\Big|^n_{i,\, J_0+1/2,\, k+1/2} = C_{x,y} H_{z,\text{inc}}\Big|^n_{i,\, J_0,\, k+1/2} = C_{x,y} H_{z,\text{inc}}\Big|^n_{i_r+m_z/2} \tag{3.59a}$$

$$C_{z,b} J_{z,\text{inc}}\Big|^n_{i+1/2,\, J_0+1/2,\, k} = -C_{z,y} H_{x,\text{inc}}\Big|^n_{i+1/2,\, J_0,\, k} = -C_{z,y} H_{x,\text{inc}}\Big|^n_{i_r+m_x/2} \tag{3.59b}$$

At $j = J_1$:

$$C_{x,b} J_{x,\text{inc}}\Big|^{n}_{i,J_1+1/2,k+1/2} = -C_{x,y} H_{z,\text{inc}}\Big|^{n}_{i,J_1+1,k+1/2} = -C_{x,y} H_{z,\text{inc}}\Big|^{n}_{i_r+m_y+m_z/2} \tag{3.59c}$$

$$C_{z,b} J_{z,\text{inc}}\Big|^{n}_{i+1/2,J_1+1/2,k} = C_{z,y} H_{x,\text{inc}}\Big|^{n}_{i+1/2,J_1+1,k} = C_{z,y} H_{x,\text{inc}}\Big|^{n}_{i_r+m_y+m_x/2} \tag{3.59d}$$

x-y faces $(I_0 \le i \le I_1$ and $J_0 \le j \le J_1)$

At $k = K_0$:

$$C_{x,b} J_{x,\text{inc}}\Big|^{n}_{i,j+1/2,K_0+1/2} = -C_{x,z} H_{y,\text{inc}}\Big|^{n}_{i,j+1/2,K_0} = -C_{x,z} H_{y,\text{inc}}\Big|^{n}_{i_r+m_y/2} \tag{3.60a}$$

$$C_{y,b} J_{y,\text{inc}}\Big|^{n}_{i+1/2,j,K_0+1/2} = C_{y,z} H_{x,\text{inc}}\Big|^{n}_{i+1/2,j,K_0} = C_{y,z} H_{x,\text{inc}}\Big|^{n}_{i_r+m_x/2} \tag{3.60b}$$

At $k = K_1$:

$$C_{x,b} J_{x,\text{inc}}\Big|^{n}_{i,j+1/2,K_1+1/2} = C_{x,z} H_{y,\text{inc}}\Big|^{n}_{i,j+1/2,K_1+1} = C_{x,z} H_{y,\text{inc}}\Big|^{n}_{i_r+m_y/2+m_z} \tag{3.60c}$$

$$C_{y,b} J_{y,\text{inc}}\Big|^{n}_{i+1/2,j,K_1+1/2} = -C_{y,z} H_{x,\text{inc}}\Big|^{n}_{i+1/2,j,K_1+1} = -C_{y,z} H_{x,\text{inc}}\Big|^{n}_{i_r+m_x/2+m_z} \tag{3.60d}$$

At each of the six faces of the TF/SF interface, the required surface magnetic current densities to be used in (3.19), (3.21), and (3.23) are:

y-z faces $(J_0 \le j \le J_1$ and $K_0 \le k \le K_1)$

At $i = I_0$:

$$D_{z,b} M_{z,\text{inc}}\Big|^{n+1/2}_{I_0,j,k+1/2} = D_{z,x} E_{y,\text{inc}}\Big|^{n+1/2}_{I_0+1/2,j,k+1/2} = D_{z,x} E_{y,\text{inc}}\Big|^{n+1/2}_{i_r+m_x/2+m_z/2} \tag{3.61a}$$

$$D_{y,b} M_{y,\text{inc}}\Big|^{n+1/2}_{I_0,j+1/2,k} = -D_{y,x} E_{z,\text{inc}}\Big|^{n+1/2}_{I_0+1/2,j+1/2,k} = -D_{y,x} E_{z,\text{inc}}\Big|^{n+1/2}_{i_r+m_x/2+m_y/2} \tag{3.61b}$$

At $i = I_1$:

$$D_{z,b} M_{z,\text{inc}}\Big|^{n+1/2}_{I_1+1,j,k+1/2} = -D_{z,x} E_{y,\text{inc}}\Big|^{n+1/2}_{I_1+1/2,j,k+1/2} = -D_{z,x} E_{y,\text{inc}}\Big|^{n+1/2}_{i_r+m_x/2+m_z/2} \tag{3.61c}$$

$$D_{y,b} M_{y,\text{inc}}\Big|^{n+1/2}_{I_1+1,j+1/2,k} = D_{y,x} E_{z,\text{inc}}\Big|^{n+1/2}_{I_1+1/2,j+1/2,k} = D_{y,x} E_{z,\text{inc}}\Big|^{n+1/2}_{i_r+m_x/2+m_y/2} \tag{3.61d}$$

x-z faces $(I_0 \le i \le I_1$ and $K_0 \le k \le K_1)$

At $j = J_0$:

$$D_{z,b} M_{z,\text{inc}}\Big|^{n+1/2}_{i,J_0,k+1/2} = -D_{z,y} E_{x,\text{inc}}\Big|^{n+1/2}_{i,J_0+1/2,k+1/2} = -D_{z,y} E_{x,\text{inc}}\Big|^{n+1/2}_{i_r+m_y/2+m_z/2} \tag{3.62a}$$

$$D_{x,b} M_{x,\text{inc}}\Big|^{n+1/2}_{i+1/2,J_0,k} = D_{x,y} E_{z,\text{inc}}\Big|^{n+1/2}_{i+1/2,J_0+1/2,k} = D_{x,y} E_{z,\text{inc}}\Big|^{n+1/2}_{i_r+m_x/2+m_y/2} \tag{3.62b}$$

At $j = J_1$:

$$D_{z,b}\, M_{z,\mathrm{inc}}\Big|_{i,\,J_1+1,\,k+1/2}^{n+1/2} = D_{z,y} E_{x,\mathrm{inc}}\Big|_{i,\,J_1+1/2,\,k+1/2}^{n+1/2} = D_{z,y} E_{x,\mathrm{inc}}\Big|_{i_r+m_y/2+m_z/2}^{n+1/2} \tag{3.62c}$$

$$D_{x,b}\, M_{x,\mathrm{inc}}\Big|_{i+1/2,\,J_1+1,\,k}^{n+1/2} = -D_{x,y} E_{z,\mathrm{inc}}\Big|_{i+1/2,\,J_1+1/2,\,k}^{n+1/2} = -D_{x,y} E_{z,\mathrm{inc}}\Big|_{i_r+m_x/2+m_y/2}^{n+1/2} \tag{3.62d}$$

x-y faces $(I_0 \le i \le I_1$ and $J_0 \le j \le J_1)$

At $k = K_0$:

$$D_{y,b}\, M_{y,\mathrm{inc}}\Big|_{i,\,j+1/2,\,K_0}^{n+1/2} = D_{y,z} E_{x,\mathrm{inc}}\Big|_{i,\,j+1/2,\,K_0+1/2}^{n+1/2} = D_{y,z} E_{x,\mathrm{inc}}\Big|_{i_r+m_y/2+m_z/2}^{n+1/2} \tag{3.63a}$$

$$D_{x,b}\, M_{x,\mathrm{inc}}\Big|_{i+1/2,\,j,\,K_0}^{n+1/2} = -D_{x,z} E_{y,\mathrm{inc}}\Big|_{i+1/2,\,j,\,K_0+1/2}^{n+1/2} = -D_{x,z} E_{y,\mathrm{inc}}\Big|_{i_r+m_x/2+m_z/2}^{n+1/2} \tag{3.63b}$$

At $k = K_1$:

$$D_{y,b}\, M_{y,\mathrm{inc}}\Big|_{i,\,j+1/2,\,K_1+1}^{n+1/2} = -D_{y,z} E_{x,\mathrm{inc}}\Big|_{i,\,j+1/2,\,K_1+1/2}^{n+1/2} = -D_{y,z} E_{x,\mathrm{inc}}\Big|_{i_r+m_y/2+m_z/2}^{n+1/2} \tag{3.63c}$$

$$D_{x,b}\, M_{x,\mathrm{inc}}\Big|_{i+1/2,\,j,\,K_1+1}^{n+1/2} = D_{x,z} E_{y,\mathrm{inc}}\Big|_{i+1/2,\,j,\,K_1+1/2}^{n+1/2} = D_{x,z} E_{y,\mathrm{inc}}\Big|_{i_r+m_x/2+m_z/2}^{n+1/2} \tag{3.63d}$$

3.12 SUMMARY OF METHOD

The building blocks for implementing an exact FDTD TF/SF plane-wave source are now complete. To summarize, the following steps are needed to implement this formulation:

Setting up the Program. Establish a regular 3-D FDTD computational domain in the normal fashion. Identify grid locations (indices) for the TF/SF interface.

Identify the angle of incident plane-wave propagation and map this to integers $[m_x, m_y, m_z]$ via (3.39) and (3.40). Establish six incident plane-wave 1-D FDTD grids based on these integers and (3.44)–(3.49). Establish the source fields for these 1-D grids based on (3.56).

During Runtime. Perform normal $\boldsymbol{E}$ and $\boldsymbol{H}$ updates in the primary 3-D FDTD grid according to (3.13), (3.15), (3.17), (3.19), (3.21), and (3.23).

Concurrently, perform normal field updates in the six auxiliary 1-D FDTD incident plane-wave grids to generate the required $\boldsymbol{J}_{\mathrm{inc}}$ and $\boldsymbol{M}_{\mathrm{inc}}$ source terms along the TF/SF interface of the primary 3-D FDTD grid, as specified by (3.58) – (3.63).

3.13 MODELING EXAMPLES

The methodology developed above is completely independent of the temporal source function used to initiate the plane-wave source. Whether the source is a Gaussian pulse, a square wave, or a completely random set of values, the set of six auxiliary 1-D FDTD grids faithfully propagates a plane wave based on that source with *absolutely zero* nonphysical scattered fields occurring at the TF/SF interface of the primary 3-D FDTD grid. This is because the numerical dispersion characteristics of the plane-wave source are *identical* to those of the primary grid.

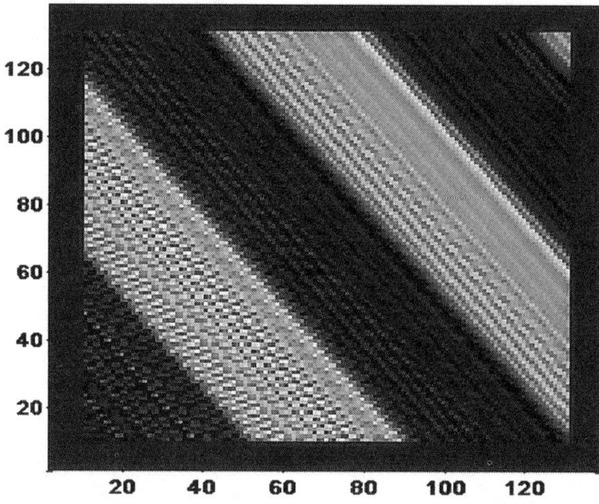

Fig. 3.3 Visualization of the FDTD-computed E_x field along a y-z planar cross-section of a 3-D TF/SF grid for an obliquely propagating square wave within the total-field region. There is zero leakage into the scattered-field region to machine precision (~10^{-15}, or –300 dB). *Source:* Tan and Potter, *IEEE Trans. Antennas and Propagation*, Vol. 58, 2010, pp. 2641–2648, ©2010 IEEE.

Figure 3.3 illustrates an example of employing the exact TF/SF plane-wave source [14]. This figure visualizes the FDTD-computed E_x field after 1500 time-steps along a y-z planar cross-section of a 3-D TF/SF grid. Here, the incident wave was assumed to have an ideal square-wave temporal dependence and propagate obliquely within the grid at $\varphi \approx 59°$ and $\theta \approx 39.8°$, or equivalently $[m_x, m_y, m_z] = [3, 5, 7]$. The free-space total-field region spanned $120 \times 120 \times 120$ cells, and was surrounded on all sides by a 10-cell scattered-field region, the latter being terminated with a perfect electric conductor. Leakage into the scattered-field region was zero to machine precision (~10^{-15}, or –300 dB) despite substantial dispersion of high-frequency wave components at the lower left corner of the total-field region (at the trailing part of the wave).

Figure 3.4 illustrates a second example of employing the exact TF/SF plane-wave source [14], here for a 160 × 160 × 160-cell free-space total-field region in the primary grid surrounded on all sides by a 10-cell scattered-field region. The incident wave had a sinusoidally modulated Gaussian pulse temporal dependence, $f(n) = \sin(n\pi/4)\exp[-(n-68)^2/125]$, and propagated obliquely within the total-field region of the primary grid at $\varphi \approx 63.4°$ and $\theta \approx 36.7°$, or equivalently $[m_x, m_y, m_z] = [1, 2, 3]$.

Figure 3.4(a) depicts two time-waveforms of the FDTD-computed E_x field in the total-field region of the primary grid. Waveform Q_{T1} was recorded near the source just inside the total-field region. Waveform Q_{T2} was recorded after the incident wave propagated ~270 cells away from the source to the corner of the total-field region furthest from the source. Significant distortion due to phase-error accumulation prompted by numerical dispersion is apparent when these two waveforms are compared. Fig. 3.4(b) depicts time-waveforms of the FDTD-computed E_x field in the scattered-field region of the primary grid. Waveform Q_{S1} was recorded immediately adjacent to the monitor point of Q_{T1}, but just outside in the scattered-field region. Similarly, waveform Q_{S2} was recorded immediately adjacent to the monitor point of Q_{T2}, but just outside in the scattered-field region. We see that, despite the numerical dispersion artifacts accumulating during pulse propagation, the amplitude levels of the scattered fields adjacent to both the near and far corners of the total-field region were zero to machine precision (~10^{-15}).

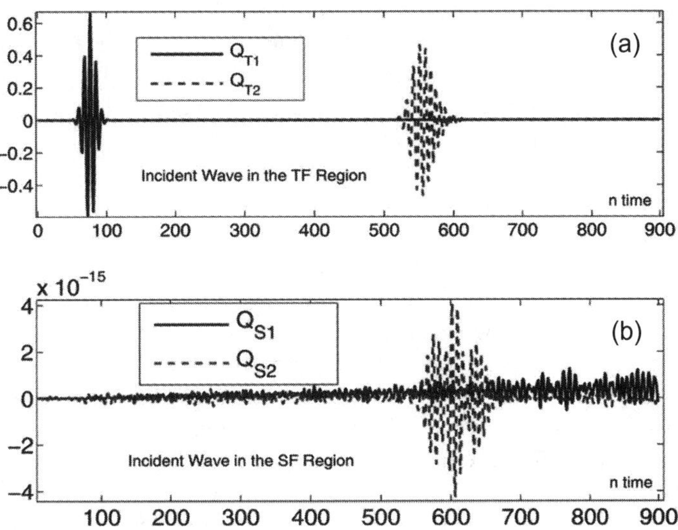

Fig. 3.4 FDTD-computed E_x field at the corners of the total-field region of a 3-D TF/SF grid for a propagating incident plane wave having a modulated Gaussian pulse time-waveform: (a) monitor points located just inside the total-field region; (b) monitor points located just outside in the scattered-field region. Solid lines—monitor points near the source; dashed lines—monitor points at the corner of the total-field region furthest from the source. Despite an accumulation of phase distortion during pulse propagation (compared with an analytical plane wave), the leakage into the scattered-field region is zero to machine precision. *Source:* Tan and Potter, *IEEE Trans. Antennas and Propagation*, Vol. 58, 2010, pp. 2641–2648, ©2010 IEEE.

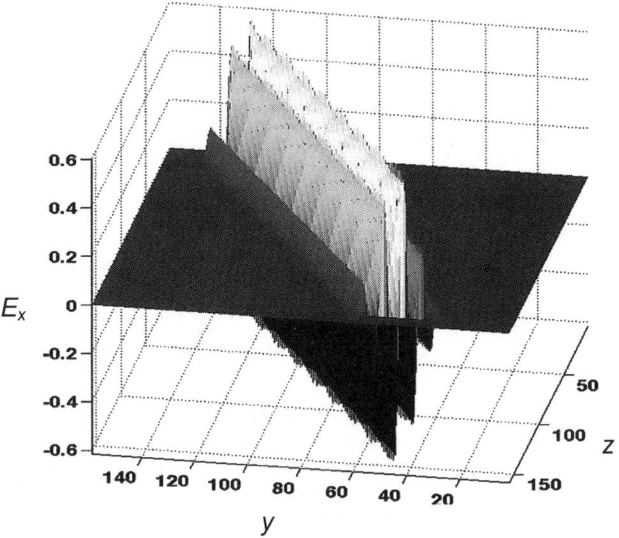

Fig. 3.5 Snapshot of the FDTD-computed E_x field in a y-z plane of the TF/SF grid used in the example of Fig. 3.4. The field levels in the scattered-field region (the small band around the outside of the plane) are all zero to machine-precision levels. *Source:* Tan and Potter, *IEEE Trans. Antennas and Propagation*, Vol. 58, 2010, pp. 2641–2648, ©2010 IEEE.

Figure 3.5 shows a snapshot in time of the modulated Gaussian pulse propagating across the domain used in the example of Fig. 3.4 [14]. This figure depicts the FDTD-computed E_x field in a complete y-z plane of the TF/SF grid, including both the total-field region and the scattered-field region (the thin band at the periphery of the plane). The field values at all points within the scattered-field region are zero to machine-precision levels.

3.14 DISCUSSION

The exact FDTD TF/SF plane-wave source has been tested for arbitrary propagation and polarization angles, stability factors, and non-unity aspect ratios. In each case studied, leakage of the incident plane wave into the scattered-field region has been found to be zero to machine-precision levels. The price one pays for this achievement is a more complicated programming than is needed for the standard incident field array reported in [1]. Otherwise, even though implementing the exact TF/SF plane-wave source requires setting up and processing six auxiliary 1-D grids, the incremental computer memory and execution time are negligible compared to the computer burdens of the primary 3-D grid, unless extremely fine angular resolution is needed.

In developing the exact TF/SF plane-wave source, geometric arguments regarding projecting field components from a 3-D space to a 1-D space (and hence, the six coupled incident field arrays) require no assumptions about the computational stencil utilized in the primary FDTD grid. Hence, this technique is not limited to the standard spatially interleaved FDTD grid, nor is it limited to the standard second-order-accurate computational stencils. For instance, in [16, 17], the exact TF/SF plane-wave source technique was modified to be utilized in a split-field formulation and with the higher-order M24 stencil; in [18], it was adapted for a high-order symplectic FDTD algorithm; and in [19] it was adapted for an unconditionally stable Crank-Nicholson time-stepping scheme.

REFERENCES

[1] Taflove, A., and S. C. Hagness, *Computational Electrodynamics: The Finite-Difference Time-Domain Method,* 3rd ed., Norwood, MA: Artech House, 2005.

[2] Stratton, J. A., *Electromagnetic Theory*, New York: McGraw Hill, 1941.

[3] Harrington, R. F., *Time-Harmonic Electromagnetic Fields*, New York: IEEE / Wiley, 2001.

[4] Balanis, C. A., *Advanced Engineering Electromagnetics, 2nd ed.*, New York: Wiley, 2012.

[5] Merewether, D., et al., "On implementing a numeric Huygen's source scheme in a finite difference program to illuminate scattering bodies," *IEEE Trans. Nuclear Science*, Vol. 27, 1980, pp. 1829–1833.

[6] Umashankar, K. R., and A. Taflove, "A novel method to analyze electromagnetic scattering of complex objects," *IEEE Trans. Electromagnetic Compatibility*, Vol. 24, 1982, pp. 397-405.

[7] Guiffaut, C., and K. Mahdjoubi, "A perfect wideband plane wave injector for FDTD method," in *Proc. IEEE APS International Symposium*, Vol. 1, Salt Lake City, UT, 2000, pp. 236–239.

[8] Oguz, U., and L. Gurel, "Interpolation techniques to improve the accuracy of the plane wave excitations in the finite difference time domain method," *Radio Science*, Vol. 32, 1997, pp. 2189–2199.

[9] Oguz, U., and L. Gurel, "An efficient and accurate technique for the incident-wave excitations in the FDTD method," *IEEE Trans. Microwave Theory and Techniques*, Vol. 46, 1998, pp. 869–882.

[10] Moss, C. D., F. L Teixeira, and J. A. Kong, "Analysis and compensation of numerical dispersion in the FDTD method for layered anisotropic media," *IEEE Trans. Antennas and Propagation*, Vol. 50, 2002, pp. 1174–1184.

[11] Schneider, J. B., "Plane waves in FDTD simulations and a nearly perfect total-field/scattered-field boundary," *IEEE Trans. Antennas and Propagation*, Vol. 52, 2004, pp. 3280–3287.

[12] Tan, T., and M. Potter, "1-D multipoint auxiliary source propagator for the total-field/scattered-field FDTD formulation," *IEEE Antennas and Wireless Propagation Lett.*, Vol. 6, 2007, pp. 144–148.

[13] Tan, T., and M. Potter, "On the nature of numerical plane waves in FDTD," *IEEE Antennas and Wireless Propagation Lett.*, Vol. 8, 2009, pp. 505–508.

[14] Tan, T., and M. Potter, "FDTD discrete planewave (FDTD-DPW) formulation for a perfectly matched source in TFSF simulations," *IEEE Trans. Antennas and Propagation*, Vol. 58, 2010, pp. 2641–2648.

[15] Tan, T., and M. Potter, "Optimized analytic field propagator (O-AFP) for plane wave injection in FDTD simulations," *IEEE Trans. Antennas and Propagation*, Vol. 58, 2010, pp. 824–831.

[16] Hadi, M. F., "A versatile split-field 1-D propagator for perfect FDTD plane wave injection," *IEEE Trans. Antennas and Propagation*, Vol. 57, 2009, pp. 2691–2697.

[17] Hui, W., et al., "Perfect plane wave injection into 3D FDTD (2,4) scheme," in *Proc. Cross Strait Quad-Regional Radio Science and Wireless Technology Conference (CSQRWC) 2011*, Harbin, China, 2011, pp. 40–43.

[18] Hui, W., et al., "Perfect plane-wave source for a high-order symplectic finite-difference time-domain scheme," *Chinese Physics B*, Vol. 20, 2011, 114701, doi:10.1088/1674-1056/20/11/114701.

[19] Huang, Z., G. Pan, and H. K. Pan, "Perfect plane wave injection for Crank-Nicholson time-domain method," *IET Microwaves, Antennas and Propagation*, Vol. 4, 2010, pp. 1855–1862.

Chapter 4

Electromagnetic Wave Source Conditions[1]

Ardavan Oskooi and Steven G. Johnson

4.1 OVERVIEW

This chapter discusses the relationships between current sources and the resulting electromagnetic waves in FDTD simulations. First, the "total-field/scattered-field" approach to creating incident plane waves is reviewed and seen to be a special case of the well-known *principle of equivalence* in electromagnetism: this can be used to construct "equivalent" current sources for any desired incident field, including waveguide modes. The effects of dispersion and discretization are discussed, and a simple technique to separate incident and scattered fields is described in order to compensate for imperfect equivalent currents. The important concept of the *local density of states* (LDOS) is reviewed, which elucidates the relationship between current sources and the resulting fields, including enhancement of the LDOS via mode cutoffs (Van Hove singularities) and resonant cavities (Purcell enhancement). We also address various other source techniques such as covering a wide range of frequencies and incident angles in a small number of simulations for waves incident on a periodic surface, sources to excite eigenmodes in rectangular supercells of periodic systems, moving sources, and thermal sources via a Monte Carlo/Langevin approach.

4.2 INCIDENT FIELDS AND EQUIVALENT CURRENTS

A common problem in FDTD simulations is to compute the interaction of a given incident wave with some geometry and materials in a localized region. For example, the incident wave could be a plane wave in vacuum (e.g., to compute the scattered and absorbed field for some isolated object), or the incident wave could be a propagating mode of a waveguide (e.g., to compute the transmission or reflection around a waveguide bend or through some other device). One cannot simply specify the fields near the boundary of a simulation — this would be a "hard" source that nonphysically scatters waves (e.g., reflected waves) that impinge on the source [1]. Instead, it is desirable to specify *equivalent electric and magnetic currents* (also called "transparent sources" [1]) that produce the desired incident wave, but which are transparent to other waves by the linearity of the Maxwell equations (at least in a linear incident medium).

---

[1]In this chapter the symbol i is used to designate $\sqrt{-1}$, rather than the symbol j; and a phasor is denoted as $e^{-i\omega t}$.

In fact, there is a simple general prescription for deriving such currents from *any* incident field based on one of the fundamental theorems of electromagnetism: the *principle of equivalence* [2–4], a precise formalization of *Huygens' principle* that fields on a wavefront can be treated as sources. In the context of FDTD, these equivalent currents enable the "total-field/scattered-field" approach [1], but many other applications are possible.

4.2.1 The Principle of Equivalence

The derivation in this section is essentially equivalent to the usual derivation of the total-field/scattered-field approach, but instead of first discretizing the equations and then writing out the FDTD component equations individually, we start with the analytical Maxwell's equations and employ a compact notation that allows us to look at all the equations together. Not only does this shorten the derivation, but it also makes clear the applicability of the approach to any linear incident medium (such as a waveguide, including inhomogeneous and bianisotropic media [4]), and highlights the explicit identification of the source terms as the electric and magnetic currents of the principle of equivalence.

It is convenient to write Maxwell's equations compactly in terms of the six-component electric ($\boldsymbol{E}$) and magnetic ($\boldsymbol{H}$) field vector, ψ, and the six-component electric ($\boldsymbol{J}$) and magnetic ($\boldsymbol{K}$) current vector, ξ, in which case the equations in linear media can be written as:

$$\underbrace{\begin{bmatrix} & \nabla\times \\ -\nabla\times & \end{bmatrix}}_{\mathbf{M}} \underbrace{\begin{bmatrix} \boldsymbol{E} \\ \boldsymbol{H} \end{bmatrix}}_{\psi} = \frac{\partial}{\partial t}(\psi + \chi * \psi) + \underbrace{\begin{bmatrix} \boldsymbol{J} \\ \boldsymbol{K} \end{bmatrix}}_{\xi} \tag{4.1}$$

where we have chosen natural units in which $\varepsilon_0 = 1$ and $\mu_0 = 1$, and $\chi *$ denotes convolution with the 6×6 linear susceptibility tensor:

$$\chi = \begin{bmatrix} \varepsilon - 1 & \\ & \mu - 1 \end{bmatrix} \tag{4.2}$$

in ordinary dielectric/magnetic media. Suppose we have a desired incident wave, ψ_+, which solves the source-free Maxwell's equations in an infinite "incident" medium, χ_+ (e.g., vacuum or an infinite waveguide) with no scatterers: $\mathbf{M}\psi_+ = \partial(\psi_+ + \chi_+ * \psi_+)/\partial t$. Now, the question is to come up with *equivalent currents*, ξ, that generate the same incident field, ψ_+, in a *finite* domain, Ω, within our computational space, into which we can then introduce scatterers or other inhomogeneities for interaction with the incident wave.

The derivation and application of the principle of equivalence can be thought of as a three-step process, illustrated in Fig. 4.1. First (a), we consider the incident wave, ψ_+, in an infinite medium, χ_+, with space arbitrarily divided into an interior domain, Ω, and an exterior, separated by an imaginary surface, $\partial\Omega$. Second (b), keeping the *same* infinite medium, χ_+, we set $\psi = 0$ in the *exterior* and keep $\psi = \psi_+$ in the *interior* Ω, and solve for the currents, ξ, which allow this *discontinuous* field, ψ, to solve Maxwell's equations: these are the *equivalent surface currents*. Third (c), given these currents, we can then truncate space [e.g., with perfectly matched layer (PML) absorbers], and introduce inhomogeneities to χ inside Ω in order to model the interaction of ψ_+ with these inhomogeneities. One can even include nonlinearities in the interior of Ω in step (c); the only real requirement is that the *incident* medium, χ_+, be linear.

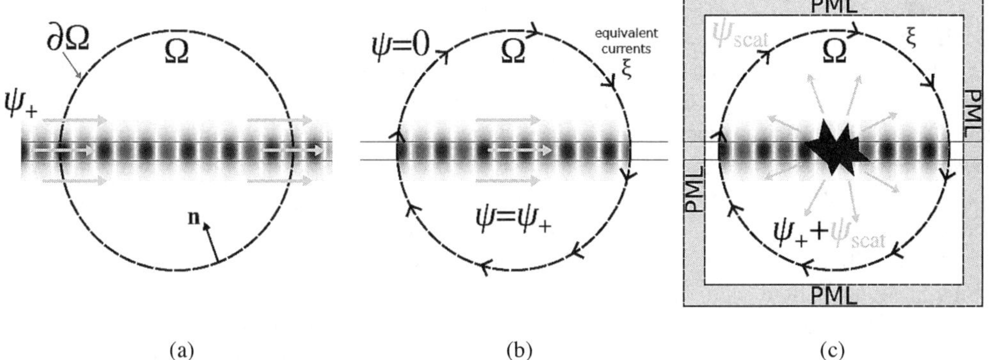

(a) (b) (c)

Fig. 4.1 Schematic construction of equivalent electric and magnetic current sources, ξ, to produce a desired incident field, ψ_+, which denotes a 6-component composite electric and magnetic field vector. (a) The field, ψ_+, in an infinite medium — here, a waveguide mode in an infinite waveguide — with space arbitrarily divided by an imaginary surface $\partial\Omega$ into an "interior" region, Ω, and an "exterior." (b) We set the field to be $\psi = 0$ in the exterior and $\psi = \psi_+$ in Ω, and construct surface currents, ξ, that allow this discontinuous ψ to satisfy Maxwell's equations. This is the *principle of equivalence*. (c) Given this ξ, we can truncate the computation space with a PML absorber and insert a scatterer or some other object into Ω. The current, ξ, produces the desired incident field, ψ_+, so that the field ψ inside Ω is the sum of ψ_+ and a scattered field, while the field outside Ω is *only* the scattered field. This is the basis of the "total-field/scattered-field" approach in FDTD.

The key is the second step, depicted in Fig. 4.1(b). If we let:

$$\psi = \begin{cases} \psi_+ & \text{inside } \Omega \\ 0 & \text{outside } \Omega \end{cases} \tag{4.3}$$

then ψ clearly solves the source-free Maxwell's equations, $\mathbf{M}\psi = \partial(\psi + \chi_+ * \psi)/\partial t$, in *both* the interior and exterior regions. Here, the only question is what happens at the surface, $\partial\Omega$. At this surface, the discontinuity of ψ has only one effect in Maxwell's equations: it produces a surface Dirac delta function, $\delta(\partial\Omega)$, in the spatial derivative, $\mathbf{M}\psi$.[2] Hence, in order to satisfy Maxwell's equations with this ψ, we must introduce a matching delta function on the right-hand side: a *surface current*, ξ. In particular, if $\boldsymbol{n}$ is the unit inward-normal vector,[3] only the normal derivative, $\boldsymbol{n}\cdot\nabla$, contains a delta function (whose amplitude is the magnitude of the discontinuity). This implies a surface current:

$$\xi = \begin{bmatrix} \boldsymbol{J} \\ \boldsymbol{K} \end{bmatrix} = \delta(\partial\Omega)\begin{bmatrix} & \boldsymbol{n}\times \\ -\boldsymbol{n}\times & \end{bmatrix}\psi_+ = \delta(\partial\Omega)\begin{bmatrix} \boldsymbol{n}\times\boldsymbol{H}_+ \\ -\boldsymbol{n}\times\boldsymbol{E}_+ \end{bmatrix} \tag{4.4}$$

---

[2]That is, $\delta(\partial\Omega)$ is the distribution such that $\iiint \phi(\boldsymbol{x})\delta(\partial\Omega) = \iint_{\partial\Omega} \phi(\boldsymbol{x})$ for any continuous test function, ϕ.

[3]For simplicity, we assume a differentiable surface, $\partial\Omega$, so that its normal, $\boldsymbol{n}$, is well defined; but a surface with corners (e.g., a cubical domain) follows as a limiting case.

That is, there is a *surface electric current* given by the surface-tangential components, $\boldsymbol{n} \times \boldsymbol{H}_{+}$, of the incident magnetic field, and a *surface magnetic current* given by the components, $-\boldsymbol{n} \times \boldsymbol{E}_{+}$, of the incident electric field. These are the *equivalent currents* of the principle of equivalence (which can also be derived in other ways, traditionally from a Green's function approach [3, 4]). This principle has a long history [5–8] and far-reaching consequences in electromagnetism. For example, not only is it useful in constructing wave sources in FDTD, but it is also central to *integral-equation* formulations of electromagnetic scattering problems and the resulting boundary-element method (BEM) [9, 10].

Essentially the same derivation applied to the discretized curl operator, **M**, yields the source terms in the total-field/scattered field approach [1], with the source currents being a discrete or "Kronecker" delta function. Here, the connection to the equivalence principle usually goes unmentioned, despite the fact that this approach was originally derived in analogy with Huygens' principle [11]. Although this technique is most commonly applied to plane waves in free space, one can just as easily apply the derivation to produce a waveguide-mode source, as shown in Fig. 4.2. In this case, the incident fields are those of a waveguide mode. Since these are *exponentially localized* to the vicinity of the waveguide, we can in practice restrict the current sources, $\boldsymbol{\xi}$, to a *short line segment crossing the waveguide*. An eigenmode current source excites only the desired rightward-propagating guided mode, whereas a simple constant-amplitude source produces a mixture of leftward- and rightward-propagating modes and radiated fields.

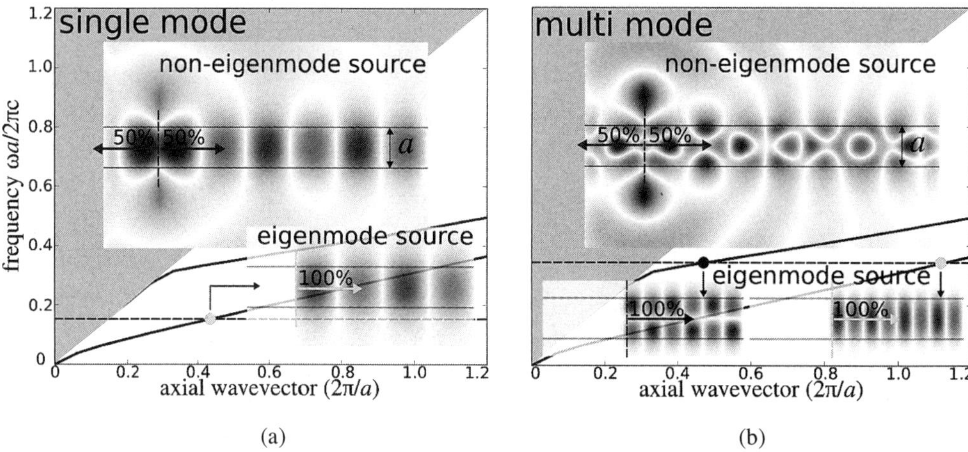

Fig. 4.2 Comparison of different current sources in a two-dimensional (2-D) dielectric waveguide ($\varepsilon/\varepsilon_0 = 12$, width a) with out-of-plane $\boldsymbol{E}$ field. The dispersion relation, $\omega(k)$, is shown in the background of each panel, and the insets visualize the FDTD-computed $\boldsymbol{E}_z$. (a) Sources at a frequency where the waveguide is single-mode, indicated by a dot on the $\omega(k)$ curve. Top inset: a simple out-of-plane electric-current source, $\boldsymbol{J}$, of constant amplitude along a transverse cross-section of the waveguide (vertical dashed line), which excites both the waveguide mode and radiating fields in both directions. Bottom inset: equivalent-current source, $\boldsymbol{\xi}$, from (4.4), having a transverse distribution consistent with the eigenmode computed in a mode solver, which excites only the right-going waveguide mode. (b) Sources at a frequency where the waveguide has two modes, indicated by two dots on the $\omega(k)$ curve. Top inset: a simple $\boldsymbol{J}$ source of constant amplitude along a transverse cross-section of the waveguide, which excites a superposition of both waveguide modes in addition to the radiating field. Bottom insets: equivalent-current source, $\boldsymbol{\xi}$, having a transverse distribution consistent with either the fundamental or the 2nd-order eigenmode; in each case, we only excite the desired right-going mode.

4.2.2 Discretization and Dispersion of Equivalent Currents

There are two practical obstacles to exact implementation of equivalent-current sources in FDTD simulations: *discretization* and *dispersion*. If these are not accounted for precisely, then the currents, ξ, produce slight errors in ψ_+ in the interior, Ω, and slightly nonzero fields in the exterior of Ω. In practice, these slight errors can often be dealt with easily by postprocessing, as described in Section 4.3, but it is still important to understand how they arise and how (at least in principle) the errors could be eliminated.

The first potential source of error is discretization. As already mentioned, for FDTD we should, in principle, apply the equivalent-currents derivation to the discretized Maxwell's equations, and in particular to the discretized curl operations, $\mathbf{M}$, in order to obtain the exact location of the surface currents as discrete delta functions on the Yee grid. This procedure is straightforward, and variations of the derivation can be found in textbook discussions of the total-field/scattered-field approach [1]. Moreover, even if the incident field ψ_+ solves the *exact* Maxwell's equations, it will not solve the *discretized* Maxwell's equations without modification. For the case of an incident plane wave in free space, the exact discretized solution for ψ_+ can be found by substituting a plane wave into FDTD's Yee discretization, and then solving for the dispersion relation, $\omega(\boldsymbol{k})$, and the corresponding $\boldsymbol{E}$ and $\boldsymbol{H}$ fields (which are not quite perpendicular to $\boldsymbol{k}$ due to numerical anisotropy) [12]. An efficient recent alternative, described in detail in Chapter 3, accomplishes this "on the fly" as FDTD time-stepping progresses.

In contrast, obtaining ψ_+ for geometries like dielectric waveguides must, in general, be done numerically, especially in three dimensions. Suppose one uses a ψ_+ solution from an eigenmode solver that is sufficiently converged as to be considered exact. In this case, small wave-source errors, which vanish with increasing FDTD grid resolution, are introduced in FDTD because of the mismatch between the exact ψ_+ and the FDTD-discretized ψ_+ solutions. This is precisely what was observed in the modeling results shown in Fig. 4.2, where the eigenmode solver used the free plane-wave expansion code, MPB [13], at a high resolution, followed by a simple interpolation onto $\partial\Omega$, the line-segment waveguide cross-section in the FDTD grid. As a result, a small backwards (leftward-propagating) wave artifact was generated within the FDTD waveguide model by the eigenmode current source, carrying approximately 10^{-5} of the power of the desired rightward-propagating mode for an FDTD grid resolution of about 27 pixels per wavelength. To eliminate this artifact, one could use the "bootstrap" technique described in [1] where a preliminary FDTD model of the infinitely long waveguide is run to obtain ψ_+, using exactly the same spatial and temporal resolutions as the subsequent "working" FDTD run. Of course, this would increase the required computational resources, but once the discretized ψ_+ is obtained and stored, it could be used in all subsequent models of the waveguide in question.

A second source of error, which is more difficult to deal with in practice, is dispersion: any nontrivial frequency dependence of the incident-wave solution, ψ_+. For a plane-wave $\sim\exp(i\boldsymbol{k}\cdot\boldsymbol{x} - i\omega t)$, the exact vacuum dispersion relation is simply $\omega = c|\boldsymbol{k}|$. However, much more complicated $\omega(\boldsymbol{k})$ dispersion relations can arise. First, there are numerical dispersion artifacts due to the FDTD spatial and temporal discretizations [1]. Second, a variety of physical dispersions are caused by frequency-dependent material dielectric properties and structural geometric effects such a waveguide cutoff phenomena. In general, suppose that we have solved for the desired incident fields (e.g., plane waves or waveguide modes), $\hat{\psi}_+(\boldsymbol{x},\omega)$, at *each* frequency, with a *frequency-independent* normalization (e.g., normalized to unit input power). Furthermore, suppose that we wish to inject a wave *pulse* into our FDTD simulation, with some pulse profile, $p(t)$ (e.g., a Gaussian pulse), and denote its Fourier transform by $\hat{p}(\omega)$. It follows that the desired equivalent currents are given by the following Fourier transform pair:

$$\hat{\boldsymbol{\xi}}(\boldsymbol{x}, \omega) = \delta(\partial\Omega)\begin{bmatrix} & \boldsymbol{n}\times \\ -\boldsymbol{n}\times & \end{bmatrix}\hat{\boldsymbol{\psi}}_{+}(\boldsymbol{x}, \omega)\,\hat{p}(\omega) \tag{4.5a}$$

$$\boldsymbol{\xi}(\boldsymbol{x}, t) = \delta(\partial\Omega)\begin{bmatrix} & \boldsymbol{n}\times \\ -\boldsymbol{n}\times & \end{bmatrix}\boldsymbol{\psi}_{+}(\boldsymbol{x}, t) * p(t) \tag{4.5b}$$

where $*$ denotes a convolution of the time-domain fields, $\boldsymbol{\psi}_+$ (the inverse Fourier transform of $\hat{\boldsymbol{\psi}}_+$), with the pulse shape $p(t)$. In any time-domain method such as FDTD, of course, we need the time-domain currents, $\boldsymbol{\xi}$. The difficulty is that these convolutions (a *different* convolution at every point $\boldsymbol{x}$ on the surface, $\partial\Omega$) can be cumbersome to perform in general. Several options are:

- Precompute $\boldsymbol{\xi}(\boldsymbol{x}, t)$ via inverse Fourier transformation: Compute $\hat{\boldsymbol{\psi}}_+(\boldsymbol{x}, \omega)$ and $\hat{p}(\omega)$ at a set of discrete ω (assuming a bandlimited pulse); multiply them; and then perform an inverse fast Fourier transform (FFT) [12, 14]. Unfortunately, this can require a large amount of storage if $\boldsymbol{\xi}(\boldsymbol{x}, t)$ is nonzero over a large surface, $\partial\Omega$.

- Precompute $\boldsymbol{\xi}(\boldsymbol{x}, t)$ via FDTD "bootstrapping" [1]: If $\boldsymbol{\psi}_+(\boldsymbol{x}, t)$ is the field generated by a known current source (e.g., a point source in an infinite waveguide) lying outside Ω, precompute these fields by an FDTD simulation in the incident medium $\boldsymbol{\chi}_+$ (i.e., with no scatterers, etc., using PML to absorb outgoing waves); store their values on $\partial\Omega$; and then convolve them if needed with any auxiliary pulse, $p(t)$, to obtain $\boldsymbol{\xi}(\boldsymbol{x}, t)$. An analogous approach can also be used to convert hard sources into equivalent transparent currents [15]. Again, this can be storage intensive.

- As first reported in [16] and described in detail in Chapter 3, for the special case of an incident plane wave in free space, the computation of $\boldsymbol{\xi}(\boldsymbol{x}, t)$ can be reduced to a *one-dimensional* (1-D) FDTD problem that is co-evolved with the main simulation along the incident wavevector, $\boldsymbol{k}$. This greatly reduces the storage requirements because all points in the same phase plane are redundant.

- Given a bandwidth of interest (since $|\hat{p}|$ is typically small outside some bandwidth), one could apply standard filter-design techniques from digital signal processing [17] to approximate $\hat{\boldsymbol{\psi}}_+(\boldsymbol{x}, \omega)$ in that bandwidth by a simple rational function of $e^{j\omega}$. This would represent a "finite impulse-response" (FIR) or recursive "infinite impulse-response" (IIR) filter with a small amount of storage (the filter "tap coefficients") for each $\boldsymbol{x}$. In this way, the $\boldsymbol{\psi}_+ * p$ convolution could be computed during the FDTD simulation with minimal storage per $\boldsymbol{x}$ (especially if $\hat{\boldsymbol{\psi}}_+$ is a slowly varying function of ω that is well approximated by a low-degree polynomial in the desired bandwidth), at the expense of greater software complexity [18].

In the absence of one of these techniques, a simple work-around is given in the next section.

4.3 SEPARATING INCIDENT AND SCATTERED FIELDS

In simulations where one is computing the scattered, reflected, or similar fields, it is necessary to distinguish the scattered fields from the incident fields. If the equivalent currents, $\boldsymbol{\xi}$, from the previous section are applied exactly, taking into account all discretization and dispersion effects, then the currents produce *zero* fields outside Ω (to nearly machine precision), so that any fields outside Ω are *only* the scattered fields. However, as noted above, exactly accounting for discretization and dispersion effects can be cumbersome. Moreover, in many cases, it is possible and convenient to use a much simpler method than the equivalent-currents prescription:

- In a single-mode waveguide as in Fig. 4.2(a), a point-dipole (or line-segment) current source is sufficient to excite the waveguide mode, as long as the source is far enough away from the region of interest that any other (non-guided) fields radiate away.
- To create an incident plane wave on a *periodic* surface, as described in Section 4.5, only the symmetry of the source matters. Any planar current source with the correct $e^{ik\cdot x}$ phase relationship and the desired polarization produces a plane wave, but generally traveling in *both* directions away from the source.

It turns out that there is a very simple and efficient way to exactly separate the incident and scattered fields, even for sources that are not the exact equivalent currents of a total-field/scattered-field approach: we simply run two FDTD simulations and subtract them. More precisely, to avoid large storage requirements, we subtract the Fourier transforms of the fields, as discussed next.

A typical use of FDTD simulations is to inject an incident electromagnetic pulse into a problem geometry; Fourier-transform the desired response (e.g., the transmitted, reflected, or scattered fields); compute the corresponding power or energy at each frequency; and thereby obtain the entire *spectrum* of the response (e.g., a transmission or absorption spectrum) in a single simulation. That is, for a field, $f(t)$, in response to the exciting pulse, one computes the Fourier transform, $\hat{f}(\omega)$, approximated by a discrete-time Fourier transform (DTFT) of the discrete-time field, $f(n\Delta t)$:

$$\hat{f}(\omega) = \frac{1}{\sqrt{2\pi}}\int_{-\infty}^{\infty} f(t)e^{i\omega t}dt \approx \frac{\Delta t}{\sqrt{2\pi}}\sum_{n=-\infty}^{\infty} f(n\Delta t)e^{i\omega n\Delta t} \tag{4.6}$$

where, for a pulsed field, the sum over n can be truncated when $|f|$ becomes sufficiently small. This computation would be performed for a set of frequencies covering the desired bandwidth. In practice, it is more storage efficient to accumulate the $\hat{f}$ summations as the FDTD simulation progresses, rather than storing $f(n\Delta t)$ for all n and computing $\hat{f}$ in postprocessing, especially if the fields at many spatial points are required [1]. For example, to compute the flux through some surface S, one would apply this procedure to obtain $\hat{\boldsymbol{E}}(\boldsymbol{x},\omega)$ and $\hat{\boldsymbol{H}}(\boldsymbol{x},\omega)$ for $\boldsymbol{x} \in S$, and then compute the flux spectrum, $P(\omega) = \frac{1}{2}\mathrm{Re}\iint_S(\hat{\boldsymbol{E}}^* \times \hat{\boldsymbol{H}})\cdot d\boldsymbol{S}$.

The procedure to separate the incident and scattered fields is then straightforward. First, perform a simulation with the desired current sources, $\boldsymbol{\xi}$, in the incident medium, $\boldsymbol{\chi}_+$, and compute the Fourier-transformed incident field, $\hat{\boldsymbol{\psi}}_+$, on the flux surface, S (storage proportional to the number of points on S multiplied by the number of desired frequencies). Repeat the calculation with the full medium, $\boldsymbol{\chi}$, including any scatterers or other devices, to obtain the Fourier-transformed *total* fields, $\hat{\boldsymbol{\psi}}$, and then subtract them to obtain the Fourier-transformed scattered fields, $\hat{\boldsymbol{\psi}}_- = \hat{\boldsymbol{\psi}} - \hat{\boldsymbol{\psi}}_+$. One can then compute the scattered power spectrum and so on as desired. As an additional benefit of this procedure, one can use the first $\boldsymbol{\psi}_+$ simulation to compute the exact incident power for normalization purposes.

Figure 4.3 illustrates an example of this procedure to compute the transmission and reflection of a single-mode dielectric waveguide through a 90° bend. Using a current source located a distance, L, from the bend, we excite a waveguide mode that is incident on the bend.

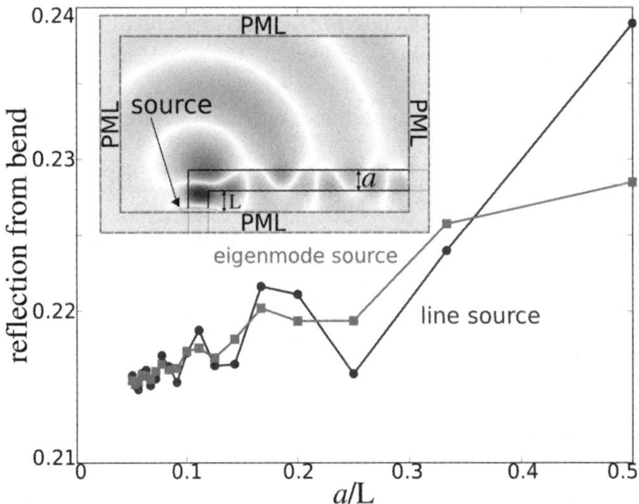

Fig. 4.3 FDTD-computed reflection coefficient in the 2-D dielectric waveguide of Fig. 4.2 ($\varepsilon/\varepsilon_0 = 12$, width a) with a 90° bend, at the single-mode frequency from Fig. 4.2(a), plotted vs. a/L (the normalized inverse of the distance between the source and the bend). Line plot with square dots: eigenmode-derived $\boldsymbol{J}$ source that excites only the propagating waveguide mode. Line plot with round dots: constant-amplitude $\boldsymbol{J}$ source along a transverse cross-section of the waveguide that excites both the waveguide mode and radiative fields. Inset: visualization of the steady-state E_z field, showing a large radiative-loss component at the bend. The reflection coefficient includes an error term which scales ~$1/L$ (hence approaching a straight line in this plot). This is due to the backscattered radiative field contributing to the measured backwards Poynting flux in the waveguide.

We compute the Fourier transform of the resulting field at two places: $\hat{\boldsymbol{\psi}}^{(1)}$ at a distance, L, before the bend (at the source location); and $\hat{\boldsymbol{\psi}}^{(2)}$ at a distance, L, after the bend. We also perform a second simulation with a *straight* waveguide, and compute the field Fourier transforms, $\hat{\boldsymbol{\psi}}_0^{(1)}$, at the source location, and $\hat{\boldsymbol{\psi}}_0^{(2)}$, after a distance, $2L$. $\hat{\boldsymbol{\psi}}_0^{(2)}$ is used to obtain the incident power, $P_+(\omega)$, and $\hat{\boldsymbol{\psi}}^{(2)}$ is used to obtain the transmitted power, $P_T(\omega)$. Hence, the normalized transmission is $T(\omega) = P_T(\omega)/P_+(\omega)$. $\hat{\boldsymbol{\psi}}_- = \hat{\boldsymbol{\psi}}^{(1)} - \hat{\boldsymbol{\psi}}_0^{(1)}$ is the reflected field, which is used to obtain the reflected power, $P_R(\omega)$. From this, we obtain the normalized reflection, $R(\omega) = P_R(\omega)/P_+(\omega)$, and also the radiative bend loss, $1 - T(\omega) - R(\omega)$.

The key here is to understand the role of the distance, L, between the source and the bend. To this end, Figure 4.3 plots the normalized reflection, R, versus a/L at the excitation frequency. For comparison, we consider two sources: an eigenmode-derived $\boldsymbol{J}$ source that excites only the propagating waveguide mode, and a constant-amplitude $\boldsymbol{J}$ source along a transverse cross-section of the waveguide that excites both the waveguide mode and radiative fields. (Because of the subtraction procedure, the fact that the latter source excites waveguide modes traveling in both directions is irrelevant.) In Fig. 4.3, the observed L variation of R is due to two factors. First, the simple constant-amplitude $\boldsymbol{J}$ source (but not the eigenmode-derived $\boldsymbol{J}$ source) generates some radiating waves in addition to the guided mode, and a portion of these radiating waves backscatter from the waveguide bend and contribute to $\hat{\boldsymbol{\psi}}^{(1)}$ for any finite L. Second, for *both* current sources, when the guided mode hits the waveguide bend, it scatters into radiating fields *in addition* to a backscattered guided mode, and a portion of this scattered radiation contributes to $\hat{\boldsymbol{\psi}}^{(1)}$ for a finite L.

However, at the observation point located a distance, L, before the waveguide bend, at which the reflection from the bend is computed, the scattering and backscattering radiative effects associated with the bend decrease as $1/L$ in 2-D (and $1/L^2$ in 3-D), the rate at which the radiative field intensity decreases with distance. Thus, as shown in Fig. 4.3, for both current sources, the computed results for reflection from the bend converge to the same value, and at approximately the same rate, as $L \to \infty$ ($a/L \to 0$). We see that there is no great advantage (at least for a single-mode waveguide) in using an equivalent-current source that generates only the waveguide mode, since this procedure does not reduce the contribution of the scattered and backscattered radiating fields.

We close this section by remarking that it is possible to exploit the orthogonality of modes in order to separate the contribution of the backscattered guided mode from any other radiative scattering, and more generally to compute the power scattered into *each* mode of a multimode waveguide. Suppose that one has derived from the FDTD results the Fourier-transformed scattered fields, $\hat{\boldsymbol{E}}$ and $\hat{\boldsymbol{H}}$, at frequency, ω, on a plane, A, perpendicular to the waveguide. Suppose also that one has computed (using a mode solver) the fields $\hat{\boldsymbol{E}}_m$ and $\hat{\boldsymbol{H}}_m$ of a particular waveguide mode, m, at the same frequency. If this waveguide mode is normalized to unit power, $\frac{1}{2}\mathrm{Re}\iint_A (\hat{\boldsymbol{E}}_m^* \times \hat{\boldsymbol{H}}_m)\cdot d\boldsymbol{A} = 1$, then the power, P_m, carried in $\hat{\boldsymbol{E}}$ and $\hat{\boldsymbol{H}}$ by *only this mode* is given by:

$$P_m = \left| \frac{1}{4}\iint_A \left(\hat{\boldsymbol{E}}_m^* \times \hat{\boldsymbol{H}} + \hat{\boldsymbol{E}} \times \hat{\boldsymbol{H}}_m^* \right)\cdot d\boldsymbol{A} \right|^2 \tag{4.7}$$

due to the well-known orthogonality properties of modes at a fixed ω [19–21].[4] Note that this calculation circumvents the need for any space-inefficient convolution because it is performed entirely in the frequency domain, and one only needs to invoke the mode solver at the frequencies where $\hat{\boldsymbol{E}}$ and $\hat{\boldsymbol{H}}$ are desired. A variation on this orthogonality relationship is available for Bloch modes in periodic waveguides or other periodic systems [21]. Some care is required with the normalization and orthogonality of evanescent modes and modes in lossy materials [21].

4.4 CURRENTS AND FIELDS: THE LOCAL DENSITY OF STATES

A key point in understanding the relationship between current sources and the resulting fields is this: *the same current radiates a different amount of power depending on the surrounding geometry.* This well-known fact takes on many forms. In antenna theory, the ratio of radiated power, P, to the mean-squared current, I^2, is known as the *radiation resistance*, $R = P/I^2$, and its dependence on the surrounding geometry can be illustrated analytically by, for example, the

[4]The field orthogonality relationships are often written without the complex conjugations [19], and it is easy to become confused on this point. For propagating modes in a lossless waveguide, one can do this because the field components transverse to the plane can be chosen to be purely real [21]. However, including the conjugation is convenient for mode solvers that do not enforce this phase choice. For lossy waveguides, dropping the conjugation is essential because the waveguide eigenproblem is complex-symmetric rather than Hermitian [21]. In this case, however, the unit-power normalization procedure is less clear. For evanescent modes, dropping the conjugation is equivalent to a time-reversal procedure that is required in such cases [21]. For periodic systems (and magneto-optic materials), matters are more complicated, but for lossless propagating modes the conjugation is essential [21].

method of images for a dipole current ("antenna") above a conducting ground plane [22]. A dipole source in a hollow metallic waveguide emits a power that diverges as a mode cutoff frequency is approached, an effect that has been explained in terms of the frequency-dependent waveguide impedance [23], but which can also be viewed as an example of a much more general phenomenon: *a Van Hove singularity in the density of states*. More precisely, the power radiated by a point-dipole source turns out to be exactly proportional to a fundamental physical quantity: the *local density of states* (LDOS), a measure of how much the harmonic modes of a system overlap with the source point. Equivalently, we can say that the LDOS is proportional to the radiation resistance of a dipole antenna.

In this section, we will review the derivation of this relationship, and the definition of the LDOS. The LDOS is of central importance not only for understanding classical dipole sources, but also in many physical phenomena that can be understood semiclassically in terms of dipole currents. For example, the spontaneous emission rate of atoms (key to fluorescence and lasing phenomena) is proportional to the LDOS [24–26], and enhancement of the LDOS by a microcavity is known as a *Purcell effect* [27]. A similar enhancement also occurs for nonlinear optical effects [28]. In understanding complex systems, the LDOS also gives a localized measure of the spectrum of eigenfrequencies present in a system, which is extremely useful in assessing a finite structure (e.g., portions of waveguides or periodic media) in which the concept of a dispersion relation (applicable to infinite waveguides or infinite periodic media) does not directly apply. The relationship of LDOS to the power exerted by a current source also makes it easier to generalize the concept of eigenfrequency spectra to lossy or leaky systems, in which the eigenproblem is non-Hermitian and more difficult to analyze directly. We will review the computation of the LDOS in FDTD by inserting dipole sources and performing appropriate Fourier transforms [29–34], or alternatively by computing resonant modes and their Purcell factors.

4.4.1 The Maxwell Eigenproblem and the Density of States

The central question is the relationship of the power radiated by a current source to the harmonic modes (eigenmodes) of Maxwell's equations in a given geometry. To illuminate this relationship, we follow the standard approach of expanding in an orthonormal basis of eigensolutions to Maxwell's equations [26, 31]. Let us begin by summarizing the Maxwell eigenproblem, which is discussed in detail elsewhere [35], considering lossless and dispersionless linear materials, $\varepsilon(\boldsymbol{x}) > 0$ and $\mu(\boldsymbol{x}) > 0$, for simplicity. For time-harmonic fields ($\sim e^{-i\omega t}$) in the absence of current sources, Maxwell's equations become $\nabla \times \boldsymbol{E} = i\omega\mu\boldsymbol{H}$ and $\nabla \times \boldsymbol{H} = -i\omega\varepsilon\boldsymbol{E}$. This can be combined into the linear eigenproblem:

$$\frac{1}{\varepsilon}\nabla \times \frac{1}{\mu}\nabla \times \boldsymbol{E} \triangleq \Theta\boldsymbol{E} = \omega^2\boldsymbol{E} \tag{4.8}$$

where we have defined the linear operator, $\Theta = (1/\varepsilon)\nabla \times (1/\mu)\nabla \times$ [35]. For appropriate boundary conditions, this operator is *Hermitian* ($\langle \boldsymbol{E}, \Theta\boldsymbol{E}'\rangle = \langle \Theta\boldsymbol{E}, \boldsymbol{E}'\rangle$ for any $\boldsymbol{E}$ and $\boldsymbol{E}'$) and semidefinite ($\langle \boldsymbol{E}, \Theta\boldsymbol{E}\rangle \geq 0$ for any $\boldsymbol{E}$) under the inner product, $\langle \boldsymbol{E}, \boldsymbol{E}'\rangle = \int \boldsymbol{E}^* \cdot \varepsilon\boldsymbol{E}'$. As a consequence, the solutions of (4.8) have real eigenfrequencies, $\omega^{(n)}$, and orthogonal eigenfields, $\boldsymbol{E}^{(n)}$ [35]. For simplicity, we imagine a problem contained in a finite box so that the eigenfrequencies form a discrete set, ω_n, for $n = 1, 2, \ldots$, taking a limit later on to obtain infinite systems. The eigensolutions, $\boldsymbol{E}^{(n)}$, can be chosen to be purely real-valued, if desired.

The density of states (DOS) is then defined by [36, 37]:

$$\text{DOS}(\omega) = \sum_n \delta\left(\omega - \omega^{(n)}\right) \tag{4.9}$$

so that $\int \text{DOS}(\omega)d\omega$ is simply a count of the number of modes in the integration interval. If we normalize the eigenfields such that $\left\langle \boldsymbol{E}^{(n)}, \boldsymbol{E}^{(n)} \right\rangle = 1$, then we can define a (per-polarization) *local* density of states by [36, 37][5]:

$$\text{LDOS}_\ell(\boldsymbol{x}, \omega) = \sum_n \delta\left(\omega - \omega^{(n)}\right) \varepsilon(\boldsymbol{x}) \left| E_\ell^{(n)}(\boldsymbol{x}) \right|^2 \tag{4.10}$$

so that

$$\text{DOS}(\omega) = \sum_{\ell=1}^{3} \int \text{LDOS}_\ell(\boldsymbol{x}, \omega)\, d^3\boldsymbol{x} \tag{4.11}$$

That is, $\text{LDOS}_\ell(\boldsymbol{x}, \omega)$ gives a measure of $\text{DOS}(\omega)$ weighted by how much of the energy density of each mode's electric field is at position $\boldsymbol{x}$ in direction ℓ. (Some applications, modeled by dipoles with random or fluctuating orientations, may instead need the polarization-independent LDOS, which is equal to $\sum_\ell \text{LDOS}$.) Remarkably, we will find that $\text{LDOS}_\ell(\boldsymbol{x}, \omega)$ is exactly proportional to the power radiated by an ℓ-oriented point-dipole current located at $\boldsymbol{x}$. In fact, that radiated-power (Green's function) version of the LDOS formula is often regarded as the more fundamental (albeit less obvious) form of the definition because it is more easily generalized to systems with loss.

Sometimes, authors say "DOS" when they are really referring to an LDOS. For an infinite system with translational invariance (or periodicity), the DOS is usually reported per unit length (or per period) in each direction of invariance (or periodicity). In a homogeneous medium (where the LDOS is position invariant), the DOS per unit distance in each direction is then precisely the LDOS.

4.4.2 Radiated Power and the Harmonic Modes

For any time-harmonic electric current source, $\boldsymbol{J}(\boldsymbol{x})e^{-i\omega t}$, we can combine Maxwell's equations, $\nabla \times \boldsymbol{E} = i\omega\mu\boldsymbol{H}$ and $\nabla \times \boldsymbol{H} = -i\omega\varepsilon\boldsymbol{E} + \boldsymbol{J}$, to obtain the equation:

$$\left(\nabla \times \frac{1}{\mu} \nabla \times - \omega^2 \varepsilon \right) \boldsymbol{E} = i\omega \boldsymbol{J} \tag{4.12}$$

---

[5]This is not the only way to define an LDOS. For example, one could define an LDOS weighted by the magnetic field energy, $\mu|\boldsymbol{H}^{(n)}|^2$ (normalized to $\int \mu|\boldsymbol{H}^{(n)}|^2 = 1$), instead of the electric field, which would be proportional to the power radiated by a *magnetic* dipole current. However, because so many physical phenomena are related to electric dipole sources, the electric-field LDOS seems to be more commonly used. Alternatively, in some circumstances (e.g., in considering the total energy of thermal fluctuations), it is useful to have an LDOS weighted by the total electromagnetic energy density, $\frac{1}{2}\varepsilon\left|\boldsymbol{E}^{(n)}(\boldsymbol{x}_0)\right|^2 + \frac{1}{2}\mu\left|\boldsymbol{H}^{(n)}(\boldsymbol{x}_0)\right|^2$ [38], which corresponds to an average of electric and magnetic dipole powers.

Equivalently, $\boldsymbol{E} = i\omega(\Theta - \omega^2)^{-1}\varepsilon^{-1}\boldsymbol{J}$ in terms of Θ from Section 4.4.1. Applying Poynting's theorem [39], the total radiated power, P, plus absorbed power in a lossy time-harmonic system, is equal and opposite to the time-average work done by the electric field on the electric current:

$$P = -\tfrac{1}{2}\mathrm{Re}\int \boldsymbol{E}^{*}\cdot\boldsymbol{J}\,d^3\boldsymbol{x} = -\tfrac{1}{2}\mathrm{Re}\langle \boldsymbol{E}, \varepsilon^{-1}\boldsymbol{J}\rangle \tag{4.13}$$

We can relate P to the spectrum of eigenfrequencies, $\omega^{(n)}$, by expanding $\varepsilon^{-1}\boldsymbol{J}$ in the basis of $\boldsymbol{E}^{(n)}$, using the orthogonality of the modes:

$$\varepsilon^{-1}\boldsymbol{J} = \sum_n \boldsymbol{E}^{(n)}\langle \boldsymbol{E}^{(n)}, \varepsilon^{-1}\boldsymbol{J}\rangle \tag{4.14}$$

Now, we can substitute the eigenequation, $\Theta\boldsymbol{E}^{(n)} = (\omega^{(n)})^2\boldsymbol{E}^{(n)}$, into $\boldsymbol{E} = i\omega(\Theta - \omega^2)^{-1}\varepsilon^{-1}\boldsymbol{J}$ to obtain:

$$\boldsymbol{E}(\boldsymbol{x}) = \sum_n \frac{i\omega\boldsymbol{E}^{(n)}(\boldsymbol{x})\langle \boldsymbol{E}^{(n)}, \varepsilon^{-1}\boldsymbol{J}\rangle}{(\omega^{(n)})^2 - \omega^2} \tag{4.15}$$

However, some care is required to substitute this expression into P. On the one hand, the numerator of $\langle \boldsymbol{E}, \varepsilon^{-1}\boldsymbol{J}\rangle$ is then $i\omega\left|\langle \boldsymbol{E}^{(n)}, \varepsilon^{-1}\boldsymbol{J}\rangle\right|^2$, which is purely imaginary, making it seem naively as if we will obtain $-\tfrac{1}{2}\mathrm{Re}\langle \boldsymbol{E}, \varepsilon^{-1}\boldsymbol{J}\rangle = 0$. On the other hand, this $\boldsymbol{E}(\boldsymbol{x})$ expression diverges when $\omega = \omega^{(n)}$, the physical result of driving a lossless oscillator at resonance for an infinite time. A basic problem here is that the equations are singular at $\omega = \omega^{(n)}$. At this frequency, the solutions are not unique since we can add any multiple of $\boldsymbol{E}^{(n)}$. Some careful regularization is required in order to treat this singularity properly to obtain the delta functions in the LDOS. In particular, the correct approach is to consider a *lossy* system (complex ε), for which the equations are nonsingular at all real ω, and then to take the *limit* as the losses go to zero ($\mathrm{Im}\,\varepsilon \to 0^+$). In effect, add an "infinitesimal loss." Physically, every material except vacuum has some loss. In an infinite system, adding an infinitesimal loss in this manner is equivalent to applying a radiation boundary condition to obtain a unique solution, since loss eliminates incoming waves from infinity. Adding an infinitesimal loss is also sometimes thought of as "enforcing causality" [26], which requires the Green's function to be analytic in the upper-half complex-ω plane [39].

Although this infinitesimal loss is often written formally as eigenfrequencies, "$\omega^{(n)} - i0^+$" or similar [24, 36], it is useful to go through this limiting process step by step [26, 31]. Suppose that we add a small positive imaginary part to ε, corresponding to an absorption loss. The operator, Θ, is no longer Hermitian, but is still complex symmetric under the unconjugated "inner product," $\langle \boldsymbol{E}, \boldsymbol{E}'\rangle = \int \boldsymbol{E}\cdot\varepsilon\boldsymbol{E}'$. Orthogonality of modes still follows, but we now have complex eigenfrequencies, $\omega_c^{(n)} = \omega^{(n)} - i\gamma^{(n)}$, with $\gamma^{(n)} > 0$ (exponential decay). (For a small $\mathrm{Im}\,\varepsilon$, one can show from perturbation theory that the real part of the frequency, $\omega^{(n)}$, is unchanged to first order, compared to the lossless system [35].) Now, substituting the above equations into P, we find:

$$P = \frac{\omega}{2}\,\mathrm{Im}\sum_n \frac{\langle \boldsymbol{E}^{(n)}, \varepsilon^{-1}\boldsymbol{J}\rangle\langle \boldsymbol{E}^{(n)}, \varepsilon^{-1}\boldsymbol{J}^{*}\rangle}{(\omega^{(n)} - i\gamma^{(n)})^2 - \omega^2} \tag{4.16}$$

Now, suppose that the loss is small, so that $\gamma^{(n)} \ll \omega^{(n)}$ for all modes. For this case, we note that $\gamma^{(n)}/\omega^{(n)}$ is proportional to $\operatorname{Im}\varepsilon / \operatorname{Re}\varepsilon$ [35], and $\boldsymbol{E}^{(n)}$ is approximately the real (lossless) eigenmode plus a correction of order $\operatorname{Im}\varepsilon / \operatorname{Re}\varepsilon$. Things simplify even further because the contribution of each term is negligible except when ω is close to $\omega^{(n)}$, making the denominator approximately 0. Altogether, to lowest order in $\operatorname{Im}\varepsilon / \operatorname{Re}\varepsilon$, one obtains:

$$P \approx \frac{1}{4}\sum_n \frac{\gamma^{(n)}\left|\left\langle \boldsymbol{E}^{(n)}, \varepsilon^{-1}\boldsymbol{J}\right\rangle\right|^2}{(\omega-\omega^{(n)})^2 + (\gamma^{(n)})^2} \tag{4.17}$$

which is simply a sum of *Lorentzian peaks* for each lossy mode. That is, each lossy mode contributes a Lorentzian peak to the radiated power that is proportional to the overlap integral, $\left|\int \boldsymbol{J}^* \cdot \boldsymbol{E}^{(n)}\right|^2$, of the current with that mode. This approximation becomes exact in the limit $\operatorname{Im}\varepsilon \to 0^+$. Importantly, in this $\gamma \to 0$ limit, the Lorentzians approach delta functions: $\lim_{\gamma\to 0} \gamma/(\Delta^2+\gamma^2) = \pi\,\delta(\Delta)$. One therefore obtains, in the lossless limit:

$$P = \frac{\pi}{4}\sum_n \left|\left\langle \boldsymbol{E}^{(n)}, \varepsilon^{-1}\boldsymbol{J}\right\rangle\right|^2 \delta(\omega-\omega^{(n)}) \tag{4.18}$$

In some treatments, this whole limiting process is circumvented by employing the complex identity, $\operatorname{Im}[1/(\Delta - i0^+)] = \pi\,\delta(\Delta)$ [24, 36]. However, the finite-loss result that lossy modes contribute Lorentzian peaks is useful in its own right. We will return to this result in Section 4.4.6 to discuss Purcell enhancement by resonant cavities.

The fact that the power in a lossless system exhibits a delta function peak at each eigenfrequency, $\omega^{(n)}$, assuming nonzero overlap of $\boldsymbol{E}^{(n)}$ with $\boldsymbol{J}$, requires some care to interpret physically. Since the delta function is a distribution rather than a classical function, it is not really valid to evaluate a delta function "at" $\omega = \omega^{(n)}$ to obtain "infinity." Instead, one should always consider a current consisting of a continuous spread of frequencies (a "test function" in distribution-theory parlance [40]). For example, consider a current source, $\boldsymbol{J}(\boldsymbol{x})p(t)$, where $p(t)$ is a pulse source with a continuous Fourier transform, $\hat{p}(\omega)$, representing its frequency spectrum. In computing the *time-integrated* power (i.e., the total work by the current), we can apply Parseval's theorem:

$$\int\left|\left\langle \boldsymbol{E}^{(n)}, \varepsilon^{-1}\boldsymbol{J}p\right\rangle\right|^2 dt = \left|\left\langle \boldsymbol{E}^{(n)}, \varepsilon^{-1}\boldsymbol{J}\right\rangle\right|^2 \int |p(t)|^2 dt = \left|\left\langle \boldsymbol{E}^{(n)}, \varepsilon^{-1}\boldsymbol{J}\right\rangle\right|^2 \int |\hat{p}(\omega)|^2 d\omega \tag{4.19}$$

Equation (4.18) becomes:

$$\int P\,dt = \frac{\pi}{4}\sum_n \left|\left\langle \boldsymbol{E}^{(n)}, \varepsilon^{-1}\boldsymbol{J}p\right\rangle\right|^2 \left|\hat{p}(\omega^{(n)})\right|^2 \tag{4.20}$$

This is a sum of *finite* contributions from each mode weighted by the Fourier amplitudes, $\left|\hat{p}(\omega^{(n)})\right|^2$, at the mode frequencies. Physically, injecting a pulse source into a lossless resonant system does a *finite* amount of work, leaving some superposition of resonant modes oscillating losslessly after the pulse source has returned to zero. Our summation gives the energy deposited by the pulse into each mode. (Another subtlety arises in an infinite system, because the number of modes increases with system size, but the overlap of each mode with any localized $\boldsymbol{J}$ decreases, so that in the limit of an infinite system, a current source still expends finite power.)

4.4.3 Radiated Power and the LDOS

Given (4.18), deriving the connection between radiated power and the LDOS is straightforward. We merely consider the case of a dipole current source, $\boldsymbol{J}(\boldsymbol{x}) = \boldsymbol{e}_\ell\,\delta(\boldsymbol{x}-\boldsymbol{x}_0)$, located at $\boldsymbol{x}_0$, where $\boldsymbol{e}_\ell$ is the unit vector in the direction, $\ell \in \{1, 2, 3\}$. In this case, $\left|\left\langle \boldsymbol{E}^{(n)}, \varepsilon^{-1}\boldsymbol{J}\right\rangle\right|^2 = \left|E_\ell^{(n)}(\boldsymbol{x}_0)\right|^2$. Comparing (4.10) and (4.18), we immediately see that:

$$\mathrm{LDOS}_\ell(\boldsymbol{x}_0, \omega) = \frac{4}{\pi}\varepsilon(\boldsymbol{x}_0)P_\ell(\boldsymbol{x}_0, \omega) \tag{4.21}$$

where $P_\ell(\boldsymbol{x}_0, \omega)$ is the power radiated by $\boldsymbol{J} = \boldsymbol{e}_\ell\,\delta(\boldsymbol{x}-\boldsymbol{x}_0)e^{-i\omega t}$.[6] Hence, the LDOS is exactly proportional to the power radiated by a dipole source, differing only by a factor of $4\varepsilon/\pi$.

The LDOS is often described as the imaginary part of the diagonal of the Green's function, similar to an analogous formula in quantum mechanics [36]. This refers to a definition of the dyadic "photon" Green's function, $\mathbf{G}_\ell$, as solving $(\omega^2 - \Theta)\mathbf{G}_\ell(\boldsymbol{x}, \boldsymbol{x}_0) = \boldsymbol{e}_\ell\,\delta(\boldsymbol{x}-\boldsymbol{x}_0)$, in which a factor, $-i\omega\varepsilon^{-1}$, is missing from the right-hand side compared to $\boldsymbol{E} = i\omega(\Theta - \omega^2)^{-1}\varepsilon^{-1}\boldsymbol{J}$, so that $\boldsymbol{E} = -i\omega\varepsilon^{-1}\mathbf{G}_\ell$, $P_\ell = -\tfrac{1}{2}\int \mathrm{Re}(\boldsymbol{E}*\boldsymbol{J}) = -\tfrac{1}{2}\omega\varepsilon^{-1}\,\mathrm{Im}\,G_{\ell\ell}$, and $\mathrm{LDOS}_\ell = -(2\omega/\pi)\,\mathrm{Im}\,G_{\ell\ell}$, matching LDOS definitions given elsewhere [24, 37].

There is an interesting subtlety in the application of (4.21) to spontaneous emission. It has been argued that the spontaneous emission rate is proportional not to the LDOS, but rather to $\mathrm{LDOS}/\varepsilon$ [24, 41] (the "radiative" LDOS). From (4.21), this implies that the spontaneous emission rate is exactly proportional to the power radiated by a dipole, with *no* ε factor. This is reminiscent of the semiclassical model of spontaneous emission as energy radiated by a classical dipole source [42], and the exact equivalence to the quantum picture can be demonstrated explicitly [25]. However, as a practical matter, one is most often interested in the *relative* enhancement or suppression of spontaneous emission by one structure relative to another, with the emitting atom embedded in the same material in both cases. In such a comparison, the presence or absence of an ε factor has no effect.

4.4.4 Computation of LDOS in FDTD

Relationship (4.21) between the LDOS and the power radiated by a dipole source makes it easy to compute the LDOS in an FDTD model. One can apply the usual technique of injecting an impulsive point-dipole source, $\boldsymbol{e}_\ell\,\delta(\boldsymbol{x}-\boldsymbol{x}_0)\,p(t)$, and accumulating the Fourier transforms, $\hat{E}_\ell(\boldsymbol{x}_0, \omega)$, of the field at $\boldsymbol{x}_0$. This yields the complete LDOS spectrum at $\boldsymbol{x}_0$ in a single calculation [30, 33]. The LDOS is then:

$$\mathrm{LDOS}_\ell(\boldsymbol{x}_0, \omega) = -\frac{2}{\pi}\varepsilon(\boldsymbol{x}_0)\frac{\mathrm{Re}\left[\hat{E}_\ell(\boldsymbol{x}_0, \omega)\,\hat{p}(\omega)^*\right]}{\left|\hat{p}(\omega)\right|^2} \tag{4.22}$$

[6] There is a subtlety in applying $P = -\tfrac{1}{2}\mathrm{Re}\left\langle \boldsymbol{E}, \varepsilon^{-1}\boldsymbol{J}\right\rangle$ to a delta-function current in 2-D or 3-D, because in this case, $\boldsymbol{E}$ diverges at the location of the current. However, the *real part* of $\boldsymbol{E}$ does *not* diverge for $\boldsymbol{J}(\boldsymbol{x}) = \boldsymbol{e}_\ell\,\delta(\boldsymbol{x}-\boldsymbol{x}_0)$ in the case of a lossless medium at $\boldsymbol{x}_0$. It is easily verified that $-\tfrac{1}{2}\mathrm{Re}\,E_\ell(\boldsymbol{x}_0)$ is then equal to the radiated power as Poynting's theorem demands (or $\tfrac{1}{2}\omega\,\mathrm{Im}\,E_\ell$ in the common convention where the dipole, $\boldsymbol{J}$, is multiplied by $-i\omega$ [39]). Matters are more subtle for lossy materials at $\boldsymbol{x}_0$, but it has been argued that the discretization in FDTD is equivalent to standard analytical regularizations [34].

Here, we need to normalize by the pulse spectrum, $|\hat{p}(\omega)|^2$, in order to obtain the power exerted by a unit-amplitude dipole (assuming linear materials), equivalent to computing the radiation resistance. Most commonly, we are not interested in the LDOS as an absolute quantity, but rather the relative enhancement (or suppression) of LDOS in one system relative to another — for example, the LDOS in some structure relative to the LDOS in a homogeneous medium. Again, this is straightforward: just compute the LDOS twice, once in the reference system (e.g., vacuum) and once in another geometry, and divide the two. This has the useful side effect of canceling any normalization factors (including $|\hat{p}(\omega)|^2$, but *not* the $\hat{p}^*$ factor inside Re) or choices of units. As will be noted in Section 4.10, when computing the LDOS of a resonant cavity with a long lifetime, it can be more efficient to employ an alternate "Purcell" formula for the LDOS following a resonant-mode calculation, unless more sophisticated Padé extrapolation techniques [43–45] are used to compute the Fourier transform of the slowly decaying fields.

Figure 4.4 illustrates an example of such a calculation. Here, we compute the LDOS spectrum at a point inside a finite *photonic crystal* comprised of a 2-D square lattice of dielectric rods ($\varepsilon = 12\varepsilon_0$) in air. For the assumed out-of-plane point-dipole current source, an infinite periodic system of this type would exhibit "photonic bandgaps," i.e., ranges of ω in which there are no electromagnetic waves propagating in any direction in the crystal [35]. In such an infinite system, (4.10) tells us that the current source would radiate zero power at any ω in the bandgaps, since there are no modes, $\omega^{(n)}$, in these gaps to make the delta function nonzero.

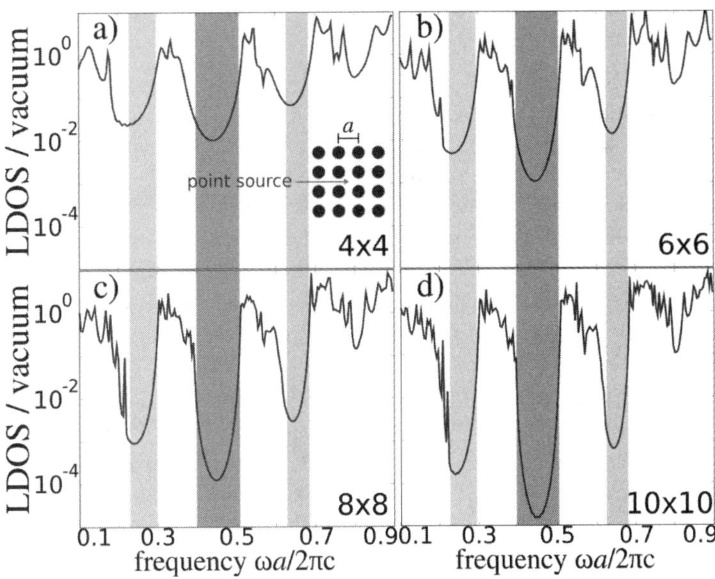

Fig. 4.4 Radiated power spectrum — proportional to the local density of states (LDOS) — for an out-of-plane point-dipole current source radiating in the center of a finite 2-D photonic crystal (inset: period = a square lattice of dielectric rods of radius = $0.2a$ and refractive index = $\sqrt{12}$, in air), normalized by a calculation for the same dipole source radiating in free space. The spectrum is drastically changed depending on how many rods (from 4×4 to 10×10) surround the source. The LDOS in the photonic bandgaps (shaded regions) decreases exponentially with an increasing number of rods. The LDOS experiences Fabry-Perot oscillations at other frequencies due to the partial reflections at the edges of the photonic crystal, and approaches a discontinuity (Van Hove singularity) at the edges of the bandgaps.

For a finite photonic crystal surrounded by an infinite air region (truncated in FDTD by PML absorbers), the LDOS in the bandgap is nonzero. This is because an evanescent tail of the field produced by the dipole source "leaks" through the photonic crystal and radiates into the air. However, this effect decreases *exponentially* as the crystal size increases. Furthermore, at the edges of the bandgaps, as ω approaches the cutoff of a band of propagating modes in the crystal, the LDOS becomes more and more singular (here, discontinuous) as the size of the crystal increases. This is called a *Van Hove singularity*, and is discussed in Section 4.4.5.

For the case of spontaneous emission for an atomic transition with a finite Lorentzian linewidth, γ_0, around a transition frequency, ω_0, the correct enhancement factor is not simply the LDOS at ω_0. In fact, it is an average of the LDOS over the transition linewidth. In particular, the emission rate for such an ℓ-directed dipole transition at $\boldsymbol{x}_0$ is proportional to [25, 32]:

$$\int \frac{1}{\varepsilon(\boldsymbol{x}_0)} \mathrm{LDOS}_\ell(\boldsymbol{x}_0, \omega) \frac{\gamma_0 / \pi}{(\omega - \omega_0)^2 + \gamma_0^2} \, d\omega \tag{4.23}$$

This converges to $\mathrm{LDOS}_\ell(\boldsymbol{x}_0, \omega_0)/\varepsilon(\boldsymbol{x}_0)$ as $\gamma_0 \to 0$. Therefore, it is reasonable to simply use $\mathrm{LDOS}_\ell(\boldsymbol{x}_0, \omega_0)$ as long as γ_0 is much smaller than any feature sizes in the LDOS spectrum. Taking the Fourier transform to the time domain, expression (4.23) is equivalent to computing the total work done by a *decaying* dipole source, $\boldsymbol{J}(\boldsymbol{x}, t) \sim \boldsymbol{e}_\ell \, \delta(\boldsymbol{x} - \boldsymbol{x}_0) \exp(-i\omega_0 t - \gamma_0 t/2)$, for $t > 0$ [32]. However, this is more efficient (i.e., a shorter-time simulation) than postprocessing $\mathrm{LDOS}_\ell(\boldsymbol{x}_0, \omega)$ only if γ_0 is large (faster decay) compared to the decay rates of the optical modes, and is less efficient if γ_0 is small.

4.4.5 Van Hove Singularities in the LDOS

The LDOS — again, just the power emitted by a dipole source — exhibits singularities in a waveguide or a periodic system whenever a cutoff frequency is approached. These are known (from solid-state physics) as *Van Hove singularities* [46]. These singularities have many important physical consequences. For example, they can be understood as the feedback mechanism in distributed feedback (DFB) lasers. More generally, it is important to understand the relationship between the dispersion characteristic, $\omega(\boldsymbol{k})$, and the LDOS or DOS, in order to understand the effect of current sources in such structures.

For simplicity, consider a system (e.g., a waveguide) that is invariant in a single direction (say z), so that the fields can be written in the separable form, $\boldsymbol{E}_k(x, y) \exp[i(kz - \omega t)]$, with some modal dispersion relation(s), $\omega_n(\boldsymbol{k})$. To understand the DOS of this system, start with a *finite* system in the z-direction of length, L, with *periodic* boundary conditions in z. In this case, the only allowed solutions are the modes where $k = 2\pi m/L$ is an integer multiple, m, of $2\pi/L$. From (4.9), the DOS per unit length, L, is then:

$$\frac{\mathrm{DOS}(\omega)}{L} = \frac{1}{L} \sum_n \sum_m \delta[\omega - \omega_n(2\pi m / L)] = \frac{1}{2\pi} \sum_n \left\{ \sum_m \delta[\omega - \omega_n(2\pi m / L)] \frac{2\pi}{L} \right\} \tag{4.24}$$

Now, if we take the $L \to \infty$ limit, it is clear that $\Sigma_m \to \int dk$, with $(2\pi/L) = \Delta k \to dk$. Hence:

$$\frac{\mathrm{DOS}}{\text{per length}}(\omega) = \frac{1}{2\pi} \sum_n \int_{-\infty}^{\infty} \delta[\omega - \omega_n(k)] \, dk \tag{4.25}$$

Even though the DOS of a *finite* system is "spiky" (a sum of δ functions), the DOS/length of an infinite system is (mostly) continuous, due to this $\int dk$. A similar argument applies to discrete periodic systems (photonic crystals), in which case one integrates the dispersion relations of the Bloch modes. The LDOS is similar to the DOS, except that, like (4.10), there is an additional factor of $\varepsilon|E_\ell|^2$ weighting the integral.

The only places that singularities arise are at cutoffs (or in general, at points of zero group velocity, $d\omega_n/dk = 0$). For example, suppose that $\omega_1(k)$ has a cutoff at $k = 0$ for a frequency, ω_c (e.g., for a hollow metal waveguide). Let us apply a Taylor expansion, $\omega_1(k) \approx \omega_c + \alpha k^2$, near $k = 0$, with $\alpha = \frac{1}{2} d^2\omega_1/dk^2\big|_{k=0} > 0$. In this case, the $\int dk$ in the DOS becomes (near ω_c):

$$\int \delta[\omega - \omega_1(k)]\, dk \approx \int \delta\left(\omega - \omega_c - \alpha k^2\right) dk = \begin{cases} 0 & \omega < \omega_c \\ 0.5\left[\alpha(\omega - \omega_c)\right]^{-1/2} & \omega > \omega_c \end{cases} \tag{4.26}$$

with the $(\omega - \omega_c)^{-1/2}$ term arising from a Jacobian factor, $1/2\alpha k$. (Note that the δ function is only nonzero when $\omega = \omega_c + \alpha k^2$, and hence, $k = [(\omega - \omega_c)/\alpha]^{1/2}$). Thus, the DOS *diverges* with an integrable $(\omega - \omega_c)^{-1/2}$ singularity as ω approaches the cutoff frequency from above. This is true for any quadratic band edge in *one dimension* of translational symmetry. A similar conclusion is reached for the LDOS, of course. With two dimensions of translational symmetry (e.g., a planar waveguide), one obtains a discontinuity in the DOS (and LDOS) at a band edge; and in three dimensions, there is a continuous $(\omega - \omega_c)^{1/2}$ singularity.[7]

Since the LDOS is proportional to the spontaneous emission rate of excited atoms (semiclassical dipoles), these singularities in the LDOS at band edges (cutoffs) can lead to lasing via "distributed feedback." In an FDTD simulation of a hollow metal waveguide, the $(\omega - \omega_c)^{-1/2}$ singularities lead to a "spectral distortion" of the output of a dipole source, where the dipole source emits more power as the cutoff is approached. This effect was explained by other authors in the language of the waveguide "impedance" [23]. (Of course, in a *finite-length* waveguide surrounded by vacuum, the LDOS is finite — a current radiates finite power — essentially because the zero group-velocity band-edge solutions can escape from the ends of the waveguide.)

4.4.6 Resonant Cavities and Purcell Enhancement

A lossless localized mode yields a δ-function spike in the LDOS, whereas a *lossy* localized mode — a *resonant cavity mode* — leads to a Lorentzian peak. This was shown in (4.17) for losses due to a small absorption, but is generically true for any loss mechanism, including "leaky" modes with radiative losses.[8] The large enhancement in the LDOS at the resonant peak is known as a *Purcell effect*, named after Purcell's proposal for enhancing spontaneous emission of an atom in a cavity (by analogy with a microwave antenna resonating in a metal box) [27]. There is a famous formula for this enhancement factor arising from the LDOS derivation above, as will be reviewed next.

---

[7]To understand this dependence on the problem dimension, consider an *isotropic* band-edge shape, $\omega_1(\boldsymbol{k}) \approx \omega_c + \alpha|\boldsymbol{k}|^2$. Performing the $\boldsymbol{k}$ integration in cylindrical or spherical coordinates then yields a $\int 2\pi k\, dk$ or $\int 4\pi k^2 dk$ integral, respectively. The additional k or k^2 factors multiply the integration result by the terms, $(\omega - \omega_c)^{1/2}$ or $(\omega - \omega_c)$.

[8]The theory of leaky "modes" is somewhat subtle because the modes are not strictly eigenfunctions [47]. They can be defined as poles in the Green's function that are close to the real-ω axis (slightly *below* it due to causality), which consequently contribute a Lorentzian peak to $\mathrm{Re}(\boldsymbol{E}^* \cdot \boldsymbol{J})$.

Typically, one characterizes the lifetime of a lossy mode by a dimensionless *quality factor*, $Q^{(n)} = \omega^{(n)}/2\gamma^{(n)}$ [35]. In terms of $Q^{(n)}$, the contribution to the LDOS at a resonant peak, $\omega = \omega^{(n)}$, from (4.17) is given by:

$$\frac{1}{\pi}\varepsilon(\boldsymbol{x}_0)\left|E_\ell^{(n)}(\boldsymbol{x}_0)\right|^2 \frac{2Q^{(n)}}{\omega^{(n)}} \tag{4.27}$$

Suppose that we choose $\boldsymbol{x}_0$ to be the point where $\varepsilon|\boldsymbol{E}^{(n)}|^2$ is *maximum*, and we consider the polarization-averaged $\text{LDOS}(\boldsymbol{x}_0) = \sum_\ell \text{LDOS}_\ell(\boldsymbol{x}_0)$ (e.g., for dipoles with random orientation). Then, because we normalized $\int\varepsilon|\boldsymbol{E}^{(n)}|^2 = 1$, the $\varepsilon(\boldsymbol{x}_0)\left|\boldsymbol{E}^{(n)}(\boldsymbol{x}_0)\right|^2$ term has units of inverse volume, the *modal volume*, $V^{(n)}$ [27, 35]:

$$V^{(n)} = \frac{\int\varepsilon|\boldsymbol{E}^{(n)}|^2}{\max\varepsilon|\boldsymbol{E}^{(n)}|^2} \tag{4.28}$$

This is essentially the volume in which $\varepsilon|\boldsymbol{E}^{(n)}|^2$ is not small. In this case, the resonant mode's contribution to the LDOS at $\omega^{(n)}$ is given by:

$$\text{resonant LDOS} \approx \frac{2}{\pi\omega^{(n)}}\frac{Q^{(n)}}{V^{(n)}} \tag{4.29}$$

If $Q^{(n)}$ is large enough, all other contributions to the LDOS (from the other modes) are negligible at $\omega^{(n)}$. Hence, for a given frequency $\omega^{(n)}$, we obtain an approximate Q/V enhancement in the LDOS at the point of the peak resonant field. This Q/V enhancement is widely known as a *Purcell factor*. Note that, upon examining (4.23), we see that this enhancement factor only applies to spontaneous emission if the atomic transition linewidth, γ_0, is much smaller than the microcavity linewidth, $\gamma^{(n)}$. If $\gamma_0 >> \gamma^{(n)}$, then the contribution from the cavity is reduced by a factor, $\gamma^{(n)}/\gamma_0$.

Figure 4.5 illustrates an example of Purcell enhancement of the LDOS. Here, we consider a 2-D perfect-metallic $a \times a$ cavity of finite wall thickness, $0.1a$. One sidewall is assumed to have a small notch of width, w, which allows the cavity modes to escape to the surrounding free-space region, as shown in the inset. In the absence of the notch, the lowest-frequency mode with out-of-plane polarization is $E_z^{(1)} = (4/a^2)\sin(\pi x/a)\sin(\pi y/a)$, with a frequency, $\omega^{(1)} = \sqrt{2}c\pi/a$, and a modal volume, $V^{(1)} = a^2/4$. While the notch slightly perturbs this solution, more importantly, it allows radiation into the surrounding region, yielding a finite Q. For $w << a$, this radiative escape occurs via an evanescent (sub-cutoff) mode of the channel waveguide formed by the notch. It follows from inspection of the evanescent decay rate, $[(\pi/w)^2 - (\omega^{(1)})^2]^{1/2}/c$, that the lifetime scales asymptotically as $Q^{(1)} \sim e^{\#/w}$ for some coefficient #.

The results of Fig. 4.5 validate both this prediction and the LDOS calculations described in this section. Here, the LDOS at the center of the cavity (the point of peak $|\boldsymbol{E}|$) is computed in two ways. The first is via the exact dipole-power expression of (4.21). The second is via the Purcell approximation, (4.29), where the cavity mode and its lifetime Q are obtained using the filter-diagonalization technique. The latter approach is much more efficient for high Q (small w), since one must otherwise run the FDTD simulation for a very long time to directly accumulate the Fourier transform of a slowly decaying mode. In Fig. 4.5, the results of the two approaches agree to within the discretization error. Furthermore, the Purcell Q/V approximation is asymptotically linear on a semilog scale versus $1/w$, as predicted. This verifies the LDOS analyses of this section.

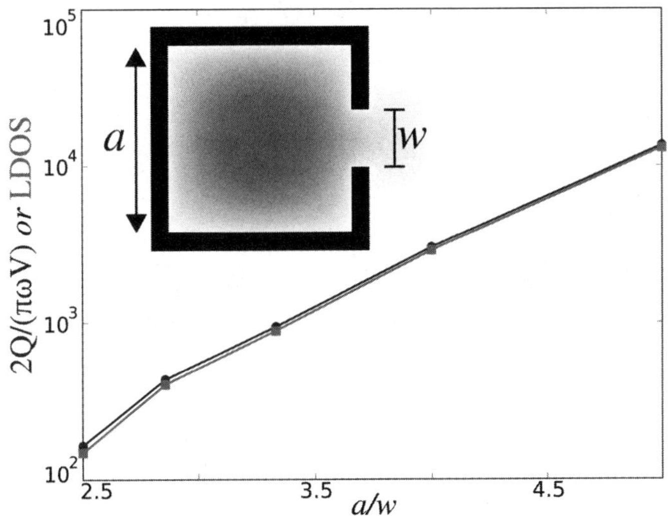

Fig. 4.5 FDTD-computed Purcell enhancement of the LDOS in a 2-D perfect-metallic $a \times a$ cavity of finite wall thickness, $0.1a$. As shown in the inset, one sidewall of the cavity is assumed to have a small notch of width, w, which allows the cavity modes to escape to the surrounding free-space region with some lifetime, Q. We consider the fundamental mode with out-of-plane (E_z) polarization. Square dots show the LDOS computed by the exact dipole-power formula of (4.21); round dots show the results of the Purcell approximation of (4.29). Agreement of the two approaches is within the computational accuracy.

4.5 EFFICIENT FREQUENCY-ANGLE COVERAGE

A common problem in nanophotonics is to evaluate the reflection, transmission, or absorption spectrum of a periodically patterned surface as a function of the incidence angle of light, as depicted schematically in Fig. 4.6. For example, calculating (and optimizing [48]) the absorption of periodically patterned surfaces is of key interest in designing efficient thin-film photovoltaic cells [49]. In this section, we review some basic properties of this problem and describe how it can be efficiently solved in FDTD, including a technique to map out the spectra as a function of *both* frequency and incidence angle by using a minimal number of FDTD simulations with simple pulse (broadband) line-current sources [50 – 52].

Such problems can be solved efficiently in FDTD because they can be reduced to a problem in the *unit cell* of the periodicity (dashed boxes in Fig. 4.6), which is finite in the plane of the surface and can be truncated by PML absorbers in the direction perpendicular to the surface. The key is to impose the correct boundary conditions. Periodicity means that the *structure* is invariant under translation by lattice vectors $\boldsymbol{R}$ [e.g., $\boldsymbol{R} = (a, 0)$ is the primitive lattice vector in the 2-D example of Fig. 4.6(b)]. However, this does *not* mean that the *fields* are periodic or that we can impose periodic boundary conditions. For an incident plane wave $\sim e^{i(\boldsymbol{k}\cdot\boldsymbol{x}-\omega t)}$ with wavevector $\boldsymbol{k}$, the incident fields are *Bloch periodic*:

$$\text{fields}(\boldsymbol{x}+\boldsymbol{R}) = \text{fields}(\boldsymbol{x})\, e^{i\boldsymbol{k}\cdot\boldsymbol{R}} \tag{4.30}$$

That is, the fields are periodic up to a phase shift depending on the components of $\boldsymbol{k}$ in the plane of periodicity.

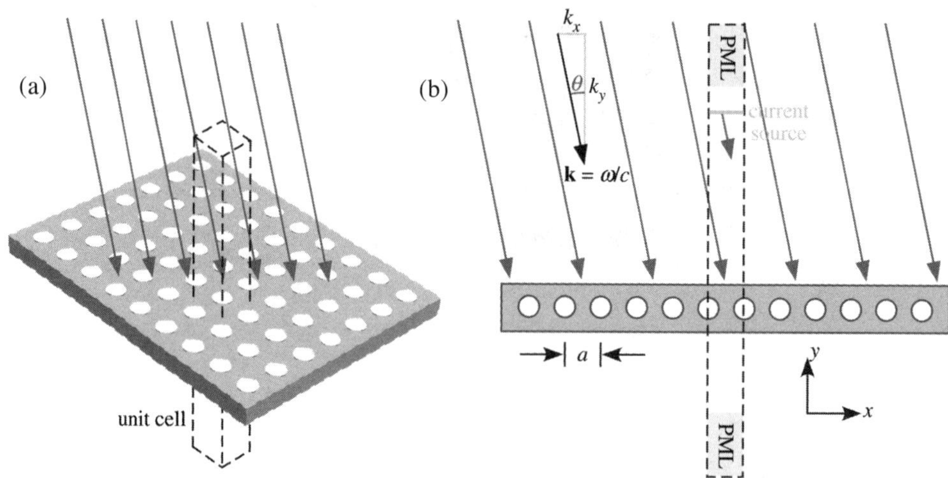

Fig. 4.6 Plane wave obliquely incident on periodic structures. (a) For a periodically patterned surface in 3-D, the problem can be reduced to a simulation within a unit cell (the dashed parallelepiped) having Bloch-periodic boundaries. (b) Analogous 2-D geometry, a holey waveguide comprised of a dielectric strip perforated by a linear array of holes of period, a. Here, the unit cell of the computation is shown as a dashed box, with PML absorbers in the y-direction and Bloch-periodic (phase $e^{ik_x a}$) boundary conditions in the x-direction, where k_x is the surface-parallel component of the incident wavevector, $\boldsymbol{k}$. In FDTD, the incident wave can be excited with a current source of appropriate phase and polarization extending transversely across the computational cell.

The crucial fact is that, in a periodic system with a Bloch-periodic incident wave, the Bloch periodicity (the surface-parallel component of $\boldsymbol{k}$) is *conserved* for linear materials. The total field (including scattering, etc.) is Bloch periodic with the *same* in-plane $\boldsymbol{k}$ (a kind of conserved "momentum") [35]. This means that we simply need to impose Bloch-periodic boundary conditions in our FDTD simulation in order to reduce the computational domain to the unit cell of the problem. Note that this means that our fields must be simulated as *complex numbers*, which poses no particular difficulty. Because Maxwell equations are unmodified, the real and imaginary parts of the fields only couple via the boundary conditions.

Even though the scattered fields are also Bloch periodic, this does not mean that they are all plane waves with the same $\boldsymbol{k}$. Consider the 2-D problem of Fig. 4.6(b), in which the incident wavevector is $\boldsymbol{k} = (k_x, k_y)$. From (4.30), $\text{fields}(x+a) = \text{fields}(x)e^{ik_x a}$, so only k_x is "conserved" and not k_y. Moreover, even k_x is only conserved up to addition of multiples of $2\pi/a$ ("reciprocal lattice vectors") [35]. For example, a reflected plane wave with $\boldsymbol{k}' = (k_x + 2n\pi/a, k_y')$ is *also* Bloch periodic for any integer, n, since $e^{i(2n\pi/a)a} = 1$. These are *diffracted* waves. Summing all such waves for all n yields a Fourier series, and we can equivalently say that the solutions are in general *Bloch waves*: periodic functions of x multiplied by $e^{ik_x x}$ [35]. Furthermore, since the solutions in air must satisfy the dispersion relationship, $\omega = c|\boldsymbol{k}'|$ (frequency is conserved in a linear system), the scattered waves satisfy $k_y' = \pm[(\omega/c)^2 - (k_x + 2n\pi/a)^2]^{1/2}$. For $n = 0$, we obtain $k_y' = \pm k_y$, the "law of equal angles" for the "specular" reflected and transmitted waves. In addition, for any ω, k_y' becomes *imaginary* for sufficiently large $|n|$, corresponding to decaying (evanescent) waves. Hence, there are a finite number of propagating diffracted waves for any finite ω; and for sufficiently small ω, there are *no* diffracted waves except for the specular $n = 0$ waves [35].

Suppose that our incident plane wave is given by fields $\boldsymbol{E}_0 e^{i(\boldsymbol{k}\cdot\boldsymbol{x}-\omega t)}$ and $\boldsymbol{H}_0 e^{i(\boldsymbol{k}\cdot\boldsymbol{x}-\omega t)}$. For a planar current source parallel to the surface [e.g. the source in Fig. 4.6(b)], the equivalent-currents prescription of Section 4.2 would specify the electric and magnetic currents, $\boldsymbol{J} = \boldsymbol{n} \times \boldsymbol{H}_0 e^{i(\boldsymbol{k}\cdot\boldsymbol{x}-\omega t)}\delta(\cdots)$ and $\boldsymbol{K} = -\boldsymbol{n} \times \boldsymbol{E}_0 e^{i(\boldsymbol{k}\cdot\boldsymbol{x}-\omega t)}\delta(\cdots)$. However, one can simplify this in several ways. First, we can use an electric current, $\boldsymbol{J}$, alone, or a magnetic current, $\boldsymbol{K}$, alone. By the equivalence principle, this corresponds to the incident wave plus (or minus) its mirror flip. That is, it generates incident waves propagating both towards and away from the surface, but the latter can be eliminated by the subtraction technique of Section 4.3. Second, it is not necessary to correct for discretization or dispersion effects. It follows from the translational symmetry of the air region (giving, again, conservation of $\boldsymbol{k}$) that an $\sim e^{i\boldsymbol{k}\cdot\boldsymbol{x}}$ current source in air produces a plane wave (in this case, a discrete-space plane wave) with the same in-plane wavevector component. Therefore, all we have to do is insert a planar current source, $\boldsymbol{J}$ or $\boldsymbol{K}$, with the correct in-plane phase relationship and the desired polarization, and perform a second simulation with only air (no surface) for normalization and for subtracting the incident field (or rather, its Fourier transform), as discussed in Section 4.3.

Figure 4.7 illustrates the final step of the analysis: efficiently mapping the reflection or transmission spectrum to its angular response. In this example, the reflection spectrum, $R(\omega,\theta)$, is obtained at multiple angles for the 2-D holey waveguide shown in Fig. 4.6(b).

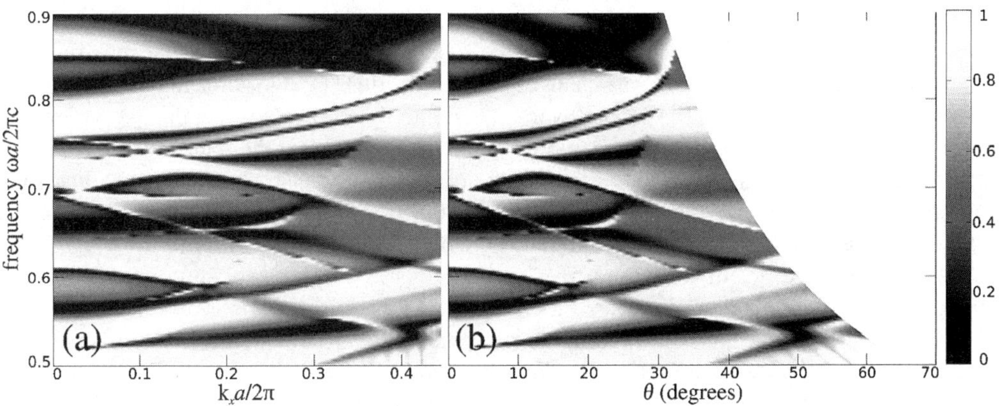

Fig. 4.7 FDTD-computed reflection spectra, R = reflected power / incident power, for oblique plane-wave incidence on the 2-D holey waveguide of Fig. 4.6(b), characterized by $\varepsilon/\varepsilon_0 = 13$, width = $1.2a$, and hole radius = $0.36a$. (a) $R(\omega, k_x)$, computed for a finite number of fixed k_x values and then interpolated between these vertical cuts to fill the rectangular $\omega \times k_x$ plot; (b) $R(\omega,\theta)$, derived from the data of (a) by first mapping the spectral data computed for each fixed k_x value onto the corresponding curve, $\theta = \sin^{-1}(ck_x/\omega)$, and then interpolating between these curves to fill the remainder of the $\omega \times \theta$ plot.

As usual, we obtain the entire spectrum at once using FDTD by injecting a short (broad-bandwidth) pulse source. We then Fourier transform the result to obtain a flux spectrum normalized by the incident-flux spectrum derived from an empty-space simulation. However, there is one subtlety. When we impose Bloch-periodic boundary conditions in FDTD, we set the fields at one side equal to $e^{ik_x a}$ times the fields at the other side, which gives the *same* k_x for *all* the frequency components, ω. That is, each simulation really computes $R(\omega, k_x)$ for a *fixed* k_x.

Note that a fixed k_x does *not* correspond to a fixed angle, θ, except for normal incidence, $k_x = 0$. This is because θ depends on ω as well as k_x: $\theta(\omega, k_x) = \sin^{-1}(ck_x/\omega)$. Of course, $R(\omega, k_x)$ is useful in its own right, and is plotted for the holey waveguide structure of Fig. 4.6(b) in Fig. 4.7(a). In fact, $R(\omega, k_x)$ is closely related to this structure's dispersion relation (band diagram), $\omega(k_x)$. The leaky modes above the light line, $\omega > ck_x$ [35], yield *Fano resonances*, adjacent peaks and dips in the reflection spectra [53] that can be observed in Fig. 4.7(a).

However, given the FDTD-computed $R(\omega, k_x)$ data, it is straightforward to convert to an $R(\omega,\theta)$ plot by making the change of variables to $\theta(\omega, k_x)$ for each k_x. By this procedure, each frequency spectrum at a fixed k_x maps onto its corresponding $\theta(\omega, k_x)$ curve in the $R(\omega,\theta)$ plot. Readily available software is then exercised to interpolate $R(\omega,\theta)$ between these curves, and thereby fill the desired ω, θ parameter space. The result is shown in Fig. 4.7(b). Here, note that the range of θ depends on ω. [Of course, we could obtain a larger range of θ by going to larger k_x values in the FDTD simulations. We could also display a "normal" looking $R(\omega,\theta)$ plot simply by cropping the ω, θ data to a rectangular region.]

4.6 SOURCES IN SUPERCELLS

A common problem that arises in FDTD simulations of periodic systems is that FDTD is usually formulated using orthogonal Cartesian grids [1], whereas periodic systems commonly have nonorthogonal (parallelogram/parallelepiped) unit cells [35]. This mismatch can be dealt with in a variety of ways. FDTD can be reformulated to use a nonorthogonal grid that represents the desired unit cell directly [1, 54], but changes to the core FDTD algorithm may require major revision of one's existing FDTD software. Essentially equivalent to this, a coordinate transformation can be used to map the nonorthogonal lattice into an orthogonal lattice. However, this requires appropriately specifying anisotropic materials and transformations of the fields and sources in order to relate them to the original problem [55]. Another option is to use a "staircased" version of the nonorthogonal unit cell with an orthogonal grid [56].

A simple and efficient technique to model a nonorthogonal unit cell in FDTD, without changing the core FDTD algorithm and without changing the materials or fields, is to modify the boundary conditions to employ an orthogonal computational cell. Here, instead of Bloch-periodic, one can use "skewed" Bloch-periodic boundary conditions in which the fields at one side of the computational cell are related to the fields at the other side plus a lateral shift [57, 58].

However, it is tempting to use FDTD *with no modifications whatsoever* in the common case where the nonorthogonal periodic lattice can be described with an orthogonal *supercell*. This is a periodic unit cell that is larger than the primitive unit cell, albeit with an increase in the computational volume. For example, this is true for triangular/hexagonal lattices in 2-D, and fcc/bcc/fct lattices in 3-D [35]. In this section we review the fact that the simplicity of a supercell comes at the price of introducing unwanted additional solutions, due to a *band-folding* phenomenon. Fortunately, there is a simple modification of the *source terms* in FDTD (and, optionally, of the postprocessing as well) that mostly eliminates this problem.

Band folding is easiest to describe in a 1-D example, as illustrated in Fig. 4.8. A typical goal in analyzing a periodic system (e.g., a photonic crystal) is to map out the dispersion relation, $\omega_n(\boldsymbol{k})$, of the solutions propagating within the crystal, which take the form of *Bloch waves*: $\boldsymbol{E}, \boldsymbol{H} \sim$ (periodic function)$e^{i(\boldsymbol{k}\cdot\boldsymbol{x}-\omega_n t)}$, i.e., plane waves multiplied by a periodic envelope [35]. In 1-D with period a, this means that the field components are of the form $p_k(x)e^{ikx}$, where $p_k(x+a) = p_k(x)$. An example of such bands, $\omega_n(k)$, for $n = 1, 2, 3$ is shown for a simple multilayer film in Fig. 4.8.

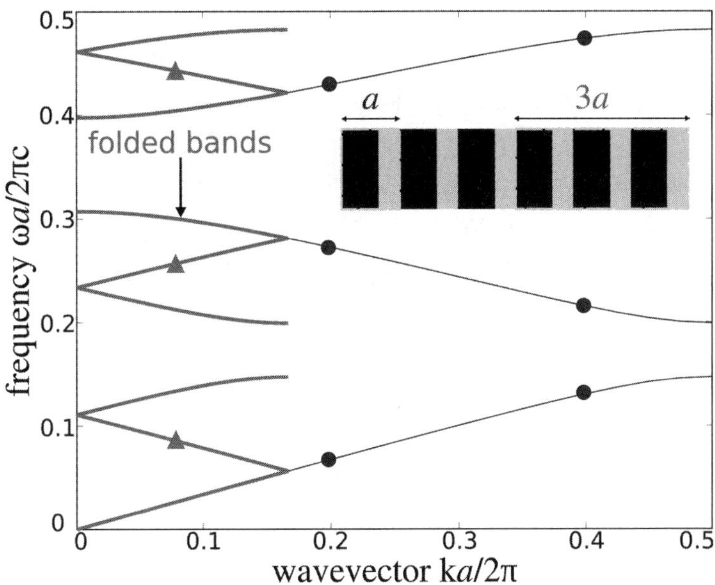

Fig. 4.8 1-D example of band folding within a supercell for a multilayer film of period = a, alternating between $\varepsilon/\varepsilon_0 = 1$ (thickness = $0.3a$) and $\varepsilon/\varepsilon_0 = 12$ (thickness = $0.7a$) [35] (inset). The curves of the dispersion relation, $\omega_n(k)$, are marked by round dots. There are three bands in this frequency range, separated by photonic bandgaps in which no propagating solutions exist. However, if we solve the same problem in a supercell of size, $3a$, the solutions are the bands "folded" three times (curves marked by triangular dots). These are the same solutions, but the meaning of the k label has changed.

In Fig. 4.8, we only plot for $k \in [0, \pi/a]$ the *irreducible Brillouin zone* [35]. Here, $\omega_n(-k) = \omega_n(+k)$ by either mirror symmetry or time-reversal symmetry. Furthermore, $\omega_n(k+2\pi/a) = \omega_n(k)$ because $p_{k+2\pi/a}(x)\, e^{i(k+2\pi/a)x} = [p_{k+2\pi/a}(x)\, e^{i2\pi x/a}]e^{ikx} = [\text{periodic}]e^{ikx}$ is a Bloch solution at k as well as at $k+2\pi/a$. Now, suppose that we solve the system using a three-period (3a) supercell. This obviously describes the same structure, and hence, would seem to imply the same solutions. In fact, we do get the same solutions in a sense, but they are mixed up because the meaning of k has changed! That is, suppose we impose Bloch-periodic boundary conditions, so that we are asking for field solutions of the form $f(x)$ such that $f(x+3a) = e^{ik(3a)}f(x)$. This is equivalent to requiring $f(x) = \tilde{p}(x)e^{ikx}$, where $\tilde{p}(x+3a) = \tilde{p}(x)$ is periodic with the period, $3a$. If we compare to our original solution, we find that $p_k(x)e^{ikx}$ is indeed a solution with the new boundary conditions. The problem is that we now get *new* solutions $p_{k\pm 2\pi/3a}(x)\, e^{i(k\pm 2\pi/3a)x}$ from $k \pm 2\pi/3a$, since $e^{\pm i(2\pi/3a)3a} = 1$. This is shown by the curves marked by the triangles in Fig. 4.8, where the original bands are "folded" onto the new irreducible Brillouin zone, $[0, \pi/3a]$, of the supercell [35]. Of course, these still solve Maxwell's equations, but understanding the dispersion relation has been complicated.

This band-folding effect is even more difficult to disentangle for supercells in 2-D or 3-D. For example, consider the case of a triangular lattice (period a) of dielectric rods in air, shown in Fig. 4.9(a). This geometry admits a rectangular $a \times \sqrt{3}a$ supercell that has twice the area of the rhombus unit cell. We wish to compute the bands, $\omega_n(\boldsymbol{k})$, around the boundaries (Γ–M–K) of its irreducible Brillouin zone of Fig. 4.9(b) [35], compared to an alternate, plane-wave method [13].

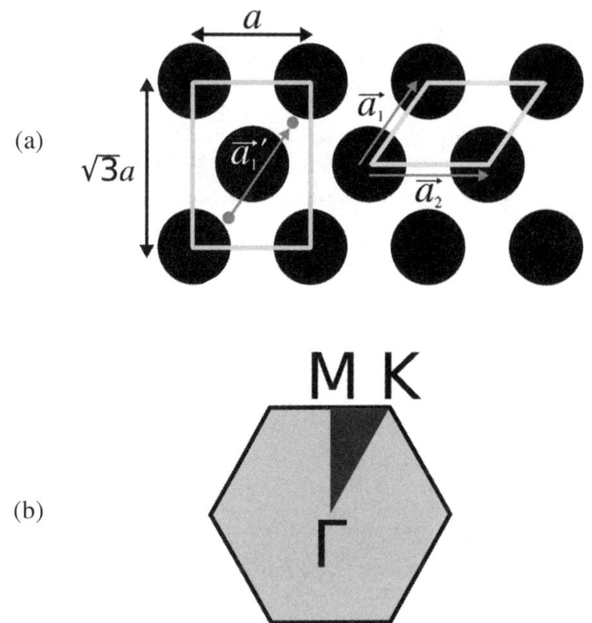

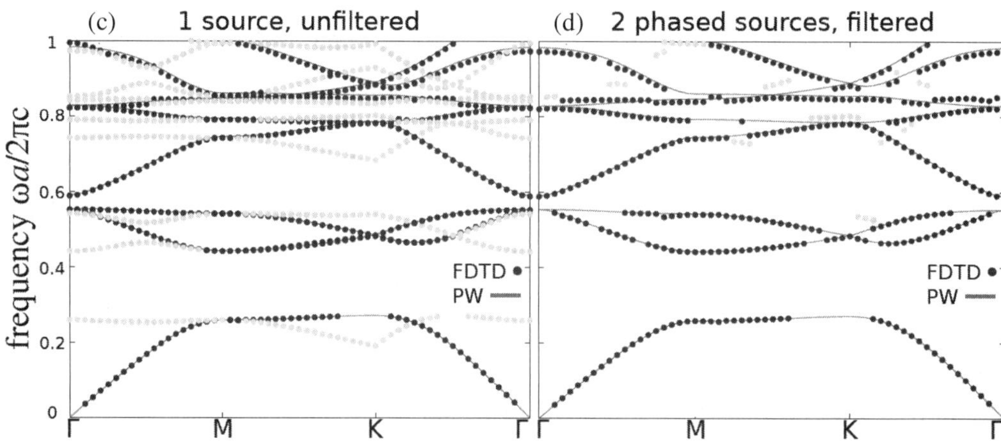

Fig. 4.9 Band-folding effects on the dispersion relation, $\omega_n(\boldsymbol{k})$, computed for a triangular lattice of dielectric rods in air. (a) Geometry of the lattice of dielectric rods, comparing the rectangular supercell of the lattice to the rhombus unit cell. (b) Boundaries of the irreducible Brillouin zone (Γ–M–K) in $\boldsymbol{k}$-space, for out-of-plane $\boldsymbol{E}$ excitation. (c) Band diagram: solid lines—correct curves computed by a plane-wave method [13]; dark dots—correct results of an FDTD calculation using the rectangular supercell of (a), where the eigenfrequencies, $\omega_n(\boldsymbol{k})$, are computed by filter diagonalization of the response to a pulsed point dipole; light dots—additional spurious "folded" bands arising from the FDTD calculation. (d) Improved FDTD band diagram results using the same rectangular supercell as in (c), but now most of the spurious solutions are eliminated by using two correctly phased sources separated by the lattice vector, $\vec{a}_1'$, in (a) to account for the underlying periodicity, and filtering the results to those with the correct phase relation. A few spurious solutions (light dots) remain due to discretization errors and signal-processing difficulty in dealing with closely spaced bands. These are easily eliminated by hand.

To obtain the FDTD band diagram results shown in Fig. 4.9(c), we excite the modes for the rectangular supercell of Fig. 4.9(a) with a broadband pulsed dipole, and then apply the filter-diagonalization algorithm to the resulting fields to extract the (lossless) resonant frequencies. This procedure yields a combination of correct results for the desired bands (dark dots) that coincide with the plane-wave results (solid lines), as well as spurious "folded" bands (light dots) due to the supercell.

The solution is conceptually simple. If our computational space were a *single* unit cell, our single point source would correspond to a source in *every* unit cell multiplied by the corresponding $e^{ik \cdot x}$ phase factor. Since we are trying to model this situation with a supercell of *two* unit cells, we should have *two* sources [shown as light dots in Fig. 4.9(a)] separated by a lattice vector, $\vec{a}_1'$, of the underlying periodicity, and differing by an $e^{ik \cdot x}$ phase. In the exact Maxwell's equations, such a two-point source would excite *only* the correct solutions at $\boldsymbol{k}$, without any possibility of spurious folded bands.

However, the discretization errors inherent in FDTD inevitably cause some spurious folded bands to be excited using this two-point source approach. While these spurious bands are of small amplitude, they still appear upon implementing filter diagonalization. Fortunately, we can filter most of them out of the results, because filter diagonalization provides both the frequency *and* the complex amplitude of the resonances. If we apply filter diagonalization at one point, we can discard modes having a small amplitude. If we apply filter diagonalization at the two source points (or any two points separated by a lattice vector), we can discard any modes whose amplitudes do not have the $e^{ik \cdot x}$ phase relationship.

Figure 4.9(d) shows the results of this two-source plus filtering algorithm. Here, we see that most of the spurious solutions (light dots) have been eliminated. There are still a few remnants of the folded bands, because the filter diagonalization technique is not magic – there is still some difficulty in the signal processing in separating the amplitudes of closely spaced modes. In addition, a few real FDTD solutions (dark dots) are also missing. Again, the signal processing is imperfect and occasionally misses resonances, especially if, by bad luck, they are excited with low amplitude.

It is certainly possible to mitigate these problems. For example, we know that the actual bands, $\omega_n(\boldsymbol{k})$, are continuous, so any gaps must be missing resonances, and any isolated dots must be spurious ones. We could also look more closely at the problematic regions with a narrow-band source, run FDTD for a longer time, or be more clever in the filtering criteria. However, the overall supercell technique described above is generally sufficiently robust that the few remaining errors in the band diagram can be spotted visually and removed manually.

Reiterating, it is more efficient to use the true unit cell via skewed boundary conditions [57, 58]. This also eliminates all spurious modes, but not all signal-processing difficulties for missed resonances or closely spaced modes. However, the use of a supercell with phased sources is a useful work-around that allows an existing FDTD program to be used *with no modifications*.

4.7 MOVING SOURCES

An interesting example of a source term is a source whose position is *moving* over time, for example to describe the radiation emitted by a particle as it moves through a medium. A moving *charged* particle can generate *Cherenkov radiation* [59], and a moving dipole antenna (e.g., a moving atom generating spontaneous emission) exhibits Doppler shifts in the frequency of its radiated waves [39].

In a homogeneous medium, the Cherenkov-radiation and Doppler-shift phenomena related to moving sources can be described analytically. However, both phenomena can be greatly modified when the source passes through an inhomogeneous medium. For example, in a homogeneous medium, a moving charge only produces Cherenkov radiation when it exceeds the phase velocity of light in the surrounding medium [59]. In contrast, Cherenkov radiation can be produced at any particle velocity in a periodic medium (the *Smith-Purcell effect* [60, 61]), in addition to a number of other anomalous effects [62, 63]. Similarly, the usual direction of the Doppler shift can be reversed in an inhomogeneous medium, among other unusual effects [64, 65]. FDTD provides a powerful tool to model these effects in complicated media [62, 64].

For example, consider the case of a point charge, q, moving with a velocity $\boldsymbol{v}$, which is described by a free-charge density, $\rho = q\delta(\boldsymbol{x}-\boldsymbol{v}t)$. To obtain the equivalent current density, $\boldsymbol{J}$, we turn to the continuity equation (describing conservation of charge) [39]: $\partial\rho/\partial t = -\nabla\cdot\boldsymbol{J}$. Since $\partial\rho/\partial t = -q\boldsymbol{v}\cdot\nabla\delta(\boldsymbol{x}-\boldsymbol{v}t)$, we immediately find that the continuity equation is satisfied by $\boldsymbol{J}(\boldsymbol{x},t) = q\boldsymbol{v}\delta(\boldsymbol{x}-\boldsymbol{v}t)$. This is a *moving dipole current* oriented in the $q\boldsymbol{v}$ direction.[9,10]

Figure 4.10 illustrates the simulation of such a moving current in FDTD. Here, the current, which is treated as an ordinary $\boldsymbol{J}$ source term in Maxwell's equations, is assumed to move to the right with a superluminal velocity, $v = 1.05\,c/n$, where n is the refractive index of the medium. This is equivalent to 0.35 pixel per time-step, Δt. At every time-step, *we simply change the location of* $\boldsymbol{J}$ *in the grid* in accordance with the assumed velocity [62]. An immediate complication arises: Because the computation space is discretized in pixels of size Δx, the successive locations of $\boldsymbol{J}$ would fall exactly on grid-points only if Δx is selected to be an integer multiple of $v\Delta t$. This is not the case in this example, and cannot be expected, in general.

One option to resolve this issue is to simply round the location of $\boldsymbol{J}$ to the nearest grid-point at every time-step, as shown in Fig. 4.10(b). However, this results in a jerky, discretized motion of $\boldsymbol{J}$ that generates spurious high-frequency components clearly visible in the radiated field. Instead, in Fig. 4.10(a), we use the interpolation scheme of Chapter 20, Section 20.3.2, to distribute the point-dipole, $\boldsymbol{J}$, to its neighboring grid-points at each time-step, with weights that change continuously with position (and with the correct total $\boldsymbol{J}$). This results in significantly fewer artifacts in the radiated field, although the effects of numerical dispersion are visible in the high-frequency components.

Doppler radiation from a moving dipole source (e.g., to model spontaneous emission from a moving atom) is even more straightforward to implement. A stationary dipole, $\boldsymbol{p}$, oscillating with frequency, ω, is a current, $\boldsymbol{J} = -i\omega\boldsymbol{p}\delta(\boldsymbol{x})e^{-i\omega t}$. The same source moving with velocity, $\boldsymbol{v}$, is $\boldsymbol{J} = -i\omega\boldsymbol{p}\delta(\boldsymbol{x}-\boldsymbol{v}t)e^{-i\omega t}$. [Again, this is nonrelativistic; the exact relativistic formula multiplies the component of $\boldsymbol{J}$ in the direction of $\boldsymbol{v}$ by $(1-v^2/c^2)^{-1/2}$ [39].] Therefore, as above, one simply has an ordinary dipole source in FDTD whose location changes every time-step [64], using interpolation to achieve continuous "motion." Figure 4.11 illustrates this for the simple case of an oscillating dipole moving at $|\boldsymbol{v}| = 0.3c$ in vacuum. The Doppler shifts (shorter wavelength in front of the source and longer wavelength behind it) are apparent in the FDTD results.

---

[9] Since $\delta(\boldsymbol{x}-\boldsymbol{v}t) = (2\pi)^{-3}\int e^{i\boldsymbol{k}\cdot(\boldsymbol{x}-\boldsymbol{v}t)}d^3\boldsymbol{k}$, upon taking the Fourier transform of this current, one immediately obtains a phase-matching condition, $\omega(\boldsymbol{k}) = \boldsymbol{k}\cdot\boldsymbol{v}$, that must be satisfied by the dispersion relation, $\omega(\boldsymbol{k})$, of any radiated field [59].

[10] Technically, this is a nonrelativistic approximation. The exact relativistic formula multiplies both the rest charge, q, and our current, $\boldsymbol{J}$, by $(1-v^2/c^2)^{-1/2}$ [39].

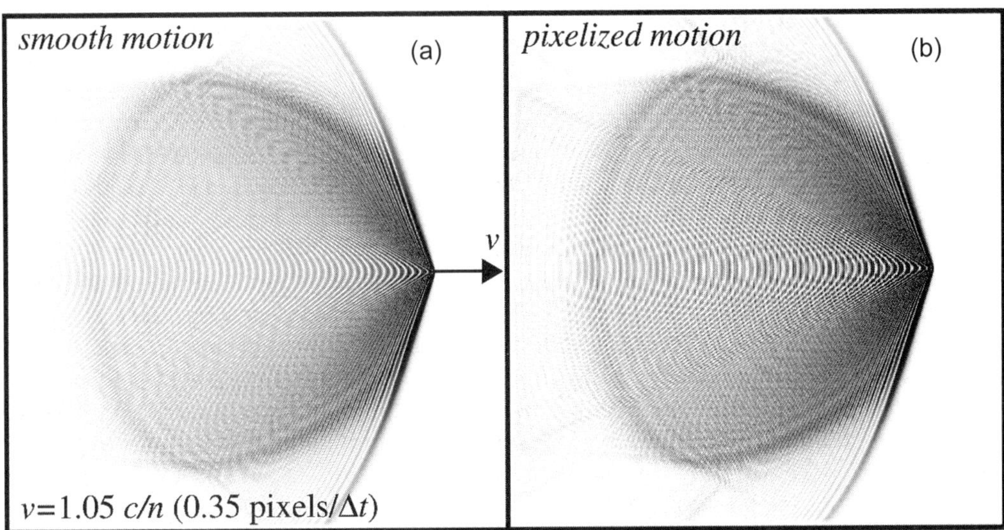

Fig. 4.10 Visualization of the FDTD-computed Cherenkov radiation (H_z field) from a point charge, q, moving to the right with a superluminal velocity, $v = 1.05\,c/n$, in a homogeneous medium, where n is the refractive index of the medium. This is simulated in FDTD by a moving current source, $\boldsymbol{J}(\boldsymbol{x}, t) = q\boldsymbol{v}\delta(\boldsymbol{x} - \boldsymbol{v}t)$, where v is equivalent to 0.35 pixel per time-step. (a) Continuous interpolation of the current source onto the FDTD grid, using the technique of Chapter 20, Section 20.3.2. (b) Pixelized motion, in which the location of $\boldsymbol{J}$ is rounded to the nearest FDTD grid-point, leading to visible high-frequency artifacts in the radiated field.

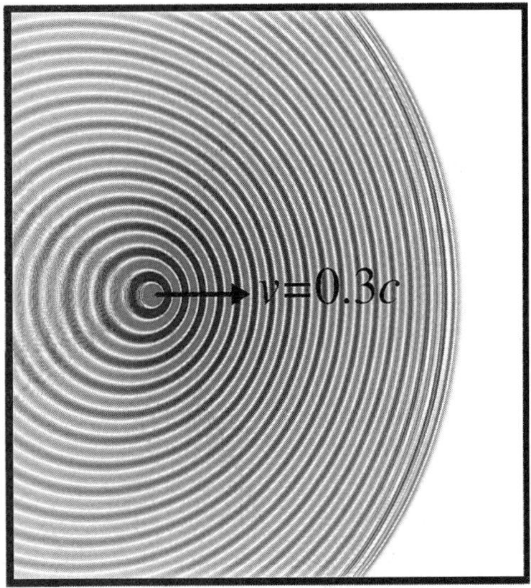

Fig. 4.11 Visualization of the FDTD-computed Doppler-shifted radiation (out-of-plane electric field) from an oscillating dipole moving to the right at $|v| = 0.3c$ in vacuum, polarized out-of-plane. The waves in front of the source are Doppler-shifted to a higher frequency (shorter wavelength), and the waves behind the source are Doppler-shifted to a lower frequency (longer wavelength).

4.8 THERMAL SOURCES

Although sources in FDTD simulations are usually deterministic, there are important cases in which the *physical* source terms in Maxwell's equations are *nondeterministic* currents resulting from thermal and quantum fluctuations. Examples include thermal radiation (radiation from thermal fluctuations in polarizable matter [66–68]), van der Waals and Casimir forces (net forces resulting from imbalances in thermal and quantum radiation pressure [69, 70]), and fluorescence (spontaneous emission from excited particles).

Because one is generally interested only in the time-average result of these fluctuations, there are usually clever ways to employ FDTD (or another Maxwell's equations solver) to obtain the averaged result without modeling the fluctuations directly. For example, far-field thermal radiation obeys *Kirchhoff's law*, which says that the thermal radiation from a body is equal to the known radiation of an ideal "black body" multiplied by the absorptivity (fraction of absorbed power) of the body at each frequency [68]. In fact, the fraction of absorbed power from an incident plane wave is easily computed in FDTD (e.g., by the techniques of Sec. 4.5). More generally (e.g., in the near field), one can employ techniques derived from electromagnetic reciprocity and related principles [71–74]. For example, the rate and extraction efficiency of spontaneous emission can be computed by the power radiated from a dipole [24–26], as described in Section 4.4. Furthermore, as discussed in detail in Chapter 19, time-average Casimir forces can be computed efficiently in FDTD using reciprocity and other techniques [75, 76]. It is also possible to efficiently model time-average blackbody radiation and electromagnetic fluctuations in dissipative open systems using FDTD, as discussed in detail in Chapter 18.

Nevertheless, directly modeling fluctuating currents in FDTD — a type of *Monte Carlo method* or *Langevin model* — has the virtues of simplicity and generality. This approach is attractive when studying new problems where more sophisticated methods are not yet implemented. For example, the use of fluctuating sources in FDTD was the first method successfully employed in studying *near-field* thermal radiation (radiative heat transport between bodies at such small separations that evanescent interactions become important [77]) for any geometry other than spheres or planes [71]. Moreover, the elementary nature of the fluctuating-currents picture may make it more approachable than more sophisticated formulations.

Consider a material with a complex permittivity tensor, $\boldsymbol{\varepsilon}(\omega)$, at a temperature, T. Physically, $\boldsymbol{\varepsilon} \neq \varepsilon_0$ corresponds to a *polarizable* medium, in which microscopic dipoles can be aligned (or created) in response to an applied electric field [39]. However, even in the absence of an applied electric field, thermal and quantum fluctuations induce *spontaneous* microscopic polarizations in the material, which rapidly fluctuate in orientation and magnitude. This fluctuating polarization density, $\boldsymbol{P}$, corresponds to a fluctuating *current* density, $\boldsymbol{J} = \partial \boldsymbol{P} / \partial t$ [39], with zero mean, $\langle \boldsymbol{J} \rangle$, and nonzero mean-square, $\langle |\boldsymbol{J}|^2 \rangle$. As a result, fluctuating electromagnetic fields are generated, as depicted in Fig. 4.12(b).

There is an important consequence if we Fourier transform these current-density fluctuations to obtain $\hat{\boldsymbol{J}}(\omega)$. Namely, the mean-squared current, $\langle |\hat{\boldsymbol{J}}|^2 \rangle$, is proportional to the absorption coefficient, $\mathrm{Im}\,\boldsymbol{\varepsilon}$, of the material. This is a result of a profound and far-reaching principle of statistical physics called the *fluctuation–dissipation theorem* [67]. In particular, the statistics of the fluctuation in $\boldsymbol{J}$ are described by [67]:

$$\left\langle \hat{J}_\ell(\omega, \boldsymbol{x})\, \hat{J}_m(\omega, \boldsymbol{x}')^* \right\rangle = \frac{1}{\pi}\, \delta(\boldsymbol{x} - \boldsymbol{x}') \left[\frac{\hbar\omega}{2} \coth\left(\frac{\hbar\omega}{2kT} \right) \right] \omega\, \mathrm{Im}\, \varepsilon_{\ell m}(\omega, \boldsymbol{x}) \tag{4.31}$$

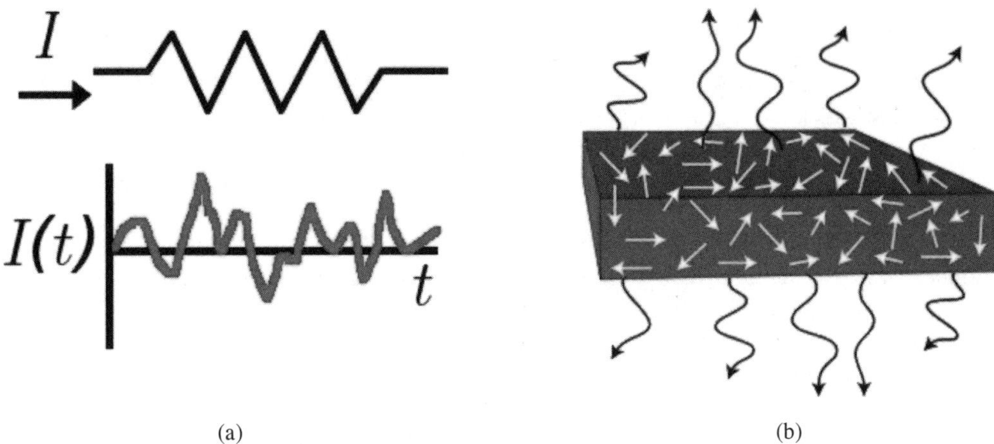

(a) (b)

Fig. 4.12 Schematic effects of random thermal and quantum fluctuations in polarizable matter. (a) At a macroscopic level, these fluctuations produce Johnson–Nyquist noise: random current fluctuations through a resistor. (b) At a microscopic level, the fluctuations can be described semiclassically as a fluctuating current density, $\boldsymbol{J}$, whose statistics are given by (4.31) from the fluctuation-dissipation theorem. This current produces thermal radiation, van der Waals forces, and other phenomena.

where $\hbar$ is Planck's constant and k is Boltzmann's constant. In signal-processing language, the right-hand side of (4.31) is the power spectral density of a "colored noise" process, i.e., the Fourier transform of the correlation function of the currents in time.

It is instructive to examine the right-hand side of (4.31) term by term. The first term, $\delta(\boldsymbol{x}-\boldsymbol{x}')$, means that the fluctuations are *uncorrelated in space*.[11] The second term (in $[\cdot\cdot\cdot]$) is the spectrum of the thermal and quantum fluctuations. For low frequencies or high temperatures ($\hbar\omega << 2kT$), this spectrum becomes simply kT, independent of frequency, since $\coth x \approx 1/x$ for small x. In fact, this is the *classical* limit. Alternatively, at high frequencies or low temperatures, $\coth \approx 1$, and the spectrum approaches $\hbar\omega/2$. This is the interesting regime of the purely quantum-mechanical phenomenon of *zero-point fluctuations* [67]. Finally, the third term, $\omega \operatorname{Im} \varepsilon$, corresponds to the *conductivity* of a conducting material [39].

This analysis leads us to a famous result in electrical engineering, the *Johnson–Nyquist* noise formula [79]. If we first integrate the current density over the cross-section of a wire to obtain the mean-square current, $\langle \tilde{I}^2 \rangle$, then integrate over a frequency bandwidth, $\Delta\omega = 2\pi\Delta f$ (summing over both positive and negative frequencies), and finally consider the classical limit, $[\cdots] \approx kT$, we arrive at the famous formula, $\langle I^2 \rangle = 4kT\Delta f/R$, where R is the wire's resistance. Thus, (4.31) is simply the microscopic generalization, including quantum effects, of Johnson–Nyquist noise, which produces a fluctuating current in any resistor as depicted in Fig. 4.12(a).

[11] Technically, the fluctuations are uncorrelated in space for the usual case of a material with a *local* dielectric response. Materials having a *nonlocal* dielectric response, in which a field at one point produces a polarization at a different point, produce correlated fluctuations. This property can be modeled in FDTD [78], as described in detail in Chapter 9.

To implement such a fluctuating current source in FDTD, one could in principle employ sources described exactly by (4.31). However, there is a useful technique that greatly simplifies the algorithm and improves its performance [71, 80, 81]. Physically, (4.31) simply describes a simulation in which there is an independent random current source at every point in space, or at least, every point where $\mathrm{Im}\,\varepsilon > 0$. From the resulting fields, we can compute the time-average energy flux (Poynting vector), momentum flux (stress tensor), or any other desired quantities. The biggest computational difficulty is that, while the sources are uncorrelated in space, they are correlated in time with power-spectrum given by (4.31). There are various ways to generate correlated random numbers (colored noise) by filtering techniques from signal processing, but they involve substantial additional storage or computation, or both [82].

Instead, we can exploit the *linearity* of electromagnetism, assuming linear materials. Namely, we can instead compute the fields due to *white-noise* currents (uncorrelated in time and space), and then only *after* the simulation, multiply the Fourier-transformed fields by the ω-dependent terms of (4.31). We begin by considering the common case in which there is only *one* absorbing material, so that $\mathrm{Im}\,\varepsilon$ is either zero (e.g., in air regions) or some function of frequency, $\varepsilon''(\omega)$, in the absorbing material. Then, we inject white noise currents in the absorbing material. Here, each component of $\boldsymbol{J}$ in the absorber is an uncorrelated random number generated at each FDTD time-step, with mean, 0, and mean-square, 1. Thanks to the central-limit theorem, the precise random-number distribution is irrelevant, e.g., Gaussian or uniform. As long as the distribution has the correct mean and mean-square, the net effect is the same when averaged over a long time, or over many simulations.

One then computes the desired Fourier-transformed fields (e.g., for the Poynting flux). Because one usually wants only the average of these quantities, one should compute the "ensemble average" over many such simulations, or over a long time. (Technically, such an average is called a periodogram; there are various windowing algorithms to speed convergence [17]). *After* this is done, the Poynting spectrum (or any quantity proportional to the squared Fourier amplitudes) is multiplied by the $[(\hbar\omega/2)\coth(\hbar\omega/2kT)]\,\omega\varepsilon''(\omega)$ frequency dependence to obtain the correct "thermal" spectrum.

Figure 4.13 illustrates the type of result that may be obtained using this method, excerpted from more detailed results described elsewhere [71]. The solid curve in the figure plots the spectral density of the near-field power flux between two one-dimensionally periodic photonic crystals of silicon carbide (SiC) separated by a short distance, d (inset). The dashed curve plots the power flux between *unpatterned* SiC slabs. In both cases, the power flux is normalized by the flux between the same structures at infinite separation, $d \to \infty$. The patterning of the slabs drastically modifies the flux spectrum as compared to the unpatterned case. In this case, however, there is a complication. Namely, we have two bodies at *different* temperatures, T. This would seem to conflict with our technique of multiplying by the $\coth(\hbar\omega/2kT)$ temperature dependence only after the simulation.

There are two possible solutions. Because the currents are uncorrelated in space, we could simulate them *separately* (the cross-terms in the resulting fields average to zero). This would mean two simulations, one with sources in each body, with the results of each simulation multiplied by its respective coth factor and summed. However, as a consequence of electromagnetic reciprocity [71], the energy flux into body 1 in response to white-noise sources in body 2 is *identical* to the energy flux into body 2 from white-noise sources in body 1. Therefore, it suffices to perform *one* simulation, with white-noise sources in only *one* body, and then multiply the results by the *difference* of the two coth factors to obtain the net energy flux [71].

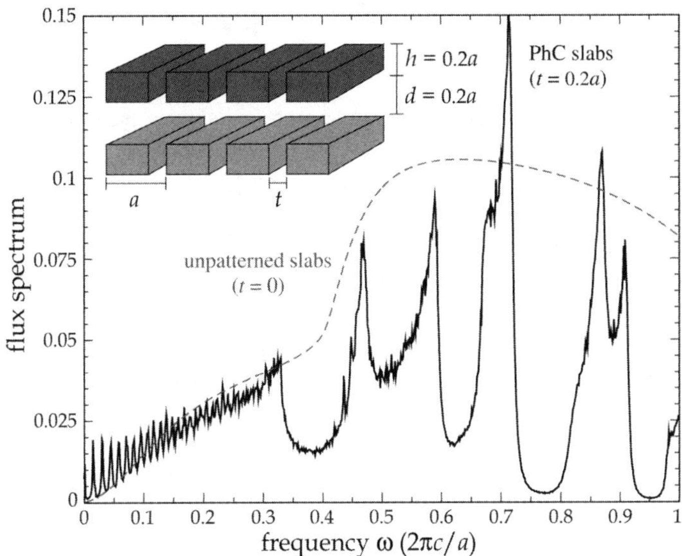

Fig. 4.13 FDTD-computed radiative heat transfer between patterned and unpatterned SiC slabs, operating in the *near field* where the slabs are close enough to couple evanescently as well as radiatively [71]. The solid curve plots the spectral density of the power flux between SiC photonic crystals (inset) maintained at unequal temperatures and surface-to-surface separation, d. The dashed curve plots the power flux between unpatterned SiC slabs. In both cases, the power flux is normalized by the power flux that would exist between the same structures at infinite separation, $d \to \infty$.

There is one additional technique that is required if there are *multiple* absorbing materials in a single simulation, each with its own absorption spectrum, $\operatorname{Im}\varepsilon$. In this case, we cannot use white-noise sources in all absorbing materials at once and multiply by the frequency-dependence *a posteriori*, since the fields from different absorbing materials are mixed together.

A potential approach to this problem would be to perform a separate simulation for each material (using white-noise sources in one material at a time), and summing them afterwards. This would work because of the lack of spatial correlation. However, there is another possibility that would use just a *single* simulation, based on the way that a frequency-dependent ε is modeled in FDTD. In particular, a frequency-dependent permittivity, $\varepsilon(\omega) = \varepsilon_0[1+\chi(\omega)]$, is typically implemented in FDTD by integrating an *auxiliary differential equation* [1]. By inserting the white-noise source into this auxiliary equation *instead* of directly into $\boldsymbol{J}$, one can build the desired $\operatorname{Im}\varepsilon$ frequency dependence directly into the current, up to material-independent factors of ω that can be multiplied *a posteriori* [80, 81].

4.9 SUMMARY

This chapter has provided a tutorial discussion of the state of knowledge regarding the relationships between current sources and the resulting electromagnetic waves in FDTD simulations. The techniques presented here are suitable for a use in a wide range of FDTD modeling applications, from classical radiation, scattering, and waveguiding problems to nanoscale material structures interacting with thermal and quantum fluctuations.

The chapter commenced with a discussion of incident fields and equivalent currents, examining in detail the principle of equivalence and the discretization and dispersion of equivalent currents in FDTD models. This was followed by a review of means to separate incident and scattered fields, whether in the context of scatterers, waveguides, or periodic structures. The next major topic was the relationship between current sources and the resulting local density of states. Here, key sub-topics included the Maxwell eigenproblem and the density of states, radiated power and the harmonic modes, radiated power and the local density of states, computation of the local density of states in FDTD, Van Hove singularities in the local density of states, and resonant cavities and Purcell enhancement. Subsequent major topics included source techniques that enable covering a wide range of frequencies and incident angles in a small number of simulations for waves incident on a periodic surface; sources to efficiently excite eigenmodes in rectangular supercells of periodic systems; moving sources to enable modeling of Cherenkov radiation and Doppler-shifted radiation; and finally thermal sources via a Monte Carlo/Langevin approach to enable modeling radiative heat transfer between complex-shaped material objects in the near field.

REFERENCES

[1] Taflove, A., and S. C. Hagness, *Computational Electrodynamics: The Finite-Difference Time-Domain Method*, 3rd ed., Norwood, MA: Artech, 2005.

[2] Harrington, R. F., *Time-Harmonic Electromagnetic Fields*, 2nd ed., Piscataway, NJ: Wiley-IEEE Press, 2001.

[3] Chen, K., "A mathematical formulation of the equivalence principle," *IEEE Trans. Microwave Theory and Techniques*, Vol. 37, 1989, pp. 1576–1581.

[4] Monzon, J. C., "On surface integral representations: Validity of Huygens' principle and the equivalence principle in inhomogeneous bianisotropic media," *IEEE Trans. Microwave Theory and Techniques*, Vol. 41, 1993, pp. 1995–2001.

[5] Love, A. E. H., "The integration of equations of propagation of electric waves," *Philosophical Trans. of the Royal Society of London A*, Vol. 197, 1901, pp. 1–45.

[6] Schelkunoff, S. A., "Some equivalence theorems of electromagnetic waves," *Bell System Technical J.*, Vol. 15, 1936, pp. 92–112.

[7] Stratton, J. A., and L. J. Chu, "Diffraction theory of electromagnetic waves," *Physical Review*, Vol. 56, 1939, pp. 99–107.

[8] Rengarajan, S. R., and Y. Rahmat-Samii, "The field equivalence principle: Illustration of the establishment of the non-intuitive null fields," *IEEE Antennas and Propagation Magazine*, Vol. 42, 2000, pp. 122–128.

[9] Harrington, R. F., "Boundary integral formulations for homogeneous material bodies," *J. Electromagnetic Waves and Applications*, Vol. 3, 1989, pp. 1–15.

[10] Chew, W. C., J.-M. Jin, E. Michielssen, and J. Song, eds., *Fast and Efficient Algorithms in Computational Electromagnetics*, Norwood, MA: Artech, 2001.

[11] Merewether, D. E., R. Fisher, and F. W. Smith, "On implementing a numeric Huygens' source scheme in a finite difference program to illuminate scattering bodies," *IEEE Trans. Nuclear Science*, Vol. 27, 1980, pp. 1829–1833.

[12] Schneider, J. B., "Plane waves in FDTD simulations and a nearly perfect total-field/scattered-field boundary," *IEEE Trans. Antennas and Propagation*, Vol. 52, 2004, pp. 3280–3287.

[13] Johnson, S., and J. Joannopoulos, "Block-iterative frequency-domain methods for Maxwell's equations in a plane wave basis," *Optics Express*, Vol. 8, 2001, pp. 173–190.

[14] Moss, C. D., F. L. Teixeira, and J. A. Kong, "Analysis and compensation of numerical dispersion in the FDTD method for layered, anisotropic media," *IEEE Trans. Antennas and Propagation*, Vol. 50, 2002, pp. 1174–1184.

[15] Schneider, J. B., C. L. Wagner, and O. M. Ramahi, "Implementation of transparent sources in FDTD simulations," *IEEE Trans. Antennas and Propagation*, Vol. 46, 1998, pp. 1159–1168.

[16] Tan, T., "FDTD discrete plane wave (FDTD-DPW) formulation for a perfectly matched source in TF/SF simulations," *IEEE Trans. Antennas and Propagation*, Vol. 58, 2010, pp. 2641–2648.

[17] Oppenheim, A. V., R. W. Schafer, and J. R. Buck, *Discrete-Time Signal Processing*, 2nd ed., Englewood Cliffs, NJ: Prentice-Hall, 1999.

[18] Oguz, U., and L. Gürel, "Application of signal-processing techniques to dipole excitations in the finite-difference time-domain method," *J. Electromagnetic Waves and Applications*, Vol. 16, 2002, pp. 671–687.

[19] Marcuse, D., *Theory of Dielectric Optical Waveguides*, 2nd ed., San Diego: Academic Press, 1991.

[20] Skorobogatiy, M., and J. Yang, *Fundamentals of Photonic Crystal Guiding*, Cambridge, England: Cambridge University Press, 2009.

[21] Johnson, S. G., P. Bienstman, M. A. Skorobogatiy, M. Ibanescu, E. Lidorikis, and J. D. Joannopoulos, "Adiabatic theorem and continuous coupled-mode theory for efficient taper transitions in photonic crystals," *Physical Review E*, Vol. 66, 2002, 066608.

[22] Elliott, R. S., *Antenna Theory and Design*, rev. ed., Piscataway, NJ: IEEE Press, 2003.

[23] Wang, S., and F. L. Teixeira, "An equivalent electric field source for wideband FDTD simulations of waveguide discontinuities," *IEEE Microwave and Wireless Components Lett.*, Vol. 13, 2003, pp. 27–29.

[24] Wijnands, F., J. B. Pendry, F. J. Garcia-Vidal, P. M. Bell, P. J. Roberts, and L. M. Moreno, "Green's functions for Maxwell's equations: Application to spontaneous emission," *Optical and Quantum Electronics*, Vol. 29, 1997, pp. 199–216.

[25] Xu, Y., R. K. Lee, and A. Yariv, "Quantum analysis and the classical analysis of spontaneous emission in a microcavity," *Physical Review A*, Vol. 61, 2000, 033807.

[26] Sakoda, K., *Optical Properties of Photonic Crystals*, Berlin, Germany: Springer, 2001.

[27] Purcell, E. M., "Spontaneous emission probabilities at radio frequencies," *Physical Review*, Vol. 69, 1946, p. 681.

[28] Hamam, R. E., M. Ibanescu, E. J. Reed, P. Bermel, S. G. Johnson, E. Ippen, J. D. Joannopoulos, and M. Soljacic, "Purcell effect in nonlinear photonic structures: A coupled mode theory analysis," *Optics Express*, Vol. 16, 2008, pp. 12523–12537.

[29] Fan, S., P. Villeneuve, and J. D. Joannopoulos, "High extraction efficiency of spontaneous emission from slabs of photonic crystals," *Physical Review Lett.*, Vol. 78, 1997, pp. 3294–3297.

[30] Ward, A. J., and J. B. Pendry, "Calculating photonic Green's functions using a nonorthogonal finite-difference time-domain method," *Physical Review B*, Vol. 58, 1998, pp. 7252–7259.

[31] Xu, Y., J. S. Vuckovic, R. K. Lee, O. J. Painter, A. Scherer, and A. Yariv, "Finite-difference time-domain calculation of spontaneous emission lifetime in a microcavity," *J. Optical Society of America B*, Vol. 16, 1999, pp. 465–474.

[32] Xu, Y., R. K. Lee, and A. Yariv, "Finite-difference time-domain analysis of spontaneous emission in a microdisk cavity," *Physical Review A*, Vol. 61, 2000, 033808.

[33] Shen, C., K. Michielsen, and H. De Raedt, "Spontaneous-emission rate in microcavities: Application to two-dimensional photonic crystals," *Physical Review Lett.*, Vol. 96, 2006, 120401.

[34] Van Vlack, C., and S. Hughes, "Finite-difference time-domain technique as an efficient tool for calculating the regularized Green function: Applications to the local-field problem in quantum optics for inhomogeneous lossy materials," *Optics Lett.*, Vol. 37, 2012, pp. 2880–2882.

[35] Joannopoulos, J. D., S. G. Johnson, J. N. Winn, and R. D. Meade, *Photonic Crystals: Molding the Flow of Light*, 2nd ed., Princeton University Press, 2008.

[36] Economou, E. N., *Green's Functions in Quantum Physics*, 3rd ed., Berlin, Germany: Springer, 2010.

[37] Lagendijk, A., and B. A. van Tiggelen, "Resonant multiple scattering of light," *Physics Reports*, Vol. 270, 1996, pp. 143–215.

[38] Joulain, K., R. Carminati, J.-P. Mulet, and J.-J. Greffet, "Definition and measurement of the local density of electromagnetic states close to an interface," *Physical Review B*, Vol. 68, 2003, 245405.

[39] Jackson, J. D., *Classical Electrodynamics*, 3rd ed., New York: Wiley, 1998.

[40] Strichartz, R. S., *A Guide to Distribution Theory and Fourier Transforms*, Boca Raton, FL: CRC Press, 1994.

[41] de Vries, P., D. V. van Coevorden, and A. Lagendijk, "Point scatterers for classical waves," *Review of Modern Physics*, Vol. 70, 1998, pp. 447–466.

[42] Milonni, P. W., "Semiclassical and quantum-electrodynamical approaches in nonrelativistic radiation theory," *Physics Reports*, Vol. 25, 1976, pp. 1–81.

[43] Dey, S., and R. Mittra, "Efficient computation of resonant frequencies and quality factors of cavities via a combination of the finite-difference time-domain technique and the Padé approximation," *IEEE Microwave and Guided Wave Lett.*, Vol. 8, 1998, pp. 415–417.

[44] Guo, W.-H., W.-J. Li, and Y.-Z. Huang, "Computation of resonant frequencies and quality factors of cavities by FDTD technique and Padé approximation," *IEEE Microwave and Wireless Components Lett.*, Vol. 11, 2001, pp. 223–225.

[45] Zhang, Y., W. Zheng, M. Xing, G. Ren, H. Wang, and L. Chen, "Application of fast Padé approximation in simulating photonic crystal nanocavities by FDTD technology," *Optics Communications,* Vol. 281, 2008, pp. 2774–2778.

[46] Ashcroft, N. W., and N. D. Mermin, *Solid State Physics*, Philadelphia, PA: Holt Saunders, 1976.

[47] Snyder, A. W., and J. D. Love, *Optical Waveguide Theory*, London, England: Chapman and Hall, 1983.

[48] Sheng, X., S. G. Johnson, J. Michel, and L. C. Kimerling, "Optimization-based design of surface textures for thin-film Si solar cells," *Optics Express*, Vol. 19, 2011, pp. A841–A850.

[49] Peters, M., A. Bielawny, B. Bläsi, R. Carius, S. W. Glunz, J. C. Goldschmidt, H. Hauser, M. Hermle, T. Kirchartz, P. Löper, J. Üpping, R. Wehrspohn, G. Willeke, and A. Ludwigs, "Photonic concepts for solar cells," Chapter 1 (pp. 1–41) in *Physics of Nanostructured Solar Cells*, V. Badescu and M. Paulescu, eds., Hauppauge, NY: Nova Science Publishers, 2010.

[50] Aminian, A., and Y. Rahmat-Samii, "Spectral FDTD: A novel technique for the analysis of oblique incident plane wave on periodic structures," *IEEE Trans. Antennas and Propagation*, Vol. 54, 2006, pp. 1818–1825.

[51] Yang, F., J. Chen, R. Qiang, and A. Elsherbeni, "A simple and efficient FDTD/PBC algorithm for periodic structure analysis," *Radio Science*, Vol. 42, 2007, RS4004.

[52] Zhou, Y. J., X. Y. Zhou, and T. J. Cui, "Efficient simulations of periodic structures with oblique incidence using direct spectral FDTD method," *Progress in Electromagnetics Research M*, Vol. 17, 2011, pp. 101–111.

[53] Fan, S., W. Suh, and J. D. Joannopoulos, "Temporal coupled-mode theory for the Fano resonance in optical resonators," *J. Optical Society of America A*, Vol. 20, 2003, pp. 569–572.

[54] Qiu, M., and S. He, "A nonorthogonal finite-difference time-domain method for computing the band structure of a two-dimensional photonic crystal with dielectric and metallic inclusions," *J. Applied Physics*, Vol. 87, 2010, pp. 8268–8275.

[55] Ward, A. J., and J. B. Pendry, "Refraction and geometry in Maxwell's equations," *J. Modern Optics*, Vol. 43, 1996, pp. 773–793, 1996.

[56] Kuang, W., W. J. Kim, and J. D. O'Brien, "Finite difference time-domain method for nonorthogonal unit-cell two-dimensional photonic crystals," *J. Lightwave Technology*, Vol. 25, 2007, pp. 2612–2617.

[57] Ma, Z., and K. Ogusu, "FDTD analysis of 2d triangular-lattice photonic crystals with arbitrary-shape inclusions based on unit-cell transformation," *Optics Commun.*, Vol. 282, 2009, pp. 1322–1325.

[58] Umenyi, A. V., K. Miura, and O. Hanaizumi, "Modified finite-difference time-domain method for triangular lattice photonic crystals," *J. Lightwave Technology*, Vol. 27, 2009, pp. 4995–5001.

[59] Jelley, J. V., *Cerenkov Radiation and Its Applications*, London, England: Pergamon, 1958.

[60] Smith, S. J., and E. M. Purcell, "Visible light from localized surface charges moving across a grating," *Physical Review*, Vol. 92, 1953, p. 1069.

[61] Potylitsyn, A. P., *Electromagnetic Radiation of Electrons in Periodic Structures*, Berlin, Germany: Springer, 2011.

[62] Luo, C., M. Ibanescu, S. G. Johnson, and J. D. Joannopoulos, "Cherenkov radiation in photonic crystals," *Science*, Vol. 299, 2003, pp. 368–371.

[63] Kramers, C., D. N. Chigrin, and J. Kroha, "Theory of Cherenkov radiation in periodic dielectric media: Emission spectrum," *Physical Review A*, Vol. 79, 2009, 013829.

[64] Luo, C., M. Ibanescu, E. J. Reed, S. G. Johnson, and J. D. Joannopoulos, "Doppler radiation emitted by an oscillating dipole moving inside a photonic band-gap crystal," *Physical Review Lett.*, Vol. 96, 2006, 043903.

[65] Reed, E. J., "Physical optics: Backwards Doppler shifts," *Nature Photonics*, Vol. 5, 2011, pp. 199–200.

[66] Rytov, S., *Theory of Electric Fluctuations and Thermal Radiation*, Electronics Research Directorate, Air Force Cambridge Research Center, Air Research and Development Command, U.S. Air Force, 1959.

[67] Eckhardt, W., "Macroscopic theory of electromagnetic fluctuations and stationary radiative heat transfer," *Physical Review A*, Vol. 29, 1984, pp. 1991–2003.

[68] Howell, J. R., R. Siegel, and M. P. Menguc, *Thermal Radiation Heat Transfer*, 5th ed., Boca Raton, FL: CRC Press, 2010.

[69] Parsegian, A. V., *Van der Waals Forces: A Handbook for Biologists, Chemists, Engineers, and Physicists*, Cambridge University Press, 2006.

[70] Rodriguez, A. W., F. Capasso, and S. G. Johnson, "The Casimir effect in microstructured geometries," *Nature Photonics*, Vol. 5, 2011, pp. 211–221.

[71] Rodriguez, A. W., O. Ilic, P. Bermel, I. Celanovic, J. D. Joannopoulos, M. Soljacic, and S. G. Johnson, "Frequency-selective near-field radiative heat transfer between photonic crystal slabs: A computational approach for arbitrary geometries and materials," *Physical Review Lett.*, Vol. 107, 2011, 114302.

[72] McCauley, A. P., M. T. H. Reid, M. Krüger, and S. G. Johnson, "Modeling near-field radiative heat transfer from sharp objects using a general three-dimensional numerical scattering technique," *Physical Review B*, Vol. 85, 2012, 165104.

[73] Rodriguez, A. W., M. T. H. Reid, and S. G. Johnson, "Fluctuating surface current formulation of radiative heat transfer for arbitrary geometries," Cornell University Library arXiv:1206.1772 [cond-mat.mtrl-sci], 2012, Online: http://arxiv.org/abs/1206.1772

[74] Guérout, R., J. Lussange, F. S. S. Rosa, J.-P. Hugonin, D. A. R. Dalvit, J.-J. Greffet, A. Lambrecht, and S. Reynaud, "Enhanced radiative heat transfer between nanostructured gold plates," Cornell University Library arXiv:1203.1496 [physics.optics], 2012, Online: http://arxiv.org/abs/1203.1496

[75] Rodriguez, A. W., A. P. McCauley, J. D. Joannopoulos, and S. G. Johnson, "Casimir forces in the time domain: Theory," *Physical Review A*, Vol. 80, 2009, p. 012115.

[76] McCauley, A. P., A. W. Rodriguez, J. D. Joannopoulos, and S. G. Johnson, "Casimir forces in the time domain: Applications," *Physical Review A*, Vol. 81, 2010, 012119.

[77] Volokitin, A. I., and B. N. J. Persson, "Near-field radiative heat transfer and noncontact friction," *Review of Modern Physics*, Vol. 79, 2007, pp. 1291–1329.

[78] McMahon, J. M., S. K. Gray, and G. C. Schatz, "Calculating nonlocal optical properties of structures with arbitrary shape," *Physical Review B*, Vol. 82, 2010, 035423.

[79] Gray, P. R., P. J. Hurst, S. H. Lewis, and R. G. Meyer, *Analysis and Design of Analog Integrated Circuits*, 5th ed., New York: Wiley, 2009.

[80] Luo, C., A. Narayanaswamy, G. Chen, and J. D. Joannopoulos, "Thermal radiation from photonic crystals: A direct calculation," *Physical Review Lett.*, Vol. 93, 2004, pp. 213905–213908.

[81] Chan, D. L., M. Soljacic, and J. D. Joannopoulos, "Direct calculation of thermal emission for three-dimensionally periodic photonic-crystal slabs," *Physical Review E*, Vol. 74, 2006, 036615.

[82] Rodriguez, A., and S. G. Johnson, "Efficient generation of correlated random numbers using Chebyshev-optimal magnitude-only IIR filters," Cornell University Library arXiv:physics/0703152 [physics.comp-ph], 2007, Online: http://arxiv.org/abs/physics/0703152

Chapter 5

Rigorous PML Validation and a Corrected Unsplit PML for Anisotropic Dispersive Media[1]

Ardavan Oskooi and Steven G. Johnson

5.1 INTRODUCTION

A perfectly matched layer (PML) is an artificial medium that is commonly used as an absorbing boundary condition (ABC) to truncate computational grids for simulating wave equations (e.g., Maxwell's equations). The PML is designed to have the property that interfaces between it and adjacent media are reflectionless in the exact wave equation [1, 2].

Previously, we found certain instances where proposed PML ABCs [3–7] for inhomogeneous media (such as photonic crystals [8]) were not, in fact, reflectionless in the exact wave equation [9]. In [10], similar failures were demonstrated for a proposed unsplit-field PML ABC for anisotropic media [11], and proposed PML ABCs for oblique waveguides [12, 13]. Errors in constructing PML ABCs are easily overlooked because all proposed implementations use an absorption that gradually increases with depth from the PML surface, in order to mitigate discretization effects [2]. Such an *adiabatic absorber* performs arbitrarily well, even if it is not a true PML, as long as the absorber is sufficiently thick [9]. Although many previous papers claim to validate their PML proposals by checking whether the numerical wave reflection is sufficiently low for a given grid resolution [14–19], and/or decreases with PML thickness [1, 20], the true test of a PML ABC is to verify that wave reflections systematically vanish as the grid resolution is progressively refined, thereby approaching the exact wave equation. However, separating reflected and incident waves can be cumbersome in complex inhomogeneous and/or anisotropic media.

In this chapter, we review the testing procedure for PML ABCs reported in [10] that involves computing the difference between two PML simulations as a function of grid resolution. The mathematical foundations for this procedure were derived from previous work in [9] that studied the effect of smoothness rather than resolution and correctness. (Reference [9] identified a failure of PML in periodic media by a different method, exact reflection computations, that is not as easily generalized as the method described in [10].) In particular, the validation procedure of [10] revealed a problem with a previously proposed unsplit PML for anisotropic media [11].

---

[1]This chapter is adapted and expanded from Ref. [10], A. Oskooi and S. G. Johnson, "Distinguishing correct from incorrect PML proposals and a corrected unsplit PML for anisotropic, dispersive media," *J. Computational Physics*, Vol. 230, 2011, pp. 2369–2377, ©2011 Elsevier. For consistency of notation with [10], the symbol i is used to designate $\sqrt{-1}$, rather than the symbol j; and a phasor is denoted as $e^{-i\omega t}$.

We also review the complex coordinate-stretching formulation of the corrected unsplit-field uniaxial PML as reported in [10] for terminating arbitrary anisotropic, dispersive, and conducting media. Modeling of anisotropic media is becoming increasingly important in computational electrodynamics, both for simulating metamaterials [21–23] and for accurately treating curved boundaries of isotropic media via subpixel smoothing [24–26]. Although other correct alternatives for PML termination of anisotropic media exist [14–17, 19, 20, 27–31] (including correct split-field proposals [14, 16, 28] by the same authors of later incorrect unsplit-field formulations), the unsplit PML formulation of [10] has the appeal of a simple correction to previous UPML-like proposals that were correct for isotropic media [11, 32, 33].

This chapter also reviews the demonstration of the PML formulation of [10] for both a plane-wave method in the frequency domain and for a finite-difference time-domain (FDTD) method (with a free-software implementation [34]). In the time domain, the frequency dependence of the PML response is well known to require additional storage (e.g., auxiliary fields) [2], but the scheme of [10] involves storage requirements at least as good as previous correct split-field PML proposals for anisotropic media [14–17, 27–30].

Finally, we review the demonstration in [10] of the inapplicability of standard PML to terminate obliquely incident dielectric waveguides [9]. Previous suggested PMLs for this case [12, 13] are invalid because the material function (and hence Maxwell's equations) is not invariant along the PML direction. While this possibility was noted in [9], Ref. [10] explicitly demonstrated the PML failure and suggested a workaround using a pseudo-PML [9].

5.2 BACKGROUND

There are several nearly equivalent formulations of PMLs. Berenger's seminal original formulation [1] split the wave solution into the sum of two new artificial field components. Subsequently, a more physically intuitive formulation was proposed [32, 33] whereby an equivalent effect can be obtained using a special uniaxial anisotropic medium ("UPML") introduced into the ordinary ("unsplit") Maxwell's equations.

Although these PML derivations were for homogeneous media in which reflections were explicitly computed and set to zero, a more elegant and general derivation was proposed in [35, 36]. Here, PML action is obtained by the analytic continuation of Maxwell's equations to complex coordinates, thereby transforming oscillating solutions into exponential decays. This coordinate-stretching viewpoint is equivalent to the split-field approach [35]. It is also equivalent to the anisotropic-media PML formulation [37], which can be viewed as a special case of Ward-Pendry transformation optics [21]. The coordinate-stretching concept is even applicable to deriving PML in inhomogeneous media, as long as the media are invariant along the PML direction [9], although there are certain backward-wave inhomogeneous media where PML fails for other reasons [38]. Subsequent refinements included the incorporation of real as well as imaginary coordinate stretching in order to attenuate evanescent waves [39]; shifting the zero-frequency pole in order to better attenuate low-frequency waves (CFS-PML) [2, 18, 40]; and implementation of the latter by a recursive-convolution technique (CPML) [19].

The same coordinate-stretching approach is also applicable for deriving reflectionless PMLs for arbitrary anisotropic media. Several such proposals employed the split-field approach [14–17, 27–30] or the convolutional approach [19, 20, 31] to obtain time-domain equations. Split-field PMLs were also derived for homogeneous anisotropic media by directly computing the reflection coefficients at the PML interface [15, 27, 30].

Transformation optics also leads to a simple uniaxial anisotropic PML medium in the frequency domain [41], and yields a straightforward time-domain PML for anisotropic and dispersive media (although in the time domain, a new factorization is required in order to efficiently implement the frequency dependence [10].)

5.3 COMPLEX COORDINATE STRETCHING BASIS OF PML

A PML medium is simplest to derive in the frequency domain for linear media with arbitrary dispersion and anisotropy. Assuming an $e^{-i\omega t}$ time dependence for all fields, an ordinary PML in Cartesian coordinates can be derived by a complex coordinate stretching, where each coordinate is stretched by a factor:

$$s_{x,y,z} = \kappa_{x,y,z} + i\left(\frac{\sigma_{x,y,z}}{\omega}\right) \tag{5.1}$$

where $\kappa \geq 1$ represents a real coordinate stretching used to attenuate evanescent waves [39], and $\sigma > 0$ is the PML "conductivity." (See Appendix 5A for a tutorial discussion of the coordinate-stretching technique for PML.) Note that the CFS variant of PML replaces ω with $\omega + i\alpha$ in order to shift the $\omega = 0$ pole below the real axis [2, 18, 19, 40]. For example, to terminate the computation space in the x-direction, only σ_x is not unity. These coordinate stretchings can be absorbed into Maxwell's equations as a change in the frequency-dependent constitutive tensors $[\varepsilon]$ and $[\mu]$, thanks to the general transformation optics principle [21]. In particular, the original dielectric permittivity tensor $[\varepsilon]$ in the PML region is replaced by an effective tensor $[\tilde{\varepsilon}]$ given by:

$$[\tilde{\varepsilon}] = \frac{\mathbf{J}\,[\varepsilon]\,\mathbf{J}^{\mathrm{T}}}{\det \mathbf{J}} \tag{5.2a}$$

where

$$\mathbf{J} = \mathrm{diag}\left(s_x^{-1},\, s_y^{-1},\, s_z^{-1}\right) \tag{5.2b}$$

is the Jacobian matrix of the coordinate stretching. There is also an identical transformation of the magnetic permeability tensor $[\mu]$, but for simplicity we focus here on $[\varepsilon]$. Equation (5.2a, b) is derived for the general anisotropic case in [25], and is also derived for the specific case of PML with a diagonal coordinate stretching in [41].

An *isotropic* tensor $[\varepsilon]$ commutes with $\mathbf{J}$. As a result, one can simply multiply $[\varepsilon]$ by:

$$\frac{\mathbf{J}\,\mathbf{J}^{\mathrm{T}}}{\det \mathbf{J}} = \mathrm{diag}\left(\frac{s_y s_z}{s_x},\, \frac{s_x s_z}{s_y},\, \frac{s_x s_y}{s_z}\right) \equiv \mathbf{\Sigma} \tag{5.3}$$

to obtain the equivalent PML "conductivity" tensor, $\mathbf{\Sigma}$. This permits Ampere's law:

$$\nabla \times \boldsymbol{H} = -i\omega[\tilde{\varepsilon}]\boldsymbol{E} \tag{5.4}$$

to be separated into two parts. The first is a time-derivative part:

$$\nabla \times \boldsymbol{H} = -i\omega \boldsymbol{\Sigma} \boldsymbol{D} \tag{5.5a}$$

The second part is the "unmodified" constitutive relation:

$$\boldsymbol{D} = [\varepsilon]\boldsymbol{E} \tag{5.5b}$$

In this way, the PML-dependent parts are separated from the material properties $[\varepsilon]$ derived from the non-PML regions. This approach was called a "material-independent PML" in [11] with an unsplit-field formulation (where $\boldsymbol{\Sigma}$ was denoted by $\hat{\varepsilon}^D$).

However, it is clear in (5.2a, b) that an arbitrary anisotropic $[\varepsilon]$ (with off-diagonal entries) cannot, in general, be commuted with $\mathbf{J}$. Therefore, the transformation to $\boldsymbol{\Sigma}[\varepsilon]$ is *no longer equivalent* to a complex coordinate stretching. Hence, one would not expect the PML proposed in [11] to be a true reflectionless PML for an arbitrary anisotropic $[\varepsilon]$ in the limit of a very finely resolved computation approaching the exact wave-equation solution. (However, if the anisotropy is sufficiently weak, residual reflections are not significant, perhaps explaining why this problem was not obvious in the previous work.)

Reference [10] considered this issue, and concluded that, rigorously, one should use the uncommuted form of (5.2a, b) [41] for an arbitrary anisotropic $[\varepsilon]$. We review below the time-domain technique reported in [10] to achieve this goal in one subcomputation. This technique uses auxiliary fields to capture the frequency dependence of the media while retaining the UPML-like unsplit-field structure and the "material-independent" property of an unmodified constitutive relation.

5.4 ADIABATIC ABSORBERS AND PML REFLECTIONS

All adiabatic absorbers, whether PML or not, are characterized by a continuous conductivity or absorption-strength profile $\sigma(x) > 0$ that becomes more gradual as the absorber is made thicker. Outside the absorber, where $\sigma = 0$, the wave equation (and thus, its solution) is unchanged — the solution becomes exponentially decaying only inside the absorber. In the exact wave equation for a true PML, this transition occurs with zero reflection, no matter how fast σ changes, even if σ changes discontinuously. In practice, even for a true PML, reflections occur due to numerical discretization effects whenever σ changes abruptly. This artifact is mitigated by smoothly increasing σ with depth from the surface of the absorber [2]. It turns out that the reflection characteristics of any such smoothly increasing σ, whether in a PML or not, have certain universal properties determined by the smoothness of $\sigma(x)$ [9]. These properties are central to understanding absorbing-layer behaviors and validation.

Reflections from both PML and non-PML absorbers consist of two parts: round-trip and transition reflections. The first component is the round-trip reflection R_{RT}, also called the "theoretical reflection" in the case of a PML [1]. This arises after a plane numerical wave enters the absorber from the adjacent interior "working" portion of the computation space at $x = 0$. This wave decays exponentially within the absorber during its propagation to the outer boundary of the overall computation space at $x = L$, where a hard-wall (e.g., Dirichlet) condition is enforced. After reflecting from the outer boundary, the numerical wave undergoes additional exponential decay during its propagation back toward the surface of the absorber at $x = 0$. R_{RT} is a measure of how much of the original incident wave remains after its round-trip through the absorber, and is exponentially small:

$$R_{\mathrm{RT}} \sim \exp\left[-4\frac{k_x}{\omega}\int_0^L \sigma(x')dx'\right] \tag{5.6}$$

where k_x is the propagation constant in the x-direction, and the factor of 4 arises because the reflection is proportional to the round-trip ($2L$) field squared. The absorption-strength profile $\sigma(x)$ is defined between $x = 0$ and $x = L$, and can be conveniently expressed as:

$$\sigma(x) = \sigma_0\, s(x/L) \tag{5.7}$$

where $s(u)$: $[0, 1] \rightarrow [0, 1]$ is a dimensionless profile function, and σ_0 is an overall amplitude that is selected to achieve some theoretical round-trip absorption according to (5.6). We note that, as the absorber thickness L is increased for a fixed $s(u)$, $\sigma(x)$ rises more gradually because $s(u)$ stretches out and σ_0 decreases. Following [10], throughout the remainder of this chapter, we shall set R_{RT} to be negligibly small (less than 10^{-25}).

The second, and much more significant component of the overall reflection, is the transition reflection R_{T} occurring at $x = 0$, the interface of the absorber and the adjacent interior working portion of the computation space. For a finite resolution, R_{T} is nonzero even for a correct PML. Applying coupled-mode theory, Ref. [9] demonstrated that, for *any* adiabatic absorber, there is a fundamental connection between the smoothness of $s(u)$ and the rate of decrease of R_{T} with L. For example, it was shown in [9] that $R_{\mathrm{T}}(L) \sim 1/L^{2d+2}$ for the monomial profile, $s(u) = u^d$, for *either* a PML or a non-PML absorber.

5.5 DISTINGUISHING CORRECT FROM INCORRECT PML PROPOSALS

Given the results of [9], we must ask the question: "Then, how can we distinguish a correct from an incorrect proposed PML, or even from a conventional absorber?" If sufficiently thick with an appropriate absorption-strength profile, all three absorbers could effectively terminate a computation space. In other words, the failure of a putative PML may be obscured by the use of a sufficiently thick absorbing region.

There are two components to the answer. First, a PML provides a potentially smaller multiplicative coefficient in the $R_{\mathrm{T}}(L)$ relation. Second, for a PML, this coefficient *vanishes* for infinitely fine resolution [9]. The latter answer suggests a straightforward way to distinguish a correct PML proposal from an incorrect PML proposal. Namely, test the proposed PML by observing the *difference* in reflections $R(L_1) - R(L_2)$ for two simulations with identical interior working regions but different PML thicknesses L_1 and L_2 (with the same R_{RT}). For a correct PML proposal, this difference must effectively vanish in the limit of infinitely fine resolution, assuming that R_{RT} is negligibly small (10^{-25} suffices). In contrast, for an incorrect PML proposal, $R(L_1) - R(L_2)$ converges at infinitely fine resolution to some nonzero value determined by the transition reflections in the exact wave equation, independent of (and much larger than) R_{RT}.

Moreover, we need not compute the reflections explicitly to distinguish a correct from an incorrect proposed PML. If we simply subtract the field arising from the two simulations at a monitor point within the interior working region of the computation space, the difference is due entirely to the reflections. For a correct PML, this difference must vanish as the resolution is refined. Note that, to perform as rigorous a test as possible of a proposed PML, it is desirable to employ a simulation involving waves incident on the PML with a wide range of angles. This can be achieved using a point-dipole source.

5.6 VALIDATION OF ANISOTROPIC PML PROPOSALS

Reference [10] applied the validation technique summarized in Section 5.5 to illustrate the failure of incorrect PML formulations in anisotropic media. Two-dimensional (2-D) simulations (E_z, H_x, H_y) were conducted to obtain the response of proposed PMLs to impinging numerical waves radiated by a point-dipole sinusoidal current source [8, Appendix D]. Numerical solutions were obtained with a plane-wave frequency-domain (PWFD) expansion method employing an iterative biconjugate gradient algorithm [42].

For each proposed PML, two sets of simulations were performed at progressively refined discretization resolutions. For the first set, the absorber thickness was L, and for the second set, the absorber thickness was $L + \Delta L$. The difference of the electric field $\boldsymbol{E}$ at the same (arbitrarily chosen) point $\boldsymbol{x}$ in the interior of both simulations was computed, defining a *field convergence factor*:

$$\frac{|\boldsymbol{E}_{L+\Delta L}(\boldsymbol{x}) - \boldsymbol{E}_L(\boldsymbol{x})|^2}{|\boldsymbol{E}_L(\boldsymbol{x})|^2} \tag{5.8}$$

As long as $\boldsymbol{E}(\boldsymbol{x})$ was not identically zero, it was found that observation of $\boldsymbol{E}$ at a single point sufficed for this purpose, as opposed to a more complex procedure using some overall norm of the difference in the interior field of the computation space [10].

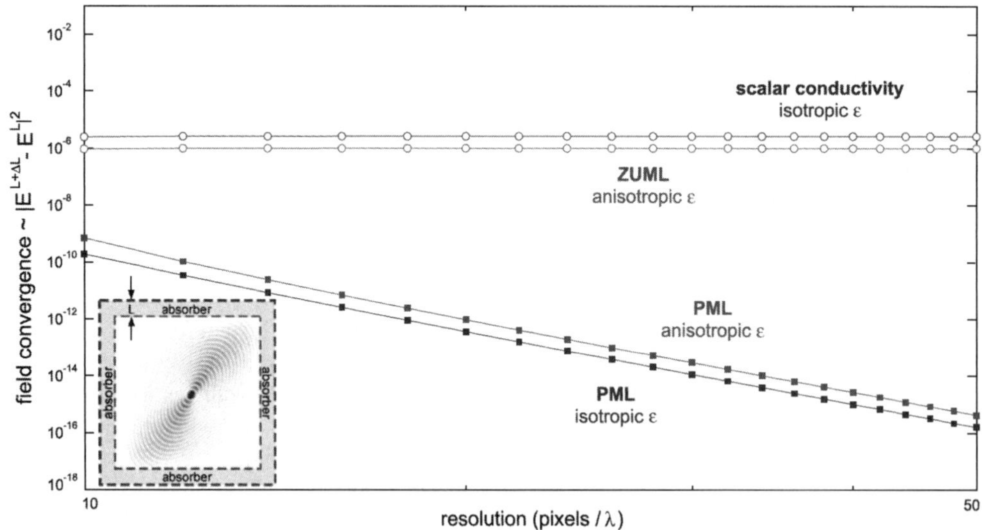

Fig. 5.1 PWFD-computed field convergence factor of (5.8) for $\Delta L = \lambda$ vs. discretization resolution for both isotropic and anisotropic internal media with PML and non-PML (scalar conductivity) absorbing boundaries. The unsplit-field PML proposal by Zhao [11] is denoted "ZUML," and is seen not to be a true PML for the anisotropic medium where it fails to be reflectionless even in the limit of high resolution, similar to the scalar conductivity for an isotropic medium. Inset: Re(E_z) field profile of a point-dipole source located at the center of the anisotropic medium computational space terminated by an absorber. *Source:* A. Oskooi and S. G. Johnson, *J. Computational Physics*, Vol. 230, 2011, pp. 2369–2377, ©2011 Elsevier.

Figure 5.1 shows the PWFD-computed field convergence factor of (5.8) as a function of discretization resolution for four cases [10]: (1) a simple non-PML absorber, consisting of just a scalar electrical conductivity, terminating a vacuum interior medium; (2) the unsplit-field PML proposed in [11] (denoted "ZUML") terminating an anisotropic interior medium with $[\varepsilon]$ formed from an isotropic diagonal tensor with eigenvalues $\{12, 1, 12\}$ and principal axes rotated 45° about the y and z axes; (3) the correct PML formulation of Section 5.3 terminating the same anisotropic interior medium; and (4) the correct PML formulation of Section 5.3 terminating the vacuum interior medium. For all of these results, the absorber thickness perturbation was $\Delta L = \lambda$, the vacuum wavelength of the point-dipole source.

From Fig. 5.1, we observe that the differences between a correct PML, the ZUML, and a non-PML absorber are stark. In the correct PML formulation, the field convergence factor of (5.8) diminishes exponentially (and nearly identically) for both the vacuum and anisotropic interior media cases as the resolution in grid cells per wavelength increases linearly toward the continuum limit. On the other hand, both the simple non-PML absorber terminating a vacuum interior medium, and the proposed ZUML terminating the anisotropic interior medium, yield field convergence factors that saturate to constant nonzero values (in fact, derivable from the corresponding Fresnel reflection coefficients from the exact Maxwell's equations). The ZUML is clearly a non-PML absorber for anisotropic media, with nonzero reflection in the continuum limit as predicted from the fact that its derivation is not equivalent to an analytic continuation of Maxwell's equations.

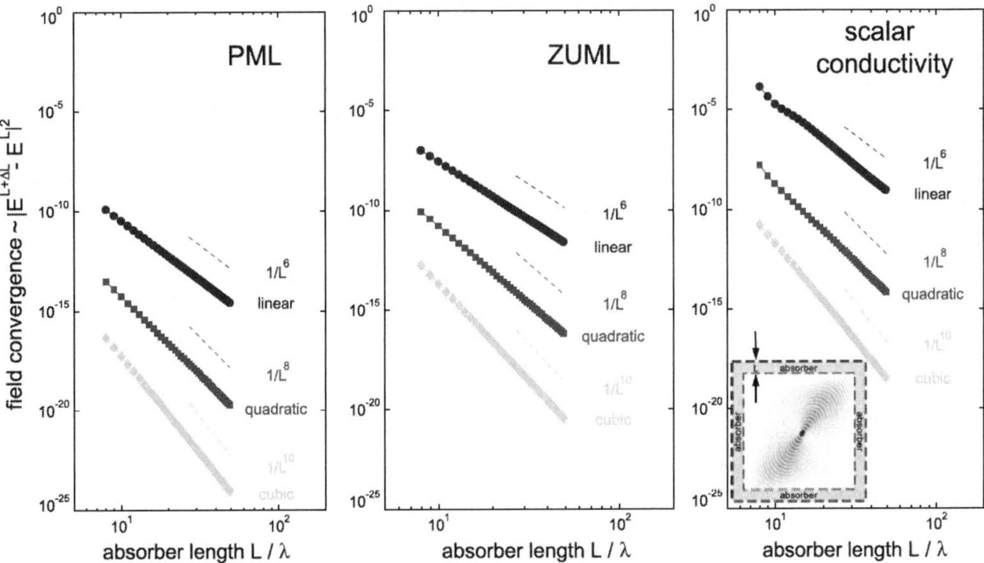

Fig. 5.2 PWFD-computed field convergence factor of (5.8) for $\Delta L = \lambda$ vs. absorber thickness L in anisotropic media at a coarse resolution of 9 pixels/λ for various monomial absorber functions $s(u)$ ranging from linear to cubic. For comparison, the predicted asymptotic power laws are shown as dashed lines. Left plot – PML; middle plot – ZUML; and right plot – scalar electric conductivity absorber. Inset: Re(E_z) field profile of a point-dipole source located at the center of the anisotropic medium computational space terminated by an absorber. *Source:* A. Oskooi and S. G. Johnson, *J. Computational Physics*, Vol. 230, 2011, pp. 2369–2377, ©2011 Elsevier.

Figure 5.2 compares the variations of the PWFD-computed field convergence factor of (5.8) vs. absorber thickness L for the PML, ZUML, and scalar electric conductivity absorber terminations of the anisotropic medium of Fig. 5.1 [10]. In this study, while $\Delta L = \lambda$ was again assumed, the resolution was instead fixed at a coarse 9 pixels/λ, and three different monomial absorber functions $s(u)$ (linear, quadratic, and cubic) were tested for each candidate. For comparison, the predicted asymptotic power laws [9] are shown as dashed lines.

From Fig. 5.2, we observe that, as discussed in Section 5.4, all absorption profiles (whether PML or not) exhibit the same characteristic power-law scaling with L, determined only by the differentiability d of the profile. The impact of a true PML is to improve the constant multiplicative factor in this relationship, although ZUML actually fares quite well in this respect at the assumed coarse resolution, and gets even better for milder anisotropy. This illustrates why merely checking whether the numerical wave reflection is low for a given resolution [14–19], and/or decreases with absorber thickness [1, 20], can be misleading tests for the validity of a PML proposal.

5.7 TIME-DOMAIN PML FORMULATION FOR TERMINATING ANISOTROPIC DISPERSIVE MEDIA

The coordinate-stretching derivation of PML, reviewed in Section 5.3, is straightforward in the frequency domain because the transformations and materials of PML are frequency dependent. To express these dependencies in the time domain involves the evolution of appropriate auxiliary differential equations. This complication did not arise in previous frequency-domain unsplit PML proposals for anisotropic media [41], and the anisotropy means that previous UPML time-domain schemes for isotropic media [11, 33] are not directly applicable, so a new reformulation is required. (Equivalently, multiplication by a frequency-dependent susceptibility corresponds in the time domain to a convolution, leading to "convolutional PML" formulations [19]. Because the frequency dependence is in the form of a polynomial fraction in ω, it can be implemented with finite auxiliary storage by a recursive "IIR" filter [43].) The emphasis is on keeping the number of auxiliary differential equations (and the resulting memory and computational costs) to a minimum, while not making the PML region too complicated compared to the non-PML regions. Because the treatment of $[\varepsilon]$ and $[\mu]$ is identical except for interchange of $\boldsymbol{D}$ with $\boldsymbol{B}$ and $\boldsymbol{E}$ with $\boldsymbol{H}$ (and a sign flip from Ampere's to Faraday's law), we only describe the $[\varepsilon]$ and $\partial \boldsymbol{D}/\partial t$ equations (Ampere's law) here.

First-order ordinary differential equations (ODEs) are the most convenient to discretize. For example, consider the frequency-domain relation $a = s \cdot b = (1 + i\sigma/\omega) \cdot b$. Multiplication of both sides by $-i\omega$ gives $-i\omega a = -i\omega b + \sigma b$, which upon inverse Fourier transformation yields the first-order time-domain ODE, $da/dt = db/dt + \sigma b$. For a more general scaling factor, s, such as $s = \kappa + \sigma/(-i\omega + \alpha)$ that appears in shifted-pole PMLs [2, 18, 40], $a = s \cdot b$ still leads to a simple first-order time-domain ODE, $da/dt + \alpha a = \kappa db/dt + (\sigma + \kappa\alpha)b$. In the language of digital signal processing, this factorization into first-order ODEs is equivalent to the "cascade" form of a recursive/IIR filter [43].

Therefore, starting in the frequency domain, our strategy is to factorize $\nabla \times \boldsymbol{H} = -i\omega\tilde{\boldsymbol{D}}$ (where $\tilde{\boldsymbol{D}} = [\tilde{\varepsilon}(\omega)]\boldsymbol{E}$) into terms with only *one* factor of s (or σ/ω) each, or ratios of single s factors. In doing so, we are free to change the definition of $\boldsymbol{D}$ and introduce new auxiliary fields as desired, since the fields in the PML region are not physical. A key means to facilitate our development is the following factorization:

$$\frac{\mathbf{J}}{\det \mathbf{J}} = s_x s_y s_z \begin{bmatrix} s_x^{-1} & & \\ & s_y^{-1} & \\ & & s_z^{-1} \end{bmatrix} = \begin{bmatrix} s_y & & \\ & s_z & \\ & & s_x \end{bmatrix} \begin{bmatrix} s_z & & \\ & s_x & \\ & & s_y \end{bmatrix} \equiv \mathbf{S}_1 \mathbf{S}_2 \qquad (5.9)$$

Using this factorization, and defining the auxiliary vector fields, $\boldsymbol{U}$ and $\boldsymbol{W}$, we can convert the frequency-domain expression (5.2a, b) into the following equivalent form, giving the same relationship between $\boldsymbol{H}$ and $\boldsymbol{E}$:

$$\tilde{\boldsymbol{D}} = \underbrace{\mathbf{S}_1 \underbrace{\mathbf{S}_2 \underbrace{[\varepsilon(\omega)] \underbrace{\mathbf{J}^{\mathrm{T}} \boldsymbol{E}}_{\boldsymbol{W}}}_{\boldsymbol{U}}}_{\boldsymbol{D}}} \qquad (5.10)$$

where:

$$\nabla \times \boldsymbol{H} = -i\omega \mathbf{S}_1 \boldsymbol{D} \qquad (5.11a)$$

$$\boldsymbol{D} = \mathbf{S}_2 \boldsymbol{U} \qquad (5.11b)$$

$$\boldsymbol{U} = [\varepsilon(\omega)]\, \boldsymbol{W} \qquad (5.11c)$$

$$\boldsymbol{W} = \mathbf{J}^{\mathrm{T}} \boldsymbol{E} \qquad (5.11d)$$

Upon inverse Fourier transformation, (5.11a) – (5.11d) each yields a first-order ODE in time. This system of equations is straightforward to implement in the FDTD context, using the following sequence:

Update $\boldsymbol{D}^{n+1}$ from $\nabla \times \boldsymbol{H}^{n+1/2}$.

Update $\boldsymbol{U}^{n+1}$ from $\boldsymbol{D}^{n+1}$.

Update $\boldsymbol{W}^{n+1}$ from $\boldsymbol{U}^{n+1}$.

Update $\boldsymbol{E}^{n+1}$ from $\boldsymbol{W}^{n+1}$.

We note that the computation of $\boldsymbol{W}$ from $\boldsymbol{U}$ is identical to the computation of $\boldsymbol{E}$ from $\boldsymbol{D}$ in the non-PML regions, including any auxiliary fields needed to implement a dispersive ε by standard techniques [2].

Figure 5.3 displays the results of a validation study where a PML formulated according to the above approach terminated a 2-D grid filled with a dispersive anisotropic medium. This model was implemented using the free-software FDTD package, *Meep* [34], described in Chapter 20. Here, the field-convergence factor defined in the frequency domain in (5.8) was calculated via a discrete-time Fourier transform of the field observed at a monitor point in the working region of the grid in response to a Gaussian-pulse dipole source, with λ assumed to be equal to the vacuum wavelength at the center frequency of the pulse. In particular, the inset of Fig. 5.3 shows that the field-convergence factor properly approaches zero with increasing grid resolution. This is the behavior of a valid PML. The main graph of Fig. 5.3 reaffirms the expected scalings with absorber thickness L for linear, quadratic, and cubic absorption profiles of the PML.

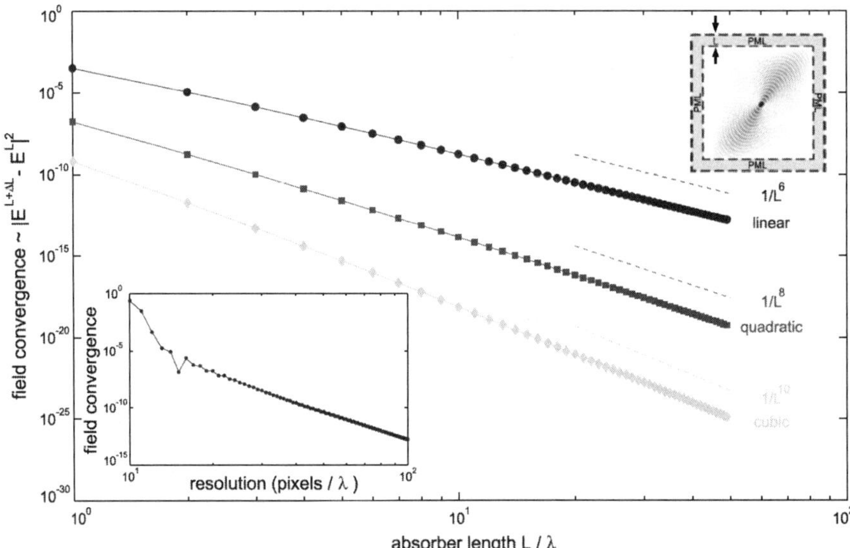

Fig. 5.3 FDTD-computed field convergence factor of (5.8) with $\Delta L = \lambda$ for a PML formulated according to Section 5.7 terminating a dispersive anisotropic medium. Main graph: field convergence factor vs. absorber thickness L at a coarse grid resolution for various monomial absorber functions ranging from linear to cubic. For comparison, the predicted asymptotic power laws are shown as dashed lines. Lower left inset: field convergence factor vs. grid resolution, demonstrating proper convergence to zero as a valid PML. Upper right inset: Re(E_z) field profile of a point-dipole source located at the center of the anisotropic medium filling the working region of the grid. *Source:* A. Oskooi and S. G. Johnson, *J. Computational Physics*, Vol. 230, 2011, pp. 2369–2377, ©2011 Elsevier.

Other valid PML alternatives include the split-field approach, which has been correctly derived for anisotropic media without invalid commutations [14–17, 27–30], or in general any coordinate-transformation approach that is not absorbed into the materials but is left as a transformation of the derivatives [19]. In terms of the number of required auxiliary variables, the computational expense of the time-domain unsplit formulation of [10] reviewed above is at least as good as that of the correct split-field methods (see Appendix 5B). Furthermore, the viewpoint in [10] of simply factorizing transformed materials has a certain aesthetic appeal, and appears to be the simplest correction to the standard UPML approaches.

5.8 PML FAILURE FOR OBLIQUE WAVEGUIDES

Reference [9] suggested that PML is inapplicable to the case of a waveguide entering the PML at an oblique angle, since the material function is not analytic in the direction of the PML. This opposed claims that ordinary PML formulations are valid for such circumstances [12, 13], when in fact the claimed PMLs were apparently adiabatic non-PML absorbers with sufficient thicknesses to mitigate the reflections. The use of unmodified "PML" equations in cases where they are inapplicable, relying on adiabatic absorption to compensate for failure of the PML, has been termed a "pseudo-PML" or pPML [9]. Using the resolution test reviewed in Section 5.6, Ref. [10] explicitly demonstrated the failure of PML for an oblique waveguide.

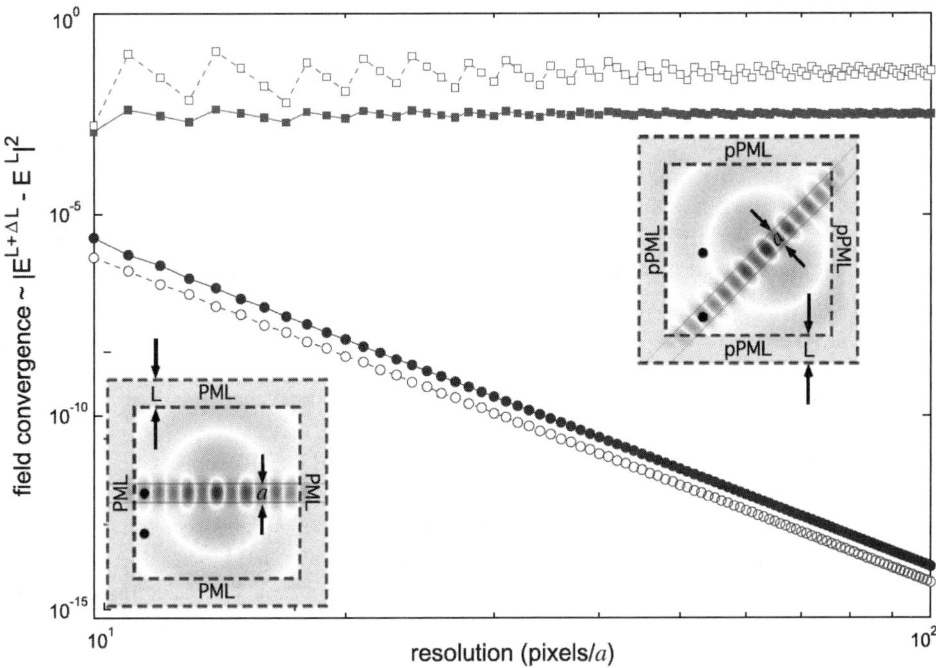

Fig. 5.4 FDTD results for a dielectric waveguide of width a penetrating PML at 45° [upper inset – Re(E_z) profile; upper pair of lines], and the same waveguide penetrating the PML at 0° [lower inset – Re(E_z) profile; lower pair of lines]. An E_z point-dipole source centered within the waveguide provides the excitation. Field monitor points are located inside and outside each waveguide, with positions indicated by black dots in the insets. For each case, the field convergence factor of (5.8) is plotted vs. grid resolution for $\Delta L = a$, with the filled and open symbols corresponding to the field monitor points inside and outside the waveguide, respectively. *Source:* A. Oskooi and S. G. Johnson, *J. Computational Physics*, Vol. 230, 2011, pp. 2369–2377, ©2011 Elsevier.

Figure 5.4 shows the FDTD-computed results reported in [10] of dielectric waveguides penetrating PML at 45° and at 0° for the 2-D $\{E_z, H_x, H_y\}$ polarization case. For the oblique waveguide, the field convergence factor is seen to saturate in the continuum limit as the grid resolution increases, indicative of a pseudo-PML absorber. On the other hand, the field convergence factor for the normally oriented waveguide decreases approximately exponentially with increased grid resolution, indicative of a correct PML.

Reference [10] noted that the exact position of the monitor point made little difference (a small constant factor), since the failure of PML generally produces reflected numerical waves throughout the computational space. It was also noted that, in problems where a waveguide channel cannot be rotated to an orientation normal to the PML, it is possible that some additional (real) coordinate transformation could be used to *locally* rotate the waveguide into the correct direction, absorbing the coordinate transformation into the materials [ε] and [μ] [21]. However, as noted in [10], care must be taken that such a coordinate transformation not introduce new variation along the PML direction that itself spoils the PML. Such a problem-dependent solution was outside the scope of [10] and this chapter.

5.9 SUMMARY AND CONCLUSION

This chapter reviewed the straightforward technique reported in [10] to verify the correctness of any proposed PML formulation, irrespective of its implementation. The PML validation method of [10] involves simply computing the difference between two computations as a function of the discretization resolution. Such a validation procedure is useful since any adiabatic absorber (PML or not) could be made to have sufficiently low reflection so that its deviation from a true PML would be overlooked. In fact, several published claims of working PMLs for anisotropic media, periodic media, and oblique waveguides appear to be just instances of adiabatic pseudo-PML absorbers.

This chapter also reviewed the corrected unsplit PML formulation of [10] for anisotropic dispersive media, involving a simple refactorization of typical UPML proposals. The computational expense of this formulation is at least as good as that of the split-field methods. In addition, its viewpoint of simply factorizing transformed materials has a certain aesthetic appeal, and appears to be the simplest correction to the standard UPML approaches.

Finally, the discussion in [10] of waveguides penetrating obliquely into PML was reviewed. Here, a possible brute-force workaround is the use of a pseudo-PML absorber, sacrificing two or three orders-of-magnitude in effectiveness compared to a true PML, necessarily being compensated by requiring a thicker absorber. A true PML solution for the oblique-waveguide problem, especially one that handles multiple waveguides incident at different angles on the same PML boundary, may be an interesting problem for future research.

Overall, the rigorous PML validation scheme reported in [10] and reviewed here is not fooled by reflections that are simply low. Reflections must converge to zero with increasing resolution. Such validation seems increasingly relevant as the PML concept is applied to more and more geometries, material systems, wave equations, and computational schemes.

ACKNOWLEDGMENTS

This work was supported in part by the Materials Research Science and Engineering Center program of the National Science Foundation under Grant Nos. DMR-9400334 and DMR-0819762, and by the Army Research Office through the Institute for Soldier Nanotechnologies under contract DAAD-19-02-D0002.

APPENDIX 5A: TUTORIAL ON THE COMPLEX COORDINATE-STRETCHING BASIS OF PML

There are several equivalent formulations of PML. Berenger's original formulation [1] is called the *split-field* PML, because he artificially split the wave solutions into the sum of two new artificial field components. Nowadays, a more common formulation is the *uniaxial* PML or *UPML*, which expresses the PML region as the ordinary wave equation with a combination of artificial *anisotropic* absorbing materials [32]. Both of these formulations were originally derived by laboriously computing the solution for a wave incident on the absorber interface at an arbitrary angle (and polarization, for vector waves), and then solving for the conditions in which the reflection is always zero. This technique, however, is labor intensive to extend to other wave equations and other coordinate systems (e.g., cylindrical or spherical rather than Cartesian). It also misses an important fact: PML still works (i.e., can be made theoretically reflectionless) for *inhomogeneous* media, such as waveguides, as long as the medium is homogeneous in the direction perpendicular to the boundary, even though the wave solutions for such media cannot generally be found analytically. It turns out, however, that *both* the split-field and UPML formulations can be derived in a much more elegant and general way by viewing them as the result of *complex coordinate stretching* [35, 36, 41].[2] It is this complex-coordinate approach, which is essentially based on *analytic continuation* of Maxwell's equations into complex spatial coordinates where the fields are exponentially decaying, that we reviewed in the first part of this chapter.

In this appendix, we first briefly remind the reader what a wave equation is, focusing on the simple case of the scalar wave equation but also giving a general definition. We then derive PML as a combination of two steps: analytic continuation into complex coordinates, and then a coordinate transformation back to real coordinates. Finally, we discuss some limitations of PML, most notably the fact that it is no longer reflectionless once the wave equation is discretized, and common workarounds for these limitations.

5A.1 Wave Equations

There are many formulations of waves and wave equations in the physical sciences. The prototypical example is the (source-free) scalar wave equation:

$$\nabla\cdot(a\nabla u) = \frac{1}{b}\frac{\partial^2 u}{\partial t^2} \triangleq \frac{\ddot{u}}{b} \tag{5A.1}$$

where $u(x, t)$ is the scalar wave amplitude, and $c = \sqrt{ab}$ is the phase velocity of the wave for some parameters, $a(x)$ and $b(x)$, of the (possibly inhomogeneous) medium. For lossless, propagating waves, a and b should be real and positive.

Both for computational convenience (in order to use a staggered-grid leapfrog discretization) and for analytical purposes, it is more convenient to split this second-order equation into two coupled first-order equations, by introducing an auxiliary field, $\mathbf{v}(x, t)$:

$$\frac{\partial u}{\partial t} = b\nabla\cdot\mathbf{v} \tag{5A.2}$$

$$\frac{\partial \mathbf{v}}{\partial t} = a\nabla u \tag{5A.3}$$

These equations are easily seen to be equivalent to (5A.1). Equations (5A.2) and (5A.3) can be written more abstractly as:

$$\frac{\partial \mathbf{w}}{\partial t} = \frac{\partial}{\partial t}\begin{bmatrix} u \\ \mathbf{v} \end{bmatrix} = \begin{bmatrix} & b\nabla\cdot \\ a\nabla & \end{bmatrix}\begin{bmatrix} u \\ \mathbf{v} \end{bmatrix} \triangleq \hat{\mathbf{D}}\mathbf{w} \tag{5A.4}$$

for a 4×4 linear operator, $\hat{\mathbf{D}}$, and a 4-component vector, $\mathbf{w} = [u, \mathbf{v}]$ (in three dimensions). The key property that makes this a "wave equation" is that $\hat{\mathbf{D}}$ is *an anti-Hermitian* operator in a

[2] It is sometimes implied that only the split-field PML can be derived via the stretched-coordinate approach [2], but the UPML media can be derived in this way as well [41].

proper choice of inner product, which leads to oscillating solutions, conservation of energy, and other "wavelike" phenomena. Every common wave equation, from scalar waves to Maxwell's equations (electromagnetism) to Schrödinger's equation (quantum mechanics) to the Lamé-Navier equations for elastic waves in solids, can be written in the abstract form, $\partial \boldsymbol{w}/\partial t = \hat{\mathbf{D}}\boldsymbol{w}$, for some wavefunction, $\boldsymbol{w}(\boldsymbol{x}, t)$, and some anti-Hermitian operator, $\hat{\mathbf{D}}$.[3] The same PML ideas apply equally well in all of these cases, although PML is most commonly applied to Maxwell's equations for computational electromagnetics.

5A.2 Complex Coordinate Stretching

Let us start with the solution, $\boldsymbol{w}(\boldsymbol{x}, t)$, of some wave equation in an unbounded region. We have a region of interest near the origin, $\boldsymbol{x} = 0$, and want to truncate space outside this region in such a way as to absorb radiating waves. In particular, we will focus on truncating the problem in the $+x$-direction (the other directions will follow by the same technique). This truncation occurs in three conceptual steps, summarized as follows:

1. In infinite space, *analytically continue* the solutions and equations to a *complex-x* contour, which changes oscillating waves into *exponentially decaying* waves outside the region of interest *without* reflections.
2. Still in infinite space, perform a *coordinate transformation* to express the complex x as a function of a real coordinate. Now, we have *real coordinates* and *complex materials*.
3. Truncate the domain of this new real coordinate inside the complex-material region. Since the solution is decaying there, if we truncate it after a sufficiently long distance (where the exponential tails are small), it won't matter what boundary condition we use (hard-wall truncations are fine).

For now, we will make two simplifications:

- We assume that the space far from the region of interest is homogeneous (deferring the inhomogeneous case until later).
- We assume that the space far from the region of interest is linear and time invariant.

Under these assumptions, the radiating solutions in infinite space must take the form of a superposition of *plane waves*:

$$\boldsymbol{w}(\boldsymbol{x}, t) = \int_{\boldsymbol{k}, \omega} \boldsymbol{W}_{\boldsymbol{k},\omega} \, e^{i(\boldsymbol{k}\cdot\boldsymbol{x} - \omega t)} \tag{5A.5}$$

for some constant amplitudes, $\boldsymbol{W}_{\boldsymbol{k},\omega}$, where ω is the (angular) frequency and $\boldsymbol{k}$ is the wavevector. [In an isotropic medium, ω and $\boldsymbol{k}$ are related by $\omega = c|\boldsymbol{k}|$, where $c(\omega)$ is some phase velocity, but we don't need to assume that here.] In particular, the key fact is that the radiating solutions can be decomposed into functions of the form:

$$\boldsymbol{W}(y, z) \, e^{i(kx - \omega t)} \tag{5A.6}$$

---

[3] See Reference [45], for example.

The ratio ω / k is the phase velocity, which can be different from the group velocity, $d\omega / dk$ (the velocity of energy transport, in lossless media). For waves propagating in the $+x$-direction, the group velocity is positive. Except in very unusual cases, the phase velocity has the same sign as the group velocity in a homogeneous medium,[4] so we will assume that k is positive.

Analytic Continuation

We note that (5A.6) is an *analytic function* of x. That means that we can freely *analytically continue* it, evaluating the solution at complex values of x. As shown in the top panels of Fig. 5A.1, the original problem corresponds to x along the real axis, which gives an oscillating e^{ikx} solution. However, instead of evaluating x along the real axis, consider what happens if we evaluate it along the contour shown in the bottom-left panel of Fig. 5A.1, where for $\operatorname{Re} x > 5$, we have added a linearly growing imaginary part.

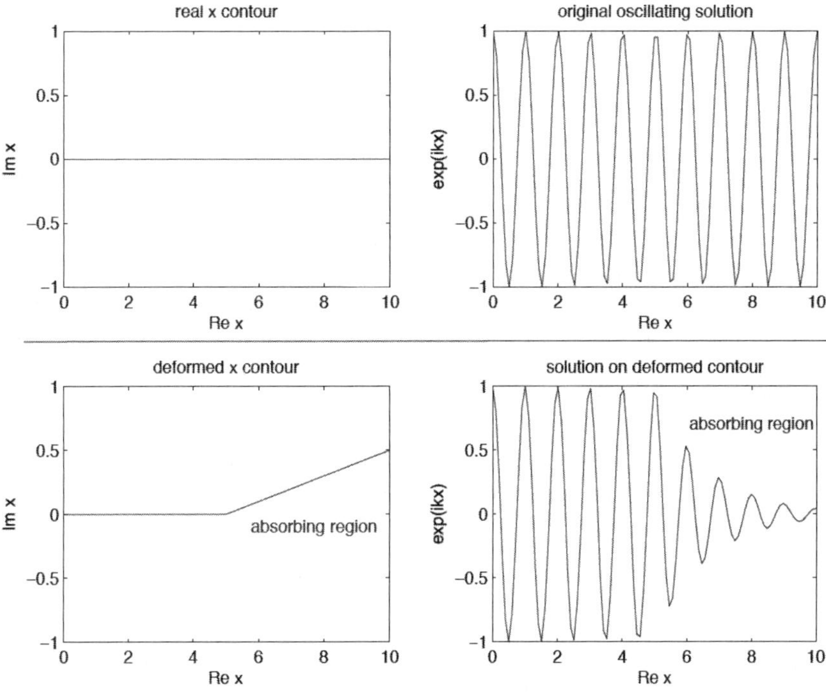

Fig. 5A.1 Top two panels: The real part of the original oscillating solution, e^{ikx} (right panel), corresponds to an evaluation contour along the real axis in the complex-x plane (left panel). Bottom two panels: We can instead evaluate the analytic function, e^{ikx}, along the *deformed* contour in the complex plane shown in the left panel, which linearly shifts in the positive imaginary direction for $\operatorname{Re} x > 5$. The resulting e^{ikx} solution (right panel) is unchanged for $\operatorname{Re} x < 5$, but exponentially decays for $\operatorname{Re} x > 5$ where the contour had been deformed, corresponding to an "absorbing" region.

[4] The formulation of PML absorbers is somewhat more involved when the phase velocity has a sign opposite to the group velocity, for example, in the "left-handed media" of electromagnetism [46, 47]. Matters are even worse in certain waveguides with both signs of group velocity at the same ω [38].

Upon adding a linearly growing positive imaginary part for $\operatorname{Re} x > 5$, as shown in the bottom-left panel of Fig. 5A.1, the solution *exponentially decays* for $\operatorname{Re} x > 5$, as shown in the bottom-right panel of this figure. This is because $e^{ik(\operatorname{Re} x + i \operatorname{Im} x)} = e^{ik \operatorname{Re} x} e^{-k \operatorname{Im} x}$ exponentially decays (for $k > 0$) as $\operatorname{Im} x$ increases. In effect, the wave solution in this region behaves like a wave propagating in an *absorbing material.*

However, there is one crucial difference here from an ordinary absorbing material: the solution is *not changed* for $\operatorname{Re} x < 5$, where x is the same as before. Hence, it not only acts like an absorbing material, it acts like a *reflectionless absorbing material*, a PML.

The key point is that the analytically continued solution satisfies the *same* differential equation. We assumed that the differential equation was x-invariant in this region, so x only appeared in derivatives like $\partial / \partial x$, and the derivative of an analytic function is the same along any dx direction in the complex plane. Therefore, we have succeeded in transforming our original wave equation into one in which the radiating solutions (in the large-x region) decay exponentially, while the solutions in the region we care about (small x) are unchanged. The only problem is that solving differential equations along contours in the complex plane is rather unfamiliar and inconvenient. This difficulty is easily fixed.

Coordinate Transformation Back to Real x

For convenience, let's denote the complex x-contour by $\tilde{x}$, and reserve the letter, x, for the real part. Thus, we have $\tilde{x}(x) = x + if(x)$, where $f(x)$ is some function indicating how we've deformed our contour in the positive imaginary direction. Since the complex coordinate $\tilde{x}$ is inconvenient, we will just *change variables* to write the equations in terms of x, the real part!

Changing variables is easy. Where our original equation had $\partial \tilde{x}$ (the differential along the deformed contour, $\tilde{x}$), we now have $\partial \tilde{x} = (1 + i\, df/dx)\, \partial x$. That's it! Since our original wave equation was assumed x-invariant (at least in the large-x regions where $f \neq 0$), we have no other substitutions to make. As we shall soon see, it will be convenient to denote $df/dx = \sigma_x(x)/\omega$ for some function, $\sigma_x(x)$, having units of $\sec^{-1}$. [For example, in the bottom panels of Fig. 5A.1, we chose $\sigma_x(x)$ to be a step function: zero for $x \leq 5$, and a positive constant for $x > 5$, which gave us exponential decay.] In terms of σ_x, the entire process of PML can be conceptually summed up by a single transformation of our original differential equation:

$$\frac{\partial}{\partial x} \rightarrow \frac{1}{1 + i\sigma_x(x)/\omega} \frac{\partial}{\partial x} \tag{5A.7}$$

In the PML regions where $\sigma_x > 0$, our oscillating solutions turn into exponentially decaying ones. In the regions where $\sigma_x = 0$, our wave equation is unchanged, and the solution is *unchanged*. There are no reflections because this is only an analytic continuation of the original solution from x to $\tilde{x}$, and where $\tilde{x} = x$, the solution cannot change.

Why did we choose σ_x/ω, as opposed to just σ_x? The answer becomes apparent if we look at what happens to e^{ikx}. In the new coordinates, this transforms to the decaying wave:

$$e^{ikx} \exp\left[-(k_x/\omega) \int_x \sigma_x(x')\, dx'\right] \tag{5A.8}$$

The factor k_x/ω in (5A.8) represents the inverse of the phase velocity, c_x, in the x-direction. For a fixed propagation angle in a dispersionless medium, c_x (and thus, the attenuation rate in the PML), is *independent of* ω, and all wavelengths decay at the same rate. In contrast, omitting $1/\omega$

would cause shorter wavelengths to decay faster than longer wavelengths. Note, however, that the PML attenuation rate varies with the wave's propagation angle, which obviously impacts c_x.

Truncating the Computational Region

Once we have performed the PML transformation (5A.7) of our wave equation, the solutions are unchanged in our small-x region of interest, but exponentially decay in the large-x outer region. This means that we can *truncate* the computational region at some sufficiently large x, perhaps by putting a hard wall (Dirichlet boundary condition). Because only exponentially small tails "see" this hard wall and reflect from it, and even they attenuate on the way back toward the region of interest, the effect on the solutions in our region of interest is exponentially small.

In principle, we can make the PML region as thin as we want by increasing σ_x, thanks to the fact that the decay rate is independent of ω. However, in practice, using a large σ_x can cause "numerical reflections" once we discretize the problem onto a grid. To minimize these reflections, $\sigma_x(x)$ is usually gradually increased from zero with a power-law dependence upon depth within the PML, as discussed in [2].

PML Boundaries in Other Directions

So far, we've seen how to truncate our computational region with a PML in the +x-direction. What about other directions? The most important case to consider is the –x-direction. Here, we do *exactly* the same thing: apply the PML transformation (5A.7) with $\sigma_x > 0$ at a sufficiently large negative x, and then truncate the computational space. This works because, for $x < 0$, the radiating waves are propagating in the –x-direction with $k < 0$ (negative phase velocity). This makes our PML solutions (5A.8) decay in the opposite direction (exponentially decaying as $x \to -\infty$) for the *same* positive σ_x.

Now that we have dealt with the $\pm x$-directions, the $\pm y$- and $\pm z$-directions are easy: we implement the same transformation, except to $\partial/\partial y$ and $\partial/\partial z$, respectively, using functions, $\sigma_y(y)$ and $\sigma_z(z)$, that are nonzero in the $\pm y$ and $\pm z$ PML regions. At the corners of the computational space, we will have regions that are PML along two or three directions simultaneously (i.e., two or three σs are nonzero), but that doesn't change anything.

Coordinate Transformations and Materials

We will see below that, in the context of the scalar wave equation, the $1 + i\sigma/\omega$ term from the PML coordinate transformation has the effect of an *artificial anisotropic absorbing material* in the wave equation. This changes a and b of (5A.1)–(5A.4) to complex numbers, and a tensor in the case of a. In the case of Maxwell's equations, this is a specific instance of a more general theorem: Maxwell's equations under *any* coordinate transformation can be expressed as Maxwell's equations in Cartesian coordinates with *transformed materials*.[5] That is, the coordinate transform is "absorbed" into a change of the permittivity, ε, and the permeability, μ, generally into anisotropic tensors. This is the reason why UPML, which constructs reflectionless anisotropic absorbers, is equivalent to complex coordinate stretching: it is just absorbing the coordinate stretching into the material tensors.

---

[5] This theorem appears to have been first clearly stated and derived by Ward and Pendry [21], and is summarized in a compact general form in [25] and [48].

5A.3 PML Examples

As we have seen, in the frequency domain, when we are solving for solutions with time-dependence $e^{-i\omega t}$, PML is almost trivial. In the x-direction, we apply PML transformation (5A.7) to every $\partial/\partial x$ derivative in our wave equation, and apply analogous transformations to the $\partial/\partial y$ and $\partial/\partial z$ derivatives to obtain PML boundaries in those directions.

In the time domain, however, things are more complicated. We chose our transformation to be of the form $1+i\sigma/\omega$, i.e., a *frequency-dependent* complex "stretch" factor, so that the attenuation rate would be *frequency independent*. But how do we express a $1/\omega$ dependence in the time domain, where we don't have ω (since the time-domain wavefunction can superimpose multiple frequencies at once)? A very limited solution is to use a stretch factor, $1+i\sigma/\omega_0$, for some constant frequency, ω_0, that is typical of our problem. Here, as long as the bandwidth is sufficiently narrow, the PML attenuation rate (and thus the truncation error) is fairly constant. However, it turns out that there is a way to implement the ideal $1/\omega$ dependence *directly* in the time domain—the *auxiliary differential equation* (ADE) approach [2].

The ADE approach is best illustrated by example, so we will consider PML absorbers in the x-direction for the scalar wave equation in 1-D and 2-D. (While an ADE is not required in 1-D, studying the 1-D case permits establishment of a conceptual foundation for the 2-D case.)

Scalar Waves in 1-D

Consider the 1-D version of the scalar wave equation (5A.2), (5A.3):

$$\frac{\partial u}{\partial t} = b\frac{\partial v}{\partial x} = -i\omega u \tag{5A.9}$$

$$\frac{\partial v}{\partial t} = a\frac{\partial u}{\partial x} = -i\omega v \tag{5A.10}$$

where we have substituted an $e^{-i\omega t}$ time dependence. Now, if we perform PML transformation (5A.7), and multiply both sides by $1+i\sigma_x/\omega$, we obtain:

$$b\frac{\partial v}{\partial x} = -i\omega u + \sigma_x u \tag{5A.11}$$

$$a\frac{\partial u}{\partial x} = -i\omega v + \sigma_x v \tag{5A.12}$$

Note that the $1/\omega$ terms have cancelled. Upon applying the inverse Fourier transform to (5A.11) and (5A.12), we obtain their corresponding time-domain forms:

$$\frac{\partial u}{\partial t} = b\frac{\partial v}{\partial x} - \sigma_x u \tag{5A.13}$$

$$\frac{\partial v}{\partial t} = a\frac{\partial u}{\partial x} - \sigma_x v \tag{5A.14}$$

Note that, for $\sigma_x > 0$, the decay terms have exactly the correct sign to make the solutions decay in *time* if u and v are constants in space. Similarly, they have the correct sign to make the solutions decay in space where $\sigma_x > 0$. In fact, this is a true PML: there are *zero* reflections from any boundary where we change σ_x, even if we change σ_x discontinuously (not including the discretization problems mentioned above).

Equations (5A.13) and (5A.14) reveal why we use the symbol σ for the PML absorption parameter. If these equations are interpreted as being equivalent to the 1-D time-dependent Maxwell's curl equations for the electric field (here, denoted by u) and the magnetic field (here, denoted by v), then σ plays the role of a *conductivity*, and conductivity is traditionally denoted by σ. However, unlike the usual electrical conductivity, in PML we have both an electric *and* a magnetic conductivity, since we have terms corresponding to currents of electric and magnetic charges. There is no reason that we need to be limited to physical materials to construct PML for our computer simulations!

Scalar Waves in 2-D

The 1-D case reviewed above is much too simple to provide the full flavor of how PML works. Hence, we now consider the 2-D scalar wave equation (again assuming an $e^{-i\omega t}$ time dependence):

$$\frac{\partial u}{\partial t} = b\nabla \cdot \boldsymbol{v} = b\frac{\partial v_x}{\partial x} + b\frac{\partial v_y}{\partial y} = -i\omega u \tag{5A.15}$$

$$\frac{\partial v_x}{\partial t} = a\frac{\partial u}{\partial x} = -i\omega v_x \tag{5A.16}$$

$$\frac{\partial v_y}{\partial t} = a\frac{\partial u}{\partial y} = -i\omega v_y \tag{5A.17}$$

After performing the PML transformation (5A.7) of $\partial/\partial x$ in (5A.15) and (5A.16), and multiplying both sides by $1 + i\sigma_x/\omega$, we obtain:

$$b\frac{\partial v_x}{\partial x} + b\frac{\partial v_y}{\partial y}\left(1 + i\frac{\sigma_x}{\omega}\right) = -i\omega u + \sigma_x u \tag{5A.18}$$

$$a\frac{\partial u}{\partial x} = -i\omega v_x + \sigma_x v_x \tag{5A.19}$$

Equation (5A.19) is easy to inverse-Fourier-transform to the time domain since $-i\omega$ becomes a time derivative, just like for the 1-D scalar-wave equation. However, (5A.18) poses a problem because we have an extra $(ib\sigma_x/\omega)\,\partial v_y/\partial y$ term with an explicit $1/\omega$ factor. What do we do with this?

In a Fourier transform, $-i\omega$ corresponds to differentiation; therefore, i/ω corresponds to *integration*. Hence, our problematic $1/\omega$ term is the *integral* of another quantity. In particular, we introduce a new *auxiliary* field variable ψ, satisfying:

$$-i\omega\psi = b\sigma_x \frac{\partial v_y}{\partial y} \tag{5A.20}$$

in which case:

$$b\frac{\partial v_x}{\partial x} + b\frac{\partial v_y}{\partial y} + \psi = -i\omega u + \sigma_x u \tag{5A.21}$$

Now, we can inverse-Fourier-transform everything to the time domain. This yields a set of *four* time-domain equations with PML absorbing boundaries in the x-direction that we can solve by our favorite discretization scheme:

$$\frac{\partial u}{\partial t} = b\nabla \cdot \mathbf{v} - \sigma_x u + \psi \tag{5A.22}$$

$$\frac{\partial v_x}{\partial t} = a\frac{\partial u}{\partial x} - \sigma_x v_x \tag{5A.23}$$

$$\frac{\partial v_y}{\partial t} = a\frac{\partial u}{\partial y} \tag{5A.24}$$

$$\frac{\partial \psi}{\partial t} = b\sigma_x \frac{\partial v_y}{\partial y} \tag{5A.25}$$

Equation (5A.25) is an *auxiliary differential equation* for ψ (with the initial condition, $\psi = 0$). Note that we have σ_x absorption terms in the u and v_x equations, but not for v_y. Hence, the PML corresponds to an *anisotropic absorber*, as if a were replaced by the 2×2 complex tensor:

$$\frac{1}{a}\begin{bmatrix} 1 + i\sigma_x/\omega & \\ & 1 \end{bmatrix}^{-1} \tag{5A.26}$$

This is an example of the general theorem alluded to in the subsection, *Coordinate Transformations and Materials*, in Section 5A.2.

5A.4 PML in Inhomogeneous Media

Media Having Translational Invariance in the PML-Normal Direction

The above derivation for a PML termination in the x-direction does *not* depend on the medium being homogenous in the y- and z-directions. We only require that the medium (and hence the wave equation and its solutions) have *translational invariance* in the x-direction for sufficiently large x.

For example, a PML can terminate a dielectric waveguide oriented in the x-direction and having an x-invariant y-z cross-section. In this example, regardless of the y-z dependence, translational invariance implies that radiating solutions can be decomposed into a sum of functions of the form of (5A.6), $\boldsymbol{W}(y, z)e^{i(kx-\omega t)}$. Note that these solutions, $\boldsymbol{W}$, are no longer plane waves. Instead, they are the *normal modes* of the x-invariant structure, and k is the *propagation constant*. Normal modes are the subject of waveguide theory in electromagnetism, a subject treated extensively elsewhere [49, 50]. Here, the bottom line is this: since the wave equation (and its solution) are still analytic in x, the PML is reflectionless.

While this "bottom line" idea is appealing, certain unusual x-invariant material inhomogeneities in the y-z plane present difficulties in practice. Perhaps the best-known example is the "backward-wave" coaxial waveguide [51–54]. At the same frequency, ω, this structure can simultaneously propagate both "right-handed" modes (where the phase and group velocities in the x-direction have the same sign) and "left-handed" modes (where the phase velocity is *opposite* to the group velocity). In this case, whatever sign one selects for the PML conductivity, σ, either the left-handed or right-handed modes grow exponentially, and the PML fails in a spectacular instability [38]. There is a subtle relationship of this failure to the orientation of the fields for a left-handed mode and the anisotropy of the PML [38]. In this case, one must abandon PML absorbers and use a different technique, such as a scalar conductivity that is turned on sufficiently gradually to adiabatically absorb outgoing waves [9].

Media Lacking Translational Invariance in the PML-Normal Direction

PML fails completely in the case where the medium is not x-invariant (for an x-boundary) [9]. One might ask: why should we care about such cases? At first glance, it appears that if the medium varies in the x-direction, then we surely observe reflections (from the variation) anyway, PML or no PML. However, this is not necessarily true.

There are several important cases of x-varying media that support reflectionless, propagating waves in an infinite, unbounded system. As discussed in Section 5.8, perhaps the simplest example is a dielectric waveguide that obliquely intersects the outer boundary of the computational space. Another, more challenging, case is that of a photonic crystal. For such a periodic medium, there are wave solutions (Bloch waves) that propagate without scattering, and can have very interesting properties that are unattainable in a uniform medium [8].

For any such case, PML seems to be irrevocably spoiled. The central idea behind PML is that the wave equations and their solutions are analytic functions in the direction perpendicular to the outer grid boundary. Hence, they can be analytically continued into the complex coordinate plane. If the medium is varying in the x-direction, it is most likely varying discontinuously, and thus analytic continuation has no foundation.

What can be done in such a case? Conventional analytical ABCs don't work here either, since they are typically designed for homogeneous media. The only fallback is the adiabatic theorem [55]: even a non-PML absorber approaches a reflectionless limit if turned on sufficiently gradually and smoothly. Then, the issue becomes how gradual is gradual enough, yet allowing for a tractable thickness of the absorber [9]. See Appendix 5C for an extensive discussion.

5A.5 PML for Evanescent Waves

The discussion in this appendix considered waves of the form, e^{ikx}, and showed that they become exponentially decaying if we replace x by $x(1 + i\sigma_x/\omega)$, assuming that $\sigma > 0$ and k is real valued (and positive). More generally, the wave equation in 2-D or 3-D allows *evanescent* solutions where k is complex valued; in fact, most commonly where k is purely imaginary.

Consider what happens to a decaying, imaginary-k evanescent wave in the PML medium. Let $k = i\kappa$. Then, in the PML:

$$e^{-\kappa x} \rightarrow e^{-\kappa x - i(\sigma_x/\omega)x} \tag{5A.27}$$

We see that the PML adds an oscillation to the evanescent wave, but does *not* increase its decay rate. While the PML is still reflectionless, it does not help terminate the wave.

Of course, one might object that an evanescent wave is decaying anyway, so we hardly need a PML — we just need to make the computational region large enough and it will vanish on its own. This is true, but it would be nice to accelerate the process, especially if $\kappa = \mathrm{Im}\,\kappa$ is relatively small, and a large grid is needed for the wave to decay sufficiently. Fortunately, this acceleration is possible because nothing in our analysis requires σ_x to be real valued. We can just as easily make σ_x *complex valued*, where $\mathrm{Im}\,\sigma_x < 0$ corresponds to a *real* coordinate stretching. That is, the imaginary part of σ_x can accelerate the decay of evanescent waves in (5A.9) without creating any reflections.

However, adding an imaginary part to σ_x does come at a price: now, *propagating* (real-k) waves in the PML oscillate *faster*. This requires a finer spatial discretization in the PML to limit numerical reflections at the PML surface. In short, everything in moderation.

APPENDIX 5B: REQUIRED AUXILIARY VARIABLES

This appendix reprises that of Ref. [10], which compared the computational cost (in terms of only the number of required auxiliary variables) of the time-domain UPML formulation of Section 5.7 for anisotropic dispersive media to that of other correct PML implementations based on the split-field approach [15, 29]. The convolutional PML (CPML) formulation is omitted from this comparison since its shifted pole requires an extra auxiliary variable [19].

Following [10], we consider for nondispersive, anisotropic media the example of a PML in a single direction, x, neglecting the case of corners of the computational grid where multiple PML directions overlap. We consider only the $\boldsymbol{E}$ and $\boldsymbol{D}$ updates since the $\boldsymbol{H}$ and $\boldsymbol{B}$ updates have an identical form, exactly doubling the storage requirements. The UPML formulation of Section 5.7 nominally requires two additional auxiliary vector fields, $\boldsymbol{U}$ and $\boldsymbol{W}$, amounting to an extra six individual field components in the PML. (Note that, for anisotropic media, separate $\boldsymbol{D}$ and $\boldsymbol{E}$ fields must be stored even in non-PML regions [34] because the $\boldsymbol{E} = [\varepsilon]^{-1}\boldsymbol{D}$ update must be non-local on the Yee grid for stability [44].) However, in a unidirectional PML region, only s_x is non-unity. Hence, we need only store two auxiliary field components, W_x and U_y, from (5.10), whereas $W_{y,z} = E_{y,z}$ and $U_{x,z} = D_{x,z}$ and need not be stored explicitly.

Exactly the same amount of auxiliary storage is required in a split-field PML formulation, as long as appropriate simplifications are made. To review, the split-field formulation can be derived from coordinate stretching in which ∇ is replaced by $\tilde{\nabla}$, where $\tilde{\nabla}_k = (1/s_k)\,\partial/\partial x_k$. Then, the frequency-domain modified Ampere's law, $-i\omega\boldsymbol{D} = \tilde{\nabla}\times\boldsymbol{H}$, is split into three equations:

$$-i\omega s_k \boldsymbol{D}^{(k)} = (\nabla\times\boldsymbol{H})^{(k)} \tag{5B.1}$$

where $\boldsymbol{D} = \Sigma_k \boldsymbol{D}^{(k)}$ and $(\nabla\times\boldsymbol{H})^{(k)}$ denotes the terms of $\nabla\times\boldsymbol{H}$ that only have $\partial/\partial x_k$ derivatives. With the usual stretch factor $s_k = 1 + i\sigma_k/\omega$, Equation (5B.1) contains only first-order terms in $-i\omega$, so that, after inverse Fourier transformation, it yields a first-order ODE in the time domain. The electric field is given by the unmodified constitutive relation $\boldsymbol{E} = [\varepsilon]^{-1}\boldsymbol{D}$. Just as in our unsplit-field PML, this ostensibly requires two auxiliary fields since $\boldsymbol{D}$ is split into three fields.

However, several simplifications occur. First, the $D_k^{(k)}$ component of each auxiliary field can be omitted since $(\nabla \times \boldsymbol{H})_k$ has no x_k derivative. Since only two components of each auxiliary field are stored, this corresponds to a total of six components. Furthermore, if only s_1 is nonunity, then the $\boldsymbol{D}^{(2)}$ and $\boldsymbol{D}^{(3)}$ auxiliary fields can be combined into a single equation for $\boldsymbol{D}^{(2)} + \boldsymbol{D}^{(3)}$, leading to a total of five components that need to be stored, or two additional components, just as for our unsplit formulation above.

One complication should be noted, however. In counting the storage requirements, we have assumed that $\boldsymbol{D}$ is not stored explicitly in the PML regions, but is rather computed as needed from $\boldsymbol{D}^{(k)}$. In this case, the implementation of the $\boldsymbol{E} = [\varepsilon]^{-1}\boldsymbol{D}$ update must be modified in the PML region to account for the splitting, noting that the splitting is different in each PML region if we exploit the optimizations above. This complicates the implementation compared to our unsplit-field formulation, in which the same implementation of the constitutive relations can be used in the PML and non-PML relations (simply passing different arrays for the components to swap $\boldsymbol{W}$ for $\boldsymbol{E}$, and so on).

APPENDIX 5C: PML IN PHOTONIC CRYSTALS

Previous proposed techniques to terminate photonic crystals of infinite extent by simply overlapping a "PML" anisotropic absorber with a crystal's periodic dielectric function [3–7] (including a similar suggestion for integral-equation methods [56]) did not attain the desired effect of a "true" PML medium. This is because the resulting wave reflection does *not* approach zero, even in the limit of infinitely fine grid resolution. We refer to such an absorbing layer as a pseudo-PML (pPML).[6] Nevertheless, previous authors were able to observe acceptably small reflections by overlapping the pPML with many periods of the photonic crystal, and increasing the loss, σ, of the pPML very gradually over this span.

Such absorbing layers are more properly understood as adiabatic absorbers rather than PML media [9]. Indeed, the "PML" property only improves the constant factor in the long-wavelength limit of an effective homogeneous medium, or in any case where there are large homogeneous-material regions compared to the wavelength. Moreover, as described in this appendix, the reflections worsen rapidly as the group velocity decreases (e.g., as a band edge is approached).

5C.1 Conductivity Profile of the pPML

Without loss of generality, let us first define $\sigma(x)$ in a pPML ($x \in [0, L]$) used to terminate a photonic crystal by a *shape function*, $s(u) \in [0,1]$:

---

[6] In the special case of an effectively 1-D medium where there is only a single propagating mode, such as a single-mode waveguide surrounded by a complete-bandgap medium, it is possible to arrange an "impedance-matched" absorber to approximately cancel that one mode [57], or alternatively to specify analytical boundary conditions of zero reflection for that one mode [58]. More generally, in a transfer-matrix or scattering-matrix method where one explicitly computes all propagating modes and expands the fields in that basis, it is possible to impose analytically reflectionless boundary conditions [59, 60], but such methods become very expensive in 3-D.

$$\sigma(x) = \sigma_0 \, s(x/L) \tag{5C.1}$$

where the argument of $s(u)$ is the rescaled coordinate, $u = x/L \in [0,1]$, and σ_0 is an overall amplitude set to achieve some theoretical round-trip reflection, R_0, for normally incident plane waves in a medium of index, n. We can define σ_0 in terms of R_0 and $s(u)$ by:

$$\sigma_0 = \frac{-\ln R_0}{4nL\int_0^1 s(u')du'} \tag{5C.2}$$

For $x < 0$ (outside the pPML), $\sigma = 0$, i.e., $s(u < 0) = 0$. As L is made longer for a fixed $s(u)$, the pPML profile, $\sigma(x)$, turns on more and more gradually, both because $s(u)$ is stretched out and σ_0 decreases.

5C.2 Coupled-Mode Theory

A natural way to analyze waves propagating along a medium that is slowly varying in the propagation direction (say, x) is *coupled-mode theory* (or coupled-wave theory) [50, 55]. Although coupled-mode theory was originally developed for media that are slowly varying in the propagation direction [50], it has been generalized to periodic media with a slowly varying unit cell [55], in which very similar coupled-mode equations appear.

In coupled-mode theory, at each x, one expands the fields in the basis of the eigenmodes (indexed by ℓ) of a uniform structure with that cross-section in terms of expansion coefficients, $c_\ell(x)$. (The eigenmodes have the x-dependence, $e^{i\beta_\ell x}$, for some propagation constants, β_ℓ.) The expansion coefficients, c_ℓ, in this basis are then determined by a set of ordinary differential equations for dc_ℓ/dx coupling the different modes, where the coupling coefficient is proportional to the rate of change [here, the derivative, $s'(x/L)$]. In the limit where the structure varies more and more slowly, the solution approaches an "adiabatic" limit in which the c_ℓ are nearly constant (i.e., no scattering between modes). A similar adiabatic limit has also been derived for slowly varying discrete systems. Using coupled-mode theory, one can derive a universal relationship between the smoothness of the rate of change, $s'(u)$, and the asymptotic rate of convergence to the adiabatic limit. This relationship analytically predicts the convergence rates of the reflection with absorber length.

In this appendix, we omit the derivation of the coupled-mode equations; their general form is considered in detail elsewhere [50, 55]. We simply quote the result: in the limit of slow variation (large L), the equations can be solved to lowest order in $1/L$ in terms of a simple integral. In particular, if the structure is smoothly parameterized by a shape function, $s(x/L)$ (e.g., the absorption profile as given here), then the amplitude, c_r, of a reflected mode (corresponding to a reflected power, $|c_r|^2$) is given to lowest order (for large L) by [55]:

$$c_r(L) = \int_0^1 s'(u)\frac{M[s(u)]}{\Delta\beta[s(u)]} \exp\left\{iL\int_0^u \Delta\beta[s(u')]du'\right\}du \tag{5C.3}$$

Here, M is a coupling coefficient depending on the mode overlap between the incident and reflected field (in the changing part of the structure), and $\Delta\beta \neq 0$ is the difference, $\beta_i - \beta_r$, between the propagation constants of the incident and reflected modes. Both of these are some analytic functions of the shape, $s(u)$. In general, there may be more than one reflected mode.

Furthermore, in a periodic structure, the coefficient for even a single reflected mode is a sum of contributions of the above form from the different Brillouin zones [55]. However, it suffices to analyze the rate of convergence of a single such integral with L. The basic reason for the adiabatic limit is that, as L grows, the phase term oscillates faster and faster, and the integral of this oscillating quantity goes to zero.

5C.3 Convergence Analysis

Many standard methods are available for analyzing the asymptotic (large-L) properties of (5C.3). In particular, we apply a technique that is commonly used to analyze the convergence rate of Fourier series. Here, one simply integrates by parts repeatedly until a nonzero boundary term is obtained [61, 62]. Each integration by parts integrates the $\exp(iL\int\Delta\beta)$ term, dividing the integrand by $iL\Delta\beta(u)$, and differentiates the $s'M/\Delta\beta$ term. If $\Delta\beta$ is real, as in the case of waveguides but not absorbers, then we can turn this expression into a Fourier transform; otherwise, we have to evaluate the expression explicitly as shown next.

After integrating by parts d times, the boundary term at $u = 0$ is zero if the corresponding derivative, $s^{(d)}(0^+)$, is zero. On the other hand, the boundary term at $u = 1$ is always negligible because of the absorption (leading to a complex $\Delta\beta$ and exponential decay), assuming a small round-trip reflection, R_0. The dominant asymptotic term is the *first* (lowest-d) $u = 0$ boundary term that is nonzero, since all subsequent integrations by parts have an additional factor of $1/L$. Here, we have assumed that s is a smooth function in $(0, 1)$, so that there are never delta-function contributions from the interior. In systems with purely decaying solutions (e.g., elliptic equations), mapping the domain, $[0, \infty]$, to $[0, 1]$ requires some care since large reflections arise if the wave oscillations vary too rapidly, exceeding the discrete grid's Nyquist limit [62].

The result is the following asymptotic form for (5C.3), independent of the particular details of the geometry or the modes:

$$c_r(L) = s^{(d)}(0^+)\frac{M(0^+)}{\Delta\beta(0^+)}[-iL\Delta\beta(0^+)]^{-d} + O(L^{-d-1}) \tag{5C.4}$$

where $s^{(d)}(0^+)$ is the first nonzero derivative of $s(u)$ at $u = 0^+$, and integrating by parts d times yielded a division by $(-iL\Delta\beta)^d$ (flipping the sign each time). This result corresponds to what is sometimes called "Darboux's principle," where the convergence is dominated by the lowest-order singularity [62]. Here, this is the first discontinuity in the rate of change, $s'(u)$, at $u = 0$. A similar result applies to the convergence rate of a Fourier series. Namely, a function that has a discontinuity in the dth derivative has a Fourier series whose coefficients, c_n, decrease asymptotically as $1/n^{(d+1)}$ [61, 62] (where $d+1$ appears as the exponent of n instead of d due to the fact that our integral starts with s').

Equation (5C.4) would seem to imply that the reflection $|c_r|^2 \sim O(L^{-2d})$, but this is not the case. This is because there is a hidden $1/L$ factor in the coupling coefficient, M, thanks to the $1/L$ dependence of σ_0 in (5C.2). In fact, M is a matrix element proportional to the rate of change of the materials [55], which in this case is $\partial\sigma/\partial u = s'(u)\sigma_0 \sim 1/L$. Therefore, the reflection scales as $|M|^2/L^{2d} \sim O(L^{-2d-2})$.

Other useful results can be obtained from (5C.4). In particular, one can show that the reflections due to nonuniformity worsen in a periodic structure as a flat band edge, (β_0, ω_0), is approached [63]. As a quadratic-shaped band edge, $\omega - \omega_0 \sim (\beta - \beta_0)^2$, is approached, the group velocity, $v_g = d\omega/d\beta$, scales proportional to $\beta - \beta_0$, while the $\Delta\beta$ between the forward and reflected modes is $2(\beta - \beta_0) \sim v_g$. Also, coupling coefficient M is proportional to $1/v_g$ because of the constant-power normalizations of the incident and reflected modes [55, 63]. Hence, by inspection of (5C.4), $|c_r|^2 \sim O(v_g^{-2d-4})$. For example, the reflection is $O(v_g^{-6})$ for a linear taper, $s(u) = u$ [63]. Because of this unfavorable scaling, an imperfect absorbing layer such as a pPML is most challenging in periodic structures when operating close to a band edge where there are slow-light modes (in the same way that other taper transitions are challenging in this regime [63]).

5C.4 Adiabatic Theorems in Discrete Systems

There is one thing missing from the above analysis, and that is the discretized-space adiabatic case. In a slowly varying discrete system [i.e., sampling some slow change, $s_m = s(m\Delta x/L)$, as L grows larger], there is still a proof of the adiabatic theorem ($c_r \rightarrow 0$), but the only published proof is currently for the lossless case (unitary evolution) [64]. Also, an analogous integral form of the lowest-order reflection has not been presented, nor has the rate of convergence to the adiabatic limit been analyzed in the discrete case. Therefore, our prediction of the asymptotic convergence rate is rigorously proven only for the case of continuous-space wave propagation. However, numerical experiments demonstrate that a slowly changing discretized system exhibits *exactly* the same scaling (e.g., in the PML case for uniform media, where the *only* reflections are due to discretization). This seems analogous to the fact that the discretization error of a discrete Fourier transform converges at the same rate as the decay of the coefficients of the continuous-space Fourier series [62].)

5C.5 Toward Better Absorbers

From the previous section, we observe that a close relationship exists between the smoothness of the absorption profile, $s(u)$, and the asymptotic convergence rate of the reflections, $R(L)$, as a function of the absorber thickness, L. If $s(u)$ has a discontinuity in the dth derivative (e.g., for $s = u^d$), then the reflection coefficient goes as $1/L^{2d+2}$ for a fixed round-trip reflection. This result raises several interesting questions. Can one do better than polynomial convergence? What is the optimal shape, $s(u)$? And what if the round-trip reflection is not fixed?

Smoothness and C_∞ Functions

The above result relating smoothness and convergence has a natural corollary: if $s(u)$ is C_∞, i.e., all of its derivatives are continuous, then the reflection goes to zero faster than any polynomial in $1/L$. This is similar to a well-known result for the convergence of Fourier series of C_∞ functions [62]. The exact rate of faster-than-polynomial convergence again depends on the strongest singularity in $s(u)$. For example, for $s(u) = [\tanh(u)+1]/2$, which goes exponentially to zero as $u \rightarrow -\infty$ and to one as $u \rightarrow +\infty$, the reflection should decrease exponentially with L, as determined by contour integration from the residue of the pole at $u = \pm i\pi/2$ that is closest to the real axis (similar to the analysis for the convergence of a Fourier series for an analytic function [62, 65]). However, such an absorption taper would require an infinitely thick absorber in order to avoid discontinuously truncating the exponential tail of tanh(u).

To have a C_∞ function with a finite absorber, with $s(u) = 0$ for $u \leq 0$, the $s(u)$ function must be non-analytic. A standard example of such a function is $s(u) = \exp(1 - 1/u)$ for $u > 0$, all of whose derivatives go to zero as $u \rightarrow 0^+$, where there is an essential singularity. Because this function is C_∞, its reflection, $R(L)$, must decrease faster than any polynomial. Exactly how much faster than a polynomial is determined by asymptotically evaluating the integral of (5C.3) by a saddle-point method [66, 67]. The result is that $R(L)$ decays asymptotically as $\exp(-\alpha\sqrt{L})$ for some constant $\alpha > 0$ [66]. This has been confirmed in numerical experiments for 1-D uniform and periodic cases.

Although $s(u) = \exp(1 - 1/u)$ yields an exponential decay of $R(L)$ with L, the constant factor and the exponential rate are almost certainly suboptimal for this arbitrary choice of a C_∞ function. Numerical experiments show that this particular choice of C_∞ $s(u)$ is superior to a polynomial $s(u)$ for the periodic case where PML is not perfect, but is inferior for the uniform case until the reflection becomes inconsequential ($\sim 10^{-20}$). This is still a useful result in the sense that one mainly needs to improve pPML for the periodic case, whereas PML is already good enough for uniform media. However, one would ideally prefer a shape function that is consistently better than the polynomial $s(u)$, regardless of the dielectric function. Thus, further exploration of the space of possible absorption profiles seems warranted.

A possibility here would be to design a custom absorber profile with optimal performance that combines the superior constant factor of quadratic absorbers at short taper lengths with the exponential convergence of a C_∞ absorber at long taper lengths. An example of such a function is $s(u) = u(e^{\beta u} - 1)/(e^{\beta} - 1)$, where $\beta = 2L^{0.5}$. In numerical experiments for 1-D uniform, 1-D periodic, and 2-D periodic media in comparison to the simple C_∞ profile, $s(u) = u^{(1 - 1/u)}$, the constant factor of the more complicated custom absorber profile was determined to be lower for small taper lengths (since it approximates a simple quadratic profile in this regime), while both absorber profiles clearly demonstrated exponential convergence as the taper length was increased. This is just a simple demonstration of the utility of custom absorber profiles for adiabatic tapers, and we hope that further research will continue to improve their properties.

Balancing Round-Trip and Transition Reflections

In the above analysis, we fixed the round-trip reflection, R_0, to approximately 10^{-25} in order for our calculations to isolate the effect of the transition reflection. Obviously, in a real application, one is unlikely to require such low reflections and would set R_0 to a larger value, corresponding to a larger $\sigma_0 \sim \ln R_0$ in (5C.2). This would also reduce the transition reflection [as seen from (5C.4)], but only by a logarithmic constant factor.

In principle, the best choice to minimize reflection for a given absorber length is to set R_0 to be roughly equal to the transition reflection for that length. (Another reason to make them equal is the possibility of destructive interference between the round-trip and transition reflections [68], but because such destructive interference is inherently restricted to narrow bandwidths and ranges of incident angles, we will not concern ourselves with this situation.) In order to make them roughly equal, one needs an estimate of the transition reflection. For example, one could simply numerically fit the power law of (5C.4). Numerical experiments in 1-D uniform media show that the overall reflection is reduced by a factor of 300 to 400 compared to a fixed $R_0 = 10^{-16}$. This is a significant reduction, but is not overwhelming (especially for smaller L), and changes the asymptotic convergence rate of (5C.4) only by a factor of $\ln R_0 \sim \ln L$. The drawback of this optimization is that it is difficult to determine the transition reflection analytically for inhomogeneous media. Hence, one is generally forced to make a conservative estimate of R_0, which reduces the advantage gained.

REFERENCES

[1] Bérenger, J.-P., "A perfectly matched layer for the absorption of electromagnetic waves," *J. Computational Physics*, Vol. 114, 1994, pp. 185–200.

[2] Taflove, A., and S. C. Hagness, *Computational Electrodynamics: The Finite-Difference Time-Domain Method*, 3rd ed., Norwood, MA: Artech, 2005.

[3] Koshiba, M., Y. Tsuji, and S. Sasaki, "High-performance absorbing boundary conditions for photonic crystal waveguide simulations," *IEEE Microwave and Wireless Components Lett.*, Vol. 11, 2001, pp. 152–154.

[4] Tsuji, Y., and M. Koshiba, "Finite element method using port truncation by perfectly matched layer boundary conditions for optical waveguide discontinuity problems," *J. Lightwave Technology*, Vol. 20, 2002, pp. 463–468.

[5] Kosmidou, E. P., T. I. Kosmani, and T. D. Tsiboukis, "A comparative FDTD study of various PML configurations for the termination of nonlinear photonic bandgap waveguide structures," *IEEE Trans. Magnetics*, Vol. 39, 2003, pp. 1191–1194.

[6] Weily, A. R., L. Horvath, K. P. Esselle, and B. C. Sanders, "Performance of PML absorbing boundary conditions in 3-D photonic crystal waveguides," *Microwave and Optical Technology Lett.*, Vol. 40, 2004, pp. 1–3.

[7] Kono, N., and M. Koshiba, "General finite-element modeling of 2-D magnetophotonic crystal waveguides," *IEEE Photonics Technology Lett.*, Vol. 17, 2005, pp. 1432–1434.

[8] Joannopoulos, J. D., S. G. Johnson, R. D. Meade, and J. N. Winn, *Photonic Crystals: Molding the Flow of Light*, 2nd ed., Princeton, NJ: Princeton University Press, 2008.

[9] Oskooi, A., L. Zhang, Y. Avniel, and S. G. Johnson, "The failure of perfectly matched layers, and towards their redemption by adiabatic absorbers," *Optics Express*, Vol. 16, 2008, pp. 11376–11392.

[10] Oskooi, A., and S. G. Johnson, "Distinguishing correct from incorrect PML proposals and a corrected unsplit PML for anisotropic, dispersive media," *J. Computational Physics*, Vol. 230, 2011, pp. 2369–2377.

[11] Zhao, A. P., "The limitations of the perfectly matched layers based on E-H fields for arbitrary anisotropic media – Reply to 'Comment on "On the matching conditions of different PML schemes applied to multilayer isotropic dielectric media,"' " *Microwave and Optical Technology Lett.*, Vol. 32, 2002, pp. 237–241.

[12] Li, L., G. P. Nordin, J. M. English, and J. Jiang, "Small-area bends and beamsplitters for low-index-contrast waveguides," *Optics Express*, Vol. 11, 2003, pp. 282–290.

[13] Tai, C. Y., S. H. Chang, and T. C. Chiu, "Design and analysis of an ultra-compact and ultra-wideband polarization beam splitter based on coupled plasmonic waveguide arrays," *IEEE Photonics Technology Lett.*, Vol. 19, 2007, pp. 1448–1450.

[14] Zhao, A. P., "Generalized-material-independent PML absorbers used for the FDTD simulation of electromagnetic waves in 3-D arbitrary anisotropic dielectric and magnetic media," *IEEE Trans. Microwave Theory and Techniques*, Vol. 46, 1998, pp. 1511–1513.

[15] García, S. G., I. V. Pérez, R. G. Martín, and B. G. Olmedo, "Extension of Berenger's PML for bi-isotropic media," *IEEE Microwave and Guided Wave Lett.*, Vol. 8, 1998, pp. 297–299.

[16] Renko, A., and A. P. Zhao, "Extension of the material-independent PML absorbers to arbitrary lossy anisotropic dielectric media," *Microwave and Optical Technology Lett.*, Vol. 20, 1999, pp. 245–249.

[17] Liu, Q. H., "PML and PSTD algorithm for arbitrary lossy anisotropic media," *IEEE Microwave and Guided Wave Lett.*, Vol. 9, 1999, pp. 48–50.

[18] Tong, M. S., Y. Chen, M. Kuzuoglu, and R. Mittra, "A new anisotropic perfectly matched layer medium for mesh truncation in finite difference time domain analysis," *International J. Electronics*, Vol. 86, 1999, pp. 1085–1095.

[19] Roden, J. A., and S. D. Gedney, "Convolution PML (CPML): An efficient FDTD implementation of the CFS-PML for arbitrary media," *Microwave and Optical Technology Lett.*, Vol. 27, 2000, pp. 334–339.

[20] Ramadan, O., and A. Y. Oztoprak, "Z-transform implementation of the perfectly matched layer for truncating FDTD domains," *IEEE Microwave and Wireless Components Lett.*, Vol. 13, 2003, pp. 402–404.

[21] Ward, A. J., and J. B. Pendry, "Refraction and geometry in Maxwell's equations," *J. Modern Optics*, Vol. 43, 1996, pp. 773–793.

[22] Engheta, N., and R. W. Ziolkowski, eds., *Metamaterials: Physics and Engineering Exploration*, IEEE Press, 2006.

[23] Marqués, R., F. Martín, and M. Sorolla, *Metamaterials with Negative Parameters: Theory, Design and Microwave Applications*, New York: Wiley, 2008.

[24] Farjadpour, A., D. Roundy, A. Rodriguez, M. Ibanescu, P. Bermel, J. Joannopoulos, S. Johnson, and G. Burr, "Improving accuracy by sub-pixel smoothing in the finite-difference time domain," *Optics Lett.*, Vol. 31, 2006, pp. 2972–2974.

[25] Kottke, C., A. Farjadpour, and S. G. Johnson, "Perturbation theory for anisotropic dielectric interfaces, and application to sub-pixel smoothing of discretized numerical methods," *Physical Review E*, Vol. 77, 2008, 036611.

[26] Oskooi, A., C. Kottke, and S. G. Johnson, "Accurate finite-difference time-domain simulation of anisotropic media by subpixel smoothing," *Optics Lett.*, Vol. 34, 2009, pp. 2778–2780.

[27] Perez, I. V., S. G. Garcia, R. G. Martin, and B. G. Olmedo, "Extension of Berenger's absorbing boundary conditions to match dielectric anisotropic media," *IEEE Microwave and Guided Wave Lett.*, Vol. 7, 1997, pp. 302–304.

[28] Zhao, A. P., J. Juntunen, and A. V. Raisanen, "Material independent PML absorbers for arbitrary anisotropic dielectric media," *Electronics Lett.*, Vol. 33, 1997, pp. 1535–1536.

[29] Teixeira, F. L., and W. C. Chew, "A general approach to extend Berenger's absorbing boundary condition to anisotropic and dispersive media," *IEEE Trans. Antennas and Propagation*, Vol. 46, 1998, pp. 1386–1387.

[30] Garcia, S. G., J. Juntunen, R. G. Martin, A. P. Zhao, B. G. Olmedo, and A. Raisanen, "A unified look at Berenger's PML for general anisotropic media," *Microwave and Optical Technology Lett.*, Vol. 28, 2001, pp. 302–304.

[31] Li, J., and J. Dai, "Z-transform implementation of the CFS-PML for arbitrary media," *IEEE Microwave and Wireless Components Lett.*, Vol. 16, 2006, pp. 437–439.

[32] Sacks, Z. S., D. M. Kingsland, R. Lee, and J. F. Lee, "A perfectly matched anisotropic absorber for use as an absorbing boundary condition," *IEEE Trans. Antennas and Propagation*, Vol. 43, 1995, pp. 1460–1463.

[33] Gedney, S. D., "An anisotropic perfectly matched layer-absorbing medium for the truncation of FDTD lattices," *IEEE Trans. Antennas and Propagation*, Vol. 44, 1996, pp. 1630–1639.

[34] Oskooi, A., D. Roundy, M. Ibanescu, P. Bermel, J. D. Joannopoulos, and S. G. Johnson, "Meep: A flexible free-software package for electromagnetic simulations by the FDTD method," *Comput. Physics Commun.*, Vol. 181, 2010, pp. 687–702.

[35] Chew, W. C., and W. H. Wheedon, "A 3D perfectly matched medium from modified Maxwell's equations with stretched coordinates," *Microwave and Optical Technology Lett.*, Vol. 7, 1994, pp. 599–604.

[36] Rappaport, C. M., "Perfectly matched absorbing boundary conditions based on anisotropic lossy mapping of space," *IEEE Microwave and Guided Wave Lett.*, Vol. 5, 1995, pp. 90–92.

[37] Zhao, L., and A. C. Cangellaris, "A general approach for the development of unsplit-field time-domain implementations of perfectly matched layers for FDTD grid truncation," *IEEE Microwave and Guided Wave Lett.*, Vol. 6, 1996, pp. 209–211.

[38] Loh, P.-R., A. F. Oskooi, M. Ibanescu, M. Skorobogatiy, and S. G. Johnson, "Fundamental relation between phase and group velocity, and application to the failure of perfectly matched layers in backward-wave structures," *Physical Review E*, Vol. 79, 2009, 065601(R).

[39] Fang, J., and Z. Wu, "Generalized perfectly matched layer for the absorption of propagating and evanescent waves in lossless and lossy media," *IEEE Trans. Microwave Theory and Techniques*, Vol. 44, 1998, pp. 2216–2222.

[40] Kuzuoglu, M., and R. Mittra, "Frequency dependence of the constitutive parameters of causal perfectly matched anisotropic absorbers," *IEEE Microwave and Guided Wave Lett.*, Vol. 6, 1996, pp. 447–449.

[41] Teixeira, F. L., and W. C. Chew, "General closed-form PML constitutive tensors to match arbitrary bianisotropic and dispersive linear media," *IEEE Microwave and Guided Wave Lett.*, Vol. 8, 1998, pp. 223–225.

[42] Barrett, R., M. Berry, T. Chan, J. Demmel, J. Donato, J. Dongarra, V. Eijkhout, R. Pozo, C. Romine, and H.V. der Vorst, *Templates for the Solution of Linear Systems: Building Blocks for Iterative Methods*, Philadelphia, PA: SIAM, 1994.

[43] Oppenheim, A. V., R. W. Schafer, and J. R. Buck, *Discrete-Time Signal Processing*, 2nd ed., Upper Saddle River, NJ: Prentice-Hall, 1999.

[44] Werner, G., and J. Cary, "A stable FDTD algorithm for non-diagonal anisotropic dielectrics," *J. Computational Physics*, Vol. 226, 2007, pp. 1085–1101.

[45] Johnson, S. G., "Notes on the algebraic structure of wave equations," 2007. Online: http://math.mit.edu/~stevenj/18.369/wave-equations.pdf

[46] Cummer, S. A., "Perfectly matched layer behavior in negative refractive index materials," *IEEE Antennas and Wireless Propagation Lett.*, Vol. 3, 2004, pp. 172–175.

[47] Dong, X. T., X. S. Rao, Y. B. Gan, B. Guo, and W. Y. Yin, "Perfectly matched layer-absorbing boundary condition for left-handed materials," *IEEE Microwave and Wireless Components Lett.*, Vol. 14, 2004, pp. 301–303.

[48] Johnson, S. G., "Coordinate transformation and invariance in electromagnetism: Notes for Course 18.369 at MIT," 2007. Online: http://math.mit. edu/~stevenj/18.369/coordinate-transform.pdf

[49] Snyder, A. W., "Radiation losses due to variations of radius on dielectric or optical fibers," *IEEE Trans. Microwave Theory and Techniques*, Vol. 18, 1970, pp. 608–615.

[50] Marcuse, D., *Theory of Dielectric Optical Waveguides*, 2nd ed., San Diego, CA: Academic Press, 1991.

[51] Clarricoats, P. J. B., and R. A. Waldron, "Non-periodic slow-wave and backward-wave structures," *J. Electronic Contr.*, Vol. 8, 1960, pp. 455–458.

[52] Waldron, R. A., "Theory and potential applications of backward waves in nonperiodic inhomogeneous waveguides," *Proc. IEE*, Vol. 111, 1964, pp. 1659–1667.

[53] Omar, A. S., and K. F. Schunemann, "Complex and backward-wave modes in inhomogeneously and anisotropically filled waveguides," *IEEE Trans. Microwave Theory and Techniques*, Vol. 35, 1987, pp. 268–275.

[54] Ibanescu, M., S. G. Johnson, D. Roundy, C. Luo, Y. Fink, and J. D. Joannopoulos, "Anomalous dispersion relations by symmetry breaking in axially uniform waveguides," *Physical Review Lett.*, Vol. 92, 2004, 063903.

[55] Johnson, S. G., P. Bienstman, M. Skorobogatiy, M. Ibanescu, E. Lidorikis, and J. D. Joannopoulos, "Adiabatic theorem and continuous coupled-mode theory for efficient taper transitions in photonic crystals," *Physical Review E*, Vol. 66, 2002, 066608.

[56] Pissoort, D., and F. Olyslager, "Termination of periodic waveguides by PMLs in time-harmonic integral equation-like techniques," *IEEE Antennas and Wireless Propagation Lett.*, Vol. 2, 2003, pp. 281–284.

[57] Mekis, A., S. Fan, and J. D. Joannopoulos, "Absorbing boundary conditions for FDTD simulations of photonic crystal waveguides," *IEEE Microwave and Guided Wave Lett.*, Vol. 9, 1999, pp. 502–504.

[58] Moreno, E., D. Erni, and C. Hafner, "Modeling of discontinuities in photonic crystal waveguides with the multiple multipole method," *Physical Review E*, Vol. 66, 2002, 036618.

[59] Li, Z.-Y., and K.-M. Ho, "Light propagation in semi-infinite photonic crystals and related waveguide structures," *Physical Review B*, Vol. 68, 2003, 155101.

[60] Pissoort, D., B. Denecker, P. Bienstman, F. Olyslager, and D. De Zutter, "Comparative study of three methods for the simulation of two-dimensional photonic crystals," *J. Optical Society of America A*, Vol. 21, 2004, pp. 2186–2195.

[61] Mead, K. O., and L. M. Delves, "On the convergence rate of generalized Fourier expansions," *IMA J. Applied Mathematics*, Vol. 12, 1973, pp. 247–259.

[62] Boyd, J. P., *Chebyshev and Fourier Spectral Methods,* 2nd ed., Springer, 1989.

[63] Povinelli, M., S. Johnson, and J. Joannopoulos, "Slow-light, band-edge waveguides for tunable time delays," *Optics Express*, Vol. 13, 2005, pp. 7145–7159.

[64] Dranov, A., J. Kellendonk, and R. Seller, "Discrete time adiabatic theorems for quantum mechanical systems," *J. Mathematical Physics*, Vol. 39, 1998, pp. 1340–1349.

[65] Elliott, D., "The evaluation and estimation of the coefficients in the Chebyshev series expansion of a function," *Mathematics of Computation*, Vol. 18, 1964, pp. 274–284.

[66] Boyd, J. P., "The optimization of convergence for Chebyshev polynomial methods in an unbounded domain," *J. Computational Physics*, Vol. 45, 1982, pp. 43–79.

[67] Cheng, H., *Advanced Analytic Methods in Applied Mathematics, Science and Engineering*, Boston: Luban Press, 2006.

[68] Juntunen, J. S., N. V. Kantartzis, and T. D. Tsiboukis, "Zero reflection coefficient in discretized PML," *IEEE Microwave and Wireless Components Lett.*, Vol. 11, 2001, pp. 155–157.

SELECTED BIBLIOGRAPHY

A tutorial discussion along the lines presented in Appendix 5A.2 has also been provided in the following reference:

Teixeira, F. L., and W. C. Chew, "Advances in the theory of perfectly matched layers," pp. 283–346 in *Fast and Efficient Algorithms in Computational Electromagnetics*, W. C. Chew, J.-M. Jin, E. Michielssen, and J. Song, eds., Norwood, MA: Artech House, 2001.

This reference discusses in considerable detail the connection between the PML and the foundation of what is now termed "transformation optics."

Regarding the issue of PML reflections considered in the appendices, the following paper should also be noted:

Chew, W. C., and J. M. Jin, "Perfectly matched layers in discretized space: An analysis and optimization," *Electromagnetics*, Vol. 16, 1996, pp. 325–340.

This paper provides a detailed analysis of PML reflections in the discretized space, including an explicit expression for the single-interface reflection coefficient.

Chapter 6

Accurate FDTD Simulation of Discontinuous Materials by Subpixel Smoothing[1]

Ardavan Oskooi and Steven G. Johnson

6.1 INTRODUCTION

The standard FDTD method discretizes Maxwell's equations on a uniform Cartesian grid in space and employs leapfrog stepping in time [1]. Because FDTD is robust and computationally efficient, especially when run on parallel-processing computers, it has become a popular numerical simulation tool for photonics. FDTD enjoys significant advantages relative to competing numerical techniques for simulating broadband device characteristics and the dynamics of propagating waves in the presence of dispersive and nonlinear materials.

However, standard FDTD has a key drawback: it aliases material interfaces mapped onto the Cartesian grid. That is, suppose we are discretizing the permittivity, ε (or permeability, μ), on a two-dimensional (2-D) Yee grid of "pixels" or a three-dimensional (3-D) Yee lattice of "voxels" with uniform spacing, Δx. In the common case where ε is piecewise constant, what values of ε do we assign to pixels that cross a sharp interface between ε^a and ε^b? Simply assigning the values of ε observed at the centers of these pixels leads to an irregular "staircased" approximation of a slanted or curved interface. In addition, serious problems can arise even for interfaces aligned with the grid (and hence discretized as flat interfaces with no staircasing).

We will see in this chapter that, in general, discretizing a sharp material interface *degrades* the accuracy of FDTD. Nominally, the central-difference approximations of standard FDTD have errors that decrease as $O(\Delta x)^2$ (second-order accuracy) [1]. However, naively discretizing a discontinuous ε leads to mean-square errors that decrease only as $O(\Delta x)$ (first-order accuracy) [2]. The basic problem is that the derivation of the second-order accuracy relies on Taylor-series expansions of the fields [1]. Unfortunately, a discontinuity in the materials yields discontinuities in the fields (e.g., $\boldsymbol{E}_\perp$ is discontinuous at an ε interface) that spoil the Taylor expansions (unless we happen to have a polarization everywhere parallel to the interfaces).

---

[1] This chapter is a synthesis derived from Refs. [6–8]: (1) A. Farjadpour, D. Roundy, A. Rodriguez, M. Ibanescu, P. Bermel, J. Joannopoulos, S. Johnson, and G. Burr, "Improving accuracy by sub-pixel smoothing in the finite-difference time domain," *Optics Lett.*, Vol. 31, 2006, pp. 2972–2974, ©2006 The Optical Society of America; (2) C. Kottke, A. Farjadpour, and S. G. Johnson, "Perturbation theory for anisotropic dielectric interfaces, and application to sub-pixel smoothing of discretized numerical methods," *Physical Review E*, Vol. 77, 2008, 036611, ©2008 The American Physical Society; and (3) A. Oskooi, C. Kottke, and S. G. Johnson, "Accurate finite-difference time-domain simulation of anisotropic media by subpixel smoothing," *Optics Lett.*, Vol. 34, 2009, pp. 2778–2780, ©2009 The Optical Society of America.

To mitigate these problems in the FDTD context, a number of schemes employing smoothed permittivities at material interfaces have been proposed [3–5]. These schemes assign to each pixel some effective ε based on the materials and interfaces in and around the pixel. The effective ε can then vary continuously with geometry. However, such smoothing perturbs the problem being solved, changing the original discontinuous geometry to a smoothed geometry. Thus, smoothing could actually increase the error.

In this chapter, we review a *subpixel smoothing* technique that (in the absence of sharp corners of the modeled structure) recovers second-order accuracy for arbitrary, discontinuous, possibly even anisotropic ε and μ distributions [6–8]. Adapted from spectral methods [9–11], subpixel smoothing has two additional benefits. Namely, it allows the simulation to respond continuously to changes in the geometry, such as during optimization or parameter studies, rather than changing in discontinuous jumps as material interfaces cross pixel boundaries. In addition, this technique yields much smoother convergence of the error with grid resolution, which makes it easier to evaluate the accuracy of a simulation, and furthermore enables the possibility of extrapolation to gain another order of accuracy. Unlike approaches that require complicated position-dependent difference equations, subpixel smoothing uses standard FDTD central differences and is easy to implement by simply preprocessing the ε and μ arrays. The subpixel-smoothing techniques of [6–8] consistently achieve the smallest errors compared with previous smoothing schemes for FDTD [3–5], and free code is available (see [12] and Chapter 20).

6.2 DIELECTRIC INTERFACE GEOMETRY

Consider an interface between two dielectric materials characterized by the permittivities ε^a and ε^b crossing a pixel in a classic (Yee-cell) FDTD grid. This interface is assumed to be locally flat for small pixels, deferring the question of corners until later. Reference [8] defined a coordinate frame relative to the interface as shown in Fig. 6.1. Here, the first component "1" is the direction normal to the interface, which also defines the direction of the unit-normal vector $\hat{\boldsymbol{n}}$.

6.3 PERMITTIVITY SMOOTHING RELATION, ISOTROPIC INTERFACE CASE

For an interface between two isotropic dielectrics, Ref. [6] reported that the proper smoothed permittivity at each point in the coordinate frame defined in Fig. 6.1 is given by:

$$\tilde{\boldsymbol{\varepsilon}} = \begin{bmatrix} \langle \varepsilon^{-1} \rangle^{-1} & 0 & 0 \\ 0 & \langle \varepsilon \rangle & 0 \\ 0 & 0 & \langle \varepsilon \rangle \end{bmatrix} \tag{6.1}$$

where $\langle \cdots \rangle$ denotes an average over the voxel $s\Delta x \times s\Delta y \times s\Delta z$ (in 3-D) surrounding the grid point in question, and s is a smoothing diameter in units of the grid spacing ($s = 1$ except where noted). This smoothing uses the mean $\langle \varepsilon \rangle$ for the surface-parallel $\boldsymbol{E}$ components, and the harmonic mean $\langle \varepsilon^{-1} \rangle^{-1}$ for the surface-perpendicular component. It has the nice property of being Hermitian for real scalar ε, and equals ε for homogeneous pixels. Rotation of $\tilde{\boldsymbol{\varepsilon}}$ of (6.1) to an $\boldsymbol{\varepsilon}$ tensor that is specified in the Cartesian coordinates of the FDTD grid permits the FDTD components of $\boldsymbol{E}$ to be obtained via $\boldsymbol{E} = \boldsymbol{\varepsilon}^{-1}\boldsymbol{D}$.

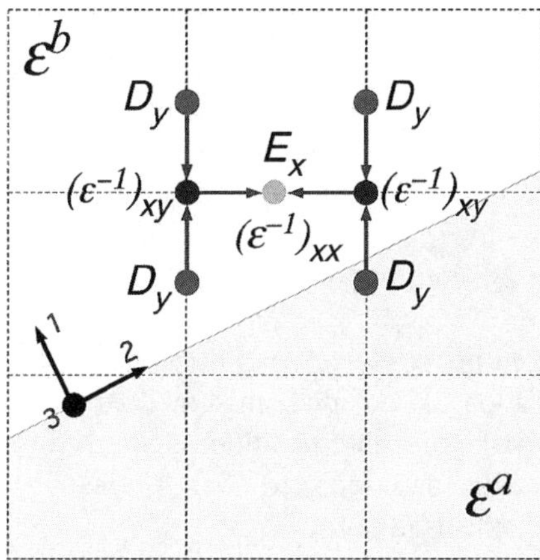

Fig. 6.1 Schematic of a 2-D Yee-grid FDTD discretization near the interface of two dielectric materials characterized by the permittivities ε^a and ε^b. This illustrates a numerically stable method used to compute the part of E_x that comes from the adjacent D_y components, and the locations where various ε^{-1} components are required. *Source:* A. Oskooi, C. Kottke, and S. G. Johnson, *Optics Lett.*, Vol. 34, 2009, pp. 2778–2780, ©2009 The Optical Society of America.

Equation (6.1) corresponds to discretizing a smoothed version of Maxwell's equations, where ε or its inverse has been anisotropically convolved with a boxlike smoothing kernel. It was proposed for use with a plane-wave method [9–11], based on effective-medium considerations, but this and other schemes can be evaluated by a simple criterion from perturbation theory (Appendix 6A). In particular, even before the problem is discretized, smoothing ε causes the solution to be perturbed, and this perturbation can be analyzed via methods recently developed for high-contrast interfaces [13, 14]. To minimize this smoothing error, one can simply require that the error be zero to first order in the smoothing diameter, s, as $s \to 0$. The remaining smoothing errors will be quadratic in the resolution (except in singular cases as discussed below).

Near a flat interface, the effect of a perturbation, $\Delta\varepsilon$, on computed quantities such as eigenfrequencies [13] or scattered powers [14] is proportional to $\Delta\varepsilon|\boldsymbol{E}_{\parallel}|^2 - \Delta(\varepsilon^{-1})|\boldsymbol{D}_{\perp}|^2$, where $\boldsymbol{E}_{\parallel}$ and $\boldsymbol{D}_{\perp}$ are the continuous surface-parallel and surface-perpendicular components of $\boldsymbol{E}$ and $\boldsymbol{D}$, respectively. For the first-order change to be zero, $\Delta\varepsilon$ must be a tensor such that $\Delta\varepsilon_{\parallel}$ and $\Delta\varepsilon_{\perp}^{-1}$ both integrate to zero (e.g., are equal and opposite on the two sides of the interface). Equation (6.1) is a simple choice satisfying this condition.

6.4 FIELD COMPONENT INTERPOLATION FOR NUMERICAL STABILITY

The spatially staggered location of field components in a Yee FDTD grid poses a challenge in implementing (6.1) for isotropic dielectric interfaces, as well as for the corresponding smoothing relation in Section 6.6 for anisotropic dielectric interfaces. At each time-step, $\boldsymbol{E} = \boldsymbol{\varepsilon}^{-1}\boldsymbol{D}$ must be computed, where $\boldsymbol{\varepsilon}$ is the $\tilde{\boldsymbol{\varepsilon}}$ of (6.1) rotated to the Cartesian FDTD grid coordinates. However, any off-diagonal elements of $\boldsymbol{\varepsilon}$ couple components of $\boldsymbol{E}$ stored at different grid locations. For example, a nonzero $(\varepsilon_{xy})^{-1}$ means that computing E_x requires D_y, but D_y is not available at the same point as E_x, as illustrated in Fig. 6.1. This mandates the use of some interpolation scheme.

Continuing in this example, one approach is to average the four adjacent D_y values and use them in updating E_x, along with $(\varepsilon_{xy})^{-1}$ at the E_x point [6]. However, Ref. [15] showed that this approach is theoretically unstable and leads to divergences for long-running simulations. Instead, Ref. [15] reported a modified interpolation technique that satisfies a necessary condition for stability with Hermitian $\boldsymbol{\varepsilon}$. Using this technique, as depicted in Fig. 6.1, one first averages D_y at $[i, j\pm0.5, k]$ and multiplies by $(\varepsilon_{xy})^{-1}$ at $[i, j, k]$; then averages D_y at $[i+1, j\pm0.5, k]$ and multiplies by $(\varepsilon_{xy})^{-1}$ at $[i+1, j, k]$; and finally averages the two results at $[i, j, k]$ and $[i+1, j, k]$ to update E_x at $[i+0.5, j, k]$.

Although Ref. [15] derived no *sufficient* condition for stability with inhomogeneous media once the Yee time discretization is included, this method has been stable in all numerical experiments to date. An analogous interpolation is used for the E_y and E_z components. Reference [8] used this scheme, and reported that it greatly improved numerical stability compared with the simpler scheme used in [6].

6.5 CONVERGENCE STUDY, ISOTROPIC INTERFACE CASE

As noted in Section 6.3, near a flat interface, to ensure that a perturbation, $\Delta\varepsilon$, yields zero first-order change on computed quantities such as eigenfrequencies or scattered powers, $\Delta\varepsilon$ must be a tensor such that $\Delta\varepsilon_{\parallel}$ and $\Delta\varepsilon_{\perp}^{-1}$ both integrate to zero (e.g., are equal and opposite on the two sides of the interface). While (6.1) satisfies this condition, previous smoothing schemes do not, and are therefore expected to have only linear convergence, in general. In fact, they may even have worse errors than unsmoothed FDTD.

In particular, Ref. [6] compared the convergence of (6.1) with that of three other proposed smoothing approaches: (1) the simplest, where the scalar mean, $\langle\varepsilon\rangle$, is used for all components [4]; (2) the anisotropic smoothing of Kaneda *et al.* [3]; and (3) the VP-EP scheme [5]. Using $\langle\varepsilon\rangle$ for all components immediately leads to an incorrect result for the surface-normal fields. Furthermore, while both the Kaneda and VP-EP schemes are equivalent to (6.1) for flat interfaces oriented along the grid's major (x, y, or z) directions, these schemes do not satisfy the perturbation criterion for diagonal interfaces. Yet another method [16] was found to be numerically unstable [6], which prevented the evaluation of its convergence. However, it is equivalent to (6.1) only for flat $x/y/z$ interfaces. Other schemes, not considered in [6], were developed for perfect conductors [1, 17] or for non-Yee lattices in 2-D [18].

Figure 6.2 illustrates the geometry and results of the first convergence study reported in [6], involving a 2-D photonic crystal excited with $\boldsymbol{E}$ in the plane of the crystal [19]. This was selected to challenge the Kaneda and VP-EP schemes, which are equivalent to (6.1) only for grid-parallel interfaces (with quadratic convergence obtained for all these methods). Here, as shown in the inset of Fig. 6.2, the photonic crystal was comprised of a periodic square lattice of elliptical air holes in a background medium of relative permittivity $\varepsilon_r = 12$.

To evaluate the discretization error, Ref. [6] computed the smallest eigenfrequency, ω, for an arbitrarily chosen Bloch wavevector, $\boldsymbol{k}$ (not aligned with the grid), so that the wavelength was comparable with the feature sizes. FDTD simulations were performed with Bloch-periodic boundaries and a Gaussian pulse source, and the responses were analyzed with a filter-diagonalization method [20]. This was compared with the exact ω_0 from a plane-wave calculation [11] at a very high resolution, to obtain the relative error, $|\omega-\omega_0|/\omega_0$, versus grid resolution. The term ω is a good proxy for other common computations, because both the change in the frequency and the scattered power for a small $\Delta\varepsilon$ vary as $\Delta\varepsilon|\boldsymbol{E}|^2$ to lowest order [15].

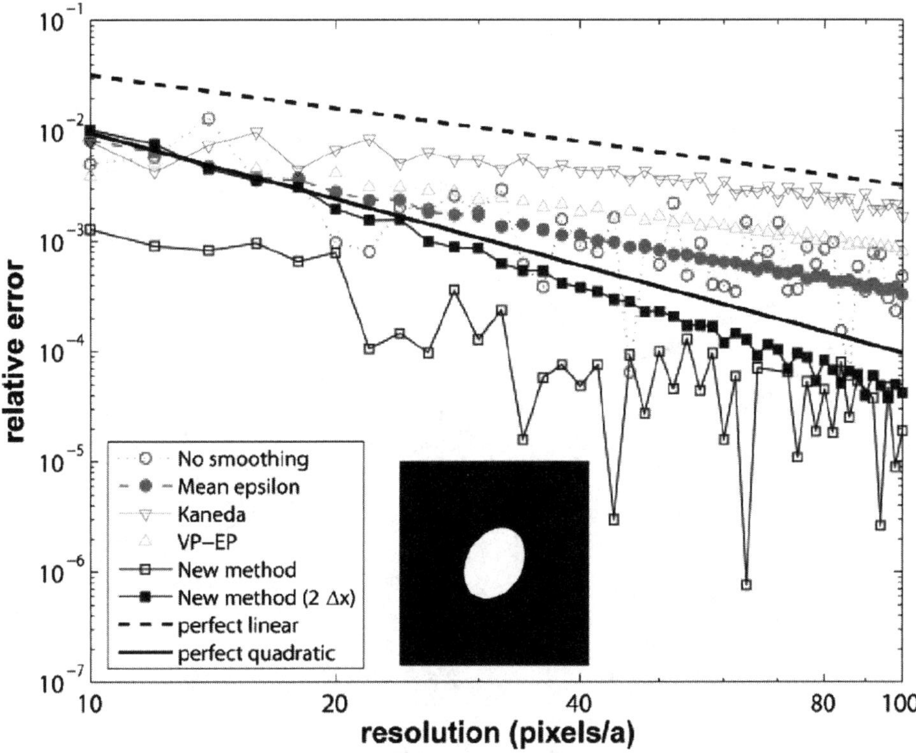

Fig. 6.2 Convergence of the FDTD-computed eigenfrequency error vs. grid resolution for a square 2-D lattice of elliptical air holes in an $\varepsilon_r = 12$ background medium (inset) with $\boldsymbol{E}$ in the 2-D plane for several proposed dielectric-interface smoothing techniques. *Source:* A. Farjadpour, D. Roundy, A. Rodriguez, M. Ibanescu, P. Bermel, J. Joannopoulos, S. Johnson, and G. Burr, *Optics Letters,* Vol. 31, 2006, pp. 2972–2974, ©2006 The Optical Society of America.

Figure 6.2 shows that the new smoothing method of (6.1) (hollow squares) reported in [6] provided the smallest errors by a large margin, while the Kaneda and VP-EP methods were actually worse than no smoothing. All methods except that of (6.1) converged linearly, whereas the method of (6.1) was found to be asymptotically quadratic. As a device to make this quadratic convergence more apparent, results were included in Fig. 6.2 for a doubling of the smoothing diameter to $s = 2$ (filled squares), at the expense of increasing the absolute error [6].

Reference [6] noted that, for the 2-D photonic crystal of Fig. 6.2 excited with $\boldsymbol{E}$ out of its plane, all of the smoothing methods became equivalent to the simple mean ε. For this case, all of the smoothing methods decreased the error compared with no smoothing, and all of the methods (including no smoothing) exhibited quadratic convergence. Since $\boldsymbol{E}$ is everywhere continuous for this polarization, numerical computations (and perturbative methods [13, 14]) are easier than for the in-plane $\boldsymbol{E}$ case.

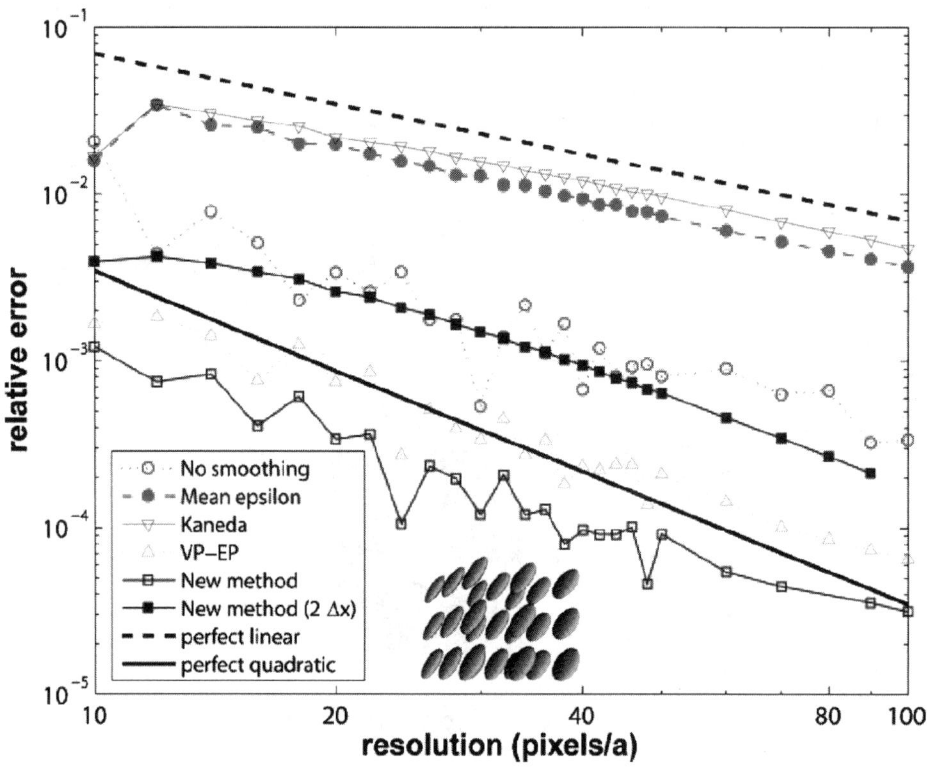

Fig. 6.3 Convergence of the FDTD-computed eigenfrequency error vs. grid resolution for a cubic 3-D lattice of $\varepsilon_r = 12$ ellipsoids in air (inset) for several proposed dielectric-interface smoothing techniques. *Source:* A. Farjadpour, D. Roundy, A. Rodriguez, M. Ibanescu, P. Bermel, J. Joannopoulos, S. Johnson, and G. Burr, *Optics Letters,* Vol. 31, 2006, pp. 2972–2974, ©2006 The Optical Society of America.

Figure 6.3 illustrates the geometry and results of the second convergence study reported in [6], involving (as shown in the inset of this figure) a 3-D cubic lattice of $\varepsilon_r = 12$ ellipsoids with an arbitrary orientation in air. The results in Fig. 6.3 show that the new smoothing method of (6.1) (hollow squares) reported in [6] again yielded the smallest error, and again converged quadratically with grid resolution. Reference [6] noted that the ordering of the other methods changed from that of Fig. 6.2, and in general provided erratic accuracy.

Figure 6.4 illustrates the geometry and results of the final convergence study reported in [6], involving a square 2-D lattice of tilted-square air holes in an $\varepsilon_r = 12$ background medium. For this case, none of the smoothing methods being studied satisfied the zero-perturbation criterion of [6]. This was a consequence of the field singularities generated by the sharp corners of the air holes. Because the new method of (6.1) reported in [6] at least properly modeled the flat edges of the air holes, it still had a lower error than the other smoothing schemes, although its suboptimal handling of the corners limited the differences. Fits of the computed data indicated that the method of (6.1) converged as $\Delta x^{1.4}$, which could be predicted analytically.

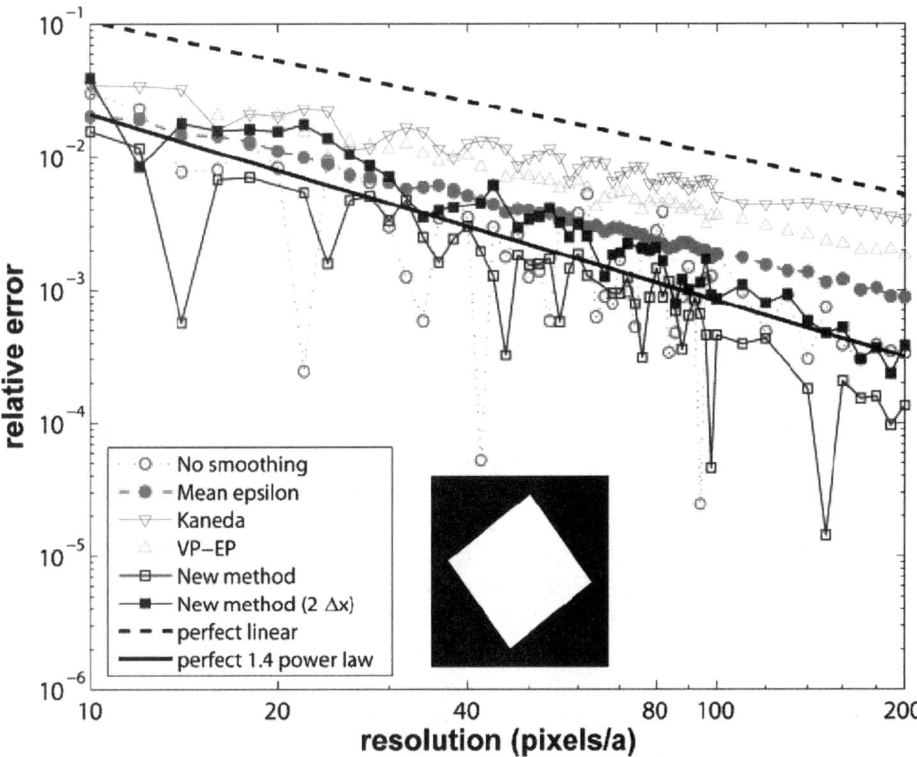

Fig. 6.4 Convergence of the FDTD-computed eigenfrequency error vs. grid resolution for a square 2-D lattice of tilted-square air holes in an $\varepsilon_r = 12$ background medium (inset) with $\boldsymbol{E}$ in the 2-D plane for several proposed dielectric-interface smoothing techniques. *Source:* A. Farjadpour, D. Roundy, A. Rodriguez, M. Ibanescu, P. Bermel, J. Joannopoulos, S. Johnson, and G. Burr, *Optics Letters,* Vol. 31, 2006, pp. 2972–2974, ©2006 The Optical Society of America.

Quite generally, any corner leads to a singularity where $\boldsymbol{E}$ diverges as r^{p-1} for a radius, r, from the corner, with p given by a transcendental equation in the corner angle and ε values (in [6], $p \approx 0.702$ [21]). This led to a perturbation in the frequency $\sim \int_\Delta |\boldsymbol{E}|^2\, r dr \sim \Delta r^{2p} \approx \Delta r^{1.404}$, where Δr was the size of the perturbation (the pixel). In contrast, the other smoothing schemes were limited by the linear error from the flat interfaces.

6.6 PERMITTIVITY SMOOTHING RELATION, ANISOTROPIC INTERFACE CASE

For an interface between anisotropic materials, Ref. [7] rigorously derived an appropriate subpixel-smoothing scheme having zero first-order perturbation. Referring to the dielectric interface geometry depicted in Fig. 6.1, this scheme is given by:

$$\tilde{\boldsymbol{\varepsilon}} = \boldsymbol{\tau}^{-1}\left[\langle \boldsymbol{\tau}(\boldsymbol{\varepsilon}) \rangle\right] \tag{6.2}$$

where $\boldsymbol{\tau}(\boldsymbol{\varepsilon})$ and its inverse are defined by:

$$\boldsymbol{\tau}(\boldsymbol{\varepsilon}) = \frac{1}{\varepsilon_{11}} \begin{bmatrix} -1 & \varepsilon_{12} & \varepsilon_{13} \\ \varepsilon_{21} & \varepsilon_{11}\varepsilon_{22} - \varepsilon_{21}\varepsilon_{12} & \varepsilon_{11}\varepsilon_{23} - \varepsilon_{21}\varepsilon_{13} \\ \varepsilon_{31} & \varepsilon_{11}\varepsilon_{32} - \varepsilon_{31}\varepsilon_{12} & \varepsilon_{11}\varepsilon_{33} - \varepsilon_{31}\varepsilon_{13} \end{bmatrix} \tag{6.3a}$$

$$\boldsymbol{\tau}^{-1}[\boldsymbol{\tau}] = \frac{1}{\tau_{11}} \begin{bmatrix} -1 & -\tau_{12} & -\tau_{13} \\ -\tau_{21} & \tau_{11}\tau_{22} - \tau_{21}\tau_{12} & \tau_{11}\tau_{23} - \tau_{21}\tau_{13} \\ -\tau_{31} & \tau_{11}\tau_{32} - \tau_{31}\tau_{12} & \tau_{11}\tau_{33} - \tau_{31}\tau_{13} \end{bmatrix} \tag{6.3b}$$

Appendix 6A provides a brief overview of the derivation of this result. Two key points should be noted here. First, smoothing relation (6.1) is obtained from (6.2) as a special case for isotropic ε. Second, just as for the isotropic case, at each field point of interest, the smoothed $\tilde{\boldsymbol{\varepsilon}}$ tensor obtained from (6.2) must be rotated from the interface-normal coordinate system shown in Fig. 6.1 to yield a tensor $\boldsymbol{\varepsilon}$ oriented in the local Cartesian coordinates of the FDTD grid [8].

For each Yee cell in three dimensions, subpixel averaging is performed four times, obtaining $(\boldsymbol{\varepsilon}^{-1})_{xx}$ at $[i+0.5, j, k]$, $(\boldsymbol{\varepsilon}^{-1})_{yy}$ at $[i, j+0.5, k]$, $(\boldsymbol{\varepsilon}^{-1})_{zz}$ at $[i, j, k+0.5]$, and all of the off-diagonal components of $\boldsymbol{\varepsilon}$ at $[i, j, k]$. In other words, the averaging procedure of (6.2) is applied to pixels centered about different points/corners in the Yee cell, and then for each point, only the components of $\boldsymbol{\varepsilon}^{-1}$ necessary for that point are stored [8]. Hence, each component of $\boldsymbol{\varepsilon}^{-1}$ need only be stored at most once per Yee cell, so no additional storage is required compared to other anisotropic FDTD schemes. Furthermore, this permits the field component interpolation required for numerical stability, reviewed in Section 6.4. After smoothing is completed, the regular anisotropic FDTD time-stepping scheme proceeds without modification [8].

As a specific example, at the E_x point $[i+0.5, j, k]$ (the orange dot in Fig. 6.1), the smoothed $\tilde{\boldsymbol{\varepsilon}}$ is computed with (6.2) by averaging over the pixel centered at that point. Then, this $\tilde{\boldsymbol{\varepsilon}}$ is rotated to the local FDTD grid coordinates, and inverted to obtain $(\boldsymbol{\varepsilon}^{-1})_{xx}$, which is stored at $[i+0.5, j, k]$. Similar subpixel averaging is performed for a pixel centered at the $[i, j, k]$ point (blue dot) located halfway between two D_y points (red dots), and $(\boldsymbol{\varepsilon}^{-1})_{xy}$ is computed and stored at that point. This is repeated for the other field components [8].

6.7 CONVERGENCE STUDY, ANISOTROPIC INTERFACE CASE

Reference [8] reported the study shown in Figs. 6.5 and 6.6 of the convergence of the discretization error of the anisotropic subpixel-smoothing scheme of (6.2) vs. FDTD grid resolution. This study involved computing an eigenfrequency, ω, of a periodic 2-D square or 3-D cubic lattice (period = a) of dielectric ellipsoids of permittivity, $\boldsymbol{\varepsilon}^a$, surrounded by $\boldsymbol{\varepsilon}^b$, i.e., a photonic crystal [19]. Here, the tensor permittivities, $\boldsymbol{\varepsilon}^a$ and $\boldsymbol{\varepsilon}^b$, were chosen to be random positive-definite symmetric matrices. Matrix $\boldsymbol{\varepsilon}^a$ had the random eigenvalues 1.45, 2.81, and 4.98 in the interval {1, 5}; and matrix $\boldsymbol{\varepsilon}^b$ had the random eigenvalues 8.49, 8.78, and 11.52 in the interval {9, 13}. The lowest eigenfrequency, ω, was computed for an arbitrary Bloch wavevector, $\boldsymbol{k} = [0.4, 0.2, 0.3]\, 2\pi/a$, giving wavelengths comparable to the feature sizes.

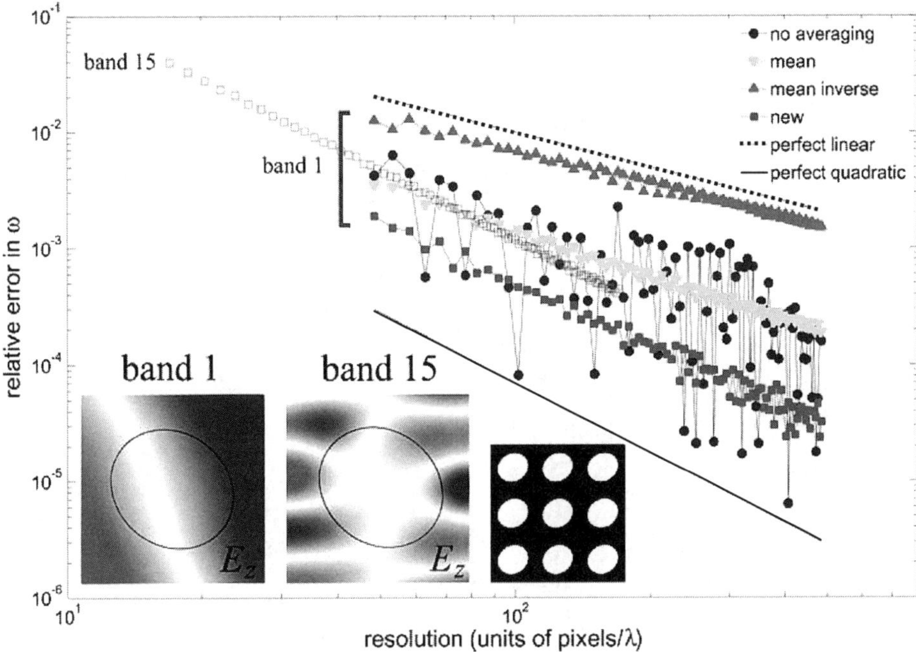

Fig. 6.5 Comparative FDTD-computed relative error $\Delta\omega/\omega$ for eigenmodes of a square lattice (period a) of 2-D anisotropic ellipsoids (right inset) versus grid resolution (pixels per vacuum wavelength λ). Straight lines indicating perfect linear (dashed) and perfect quadratic (solid) convergence are shown for reference. Most curves are for the first eigenvalue band (left inset shows E_z in a unit cell), with vacuum wavelength $\lambda = 4.85a$. Hollow squares show new method of (6.2) for band 15 (middle inset), with $\lambda = 1.7a$. Maximum resolution for all curves is 100 pixels/a. *Source:* A. Oskooi, C. Kottke, and S. G. Johnson, *Optics Lett.*, Vol. 34, 2009, pp. 2778–2780, ©2009 The Optical Society of America.

Each FDTD simulation in Figs. 6.5 and 6.6 employed Bloch-periodic boundaries and a Gaussian pulse source [8]. FDTD-computed time-waveforms were analyzed with a filter-diagonalization method [20] to obtain the eigenfrequency, ω. This yielded the relative error, $|\omega - \omega_0| / \omega_0$, by comparison with the "exact" ω_0 from a plane-wave calculation [11] at a high resolution. To counter suggestions that subpixel smoothing could perform poorly for higher eigenvalue bands [15], Ref. [8] examined eigenvalue bands 1 and 15 in 2-D, and 1 and 13 in 3-D. In each case, the higher band provided clearly non-plane-wave-like behavior.

Figures 6.5 and 6.6, respectively, show the results of the 2-D and 3-D convergence studies reported in [8]. These figures compare the convergence with grid resolution of the new smoothing technique of (6.2) with the convergence behavior of three simple alternatives: no smoothing; using the mean $\langle \varepsilon \rangle$ [4]; and using the harmonic mean $\langle \varepsilon^{-1} \rangle^{-1}$. In both of these case studies, similar to the results in Section 6.5 for isotropic materials [6], the new smoothing algorithm had the lowest error, often by one order of magnitude or more. Furthermore, it was the only technique that appeared to give second-order accuracy in the limit of fine grid resolution. (Interestingly, the simple mean $\langle \varepsilon \rangle$ performed better than the harmonic mean $\langle \varepsilon^{-1} \rangle^{-1}$, probably because it treated roughly two of the three field components correctly [6].)

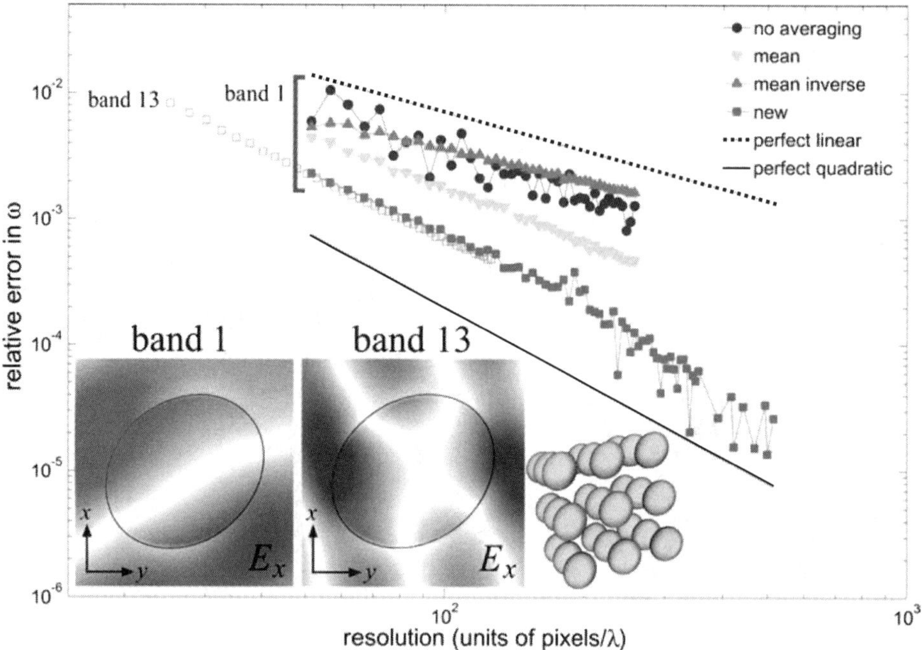

Fig. 6.6 Comparative FDTD-computed relative error $\Delta\omega/\omega$ for eigenmodes of a cubic lattice (period a) of 3-D anisotropic ellipsoids (right inset) versus grid resolution (pixels per vacuum wavelength λ). Straight lines indicating perfect linear (dashed) and perfect quadratic (solid) convergence are shown for reference. Most curves are for the first eigenvalue band (left inset shows E_x in x-y cross-section of a unit cell), with vacuum wavelength $\lambda = 5.15a$. Hollow squares show new method of (6.2) for band 13 (middle inset), with $\lambda = 2.52a$. Results of the new method for bands 1 and 13 are shown for resolution up to 100 pixels/a. *Source:* A. Oskooi, C. Kottke, and S. G. Johnson, *Optics Lett.*, Vol. 34, 2009, pp. 2778–2780, ©2009 The Optical Society of America.

In viewing the results shown in Figs. 6.5 and 6.6, it is noteworthy that similar accuracy was obtained for both the lower and higher (non-plane-wave-like) eigenvalue bands at comparable resolutions per wavelength, although the higher bands require greater absolute resolution per period, a, because their wavelengths are smaller. While Ref. [8] noted that apparent quadratic convergence obtained in a single structure can sometimes be fortuitous, it also expressed confidence in its reported results. These were obtained in multiple settings, and were backed by a clear theory rather than an *ad hoc* heuristic.

6.8 CONCLUSIONS

The subpixel smoothing schemes of (6.1) and (6.2), reported in [6] and [8], respectively, for isotropic and anisotropic interfaces, are attractive for FDTD modeling applications. These schemes are backed by a rigorous perturbation theory, rather than an *ad hoc* heuristic. Furthermore, they provide greatly improved accuracy without increasing computer storage or running time, other than a one-time preprocessing step. As noted in [8], a remaining challenge is to accurately model dielectric objects having sharp corners, where the resulting field singularities are known to degrade the accuracy of (6.1) and (6.2) to between first and second order, once the smoothing eliminates the first-order error [6]. In principle, an accurate smoothing can be developed for such corners once the corresponding perturbation theory is derived.

We also note that additional developments are required to treat interfaces with metals [22, 23] and dispersive materials [24]. Here, it is possible that a similar derivation strategy can be pursued using the unconjugated form of perturbation theory for the complex-symmetric Maxwell equations in reciprocal media with losses [25].

Overall, subpixel smoothing at the material interfaces of arbitrary-shaped dielectric objects results in a significant improvement (to global second-order accuracy) in FDTD computations of (a) resonant frequencies and eigenfrequencies; (b) scattered flux integrated over a virtual surface located away from the dielectric interfaces; and (c) other integrals of fields and energies to which the interfaces contribute only $O(\Delta x)$ of the integration domain. This is because the residual localized error currents at the interfaces due to subpixel smoothing radiate zero power to first order [7, 14]. Computational resources are identical to those of the classic unsmoothed (staircased) FDTD model, except for a relatively small preprocessing requirement.

ACKNOWLEDGMENT

This work was supported in part by the Materials Research Science and Engineering Center program of the National Science Foundation under awards DMR-9400334 and DMR-0819762; by the Office of Naval Research under award N00014-05-1-0700; and by the Army Research Office through the Institute for Soldier Nanotechnologies under contract DAAD-19-02-D0002.

APPENDIX 6A: OVERVIEW OF THE PERTURBATION TECHNIQUE USED TO DERIVE SUBPIXEL SMOOTHING[2]

There are many ways to formulate perturbation techniques in electromagnetism. One common formulation is analogous to "time-independent perturbation theory" in quantum mechanics [26]. This expresses Maxwell's equations as a generalized Hermitian eigenproblem, $\nabla \times \nabla \times \boldsymbol{E} = \omega^2 \varepsilon \boldsymbol{E}$, in terms of the frequency, ω, and the electric field, $\boldsymbol{E}$ (or equivalent formulations in terms of the magnetic field, $\boldsymbol{H}$) [19]. Then, one considers the first-order change, $\Delta\omega$, in the frequency resulting from a small change, $\Delta\varepsilon$, in the dielectric function, $\varepsilon(\boldsymbol{x})$, which is assumed real and positive. This yields the expression [19]:

$$\frac{\Delta\omega}{\omega} = -\frac{\int \boldsymbol{E}^* \cdot \Delta\varepsilon \boldsymbol{E}\, d^3\boldsymbol{x}}{2\int \boldsymbol{E}^* \cdot \varepsilon \boldsymbol{E}\, d^3\boldsymbol{x}} + O[(\Delta\varepsilon)^2] \tag{6A.1}$$

where $\boldsymbol{E}$ and ω are, respectively, the electric field and eigenfrequency of the *unperturbed* structure ε, and * denotes complex conjugation.

[2]See Ref. [7] for the complete derivation.

The key part of (6A.1) is the numerator on the right-hand side, which is what expresses the effect of the perturbation. This same numerator appears in a nearly identical form for many different perturbation techniques. For example, one obtains a similar expression in finding the perturbation, $\Delta\beta$, in the propagation constant, β, of a waveguide mode [27]; the coupling coefficient, $\sim\int \boldsymbol{E}^{*}\cdot\Delta\varepsilon \boldsymbol{E}'$, between two modes, $\boldsymbol{E}$ and $\boldsymbol{E}'$, in coupled-wave theory [28–30]; or the scattering current, $\boldsymbol{J} \sim \Delta\varepsilon \boldsymbol{E}$, and the scattered power, $\sim\int \boldsymbol{J}^{*}\cdot \boldsymbol{E}$, in the "volume-current" method (equivalent to the first Born approximation) [14, 31–33].

Equation (6A.1) also corresponds to an *exact* result for the *derivative* of ω with respect to any parameter, p, of ε, since if we write $\Delta\varepsilon = (\partial\varepsilon/\partial p)\Delta p + O[(\Delta p)^2]$, we can divide both sides by Δp and take the limit $\Delta p \rightarrow 0$. This result is equivalent to the Hellmann-Feynman theorem of quantum mechanics [13, 26]. In cases where the unperturbed ε is not real, corresponding to absorption or gain, or when one is considering "leaky modes," the eigenproblem typically becomes complex symmetric rather than Hermitian, and one obtains a similar formula but without the complex conjugation [25]. Therefore, any modification to the form of the numerator in (6A.1) for the frequency-perturbation theory immediately leads to corresponding modified formulas in many other perturbative techniques. It is sufficient for our purposes to consider frequency-perturbation theory only.

As shown in [13], Equation (6A.1) is not valid when $\Delta\varepsilon$ is due to a small change in the position of an interface between two dielectric materials, except in the limit of low dielectric contrast. However, a simple correction is possible. In particular, let us consider situations like the one illustrated in Fig. 6A.1, where the interface between two dielectric materials, ε^a and ε^b, is shifted by some small displacement, h, which may be a function of position.

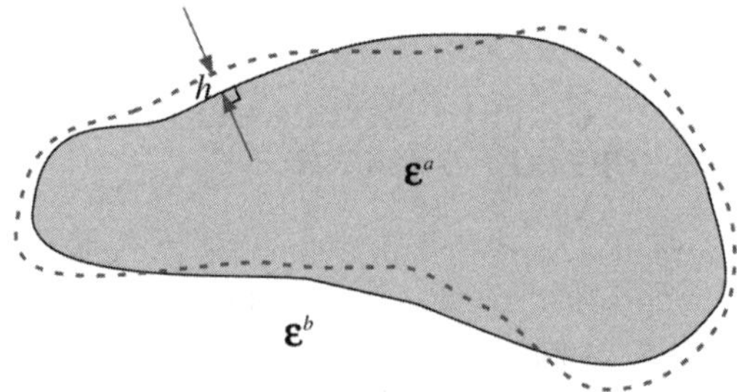

Fig. 6A.1 Schematic of an interface perturbation. Here, the interface between two possibly anisotropic dielectric materials, ε^a and ε^b, is shifted by some small position-dependent displacement, h. *Source:* C. Kottke, A. Farjadpour, and S. G. Johnson, *Physical Review E*, Vol. 77, 2008, 036611, ©2008 The American Physical Society.

Referring to Fig. 6A.1, directly applying (6A.1) with $\Delta\varepsilon = \pm(\varepsilon^a - \varepsilon^b)$ in the regions where the material has changed, gives an incorrect result. Here, $\Delta\omega/h$, which should ideally approach the exact derivative, $d\omega/dh$, is incorrect even for $h \rightarrow 0$. This is because $\boldsymbol{E}$ is discontinuous at the interface. Hence, in the limit $h \rightarrow 0$, the standard method leads to an ill-defined surface integral of $\boldsymbol{E}$ over the interface. For *isotropic* materials, corresponding to scalar ε^a and ε^b, the correct numerator in (6A.1) is instead the following surface integral over the interface [13]:

$$\int \boldsymbol{E}^{*}\cdot\Delta\varepsilon \boldsymbol{E}\, d^3\boldsymbol{x} \;\rightarrow\; \iint \left[(\varepsilon^a - \varepsilon^b)|\boldsymbol{E}_{\|}|^2 - \left(\frac{1}{\varepsilon^a} - \frac{1}{\varepsilon^b}\right)|D_{\perp}|^2 \right] \boldsymbol{h}\cdot d\boldsymbol{A} \qquad (6A.2)$$

where $E_{\|}$ and $D_{\perp}$ are, respectively, the (continuous) components of $\boldsymbol{E}$ and $\boldsymbol{D} = \varepsilon\boldsymbol{E}$ parallel and perpendicular to the interface; $d\boldsymbol{A}$ points toward ε^b; and $\boldsymbol{h}$ is the displacement of the interface from ε^a toward ε^b.

Reference [7] generalized (6A.2) to deal with the case where the two materials are *anisotropic*. This corresponds to arbitrary 3×3 tensors, $\boldsymbol{\varepsilon}^a$ and $\boldsymbol{\varepsilon}^b$, which were assumed to be Hermitian and positive-definite to obtain a well-behaved Hermitian eigenproblem. Reference [7] defined a local coordinate frame, (x_1, x_2, x_3), at each point on the interface between $\boldsymbol{\varepsilon}^a$ and $\boldsymbol{\varepsilon}^b$, where the x_1 direction was orthogonal to the interface, and the (x_2, x_3) directions were parallel. A continuous field "vector," $\boldsymbol{F} = [D_1, E_2, E_3]$, was also defined such that $F_1 = D_{\perp}$ and $\boldsymbol{F}_{2,3} = \boldsymbol{E}_{\|}$. After a rigorous derivation, the resulting numerator of (6A.1), representing the generalization of (6A.2), was determined to be:

$$\iint \boldsymbol{F}^* \cdot \left[\boldsymbol{\tau}(\boldsymbol{\varepsilon}^a) - \boldsymbol{\tau}(\boldsymbol{\varepsilon}^b)\right] \boldsymbol{F}\, \boldsymbol{h} \cdot d\boldsymbol{A} \tag{6A.3}$$

where $\boldsymbol{\tau}(\boldsymbol{\varepsilon})$ is given by the following 3×3 matrix [denoted in Section 6.6 as (6.3a)]:

$$\boldsymbol{\tau}(\boldsymbol{\varepsilon}) = \frac{1}{\varepsilon_{11}} \begin{bmatrix} -1 & \varepsilon_{12} & \varepsilon_{13} \\ \varepsilon_{21} & \varepsilon_{11}\varepsilon_{22} - \varepsilon_{21}\varepsilon_{12} & \varepsilon_{11}\varepsilon_{23} - \varepsilon_{21}\varepsilon_{13} \\ \varepsilon_{31} & \varepsilon_{11}\varepsilon_{32} - \varepsilon_{31}\varepsilon_{12} & \varepsilon_{11}\varepsilon_{33} - \varepsilon_{31}\varepsilon_{13} \end{bmatrix} \tag{6A.4}$$

Note that (6A.4) reduces to (6A.2) when $\boldsymbol{\varepsilon}$ is a scalar multiple, ε, of the identity matrix. (The assumption that $\boldsymbol{\varepsilon}$ is positive definite guarantees that $\varepsilon_{11} > 0$.)

Note also a key restriction: (6A.2) and (6A.3) require that the radius of curvature of the dielectric interface be much larger than h, except possibly on a set of measure zero such as at isolated corners or edges. Otherwise, more complicated methods must be employed [14]. For example, one cannot apply the above equations to the case of a hemispherical "bump" of radius, h, on the unperturbed surface. In this case, the lowest-order perturbation is $\Delta\omega \sim O(h^3)$, and requires a numerical computation of the polarizability of the hemisphere [14].

Next, we analyze how perturbation theory leads to a subpixel smoothing scheme. Suppose that we smooth the underlying dielectric tensor, $\boldsymbol{\varepsilon}(\boldsymbol{x})$, into some locally averaged tensor, $\bar{\boldsymbol{\varepsilon}}(\boldsymbol{x})$, by some method to be determined. This involves a change, $\Delta\boldsymbol{\varepsilon} = \bar{\boldsymbol{\varepsilon}} - \boldsymbol{\varepsilon}$, which is likely to be large near points where ε is discontinuous (and, conversely, is zero well inside regions where ε is constant). In particular, suppose that we employ a smoothing radius (defined more precisely below) that is proportional to the spatial resolution, Δx, so that $\Delta\boldsymbol{\varepsilon}$ is zero, or at most $O[(\Delta x)^2]$, except within a distance $\sim\Delta x$ of the discontinuous interface. To evaluate the effect of this large perturbation near the interface, we must employ an equivalent reformulation of (6A.3):

$$\Delta\omega \sim \int \boldsymbol{F}^* \cdot \Delta\boldsymbol{\tau}\, \boldsymbol{F}\, d^3\boldsymbol{x} \tag{6A.5}$$

where $\Delta\boldsymbol{\tau} = \boldsymbol{\tau}(\bar{\boldsymbol{\varepsilon}}) - \boldsymbol{\tau}(\boldsymbol{\varepsilon})$. It is sufficient to look at the perturbation in ω because, as remarked above, the same integral appears in the perturbation theory for many other quantities such as scattered power. With x_1 denoting the local coordinate orthogonal to the interface, then the x_1 integral is simply proportional to $\int \Delta\boldsymbol{\tau}\, dx_1 + O[(\Delta x)^2]$. Because $\boldsymbol{F}$ is continuous and $\Delta\boldsymbol{\tau} = 0$ except near the interface, we can pull $\boldsymbol{F}$ out of the x_1 integral to lowest order. Hence, it is sufficient to have $\int \Delta\boldsymbol{\tau}\, dx_1 = 0$ so that the first-order perturbation *equals zero* for all $\boldsymbol{F}$. This is achieved by averaging $\boldsymbol{\tau}$, as per the following discussion.

The most straightforward interpretation of subpixel smoothing would be to convolve $\boldsymbol{\varepsilon}$ with some localized kernel, $s(\boldsymbol{x})$, where $\int s(\boldsymbol{x})\, d^3\boldsymbol{x} = 1$ and $s(\boldsymbol{x}) = 0$ for $|\boldsymbol{x}|$ greater than some smoothing radius (the support radius) proportional to the resolution, Δx. That is:

$$\bar{\boldsymbol{\varepsilon}}(\boldsymbol{x}) = \boldsymbol{\varepsilon} * s = \int \boldsymbol{\varepsilon}(\boldsymbol{x}')\, s(\boldsymbol{x} - \boldsymbol{x}')\, d^3\boldsymbol{x}' \tag{6A.6}$$

In this interpretation, the simplest subpixel smoothing would merely involve computing the average of $\boldsymbol{\varepsilon}$ over each pixel. This would correspond to $s = 1$ inside a pixel at the origin, and $s = 0$ elsewhere. However, this would not lead to the desired $\int \Delta\boldsymbol{\tau}\, dx_1 = 0$ condition that is necessary to obtain second-order accuracy. Instead, we employ:

$$\bar{\boldsymbol{\varepsilon}}(\boldsymbol{x}) = \boldsymbol{\tau}^{-1}[\boldsymbol{\tau}(\boldsymbol{\varepsilon}) * s] = \boldsymbol{\tau}^{-1}\left(\int \boldsymbol{\tau}[\boldsymbol{\varepsilon}(\boldsymbol{x}')]\, s(\boldsymbol{x} - \boldsymbol{x}')\, d^3\boldsymbol{x}'\right) \tag{6A.7}$$

where $\boldsymbol{\tau}^{-1}$ is given by the following 3×3 matrix [denoted in Section 6.6 as (6.3b)], which is the inverse of the $\boldsymbol{\tau}(\boldsymbol{\varepsilon})$ mapping:

$$\boldsymbol{\tau}^{-1}[\boldsymbol{\tau}] = \frac{1}{\tau_{11}} \begin{bmatrix} -1 & -\tau_{12} & -\tau_{13} \\ -\tau_{21} & \tau_{11}\tau_{22} - \tau_{21}\tau_{12} & \tau_{11}\tau_{23} - \tau_{21}\tau_{13} \\ -\tau_{31} & \tau_{11}\tau_{32} - \tau_{31}\tau_{12} & \tau_{11}\tau_{33} - \tau_{31}\tau_{13} \end{bmatrix} \tag{6A.8}$$

The reason why (6A.7) works, regardless of the smoothing kernel, $s(\boldsymbol{x})$, is that:

$$\begin{aligned} \int \Delta\boldsymbol{\tau}\, d^3\boldsymbol{x} &= \int \left\{ \int \boldsymbol{\tau}[\boldsymbol{\varepsilon}(\boldsymbol{x}')]\, s(\boldsymbol{x} - \boldsymbol{x}')\, d^3\boldsymbol{x}' - \boldsymbol{\varepsilon}(\boldsymbol{x}) \right\} d^3\boldsymbol{x} \\ &= \int \boldsymbol{\tau}[\boldsymbol{\varepsilon}(\boldsymbol{x}')] \left\{ \int s(\boldsymbol{x} - \boldsymbol{x}')\, d^3\boldsymbol{x}' - 1 \right\} d^3\boldsymbol{x}' = 0 \end{aligned} \tag{6A.9}$$

Equation (6A.9) guarantees that the integral of $\Delta\boldsymbol{\tau}$ is zero over all space. However, it is not immediately obvious that this equation also implies the sufficiency condition that the local interface-perpendicular integral, $\int \Delta\boldsymbol{\tau}\, dx_1$, equals zero, at least to first order. In this regard, a key observation is that, within a pixel (where thc interface is locally flat to first order in the smoothing radius), $\Delta\boldsymbol{\tau}$ *must* be only a function of x_1 by translational symmetry. Hence, it becomes clear that (6A.9) implies $\int \Delta\boldsymbol{\tau}\, dx_1 = 0$ *by itself.*

Although the above convolution formulas may look complicated, for the simplest smoothing kernel, $s(\boldsymbol{x})$, the procedure is quite simple. Namely, in each pixel, first average $\boldsymbol{\tau}(\boldsymbol{\varepsilon})$ in the pixel, and then apply $\boldsymbol{\tau}^{-1}$ to the result. This is as straightforward to apply as the procedure reported in [11], for example.

Strictly speaking, the use of this smoothing does not always guarantee global second-order accuracy, even though FDTD is nominally second-order accurate. Reduced accuracy occurs when modeling a dielectric object that has sharp corners and associated local field singularities, as discussed in [13]. Although the first-order error is canceled by applying smoothing, the next-order correction might not be second order. In this case, smoothing leads to a convergence rate between first order (which would be obtained with no smoothing) and second order, with the exponent determined by the nature of the local field singularities at the corners.

REFERENCES

[1] Taflove, A., and S. C. Hagness, *Computational Electrodynamics: The Finite-Difference Time-Domain Method*, 3rd ed., Norwood, MA: Artech, 2005.

[2] Ditkowski, A., K. Dridi, and J. S. Hesthaven, "Convergent Cartesian grid methods for Maxwell's equations in complex geometries," *J. Computational Physics*, Vol. 170, 2001, pp. 39–80.

[3] Kaneda, N., B. Houshmand, and T. Itoh, "FDTD analysis of dielectric resonators with curved surfaces," *IEEE Trans. Microwave Theory and Techniques*, Vol. 45, 1997, pp. 1645–1649.

[4] Dey, S., and R. Mittra, "A conformal finite-difference time-domain technique for modeling cylindrical dielectric resonators," *IEEE Trans. Microwave Theory and Techniques*, Vol. 47, 1999, pp. 1737–1739.

[5] Mohammadi, A., H. Nadgaran, and M. Agio, "Contour-path effective permittivities for the two-dimensional finite-difference time-domain method," *Optics Express*, Vol. 13, 2005, pp. 10367–10381.

[6] Farjadpour, A., D. Roundy, A. Rodriguez, M. Ibanescu, P. Bermel, J. Joannopoulos, S. Johnson, and G. Burr, "Improving accuracy by sub-pixel smoothing in the finite-difference time domain," *Optics Letters,* Vol. 31, 2006, pp. 2972–2974.

[7] Kottke, C., A. Farjadpour, and S. G. Johnson, "Perturbation theory for anisotropic dielectric interfaces, and application to sub-pixel smoothing of discretized numerical methods," *Physical Review E*, Vol. 77, 2008, 036611.

[8] Oskooi, A., C. Kottke, and S. G. Johnson, "Accurate finite-difference time-domain simulation of anisotropic media by subpixel smoothing," *Optics Letters,* Vol. 34, 2009, pp. 2778–2780.

[9] Meade, R. D., A. M. Rappe, K. D. Brommer, J. D. Joannopoulos, and O. L. Alerhand, "Accurate theoretical analysis of photonic band-gap materials," *Physical Review B*, Vol. 48, 1993, pp. 8434–8437.

[10] Meade, R. D., A. M. Rappe, K. D. Brommer, J. D. Joannopoulos, and O. L. Alerhand, "Erratum: Accurate theoretical analysis of photonic band-gap materials [Phys. Rev. B. 48, 8434 (1993)]," *Physical Review B*, Vol. 55, 1997, pp. 15942–15942.

[11] Johnson, S. G., and J. D. Joannopoulos, "Block-iterative frequency-domain methods for Maxwell's equations in a planewave basis," *Optics Express*, Vol. 8, 2001, pp. 173–190.

[12] Meep (free FDTD simulation software package developed at MIT to model electromagnetic systems), Online: http://ab-initio.mit.edu/wiki/index.php/Meep

[13] Johnson, S. G., M. Ibanescu, M. A. Skorobogatiy, O. Weisberg, J. D. Joannopoulos, and Y. Fink, "Perturbation theory for Maxwell's equations with shifting material boundaries," *Physical Review E*, Vol. 65, 2002, 066611.

[14] Johnson, S. G., M. L. Povinelli, M. Soljacic, A. Karalis, S. Jacobs, and J. D. Joannopoulos, "Roughness losses and volume-current methods in photonic-crystal waveguides," *Applied Physics B*, Vol. 81, 2005, pp. 283–293.

[15] Werner, G., and J. Cary, "A stable FDTD algorithm for non-diagonal, anisotropic dielectrics," *J. Computational Physics*, Vol. 226, 2007, pp. 1085–1101.

[16] Nadobny, J., D. Sullivan, W. Wlodarczyk, P. Deuflhard, and P. Wust, "A 3-D tensor FDTD formulation for treatment of sloped interfaces in electrically inhomogeneous media," *IEEE Trans. Antennas and Propagation*, Vol. 51, 2003, pp. 1760–1770.

[17] Zagorodnov, I. A., R. Schuhmann, and T. Weiland, "A uniformly stable conformal FDTD method in Cartesian grids," *International J. Numerical Modeling*, Vol. 16, 2003, pp. 127–141.

[18] Moskow, S., V. Druskin, T. Habashy, P. Lee, and S. Davidycheva, "A finite difference scheme for elliptic equations with rough coefficients using a Cartesian grid nonconforming to interfaces," *SIAM J. Numerical Analysis*, Vol. 36, 1999, 442–464.

[19] Joannopoulos, J. D., R. D. Meade, and J. N. Winn, *Photonic Crystals: Molding the Flow of Light*, Princeton, NJ: Princeton University Press, 1995.

[20] Mandelshtam, V. A., and H. S. Taylor, "Harmonic inversion of time signals and its applications," *J. Chemical Physics*, Vol. 107, 1997, 6756.

[21] Andersen, J. B., and V. Solodukhov, "Field behavior near a dielectric wedge," *IEEE Trans. Antennas and Propagation*, Vol. 26, 1978, pp. 598–602.

[22] Mezzanotte, P., L. Roselli, and R. Sorrentino, "A simple way to model curved metal boundaries in FDTD algorithm avoiding staircase approximation," *IEEE Microwave and Guided Wave Lett.*, Vol. 5, 1995, pp. 267–269.

[23] Anderson, J., M. Okoniewski, and S. S. Stuchly, "Practical 3-D contour/staircase treatment of metals in FDTD," *IEEE Microwave and Guided Wave Lett.*, Vol. 6, 1996, pp. 146–148.

[24] Deinega, A., and I. Valuev, "Subpixel smoothing for conductive and dispersive media in the finite-difference time-domain method," *Optics Lett.*, Vol. 32, 2007, pp. 3429–3431.

[25] Leung, P., S. Liu, and K. Young, "Completeness and time-independent perturbation of the quasinormal modes of an absorptive and leaky cavity," *Physical Review A*, Vol. 49, 1994, pp. 3982–3989.

[26] Cohen-Tannoudji, C., B. Din, and F. Laloë, *Quantum Mechanics*, Paris: Hermann, 1977.

[27] Johnson, S. G., M. Ibanescu, M. Skorobogatiy, O. Weisberg, T. D. Engeness, M. Soljačić, S. A. Jacobs, J. D. Joannopoulos, and Y. Fink, "Low-loss asymptotically single-mode propagation in large-core OmniGuide fibers," *Optics Express*, Vol. 9, 2001, pp. 748–779.

[28] Marcuse, D., *Theory of Dielectric Optical Waveguides*, 2nd ed., San Diego, CA: Academic Press, 1991.

[29] Katsenelenbaum, B. Z., L. Mercader del Río, M. Pereyaslavets, M. Sorolla Ayza, and M. Thumm, *Theory of Nonuniform Waveguides: The Cross-Section Method*, London: Institute of Electrical Engineers, 1998.

[30] Johnson, S. G., P. Bienstman, M. A. Skorobogatiy, M. Ibanescu, E. Lidorikis, and J. D. Joannopoulos, "Adiabatic theorem and continuous coupled-mode theory for efficient taper transitions in photonic crystals," *Physical Review E*, Vol. 66, 2002, 066608.

[31] Snyder, A. W., and J. D. Love, *Optical Waveguide Theory*, London: Chapman and Hall, 1983.

[32] Kuznetsov, M., and H. A. Haus, "Radiation loss in dielectric waveguide structures by the volume current method," *IEEE J. Quantum Electronics*, Vol. 19, 1983, pp. 1505–1514.

[33] Chew, W. C., *Waves and Fields in Inhomogeneous Media*, New York: IEEE Press, 1995.

Chapter 7

Stochastic FDTD for Analysis of Statistical Variation in Electromagnetic Fields[1]

Steven M. Smith and Cynthia M. Furse

7.1 INTRODUCTION

The finite-difference time-domain (FDTD) method is commonly used to evaluate the electromagnetic fields in numerous applications including, for example, bioelectromagnetics [1–3], biophotonics [4, 5], geophysical prospecting [6], and atmospheric studies [7, 8]. As with all electromagnetic simulations, the computed fields are controlled by the geometry of the model and the source, and by the electrical properties of the materials in the model.

An FDTD practitioner working with applications such as those of [1–8] immediately confronts a fundamental issue. Namely, whereas a traditional FDTD model assigns a source and an interaction structure in a deterministic manner in space-time, the properties of the source and the interaction structure in [1–8] (and similar applications) vary randomly. At best, only a statistical description of this random behavior is available. The most common such randomness involves variations of the structure's electrical properties with respect to spatial position within a particular model, or variations of the structure's geometry and electrical properties from one complete realization of the model to another.

While it is expected that the fields in an electromagnetic wave interaction model having random, statistically characterized material and geometry properties must also vary statistically, traditional deterministic numerical simulations do not provide this information. In such circumstances, FDTD practitioners generally use average values of the spatial configuration and electrical properties arising from the presumed statistical variations, and consequently compute the average fields produced in the model.

If an estimate of the field variability is required, multiple FDTD simulations can be conducted, with the properties for each simulation selected at "random" according to their statistics, emulating the *Monte Carlo* method [9, 10]. However, given the complexity and size of many two-dimensional (2-D) and three-dimensional (3-D) electromagnetic wave interaction problems, the full range of field variability is likely not deduced by this procedure, nor are useful statistical properties such as variance, standard deviation, and 90% confidence intervals. This is because of the impractically large number of required FDTD simulations.

[1]This chapter is adapted from Ref. [11], S. M. Smith and C. M. Furse, "Stochastic FDTD for analysis of statistical variation in electromagnetic fields," *IEEE Trans. Antennas and Propagation*, Vol. 60, 2012, pp. 3343–3350, ©2012 IEEE.

This chapter reviews in detail the formulation of the FDTD technique reported in [11] that treats electrical properties and fields as stochastic variables. These variables are iteratively stepped through time in the typical FDTD fashion at each grid point. This *stochastic FDTD* (S-FDTD) method allows the direct evaluation of the statistical mean and variance of the fields at every point in the grid and at each time-step, due to the measured or estimated variance in the electrical properties. Hence, a *single* FDTD modeling run, rather than scores or even hundreds of runs, suffices to evaluate the approximate statistical parameters of the electromagnetic fields within the randomly constituted structure of interest.

The S-FDTD technique of [11] arose from an application of perturbation theory [12] similar in concept to a technique previously applied to find the stochastic properties of mechanical systems in finite-element simulations [13]. This technique, also called the *delta method* [14], assumes that the solution has a Taylor-series expansion. Suitably truncated, this series is substituted into the equation being approximated; the equation is expanded; and the coefficients of the Taylor series are determined via linear algebra. In the case of [11], the solution involved electric (E) and magnetic (H) field variables, and Taylor series for these variables were substituted into the FDTD numerical approximation of Maxwell's time-dependent curl equations in one dimension (1-D). Generalization to 2-D and 3-D models was proposed in [15, 16].

Note that the S-FDTD technique reported in [11] and reviewed in this chapter works best only for small perturbations since the Taylor series is truncated to low order. If larger perturbations are assumed, higher-order terms would need to be retained, and the S-FDTD update equations would not apply as is.

Sections 7.2 and 7.3 review the basic formulation of the delta-method derivation of the mean and variance, respectively, of a generic function of multiple random variables. Section 7.4 specifies the 1-D FDTD field equations to be analyzed. Then, Sections 7.5 and 7.6 apply the results of Sections 7.2 and 7.3 to derive approximate expressions for the mean and variance of the FDTD time-stepped E and H fields. This yields the 1-D S-FDTD algorithm, which is summarized in Section 7.7 and subsequently applied in Section 7.8 to a layered biological tissue model in order to evaluate its accuracy relative to a Monte Carlo benchmark. Finally, Section 7.9 concludes with a summary and assessment of future research in this topical area.

7.2 DELTA METHOD: MEAN OF A GENERIC MULTIVARIABLE FUNCTION

Reference [11] reported applying the delta method [14] to derive the mean and variance of the FDTD-computed fields in a 1-D electromagnetic wave interaction model. Assuming wave propagation in the $\pm z$-directions, the FDTD time-stepping algorithm for this case involved four random variables: electric field E_x, magnetic field H_y, relative permittivity ε_r, and electrical conductivity $\hat{\sigma}$.

This section and Section 7.3 provide the foundation needed for these delta-method derivations, which are described in detail in Sections 7.5 and 7.6. Following [11], we begin with the Taylor's series expansion of a generic function, g, of stochastic variables $x_1, x_2, \ldots, x_n$:

$$\begin{aligned} g(x_1, x_2, x_3, \ldots, x_n) &= g\left(\langle x_1 \rangle, \langle x_2 \rangle, \langle x_3 \rangle, \ldots, \langle x_n \rangle\right) \\ &+ \sum_{i=1}^{n} \frac{\partial g}{\partial x_i}\bigg|_{\langle x_1 \rangle, \langle x_2 \rangle, \ldots, \langle x_n \rangle} \left(x_i - \langle x_i \rangle\right) + \frac{1}{2!}\sum_{i=1}^{n}\sum_{j=1}^{n} \frac{\partial^2 g}{\partial x_i x_j}\bigg|_{\langle x_1 \rangle, \langle x_2 \rangle, \ldots, \langle x_n \rangle} \left(x_i - \langle x_i \rangle\right)\left(x_j - \langle x_j \rangle\right) + \ldots \end{aligned} \tag{7.1}$$

Here, the mean of the ith stochastic variable is denoted as $\langle x_i \rangle$, and in Sections 7.4 and 7.5, the function, g, will be obtained from the FDTD time-stepping expressions for E_x and H_y. Now, taking the expectation of (7.1) and applying the linearity of the expectation operator gives:

$$\begin{aligned} \langle g(x_1, x_2, x_3, \dots, x_n) \rangle &= \langle g(\langle x_1 \rangle, \langle x_2 \rangle, \langle x_3 \rangle, \dots, \langle x_n \rangle) \rangle \\ &+ \left\langle \sum_{i=1}^{n} \frac{\partial g}{\partial x_i}\bigg|_{\langle x_1 \rangle, \langle x_2 \rangle, \dots, \langle x_n \rangle} (x_i - \langle x_i \rangle) \right\rangle + \left\langle \frac{1}{2!} \sum_{i=1}^{n} \sum_{j=1}^{n} \frac{\partial^2 g}{\partial x_i x_j}\bigg|_{\langle x_1 \rangle, \langle x_2 \rangle, \dots, \langle x_n \rangle} (x_i - \langle x_i \rangle)(x_j - \langle x_j \rangle) \right\rangle + \dots \end{aligned} \tag{7.2}$$

Several terms in (7.2) go to zero. For example, $\langle x_i - \langle x_i \rangle \rangle = \langle x_i \rangle - \langle x_i \rangle = 0$, recognizing that expectation is a linear operator having the distributive property, and the expectation of a constant is a constant [17, 18]. Noting also that $\langle aX \rangle = a \langle X \rangle$ [17, 18], (7.2) can now be simplified to:

$$\begin{aligned} \langle g(x_1, x_2, x_3, \dots, x_n) \rangle &= g(\langle x_1 \rangle, \langle x_2 \rangle, \langle x_3 \rangle, \dots, \langle x_n \rangle) \\ &+ \underbrace{\sum_{i=1}^{n} \frac{\partial g}{\partial x_i}\bigg|_{\langle x_1 \rangle, \langle x_2 \rangle, \dots, \langle x_n \rangle} \langle x_i - \langle x_i \rangle \rangle}_{0} + \frac{1}{2!} \sum_{i=1}^{n} \sum_{j=1}^{n} \frac{\partial^2 g}{\partial x_i x_j}\bigg|_{\langle x_1 \rangle, \langle x_2 \rangle, \dots, \langle x_n \rangle} \langle (x_i - \langle x_i \rangle)(x_j - \langle x_j \rangle) \rangle + \dots \end{aligned} \tag{7.3}$$

Neglecting higher-order terms removes the double sum in (7.3), thus yielding:

$$\langle g(x_1, x_2, x_3, \dots, x_n) \rangle \approx g(\langle x_1 \rangle, \langle x_2 \rangle, \langle x_3 \rangle, \dots, \langle x_n \rangle) \tag{7.4}$$

Equation (7.4) is the mathematical verification of the traditional FDTD approach—namely, the average (or expected) fields on the left-hand side of (7.4) can be found by solving the field equations using the means or averages of the variables on the right-hand side of (7.4). Thus, the equations for the mean values of the fields in the S-FDTD method are the traditional FDTD field equation updates. The usual FDTD field values are now recognized to be the mean field values, and can be found by inputting the mean electrical properties, ε_r and $\hat{\sigma}$.

7.3 DELTA METHOD: VARIANCE OF A GENERIC MULTIVARIABLE FUNCTION

Following [11], we now turn our attention to finding the variance of a generic multivariable function, $\sigma^2\{g\}$, defined as [17, 18]:

$$\sigma^2 \{g(x_1, x_2, \dots, x_n)\} = \langle [g(x_1, x_2, \dots, x_n)]^2 \rangle - \langle g(x_1, x_2, \dots, x_n) \rangle^2 \tag{7.5}$$

To obtain the first term on the right-hand side, $\langle [g(x_1, x_2, \dots, x_n)]^2 \rangle$, we first find $[g(x_1, x_2, \dots, x_n)]^2$ by squaring (7.1), the Taylor's series expansion of $g(x_1, x_2, \dots, x_n)$. Retaining only terms through second order, this yields:

$$
\begin{aligned}
\left[g(x_1, x_2, x_3, \ldots, x_n)\right]^2 &= \left[g\left(\langle x_1\rangle, \langle x_2\rangle, \langle x_3\rangle, \ldots, \langle x_n\rangle\right)\right]^2 \\
&+ 2g\left(\langle x_1\rangle, \langle x_2\rangle, \langle x_3\rangle, \ldots, \langle x_n\rangle\right) \sum_{i=1}^{n} \left.\frac{\partial g}{\partial x_i}\right|_{\langle x_1\rangle, \langle x_2\rangle, \ldots, \langle x_n\rangle} \left(x_i - \langle x_i\rangle\right) \\
&+ \sum_{i=1}^{n}\sum_{j=1}^{n} \left.\frac{\partial g}{\partial x_i}\frac{\partial g}{\partial x_j}\right|_{\langle x_1\rangle, \langle x_2\rangle, \ldots, \langle x_n\rangle} \left(x_i - \langle x_i\rangle\right)\left(x_j - \langle x_j\rangle\right) + \\
&+ \frac{2}{2!} g\left(\langle x_1\rangle, \langle x_2\rangle, \langle x_3\rangle, \ldots, \langle x_n\rangle\right) \sum_{i=1}^{n}\sum_{j=1}^{n} \left.\frac{\partial^2 g}{\partial x_i x_j}\right|_{\langle x_1\rangle, \langle x_2\rangle, \ldots, \langle x_n\rangle} \left(x_i - \langle x_i\rangle\right)\left(x_j - \langle x_j\rangle\right) + \ldots
\end{aligned}
\tag{7.6}
$$

Taking the expectation of this equation yields:

$$
\begin{aligned}
\left\langle \left[g(x_1, x_2, x_3, \ldots, x_n)\right]^2 \right\rangle &= \left[g\left(\langle x_1\rangle, \langle x_2\rangle, \langle x_3\rangle, \ldots, \langle x_n\rangle\right)\right]^2 \\
&+ 2g\left(\langle x_1\rangle, \langle x_2\rangle, \langle x_3\rangle, \ldots, \langle x_n\rangle\right) \sum_{i=1}^{n} \left.\frac{\partial g}{\partial x_i}\right|_{\langle x_1\rangle, \langle x_2\rangle, \ldots, \langle x_n\rangle} \left\langle\left(x_i - \langle x_i\rangle\right)\right\rangle \\
&+ \sum_{i=1}^{n}\sum_{j=1}^{n} \left.\frac{\partial g}{\partial x_i}\frac{\partial g}{\partial x_j}\right|_{\langle x_1\rangle, \langle x_2\rangle, \ldots, \langle x_n\rangle} \left\langle\left(x_i - \langle x_i\rangle\right)\left(x_j - \langle x_j\rangle\right)\right\rangle \\
&+ \frac{2}{2!} g\left(\langle x_1\rangle, \langle x_2\rangle, \langle x_3\rangle, \ldots, \langle x_n\rangle\right) \sum_{i=1}^{n}\sum_{j=1}^{n} \left.\frac{\partial^2 g}{\partial x_i x_j}\right|_{\langle x_1\rangle, \langle x_2\rangle, \ldots, \langle x_n\rangle} \left\langle\left(x_i - \langle x_i\rangle\right)\left(x_j - \langle x_j\rangle\right)\right\rangle + \ldots
\end{aligned}
\tag{7.7}
$$

Terms containing expressions such as $\left\langle\left(x_i - \langle x_i\rangle\right)\right\rangle$ go to zero as discussed previously, leaving the following equation:

$$
\begin{aligned}
\left\langle \left[g(x_1, x_2, x_3, \ldots, x_n)\right]^2 \right\rangle &= \left[g\left(\langle x_1\rangle, \langle x_2\rangle, \langle x_3\rangle, \ldots, \langle x_n\rangle\right)\right]^2 \\
&+ \sum_{i=1}^{n}\sum_{j=1}^{n} \left.\frac{\partial g}{\partial x_i}\frac{\partial g}{\partial x_j}\right|_{\langle x_1\rangle, \langle x_2\rangle, \ldots, \langle x_n\rangle} \left\langle\left(x_i - \langle x_i\rangle\right)\left(x_j - \langle x_j\rangle\right)\right\rangle + \ldots
\end{aligned}
\tag{7.8}
$$

Next, to obtain the second term on the right-hand side of (7.5), we square (7.3). This yields:

$$\langle g(x_1, x_2, x_3, \ldots, x_n)\rangle^2 = \left[g(\langle x_1\rangle, \langle x_2\rangle, \langle x_3\rangle, \ldots, \langle x_n\rangle)\right]^2$$

$$+ \frac{2}{2!} g(\langle x_1\rangle, \langle x_2\rangle, \langle x_3\rangle, \ldots, \langle x_n\rangle) \sum_{i=1}^{n}\sum_{j=1}^{n} \left.\frac{\partial^2 g}{\partial x_i x_j}\right|_{\langle x_1\rangle, \langle x_2\rangle, \ldots, \langle x_n\rangle} \left\langle \left(x_i - \langle x_i\rangle\right)\left(x_j - \langle x_j\rangle\right)\right\rangle \tag{7.9}$$

$$+ \left\{ \frac{1}{2!} \sum_{i=1}^{n}\sum_{j=1}^{n} \left.\frac{\partial^2 g}{\partial x_i x_j}\right|_{\langle x_1\rangle, \langle x_2\rangle, \ldots, \langle x_n\rangle} \left\langle \left(x_i - \langle x_i\rangle\right)\left(x_j - \langle x_j\rangle\right)\right\rangle \right\}^2 + \ldots$$

Upon implementing $\left\langle \left[g(x_1, x_2, \ldots, x_n)\right]^2\right\rangle - \langle g(x_1, x_2, \ldots, x_n)\rangle^2$, i.e., subtracting (7.9) from (7.8), and removing the higher order terms, we obtain an approximation for the variance of g:

$$\sigma^2\left\{g(x_1, x_2, \ldots, x_n)\right\} \approx \sum_{i=1}^{n}\sum_{j=1}^{n} \left.\frac{\partial g}{\partial x_i}\frac{\partial g}{\partial x_j}\right|_{\langle x_1\rangle, \langle x_2\rangle, \ldots, \langle x_n\rangle} \left\langle \left(x_i - \langle x_i\rangle\right)\left(x_j - \langle x_j\rangle\right)\right\rangle \tag{7.10}$$

Putting this equation in terms of the covariance:

$$\sigma^2\left\{g(x_1, x_2, \ldots, x_n)\right\} \approx \sum_{i=1}^{n}\sum_{j=1}^{n} \left.\frac{\partial g}{\partial x_i}\frac{\partial g}{\partial x_j}\right|_{\langle x_1\rangle, \langle x_2\rangle, \ldots, \langle x_n\rangle} \mathrm{Cov}\left\{x_i, x_j\right\} \tag{7.11}$$

Following [11], at this point we have derived two important approximate relations: (7.4) for the mean, and (7.11) for the variance, of a generic function g of random variables $x_1, x_2, \ldots, x_n$, all based on truncated Taylor-series expansions. Before applying these relations to develop the S-FDTD algorithm for 1-D electromagnetic wave interactions, it is useful to recall three fundamental identities regarding the random variables, X and Y. First, the variance of $X \pm Y$:

$$\sigma^2\left\{X \pm Y\right\} = \sigma_X^2 + \sigma_Y^2 \pm 2\mathrm{Cov}\left\{X, Y\right\} \tag{7.12}$$

Second, the variance of X scaled by the constant, a:

$$\sigma^2\left\{aX\right\} = a^2\sigma^2\left\{X\right\} \tag{7.13}$$

And third, the covariance identity:

$$\mathrm{Cov}\left\{X, Y\right\} = \rho\left\{X, Y\right\}\sigma\left\{X\right\}\sigma\left\{Y\right\} \tag{7.14}$$

In the covariance identity, the two terms in the form of $\sigma\{\,\}$ are the standard deviations of X and Y, and $\rho\{X, Y\}$ is the correlation coefficient of these two random variables. This correlation is bounded between -1 and 1.

7.4 FIELD EQUATIONS

The S-FDTD algorithm reported in [11] was developed for $\pm z$-directed 1-D electromagnetic wave interactions, assuming field components E_x and H_y. For this case, Faraday's Law is given by:

$$\mu \frac{\partial H_y}{\partial t} = -\frac{\partial E_x}{\partial z} \tag{7.15}$$

and the resulting classic FDTD field update for H_y [19] can be written as either:

$$H_y\Big|_{k+1/2}^{n+1/2} = H_y\Big|_{k+1/2}^{n-1/2} - \frac{\Delta t}{\mu_{k+1/2}\Delta z}\left(E_x\Big|_{k+1}^{n} - E_x\Big|_{k}^{n}\right) \tag{7.16a}$$

or

$$H_y\Big|_{k+1/2}^{n+1/2} - H_y\Big|_{k+1/2}^{n-1/2} = -\frac{\Delta t}{\mu_{k+1/2}\Delta z}\left(E_x\Big|_{k+1}^{n} - E_x\Big|_{k}^{n}\right) \tag{7.16b}$$

where $H_y\big|_{k+1/2}^{n+1/2}$ denotes the y-component of $\boldsymbol{H}$ evaluated at grid-point $(k+1/2)\Delta z$ and time-step $(n+1/2)\Delta t$, and similarly $E_x\big|_{k}^{n}$ denotes the x-component of $\boldsymbol{E}$ evaluated at grid-point $k\Delta z$ and time-step $n\Delta t$. The magnetic permeability at grid-point $(k+1/2)\Delta z$, denoted here as $\mu_{k+1/2}$, is henceforth assumed to equal the vacuum permeability, μ_0.

Subject to these assumptions, Ampere's law is given by:

$$\varepsilon \frac{\partial E_x}{\partial t} = -\frac{\partial H_y}{\partial z} - \hat{\sigma} E_x \tag{7.17}$$

and the resulting classic FDTD field update for E_x [19] can be written as either:

$$\begin{aligned} E_x\Big|_{k}^{n+1} &= \left(\frac{\dfrac{\varepsilon_r\big|_k \varepsilon_0}{\Delta t} - \dfrac{\hat{\sigma}_k}{2}}{\dfrac{\varepsilon_r\big|_k \varepsilon_0}{\Delta t} + \dfrac{\hat{\sigma}_k}{2}}\right) E_x\Big|_{k}^{n} - \frac{1}{\left(\dfrac{\varepsilon_r\big|_k \varepsilon_0}{\Delta t} + \dfrac{\hat{\sigma}_k}{2}\right)\Delta z}\left(H_y\Big|_{k+1/2}^{n+1/2} - H_y\Big|_{k-1/2}^{n+1/2}\right) \\ &\triangleq C_a\big|_k E_x\Big|_{k}^{n} - C_b\big|_k\left(H_y\Big|_{k+1/2}^{n+1/2} - H_y\Big|_{k-1/2}^{n+1/2}\right) \end{aligned} \tag{7.18a}$$

or

$$E_x\Big|_{k}^{n+1} - C_a\big|_k E_x\Big|_{k}^{n} = -C_b\big|_k\left(H_y\Big|_{k+1/2}^{n+1/2} - H_y\Big|_{k-1/2}^{n+1/2}\right) \tag{7.18b}$$

where, at grid-point $k\Delta z$, the relative permittivity is denoted as $\varepsilon_r\big|_k$, the electrical conductivity is denoted as $\hat{\sigma}_k$ (to distinguish it from the statistical term, σ), and the updating coefficients, $C_a\big|_k$ and $C_b\big|_k$, are given by:

$$C_a\big|_k = \frac{\dfrac{\varepsilon_r\big|_k \varepsilon_0}{\Delta t} - \dfrac{\hat{\sigma}_k}{2}}{\dfrac{\varepsilon_r\big|_k \varepsilon_0}{\Delta t} + \dfrac{\hat{\sigma}_k}{2}}, \qquad C_b\big|_k = \frac{1}{\left(\dfrac{\varepsilon_r\big|_k \varepsilon_0}{\Delta t} + \dfrac{\hat{\sigma}_k}{2}\right)\Delta z} \tag{7.19a, b}$$

7.5 FIELD EQUATIONS: MEAN APPROXIMATION

Following [11], applying the expectation operator to the FDTD time-stepping equations allows determining the mean of $H_y\big|_{k+1/2}^{n+1/2}$ and $E_x\big|_k^{n+1}$. For example, taking the mean of both sides of the Faraday's law update of (7.16a) yields:

$$\left\langle H_y\big|_{k+1/2}^{n+1/2}\right\rangle = \left\langle H_y\big|_{k+1/2}^{n-1/2} - \frac{\Delta t}{\mu_0 \Delta z}\left(E_x\big|_{k+1}^{n} - E_x\big|_k^{n}\right)\right\rangle \tag{7.20}$$

Because expectation is a linear operator (i.e., $\langle aX + bY\rangle = a\langle X\rangle + b\langle Y\rangle$ for constants a and b), it can be distributed within the bracket on the right-hand side of (7.20), leading to:

$$\left\langle H_y\big|_{k+1/2}^{n+1/2}\right\rangle = \left\langle H_y\big|_{k+1/2}^{n-1/2}\right\rangle - \frac{\Delta t}{\mu_0 \Delta z}\left(\left\langle E_x\big|_{k+1}^{n}\right\rangle - \left\langle E_x\big|_k^{n}\right\rangle\right) \tag{7.21}$$

This is the just the classic FDTD updating equation for Faraday's law in 1-D.

Taking the mean of both sides of the Ampere's law update of (7.18a) yields:

$$\left\langle E_x\big|_k^{n+1}\right\rangle = \left\langle \left(\frac{\dfrac{\varepsilon_r\big|_k \varepsilon_0}{\Delta t} - \dfrac{\hat{\sigma}_k}{2}}{\dfrac{\varepsilon_r\big|_k \varepsilon_0}{\Delta t} + \dfrac{\hat{\sigma}_k}{2}}\right) E_x\big|_k^{n}\right\rangle - \frac{1}{\Delta z}\left\langle \frac{H_y\big|_{k+1/2}^{n+1/2} - H_y\big|_{k-1/2}^{n+1/2}}{\dfrac{\varepsilon_r\big|_k \varepsilon_0}{\Delta t} + \dfrac{\hat{\sigma}_k}{2}}\right\rangle \tag{7.22}$$

Here, it is difficult to separate $E_x\big|_k^n$ from the random variables $\varepsilon_r\big|_k$ and $\hat{\sigma}_k$, because they are not constant. The field $E_x\big|_k^{n+1}$ is clearly a multivariable stochastic function. Applying the approximation developed in (7.4) to this function yields:

$$\left\langle E_x\big|_k^{n+1}\right\rangle \approx \left(\frac{\dfrac{\left\langle\varepsilon_r\big|_k\right\rangle \varepsilon_0}{\Delta t} - \dfrac{\left\langle\hat{\sigma}_k\right\rangle}{2}}{\dfrac{\left\langle\varepsilon_r\big|_k\right\rangle \varepsilon_0}{\Delta t} + \dfrac{\left\langle\hat{\sigma}_k\right\rangle}{2}}\right)\left\langle E_x\big|_k^{n}\right\rangle - \frac{\left\langle H_y\big|_{k+1/2}^{n+1/2}\right\rangle - \left\langle H_y\big|_{k-1/2}^{n+1/2}\right\rangle}{\left(\dfrac{\left\langle\varepsilon_r\big|_k\right\rangle \varepsilon_0}{\Delta t} + \dfrac{\left\langle\hat{\sigma}_k\right\rangle}{2}\right)\Delta z} \tag{7.23}$$

This is equivalent to simply applying (7.18a), the original FDTD update equation, using just the mean of the electrical permittivity and conductivity at each grid point. This seems intuitive — that the mean electric field at each grid point would result from assuming mean values of the electrical parameters at each grid point.

7.6 FIELD EQUATIONS: VARIANCE APPROXIMATION

7.6.1 Variance of the *H*-Fields

Following [11], applying the variance operator, σ^2, to (7.16b) yields:

$$\sigma^2\left\{ H_y\Big|_{k+1/2}^{n+1/2} - H_y\Big|_{k+1/2}^{n-1/2} \right\} = \left(\frac{\Delta t}{\mu_0 \Delta z}\right)^2 \sigma^2\left\{ E_x\Big|_{k+1}^{n} - E_x\Big|_{k}^{n} \right\} \tag{7.24}$$

Applying (7.12) and rearranging, we arrive at:

$$\sigma^2\left\{ H_y\Big|_{k+1/2}^{n+1/2} \right\} + \sigma^2\left\{ H_y\Big|_{k+1/2}^{n-1/2} \right\} - 2\rho\left\{ H_y\Big|_{k+1/2}^{n+1/2}, H_y\Big|_{k+1/2}^{n-1/2} \right\} \sigma\left\{ H_y\Big|_{k+1/2}^{n+1/2} \right\} \sigma\left\{ H_y\Big|_{k+1/2}^{n-1/2} \right\}$$
$$= \left(\frac{\Delta t}{\mu_0 \Delta z}\right)^2 \begin{pmatrix} \sigma^2\left\{ E_x\Big|_{k+1}^{n} \right\} + \sigma^2\left\{ E_x\Big|_{k}^{n} \right\} \\ - 2\rho\left\{ E_x\Big|_{k+1}^{n}, E_x\Big|_{k}^{n} \right\} \sigma\left\{ E_x\Big|_{k+1}^{n} \right\} \sigma\left\{ E_x\Big|_{k}^{n} \right\} \end{pmatrix} \tag{7.25}$$

Now, critical approximations for the correlation coefficients, ρ, must be made. These are bounded between -1 and 1 [17, 18]. For example, $\rho\left\{ H_y\big|_{k+1/2}^{n+1/2}, H_y\big|_{k+1/2}^{n-1/2} \right\}$ represents the correlation between two H-fields at time-steps $n+1/2$ and $n-1/2$. Since these two fields are separated by only one time-step, Δt, they should be highly correlated, and hence merit a ρ-value of ~1 [15]. On the right side of (7.25), $\rho\left\{ E_x\big|_{k+1}^{n}, E_x\big|_{k}^{n} \right\}$ represents the correlation between two E-fields at grid-points k+1 and k, i.e., separated by only a single space cell, Δz. In a uniform medium, these are highly correlated, and also merit a ρ-value of ~1 [15]. For nonuniform media, better approximations may be derived. Using the approximation $\rho = 1$ for these two correlation coefficients and rearranging terms gives:

$$\sigma^2\left\{ H_y\Big|_{k+1/2}^{n+1/2} \right\} - 2\sigma\left\{ H_y\Big|_{k+1/2}^{n+1/2} \right\} \sigma\left\{ H_y\Big|_{k+1/2}^{n-1/2} \right\} + \sigma^2\left\{ H_y\Big|_{k+1/2}^{n-1/2} \right\}$$
$$\approx \left(\frac{\Delta t}{\mu_0 \Delta z}\right)^2 \left(\sigma^2\left\{ E_x\Big|_{k+1}^{n} \right\} - 2\sigma\left\{ E_x\Big|_{k+1}^{n} \right\} \sigma\left\{ E_x\Big|_{k}^{n} \right\} + \sigma^2\left\{ E_x\Big|_{k}^{n} \right\} \right) \tag{7.26}$$

This equation has perfectly squared terms on both sides. Hence, taking the square root of each side yields an expression for the deviation, σ, from the mean:

$$\sigma\left\{ H_y\Big|_{k+1/2}^{n+1/2} \right\} - \sigma\left\{ H_y\Big|_{k+1/2}^{n-1/2} \right\} \approx \left(\frac{\Delta t}{\mu_0 \Delta z}\right) \left(\sigma\left\{ E_x\Big|_{k+1}^{n} \right\} - \sigma\left\{ E_x\Big|_{k}^{n} \right\} \right) \tag{7.27}$$

Importantly, completing the square and taking the square root of the result *preserves the sign* of σ, which can be either instantaneously negative (smaller than the mean) or instantaneously positive (larger than the mean). Since, by definition, the standard deviation, $\sqrt{\sigma^2(\,)}$, is the positive square root of the variance, the σ of (7.27) is *not* the standard deviation of the H-field. In fact, upon rearranging (7.27), we see that σ appears as a wave that can be marched in time:

$$\sigma\left\{ H_y\Big|_{k+½}^{n+½} \right\} \approx \sigma\left\{ H_y\Big|_{k+½}^{n-½} \right\} + \left(\frac{\Delta t}{\mu_0 \Delta z}\right)\left(\sigma\left\{ E_x\Big|_{k+1}^{n} \right\} - \sigma\left\{ E_x\Big|_{k}^{n} \right\} \right) \tag{7.28}$$

This time-marching of σ, in a manner analogous to the E- and H-fields in the FDTD simulation, is key to the S-FDTD technique.

7.6.2 Variance of the *E*-Fields

Following [11], the variance of the E-fields is found by taking the variance of both sides of (7.18b), yielding:

$$\sigma^2\left\{ E_x\Big|_k^{n+1} - C_a\Big|_k E_x\Big|_k^n \right\} = \sigma^2\left\{ -C_b\Big|_k \left(H_y\Big|_{k+½}^{n+½} - H_y\Big|_{k-½}^{n+½} \right) \right\} \tag{7.29}$$

Reduction of the Left-Hand Side of (7.29)

We first expand the left-hand side of (7.29) using (7.12), and then separate the compound terms containing $C_a\big|_k$ and E using the delta approximation (7.11). This yields:

$$\begin{aligned} \sigma^2\left\{ E_x\Big|_k^{n+1} - C_a\Big|_k E_x\Big|_k^n \right\} &= \sigma^2\left\{ E_x\Big|_k^{n+1} \right\} + \sigma^2\left\{ C_a\Big|_k E_x\Big|_k^n \right\} \\ &\quad - 2\,\mathrm{Cov}\left\{ E_x\Big|_k^{n+1},\ C_a\Big|_k E_x\Big|_k^n \right\} \end{aligned} \tag{7.30}$$

Upon applying the identity:

$$\mathrm{Cov}\{X, Y\} = \rho\{X, Y\}\sqrt{\sigma^2\{X\}\,\sigma^2\{Y\}} = \rho\{X, Y\}\sigma\{X\}\sigma\{Y\} \tag{7.31}$$

we obtain:

$$\begin{aligned} \sigma^2\left\{ E_x\Big|_k^{n+1} - C_a\Big|_k E_x\Big|_k^n \right\} &= \sigma^2\left\{ E_x\Big|_k^{n+1} \right\} + \sigma^2\left\{ C_a\Big|_k E_x\Big|_k^n \right\} \\ &\quad - 2\rho\left\{ E_x\Big|_k^{n+1},\ C_a\Big|_k E_x\Big|_k^n \right\} \sigma\left\{ E_x\Big|_k^{n+1} \right\} \sigma\left\{ C_a\Big|_k E_x\Big|_k^n \right\} \end{aligned} \tag{7.32}$$

We will next complete the square of (7.32). As noted in Section 7.6.1, this step is very important because it preserves the signs of the variables in the equation. This allows a wavelike function to exist, which in turn allows the use of typical FDTD time-stepping and boundary conditions. Upon completing the square of (7.32), combining terms, and simplifying, we obtain:

$$\begin{aligned} \sigma^2\left\{ E_x\Big|_k^{n+1} - C_a\Big|_k E_x\Big|_k^n \right\} &= \left(\sigma\left\{ E_x\Big|_k^{n+1} \right\} - \rho\left\{ E_x\Big|_k^{n+1},\ C_a\Big|_k E_x\Big|_k^n \right\} \sigma\left\{ C_a\Big|_k E_x\Big|_k^n \right\} \right)^2 \\ &\quad + \left[1 - \left(\rho\left\{ E_x\Big|_k^{n+1},\ C_a\Big|_k E_x\Big|_k^n \right\} \right)^2 \right] \sigma^2\left\{ C_a\Big|_k E_x\Big|_k^n \right\} \end{aligned} \tag{7.33}$$

Again using the approximation that E_x fields are highly correlated to each other in time, the correlation coefficient, $\rho\left\{ E_x\big|_k^{n+1},\ C_a\big|_k E_x\big|_k^n \right\}$, is approximated as 1, yielding:

$$\sigma^2\left\{ E_x\big|_k^{n+1} - C_a\big|_k E_x\big|_k^n \right\} \approx \left(\sigma\left\{ E_x\big|_k^{n+1} \right\} - \sigma\left\{ C_a\big|_k E_x\big|_k^n \right\} \right)^2 \tag{7.34}$$

The $\sigma\left\{ C_a\big|_k E_x\big|_k^n \right\}$ term in (7.34) can be derived as follows. First, we apply the approximation found in (7.11) to the *square* of this term:

$$\begin{aligned}
\left(\sigma\left\{ C_a\big|_k E_x\big|_k^n \right\}\right)^2 \approx & \left(\frac{2\varepsilon_0\left\langle \varepsilon_r\big|_k \right\rangle - \left\langle \hat{\sigma}\big|_k \right\rangle \Delta t}{2\varepsilon_0\left\langle \varepsilon_r\big|_k \right\rangle + \left\langle \hat{\sigma}\big|_k \right\rangle \Delta t} \right)^2 \sigma^2\left\{ E_x\big|_k^n \right\} \\
& + \frac{8\varepsilon_0 \Delta t\, \sigma\left\{ E_x\big|_k^n \right\}\left\langle E_x\big|_k^n \right\rangle\left(2\varepsilon_0\left\langle \varepsilon_r\big|_k \right\rangle - \left\langle \hat{\sigma}\big|_k \right\rangle \Delta t\right)}{\left(2\varepsilon_0\left\langle \varepsilon_r\big|_k \right\rangle + \left\langle \hat{\sigma}\big|_k \right\rangle \Delta t\right)^3} \cdot \begin{pmatrix} \left\langle \hat{\sigma}\big|_k \right\rangle \rho\left\{ \varepsilon_r\big|_k, E_x\big|_k^n \right\} \sigma\left\{ \varepsilon_r\big|_k \right\} \\ -\left\langle \varepsilon_r\big|_k \right\rangle \rho\left\{ \hat{\sigma}\big|_k, E_x\big|_k^n \right\} \sigma\left\{ \hat{\sigma}\big|_k \right\} \end{pmatrix} \\
& + \frac{\left(4\varepsilon_0 \Delta t \left\langle E_x\big|_k^n \right\rangle\right)^2}{\left(2\varepsilon_0\left\langle \varepsilon_r\big|_k \right\rangle + \left\langle \hat{\sigma}\big|_k \right\rangle \Delta t\right)^4} \begin{pmatrix} \left(\left\langle \hat{\sigma}\big|_k \right\rangle \sigma\left\{ \varepsilon_r\big|_k \right\}\right)^2 + \left(\left\langle \varepsilon_r\big|_k \right\rangle \sigma\left\{ \hat{\sigma}\big|_k \right\}\right)^2 \\ -2\rho\left\{ \hat{\sigma}\big|_k, \varepsilon_r\big|_k \right\} \left\langle \varepsilon_r\big|_k \right\rangle \left\langle \hat{\sigma}\big|_k \right\rangle \sigma\left\{ \varepsilon_r\big|_k \right\} \sigma\left\{ \hat{\sigma}\big|_k \right\} \end{pmatrix}
\end{aligned} \tag{7.35}$$

After considerable manipulation to complete the square, we obtain:

$$\begin{aligned}
\left(\sigma\left\{ C_a\big|_k E_x\big|_k^n \right\}\right)^2 \approx & \left[\begin{matrix} \dfrac{2\varepsilon_0\left\langle \varepsilon_r\big|_k \right\rangle - \left\langle \hat{\sigma}\big|_k \right\rangle \Delta t}{2\varepsilon_0\left\langle \varepsilon_r\big|_k \right\rangle + \left\langle \hat{\sigma}\big|_k \right\rangle \Delta t} \cdot \sigma\left\{ E_x\big|_k^n \right\} \\ + \dfrac{4\varepsilon_0 \Delta t \left\langle E_x\big|_k^n \right\rangle}{\left(2\varepsilon_0\left\langle \varepsilon_r\big|_k \right\rangle + \left\langle \hat{\sigma}\big|_k \right\rangle \Delta t\right)^2} \begin{pmatrix} \left\langle \hat{\sigma}\big|_k \right\rangle \rho\left\{ \varepsilon_r\big|_k, E_x\big|_k^n \right\} \sigma\left\{ \varepsilon_r\big|_k \right\} \\ -\left\langle \varepsilon_r\big|_k \right\rangle \rho\left\{ \hat{\sigma}\big|_k, E_x\big|_k^n \right\} \sigma\left\{ \hat{\sigma}\big|_k \right\} \end{pmatrix} \end{matrix} \right]^2 \cdot \\
& \left[1 + \underbrace{\frac{\left(4\varepsilon_0 \Delta t \left\langle E_x\big|_k^n \right\rangle\right)^2}{\left(2\varepsilon_0\left\langle \varepsilon_r\big|_k \right\rangle + \left\langle \hat{\sigma}\big|_k \right\rangle \Delta t\right)^4}}_{\text{prefactor}} \cdot \mathit{function} \begin{pmatrix} \left\langle \varepsilon_r\big|_k \right\rangle, \left\langle \hat{\sigma}\big|_k \right\rangle, \sigma\left\{ \varepsilon_r\big|_k \right\}, \sigma\left\{ \hat{\sigma}\big|_k \right\}, \\ \rho\left\{ \varepsilon_r\big|_k, E_x\big|_k^n \right\}, \rho\left\{ \hat{\sigma}\big|_k, E_x\big|_k^n \right\} \end{pmatrix} \right]
\end{aligned} \tag{7.36}$$

Note that:

$$\text{prefactor} \sim \begin{cases} O\left(\Delta t/\varepsilon_0\right)^2 & \Delta t \ll \varepsilon_0 \\ O\left(\varepsilon_0/\Delta t\right)^2 & \Delta t \gg \varepsilon_0 \end{cases} \tag{7.37}$$

Then, for many common FDTD models, the prefactor term is negligible, and (7.36) reduces to the following expression upon taking the square root:

$$\sigma\left\{ C_a\big|_k E_x\big|_k^n \right\} \approx \left(\frac{2\varepsilon_0 \langle \varepsilon_r|_k \rangle - \langle \hat{\sigma}|_k \rangle \Delta t}{2\varepsilon_0 \langle \varepsilon_r|_k \rangle + \langle \hat{\sigma}|_k \rangle \Delta t} \right) \sigma\left\{ E_x\big|_k^n \right\}$$
$$+ \; 4\varepsilon_0 \Delta t \left[\frac{\langle \hat{\sigma}|_k \rangle \, \rho\left\{ \varepsilon_r|_k, E_x|_k^n \right\} \sigma\left\{ \varepsilon_r|_k \right\} \; - \; \langle \varepsilon_r|_k \rangle \, \rho\left\{ \hat{\sigma}|_k, E_x|_k^n \right\} \sigma\left\{ \hat{\sigma}|_k \right\}}{\left(2\varepsilon_0 \langle \varepsilon_r|_k \rangle + \langle \hat{\sigma}|_k \rangle \Delta t \right)^2} \right] \left\langle E_x\big|_k^n \right\rangle \tag{7.38}$$

Substituting (7.38) into (7.34) and taking the square root yields the reduction of the left-hand side (LHS) of (7.29):

$$\text{Reduced LHS of (7.29)} \;\approx\; \sigma\left\{ E_x\big|_k^{n+1} \right\} \;-\; \left(\frac{2\varepsilon_0 \langle \varepsilon_r|_k \rangle - \langle \hat{\sigma}|_k \rangle \Delta t}{2\varepsilon_0 \langle \varepsilon_r|_k \rangle + \langle \hat{\sigma}|_k \rangle \Delta t} \right) \sigma\left\{ E_x\big|_k^n \right\}$$
$$- \; 4\varepsilon_0 \Delta t \left[\frac{\langle \hat{\sigma}|_k \rangle \, \rho\left\{ \varepsilon_r|_k, E_x|_k^n \right\} \sigma\left\{ \varepsilon_r|_k \right\} \; - \; \langle \varepsilon_r|_k \rangle \, \rho\left\{ \hat{\sigma}|_k, E_x|_k^n \right\} \sigma\left\{ \hat{\sigma}|_k \right\}}{\left(2\varepsilon_0 \langle \varepsilon_r|_k \rangle + \langle \hat{\sigma}|_k \rangle \Delta t \right)^2} \right] \left\langle E_x\big|_k^n \right\rangle \tag{7.39}$$

Reduction of the Right-Hand Side of (7.29)

Following [11], the right-hand side (RHS) of (7.29) is processed in the same manner as the left-hand side, subject to the following approximations:

$$\rho\left\{ H_y\big|_{k+1/2}^{n+1/2},\; H_y\big|_{k-1/2}^{n+1/2} \right\} \approx 1 \tag{7.40a}$$

$$\rho\left\{ \varepsilon_r\big|_k,\; H_y\big|_{k+1/2}^{n+1/2} \right\} \approx \rho\left\{ \varepsilon_r\big|_k,\; H_y\big|_{k-1/2}^{n+1/2} \right\} \tag{7.40b}$$

$$\rho\left\{ \hat{\sigma}\big|_k,\; H_y\big|_{k+1/2}^{n+1/2} \right\} \approx \rho\left\{ \hat{\sigma}\big|_k,\; H_y\big|_{k-1/2}^{n+1/2} \right\} \tag{7.40c}$$

This yields:

$$\text{Reduced RHS of (7.29)} \;\approx\; \sigma\left\{ -C_b\big|_k \left(H_y\big|_{k+1/2}^{n+1/2} - H_y\big|_{k-1/2}^{n+1/2} \right) \right\}$$
$$\approx \; \frac{2\Delta t}{\left(2\varepsilon_0 \langle \varepsilon_r|_k \rangle + \langle \hat{\sigma}|_k \rangle \Delta t \right) \Delta z} \cdot$$
$$\left[\begin{array}{l} \sigma\left\{ H_y\big|_{k+1/2}^{n+1/2} \right\} - \sigma\left\{ H_y\big|_{k-1/2}^{n+1/2} \right\} + \left(\left\langle H_y\big|_{k+1/2}^{n+1/2} \right\rangle - \left\langle H_y\big|_{k-1/2}^{n+1/2} \right\rangle \right) \cdot \\ \left(\dfrac{2\varepsilon_0 \sigma\left\{ \varepsilon_r|_k \right\} \rho\left\{ \varepsilon_r|_k, H_y|_{k\pm 1/2}^{n+1/2} \right\} + \sigma\left\{ \hat{\sigma}|_k \right\} \rho\left\{ \hat{\sigma}|_k, H_y|_{k\pm 1/2}^{n+1/2} \right\} \Delta t}{2\varepsilon_0 \langle \varepsilon_r|_k \rangle + \langle \hat{\sigma}|_k \rangle \Delta t} \right) \end{array} \right] \tag{7.41}$$

Final Expression for the Statistical Deviation of the E-Field from the Mean

Following [11], upon equating the reduced left-hand and right-hand sides of (7.29) and solving for $\sigma\left\{ E_x\big|_k^{n+1} \right\}$, we obtain the final expression for the deviation, σ, of the E-field from the mean:

$$
\begin{aligned}
\sigma\left\{ E_x\big|_k^{n+1} \right\} \approx\ & \frac{2\varepsilon_0\left\langle \varepsilon_r\big|_k \right\rangle - \left\langle \hat{\sigma}\big|_k \right\rangle \Delta t}{2\varepsilon_0\left\langle \varepsilon_r\big|_k \right\rangle + \left\langle \hat{\sigma}\big|_k \right\rangle \Delta t} \cdot \sigma\left\{ E_x\big|_k^{n} \right\} + \frac{2\Delta t}{\left(2\varepsilon_0\left\langle \varepsilon_r\big|_k \right\rangle + \left\langle \hat{\sigma}\big|_k \right\rangle \Delta t\right)\Delta z} \cdot \\
& \left[\begin{array}{l} \sigma\left\{ H_y\big|_{k+1/2}^{n+1/2} \right\} - \sigma\left\{ H_y\big|_{k-1/2}^{n+1/2} \right\} + \left(\left\langle H_y\big|_{k+1/2}^{n+1/2} \right\rangle - \left\langle H_y\big|_{k-1/2}^{n+1/2} \right\rangle \right) \cdot \\ \left(\dfrac{2\varepsilon_0 \sigma\left\{ \varepsilon_r\big|_k \right\} \rho\left\{ \varepsilon_r\big|_k, H_y\big|_{k\pm 1/2}^{n+1/2} \right\} + \sigma\left\{ \hat{\sigma}\big|_k \right\} \rho\left\{ \hat{\sigma}\big|_k, H_y\big|_{k\pm 1/2}^{n+1/2} \right\} \Delta t}{2\varepsilon_0\left\langle \varepsilon_r\big|_k \right\rangle + \left\langle \hat{\sigma}\big|_k \right\rangle \Delta t} \right) \end{array} \right] \\
& + 4\varepsilon_0 \Delta t \, \frac{\left\langle \hat{\sigma}\big|_k \right\rangle \rho\left\{ \varepsilon_r\big|_k, E_x\big|_k^{n} \right\} \sigma\left\{ \varepsilon_r\big|_k \right\} - \left\langle \varepsilon_r\big|_k \right\rangle \rho\left\{ \hat{\sigma}\big|_k, E_x\big|_k^{n} \right\} \sigma\left\{ \hat{\sigma}\big|_k \right\}}{\left(2\varepsilon_0\left\langle \varepsilon_r\big|_k \right\rangle + \left\langle \hat{\sigma}\big|_k \right\rangle \Delta t\right)^2} \left\langle E_x\big|_k^{n} \right\rangle
\end{aligned}
\tag{7.42}
$$

Equation (7.42) provides a time-stepping expression for the E-field σ, just as (7.28) does for the H-field σ. Both the E-field σ and the H-field σ can assume positive and negative values, and both behave much like waves in the FDTD grid. In effect, the S-FDTD technique time-marches (in a single modeling run) the usual E- and H-fields concurrently with the statistics of these fields, thereby avoiding the need to conduct numerous runs to deduce the field statistics as per brute-force Monte Carlo approaches.

7.7 SEQUENCE OF THE FIELD AND σ UPDATES

Figure 7.1 is a flowchart that shows the sequence of the field and σ updates implemented in the S-FDTD technique [11]. Here, the white boxes designate the classic FDTD time-stepping operations for the 1-D E- and H-fields specified by (7.23) and (7.21), respectively. In the context of the S-FDTD scheme, these are considered to be mean field values. The gray boxes designate the corresponding σ updates for these fields specified by (7.42) and (7.28).

Here, it is important to note that all E-field components are collocated in space and time with their σ counterparts. Similarly, all H-field components are collocated in space and time with their σ counterparts. Furthermore, standard leapfrog time-stepping is applied to all E- and H-field components and their σ counterparts, so that all are marched in time in a synchronous manner. A standard absorbing boundary condition (ABC) is applied to the E-field components at the outer grid points; the identical ABC is applied to their σ counterparts at the same points. The latter exploits the wavelike nature of the σ terms derived in the construction of the S-FDTD algorithm.

It should be clear from this discussion that implementing the S-FDTD technique requires a doubling of the classic FDTD computer storage, due to the mirroring of each field component with its σ counterpart. The computer running time is slightly more than doubled due to the additional operations involved in implementing (7.42) and (7.28). These additional requirements pale in comparison to those involved in implementing a full-scale Monte Carlo series of runs.

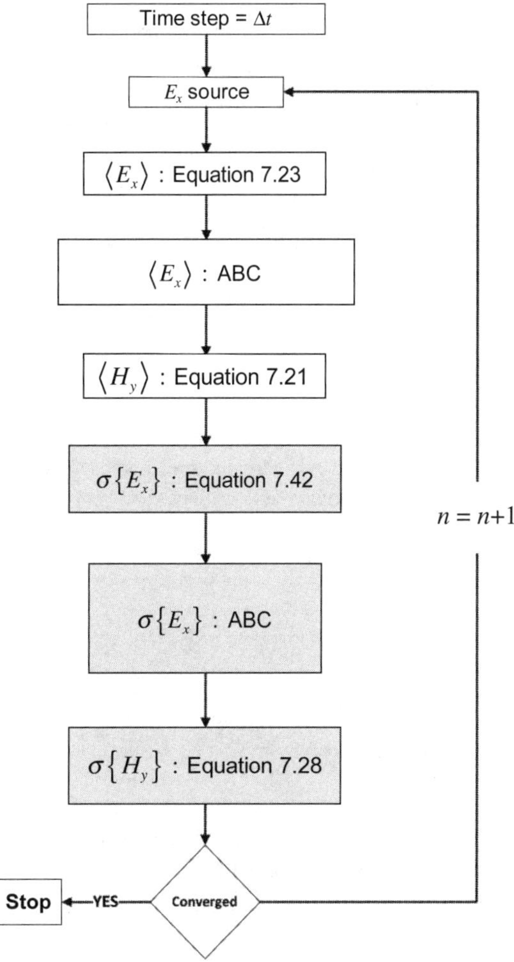

Fig. 7.1 Flow chart for the 1-D S-FDTD algorithm. *Adapted from:* S. M. Smith and C. M. Furse, *IEEE Trans. Antennas and Propagation*, Vol. 60, 2012, pp. 3343–3350, ©2012 IEEE.

7.8 LAYERED BIOLOGICAL TISSUE EXAMPLE

To evaluate the accuracy of the S-FDTD method relative to the Monte Carlo technique, Ref. [11] reported modeling the 1-D layered biological tissue model shown in Fig. 7.2. The relative permittivity (ε_r) and electrical conductivity ($\hat{\sigma}$) of each layer were assumed to be normally distributed random variables having their mean and variance specified as shown [20].

In Fig. 7.2, a 1 V/m 2-GHz sinusoidal plane wave impinging from the left was used as the source. The space-cell size was $\Delta z = 0.51$ mm, equivalent to 1/40th of the smallest wavelength within the dielectric media of the model. The time-step was $\Delta t = \Delta z / 2c \cong 8.5 \times 10^{-13}$ s, which satisfied the requirement posed by (7.37). Standard Mur absorbing boundary conditions [21], which are theoretically exact in 1-D, were applied to both E_x and σ at the outermost grid points. For example, at the left outermost grid point ($z = 0$), these were given by:

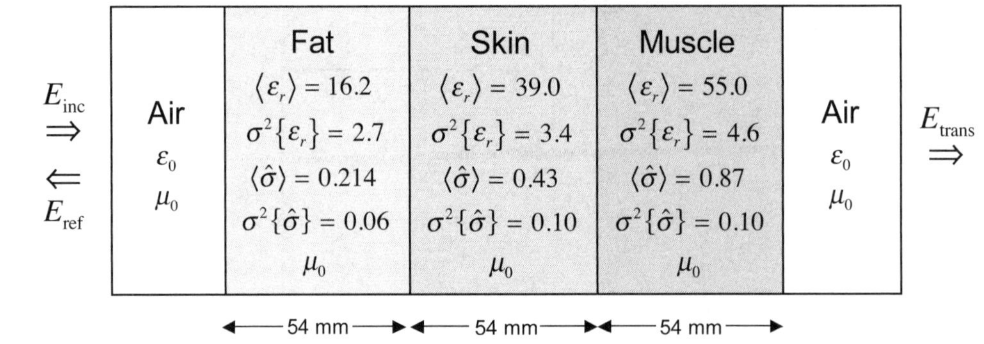

E_{inc} ⇒ ⇐ E_{ref}	Air ε_0 μ_0	Fat	Skin	Muscle	Air ε_0 μ_0	E_{trans} ⇒
		$\langle\varepsilon_r\rangle = 16.2$	$\langle\varepsilon_r\rangle = 39.0$	$\langle\varepsilon_r\rangle = 55.0$		
		$\sigma^2\{\varepsilon_r\} = 2.7$	$\sigma^2\{\varepsilon_r\} = 3.4$	$\sigma^2\{\varepsilon_r\} = 4.6$		
		$\langle\hat{\sigma}\rangle = 0.214$	$\langle\hat{\sigma}\rangle = 0.43$	$\langle\hat{\sigma}\rangle = 0.87$		
		$\sigma^2\{\hat{\sigma}\} = 0.06$	$\sigma^2\{\hat{\sigma}\} = 0.10$	$\sigma^2\{\hat{\sigma}\} = 0.10$		
		μ_0	μ_0	μ_0		
		← 54 mm →	← 54 mm →	← 54 mm →		

Fig. 7.2 1-D layered biological tissue model for testing the S-FDTD algorithm. The relative permittivity (ε_r) and electrical conductivity ($\hat{\sigma}$) of each layer were assumed to be normally distributed random variables having their mean and variance specified as shown. *Adapted from:* S. M. Smith and C. M. Furse, *IEEE Trans. Antennas and Propagation*, Vol. 60, 2012, pp. 3343–3350, ©2012 IEEE.

$$E_x\big|_0^{n+1} = E_x\big|_1^n + \frac{c\Delta t - \Delta z}{c\Delta t + \Delta z}\cdot\left(E_x\big|_1^{n+1} - E_x\big|_0^n\right) \tag{7.43a}$$

$$\sigma\left\{E_x\big|_0^{n+1}\right\} = \sigma\left\{E_x\big|_1^n\right\} + \frac{c\Delta t - \Delta z}{c\Delta t + \Delta z}\cdot\left(\sigma\left\{E_x\big|_1^{n+1}\right\} - \sigma\left\{E_x\big|_0^n\right\}\right) \tag{7.43b}$$

A Monte Carlo analysis was used to determine the actual mean and variance of the fields in the model of Fig. 7.2. For this purpose, 10,000 FDTD simulations were done, after which the mean and variance of the fields were calculated from their outputs.

Figure 7.3 compares the S-FDTD and Monte Carlo results for the mean of the electric field within the 1-D layered biological tissue model of Fig. 7.2 [11]. Here, the mean values computed using the S-FDTD method were all within approximately 0.02 V/m of the Monte Carlo mean values for the 1 V/m 2-GHz incident plane wave. (Recall that the S-FDTD mean-value time-stepping equations are the same as the classic FDTD equations using mean values of the stochastic electrical parameters of the materials in the simulations.) The comparison shown in Fig. 7.3 verifies that this mean approximation was sufficiently accurate for this application.

Figure 7.4 compares the S-FDTD and Monte Carlo results for the variance of the electric field within the 1-D layered tissue model of Fig. 7.2 [11]. Recognizing that implementing (7.42) requires inputting the correlation coefficients $\rho\left\{\varepsilon_r\big|_k, E_x\big|_k^n\right\}$, $\rho\left\{\hat{\sigma}\big|_k, E_x\big|_k^n\right\}$, $\rho\left\{\varepsilon_r\big|_k, H_y\big|_{k\pm 1/2}^{n+1/2}\right\}$, and $\rho\left\{\hat{\sigma}\big|_k, H_y\big|_{k\pm 1/2}^{n+1/2}\right\}$, and that these coefficients are *a priori* unknown, this figure shows S-FDTD results for two approximations of the correlation coefficients: (a) each ρ value set equal to 1; and (b) each ρ value set equal to the reflection coefficient due to the nearest dielectric interface. As seen in Fig. 7.4, relative to the Monte Carlo benchmark, the former choice overestimated the variance at most points in the model, whereas the latter choice underestimated the variance.

This characteristic was seen in numerous similar examples, only one of which was reported in [11]. It was concluded that it is possible to *bracket* the actual field variances within one order of magnitude by using two simple S-FDTD runs. More research to develop better approximations for the correlation coefficients would be expected to improve the accuracy of these results.

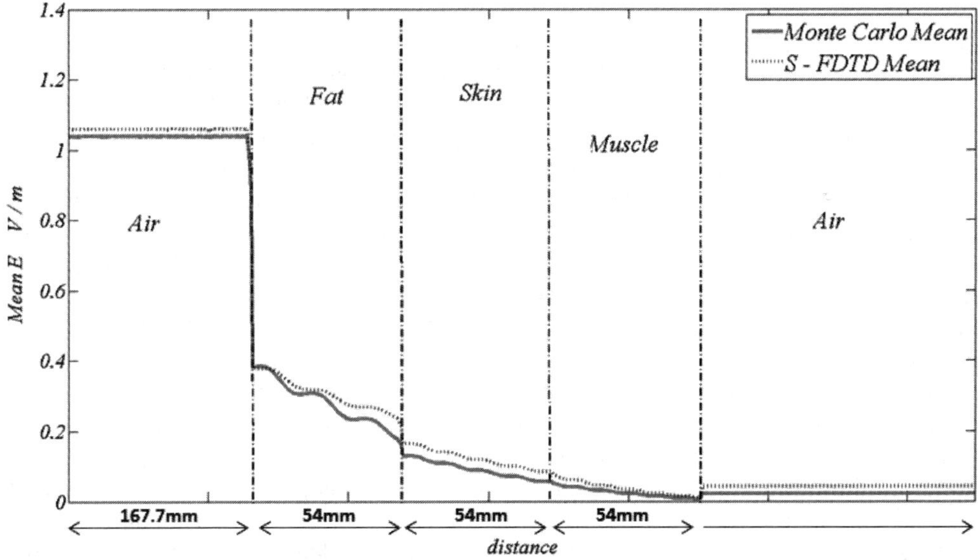

Fig. 7.3 Comparison of the S-FDTD and Monte Carlo results for the mean of the electric field within the 1-D layered biological tissue model of Fig. 7.2. *Adapted from:* S. M. Smith and C. M. Furse, *IEEE Trans. Antennas and Propagation*, Vol. 60, 2012, pp. 3343–3350, ©2012 IEEE.

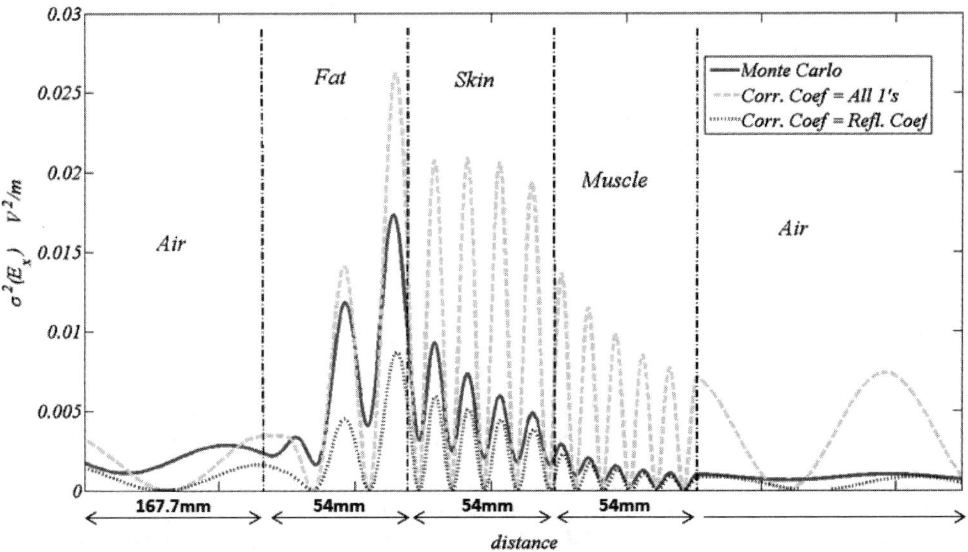

Fig. 7.4 Comparison of the S-FDTD and Monte Carlo results for the variance of the electric field within the 1-D layered biological tissue model of Fig. 7.2. *Adapted from:* S. M. Smith and C. M. Furse, *IEEE Trans. Antennas and Propagation*, Vol. 60, 2012, pp. 3343–3350, ©2012 IEEE.

7.9 SUMMARY AND CONCLUSIONS

This chapter reviewed a new stochastic FDTD (S-FDTD) method originally reported in [11] that provides an efficient way to evaluate statistical variations in numerical simulations of electromagnetic wave interactions caused by random variations of the electrical properties of the model. The statistics of these variations are incorporated directly into FDTD, which computes an estimate of the resulting mean and variance of the fields at every point in space and time.

As an example of the S-FDTD method, the 1-D bioelectromagnetics example reported in [11] (sinusoidal plane-wave illumination of three tissue layers) was reviewed. Field variances computed using S-FDTD were compared to those computed using the brute-force Monte Carlo technique. It was found possible to bracket the Monte Carlo result for the field variance at each point within the model within one order of magnitude, using only two S-FDTD runs based on simple estimates for the cross-correlations of the fields and electrical properties of the materials. Additional research is needed to improve the estimates for these correlations, which would result in tightening this bracket. While this example involved a simple 1-D geometry, the S-FDTD method is not limited to 1-D, having already been extended to 3-D in [15].

The total additional memory for the S-FDTD method is double that for the traditional FDTD method, since storage must be provided for the variance of the fields in addition to the mean fields. Simulation time per modeling run is slightly more than doubled, since S-FDTD has twice as many equations to compute as FDTD, and the variance time-stepping equation (7.42) is somewhat more complicated than the others. Still, when compared with having to run hundreds, thousands, or more of Monte Carlo simulations, the S-FDTD method offers a huge savings in computation time. This opens up the possibility of additional assessment of confidence intervals, expected variations, and other statistical parameters for applications in bioelectromagnetics, biophotonics, and geophysics where the electrical properties have uncertainty or variability.

ACKNOWLEDGMENT

The authors would like to thank Dr. Camelia Gabriel for providing the electrical properties of human tissues used in the simulations.

REFERENCES

[1] Taflove, A., and M. E. Brodwin, "Computation of the electromagnetic fields and induced temperatures within a model of the microwave-irradiated human eye," *IEEE Trans. Microwave Theory and Techniques*, Vol. 23, 1975, pp. 888–896.

[2] Sullivan, D. M., O. P. Gandhi, and A. Taflove, "Use of the finite-difference time-domain method in calculating EM absorption in man models," *IEEE Trans. Biomedical Engineering*, Vol. 35, 1988, pp. 179–186.

[3] Gandhi, O. P., G. Lazzi, and C. Furse, "Electromagnetic absorption in the human head and neck for mobile telephones at 835 and 1900 MHz," *IEEE Trans. Microwave Theory and Techniques*, Vol. 44, 1996, pp. 1884–1897.

[4] Piket-May, M. J., A. Taflove, and J. B. Troy, "Electrodynamics of visible light interactions with the vertebrate retinal rod," *Optics Lett.*, Vol. 18, 1993, pp. 568–570.

[5] Dunn, A., and R. Richards-Kortum, "Three-dimensional computation of light scattering from cells," *IEEE J. Selected Topics in Quantum Electronics*, Vol. 2, 1996, pp. 898–905.

[6] Johnson, D., E. Cherkaev, C. Furse, and A. C. Tripp, "Cross-borehole delineation of a conductive ore deposit in a resistive host – Experimental design," *Geophysics*, Vol. 66, 2001, pp. 824–835.

[7] Ward, J., C. Swenson, and C. Furse, "The impedance of a short dipole antenna in a magnetized plasma via a FDTD model," *IEEE Trans. Antennas and Propagation*, Vol. 53, 2005, pp. 2711–2718.

[8] Simpson, J. J., and A. Taflove, "A review of progress in FDTD Maxwell's equations modeling of impulsive subionospheric propagation below 300 kHz," *IEEE Trans. Antennas and Propagation*, Vol. 55, 2007, pp. 1582–1590.

[9] Sadiku, M. N. O., *Numerical Techniques in Electromagnetics*, 2nd ed., New York: CRC Press, 2001.

[10] Borcherds, P. H., "Importance sampling: An illustrative introduction," *European J. Physics*, Vol. 21, 2000, pp. 405–411.

[11] Smith, S. M., and C. M. Furse, "Stochastic FDTD for analysis of statistical variation in electromagnetic fields," *IEEE Trans. Antennas and Propagation*, Vol. 60, 2012, pp. 3343–3350.

[12] Nayfeh, A., *Perturbation Methods*, 1st ed., New York: Wiley, 1973.

[13] Kleiber, M., and T. D. Hein, *The Stochastic Finite Element Method: Basic Perturbation Technique and Computer Implementation*, 1st ed., New York: Wiley, 1992.

[14] George Casella, R. L. B., *Statistical Inference*, 2nd ed., Pacific Grove, CA: Duxbury, Thomson Learning, 2002.

[15] Smith, S. M., *Stochastic FDTD*, Ph.D. dissertation, University of Utah, Salt Lake City, 2011.

[16] Smith, S. M., and C. Furse, "A stochastic FDTD method for statistically varying biological tissues," *Proc. 2011 IEEE AP-S International Symposium on Antennas and Propagation and USNC/CNC/URSI*, Spokane, WA, 2011.

[17] Krishnan, V., *Probability and Random Process*, 1st ed., New York: Wiley, 2006.

[18] Walpole, R. H. M., S. L. Myers, and K. Ye, *Probability and Statistics for Engineers and Scientists*, Upper Saddle River, NJ: Pearson Prentice-Hall, 2007, p. 816.

[19] Taflove, A., and S. C. Hagness, *Computational Electrodynamics: The Finite-Difference Time-Domain Method*, 3rd ed., Norwood, MA: Artech, 2005.

[20] Gabriel, C., personal communication, 2006.

[21] Mur, G., "Absorbing boundary conditions for the finite-difference approximation of the time-domain electromagnetic-field equations," *IEEE Trans. Electromagnetic Compatibility*, Vol. 23, 1981, pp. 377–382.

Chapter 8

FDTD Modeling of Active Plasmonics[1]

Iftikhar Ahmed, Eng Huat Khoo, and Er Ping Li

8.1 INTRODUCTION

Until a few years ago, the miniaturization of photonic devices was a challenge due to the diffraction limit, which restricted the minimum size of a component to approximately one-half of the optical wavelength ($\lambda/2$). However, recent developments in plasmonics technologies have enabled photonic device features to have key characteristic dimensions smaller than the diffraction limit. Plasmonics technologies exploit optical electromagnetic wave propagation that is locally confined to a metal–dielectric interface, despite the potentially nanoscale transverse dimensions of the metal component. Several types of plasmonics structures capable of guiding, manipulating, and radiating electromagnetic signals have been reported, including nano-antennas, lenses, resonators, sensors, and waveguides [1–4].

While most investigations in this area to date have been concerned with passive plasmonics phenomena and applications [1–3], active plasmonics technologies are receiving increased attention owing to their potential for increased flexibility in manipulating light [4–7]. At the same time, complementary metal-oxide-semiconductor (CMOS) electronics technology is reaching limiting values in terms of size and speed. It is difficult to abandon CMOS technology due to its numerous applications, inexpensive processes, and mature fabrication technologies. However, it may be possible to overcome the limitations of CMOS by interfacing it with plasmonics. Active plasmonics is believed to be an excellent candidate for this purpose because the interface between both technologies (which use similar semiconductor materials) is easier to realize than with passive plasmonics.

This chapter incorporates a nonlinear, multi-level, multi-electron, quantum system formulation into Maxwell's time-dependent curl equations to simulate the solid-state part of an active plasmonics structure. Here, electron dynamics are governed by the Pauli exclusion principle, state filling, and dynamical Fermi–Dirac thermalization. As reported in the recent literature, this approach has been used to model molecular or atomic media [8], as well as active photonics devices such as lasers and optical switches [9–11]. To simulate the metallic part of an active plasmonics structure, we incorporate a linear Lorentz–Drude dielectric dispersion model [12, 13] into Maxwell's curl equations. The Lorentz–Drude model deals with free electrons (intraband effects) and bound electrons (interband effects) in metals.

---

[1]This chapter is adapted from Ref. [16], I. Ahmed, E. H. Khoo, O. Kurniawan, and E. P. Li, "Modeling and simulation of active plasmonics with the FDTD method by using solid state and Lorentz–Drude dispersive model," *J. Optical Society of America B*, Vol. 28, 2011, pp. 352–359, ©2011 The Optical Society of America.

We apply the finite-difference time-domain (FDTD) method to temporally integrate the resulting set of linear and nonlinear equations, exploiting the proven ability of FDTD to robustly model the physics of complex dynamic media. Related previous work used FDTD to integrate Maxwell's curl equations concurrently with either a stochastic ensemble Monte Carlo approach to simulate carrier transport [14], or the Schrödinger equation to simulate carrier dynamics [15]. Relative to these previous approaches, the technique reviewed in this chapter (originally presented in [16]) has two important modeling capabilities that allow more realistic simulations of active plasmonics devices: (a) intraband and interband electron transitions (i.e., transitions from one energy level to another, transitions from the valance band to the conduction band, and vice versa); and (b) stimulated emission. This appears to be the first time-domain approach for simulating active plasmonics devices to employ a realistic solid-state physics model combined with a Lorentz–Drude dispersive metallic model. In subsequent sections, we shall detail the numerical methodology, show the results of illustrative simulations, and draw conclusions.

8.2 OVERVIEW OF THE COMPUTATIONAL MODEL

Interactions between electrons and photons comprise the fundamental physics basis of electro-optical systems. Photon absorption causes an electron to transition from a lower energy state to a higher energy state. In the inverse process, photon emission occurs when an electron moves from the conduction band to the valence band.

This chapter reviews a semiclassical computational model [16] wherein the macroscopic time-domain Maxwell's curl equations representing Ampere's law and Faraday's law:

$$\nabla \times \boldsymbol{H} = \frac{\partial \boldsymbol{D}}{\partial t} \tag{8.1}$$

$$\nabla \times \boldsymbol{E} = -\frac{\partial \boldsymbol{B}}{\partial t} \tag{8.2}$$

for $\boldsymbol{D} = \varepsilon_0 \varepsilon_r \boldsymbol{E}$ and $\boldsymbol{B} = \mu_0 \mu_r \boldsymbol{H}$, are integrated in time concurrently with partial differential equations that model intraband (Drude) and interband (Lorentz) electron transitions in metal, as well as transient intraband and interband electron dynamics and the carrier thermal equilibrium process for direct-bandgap semiconductors. We note that the semiconductor model is formulated to automatically incorporate the band-filling and nonlinear optical effects associated with carrier dynamics. Although the overall formulation developed to simulate active plasmonics devices is complicated, it efficiently covers the primary physical phenomena of interest.

8.3 LORENTZ–DRUDE MODEL FOR METALS

At near-infrared and visible optical wavelengths, the frequency-dependent relative permittivity $\varepsilon_r(\omega)$ of numerous metals can be described by a Lorentz–Drude model written as [12, 16]:

$$\varepsilon_r(\omega) = \varepsilon_\infty - \frac{\omega_{p\mathrm{D}}^2}{\omega^2 - j\omega\Gamma_\mathrm{D}} + \frac{\omega_{p\mathrm{L}}^2 \Delta\varepsilon_\mathrm{L}}{\omega_\mathrm{L}^2 - \omega^2 + j\omega\Gamma_\mathrm{L}} \tag{8.3}$$

where $\omega_{p\mathrm{D}}$ is the plasma frequency and Γ_D is the damping coefficient for the Drude model; and $\Delta\varepsilon_\mathrm{L}$ is a weighting factor, $\omega_{p\mathrm{L}}$ is the resonance frequency, and Γ_L is the spectral width for the Lorentz model. While a number of oscillators can be considered for the Lorentz model, for simplicity we consider only one oscillator having the frequency ω_L.

Upon inserting the frequency-dependent Lorentz model of (8.3) into the Fourier transform of (8.1), and then performing the inverse Fourier transform, after some simplification we obtain the following system of time-dependent partial differential equations for the Maxwell–Ampere law:

$$\nabla \times \boldsymbol{H} = \varepsilon_0 \varepsilon_\infty \frac{\partial \boldsymbol{E}}{\partial t} + \boldsymbol{J}_\mathrm{D} + \varepsilon_0 \frac{\partial \boldsymbol{Q}_\mathrm{L}}{\partial t} \tag{8.4}$$

$$\omega_{p\mathrm{D}}^2 \varepsilon_0 \boldsymbol{E} = \frac{\partial \boldsymbol{J}_\mathrm{D}}{\partial t} + \Gamma_\mathrm{D} \boldsymbol{J}_\mathrm{D} \tag{8.5}$$

$$\Delta\varepsilon_\mathrm{L} \omega_{p\mathrm{L}}^2 \boldsymbol{E} = \frac{\partial^2 \boldsymbol{Q}_\mathrm{L}}{\partial t^2} + \Gamma_\mathrm{L} \frac{\partial \boldsymbol{Q}_\mathrm{L}}{\partial t} + \omega_\mathrm{L}^2 \boldsymbol{Q}_\mathrm{L} \tag{8.6}$$

where the Maxwell–Ampere curl equation containing both Drude $\boldsymbol{J}_\mathrm{D}$ and Lorentz $\varepsilon_0 \partial \boldsymbol{Q}_\mathrm{L} / \partial t$ terms is given by (8.4), and the time evolution of the Drude and Lorentz terms is governed by (8.5) and (8.6), respectively.

After incorporating the Drude and Lorentz terms, Maxwell's equations (8.4) and (8.2) are written in Cartesian coordinates as the following six coupled scalar equations, suitable for FDTD time marching:

$$\frac{\partial E_x}{\partial t} = \frac{1}{\varepsilon_0 \varepsilon_\infty}\left(\frac{\partial H_z}{\partial y} - \frac{\partial H_y}{\partial z}\right) - \frac{1}{\varepsilon_0 \varepsilon_\infty} J_{\mathrm{D}_x} - \frac{1}{\varepsilon_\infty}\frac{\partial Q_{\mathrm{L}_x}}{\partial t} \tag{8.7}$$

$$\frac{\partial E_y}{\partial t} = \frac{1}{\varepsilon_0 \varepsilon_\infty}\left(\frac{\partial H_x}{\partial z} - \frac{\partial H_z}{\partial x}\right) - \frac{1}{\varepsilon_0 \varepsilon_\infty} J_{\mathrm{D}_y} - \frac{1}{\varepsilon_\infty}\frac{\partial Q_{\mathrm{L}_y}}{\partial t} \tag{8.8}$$

$$\frac{\partial E_z}{\partial t} = \frac{1}{\varepsilon_0 \varepsilon_\infty}\left(\frac{\partial H_y}{\partial x} - \frac{\partial H_x}{\partial y}\right) - \frac{1}{\varepsilon_0 \varepsilon_\infty} J_{\mathrm{D}_z} - \frac{1}{\varepsilon_\infty}\frac{\partial Q_{\mathrm{L}_z}}{\partial t} \tag{8.9}$$

$$\frac{\partial H_x}{\partial t} = \frac{1}{\mu_r \mu_0}\left(\frac{\partial E_y}{\partial z} - \frac{\partial E_z}{\partial y}\right) \tag{8.10}$$

$$\frac{\partial H_y}{\partial t} = \frac{1}{\mu_r \mu_0}\left(\frac{\partial E_z}{\partial x} - \frac{\partial E_x}{\partial z}\right) \tag{8.11}$$

$$\frac{\partial H_z}{\partial t} = \frac{1}{\mu_r \mu_0}\left(\frac{\partial E_x}{\partial y} - \frac{\partial E_y}{\partial x}\right) \tag{8.12}$$

For example, applying standard FDTD central differencing in time to (8.7), (8.5), and (8.6) yields, respectively, the following time-stepping relations:

$$E_x^{n+1} = \frac{1}{\Omega_x} E_x^n + \frac{\Delta t}{\Omega_x \varepsilon_0 \varepsilon_\infty}\left(\frac{\partial H_z^{n+1/2}}{\partial y} - \frac{\partial H_y^{n+1/2}}{\partial z}\right) - \frac{\Delta t}{2\Omega_x \varepsilon_0 \varepsilon_\infty}\left[(\alpha_x + 1)J_{\mathrm{D}_x}^n + \beta_x E_x^n\right] - \frac{1}{\Omega_x \varepsilon_\infty}\left[\varsigma_x E_x^n + (\tau_x - 1)Q_{\mathrm{L}_x}^n - \rho_x Q_{\mathrm{L}_x}^{n-1}\right] \tag{8.13}$$

$$J_{\mathrm{D}_x}^{n+1} = \alpha_x J_{\mathrm{D}_x}^n + \beta_x\left(E_x^{n+1} + E_x^n\right) \tag{8.14}$$

$$Q_{\mathrm{L}_x}^{n+1} = \tau_x Q_{\mathrm{L}_x}^n - \varsigma_x Q_{\mathrm{L}_x}^{n-1} + \varsigma_x\left(E_x^{n+1} + E_x^n\right) \tag{8.15}$$

where

$$\alpha_x = \frac{1 - \Gamma_\mathrm{D}\Delta t/2}{1 + \Gamma_\mathrm{D}\Delta t/2}, \qquad \beta_x = \frac{\omega_{p\mathrm{D}}^2 \varepsilon_0 \Delta t/2}{1 + \Gamma_\mathrm{D}\Delta t/2} \tag{8.16a}$$

$$\varsigma_x = \frac{(\omega_{p\mathrm{L}}\Delta t)^2 \Delta\varepsilon_\mathrm{L}/2}{1 + \Gamma_\mathrm{L}\Delta t + (\omega_\mathrm{L}\Delta t)^2/2}, \qquad \tau_x = \frac{2 + \Gamma_\mathrm{L}\Delta t - (\omega_\mathrm{L}\Delta t)^2/2}{1 + \Gamma_\mathrm{L}\Delta t + (\omega_\mathrm{L}\Delta t)^2/2} \tag{8.16b}$$

$$\rho_x = \frac{1}{1 + \Gamma_\mathrm{L}\Delta t + (\omega_\mathrm{L}\Delta t)^2/2}, \qquad \Omega_x = 1 + \frac{\varsigma_x}{\varepsilon_\infty} + \frac{\beta_x \Delta t}{2\varepsilon_0 \varepsilon_\infty} \tag{8.16c}$$

Analogous time-stepping relations can be obtained for the E_y and E_z field components.

8.4 DIRECT-BANDGAP SEMICONDUCTOR MODEL

The starting point for the solid-state materials model, as reported in [16], is the Maxwell–Ampere law (8.1), which can be written for this case as:

$$\nabla \times \boldsymbol{H} = \varepsilon_0 \varepsilon_\infty \frac{\partial \boldsymbol{E}}{\partial t} + \frac{\partial \boldsymbol{P}}{\partial t} \tag{8.17}$$

Here, $\boldsymbol{P}$ is the macroscopic polarization density representing the total dipole moment per unit volume:

$$\boldsymbol{P}(\boldsymbol{r}, t) = \boldsymbol{U}_\mathrm{m}(t) \sum_h N_{\mathrm{dip},h}(\boldsymbol{r}) \tag{8.18}$$

where $\boldsymbol{U}_\mathrm{m}(t)$ is the atomic dipole moment, and $N_{\mathrm{dip},h}(\boldsymbol{r})$ is the dipole volume density for level h within energy width δE, specified by the number of dipoles N_0 divided by the unit volume δV.

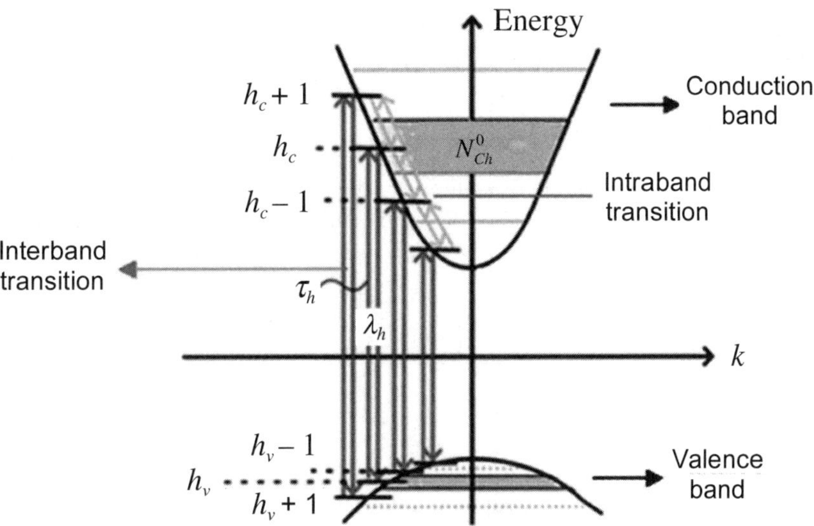

Fig. 8.1 Discretization of the conduction and valence bands for a multi-level multi-electron model of a direct-bandgap semiconductor for simulation with the FDTD method. *Source:* Ahmed, Khoo, Kurniawan, and Li, *J. Optical Society of America B*, Vol. 28, 2011, pp. 352–359, ©2011 The Optical Society of America.

In this chapter, following [16], we model electron dynamics in a direct-bandgap semiconductor by examining electron transitions between (and among) 10 discrete energy levels in the conduction band and 10 discrete energy levels in the valence band. Figure 8.1 illustrates this approach, denoting representative valence-band energy levels as h_v and representative conduction-band energy levels as h_c. In this figure, an arrow drawn between a conduction-band energy level and a valence-band energy level (or vice versa) depicts an interband transition, whereas an arrow drawn between different energy levels of the same band depicts an intraband transition. The occupation probability with respect to time for each energy level in the conduction and valence bands, and the effect of carrier densities for different level pairs, has been studied in [8]. It was observed that the electrons' relaxation time to equilibrium is much slower in the conduction band than in the valence band. From these data and knowledge of our assumed energy levels, we can calculate intraband and interband transition times τ. Specification of the energy levels is also helpful in finding Fermi–Dirac thermalization dynamics in the semiconductor, discussed later in this section.

After a mathematical derivation, we find that the dynamics of the vector components of the polarization $\boldsymbol{P}$ in (8.18) are governed by second-order differential equations. For example, we have the following differential equation for the x-component, $P_{h,x}$:

$$\frac{\partial^2 P_{h,x}(\boldsymbol{r},t)}{\partial t^2} + \Gamma_h \frac{\partial P_{h,x}(\boldsymbol{r},t)}{\partial t} + \left\{ \omega_{ah} + \left(\frac{2\omega_{ah}}{\hbar}\right)^2 |U_h|^2 \left[A_x(\boldsymbol{r},t)\right]^2 \right\} P_{h,x}(\boldsymbol{r},t)$$
$$= \frac{2\omega_{ah}}{\hbar}|U_h|^2 \left[\frac{N_{\text{dip},h}(\boldsymbol{r})}{N^0_{Vh}(\boldsymbol{r})} N_{Vh}(\boldsymbol{r},t) - \frac{N_{\text{dip},h}(\boldsymbol{r})}{N^0_{Ch}(\boldsymbol{r})} N_{Ch}(\boldsymbol{r},t) \right] E_x(\boldsymbol{r},t) \quad (8.19)$$

where Γ_h represents the dephasing rate at the hth energy level after excitation; ω_{ah} is the interband transition frequency; $\hbar$ is Planck's constant divided by 2π; $|U_h|^2 = 3\pi\hbar\varepsilon_0 c^3 / \omega_{ah}^3 \tau_h$, where c is the free-space speed of light and τ_h is the interband transition time; A_x is the x-component of the vector potential $\boldsymbol{A}$; N_{Vh}^0 and N_{Ch}^0 are the initial values of the volume densities of the energy states at level h in the valence and conduction bands, respectively; and N_{Vh} and N_{Ch} are the later values of these volume densities.

After further simplification of (8.19) and application of standard FDTD temporal differencing, we obtain the following updating expression for $P_{h,x}$ at grid-point $(i+1/2, j, k)$ and time-step n:

$$\begin{aligned} P_{h,x}\Big|_{i+1/2,j,k}^{n+1} &= \frac{1}{\Gamma_h \Delta t + 2}\left\{4 - 2(\Delta t)^2\left[\omega_{ah}^2 + \left(\frac{2\omega_{ah}|U_h|A_x\big|_{i+1/2,j,k}^{n}}{\hbar}\right)^2\right]\right\} P_{h,x}\Big|_{i+1/2,j,k}^{n} \\ &+ \frac{\Gamma_h \Delta t - 2}{\Gamma_h \Delta t + 2} P_{h,x}\Big|_{i+1/2,j,k}^{n-1} - \frac{4(\Delta t)^2\omega_{ah}}{\hbar(\Gamma_h \Delta t + 2)}|U_h|^2\left(N_{Ch}\Big|_{i+1/2,j,k}^{n} - N_{Vh}\Big|_{i+1/2,j,k}^{n}\right)E_x\Big|_{i+1/2,j,k}^{n} \end{aligned} \tag{8.20}$$

where the vector-potential component A_x is obtained via the recursive sum:

$$A_x\Big|_{i+1/2,j,k}^{n} = A_x\Big|_{i+1/2,j,k}^{n-1} - E_x\Big|_{i+1/2,j,k}^{n}\Delta t \tag{8.21}$$

and N_{Ch} and N_{Vh} are obtained via time-stepped rate equations, discussed next. Note that expressions analogous to (8.20) and (8.21) can be written for $P_{h,y}$ and $P_{h,z}$.

To develop rate equations for N_{Ch} and N_{Vh} in (8.20), we begin by examining electron transitions between energy levels in the semiconductor. First, we consider the interband transition and spontaneous decay of electrons from conduction-band energy level h_c to valence-band energy level h_v. Let ΔN_h denote the number of electrons per unit volume transferred in this manner. Then, at grid-point $(i+1/2, j, k)$ and time-step n, we have:

$$\begin{aligned} \Delta N_h\Big|_{i+1/2,j,k}^{n} &= \frac{\omega_{ah}}{\hbar}\left(\begin{aligned} &A_x\Big|_{i+1/2,j,k}^{n} P_{h,x}\Big|_{i+1/2,j,k}^{n} + A_y\Big|_{i+1/2,j,k}^{n} P_{h,y}\Big|_{i+1/2,j,k}^{n} \\ &+ A_z\Big|_{i+1/2,j,k}^{n} P_{h,z}\Big|_{i+1/2,j,k}^{n} \end{aligned}\right) \\ &+ \frac{1}{\tau_h} N_{Ch}\Big|_{i+1/2,j,k}^{n}\left(1 - N_{Vh}\Big|_{i+1/2,j,k}^{n} \Big/ N_{Vh}^0\Big|_{i+1/2,j,k}^{n}\right) \end{aligned} \tag{8.22}$$

Similarly, relations for intraband electron transitions within the conduction and valence bands can be derived. For example, the number of electrons per unit volume $\Delta N_{C(h,h-1)}$ transferred within the conduction band from energy level h_c to $h_c - 1$ is given by:

$$\begin{aligned} \Delta N_{C(h,h-1)}\Big|_{i+1/2,j,k}^{n} &= \frac{1}{\tau_{C(h,h-1)}} N_{Ch}\Big|_{i+1/2,j,k}^{n}\left(1 - N_{C(h-1)}\Big|_{i+1/2,j,k}^{n} \Big/ N_{C(h-1)}^0\Big|_{i+1/2,j,k}^{n}\right) \\ &- \frac{1}{\tau_{C(h-1,h)}} N_{C(h-1)}\Big|_{i+1/2,j,k}^{n}\left(1 - N_{Ch}\Big|_{i+1/2,j,k}^{n} \Big/ N_{Ch}^0\Big|_{i+1/2,j,k}^{n}\right) \end{aligned} \tag{8.23}$$

where $\tau_{C(h,h-1)}$ is the downward transition rate and $\tau_{C(h-1,h)}$ is the upward transition rate between levels h_c and $h_c - 1$ in the conduction band. By analogy, the number of electrons per unit volume $\Delta N_{V(h,h-1)}$ transferred within the valence band from energy level h_v to $h_v - 1$ is given by:

$$\begin{aligned}\Delta N_{V(h,h-1)}\Big|_{i+1/2,j,k}^{n} &= \frac{1}{\tau_{V(h,h-1)}} N_{Vh}\Big|_{i+1/2,j,k}^{n} \left(1 - N_{V(h-1)}\Big|_{i+1/2,j,k}^{n} \Big/ N_{V(h-1)}^{0}\Big|_{i+1/2,j,k}^{n}\right) \\ &- \frac{1}{\tau_{V(h-1,h)}} N_{V(h-1)}\Big|_{i+1/2,j,k}^{n} \left(1 - N_{Vh}\Big|_{i+1/2,j,k}^{n} \Big/ N_{Vh}^{0}\Big|_{i+1/2,j,k}^{n}\right)\end{aligned} \tag{8.24}$$

where $\tau_{V(h,h-1)}$ is the downward transition rate and $\tau_{V(h-1,h)}$ is the upward transition rate between levels h_v and $h_v - 1$ in the valence band.

Using the results for ΔN_h, $\Delta N_{C(h,h-1)}$, and $\Delta N_{V(h,h-1)}$ at time-step n given in (8.22)–(8.24), the following time-marched rate equations can be formulated to provide the N_{Ch} and N_{Vh} terms required in (8.20):

$$N_{Ch}\Big|_{i+1/2,j,k}^{n+1} = N_{Ch}\Big|_{i+1/2,j,k}^{n-1} + 2\Delta t \left(\begin{aligned} &-\Delta N_h\Big|_{i+1/2,j,k}^{n} - \Delta N_{C(h,h-1)}\Big|_{i+1/2,j,k}^{n} \\ &+ \Delta N_{C(h+1,h)}\Big|_{i+1/2,j,k}^{n} + W_{\text{pump}}\Big|_{i+1/2,j,k}^{n}\end{aligned}\right) \tag{8.25}$$

$$N_{Vh}\Big|_{i+1/2,j,k}^{n+1} = N_{Vh}\Big|_{i+1/2,j,k}^{n-1} + 2\Delta t \left(\begin{aligned} &-\Delta N_h\Big|_{i+1/2,j,k}^{n} - \Delta N_{V(h,h-1)}\Big|_{i+1/2,j,k}^{n} \\ &+ \Delta N_{V(h+1,h)}\Big|_{i+1/2,j,k}^{n} - W_{\text{pump}}\Big|_{i+1/2,j,k}^{n}\end{aligned}\right) \tag{8.26}$$

where W_{pump} denotes an electrical pumping term.

Now, all of the terms required to time-step the polarization in (8.20) are available, and the Maxwell–Ampere law yields the following FDTD updating relation for the electric-field vector component E_x:

$$\begin{aligned}E_x\Big|_{i+1/2,j,k}^{n+1} &= E_x\Big|_{i+1/2,j,k}^{n} + \frac{\Delta t}{\varepsilon_{i+1/2,j,k}\Delta y}\left(H_z\Big|_{i+1/2,j+1/2,k}^{n+1/2} - H_z\Big|_{i+1/2,j-1/2,k}^{n+1/2}\right) \\ &+ \frac{\Delta t}{\varepsilon_{i+1/2,j,k}\Delta z}\left(H_y\Big|_{i+1/2,j,k+1/2}^{n+1/2} - H_y\Big|_{i+1/2,j,k-1/2}^{n+1/2}\right) \\ &- \frac{1}{\varepsilon_{i+1/2,j,k}} \sum_{h=1}^{M}\left(P_{h,x}\Big|_{i+1/2,j,k}^{n+1} - P_{h,x}\Big|_{i+1/2,j,k}^{n}\right)\end{aligned} \tag{8.27}$$

where M is the total number of energy states modeled. Analogous time-stepping relations can be formulated for E_y and E_z.

We note that the Fermi–Dirac thermalization dynamics in the semiconductor are obtained by taking the ratio between the upward and downward intraband transitions for two neighboring energy levels. The relation between two energy levels can be obtained by the intraband transition rate equations. For example, using (8.24) for the valence band, we obtain:

$$\frac{\tau_{V(h-1,h)}}{\tau_{V(h,h-1)}} = \frac{N^0_{V(h-1)}(\boldsymbol{r})}{N^0_{Vh}(\boldsymbol{r})} \exp\left[\frac{E_{Vh} - E_{V(h-1)}}{k_B T}\right] \tag{8.28}$$

where k_B is the Boltzmann constant, T is the absolute temperature, and E_{Vh} and $E_{V(h-1)}$ are energies at levels h_v and $h_v - 1$, respectively. A similar relation for the conduction band can be obtained. While the approach adopted for intraband thermalization can also be applied to the interband transition, the contribution is negligible due to a large energy gap. Therefore, in this chapter, the thermalization effect is only considered for the intraband transition.

8.5 NUMERICAL RESULTS

Previous work has demonstrated the applicability and accuracy of the Lorentz–Drude model of metals of Section 8.3 and the direct-bandgap semiconductor model of Section 8.4, when each model was applied individually. Specifically, the Lorentz–Drude model was found to accurately simulate interactions between magnetic and nonmagnetic metals for plasmonics applications [13]. The direct-bandgap semiconductor model was found to accurately quantify (for switching applications) the enhanced extraction of light energy from an elliptical microcavity using an external magnetic field [9], and the extraction of light energy from the minor arc of an electrically pumped elliptical microcavity laser [10].

This chapter reviews the two examples reported in [16] wherein both the Lorentz–Drude model of metals and the direct-bandgap semiconductor model were run concurrently to simulate the plasmonic effects of a metal structure strongly coupled to a semiconductor medium. The first example involved a gold parallel-plate waveguide wherein the guided optical mode can be amplified by electrically pumping the thin gallium-arsenide (GaAs) slab separating the gold plates. The second example investigated the resonance shift and radiation from a GaAs microcavity resonator with embedded gold nanocylinders.

In each of these examples, gold was assumed to be characterized by the following Lorentz–Drude model parameters in (8.3) [12]: $\varepsilon_\infty = 1$; $\omega_{pD} = 1.196 \times 10^{16}$ rad/s; $\Gamma_D = 1.28 \times 10^{13}$ s^{-1}; $\omega_{pL} = 1.37 \times 10^{16}$ rad/s; $\Delta\varepsilon_L = 0.024$; $\omega_L = 6.30 \times 10^{14}$ rad/s; and $\Gamma_L = 5.83 \times 10^{13}$ s^{-1}. GaAs was assumed to be characterized by the following direct-bandgap semiconductor model parameters: refractive index equal to 3.54; carrier density equal to 3×10^{22} m^{-3}; and effective masses of electrons and holes in the conduction and valence bands equal to $0.047 m_e$ and $0.36 m_e$, respectively, where m_e is the mass of a free electron. Ten energy levels were assumed for both the conduction and valence bands of GaAs, with transition-time parameters taken from [17].

8.5.1 Amplification of a 175-fs Optical Pulse in a Pumped Parallel-Plate Waveguide

Figure 8.2 illustrates the geometry of the first example reported in [16], a parallel-plate waveguide comprised of two 100-nm-thick gold plates separated by a 50-nm-thick GaAs slab. The waveguide "sandwich" was assumed to be 4 μm long and 100 nm wide.

In the FDTD simulation, a 175-fs full-width at half-maximum (FWHM) Gaussian pulse with a carrier wavelength of 800 nm was injected as a source at the center of the GaAs slab. This generated a mode that propagated equally toward both ends of the waveguide. The dimensions of the waveguide were selected to allow only single-mode propagation at this carrier wavelength.

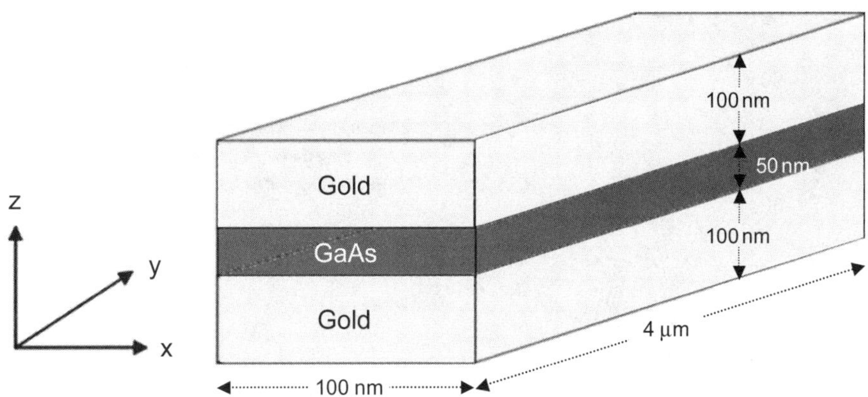

Fig. 8.2 Geometry (not to scale) of a parallel-plate waveguide comprised of two 100-nm-thick gold plates separated by a 50-nm-thick gallium-arsenide slab. The waveguide "sandwich" was 4 µm long and 100 nm wide. *Adapted from:* Ahmed, Khoo, Kurniawan, and Li, *J. Optical Society of America B*, Vol. 28, 2011, pp. 352–359, ©2011 The Optical Society of America.

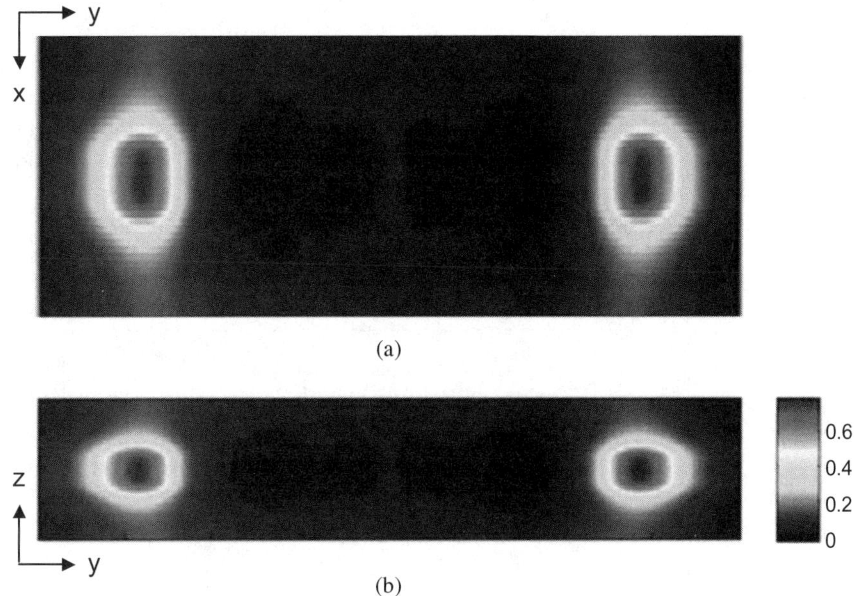

Fig. 8.3 (a) Top view and (b) side view visualizations of a snapshot of the FDTD-computed electric field intensity within the parallel-plate waveguide of Fig. 8.2. *Adapted from:* Ahmed, Khoo, Kurniawan, and Li, *J. Optical Society of America B*, Vol. 28, 2011, pp. 352–359, ©2011 The Optical Society of America.

Figure 8.3 shows top view and side view visualizations of a snapshot of the FDTD-computed electric field intensity distribution within the waveguide structure of Fig. 8.2 at a point in time when the pulsed oppositely propagating waves neared the two ends of the waveguide. The single-mode nature of the propagation is evident.

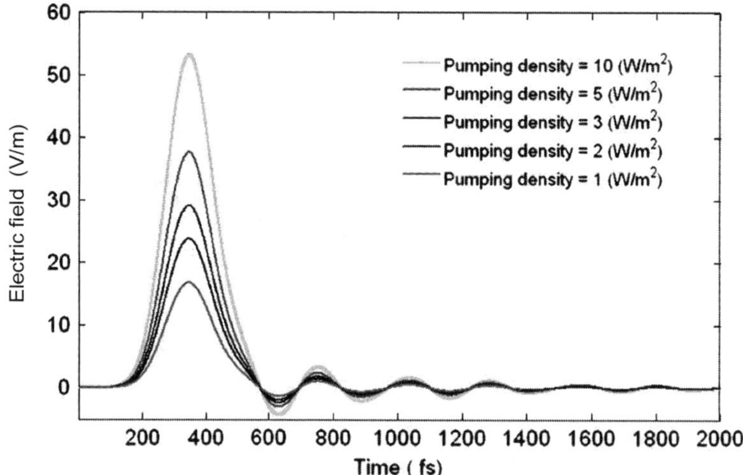

Fig. 8.4 FDTD-computed E-field intensity with respect to time for different pumping densities within the parallel-plate waveguide of Fig. 8.2. *Source:* Ahmed, Khoo, Kurniawan, and Li, *J. Optical Society of America B*, Vol. 28, 2011, pp. 352–359, ©2011 The Optical Society of America.

Figure 8.4 depicts FDTD-computed electric field intensities with respect to time for different pumping densities within the parallel-plate waveguide of Fig. 8.2 [16]. This figure shows that computed propagated field intensities at the observation point increased as the pumping density was elevated. This effective gain phenomenon was due to the simulation of plasmonic effects in the vicinity of the two gold-GaAs interfaces that coupled additional electrons and holes to the adjacent GaAs slab. A similar approach to enhance wave propagation distance by incorporating quantum dots in the dielectric medium was studied in [5] for splitters and interferometers.

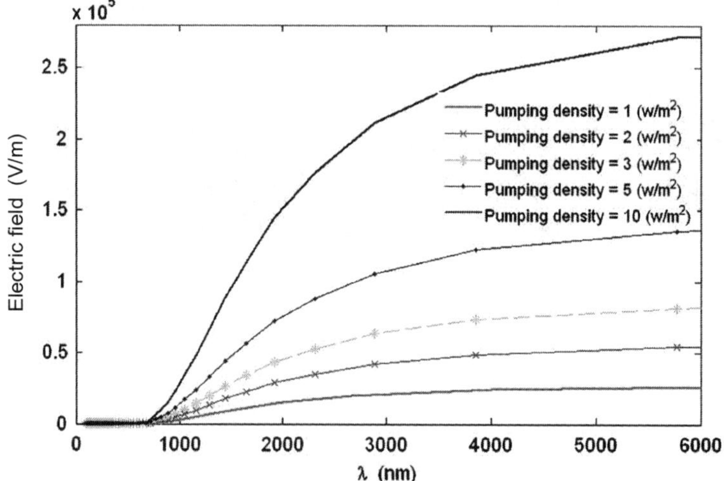

Fig. 8.5 FDTD-computed E-field intensity with respect to wavelength for different pumping densities within the parallel-plate waveguide of Fig. 8.2. *Source:* Ahmed, Khoo, Kurniawan, and Li, *J. Optical Society of America B*, Vol. 28, 2011, pp. 352–359, ©2011 The Optical Society of America.

Figure 8.5 depicts FDTD-computed electric field intensities with respect to wavelength for different pumping densities within the parallel-plate waveguide of Fig. 8.2 [16]. This figure shows that, despite pumping, computed propagated field intensities at the observation point were negligible for carrier wavelengths shorter than ~800 nm. For wavelengths longer than this gain cutoff, simulated field intensities increased rapidly, eventually saturating at maximum values having approximately a linear dependence on the pumping density. Such a structure could provide nearly wavelength-independent linear amplification in the infrared spectrum.

8.5.2 Resonance Shift and Radiation from a Passive Disk-Shaped GaAs Microcavity with Embedded Gold Nanocylinders

In its second example, Ref. [16] reported the FDTD simulation of a disk-shaped GaAs microcavity of radius $R = 700$ nm and thickness $t = 196$ nm, shown in Fig. 8.6(a).

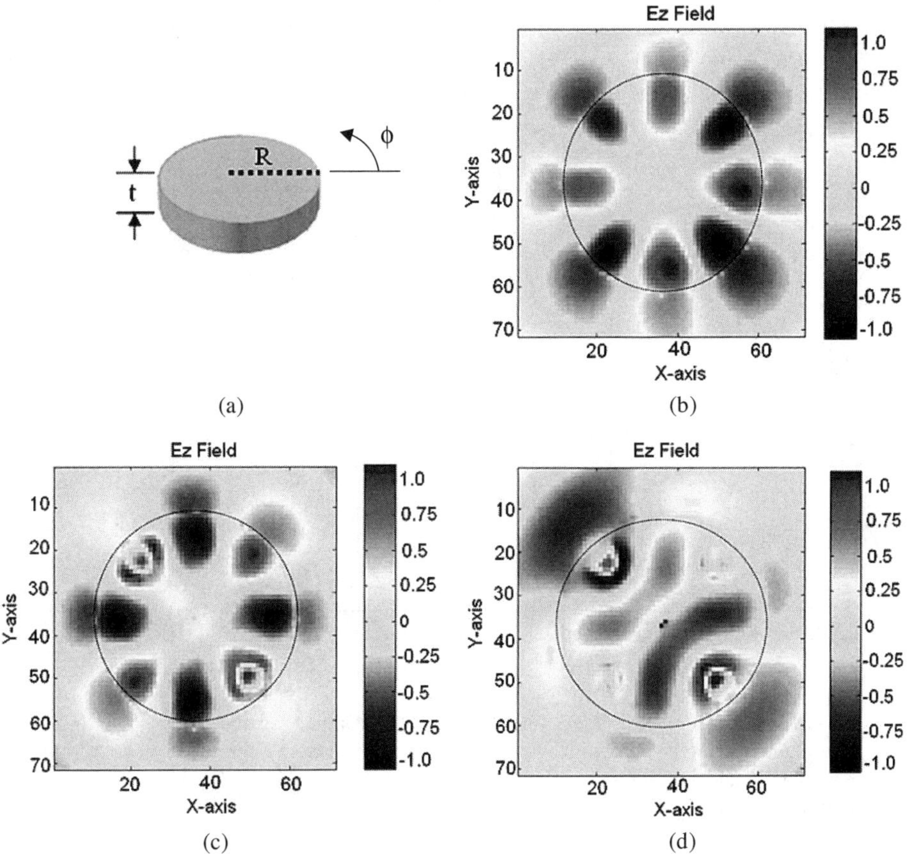

Fig. 8.6 Geometry and FDTD-computed results for the passive disk-shaped GaAs microcavity: (a) 3-D view; (b) unperturbed E-field distribution at $\lambda_{res} = 642.8$ nm; (c) E-field distribution with two embedded gold nanocylinders at $\lambda_{res} = 640.2$ nm; (d) E-field distribution with four embedded gold nanocylinders at $\lambda_{res} = 661.2$ nm. *Source:* Ahmed, Khoo, Kurniawan, and Li, *J. Optical Society of America B*, Vol. 28, 2011, pp. 352–359, ©2011 The Optical Society of America.

Figure 8.6(b) depicts a snapshot of the FDTD-computed electric field distribution within the GaAs disk microcavity at its resonant wavelength of 642.8 nm [16]. We observe an internal field distribution that is highly symmetrical with four peaks and nodes positioned regularly along the circumference of the disk.

Subsequently, two gold nanocylinders of radius of 140 nm and thickness 224 nm were embedded at opposite ends of a diameter of the GaAs disk model oriented at $\phi = 135^\circ$ and $\phi = -45^\circ$. This dropped the FDTD-computed resonant wavelength of the microcavity to 640.2 nm. Figure 8.6(c) depicts a snapshot of the FDTD-computed electric field distribution at this wavelength [16]. This snapshot shows significant field enhancement in the vicinity of each of the embedded gold nanocylinders. We also observe a mirror-image symmetry of the overall field distribution about the diameter of the disk connecting the pair of gold nanocylinders.

Finally, a second pair of gold nanocylinders of radius 140 nm and thickness 224 nm was added to the GaAs disk model. This pair was embedded at opposite ends of a diameter of the disk oriented at $\phi = 45^\circ$ and $\phi = -135^\circ$, i.e., perpendicular to the diameter orientation of the first pair of gold nanocylinders. Notably, this increased the FDTD-computed resonant wavelength of the microcavity to 661.2 nm. Furthermore, from the FDTD-computed electric field snapshot shown in Fig. 8.6(d), we observe significant field oscillation and radiation along the $\phi = 135^\circ$ and $\phi = -45^\circ$ directions, and suppression of the field in other directions [16]. This demonstrates a dipole antenna-like behavior of the GaAs disk microcavity.

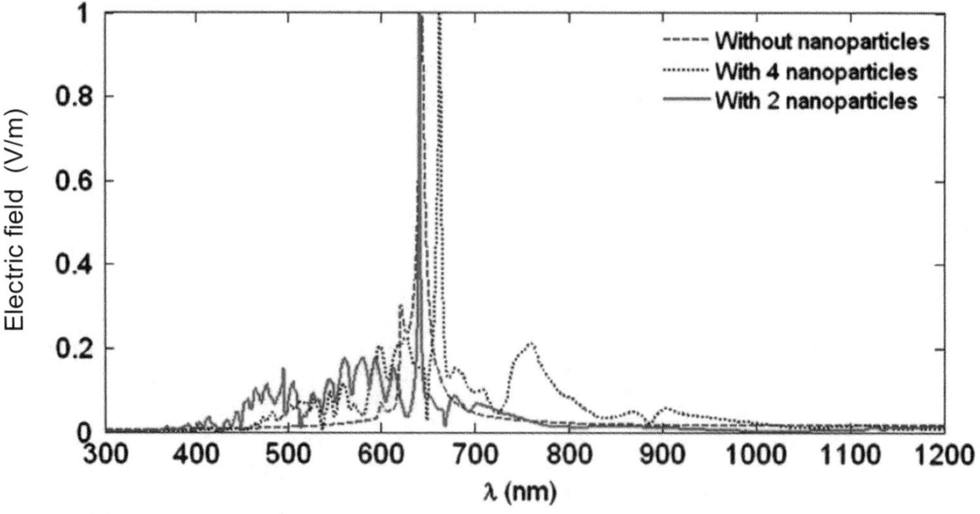

Fig. 8.7 FDTD-computed electric field intensity vs. wavelength for the GaAs disk microcavity of Fig. 8.6(a) without and with the embedded gold nanocylinders. *Source:* Ahmed, Khoo, Kurniawan, and Li, *J. Optical Society of America B*, Vol. 28, 2011, pp. 352–359, ©2011 The Optical Society of America.

Figure 8.7 depicts the FDTD-computed resonance behavior of the GaAs disk microcavity of Fig. 8.6(a) without and with the embedded gold nanocylinders [16]. Combined with the results shown in Fig. 8.6, we see that adding gold nanoparticles to an open semiconductor microcavity allows control of both its resonance and radiation characteristics.

8.6 SUMMARY

This chapter has reviewed a recently reported FDTD technique [16] for modeling active plasmonics devices. The new technique integrates a Lorentz–Drude model to simulate the metal components of the device with a multi-level multi-electron quantum model of the semiconductor component. Two examples of applications were summarized: amplification of a 175-fs optical pulse propagating in a thin, electrically pumped GaAs medium between two gold plates; and resonance shift and radiation from a passive GaAs microcavity resonator with embedded gold nanocylinders. The new approach is promising for stimulating progress in developing new applications in the field of active plasmonics, which is still in its development stage.

APPENDIX 8A: CRITICAL POINTS MODEL FOR METAL OPTICAL PROPERTIES

The Lorentz–Drude model used in this chapter to simulate the dispersive dielectric behavior of the gold components of active plasmonics devices is sufficiently accurate for the range of optical wavelengths (>800 nm) studied herein. However, FDTD simulations of such devices at shorter wavelengths may require a more precise dielectric model of gold (and other metals, if used). Regarding gold, Etchegoin *et al.* stated in their 2006 paper that [18]:

> Compared to silver (Ag), the optical properties of gold (Au) are more difficult to represent in the visible/near-uv region with an analytic model. The reason for that is the more important role in the latter played by interband transitions in the violet/near-uv region. Au has at least two interband transitions at $\lambda \sim 470$ and ~ 330 nm that do play an important role and must be included explicitly if a realistic analytic model for $\varepsilon(\omega)$ is sought. Their line shapes (coming from the joint density of states of the interband transitions) are not very well accounted for by a simple Lorentz oscillator (as in a simple molecular transition), and attempts to add simple Lorentz oscillators to a Drude term to account for the interband transitions very rapidly face limitations.
>
> A different type of analytic model for the two interband transitions in gold in the violet/near-uv region has to be included to achieve a reasonable representation of $\varepsilon(\omega)$ with a minimum set of parameters. It is possible to include a family of analytical models (called critical points) for transitions in solids, which satisfy a set of minimum requirements (like Kramers–Kronig consistency) and reproduce most of the line shapes in $\varepsilon(\omega)$ observed experimentally.

Subsequently, this "critical points model" proposed by Etchegoin *et al.* [18, 19] was delineated by Vial *et al.* [20] in the following form suitable for FDTD simulations:

$$\varepsilon(\omega) = \varepsilon_\infty - \frac{\omega_\mathrm{D}^2}{\omega^2 - j\omega\Gamma_\mathrm{D}} + \sum_l A_l \Omega_l \left(\frac{e^{j\phi_l}}{\Omega_l - \omega - j\Gamma_l} + \frac{e^{-j\phi_l}}{\Omega_l + \omega + j\Gamma_l} \right) \tag{8A.1}$$

The first and second terms of (8A.1) comprise the standard permittivity contribution of a Drude model, and the remaining terms are contributions from interband transitions [18]. Table 8A.1 provides values for the various parameters in (8A.1) for gold, silver, aluminum, and chromium with $l = 1, 2$ [20]. Over the wavelength range $400 < \lambda < 1000$ nm, these parameters fit experimental data for gold from [21], and for silver, aluminum, and chromium from [22].

TABLE 8A.1

Parameters for Drude + 2 Critical Points Fit of (8A.1) for Au, Ag, Al, and Cr over the Wavelength Range $400 < \lambda < 1000$ nm. *Adapted from:* Vial *et al.*, *Applied Physics A*, 2011, Vol. 103, pp. 849–853, ©2011 Springer.

Gold

$$\varepsilon_\infty = 1.1431,\ \omega_D = 1.3202 \times 10^{16} \text{ rads/s},\ \Gamma_D = 1.0805 \times 10^{14} \text{ s}^{-1}$$
$$A_1 = 0.26698,\ \phi_1 = -1.2371,\ \Omega_1 = 3.8711 \times 10^{15} \text{ rads/s},\ \Gamma_1 = 4.4642 \times 10^{14} \text{ s}^{-1}$$
$$A_2 = 3.0834,\ \phi_2 = -1.0968,\ \Omega_2 = 4.1684 \times 10^{15} \text{ rads/s},\ \Gamma_2 = 2.3555 \times 10^{15} \text{ s}^{-1}$$

Silver

$$\varepsilon_\infty = 1.4447,\ \omega_D = 1.3280 \times 10^{16} \text{ rads/s},\ \Gamma_D = 9.1269 \times 10^{13} \text{ s}^{-1}$$
$$A_1 = -1.5951,\ \phi_1 = 3.1288,\ \Omega_1 = 8.2749 \times 10^{15} \text{ rads/s},\ \Gamma_1 = 5.1770 \times 10^{15} \text{ s}^{-1}$$
$$A_2 = 0.25261,\ \phi_2 = -1.5066,\ \Omega_2 = 6.1998 \times 10^{15} \text{ rads/s},\ \Gamma_2 = 5.4126 \times 10^{14} \text{ s}^{-1}$$

Aluminum

$$\varepsilon_\infty = 1.0000,\ \omega_D = 2.0598 \times 10^{16} \text{ rads/s},\ \Gamma_D = 2.2876 \times 10^{14} \text{ s}^{-1}$$
$$A_1 = 5.2306,\ \phi_1 = -0.51202,\ \Omega_1 = 2.2694 \times 10^{15} \text{ rads/s},\ \Gamma_1 = 3.2867 \times 10^{14} \text{ s}^{-1}$$
$$A_2 = 5.2704,\ \phi_2 = 0.42503,\ \Omega_2 = 2.4668 \times 10^{15} \text{ rads/s},\ \Gamma_2 = 1.7731 \times 10^{15} \text{ s}^{-1}$$

Chromium

$$\varepsilon_\infty = 1.1297,\ \omega_D = 8.8128 \times 10^{15} \text{ rads/s},\ \Gamma_D = 3.8828 \times 10^{14} \text{ s}^{-1}$$
$$A_1 = 33.086,\ \phi_1 = -0.25722,\ \Omega_1 = 1.7398 \times 10^{15} \text{ rads/s},\ \Gamma_1 = 1.6329 \times 10^{15} \text{ s}^{-1}$$
$$A_2 = 1.6592,\ \phi_2 = 0.83533,\ \Omega_2 = 3.7925 \times 10^{15} \text{ rads/s},\ \Gamma_2 = 7.3567 \times 10^{14} \text{ s}^{-1}$$

Shibayama *et al.* reported an efficient frequency-dependent FDTD algorithm for the Drude critical points model of (8A.1) employing a trapezoidal recursive convolution technique [23]. Retaining the notation of (8A.1), the explicit time-stepping algorithm of Shibayama *et al.* is given by:

$$\boldsymbol{E}^{n+1} = \left(\frac{\varepsilon_\infty - \chi^0/2}{\varepsilon_\infty + \chi^0/2}\right)\boldsymbol{E}^n + \left(\frac{1}{\varepsilon_\infty + \chi^0/2}\right)\xi^n + \left[\frac{\Delta t}{\varepsilon_0(\varepsilon_\infty + \chi^0/2)}\right]\nabla\times\boldsymbol{H}^{n+1/2} \tag{8A.2}$$

In (8A.2), the scalar factor χ^0 is given by:

$$\chi^0 = \chi_D^0 + \sum_l \text{Re}\left(\chi_l^0\right) \tag{8A.3}$$

for:

$$\chi_D^0 = \frac{\omega_D^2}{\Gamma_D}\left[\Delta t - \frac{1}{\Gamma_D}\left(1 - e^{-\Gamma_D \Delta t}\right)\right], \quad \chi_l^0 = \frac{2jA_l\Omega_l e^{-j\phi_l}}{j\Omega_l - \Gamma_l}\left[1 - e^{(-\Gamma_l + j\Omega_l)\Delta t}\right] \tag{8A.4}$$

Also in (8A.2), the vector factor $\boldsymbol{\xi}^n$ is given by:

$$\boldsymbol{\xi}^n = \boldsymbol{\xi}_{\mathrm{D}}^n + \sum_l \mathrm{Re}\left(\boldsymbol{\xi}_l^n\right) \tag{8A.5}$$

where $\boldsymbol{\xi}_{\mathrm{D}}^n$ and $\boldsymbol{\xi}_l^n$ are obtained via the recursion relations:

$$\boldsymbol{\xi}_{\mathrm{D}}^n = \left(\frac{\boldsymbol{E}^n + \boldsymbol{E}^{n-1}}{2}\right)\Delta\chi_{\mathrm{D}}^0 + e^{-\Gamma_{\mathrm{D}}\Delta t}\boldsymbol{\xi}_{\mathrm{D}}^{n-1} \tag{8A.6a}$$

$$\boldsymbol{\xi}_l^n = \left(\frac{\boldsymbol{E}^n + \boldsymbol{E}^{n-1}}{2}\right)\Delta\chi_l^0 + e^{(-\Gamma_l + j\Omega_l)\Delta t}\boldsymbol{\xi}_l^{n-1} \tag{8A.6b}$$

for:

$$\Delta\chi_{\mathrm{D}}^0 = -\left[\frac{\omega_{\mathrm{D}}\left(1 - e^{-\Gamma_{\mathrm{D}}\Delta t}\right)}{\Gamma_{\mathrm{D}}}\right]^2, \qquad \Delta\chi_l^0 = \chi_l^0\left[1 - e^{(-\Gamma_l + j\Omega_l)\Delta t}\right] \tag{8A.6c}$$

APPENDIX 8B: OPTIMIZED STAIRCASING FOR CURVED PLASMONIC SURFACES

Okada and Cole reported what amounts to be an optimized staircasing approach [24] to represent curved plasmonic surfaces in 2-D and 3-D Cartesian-cell FDTD grids. They asserted that the use of interpolated (i.e., effective) permittivities for $\boldsymbol{E}$-component updates adjacent to a plasmonic surface is not appropriate. This is because the surface plasmon resonance condition determined by the original permittivity differs from those of the interpolated permittivity values. Okada and Cole termed their approach the *staircased effective permittivity* (S-EP) model.

Consider an arbitrarily shaped plasmonic medium of permittivity ε_2 embedded within a dielectric region of permittivity ε_1. Upon performing a series of numerical experiments, Okada and Cole developed the following simple rule to assign permittivity values to $\boldsymbol{E}$ components of any orientation near the surface of the plasmonic medium [24]. If the $\boldsymbol{E}$ component is located within the plasmonic medium, or even *outside* the plasmonic medium, but at a distance of *less than one-half grid cell* from its surface, then the $\boldsymbol{E}$ component is assigned the plasmonic medium permittivity ε_2. Else, the $\boldsymbol{E}$ component is assigned the exterior dielectric region permittivity ε_1.

Using this S-EP approach, Okada and Cole reported FDTD simulations of plane-wave scattering by an infinitely long silver microcylinder and a gold microsphere at optical wavelengths [24]. Numerical results were compared with Mie theory in the surface plasmon waveband, and the S-EP model provided higher accuracy than conventional staircase and effective-permittivity models, even on a coarse grid.

Okada and Cole stated that the S-EP approach has achieved the same performance improvement when applied to other material dispersion models (specifically, the Lorentz model and the Drude with critical points), as well as other FDTD treatments of dispersion (recursive convolution and piecewise linear recursive convolution) [24]. Furthermore, they asserted that the S-EP method is robust relative to changes of material parameters. All of these desirable attributes result because the S-EP method is implemented only by changing how permittivity values are assigned to $\boldsymbol{E}$ components located near the surface of a plasmonic medium.

REFERENCES

[1] Maier, S. A., *Plasmonics: Fundamentals and Applications*, Berlin: Springer-Verlag, 2007.

[2] Brongersma, M. L., and P. G. Kik, *Surface Plasmon Nanophotonics*, Berlin: Springer, 2007.

[3] Ahmed, I., C. E. Png, E. P. Li, and R. Vahldieck, "Electromagnetic propagation in a novel Ag nanoparticle based plasmonic structure," *Optics Express*, Vol. 17, 2009, pp. 337–345.

[4] MacDonald, K. F., Z. L. Samson, M. I. Stockman, and N. I. Zheludev, "Ultrafast active plasmonics," *Nature Photonics*, Vol. 3, 2008, pp. 55–58.

[5] Krasavin, A. V., and A. V. Zayats, "Three-dimensional numerical modeling of photonics integration with dielectric loaded SPP waveguides," *Physical Review B*, Vol. 78, 2008, 045425.

[6] Dionne, J. A., K. Diest, L. A. Sweatlock, and H. A. Atwater, "PlasMOStor: A metal-oxide-Si field effect plasmonic modulator," *Nano Letters*, Vol. 9, 2009, pp. 897–902.

[7] Hill, M. T., M. Marell, E. S. P. Leong, B. Smalbrugge, Y. Zhu, M. Sun, P. J. van Veldhoven, E. J. Geluk, F. Karouta, Y.-S. Oei, R. Nötzel, C.-Z. Ning, and M. K. Smit, "Lasing in metal-insulator-metal sub-wavelength plasmonic waveguides," *Optics Express*, Vol. 17, 2009, pp. 11107–11112.

[8] Huang, Y., and S. T. Ho, "Computational model of solid state, molecular, or atomic media for FDTD simulation based on a multilevel multi-electron system governed by Pauli exclusion and Fermi-Dirac thermalization with application to semiconductor photonics," *Optics Express*, Vol. 14, 2006, pp. 3569–3587.

[9] Khoo, E. H., I. Ahmed, and E. P. Li, "Enhancement of light energy extraction from elliptical microcavity using external magnetic field for switching applications," *Applied Physics Letters*, Vol. 95, 2009, pp. 121104–121106.

[10] Khoo, E. H., S. T. Ho, I. Ahmed, E. P. Li, and Y. Huang, "Light energy extraction from the minor surface arc of an electrically pumped elliptical microcavity laser," *IEEE J. Quantum Electronics*, Vol. 46, 2010, pp. 128–136.

[11] Huang, Y., and S. T. Ho, "Simulation of electrically-pumped nanophotonic lasers using dynamical semiconductor medium FDTD method," in *Proc. 2nd IEEE International Nanoelectronics Conf.*, 2008, pp. 202–205.

[12] Rakic, D., A. B. Djurisic, J. M. Elazar, and M. L. Majewski, "Optical properties of metallic films for vertical-cavity optoelectronic devices," *Applied Optics*, Vol. 37, 1998, pp. 5271–5283.

[13] Ahmed, I., E. P. Li, and E. H. Khoo, "Interactions between magnetic and non-magnetic materials for plasmonics," in *Proc. International Conf. on Materials and Advanced Technologies*, Materials Research Society of Singapore, 2009.

[14] Willis, K. J., J. S. Ayubi-Moak, S. C. Hagness, and I. Knezevic, "Global modeling of carrier-field dynamics in semiconductor using EMC-FDTD," *J. Computational Electronics*, Vol. 8, 2009, pp. 153–171.

[15] Ahmed, I., E. H. Khoo, E. P. Li, and R. Mittra, "A hybrid approach for solving coupled Maxwell and Schrödinger equations arising in the simulation of nano-devices," *IEEE Antennas & Wireless Propagation Lett.*, Vol. 9, 2010, pp. 914–917.

[16] Ahmed, I., E. H. Khoo, O. Kurniawan, and E. P. Li, "Modeling and simulation of active plasmonics with the FDTD method by using solid state and Lorentz–Drude dispersive model," *J. Optical Society of America B*, Vol. 28, 2011, pp. 352–359.

[17] Marrin, S., B. Deveaud, F. Clerot, K. Fuliwara, and K. Mitsunaga, "Capture of photoexcited carriers in a single quantum well with different confinement structures," *IEEE J. Quantum Electronics*, Vol. 27, 1991, pp. 1669–1675.

[18] Etchegoin, P. G., E. C. Le Ru, and M. Meyer, "An analytic model for the optical properties of gold," *J. Chemical Physics*, Vol. 125, 2006, 164705.

[19] Etchegoin, P. G., E. C. Le Ru, and M. Meyer, "Erratum: An analytic model for the optical properties of gold," *J. Chemical Physics*, Vol. 127, 2007, 189901.

[20] Vial, A., T. Laroche, M. Dridi, and L. Le Cunff, "A new model of dispersion for metals leading to a more accurate modeling of plasmonic structures using the FDTD method," *Applied Physics A*, Vol. 103, 2011, pp. 849–853.

[21] Johnson, P. B., and R. W. Christy, "Optical constants of the noble metals," *Physical Review B*, Vol. 6, 1972, pp. 4370–4379.

[22] Palik, E. D., ed., *Handbook of Optical Constants of Solids*, San Diego, CA: Academic Press, 1985.

[23] Shibayama, J., K. Watanabe, R. Ando, J. Yamauchi, and H. Nakano, "Simple frequency-dependent FDTD algorithm for a Drude-critical points model," *Proc. Asia-Pacific Microwave Conf. 2010*, WE1D-4, 2010, pp. 73–75.

[24] Okada, N., and J. B. Cole, "Effective permittivity for FDTD calculation of plasmonic materials," *Micromachines*, Vol. 3, 2012, pp. 168–179.

SELECTED BIBLIOGRAPHY

Heltzel, S. Theppakuttai, S. C. Chen, and J. R. Howell, "Surface plasmon-based nanopatterning assisted by gold nanospheres," *Nanotechnology*, Vol. 19, 2008, 025305.

Jung, K.-Y., and F. L. Teixeira, "Multispecies ADI-FDTD algorithm for nanoscale three-dimensional photonic metallic structures," *IEEE Photonics Technology Lett.*, Vol. 19, 2007, pp. 586–588.

Oubre, C., and P. Nordlander, "Optical properties of metallodielectric nanostructures calculated using the FDTD method," *J. Physical Chemistry*, Vol. 108, 2004, pp. 17740–17747.

Shibayama, J., A. Nomura, R. Ando, J. Yamauchi, and H. Nakano, "A frequency-dependent LOD-FDTD method and its applications to the analyses of plasmonic waveguide devices," *IEEE J. Quantum Electronics*, Vol. 46, 2010, pp. 40–49.

Shibayama, J., R. Takahashi, J. Yamauchi, and H. Nakano, "Frequency-dependent locally one-dimensional FDTD implementation with a combined dispersion model for the analysis of surface plasmon waveguides," *IEEE Photonics Technology Lett.*, Vol. 20, 2008, pp. 824–826.

Zhou, J., M. Hu, Y. Zhang, P. Zhang, W. Liu, and S. Liu, "Numerical analysis of electron-induced surface plasmon excitation using the FDTD method," *J. Optics*, Vol. 13, 2011, 035003.

Chapter 9

FDTD Computation of the Nonlocal Optical Properties of Arbitrarily Shaped Nanostructures[1]

Jeffrey M. McMahon, Stephen K. Gray, and George C. Schatz

9.1 INTRODUCTION

Interest in the optical properties of metallic nanostructures has been steadily increasing as experimental techniques for their fabrication and investigation have become more sophisticated [1]. One of the main driving forces of this is their potential utility in sensing, photonic, and optoelectronics applications [1–3]. However, there can also be interesting fundamental issues to consider, particularly as very small length scales are approached (approximately less than 10 nm). In this limit, quantum-mechanical effects can lead to unusual optical properties relative to predictions based on classical electrodynamics applied with bulk, local dielectric values for the metal [4]. In isolated spherical nanoparticles, for example, localized surface-plasmon resonances (LSPRs) are found to be blue-shifted relative to Mie theory predictions [5], and in thin metal films anomalous absorption is observed [6, 7].

Roughly speaking, when light interacts with a structure of size d (e.g., a nanoparticle size or junction gap distance), wavevector components $\boldsymbol{k}$, which are related to the momentum $\boldsymbol{p}$ by $\boldsymbol{p} = \hbar\boldsymbol{k}$, where $\hbar$ is the Planck constant, are generated with magnitude $k = 2\pi/d$. These, in turn, impart an energy of $E = (\hbar k)^2/2m_e$, where m_e is the mass of an electron, to (relatively) free electrons in the metal. For small d, these energies can correspond to the optical range (1–6 eV). This analysis suggests that such effects should come into play for d less than approximately 2 nm. In metals, however, somewhat larger d values also exhibit these effects, because electrons in motion at the Fermi velocity can be excited by the same energy with a smaller momentum increase, due to dispersion effects.

A full quantum-mechanical treatment of such structures would of course be best, but this is not practical for these sizes. However, it is possible to incorporate some "quantum effects" within classical electrodynamics via use of a different dielectric model than that for the bulk metal. At least four such effects can be addressed in this way: electron scattering, electron spill-out, quantum-size effects, and spatial nonlocality of the material polarization.

---

[1]This chapter is adapted from Ref. [24], J. M. McMahon, S. K. Gray, and G. C. Schatz, "Calculating nonlocal optical properties of structures with arbitrary shape," *Physical Review B*, Vol. 82, 2010, 035423,

For consistency of notation relative to this source paper, in this chapter the symbol i is used to designate $\sqrt{-1}$, rather than the symbol j; and a phasor is denoted as $e^{-i\omega t}$.

The additional losses due to increased electron scattering at the metal surface can be described by a size-dependent damping term [8] that effectively broadens spectral peaks [9]. Electron spill-out from the metal into the medium, due to the electron density varying smoothly, can be partially accounted for by a dielectric layer model. The effect of this is varied, and depends on a number of details, including the surface chemistry of the structure [10] and its local dielectric environment. Quantum-size effects due to discrete electronic energy levels can lead to a size- and shape-dependent conductivity. At least for metal films [11], this quantity is reduced relative to the bulk and exhibits peaks for certain film thicknesses. Such effects can be incorporated directly into classical calculations for some simple systems, based on rigorous theory. Although, because this effect and electron spill-out are both highly dependent on system specifics, incorporating them into a general framework is not straightforward. Therefore, they will not be considered in this chapter.

The fourth quantum effect, and the one that is of main interest here, is the need for a dielectric model which considers that the material polarization at a point $\boldsymbol{x}$ in space depends not only on the local electric field but also that in its neighborhood [12, 13]. This contrasts with the undergraduate perspective, wherein materials are described through a dielectric function, ε, that relates the electric displacement field, $\boldsymbol{D}$, to the local electric field, $\boldsymbol{E}$, at a given frequency, ω. Whereas this simplified relationship assumes that the polarizability of a material at $\boldsymbol{x}$ depends only on $\boldsymbol{E}$ at $\boldsymbol{x}$, the more general case can be characterized by:

$$\boldsymbol{D}(x, \omega) = \varepsilon_0 \int \varepsilon(\boldsymbol{x}, \boldsymbol{x}', \omega) \boldsymbol{E}(\boldsymbol{x}', \omega) \, d\boldsymbol{x}' \tag{9.1}$$

where $\varepsilon(\boldsymbol{x}, \boldsymbol{x}', \omega)$ is a spatially dependent (nonlocal) and frequency-dispersive relative dielectric function. In a homogeneous environment (an approximation that is made for the finite, arbitrarily shaped structures considered here), $\varepsilon(\boldsymbol{x}, \boldsymbol{x}', \omega)$ only spatially depends on $|\boldsymbol{x} - \boldsymbol{x}'|$. Therefore, in $\boldsymbol{k}$ space, (9.1) is more simply expressed as:

$$\boldsymbol{D}(\boldsymbol{k}, \omega) = \varepsilon_0 \, \varepsilon(\boldsymbol{k}, \omega) \, \boldsymbol{E}(\boldsymbol{k}, \omega) \tag{9.2}$$

Since the early formulations of nonlocal electromagnetics [12, 13], applications of $\boldsymbol{k}$-dependent dielectric functions have remained limited to simple systems such as spherical structures [14, 15], aggregates of spherical structures [16–19], and planar surfaces [20]. Nonetheless, this $\boldsymbol{k}$ dependence has been found experimentally [6, 7] and proven theoretically [14, 20] to have important consequences. For example, such dependence is responsible for the aforementioned anomalous absorption and LSPR blue-shifting.

Reference [21] outlined a method by which the optical properties of arbitrarily shaped structures with a nonlocal dielectric function can easily be calculated. This was done by deriving an equation of motion for the current associated with the hydrodynamic Drude model [22] that was solved within the framework of the finite-difference time-domain (FDTD) method [23]. The advantage of this approach is that it can describe the dynamical optical response of structures that are too large to treat using quantum mechanics, yet small enough such that the application of local continuum electrodynamics becomes questionable. Reference [24], reviewed in this chapter, expanded on the work reported in [21], and detailed the full method. This review includes an analytical verification of the results for the cylindrical gold (Au) nanowires considered in [21], as well as new examples of calculations of the optical properties of one-dimensional (1-D), two-dimensional (2-D), and three-dimensional (3-D) Au nanostructures.

9.2 THEORETICAL APPROACH

The interaction of light with matter in the classical continuum limit (i.e., many hundreds of atoms or more) is described by Maxwell's equations:

$$\frac{\partial}{\partial t}\boldsymbol{D}(\boldsymbol{x},t) + \boldsymbol{J}(\boldsymbol{x},t) = \nabla \times \boldsymbol{H}(\boldsymbol{x},t) \tag{9.3}$$

$$\frac{\partial}{\partial t}\boldsymbol{B}(\boldsymbol{x},t) = -\nabla \times \boldsymbol{E}(\boldsymbol{x},t) \tag{9.4}$$

$$\nabla \cdot \boldsymbol{D}(\boldsymbol{x},t) = \rho(\boldsymbol{x},t) \tag{9.5}$$

$$\nabla \cdot \boldsymbol{B}(\boldsymbol{x},t) = 0 \tag{9.6}$$

where $\boldsymbol{H}(\boldsymbol{x},t)$ and $\boldsymbol{B}(\boldsymbol{x},t)$ are, respectively, the magnetic field intensity and the magnetic flux density; and $\boldsymbol{J}(\boldsymbol{x},t)$ and $\rho(\boldsymbol{x},t)$ are, respectively, the electric current and electric charge density. Except for the most simple systems, such as spheres or metal films, analytical solutions or simplifying approximations to (9.3) – (9.6) do not exist. Therefore, computational methods are often used to solve them, one of the most popular being FDTD [23]. For dynamic fields, (9.3) and (9.4) are explicitly solved, while (9.5) and (9.6) are considered to be implied by the solution to (9.3) and (9.4), if properly posed.

However, before Maxwell's equations can be solved, an explicit form for $\varepsilon(\boldsymbol{k},\omega)$ in the constitutive relationship between $\boldsymbol{D}(\boldsymbol{k},\omega)$ and $\boldsymbol{E}(\boldsymbol{k},\omega)$ in (9.2) must be specified. Note that (9.3) – (9.6) are in terms of $\boldsymbol{x}$ and t, but material properties are often dependent on $\boldsymbol{k}$ and ω, which are related to the former via Fourier transforms. Also note that no magnetic materials are assumed present. Thus, the magnetic field constitutive relationship is $\boldsymbol{B}(\boldsymbol{x},\omega) = \mu_0 \boldsymbol{H}(\boldsymbol{x},\omega)$, where μ_0 is the vacuum permeability. Returning to the current discussion, the permittivity of a metal like Au is well described in the classical continuum limit by three separate components:

$$\varepsilon(\boldsymbol{k},\omega) = \varepsilon_\infty + \varepsilon_{\text{inter}}(\omega) + \varepsilon_{\text{intra}}(\boldsymbol{k},\omega) \tag{9.7}$$

Here, ε_∞ is the permittivity value as $\omega \to \infty$; $\varepsilon_{\text{inter}}(\omega)$ is the permittivity contribution of the interband electron transitions from the *d*-band to the *sp*-band (conduction-band); and $\varepsilon_{\text{intra}}(\boldsymbol{k},\omega)$ is the contribution due to *sp*-band electron excitations. The notation in (9.7) highlights the $\boldsymbol{k}$ and ω dependencies.

Permittivity $\varepsilon_{\text{inter}}(\omega)$ can be physically described using a multipole Lorentz oscillator model [25]:

$$\varepsilon_{\text{inter}}(\omega) = \sum_j \frac{\Delta\varepsilon_{\text{L}j}\,\omega_{\text{L}j}^2}{\omega_{\text{L}j}^2 - \omega\left(\omega + i2\delta_{\text{L}j}\right)} \tag{9.8}$$

where j is an index labeling the individual *d*-band to *sp*-band electron transitions occurring at $\omega_{\text{L}j}$, $\Delta\varepsilon_{\text{L}j}$ is the shift in relative permittivity at the transition, and $\delta_{\text{L}j}$ is the electron dephasing rate. Because there are two interband transitions in Au at optical frequencies near 3 and 4 eV [26], $j = 2$ is assumed in this work.

Permittivity $\varepsilon_{\text{intra}}(\boldsymbol{k},\omega)$ is responsible for both the plasmonic optical response of metals and nonlocal effects. Both of these can be described by the hydrodynamic Drude model [22], which reduces to the local Drude expression for electron motion if $\boldsymbol{k} \rightarrow 0$ [25]:

$$\varepsilon_{\text{intra}}(\boldsymbol{k},\omega) = -\frac{\omega_{\text{D}}^2}{\omega(\omega + i\gamma) - \beta^2 k^2} \tag{9.9}$$

where ω_{D} is the plasma frequency; γ is the collision frequency; and $\beta^2 = C v_{\text{F}}^2 / D$ for a free-electron gas (i.e., one with only kinetic energy), where v_{F} is the Fermi velocity (1.39×10^6 m/s for Au), D is the dimension of the system, and $C = 1$ at low frequencies and $3D/(D + 2)$ at high frequencies [27]. (The low-frequency 2-D value of β^2 was used in the work reported in [21].) Other analytical forms for $\varepsilon_{\text{intra}}(\boldsymbol{k},\omega)$ could also be used, such as those inferred from representative quantum-mechanical electronic structure calculations [28].

Using (9.8) and (9.9), the insertion of (9.2) and (9.7) into the Maxwell–Ampere law in $\boldsymbol{k}$-space for a time-harmonic field, $-i\omega\boldsymbol{D}(\boldsymbol{k},\omega) = i\boldsymbol{k} \times \boldsymbol{H}(\boldsymbol{k},\omega)$, leads to:

$$-i\omega\varepsilon_0\varepsilon_\infty \boldsymbol{E}(\boldsymbol{k},\omega) + \sum_j \boldsymbol{J}_{\text{L}j}(\boldsymbol{k},\omega) + \boldsymbol{J}_{\text{HD}}(\boldsymbol{k},\omega) = i\boldsymbol{k} \times \boldsymbol{H}(\boldsymbol{k},\omega) \tag{9.10}$$

where the $\boldsymbol{J}_{\text{L}j}(\boldsymbol{k},\omega)$ are polarization currents associated with (9.8):

$$\boldsymbol{J}_{\text{L}j}(\boldsymbol{k},\omega) = -i\omega\varepsilon_0 \frac{\Delta\varepsilon_{\text{L}j}\,\omega_{\text{L}j}^2}{\omega_{\text{L}j}^2 - \omega(\omega + i2\delta_{\text{L}j})} \boldsymbol{E}(\boldsymbol{k},\omega) \tag{9.11}$$

and $\boldsymbol{J}_{\text{HD}}(\boldsymbol{k},\omega)$ is a nonlocal polarization current associated with (9.9):

$$\boldsymbol{J}_{\text{HD}}(\boldsymbol{k},\omega) = i\omega\varepsilon_0 \frac{\omega_{\text{D}}^2}{\omega(\omega + i\gamma) - \beta^2 k^2} \boldsymbol{E}(\boldsymbol{k},\omega) \tag{9.12}$$

Note that the current densities in (9.11) and (9.12) are unrelated to the external current density $\boldsymbol{J}(\boldsymbol{x},t)$ in (9.3), which has been assumed to be zero. Now, equations of motion (i.e., partial differential equations in terms of $\boldsymbol{x}$ and t) for the currents in (9.11) and (9.12) can be obtained by multiplying through each equation by the appropriate denominator, and then applying the inverse Fourier transformations $i\boldsymbol{k} \rightarrow \nabla$ and $-i\omega \rightarrow \partial/\partial t$:

$$\frac{\partial^2}{\partial t^2}\boldsymbol{J}_{\text{L}j}(\boldsymbol{x},t) + 2\delta_{\text{L}j}\frac{\partial}{\partial t}\boldsymbol{J}_{\text{L}j}(\boldsymbol{x},t) + \omega_{\text{L}j}^2\boldsymbol{J}_{\text{L}j}(\boldsymbol{x},t) = \varepsilon_0\Delta\varepsilon_{\text{L}j}\,\omega_{\text{L}j}^2\frac{\partial}{\partial t}\boldsymbol{E}(\boldsymbol{x},t) \tag{9.13}$$

$$\frac{\partial^2}{\partial t^2}\boldsymbol{J}_{\text{HD}}(\boldsymbol{x},t) + \gamma\frac{\partial}{\partial t}\boldsymbol{J}_{\text{HD}}(\boldsymbol{x},t) - \beta^2\nabla^2\boldsymbol{J}_{\text{HD}}(\boldsymbol{x},t) = \varepsilon_0\,\omega_{\text{D}}^2\frac{\partial}{\partial t}\boldsymbol{E}(\boldsymbol{x},t) \tag{9.14}$$

Because of the spatial derivatives in (9.14), the equation of motion for the hydrodynamic Drude model is second order, unlike the normal Drude model that is first order [23].

Equations (9.13) and (9.14) can be solved self-consistently with (9.4) and the inverse Fourier-transformed form of (9.10):

$$\varepsilon_0 \varepsilon_\infty \frac{\partial}{\partial t} \boldsymbol{E}(\boldsymbol{x},t) + \sum_j \boldsymbol{J}_{\mathrm{L}j}(\boldsymbol{x},t) + \boldsymbol{J}_{\mathrm{HD}}(\boldsymbol{x},t) = \nabla \times \boldsymbol{H}(\boldsymbol{x},t) \tag{9.15}$$

along with the requirement that (9.5) and (9.6) are satisfied. The implementation of (9.4) and (9.13) – (9.15) using standard FDTD techniques, as reported in [24], is outlined in Appendix 9A.

9.3 GOLD DIELECTRIC FUNCTION

To model Au nanostructures using the approach outlined in Section 9.2, (9.7) must first be fit to the experimentally determined dielectric data of bulk Au [26]. This is done in the limit of $\boldsymbol{k} \to \boldsymbol{0}$ in (9.9), which is valid for the relatively large structures used to obtain such experimental data. To make sure that the separate terms in (9.7) accurately capture the physics of the problem, it is necessary to fit (9.8) and (9.9) over the appropriate energy ranges separately. Using simulated annealing, Ref. [24] first fit (9.9) (also incorporating ε_∞) over the range 1.0 – 1.8 eV, where $\varepsilon(\boldsymbol{0}, \omega)$ is dominated by *sp*-band electron motion. Then, keeping the parameters in (9.9) constant (but not ε_∞), the entire dielectric function in (9.7) was fit over the full range of interest, 1.0 – 6.0 eV. The resulting parameters were $\varepsilon_\infty = 3.559$, $\omega_{\mathrm{D}} = 8.812$ eV, $\gamma = 0.0752$ eV, $\Delta\varepsilon_{\mathrm{L1}} = 2.912$, $\omega_{\mathrm{L1}} = 4.693$ eV, $\delta_{\mathrm{L1}} = 1.541$ eV, $\Delta\varepsilon_{\mathrm{L2}} = 1.272$, $\omega_{\mathrm{L2}} = 3.112$ eV, and $\delta_{\mathrm{L2}} = 0.525$ eV. Figure 9.1 compares the resulting fitted dielectric data to the experimental values [24].

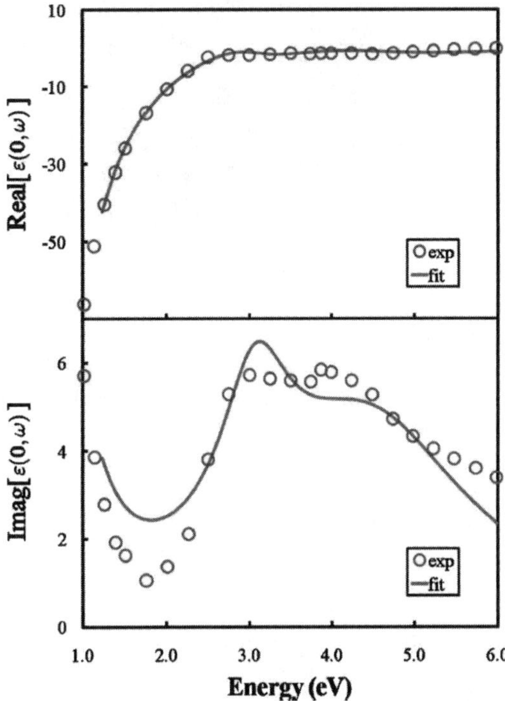

Fig. 9.1 Fitted dielectric data for bulk Au (red lines) compared to experimental values (blue circles) [26]. *Source:* J. M. McMahon, S. K. Gray, and G. C. Schatz, *Physical Review B*, Vol. 82, 2010, 035423, ©2010 The American Physical Society.

We can see from Fig. 9.1 that the dielectric function fit reported in [24] was reasonably good, given the simple form of (9.7). For example, features of the two interband transitions were captured near 3.15 and 4.30 eV, as evident in imag[$\varepsilon(\mathbf{0}, \omega)$]. Note that ω_{L1} and ω_{L2} were also close to these values, as expected based on the discussion in Section 9.2. While this fit was not as good as could be achieved with a more flexible function, such as an unrestricted fit, the fitting scheme of [24] led to parameters that were more physically realistic. This was essential given that Ref. [24] used these local ($\boldsymbol{k} = \mathbf{0}$) parameters in the nonlocal ($\boldsymbol{k} \neq \mathbf{0}$) expression. One consequence of this fit was that the minimum value of imag[$\varepsilon(\mathbf{0}, \omega)$] near 1.85 eV was not as small as the experimental one, which ended up giving broader plasmon resonances near this energy than expected. Fortunately, such differences did not play a significant role in the results reported in [24].

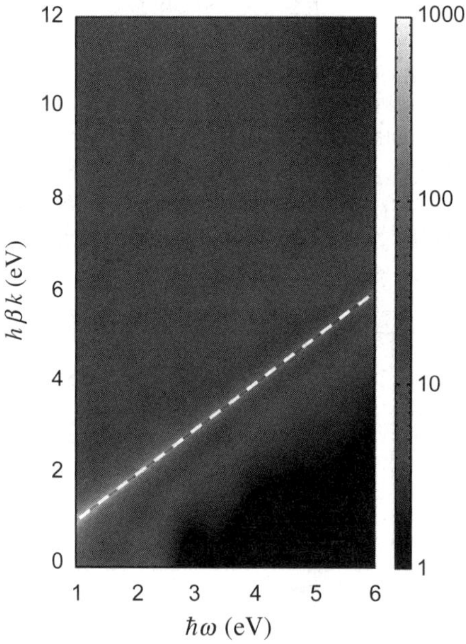

Fig. 9.2 $|\text{Real}[\varepsilon(\boldsymbol{k},\omega)]|$ of Au as a function of both k and ω. Below $\beta k \approx \omega$, $\varepsilon(\boldsymbol{k},\omega) < 0$; and above, $\varepsilon(\boldsymbol{k},\omega) > 0$. No specific value is attached to β^2. The condition $\beta k = \omega$ is shown using a dashed white line. *Adapted from:* J. M. McMahon, S. K. Gray, and G. C. Schatz, *Physical Review B*, Vol. 82, 2010, 035423, ©2010 The American Physical Society.

Figure 9.2 shows $|\text{Real}[\varepsilon(\boldsymbol{k},\omega)]|$ calculated using (9.7) as a function of both k and ω [24]. Here, a slice through $k = 0$ gives the local dielectric data shown in Fig. 9.1. In Fig. 9.2, when $\beta k << \omega$, $\varepsilon(\boldsymbol{k}, \omega)$ is relatively constant for a given ω, i.e., it remains close to the local value. However, as βk approaches ω from below, $\varepsilon(\boldsymbol{k}, \omega)$ quickly becomes very negative, and then increases rapidly and changes sign as it passes through $\beta k \approx \omega$, after which $\varepsilon(\boldsymbol{k}, \omega)$ is no longer plasmonic. We note that absorption of light by a system is related to the value of $\varepsilon(\boldsymbol{k}, \omega)$ and the structure under consideration. (For example, for a small spherical particle in air, the maximum absorption occurs when $\text{Real}[\varepsilon(\boldsymbol{k},\omega)] = -2$ [25].) Figure 9.2 therefore indicates that, in addition to the local absorption, additional (anomalous) absorption occurs when $\beta k \approx \omega$, i.e., when the rapid variation in $\text{Real}[\varepsilon(\boldsymbol{k},\omega)]$ occurs.

Nonlocal effects are very prominent for nanostructures [21]. In such systems, it is necessary to consider the reduced mean-free path of the *sp*-band electrons due to electron-interface scattering. As was briefly discussed in the introduction, this can be taken into account by using a modified collision frequency in (9.9), $\gamma' = \gamma + Av_F / L_{eff}$ [8]. In this expression, the effective mean-free path is $L_{eff} = 4V/S$ in 3-D and $\pi S/P$ in 2-D, where V is the volume of the structure having the surface area S with perimeter P, and A can be considered the proportion of electron-interface collisions that are totally inelastic. Such scattering can also be considered a nonlocal effect [16, 29].

In a formal sense, A is related to the translational invariance at the surface, the full description of which depends on the geometry of the structure, its local dielectric environment [30], and the dielectric function of the material, which is ultimately nonlocal in character. Because of these complex details, correctly choosing the value of A can be challenging, and large values can have a significant effect [9]. Although the general magnitude of A can be obtained in the local limit in a variety of ways [4], for simplicity and consistency with [21], Ref. [24] took $A = 0.1$ for its calculations, which are reviewed later in this chapter.

9.4 COMPUTATIONAL CONSIDERATIONS

The FDTD computational domains in [24] were discretized using a Yee spatial lattice [23, 31], as outlined in Appendix 9A. The outer edges of these domains were truncated using convolutional perfectly matched layers (CPMLs) [23, 32].

Optical responses in [24] were determined by calculating extinction cross-sections [25], i.e., the total amount of power absorbed and scattered relative to the incident light. These calculations were implemented by integrating the normal component of the Poynting vector around surfaces enclosing the particles [33]. It was found that an overall increase in the cross-section could occur by performing the integration too close to the structure. In [21], this resulted in the cross-sections for the 2-nm cylindrical nanowire and the 5-nm triangular nanowire to approach 0.1 at low energies, rather than 0.

To obtain the accurate Fourier-transformed fields needed for these calculations, as well as field intensity profiles, incident Gaussian damped sinusoidal pulses with frequency content over the range of interest (1 – 6 eV) were introduced into the FDTD computational domains using the total-field/scattered-field technique [23, 34]. Furthermore, all simulations were carried out to at least 100 fs [24]. Numerical instabilities were encountered in some of the 3-D calculations. For example, simulations of 1.0-nm Au nanoparticles became unstable when grid spacings of 0.05 nm were used. In 2-D, such instabilities did not seem to exist. This issue remained under investigation as of the publication of [24].

9.5 NUMERICAL VALIDATION

One way to determine the accuracy of the method reported in [24], and reviewed in Section 9.2 and Appendix 9A, is to compare results of the computational model to exact analytical ones, where available. Such comparisons are possible for metal nanofilms [20], cylindrical nanowires [35], and spherical nanoparticles [36]. This section reviews the comparison made in [24] for a 2-nm radius cylindrical nanowire, an example that had previously been considered in [21]. For these calculations, variable grid spacings were used for the discretizations, as well as the low-frequency 2-D value of β^2 for consistency with [21].

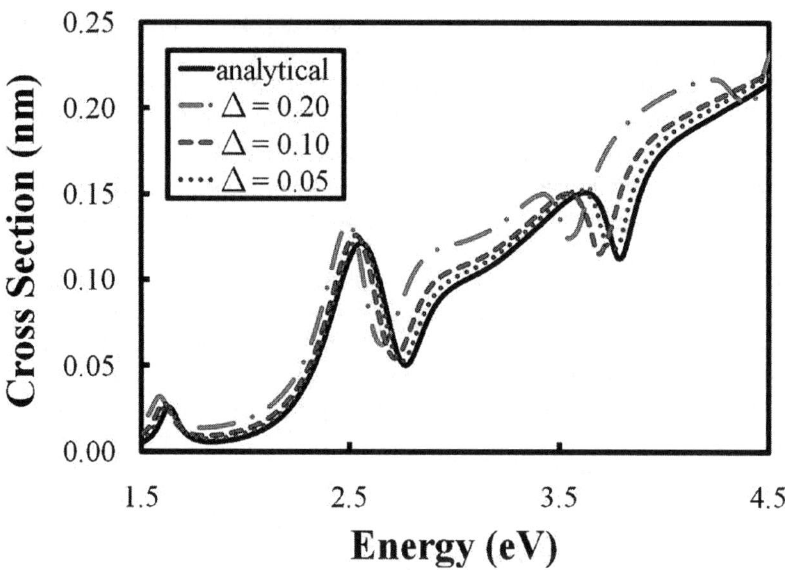

Fig. 9.3 Convergence of the FDTD-calculated extinction cross-section (color dashed and dotted lines) to the analytical results of [35] (black line) for a 2-nm radius cylindrical Au nanowire, with respect to the FDTD grid cell size, Δ, in nm. *Source:* J. M. McMahon, S. K. Gray, and G. C. Schatz, *Physical Review B*, Vol. 82, 2010, 035423, ©2010 The American Physical Society.

Figure 9.3 displays the convergence with grid resolution of the FDTD-calculated extinction cross-section of a 2-nm radius cylindrical Au nanowire, as reported in [24]. Here, the nanowire's optical response was calculated using a uniform grid cell size, Δ, in both x and y of 0.2, 0.1, and 0.05 nm. These were compared to the available analytical results [35]. Several peaks and valleys are seen in all of these results, corresponding to the dipolar LSPR near 2.55 eV and anomalous absorption near 1.61, 2.75, and 3.78 eV [20].

We can see from Fig. 9.3 that the FDTD calculations and the analytical results agreed both qualitatively and quantitatively. Decreasing Δ led to significantly better agreement, especially for the higher energy peaks. For example, the peak near 3.78 eV converged from 3.53 to 3.68 to 3.72 eV as Δ was reduced from 0.2 to 0.1 to 0.05 nm, respectively. Such convergence is understandable because, for a given grid spacing, there was an uncertainty in the nanowire radius r of $\pm\Delta$. Since the results appear to red-shift with increasing Δ, it was inferred that this radius was approximately $r + \Delta$. The convergence of these effects was much slower than those in local electrodynamics, where even $\Delta = 0.2$ nm was sufficient in the latter case (not shown). These results demonstrate the sensitivity of nonlocal effects to even minor geometric features.

Grid cell sizes of 0.05 or 0.1 nm are impractical for most calculations because of the resulting computational effort. At first this appears troublesome, given that nonlocal effects are so sensitive to this parameter. However, the only real downside is that an uncertainty of $\pm\Delta$ in the geometry must be accepted (which is also the case in local electrodynamics, but is less important). This is because it is found that, for a given grid cell size, the calculated results always fall between the analytical ones that incorporate the $\pm\Delta$ tolerance. For example, in the case of the $r = 2$ nm cylindrical nanowire in Fig. 9.4, the calculated results with $\Delta = 0.2$ nm were constrained by the analytical ones with $r = 1.8$ nm and $r = 2.2$ nm. Results for $\Delta = 0.1$ nm and $\Delta = 0.05$ nm are shown in Fig. 9.4 as well and are also consistent with this analysis.

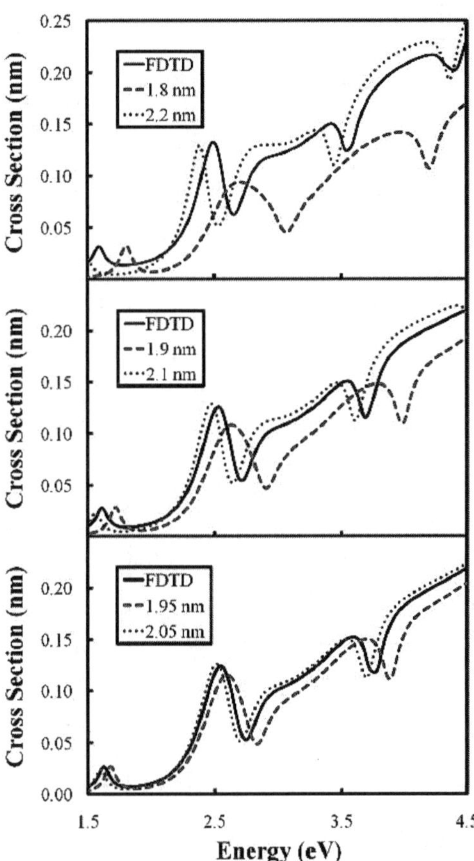

Fig. 9.4 Extinction cross-section of a 2-nm radius cylindrical Au nanowire, calculated using grid cell sizes of $\Delta = 0.2$ nm (top), $\Delta = 0.1$ nm (middle), and $\Delta = 0.05$ nm (bottom). The results show that the FDTD calculations (solid black lines), were always constrained by the analytical results (color dotted and dashed lines) when tolerances for the grid cell sizes were considered. *Source:* J. M. McMahon, S. K. Gray, and G. C. Schatz, *Physical Review B*, Vol. 82, 2010, 035423, ©2010 The American Physical Society.

9.6 APPLICATION TO GOLD NANOFILMS (1-D SYSTEMS)

This section reviews the study reported in [24] of the FDTD-computed transmission, reflection, and absorption of light by Au nanofilms illuminated at normal incidence, i.e., systems having an effective dimension of one. For simplicity, the surrounding medium was assumed to be air, although introducing other dielectric layers would have been straightforward. Because wavevector components were assumed to exist only in the direction normal to the surface of each film, these systems were ideal for studying and qualitatively highlighting nonlocal effects. Furthermore, they allowed drawing some connections with related experimental results [6, 7].

Figures 9.5(a), 9.5(b), and 9.6 show, respectively, the FDTD modeling results for Au nanofilms of thickness 2, 10, and 20 nm [24]. In these models, the high-frequency 2-D value of β^2 was assumed, with a grid cell size $\Delta = 0.1$ nm used for the 2-nm film, and $\Delta = 0.2$ nm used for the 10- and 20-nm films.

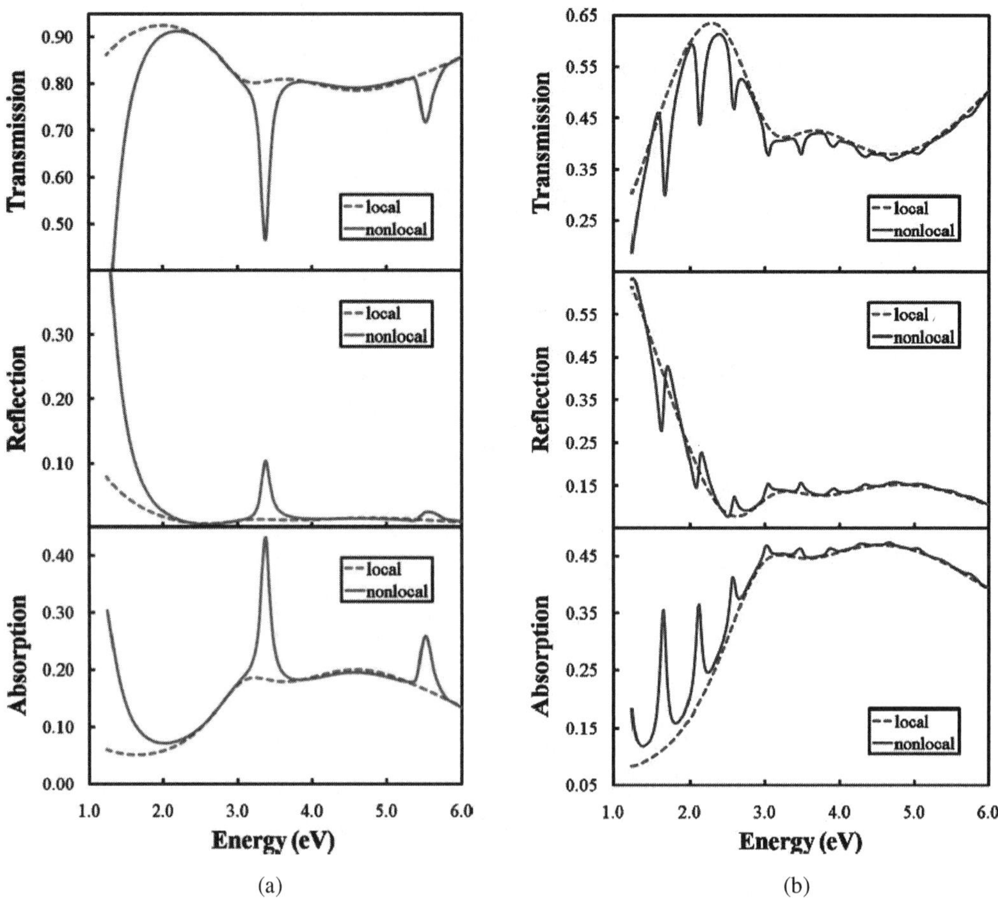

Fig. 9.5 FDTD-computed spectra for a gold nanofilm illuminated at normal incidence: (a) 2-nm-thick film; (b) 10-nm-thick film. Top panels – transmission; middle panels – reflection; bottom panels – absorption. Red dashed lines – local dielectric function; blue solid lines – nonlocal dielectric function. *Source:* J. M. McMahon, S. K. Gray, and G. C. Schatz, *Physical Review B*, Vol. 82, 2010, 035423, ©2010 The American Physical Society.

Especially for the 2- and 10-nm-thick Au films shown in Fig. 9.5, significant differences were found in the FDTD-computed transmission, reflection, and absorption of light depending on whether the local or nonlocal dielectric function was modeled [24]. Specifically, narrow additional (anomalous) absorption peaks in the absorption spectra occurred in the nonlocal results, relative to the local ones. The appearance of these peaks is identical to theoretical predictions [20] and experimental observations [6, 7] on other thin metal films, where they were the result of optically excited longitudinal (or volume) plasmons. Unlike surface plasmons, which propagate along a metal-dielectric interface, these waves are longitudinal to $\boldsymbol{k}$ and are contained within the volume of the nanostructure. Not surprisingly, at the anomalous absorption energies, there was a corresponding decrease in the transmission. However, contrary to an initial expectation of a decrease in reflection, there was found either an increase or a decrease, depending on if the corresponding absorption occurred well above (giving an increase) or below (giving a decrease) the surface-plasmon energy, for example ~2.65 eV for the 10-nm Au film.

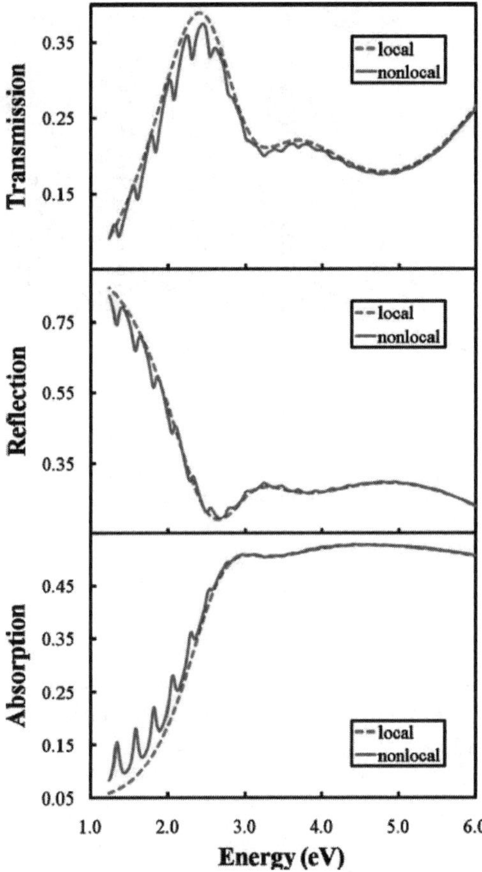

Fig. 9.6 FDTD-computed spectra for a 20-nm-thick Au nanofilm illuminated at normal incidence. Top panel – transmission; middle panel – reflection; bottom panel – absorption. Red dashed lines – local dielectric function; blue solid lines – nonlocal dielectric function. *Source:* J. M. McMahon, S. K. Gray, and G. C. Schatz, *Physical Review B*, Vol. 82, 2010, 035423, ©2010 The American Physical Society.

Although a little hard to discern from Figs. 9.5 and 9.6, but can be inferred from previous results [21], the anomalous absorption peaks were red-shifted to longer wavelengths as the Au film thickness was increased [24]. This caused more such peaks, occurring initially at higher energies, to appear in the optical range. For example, there were only three peaks computed for the 2-nm film of Fig. 9.5(a), but 12 peaks for the 10-nm film of Fig. 9.5(b). In addition, their intensities decreased quickly, so that for the 20-nm film, the nonlocal dielectric results were almost converged to the local ones, as shown in Fig. 9.6.

We wanted to determine whether the anomalous absorption in this model resulted from the excitation of longitudinal plasmons. To this end, intensity profiles of $|\boldsymbol{D}|^2$ were obtained using FDTD at the three anomalous absorption energies for the 2-nm film shown in Fig. 9.5(a): 1.14, 3.36, and 5.54 eV. These profiles are visualized in Fig. 9.7 [24]. Well-defined standing-wave patterns of $|\boldsymbol{D}|^2$ are seen inside the films longitudinal to $\boldsymbol{k}$, confirming the assumption of their nature. The wavelengths of these standing waves were found to satisfy the condition:

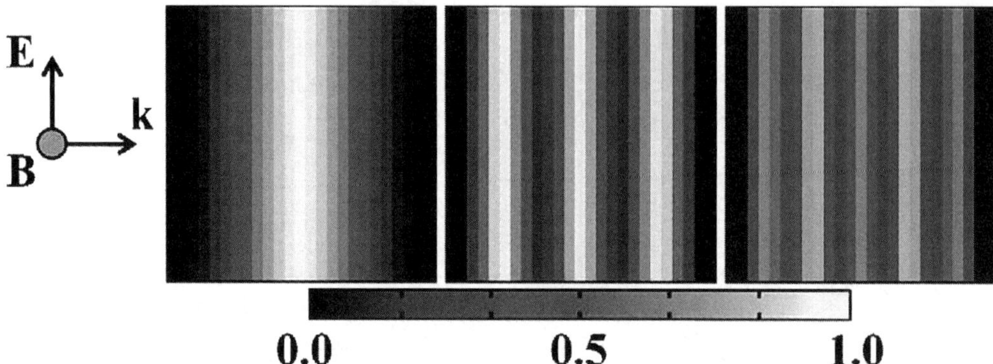

Fig. 9.7 Normalized $|\boldsymbol{D}|^2$ intensity profiles inside a 2-nm-thick Au film at energies of (left) 1.14 eV; (middle) 3.36 eV; and (right) 5.54 eV. The polarization and direction of the incident light are indicated. In each image, the sides of the metal film are padded on the left and right using solid black lines. *Source:* J. M. McMahon, S. K. Gray, and G. C. Schatz, *Physical Review B*, Vol. 82, 2010, 035423, ©2010 The American Physical Society.

$$\lambda_{\mathrm{L}} = 2d/m \tag{9.16}$$

where d is the film thickness and $m = 1, 3, 5, \ldots$. This means that odd numbers of half-wavelengths fit longitudinally into the film. The wavelengths defined by (9.16) will hereafter be referred to as "modes" characterized by m.

Figure 9.7 explicitly shows the m = 1, 3, and 5 modes [24]. These results were significantly different from those computed assuming a local dielectric function, where relatively uniform $|\boldsymbol{D}|^2$ patterns were found at all energies (not shown). Nonetheless, as stated previously, this analysis agreed with previous results on analogous systems [6, 7, 20], providing further support for the validity of the method reported in [24]. Based on the observations in Fig. 9.7 and this analysis, it makes sense that the anomalous absorption features should be red-shifted with increasing film thickness, and that their intensity should decrease with increasing m.

From the above discussion and that in Section 9.3, the approximate anomalous absorption energies can be predicted analytically. From (9.9), it is seen that rapid variations in $\varepsilon(\boldsymbol{k}, \omega)$ occur when $\omega \approx \beta k$, which likely leads to an absorption condition. From (9.16), it is seen that longitudinal plasmons with wavelength λ_{L} are excited inside a gold film of thickness, d, which results in momentum states of magnitude, $k = 2\pi/\lambda_{\mathrm{L}}$. Thus, everything needed to predict the approximate anomalous absorption energies is known: $\hbar\omega = m\beta\pi/d$.

Using the 2-nm film as an example, this analysis predicts anomalous absorption at energies of approximately $\hbar\omega = m \times 1.44$ eV. For the first three m modes, these are 1.44, 4.31, and 7.19 eV, while those rigorously calculated using FDTD were, respectively, 1.14, 3.36, and 5.54 eV, per Fig. 9.5(a). While differing somewhat from the FDTD results, the approximations are reasonably close. Part of the differences can be attributed to the grid spacing uncertainty, as outlined in Section 9.5, which leads in this case to an uncertainty of ± 0.2 nm in the film thickness, d. While this approximate analysis could also be applied to related experimental results [6], it is important to keep in mind that this analytical approximation is based on simple considerations. If appropriate, more accurate values should be obtained from full FDTD computational models or, if available for simple geometries, by using rigorous theory [20].

9.7 APPLICATION TO GOLD NANOWIRES (2-D SYSTEMS)

References [21] and [24] demonstrated that nonlocal dielectric effects are particularly important in structures with apex features, such as nanowires of triangular cross-section. In fact, much larger structures of this type have optical responses affected by nonlocal dielectric phenomena than structures with smooth geometries, such as cylindrical nanowires. Additionally, near-field properties (such as $|\boldsymbol{E}|^2$ enhancements) of structures having nonlocal dielectric effects do not converge at large sizes to those of structures having only local dielectric effects.

We now review the 2-D FDTD modeling results reported in [24] obtained by applying the nonlocal electrodynamics technique of Section 9.2 and Appendix 9A to model nanowires with apex features. For these models, grid spacings of 0.25 nm were used, along with the high-frequency value of β^2. Figures 9.8, 9.9(a), and 9.9(b) show, respectively, the computed extinction cross-sections as well as $|\boldsymbol{D}|^2$ and $|\boldsymbol{E}|^2$ profiles of isolated Au nanowires of circular, square, and triangular cross-section having 50-nm diameters or side lengths (common sizes used in experimental and theoretical studies). While electric-field enhancements of such structures have been studied in the past [37, 38], it appears that all previous studies were carried out using local electrodynamics (at least for non-circular cross-section structures), except for the work reported in [21] and [24].

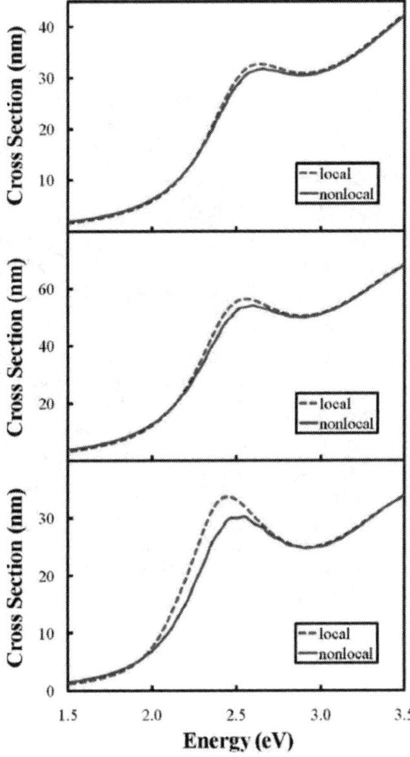

Fig. 9.8 FDTD-computed extinction cross-sections of Au nanowires of (top) circular, (middle) square, and (bottom) triangular cross-sections with 50-nm diameters or side lengths. Red dashed lines – local dielectric function; solid blue lines – nonlocal. The illuminating electric field was assumed to parallel the longest axis of each cross-section. *Source:* J. M. McMahon, S. K. Gray, and G. C. Schatz, *Physical Review B*, Vol. 82, 2010, 035423, ©2010 The American Physical Society.

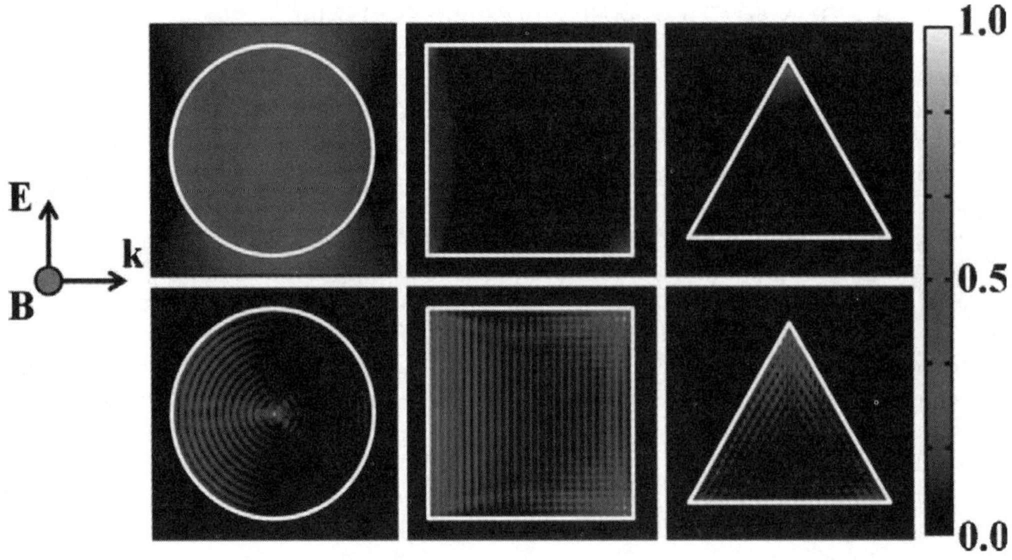

(a) Normalized $|\boldsymbol{D}|^2$ intensity profiles.

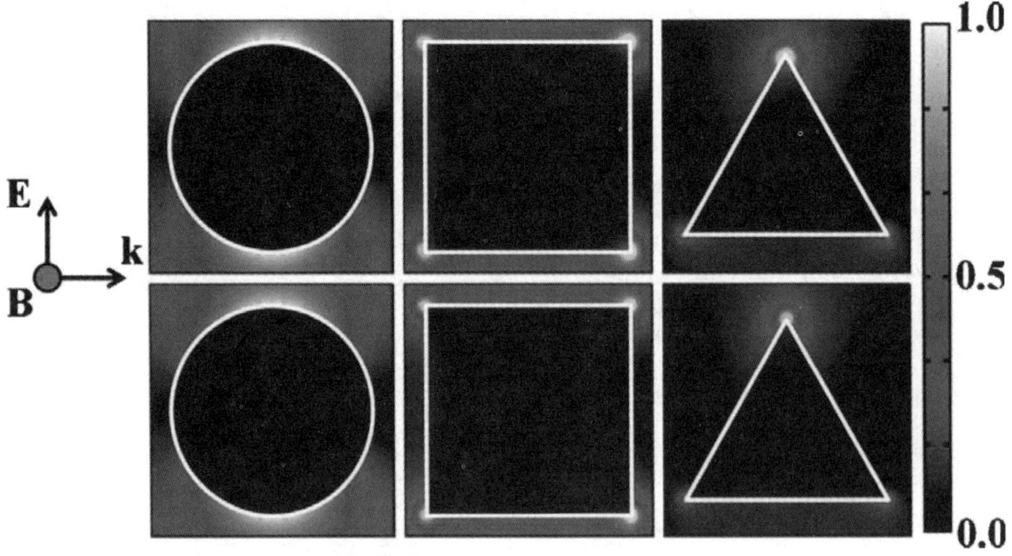

(b) Normalized $|\boldsymbol{E}|^2$ intensity profiles.

Fig. 9.9 Visualization of the FDTD-computed normalized $|\boldsymbol{D}|^2$ and $|\boldsymbol{E}|^2$ intensity profiles at the LSPR energies in and around isolated Au nanowires of (left) circular, (middle) square, and (right) triangular cross-sections with 50-nm diameters or side lengths. Top panels – local dielectric function; bottom panels – nonlocal dielectric function. The polarization and direction of the incident light are indicated, and the perimeters of the nanowires are outlined in white. *Source:* J. M. McMahon, S. K. Gray, and G. C. Schatz, *Physical Review B*, Vol. 82, 2010, 035423, ©2010 The American Physical Society.

Because of their relatively large sizes, these structures did not exhibit distinct anomalous absorption upon modeling. Nonetheless, many closely spaced longitudinal plasmon modes did exist (*vide infra*), which led to very minor, closely spaced "bumps" in the nonlocal dielectric function results, as well as LSPR blue-shifting [24].

These modes can again be confirmed by looking at FDTD-computed intensity profiles of $|\mathbf{D}|^2$ in Fig. 9.9(a). Unlike the results in Fig. 9.7, the longitudinal plasmons in Fig. 9.9(a) formed much more complex patterns, attributed by Ref. [24] to two related effects. First, the span of the nanowire cross-section along the longitudinal direction of the incident field was not the same at all positions, except for the square nanowire. Therefore, for a given energy, modes of different order were sustained at multiple positions along the cross-section at each place where (9.16) was satisfied [21]. This was also one of the reasons why nonlocal effects were observed to be so strong in structures with apex features, and why they could remain important in such structures for arbitrarily large sizes. In other words, low-order longitudinal plasmon modes could always be sustained near the apex. And second, scattering of the incident field from a curved nanowire surface generated many $\mathbf{k}$ components, which excited longitudinal plasmons along directions other than that of the incident $\mathbf{k}$, creating an interference pattern. This effect also led to the dephasing of longitudinal plasmons.

Reference [24] also commented on the FDTD-computed intensity profiles of $|\mathbf{E}|^2$ at the LSPR energies in Fig. 9.9(b), near where this quantity was expected to be maximized [39]. These energies were slightly different in the local and nonlocal results due to LSPR blue-shifting. In this figure, the $|\mathbf{E}|^2$ profiles were normalized for each nanowire cross-section geometry so that relative comparisons could be made between the local and nonlocal results for each nanowire. However, because of this normalization, Fig. 9.9(b) was not appropriate for comparing the results of different nanowire cross-section geometries.

To provide this comparison, Ref. [24] provided a table, here shown as Table 9.1, which listed the maximum and average FDTD-computed $|\mathbf{E}|^2$ enhancements at the LSPR energies around the isolated Au nanowires of circular, square, and triangular cross-section having 50-nm diameters or side lengths. The average values refer to fields averaged over certain distances from the nanowire surfaces.

TABLE 9.1

FDTD-Computed Maximum and Average $|\mathbf{E}|^2$ Enhancements at the LSPR Energies around Au Nanowires of Circular, Square, and Triangular Cross-Sections with Diameters or Side Lengths of 50 nm.

Nanowire cross-section	*Maximum*	*Average at 0.5 nm*	*Average at 1.0 nm*	*Average at 2.0 nm*
Circular (local)	8.64	2.42	2.47	2.40
Circular (nonlocal)	7.85	2.32	2.39	2.34
Square (local)	60.58	3.54	3.33	3.01
Square (nonlocal)	39.79	3.02	2.91	2.69
Triangular (local)	145.77	5.49	4.90	4.18
Triangular (nonlocal)	71.40	3.42	3.30	3.01

Source: J. M. McMahon, S. K. Gray, and G. C. Schatz, *Physical Review B*, Vol. 82, 2010, 035423, ©2010 The American Physical Society.

In all cases, decreases in both quantities were obtained in the nonlocal results. However, for the nanowires of circular cross-section, these decreases were negligible. It is also interesting to note that the average enhancements were higher 1.0 nm away from the surfaces than they were at 0.5 nm. For the square cross-section nanowires, the decreases were noticeably larger. There was approximately a 13% difference at 1.0 nm, and an 11% difference at 2.0 nm in the average values. The decrease in the difference between average enhancements at a further distance from the nanowire surface was expected, since the contributing near fields decay exponentially. Decreases in $|\boldsymbol{E}|^2$ enhancements for the triangular cross-section nanowires were strikingly larger than for the other geometries. For example, decreases of 51% and 38% in the maximum and average values at 0.5 nm were obtained, respectively.

Considering that some physical processes are dependent on $|\boldsymbol{E}|^4$ enhancements [1], such as surface-enhanced Raman scattering (SERS), the differences between local and nonlocal electrodynamics could have significant implications for the interpretation of results. This statement is based on the fact that the nonlocal calculations are, in principle, more rigorous than the local ones. For example, if the actual electromagnetic contribution to SERS is smaller than expected on the basis of local theory, it is possible that chemical effects play a more important role than has been considered in the past [40]. Such results are also likely to play a large role in the accurate interpretation of electron energy-loss measurements for anisotropic nanoparticle structures, which have recently received attention within the framework of local electrodynamics [41].

9.8 APPLICATION TO SPHERICAL GOLD NANOPARTICLES (3-D SYSTEMS)

This section reviews the FDTD modeling results reported in [24] obtained by applying the full 3-D nonlocal electrodynamics technique of Section 9.2 and Appendix 9A to model spherical gold nanoparticles of diameters 4, 7, and 15 nm. For these models, grid spacings of 0.2 nm were used for the 4- and 7-nm nanoparticles, and 0.5 nm for the 15-nm nanoparticle, along with the high-frequency value of β^2. The polarization and direction of the incident light were irrelevant for these calculations.

Figure 9.10 shows the FDTD-computed extinction cross-sections of the spherical nanoparticles modeled in [24]. Note that, for small nanoparticles such as these, the computed optical responses are predominately absorption, because scattering does not play a significant role for particle sizes less than approximately 20 nm. From this figure, it is seen that including nonlocal effects resulted in significant anomalous absorption and LSPR blue-shifting for both the 4- and 7-nm nanoparticles. In fact, these effects were so large that the main LSPRs were barely distinguishable.

As reviewed in Section 9.6, the appearance of anomalous absorption peaks in Au nanofilms could be attributed to the excitation of longitudinal plasmons [24]. However, this effect was found to diminish much faster for Au nanospheres, relative to Au nanofilms, as their characteristic sizes were increased. In fact, for a nanosphere diameter of 15 nm, the anomalous peaks showed up only as slight indents on the main LSPR. This difference was attributed to two effects. First, for a spherical nanoparticle, external scattering of the incident light from the surface of the nanoparticle generates many $\boldsymbol{k}$ components that can interact and dephase one another, especially for the high-order m modes with multiple nodes. Second, internal scattering of the conduction electrons (that compose the longitudinal plasmons) from the surface of the nanoparticle could also lead to dephasing. Each of these processes causes nonlocal effects to diminish at much smaller distances in spherical Au nanoparticles, relative to Au nanofilms.

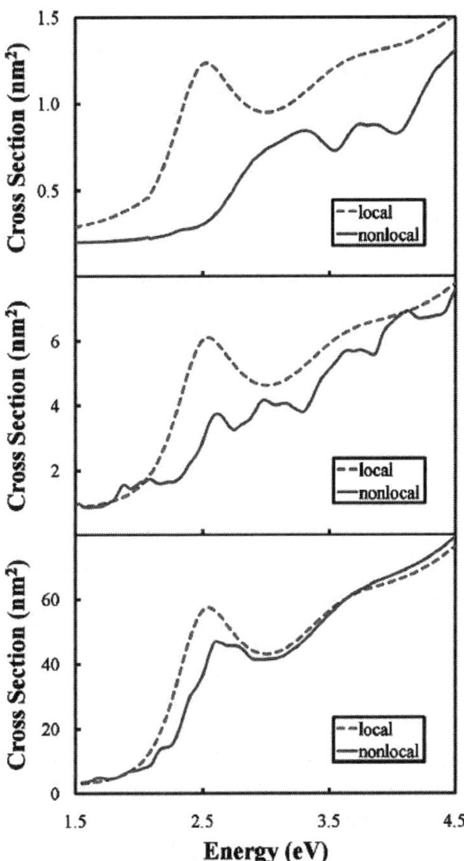

Fig. 9.10 FDTD-computed extinction cross-sections of spherical Au nanoparticles of diameter (top) – 4 nm; (middle) – 7 nm; and (bottom) – 15 nm. Red dashed lines – local dielectric function; blue solid lines – nonlocal dielectric function. *Source:* J. M. McMahon, S. K. Gray, and G. C. Schatz, *Physical Review B*, Vol. 82, 2010, 035423, ©2010 The American Physical Society.

The LSPR blue-shifting was most apparent for the 15-nm nanoparticle [24]. Here, the local electrodynamics LSPR was at 2.57 eV, whereas the nonlocal electrodynamics LSPR was at 2.71 eV. This was attributed to a low anomalous absorption that allowed this peak to be clearly identified, an effect that can be understood by looking at the form of (9.9). When nonlocal effects are included, the interplay between ω and $\boldsymbol{k}$ causes all effects (e.g., the absorption condition of Real$[\varepsilon(\boldsymbol{k}, \omega)] = -2$, in this case) to appear at higher energies compared to $\boldsymbol{k} = \mathbf{0}$.

Based on the results in Fig. 9.10, one might wonder why such strong nonlocal effects have not yet been experimentally observed in such nanoparticle systems, whereas these effects are important and have been observed in other cases [6, 7]. The most probable reason is that experimental measurements are often made on heterogeneous collections of nanoparticles [24]. Given that nonlocal effects are very sensitive to nanoparticle dimensions (as reviewed in Section 9.5), slight heterogeneity could effectively average them away. Support for this claim comes from an experimental study of isolated Au nanoparticles [5] that clearly demonstrated the LSPR blue-shift and possibly anomalous absorption features [21].

There may be two additional possible explanations [24]. The first possibility is a non-optimal choice of β^2, which is directly related to the strength of nonlocal effects, as recently argued for metallic nanoshells [9]. The hydrodynamic Drude model neglects quantum-mechanical exchange and correlation effects, which in a local-density approximation would decrease β^2. Another possibility is that the damping parameter, A, was set too low. Increasing A would smooth all spectral features, and the anomalous absorption would not appear to be as strong. Support for this arises from a combined theoretical (local electrodynamics) and experimental study of metallic nanoshells, where values of A greater than 1.0 were needed to describe the results. Note that this corresponds to L_{eff} reduced below that based on geometric considerations alone [42].

9.9 SUMMARY AND OUTLOOK

This chapter reviewed the electrodynamics technique originally reported in [21] and [24] to calculate the optical response of an arbitrarily shaped structure described by a spatially nonlocal dielectric function. This technique was based on converting the hydrodynamic Drude model into an equation of motion for the conduction electrons, which then served as a current field in the Maxwell–Ampere law. By discretizing this equation using standard finite-difference techniques, it was incorporated into a self-consistent computational scheme along with the standard equations used in the FDTD method.

In this chapter, the full theory, FDTD algorithm, and modeling examples reported in [24] were reviewed. The review commenced with the validation study for cylindrical nanowires presented in [24], which demonstrated the accuracy of the technique via comparison to analytical theory. Then, the review continued with models of the optical properties of Au nanostructures in one, two, and three dimensions presented in [24]. The first models involved 1-D calculations of the transmission, reflection, and absorption spectra of Au nanofilms. Because of their simplicity, these systems demonstrated clearly the longitudinal (or volume) plasmons characteristic of nonlocal effects, which result in anomalous absorption and LSPR blue-shifting. The next models involved 2-D calculations of the maximum and average electric field enhancements around Au nanowires of various shapes, which were compared to local-electrodynamics theory predictions. The final models involved 3-D calculations of the optical properties of spherical Au nanoparticles, which again showed the significant impact of nonlocal dielectric effects.

The work of [24] reviewed in this chapter demonstrates the importance of including nonlocal dielectric effects when describing metal–light interactions at the nanometer length scale. At present, it is difficult to directly compare the results of applying the techniques of [24] with experimental studies. This is because most experiments currently involve heterogeneous collections of particles or noncontinuous systems, which in the small-size limit tend to average over nonlocal electrodynamics effects. Consequently, it anticipated that the results reported in [24] and reviewed here will motivate new and more precise experimental studies, in particular involving isolated nanostructures where nonlocal effects are likely to play a large role.

An important future research direction involves deriving more accurate expressions than the hydrodynamic Drude model. As stated, this model neglects quantum-mechanical exchange and correlation effects. By incorporating these effects, nonlocal electrodynamics calculations could be compared directly with quantum-mechanical approaches, such as electronic structure theory. Such an improved model would allow even more accurate descriptions of nonlocal optical phenomena than were presented in [24] and reviewed in this chapter.

ACKNOWLEDGMENTS

The work reported in [21] and [24], which was reviewed in this chapter, was supported in part by AFOSR/DARPA Project BAA07-61 (Grant No. FA9550-08-1-0221) and by the NSF MRSEC (Grant No. DMR-0520513) at the Materials Research Center of Northwestern University. Use of the Center for Nanoscale Materials was supported by the U.S. Department of Energy, Office of Science, Office of Basic Energy Sciences, under Contract No. DE-AC02-06CH11357.

APPENDIX 9A: NONLOCAL FDTD ALGORITHM

This appendix reviews the derivation of the 3-D FDTD algorithm presented in [24] for modeling nonlocal dielectric effects. This algorithm operates within the framework of the classic Yee space-time discretization [23, 31] of the components of $\boldsymbol{E}(\boldsymbol{x})$ and $\boldsymbol{H}(\boldsymbol{x})$ (i.e., 0.5 grid-cell spatial offsets, mutual circulations about the other, and 0.5 time-step temporal offsets). All spatial and temporal derivatives are approximated using second-order accurate central finite differences.

In the expressions to follow, the $\boldsymbol{J}_{\mathrm{L}j}(\boldsymbol{x})$ and $\boldsymbol{J}_{\mathrm{HD}}(\boldsymbol{x})$ components are centered at the same spatial locations as the corresponding $\boldsymbol{E}(\boldsymbol{x})$ components. To model an arbitrarily shaped structure, these $\boldsymbol{J}$ components exist only at the grid positions of the corresponding nonlocal material. By not updating the currents outside of the structure, the additional boundary condition of Pekar is imposed [43] — i.e., the total nonlocal polarization current vanishes outside of the structure.

First, the temporal derivatives in (9.4) and (9.15) are discretized using a leapfrog algorithm [23]:

$$\mu_0\left(\frac{\boldsymbol{H}(\boldsymbol{x})^{n+1/2} - \boldsymbol{H}(\boldsymbol{x})^{n-1/2}}{\Delta t}\right) = -\nabla\times\boldsymbol{E}(\boldsymbol{x})^n \tag{9A.1}$$

$$\varepsilon_\infty\varepsilon_0\left(\frac{\boldsymbol{E}(\boldsymbol{x})^{n+1} - \boldsymbol{E}(\boldsymbol{x})^{n}}{\Delta t}\right) + \sum_j \boldsymbol{J}_{\mathrm{L}j}(\boldsymbol{x})^{n+1/2} + \boldsymbol{J}_{\mathrm{HD}}(\boldsymbol{x})^{n+1/2} = \nabla\times\boldsymbol{H}(\boldsymbol{x})^{n+1/2} \tag{9A.2}$$

where the superscript n denotes a discrete time-step. Rearrangement of (9A.1) immediately provides the updating expression for $\boldsymbol{H}(\boldsymbol{x})$:

$$\boldsymbol{H}(\boldsymbol{x})^{n+1/2} = \boldsymbol{H}(\boldsymbol{x})^{n-1/2} - \left(\frac{\Delta t}{\mu_0}\right)\nabla\times\boldsymbol{E}(\boldsymbol{x})^n \tag{9A.3}$$

Next, (9.13) and (9.14) are discretized using central finite differences centered at time-step n:

$$\begin{aligned}\frac{\boldsymbol{J}_{\mathrm{L}j}(\boldsymbol{x})^{n+1} - 2\boldsymbol{J}_{\mathrm{L}j}(\boldsymbol{x})^{n} + \boldsymbol{J}_{\mathrm{L}j}(\boldsymbol{x})^{n-1}}{(\Delta t)^2} &+ 2\delta_{\mathrm{L}j}\left(\frac{\boldsymbol{J}_{\mathrm{L}j}(\boldsymbol{x})^{n+1} - \boldsymbol{J}_{\mathrm{L}j}(\boldsymbol{x})^{n-1}}{2\Delta t}\right) \\ + \omega_{\mathrm{L}j}^2\boldsymbol{J}_{\mathrm{L}j}(\boldsymbol{x})^n &= \varepsilon_0\,\Delta\varepsilon_{\mathrm{L}j}\,\omega_{\mathrm{L}j}^2\left(\frac{\boldsymbol{E}(\boldsymbol{x})^{n+1} - \boldsymbol{E}(\boldsymbol{x})^{n-1}}{2\Delta t}\right)\end{aligned} \tag{9A.4}$$

$$\frac{\boldsymbol{J}_{\mathrm{HD}}(\boldsymbol{x})^{n+1} - 2\boldsymbol{J}_{\mathrm{HD}}(\boldsymbol{x})^{n} + \boldsymbol{J}_{\mathrm{HD}}(\boldsymbol{x})^{n-1}}{(\Delta t)^2} + \gamma\left(\frac{\boldsymbol{J}_{\mathrm{HD}}(\boldsymbol{x})^{n+1} - \boldsymbol{J}_{\mathrm{HD}}(\boldsymbol{x})^{n-1}}{2\Delta t}\right) - \beta^2\nabla^2\boldsymbol{J}_{\mathrm{HD}}(\boldsymbol{x})^{n} = \varepsilon_0\omega_{\mathrm{D}}^2\left(\frac{\boldsymbol{E}(\boldsymbol{x})^{n+1} - \boldsymbol{E}(\boldsymbol{x})^{n-1}}{2\Delta t}\right) \tag{9A.5}$$

The updating equation for $\boldsymbol{J}_{\mathrm{L}j}(\boldsymbol{x})$ is obtained by rearranging (9A.4):

$$\boldsymbol{J}_{\mathrm{L}j}(\boldsymbol{x})^{n+1} = \alpha_{\mathrm{L}j}\boldsymbol{J}_{\mathrm{L}j}(\boldsymbol{x})^{n} + \xi_{\mathrm{L}j}\boldsymbol{J}_{\mathrm{L}j}(\boldsymbol{x})^{n-1} + \eta_{\mathrm{L}j}\left(\frac{\boldsymbol{E}(\boldsymbol{x})^{n+1} - \boldsymbol{E}(\boldsymbol{x})^{n-1}}{2\Delta t}\right) \tag{9A.6}$$

where

$$\alpha_{\mathrm{L}j} = \frac{2 - \omega_{\mathrm{L}j}^2(\Delta t)^2}{1 + \delta_{\mathrm{L}j}\Delta t} \tag{9A.7a}$$

$$\xi_{\mathrm{L}j} = -\frac{1 - \delta_{\mathrm{L}j}\Delta t}{1 + \delta_{\mathrm{L}j}\Delta t} \tag{9A.7b}$$

$$\eta_{\mathrm{L}j} = \frac{\varepsilon_0\,\Delta\varepsilon_{\mathrm{L}j}\,\omega_{\mathrm{L}j}^2(\Delta t)^2}{1 + \delta_{\mathrm{L}j}\Delta t} \tag{9A.7c}$$

Similarly, the updating equation for $\boldsymbol{J}_{\mathrm{HD}}(\boldsymbol{x})$ is obtained by rearranging (9A.5):

$$\boldsymbol{J}_{\mathrm{HD}}(\boldsymbol{x})^{n+1} = \alpha_{\mathrm{HD}}\boldsymbol{J}_{\mathrm{HD}}(\boldsymbol{x})^{n} + \xi_{\mathrm{HD}}\boldsymbol{J}_{\mathrm{HD}}(\boldsymbol{x})^{n-1} + \eta_{\mathrm{HD}}\left(\frac{\boldsymbol{E}(\boldsymbol{x})^{n+1} - \boldsymbol{E}(\boldsymbol{x})^{n-1}}{2\Delta t}\right) \tag{9A.8}$$

where

$$\alpha_{\mathrm{HD}} = \frac{4 + 2(\Delta t)^2\beta^2\nabla^2}{2 + \gamma\Delta t} \tag{9A.9a}$$

$$\xi_{\mathrm{HD}} = -\frac{2 - \gamma\Delta t}{2 + \gamma\Delta t} \tag{9A.9b}$$

$$\eta_{\mathrm{HD}} = \frac{2\varepsilon_0\omega_{\mathrm{D}}^2(\Delta t)^2}{2 + \gamma\Delta t} \tag{9A.9c}$$

Note that α_{HD} is an operator, rather than a simple coefficient. To use (9A.6) and (9A.8) in (9A.2), $\boldsymbol{J}_{\mathrm{L}j}(\boldsymbol{x})$ and $\boldsymbol{J}_{\mathrm{HD}}(\boldsymbol{x})$ are centered at time-step n+1/2 by averaging:

$$\boldsymbol{J}_{\mathrm{L}j}(\boldsymbol{x})^{n+1/2} = \frac{\boldsymbol{J}_{\mathrm{L}j}(\boldsymbol{x})^{n+1} + \boldsymbol{J}_{\mathrm{L}j}(\boldsymbol{x})^{n}}{2} \tag{9A.10}$$

$$\boldsymbol{J}_{\text{HD}}(\boldsymbol{x})^{n+1/2} = \frac{\boldsymbol{J}_{\text{HD}}(\boldsymbol{x})^{n+1} + \boldsymbol{J}_{\text{HD}}(\boldsymbol{x})^{n}}{2} \tag{9A.11}$$

Equations (9A.2), (9A.6), and (9A.8) all contain $\boldsymbol{E}(\boldsymbol{x})^{n+1}$. To obtain a consistent updating, (9A.10) and (9A.11) [using (9A.6) and (9A.8)] are inserted into (9A.2) and rearranged:

$$\boldsymbol{E}(\boldsymbol{x})^{n+1} = \left(\frac{1}{\zeta_1 + \zeta_2}\right)\left[\zeta_1 \boldsymbol{E}(\boldsymbol{x})^{n} + \zeta_2 \boldsymbol{E}(\boldsymbol{x})^{n-1} + \nabla \times \boldsymbol{H}(\boldsymbol{x})^{n+1/2} - \boldsymbol{J}_{\text{T}}(\boldsymbol{x})^{n,n-1}\right] \tag{9A.12}$$

where

$$\zeta_1 = \frac{\varepsilon_\infty \varepsilon_0}{\Delta t} \tag{9A.13a}$$

$$\zeta_2 = \frac{1}{4\Delta t}\left(\sum_j \eta_{\text{L}j} + \eta_{\text{HD}}\right) \tag{9A.13b}$$

$$\boldsymbol{J}_{\text{T}}(\boldsymbol{x})^{n,n-1} = \frac{1}{2}\left\{\begin{array}{l} \sum_j \left[\left(\alpha_{\text{L}j} + 1\right)\boldsymbol{J}_{\text{L}j}(\boldsymbol{x})^{n} + \xi_{\text{L}j}\,\boldsymbol{J}_{\text{L}j}(\boldsymbol{x})^{n-1}\right] \\ + \left(\alpha_{\text{HD}} + 1\right)\boldsymbol{J}_{\text{HD}}(\boldsymbol{x})^{n} + \xi_{\text{HD}}\,\boldsymbol{J}_{\text{HD}}(\boldsymbol{x})^{n-1} \end{array}\right\} \tag{9A.13c}$$

Equations (9A.3), (9A.6), (9A.8), and (9A.12) form the complete and consistent set necessary to solve (9.3) – (9.6) for materials described by the constitutive relationship in (9.2) with the dielectric function given in (9.7) – (9.9).

REFERENCES

[1] Willets, K. A., and R. P. Van Duyne, "Localized surface plasmon resonance spectroscopy and sensing," *Annual Review of Physical Chemistry*, Vol. 58, 2007, pp. 267–297.

[2] Stewart, M. E., C. R. Anderton, L. B. Thompson, J. Maria, S. K. Gray, J. A. Rogers, and R. G. Nuzzo, "Nanostructured plasmonic sensors," *Chemical Review*, Vol. 108, 2008, pp. 494–521.

[3] Ozbay, E., "Plasmonics: Merging photonics and electronics at nanoscale dimensions," *Science*, Vol. 311, 2006, pp. 189–193.

[4] Kreibig, U., and M. Vollmer, *Optical Properties of Metal Clusters*, Berlin: Springer, 1995.

[5] Palomba, S., L. Novotny, and R. E. Palmer, "Blue-shifted plasmon resonance of individual size-selected gold nanoparticles," *Optics Communications*, Vol. 281, 2008, pp. 480–483.

[6] Anderegg, M., B. Feuerbacher, and B. Fitton, "Optically excited longitudinal plasmons in potassium," *Physical Review Lett.*, Vol. 27, 1971, pp. 1565–1568.

[7] Lindau, I., and P. O. Nilsson, "Experimental evidence for excitation of longitudinal plasmons by photons," *Physics Lett. A*, Vol. 31, 1970, pp. 352–353.

[8] Coronado, E. A., and G. C. Schatz, "Surface plasmon broadening for arbitrary shape nanoparticles: A geometrical probability approach," *J. Chemical Physics*, Vol. 119, 2003, 3926.

[9] McMahon, J. M., S. K. Gray, and G. C. Schatz, "Nonlocal dielectric effects in core–shell nanowires," *J. Physical Chemistry C*, Vol. 114, 2010, pp. 15903–15908.

[10] Peng, S., J. M. McMahon, G. C. Schatz, S. K. Gray, and Y. Sun, "Reversing the size-dependence of surface plasmon resonances," *Proc. National Academy of Sciences USA*, Vol. 107, 2010, pp. 14530–14534.

[11] Trivedi, N., and N. W. Ashcroft, "Quantum size effects in transport properties of metallic films," *Physical Review B*, Vol. 38, 1988, pp. 12298–12309.

[12] Agarwal, G. S., D. N. Pattanayak, and E. Wolf, "Electromagnetic fields in spatially dispersive media," *Physical Review B*, Vol. 10, 1974, pp. 1447–1475.

[13] Landau, L. D., E. M. Lifshitz, and L. P. Pitaevskii, *Electrodynamics of Continuous Media*, 2nd ed., Oxford: Butterworth-Heinemann, 1984.

[14] Dasgupta, B. B., and R. Fuchs, "Polarizability of a small sphere including nonlocal effects," *Physical Review B*, Vol. 24, 1981, pp. 554–561.

[15] Chang, R., and P. T. Leung, "Nonlocal effects on optical and molecular interactions with metallic nanoshells," *Physical Review B*, Vol. 73, 2006, 125438.

[16] García de Abajo, F. J., "Nonlocal effects in the plasmons of strongly interacting nanoparticles, dimers, and waveguides," *J. Physical Chemistry C*, Vol. 112, 2008, pp. 17983–17987.

[17] Tserkezis, C., G. Gantzounis, and N. Stefanou, "Collective plasmonic modes in ordered assemblies of metallic nanoshells," *J. Physics: Condensed Matter*, Vol. 20, 2008, 075232.

[18] Pack, A., M. Hietschold, and R. Wannemacher, "Failure of local Mie theory: Optical spectra of colloidal aggregates," *Optics Communications*, Vol. 194, 2001, pp. 277–287.

[19] Yannopapas, V., "Non-local optical response of two-dimensional arrays of metallic nanoparticles," *J. Physics: Condensed Matter*, Vol. 20, 2008, 325211.

[20] Jones, W. E., K. L. Kliewer, and R. Fuchs, "Nonlocal theory of the optical properties of thin metallic films," *Physical Review*, Vol. 178, 1969, pp. 1201–1203.

[21] McMahon, J. M., S. K. Gray, and G. C. Schatz, "Nonlocal optical response of metal nanostructures with arbitrary shape," *Physical Review Lett.*, Vol. 103, 2009, 097403.

[22] Boardman, A. D., ed., *Electromagnetic Surface Modes*, New York: Wiley, 1982.

[23] Taflove, A., and S. C. Hagness, *Computational Electrodynamics: The Finite-Difference Time-Domain Method*, 3rd ed., Norwood, MA: Artech House, 2005.

[24] McMahon, J. M., S. K. Gray, and G. C. Schatz, "Calculating nonlocal optical properties of structures with arbitrary shape," *Physical Review B*, Vol. 82, 2010, 035423.

[25] Bohren, C. F., and D. R. Huffman, *Absorption and Scattering of Light by Small Particles*, New York: Wiley, 1983.

[26] Johnson, P. B., and R. W. Christy, "Optical constants of the noble metals," *Physical Review B*, Vol. 6, 1972, pp. 4370–4379.

[27] Fetter, A. L., "Electrodynamics of a layered electron gas. I. Single layer," *Annals of Physics*, Vol. 81, 1973, pp. 367–393.

[28] Marinopoulos, A. G., L. Reining, and A. Rubio, "Ab initio study of the dielectric response of crystalline ropes of metallic single-walled carbon nanotubes: Tube-diameter and helicity effects," *Physical Review B*, Vol. 78, 2008, 235428.

[29] Apell, P., and D. R. Penn, "Optical properties of small metal spheres: Surface effects," *Physical Review Lett.*, Vol. 50, 1983, pp. 1316–1319.

[30] Pinchuk, A., U. Kreibig, and A. Hilger, "Optical properties of metallic nanoparticles: Influence of interface effects and interband transitions," *Surface Science*, Vol. 557, 2004, pp. 269–280.

[31] Yee, K. S., "Numerical solution of initial boundary value problems involving Maxwell's equations in isotropic media," *IEEE Trans. Antennas and Propagation*, Vol. 14, 1966, pp. 302–307.

[32] Roden, J. A., and S. D. Gedney, "Convolution PML (CPML): An efficient FDTD implementation of the CFS–PML for arbitrary media," *Microwave and Optical Technology Lett.*, Vol. 27, 2000, pp. 334–339.

[33] Gray, S. K., and T. Kupka, "Propagation of light in metallic nanowire arrays: Finite-difference time-domain studies of silver cylinders," *Physical Review B*, Vol. 68, 2003, 045415.

[34] Umashankar, K. R., and A. Taflove, "A novel method to analyze electromagnetic scattering of complex objects," *IEEE Trans. Electromagnetic Compatibility*, Vol. 24, 1982, pp. 397–405.

[35] Ruppin, R., "Optical properties of a spatially dispersive cylinder," *J. Optical Society of America B*, Vol. 6, 1989, pp. 1559–1563.

[36] Ruppin, R., "Optical properties of spatially dispersive dielectric spheres," *J. Optical Society of America*, Vol. 71, 1981, pp. 755–758.

[37] Kottmann, J. P., O. J. F. Martin, D. R. Smith, and S. Schultz, "Spectral response of plasmon resonant nanoparticles with a non-regular shape," *Optics Express*, Vol. 6, 2000, pp. 213–219.

[38] Kottmann, J. P., O. J. F. Martin, D. R. Smith, and S. Schultz, "Plasmon resonances of silver nanowires with a nonregular cross section," *Physical Review B*, Vol. 64, 2001, 235402.

[39] Camden, J. P., J. A. Dieringer, Y. Wang, D. J. Masiello, L. D. Marks, G. C. Schatz, and R. P. Van Duyne, "Probing the structure of single-molecule surface-enhanced Raman scattering hot spots," *J. American Chemical Society*, Vol. 130, 2008, pp. 12616–12617.

[40] Qian, X.-M., and S. M. Nie, "Single-molecule and single-nanoparticle SERS: From fundamental mechanisms to biomedical applications," *Chemical Society Review*, Vol. 37, 2008, pp. 912–920.

[41] N'Gom, M., S. Li, G. Schatz, R. Erni, A. Agarwal, N. Kotov, and T. B. Norris, "Electron-beam mapping of plasmon resonances in electromagnetically interacting gold nanorods," *Physical Review B*, Vol. 80, 2009, 113411.

[42] Westcott, S. L., J. B. Jackson, C. Radloff, and N. J. Halas, "Relative contributions to the plasmon line shape of metal nanoshells," *Physical Review B*, Vol. 66, 2002, 155431.

[43] Halevi, P., and R. Fuchs, "Generalised additional boundary condition for non-local dielectrics. I. Reflectivity," *J. Physics C*, Vol. 17, 1984, 3869.

Chapter 10

Classical Electrodynamics Coupled to Quantum Mechanics for Calculation of Molecular Optical Properties: An RT-TDDFT/FDTD Approach[1]

Hanning Chen, Jeffrey M. McMahon, Mark A. Ratner, and George C. Schatz

10.1 INTRODUCTION

Optical response is one of the fundamental characteristics of any physical system, usually providing a measure of the charge redistribution induced by an applied radiation field. The perturbation induced by light on a microscopic charge distribution is externally reflected in such macroscopic electromagnetic phenomena as absorption, refraction, luminescence, and scattering of light. In general, no two physical objects exhibit the same optical properties unless they are identical to each other, making the optical spectrum a powerful tool to detect, identify, and measure chemical substances.

Among the many optical techniques available nowadays, absorption spectroscopy [1] and Raman spectroscopy [2] are widely used as a result of a number of technological advances, including the development of highly coherent and narrowly diverging monochromatic lasers [3]. A major challenge of Raman spectroscopy arises from its feeble sensitivity. However, the amplification of this signal when molecules are adsorbed on silver nanoparticle substrates provides an important technique for circumventing this limitation [4]. Most of this amplification is now considered to arise from local field enhancement that results from plasmon excitation in the silver particles [5], although chemical contributions to the enhancement factor likely also exist.

Other examples of photophysical phenomena associated with light interacting with a system composed of plasmonic metal particles and molecules are also of interest. For example, photo-induced electron transfer in a single-molecule junction (SMJ) has recently gained attention. Here, it was demonstrated that switching from conducting to insulating states in a photochromic molecule anchored between two gold electrodes occurs under visible light irradiation [6].

---

[1]This chapter is adapted from Ref. [21], H. Chen, J. M. McMahon, M. A. Ratner, and G. C. Schatz, "Classical electrodynamics coupled to quantum mechanics for calculation of molecular optical properties: An RT-TDDFT/FDTD approach," *J. Physical Chemistry C*, Vol. 114, 2010, pp. 14384–14392, ©2010 American Chemical Society.

For consistency of notation relative to this source paper, in this chapter the symbol i is used to designate $\sqrt{-1}$, rather than the symbol j; and a phasor is denoted as $e^{-i\omega t}$.

Although the unambiguous observation of photoconductance is rather hard to prove due to associated thermal expansion [7] and charge trapping [8], it has been argued that the incident light in resonance with electronic transitions between molecule and electrodes can amplify the photocurrent by orders of magnitude when the Fermi energy of the electrode lies between the molecule's highest occupied molecular orbital (HOMO) and lowest unoccupied molecular orbital (LUMO) [9, 10]. The nonequilibrium Green's function (NEGF) formalism [11] has been generalized for both light absorption [12] and Raman scattering [13], providing a rationale for the strong mediation between bridge molecule and metal electrodes that arises from electronic and vibrational couplings. Plasmonic enhancement has also been applied to dye-sensitized solar cells (DSSC) [14], where it has recently been shown that photocurrent can be enhanced by nearly a factor of 10 when the thickness of a TiO_2 layer on the silver particles is reduced from 4.8 to 2.0 nm for a low-efficiency cell [15].

It is very challenging to develop a self-consistent theory for processes that couple light simultaneously to nanoparticles spanning 10 to 100 nm and molecules spanning <1 nm. Most past work treated the nanoparticle with classical electrodynamics in the absence of the molecule. The field arising from the particle's plasmonic excitation was assumed to be applied to the molecule as an external constant field [16]. Some studies treated both the molecule and the particle with quantum mechanics, but these were limited to particles containing fewer than ~100 atoms [17]. Hybrid approaches have been proposed in which classical electrodynamics in the particle were explicitly coupled to electronic structure calculations in the molecule. These include work by Corni and Tomasi, who described the metal particle polarization effects in the frequency domain by effective charges that were included in the molecule's Hamiltonian under the quasi-static approximation [18]. Also, Neuhauser developed a localized two-level random-phase approximation (RPA) model to evaluate the molecule's population transfer rate in the presence of surface plasmons by means of a density-matrix evolution [19]. Most recently Masiello and Schatz applied a many-body Green's function method to plasmon-enhanced molecular absorption [20].

This chapter reviews the new formalism reported in [21] that couples classical electrodynamics for the nanoparticle (as described using the finite-difference time-domain (FDTD) method [22]) with electronic structure theory for a nearby molecule [as described using real-time time-dependent density functional theory (RT-TDDFT)]. In this formalism, the disparate spatial and time scales needed to describe the optical response of the nanoparticle and molecule are such that the calculations are done sequentially. While Ref. [21] neglected the "backcoupling" of the molecule on the particle [20], its approach was otherwise completely general. It provided the capability to determine local field-enhancement effects on absorption and scattering that include the wavevector dependence of the incident light. In addition, the influence of the polarization of the electromagnetic field near the particle surface, and its coupling with transition moments associated with excitations in the molecule, could be automatically accounted.

Sections 10.2 through 10.5 of this chapter first briefly describe the RT-TDDFT and FDTD methods, and then discuss the coupling between these methods through a scattering response function. Sections 10.6 through 10.8 review the application of the resulting hybrid quantum mechanics / classical electrodynamics (QM/ED) computational technique to study: (1) surface-enhanced absorption in a system that includes the DSSC ruthenium-based dye molecule N3; and (2) the surface-enhanced Raman scattering (SERS) spectrum of pyridine. Section 10.9 concludes with a discussion of the applicability of the new QM/ED method to model linear optical properties, and its possible extension to nonlinear optics.

10.2 REAL-TIME TIME-DEPENDENT DENSITY FUNCTION THEORY

For a molecule exposed to a time-dependent external electric field, E_i, along axis i, the dipole moment, P_j, along axis j, in a first-order (linear) approximation is given by [21]:

$$P_j = P_{j0} + \alpha_{ij} E_i \tag{10.1}$$

where P_{j0} is the permanent dipole moment and α_{ij} is the linear polarizability tensor. The Einstein summation convention is used in this expression and throughout this chapter, when appropriate. In the time domain, (10.1) can be written as:

$$P_j(t) = P_{j0} + \int \alpha_{ij}(t-t_1) E_i(t_1) dt_1 \tag{10.2}$$

where $\alpha_{ij}(t-t_1)$ is related to the frequency-domain polarizability via:

$$\alpha_{ij}(t-t_1) = \frac{1}{2\pi} \int \alpha_{ij}(\omega)\, e^{-i\omega(t-t_1)} d\omega \tag{10.3}$$

Combining (10.2) and (10.3) yields:

$$\begin{aligned} P_j(t) &= P_{j0} + \frac{1}{2\pi} \int dt_1 \int \alpha_{ij}(\omega)\, E_i(t_1) e^{-i\omega(t-t_1)} d\omega \\ &= P_{j0} + \frac{1}{2\pi} \int \alpha_{ij}(\omega)\, E_i(\omega) e^{-i\omega t} d\omega \end{aligned} \tag{10.4}$$

If the induced dipole, $P_j^1(t)$, is defined as:

$$P_j^1(t) = P_j(t) - P_{j0} \tag{10.5}$$

its frequency-domain form is recognized as [21]:

$$P_j^1(\omega) = \alpha_{ij}(\omega)\, E_i(\omega) \tag{10.6}$$

where

$$\alpha_{ij}(\omega) = \frac{P_j^1(\omega)}{E_i(\omega)} = \frac{\int P_j^1(t) e^{-\Gamma t} e^{i\omega t} dt}{\int E_i(t) e^{i\omega t} dt} \tag{10.7}$$

Equation (10.7) relates a molecule's frequency-dependent polarizability tensor, $\alpha_{ij}(\omega)$, to the evolution of its induced dipole moment, $P_j^1(t)$, under a time-dependent external electric field, $E_i(t)$. Note that a damping factor Γ has been added to the numerator term of (10.7) to reflect the finite lifetime of excited electronic states due to quantum dephasing and vibronic coupling. This *ad hoc* procedure allows incorporating the effect of coupling to the metal nanoparticle on the excited state dynamics of the molecule. A more rigorous method for introducing this effect had been recently described [20], but was not implemented in [21].

This *ad hoc* method was consistent with earlier work using pure QM methods to describe resonance Raman and SERS [17]. It allows using a relatively short time integration to evaluate optical properties. The commonly used value of 0.1 eV was chosen for Γ in the applications presented in [21] and reviewed in this chapter. Note that damping was not applied to the denominator term in (10.7) since this represented the applied field, rather than the polarization response to the applied field.

Within the framework of density functional theory (DFT), $\boldsymbol{P}(t)$ can be calculated from the perturbed electron density that arises when the system is subjected to an applied field, $\boldsymbol{E}_0(t)$, by using the time-dependent Schrödinger equation (TDSE) [23]:

$$i\frac{\partial}{\partial t}\varphi(r,t) = \left\{-\frac{1}{2}\nabla^2 + \int\frac{\rho(r',t)}{|r-r'|}dr' + \frac{\delta E_{\mathrm{xc}}[\rho(r,t)]}{\delta\rho(r,t)} - \boldsymbol{E}_0\cdot\boldsymbol{r}\right\}\varphi(r,t) \tag{10.8}$$

Here, the four operators within the brackets on the right-hand side correspond to the kinetic energy, the Coulomb repulsion, the exchange-correlation energy, and the external electric field, respectively. The coupling Hamiltonian between the external electric field and the molecule is given by:

$$-\int\varphi^*(r)\,\boldsymbol{E}_0\cdot\boldsymbol{r}\,\varphi(r)dr = -\boldsymbol{E}_0\cdot\int\varphi^*(r)\,\boldsymbol{r}\,\varphi(r)dr = -\boldsymbol{E}_0\cdot\boldsymbol{P} \tag{10.9}$$

where the asterisk indicates the complex conjugate operator. Although an analytical solution is typically not available for the TDSE, it can be propagated by numerical integration schemes such as the first-order Crank–Nicholson approximation [24] or the enforced time reversible symmetry (ETRS) algorithm [25]. For an isolated and freely rotated molecule, the absorption cross-section $\sigma(\omega)$ can be obtained from [26]:

$$\sigma(\omega) = \frac{4\pi\omega}{c}\left\langle\frac{1}{3}\left[\alpha_{ii}(\omega) + \alpha_{jj}(\omega) + \alpha_{kk}(\omega)\right]\right\rangle_{\mathrm{imag}} \tag{10.10}$$

where $\langle\,\rangle_{\mathrm{imag}}$ denotes the imaginary part, and c is the free-space speed of light. In addition, the Raman differential cross-section for a given vibrational normal mode, p, is provided by the following expression [27]:

$$\frac{d\sigma}{d\Omega} = \frac{\pi^2}{\varepsilon_0^2}\left(\tilde{\nu}_{in} - \tilde{\nu}_p\right)^4\frac{h}{8\pi^2 c\,\tilde{\nu}_p}\left(\frac{45\left|\alpha_p\right|^2 + 7\gamma_p^2}{45}\right)\cdot\left(\frac{1}{1 - e^{-hc\tilde{\nu}_p/k_{\mathrm{B}}T}}\right) \tag{10.11}$$

where $|\ |$ denotes the complex modulus, ε_0 is the vacuum permittivity, $\tilde{\nu}_{in}$ is the wavenumber of the incident light, $\tilde{\nu}_p$ is the wavenumber of the normal mode, h is Planck's constant, k_{B} is the Boltzmann constant, and T is temperature. In addition, α_p and γ_p are the isotropic and anisotropic polarizability derivatives, respectively:

$$\alpha_p = \frac{1}{3}\left(\frac{\partial\alpha_{ii}}{\partial p} + \frac{\partial\alpha_{jj}}{\partial p} + \frac{\partial\alpha_{kk}}{\partial p}\right) \tag{10.12}$$

$$\gamma_p^2 = \frac{1}{2}\left(\left|\frac{\partial\alpha_{ii}}{\partial p} - \frac{\partial\alpha_{jj}}{\partial p}\right|^2 + \left|\frac{\partial\alpha_{ii}}{\partial p} - \frac{\partial\alpha_{kk}}{\partial p}\right|^2 + \left|\frac{\partial\alpha_{jj}}{\partial p} - \frac{\partial\alpha_{kk}}{\partial p}\right|^2\right)$$
$$+ 3\left(\left|\frac{\partial\alpha_{ii}}{\partial p}\right|^2 + \left|\frac{\partial\alpha_{jj}}{\partial p}\right|^2 + \left|\frac{\partial\alpha_{kk}}{\partial p}\right|^2\right) \quad (10.13)$$

10.3 BASIC FDTD CONSIDERATIONS

In FDTD simulations, light is assumed incident on a system that is discretized into many small grid cells, each characterized by a dielectric permittivity, $\varepsilon(r)$, and a magnetic permeability, $\mu(r)$. Then Maxwell equations [28]:

$$\varepsilon(r)\frac{\partial}{\partial t}\boldsymbol{E}(r,t) = \nabla\times\boldsymbol{H}(r,t) - \boldsymbol{J}(r,t) \quad (10.14)$$

$$\mu(r)\frac{\partial}{\partial t}\boldsymbol{H}(r,t) = -\nabla\times\boldsymbol{E}(r,t) \quad (10.15)$$

are solved in the real time domain to obtain the evolution of the electric field, $\boldsymbol{E}(r,t)$, the magnetic field, $\boldsymbol{H}(r,t)$, and the electric current density, $\boldsymbol{J}(r,t)$.

The properties of the electromagnetic field can also be determined in the frequency domain via Fourier transformation of the FDTD-computed fields. To study a broad spectral range, an impulsive (finite-duration) wave having zero initial value and zero average value is typically chosen as the incident field [22]. Since the total electric field, $\boldsymbol{E}_{\text{total}}(r,\omega)$, at a given observation point r is the sum of the scattered field, $\boldsymbol{E}_{\text{sca}}(r,\omega)$, and the incident field, $\boldsymbol{E}_0(r,\omega)$:

$$\boldsymbol{E}_{\text{total}}(r,\omega) = \boldsymbol{E}_{\text{sca}}(r,\omega) + \boldsymbol{E}_0(r,\omega) \quad (10.16)$$

a scattering response function (SRF), $\lambda(r,\omega)$, can be defined as:

$$\lambda_{ij}(r,\omega) = \frac{E_{i,\text{total}}(r,\omega)}{E_{j0}(r,\omega)} - \delta_{ij} \quad (10.17)$$

Note that $\lambda(r,\omega)$, which provides a measure of local field enhancement, is a complex tensor that depends on the propagation and polarization directions of the incident light. Normally, the light propagation direction is irrelevant in quantum chemistry due to the small size of molecular systems compared to the wavelength of light. However, in the context of a metal particle that is coupled to a molecule, it can play a role [21], as will be reviewed later in this chapter.

10.4 HYBRID QUANTUM MECHANICS/CLASSICAL ELECTRODYNAMICS

Under the assumption of a uniform scattered electric field inside an FDTD grid cell where a dye molecule of interest is located, the Hamiltonian operator of the dye molecule in the presence of an incident field, $\boldsymbol{E}_0(t)$, can be rewritten as:

$$\hat{H}(t) = -\frac{1}{2}\nabla^2 + \int \frac{\rho(r',t)}{|r-r'|}dr' + \frac{\delta E_{xc}[\rho(r,t)]}{\delta\rho(r,t)} - \boldsymbol{E}_0(t)\cdot\boldsymbol{r} - \boldsymbol{E}_{sca}(t)\cdot\boldsymbol{r} \tag{10.18}$$

where the electric field, $\boldsymbol{E}_{sca}(t)$, imposed by the polarized nanoparticle is included. Using the definition of $\lambda(r,\omega)$, $\boldsymbol{E}_{sca}(t)$ can be expressed as a two-dimensional (2-D) inverse Fourier transform of $\boldsymbol{E}_0(t)$ and $\lambda(\omega)$:

$$\begin{aligned} E_{i,sca}(t) &= \frac{1}{2\pi}\int E_{i,sca}(\omega)\,e^{-i\omega t}d\omega = \frac{1}{2\pi}\int e^{-i\omega t}d\omega \sum_j \lambda_{ij}(\omega)E_{j0}(\omega) \\ &= \frac{1}{2\pi}\int e^{-i\omega t}d\omega \sum_j \lambda_{ij}(\omega)\int E_{j0}(t_1)e^{i\omega t_1}dt_1 \\ &= \frac{1}{2\pi}\sum_j \int\int \lambda_{ij}(\omega)\,E_{j0}(t_1)\,e^{-i\omega(t-t_1)}d\omega\,dt_1 \end{aligned} \tag{10.19}$$

If the incident light is a stepwise pulse with a short duration, ΔT, where:

$$E_{j0}(t) = \begin{cases} E_{j0} & 0 < t < \Delta T \\ 0 & \text{otherwise} \end{cases} \tag{10.20}$$

then $\boldsymbol{E}_{sca}(t)$ can be reduced to the one-dimensional (1-D) inverse Fourier transform of $\lambda(\omega)$:

$$E_{i,sca}(t) \approx \frac{1}{2\pi}\sum_j E_{j0}\,\Delta T\int \lambda_{ij}(\omega)e^{-i\omega t}d\omega \tag{10.21}$$

10.5 OPTICAL PROPERTY EVALUATION FOR A PARTICLE-COUPLED DYE MOLECULE FOR RANDOMLY DISTRIBUTED INCIDENT POLARIZATION

As reflected in (10.17) and (10.19), the optical response of a dye molecule bound to a metal particle is a function of the light propagation direction, z, and polarization direction, x, through $\lambda(r,\omega)$. In most experimental settings, the relative orientation of the dye molecule with respect to the metal particle is fixed. Without loss of generality, the dye's molecular frame, denoted by three Cartesian axes $\{i,j,k\}$, was assumed in [21] to have its k-axis overlap with the light propagation axis, z, of the experimental frame denoted by $\{x,y,z\}$. Therefore, under the condition of randomly distributed polarization direction of the illumination, the spatial average of the polarizability tensor diagonal component, $\bar{\alpha}_{xx}$, is given by:

$$\bar{\alpha}_{xx} = \sum_a\sum_b \bar{\alpha}_{ab}\cos(xa)\cos(xb) \tag{10.22}$$

where $(a,b) \in (i,j,k)$ and xa is the angle between axis x and axis a. Following [21], since:

$$\cos(xk) = \cos(\pi/2) = 0 \tag{10.23}$$

$\bar{\alpha}_{xx}$ can be reduced to:

$$\begin{aligned}\bar{\alpha}_{xx} &= \alpha_{ii}\overline{\cos(ix)\cos(ix)} + \alpha_{ij}\overline{\cos(ix)\cos(jx)} \\ &+ \alpha_{ji}\overline{\cos(jx)\cos(ix)} + \alpha_{jj}\overline{\cos(jx)\cos(jx)}\end{aligned} \tag{10.24}$$

Averaging over the rotation angles between the coordinate axes yields:

$$\bar{\alpha}_{xx} = \frac{1}{2}\left(\alpha_{ii} + \alpha_{jj}\right) = \alpha_k \tag{10.25}$$

Similarly, the average value of λ_{xx} is given by:

$$\bar{\lambda}_{xx} = \frac{1}{2}\left(\lambda_{ii} + \lambda_{jj}\right) = \lambda_k \tag{10.26}$$

Under steady-state conditions, the absorption cross-section, $\sigma(\omega)$, of the particle-bound dye molecule is related to its stimulated transition rate, R, and the incident photon flux density, I, by [21]:

$$\sigma(\omega) = \frac{R}{I} = \frac{\dfrac{4\pi E_x}{h\omega}\dfrac{dP_x}{dt}}{\dfrac{c}{2h\omega}\left|E_x\right|^2} = \frac{4\pi}{c}\left(\frac{2}{\left|E_x\right|^2}E_x\frac{dP_x}{dt}\right) \tag{10.27}$$

where E_x is the electric field imposed on the dye molecule, and P_x is its corresponding dipole moment component. For plane-wave incident illumination, $E_{x0}\cos(\omega t)$:

$$E_x = E_{x0}\cos(\omega t) + E_{x0}\left|\lambda_{xx}\right|\cos(\omega t - \varphi) \tag{10.28}$$

$$P_x = P_{x0} + \left|\alpha_{xx}\right|E_{x0}\cos(\omega t - \theta) \tag{10.29}$$

where φ and θ represent the phase shift of the scattering response function, λ, and the polarizability, α, relative to the incident light, respectively. After some mathematical manipulations, this yields [21]:

$$\begin{aligned}\sigma(\omega) &= \frac{4\pi\omega}{c}\left(\left|\alpha_{xx}\right|\sin\theta + \left|\alpha_{xx}\right|\left|\lambda_{xx}\right|\sin\theta\cos\varphi - \left|\alpha_{xx}\right|\left|\lambda_{xx}\right|\cos\theta\sin\varphi\right) \\ &= \frac{4\pi\omega}{c}\left(\left\langle\alpha_{xx}\right\rangle_{\text{imag}} + \left\langle\alpha_{xx}\lambda^*_{xx}\right\rangle_{\text{imag}}\right) = \frac{4\pi\omega}{c}\left\langle\alpha_{xx}\left(1 + \lambda^*_{xx}\right)\right\rangle_{\text{imag}}\end{aligned} \tag{10.30}$$

In the molecular frame, the absorption cross-section, $\sigma_k(\omega)$, for a bound dye molecule illuminated by light with a fixed propagation direction, k, is given by:

$$\sigma_k(\omega) = \frac{4\pi\omega}{c}\left\langle \alpha_k\left(1 + \lambda_k^*\right) + \gamma_k/8\right\rangle_{\text{imag}} \tag{10.31}$$

where the anisotropic polarizability is:

$$\gamma_k = \left(\alpha_{ii} - \alpha_{jj}\right)\left(\lambda_{ii}^* - \lambda_{jj}^*\right) + \left(\alpha_{ij} + \alpha_{ji}\right)\left(\lambda_{ij}^* + \lambda_{ji}^*\right) \tag{10.32}$$

Following a similar procedure, the Raman differential cross-section, $d\sigma/d\Omega$, of such a bound dye molecule can be inferred following Long [29]:

$$\frac{d\sigma}{d\Omega} = \frac{\pi^2}{\varepsilon_0^2}\left(\tilde{v}_{in} - \tilde{v}_p\right)^4\left(\frac{h}{8\pi^2 c\,\tilde{v}_p}\right)\left|\frac{8\alpha_{k,p} + \gamma_{k,p}}{8}\right|^2\left(\frac{1}{1 - e^{-hc\tilde{v}_p/k_B T}}\right) \tag{10.33}$$

where the isotropic polarizability derivative, $\alpha_{k,p}$, and the anisotropic polarizability derivative, $\gamma_{k,p}$, are defined as:

$$\alpha_{k,p} = \left(1 + \lambda_k\right)\frac{\partial \alpha_k}{\partial p} \tag{10.34a}$$

$$\gamma_{k,p} = \left(\frac{\partial \alpha_{ii}}{\partial p} - \frac{\partial \alpha_{jj}}{\partial p}\right)\left(\lambda_{ii} - \lambda_{jj}\right) + \left(\frac{\partial \alpha_{ij}}{\partial p} + \frac{\partial \alpha_{ji}}{\partial p}\right)\left(\lambda_{ij} + \lambda_{ji}\right) \tag{10.34b}$$

Note that the $\left(1 + \lambda_k\right)$ term in (10.34a) is evaluated at the Stoke's frequency, while the polarizability derivative term is at the incident frequency. In this evaluation, the zero Stoke's shift limit of this expression was assumed [21].

10.6 NUMERICAL RESULTS 1: SCATTERING RESPONSE FUNCTION OF A 20-nm-DIAMETER SILVER NANOSPHERE

The first application reported in [21] studied the effect of polarization on the FDTD-computed field enhancement factor, $\lambda(r, \omega)$, as defined in (10.17), for the scattering of light from a 20-nm-diameter silver nanosphere. All modeling results were obtained using the JFDTD3D package [30].

The nanosphere was centered at $(0, 0, 0)$ within a $160 \times 160 \times 160$-cell Cartesian FDTD grid containing 4,096,000 cubic cells of uniform size 0.25 nm. Because the incident light was assumed to propagate along the $+z$-axis, perfectly matched layer (PML) absorbing boundary conditions (ABCs) were implemented adjacent to the outer grid planes at $z = \pm 20$ nm to remove reflections, while periodic boundary conditions were applied at the outer grid planes at $x = \pm 20$ nm and $y = \pm 20$ nm.

Each FDTD grid cell was characterized by ε and μ values based on its distance to the center of the silver nanosphere. The values of ε and μ for silver were determined from experiment [31], while ε_0 and μ_0 were assigned to free space.

Incident light was injected impulsively at grid plane $z = -16$ nm using the functional form:

$$E_0(t) = e^{-(t-t_0)^2/\tau^2}\sin[\omega_0(t - t_0)] \tag{10.35}$$

where $t_0 = 10.0$ fs, $\tau = 0.7$ fs, and $\omega_0 = 3.14 \times 10^{15}$ sec^{-1} ($\lambda_0 = 600.0$ nm). These parameters were chosen to fully cover the visible spectrum from 300 to 800 nm. A time-step of 4.57×10^{-4} fs and a total simulation time of 100 fs were employed so that the value of each FDTD-computed scattering response function, $\lambda(r, \omega)$, converged to $\pm 0.2\%$.

Five observation points along the x-axis, x = 11, 12, 13, 14, and 15 nm, were selected for computing $\lambda(r, \omega)$. These were, respectively, 1, 2, 3, 4, and 5 nm from the nanosphere's surface. At each of these points, two separate FDTD calculations were carried out for different polarization directions of the incident light: one along the x-axis, and the other along the y-axis. During these calculations, the E-field vectors at the observation points were saved every 10 time-steps for subsequent Fourier transformation to determine $\lambda(r, \omega)$ according to (10.17).

Figure 10.1 shows the FDTD-computed tensor components of the scattering function, $\lambda(r, \omega)$, at the five observation points [21]. Only the diagonal terms in $\lambda(r, \omega)$ are shown because the scattered field along the pulse polarization direction was found to be much stronger than that along the two other directions by at least two orders of magnitude. Hence, the off-diagonal terms in $\lambda(r, \omega)$ were nearly negligible compared to the diagonal terms, which were subsequently used to evaluate the time-dependent scattered field in the RT-TDDFT calculations given by (10.19).

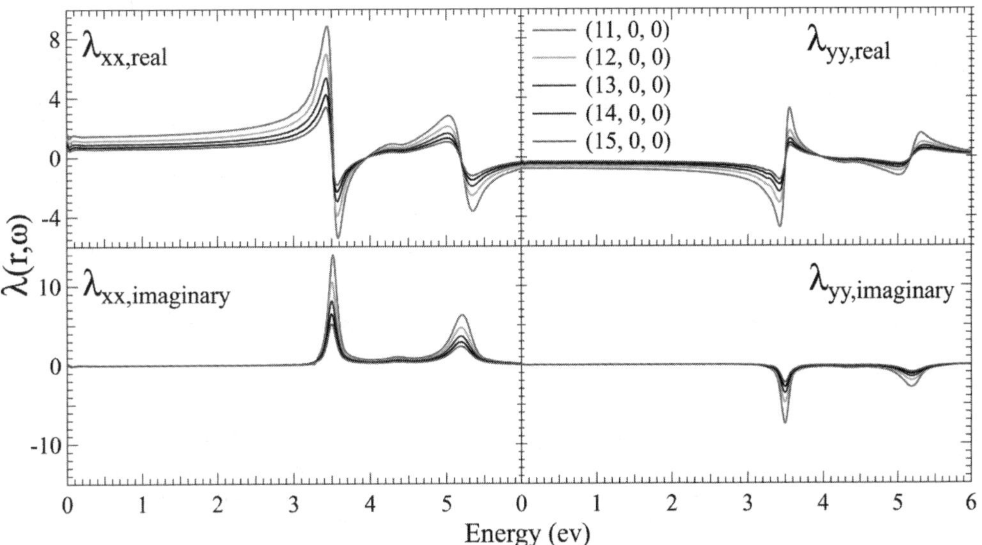

Fig. 10.1 FDTD-computed tensor components of the scattering function, $\lambda(r, \omega)$, at five observation points along the x-axis adjacent to a 20-nm-diameter silver nanosphere centered at (0, 0, 0) nm. *Source:* H. Chen, J. M. McMahon, M. A. Ratner, and G. C. Schatz, *J. Physical Chemistry C*, Vol. 114, 2010, pp. 14384–14392, ©2010 American Chemical Society.

From Fig. 10.1, it is apparent that the real parts of λ_{xx} and λ_{yy} were found to exhibit two dip-peak features centered at ~3.4 and ~5.2 eV, while their imaginary parts displayed two peaks at the same locations. For the imaginary parts, which are pertinent to light absorption, the sharper and stronger peaks at 3.4 eV could be ascribed to an intraband transition (plasmon excitation), whereas the broader and weaker peaks at 5.2 eV were induced by an interband transition [21].

Given the geometric symmetry of the silver nanosphere, with the total induced dipole moment parallel to the incident light polarization direction, it is not surprising that λ_{xx} was found to be stronger than λ_{yy} at each of the observation points along the x-axis. Nevertheless, λ_{yy} was not negligibly small compared to λ_{xx}, particularly when the observation point was within 2.0 nm of the nanosphere's surface.

Interestingly, the sign of $\lambda_{xx,\text{imag}}$ was determined to be opposite to that of $\lambda_{yy,\text{imag}}$ for both peaks. The positive value of $\lambda_{xx,\text{imag}}$ indicates that the corresponding scattered field, $E_{\text{sca},xx}$, lagged behind the incident light in phase. On the other hand, the negative value of $\lambda_{yy,\text{imag}}$ suggests that $E_{\text{sca},yy}$ led the incident light. Outside the two peak regions, the values of $\lambda_{xx,\text{imag}}$ and $\lambda_{yy,\text{imag}}$ were close to zero, indicating that there was neither energy loss nor phase change of the incident light at off-resonance frequencies. Similarly, $\lambda_{xx,\text{real}}$ and $\lambda_{yy,\text{real}}$ exhibited opposite peak patterns at the two resonance frequencies.

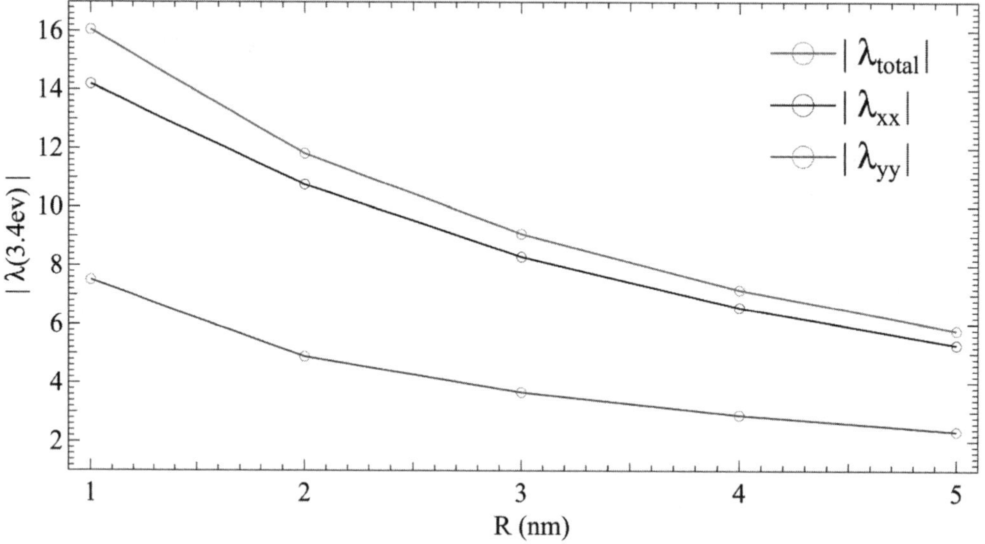

Fig. 10.2 Strength of the FDTD-computed on-resonance scattering response function, $\lambda(3.4\,\text{eV})$, as a function of R, the distance from the observation point to the nanosphere's surface. *Source:* H. Chen, J. M. McMahon, M. A. Ratner, and G. C. Schatz, *J. Physical Chemistry C*, Vol. 114, 2010, pp. 14384–14392, ©2010 American Chemical Society.

Figure 10.2 shows the decay of the scattering response function, λ, at 3.4 eV (the plasmon frequency) with increasing distance, R, from the observation point to the nanosphere's surface [21]. As R increased from 1 to 5 nm, in addition to a 50% drop in $|\lambda_{\text{total}}|$ defined by:

$$|\lambda_{\text{total}}| = \sqrt{\lambda_{xx}^2 + \lambda_{yy}^2} \qquad (10.36)$$

$|\lambda_{xx}|$ was observed to decay slightly faster than $|\lambda_{yy}|$. Moreover, the imaginary part was shown to contribute more significantly to either $|\lambda_{xx}|$ or $|\lambda_{yy}|$ than the real part. Since $|\lambda_{\text{total}}|$ was still as large as ~6 even for $R = 5\,\text{nm}$, the 20-nm silver nanosphere was an ideal system to investigate plasmonic enhancement for absorption and Raman spectra, as reviewed in the next two sections.

10.7 NUMERICAL RESULTS 2: OPTICAL ABSORPTION SPECTRA OF THE N3 DYE MOLECULE

Reference [21] reported numerical simulations of the optical absorption spectra of the *cis*-Bis(isothiocyanato)-bis(2,2′-bipyridyl-4,4′-dicarboxylato)-ruthenium(II) molecule, known as N3 [32]. This molecule, shown in the inset of Fig. 10.3, is one of the most widely used charge-transfer sensitizers in dye-sensitized solar cells due to its high extinction coefficient and extraordinary chemical stability [33]. More impressively, the N3 dye molecule can transfer its excited electrons to a TiO_2 layer within femtoseconds, much faster than other competing deactivation processes [34]. Recent work [15] revealed a strong dependence of the N3 molecule's incident photon conversion efficiency on the thickness of a TiO_2 layer separating it from an adjacent silver nanoparticle. This prompted the study reported in [21], reviewed next, of improving the N3 molecule's light absorption through plasmonic enhancement.

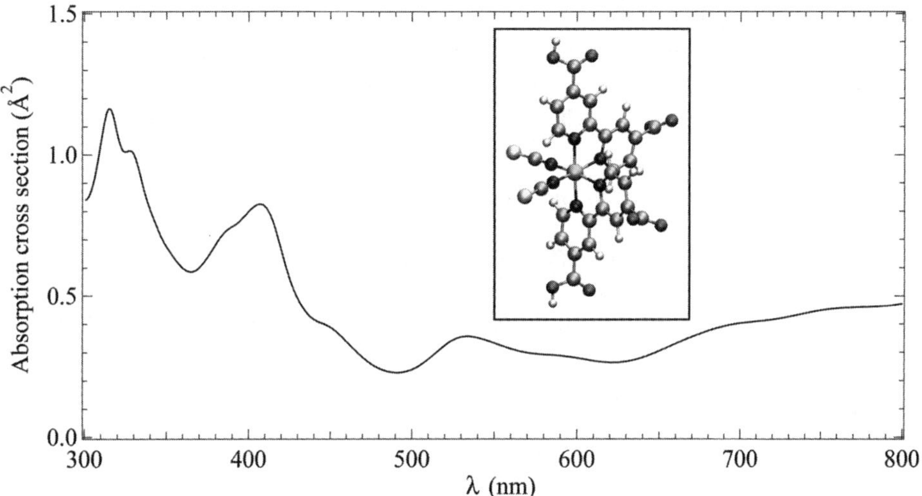

Fig. 10.3 Calculated optical absorption cross-section spectrum of an isolated N3 dye molecule (illustrated in the inset). *Source:* H. Chen, J. M. McMahon, M. A. Ratner, and G. C. Schatz, *J. Physical Chemistry C*, Vol. 114, 2010, pp. 14384–14392, ©2010 American Chemical Society.

10.7.1 Isolated N3 Dye Molecule

Following [21], the analysis commenced with modeling the isolated N3 molecule. Here, the first step was to optimize the structure of this molecule with a neutral charge by a ground-state DFT calculation using the Goedecker–Teter–Hutter (GTH) dual-space Gaussian pseudo-potential [35], parameterized with the Perdew–Burke–Ernzerhof (PBE) exchange correlation functional [36]. A polarized-valence-double-zeta (PVDZ) [37] basis set was chosen, and the wavelet Poisson solver [38] was applied to treat electrostatic interactions under non-periodic boundary conditions. All DFT simulations were performed using the CP2K molecular simulation package [39].

After geometry optimization, three separate RT-TDDFT simulations were conducted. Each simulation employed a rectangular electric field pulse [as per (10.20)] of 0.02745 V/nm magnitude and 0.0121-fs duration, applied along one of the three coordinate axes. The field was evaluated at the center of mass of the molecule for each RT-TDDFT calculation. Field gradient effects were found to be negligible for this case. The wavefunction of the system was propagated for 4000 steps according to the time-dependent Schrödinger equation using the ETRS algorithm [25] with a time-step of 0.0121 fs. Then, the polarizability tensor was calculated using (10.7), and the optical absorption cross-section was determined according to (10.10).

Figure 10.3 shows the calculated optical absorption cross-section spectrum of the isolated N3 dye molecule [21]. From this figure, it is apparent that the isolated N3 molecule was found to be a wide-spectrum light harvester, covering the whole visible range. The maximum cross-section of 1.17 Å^2 occurred for the sharp peak at 315 nm, which also exhibited a small shoulder at 328 nm. In addition, there was a moderately broad peak at 407 nm with a maximum value of 0.83 Å^2. The broadest absorption band, spanning 530 to 800 nm, had an average absorption cross-section of 0.4 Å^2. Compared to experimental measurements [15], this calculated absorption profile was shifted toward the red by 20 nm on average, primarily due to the incorrect asymptotic behavior of the short-range DFT functional. Although the underestimated HOMO-LUMO gap typically encountered in DFT calculations [40] can be partially improved by adding long-range corrections, such as via the statistical averaging of orbital potentials (SAOP) [41], the associated computational cost is prohibitively expensive for RT-TDDFT. The same can be said about improvements on the PVDZ basis. However, in light of the reasonably good results shown in Fig. 10.3, no effort was invested in achieving better Kohn–Sham orbital energies [21].

10.7.2 N3 Dye Molecule Bound to an Adjacent 20-nm Silver Nanosphere

Reference [21] then reported the results of modeling a system consisting of the N3 dye molecule of Section 10.7.1 and Fig. 10.3 located at various distances, R, from the 20-nm-diameter silver nanosphere of Section 10.6. The geometry of this system is illustrated in the inset of Fig. 10.4.

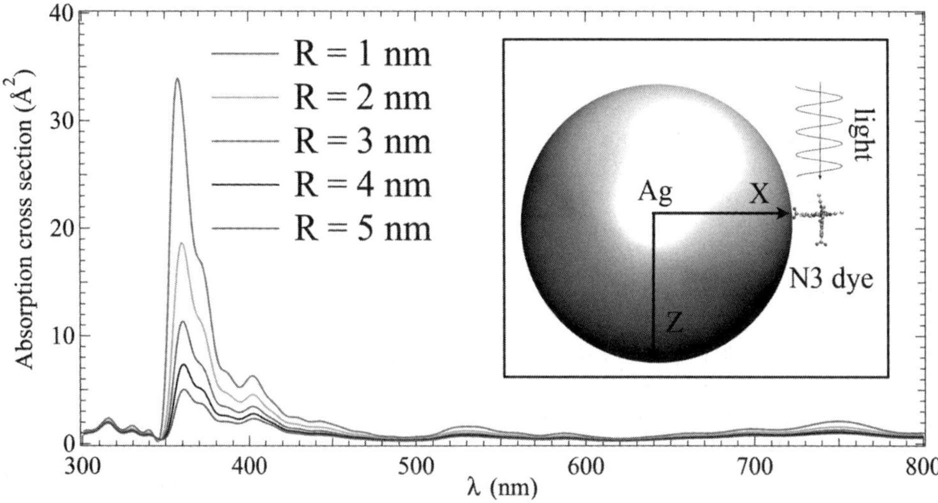

Fig. 10.4 Calculated optical absorption cross-section spectra of an N3 dye molecule located at various distances, R, from an adjacent 20-nm silver nanosphere. The system geometry is illustrated in the inset. *Source:* H. Chen, J. M. McMahon, M. A. Ratner, and G. C. Schatz, *J. Physical Chemistry C*, Vol. 114, 2010, pp. 14384–14392, ©2010 American Chemical Society.

The system of Fig. 10.4 was set up such that one of the four carboxyl groups of the N3 molecule pointed toward the center of the silver nanosphere to mimic the binding orientation of this molecule on a semiconductor surface, as revealed by earlier studies [42]. In addition, the aromatic plane of the N3 molecule's binding bipyridine group was assumed to be normal to the z-axis. The ruthenium atom, which was assumed to be the molecule's center of mass, was placed at the observation points discussed in Section 10.6: 1, 2, 3, 4, and 5 nm from the surface of the silver nanosphere along the x-axis.

Since even the long axis of the N3 dye molecule is only 0.8 nm, the scattered field, E_{sca}, imposed on it was regarded as spatially homogeneous in the model of [21]. For a given observation point, E_{sca} was determined according to (10.21) using the same E_0 applied to the isolated N3 system together with the corresponding $\lambda(r, \omega)$, as discussed in Section 10.6. Subsequently, two independent RT-TDDFT simulations were performed with the added E_{sca}. The polarization direction associated with E_0 was along the x-axis and y-axis in the first and second simulations, respectively. In a similar manner to the isolated N3 dye molecule, the system's induced dipole was saved before it was used to evaluate the absorption cross-section using (10.31).

Figure 10.4 shows that the formation of a sharp peak at 357 nm (equivalent to 3.48 eV) was found to be the primary consequence of plasmon enhancement on light absorption [21]. However, the absorption cross-section was found to be significantly increased at nearly all visible wavelengths. The width of the 357-nm peak was approximately 50 nm, consistent with the peak width of $\lambda(r, \omega)$ as depicted in Fig. 10.1. Therefore, the primary absorption peak was ascribed to strong coupling between the electronic transitions of the N3 dye molecule and the plasmon oscillations of the silver nanosphere [21]. More interestingly, binding N3 to the nanosphere not only increased the absolute values of absorption cross-section, but also changed the relative height of the different absorption peaks. For example, the peak at 311 nm, which was at least 50% stronger than other peaks for the isolated N3 molecule, was now nearly as intense as the broad absorption band ranging from 500 to 800 nm. In addition, the absorption peak at 407 nm was significantly amplified, and was now second only to the 357-nm resonance peak in the presence of the silver nanosphere.

Regardless of the underlying physical mechanisms, the absorption enhancement effect can be quantified by calculating the ratio of $\sigma_{bound}(\omega)$ and $\sigma_{isolated}(\omega)$, the absorption cross-sections for the silver-bound and isolated N3 dye molecule, respectively. As shown in Fig. 10.5, this enhancement was found to be strongest at the plasmon resonance wavelength of 360 nm. When the N3 dye molecule was located only 1 nm away from the silver nanosphere's surface, the on-resonance enhancement ratio reached a maximum value of ~55. This value was roughly proportional to $|\lambda + 1|^2$, although there was a prefactor that reflected polarization effects. The corresponding maximum absorption cross-section of ~34Å^2 was not easily attainable by any bare organic sensitizer with a molecular size similar to N3. On the red side of the plasmon frequency, the absorption enhancement ratio exhibited peaks at 407, 440, 530, and 750 nm, leading to the enhanced absorption peaks at these four wavelengths. Interestingly, there was little (or weak at best) absorption enhancement on the blue side of the plasmon frequency where the enhancement ratio was close to 1. Nevertheless, even when the N3 molecule and the silver nanosphere were located 5 nm apart, the corresponding on-resonance enhancement ratio at 360 nm was larger than 8. As anticipated, both the absorption profile and the enhancement ratio were monotonically decreasing functions of the molecule-nanoparticle separation over the entire visible range, demonstrating a distinct advantage of diminishing the protective layer thickness in plasmonic dye-sensitized solar cells [15].

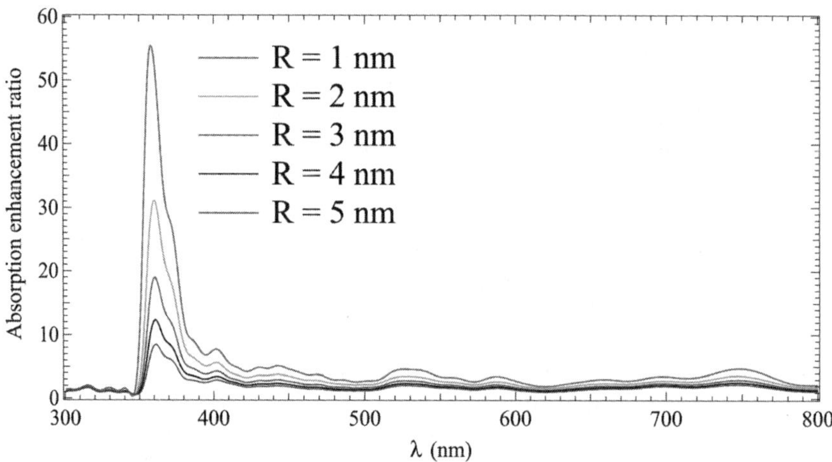

Fig. 10.5 Dependence of the computed absorption enhancement ratio on the separation, R, between the N3 dye molecule and the 20-nm silver nanosphere. *Source:* H. Chen, J. M. McMahon, M. A. Ratner, and G. C. Schatz, *J. Physical Chemistry C*, Vol. 114, 2010, pp. 14384–14392, ©2010 American Chemical Society.

10.8 NUMERICAL RESULTS 3: RAMAN SPECTRA OF THE PYRIDINE MOLECULE

Reference [21] reported numerical simulations of the Raman spectra of the pyridine molecule. This molecule, shown in the inset of Fig. 10.6, has been used as a model system to study the interaction between an analyte molecule and a plasmonic surface since the early development of surface-enhanced Raman scattering (SERS), due to its easily assignable vibrational modes and considerable signal amplification [43].

10.8.1 Isolated Pyridine Molecule

Following [21], the analysis commenced with modeling of the isolated pyridine molecule. Here, the first step was to perform a vibrational mode analysis on the optimized geometry of this molecule to determine its normal mode vectors and frequencies by diagonalizing the Hessian matrix constructed after geometry optimization. A total of 27 normal vectors were ascertained. Then, an RT-TDDFT simulation was carried out on each of the 27 perturbed pyridine geometries, which were built from the optimized geometry by adding the corresponding normal mode vectors. A rectangular electric field pulse of magnitude 0.02745 V/nm and duration 0.0242 fs was applied at the beginning of each simulation to trigger the wave function propagation. The field was evaluated at the center of mass of the molecule. Field gradient effects were found not to be important in this application, so they were neglected. Each RT-TDDFT trajectory was 48.36 fs long, and a time-step of 0.0242 fs was used. Once again, the GTH pseudopotential [35] and PBE functional [36] were chosen to model the core-valence and exchange-correlation interactions, respectively. The derivatives of the polarizability tensor with respect to the normal coordinates were evaluated through numerical differentiation prior to the calculation of differential Raman cross-section for isolated pyridine according to (10.11). The wavelength of the incident light was chosen to be 514 nm [4].

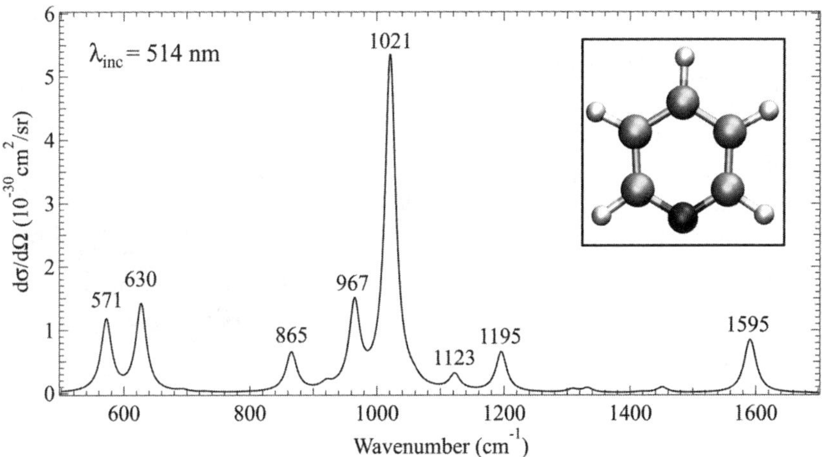

Fig. 10.6 Calculated normal Raman spectrum of the isolated pyridine molecule. All peaks were broadened by a Lorentzian function of width 10 cm^{-1}. *Source:* H. Chen, J. M. McMahon, M. A. Ratner, and G. C. Schatz, *J. Physical Chemistry C*, Vol. 114, 2010, pp. 14384–14392, ©2010 American Chemical Society.

Figure 10.6 shows the calculated normal Raman spectrum of the isolated pyridine molecule [21]. A pronounced peak is seen at 1021 cm^{-1} along with seven modest peaks at 571, 630, 865, 967, 1123, 1195, and 1595 cm^{-1}. The locations of the primary peak and its closest secondary peak attributed to ring breathing were very close to the experimental values [4] of 1030 and 991 cm^{-1} (slightly variable through solvent effects [44].) However, the height of the 967-cm^{-1} peak was underestimated compared to the experimental result in which it was nearly as strong as the 1021-cm^{-1} peak. This underestimate was likely due to limited accuracy of the underlying exchange-correlation functionals and core-valence pseudopotentials. Nevertheless, these simulations properly exhibited stronger intensities for the two ring-breathing vibrational modes than the others. The cross-section values of ~10^{-30} cm^2/sr were similar to previous estimates, and illustrated why normal Raman is not suitable for ultra-trace or single-molecule detection [45].

10.8.2 Pyridine Molecule Adjacent to a 20-nm Silver Nanosphere

Reference [21] then reported the results of modeling a system consisting of the pyridine molecule of Section 10.8.1 and Fig. 10.6 located at various distances, R, from the 20-nm-diameter silver nanosphere of Section 10.6. The geometry of this system, illustrated in the inset of Fig. 10.7, was configured so that the center of the nanosphere, the nitrogen, and the para-carbon atoms of pyridine were all along the x-axis. The normal vector to the pyridine ring was along the z-axis, which was also the propagation direction of the incident light. Distance R was measured between the surface of the silver nanosphere and the center of the pyridine ring.

A total of five simulations with uniformly spaced values of R from 1 to 5 nm were conducted. Each simulation followed the same procedure as that for the isolated pyridine except for the presence of an external scattered field, E_{sca}, which was determined according to (10.21) using the values of $\lambda(r, \omega)$, as discussed in Section 10.6. The incident light was polarized along the x- or y-axis, and the differential Raman cross-section was evaluated by (10.33) assuming a uniformly distributed light polarization direction on the x-y plane of the molecular frame.

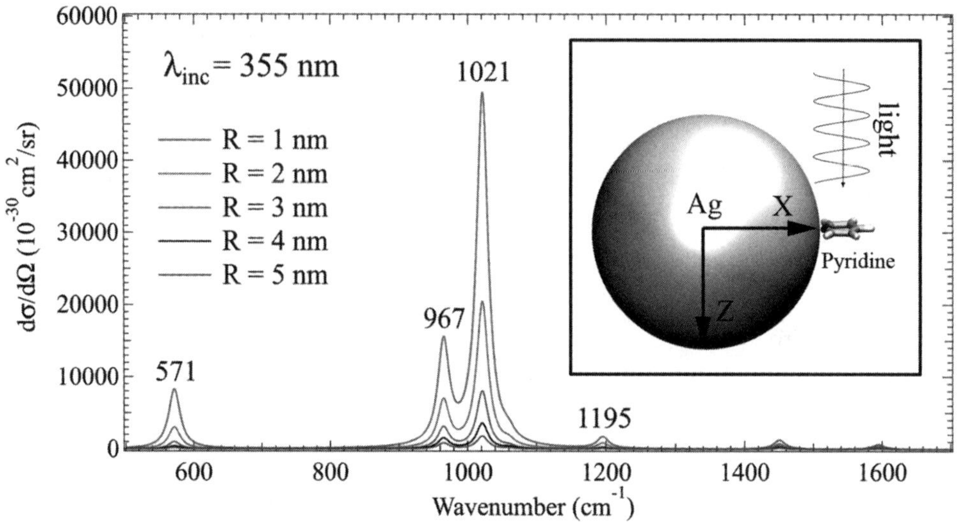

Fig. 10.7 Calculated surface-enhanced Raman scattering (SERS) spectrum of a pyridine molecule located at various distances, R, from an adjacent 20-nm silver nanosphere. The system geometry is illustrated in the inset. *Source:* H. Chen, J. M. McMahon, M. A. Ratner, and G. C. Schatz, *J. Physical Chemistry C*, Vol. 114, 2010, pp. 14384–14392, ©2010 American Chemical Society.

It was determined that the Raman signals of all five simulations were maximally amplified when $\lambda_{\text{inc}} = 355$ nm, coincident with the plasmon wavelength of the silver nanosphere. Hence, the SERS spectrum of pyridine was calculated at that incident wavelength. The results of this model are shown in Fig. 10.7 [21].

Comparison of the results in Fig. 10.7 to those in Fig. 10.6 indicates that the QM/ED modeling technique of [21] yielded a very large predicted enhancement of the SERS response at several peaks, most prominently the central peak at 1021 cm^{-1}. Specifically, at $R = 1$ nm, the closest separation between the pyridine molecule and the silver nanosphere, this peak exceeded $50{,}000 \times 10^{-30}$ cm^2/sr, representing a 10,000:1 enhancement over the isolated-molecule Raman result of Fig. 10.6 (roughly proportional to $|\lambda_k + 1|^4$, considered to be reasonable for a silver nanosphere at 355 nm [46]). As expected, this plasmon-enhanced SERS response decreased rapidly with increasing R, dropping to $\sim 20{,}000 \times 10^{-30}$ cm^2/sr at $R = 2$ nm, and $\sim 1800 \times 10^{-30}$ cm^2/sr at $R = 5$ nm.

Note that the 10,000:1 enhancement for the 1021-cm^{-1} peak at $R = 1$ nm could be further increased by changing the local chemical environment of the pyridine molecule through chemical bonding and intermolecular charge transfer. Within the framework of the QM/ED technique of [21], these effects could be modeled with moderate additional computational cost by including the binding site of the silver nanosphere into the QM level.

Figure 10.7 also shows significant enhancement of the SERS response at two secondary peaks, 571 and 1195 cm^{-1}. These arose from ring-deformation and ring-stretch modes, respectively, as illustrated in Fig. 10.8, and were consistent with experimental results [47]. The large amplification of the 571-cm^{-1} mode could be partially ascribed to the nitrogen and para-carbon atoms vibrating towards the silver nanosphere to couple with the induced plasmon field, whereas these two atoms were nearly stationary in the 1195-cm^{-1} mode. The strong coupling was also reflected in the dependence of the peak heights on R.

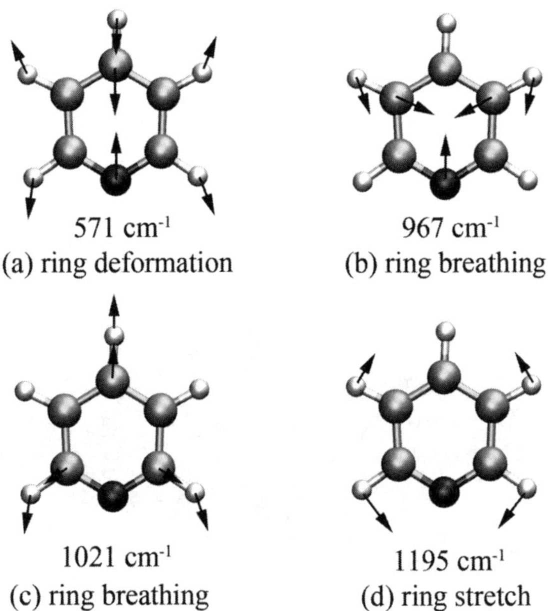

Fig. 10.8 Plasmon-enhanced vibrational normal modes of the pyridine molecule. *Source:* H. Chen, J. M. McMahon, M. A. Ratner, and G. C. Schatz, *J. Physical Chemistry C*, Vol. 114, 2010, pp. 14384–14392, ©2010 American Chemical Society.

Figure 10.9 displays three-dimensional (3-D) surface plots of the calculated SERS response as a function of both the incident wavelength and the vibrational wavenumber [21]. Despite the distinct absolute values of the cross-sections, these plots exhibit a similar pattern having only a countable number of vibrational modes, all with peaks at 3.49 eV. Moreover, the computed plasmonic enhancement was highly selective, nearly disappearing when the incident photon energy was more than 0.5 eV off the plasmon frequency. In fact, the strength of the Raman signal typically retained more than half of its maximum value within a narrow window of 0.2 eV around the plasmon maximum.

Two final points should be considered in assessing these results. First, to examine the validity of assuming a uniform scattered E-field across a small molecule such as pyridine (and N3), a justification calculation was performed on the SERS of the pyridine molecule by including the coupling between its quadrupole moment and the E-field gradient. In this calculation, a molecule-particle separation of 1.0 nm was chosen to realize the large E-field gradient near the silver nanosphere surface. Importantly, the resulting computed SERS differential cross-section changed by less than 1% from the uniform E-field case, indicating that the effect of the E-field gradient could be safely ignored in the study reviewed here.

Second, chemical contributions to the SERS enhancement factor were also ignored in this modeling work. In general, such effects play a role in the SERS spectrum of pyridine, as has been discussed in the past [17, 48, 49]. Furthermore, if required, these effects can be included in the QM/ED technique of [21] by adding silver atoms at the point where the pyridine molecule is adsorbed on the surface of the silver nanoparticle. However, such chemical effects should not have played a significant role in the study reviewed here. This is because the pyridine molecule was always assumed to be separated from the surface of the silver nanosphere by at least 1 nm.

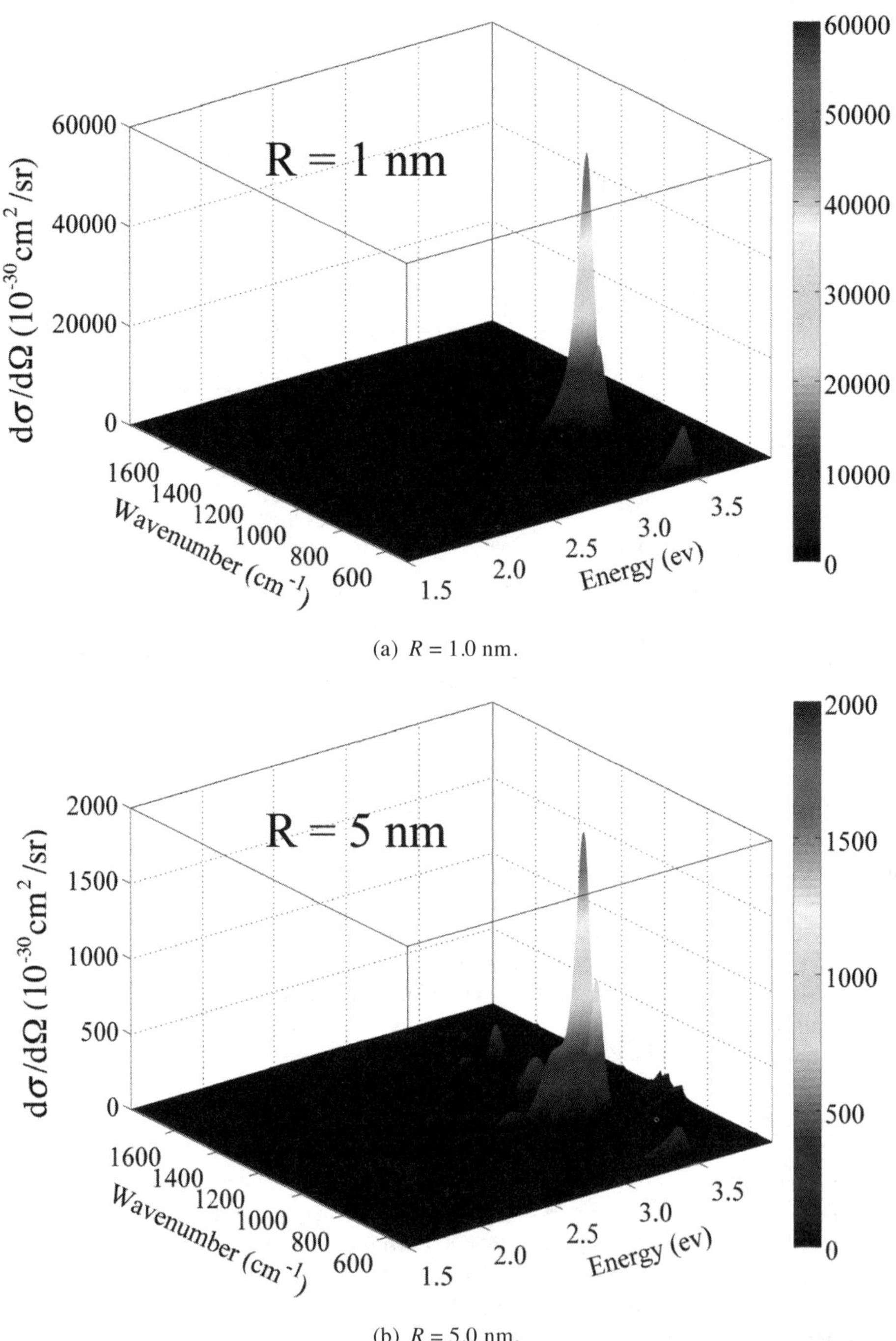

(a) $R = 1.0$ nm.

(b) $R = 5.0$ nm.

Fig. 10.9 Surface plots of the calculated differential Raman cross-section of the pyridine molecule/silver nanosphere system of Fig. 10.7 as a function of the incident light energy and the vibrational mode wavenumber for two separations, R, between the molecule and the nanosphere. *Source:* H. Chen, J. M. McMahon, M. A. Ratner, and G. C. Schatz, *J. Physical Chemistry C*, Vol. 114, 2010, pp. 14384–14392, ©2010 American Chemical Society.

10.9 SUMMARY AND DISCUSSION

This chapter reviewed a new multiscale computational methodology reported in [21] to incorporate the scattered *E*-field of a plasmonic nanoparticle into a quantum-mechanical (QM) optical property calculation for a nearby dye molecule. For a given location of the molecule with respect to the nanoparticle, a frequency-dependent scattering response function was first computed by the classical electrodynamics (ED) FDTD method. Subsequently, the time-dependent scattered *E*-field at the dye molecule was calculated using this response function through a multidimensional Fourier transform to reflect the effect of polarization of the nanoparticle on the local field at the molecule. Finally, a real-time time-dependent density function theory (RT-TDDFT) approach was employed to obtain the desired optical property (such as absorption cross-section) of the dye molecule in the presence of the nanoparticle's scattered *E*-field.

The review of the hybrid QM/ED technique of [21] continued with demonstrations of its application to calculate absorption spectra of the N3 dye molecule and Raman spectra of the pyridine molecule. Both molecules' spectra were shown to be significantly enhanced by a closely located 20-nm-diameter silver nanosphere. In contrast to traditional QM optical calculations in which the field at the molecule is entirely determined by the intensity and polarization direction of the incident light, this work showed that the light propagation direction as well as polarization and intensity are important to the response of systems comprised of a dye molecule bound to a metal nanoparticle. At no additional computation cost compared to conventional electrodynamics and quantum-mechanics calculations, this technique provided a reliable way to couple the response of the dye molecule's individual electrons to the collective dielectric response of the nearby nanoparticle.

Overall, the hybrid QM/ED method was demonstrated to provide a bridge spanning the wide gap between quantum mechanics and classical electrodynamics with respect to both length and time scales. This was manifested by the successful calculation of absorption spectra of the N3 dye molecule and Raman spectra of the pyridine molecule, for each molecule located within a few nanometers of a 20-nm silver nanosphere. In spite of the distinct underlying physical mechanisms of these two spectra, experimentally consistent enhancements in the simulated cross-sections were calculated. Fundamentally, the computed signal amplifications reflected the strong coupling between the wavelike response behavior of each dye molecule's individual electrons and the particle-like collective motion of the adjacent silver nanosphere's dielectric medium. This coupling was sensitive to both their relative location and orientation, in addition to the propagation and polarization directions of the incident light.

This study showed that the QM/ED method is well suited to study the optical properties of one or several dye molecules near plasmonic nanoparticles. The conventional frequency-domain TDDFT (FD-TDDFT) requires diagonalization of a response matrix that is constructed from the occupied-unoccupied orbital pairs, and thus grows rapidly with system size [50]. In contrast, RT-TDDFT provides significant savings in computer time and memory allocation through the use of one-particle wavefunction propagations starting from occupied orbitals [51]. In addition, the time scale of the propagation is relatively short due to the assumed damping of the molecular excited state. As a result, the size of a system that can be treated at the QM level in the QM/ED calculations can be expanded to a few hundred atoms, large enough to cover most organic dye molecules of interest.

The FDTD method was selected as the counterpart of RT-TDDFT at the ED level, because it proceeds in the real time domain and thus is able to generate the frequency-dependent field enhancement factor in a single simulation through Fourier transformation. Note, however, that it is convenient to express the coupling between ED and QM via in the frequency domain due to the different ways that time is used in FDTD and RT-TDDFT. In FDTD, time starts when the incident wave leaves its source. On the other hand, the timer in RT-TDDFT is triggered when the incident light actually reaches the molecule. Since the frequency domain is used to interconnect ED and QM, ED methods that work in the frequency domain could be used directly instead of FDTD. For example, the finite-element method (FEM) [52] might be used to advantage here, because it can circumvent some of thc errors in FDTD, such as the staircasing treatment of curved surfaces.

Finally, in addition to the flexible coupling scheme between the QM and ED levels of theory, QM/ED allows for an arbitrary number of incident light pulses with any choice of propagation and polarization directions, paving the way to the investigation of time-resolved optical phenomena. Also, following the successful application of QM/ED in [21] for linear optical properties such as absorption and Raman spectra, its extension to nonlinear optical materials [53] is anticipated. Other generalizations of this work could arise in the treatment of layers of molecules near plasmonic nanoparticles where the self-consistent coupling of the dielectric properties of the molecules to the optical response of the nanoparticles is important. In this case it would be necessary to use the fields from the QM calculations as input to the ED calculations.

ACKNOWLEDGMENT

The research reported in [21] and reviewed in this chapter was supported by the ANSER Energy Frontier Research Center (DE-SC0001785) funded by the U.S. Department of Energy, Office of Science, Office of Basic Energy Sciences. The computational resources utilized in this research were provided by Quest cluster system administered by Northwestern University Information Technology (NUIT).

REFERENCES

[1] Kirchhoff, G., and R. Bunsen, "Chemische analyse durch spectralbeobachtungen," *Annalen der Physik und Chemie*, Vol. 186, 1860, pp. 161–189.

[2] Raman, C. V., and K. S. Krishnan, "A new type of secondary radiation," *Nature*, Vol. 121, 1928, pp. 501–502.

[3] Oulton, R. F., V. J. Sorger, T. Zentgraf, R.-M. Ma, C. Gladden, L. Dai, G. Bartal, and X. Zhang, "Plasmon lasers at deep subwavelength scale," *Nature*, Vol. 461, 2009, pp. 629–632.

[4] Jeanmaire, D. L., and R. P. Van Duyne, "Surface Raman spectroelectrochemistry: Part I. Heterocyclic, aromatic, and aliphatic amines adsorbed on the anodized silver electrode," *J. Electroanalytical Chemistry*, 1977, Vol. 84, pp. 1–20.

[5] King, F. W., R. P. Van Duyne, and G. C. Schatz, "Theory of Raman scattering by molecules adsorbed on electrode surfaces," *J. Chemical Physics*, Vol. 69, 1978, pp. 4472–4481.

[6] Dulić, D., S. J. van der Molen, T. Kudernac, H. T. Jonkman, J. J. D. de Jong, T. N. Bowden, J. van Esch, B. L. Feringa, and B. J. van Wees, "One-way optoelectronic switching of photochromic molecules on gold," *Physical Review Lett.*, Vol. 91, 2003, 207402.

[7] Grafstrom, S., "Photoassisted scanning tunneling microscopy," *J. Applied Physics*, Vol. 91, 2002, pp. 1717–1753.

[8] Nakanishi, H., K. J. M. Bishop, B. Kowalczyk, A. Nitzan, E. A. Weiss, K. V. Tretiakov, M. M. Apodaca, R. Klajn, J. F. Stoddart, and B. A. Grzybowski, "Photoconductance and inverse photoconductance in films of functionalized metal nanoparticles," *Nature*, Vol. 460, 2009, pp. 371–375.

[9] Viljas, J. K., F. Pauly, and J. C. Cuevas, "Photoconductance of organic single-molecule contacts," *Physical Review B*, Vol. 76, 2007, 033403.

[10] Galperin, M., and A. Nitzan, A. "Optical properties of current carrying molecular wires," *J. Chemical Physics*, Vol. 124, 2006, 234709-17.

[11] Galperin, M., and A. Nitzan, "Current-induced light emission and light-induced current in molecular-tunneling junctions," *Physical Review Lett.*, Vol. 95, 2005, 206802.

[12] Galperin, M., and S. Tretiak, "Linear optical response of current-carrying molecular junction: A nonequilibrium Green's function–time-dependent density functional theory approach," *J. Chemical Physics*, Vol. 128, 2008, 124705-9.

[13] Galperin, M., M. A. Ratner, and A. Nitzan, "Raman scattering from nonequilibrium molecular conduction junctions," *Nano Lett.*, Vol. 9, 2009, pp. 758–762.

[14] Zhao, G., H. Kozuka, and T. Yoko, "Effects of the incorporation of silver and gold nanoparticles on the photoanodic properties of rose bengal sensitized TiO_2 film electrodes prepared by sol-gel method," *Solar Energy Materials Solar Cells*, Vol. 46, 1997, pp. 219–231.

[15] Standridge, S. D., G. C. Schatz, and J. T. Hupp, "Distance dependence of plasmon-enhanced photocurrent in dye-sensitized solar cells," *J. American Chemical Society*, Vol. 131, 2009, pp. 8407–8409.

[16] Zhao, J., A. O. Pinchuk, J. M. McMahon, S. Li, L. K. Ausman, A. L. Atkinson, and G. C. Schatz, "Methods for describing the electromagnetic properties of silver and gold nanoparticles," *Accts. Chemical Research*, Vol. 41, 2008, pp. 1710–1720.

[17] Jensen, L., C. M. Aikens, and G. C. Schatz, "Electronic structure methods for studying surface-enhanced Raman scattering," *Chemical Society Reviews*, Vol. 37, 2008, pp. 1061–1073.

[18] Corni, S., and J. Tomasi, "Enhanced response properties of a chromophore physisorbed on a metal particle," *J. Chemical Physics*, Vol. 114, 2001, pp. 3739–3751.

[19] Lopata, K., and D. Neuhauser, "Multiscale Maxwell–Schrödinger modeling: A split field finite-difference time-domain approach to molecular nanopolaritonics," *J. Chemical Physics*, Vol. 130, 2009, 104707-7.

[20] Masiello, D., and G. C. Schatz, "On the linear response and scattering of an interacting molecule-metal system," *J. Chemical Physics*, Vol. 132, 2010, 064102.

[21] Chen, H., J. M. McMahon, M. A. Ratner, and G. C. Schatz, "Classical electrodynamics coupled to quantum mechanics for calculation of molecular optical properties: An RT-TDDFT/FDTD approach," *J. Physical Chemistry C*, Vol. 114, 2010, pp. 14384–14392.

[22] Taflove, A., and S. C. Hagness, *Computational Electrodynamics: The Finite-Difference Time-Domain Method*, 3rd ed., Norwood, MA: Artech, 2005.

[23] Schrödinger, E., "An undulatory theory of the mechanics of atoms and molecules," *Physical Review*, Vol. 28, 1926, p. 1049.

[24] Crank, J., and P. Nicolson, "A practical method for numerical evaluation of solutions of partial differential equations of the heat-conduction type," *Advances in Computational Mathematics*, Vol. 6, 1996, pp. 207–226.

[25] Castro, A., M. A. L. Marques, and A. Rubio, "Propagators for the time-dependent Kohn–Sham equations," *J. Chemical Physics*, Vol. 121, 2004, pp. 3425–3433.

[26] Castro, A., H. Appel, M. Oliveira, C. A. Rozzi, X. Andrade, F. Lorenzen, M. A. L. Marques, E. K. U. Gross, and A. Rubio, "Octopus: A tool for the application of time-dependent density functional theory," *Phys. Status Solidi B*, Vol. 243, 2006, pp. 2465–2488.

[27] Neugebauer, J., M. Reiher, C. Kind, and B. A. Hess, "Quantum chemical calculation of vibrational spectra of large molecules: Raman and IR spectra for Buckminsterfullerene," *J. Computational Chemistry*, Vol. 23, 2002, pp. 895–910.

[28] Maxwell, J. C., "A dynamical theory of the electromagnetic field," *Philosophical Trans. of the Royal Society of London*, Vol. 155, 1865, pp. 459–512.

[29] Long, D. A., *Raman Spectroscopy*, New York: McGraw-Hill, 1977.

[30] McMahon, J. M., Y. Wang, L. J. Sherry, R. P. Van Duyne, L. D. Marks, S. K. Gray, and G. C. Schatz, "Correlating the structure, optical spectra, and electrodynamics of single silver nanocubes," *J. Physical Chemistry C*, Vol. 113, 2009, pp. 2731–2735.

[31] Johnson, P. B., and R. W. Christy, "Optical constants of the noble metals," *Physical Review B*, Vol. 6, 1972, p. 4370.

[32] Grätzel, M., "Dye-sensitized solar cells," *J. Photochem. Photobiol. C*, Vol. 4, 2003, pp. 145–153.

[33] Nazeeruddin, M. K., A. Kay, I. Rodicio, R. Humphry-Baker, E. Mueller, P. Liska, N. Vlachopoulos, and M. Graetzel, "Conversion of light to electricity by cis-X2bis(2,2'-bipyridyl-4,4'-dicarboxylate) ruthenium(II) charge-transfer sensitizers (X = Cl–, Br–, I–, CN–, and SCN–) on nanocrystalline titanium dioxide electrodes," *J. American Chemical Society*, Vol. 115, 1993, pp. 6382–6390.

[34] Hannappel, T., B. Burfeindt, W. Storck, and F. Willig, "Measurement of ultrafast photoinduced electron transfer from chemically anchored Ru-dye molecules into empty electronic states in a colloidal anatase TiO_2 film," *J. Physical Chemistry B*, Vol. 101, 1997, pp. 6799–6802.

[35] Krack, M., "Pseudopotentials for H to Kr optimized for gradient-corrected exchange-correlation functionals," *Theor. Chem. Acc.*, Vol. 114, 2005, pp. 145–152.

[36] Perdew, J. P., K. Burke, and M. Ernzerhof, "Generalized gradient approximation made simple," *Physical Review Lett.*, Vol. 77, 1996, p. 3865.

[37] Woon, D. E., and J. T. H. Dunning, "Gaussian basis sets for use in correlated molecular calculations. IV. Calculation of static electrical response properties," *J. Chemical Physics*, Vol. 100, 1994, pp. 2975–2988.

[38] Genovese, L., T. Deutsch, A. Neelov, S. Goedecker, and G. Beylkin, "Efficient solution of Poisson's equation with free boundary conditions," *J. Chemical Physics*, Vol. 125, 2006, 074105-5.

[39] VandeVondele, J., M. Krack, F. Mohamed, M. Parrinello, T. Chassaing, and J. R. Hutter, "Quickstep: Fast and accurate density functional calculations using a mixed Gaussian and plane waves approach," *Comput. Phys. Commun.*, Vol. 167, 2005, pp. 103–128.

[40] Gritsenko, O., and E. J. Baerends, "Asymptotic correction of the exchange–correlation kernel of time-dependent density functional theory for long-range charge-transfer excitations," *J. Chemical Physics*, Vol. 121, 2004, pp. 655–660.

[41] Gritsenko, O. V., P. R. T. Schipper, and E. J. Baerends. "Approximation of the exchange-correlation Kohn-Sham potential with a statistical average of different orbital model potentials," *Chemical Physics Lett.*, Vol. 302, 1999, pp. 199–207.

[42] Duncan, W. R., and O. V. Prezhdo, "Theoretical studies of photoinduced electron transfer in dye-sensitized TiO_2," *Annual Rev. Physical Chemistry*, Vol. 58, 2007, pp. 143–184.

[43] Creighton, J. A., "Contributions to the early development of surface-enhanced Raman spectroscopy," *Notes and Records of the Royal Society*, Vol. 64, 2010, pp. 175–183.

[44] Johnson, A. E., and A. B. Myers, "Solvent effects in the Raman spectra of the triiodide ion: Observation of dynamic symmetry breaking and solvent degrees of freedom," *J. Physical Chemistry*, Vol. 100, 1996, pp. 7778–7788.

[45] Nie, S., and S. R. Emory, "Probing single molecules and single nanoparticles by surface-enhanced Raman scattering," *Science*, Vol. 275, 1997, pp. 1102–1106.

[46] Kelly, K. L., E. Coronado, L. L. Zhao, and G. C. Schatz, "The optical properties of metal nanoparticles: The influence of size, shape, and dielectric environment," *J. Physical Chemistry B*, Vol. 107, 2003, pp. 668–677.

[47] Arenas, J. F., I. Lopez Tocon, J. C. Otero, and J. I. Marcos, "Charge transfer processes in surface-enhanced Raman scattering. Franck-Condon active vibrations of pyridine," *J. Physical Chemistry*, Vol. 100, 1996, pp. 9254–9261.

[48] Lombardi, J. R., and R. L. Birke, "A unified approach to surface-enhanced Raman spectroscopy," *J. Physical Chemistry C*, Vol. 112, 2008, pp. 5605–5617.

[49] Morton, S. M., and L. Jensen, "Understanding the molecule-surface chemical coupling in SERS," *J. American Chemical Society*, Vol. 131, 2009, pp. 4090–4098.

[50] Casida, M. E., C. Jamorski, K. C. Casida, and D. R. Salahub, "Molecular excitation energies to high-lying bound states from time-dependent density-functional response theory: Characterization and correction of the time-dependent local density approximation ionization threshold," *J. Chemical Physics*, Vol. 108, 1998, pp. 4439–4449.

[51] Takimoto, Y., F. D. Vila, and J. J. Rehr, "Real-time time-dependent density functional theory approach for frequency-dependent nonlinear optical response in photonic molecules," *J. Chemical Physics*, Vol. 127, 2007, 154114-10.

[52] Coccioli, R., T. Itoh, G. Pelosi, and P. P. Silvester, "Finite-element methods in microwaves: A selected bibliography," *IEEE. Trans. Antennas and Propagation*, Vol. 38, 1996, pp. 34–48.

[53] Eaton, D. F., "Nonlinear optical materials," *Science*, Vol. 253, 1991, pp. 281–287.

Chapter 11

Transformation Electromagnetics Inspired Advances in FDTD Methods

Roberto B. Armenta and Costas D. Sarris

11.1 INTRODUCTION

Transformation electromagnetics is an exciting new research area that employs the coordinate-invariance property of Maxwell's equations as a tool to synthesize the permeability and permittivity tensors of artificial materials that guide electromagnetic waves in specified manners [1, 2]. The origins of this new field are closely linked to the modeling of curved material boundaries using FDTD techniques [3, 4]. In fact, the idea of employing a coordinate-invariant representation of Maxwell's equations to incorporate curved material boundaries well preceded the idea of exploiting the same representation to design artificial materials (see the Selected Bibliography). However, the recent explosion of research in artificial materials, especially to achieve functional invisibility, has sparked renewed interest in exploring both old and new applications of the coordinate-invariance property of Maxwell's equations within the context of FDTD numerical techniques.

The coordinate-invariance property of Maxwell's equations is one aspect of a more general physical principle — known as the principle of invariance — that requires all physical laws to be formulated in a manner that is independent of the particular properties of the coordinate system being employed. This principle was born out of the need for a clear conceptual separation between the laws of physics and the mathematical tools that are used to describe them.

A second, stronger property that is a peculiarity of Maxwell's equations is that deformations on the spatial metric can be incorporated through an equivalent set of constitutive tensors whereby the generic form of the equations does not change. The second property is explicitly manifested when Maxwell's equations are written in the language of differential forms (exterior calculus) instead of vector calculus.

The goals of this chapter are first, to present the foundation of coordinate transformations of Maxwell's equations in the context of the FDTD method, and second, to describe emerging applications, including the design of artificial materials. The remainder of this chapter is divided into seven sections. First, two sections introduce the invariance and relativity principles in the context of FDTD techniques. Then, the three sections that follow explain how to use coordinate transformations to incorporate curved material boundaries into FDTD discretizations of Maxwell's equations in classical form. A key component of the discussion includes the development of a projection of Maxwell's equations onto the covariant-contravariant vector bases associated with a general curvilinear coordinate system. Subsequently, the next section describes how coordinate transformations can be used for the dual purposes of designing an artificial material and creating an FDTD discretization. Finally, the last section provides an overview of how to exploit the general relativistic form of Maxwell's equations to create a time-dependent discretization.

11.2 INVARIANCE PRINCIPLE IN THE CONTEXT OF FDTD TECHNIQUES

In the past, FDTD methods were often viewed as inherently poorly suited to model objects with curved material boundaries. This view stemmed from the fact that most FDTD discretizations of Maxwell's equations have been derived using the Cartesian coordinate system, and this coordinate system is not well equipped to describe curved material boundaries. In any coordinate system, the simplest way to describe material boundaries is by using coordinate surfaces defined by equations of the form:

$$x^i = const. \quad \text{for } i = 1, 2, 3 \tag{11.1}$$

Assuming that the x^i set of coordinates represents the three Cartesian coordinates, the surfaces described in (11.1) are flat sheets that intersect each other at right angles. For this reason, the Cartesian coordinate system is well suited to model objects with rectangular material boundaries, and poorly suited to model objects with curved material boundaries. In spite of this drawback, many methods have been developed to enforce different kinds of boundary conditions on curved material boundaries using the Cartesian coordinate system [5–15].

However, there is no compelling reason to always use Cartesian coordinates. The ability to model curved material boundaries can be more easily and systematically integrated into an FDTD method if the method's derivation is carried out using a set of general curvilinear coordinates — denoted here by u^j for $j = 1, 2, 3$ — rather than Cartesian coordinates. To achieve the simplicity of (11.1), the u^j set of coordinates must be constructed so that all the boundaries of a given object are described by coordinate surfaces defined by equations of the form:

$$u^j = const. \quad \text{for } j = 1, 2, 3 \tag{11.2}$$

Enforcing boundary conditions on the above coordinate surfaces requires a vector component representation of Maxwell's equations derived using the covariant and contravariant vector bases associated with the general curvilinear coordinate system rather than the Cartesian basis vectors. In other words, the starting point for any FDTD discretization must be a covariant-contravariant representation of Maxwell's equations that does not place any restrictions on how the u^j coordinates are constructed.

The important role that the invariance principle plays in the derivation of FDTD methods was first recognized by Holland [16], inspired by a passage in Stratton's book [17]. Since Holland's work appeared, a number of FDTD methods have been proposed that seek to exploit general curvilinear coordinate systems and their associated covariant and contravariant basis vectors [4, 18–26]. The earliest of these methods [18–23] resorted to crude polyhedral approximations of the material boundaries to generate approximations for the needed covariant and contravariant basis vectors. Because these early methods tied the accuracy with which the covariant and contravariant basis vectors were generated to the size of the grid cells used to discretize Maxwell's equations, it was impossible to refine the boundary representations without incurring a large computational cost.

Subsequently, it was realized that the accuracy with which the covariant and contravariant basis vectors are constructed does not have to be tied to the accuracy with which Maxwell's equations are discretized. This point was first made by Ward and Pendry [3, 4] who explained that, if all the material boundaries of an object are described by analytically constructing an invertible coordinate transformation:

$$x^i = x^i(u^j) \quad \text{for} \quad i, j = 1, 2, 3 \tag{11.3}$$

then the boundary representation is exact and the covariant and contravariant basis vectors can be constructed exactly. When an exact boundary representation is available, the global error is dominated by the local truncation error of the finite-difference approximations used to discretize Maxwell's equations rather than the crudeness of the boundary representations. As pointed out in [25, 26], this opened up the possibility of exploiting high-order finite differences.

The introduction of curved material boundaries is not the only possible application of the invariance principle in the context of FDTD numerical techniques. A very different and interesting application is introduced next: the creation of time-dependent FDTD discretizations.

11.3 RELATIVITY PRINCIPLE IN THE CONTEXT OF FDTD TECHNIQUES

When Fitzgerald, Heaviside, Hertz, Lodge, and others used vector algebra to rewrite Maxwell's equations in the modern classical form [27], the invariance principle was clearly embraced in the spatial sense. In other words, for any set of three spatial coordinates u^j for $j = 1, 2, 3$, Maxwell's equations always take the form:

$$\nabla \times \boldsymbol{E}(u^j, t) = -\frac{\partial \boldsymbol{B}(u^j, t)}{\partial t} \tag{11.4}$$

$$\nabla \times \boldsymbol{H}(u^j, t) = \frac{\partial \boldsymbol{D}(u^j, t)}{\partial t} + \boldsymbol{J}(u^j, t) \tag{11.5}$$

$$\boldsymbol{B}(u^j, t) = \mathcal{M}(u^j)\boldsymbol{H}(u^j, t), \qquad \boldsymbol{D}(u^j, t) = \mathcal{E}(u^j)\boldsymbol{E}(u^j, t) \tag{11.6, 7}$$

However, these equations do not embody the invariance principle in the temporal sense because they are tied to a time-orthogonal coordinate system. In essence, time orthogonality means that only one of the four independent variables (t in this case) can contain time information.

When the general theory of relativity was developed, Einstein and Minkowski sought to put space and time on an equal footing by rewriting Maxwell's equations in a mathematical framework that embodies the principle of invariance in both the temporal and spatial sense [28]. As a result, all four independent variables in the modern general relativistic form of Maxwell's equations contain both temporal and spatial information. This feature enables the introduction of a four-dimensional (4-D) coordinate transformation that intertwines space with time:

$$x^i = x^i(u^j) \quad \text{for} \quad i, j = 1, 2, 3, 4 \tag{11.8}$$

As before, the x^i set in the above transformation stands for the Cartesian or Minkowski coordinates — with $x^4 = c_0 t$ — while the u^j set stands for a set of general curvilinear coordinates.

The invariance principle in its combined temporal and spatial interpretation is also known as the relativity principle, and can be exploited to incorporate time-varying discretizations in the following manner. The coordinate transformation in (11.8) is constructed so that u^4 remains time-like, and a uniform discretization with respect to all four u^j coordinates produces a time-varying (or x^4-dependent) discretization in the x^i coordinates. By treating u^4 as a time-like coordinate, all previously known strategies to discretize Maxwell's equations in the time domain on a time-orthogonal coordinate system can be ported over to the u^j coordinate system.

A time-varying discretization can serve two different purposes: enhancing the resolution of a discretization grid through a dynamic reconfiguration process, and introducing moving material boundaries. Although a number of methods to dynamically refine grids [29, 30] as well as to introduce moving material boundaries [31–33] have been proposed over the years, the idea of exploiting the relativity principle has only recently been explored [34]. While the classical and relativistic formulations of Maxwell's equations contain the same physics when applied to a lab on the surface of the Earth, the relativistic framework offers a distinct advantage: it provides a clear path to systematically derive time-domain discretizations of Maxwell's equations on a set of continuously moving grid points whose velocity can come close to the velocity of the waves in the solution.

11.4 COMPUTATIONAL COORDINATE SYSTEM AND ITS COVARIANT AND CONTRAVARIANT VECTOR BASES

Any three nonorthogonal vectors can be used as a basis for a three-dimensional (3-D) Euclidean (or flat) space, but not all vectors are equally suited to construct physical laws. Because Maxwell's equations are closely tied to flux and path integrals, it makes sense to employ two different sets of basis vectors that are appropriate for the formulation of these two types of integrals. The covariant and contravariant sets of basis vectors were constructed with this particular consideration in mind.

11.4.1 Covariant and Contravariant Basis Vectors

The covariant basis vectors are denoted by $\boldsymbol{g}_j$ for $j = 1, 2, 3$. They can be constructed directly from the position vector $\boldsymbol{r}$ in Cartesian coordinates as:

$$\boldsymbol{g}_j = \frac{\partial \boldsymbol{r}}{\partial u^j} \quad \text{where} \quad \boldsymbol{r} = x^1\boldsymbol{i}_1 + x^2\boldsymbol{i}_2 + x^3\boldsymbol{i}_3 \tag{11.9}$$

and $\boldsymbol{i}_j$ (or $\boldsymbol{i}^j$) stands for the jth Cartesian unit vector. From the above definition, it follows that the Cartesian components of the jth covariant basis vector are:

$$\boldsymbol{g}_j \cdot \boldsymbol{i}^i = \frac{\partial x^i}{\partial u^j} \quad \text{for} \quad i, j = 1, 2, 3 \tag{11.10}$$

The contravariant basis vectors are denoted by $\boldsymbol{g}^i$ for $i = 1, 2, 3$. Their Cartesian components are given by:

$$\boldsymbol{g}^i \cdot \boldsymbol{i}_j = \frac{\partial u^i}{\partial x^j} \quad \text{for} \quad i, j = 1, 2, 3 \tag{11.11}$$

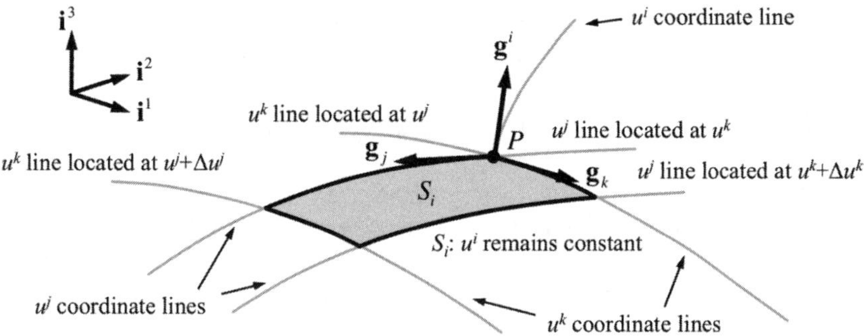

Fig. 11.1 Diagram of the covariant ($\boldsymbol{g}_j$) and contravariant ($\boldsymbol{g}^i$) basis vectors in 3-D Euclidean space. The indices $\{i, j, k\}$ must be cycled through the sets $\{1,2,3\}$, $\{3,2,1\}$, and $\{2,3,1\}$ to visualize the three covariant-contravariant combinations.

Figure 11.1 illustrates the two different geometrical roles that the covariant and contravariant basis vectors play. The jth covariant basis vector $\boldsymbol{g}_j$ is tangential to a u^j coordinate line, that is, a line traced by varying u^j while keeping u^i and u^k constant. The ith contravariant basis vector $\boldsymbol{g}^i$ is normal to the surface S_i defined by an equation of the form $u^i = \text{const}$.

11.4.2 Covariant and Contravariant Components of the Metric Tensor

From the covariant and contravariant basis vectors, a second-rank tensor **G**, which is known as the metric tensor, can be constructed. The metric tensor plays a key role in the definition of the constitutive relations, and its components can be defined in either a covariant or a contravariant form. In covariant form, the components of the metric tensor are given by:

$$g_{ij} = \boldsymbol{g}_i \cdot \boldsymbol{g}_j = \sum_{q=1}^{3} \frac{\partial x^q}{\partial u^i} \frac{\partial x^q}{\partial u^j} \quad \text{for } i, j = 1, 2, 3 \tag{11.12}$$

In contravariant form, the components of the metric tensor are given by:

$$g^{ij} = \boldsymbol{g}^i \cdot \boldsymbol{g}^j = \sum_{s=1}^{3} \frac{\partial u^i}{\partial x^s} \frac{\partial u^j}{\partial x^s} \quad \text{for } i, j = 1, 2, 3 \tag{11.13}$$

From (11.10) – (11.13), it is easy to establish that, once the coordinate transformation in (11.3) has been specified, the covariant and contravariant sets of basis vectors as well as the covariant and contravariant components of the metric tensor are also specified. A coordinate system is said to be orthogonal if the covariant and contravariant basis vectors remain perpendicular to each other. For the covariant and contravariant components of the metric tensors, orthogonality means that:

$$g_{ij} = \boldsymbol{g}_i \cdot \boldsymbol{g}_j = 0 \quad \text{and} \quad g^{ij} = \boldsymbol{g}^i \cdot \boldsymbol{g}^j = 0 \quad \text{for } i \neq j \tag{11.14}$$

In many textbooks, orthogonal coordinate systems are often assumed so that the distinction between covariant and contravariant basis vectors can be disregarded. However, this assumption severely restricts the types and combinations of boundary shapes that can be used. For the invariance principle to be fully embraced, no orthogonality assumptions must be made.

11.4.3 Covariant and Contravariant Representation of a Vector

To avoid confusion as to what vector basis is being used to project a given vector, lowercase letters are used to express vector components in terms of the vector basis of the computational coordinate system, while uppercase letters are used to express vector components in terms of the Cartesian vector basis. Using this convention, the covariant and contravariant components of $\boldsymbol{E}$ in the vector basis of the computational coordinate system are given by:

$$e_i = \boldsymbol{E} \cdot \boldsymbol{g}_i \quad \text{and} \quad e^i = \boldsymbol{E} \cdot \boldsymbol{g}^i \quad \text{for} \ \ i = 1, 2, 3 \tag{11.15}$$

The vector $\boldsymbol{E}$ can be easily reconstructed from the above components as follows:

$$\boldsymbol{E} = \sum_{i=1}^{3} e_i \boldsymbol{g}^i = \sum_{i=1}^{3} e^i \boldsymbol{g}_i \tag{11.16}$$

In a similar manner, the covariant and contravariant components of $\boldsymbol{E}$ in the vector basis of the Cartesian coordinate system are given by:

$$E_i = \boldsymbol{E} \cdot \boldsymbol{i}_i \quad \text{and} \quad E^i = \boldsymbol{E} \cdot \boldsymbol{i}^i \quad \text{for} \ \ i = 1, 2, 3 \tag{11.17a}$$

where

$$\boldsymbol{E} = \sum_{i=1}^{3} E_i \boldsymbol{i}^i = \sum_{i=1}^{3} E^i \boldsymbol{i}_i \tag{11.17b}$$

For clarity, subscripts are always used to identify covariant components while superscripts are always used to identify contravariant components. The covariant-contravariant distinction is unimportant in the Cartesian basis because $E_i = E^i$ and $\boldsymbol{i}_i = \boldsymbol{i}^i$ for $i = 1, 2, 3$. On the other hand, the covariant-contravariant distinction is extremely important when the vector basis of the computational coordinate system is used because $e_i \neq e^i$ and $\boldsymbol{g}_i \neq \boldsymbol{g}^i$ for $i = 1, 2, 3$. To convert vector components from a covariant to a contravariant representation and vice versa, the metric tensor is employed:

$$e^i = \sum_{j=1}^{3} g^{ij} e_j \quad \text{and} \quad e_i = \sum_{j=1}^{3} g_{ij} e^j \quad \text{for} \quad i = 1, 2, 3 \tag{11.18}$$

In Cartesian coordinates, the covariant and contravariant components of the metric tensor turn into Kronecker delta symbols with upper and lower indices: $G^{ij} = \delta^{ij}$ and $G_{ij} = \delta_{ij}$ for $i, j = 1, 2, 3$. Consequently, when the metric tensor is used in a Cartesian representation to convert components from a covariant to a contravariant representation, the aforementioned property is obtained:

$$E^i = \sum_{j=1}^{3} G^{ij} E_j = E_i \quad \text{and} \quad E_i = \sum_{j=1}^{3} G_{ij} E^j \quad \text{for} \quad i = 1, 2, 3 \tag{11.19}$$

For future reference, we note that the Kronecker delta is the only second-rank object whose indices can be changed arbitrarily: $\delta_{ij} = \delta^{i}{}_{j} = \delta_{j}{}^{i} = \delta^{ij}$.

11.4.4 Converting Vectors to the Cartesian Basis and Vice Versa

When Maxwell's equations are discretized and solved in the computational coordinate system, it is necessary to know how to convert vector components from the vector basis representation of the computational coordinate system to the vector basis representation of the Cartesian coordinate system. The following two relations are used for this purpose:

$$E_i = \sum_{j=1}^{3} \frac{\partial u^j}{\partial x^i} e_j \quad \text{and} \quad E^i = \sum_{j=1}^{3} \frac{\partial x^i}{\partial u^j} e^j \quad \text{for} \quad i = 1, 2, 3 \tag{11.20}$$

The reverse conversion is accomplished by flipping the partial derivatives:

$$e_i = \sum_{j=1}^{3} \frac{\partial x^j}{\partial u^i} E_j \quad \text{and} \quad e^i = \sum_{j=1}^{3} \frac{\partial u^i}{\partial x^j} E^j \quad \text{for} \quad i = 1, 2, 3 \tag{11.21}$$

To express the constitutive relations, second-rank tensors are needed in addition to vectors. The notational conventions that have been described thus far regarding the use of uppercase and lowercase letters, as well as the use of subscripts and superscripts, can be easily applied to second-rank tensors. However, to project the permeability and permittivity tensors into covariant and contravariant components, it is first necessary to construct a second-rank basis from the covariant and contravariant basis vectors.

11.4.5 Second-Rank Tensors in the Covariant and Contravariant Bases

A second-rank tensor is defined as a linear mapping from one vector field onto another. The dielectric constitutive relation is a clear example of such mapping where the permittivity tensor takes the electric field vector and maps it to the electric flux density vector. This mapping is expressed here as $\boldsymbol{D} = \boldsymbol{\mathcal{E}}\boldsymbol{E}$. When working with Cartesian coordinates, the dielectric constitutive relation can be reduced, for the case of a simple material, to a multiplication by a scalar. Because the assumption $\boldsymbol{\mathcal{E}} \rightarrow \varepsilon$ is quite common, it is easy to overlook the fact that the permittivity tensor represents a mapping of one vector field onto another. (Note that the metric tensor $\mathbf{G}$ is defined as the special mapping that takes a vector $\boldsymbol{E}$ and maps it onto itself: $\boldsymbol{E} = \mathbf{G}\boldsymbol{E}$.)

To represent arbitrary vector mappings in a systematic way, it is necessary to introduce the dyadic product between two vectors. This product, whose end result is a second-rank tensor, is denoted here by the symbol $\otimes$. The dyadic product is defined by two key properties:

$$(\boldsymbol{A} \otimes \boldsymbol{B})\boldsymbol{C} = \boldsymbol{A}(\boldsymbol{B} \cdot \boldsymbol{C}) \quad \text{and} \quad \boldsymbol{A}(\boldsymbol{B} \otimes \boldsymbol{C}) = (\boldsymbol{A} \cdot \boldsymbol{B})\boldsymbol{C} \tag{11.22}$$

where $\boldsymbol{A}$, $\boldsymbol{B}$, and $\boldsymbol{C}$ are arbitrary vectors. Although the dyadic product is linear, it is not commutative: $\boldsymbol{A} \otimes \boldsymbol{B} \neq \boldsymbol{B} \otimes \boldsymbol{A}$.

To create a second-rank basis from the covariant and contravariant basis vectors, the dyadic product between two vectors is used. A second-rank tensor basis can be created from any of the four possible dyadic product combinations of $\boldsymbol{g}_i$ and $\boldsymbol{g}^j$. For instance, the permittivity tensor $\boldsymbol{\mathcal{E}}$ has four possible representations:

$$\boldsymbol{\mathcal{E}} = \sum_{i=1}^{3}\sum_{j=1}^{3}\varepsilon^{ij}\mathbf{g}_i \otimes \mathbf{g}_j = \sum_{i=1}^{3}\sum_{j=1}^{3}\varepsilon_{ij}\mathbf{g}^i \otimes \mathbf{g}^j = \sum_{i=1}^{3}\sum_{j=1}^{3}\varepsilon^{i}{}_{j}\mathbf{g}_i \otimes \mathbf{g}^j = \sum_{i=1}^{3}\sum_{j=1}^{3}\varepsilon_{i}{}^{j}\mathbf{g}^i \otimes \mathbf{g}_j \tag{11.23}$$

In the above component expansions, $\varepsilon_{ij} = \mathbf{g}_i \boldsymbol{\mathcal{E}} \mathbf{g}_j$ is referred to as the (i, j)th covariant component and $\varepsilon^{ij} = \mathbf{g}^i \boldsymbol{\mathcal{E}} \mathbf{g}^j$ is the (i, j)th contravariant component of tensor $\boldsymbol{\mathcal{E}}$. Second-rank tensors can have mixed covariant-contravariant representations; hence $\varepsilon^{i}{}_{j} = \mathbf{g}^i \boldsymbol{\mathcal{E}} \mathbf{g}_j$ and $\varepsilon_{i}{}^{j} = \mathbf{g}_i \boldsymbol{\mathcal{E}} \mathbf{g}^j$. By convention, Greek letters in italics are used here to represent material tensors, while Latin letters are used to represent electric and magnetic fields.

When designing artificial materials, it is useful to know how to transform the components of $\boldsymbol{\mathcal{E}}$ from a general curvilinear covariant or contravariant vector basis representation to a Cartesian vector basis representation. This conversion is accomplished by using:

$$\mathcal{E}_{js} = \sum_{q=1}^{3}\sum_{p=1}^{3}\frac{\partial u^q}{\partial x^j}\frac{\partial u^p}{\partial x^s}\varepsilon_{qp} \quad \text{and} \quad \mathcal{E}^{js} = \sum_{q=1}^{3}\sum_{p=1}^{3}\frac{\partial x^j}{\partial u^q}\frac{\partial x^s}{\partial u^p}\varepsilon^{qp} \quad \text{for } j, s = 1, 2, 3 \tag{11.24}$$

The reverse conversion is accomplished by using:

$$\varepsilon_{js} = \sum_{q=1}^{3}\sum_{p=1}^{3}\frac{\partial x^q}{\partial u^j}\frac{\partial x^p}{\partial u^s}\mathcal{E}_{qp} \quad \text{and} \quad \varepsilon^{js} = \sum_{q=1}^{3}\sum_{p=1}^{3}\frac{\partial u^j}{\partial x^q}\frac{\partial u^s}{\partial x^p}\mathcal{E}^{qp} \quad \text{for } j, s = 1, 2, 3 \tag{11.25}$$

If a mixed covariant and contravariant representation is used, the partial derivatives must be flipped accordingly to obtain the appropriate conversions:

$$\varepsilon^{j}{}_{s} = \sum_{q=1}^{3}\sum_{p=1}^{3}\frac{\partial u^j}{\partial x^q}\frac{\partial x^p}{\partial u^s}\mathcal{E}^{q}{}_{p}, \qquad \varepsilon_{j}{}^{s} = \sum_{q=1}^{3}\sum_{p=1}^{3}\frac{\partial x^q}{\partial u^j}\frac{\partial u^s}{\partial x^p}\mathcal{E}_{q}{}^{p} \tag{11.26}$$

$$\mathcal{E}^{j}{}_{s} = \sum_{q=1}^{3}\sum_{p=1}^{3}\frac{\partial x^j}{\partial u^q}\frac{\partial u^p}{\partial x^s}\varepsilon^{q}{}_{p}, \qquad \mathcal{E}_{j}{}^{s} = \sum_{q=1}^{3}\sum_{p=1}^{3}\frac{\partial u^q}{\partial x^j}\frac{\partial x^s}{\partial u^p}\varepsilon_{q}{}^{p} \tag{11.27}$$

for $j, s = 1, 2, 3$. As with vectors, the metric tensor can be used in the case of second-rank tensors to switch back and forth between covariant and contravariant representations. All four possible component representations of $\boldsymbol{\mathcal{E}}$ are related to each other as follows:

$$\varepsilon_{js} = \sum_{q=1}^{3} g_{jq}\varepsilon^{q}{}_{s} = \sum_{p=1}^{3}\varepsilon_{j}{}^{p} g_{ps} = \sum_{q=1}^{3}\sum_{p=1}^{3} g_{jq}\varepsilon^{qp} g_{ps} \tag{11.28}$$

$$\varepsilon^{js} = \sum_{q=1}^{3} g^{jq}\varepsilon_{q}{}^{s} = \sum_{p=1}^{3}\varepsilon^{j}{}_{p} g^{ps} = \sum_{q=1}^{3}\sum_{p=1}^{3} g^{jq}\varepsilon_{qp} g^{ps} \tag{11.29}$$

for $j, s = 1, 2, 3$.

Lastly, for (11.28) and (11.29) to be consistent, the following properties must hold:

$$\sum_{s=1}^{3} g^{is} g_{sj} = \delta^{i}{}_{j} \quad \text{and} \quad g^{i}{}_{j} = \delta^{i}{}_{j} \quad \text{for } i, j = 1, 2, 3 \tag{11.30}$$

This can be proved by using (11.12) and (11.13) together with the chain rule. With the definitions of the covariant and contravariant vector bases and their associated metric tensor representations now in place, a mixed covariant-contravariant projection of Maxwell's equations is developed in the next section.

11.5 EXPRESSING MAXWELL'S EQUATIONS USING THE BASIS VECTORS OF THE COMPUTATIONAL COORDINATE SYSTEM

From the point of view of deriving an FDTD discretization, it is better to work with $\mathcal{M}^{-1}$ and $\mathcal{E}^{-1}$ rather than with $\mathcal{M}$ and $\mathcal{E}$. If (11.6) and (11.7) are employed, the resulting discrete equations cannot be updated explicitly. Using the inverse permeability and permittivity tensors, the constitutive relations are given by:

$$\boldsymbol{H}(u^j,t) = \mathcal{M}^{-1}(u^j)\,\boldsymbol{B}(u^j,t) \tag{11.31}$$

$$\boldsymbol{E}(u^j,t) = \mathcal{E}^{-1}(u^j)\,\boldsymbol{D}(u^j,t) \tag{11.32}$$

Applying the inner product of the above two equations with the covariant basis vector $\boldsymbol{g}_j$ yields:

$$h_j = \sum_{s=1}^{3} \mu_{js}^{-1}\, b^s \tag{11.33}$$

$$e_j = \sum_{s=1}^{3} \varepsilon_{js}^{-1}\, d^s \tag{11.34}$$

where, for conciseness, the spatial and temporal dependence of all quantities was omitted. Similarly, the inner product of both sides of (11.4) and (11.5) with the contravariant basis vector $\boldsymbol{g}^i$ leads to the following component equations:

$$\frac{1}{\sqrt{g}}\left(\frac{\partial e_k}{\partial u^j} - \frac{\partial e_j}{\partial u^k}\right) = -\frac{\partial b^i}{\partial t} \tag{11.35}$$

$$\frac{1}{\sqrt{g}}\left(\frac{\partial h_k}{\partial u^j} - \frac{\partial h_j}{\partial u^k}\right) = \frac{\partial d^i}{\partial t} + j^i \tag{11.36}$$

where $g = \det(g_{ij})$. In the above two equations, the set of indices $\{i, j, k\}$ must be cycled through $\{1,2,3\}$, $\{3,1,2\}$, and $\{2,3,1\}$ to obtain the three components of the two curl equations. To gain a better understanding of why the curl operator is able to make one contravariant component out of two covariant components, the interested reader is referred to [35].

A mixed covariant-contravariant projection of Maxwell's equations is more useful than a purely covariant or contravariant representation. From an algebraic point of view, a mixed covariant-contravariant projection avoids the presence of the metric tensor in the curl equations. If b^i and d^i are converted to a covariant representation, or if e_j and h_j are converted to a contravariant representation, then, by virtue of (11.18), the curl equations in (11.35) and (11.36) would contain the metric tensor on either side. From a geometric point of view, using contravariant components for the electric and magnetic flux densities is a better choice than using covariant components because the contravariant basis vectors are normal to the coordinate surfaces that are used here to define material boundaries. Moreover, using covariant field components for the electric and magnetic field intensities is a better choice than using contravariant components because the covariant basis vectors are tangential to the coordinate lines that bound any given coordinate surface.

The component form of Maxwell's equations given in (11.33) – (11.36) can be discretized using staggered second-order accurate finite differences in the same way that Yee [36] discretized his Cartesian counterparts. For conciseness, the resulting discrete equations have not been included here but they can be found in [26]. The discrete equations can be leapfrogged in the usual manner, and, for visualization purposes, the obtained solution can be mapped to the Cartesian basis representation by using (11.18) and (11.20). As stated in the introduction, the advantage of employing (11.33) – (11.36) rather than a Cartesian projection becomes apparent when faced with the task of enforcing boundary conditions. To illustrate this point more clearly, the next section provides a concrete example from the microwave literature.

11.6 ENFORCING BOUNDARY CONDITIONS BY USING COORDINATE SURFACES IN THE COMPUTATIONAL COORDINATE SYSTEM

Consider the perfect electric conductor (PEC) waveguide bend described in Fig. 11.2. The following coordinate transformation suitable for modeling this structure was proposed in [37]:

$$x^1 = \begin{cases} u^1 & \text{for } u^1 \le \ell \\ \ell + \dfrac{r + u^2 - w/2}{\csc\left[\left(u^1 - \ell\right)/r\right]} & \text{for } \ell < u^1 \le \ell + r\varphi \\ \ell + \dfrac{r + u^2 - w/2}{\csc\varphi} + \dfrac{u^1 - \ell - r\varphi}{\sec\varphi} & \text{for } u^1 > \ell + r\varphi \end{cases} \tag{11.37}$$

$$x^2 = \begin{cases} u^2 & \text{for } u^1 \le \ell \\ \dfrac{w}{2} - r + \dfrac{r + u^2 - w/2}{\sec\left[\left(u^1 - \ell\right)/r\right]} & \text{for } \ell < u^1 \le \ell + r\varphi \\ \dfrac{w}{2} - r + \dfrac{r + u^2 - w/2}{\sec\varphi} + \dfrac{u^1 - \ell - r\varphi}{\csc\varphi} & \text{for } u^1 > \ell + r\varphi \end{cases} \tag{11.38}$$

$$x^3 = u^3 \tag{11.39}$$

where $0 \le u^1 \le 2\ell + r\varphi$, $0 \le u^2 \le w$, and $0 \le u^3 \le h$. This transformation creates an orthogonal computational coordinate system because of the cylindrical curvature of the bend. For other examples requiring nonorthogonal coordinate transformations, the reader is referred to [26].

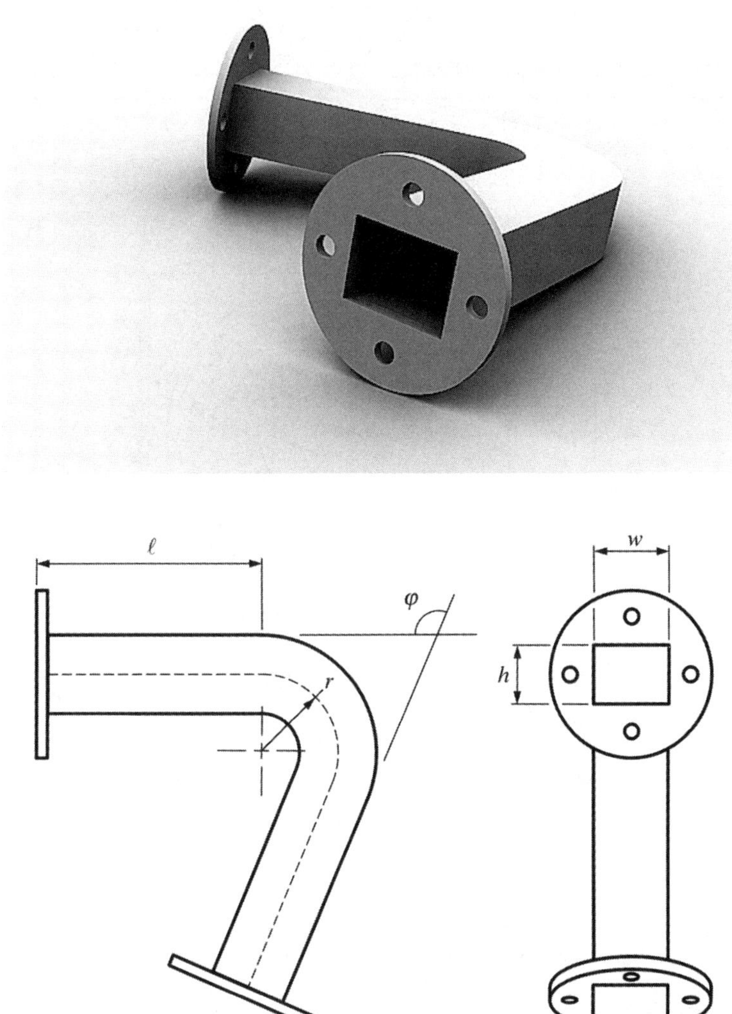

Fig. 11.2 PEC rectangular waveguide bend with the following geometric parameters: $w = 1$ cm, $h = 0.75$ cm, $r = 1$ cm, $\ell = 3$ cm, and $\varphi = 112.5°$.

Figure 11.3 illustrates the geometrical information contained in (11.37) – (11.39) via coordinate line plots for this coordinate transformation. From these two plots, it is easy to see that enforcing the PEC boundary conditions on the waveguide's sidewalls is much easier in the computational coordinate system than in the Cartesian coordinate system. In the computational coordinate system, the waveguide's sidewalls are described by $u^2 = 0$ and $u^2 = w$, and the boundary conditions on the waveguide's sidewalls are:

$$e_1(u^1, u^2 = 0, u^3) = 0 \quad \text{and} \quad e_1(u^1, u^2 = w, u^3) = 0 \tag{11.40}$$

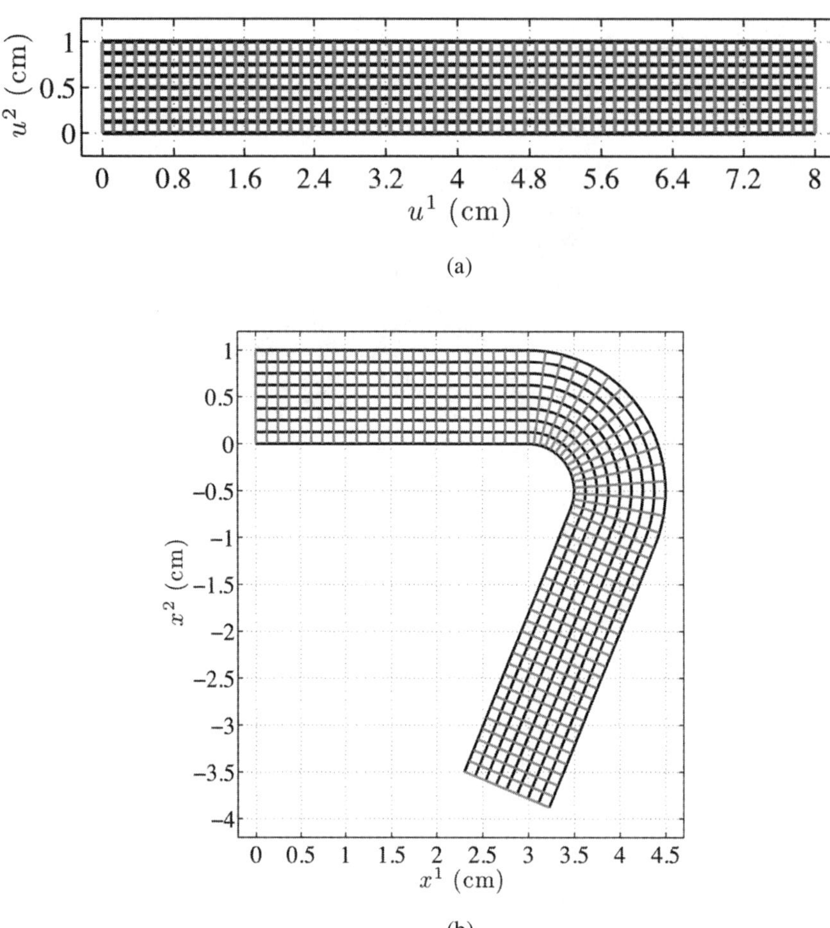

Fig. 11.3 Coordinate line plots for the coordinate transformation given in (11.37) – (11.39) to model a waveguide bend. The plots were made on a plane that cuts halfway through the height of the bend ($x^3 = u^3 = h/2$). The black and gray lines plotted in the computational coordinate system (a) map to the black and gray lines plotted in Cartesian coordinates (b).

Figure 11.4 displays sample FDTD-computed results for the waveguide bend of Fig. 11.2: snapshot visualizations of a Gaussian pulse propagating through the bend. These results were originally reported in [38] using the coordinate-transformation approach outlined here and illustrated in the coordinate line plots of Fig. 11.3.

We note that the enforcement of boundary conditions other than at a PEC surface, for example, continuity of tangential electric-field components at a dielectric interface, can also be greatly facilitated by constructing a coordinate transformation that simplifies the boundary description. Furthermore, to take advantage of the coordinate-invariant representation of Maxwell's equations given in (11.33) – (11.36), it is not necessary to construct the coordinate transformation analytically as is done here for the waveguide bend. The coordinate transformation can also be defined numerically by mapping a set of grid points in the Cartesian coordinate system to a uniformly spaced set of points in the u^j coordinate system.

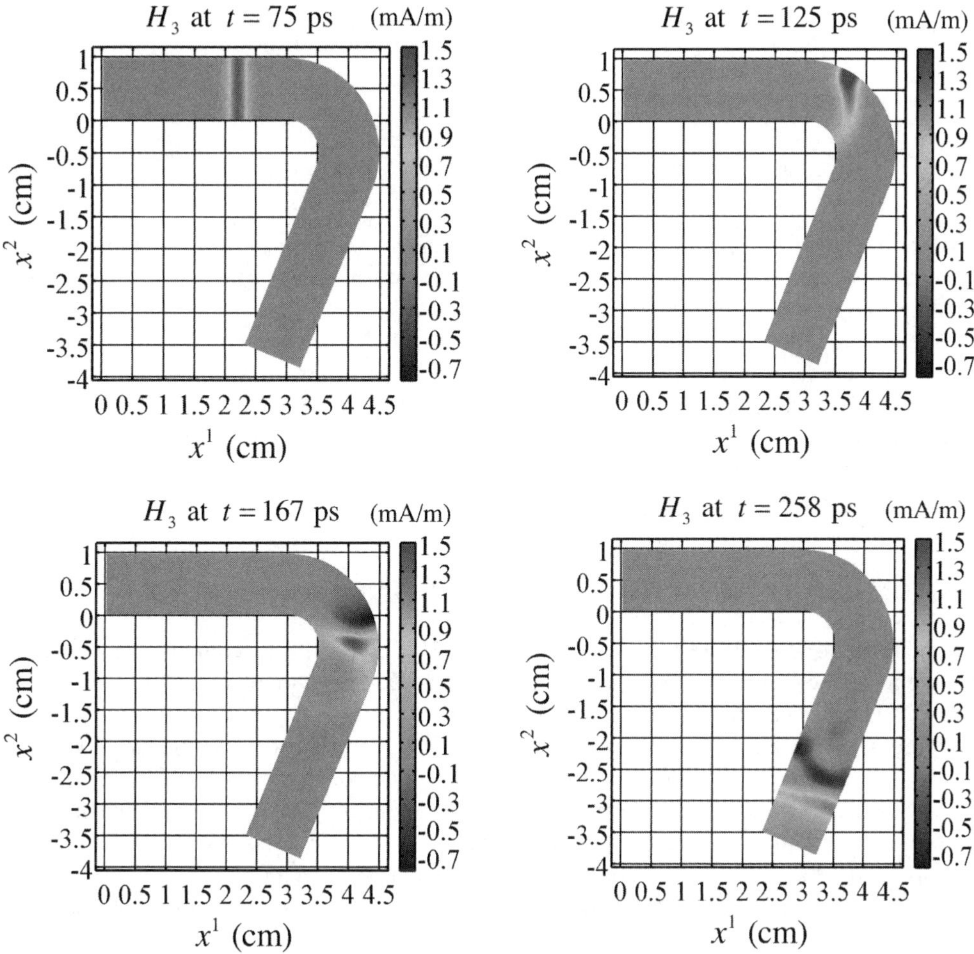

Fig. 11.4 Snapshots of an FDTD-computed Gaussian pulse propagating through the PEC waveguide bend of Fig. 11.2. The coordinate transformation of (11.37) – (11.39), visualized in Fig. 11.3, was used to model the waveguide bend.

Nevertheless, when defining a coordinate transformation numerically, an important consideration must be kept in mind. The grid used to define the coordinate transformation does not need to have the same density as the grid used to discretize Maxwell's equations. In fact, it is usually advantageous to use a dense grid with discretizations $\Delta\bar{u}^j$ to specify the coordinate transformation, and a relatively coarse grid with discretizations Δu^j for Maxwell's equations. The finite-difference approximations used to estimate the partial derivatives of the coordinate transformation are given by:

$$\frac{\partial x^i}{\partial u^j} \approx \frac{x^i(u^j + \Delta\bar{u}^j/2) - x^i(u^j - \Delta\bar{u}^j/2)}{\Delta\bar{u}^j} \quad \text{for } i, j = 1, 2, 3 \tag{11.41}$$

These must be computed only once, whereas the finite-difference approximations used to estimate the partial derivatives in Maxwell's equations, given by:

$$\frac{\partial e_i}{\partial u^j} \approx \frac{e^i(u^j + \Delta u^j/2) - e^i(u^j - \Delta u^j/2)}{\Delta u^j} \quad \text{for } i, j = 1, 2, 3 \tag{11.42}$$

and

$$\frac{\partial h_i}{\partial u^j} \approx \frac{h^i(u^j + \Delta u^j/2) - h^i(u^j - \Delta u^j/2)}{\Delta u^j} \quad \text{for } i, j = 1, 2, 3 \tag{11.43}$$

must be computed every time-step. Therefore, refining (decreasing) Δu^j is computationally much more expensive than refining $\Delta \bar{u}^j$. Setting $\Delta \bar{u}^j << \Delta u^j$ is a simple way to ensure that the available computational resources are not being wasted.

We note that the partial derivatives in (11.41), which define the metric tensor and the basis vectors, are used to describe the geometrical information of the given object. In contrast, the partial derivatives in (11.42) and (11.43) are used to describe electromagnetic wave propagation in the context of Maxwell's equations. This distinction between geometry and physics is key to understanding how to exploit high-order finite-difference operators effectively, and furthermore how to create multiphysics platforms.

11.7 CONNECTION WITH THE DESIGN OF ARTIFICIAL MATERIALS

Interestingly, the coordinate transformation in (11.37) – (11.39) was crafted with the idea of creating an artificial material [37] rather than modeling the curvature of a waveguide's sidewalls in an FDTD discretization. This suggests an interesting synergy: if constructed carefully, the coordinate transformation used to create an artificial material for a particular structure can also be used to incorporate the description of the structure's boundaries into an FDTD method.

Such dual usage of coordinate transformations has recently inspired significant interest in exploring novel applications of the invariance principle to FDTD techniques. In this context, an often-asked question is this: What do μ_{js}^{-1} and ε_{js}^{-1} in (11.33) and (11.34) mean in the context of designing artificial materials? A goal of this section is to provide a clear answer to this question.

11.7.1 Constitutive Tensors of a Simple Material

The tensor components μ_{js}^{-1} and ε_{js}^{-1} in (11.33) and (11.34) do not represent an artificial material unless they are specifically constructed to do so. Up to now, linearity is the only firm assumption that has been applied to the constitutive relations. The medium in question can still be inhomogeneous, anisotropic, or even dispersive. For the particular case of the waveguide bend of Fig. 11.2, the interior medium is vacuum. Hence, the permeability and permittivity inverse tensors are given by:

$$\mathcal{M}^{-1} = \mathbf{G}/\mu_0 \quad \text{and} \quad \mathcal{E}^{-1} = \mathbf{G}/\varepsilon_0 \tag{11.44}$$

In a covariant representation, the components of these two inverse tensors are:

$$\mu_{js}^{-1} = g_{js}/\mu_0 \quad \text{and} \quad \varepsilon_{js}^{-1} = g_{js}/\varepsilon_0 \quad \text{for } j, s = 1, 2, 3 \tag{11.45}$$

where the metric tensor components are specified by (11.37)–(11.39). The use of these material tensors enables accurate conformal FDTD modeling of the curved waveguide walls. As shown in Fig. 11.4, the transmitted pulse is a distorted version of the input.

11.7.2 Constitutive Tensors of an Artificial Material

Now, assume that the goal is to synthesize an artificial material within the waveguide bend that *allows the pulse to propagate without distortion.* In fact, this is theoretically possible, as illustrated in Fig. 11.5.

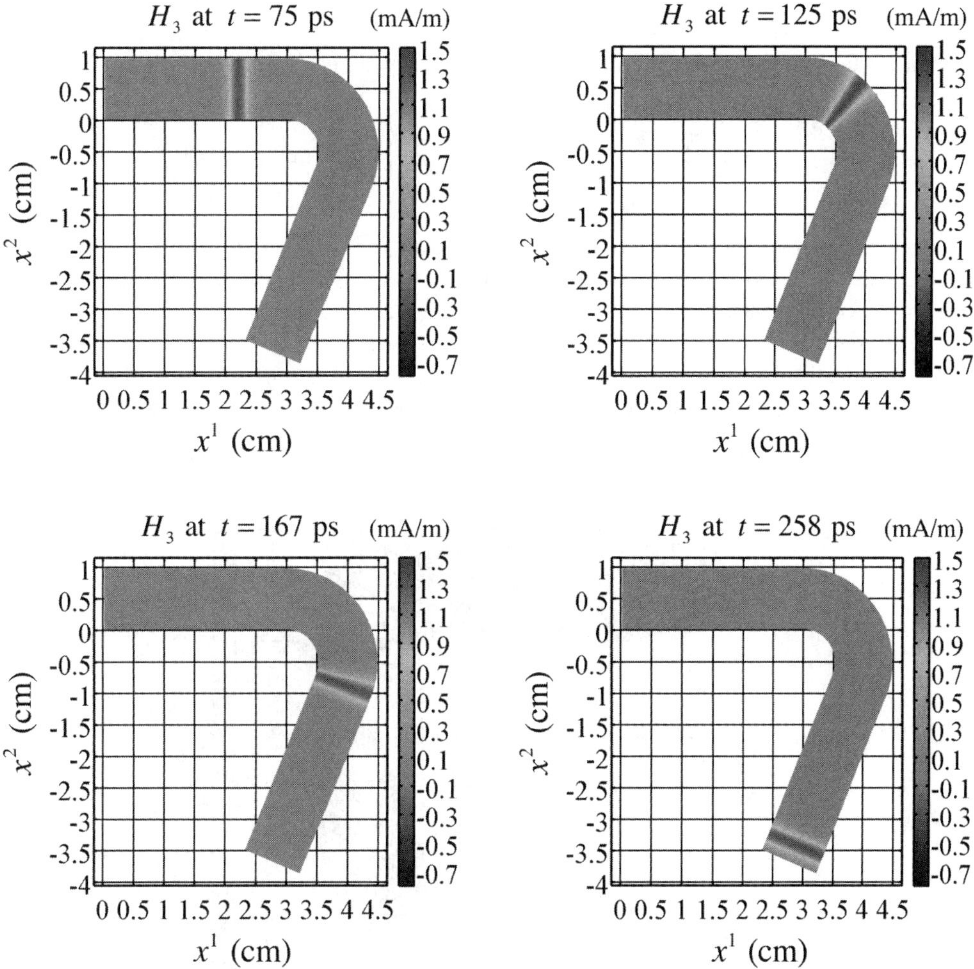

Fig. 11.5 Snapshots of an FDTD-computed Gaussian pulse propagating through the PEC waveguide bend of Fig. 11.2. Here, unlike the FDTD results shown in Fig. 11.4, the output pulse is undistorted because the artificial material specified by (11.46) together with (11.37)–(11.39) fills the bend.

To accomplish this goal for the coordinate transformation of (11.37) – (11.39), the required permeability and permittivity inverse tensors are given in covariant form by:

$$\mu_{js}^{-1} = \delta_{js} / \mu_0 \quad \text{and} \quad \varepsilon_{js}^{-1} = \delta_{js} / \varepsilon_0 \quad \text{for } j, s = 1, 2, 3 \tag{11.46}$$

The material tensors of (11.46) correspond to an inhomogeneous anisotropic medium. Even though this statement might strike the reader as being counterintuitive, it can be verified by mapping ε_{js}^{-1} to the vector basis of the Cartesian coordinate system. Applying the first relation in (11.24) to ε_{js}^{-1} yields:

$$\mathcal{E}_{pq}^{-1} = \sum_{j=1}^{3}\sum_{s=1}^{3} \frac{\partial u^j}{\partial x^p}\frac{\partial u^s}{\partial x^q}\varepsilon_{js}^{-1} \quad \text{for } p, q = 1, 2, 3 \tag{11.47}$$

Substituting the expression for ε_{js}^{-1} of (11.46) into (11.47) yields:

$$\mathcal{E}_{pq}^{-1} = \frac{1}{\varepsilon_0}\sum_{j=1}^{3}\sum_{s=1}^{3} \frac{\partial u^j}{\partial x^p}\frac{\partial u^s}{\partial x^q}\delta_{js} = \frac{1}{\varepsilon_0}\sum_{j=1}^{3} \frac{\partial u^j}{\partial x^p}\frac{\partial u^j}{\partial x^q} \quad \text{for } p, q = 1, 2, 3 \tag{11.48}$$

When the expressions for the coordinate transformation in (11.37)–(11.39) are substituted into (11.48), it can be verified that $\mathcal{E}_{pq}^{-1} \neq \delta_{pq}/\varepsilon_0$ for $p, q = 1, 2, 3$. Therefore, the covariant inverse permittivity components in (11.46) do correspond to an anisotropic material. Because the coordinate transformation in (11.37)–(11.39) is orthogonal, $\mathcal{E}_{pq}^{-1} = 0$ for $p \neq q$, and the material is only diagonally anisotropic. Of the three diagonal elements, only $\mathcal{E}_{11}^{-1}$ is inhomogeneous. A visualization of $\varepsilon_0\mathcal{E}_{11}^{-1}$ is provided in Fig. 11.6. The other two diagonal components are given by $\mathcal{E}_{22}^{-1} = \mathcal{E}_{33}^{-1} = 1/\varepsilon_0$.

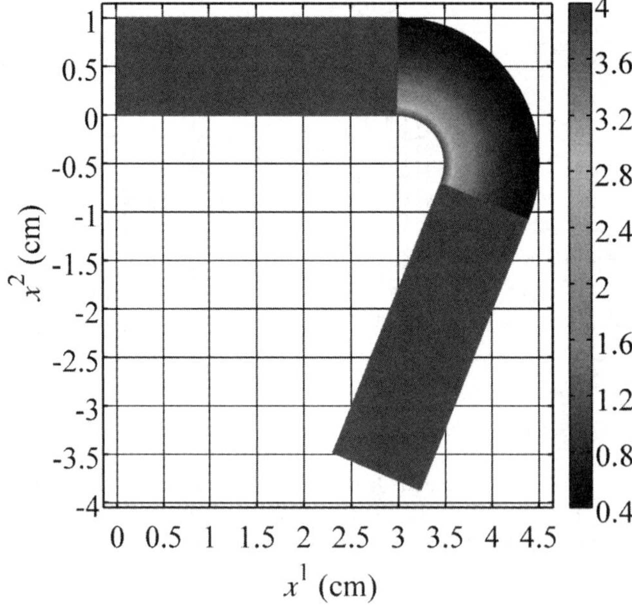

Fig. 11.6 Visualization of $\varepsilon_0\mathcal{E}_{11}^{-1}$ generated from (11.48) using the coordinate transformation in (11.37)–(11.39).

In contrast, if the expression for ε_{js}^{-1} of (11.45) is substituted into (11.47), we obtain:

$$\begin{aligned}\varepsilon_{pq}^{-1} &= \frac{1}{\varepsilon_0}\sum_{j=1}^{3}\sum_{s=1}^{3}\frac{\partial u^j}{\partial x^p}\frac{\partial u^s}{\partial x^q}g_{js} = \frac{1}{\varepsilon_0}\sum_{j=1}^{3}\sum_{s=1}^{3}\frac{\partial u^j}{\partial x^p}\frac{\partial u^s}{\partial x^q}\sum_{m=1}^{3}\frac{\partial x^m}{\partial u^j}\frac{\partial x^m}{\partial u^s} \\ &= \frac{1}{\varepsilon_0}\sum_{m=1}^{3}\sum_{j=1}^{3}\frac{\partial u^j}{\partial x^p}\frac{\partial u^m}{\partial x^j}\sum_{s=1}^{3}\frac{\partial u^s}{\partial x^q}\frac{\partial u^m}{\partial x^s} = \frac{1}{\varepsilon_0}\sum_{m=1}^{3}\delta_p^m\delta_q^m = \frac{\delta_{pq}}{\varepsilon_0} \quad \text{for } p, q = 1, 2, 3\end{aligned} \tag{11.49}$$

Whether the material represented by the permeability and permittivity inverse tensors of (11.46) can be physically (as opposed to theoretically) fabricated is beyond the scope of the present discussion. The interested reader is referred to [37].

With the connection between the two different uses of a coordinate transformation now established, the next section discusses the use of the relativity principle to introduce time-dependent discretizations for FDTD.

11.8 TIME-VARYING DISCRETIZATIONS

As stated in the introduction, the invariance principle can be exploited to incorporate time-varying discretizations into an FDTD method through a 4-D coordinate transformation [as per (11.8)] that can entangle space with time. To explain more clearly how this works, consider the 2-D coordinate transformation depicted in Fig. 11.7.

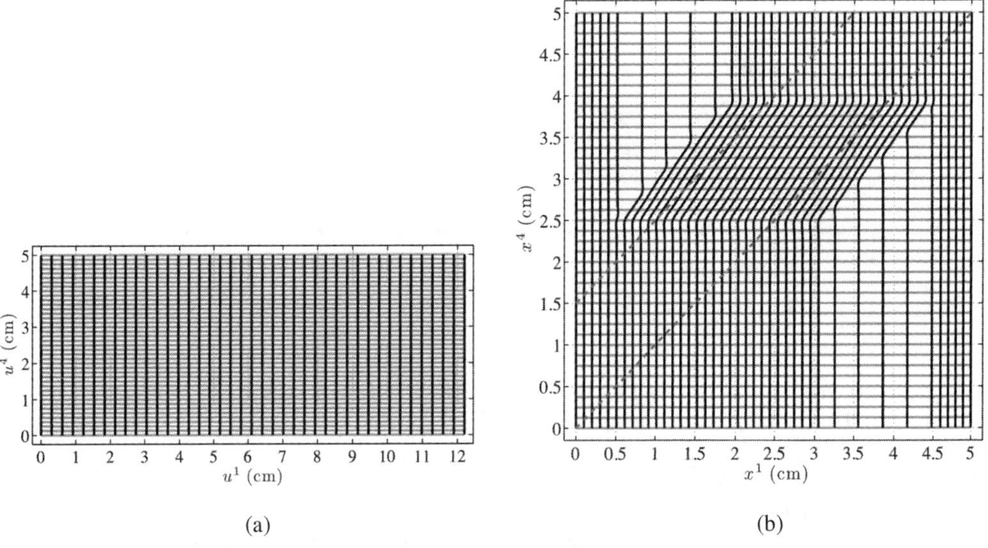

Fig. 11.7 Coordinate line plots for a 2-D time-dependent coordinate transformation that was constructed with the purpose of creating a moving set of grid points that tracks a Gaussian pulse as it propagates through free space. The solid black and gray lines plotted in the computational coordinate system (a) map to the solid black and gray lines plotted in Cartesian coordinates (b). The Gaussian pulse propagates through the window marked in (b) by the dashed lines. The units are consistent with the choice $x^4 = c_0 t$.

The coordinate transformation of Fig. 11.7 was used in [34] to generate a moving set of grid points that track a Gaussian pulse propagating in vacuum. The solid black and gray lines plotted in the computational coordinate system map to the solid black and gray lines plotted in Cartesian coordinates. Observe that the u^4 coordinate contains both time and space information. As one moves forward along a u^4 coordinate line (in solid black), one moves forward in both time and space. The classical distinction between time and space cannot be made in the computational coordinate system, but it remains valid in the Cartesian (or Minkowski) coordinate system.

By creating a uniform FDTD discretization (with respect to both u^1 and u^4) that advances forward with respect to the time-like u^4 coordinate, some of the grid points travel along with the Gaussian pulse. This point is illustrated in Fig. 11.8. As pointed out in [34], clustering the grid points in the region of the pulse enhances the resolution of the discretization grid. For more details on how the results in Fig. 11.8 were obtained, the reader is referred to [34] where a 2-D FDTD discretization of the general relativistic formulation of Maxwell's equations is provided.

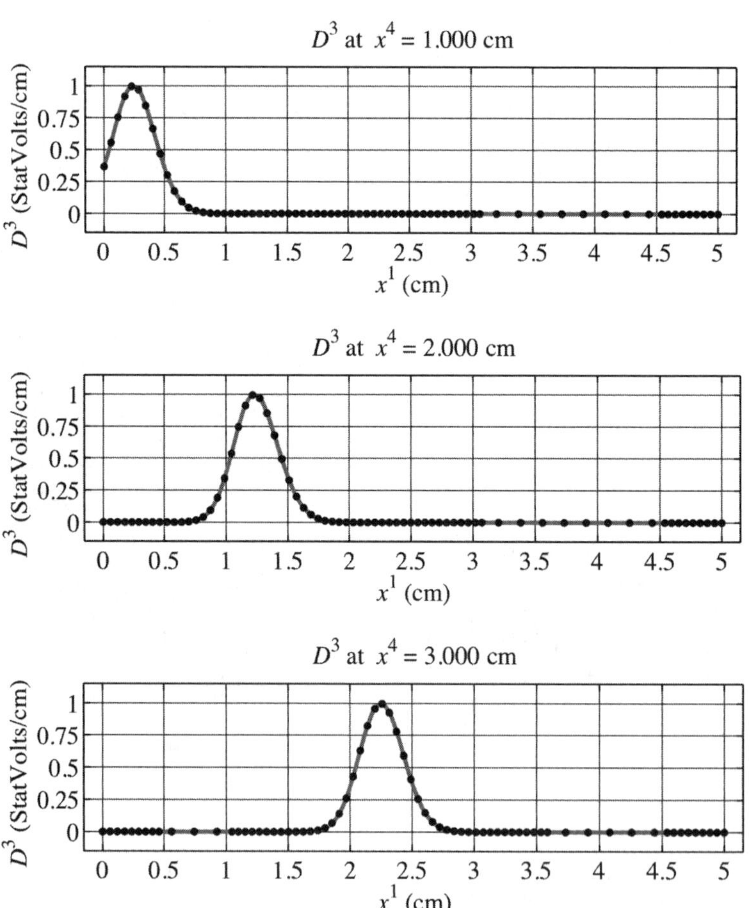

Fig. 11.8 FDTD-computed snapshots of a Gaussian pulse propagating in vacuum. The results were obtained using the coordinate transformation in Fig. 11.6 using an FDTD discretization that advances forward with respect to the u^4 coordinate. The black dots mark the positions of the grid points at a particular point in time. The units are consistent with the choice $x^4 = c_0 t$.

Finally, we observe that the coordinate transformation described in Fig. 11.7 can be combined with the coordinate transformation described earlier in Fig. 11.3 to create a time-dependent coordinate transformation that generates a dense grid region that tracks the Gaussian pulse in Fig. 11.4. Coordinate line plots for this transformation are displayed in Fig. 11.9.

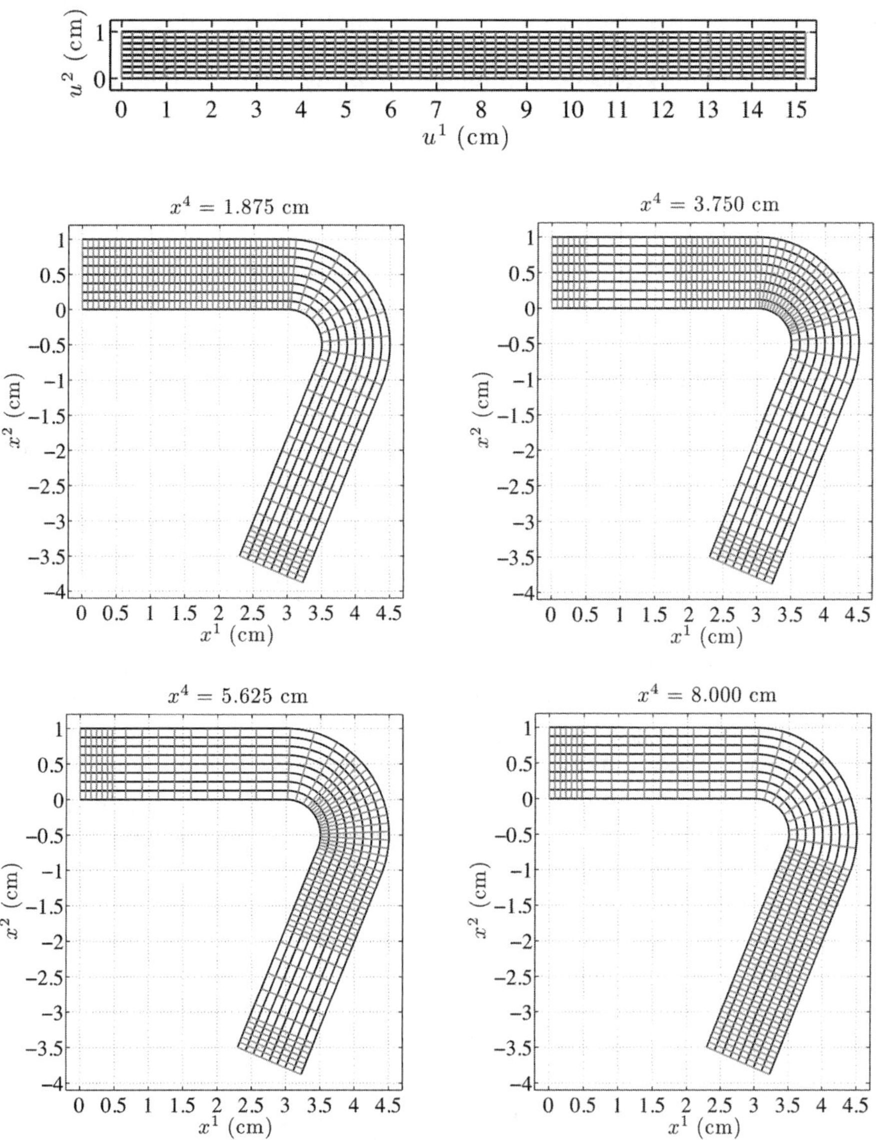

Fig. 11.9 Coordinate line plots for a time-dependent coordinate transformation used to generate a moving grid where a cluster of grid points tracks a Gaussian pulse (Fig. 11.4) that propagates in a waveguide bend (Fig. 11.2). The plots were made in a plane located midway through the height of the bend ($x^3 = u^3 = h/2$). The black and gray lines plotted in the computational coordinate system (top plot) map to the black and gray lines plotted in Cartesian coordinates (four lower plots). Note that the mapping depends on the time coordinate x^4.

11.9 CONCLUSION

Until recently, FDTD grids have been primarily constructed in Cartesian coordinates, and to a lesser degree, cylindrical and spherical coordinates, to exploit the algebraic simplifications inherent in these systems. However, as this chapter demonstrates, employing a coordinate-system-independent representation of Maxwell's equations based on the invariance principle provides powerful additional FDTD capabilities. These include conformal modeling of curved material boundaries, incorporation of artificial materials providing novel wave-propagation characteristics, and time-dependent discretizations permitting high-resolution tracking of moving electromagnetic pulses. Much more work in this area is expected in the coming years.

REFERENCES

[1] Pendry, J. B., D. Schurig, and D. R. Smith, "Controlling electromagnetic fields," *Science*, Vol. 312, 2006, pp. 1780–1782.

[2] Leonhardt, U., "Optical conformal mapping," *Science*, Vol. 312, 2006, pp. 1777–1780.

[3] Ward, A. J., and J. B. Pendry, "Refraction and geometry in Maxwell's equations," *J. Modern Optics*, Vol. 43, 1996, pp. 773–793.

[4] Ward, A. J., and J. B. Pendry, "Calculating photonic Green's functions using a nonorthogonal finite-difference time-domain method," *Physical Review Lett. B*, Vol. 58, 1998, pp. 7252–7259.

[5] Jurgens, T. G., A. Taflove, K. Umashankar, and T. G. Moore, "Finite-difference time-domain modeling of curved surfaces," *IEEE Trans. Antennas & Propagation*, Vol. 40, 1992, pp. 357–366.

[6] Railton, C. J., I. J. Craddock, and J. B. Schneider, "Improved locally distorted CP-FDTD algorithm with provable stability," *Electronics Lett.*, Vol. 31, 1995, pp. 1585–1586.

[7] Kaneda, N., B. Houshmand, and T. Itoh, "FDTD analysis of dielectric resonators with curved surfaces," *IEEE Trans. Microwave Theory & Techniques*, Vol. 45, 1997, pp. 1645–1649.

[8] Hao, Y., and C. J. Railton, "Analyzing electromagnetic structures with curved boundaries on Cartesian FDTD meshes," *IEEE Trans. Microwave Theory & Techniques*, Vol. 46, 1998, pp. 82–88.

[9] Nadobny, J., D. Sullivan, P. Wust, M. Seebass, P. Deuflhard, and R. Felix, "A high-resolution interpolation at arbitrary interfaces for the FDTD method," *IEEE Trans. Microwave Theory & Techniques*, Vol. 46, 1998, pp. 1759–1766.

[10] Dey, S., and R. Mittra, "A conformal finite-difference time-domain technique for modeling cylindrical dielectric resonators," *IEEE Trans. Microwave Theory & Techniques*, Vol. 47, 1999, pp. 1737–1739.

[11] Kosmanis, T. I., and T. D. Tsiboukis, "A systematic and topologically stable conformal finite-difference time-domain algorithm for modeling curved dielectric interfaces in three dimensions," *IEEE Trans. Microwave Theory & Techniques*, Vol. 51, 2003, pp. 839–847.

[12] Mohammadi, A., H. Nadgaran, and M. Agio, "Contour-path effective permittivities for the two-dimensional finite-difference time-domain method," *Optics Express*, Vol. 13, 2005, pp. 10367–10378.

[13] Schuhmann, R., I. A. Zagorodnov, and T. Weiland, "A simplified conformal (SC) method for modeling curved boundaries in FDTD without time step Reduction," *Proc. 2006 IEEE International Microwave Symposium*, June 2006, pp. 177–180.

[14] Zhao, S., "High order vectorial analysis of waveguides with curved dielectric interfaces," *IEEE Microwave & Wireless Components Lett.*, Vol. 19, 2009, pp. 266–268.

[15] Shyroki, M. D., "Modeling of sloped interfaces on a Yee grid," *IEEE Trans. Antennas & Propagation*, Vol. 59, 2011, pp. 3290–3295.

[16] Holland, R., "Finite-difference solution of Maxwell's equations in generalized nonorthogonal coordinates," *IEEE Trans. Nuclear Science*, Vol. 30, 1983, pp. 4589–4591.

[17] Stratton, J. A., *Electromagnetic Theory*, Piscataway, NJ: Wiley-IEEE, 2007, pp. 38–50.

[18] Fusco, M., "FDTD algorithm in curvilinear coordinates," *IEEE Trans. Antennas & Propagation*, Vol. 38, 1990, pp. 76–88.

[19] Fusco, M., M. V. Smith, and L. W. Gordon, "A three-dimensional FDTD algorithm in curvilinear coordinates," *IEEE Trans. Antennas & Propagation*, Vol. 39, 1991, pp. 1463–1471.

[20] Lee, J. F., R. Palandech, and R. Mittra, "Modeling three-dimensional discontinuities in waveguides using nonorthogonal FDTD algorithm," *IEEE Trans. Microwave Theory & Techniques*, Vol. 40, 1992, pp. 346–352.

[21] Madsen, N., "Divergence preserving discrete surface integral methods for Maxwell's equations using nonorthogonal unstructured grids," *J. Computational Physics*, Vol. 119, 1995, pp. 34–45.

[22] Schuhmann, R., and T. Weiland, "A stable interpolation technique for FDTD on non-orthogonal grids," *International. J. Numerical Modeling*, Vol. 11, 1998, pp. 299–306.

[23] Gedney, S. D., and J. A. Roden, "Numerical stability of nonorthogonal FDTD methods," *IEEE Trans. Antennas & Propagation*, Vol. 48, 2000, pp. 231–239.

[24] Russer, J. A., P. S. Sumant, and A. C. Cangellaris, "A Lagrangian approach for the handling of curved boundaries in the finite-difference time-domain method," in *Proc. 2007 IEEE International Microwave Symposium*, June 2007, pp. 717–720.

[25] Kantartzis, N. V., T. I. Kosmanis, T. V. Yioultsis, and T. D. Tsiboukis, "A nonorthogonal higher-order wavelet-oriented FDTD technique for 3-D waveguide structures on generalized curvilinear grids," *IEEE Trans. Magnetics*, Vol. 37, 2001, pp. 3264–3268.

[26] Armenta, R. B., and C. D. Sarris, "A general procedure for introducing structured nonorthogonal discretization grids into high-order finite-difference time-domain methods," *IEEE Trans. Microwave Theory & Techniques*, Vol. 58, 2010, pp. 1818–1829.

[27] Hunt, B. J., *The Maxwellians*, Ithaca, NY: Cornell University Press, 1991.

[28] Einstein, A., and H. Minkowski, *The Principle of Relativity: Original Papers by A. Einstein and H. Minkowski. Translated into English by M. N. Saha and S. N. Bose; with a Historical Introduction by P. C. Mahalanobis*, Calcutta, India: University of Calcutta, 1920.

[29] Kim, I. S., and W. J. R. Hoefer, "A local mesh refinement algorithm for the time-domain finite-difference method using Maxwell's curl equations," *IEEE Trans. Microwave Theory & Techniques*, Vol. 38, 1990, pp. 812–815.

[30] Liu, Y., and C. D. Sarris, "Efficient modeling of microwave integrated-circuit geometries via a dynamically adaptive mesh refinement FDTD technique," *IEEE Trans. Microwave Theory & Techniques*, Vol. 54, 2006, pp. 689–703.

[31] Harfoush, F., A. Taflove and G. A. Kriegsmann, "A numerical technique for analyzing electromagnetic wave scattering from moving surfaces in one and two dimensions," *IEEE Trans. Antennas & Propagation*, Vol. 37, 1989, pp. 55–63.

[32] Mueller, U., A. Beyer, and W. J. R. Hoefer, "Moving boundaries in 2-D and 3-D TLM simulations realized by recursive formulas," *IEEE Trans. Microwave Theory & Techniques*, Vol. 40, 1992, pp. 2267–2271.

[33] Russer, J. A., and A. C. Cangellaris, "An efficient methodology for the modeling of electromagnetic wave phenomena in domains with moving boundaries," *Proc. 2008 IEEE International Microwave Symposium*, June 2008, pp. 157–160.

[34] Armenta, R. B., and C. D. Sarris, "Exploiting the relativistic formulation of Maxwell's equations to introduce moving grids into finite-difference time-domain solvers," *Proc. 2010 IEEE International Microwave Symposium*, May 2010, pp. 93–96.

[35] Itskov, M., *Tensor Algebra and Tensor Analysis for Engineers*, New York: Springer, 2007, chaps. 1, 2.

[36] Yee, K. S., "Numerical solution of initial boundary value problems involving Maxwell's equations in isotropic media," *IEEE Trans. Antennas & Propagation*, Vol. AP-14, 1966, pp. 302–307.

[37] Donderici, B., and F. L. Teixeira, "Metamaterial blueprints for reflectionless waveguide bends," *IEEE Microwave & Wireless Components Lett.*, Vol. 18, 2008, pp. 233–235.

[38] Armenta, R. B., and C. D. Sarris, "A general methodology for introducing structured nonorthogonal grids into high-order finite-difference time-domain methods," *Proc. 2009 IEEE International Microwave Symposium*, June 2009, pp. 257–260.

SELECTED BIBLIOGRAPHY

The introduction states that "...the idea of employing a coordinate-invariant representation of Maxwell's equations to incorporate curved material boundaries well preceded the idea of exploiting the same representation to design artificial materials." In fact, the first papers that independently recognized this interesting property of Maxwell's equations appeared much earlier than the recent highly cited work reported in [1–4], and are listed below by date of publication:

Van Dantzig, D., "The fundamental equations of electromagnetism, independent of metrical geometry," *Proc. Cambridge Philosophical Society*, Vol. 30, 1934, pp. 421–427.

Dolin, S. L., "On a possibility of comparing three-dimensional electromagnetic systems with inhomogeneous filling," *Izv. Vyssh. Uchebn. Zaved. Radiofiz.*, Vol. 4, 1961, pp. 964–967 (in Russian).

Schonberg, M., "Electromagnetism and gravitation," *Brazilian J. Physics*, Vol. 1, 1971, pp. 91–122.

Lax, M., and D. F. Nelson, "Maxwell's equations in material form," *Physical Review B*, Vol. 13, 1976, pp. 1777–1784.

Deschamps, G., "Electromagnetics and differential forms," *Proc. IEEE*, Vol. 69, 1981, pp. 676–696.

This property has also been independently exploited for the various finite methods, including irregular-grid finite differences and finite elements, and in the design of perfectly matched layers in constitutive form for general curvilinear coordinate systems. References include:

Kotiuga, P. R., "Helicity functionals and metric invariance in three dimensions," *IEEE Trans. Magnetics*, Vol. 25, 1989, pp. 2813–2186.

Teixeira, F. L., and W. C. Chew, "Lattice electromagnetic theory from a topological viewpoint," *J. Mathematical Physics*, Vol. 40, 1999, pp. 169–187.

Teixeira, F. L., and W. C. Chew, "Differential forms, metrics, and the reflectionless absorption of electromagnetic waves," *J. Electromagnetic Waves and Applications*, Vol. 13, 1999, pp. 665–686.

Chapter 12

FDTD Modeling of Nondiagonal Anisotropic Metamaterial Cloaks[1]

Naoki Okada and James B. Cole

12.1 INTRODUCTION

Metamaterials are artificial materials that can be used to manipulate electromagnetic waves in novel ways [1, 2]. Pendry et al. developed a method to design metamaterials via coordinate transformations in what is called *transformation optics* [3, 4]. Transformation optics can be used to design metamaterials for such devices as invisibility cloaks [3, 5–14], concentrators [11], rotation coatings [15], polarization controllers [16–18], waveguides [19–23], reflectionless waveguide bends [24, 25], wave shape conversion [26], object illusions [27–29], and optical black holes [30, 31]. Some of these designs have been experimentally demonstrated [32–39].

In the literature, metamaterial designs have usually been validated by conducting frequency-domain finite-element computer simulations [5, 7–9, 11, 14–16, 19, 22, 23, 27–29]. However, this approach is inconvenient if modeling data over a wide band of frequencies are required, because only a single frequency can be considered in each finite-element simulation run.

It is well known that the finite-difference time-domain (FDTD) method naturally provides broadband data via Fourier transformation of the computed temporal waveforms [40]. This has prompted a number of reported FDTD models of metamaterials, including dispersionless cloaks [41] and diagonal (uniaxial) anisotropic cloaks [42–49]. Classic FDTD field updates can be used to model dispersionless cloaks because these have permittivity ε and permeability μ parameters that are larger than one. To model diagonal anisotropic cloaks having $\varepsilon, \mu < 1$, a suitable frequency-dependent FDTD time-stepping algorithm is applied since such materials must be dispersive due to considerations of causality [50, 51].

However, without proper care, the direct application of FDTD to simulate the broad class of metamaterials characterized by nondiagonal anisotropic parameters [7–9, 19–21, 27–29] is prone to numerical instabilities. Reference [52] reported a technique to solve this instability problem by mapping eigenvalues of ε, μ to dispersion models, thereby ensuring that the numerically derived FDTD equations are exactly symmetric. A validation study was presented in [52] wherein this technique was applied to model a two-dimensional (2-D) elliptical cloak comprised of a sample nondiagonal anisotropic metamaterial [2, 9]. This chapter reviews the basis, formulation, and validation of the technique reported in [52].

---

[1]This chapter is adapted from Ref. [52], N. Okada and J. B. Cole, "FDTD modeling of a cloak with a nondiagonal permittivity tensor," *ISRN Optics*, Vol. 2012, 536209, ©2012 Okada and Cole.

12.2 STABLE FDTD MODELING OF METAMATERIALS HAVING NONDIAGONAL PERMITTIVITY TENSORS

The numerical stability of the FDTD method is given by the Courant limit [40]. The goal of [52] was to derive a stable FDTD modeling scheme for metamaterials having nondiagonal permittivity tensors. According to transformation optics [4], anisotropic metamaterial parameters are expressed as:

$$[\varepsilon^{ij}] = [\mu^{ij}] = \pm\sqrt{g}\,[g^{ij}] \tag{12.1}$$

where $[\varepsilon^{ij}]$ is the relative permittivity, $[\mu^{ij}]$ is the relative permeability, $[g^{ij}]$ is the metric tensor, and $g = \det[g^{ij}]$. The eigenvalues λ of $[\varepsilon^{ij}]$ and $[\mu^{ij}]$ for an eigenvector $\mathbf{V}$ are defined by:

$$[\varepsilon^{ij}]\mathbf{V} = [\mu^{ij}]\mathbf{V} = \lambda\mathbf{V} \tag{12.2}$$

From (12.2), the speed of light in a metamaterial is given by $c = c_0/\lambda$, where c_0 is the speed of light in vacuum. Then, the Courant limit in the metamaterial is expressed as:

$$\Delta t \le \frac{\lambda h}{c_0\sqrt{d}} \tag{12.3}$$

where Δt is the time-step, h is the grid-cell size (assumed uniform for simplicity), and $d = 1, 2,$ or 3 dimensions. Thus, numerically stable FDTD modeling of a metamaterial depends on the eigenvalues λ.

Reference [52] determined that it is possible to construct stable FDTD models of metamaterials having nondiagonal $[\varepsilon^{ij}]$ and $[\mu^{ij}]$. A key observation is that $[\varepsilon^{ij}]$ and $[\mu^{ij}]$ are symmetric because they are constructed from the symmetric metric tensor $[g^{ij}]$. Consequently, $[\varepsilon^{ij}]$ and $[\mu^{ij}]$ have real eigenvalues with orthogonal eigenvectors, and are thus diagonalizable. Hence, the crucial first step is to find the eigenvalues λ as defined by (12.2), and *diagonalize* $[\varepsilon^{ij}]$ and $[\mu^{ij}]$. After this diagonalization, any of the previously reported FDTD algorithms for purely diagonal cases [42–49] can be applied. In these algorithms, diagonal elements having values less than unity are treated using a suitable dispersive time-stepping scheme to avoid violating causality [50, 51] and the FDTD stability limit [53, 54].

12.3 FDTD FORMULATION OF THE ELLIPTIC CYLINDRICAL CLOAK

Two designs of elliptic cylindrical cloaks have been proposed. One has diagonal $[\varepsilon^{ij}]$ and $[\mu^{ij}]$ in orthonormal elliptic cylindrical coordinates [12, 13]. In the other, $[\varepsilon^{ij}]$ and $[\mu^{ij}]$ are nondiagonal in Cartesian coordinates [2, 9]. Reference [52] derived a 2-D FDTD formulation for the latter in the field component set $\{E_x, E_y, H_z\}$.

12.3.1 Diagonalization

Figure 12.1 depicts the 2-D elliptic cylindrical cloak in (a) Cartesian coordinates, and (b) transformed coordinates. The major axis of the cloak is horizontal when $k > 1$, and vertical when $k < 1$. In the cloak region, $ka \le (x^2 + k^2y^2)^{1/2} \le kb$, the material parameters are expressed by:

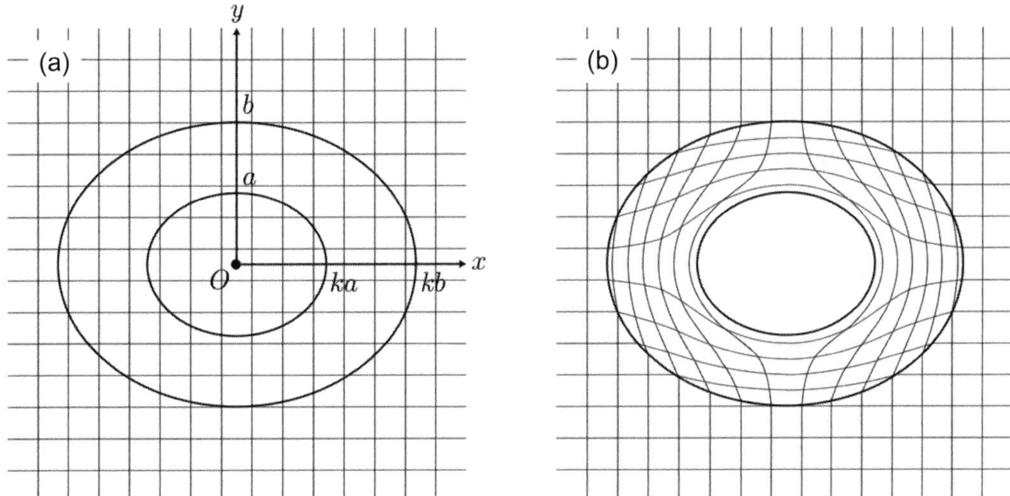

Fig. 12.1 Elliptic cylindrical cloak: (a) Cartesian coordinates; (b) transformed coordinates. *Source:* N. Okada and J. B. Cole, *ISRN Optics*, Vol. 2012, 536209, doi:10.5402/2012/536209, ©2012 Okada and Cole.

$$[\varepsilon^{ij}] = [\mu^{ij}] = \begin{bmatrix} \varepsilon_{xx} & \varepsilon_{xy} & 0 \\ \varepsilon_{xy} & \varepsilon_{yy} & 0 \\ 0 & 0 & \varepsilon_{zz} \end{bmatrix} \tag{12.4}$$

where:

$$\varepsilon_{xx} = \frac{r}{r-ka} + \left[\frac{k^2a^2R^2 - 2kar^3}{(r-ka)r^5}\right]\cdot x^2 \tag{12.5a}$$

$$\varepsilon_{xy} = \left[\frac{k^2a^2R^2 - ka(1+k^2)r^3}{(r-ka)r^5}\right]\cdot xy \tag{12.5b}$$

$$\varepsilon_{yy} = \frac{r}{r-ka} + \left[\frac{k^2a^2R^2 - 2k^3ar^3}{(r-ka)r^5}\right]\cdot y^2 \tag{12.5c}$$

$$\varepsilon_{zz} = \left(\frac{b}{b-a}\right)^2 \frac{r-ka}{r} \tag{12.5d}$$

with $r = (x^2 + k^2y^2)^{1/2}$ and $R = (x^2 + k^4y^2)^{1/2}$. From (12.4) and (12.5), we find three eigenvalues:

$$\lambda_1 = \frac{\sqrt{\alpha} - 1}{\sqrt{\alpha} + 1}, \qquad \lambda_2 = \frac{1}{\lambda_1}, \qquad \lambda_3 = \varepsilon_{zz} \tag{12.6}$$

where:

$$\alpha = \frac{\varepsilon_{xx} + \varepsilon_{yy} + 2}{\varepsilon_{xx} + \varepsilon_{yy} - 2} \tag{12.7}$$

Since $[\varepsilon^{ij}]$ is symmetric, it is diagonalized by the eigenvalue matrix $\mathbf{\Lambda}$ and its orthogonal matrix $\mathbf{P}$ as follows:

$$[\varepsilon^{ij}] = \mathbf{P\Lambda P}^{\mathrm{T}} \tag{12.8}$$

where:

$$\mathbf{\Lambda} = \begin{bmatrix} \lambda_1 & 0 & 0 \\ 0 & \lambda_2 & 0 \\ 0 & 0 & \lambda_3 \end{bmatrix} \tag{12.9a}$$

$$\mathbf{P} = \begin{bmatrix} \xi_1 & \xi_2 & 0 \\ -\xi_2 & \xi_1 & 0 \\ 0 & 0 & 1 \end{bmatrix} \tag{12.9b}$$

with $\xi_1 = (1+\beta^2)^{-1/2}$, $\xi_2 = \beta\xi_1$, and $\beta = (\lambda_2 - \varepsilon_{yy})/\varepsilon_{xy}$.

12.3.2 Mapping Eigenvalues to a Dispersion Model

From (12.6) and (12.7), λ_1 and λ_3 have values less than one in the cloak region ($ka \le r \le kb$). Based on previous FDTD modeling of diagonal anisotropic cloaks [42–49], Ref. [52] mapped these eigenvalues to the Drude dispersion model as:

$$\lambda_i = \varepsilon_{\infty_i} - \frac{\omega_{p_i}^2}{\omega^2 - j\omega\gamma_i} \qquad (i = 1, 3) \tag{12.10}$$

where ω is the angular frequency, ε_{∞_i} is the infinite-frequency permittivity, ω_{p_i} is the plasma frequency, and γ_i is the collision frequency. For simplicity, Ref. [52] considered the lossless case, $\gamma_i = 0$. Then, the plasma frequencies are given by $\omega_{p_i} = \omega(\varepsilon_{\infty_i} - \lambda_i)^{1/2}$, where $\varepsilon_{\infty_i} = \max(1, \lambda_i)$. Since $\lambda_2 \ge 1$ in the cloak region, $\omega_{p_2} = 0$ and $\varepsilon_{\infty_2} = \lambda_2$.

12.3.3 FDTD Discretization

Using the diagonalized material parameters of (12.8) and (12.9), with eigenvalues mapped to a Drude dispersion model as per (12.10), Ref. [52] could commence the derivation of a numerically stable FDTD time-stepping algorithm. As always, the starting point is the foundation of classical electrodynamics, Maxwell's time-dependent curl equations:

$$\nabla \times \boldsymbol{H} = \frac{\partial \boldsymbol{D}}{\partial t} \tag{12.11}$$

$$\nabla \times \boldsymbol{E} = -\frac{\partial \boldsymbol{B}}{\partial t} \tag{12.12}$$

where $\boldsymbol{E}$ is the electric field intensity, $\boldsymbol{D}$ is the electric flux density, $\boldsymbol{H}$ is the magnetic field intensity, and $\boldsymbol{B}$ is the magnetic flux density.

As stated earlier, Ref. [52] considered a 2-D case wherein the electromagnetic fields reduce to three nonzero components: E_x, E_y, and H_z (D_x, D_y, and B_z). Standard leapfrog FDTD time-stepping equations can be used to update D_x, D_y, and B_z [40]:

$$D_x^{n+1} = D_x^n + \frac{\Delta t}{h}\partial_y\left(H_z^{n+1/2}\right) \tag{12.13}$$

$$D_y^{n+1} = D_y^n - \frac{\Delta t}{h}\partial_x\left(H_z^{n+1/2}\right) \tag{12.14}$$

$$B_z^{n+3/2} = B_z^{n+1/2} - \frac{\Delta t}{h}\left[\partial_x\left(E_y^{n+1}\right) - \partial_y\left(E_x^{n+1}\right)\right] \tag{12.15}$$

where ∂_x and ∂_y are spatial difference operators defined by:

$$\partial_x\left[f(x,y)\right] = f(x+h/2,\, y) - f(x-h/2,\, y) \tag{12.16a}$$

$$\partial_y\left[f(x,y)\right] = f(x,\, y+h/2) - f(x,\, y-h/2) \tag{12.16b}$$

With updated values of D_x and D_y available after applying (12.13) and (12.14), one needs to find the resulting updated values of E_x and E_y. To derive an algorithm for this purpose, consider the constitutive relation:

$$\boldsymbol{D} = \varepsilon_0[\varepsilon^{ij}]\boldsymbol{E} \tag{12.17}$$

where ε_0 is the vacuum permittivity. From (12.8), this yields:

$$\varepsilon_0 \boldsymbol{E} = [\varepsilon^{ij}]^{-1}\boldsymbol{D} = \mathbf{P}\boldsymbol{\Lambda}^{-1}\mathbf{P}^{\mathrm{T}}\boldsymbol{D} \tag{12.18}$$

Substituting (12.9a) in (12.18) and multiplying both sides by $\lambda_1\lambda_2$, one obtains:

$$\varepsilon_0\lambda_1\lambda_2 E_x = (\lambda_1\xi_2^2 + \lambda_2\xi_1^2)D_x + \xi_1\xi_2(\lambda_1 - \lambda_2)D_y \tag{12.19}$$

$$\varepsilon_0\lambda_1\lambda_2 E_y = (\lambda_1\xi_1^2 + \lambda_2\xi_2^2)D_y + \xi_1\xi_2(\lambda_1 - \lambda_2)D_x \tag{12.20}$$

where ξ_1 and ξ_2 are as defined previously. Substituting the Drude model for λ_1 as shown in (12.10) and using the inverse Fourier transformation rule, $-\omega^2 \rightarrow \partial^2/\partial t^2$, Equation (12.19) becomes:

$$\varepsilon_0 \lambda_2 \left(\varepsilon_{\infty_1} \frac{\partial^2}{\partial t^2} + \omega_{p_1}^2 \right) E_x = \left[\left(\varepsilon_{\infty_1} \xi_2^2 + \lambda_2 \xi_1^2 \right) \frac{\partial^2}{\partial t^2} + \omega_{p_1}^2 \xi_2^2 \right] D_x + \xi_1 \xi_2 \left[\left(\varepsilon_{\infty_1} - \lambda_2 \right) \frac{\partial^2}{\partial t^2} + \omega_{p_1}^2 \right] D_y \tag{12.21}$$

Equation (12.21) is discretized using the following central-difference and central-average operators in time [52]:

$$\frac{\partial^2 F^n}{\partial t^2} = \frac{F^{n+1} - 2F^n + F^{n-1}}{(\Delta t)^2} \tag{12.22}$$

$$F^n = \frac{F^{n+1} + 2F^n + F^{n-1}}{4} \tag{12.23}$$

where $t = n\Delta t$ (n = integer) and $F = E_x, D_x$, or D_y. Use of the central-average operator of (12.23) improves the stability and accuracy of the overall algorithm [53, 55, 56]. This yields the following time-stepping relation for E_x [52]:

$$E_x^{n+1} = -E_x^{n-1} + 2\frac{a_1^-}{a_1^+} E_x^n + \frac{1}{\varepsilon_0 \lambda_2 a_1^+} \begin{bmatrix} b_1^+ \left(D_x^{n+1} + D_x^{n-1} \right) - 2b_1^- D_x^n + \\ c_1^+ \left(D_y^{n+1} + D_y^{n-1} \right) - 2c_1^- D_y^n \end{bmatrix} \tag{12.24}$$

where D_y^{n+1}, D_y^n, and D_y^{n-1} must be spatially interpolated to the location of the E_x component due to the staggered nature of the FDTD grid cell [42], and:

$$a_i^{\pm} = \frac{\varepsilon_{\infty_i}}{(\Delta t)^2} \pm \frac{\omega_{p_i}^2}{4} \tag{12.25a}$$

$$b_i^{\pm} = \frac{\varepsilon_{\infty_i} \xi_2^2 + \lambda_2 \xi_1^2}{(\Delta t)^2} \pm \frac{\omega_{p_i}^2 \xi_2^2}{4} \tag{12.25b}$$

$$c_i^{\pm} = \xi_1 \xi_2 \left[\frac{\varepsilon_{\infty_i} - \lambda_2}{(\Delta t)^2} \right] \pm \frac{\omega_{p_i}^2 \xi_2^2}{4} \tag{12.25c}$$

In a similar manner, the time-stepping equation for E_y is obtained by exchanging $\xi_1 \leftrightarrow \xi_2$, $E_x \leftrightarrow E_y$, and $D_x \leftrightarrow D_y$ in the above development.

To derive the FDTD time-stepping equation for H_z, one considers the relation:

$$B_z = \mu_0 \varepsilon_{zz} H_z = \mu_0 \left(\varepsilon_{\infty_3} - \frac{\omega_{p_3}^2}{\omega^2} \right) H_z \tag{12.26}$$

where μ_0 is the vacuum permeability. After inverse Fourier transformation of (12.26), the resulting time-domain differential equation is discretized using the central-difference and central-average operators of (12.22) and (12.23) applied to B_z and H_z. This yields [52]:

$$H_z^{n+3/2} = -H_z^{n-1/2} + 2\frac{a_3^-}{a_3^+}H_z^{n+1/2} + \frac{B_z^{n+3/2} - 2B_z^{n+1/2} + B_z^{n-1/2}}{\mu_0(\Delta t)^2 a_3^+} \tag{12.27}$$

In summary, the electromagnetic fields for this 2-D case are iteratively updated in the following sequence [52]:

1. Update D_x and D_y to time-step $n+1$ according to (12.13) and (12.14).
2. Update E_x and E_y to time-step $n+1$ according to the sample update for E_x given in (12.24).
3. Update B_z to time-step $n+3/2$ according to (12.15).
4. Update H_z to time-step $n+3/2$ according to (12.27). Then, cycle back to Update 1.

12.4 MODELING RESULTS FOR AN ELLIPTIC CYLINDRICAL CLOAK

This section reviews the FDTD modeling results for the 2-D elliptic cylindrical cloak reported in [52], obtained using the formulation of Section 12.3. Figure 12.2 illustrates the geometry of this simulation. The item to be cloaked was an elliptic perfect electric conductor (PEC) shell having a 500-nm minor semi-axis (a in Fig. 12.1) and a major-to-minor axis ratio of 2 (k in Fig. 12.1). The outer surface of the cloak had a 1000-nm minor semi-axis (b in Fig. 12.1) and the same major-to-minor axis ratio, $k = 2$. The PEC shell and its surrounding cloak were illuminated at the wavelength $\lambda_0 = 750$ nm by a $+x$-directed plane wave with field components E_y^{inc} and H_z^{inc}.

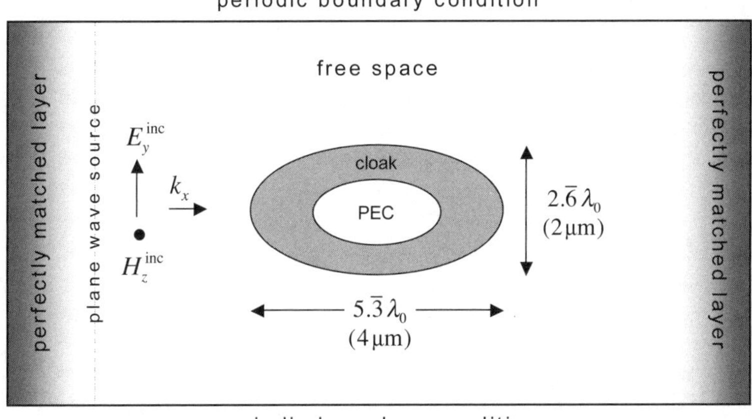

Fig. 12.2 Geometry of 2-D FDTD simulation of an elliptic cylindrical cloak covering an elliptic PEC shell. The illumination was a $+x$-directed $\lambda_0 = 750$-nm plane wave with field components E_y^{inc} and H_z^{inc}. *Adapted from:* N. Okada and J. B. Cole, *ISRN Optics*, Vol. 2012, 536209, doi:10.5402/2012/536209, ©2012 Okada and Cole.

The FDTD grid depicted in Fig. 12.2 contained 600 × 600 grid cells, each 10 × 10 nm ($\lambda_0/h = 75$). This grid was terminated with uniaxial perfectly matched layer (UPML) absorbing boundaries in the $\pm x$-directions, and a periodic boundary condition in the $\pm y$-directions [40]. Time-stepping was conducted with Δt at the Courant limit, $h/(c_0\sqrt{2})$.

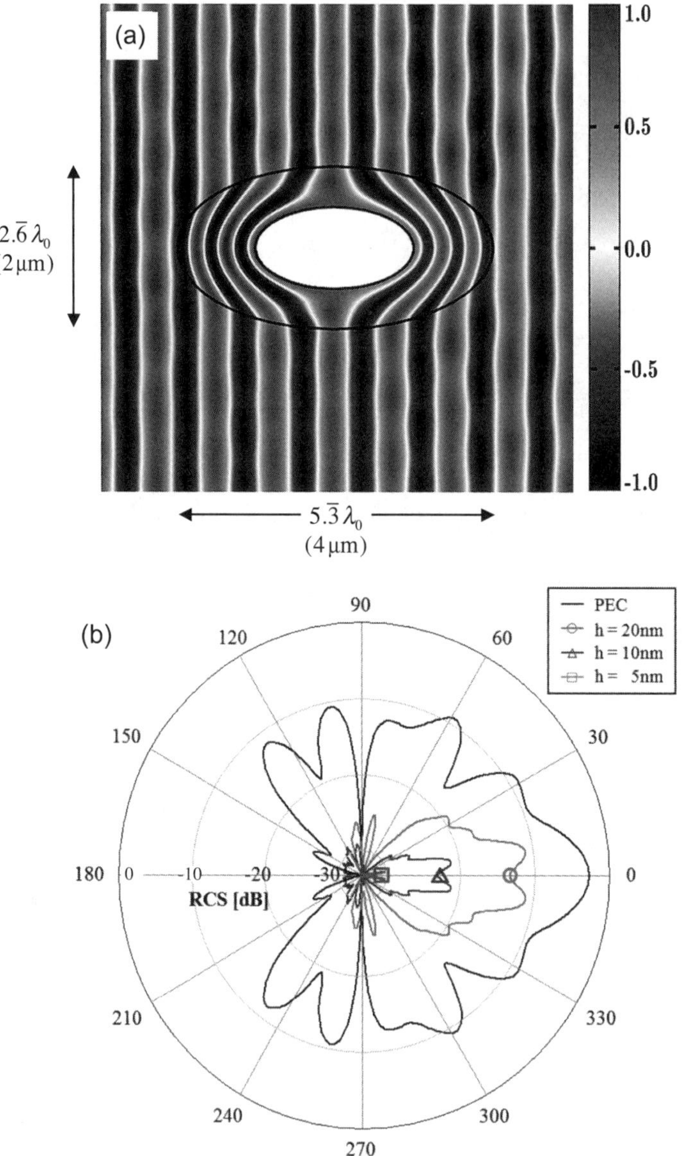

Fig. 12.3 FDTD-computed results for the 2-D elliptic cylindrical cloak model of Fig. 12.2 illuminated at $\lambda_0 = 750$ nm: (a) visualization of the H_z field distribution for 10-nm grid resolution; (b) bistatic radar cross-section without the cloak and with the cloak for 20-, 10-, and 5-nm grid resolutions. *Source:* N. Okada and J. B. Cole, *ISRN Optics*, Vol. 2012, 536209, doi:10.5402/2012/536209, ©2012 Okada and Cole.

Figure 12.3 shows the FDTD-computed results for the 2-D elliptic cylindrical cloak of Fig. 12.2 at the sinusoidal steady-state (50 wave periods at $\lambda_0 = 750$ nm) [52]. Figure 12.3(a) is a visualization of the computed H_z distribution for a 10-nm grid resolution. The simulated wave propagated without significant disturbance around the cloak, and the calculation was stable. Small ripples observed on the phase planes were purely numerical errors, and could be made to vanish by refining the grid resolution.

Figure 12.3(b) provides quantitative data that show how the performance of the elliptic cylindrical cloak at $\lambda_0 = 750$ nm improved as the FDTD grid resolution was refined [52]. This figure is a polar plot in decibels (dB) of the bistatic radar cross-section (RCS), σ, versus scattering angle, ϕ, of the cloaked PEC cylinder. In 2-D, σ is defined by [40]:

$$\sigma(\phi) = \lim_{r\to\infty} 2\pi r \frac{|E_s(\phi)|^2}{|E_0|^2} \tag{12.28}$$

where $|E_s(\phi)|^2$ is the scattered power in the far field, and $|E_0|^2$ is the incident power. If there is no significant disturbance by the object, σ approaches zero. From Fig. 12.3(b), we see that for a 20-nm grid resolution, the cloak reduced the maximum value of σ by ~10 dB relative to the bare PEC cylinder. However, when the FDTD grid was refined to use 10-nm grid cells, the σ reduction improved to ~20 dB. Finally, when the FDTD grid was further refined to use 5-nm grid cells, the σ reduction improved to ~30 dB. We note that this rapid improvement was likely due to a desirable combination of (a) decreasing errors in staircasing the surface of the cloak; and (b) decreasing errors in implementing the finite-difference field updates of (12.13)–(12.15), (12.24) and the analogous E_y update, as well as (12.27).

We next consider the broadband cloaking performance of the 2-D elliptic cylindrical cloak of Fig. 12.2, as reported in [52]. This study assumed the same cloak composition (optimized for $\lambda_0 = 750$ nm) and FDTD simulation parameters previously used to obtain the results shown in Fig. 12.3. However, leveraging the time-domain nature of FDTD modeling, a single simulation run for an impulsive plane-wave source provided $\sigma_t(\lambda)$, the total cross-section (TCS) of scattering as a function of wavelength, defined by:

$$\sigma_t(\lambda) = \int_0^{2\pi} \sigma(\lambda, \phi)\, d\phi \tag{12.29}$$

Figure 12.4(a) shows the FDTD-computed TCS spectrum, $\sigma_t(\lambda)$, of the 2-D elliptic cylindrical cloak over the wavelength band of 600 to 900 nm [52]. While the TCS was reduced by ~30 dB at the 750-nm design wavelength, it increased rapidly with wavelength shifts to either side off this optimum. In fact, the bandwidth of the effective scattering reduction was only ~30 nm, or ~4%; so narrow that it may be described as a scattering *null*.

Figure 12.4(b) compares the FDTD-computed bistatic RCS patterns corresponding to the three TCS values marked as circled dots in Fig. 12.4(a): at $\lambda_A = 730$ nm; at $\lambda_B = 750$ nm (the design wavelength); and at $\lambda_C = 830$ nm [52]. The peak RCS values (observed at $\phi = 0$) at λ_A and λ_C were, respectively, ~30 and ~45 dB larger than the peak value at the design wavelength, λ_B.

Overall, the results of this FDTD simulation demonstrate a key limitation of the transformation-based cloaking technique: its narrowband nature. At present, while of significant theoretical interest, this technique appears to have only limited practical applications.

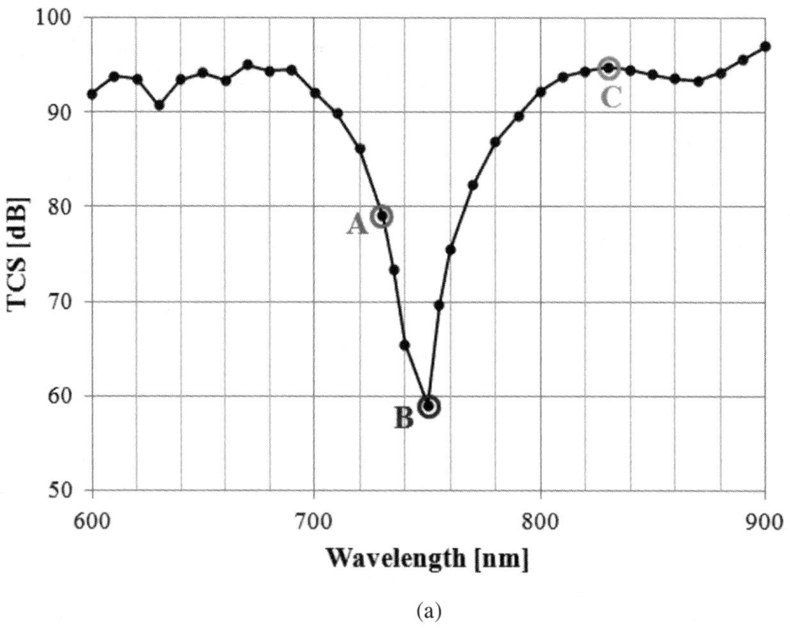

(a)

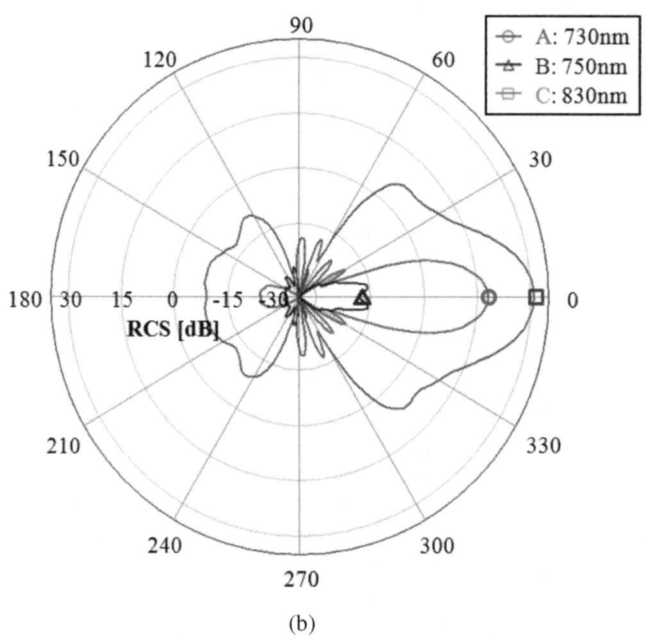

(b)

Fig. 12.4 FDTD-computed results for the 2-D elliptic cylindrical cloak model of Fig. 12.2 for 10-nm grid resolution: (a) total cross-section (TCS) of scattering vs. wavelength; (b) sample bistatic RCS patterns corresponding to the three TCS values marked as circled dots in (a) – at $\lambda_A = 730$ nm, at $\lambda_B = 750$ nm (the design wavelength), and at $\lambda_C = 830$ nm. *Source:* N. Okada and J. B. Cole, *ISRN Optics*, Vol. 2012, 536209, doi:10.5402/2012/536209, ©2012 Okada and Cole.

12.5 SUMMARY AND CONCLUSIONS

Without proper care, the direct application of FDTD to simulate transformation-based metamaterials having nondiagonal constitutive parameters, $[\varepsilon^{ij}]$ and $[\mu^{ij}]$, is prone to numerical instabilities. Reference [52] reported a technique to solve this instability problem by ensuring that the numerically derived FDTD equations are exactly symmetric.

This chapter reviewed the basis, formulation, and validation of the technique reported in [52]. The crucial step in this technique involves finding the eigenvalues and *diagonalizing* $[\varepsilon^{ij}]$ and $[\mu^{ij}]$. After this diagonalization, any of the previously reported FDTD algorithms for purely diagonal metamaterial cases can be applied. In these algorithms, diagonal elements having values less than unity are treated using a suitable dispersive time-stepping scheme to avoid violating causality and the FDTD stability limit.

The validation study of [52] reviewed in this chapter involved 2-D FDTD modeling of a transformation-based elliptical cylindrical cloak comprised of a nondiagonal anisotropic metamaterial. The cloak was found to greatly reduce both the bistatic radar cross-section and the total scattering cross-section of the enclosed elliptical PEC cylinder at the design wavelength. In fact, as the grid of the FDTD model was progressively refined, scattering by the cloaked PEC cylinder trended rapidly toward zero. However, the bandwidth of the effective scattering reduction was only ~4%; so narrow that it may be described as just a scattering *null*. This narrow bandwidth appears to limit practical applications of such cloaks.

REFERENCES

[1] Solymar, L., and E. Shamonina, *Waves in Metamaterials*, New York: Oxford University Press, 2009.

[2] Cui, T. J., D. R. Smith, and R. Liu. *Metamaterials: Theory, Design, and Applications*, Berlin: Springer-Verlag, 2009.

[3] Pendry, J. B., D. Schuring, and D. R. Smith, "Controlling electromagnetic fields," *Science*, Vol. 312, 2006, pp. 1780–1782.

[4] Leonharbt, U., and T. G. Philbin, "Transformation optics and the geometry of light," *Progress in Optics*, Vol. 53, 2009, pp. 69–152.

[5] Cummer, S. A., B. L. Popa, D. Schurig, D. R. Smith, and J. B. Pendry, "Full-wave simulations of electromagnetic cloaking structures," *Physical Review E*, Vol. 74, 2006, 036621.

[6] Li, J., and J. B. Pendry, "Hiding under the carpet: A new strategy for cloaking," *Physical Review Lett.*, Vol. 101, 2008, 203901.

[7] Jiang, W. X., T. J. Cui, Q. Cheng, J. Y. Ching, X. M. Yang, R. Liu, and D. R. Smith, "Design of arbitrarily shaped concentrators based on conformally optical transformation of nonuniform rational b-spline surfaces," *Applied Physics Lett.*, Vol. 92, 2008, 264101.

[8] Jiang, W. X., J. Y. Chin, Z. Li, Q. Cheng, R. Liu, and T. J. Cui, "Analytical design of conformally invisible cloaks for arbitrarily shaped objects," *Physical Review E*, Vol. 77, 2008, 066607.

[9] Jiang, W. X., T. J. Cui, G. X. Yu, X. Q. Lin, Q. Cheng, and J. Y. Chin, "Arbitrarily elliptical-cylindrical invisible cloaking," *J. Physics D: Applied Physics*, Vol. 41, 2008, 085504.

[10] You, Y., G. W. Kattawar, P. W. Zhai, and P. Yang, "Invisibility cloaks for irregular particles using coordinate transformations," *Optics Express*, Vol. 16, 2008, pp. 6134–6145.

[11] Rahm, M., D. Schurig, D. A. Roberts, S. A. Cummer, D. R. Smith, and J. B. Pendry, "Design of electromagnetic cloaks and concentrators using form-invariant coordinate transformations of Maxwell's equations," *Photonics and Nanostructures Fundamentals & Applications*, Vol. 6, 2008, pp. 87–95.

[12] Ma, H., S. Qu, Z. Xu, J. Zhang, B. Chen, and J. Wang, "Material parameter equation for elliptical cylindrical cloaks," *Physical Review A*, Vol. 77, 2008, 013825.

[13] Cojocaru, E., "Exact analytical approaches for elliptic cylindrical invisibility cloaks," *J. Optical Society of America B*, Vol. 26, 2009, pp. 1119–1128.

[14] Huidobro, P. A., M. L. Nesterov, L. Martin-Moreno, and F. J. Garcia-Vidal, "Transformation optics for plasmonics," *Nano Lett.*, Vol. 10, 2010, pp. 1985–1990.

[15] Chen, H., and C. T. Chan, "Transformation media that rotate electromagnetic fields," *Applied Physics Lett.*, Vol. 90, 2007, 241105.

[16] Kwon, D. H., and D. H. Werner, "Polarization splitter and polarization rotator designs based on transformation optics," *Optics Express*, Vol. 16, 2008, pp. 18731–18738.

[17] Luo, Y., J. Zhang, B. Wu, and H. Chen, "Interaction of an electromagnetic wave with a cone-shaped invisibility cloak and polarization rotator," *Physical Review B*, Vol. 78, 2008, 125108.

[18] Zhai, T., Y. Zhou, J. Zhou, and D. Liu, "Polarization controller based on embedded optical transformation," *Optics Express*, Vol. 17, 2009, pp. 17206–17213.

[19] Rahm, M., D. A. Roberts, J. B. Pendry, and D. R. Smith, "Transformation-optical design of adaptive beam bends and beam expanders," *Optics Express*, Vol. 16, 2008, pp. 11555–11567.

[20] Rahm, M., S. A. Cummer, D. Schurig, J. B. Pendry, and D. R. Smith, "Optical design of reflectionless complex media by finite embedded coordinate transformations," *Physical Review Lett.*, Vol. 100, 2008, 063903.

[21] Lin, L., W. Wang, J. Cui, C. Du, and X. Luo, "Design of electromagnetic refractor and phase transformer using coordinate transformation theory," *Optics Express*, Vol. 16, 2008, pp. 6815–6821.

[22] Han, S., Y. Xiong, D. Genov, Z. Liu, G. Bartal, and X. Zhang, "Ray optics at a deep subwavelength scale: A transformation optics approach," *Nano Lett.*, Vol. 8, 2008, pp. 4243–4247.

[23] Kwon, D. H., and D. H. Werner, "Transformation optical designs for wave collimators, flat lenses and right-angle bends," *New Journal of Physics*, Vol. 10, 2008, 115023.

[24] Donderici, B., and F. L. Teixeira, "Metamaterial blueprints for reflectionless waveguide bends," *IEEE Microwave and Wireless Components Lett.*, Vol. 18, 2008, pp. 233–235.

[25] Roberts, D. A., M. Rahm, J. B. Pendry, and D. R. Smith, "Transformation-optical design of sharp waveguide bends and corners," *Applied Physics Lett.*, Vol. 93, 2008, 251111.

[26] Jiang, W. X., T. J. Cui, H. F. Ma, X. Y. Zhou, and Q. Cheng, "Cylindrical-to-plane-wave conversion via embedded optical transformation," *Applied Physics Lett.*, Vol. 92, 2008, 261903.

[27] Lai, Y., J. Ng, H. Chen, D. Han, J. Xiao, Z. Zhang, and C. T. Chan, "Illusion optics: The optical transformation of an object into another object," *Physical Review Lett.*, Vol. 102, 2009, 253902.

[28] Jiang, W. X., and T. J. Cui, "Moving targets virtually via composite optical transformation," *Optics Express*, Vol. 18, 2010, pp. 5161–5167.

[29] Chen, H., C. T. Chan, and P. Sheng, "Transformation optics and metamaterials," *Nature Materials*, Vol. 9, 2010, pp. 387–396.

[30] Narimanov, E. E., and A. V. Kildishev, "Optical black hole: Broadband omnidirectional light absorber," *Applied Physics Lett.*, Vol. 95, 2009, 041106.

[31] Genov, D. A., S. Zhang, and X. Zhang, "Mimicking celestial mechanics in metamaterials," *Nature Physics*, Vol. 5, 2009, pp. 687–692.

[32] Schurig, D., J. J. Mock, B. J. Justice, S. A. Cummer, J. B. Pendry, A. F. Starr, and D. R. Smith, "Metamaterial electromagnetic cloak at microwave frequencies," *Science*, Vol. 314, 2006, pp. 977–980.

[33] Kante, B., D. Germain, and A. Lustrac, "Experimental demonstration of a nonmagnetic metamaterial cloak at microwave frequencies," *Physical Review B*, Vol. 80, 2009, 201104.

[34] You, Y., G. W. Kattawar, P. W. Zhai, and P. Yang, "Broadband ground-plane cloak," *Science*, Vol. 323, 2009, pp. 366–369.

[35] Ma, Y. G., C. K. Ong, T. Tyc, and U. Leonharbt, "An omnidirectional retroreflector based on the transmutation of dielectric singularities," *Nature Materials*, Vol. 8, 2009, pp. 639–642.

[36] Ma, H. F., and T. J. Cui, "Three-dimensional broadband ground-plane cloak made of metamaterials," *Nature Communications*, Vol. 1, 2010, 21.

[37] Ergin, T., N. Stenger, P. Brenner, J. B. Pendry, and M. Wegener, "Three-dimensional invisibility cloak at optical wavelengths," *Science*, Vol. 328, 2010, pp. 337–339.

[38] Chen, X., Y. Luo, J. Zhang, K. Jiang, J. B. Pendry, and S. Zhang, "Macroscopic invisibility cloaking of visible light," *Nature Communications*, Vol. 2, 2010, 176.

[39] Zhang, J., L. Liu, Y. Luo, S. Zhang, and N. A. Mortensen, "Homogeneous optical cloak constructed with uniform layered structures," *Optics Express*, Vol. 19, 2011, pp. 8625–8631.

[40] A. Taflove and S. C. Hagness, *Computational Electrodynamics: The Finite-Difference Time-Domain Method,* 3rd ed., Norwood, MA: Artech House, 2005.

[41] Kallos, E., C. Argyropoulos, and Y. Hao, "Ground-plane quasi-cloaking for free space," *Physical Review A*, Vol. 79, 2009, 063825.

[42] Zhao, Y., C. Argyropoulos, and Y. Hao, "Full-wave finite-difference time-domain simulation of electromagnetic cloaking structures," *Optics Express*, Vol. 16, 2008, pp. 6717–6730.

[43] Hao, Y., and R. Mittra, *FDTD Modeling of Metamaterials: Theory and Applications*, Norwood, MA: Artech House, 2008.

[44] Silva-Macedo, J. A., M. A. Romero, and B. H. V. Borges, "An extended FDTD method for the analysis of electromagnetic field rotations and cloaking devices," *Progress in Electromagnetics Research*, Vol. 87, 2008, pp. 183–196.

[45] Argyropoulos, C., Y. Zhao, and Y. Hao, "A radially-dependent dispersive finite-difference time-domain method for the evaluation of electromagnetic cloaks," *IEEE Trans. Antennas and Propagation*, Vol. 57, 2009, pp. 1432–1441.

[46] Argyropoulos, C., E. Kallos, Y. Zhao, and Y. Hao, "Manipulating the loss in electromagnetic cloaks for perfect wave absorption," *Optics Express*, Vol. 17, 2009, pp. 8467–8475.

[47] Argyropoulos, C., E. Kallos, and Y. Hao, "Dispersive cylindrical cloaks under non-monochromatic illumination," *Physical Review E*, Vol. 81, 2010, 016611.

[48] Argyropoulos, C., E. Kallos, and Y. Hao, "FDTD analysis of the optical black hole," *J. Optical Society of America B*, Vol. 27, 2010, pp. 2020–2025.

[49] Argyropoulos, C., Y. Zhao, and Y. Hao, "Bandwidth evaluation of dispersive transformation electromagnetics based devices," *Applied Physics A: Materials Science and Processing*, Vol. 103, 2011, pp. 715–719.

[50] Tretyakov, S. A., and S. I. Maslovski, "Veselago materials: What is possible and impossible about the dispersion of the constitutive parameters," *IEEE Trans. Antennas and Propagation*, Vol. 49, 2007, pp. 37–43.

[51] Yao, P., Z. Liang, and X. Jiang, "Limitation of the electromagnetic cloak with dispersive material," *Applied Physics Lett.*, Vol. 92, 2008, 031111.

[52] Okada, N., and J. B. Cole, "FDTD modeling of a cloak with a nondiagonal permittivity tensor," *ISRN Optics*, Vol. 2012, 536209, doi:10.5402/2012/536209.

[53] Pereda, A., L. A. Vielva, A. Vegas, and A. Prieto, "Analyzing the stability of the FDTD technique by combining the von Neumann method with the Routh-Hurwitz criterion," *IEEE Trans. Microwave Theory and Techniques*, Vol. 49, 2001, pp. 377–381.

[54] Lin, Z., and L. Thylen, "On the accuracy and stability of several widely used FDTD approaches for modeling Lorentz dielectrics," *IEEE Trans. Antennas and Propagation*, Vol. 57, 2009, pp. 3378–3381.

[55] Hildebrand, F. B., *Introduction to Numerical Analysis*, 2nd ed., Dover Publications, 1987.

[56] Zhao, Y., P. A. Belov, and Y. Hao, "Modeling of wave propagation in wire media using spatially dispersive finite-difference time-domain method: Numerical aspects," *IEEE Trans. Antennas and Propagation*, Vol. 55, 2007, pp. 1506–1513.

Chapter 13

FDTD Modeling of Metamaterial Structures

Costas D. Sarris

13.1 INTRODUCTION

In a seminal paper published in 1968 [1], Victor Veselago predicted the possibility of media simultaneously exhibiting negative dielectric permittivity, ε, and magnetic permeability, μ. In such media, the index of refraction, involving a square root of the product of ε and μ, assumes negative values. Therefore, they can be referred to as negative-refractive-index (NRI) media, as opposed to conventional positive-refractive-index (PRI) ones. As pointed out by Veselago, these media can support a number of unconventional electromagnetic wave phenomena, notably negative refraction, inverted Doppler shift, and Cherenkov radiation. The combination of positive-index and negative-index media can result in planar-lens geometries that act as "perfect lenses" achieving sub-wavelength resolution [2].

Over the course of the past decade, Veselago's theoretical predictions were experimentally validated by the engineering of artificial dielectrics ("metamaterials") that macroscopically exhibit a negative refractive index. Examples of these include the three-dimensional (3-D) split-ring-resonator (SRR) medium of [3] and the planar inductor-capacitor (L-C) grid based medium of [4]. While these concepts were initially demonstrated at microwave frequencies [3–5], several extensions to optical frequencies have also been theoretically and experimentally investigated [6, 7].

Regarding computational modeling, early studies in the frequency domain (via the finite-element method and associated commercial software packages) verified properties that had been analytically predicted by Veselago [8], the growth of evanescent waves in L-C grid based NRI media, and the dispersion properties of the latter [9]. Subsequently, time-domain modeling of metamaterials, especially using the finite-difference time-domain (FDTD) method, was motivated by the richness and non-intuitive nature of metamaterials' transient behavior, which is naturally observed in FDTD simulations [10]. For example, the causal evolution of negative refraction, which had been initially disputed, was verified in several FDTD papers that illustrated the transient development of negatively refracted wavefronts at the interface of a positive-index and a negative-index medium [11 – 14]. Currently, with even more exotic metamaterials structures being investigated (e.g., cloaking devices and electromagnetic black holes), rigorous time-domain analysis using FDTD is proving to be of exceptional importance.

This chapter provides an overview of the application of FDTD to several key classes of problems in metamaterial analysis and design, from two complementary perspectives. First, the FDTD analysis of the *transient response* of several metamaterial structures of interest is presented. These include negative-refractive-index media and the "perfect lens," an artificial transmission line supporting abnormal group velocities, and a planar anisotropic grid supporting resonance cone phenomena. Second, periodic FDTD simulations are used to characterize the *dispersion* of metamaterial structures, realized as periodic geometries. In addition to the negative-refractive-index transmission-line [4] and the plasmonic sub-wavelength photonic crystal [7], the modeling of driven periodic structures within the framework of periodic FDTD methods is discussed.

13.2 TRANSIENT RESPONSE OF A PLANAR NEGATIVE-REFRACTIVE-INDEX LENS

We shall first consider FDTD modeling of the transient (dynamic) response of a planar negative-refractive-index lens. To this end, we first briefly review how appropriate electric and magnetic field dispersions can be incorporated into the FDTD simulation to permit the stable numerical modeling of negative values of ε and μ. A detailed derivation of the *auxiliary differential equation* (ADE) technique used here is found in Section 9.4.3 of [15].

13.2.1 Auxiliary Differential Equation Formulation

Slightly adapting the notation used in [15], consider first a medium having a frequency-dependent relative permittivity, $\varepsilon_r(\omega)$, characterized by a single-pole Drude dispersion:

$$\varepsilon_r(\omega) = 1 - \frac{\omega_{\text{pe}}^2}{\omega^2 - j\omega\gamma_{\text{pe}}} \tag{13.1}$$

Then, according to [15], upon applying Ampere's law, the following explicit time-stepping relation to update the electric field, $\boldsymbol{E}$, from time-step n to time-step n+1 is derived:

$$\boldsymbol{E}^{n+1} = \left(\frac{2\varepsilon_0 - \beta_{\text{pe}}\Delta t}{2\varepsilon_0 + \beta_{\text{pe}}\Delta t}\right)\boldsymbol{E}^n + \left(\frac{2\Delta t}{2\varepsilon_0 + \beta_{\text{pe}}\Delta t}\right)\cdot\left[\nabla\times\boldsymbol{H}^{n+1/2} - 0.5\left(1+k_{\text{pe}}\right)\boldsymbol{J}_{\text{pe}}^n\right] \tag{13.2}$$

where $\boldsymbol{H}$ is the magnetic field, and the spatial derivatives of the curl operator are implemented as standard central differences on the classic interleaved Yee mesh [16]. Furthermore, the coefficients k_{pe} and β_{pe} are given by:

$$k_{\text{pe}} = \frac{1 - 0.5\gamma_{\text{pe}}\Delta t}{1 + 0.5\gamma_{\text{pe}}\Delta t}; \qquad \beta_{\text{pe}} = \frac{0.5\omega_{\text{pe}}^2\varepsilon_0\Delta t}{1 + 0.5\gamma_{\text{pe}}\Delta t} \tag{13.3}$$

and $\boldsymbol{J}$ is an auxiliary field variable having the following explicit time-stepping relation:

$$\boldsymbol{J}_{\text{pe}}^{n+1} = k_{\text{pe}}\boldsymbol{J}_{\text{pe}}^n + \beta_{\text{pe}}\left(\boldsymbol{E}^{n+1} + \boldsymbol{E}^n\right) \tag{13.4}$$

Starting with the assumed known (stored) component values of $\boldsymbol{E}^n$, $\boldsymbol{J}_{\text{pe}}^n$, and $\boldsymbol{H}^{n+1/2}$, we first calculate the complete set of $\boldsymbol{E}^{n+1}$ components using (13.2). Then, we calculate the complete set of $\boldsymbol{J}_{\text{pe}}^{n+1}$ components using (13.4) applied to the just-computed set of $\boldsymbol{E}^{n+1}$ components.

Next, assume that this medium has a frequency-dependent relative permeability, $\mu_r(\omega)$, that is similarly characterized by a single-pole Drude dispersion:

$$\mu_r(\omega) = 1 - \frac{\omega_{\text{pm}}^2}{\omega^2 - j\omega\gamma_{\text{pm}}} \tag{13.5}$$

Then, upon applying Faraday's law and following a dual procedure relative to that described in [15], the following explicit time-stepping relation to update the magnetic field, $\boldsymbol{H}$, from time-step $n+½$ to time-step $n+1½$ is derived:

$$\boldsymbol{H}^{n+1½} = \left(\frac{2\mu_0 + \beta_{\text{pm}}\Delta t}{2\mu_0 - \beta_{\text{pm}}\Delta t}\right)\boldsymbol{H}^{n+½} + \left(\frac{2\Delta t}{2\mu_0 - \beta_{\text{pm}}\Delta t}\right)\cdot\left[0.5\left(1+k_{\text{pm}}\right)\boldsymbol{M}_{\text{pm}}^{n+½} - \nabla\times\boldsymbol{E}^{n+1}\right] \tag{13.6}$$

where the spatial derivatives of the curl operator are again implemented as standard central differences on the classic interleaved Yee mesh [16]. Here, the coefficients k_{pm} and β_{pm} are given by:

$$k_{\text{pm}} = \frac{1 - 0.5\gamma_{\text{pm}}\Delta t}{1 + 0.5\gamma_{\text{pm}}\Delta t}; \qquad \beta_{\text{pm}} = \frac{0.5\omega_{\text{pm}}^2\varepsilon_0\Delta t}{1 + 0.5\gamma_{\text{pm}}\Delta t} \tag{13.7}$$

and $\boldsymbol{M}$ is an auxiliary field variable having the following explicit time-stepping relation:

$$\boldsymbol{M}_{\text{pm}}^{n+1½} = k_{\text{pm}}\boldsymbol{M}_{\text{pm}}^{n+½} + \beta_{\text{pm}}\left(\boldsymbol{H}^{n+1½} + \boldsymbol{H}^{n+½}\right) \tag{13.8}$$

Starting with the assumed known (stored) component values of $\boldsymbol{H}^{n+½}$, $\boldsymbol{M}_{\text{pm}}^{n+½}$, and (the recently computed) $\boldsymbol{E}^{n+1}$, we first calculate the complete set of $\boldsymbol{H}^{n+1½}$ components using (13.6). Then, we calculate the complete set of $\boldsymbol{M}_{\text{pm}}^{n+1½}$ components using (13.8) applied to the just-computed set of $\boldsymbol{H}^{n+1½}$ components. This completes a full time-step. Then, the entire algorithm consisting of (13.2 – 13.4) and (13.6 – 13.8) can be iterated indefinitely in the usual manner of FDTD simulations.

13.2.2 Illustrative Problem

Figure 13.1 illustrates the FDTD modeling geometry of a low-loss, two-dimensional (2-D), dispersive metamaterial lens having the negative refractive index, n, with Re(n) = –1, at 15.98 GHz. The lens is comprised of a 2-cm-thick dispersive slab placed in free space. Its electric and magnetic field material dispersions are characterized by the Drude models of (13.1) and (13.5), respectively, with $\omega_{\text{p}} = 2\pi\times 22.6$ GHz and $\gamma_{\text{p}} = 2\pi\times 100\ \text{sec}^{-1}$ used for both dispersions. The FDTD computational domain is discretized with 118 × 120 Yee cells, and is terminated in 10-cell uniaxial perfectly matched layers (UPMLs) in the $\pm x$-directions, and in periodic boundary conditions in the $\pm y$-directions. The latter also includes the negative-refractive-index region occupied by the slab. A time-step of 0.83 ps is employed. Finally, a z-directed (out-of-plane) time-harmonic current source is placed 1 cm from the first (left) interface between free space and the slab. Figure 13.2 shows snapshots of the time evolution of the FDTD-computed out-of-plane electric field, E_z, at the first and the second (right) interfaces of the slab.

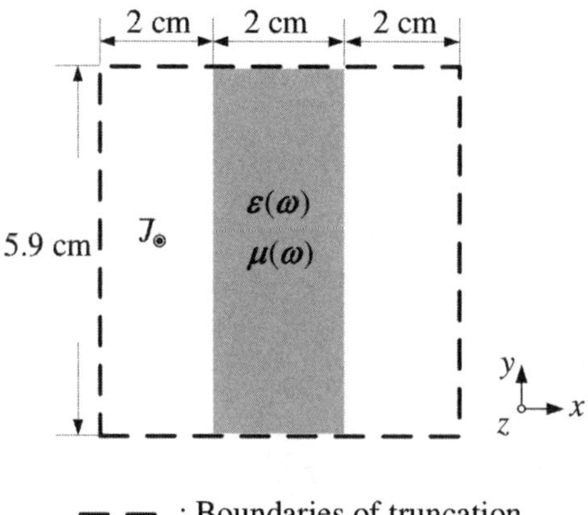

Fig. 13.1 FDTD computational space containing a model of a low-loss, 2-D, dispersive metamaterial slab in free space. The slab has a negative refractive index (NRI) of $n = -1$ at 15.98 GHz, and is sinusoidally excited by a z-directed current source, $\boldsymbol{J}$, located at $(x, y) = (1, 2.95)$ cm.

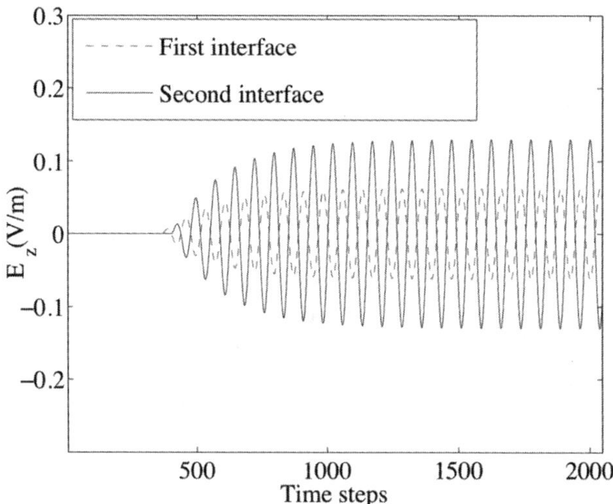

Fig. 13.2 FDTD-computed E_z time-waveforms at $y = 2.95$ cm on the first (left) and second (right) interfaces of the NRI slab of Fig. 13.1. *Source:* D. Li and C. D. Sarris, *J. Lightwave Technology*, Vol. 28, 2010, pp. 1447–1454, ©2010 IEEE.

Figure 13.2 compares the FDTD-computed time-waveforms of the E_z field at the left and right interfaces of the negative-refractive-index slab of Fig. 13.1, as observed at $y = 2.95$ cm, the y-coordinate of the sinusoidal current source [17]. The observed growth is due to multiple reflections between the two interfaces. These eventually culminate in the sinusoidal steady state after ~1500 time-steps (20 periods of the exciting source). Convergence to the sinusoidal steady state (and to the analytical solution) is further illustrated in Fig. 13.3. This shows the amplitude of the FDTD-computed E_z along an x-cut through the computational space of Fig. 13.1 at $y = 2.95$ cm after time-stepping through 9, 13, and 20 periods of the source [17, 18].

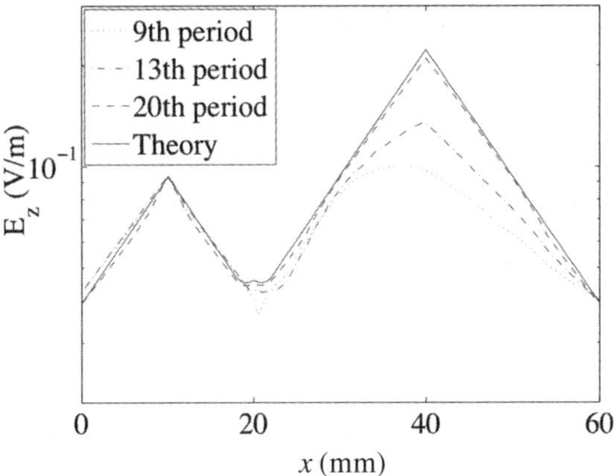

Fig. 13.3 Convergence of the amplitude of the FDTD-computed E_z to the analytical solution for the sinusoidal steady-state field. The graph displays an x-cut through the computational space of Fig. 13.1 at $y = 2.95$ cm after 9, 13, and 20 periods of the exciting current source. *Source:* D. Li and C. D. Sarris, *J. Lightwave Technology*, Vol. 28, 2010, pp. 1447–1454, ©2010 IEEE.

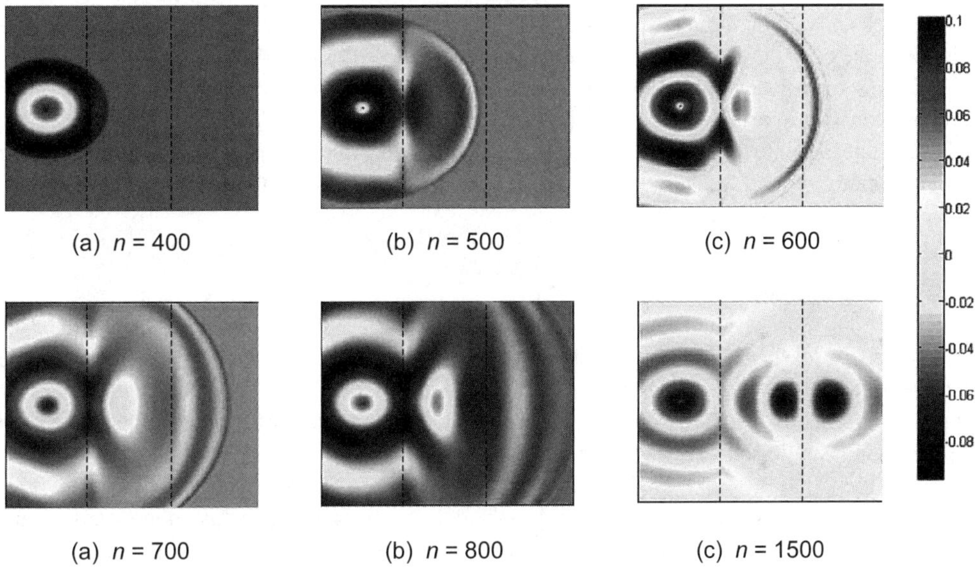

Fig. 13.4 Snapshot visualizations of the FDTD-computed E_z within the computational space of Fig. 13.1 at n = 400, 500, 600, 700, 800, and 1500 time-steps. The locations of the left and right interfaces of the NRI slab are indicated by vertical dashed lines. *Adapted from:* D. Li and C. D. Sarris, *J. Lightwave Technology*, Vol. 28, 2010, pp. 1447–1454, ©2010 IEEE.

Figure 13.4 displays snapshots of the FDTD-computed E_z within the computational domain of Fig. 13.1. After the simulation reaches the sinusoidal steady state, a focal point representing an image of the current source is observed at the right interface of the slab. The full-width at half-maximum (FWHM) diameter of this image is approximately 1/5 free-space wavelength (i.e., $\lambda_0/5$), well below the diffraction limit for a conventional lens.

13.3 TRANSIENT RESPONSE OF A LOADED TRANSMISSION LINE EXHIBITING A NEGATIVE GROUP VELOCITY

The existence of superluminal and negative group velocities was first predicted mathematically by Sommerfeld and Brillouin [19]. A superluminal group velocity signifies that the time delay between the peak of an input pulse to a device and the appearance of the peak of the resulting output pulse from the same device is less than that possible for a signal traveling at the speed of light. More extremely, a negative group velocity, or equivalently, a negative group delay, signifies that the peak of the output pulse occurs *before* the peak of its causative input pulse.

Intuitively, such group velocities appear to violate relativity and causality. Nevertheless, experimental evidence in diverse situations has shown that these peculiar velocities *can* exist. For example, electron wave packets exhibit negative delays when tunneling through potential barriers, as do electromagnetic waveguide modes tunneling through a below-cutoff section [20]. In electronic and microwave circuits, voltage pulses traveling through resonant resistor-inductor-capacitor (RLC) circuits can also exhibit negative delays [21]. In particular, several passive negative-delay circuits have been recently constructed. These consist of transmission lines loaded with series capacitors, RLC resonators, and shunt inductors. In addition to exhibiting negative group delays at and near the RLC resonance frequency, these circuits also exhibit a negative effective index. The RLC resonators produce the negative delay, and the series capacitors and shunt inductors create the negative effective index. Measurements of these circuits have been performed in the frequency and the time domain [21, 22].

This section shows how FDTD simulations can vividly reveal the dynamics of wave propagation in a one-dimensional (1-D) lossless transmission line exhibiting a negative group velocity due an embedded resonant circuit. Phenomena that, on face value, would appear to violate relativity and causality are easily seen in this manner to be perfectly natural and understandable.

13.3.1 Formulation

We assume a lossless 1-D transmission line having a uniform distributed series inductance, L henrys/m, and a uniform distributed shunt capacitance, C farads/m. We further assume that the transmission line is loaded by embedding the lumped-element circuit shown in the dashed box in Fig. 13.5 in a single short segment (unit cell) of length, Δz, of the line.

FDTD modeling of this loaded transmission line can be implemented in either of two ways: (a) by applying Kirchhoff's voltage and current laws at each individual node while leapfrog time-stepping the node voltages via $i_{\mathrm{C}} = \mathrm{C}\, dv_{\mathrm{C}}/dt$ and branch currents via $v_{\mathrm{L}} = \mathrm{L}\, di_{\mathrm{L}}/dt$; or (b) by developing a state equation [23] for the complete lumped-element circuit in the dashed box:

$$\mathbf{A}\cdot\frac{dX}{dt} = \mathbf{B}\cdot X + F \tag{13.9}$$

where vector X contains the state variables (for example, $X = [V_{\mathrm{dev1}}, I_{\mathrm{dev1}}, v_{\mathrm{C1}}, v_{\mathrm{C2}}, i_{\mathrm{L1}}, i_{\mathrm{L2}}]^T$); matrices **A** and **B** contain coefficients derived from the circuit elements based on circuit theory; and column vector F contains the source terms. Then, this state-equation for the dashed-box circuit can be linked (connected) to the transmission line on the left and right by Norton and Thevenin equivalent circuits, as illustrated in Fig. 13.6.

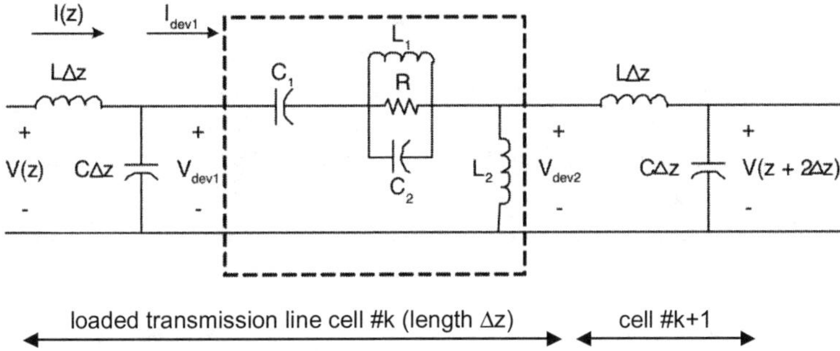

Fig. 13.5 Negative group-delay circuit of [21] inserted into a lossless transmission line. The dashed box encloses the lumped-element loading, and $C\Delta z$ and $L\Delta z$ represent the distributed capacitance and inductance, respectively, of the transmission line, as lumped elements.

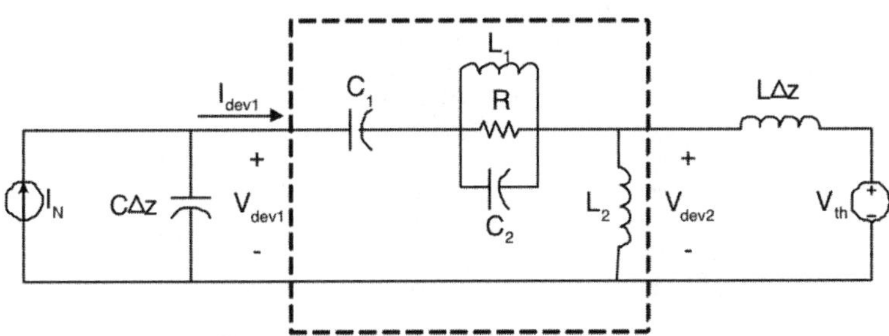

Fig. 13.6 Equivalent circuit used in the state-equation-based FDTD model of the negative-group-delay loaded transmission line of Fig. 13.5.

13.3.2 Numerical Simulation Parameters and Results

The dashed-box circuit in Fig. 13.5 was embedded in a single cell of a uniform lossless transmission line of characteristic impedance, $Z_0 = 150\,\Omega$. Component values were selected as follows: $C_1 = 1\,\text{pF}$, $C_2 = 5\,\text{pF}$, $L_1 = 2.7\,\text{nH}$, $L_2 = 11\,\text{nH}$, and $R = 300\,\Omega$. This caused a parallel resonance of L_1, C_2, and R at 1.37 GHz with a Q factor of ~13. The cell size, Δz, was set to 3 mm, corresponding to $\lambda/20$ for the maximum frequency of 2.0 GHz. The time-step, Δt, was set to 22.6 ps, corresponding to a Courant factor of 0.9, according to the 1-D stability analysis.

The source was a modulated Gaussian voltage pulse, $10\exp\{-[(t-t_0)/T_\omega]^2\}\cos\omega t$, where T_ω denotes the period of the sinusoid. This source had a 0.23-GHz bandwidth with varying center frequencies, ω, and was located $400\Delta z$ from the loaded cell. For Figs. 13.7(a) and 13.7(b), t_0 was set at $3T_\omega$ to generate a pulse rising smoothly from zero at $t = 0$, whereas for Fig. 13.8, t_0 was set at T_ω to generate a pulse rising abruptly from zero at $t = 0$. Probes of the input and output pulses were located on each side of the lumped load, ~2 cm apart. (Note that the positive delay due to this 2 cm of line was only ~0.1 ns, essentially negligible compared to the horizontal time scales in Figs. 13.7 and 13.8.) Finally, to simulate an infinite length, the transmission line was terminated at both ends with Mur's first-order absorbing boundary condition (ABC). No special boundary conditions were necessary despite the presence of backward waves in the line, in agreement with the observations in [13].

Figure 13.7 shows the input and output voltage waveforms computed with a state-equation-based FDTD model for Fig. 13.5, assuming a smoothly rising input pulse. No pulse advancement is observed in Fig. 13.7(a), where the center frequency was 255 MHz below the 1.37-GHz parallel resonance of L_1, C_2, and R. (A virtually identical result was obtained when the center frequency was 255 MHz above this resonance.) However, as seen in Fig. 13.7(b), when the excitation was centered at thc 1.37-GHz resonance frequency, the peak of the output pulse was sensed ~3 ns *before* the peak of the input pulse, i.e., exhibiting ~3 ns of *negative group delay*.

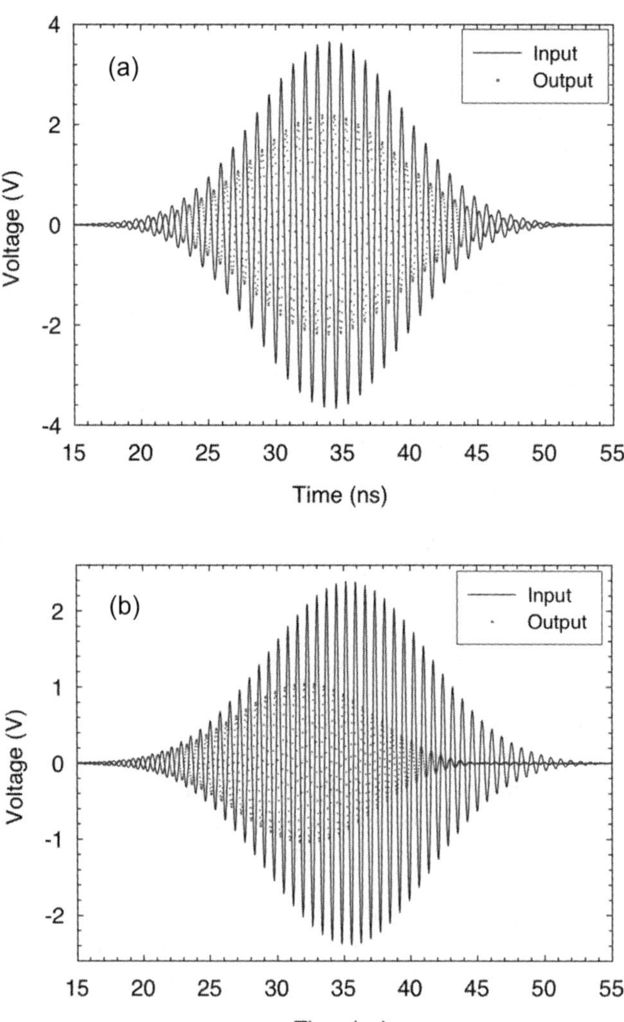

Fig. 13.7 FDTD-computed input and output voltage time-waveforms for the negative-group-delay structure of Fig. 13.5, assuming a smoothly rising input pulse: (a) No pulse advancement when the center frequency was 255 MHz below the 1.37-GHz parallel resonance of L_1, C_2, and R. A virtually identical result was obtained when the center frequency was 255 MHz above this resonance. (b) At the 1.37-GHz resonance frequency, the peak of the output pulse was sensed ~3 ns *before* the peak of the input pulse, i.e., exhibiting ~3 ns of *negative group delay*.

Figure 13.8 shows the input and output voltage waveforms for the same on-resonance case studied in Fig. 13.7(b), but here, t_0 was set equal to T_ω. Consequently, the excitation abruptly rose from zero to $10\exp[-(t_0/T_\omega)^2] = 10e^{-1} \cong 3.679$ at $t = 0$. As in Fig. 13.7(b), the peak of the output pulse was sensed ~3 ns before the peak of the input pulse. However, the *front* of the output pulse was sensed *after* the front of the input pulse. This corresponded to the expected propagation delay (less than 0.1 ns) over the 2-cm distance between the input and output probes.

In fact, the front of a pulse defines the beginning of its associated signal, *not* the position of the peak of the pulse. Abrupt discontinuities are expected to propagate with the speed of light in vacuum [19, 22]. Hence, the time-domain results of Fig. 13.8 confirm the consistency of the negative group-delay phenomenon with relativity and causality.

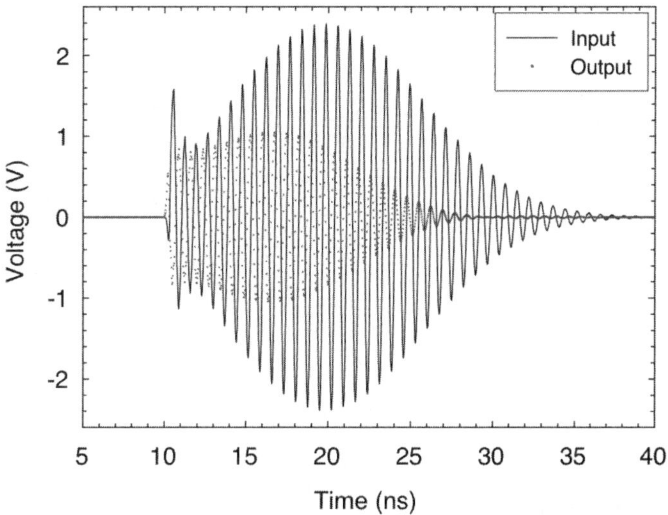

Fig. 13.8 FDTD-computed input and output voltage time-waveforms for the negative-group-delay structure of Fig. 13.5, assuming a center frequency at the 1.37-GHz resonance, and an abruptly rising input pulse. As in Fig. 13.7(b), the peak of the output pulse was sensed ~3 ns before the peak of the input pulse. However, the front of the output pulse was sensed *after* the front of the input pulse by the propagation delay time between the input and output probes.

13.4 PLANAR ANISOTROPIC METAMATERIAL GRID

Analysis of an oscillating dipole in an anisotropic plasma shows that the fields should become infinite along a cone whose axis is parallel to the static magnetic field, and whose angle is determined by the incident frequency, plasma density, and the magnetic field strength. Furthermore, the Poynting vector along the cone is singular, presenting an infinite radiation resistance to the dipole. In 1969, Fischer and Gould [24] gave experimental verification that the power flow of a short dipole antenna in an anisotropic (magnetized) plasma travels along the path of resonance cones under the condition that two of the diagonal elements of the plasma permittivity are opposite in sign. Later, Balmain and Oksiutik [25] noted that the plasma negative and positive permittivities could be interpreted, respectively, in terms of inductors and capacitors. Recently, it was observed that a grid of inductors and capacitors over ground can be viewed as a metamaterial that could exhibit resonance cone phenomena [26].

This section discusses the formulation and results of an FDTD model of the equivalent planar anisotropic metamaterial previously studied in [26]. As illustrated in Figs. 13.9 and 13.10, the FDTD model enclosed a planar grid of orthogonal lumped inductors and capacitors over ground (as reported in [26]) in a 3-D computational domain.

13.4.1 Formulation

The simulations of this section employed a 3-D FDTD code allowing for the insertion of passive lumped L, C, and R circuit elements, as initially reported in [27] and later in more detail in Section 15.9 of [15]. The ground plane and all lumped-element interconnects were modeled as perfect electric conductors (PECs). Because the L–C grid was terminated in matching resistors to reproduce the experiment described in [26], it was found sufficient to employ Mur's first-order ABC to truncate the computational domain. The following special field updates were used [15]:

Resistor R *oriented in the z-direction*

$$E_z\Big|_{i,j,k}^{n+1} = \left(\frac{1 - \dfrac{\Delta t \Delta z}{2\mathrm{R}\varepsilon_0 \Delta x \Delta y}}{1 + \dfrac{\Delta t \Delta z}{2\mathrm{R}\varepsilon_0 \Delta x \Delta y}} \right) E_z\Big|_{i,j,k}^{n} + \left(\frac{\dfrac{\Delta t}{\varepsilon_0}}{1 + \dfrac{\Delta t \Delta z}{2\mathrm{R}\varepsilon_0 \Delta x \Delta y}} \right) (\nabla \times \boldsymbol{H})_z\Big|_{i,j,k}^{n+1/2} \tag{13.10}$$

Inductor L *oriented in the x-direction*

$$E_x\Big|_{i,j,k}^{n+1} = E_x\Big|_{i,j,k}^{n} + \left(\frac{\Delta t}{\varepsilon_0} \right) (\nabla \times \boldsymbol{H})_x\Big|_{i,j,k}^{n+1/2} - \frac{\Delta x (\Delta t)^2}{\varepsilon_0 \mathrm{L} \Delta y \Delta z} \sum_{m=1}^{n} E_x\Big|_{i,j,k}^{m} \tag{13.11}$$

Capacitor C *oriented in the y-direction*

$$E_y\Big|_{i,j,k}^{n+1} = E_y\Big|_{i,j,k}^{n} + \left(\frac{\Delta t / \varepsilon_0}{1 + \dfrac{\mathrm{C}\Delta y}{\varepsilon_0 \Delta x \Delta z}} \right) (\nabla \times \boldsymbol{H})_y\Big|_{i,j,k}^{n+1/2} \tag{13.12}$$

13.4.2 Numerical Simulation Parameters and Results

Referring to Fig. 13.10, a $72 \times 72 \times 15$-cell 3-D FDTD space lattice with cubic unit cells, $\Delta = \Delta x = \Delta y = \Delta z = 0.5$ mm, was used in the simulation. The time-step was set at $\Delta t = 770$ fs ($0.8 \times$ the Courant stability limit), and the simulation was run for $10{,}240\Delta t$ (7.88 ns). A planar, orthogonal grid of lumped inductors, L, and capacitors, C, spanning $60\Delta \times 60\Delta$ (30×30 mm) was positioned in air 5Δ (2.5 mm) above the PEC ground plane, which comprised the lower boundary of the FDTD space lattice. This L–C grid was comprised of x-oriented L = 5.6 nH inductors and y-oriented C = 2 pF capacitors. Each L or C was connected with PEC filaments (E_x or E_y components set to zero) between space-lattice points in the L–C grid plane located 5Δ (2.5 mm) apart. Also, 100-kΩ resistors connected each interior node of the L–C grid to the ground plane below with PEC filaments (E_z components set to zero). In a similar manner, 50Ω resistors connected each edge node of the L–C grid to the ground plane. A 2-GHz, 1V sinusoidal source was connected between the ground plane and the L–C grid at the grid's lower left corner.

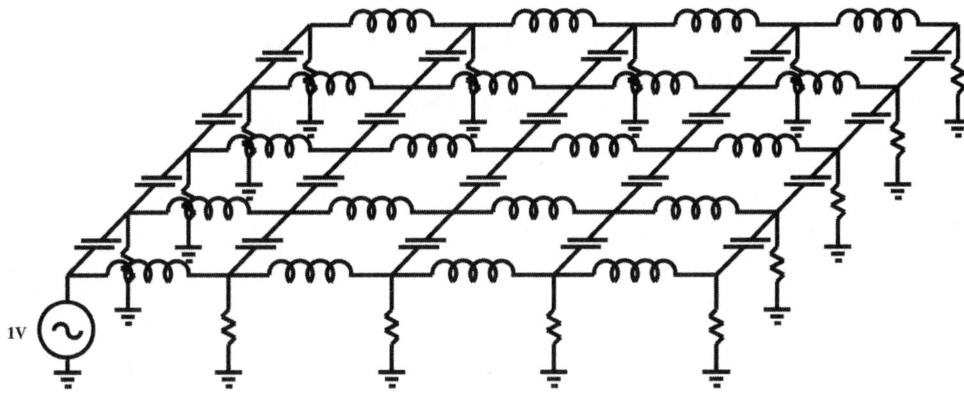

Fig. 13.9 Grid of orthogonal inductors and capacitors over a ground plane that can be viewed as a planar anisotropic metamaterial exhibiting resonance cone phenomena [26]. *Source:* K. G. Balmain, A. E. Lüttgen, and P. C. Kremer, *IEEE Trans. Antennas and Propagation*, Vol. 51, 2003, pp. 2614–2618, ©2003 IEEE.

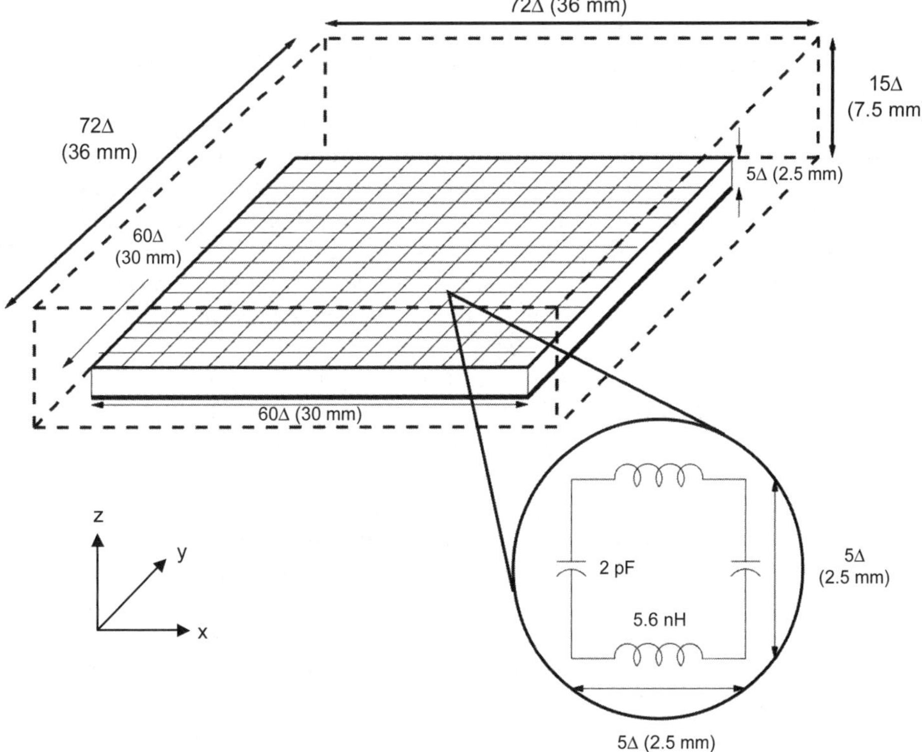

Fig. 13.10 Computational domain for the FDTD simulation of the electromagnetic wave properties of the planar grid of inductors and capacitors of Fig. 13.9.

Figure 13.11 shows FDTD-computed snapshot visualizations of the spatial distributions of $E_z(x, y)$ along an observation plane located 5 mm above the L–C grid of Fig. 13.10 at 600, 1922, and 10,240 time-steps. These visualizations depict the formation of a resonance cone as time progresses. We observe that there is little variation in amplitude along the path of the resonance cone. Using circuit theory as an analogy, the path of the resonance cone appears to be simply the path of zero reactance across the surface of the L–C grid.

Referring to Fig. 13.11(c), one can define the beam angle, θ_B, as the angle between the left edge of the L–C grid and the path of the resonance cone. Here, θ_B is seen to be approximately 30.9°. We can obtain a simple, intuitive confirmation of this result by applying the following approximate circuit-theoretic argument. Assume that a resonance path is formed by N inductors, L, in series along the x-direction, and one capacitor, C, along the y-direction. By the geometry of the L–C grid, $\theta_B = \tan^{-1}(N\Delta x/\Delta y) = \tan^{-1}(N)$, since $\Delta x = \Delta y$. Noting that the resonance frequency for the assumed path is $\omega^2 = 1/(N\text{LC})$, we see that $N = 1/(\omega^2\text{LC})$. This finally leads to:

$$\theta_B = \tan^{-1}(1/\omega^2\text{LC}) \tag{13.13}$$

For a source frequency of 2 GHz, (13.13) yields $\theta_B = 30°$, which is in good agreement with the FDTD result of Fig. 13.11(c). Evidently, FDTD captures the whole range of wave effects in this structure, including a readily apparent field localization toward the edges of the L–C grid outside the resonance cone (also apparent in the simulation results of [26]).

13.5 PERIODIC GEOMETRIES REALIZING METAMATERIAL STRUCTURES

The dispersion analysis of a periodic structure can be performed by simulating a single unit cell of the structure, terminated with periodic boundary conditions. Originally cast in the frequency domain, periodic boundary conditions can be translated into the framework of the FDTD method, hence becoming a useful tool for the analysis of periodic geometries realizing metamaterials.

In this and following sections, we focus on applying the sine-cosine FDTD method ([28] and Section 13.3.3 of [15]) to compute the dispersion diagrams of periodic structures employed in metamaterial applications. The sine-cosine method was employed in [29] to analyze a 2-D negative-refractive-index transmission line [4]. In [30], this method was extended to account for leaky-wave radiation from the same structure, thereby enabling an efficient FDTD technique to concurrently compute the attenuation and phase constants of fast waves in periodic geometries. Finally, to investigate the possibility of transferring concepts of negative-refractive-index transmission lines from the microwave regime to the optical regime, a conformal periodic FDTD analysis of plasmonic nanoparticle arrays in a mesh of triangular Yee cells was presented in [31].

Despite their several advantages, techniques based on using periodic boundary conditions suffer from the shortcoming that they cannot immediately incorporate nonperiodic sources and boundary conditions. This is an important limitation, since many practical devices employ periodic substrates or superstrates that closely interact with adjacent nonperiodic structures. This problem was first addressed in the context of the frequency-domain method of moments in [32] by invoking the array-scanning method [33]. In the context of FDTD, the interaction of nonperiodic sources with periodic structures was considered in [34–37], again by employing the array-scanning method. Moreover, Ref. [37] provided an efficient framework for incorporating nonperiodic boundary conditions into single unit-cell simulations, paving the way for modeling various guided-wave and radiating structures constructed over periodic substrates.

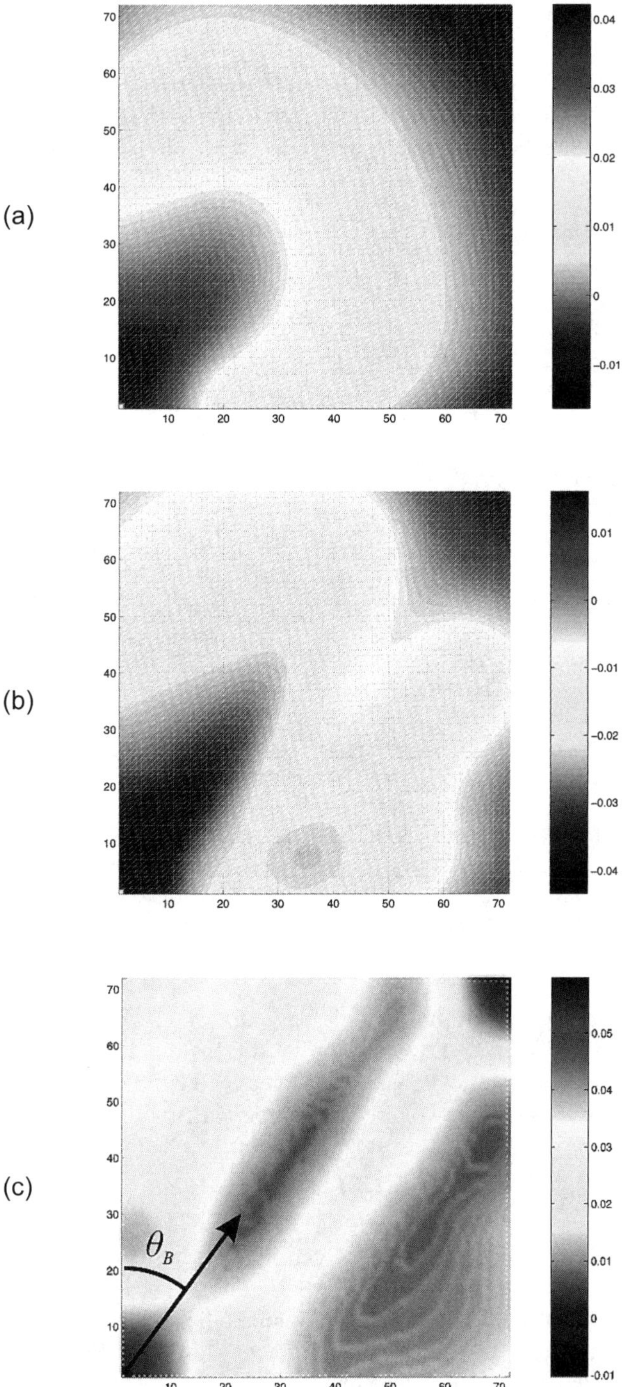

Fig. 13.11 Snapshot visualizations of the FDTD-computed distribution of E_z (V/mm) along an observation plane located 5 mm above the L–C grid of Fig. 13.10 at (a) 600; (b) 1922; and (c) 10,240 time-steps. The evolution of a beam angle (resonance cone) at 30.9° is observed.

13.6 THE SINE-COSINE METHOD

Figure 13.12 illustrates the problem under consideration in its general form. Here, a wideband nonperiodic source is shown exciting an infinite, periodic 2-D structure having a spatial periodicity, d_x, along the x-direction, and a spatial periodicity, d_y, along the y-direction.

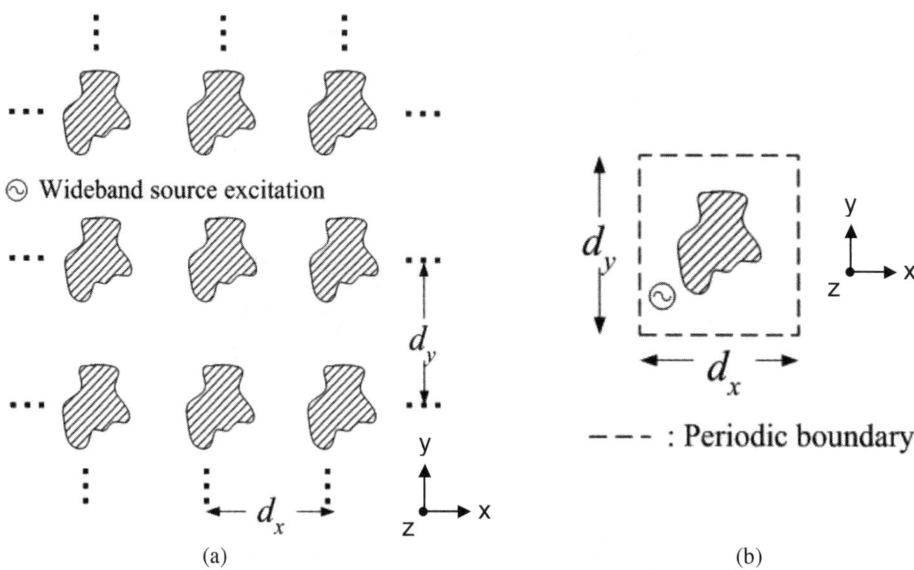

Fig. 13.12 Geometry of a wideband, possibly nonperiodic source exciting an infinite periodic 2-D structure: (a) Original geometry; (b) equivalent geometry for the sine-cosine method. *Adapted from:* D. Li and C. D. Sarris, *IEEE Trans. Microwave Theory and Techniques*, Vol. 56, 2008, pp. 1928–1937, ©2008 IEEE.

Instead of approximating the infinite periodic structure of Fig. 13.12(a) by truncating it, the computational domain of Fig. 13.12(b) can be used. In this domain, periodic boundary conditions are applied to the electric field phasors (denoted by the tilde symbol, ~) situated at the rectangular dashed-line locus along the two directions of periodicity:

$$\tilde{\boldsymbol{E}}(\boldsymbol{r}+\boldsymbol{p}) = \tilde{\boldsymbol{E}}(\boldsymbol{r})e^{-j\boldsymbol{k}_p\cdot\boldsymbol{p}} \tag{13.14}$$

where $\boldsymbol{p} = \hat{\boldsymbol{x}}d_x + \hat{\boldsymbol{y}}d_y$ is the lattice vector of the periodic structure, and $\boldsymbol{k}_p = \hat{\boldsymbol{x}}k_x + \hat{\boldsymbol{y}}k_y$ is a Floquet wavevector.

It is important to note that, if the source is nonperiodic, the computational domain of Fig. 13.12(b) does *not* conform to the original problem shown in Fig. 13.12(a). Instead, Fig. 13.12(b) models a *different* problem — the one shown in Fig. 13.13, consisting of an infinite array of phase-shifted, periodic replicas of the desired original single source. These phase shifts are given by $\phi_x = k_x d_x$ and $\phi_y = k_y d_y$. In a later section, the array-scanning technique will be used to isolate the effect of the desired original single source from the combined effect of the phased array of sources shown in Fig. 13.13.

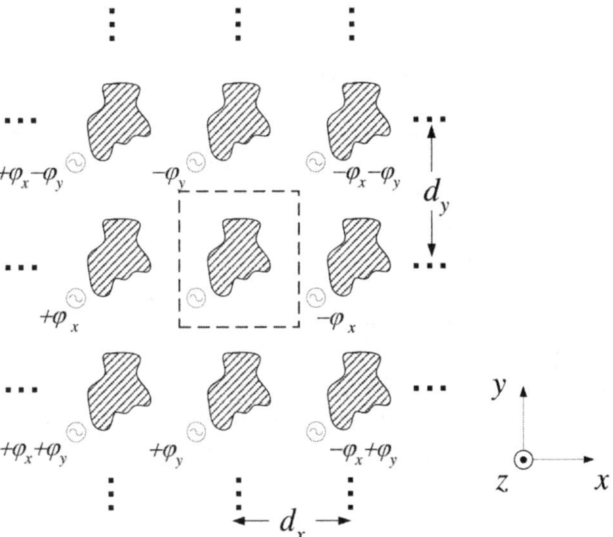

Fig. 13.13 Electromagnetic wave interaction geometry consistent with the computational domain of Fig. 13.12(b), showing an infinite array of phase-shifted, periodic replicas of the original single source to be modeled in Fig. 13.12(a). *Adapted from:* D. Li and C. D. Sarris, *IEEE Trans. Microwave Theory and Techniques*, Vol. 56, 2008, pp. 1928–1937, ©2008 IEEE.

For the moment, let us proceed to dig deeper into the sine-cosine method. Consider a field expansion in terms of Floquet modes in a periodic structure of lattice vector $\boldsymbol{p}$ inverse Fourier-transformed from the frequency domain to the time domain:

$$\begin{aligned} \boldsymbol{E}(\boldsymbol{r},t) &= \mathrm{Re}\sum_{p} e^{-j\boldsymbol{k}_p\cdot\boldsymbol{r}} \frac{1}{2\pi} \int_{\omega(\boldsymbol{k}_p)} \boldsymbol{E}(\boldsymbol{k}_p,\omega)e^{j\omega t}d\omega = \mathrm{Re}\sum_{p} e^{-j\boldsymbol{k}_p\cdot\boldsymbol{r}}\boldsymbol{E}(\boldsymbol{k}_p,t) \\ &= \mathrm{Re}\sum_{p}\left[\boldsymbol{E}_c^p(\boldsymbol{r},t) - j\boldsymbol{E}_s^p(\boldsymbol{r},t)\right] \end{aligned} \tag{13.15}$$

where $\omega(\boldsymbol{k}_p)$ is either a discrete or a continuous spectrum of frequencies corresponding to the Floquet wavevector, $\boldsymbol{k}_p$, and:

$$\boldsymbol{E}_c^p(\boldsymbol{r},t) = \cos(\boldsymbol{k}_p\cdot\boldsymbol{r})\boldsymbol{E}(\boldsymbol{k}_p,t)\,; \qquad \boldsymbol{E}_s^p(\boldsymbol{r},t) = \sin(\boldsymbol{k}_p\cdot\boldsymbol{r})\boldsymbol{E}(\boldsymbol{k}_p,t) \tag{13.16a, b}$$

Note that these two waves have identical frequency spectra because they share a common temporal dependence. Moreover:

$$\begin{aligned} \boldsymbol{E}_c^p(\boldsymbol{r}+\boldsymbol{p},\,t) &= \cos(\boldsymbol{k}_p\cdot\boldsymbol{r} + \boldsymbol{k}_p\cdot\boldsymbol{p})\,\boldsymbol{E}(\boldsymbol{k}_p,t) \\ &= \cos(\boldsymbol{k}_p\cdot\boldsymbol{p})\cos(\boldsymbol{k}_p\cdot\boldsymbol{r})\,\boldsymbol{E}(\boldsymbol{k}_p,t) - \sin(\boldsymbol{k}_p\cdot\boldsymbol{p})\sin(\boldsymbol{k}_p\cdot\boldsymbol{r})\,\boldsymbol{E}(\boldsymbol{k}_p,t) \\ &= \cos(\boldsymbol{k}_p\cdot\boldsymbol{p})\,\boldsymbol{E}_c^p(\boldsymbol{r},t) - \sin(\boldsymbol{k}_p\cdot\boldsymbol{p})\,\boldsymbol{E}_s^p(\boldsymbol{r},t) \end{aligned} \tag{13.17}$$

Similarly:

$$E_s^p(\boldsymbol{r}+\boldsymbol{p},\,t) \;=\; \sin(\boldsymbol{k}_p\cdot\boldsymbol{p})\,E_c^p(\boldsymbol{r},t) \;+\; \cos(\boldsymbol{k}_p\cdot\boldsymbol{p})\,E_s^p(\boldsymbol{r},t) \tag{13.18}$$

Therefore, the Floquet waves, $E_c^p(\boldsymbol{r},t)$ and $E_s^p(\boldsymbol{r},t)$, are shown to satisfy the "sine-cosine" boundary conditions of [28]. Note that it is straightforward to implement (13.17) and (13.18) in the discrete-time framework of FDTD, since all terms of these equations are evaluated at the same time-step.

This formulation offers some useful insights into the sine-cosine method. Clearly, these two waves are neither monochromatic nor at phase quadrature in time. In fact, our sine-/cosine waves are distinguished based on their *spatial* dependence, rather than their temporal dependence. Therefore, they can be excited by identical *broadband* sources instead of sine-/cosine modulated ones, provided that the frequency spectrum of such sources includes $\omega(\boldsymbol{k}_p)$. With $E_c^p(\boldsymbol{r},t)$ and $E_s^p(\boldsymbol{r},t)$ being excited (in their respective meshes), their spectral analysis yields all frequencies, $\omega(\boldsymbol{k}_p)$, at once. This is demonstrated through the numerical results of the following section.

13.7 DISPERSION ANALYSIS OF A PLANAR NEGATIVE-REFRACTIVE-INDEX TRANSMISSION LINE

In this section, we consider the application of the sine-cosine method of Section 13.6 to analyze the dispersion properties of a planar negative-refractive-index transmission-line structure [4, 37]. In [37], this 2-D structure was assumed to be periodic in both the x- and y-directions with the periodicity, $d_x = d_y = 8.4$ mm, and was comprised of the unit cell shown in Fig. 13.14. Each unit cell resided on a substrate of thickness $h = 1.52$ mm and relative permittivity $\varepsilon_r = 3$ over a PEC ground plane. The PEC microstrip lines had a width of $w = 0.75$ mm and a 50Ω characteristic impedance. To exhibit an equivalent negative refractive index at 1 GHz, these lines were assumed to be loaded as shown with series capacitors, C = 3.34 pF, and shunt inductors, L = 16.02 nH.

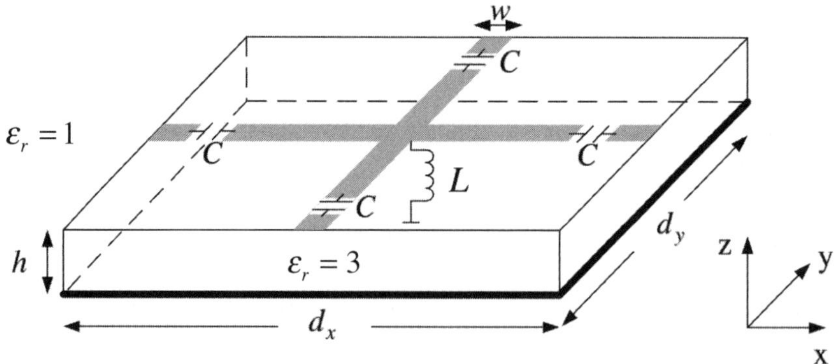

Fig. 13.14 Unit cell of the 2-D planar negative-refractive-index transmission line, where L = 16.02 nH and C = 3.34 pF. Note that by removing the shunt inductor and the four series capacitors (thereby realizing an unloaded unit cell with continuous microstrip lines), this structure becomes the unit cell of a 2-D planar *positive*-refractive-index transmission line. *Adapted from:* D. Li and C. D. Sarris, *IEEE Trans. Microwave Theory and Techniques*, Vol. 56, 2008, pp. 1928–1937, ©2008 IEEE.

The negative-refractive-index transmission-line unit cell shown in Fig. 13.14 was discretized in a $22 \times 22 \times 16$ Yee-cell FDTD mesh, wherein 3 of the 16 Yee cells in the z-direction modeled the substrate. The open-region extension of the FDTD mesh to infinity in the $+z$-direction was simulated by a UPML ABC [15]. This absorber consisted of 10 cells with a fourth-order polynomial conductivity grading, assuming a maximum conductivity value, $\sigma_{max} = 0.01194/\Delta$, with Δ being the Yee-cell size in the direction of the mesh truncation (hence, in this case $\Delta = \Delta z$). The same absorber was used to simulate open-region FDTD mesh truncations in all of the numerical simulations reviewed here.

The two sine-cosine grids were excited by a modulated Gaussian pulse that covered the frequency range, 0.5 – 5 GHz. This pulse, $\exp\{-[(t-t_0)/t_w]^2\}\sin(2\pi f_c t)$, with $t_w = 624$ ps and $t_0 = 3t_w$, was applied to three collinear E_z components inside the substrate extending from the ground plane to the microstrip line in mesh cells (6, 11, 1), (6, 11, 2), and (6, 11, 3). The time-step was set to 0.723 ps and 60,000 time-steps were performed for three cases of $k_x d_x = 0.0833\pi$, 0.167π, and 0.333π, while $k_y = 0$. Hence, all three points were along the Γ–X portion of the Brillouin diagram of the unit cell structure occupied by three transverse magnetic (TM) waves, as had been shown in previous studies [29]: a backward wave, a forward wave, and a surface wave.

For each of the three cases studied in [37] ($k_x d_x = 0.0833\pi$, 0.167π, and 0.333π), the left panels of Fig. 13.15 show the normalized magnitude of the Fourier transform of E_z within the substrate of the unit cell of Fig. 13.14, as computed by the sine-cosine FDTD method from 0 to 5 GHz. These spectra are juxtaposed with the right panels of Fig. 13.15, which display the Γ–X part of the Brillouin diagram for the same unit cell, as independently determined using Ansoft's HFSS® software. For each case, the FDTD-computed E_z field exhibited multiple resonances. Importantly, these resonances corresponded very closely with the frequencies, $\omega(k_x d_x)$, given by the intersections of the Brillouin-diagram curves with the vertical line representing the $k_x d_x$ value assumed in that particular simulation. In other words, the resonant frequencies obtained using the sine-cosine FDTD method and HFSS were in excellent agreement.

Moreover, these results clearly showed that a *single* wideband run of the sine-cosine FDTD method, with the same excitation for each grid, could simultaneously determine all of the resonant frequencies. Note that the boundary conditions, (13.17) and (13.18), which enforce the Floquet wavevector, are independent of frequency. This permits setting up an eigenvalue problem in the time domain, where *only* the modes with that given wavevector are excited. In fact, this is analogous to the way that FDTD has been used for many years to characterize cavity resonances and waveguide dispersion [38] over a broad bandwidth. This fundamental advantage of FDTD modeling (relative to frequency-domain methods) is also preserved within the context of the sine-cosine analysis of periodic structures.

13.8 COUPLING THE ARRAY-SCANNING AND SINE-COSINE METHODS

This section describes how the combination of periodic boundary conditions with a broadband source can lead to the solution of the difficulty illustrated in Fig. 13.13. There, in an attempt to apply the sine-cosine method, the electromagnetic response of a periodic structure to a single (nonperiodic) source was seen to be contaminated by the appearance of an artifact—an array of phase-shifted, periodic repetitions of the original source. The goal of the array-scanning technique is to isolate the effect of the single original source without sacrificing the computational efficiency and elegance of the sine-cosine method.

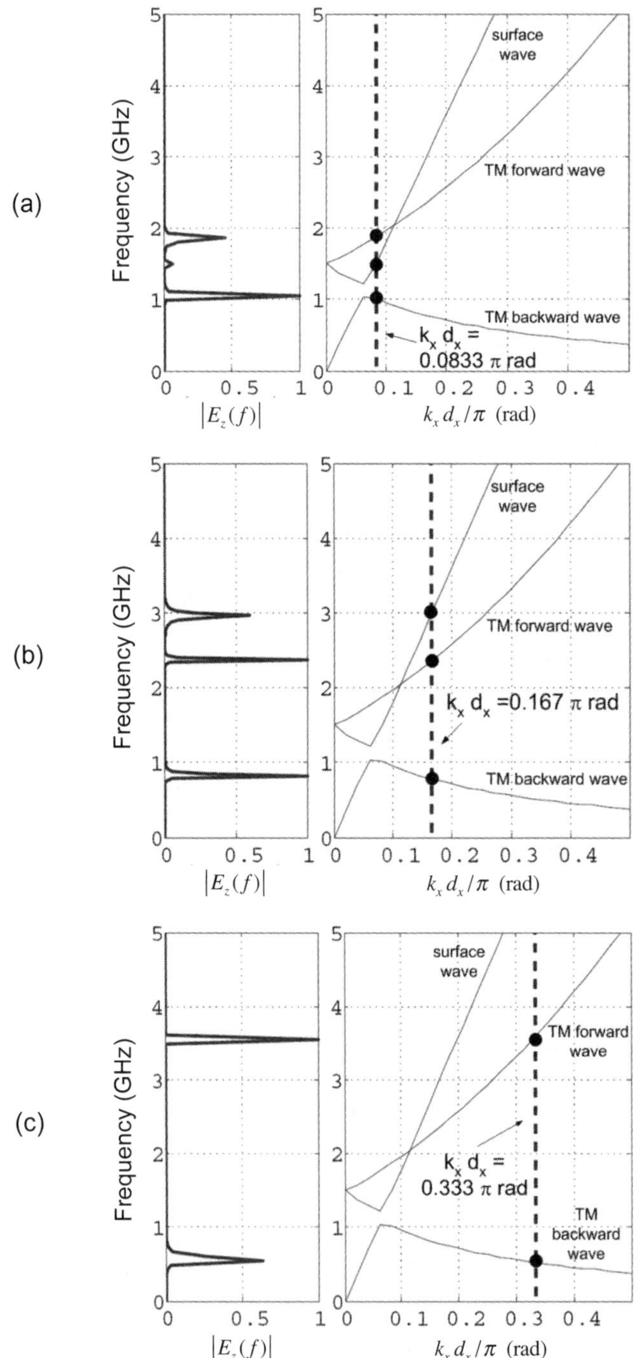

Fig. 13.15 Left panels: Normalized magnitude of Fourier transform of E_z within the substrate of the NRI unit cell of Fig. 13.14, determined by the sine-cosine FDTD method for (a) $k_x d_x = 0.0833\pi$; (b) $k_x d_x = 0.167\pi$; and (c) $k_x d_x = 0.333\pi$, while $k_y = 0$. Right panels: dispersion diagram (Γ–X) for the unit cell of Fig. 13.14, determined by HFSS. *Source:* D. Li and C. D. Sarris, *IEEE Trans. Microwave Theory and Techniques*, Vol. 56, 2008, pp. 1928–1937, ©2008 IEEE.

Let $\boldsymbol{E}_{\text{array}}(\boldsymbol{r}_0, \boldsymbol{k}_p, t)$ be the electric field determined by the sine-cosine method at a point, $\boldsymbol{r}_0$, within the unit cell, for a Floquet wavevector, $\boldsymbol{k}_p = \hat{\boldsymbol{x}}k_x + \hat{\boldsymbol{y}}k_y$, within the Brillouin zone of the structure. (Hence, $-\pi/d_x \le k_x \le \pi/d_x$ and $-\pi/d_y \le k_y \le \pi/d_y$.) The electric field, $\boldsymbol{E}_0$, at this point that is only due to the original source can be found by integrating over k_x, k_y [33]:

$$\boldsymbol{E}_0(\boldsymbol{r}_0, t) = \frac{d_x d_y}{4\pi^2} \int_{-\pi/d_x}^{\pi/d_x} \int_{-\pi/d_y}^{\pi/d_y} \boldsymbol{E}_{\text{array}}(\boldsymbol{r}_0, \boldsymbol{k}_p, t)\, dk_x\, dk_y \tag{13.19}$$

Since (13.19) is a continuous integral, whereas only N discrete k_x points and M discrete k_y points are sampled, this equation is approximated at $t = l\Delta t$ (the lth FDTD time-step) by the sum:

$$\boldsymbol{E}_0(\boldsymbol{r}_0, l\Delta t) \approx \frac{1}{NM} \sum_{n=-N/2}^{N/2} \sum_{m=-M/2}^{M/2} \boldsymbol{E}_{\text{array}}\left(\boldsymbol{r}_0,\ \hat{\boldsymbol{x}}\frac{2\pi n}{Nd_x} + \hat{\boldsymbol{y}}\frac{2\pi m}{Md_y},\ l\Delta t\right) \tag{13.20}$$

By invoking periodic boundary conditions (13.14), a modified form of (13.20) can be employed to determine the electric field at points outside the simulated unit cell. In particular:

$$\boldsymbol{E}_0(\boldsymbol{r}_0 + \boldsymbol{p}_{i,j}, l\Delta t) \approx \frac{1}{NM} \sum_{n=-N/2}^{N/2} \sum_{m=-M/2}^{M/2} \boldsymbol{E}_{\text{array}}\left(\boldsymbol{r}_0,\ \hat{\boldsymbol{x}}\frac{2\pi n}{Nd_x} + \hat{\boldsymbol{y}}\frac{2\pi m}{Md_y},\ l\Delta t\right) e^{-j\boldsymbol{k}_p \cdot \boldsymbol{p}_{i,j}} \tag{13.21}$$

with $\boldsymbol{p}_{i,j} = \hat{\boldsymbol{x}} i d_x + \hat{\boldsymbol{y}} j d_y$ for integer i, j.

The sampling theorem is the basic principal that governs the number of points, $N \times M$, that must be sampled inside the Brillouin zone. If the fields in the driven periodic structure under study are spatially limited within the area $(-W_x \le x \le W_x,\ -W_y \le y \le W_y)$, then the sampling rates, $S_x = N/(2\pi/d_x)$ and $S_y = M/(2\pi/d_y)$ [units of samples/(rad/m)], should obey the inequalities:

$$2\pi S_x \ge 2W_x\,, \qquad 2\pi S_y \ge 2W_y \tag{13.22}$$

which lead to:

$$N \ge 2W_x/d_x\,, \qquad M \ge 2W_y/d_y \tag{13.23}$$

Practically, bounds for W_x and W_y can be deduced from the physics of the problem or through a convergence study. It is noteworthy that all sine-cosine FDTD simulations for different values of k_x, k_y are independent from each other. Hence, in a parallel-computing environment, there is no additional cost as the number of sampled Floquet wavevectors increases.

13.9 APPLICATION OF THE ARRAY-SCANNING METHOD TO A POINT-SOURCED PLANAR POSITIVE-REFRACTIVE-INDEX TRANSMISSION LINE

This section reviews the application in [37] of the FDTD array-scanning method of Section 13.8 to compute the substrate electric field within a planar, positive-refractive-index transmission line having the same unit cell shown in Fig. 13.14, but *without the L–C loading*. Excitation to this 2-D periodic structure was provided by a nonperiodic E_z hard source (incompatible with the basic sine-cosine method). The goal was to investigate the convergence of the array-scanning method with respect to the number of k points used, as well as the convergence of a "brute-force" approach with respect to the transverse dimensions of the planar structure included in the model.

To this end, the computational domain used for the FDTD sine-cosine-based array-scanning method consisted of simply 18 positive-refractive-index transmission-line unit cells along the y-axis, as shown in the inset of Fig. 13.16(a). Periodic boundary conditions (PBCs) were applied on the lateral sides of each of these unit cells to simulate extension of the array to infinity in the (transverse) $\pm x$-directions. At the center of unit cell #1, the vertical electric field, E_z, within the substrate was sinusoidally hard-sourced at 1 GHz. Four simulations were conducted to investigate the effect of using an increasing number of k_x points: $N = 4$, 8, 16, and 32. Each simulation was run for 16,384 time-steps (with the same spatial and temporal discretizations as in Section 13.7) to obtain the sinusoidal steady-state E_z field within the substrate at the center of each transmission-line unit cell.

Figure 13.16(a) graphs the convergence of the E_z field distribution obtained using the sine-cosine-based array-scanning method as a function of N [37]. This figure shows convergence for $N = 16$ k_x-points, equivalent to a sampling rate of 0.125π rad/m in the wavenumber domain.

For purposes of comparison, the conventional FDTD method was also used to model this structure. Clearly, an infinite number of transmission-line unit cells in the $\pm x$-directions could not be simulated. Hence, the transverse extent of the computational domain was truncated to have only a few cells located adjacent to the $\pm x$ sides of a primary cell along the y-axis. It was reasoned that the computed results for the E_z field along the y-axis would eventually converge as the transverse extent of the structure being modeled increased sufficiently. As shown in the inset of Fig. 13.16(b), five computational domains of variable transverse dimensions were modeled in this manner, each with 18 transmission-line unit cells along the y-axis: 5 transverse unit cells (2 on each side of a y-axis cell); 9 transverse unit cells (4 on each side of a y-axis cell); 13 transverse unit cells (6 on each side of a y-axis cell); 17 transverse unit cells (8 on each side of a y-axis cell); and finally 31 transverse unit cells (15 on each side of a y-axis cell). Each of these simulations was run for 16,384 time-steps (with $\Delta t = 0.723$ ps) to permit direct comparison with the results of the sine-cosine-based array-scanning method.

Figure 13.16(b) graphs the convergence of the E_z field distribution obtained using the finite-structure conventional FDTD method as a function of the transverse dimensions of the modeled geometry [37]. This figure shows that 17 unit cells in the transverse direction were needed for convergence to within 1% of the fields obtained for the infinite periodic case.

Figure 13.16(c) compares the convergence properties of the FDTD sine-cosine-based array-scanning method and the finite-structure conventional FDTD method by graphing the error norm [37]:

$$\mathcal{E} = \sum_j \left[\frac{E_z(j) - E_z^{\text{ref}}(j)}{E_z^{\text{ref}}(j)} \right]^2 \tag{13.24}$$

Here, $E_z(j)$ is the z-component of the electric field in the middle of the substrate along the y-axis, calculated with either method [plotted in Figs. 13.16(a) and (b)], and $E_z^{\text{ref}}(j)$ is the same field calculated with a very large finite-structure conventional FDTD simulation. In Fig. 13.16(c), the error norm, $\mathcal{E}$, is plotted with respect to the number of k_x points used for the array-scanning-based field calculation, and with respect to the number of transmission-line unit cells in the transverse direction used for the finite-structure conventional FDTD calculation.

We observe that Figs. 13.16(a) and (b) show that the electric-field amplitude decays with distance from the source. As discussed in [5], this amplitude decay, and the resulting loss of the evanescent spectral components of the source, can be compensated for by introducing the negative-refractive-index transmission-line unit cells of Fig. 13.14. This is discussed in the next section in the context of a transmission-line-based microwave "perfect lens."

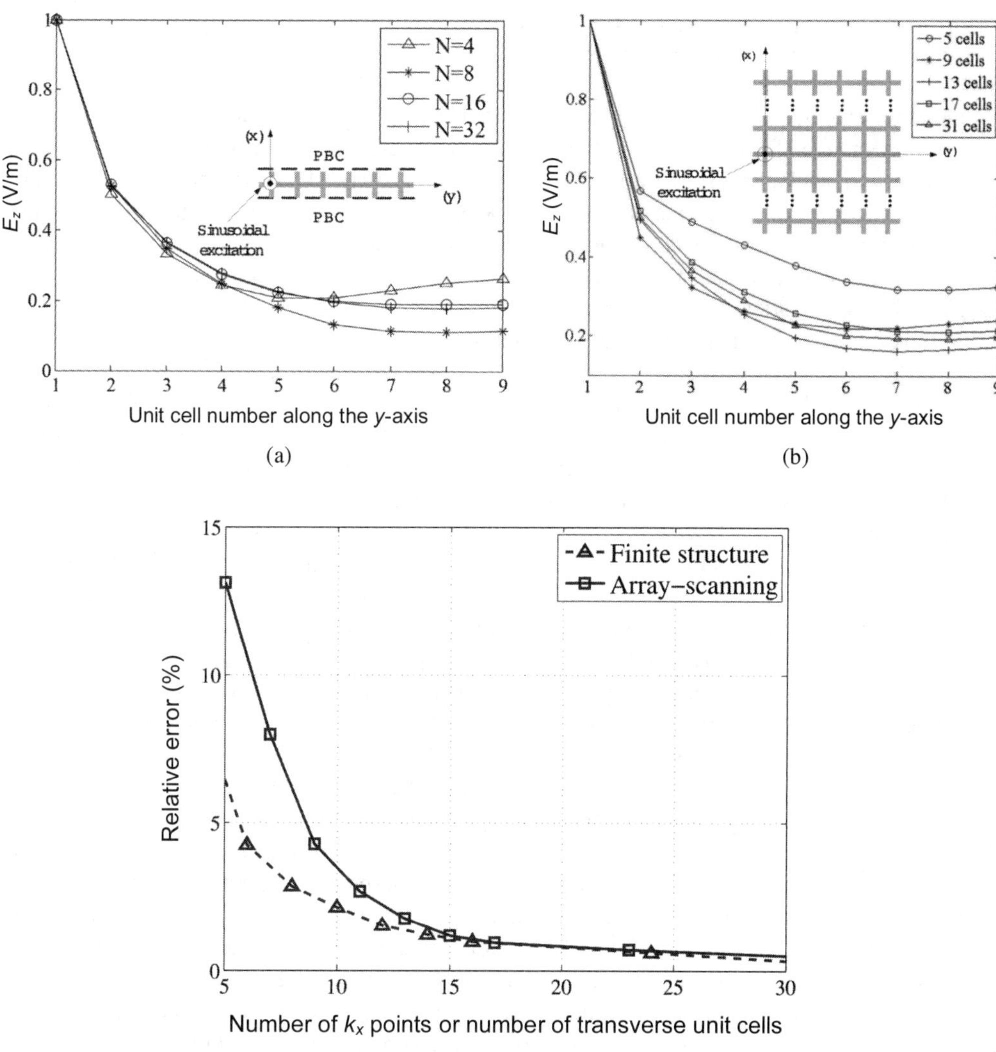

Fig. 13.16 Convergence properties of the FDTD sine-cosine-based array-scanning method and the conventional FDTD method when used to model a planar, periodic (in 2-D), positive-refractive-index transmission line having the unit cell shown in Fig. 13.14, but without the L–C loading. Sinusoidal excitation to this structure was provided by a single 1 V/m, 1-GHz, E_z hard source (incompatible with the basic sine-cosine method) located at the center of unit cell #1 on the y-axis of the transmission line. (a) Convergence of the vertical electric field, E_z, computed by the FDTD array-scanning method within the substrate along the y-axis, as a function of the number ($N = 4, 8, 16, 32$) of k_x points used, with the planar transmission line modeled as being of infinite extent in the x-direction. (b) Convergence of the vertical electric field, E_z, computed by the conventional FDTD method within the substrate along the y-axis, as a function of the total number of transmission-line unit cells ($5, 9, 13, 17, 31$) modeled in the (transverse) x-direction before arbitrary truncation of the structure. (c) Comparison of the error norm, $\mathcal{E}$, of (13.24) derived from the array-scanning-method results of (a) and the conventional-FDTD-method results of (b). *Source:* D. Li and C. D. Sarris, *IEEE Trans. Microwave Theory and Techniques*, Vol. 56, 2008, pp. 1928–1937, ©2008 IEEE.

13.10 APPLICATION OF THE ARRAY-SCANNING METHOD TO THE PLANAR MICROWAVE "PERFECT LENS"

This section reviews the application reported in [37] of the FDTD sine-cosine-based array-scanning method of Section 13.8 to compute the substrate electric field within a microwave implementation of Pendry's concept of a "perfect lens" [2], which has been experimentally demonstrated [5]. This 2-D planar structure was comprised of both negative-refractive-index transmission-line unit cells, as shown in Fig. 13.14, and identical unit cells, but without the L–C loading, which exhibit a positive refractive index. The parameters of both types of cells and of their FDTD discretization were the same as in Section 13.7.

Referring to the geometry diagram depicted in Fig. 13.17(a), the planar lens consisted of a total of 10 transmission-line unit cells in the y-direction, of which the middle 5 were provided with the L–C loading of Fig. 13.14 to exhibit a negative refractive index at 1 GHz. In the transverse (x-direction), only a single unit cell with periodic boundary conditions was required to implement the FDTD sine-cosine-based array-scanning method of Section 13.8, with 16 k_x points calculated. For purposes of comparison, 17 unit cells in the x-direction were used for the finite-structure conventional FDTD method. Excitation was provided by a 1-GHz sinusoidal E_z hard source located within the substrate on the y-axis 2.5 unit cells from the front interface of the negative-refractive-index region. The image plane was located within the substrate along a transverse locus positioned 2.5 unit cells beyond the back interface of the negative-index region.

Each FDTD model was run for 60,000 time-steps to reach the sinusoidal steady state. Importantly, using a single computer processor, the sine-cosine-based array-scanning method ran 8.36 times *faster* than the finite-structure conventional FDTD method to complete the simulation.

Figure 13.17(a) shows that the results of both FDTD modeling techniques were in excellent agreement [37]. Both techniques indicated growth of the E_z amplitude along the y-axis within the negative-refractive-index region, apparently due to resonant coupling between the surfaces of this region. However, the matching of the positive- and negative-index regions was imperfect. This was mainly due to the fringing capacitance at the microstrip gaps where the series lumped capacitors were placed, which contributed to the total gap capacitance. As a result, the E_z field growth began outside of the negative-index region, arising from the interaction of the incident and reflected waves. This was also evident in the experimental results of [5]. Because this growth occurred slowly during the time-domain simulation, there was obvious value (from a running-time perspective) in reducing the size of the computational domain by using the FDTD sine-cosine-based array-scanning method.

Figure 13.17(b) shows the E_z amplitude along the x-direction at the source and image planes, determined using both FDTD techniques [37]. For comparison, the diffraction-limited image of the source for an all positive-index space is also shown. We note that the half-power beamwidth of the "perfect lens" image extends over four unit cells, as opposed to six unit cells for the diffraction-limited image. While still better than the conventional diffraction-limited case, this represents an imperfect restoration of the source at the image plane caused by mismatch between the positive- and negative-index regions and the finite spatial period of the lens structure. Nevertheless, these results are in excellent agreement with the measurements of [5] for the same structure, and comprise the first full-wave validation of those measurements.

In Fig. 13.17(b), the array-scanning E_z values, located beyond the single simulated unit cell in the transverse direction, were calculated by means of (13.21). Note that there are significant field values up to about ±6 unit cells relative to $x = 0$. As a result, applying (13.23) with $W_x \approx 6d_x$ yields $N \geq 12$ as a limit for the number of k_x points needed for the reconstruction of the field profile in the space domain. This is an *a posteriori* verification of the bounds of (13.23).

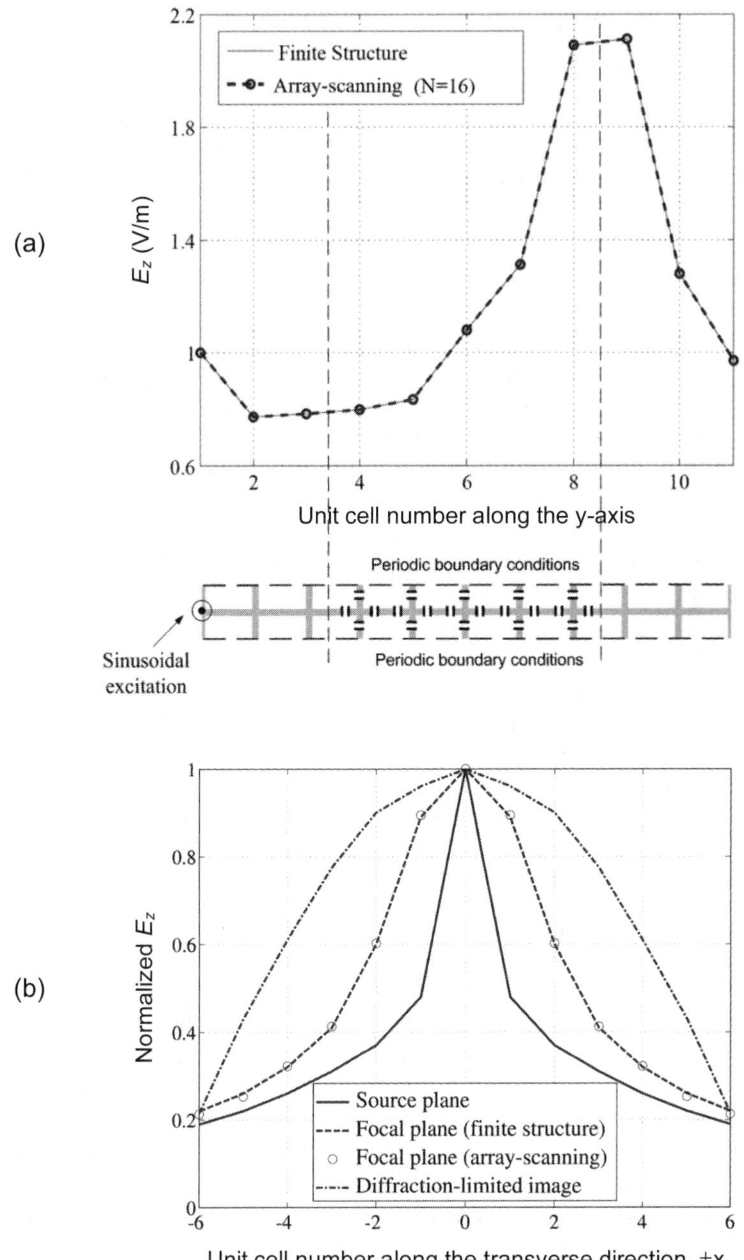

Fig. 13.17 Results of applying the FDTD sine-cosine-based array-scanning method (N=16) of Section 13.8 and the finite-structure conventional FDTD method (17 cells in the x-direction) to model a microwave implementation of a "perfect lens" [2, 5]. (a) Computed vertical electric field, E_z, along the y-axis in the middle of the substrate; (b) computed E_z along the transverse (x-direction) at the source and the image (focal) planes, with the fields normalized to their maximum amplitude. *Source:* D. Li and C. D. Sarris, *IEEE Trans. Microwave Theory and Techniques*, Vol. 56, 2008, pp. 1928–1937, ©2008 IEEE.

13.11 TRIANGULAR-MESH FDTD TECHNIQUE FOR MODELING OPTICAL METAMATERIALS WITH PLASMONIC ELEMENTS

The extension of the negative-refractive-index transmission-line (NRI-TL) concept to optical frequencies is appealing due to both its potential for broad bandwidth and its fundamental idea of developing optical analogs of lumped-circuit elements. This possibility was recently addressed in [6, 39, 40], where lattices of plasmonic spheres and ellipsoids were identified as counterparts of microwave NRI-TL structures in the infrared and visible regime. Furthermore, 1-D and 2-D arrays of silver nanoparticles have been shown to exhibit backward-wave bands, the signature property of NRI media, within their Brillouin zone [7, 41]. In this regard, Ref. [7] studied a deeply sub-wavelength unit cell of a 2-D lattice comprised of silver rods. A plasmonic backward-wave mode found in the Brillouin zone of this photonic crystal was used to realize Pendry's "perfect lens" [2], albeit with a very small bandwidth (~0.01%). Similarly challenging issues of analysis and design are arising as candidate NRI topologies are proposed at optical wavelengths.

In principle, the FDTD method is capable of rigorously analyzing the electromagnetic wave phenomena relevant to the analysis, design, and optimization of NRI metamaterials at optical wavelengths. However, when plasmonic modes are involved, numerical studies show that the staircasing errors of FDTD, which have been well known for many years [42, 43], can become quite pronounced and result in spurious resonances. In a dispersion analysis, these spurious resonances manifest themselves in mesh-dependent, poorly convergent bands. Similar effects have been observed when applying FDTD to model non-plasmonic scattering problems having significant surface-wave contribution [44].

This section discusses a triangular-mesh formulation of FDTD aimed at overcoming the artifacts associated with staircasing, especially for optical metamaterials with plasmonic elements [31]. Similar to the simple, explicit schemes proposed in [44, 45], the triangular-mesh FDTD formulation is extended to incorporate periodic boundary conditions, thereby reducing the computational domain to a single unit cell of a periodic structure via the sine-cosine method.

13.11.1 Formulation and Update Equations

A 2-D transverse electric (TE) case is considered, with E_x, E_y, and H_z field components. As shown in Fig. 13.18(a), the computational domain is divided into triangular elements, generated by a quality Delauny triangular mesh generator [46]. In each element, tangential electric field components are sampled at the center of each edge, and the perpendicular magnetic field component is sampled at a "centroid" point whose coordinates are computed by averaging the coordinates of the triangle vertices. In [44, 45], H_z was sampled at the circumcenter of each element, which may possibly lie outside a triangle containing an obtuse angle. In that case, the distance between H_z sampling points in adjacent triangles could diminish, thereby impairing the stability of the resulting FDTD time-stepping scheme. This problem is overcome by sampling H_z at the centroid instead. Note that the two approaches tend to become equivalent in the limit of a dense mesh, where the Delauny mesh generator is expected to produce nearly equilateral triangles with almost collocated centroids and circumcenters.

Upon setting up the mesh, the derivation of the field update equations proceeds as follows. For the H_z update, the integral form of Faraday's law is discretized. Considering the geometry of Fig. 13.18(b), the line integral of the electric field is computed along the triangular contour enclosing the element shown, while the surface integral of the magnetic field is approximated by the product of the H_z component at the centroid of the triangle times its area:

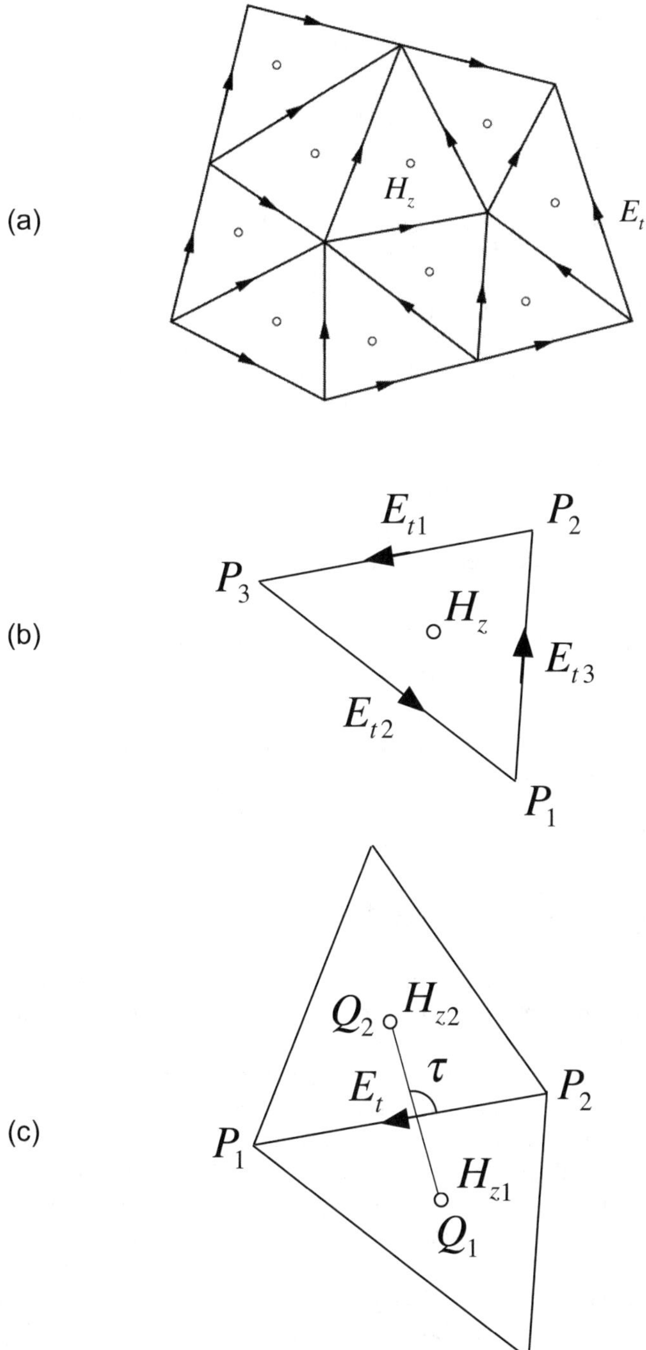

Fig. 13.18 Triangular FDTD mesh: (a) General layout for the TE case; (b) geometry illustrating the update equation for a perpendicular magnetic field component; (c) geometry illustrating the update equation for a tangential electric field component. *Source:* Y. Liu, C. D. Sarris, and G. V. Eleftheriades, *IEEE J. Lightwave Technology*, Vol. 25, 2007, pp. 938–945, ©2007 IEEE.

$$\mu \frac{\partial H_z}{\partial t} A \approx E_{t1} \left| \overline{P_2 P_3} \right| + E_{t2} \left| \overline{P_3 P_1} \right| + E_{t3} \left| \overline{P_1 P_2} \right| \tag{13.25}$$

Approximating the remaining time derivative by a second-order-accurate centered difference, the following updating equation is derived:

$$H_z^{n+1/2} = H_z^{n-1/2} + \frac{\Delta t}{\mu A} \left(E_{t1}^n \left| \overline{P_2 P_3} \right| + E_{t2}^n \left| \overline{P_3 P_1} \right| + E_{t3}^n \left| \overline{P_1 P_2} \right| \right) \tag{13.26}$$

where Δt is the time step and n is the time-step index.

Similarly, the update equation for the tangential electric field components stems from the following application of Ampere's law, using the line segment connecting the centroids of two adjacent triangles, shown in Fig. 13.18(c), as an Amperian path:

$$\frac{\partial D_t}{\partial t} \left| \overline{Q_1 Q_2} \right| \sin \tau \approx H_{z1} - H_{z2} \tag{13.27}$$

where D_t is the tangential electric flux-density component. Hence, the update equation for a D_t assumes the form:

$$D_t^{n+1} = D_t^n + \frac{\Delta t}{\left| \overline{Q_1 Q_2} \right| \sin \tau} \left(H_{z1}^{n+1/2} - H_{z2}^{n+1/2} \right) \tag{13.28}$$

The incorporation of a material dispersion characterizing a plasmonic component is straightforward. We use the constitutive relation $D_t = \varepsilon_r(\omega)\varepsilon_0 E_t$ and implement the Drude model for the dielectric permittivity specified in (13.1) via the corresponding auxiliary differential equation update for E reviewed in (13.2)–(13.4).

13.11.2 Implementation of Periodic Boundary Conditions

In a 2-D periodic structure of spatial period, d_x and d_y, along the x- and y-axes of a rectangular coordinate system, phasor field components one period away in either direction differ only by a constant attenuation and phase-shift term, $e^{-jk_x d_x}$ and $e^{-jk_y d_y}$, respectively, where $\boldsymbol{k} = \hat{\boldsymbol{x}} k_x + \hat{\boldsymbol{y}} k_y$ is a Bloch wavevector. This frequency-domain relationship is translated into the time-domain via the sine-cosine method of [28], discussed in detail earlier in Section 13.6. To determine the dispersion characteristics of the structure, we apply a modulated Gaussian point source to a unit cell with an excitation bandwidth covering the frequency range of interest. For a fixed Bloch wavevector $\boldsymbol{k}$, the resulting time-domain waveform of the field is sampled within the unit cell. Then, its Fourier transform (from the time domain to the frequency domain) reveals the resonances that represent the modal frequencies, $\omega(\boldsymbol{k})$.

The following considerations are specifically pertinent to the implementation of periodic boundaries in the triangular FDTD mesh. Consider, for example, a wave propagating along the y-axis with a Bloch wavevector, $\boldsymbol{k} = \hat{\boldsymbol{y}} k_y$, in a periodic structure of the unit cell shown in Fig. 13.19. Although the magnetic field node, $H_{z,b}$, is outside the unit cell, it is connected to the magnetic field node, $H_{z,a}$, at a distance, d_y, along the y-axis inside the cell, through the periodic boundary condition:

$$\tilde{H}_{z,b} = \tilde{H}_{z,a} e^{-jk_y d_y} \tag{13.29}$$

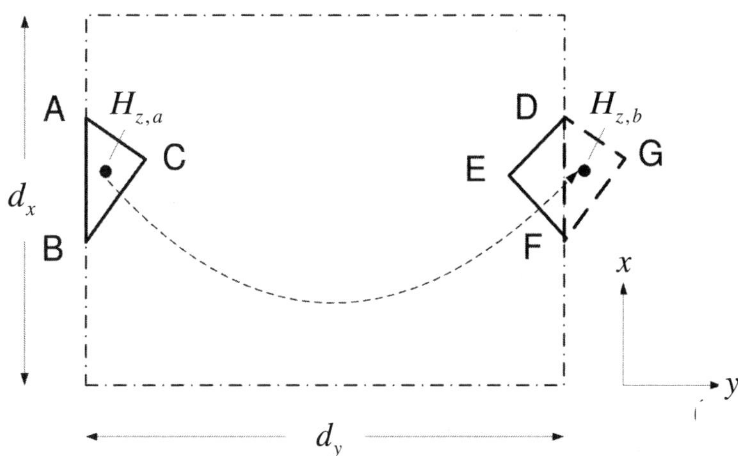

Fig. 13.19 Implementation of periodic boundary conditions (at the dot-dashed locus) in the triangular-mesh FDTD method. *Source:* Y. Liu, C. D. Sarris, and G. V. Eleftheriades, *IEEE J. Lightwave Technology*, Vol. 25, 2007, pp. 938–945, ©2007 IEEE.

Then, the tangential electric field component along DF can be updated according to the stencil of Fig. 13.18. However, the triangle enclosing the $H_{z,b}$ node has to be the same as the one enclosing the $H_{z,a}$ node for the two periodic boundaries to be identical, not just physically but numerically as well. Therefore, triangle DFG is generated by translating ABC by d_y along the y-axis. Evidently, the segmentation of the periodic boundaries is also identical. If the left boundary ($y = 0$) is divided in segments, $\overline{P_1 P_2 \ldots P_N}$, and the right boundary ($y = d_y$) is divided in segments, $Q_1 Q_2 \ldots Q_N$, then $|\overline{P_i P_{i+1}}| = |\overline{Q_i Q_{i+1}}|$ for $i = 1, 2, \ldots, N{-}1$. Similarly, the discretization of the lower and upper boundaries is identical.

Finally, modal field distributions can be obtained by Fourier transforming the sampled field values over a mesh of points. Note that the electric field at an arbitrary location within the computational domain can be interpolated by using Whitney 1-forms [47], knowing the tangential component along each edge.

13.11.3 Stability Analysis

As reported in [31], the triangular-mesh FDTD method can be proven to be conditionally stable for a Drude medium using the equivalent-circuit approach of [48]. Equations (13.25) and (13.27) are considered in the sinusoidal steady state, and are discretized in space with the triangular mesh of Fig. 13.18, but not in time. The resulting system is called semi-discretized. The geometry and nodes involved are shown in Fig. 13.20. Magnetic fields H_{z1} and H_{z2} are sampled at the centroids of triangles ABC and BCD, referred to as Q_1 and Q_2, respectively. The tangential electric field components at the centers of CA, AB, and BC are denoted as $\boldsymbol{E}_{t1}$, $\boldsymbol{E}_{t2}$, and $\boldsymbol{E}_{t3}$. The spatially discretized, sinusoidal steady-state versions of (13.25) and (13.27) involving the aforementioned field components are then:

$$-j\omega\mu\,\Delta_{\mathrm{ABC}}\,H_{z1} \;=\; \boldsymbol{E}_{t1}\cdot\overline{\mathrm{CA}} + \boldsymbol{E}_{t2}\cdot\overline{\mathrm{AB}} + \boldsymbol{E}_{t3}\cdot\overline{\mathrm{BC}} \tag{13.30}$$

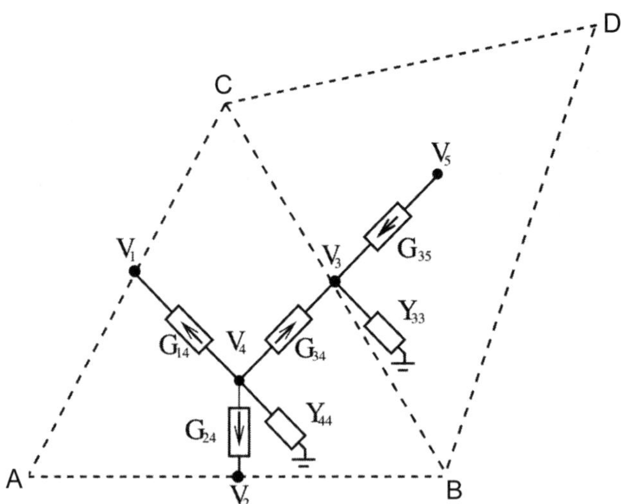

Fig. 13.20 Equivalent circuit of the triangular mesh FDTD. *Source:* Y. Liu, C. D. Sarris, and G. V. Eleftheriades, *IEEE J. Lightwave Technology*, Vol. 25, 2007, pp. 938–945, ©2007 IEEE.

$$j\omega\varepsilon(\omega)\left|\overline{Q_1 Q_2}\right|\sin\tau\, E_{t3} \;=\; H_{z1} - H_{z2} \tag{13.31}$$

where Δ_{ABC} is the area of triangle ABC. Introducing $V_1 = \boldsymbol{E}_{t1}\cdot\overline{\mathsf{CA}}$, $V_2 = \boldsymbol{E}_{t2}\cdot\overline{\mathsf{AB}}$, $V_3 = \boldsymbol{E}_{t3}\cdot\overline{\mathsf{BC}}$, $V_4 = H_{z1}$, and $V_5 = H_{z2}$, Equations (13.30) and (13.31) can be written as the linear system:

$$\begin{aligned} -Y_{44}V_4 &= G_{14}V_1 + G_{24}V_2 + G_{34}V_3 \\ Y_{33}V_3 &= G_{34}V_4 + G_{35}V_5 \end{aligned} \tag{13.32}$$

where $Y_{44} = j\omega\mu\,\Delta_{\mathsf{ABC}}$ is the admittance of a capacitor; $G_{14} = G_{24} = G_{34} = 1$ and $G_{35} = -1$ are gyrators; and $Y_{33} = j\omega\varepsilon(\omega)\left|\overline{Q_1 Q_2}\right|\sin\tau / \left|\overline{\mathsf{BC}}\right|$. If edge BC is within a lossless, nondispersive medium, Y_{33} is simply the admittance of a capacitor. If BC is within a lossless Drude medium, Y_{33} corresponds to a parallel L–C circuit; while if BC lies on the interface between two media, it can be characterized by an averaged $\varepsilon(\omega)$, which also corresponds to a parallel L–C circuit. Therefore, the circuit analog of the semi-discretized system is a lossless, passive network of gyrators, capacitors, and inductors.

As a result, the FDTD update equations arising from the time integration of this semi-discretized system are *not* associated with late-time numerical instability [48]. For stable, conventional leapfrog time-stepping, Δt is given by:

$$\Delta t \;=\; (K/c)\,\min_i l_i \tag{13.33}$$

where l_i is the length of the ith edge of the 2-D triangular mesh, c is the free-space speed of light, and K is a constant less than or equal to 0.5. All simulation results presented in the next section were obtained with K set to 0.1. No instability was observed for up to several million time-steps.

13.12 ANALYSIS OF A SUB-WAVELENGTH PLASMONIC PHOTONIC CRYSTAL USING THE TRIANGULAR-MESH FDTD TECHNIQUE

This section reviews the FDTD analysis reported in [31] of a sub-wavelength photonic crystal comprised of a 2-D array of circular silver microcylinders. Referring to Fig. 13.21, the silver cylinders were assumed to be arranged in a uniform, periodic lattice of square unit cells of size $a = c/\omega_p$, where c is the free-space speed of light and ω_p is the radian plasma frequency of silver. The diameter of each silver rod was assumed to be $d = 0.9a$, equivalent to approximately 1/7th of the free-space wavelength at the plasma frequency.

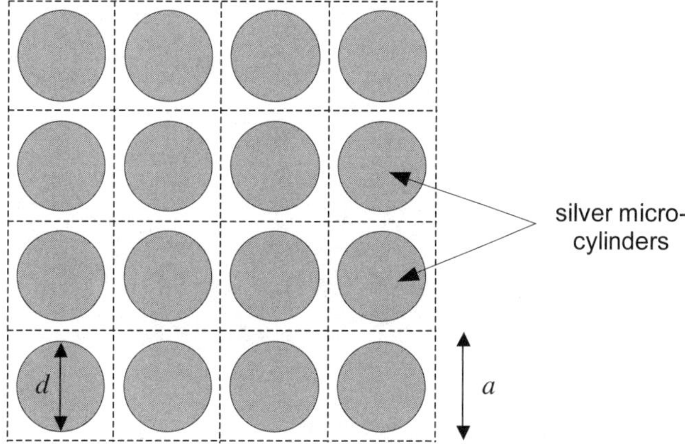

Fig. 13.21 Geometry of the 2-D plasmonic photonic crystal of [7]. *Adapted from:* Y. Liu, C. D. Sarris, and G. V. Eleftheriades, *IEEE J. Lightwave Technology*, Vol. 25, 2007, pp. 938–945, ©2007 IEEE.

The photonic crystal of Fig. 13.21 had been modeled in [7] using both quasi-static and frequency-domain full-wave (finite-element) techniques. Subsequently, Ref. [31] employed the triangular FDTD technique of Section 13.11 to implement a wideband periodic analysis of the same structure. Recall that this analysis proceeds by first fixing the phase shift between the periodic boundaries to correspond to a wavevector within the Brillouin zone, and then sampling and Fourier-transforming a field component inside the unit cell. The resonant frequencies associated with the wavevector of choice are determined by the position of the field resonances in the frequency domain.

To obtain more insights into the usefulness of the triangular mesh FDTD formulation, results obtained via the conventional (staircased) Cartesian mesh FDTD for frequencies up to $0.7\omega_p$ are presented first in Fig. 13.22 [31]. This figure shows the Fourier transform of H_z sampled inside the unit cell of the photonic crystal for several Bloch wavevectors, $\boldsymbol{k} = \hat{\boldsymbol{x}} k_x$ (along Γ–X). Although clear resonances can be seen up to $0.38\omega_p$, higher-frequency resonances present a noise-like pattern, rather than a resonant pattern. In addition, convergence of the resonant frequency peaks that appear above $0.38\omega_p$ cannot be achieved by mesh refinement. On the other hand, as shown in Fig. 13.23, the triangular-mesh FDTD method allows for a quickly convergent calculation of the resonant frequencies even at a relatively coarse mesh density [31].

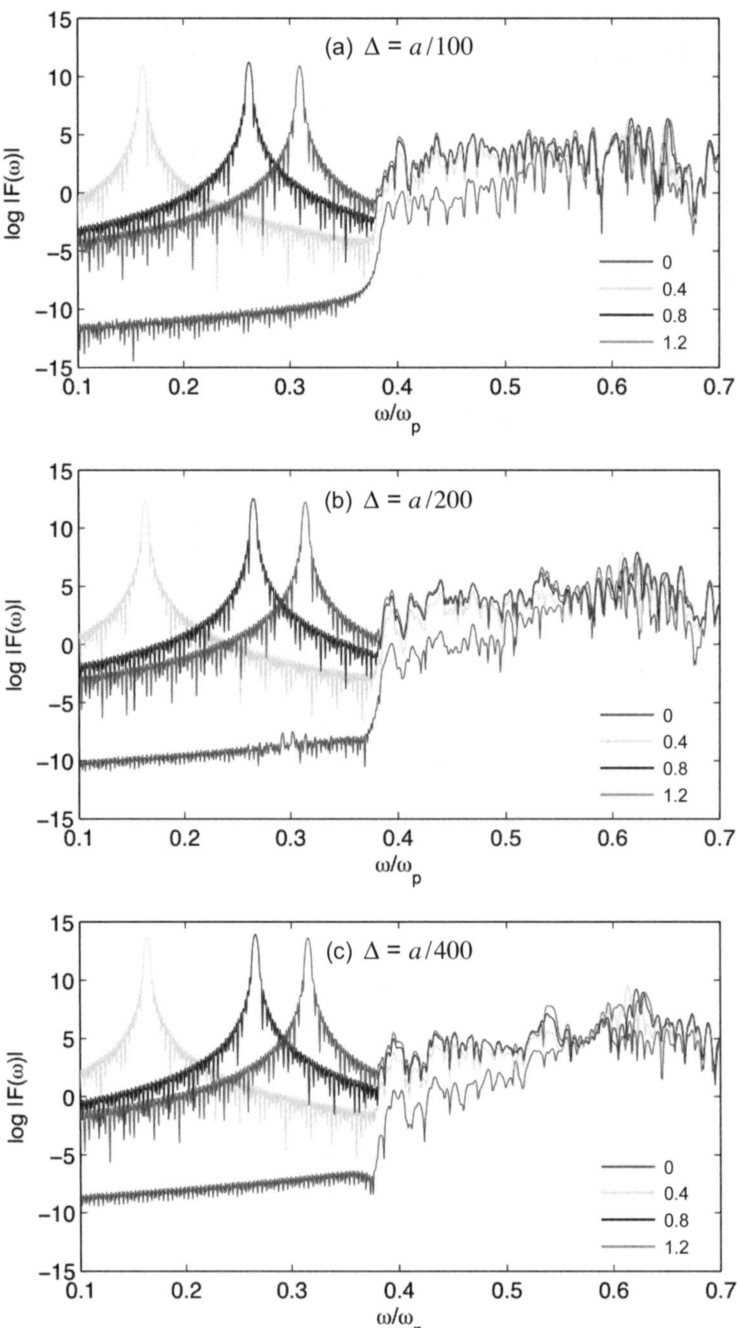

Fig. 13.22 Power spectral density of H_z in the plasmonic photonic crystal of Fig. 13.21 computed using a Cartesian (staircased) FDTD mesh for various sizes of square Yee cells, Δ, and Bloch wavevectors, $\boldsymbol{k} = \hat{\boldsymbol{x}}k_x$. The legends indicate k_xa values (a = lattice period). Although clear resonances can be seen up to $0.38\omega_p$, the higher-frequency resonances present a noise-like pattern despite very fine mesh refinement. *Adapted from:* Y. Liu, C. D. Sarris, and G. V. Eleftheriades, *IEEE J. Lightwave Technology*, Vol. 25, 2007, pp. 938–945, ©2007 IEEE.

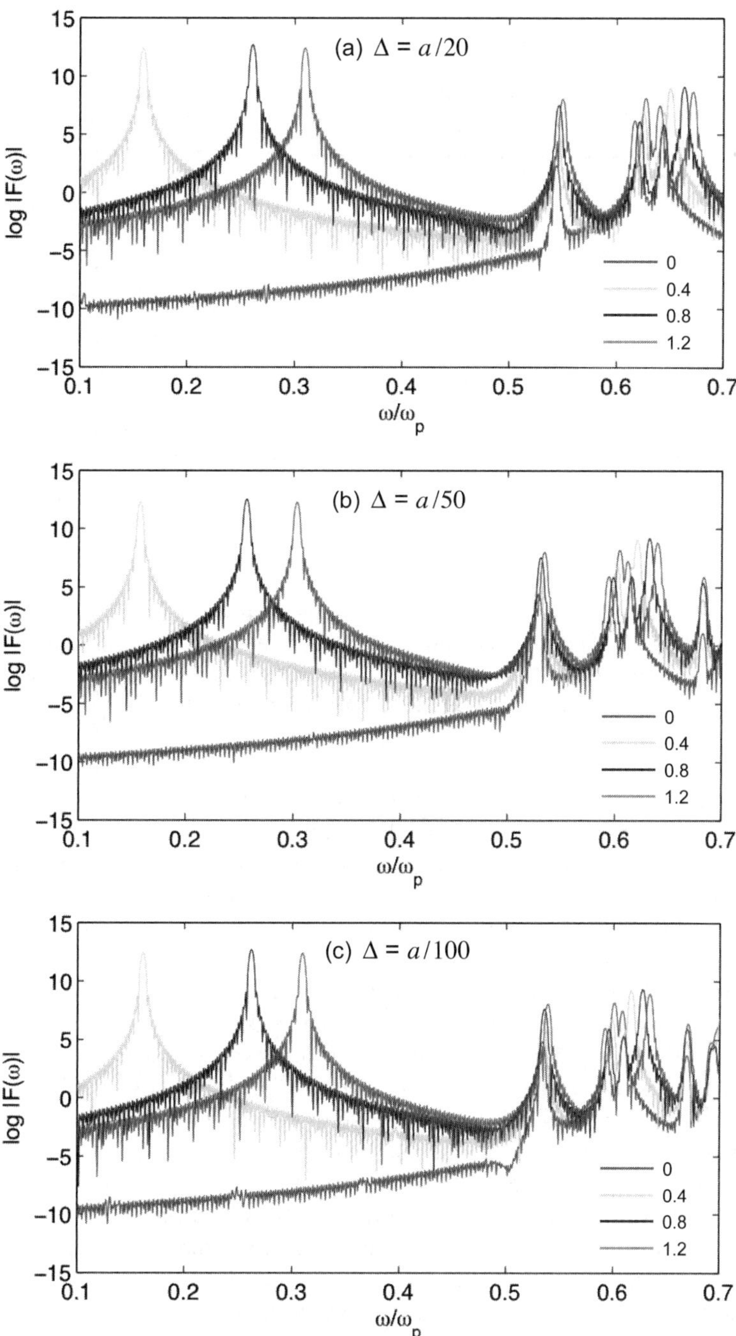

Fig. 13.23 Power spectral density of H_z in the plasmonic photonic crystal of Fig. 13.21 computed using the triangular-mesh FDTD method of Section 13.11 for various minimum sizes of triangle edges, Δ, and Bloch wavevectors, $\boldsymbol{k} = \hat{\boldsymbol{x}} k_x$. The legends indicate $k_x a$ values (a = lattice period). Unlike the staircased FDTD results in Figure 13.22, the higher-frequency resonances quickly converge even at a relatively coarse mesh. *Adapted from:* Y. Liu, C. D. Sarris, and G. V. Eleftheriades, *IEEE J. Lightwave Technology*, Vol. 25, 2007, pp. 938–945, ©2007 IEEE.

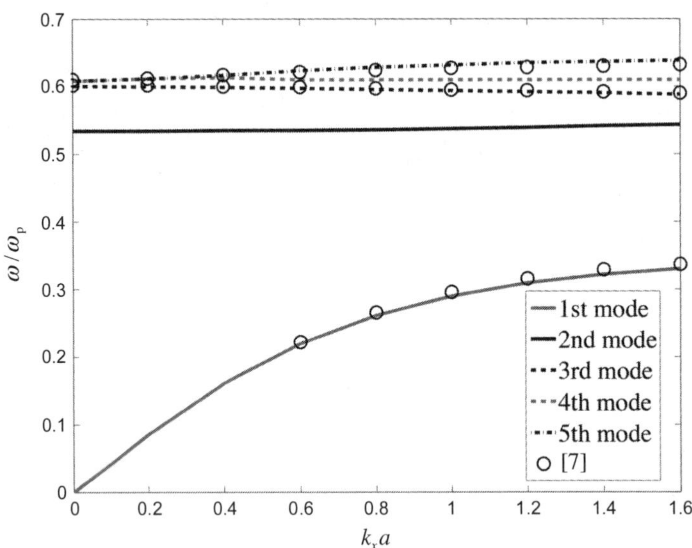

Fig. 13.24 Computed results for the Γ–X part of the Brillouin diagram of the sub-wavelength plasmonic photonic crystal of Fig. 13.21. Solid and dashed curves—triangular-mesh FDTD results; open dots—Reference [7]. *Adapted from:* Y. Liu, C. D. Sarris, and G. V. Eleftheriades, *IEEE J. Lightwave Technology*, Vol. 25, 2007, pp. 938–945, ©2007 IEEE.

Figure 13.24 displays the convergent periodic-analysis results derived by the triangular-mesh FDTD method for the Γ – X part of the Brillouin diagram of the sub-wavelength plasmonic photonic crystal of Fig. 13.21. For comparison, this figure also shows the corresponding results for the same structure reported in [7]. There is very good agreement for all of the modes.

The nature of these modes is illustrated by the visualizations of their *E*-field distributions (all corresponding to $k_x a = 1$, $k_y = 0$), shown in Fig. 13.25. The first mode (which is forward) exhibits a relatively uniform field distribution within the circular cross-section of the silver microcylinder. The other four mode patterns are characterized by strong field localizations at the microcylinder's surface, representative of their surface plasmon nature. Only the third mode is backward due to its slightly negative group velocity. This mode was used in [7] to realize a "perfect lens." However, since this mode is closely surrounded by forward modes, the lensing effect cannot be maintained over a significant bandwidth. For completeness, the full Brillouin diagram of the plasmonic photonic crystal of Fig. 13.21 is shown in Fig. 13.26.

Overall, the above comparison of FDTD results obtained using a staircased Cartesian grid and a triangular grid indicates that both accurately determined the fundamental forward mode of the sub-wavelength plasmonic photonic crystal of Fig. 13.21. In fact, the distinct Fourier transform peaks in Fig. 13.22, up to $0.38\omega_p$, correspond to this mode. However, from that point on, the Cartesian FDTD grid provided poor convergence due to the excitation of surface plasmon modes. The resonant frequencies of these modes were significantly perturbed by small errors related to staircasing approximations at the surface of the silver microcylinder, where the modal field patterns were strongly confined. These perturbations converged poorly with mesh refinement. Such issues do not accompany the triangular-mesh FDTD formulation, which can be a method of choice for studies involving surface plasmons at non-rectangular interfaces.

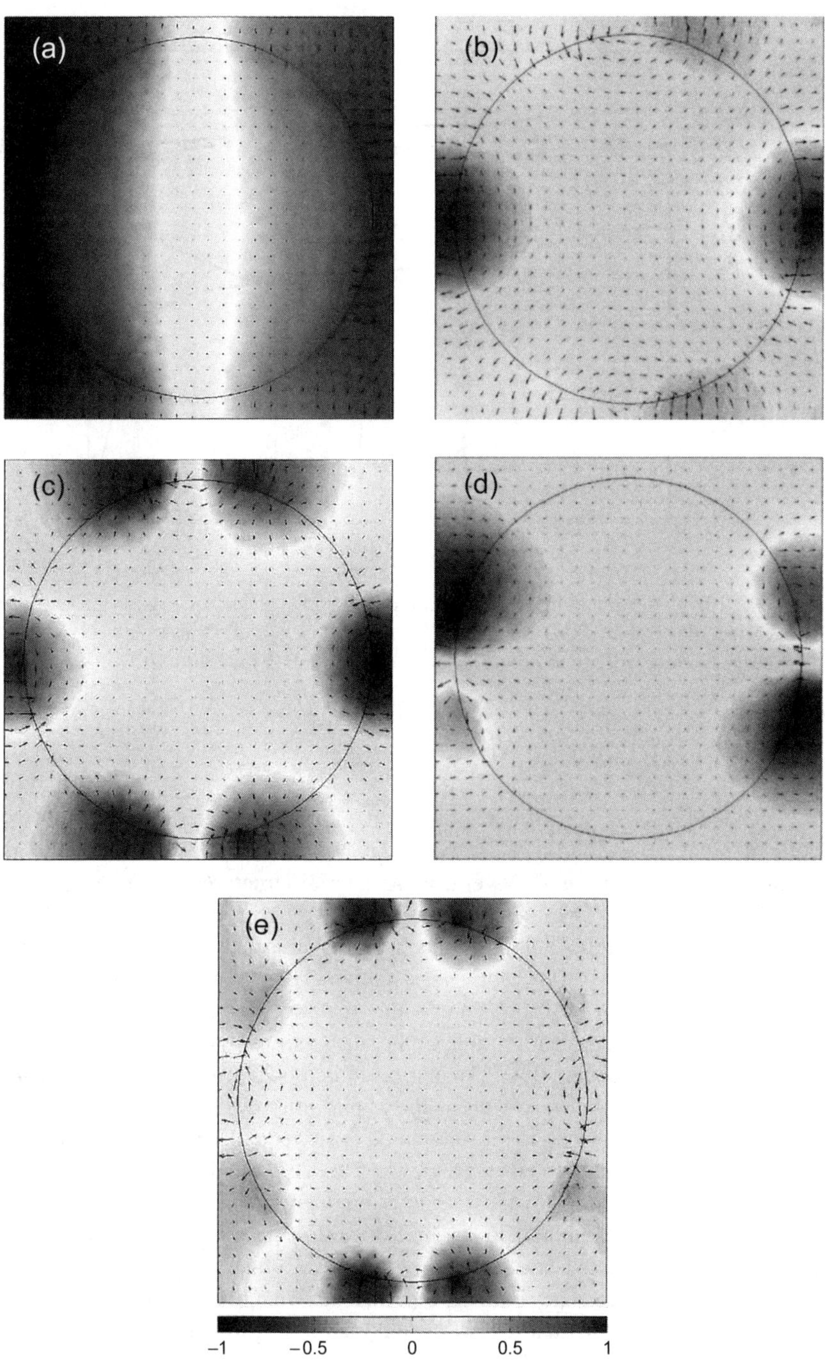

Fig. 13.25 E-field vector/magnitude plots for the plasmonic photonic crystal modes of Fig. 13.24 obtained using triangular-mesh FDTD. In all cases, $k_x a = 1$ and $k_y = 0$. (a) First mode: $\omega = 0.285\omega_p$; (b) second mode: $\omega = 0.532\omega_p$; (c) third mode: $\omega = 0.597\omega_p$; (d) fourth mode: $\omega = 0.616\omega_p$; (e) fifth mode: $\omega = 0.637\omega_p$. *Adapted from:* Y. Liu, C. D. Sarris, and G. V. Eleftheriades, *IEEE J. Lightwave Technology*, Vol. 25, 2007, pp. 938–945, ©2007 IEEE.

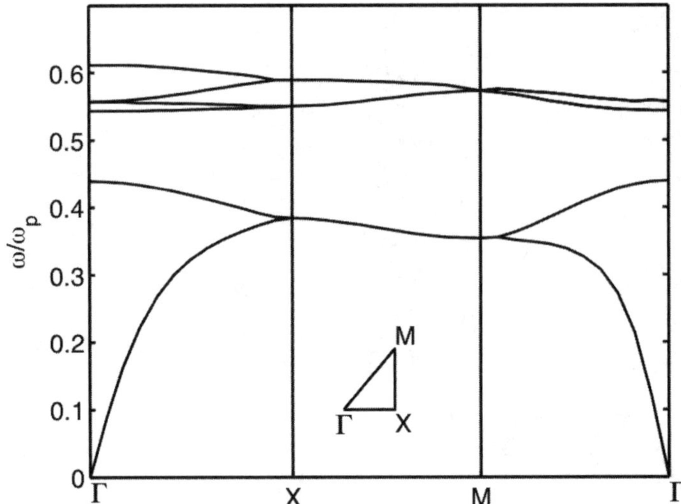

Fig. 13.26 The full Brillouin diagram of the plasmonic photonic crystal of Fig. 13.21, as determined by the triangular mesh FDTD method. *Source:* Y. Liu, C. D. Sarris, and G. V. Eleftheriades, *IEEE J. Lightwave Technology*, Vol. 25, 2007, pp. 938–945, ©2007 IEEE.

13.13 SUMMARY AND CONCLUSIONS

This chapter provided an overview of the application of FDTD to several key classes of problems in metamaterial analysis and design, from two complementary perspectives. First, FDTD analyses of the *transient response* of several metamaterial structures of interest were presented. These included negative-refractive-index media and the "perfect lens," an artificial transmission line exhibiting a negative group velocity, and a planar anisotropic grid supporting resonance cone phenomena. Second, *periodic geometries* realizing metamaterial structures were studied. The primary tool used here was the sine-cosine method, coupled with the array-scanning technique. This tool was applied to obtain the dispersion characteristics (and, as needed, the electromagnetic field) associated with planar periodic positive-refractive-index and negative-refractive-index transmission lines, as well as the planar microwave "perfect lens" comprised of sections of 2-D transmission lines exhibiting both positive and negative equivalent refractive indices. The chapter continued with a review of the triangular-mesh FDTD technique for modeling optical metamaterials with plasmonic components, and how this technique could be coupled with the sine-cosine method to analyze periodic plasmonic microstructures requiring much better modeling of slanted and curved metal surfaces than is possible using a Cartesian FDTD grid and simple staircasing. Finally, the periodic triangular-mesh FDTD technique was applied to obtain the dispersion characteristics and electromagnetic modes of a sub-wavelength plasmonic photonic crystal comprised of an array of silver microcylinders.

It was concluded that a standard Cartesian FDTD grid provides poor convergence for many of the resonant frequencies of photonic crystals comprised of nonrectangular-shaped metal elements. This is due to the excitation of surface plasmon modes, which are significantly perturbed by staircasing metal surfaces where strong modal fields are located. The triangular-mesh FDTD formulation could be a method of choice for modeling such metamaterials.

ACKNOWLEDGMENTS

The author gratefully acknowledges contributions to the research work included in this chapter made by his colleagues, Prof. George Eleftheriades and Dr. Yaxun Liu (Sections 13.11 and 13.12); by his former graduate student, Dr. Dongying Li (Sections 13.2 and 13.5–10); by Dr. Joshua Wong (Section 13.4); and by Ms. Suzanne Erickson (Section 13.3).

REFERENCES

[1] Veselago, V. G., "The electrodynamics of substances with simultaneously negative values of ε and μ," *Soviet Physics Usp.*, Vol. 10, 1968, pp. 509–514.

[2] Pendry, J. B., "Negative refraction makes a perfect lens," *Physical Review Lett.*, Vol. 85, 2000, pp. 3966–3969.

[3] Smith, D. R., W. J. Padilla, D. C. Vier, S. C. Nemat-Nasser, and S. Schultz, "Composite medium with simultaneously negative permeability and permittivity," *Physical Review Lett.*, Vol. 78, 2000, pp. 2933–2936.

[4] Eleftheriades, G. V., A. K. Iyer, and P. C. Kremer, "Planar negative refractive index media using periodically loaded transmission lines," *IEEE Trans. Microwave Theory and Techniques*, Vol. 50, 2002, pp. 2702–2712.

[5] Grbic, A., and G. V. Eleftheriades, "Overcoming the diffraction limit with a planar left-handed transmission-line lens," *Physical Review Lett.*, Vol. 92, 2004, 117403.

[6] Alú, A., and N. Engheta, "Optical nanotransmission lines: Synthesis of planar left-handed metamaterials in the infrared and visible regimes," *J. Optical Society of America B*, Vol. 23, 2006, pp. 571–583.

[7] Shvets, G., and Y. A. Urzhumov, "Engineering the electromagnetic properties of periodic nanostructures using electrostatic resonances," *Physical Review Lett.*, Vol. 93, 2004, 243904.

[8] Caloz, C., C. C. Chang, and T. Itoh, "Full-wave verification of the fundamental properties of left-handed materials in waveguide configurations," *J. Applied Physics*, Vol. 90, 2001, pp. 5483–5486.

[9] Grbic, A., and G. V. Eleftheriades, "Negative refraction, growing evanescent waves, and sub-diffraction imaging in loaded transmission-line metamaterials," *IEEE Trans. Microwave Theory and Techniques*, Vol. 51, 2003, pp. 2297–2305.

[10] Ziolkowski, R., and E. Heyman, "Wave propagation in media having negative permittivity and permeability," *Physical Review E*, Vol. 64, 2001, 056625.

[11] Foteinopoulou, S., E. N. Economou, and C. M. Soukoulis, "Refraction in media with a negative refractive index," *Physical Review Lett.*, Vol. 90, 2003, 107402.

[12] Cummer, S. A., "Dynamics of causal beam refraction in negative refractive index materials," *Applied Physica Lett.*, Vol. 82, 2003, pp. 2008–2010.

[13] Kokkinos, T., R. Islam, C. D. Sarris, and G. V. Eleftheriades, "Rigorous analysis of negative refractive index metamaterials using FDTD with embedded lumped elements," *IEEE MTT-S International Microwave Symposium Digest,* Vol. 3, 2004, pp. 1783–1786.

[14] So, P. M., H. Du, and W. J. R. Hoefer, "Modeling of metamaterials with negative refractive index using 2D-shunt and 3-D SCN TLM networks," *IEEE Trans. Microwave Theory and Techniques*, Vol. 53, 2005, pp. 1496–1504.

[15] Taflove, A., and S. C. Hagness, *Computational Electrodynamics: The Finite-Difference Time-Domain Method*, 3rd ed., Norwood, MA: Artech, 2005.

[16] Yee, K. S., "Numerical solution of initial boundary value problems involving Maxwell's equations in isotropic media," *IEEE Trans. Antennas and Propagation*, Vol. 14, 1966, pp. 302–307.

[17] Li, D., and C. D. Sarris, "A unified FDTD lattice truncation method for dispersive media based on periodic boundary conditions," *J. Lightwave Technology*, Vol. 28, 2010, pp. 1447–1454.

[18] Smith, D. R., D. Schurig, M. Rosenbluth, S. Schultz, S. A. Ramakrishna, and J. B. Pendry, "Limitations on sub-diffraction imaging with a negative refractive index slab," *Applied Physics Lett.*, Vol. 82, 2003, pp. 1506–1508.

[19] Brillouin, L., ed., *Wave Propagation and Group Velocity*, New York: Academic Press, 1960.

[20] Chiao, R. Y., and A. M. Steinberg, *Tunneling Times and Superluminality*, Amsterdam: North-Holland, 1997, chap. 37.

[21] Siddiqui, O., M. Mojahedi, and G. V. Eleftheriades, "Periodically loaded transmission line with effective negative refractive index and negative group velocity," *IEEE Trans. Antennas and Propagation*, Vol. 51, 2003, pp. 2619–2625.

[22] Siddiqui, O., S. Erickson, G. V. Eleftheriades, and M. Mojahedi, "Time-domain measurement of negative group delay in negative-refractive-index transmission-line metamaterials," *IEEE Trans. Microwave Theory and Techniques*, Vol. 52, 2004, pp. 1449–1456.

[23] Kuo, C., B. Houshmand, and T. Itoh, "Full-wave analysis of packaged microwave circuits with active and non-linear devices: An FDTD approach," *IEEE Trans. Microwave Theory and Techniques*, Vol. 45, 1997, pp. 819–826.

[24] Fischer, R. K., and R. W. Gould, "Resonance cones in the field pattern of a short antenna in an anisotropic plasma," *Physical Review Lett.*, Vol. 22, 1969, pp. 1093–1095.

[25] Balmain, K. G., and G. A. Oksiutik, *Plasma Waves in Space and in the Laboratory*, Vol. 1., Edinburgh, UK: Edinburgh University Press, 1997, pp. 247–261.

[26] Balmain, K. G., A. E. Lüttgen, and P. C. Kremer, "Power flow for resonance cone phenomena in planar anisotropic metamaterials," *IEEE Trans. Antennas and Propagation*, Vol. 51, 2003, pp. 2614–2618.

[27] Piket-May, M., A. Taflove, and J. Baron, "FD-TD modeling of digital signal propagation in 3-D circuits with passive and active loads," *IEEE Trans. Microwave Theory and Techniques*, Vol. 42, 1994, pp. 1514–1523.

[28] Harms, P., R. Mittra, and W. Ko, "Implementation of the periodic boundary condition in the finite-difference time-domain algorithm for FSS structures," *IEEE Trans. Antennas and Propagation*, Vol. 42, 1994, pp. 1317–1324.

[29] Kokkinos, T., C. D. Sarris, and G. V. Eleftheriades, "Periodic finite-difference time-domain analysis of loaded transmission-line negative-refractive-index metamaterials," *IEEE Trans. Microwave Theory and Techniques*, Vol. 53, 2005, pp. 1488–1495.

[30] Kokkinos, T., C. D. Sarris, and G. V. Eleftheriades, "Periodic FDTD analysis of leaky-wave structures and applications to the analysis of negative-refractive-index leaky-wave antennas," *IEEE Trans. Microwave Theory and Techniques*, Vol. 54, 2006, pp. 1619–1630.

[31] Liu, Y., C. D. Sarris, and G. V. Eleftheriades, "Triangular-mesh-based FDTD analysis of two-dimensional plasmonic structures supporting backward waves at optical frequencies," *IEEE J. Lightwave Technology*, Vol. 25, 2007, pp. 938–945.

[32] Yang, H.-Y. D., "Theory of microstrip lines on artificial periodic substrates," *IEEE Trans. Microwave Theory and Techniques*, Vol. 47, 1999, pp. 629–635.

[33] Munk, B., and G. A. Burrell, "Plane-wave expansion for arrays of arbitrarily oriented piecewise linear elements and its application in determining the impedance of a single linear antenna in a lossy half-space," *IEEE Trans. Antennas and Propagation*, Vol. 27, 1979, pp. 331–343.

[34] Qiang, R., J. Chen, F. Capolino, D. R. Jackson, and D. R. Wilton, "ASM-FDTD: A technique for calculating the field of a finite source in the presence of an infinite periodic artificial material," *IEEE Microwave and Wireless Components Lett.,* Vol. 17, 2007, pp. 271–273.

[35] Qiang, R., J. Chen, F. Capolino, and D. Jackson, "The array scanning method (ASM) FDTD algorithm and its application to the excitation of two-dimensional EBG materials and waveguides," *Proc. IEEE AP-S International Symposium on Antennas and Propagation*, June 2007, pp. 4457–4460.

[36] Li, D., and C. D. Sarris, "Efficient finite-difference time-domain modeling of driven periodic structures," *Proc. IEEE AP-S International Symposium on Antennas and Propagation*, June 2007, pp. 5247–5250.

[37] Li, D., and C. D. Sarris, "Efficient finite-difference time-domain modeling of driven periodic structures and related microwave circuit applications," *IEEE Trans. Microwave Theory and Techniques*, Vol. 56, 2008, pp. 1928–1937.

[38] Xiao, S., A. Vahldieck, and H. Jin, "Full-wave analysis of guided wave structures using a novel 2-D FDTD," *IEEE Microwave and Guided Wave Lett.*, Vol. 2, 1992, pp. 165–167.

[39] Engheta, N., A. Salandrino, and A. Alú, "Circuit elements at optical frequencies: nanoinductors, nanocapacitors, and nanoresistors," *Physical Review Lett.*, Vol. 95, 2005, 095504.

[40] Alú, A., A. Salandrino, and N. Engheta, "Negative effective permeability and left-handed materials at optical frequencies," *Optics Express*, Vol. 14, 2006, pp. 1557–1567.

[41] Maier, S. A., P. G. Kik, and H. A. Atwater, "Observation of coupled plasmon-polariton modes in au nanoparticle chain waveguides of different lengths: Estimation of waveguide loss," *Applied Physics Lett.*, Vol. 81, 2002, pp. 1714–1716.

[42] Holland, R., "Pitfalls of staircase meshing," *IEEE Trans. Electromagnetic Compatibility*, Vol. 35, 1993, pp. 434–439.

[43] Cangellaris, A. C., and D. B. Wright, "Analysis of the numerical error caused by the stair-stepped approximation of a conducting boundary in FDTD simulations of electromagnetic phenomena," *IEEE Trans. Antennas and Propagation*, Vol. 39, 1991, pp. 1518–1525.

[44] Lee, C. F., R. T. McCartin, R. T. Shin, and J. A. Kong, "A triangle-grid finite-difference time-domain method for electromagnetic scattering problems," *J. Electromagnetic Waves and Applications*, Vol. 8, 1994, pp. 449–470.

[45] Hano, M., and T. Itoh, "Three-dimensional time-domain method for solving Maxwell's equations based on circumcenters of elements," *IEEE Trans. Magnetics*, Vol. 32, 1996, pp. 946–949.

[46] Shewchuk, R., "Triangle: a two-dimensional quality mesh generator and Delaunay triangulator," Online: http://www.cs.cmu.edu/ quake/triangle.html

[47] Whitney, H., *Geometric Integration Theory*, Princeton, NJ: Princeton University Press, 1957.

[48] Craddock, I. J., C. J. Railton, and J. P. McGeehan, "Derivation and application of a passive equivalent circuit for the finite difference time domain algorithm," *IEEE Microwave and Guided Wave Lett.*, Vol. 6, 1996, pp. 40–42.

SELECTED BIBLIOGRAPHY

Bilotti, F., and L. Sevgi, "Metamaterials: Definitions, properties, applications, and FDTD-based modeling and simulation," *Int. J. RF and Microwave Computer-Aided Engineering*, Vol. 22, 2012, pp. 422–438.

Hao, Y., and R. Mittra, *FDTD Modeling of Metamaterials: Theory and Applications*, Norwood, MA: Artech, 2008.

Teixeira, F. L., "Time-domain finite-difference and finite-element methods for Maxwell's equations in complex media," *IEEE Trans. Antennas and Propagation*, Vol. 56, 2008, pp. 2150–2166.

Chapter 14

Computational Optical Imaging Using the Finite-Difference Time-Domain Method[1]

Ilker R. Capoglu, Jeremy D. Rogers, Allen Taflove, and Vadim Backman

14.1 INTRODUCTION

Optical imaging systems have traditionally been analyzed using well-established approximations such as ray-based geometrical optics [1] and scalar Fourier theory [2]. However, there has recently been increased interest in applying the full-vector Maxwell's equations to implement rigorous and robust first-principles numerical models of such systems. The availability of increasingly powerful computers and computational algorithms has contributed to this interest, as well as the need for improved accuracy for key scientific and engineering applications.

Although the basic principles of light scattering derived from Maxwell's equations have been known for many decades, the application of these principles to model complete optical imaging systems had to wait until the advent of computers having sufficient speed and memory. This allows modeling objects that are comparable in size to the wavelength of the illuminating light, 400 to 800 nm in the visible range. With the arrival of these capabilities, it is possible to bypass most of the traditional simplifying approximations and proceed to calculate the optical image of an arbitrary object directly from Maxwell's equations in three dimensions.

The need for computational modeling of optical imaging is presented by a number of important scientific and engineering applications that require controlling all aspects of the imaging system down to sub-wavelength precision. Early examples involved modeling photo microlithography techniques for integrated-circuit production [3–6]. More recently, there has been increasing interest in modeling optical microscopy modalities [7–11], especially regarding optical detection of early-stage nanoscale alterations in precancerous cells [12, 13].

This chapter is primarily intended as a reference for the numerical algorithms and techniques necessary for implementing a purely virtual imaging system, which we will refer to as a "microscope in a computer." Since the basic principles are also applicable to any other optical imaging system, the results here can also be consulted for modeling photolithography and metrology systems. For a more focused discussion of the finite-difference time-domain (FDTD) simulation of photolithography processes, see Chapter 15.

[1]This chapter is adapted from Ref. [137], I. R. Capoglu, J. D. Rogers, A. Taflove, and V. Backman, "The microscope in a computer: Image synthesis from three-dimensional full-vector solutions of Maxwell's equations at the nanometer scale," *Progress in Optics*, Vol. 57, 2012, pp. 1–91, ©2012 Elsevier.

In this chapter, we present a self-contained account of the numerical electromagnetic simulation of a general optical imaging system. We place special emphasis on numerical modeling issues such as discretization, sampling, and signal processing. Although most of the presentation is tailored for optics, many of the concepts and equations given in Sections 14.2 to 14.6 are applicable to a broader range of electromagnetics problems involving antennas, antenna arrays, metamaterials, and radio-frequency (RF) and microwave circuits and radars. The refocusing concept in Section 14.7, however, is a defining characteristic of an optical imaging system, with few exceptions such as focused antenna arrays in RF electromagnetics [14].

14.2 BASIC PRINCIPLES OF OPTICAL COHERENCE

In most practical situations, the excitation in an optical system (whether it is a filament or a laser source) has a certain random character. This creates randomness in the resulting optical electromagnetic field in both space and time. If this is the case, the electromagnetic field can only be represented as a *random field* that possesses certain statistical properties. Fortunately, we are almost always concerned with time averages of optical parameters such as intensity or polarization, because these are the only parameters that most optical instruments can measure. If an adequate statistical model is constructed for the random electromagnetic field, the average quantities measured at the output of the system can be inferred mathematically. The categorization and rigorous mathematical description of these matters is the subject of optical coherence [1, 15].

Although optical illumination systems almost always have a random character, the FDTD method operates on deterministic field values that are known precisely in space and time. The question arises, therefore, as to whether it is possible to compute statistical averages belonging to infinite random processes using a completely deterministic numerical electromagnetic simulation method such as FDTD. It turns out that this is possible, provided that the physical system satisfies certain conditions. One of the simplest of such situations is when the excitation is *statistically stationary* in time. In its strictest form, this means that the statistical properties of the waveforms anywhere in the system do not change in time. This is a reasonable assumption for many forms of optical sources, and will be made throughout this chapter. The study of non-stationary, spectrally partially coherent sources is outside our scope. Interested readers may consult Refs. [16–18].

The importance of a stationary process is manifested when the response of a linear system to a stationary time waveform is sought. This is the case in our analysis, because Maxwell's equations and the scattering materials are assumed to be linear. Let us consider an input waveform $x_i(t)$ exciting the system in some way and an output waveform $x_o(t)$ measured somewhere else in the system. If $x_i(t)$ is the only excitation, the relation between these is a convolution with the *impulse response* $h(t)$ of the system:

$$x_o(t) = \int_{-\infty}^{\infty} h(\tau) x_i(t-\tau) d\tau \tag{14.1}$$

The *transfer function* $H(\omega)$ is defined as the Fourier transform of $h(t)$:

$$H(\omega) = \int_{-\infty}^{\infty} h(t) e^{-j\omega t} dt \tag{14.2}$$

It can be shown that the *power spectral densities* $S_i(\omega)$ and $S_o(\omega)$ of the input and output waveforms, respectively, are related linearly by the absolute square of the transfer function [1, 15, 19, 20]:

$$S_o(\omega) = |H(\omega)|^2 \, S_i(\omega) \tag{14.3}$$

The power spectral density is an optically relevant and directly measurable quantity, defined as the power at the output of a narrowband filter centered at ω. The Wiener-Khintchine theorem [1] states that it is also the Fourier transform of the correlation function associated with the stationary waveform.

Equation (14.3) is the central result that connects random waveforms in optics with the deterministic numerical methods of electromagnetics. In a given problem, the power spectral density of the source $S_i(\omega)$ is usually known; and the power spectral density of the output $S_o(\omega)$ is desired. The necessary link is provided by the absolute square of the transfer function $H(\omega)$. A numerical electromagnetics method such as FDTD can be used to find $H(\omega)$ by sending deterministic signals through the optical system, and calculating the response. Although the majority of the results in this chapter will be given for a fixed frequency ω, the response to a broadband stationary waveform can easily be obtained by taking the temporal Fourier transform of the response to a broadband deterministic waveform.

14.3 OVERALL STRUCTURE OF THE OPTICAL IMAGING SYSTEM

An optical imaging system can be decomposed into several subsystems, wherein each performs a self-contained task that is simple enough to model theoretically. Once the theoretical underpinnings of each subsystem are laid out, the numerical computation of the actual physical parameters concerning the subsystem (transmission coefficients, far-field intensities, aberrations, etc.) becomes a matter of approximating the analytical equations in a suitable manner. We represent the optical imaging system as a combination of four subsystems: illumination, scattering, collection, and refocusing. These subsystems are drawn schematically in Fig. 14.1.

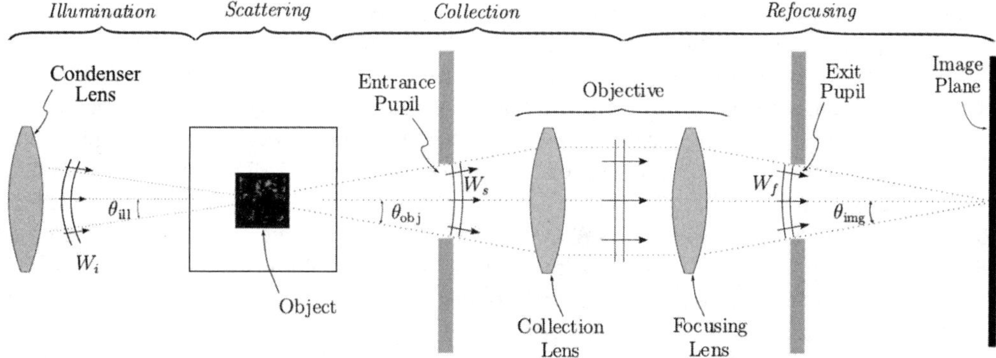

Fig. 14.1 The four subcomponents of an optical imaging system: illumination, scattering, collection, and refocusing.

14.4 ILLUMINATION SUBSYSTEM

The light source and the lens system (usually called the condenser) that focuses the light created by the source onto the object are included in this subsystem. The last lens in the condenser system is shown on the left side of Fig. 14.1, along with the wavefront incident on the object. We will base our treatment of illumination systems on whether they are *spatially coherent* or *incoherent*. Temporal coherence is a secondary concern since the sources considered here are always stationary (see Section 14.2). Once the responses to all the frequencies in the temporal spectrum of the source are found, then the synthesis of the output intensity is simply a matter of adding the intensities of the responses at each frequency.

14.4.1 Coherent Illumination

Spatially coherent illumination means that the fields at all points within the illumination are fully monochromatic with fixed amplitude and phase relations. This illumination can be created by infinitesimally small sources or by the atomic process of stimulated emission, as with lasers.

The simplest coherent illumination used in numerical modeling is the *plane wave*. Being invariant in all but one dimension, the plane wave is one of the most basic solutions to Maxwell's equations. Here, the planes of constant phase are all perpendicular to the direction of propagation $\hat{k}_{\text{inc}}$, and the electric and magnetic field vectors are perpendicular to each other and $\hat{k}_{\text{inc}}$. Individually, the plane wave can approximate a more complicated coherent illumination scheme over a very small illumination angle [21,22]. Full treatments of some of these illumination schemes in large-θ_{ill} cases have also been considered in literature, albeit with less popularity. This is primarily because non-planar coherent beams are often difficult to compute and/or implement numerically.

One of the most popular non-planar coherent illumination beams is the *Gaussian beam* [23]. Although the Gaussian beam has an approximate closed-form analytical expression that can be used in limited cases [24–26], it is often decomposed into its plane-wave components, resulting in a more accurate description than the more limited closed-form expression [27,28]. This method has the additional advantage of permitting the use of efficient and readily-available plane-wave algorithms, such as the total-field/scattered field (TF/SF) algorithm for FDTD [29]. Although the Gaussian beam is defined at a single frequency, it can be used in conjunction with an FDTD model set up for time-harmonic operation [8, 30–32].

The plane-wave spectrum (or the angular-spectrum) method can also be used to synthesize arbitrary coherent-illumination beams of non-Gaussian shape. A practical example of a coherent beam derived using the angular-spectrum method is the electromagnetic field distribution around the focal region of an aplanatic lens excited by a plane wave [33, 34]. This beam has been used to simulate the coherent illumination in scanning-type confocal or differential-interference contrast (DIC) microscopes [35, 36]. An extension of this technique to time-domain focused pulses was described in [37]. This can be used either to simulate ultrafast optical pulses [38–42] or stationary broadband systems via temporal Fourier analysis. The latter type of systems have recently become feasible with the development of white-light laser sources [43, 44].

Illumination modeling generally becomes more difficult when the object space is multilayered. To simplify this task, the TF/SF algorithm in FDTD has been generalized to deal with multilayered spaces [45–47]. This technique can be used as a means to inject arbitrary coherent beams into a multilayered space, since in principle any beam can be decomposed into a plane-wave spectrum.

14.4.2 Incoherent Illumination

The term "incoherent illumination" is traditionally used to designate a scheme that exhibits partial spatial coherence over the illumination area. Here, the light source has finite spatial extent, with every point on the source radiating in an incoherent fashion. This is an adequate model for many natural and artificial light sources such as the sun, a xenon arc lamp, or a tungsten filament. Incoherence also requires that the excitation source have a finite bandwidth, however small it may be. In fact, the converse of this requirement (strict monochromaticity) is very hard to achieve, for even the most coherent laser sources have a finite bandwidth.

Perhaps the most prominent incoherent illumination scheme in use today is called *Köhler illumination* [1, 48], named after August Köhler who introduced it in the late 1800s. One of the key advantages of this scheme is that it provides spatially uniform illumination throughout the sample, regardless of the inhomogeneities of the light source. This is accomplished by sending a collimated beam on the sample for every infinitesimal point on the light source. Figure 14.2 illustrates the details of Köhler illumination.

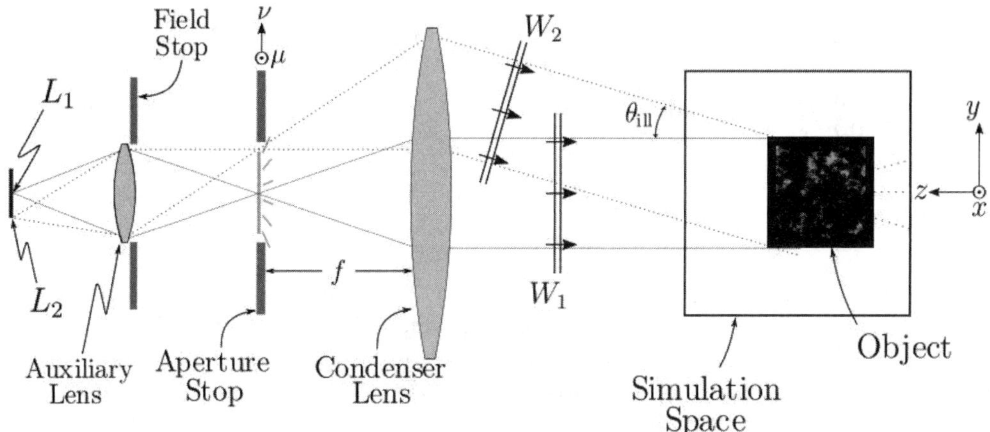

Fig. 14.2 Schematic illustration of Köhler illumination.

In Fig. 14.2, the light source on the left is imaged on the aperture stop by an auxiliary lens. The image of the light source on the aperture stop acts as a secondary source for the succeeding portion of the system. Unlike the original light source, the spatial coherence length on this secondary source is not zero. In other words, the secondary source is technically a partially coherent source. Fortunately, if the aperture stop is much larger than the size of the diffraction spot (also called the Airy disc) associated with the auxiliary lens, there is little accuracy lost if every point on this secondary source is also assumed incoherent [1]. The analysis of Köhler illumination is based on this assumption.

Two rays emanating from each of two mutually incoherent infinitesimal point sources L_1 and L_2 on the light source are shown in Fig. 14.2 by solid and dotted lines, respectively. Since the aperture stop is situated at the front focal plane of the condenser lens, every point source on the aperture stop creates a collimated beam illuminating the object from a different direction. In addition, since the secondary point sources on the aperture stop are assumed incoherent, these beams are also incoherent.

The flat wavefronts created by L_1 and L_2 are denoted by W_1 and W_2 in Fig. 14.2. The aperture stop limits the angles from which the incoherent beams hit the object within an illumination cone, defined by θ_{ill}. In general, the image of the source on the aperture stop may be inhomogeneous, and therefore the beams hitting the object may have different amplitudes. If the source is of uniform intensity, these amplitudes are also uniform. The spatial extent of the illumination, on the other hand, is controlled by the field stop in Figure 14.2.

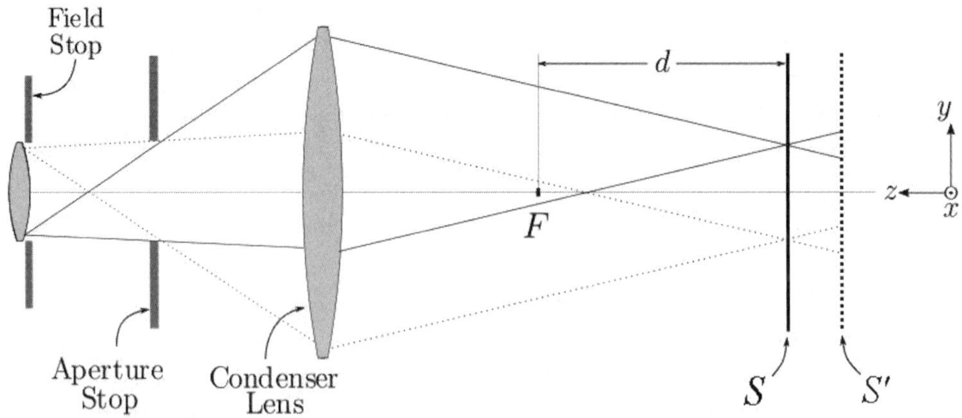

Fig. 14.3 Optimal placement of the sample for spatially uniform illumination.

As shown in Fig. 14.3, the field stop is imaged by the condenser lens at plane S at distance d from the back focal plane F. For illustration purposes, distance d in Fig. 14.3 is drawn much larger than usual. The field stop is usually at several focal lengths in front of the aperture stop, so S is usually quite close to F. It is clearly seen from Fig. 14.3 that the optimum position for the sample is at S, since any forward or backward movement of the sample will cause the elimination of some rays incident from certain directions. As the sample is moved away from focus (say, to S'), the illumination area gets larger and starts blurring at the edges. This undesirable effect is avoided by focusing the field stop sharply over the sample at all times.

If the sample is close enough to the center of the illumination area on S, the collimated beams can be well approximated by plane waves. In numerical computations, the continuum of mutually incoherent plane waves over the illumination cone has to be approximated by a finite sum. This is, in effect, a two-dimensional numerical quadrature problem, for which unfortunately no universally optimum method exists [49]. A heuristic and straightforward method that is applicable regardless of the actual shape of the source image on the aperture stop is an equally spaced arrangement of point sources, combined with the midpoint rule [49].

The corresponding placement of the plane waves incident on the sample in Fig. 14.3 can be found from geometrical optics (see [1], Section 10.6.2). Within the accuracy of Gaussian optics (small off-axis distances, small angles around the axis), every position (μ, ν) on the aperture stop corresponds to a plane wave with direction cosines $(s_x, s_y) = (\mu/f, \nu/f)$ at the back focal plane of the condenser, where f is the focal length of the condenser. The direction cosines are defined as:

$$s_x = \sin\theta\cos\phi = \cos\chi\,, \qquad s_y = \sin\theta\sin\phi = \cos\eta \tag{14.4}$$

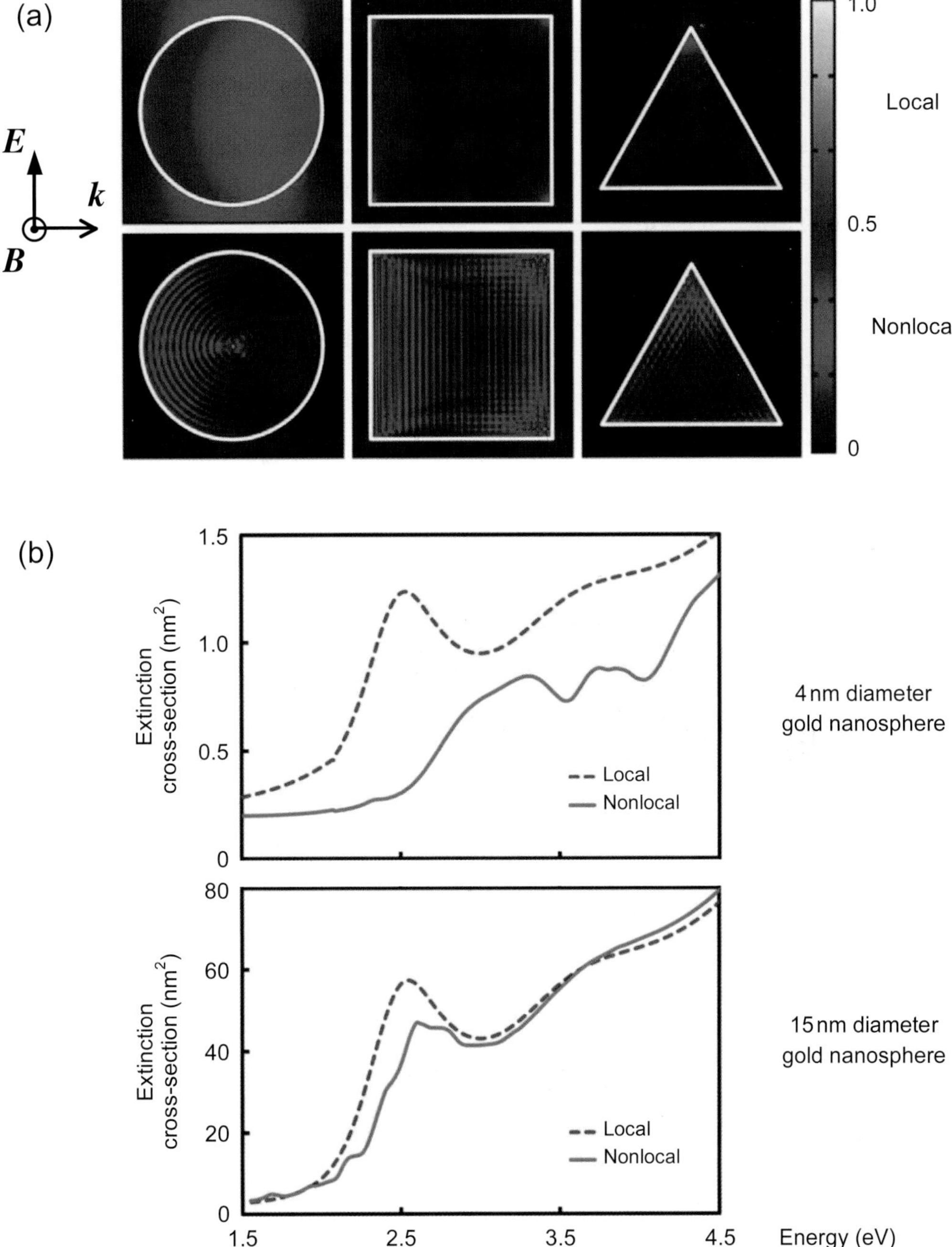

Plate 1 (a) Comparison of FDTD-computed normalized $|\boldsymbol{D}|^2$ profiles at LSPR energies in and around isolated gold nanowires of circular, square, and triangular cross-sections with 50-nm diameters or side lengths for local and nonlocal dielectric functions. (b) Comparison of FDTD-computed extinction cross-sections of gold nanospheres for local and nonlocal dielectric functions. *Adapted from:* J. M. McMahon, S. K. Gray, and G. C. Schatz, *Physical Review B*, Vol. 82, 2010, 035423, ©2010 The American Physical Society. See Chapter 9, Sections 9.7 and 9.8, for a discussion.

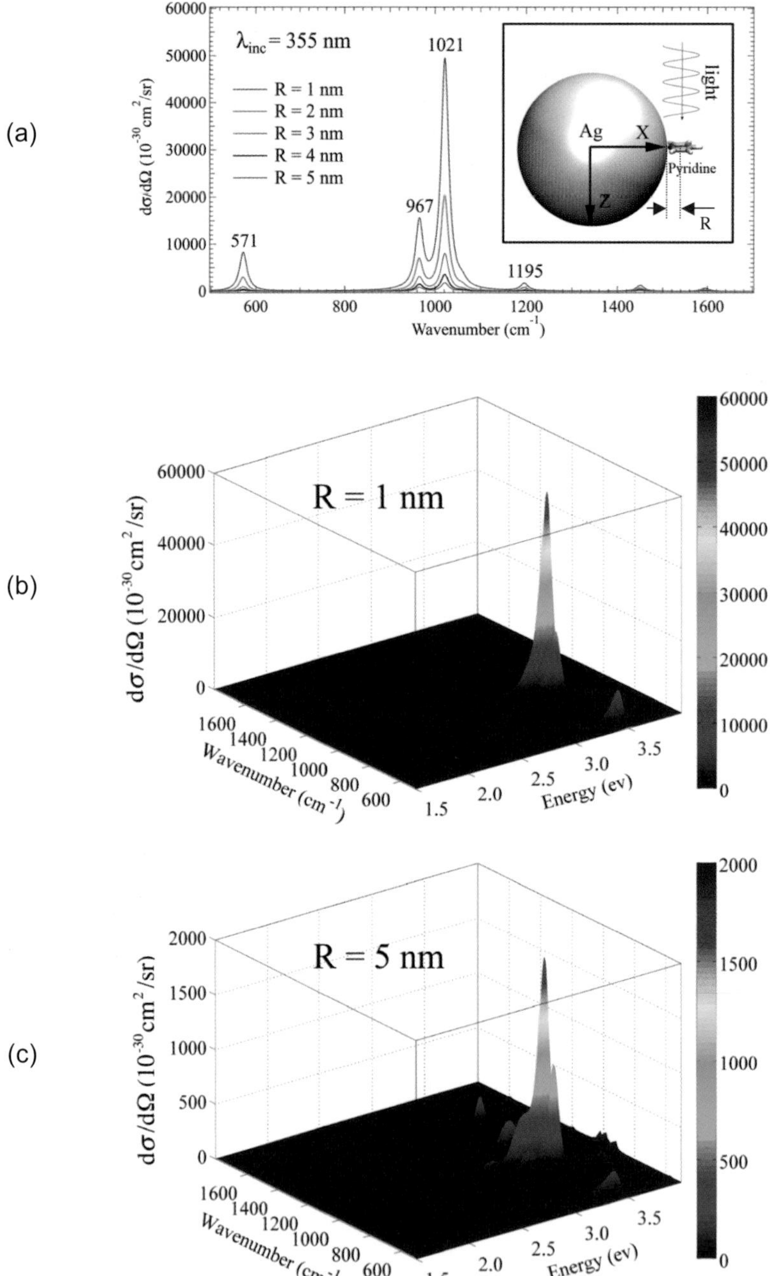

Plate 2 (a) Calculated surface-enhanced Raman scattering (SERS) spectrum of a pyridine molecule located at various distances, *R*, from an adjacent 20-nm silver nanosphere. (b, c) Surface plots of the calculated SERS spectrum of the pyridine molecule/silver nanosphere system as a function of the incident light energy and the vibrational mode wavenumber for two separations, *R*. *Adapted from:* H. Chen, J. M. McMahon, M. A. Ratner, and G. C. Schatz, *J. Physical Chemistry C*, Vol. 114, 2010, pp. 14384–14392, ©2010 American Chemical Society. See Chapter 10, Section 10.8.2, for a discussion.

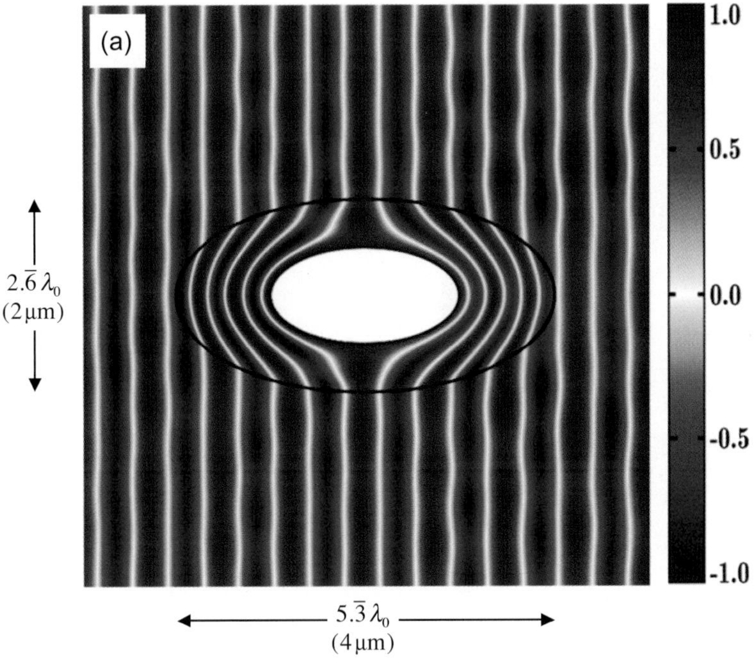

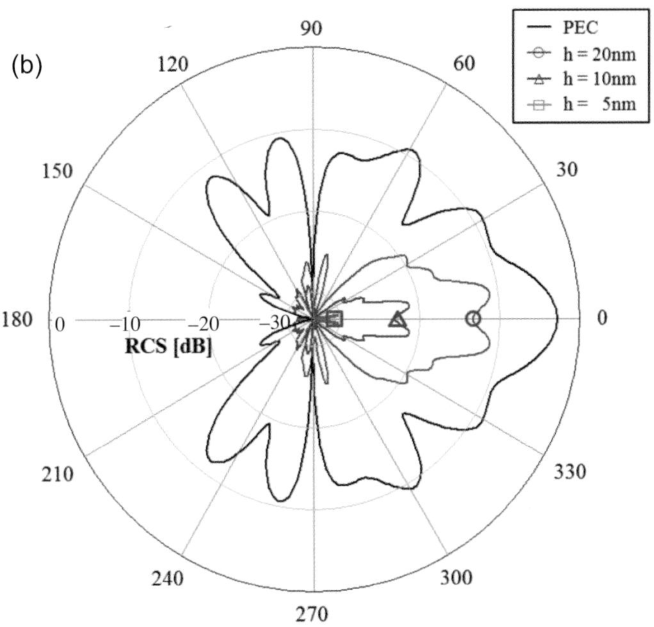

Plate 3 FDTD-computed results for a 2-D elliptic cylindrical cloak model (field components E_x, E_y, and H_z) illuminated at $\lambda_0 = 750$ nm: (a) Visualization of the H_z field distribution for 10-nm grid resolution; (b) bistatic radar cross-section without the cloak (PEC) and with the cloak for 20-, 10-, and 5-nm grid resolutions. *Source:* N. Okada and J. B. Cole, *ISRN Optics*, Vol. 2012, 536209, doi:10.5402/2012/536209, ©2012 Okada and Cole. See Chapter 12, Section 12.4, for a discussion.

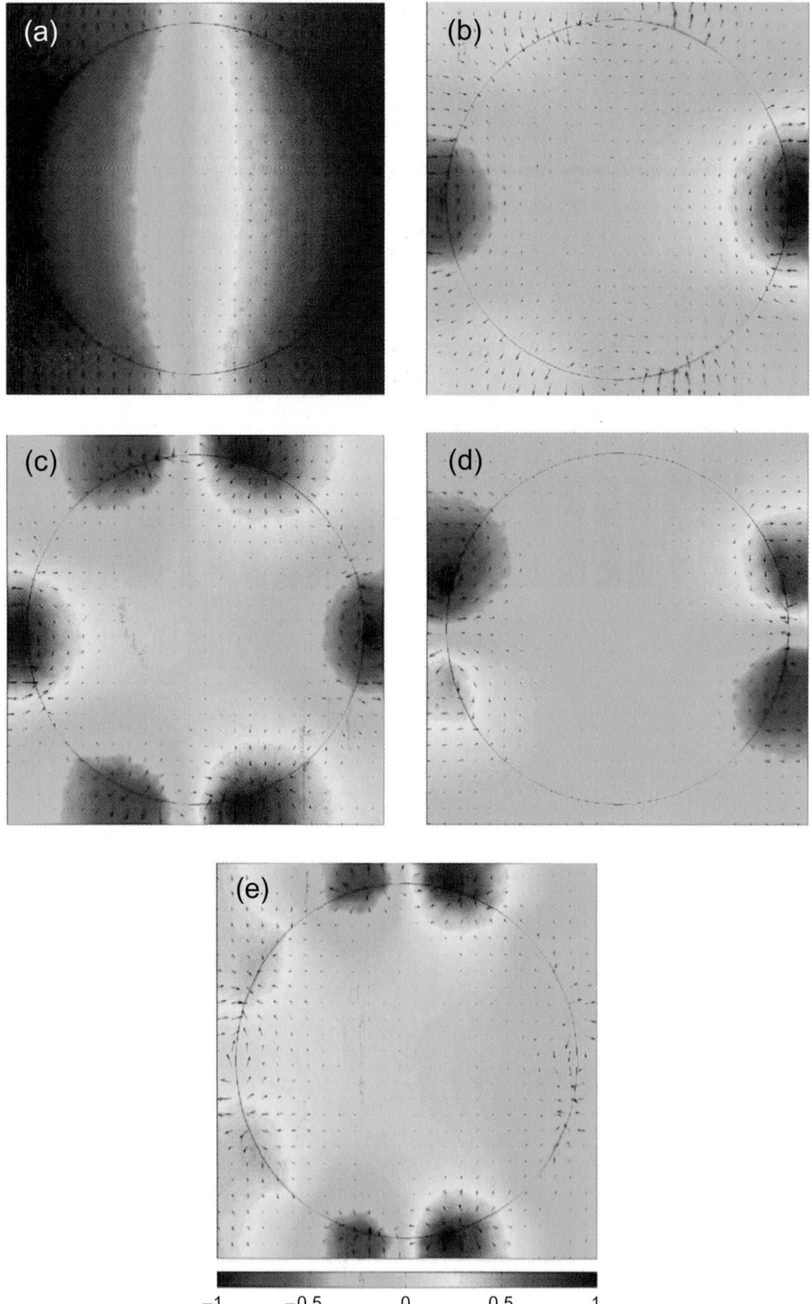

Plate 4 E-field vector and magnitude visualizations for the modes of the plasmonic photonic crystal of Fig. 13.21 obtained by the triangular-mesh FDTD method. In all cases, $k_x a = 1$ and $k_y = 0$, and ω_p denotes the plasma frequency. (a) First mode: $\omega = 0.285\omega_p$; (b) second mode: $\omega = 0.532\omega_p$; (c) third mode: $\omega = 0.597\omega_p$; (d) fourth mode: $\omega = 0.616\omega_p$; (e) fifth mode: $\omega = 0.637\omega_p$. *Adapted from:* Y. Liu, C. D. Sarris, and G. V. Eleftheriades, *IEEE J. Lightwave Technology*, Vol. 25, 2007, pp. 938–945, ©2007 IEEE. See Chapter 13, Section 13.12, for a discussion.

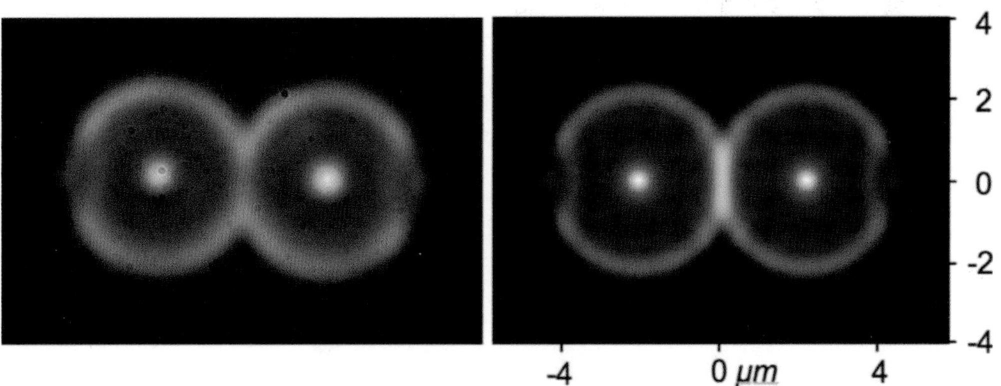

Plate 5(a) Comparison of measured (left panel) and FDTD-synthesized (right panel) bright-field grayscale microscope images of a pair of adjacent contacting 4.3-μm-diameter polystyrene microspheres in air for numerical aperture (NA) = 0.6 and wavelengths between 400 and 700 nm. See Chapter 14, Section 14.8.3, for a discussion.

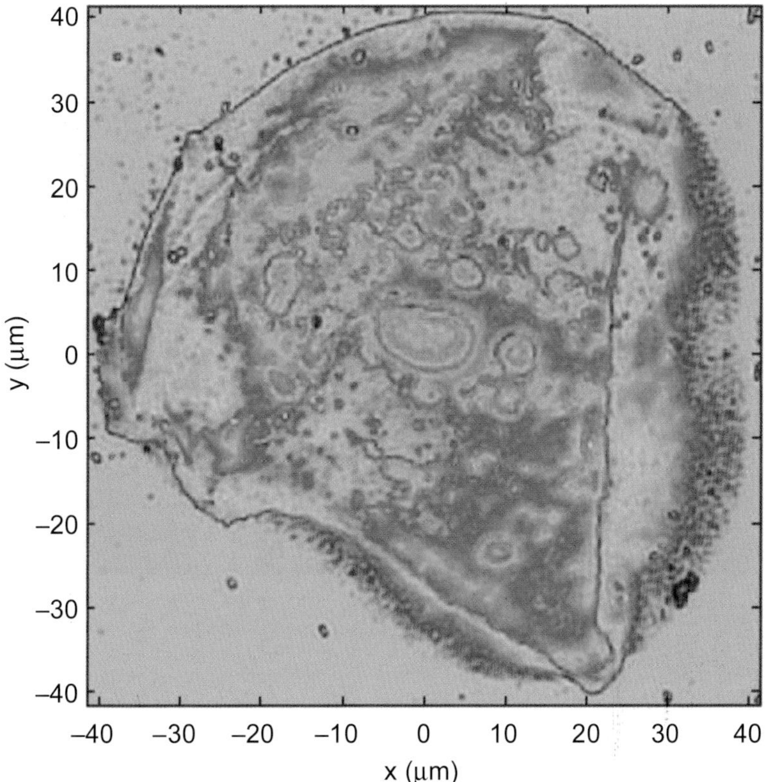

Plate 5(b) FDTD-synthesized bright-field true-color microscope image of a human cheek (buccal) cell modeled in a 3-D grid having uniform 25-nm cubic cells. See Chapter 14, Section 14.8.4, for a discussion.

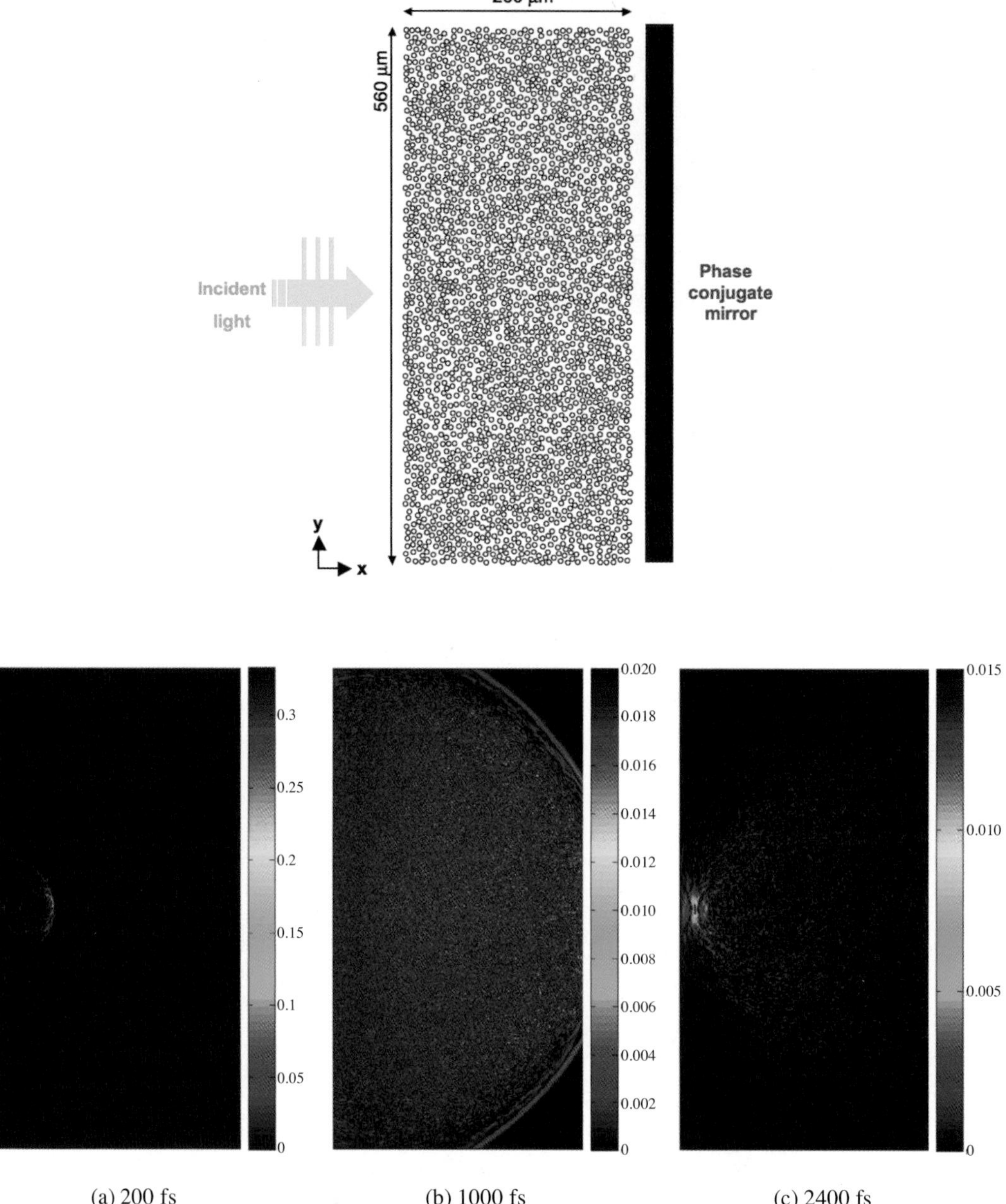

(a) 200 fs (b) 1000 fs (c) 2400 fs

Plate 6 Pseudospectral time-domain (PSTD) simulation of optical phase conjugation. Top panel: simulation geometry — a rectangular (560- × 260-μm) cluster of 2,500 randomly positioned 2.5-μm-diameter dielectric cylinders ($n = 1.2$) illuminated by a pulsed light beam. Bottom panels: visualizations of the PSTD-computed E-field. (a) Wavefront of the incident light pulse upon entering the cluster of cylinders; (b) wavefront spreading due to multiple scattering and diffraction before reaching a phase-conjugate mirror at the far right; (c) phase and propagation direction of the light is inverted by the phase-conjugate mirror, causing the light to trace back to its origination point. *Source:* S. H. Tseng and C. Yang, *Optics Express*, Vol. 15, 2007, pp. 16005–16016, ©2007 The Optical Society of America. See Chapter 16, Section 16.4.4, for a discussion.

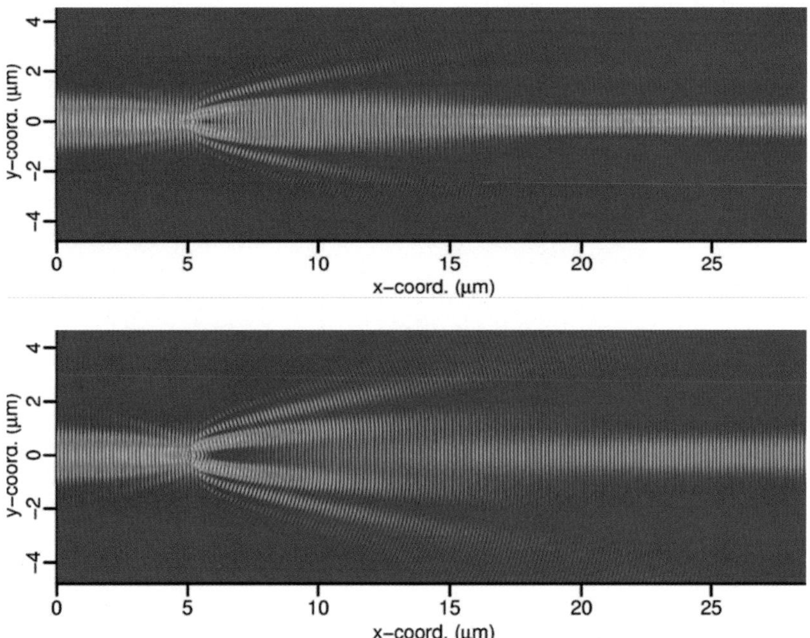

Plate 7(a) Visualization of the GVADE-FDTD-computed $|\boldsymbol{E}|$ of the scattering of a 2-D bright, narrow, overpowered spatial soliton (field components E_x, E_y, and H_z) by: top view—a square 250-nm air hole located at $x = 5$ μm; and bottom view—a square 350-nm air hole located at $x = 5$ μm. *Source:* J. H. Greene and A. Taflove, *IEEE Microwave and Wireless Components Lett.*, Vol. 17, 2007, pp. 760–762, ©2007 IEEE. See Chapter 17, Section 17.6.1, for a discussion.

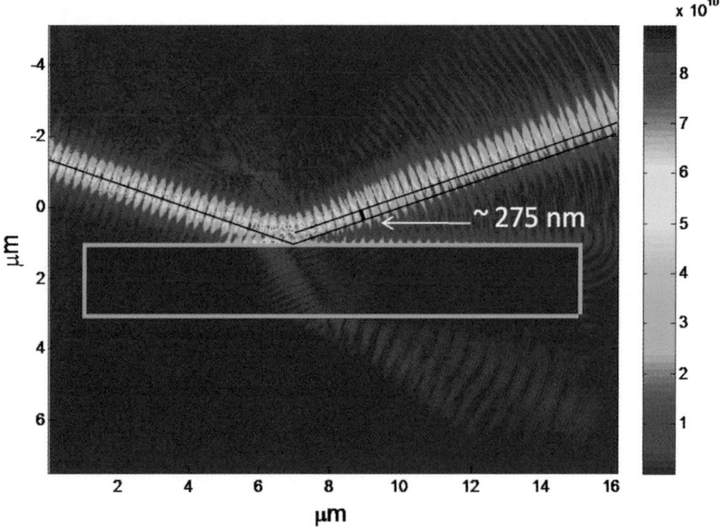

Plate 7(b) Visualization of the GVADE-FDTD-computed $|\boldsymbol{E}|$ of a bright, narrow, spatial soliton (field components E_x, E_y, and H_z) reflecting from a 2-μm-thick gold film, demonstrating an approximate 275-nm Goos-Hänchen lateral shift of the reflected beam. *Source:* Z. Lubin, J. H. Greene, and A. Taflove, *Microwave and Optical Technology Lett.*, Vol. 54, 2012, pp. 2679–2684, ©2012 Wiley Periodicals, Inc. See Chapter 17, Section 17.6.2, for a discussion.

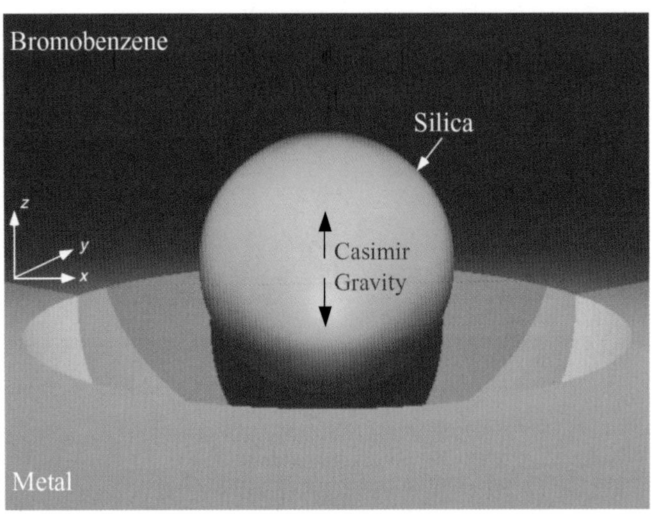

Plate 8(a) Model of the Casimir force between a silica microsphere (immersed in bromobenzene) and a perfect metal plane with a spherical indentation, showing a stable levitation of the microsphere.

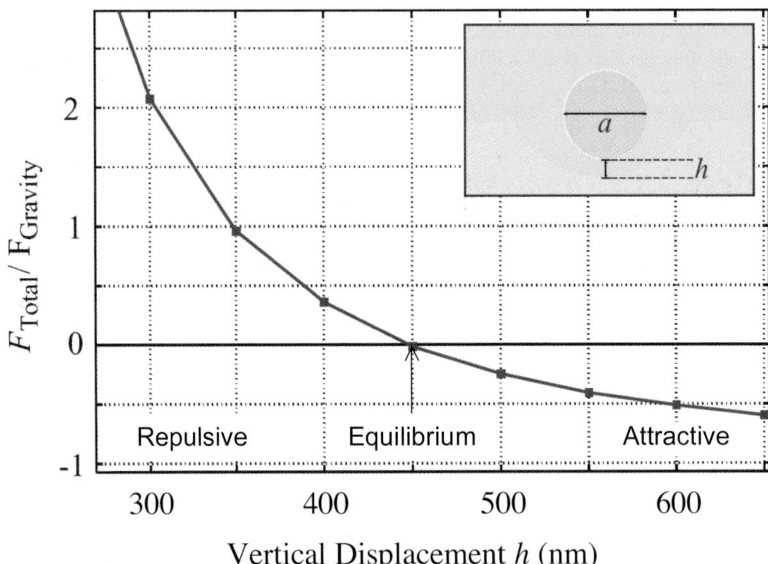

Plate 8(b) FDTD-computed total vertical force (Casimir + gravity) acting on the silica microsphere of Plate 8(a) (also see the inset) of radius 500 nm above a spherical metal indentation of radius 1 µm. As the assumed height, h, of the microsphere's surface above the bottom of the indentation was increased from zero, the initially computed net repulsive force diminished; passed through 0 at the vertical equilibrium point, h = 450 nm; and thereafter became attractive. *Adapted from:* A. P. McCauley, A. W. Rodriguez, J. D. Joannopoulos, and S. G. Johnson, *Physical Review A*, Vol. 81, 2010, 012119, ©2010 The American Physical Society. See Chapter 19, Section 19.8, for a discussion.

where angles θ, ϕ, χ, and η are as shown in Fig. 14.4. Here, angles θ and ϕ are the usual zenith and azimuthal angles in a spherical coordinate system centered about the z-axis.

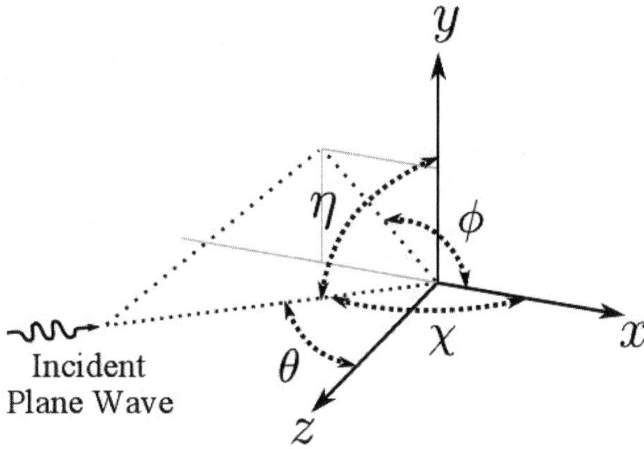

Fig. 14.4 Definitions of certain angles associated with plane-wave incidence.

An equal spacing of point sources on the aperture stop results in the equal spacing of the direction cosines (s_x, s_y) at the back focal plane of the condenser. An example of an equal-spaced arrangement of the direction cosines is shown in Fig. 14.5.

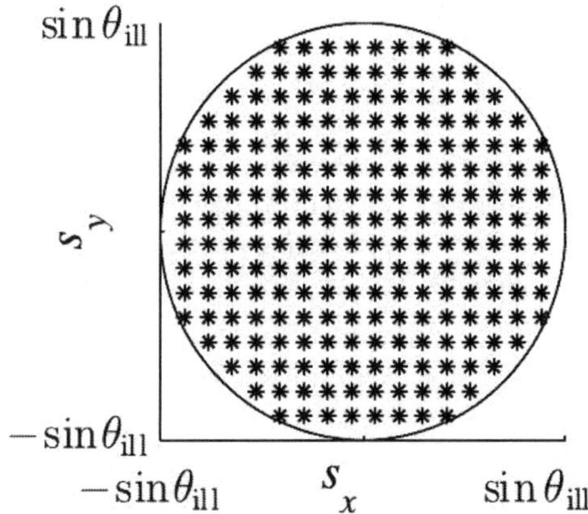

Fig. 14.5 Equal spacing of plane waves in Köhler illumination. Two orthogonal polarizations (+) and (×) are shown for each direction of incidence.

The maximum value that either s_x or s_y can attain is $\sin\theta_{\text{ill}}$, where θ_{ill} is the illumination half-angle in Fig. 14.2. We define the *illumination numerical aperture* as the quantity $\text{NA}_{\text{ill}} = n\sin\theta_{\text{ill}}$, where n is the refractive index of the medium.

As every plane wave in Fig. 14.5 propagates to the sample plane S (at distance d from F), it acquires a phase shift that will also be preserved in the scattered field due to the linearity of the system. If the intensities of the scattered field are of interest (as is the case in Köhler illumination), this extra phase shift will not have any effect on the output because of the mutual incoherence of the incident plane waves.

One can quantify the quality of the approximation that results from the discrete arrangement of the plane waves in Fig. 14.5. Let us consider quasi-monochromatic illumination with mean wavenumber $k = n k_0$, where k_0 is the mean wavenumber in free space. Let $J(x_1, y_1; x_2, y_2)$ denote the mutual coherence function at the sample plane (x, y), which quantifies the statistical correlation between two points with coordinates (x_1, y_1) and (x_2, y_2). As far as second-order quantities (intensity, two-point correlation, power-spectral density, etc.) are concerned, $J(x_1, y_1; x_2, y_2)$ completely specifies the excitation. Any illumination scheme that results in the same $J(x_1, y_1; x_2, y_2)$ will yield the same second-order quantities at the output. For the Köhler illumination scheme considered here, $J(x_1, y_1; x_2, y_2)$ is given by [1]:

$$J(x_1, y_1; x_2, y_2) = J(x_d; y_d) = \iint_{\Omega_{\text{ill}}} \exp[-jk(s_x x_d + s_y y_d)]\, d\Omega \tag{14.5}$$

where $x_d = x_1 - x_2$, $y_d = y_1 - y_2$, Ω_{ill} is the illumination solid angle bounded by $s_x^2 + s_y^2 < \sin^2\theta_{\text{ill}}$, and the differential solid angle $d\Omega$ is equal to $ds_x ds_y / \cos\theta$. Assuming moderate θ_{ill} values and neglecting the $\cos\theta$ term, this expression can also be written as:

$$J(x_d; y_d) = \int_{-\infty}^{\infty}\int_{-\infty}^{\infty} P(s_x, s_y) \exp[-jk(s_x x_d + s_y y_d)]\, ds_x ds_y \tag{14.6}$$

where $P(s_x, s_y)$ is equal to unity within the circle $s_x^2 + s_y^2 < \sin^2\theta_{\text{ill}}$ and zero elsewhere. Let us label the discrete directions in Fig. 14.5 with indices (m, n), with the direction cosines:

$$s_{x_m} = m\Delta s_x\,, \qquad s_{y_n} = n\Delta s_y \tag{14.7}$$

The indices m and n can be assumed to run from $-\infty$ to ∞. The discrete plane waves should be weighed by $(\Delta s_x \Delta s_y)^{1/2}$ (the square root of the differential area in the direction-cosine space), so that the mutual coherence function is weighted by the differential area $\Delta s_x \Delta s_y$ in the direction-cosine space. With these weights, the arrangement in Fig. 14.5 results in the following mutual coherence function:

$$J^*(x_d; y_d) = \Delta s_x \Delta s_y \sum_{m,n} P(s_{x_m}, s_{y_n}) \exp[-jk(s_{x_m} x_d + s_{y_n} y_d)] \tag{14.8}$$

Appendix 14A shows that $J^*(x_d, y_d)$ is a sum of shifted copies of the original mutual coherence function $J(x_d, y_d)$:

$$J^*(x_d; y_d) = \sum_{r=-\infty}^{\infty}\sum_{s=-\infty}^{\infty} J\left(x_d + r\frac{2\pi}{k\Delta s_x};\; y_d + s\frac{2\pi}{k\Delta s_y}\right) \tag{14.9}$$

This is called *aliasing* in signal processing [50]. For $J^*(x_d, y_d)$ to represent $J(x_d, y_d)$ in a faithful manner, the shifted copies must not overlap; i.e.:

$$\Delta s_x < \frac{2\pi}{kW_c} , \qquad \Delta s_y < \frac{2\pi}{kW_c} \tag{14.10}$$

where W_c is defined as the distance $(x_d^2 + y_d^2)^{1/2}$ at which $J(x_d, y_d)$ falls below a negligible value. Using (14.6), a closed-form expression can be found for $J(x_d, y_d)$, with a W_c value on the order of $1/(k\sin\theta_{\text{ill}}) = 1/(k_0 NA_{\text{ill}})$. If the sample dimension D is larger than W_c, then D must be substituted for W_c in (14.10). Otherwise, the mutual coherence function $J^*(x_d, y_d)$ evaluated between two most distant points on the sample will be aliased and incorrect. A more general form of the non-aliasing condition (14.10) is therefore:

$$\Delta s_x < \frac{2\pi}{k\max\{D,W_c\}} , \qquad \Delta s_y < \frac{2\pi}{k\max\{D,W_c\}} \tag{14.11}$$

For a stationary broadband excitation, the largest wavenumber k (the smallest wavelength λ) present in the illumination waveform determines the nonaliasing condition (14.11).

Summing the responses of the object to each plane wave in Fig. 14.5 is known as the *source-point* or *Abbe* integration [51–53]. Since the plane waves in Fig. 14.5 are all mutually incoherent, a separate simulation should be run for each of them. The resulting image intensities (not field values) of each simulation are then added to yield the final image intensity (see Section 14.7).

The treatment so far has been for a scalar field. It turns out that two orthogonal, mutually incoherent polarizations for the electric field of the plane wave can always be chosen for every direction (s_x, s_y), as will be seen shortly. The two polarizations for each plane wave are denoted by + and ×, and shown superposed at each (s_x, s_y) direction in Fig. 14.5. If polarization information is critical, these two polarizations should also be simulated separately, because they are mutually incoherent. The overall number of simulations is therefore twice the number of direction cosines in Fig. 14.5. This brute-force repetition of the entire simulation for incoherent illumination is a consequence of the deterministic nature of the FDTD method. A more efficient way to reduce this burden is a topic for future research.

The determination of the two orthogonal, mutually incoherent polarization states for the plane waves in Fig. 14.5 requires knowledge of the polarization properties of the sources on the aperture stop. Here, we restrict ourselves to sources that exhibit uniform polarization properties throughout the aperture stop. Assuming quasi-monochromaticity at frequency ω and denoting the coordinates on the aperture stop as (μ, ν), we can express the uniform second-order polarization properties of the source using the *cross-spectral coherency matrix* $\mathbf{J}(\mu, \nu; \omega)$ [1]:

$$\mathbf{J}(\mu,\nu;\omega) = \begin{bmatrix} \langle E_\mu^2 \rangle & \langle E_\mu E_\nu^* \rangle \\ \langle E_\mu^* E_\nu \rangle & \langle E_\nu^2 \rangle \end{bmatrix} \tag{14.12}$$

where E_μ, E_ν are the tangential components of the electric field on the aperture stop, $\langle \ldots \rangle$ denotes temporal averages (or statistical expectation values), and $\exp(j\omega t)$ dependence is implicit. Since $\mathbf{J}(\mu, \nu; \omega)$ is Hermitian, it can be represented as a weighted sum of two orthogonal coherency matrices [54, 55]:

$$\mathbf{J}(\mu,\nu;\omega) = \begin{bmatrix} A & B \\ B^* & C \end{bmatrix} + D\begin{bmatrix} C & -B \\ -B^* & A \end{bmatrix} \tag{14.13}$$

where A, B, C, D are complex scalars subject to $A, C, D \geq 0$ and $AC = |B|^2$. This corresponds to decomposing the partially polarized field on the aperture stop into two orthogonal, mutually incoherent, fully polarized fields. The directions of these polarization states coincide with the orthogonal eigenvectors of the coherency matrix $\mathbf{J}(\mu, \nu; \omega)$, and their relative weights are determined by its eigenvalues. Explicit formulas for the parameters A, B, C, D can be found in [55]. A, B, C determine the angles and the ellipticity of the two polarization states at each (μ, ν), while D determines the relative powers of these components.

Once the two orthogonal, mutually incoherent polarization states are determined, they should be treated individually in separate numerical simulations. The problem is thus reduced to fully polarized excitation, in which the electric field on the aperture stop is uniformly polarized in a certain direction. Since the general case of elliptical polarization can be handled as a complex superposition of two linearly polarized states, it suffices to treat only linear polarization. A good approximation for the polarization of the resulting plane waves in the back focal plane of the condenser can be obtained using the construction in Fig. 14.6.

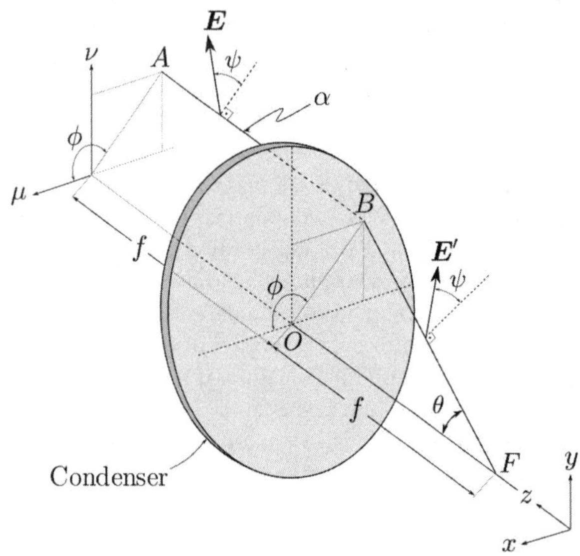

Fig. 14.6 Polarization of the plane wave created by the condenser at focus F due to an infinitesimal source at point A on the aperture stop.

In Fig. 14.6, an arbitrary point A on the plane of the aperture stop (μ, ν) is shown on the left. Let B denote the point on the lens such that $\overline{AB}$ is parallel to the optical axis OF. Let α denote the ray emanating from A, hitting the lens at B, and intersecting the optical axis at the back focal point F. The plane including the ray α and the optical axis OF is called the meridional plane; which, in our case, makes an angle ϕ with the μ axis. The key observation is thus: The vector electric field on the rays in the neighborhood of α, which are parallel to α around the focus F, will be almost the same as that of the ray α. Therefore, if the illuminated sample at the back focal plane at F is confined to a small area with dimensions $D = f$, the polarization and magnitude of the electric field on the sample at F are determined by the ray α.

The magnitude of the electric field at F follows from the intensity law of geometrical optics. Referring to Fig. 14.6, an infinitesimal source at A creates a spherical wavefront centered around A, and launches the ray α toward B. The magnitude of the electric field $\boldsymbol{E}$ at B due to this source is proportional to $|\boldsymbol{E}_s|/f$, where the vector quantity $\boldsymbol{E}_s$ is the *strength factor* of the ray α, which depends only on the magnitude and polarization of the source at A but not on f. If we consider an infinitesimal bundle of rays emanating from A and spread over an infinitesimal area around B, these rays are then collimated by the condenser into a parallel tube of rays intersecting the optical axis around the back focal point F. The infinitesimal area subtended by this parallel tube of rays is $\cos\theta$ times the infinitesimal area subtended by the ray bundle on the other side of the condenser. From the intensity law of geometrical optics [1], it follows that the magnitude of the electric field $\boldsymbol{E}'$ at the back focal point F is given by:

$$|\boldsymbol{E}'| = (\cos\theta)^{-1/2}|\boldsymbol{E}'_s|/f \qquad (14.14)$$

The polarization of $\boldsymbol{E}'$ still remains to be found. Let ψ denote the angle that $\boldsymbol{E}$ makes with the meridional plane, as shown in Fig. 14.6. If the angles of incidence at every surface of refraction through the lens are small, the angle ψ between the electric field vector on the ray and the meridional plane stays constant [1, 34]. This fact has been previously used in similar Köhler-illumination constructions [56, 57] as well as the synthesis of a coherent converging light pulse in the FDTD method [37]. Apart from the factor $(\cos\theta)^{-1/2}/f$, the electric field vector $\boldsymbol{E}'$ is then a *rotation* of $\boldsymbol{E}$ by an angle θ around an axis perpendicular to the meridional plane [56]. [For additional discussion of the strength factor of a ray, see (14.15) and its related text.]

An important special case of incoherent Köhler-style illumination is when the two orthogonal components E_μ and E_ν of the electric field on the aperture stop are of equal power and are completely uncorrelated: $|E_\mu|^2 = |E_\nu|^2$ and $\langle E_\mu E_\nu^*\rangle = \langle E_\mu^* E_\nu\rangle = 0$. The source on the aperture stop is then said to be *natural*, or *completely unpolarized*. The cross-spectral coherency matrix $\mathbf{J}(\mu, \nu)$ in (14.12) is then proportional to the identity matrix, which amounts to $D = 1$ in the decomposition (14.13). This means that the A, B, C values for the decomposition in (14.13) are not constrained by anything but the coherency condition $AC = |B|^2$. As a result, the choice of the two orthogonal polarization states + and × for each and every plane wave in Fig. 14.5 becomes *completely arbitrary*.

Note that there are other aperture shapes besides the circular shape in Fig. 14.5 employed in practice. Depending on the geometry of the aperture, the discretization scheme for the incidence directions can be slightly modified. For example, the annular aperture, a characteristic element of phase contrast microscopy [10, 58], can be accommodated using an equal spacing of the incidence angles, rather than the direction cosines.

14.5 SCATTERING SUBSYSTEM

If the scattering object has key structural details comparable in size to the wavelength of the incident light, ray-based or asymptotic methods completely fail to describe the scattering process. Some examples include biological cells, photonic crystals, and phase-shift masks in lithography. In this situation, one must seek a direct numerical solution for Maxwell's equations. The finite-difference time-domain (FDTD) method is very popular for such applications [29, 59, 60] because of its simplicity, intuitiveness, robustness, and ease of implementation. In its most basic form, the electric and magnetic fields are updated in time using a simple leapfrog updating procedure, without any matrix inversions. The FDTD method also has the advantage of yielding direct time-domain data, allowing immediate broadband analysis.

In spite of its simplicity, the Cartesian FDTD grid can be overly restrictive when local mesh refinement or conformal grids are required. Furthermore, staircase approximations have to be made for modeling curved surfaces in FDTD grids. When these geometrical constraints are too stringent, one might prefer to employ triangular meshes that allow a much finer discretization of round surfaces and much easier mesh refinement. In spite of the latitude they offer in representing different geometries, irregular grids require much more effort to ensure the consistency and stability of the numerical solution algorithm. Collectively, these algorithms are referred to as finite-element (FE) methods. For a review of the FE methods, see [61, 62].

FD and FE methods share some inherent drawbacks that are a result of their very construction. Since both methods operate on field values in a finite volume, they require auxiliary techniques for handling sources that radiate in an unbounded region. Many so-called "absorbing boundary conditions" (ABCs) have been developed for truncating FD and FE solution spaces [29]. These conditions simulate the extension of the computation grid to infinity by absorbing most of the energy incident on the outer grid boundary. The most popular ABC in use today is Berenger's "perfectly matched layer" [29, 63], which constitutes a milestone in the development of differential-equation methods. Grid dispersion and grid anisotropy are two other major problems caused by the finite size of the grid voxels and their lack of rotational symmetry. These problems can never be completely eliminated, but can be partially alleviated by choosing finer spatial steps and/or employing more rotationally symmetric discretization schemes [29].

Among the various rigorous numerical approaches for solving Maxwell's equations, the FDTD method seems to have gained wider acceptance than others primarily because it is conceptually simpler, physically more intuitive, and easier to implement. One of the earliest applications of FDTD to numerical optical imaging is the TEMPEST software of the University of California–Berkeley, which was developed for photolithography simulation [64]. Originally designed to compute two-dimensional (2-D) mask patterns, it was later generalized to three dimensions (3-D) [65–67] and then further enhanced to handle extreme-ultraviolet (EUV) photolithography simulations [68–72]. In addition to its initial purpose of simulating the scattering response of photo masks [73, 74], TEMPEST has also been used for simulating metrology [75] and alignment systems [76, 77].

The FDTD method has also found use in modeling microscopy modalities. The earliest work on this subject began with the simulation of near-field imaging modalities such as the near-field scanning optical microscope (NSOM) [78–83]. Far-field microscopy modeling was later tackled by the incorporation of ray principles and diffraction formulas from optics into the solution algorithm [7, 11, 36]. Upon suitable modification of the optical far-field data, different modalities such as differential-contrast microscopy (DIC) [35], phase-contrast microscopy [58], and confocal microscopy [8, 25] can be handled. A novel algorithm based on the extended–Nijboer–Zernike (ENZ) theory of diffraction (see Section 14.7.3) was coupled with FDTD for photo mask imaging [53, 84] and the imaging of general 3-D objects [85]. A variant of the FDTD method, called the pseudospectral time-domain (PSTD) method [86], is particularly suited to the analysis of scattering from optically large structures, such as macroscopic biological tissue [87]. Recent advances and applications of PSTD are discussed in Chapters 1 and 16.

This chapter focuses primarily on the FDTD solution of the optical imaging problem. However, it is the trade-off between accuracy requirements and resource constraints that determines the specific numerical method best suited to a problem. For guidance, one can consult studies that compare the attributes of a variety of numerical methods [6, 88–91].

14.6 COLLECTION SUBSYSTEM

After the scattering from the sample is calculated, the scattered field should be propagated to the image plane to complete the imaging process. These two steps are commonly performed by an *objective*. As shown diagrammatically in Fig. 14.1, the first task of the objective is to collect the portion of the light scattered from the sample that falls within its entrance pupil. The entrance pupil is defined as the image of the limiting aperture stop as seen from the object side. Among all the aperture stops in the system, the one that limits the angular extent of the rays emanating from the object is the limiting aperture [1]. In general, the entrance pupil can be located at any finite distance from the sample. However, it is more advantageous to place the entrance pupil at infinity whenever possible. Such a system is referred to as *object-side telecentric* [1]. In a telecentric system, the size of the blurred image does not change as the object moves out of focus. This is a very desirable feature in both microscopy and photolithography. In the former, telecentricity makes it easier to manually focus on a sample, because the image size does not change with defocus. In the latter, telecentricity is much more crucial, for it ensures that the image of the photomask is much less sensitive to positioning tolerances (see Chapter 15).

In addition to its advantages, telecentricity poses its own difficulties. First of all, it should be remembered that the limiting aperture stop should be located at the back focal plane of the optical system preceding it. If the lateral dimension of a certain lens in the optical setup is not large enough, that lens could act as the limiting aperture and destroy telecentricity. For this reason, telecentricity usually requires that the lenses have much larger lateral dimensions than the sample, resulting in more expensive equipment. This is usually not a big issue for microscopes. However, telecentricity places a more stringent constraint on photolithography projection lenses in terms of both price and performance. Almost all modern microscopes and photolithography projection lenses are telecentric on the object side. We will assume object-side telecentricity in the following analysis.

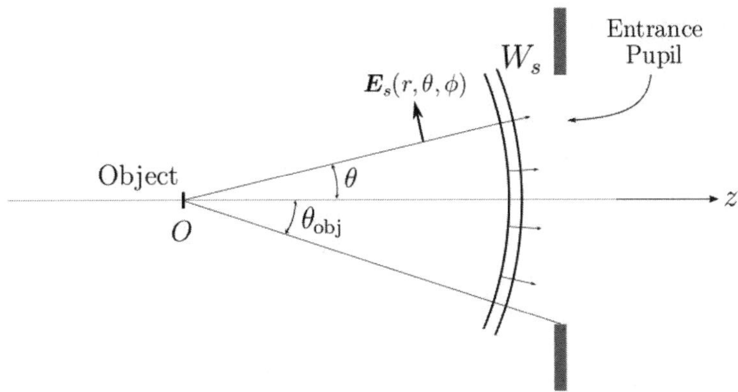

Fig. 14.7 The collection geometry for a telecentric system.

The collection geometry for an object-side telecentric system is shown in Fig. 14.7. Since the entrance pupil is at infinity, the scattering object can be regarded as a point at O, and the scattered wavefront W_s is spherical. Conventional spherical coordinates (r, θ, ϕ) are defined with respect to O and the z-axis chosen to coincide with the optical axis. The *far zone* (also called the

Fraunhofer or *radiation zone*) is defined as the region $r >> d^2/\lambda$ where d is the maximum dimension of the sample and λ is the wavelength. In the far zone, the radial dependence of the field can be factored out, and the wavefront W_s is completely specified by its angular profile [92, 93]:

$$\boldsymbol{E}_s(r,\theta,\phi) = \boldsymbol{E}_s(\theta,\phi)\frac{e^{-jkr}}{r} \tag{14.15}$$

Here, $k = nk_0$ is the wavenumber in the homogeneous space located between the object and the entrance pupil, and n is the refractive index of the same space. The vector quantity $\boldsymbol{E}_s(\theta,\phi)$ in (14.15) is called the *strength factor* of the ray associated with the far-zone direction (θ, ϕ) [34, 94]. The collection step therefore reduces to numerically calculating $\boldsymbol{E}_s(\theta,\phi)$ at various observation directions. Depending on the scattering geometry, this may be accomplished in several different ways. One common property of almost all numerical collection methods is that the near-field information is used to obtain the far-zone field using certain theorems of electromagnetics. The term *near-field-to-far-field transform* (NFFFT) is commonly used for such an algorithm that computes the far-field from the near-field. All NFFFT algorithms rely on either spatial Fourier analysis or a Green's function formalism. We will examine these two cases separately.

14.6.1 Fourier Analysis

Transformation Along an Infinite Planar Surface in the Near Field

In certain cases, the strength factor $\boldsymbol{E}_s(\theta,\phi)$ can be found using the spatial Fourier transform of the near fields around the scatterer. This near field should be given on an infinite planar surface S between the scatterer and the entrance pupil, as shown in Fig. 14.8.

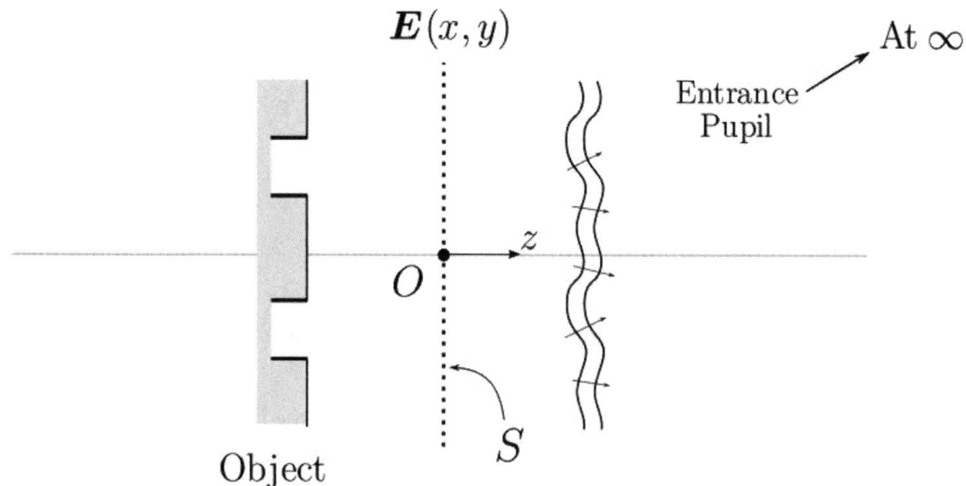

Fig. 14.8 Collection geometry for a Fourier-analysis-based near-field-to-far-field transform. The near fields should be given on an infinite planar surface S between the scatterer and the entrance pupil.

Referring to Fig. 14.8, we define the origin O on S, and denote the vector electric field on this plane as $\boldsymbol{E}(x, y)$. The far-zone field $\boldsymbol{E}(r, \theta, \phi)$ on the wavefront W_s at the entrance pupil can be found by expanding $\boldsymbol{E}(x, y)$ into its plane-wave (or angular) spectrum, and propagating it to very large r using the steepest-descent method [23]. Let us define the plane-wave spectrum of the 2-D electric-field distribution $\boldsymbol{E}(x, y)$ as the following Fourier transform operation:

$$\tilde{\boldsymbol{E}}(k_x, k_y) = \int_{-\infty}^{\infty}\int_{-\infty}^{\infty} \boldsymbol{E}(x,y)\exp[j(k_x x + k_y y)]\, dx\, dy \tag{14.16}$$

with the inverse transform (or the plane-wave representation) given by:

$$\boldsymbol{E}(x,y) = \frac{1}{(2\pi)^2}\int_{-\infty}^{\infty}\int_{-\infty}^{\infty} \tilde{\boldsymbol{E}}(k_x, k_y)\exp[-j(k_x x + k_y y)]\, dk_x\, dk_y \tag{14.17}$$

It is understood in (14.16) and (14.17) that the Fourier transform is applied to the Cartesian components of the vector integrands separately. The representation (14.17) for the vector field $\boldsymbol{E}(x, y)$ is an infinite summation of plane waves (propagating and evanescent) whose lateral propagation coefficients are (k_x, k_y).

This plane-wave representation can be used to extrapolate the electric field to the region above the plane S, i.e., $z > 0$. The following field satisfies (14.17) at $z = 0$ and Maxwell's equations in the region $z > 0$:

$$\boldsymbol{E}(x,y,z) = \frac{1}{(2\pi)^2}\int_{-\infty}^{\infty}\int_{-\infty}^{\infty} \tilde{\boldsymbol{E}}(k_x, k_y)\exp[-j(k_x x + k_y y + k_z z)]\, dk_x\, dk_y \tag{14.18}$$

where $k_z = (k^2 - k_x^2 - k_y^2)^{1/2}$ [23]. Here, k is the wavenumber ω/c in the homogeneous space between S and the entrance pupil. Only the plane-wave components in (14.17) with $k_x^2 + k_y^2 < k^2$ contribute to the far field because a complex k_z represents an evanescent plane wave decaying exponentially in z. Now, propagating the plane waves with $k_x^2 + k_y^2 < k^2$ into the space $z > 0$ and using the steepest-descent method at the far zone, one arrives at the following expression for the strength factor $\boldsymbol{E}_s(\theta,\phi)$ [23]:

$$\boldsymbol{E}_s(\theta,\phi) = E_\theta(\theta,\phi)\hat{\boldsymbol{\theta}} + E_\phi(\theta,\phi)\hat{\boldsymbol{\phi}} \tag{14.19}$$

where the θ and ϕ components are given by:

$$E_\theta(\theta,\phi) = \frac{jk}{2\pi}\left[\tilde{E}_x(\alpha,\beta)\cos\phi + \tilde{E}_y(\alpha,\beta)\sin\phi\right] \tag{14.20}$$

$$E_\phi(\theta,\phi) = \frac{jk}{2\pi}\cos\theta\left[-\tilde{E}_x(\alpha,\beta)\sin\phi + \tilde{E}_y(\alpha,\beta)\cos\phi\right] \tag{14.21}$$

in which $\tilde{E}_x(k_x, k_y)$ and $\tilde{E}_y(k_x, k_y)$ are the x and y components of the plane-wave spectrum $\tilde{\boldsymbol{E}}(k_x, k_y)$ in (14.16), and the definition:

$$(\alpha,\beta) = (k\cos\phi\sin\theta,\, k\sin\phi\sin\theta) \tag{14.22}$$

has been introduced for brevity. This expression can be put into a more compact vectorial form. Applying the zero-divergence condition of the electric field in charge-free space to (14.18) reveals that $\tilde{\boldsymbol{E}}(k_x, k_y)$ is transverse to the propagation vector $\vec{k} = k_x\hat{x} + k_y\hat{y} + k_z\hat{z}$. It follows that the vector $\tilde{\boldsymbol{E}}(\alpha, \beta)$ only possesses θ and ϕ components. Expanding $\tilde{E}_x$ and $\tilde{E}_y$ in (14.20) and (14.21) in terms of $\tilde{E}_\theta$ and $\tilde{E}_\phi$, it is found that:

$$\boldsymbol{E}_s(\theta,\phi) = \frac{jk\cos\theta}{2\pi}\tilde{\boldsymbol{E}}(\alpha,\beta) \tag{14.23}$$

Expressions (14.19) – (14.23) can be used for certain scattering geometries to calculate the strength factor $\boldsymbol{E}_s(\theta,\phi)$ numerically. The simplest such geometry results when the sample is a *phase object*, represented by a complex transmittance $T(x, y)$. This assumption is appropriate for photolithography when dealing with thin masks [95]. If the object is not sufficiently thin, the complex transmittance model can be improved by calculating the response of the object to a normally incident plane wave by some rigorous method, and then obtaining the responses to other wave incidences perturbatively from this result. However, this approach is viable only if the illumination and collection NAs are very small [96]. Such an assumption is often valid in photolithography, but not microscopy [97] (see Fig. 14.12 later in this chapter). When neither the thin-film assumption nor the perturbation assumption is valid, the scattered electric field $\boldsymbol{E}(x, y)$ on plane S and its plane-wave spectrum $\tilde{\boldsymbol{E}}(k_x, k_y)$ should be calculated using a rigorous numerical method such as FDTD.

Application of Periodic Boundary Conditions

A key difficulty in using the Fourier-analysis-based near-field-to-far-field transform in the context of FDTD is that the computational grid is always bounded in space. Since Fourier analysis requires that the near field be given on an infinite planar surface, the range of FDTD applications suitable for Fourier-based collection algorithms is limited.

However, a significant category of FDTD-modeled problems amenable to the Fourier-analysis-based near-field-to-far-field transform includes those scatterers that can be treated using *periodic boundary conditions*. In such problems, both the simulation geometry and the incident wave are constrained to be periodic along a certain direction or two orthogonal directions. The electromagnetic field on an infinite lateral plane (parallel to the direction of periodicity) is therefore determined completely by the electromagnetic field in the finite grid. This allows the use of Fourier analysis for the collection of the far-zone field. For a good review on periodic boundary conditions in FDTD, the reader could consult [29]. A more recent method for enforcing periodic boundary conditions in FDTD can be found in [98].

It is sufficient to consider the simpler 2-D case in order to illustrate the concept of periodicity. Let us consider a structure that is invariant in the y-direction and periodic with spatial period d in the x-direction. Let a monochromatic unit-amplitude transverse-electric (TE) plane wave be incident from angle θ_i with respect to the z-axis, with y component:

$$E_y^i(x,z) = \exp[-jk(\gamma_x x - \gamma_z z)] \tag{14.24}$$

in which $\gamma_x = \sin\theta_i$ and $\gamma_z = \cos\theta_i$. The periodicity of the structure requires that the electromagnetic field obey the following pseudo-periodicity condition everywhere:

$$E_y(x+d,z) = E_y(x,z)\exp(-jk\gamma_x d) \tag{14.25}$$

The linear phase term $\exp(-jk\gamma_x d)$ is enforced by the incident plane wave, and becomes unity for normal incidence. From Floquet's theorem [99], the scattered electric field on S (see Fig. 14.8) for TE illumination can be written in the following series form:

$$E_y(x) = \sum_{p=-\infty}^{\infty} R_p \exp(-j\beta_p x) \tag{14.26}$$

where $\beta_p = k\sin\theta_i - p(2\pi/d)$ are the equally spaced Floquet wavenumbers and R_p ($p = -\infty \ldots \infty$) are the reflection coefficients for the pth Floquet mode with lateral wavenumber β_p. Comparing (14.26) with (14.17), we see that the reflection coefficients R_p play the role of the plane-wave spectral coefficients $\tilde{E}(k_x, k_y)$, and the Floquet wavenumbers β_p play the role of k_x. One difference here is that R_p is a finite set of numbers unlike $\tilde{E}(k_x, k_y)$ because of the periodicity of the structure. Another difference is that R_p is defined for a 2-D geometry that is invariant in y.

If the scattering structure is periodic in both x and y with periods d_x and d_y, the scattered field can likewise be expressed as a doubly infinite sum of vector-valued Floquet modes $\boldsymbol{R}_{pq}$ [100]:

$$\boldsymbol{E}(x,y) = \sum_{p=-\infty}^{\infty}\sum_{q=-\infty}^{\infty} \boldsymbol{R}_{pq}\exp[-j(\beta_p x + \beta_q y)] \tag{14.27}$$

where now the Floquet wavenumbers:

$$\beta_p = k\cos\phi_i \sin\theta_i - p(2\pi/d_x) \tag{14.28a}$$

$$\beta_q = k\sin\phi_i \sin\theta_i - q(2\pi/d_y) \tag{14.28b}$$

play the roles of k_x and k_y, respectively. Here, ϕ_i and θ_i are the spherical incidence angles of the incident plane wave. Comparing this expression with (14.17), the plane-wave spectrum $\tilde{\boldsymbol{E}}(k_x, k_y)$ can be written in terms of $\boldsymbol{R}_{pq}$ as follows:

$$\tilde{\boldsymbol{E}}(k_x, k_y) = (2\pi)^2 \sum_{p=-\infty}^{\infty}\sum_{q=-\infty}^{\infty} \boldsymbol{R}_{pq}\,\delta(k_x - \beta_p)\,\delta(k_y - \beta_q) \tag{14.29}$$

where $\delta(\cdot)$ is the Dirac delta function. Substituting this expression in (14.23), the strength factor $\boldsymbol{E}_s(\theta,\phi)$ is obtained as:

$$\boldsymbol{E}_s(\theta,\phi) = (jk2\pi\cos\theta)\sum_{p=-\infty}^{\infty}\sum_{q=-\infty}^{\infty} \boldsymbol{R}_{pq}\,\delta(\alpha - \beta_p)\,\delta(\beta - \beta_q) \tag{14.30}$$

Carrying the term $\cos\theta = [1 - (\alpha/k)^2 - (\beta/k)^2]^{1/2}$ inside the summation and using the properties of the Dirac delta function, this can be written as:

$$\boldsymbol{E}_s(\theta,\phi) = (jk2\pi)\sum_{p=-\infty}^{\infty}\sum_{q=-\infty}^{\infty} c_{pq}\boldsymbol{R}_{pq}\,\delta(\alpha - \beta_p)\,\delta(\beta - \beta_q) \tag{14.31}$$

in which c_{pq} is the dimensionless cosine parameter:

$$c_{pq} = \sqrt{1 - (\beta_p / k)^2 - (\beta_q / k)^2} \tag{14.32}$$

We see that the far-zone field is nonzero only at a discrete set of directions. This is a direct result of the periodicity of the system and the discreteness of the Floquet wavenumbers β_p, β_q. Second, the finite range of the variables $\alpha = k\cos\phi\sin\theta$, $\beta = k\sin\phi\sin\theta$ between 0 and $k\sin\theta_{\text{obj}}$ (where $\sin\theta_{\text{obj}}$ is the collection NA in Fig. 14.8) only allows for a finite number of observation directions to be collected by the objective. It is easy to see that any Floquet mode $\boldsymbol{R}_{pq}$ with p index higher than a maximum value $p_{\max}$ will not be collected by the objective, where

$$p_{\max} = \frac{d\sin\theta_{\text{obj}}}{\lambda} \tag{14.33}$$

The same concept applies to the q index. For the best reconstruction of the scattered field at the image plane, $p_{\max}$ should be maximized. This can be accomplished by reducing the wavelength, λ, or increasing the collection NA, $\sin\theta_{\text{obj}}$. On the other extreme, if $p_{\max}$ is less than 1, only the homogeneous zeroth-order mode ($p = 0$) propagates to the objective, resulting in a uniform image (see Section 14.7.2, specifically Fig. 14.14).

Let us assume that, using FDTD in conjunction with some periodic boundary conditions, the vector field $\boldsymbol{E}(x, y)$ scattered from a 3-D object has been computed on plane S of Fig. 14.8 at equally spaced spatial points $m\Delta x$ and $n\Delta y$. This results in the discrete array $\boldsymbol{E}[m, n]$:

$$\boldsymbol{E}[m,n] = \boldsymbol{E}(m\Delta x, n\Delta y)\,, \qquad m = 0,\ldots,M-1\,; \quad n = 0,\ldots,N-1 \tag{14.34}$$

Here, the entire periods in both x and y are covered by the sampling:

$$M\Delta x = d_x\,, \qquad N\Delta y = d_y \tag{14.35}$$

We now describe how the vector amplitudes of the Floquet modes in (14.27) for this case can be obtained by using the 2-D discrete Fourier transform (DFT). Expressing the results in terms of a DFT is always advantageous, since there exists a very efficient algorithm for the evaluation of the DFT called the fast Fourier transform (FFT). We first define the phase-shifted sampled array $\bar{\boldsymbol{E}}[m,n]$ as follows:

$$\bar{\boldsymbol{E}}[m,n] = \boldsymbol{E}[m,n]\exp\left[jk\sin\theta_i\left(d_x\cos\phi_i\frac{m}{M} + d_y\sin\phi_i\frac{n}{N}\right)\right] \tag{14.36}$$

The above phase shift depends on the direction of incidence (θ_i, ϕ_i) of the impinging plane wave. This shift removes the phase condition (14.25) imposed by the incident plane wave, and simplifies the resulting expression considerably. The 2-D DFT of this modified array is conventionally defined as [50, 101]:

$$\tilde{\bar{\boldsymbol{E}}}[p,q] = \sum_{m=0}^{M-1}\sum_{n=0}^{N-1}\bar{\boldsymbol{E}}[m,n]\exp\left[-j2\pi\left(\frac{pm}{M} + \frac{qn}{N}\right)\right] \tag{14.37}$$

It can be shown (see Appendix 14B) that the DFT array $\tilde{\bar{\boldsymbol{E}}}[p,q]$ is related to the Floquet modes $\boldsymbol{R}_{pq}$ as follows:

$$\tilde{\bar{E}}[p,q] = MN \sum_{r=-\infty}^{\infty} \sum_{s=-\infty}^{\infty} \boldsymbol{R}_{p+rM,\, q+sN} \tag{14.38}$$

Equation (14.38) expresses the results of the 2-D DFT operation on the phase-shifted sampled field array $\bar{E}[m,n]$ in terms of the Floquet modes $\boldsymbol{R}_{pq}$ of the original, *continuous* field $\boldsymbol{E}(x,y)$. From (14.38), we can immediately draw some key conclusions. First, the DFT array $\tilde{\bar{E}}[p,q]$ is seen to be equal to an infinite summation of shifted copies of $\boldsymbol{R}_{pq}$ in both the p and q indices. In other words, $\tilde{\bar{E}}[p,q]$ is an aliased version of $\boldsymbol{R}_{pq}$. In order for $\tilde{\bar{E}}[p,q]$ to faithfully represent $\boldsymbol{R}_{pq}$, the shifted copies of $\boldsymbol{R}_{pq}$ should not overlap. This requires that the shifting periods M and N be larger than the effective widths W_p and W_q of $\boldsymbol{R}_{pq}$ in the p and q indices:

$$M > W_p, \qquad N > W_q \tag{14.39}$$

which, using (14.35), can also be written in terms of the sampling periods as:

$$\Delta x < d_x / W_p, \qquad \Delta y < d_y / W_q \tag{14.40}$$

If M or N is too small, shifted copies of $\boldsymbol{R}_{pq}$ overlap, and $\boldsymbol{R}_{pq}$ cannot be recovered fully from $\tilde{\bar{E}}[p,q]$. If both M and N are large enough so that neighboring replicas of $\boldsymbol{R}_{pq}$ do not overlap, then $\boldsymbol{R}_{pq}$ can be recovered from $\tilde{\bar{E}}[p,q]$ using the relationship (14.38):

$$\boldsymbol{R}_{pq} = \frac{1}{MN} \tilde{\bar{E}}[p,q] \tag{14.41}$$

for a range of p, q values around $p = q = 0$. Some simple estimates of the effective widths W_p and W_q of $\boldsymbol{R}_{pq}$ can be made in certain circumstances. For example, if the plane S is *far enough* above the periodic structure, the evanescent Floquet modes in (14.27) become negligible, and it is only necessary to consider the propagating modes. Using the propagation conditions $|\beta_p| < k$ and $|\beta_q| < k$, the following expressions are obtained for W_p and W_q:

$$W_p = 2d_x / \lambda, \qquad W_q = 2d_y / \lambda \tag{14.42}$$

Substituting these expressions into (14.40), we obtain:

$$\Delta x < \lambda / 2, \qquad \Delta y < \lambda / 2 \tag{14.43}$$

This is, in effect, a discrete version of the celebrated Nyquist sampling theorem for bandlimited signals [50]. It states that, if only propagating modes are present on the plane S, the spatial sampling periods need only be smaller than one-half wavelength.

Fourier-analysis-based NFFFTs have the advantage of not requiring separate treatments for multilayered media. This is because the sampling plane S lies above the scattering structure and any of its stratifications. However, these NFFFTs have a key limitation in that the field must be known on a laterally infinite plane S. As discussed above, this is a surmountable problem if the scattering structure is periodic in space. In addition, a Fourier-analysis-based NFFFT might still be feasible even for nonperiodic structures if both the scattered field and reflections of the incident field from the interfaces are bounded in space. Then, the collection surface S can, in principle, be made large enough to cover them. Nevertheless, the preferred method for nonperiodic structures remains the Green's function formalism explained in the next subsection.

14.6.2 Green's Function Formalism

If the scattering medium is not periodic, the Fourier-analysis-based NFFFT approaches of the previous subsection have limited use. The preferred method for a nonperiodic scatterer is usually based on a Green's function formalism, which has the geometry illustrated in Fig. 14.9.

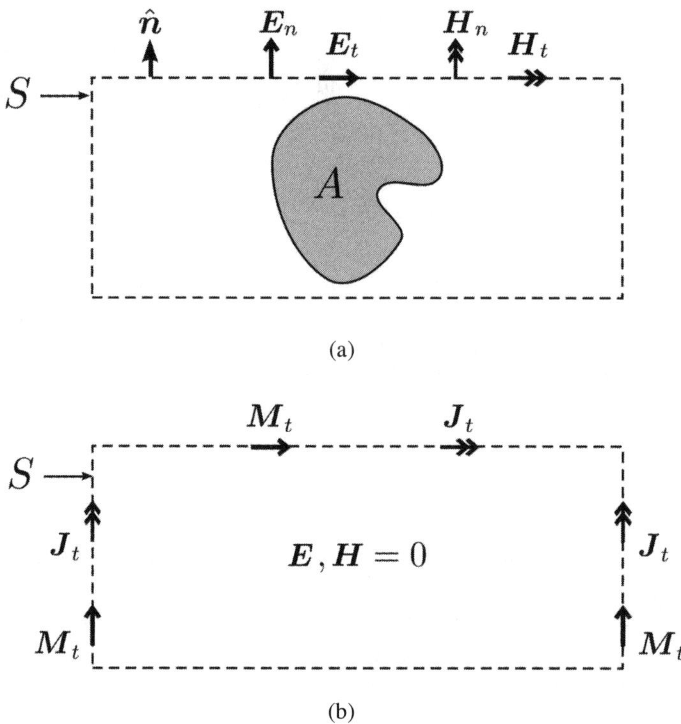

Fig. 14.9 Pictorial description of the Green's function near-field-to-far-field transformation (NFFFT). (a) Geometry of the NFFFT, wherein a scattering (or radiating) structure *A* is located within the closed virtual surface *S*. (b) Equivalent surface electric and magnetic currents on *S*, with the interior of *S* removed, which produces the same far field as that originally generated by structure A.

In Fig. 14.9, structure *A* is located within a closed virtual surface *S*, which we shall designate as the NFFFT surface. We assume that the electric and magnetic near fields scattered (or radiated) by structure *A* at the location of *S* are known by virtue of having applied a rigorous solution of Maxwell's equations. We first convert these fields to equivalent electric currents $\boldsymbol{J}_t$ and magnetic currents $\boldsymbol{M}_t$ flowing on surface *S*, as follows:

$$\boldsymbol{J}_t = \hat{\boldsymbol{n}} \times \boldsymbol{H} , \qquad \boldsymbol{M}_t = -\hat{\boldsymbol{n}} \times \boldsymbol{E} \tag{14.44}$$

where $\hat{\boldsymbol{n}}$ is the outward normal unit vector on *S* shown in Fig. 14.9(a). Note that these cross-product operations select only the *tangential* components of the electric and magnetic near fields on *S* for further processing.

The most prominent method for obtaining the scattered or radiated far-field in terms of $\boldsymbol{J}_t$ and $\boldsymbol{M}_t$ of (14.44) is a vector-potential formulation using the *surface equivalence theorem* [92, 102, 103]. Because the derivation of this theorem is quite lengthy, it will not be reproduced here [29]. However, the result of the theorem is very simple, as illustrated in Fig. 14.9(b). Namely, $\boldsymbol{J}_t$ and $\boldsymbol{M}_t$ radiating in free space, with the interior of S removed, produce the same far fields as those originally generated by structure A. (The removal of structure A is justified because $\boldsymbol{J}_t$ and $\boldsymbol{M}_t$ are defined to create a null field inside S.)

Now, the fields radiated by $\boldsymbol{J}_t$ and $\boldsymbol{M}_t$ can be found using a variety of methods. In the formulation discussed here, these currents are first inserted into certain integrals that yield intermediate quantities called the vector potentials. Among several slightly different conventions for their definitions, we will follow that of Balanis [103]. Assuming sinusoidal steady-state oscillations of the electric and magnetic fields of angular frequency ω with the corresponding wavenumber k, the vector potentials are obtained from $\boldsymbol{J}_t$ and $\boldsymbol{M}_t$ as:

$$\boldsymbol{A}(\boldsymbol{r}) = \frac{\mu_0}{4\pi}\iint_S \boldsymbol{J}_t(\boldsymbol{r}')\frac{e^{-jk|\boldsymbol{r}-\boldsymbol{r}'|}}{|\boldsymbol{r}-\boldsymbol{r}'|}d\boldsymbol{r}' \tag{14.45}$$

$$\boldsymbol{F}(\boldsymbol{r}) = \frac{\varepsilon_0}{4\pi}\iint_S \boldsymbol{M}_t(\boldsymbol{r}')\frac{e^{-jk|\boldsymbol{r}-\boldsymbol{r}'|}}{|\boldsymbol{r}-\boldsymbol{r}'|}d\boldsymbol{r}' \tag{14.46}$$

in which $\boldsymbol{A}(\boldsymbol{r})$ and $\boldsymbol{F}(\boldsymbol{r})$ are called, respectively, the electric and magnetic vector potentials. The primed coordinates $\boldsymbol{r}'$ represent the source points on S, while the unprimed coordinates $\boldsymbol{r}$ represent observation points outside of S. The electric and magnetic fields at the observation point $\boldsymbol{r}$ are obtained from the following differentiation operations on the vector potentials:

$$\boldsymbol{E}(\boldsymbol{r}) = -j\omega\left[\boldsymbol{A} + \frac{1}{k^2}\nabla(\nabla\cdot\boldsymbol{A})\right] - \frac{1}{\varepsilon_0}\nabla\times\boldsymbol{F} \tag{14.47}$$

$$\boldsymbol{H}(\boldsymbol{r}) = -j\omega\left[\boldsymbol{F} + \frac{1}{k^2}\nabla(\nabla\cdot\boldsymbol{F})\right] + \frac{1}{\mu_0}\nabla\times\boldsymbol{A} \tag{14.48}$$

In the near field outside of S, the evaluation of (14.45)–(14.48) can be complicated. However, considerable simplification occurs when the observation point $\boldsymbol{r}$ approaches infinity ($|\boldsymbol{r}| = r \to \infty$). In this *far zone*, the $|\boldsymbol{r}-\boldsymbol{r}'|$ term in the exponentials in (14.45) and (14.46) can be approximated as $(r - \hat{\boldsymbol{r}}\cdot\boldsymbol{r}')$, where $\hat{\boldsymbol{r}} = \boldsymbol{r}/r = (\cos\phi\sin\theta, \sin\phi\sin\theta, \cos\theta)$ is the unit vector in the direction of observation. In the denominators, the $|\boldsymbol{r}-\boldsymbol{r}'|$ term can be approximated as simply r. This results in the following far-zone expressions for the vector potentials:

$$\boldsymbol{A}(\boldsymbol{r}) = \mu_0\frac{e^{-jkr}}{4\pi r}\iint_S \boldsymbol{J}_t(\boldsymbol{r}')e^{jk\hat{\boldsymbol{r}}\cdot\boldsymbol{r}'}d\boldsymbol{r}' \tag{14.49}$$

$$\boldsymbol{F}(\boldsymbol{r}) = \varepsilon_0\frac{e^{-jkr}}{4\pi r}\iint_S \boldsymbol{M}_t(\boldsymbol{r}')e^{jk\hat{\boldsymbol{r}}\cdot\boldsymbol{r}'}d\boldsymbol{r}' \tag{14.50}$$

As a result of the far-zone approximation, the r-dependence in (14.49) and (14.50) is completely factored out, and the surface integrals only depend on the observation angles θ, ϕ. The differentiation relations (14.47) and (14.48) also assume simpler forms for large r if the terms that decay faster than $1/r$ are neglected. Expanding the ∇ operator in the spherical coordinates (r, θ, ϕ) and neglecting $1/r^2$ and higher-order terms, (14.47) and (14.48) simplify to:

$$E_r = 0 \tag{14.51a}$$

$$E_\theta = -j\omega\left(A_\theta + \eta_0 F_\phi\right) \tag{14.51b}$$

$$E_\phi = -j\omega\left(A_\phi - \eta_0 F_\theta\right) \tag{14.51c}$$

and

$$H_r = 0 \tag{14.52a}$$

$$H_\theta = \frac{j\omega}{\eta_0}\left(A_\phi - \eta_0 F_\theta\right) \tag{14.52b}$$

$$H_\phi = -\frac{j\omega}{\eta_0}\left(A_\theta + \eta_0 F_\phi\right) \tag{14.52c}$$

where $\eta_0 = (\mu_0/\varepsilon_0)^{1/2}$ is the wave impedance of free space. The far-zone electric and magnetic fields are transverse ($\hat{\boldsymbol{r}}\cdot\boldsymbol{E} = 0,\ \hat{\boldsymbol{r}}\cdot\boldsymbol{H} = 0$) and orthogonal to each other ($\eta_0\boldsymbol{H} = \hat{\boldsymbol{r}}\times\boldsymbol{E}$).

Within the context of finite numerical methods, the term "near-field-to-far-field transform" is usually reserved for the implementation of (14.49)–(14.52). The frequency-domain NFFFT described above was first incorporated into the FDTD method by Umashankar and Taflove [104, 105]. A time-domain version of the vector-potential NFFFT in 3-D was developed later [106, 107]. For a good review of these methods, see [29].

Near-field-to-far-field transforms based on the Green's function formalism pose a difficulty when structure A in Fig. 14.9(a) is located in the vicinity of a planar multilayered medium. In this case, the equivalent surface currents $\boldsymbol{J}_t$, $\boldsymbol{M}_t$ in Fig. 14.9(b) do not radiate in free space; therefore the free-space Green's function $G(r) = \exp(-jkr)/4\pi r$ cannot be used. Instead, the appropriate Green's functions associated with the multilayered medium should be used in (14.45) and (14.46). In the near field, obtaining exact expressions for these Green's functions can be a very complicated task [108, 109]. In the far zone, however, closed-form analytical expressions of these Green's functions can be found. Frequency-domain NFFFT algorithms for the FDTD analysis of two-layered media were introduced in [110] and [111]. A direct time-domain FDTD NFFFT was developed later for a three-layered medium [112, 113]. Spatial Fourier transform methods have also been used to obtain the far-zone field in FDTD [84, 85]. The transmission-line formulation used in [112, 113] and the spatial Fourier transform method operate on the same basic principles.

In imaging applications, the far fields of (14.51) and (14.52) have to be calculated at multiple observation directions (θ, ϕ) in order to construct the final image. The choice of these observation directions is more obvious in the Fourier-based NFFFT of the previous subsection.

If the scattering is calculated using a modal method, the Floquet modes $\boldsymbol{R}_{pq}$ in (14.27) contain all the necessary information regarding the far-zone scattered field. For a finite method such as FDTD applied to a periodic structure, the 2-D discrete Fourier transform operation of the phase-shifted sampled field $\bar{\boldsymbol{E}}[m, n]$ in (14.36) was shown to contain sufficient information regarding the far-zone scattered field, provided the sampling is fine enough to satisfy (14.40) or (14.43). In a Green's function–based NFFFT, however, the choice of the observation directions is not immediately obvious. It is evident that a discrete arrangement of observation directions (similar to that of the incoherent plane waves constituting Köhler illumination in Fig. 14.5) is needed. Two different arrangements of observation directions are shown in Fig. 14.10.

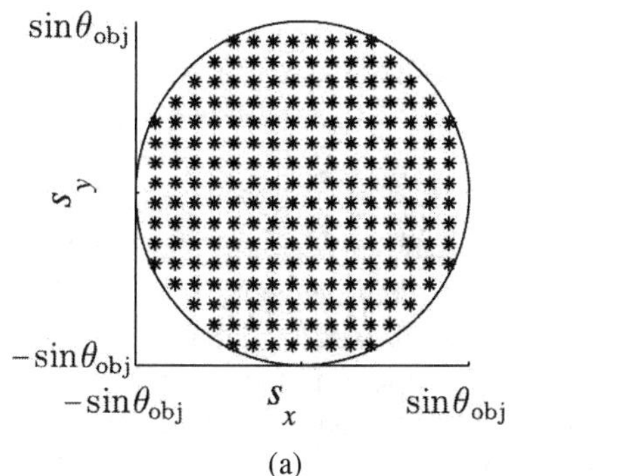

(a)

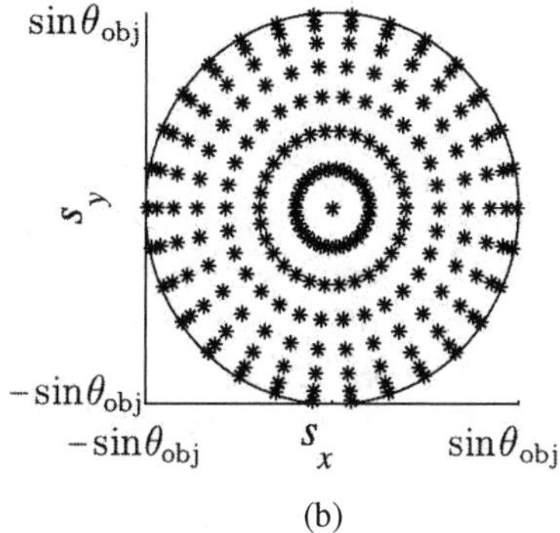

(b)

Fig. 14.10 Two types of discrete arrangements for far-zone observation directions in numerical imaging applications: (a) Equal spacing of the direction cosines $(s_x, s_y) = (\cos\phi \sin\theta, \sin\phi \sin\theta)$; (b) polar representation $(s_x, s_y) = (\rho\cos\phi, \rho\sin\phi)$, followed by Gaussian quadrature in $-\sin\theta_{\text{obj}} < \rho < \sin\theta_{\text{obj}}$ and equal spacing of $-\pi/2 < \phi < \pi/2$.

In Figure 14.10(a), the direction cosines $(s_x, s_y) = (\cos\phi \sin\theta, \sin\phi \sin\theta)$ are equally spaced, resulting in a Cartesian distribution of observation directions in the (s_x, s_y) space. The loss of rotational symmetry in ϕ can be mitigated by increasing the number of points. Alternatively, a rotationally symmetric arrangement can be obtained by parameterizing the region inside the circle $s = (s_x^2 + s_y^2)^{1/2} < \sin\theta_{\text{obj}}$ by the polar coordinates (ρ, ϕ), such that:

$$s_x = \rho\cos\phi, \qquad s_y = \rho\sin\phi \tag{14.53}$$

with the ranges

$$-\sin\theta_{\text{obj}} < \rho < \sin\theta_{\text{obj}}, \qquad -\pi/2 < \phi < \pi/2 \tag{14.54}$$

Applying Gaussian quadrature in ρ [49, 114] and maintaining equal spacing in ϕ, the discrete arrangement in Fig. 14.10(b) is obtained. Note that the rotational symmetry is preserved, but there is an inhomogeneity in the density of points inside the collection numerical aperture. In Section 14.7, the respective advantages of the arrangements in Fig. 14.10 will be seen more clearly.

For the Cartesian arrangement in Fig. 14.10(a), there is an upper limit for the spacings Δs_x, Δs_y of the direction cosines if the resulting image is to be constructed accurately. Here, we will merely note this limit, and defer its derivation until Section 14.7. Consider the scattering or radiating structure, A, in Fig. 14.9(a). It is obvious that the scattered electromagnetic field will be stronger near structure A, and will gradually decay to zero away from it. Let us define an area of dimensions W_x and W_y around structure A, outside which the scattered electromagnetic field can be assumed negligible. An area having dimensions several times the dimensions of this structure L_x and L_y will usually be sufficient. Given the dimensions W_x, W_y of the "nonzero-field" area, the condition for the image to be constructed without loss of information is:

$$\Delta s_x < \frac{2\pi}{kW_x}, \qquad \Delta s_y < \frac{2\pi}{kW_y} \tag{14.55}$$

This implies that a larger scatterer requires a finer sampling of the far-zone electromagnetic field. In a sense, this relation is dual to (14.43), which describes the condition for the reconstruction of the far-zone field from the sampled near field.

There is a subtle complication that arises in the collection step when either the incident beam or the reflection of this beam from the planar multilayers (if applicable) falls within the angular collection range of the objective. The former case may happen in transmission-mode microscopy or photolithography, where the illumination and scattering happen on opposite sides of the object. The latter case will happen even in reflection-mode microscopy, if part of the beam reflected from the layer interfaces is within the collection numerical aperture. This is usually less of a problem for a Fourier-based collection scheme because it is the *total field* that is observed on the planar surface S of Fig. 14.8, including the incident or reflected beams. The real problem arises when a Green's function–based scheme is used with near-field information on a closed surface S as in Fig. 14.9(a). Almost invariably, the near field on S is only the *scattered* field. The incident field is calculated only as an excitation term either inside the scattering regions (called the pure scattered-field formalism, used both in FDTD [29] and FEM [61]) or inside a fictitious surface surrounding the scatterer [called the total-field/scattered-field (TF/SF) formalism, used mostly in FDTD]. In the TF/SF formalism, the fictitious surface should be inside the NFFFT surface S. Otherwise, the imbalance between the magnitudes of the incident and scattered fields will cause larger numerical errors in the scattered field. For this reason, the incident or reflected beam should be treated separately from the scattered field and propagated individually through the collection and refocusing system. This will be discussed further at the end of Section 14.7.

14.7 REFOCUSING SUBSYSTEM

Since we are only concerned with real images that can be projected on a recording medium, the final step of the imaging process involves the refocusing of the rays collected from the scatterer onto an image plane. The collection and refocusing steps in Fig. 14.1 are reproduced schematically in Fig. 14.11 for convenience. The entrance and exit pupils of the system are images of each other with respect to the collection-refocusing optics in the middle. The direction cosine variables (s_x, s_y) and (s'_x, s'_y) are used to parameterize the entrance and exit pupils. The object and the image are centered around O and O', and the angles subtended by the entrance and exit pupils at O and O' are denoted by θ_{obj} and θ_{img}. The refractive indices of the object and image spaces are n and n', respectively. Allowing arbitrary n and n' can be useful for modeling liquid-immersion lenses. Two Cartesian coordinate systems are defined with respect to the origins O and O', having common z and z' axes along the optical axis OO'. The x, y and x', y' axes are anti-parallel; i.e., $x' = -x$ and $y' = -y$. In the following, unprimed and primed coordinates are used to denote variables in the object and image spaces, respectively.

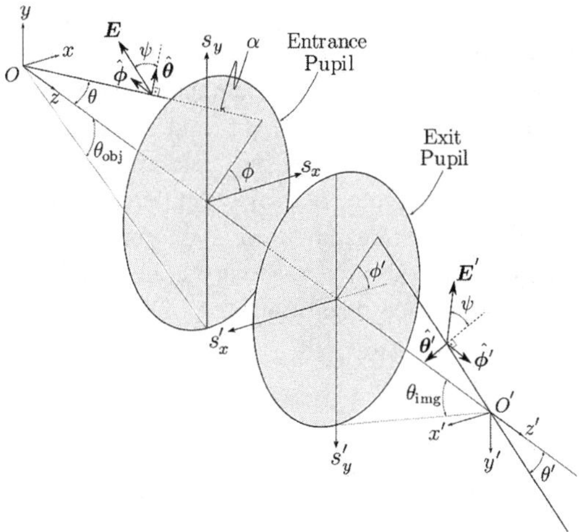

Fig. 14.11 General geometry of the collection and refocusing optics.

14.7.1 Optical Systems Satisfying the Abbe Sine Condition

In Section 14.6, the assumption of telecentricity was made in the object space; meaning that the entrance pupil is at infinity. In Fig. 14.11, a ray α is shown entering the entrance pupil at angles (θ, ϕ). This ray traverses the collection-refocusing system, leaving the exit pupil at angles (θ', ϕ'). Assuming that the collection-refocusing system is rotationally symmetric, the ray stays on the meridional plane defined by the ray α and the line OO'. This requires the azimuthal angles to be equal: $\phi' = \phi$.

We will only consider a subclass of optical systems that satisfy the *Abbe sine condition* [1, 115, 116] between the sines of the ray angles θ and θ' at the entrance and exit pupils:

$$\frac{n \sin\theta}{n' \sin\theta'} = M = \frac{NA_{\text{obj}}}{NA_{\text{img}}} \tag{14.56}$$

where M is a constant that is a characteristic of the collection-refocusing system. In (14.56), NA_{obj} and NA_{img} are, respectively, the *collection* and *imaging numerical apertures* defined as $NA_{\text{obj}} = n \sin\theta_{\text{obj}}$ and $NA_{\text{img}} = n' \sin\theta_{\text{img}}$. Up to the first order in off-axis distances, M is equal to the negative of the *lateral magnification* of the imaging system [1]. This negative scaling is a consequence of the fact that the imaging geometry in Fig. 14.11 always results in an *inverted* image. The constant M will be called "magnification" in the following, bearing in mind that the actual lateral magnification is $(-M)$. For notational convenience, we define another parameter M' representing the angular demagnification:

$$M' = \frac{n'}{n} M = \frac{\sin\theta}{\sin\theta'} \tag{14.57}$$

The Abbe sine condition (14.56) ensures that aberrations that depend on the first power of the off-axis distance of an object point (called circular coma) are absent. In other words, it is the required condition for the sharp imaging of points at small off-axis distances. This condition is usually satisfied in well-corrected optical imaging systems.

Two opposite situations regarding lateral magnification are encountered in photolithography and microscopy, as illustrated in Fig. 14.12 [97]. We consider first a general photolithography system, as illustrated in Fig. 14.12(a). Here, a demagnified image of the mask is projected on the photoresist by a projection lens, so $M < 1$ and $NA_{\text{obj}} < \text{NA}_{\text{img}}$. Demagnification in photolithography is usually specified in terms of the "reduction ratio," defined as the inverse of M and notated as $(1/M)$:1. For example, a projection lens with $M = 0.2$ is said to have a reduction ratio of 5:1. The nonlinear response of a photoresist in modern photolithography requires the projection lenses to be corrected for every conceivable aberration well beyond the requirement of simple diffraction-limited resolution (see Chapter 15).

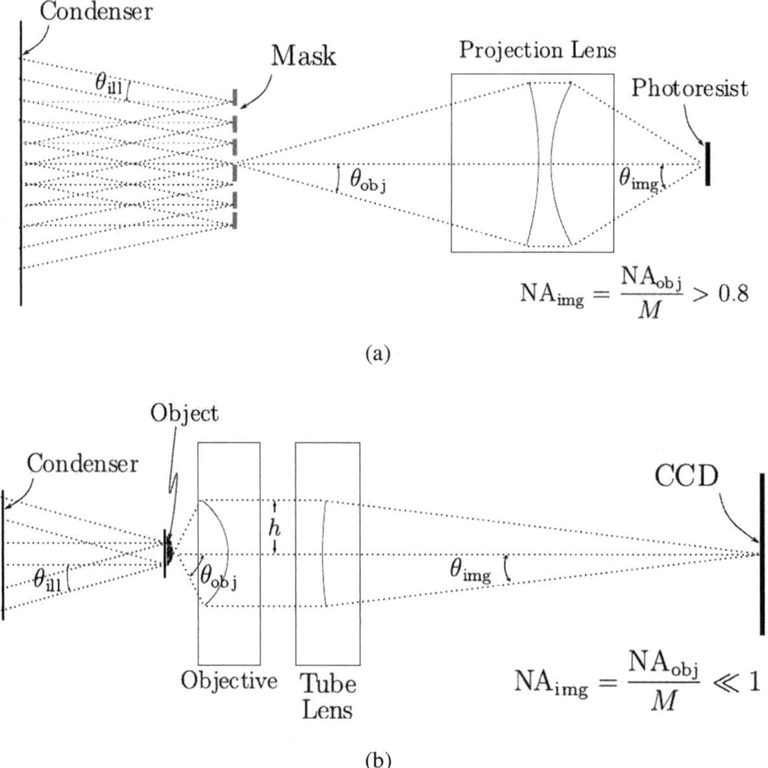

Fig. 14.12 Comparison of the collection and refocusing geometries in photolithography and microscopy. (a) In photolithography, a demagnified image of the mask is projected onto the photoresist. Typical values are $M = 0.1 - 0.25$. (b) In microscopy, a magnified image of the object is projected onto the charge-coupled device (CCD) by the objective/tube lens combination. Typical values are M = 10 – 100.

We consider next a general microscopy system, as illustrated in Fig. 14.12(b). Here, the requirements for aberration correction are less stringent than for photolithography, and most of the challenge lies with the higher NA of the objective lens. Most modern microscope objectives are infinite-conjugate, meaning that the optimum sample position for best aberration correction is at the front focal plane of the objective, resulting in an image at infinity. This image is brought to a finite position by the tube lens. The image-side numerical aperture NA_{img} of a microscope is equal to the object-side numerical aperture NA_{obj} divided by the magnification, which can be as high as $M = 100$. This results in very small incidence angles for the rays in the image space, which can be handled without much error by inexpensive tube lenses that do not require much aberration correction. On the other hand, the microscope objective is usually an expensive, well-corrected optical component. This is because the maximum object-side ray angle θ_{obj} has to be quite large for good imaging resolution. If both the objective and the tube lens satisfy the Abbe sine condition (14.56), the magnification M can also be expressed in terms of the focal lengths f_1, f_2 of the objective and the tube lens. Let us denote the height of the marginal ray between the objective and the tube lens by h. It can be shown that the Abbe sine condition for this ray takes the form [1]:

$$h = f_1 \sin\theta_{\text{obj}} = f_2 \sin\theta_{\text{img}} \tag{14.58}$$

Using (14.56), the magnification is then equal to:

$$M = \frac{n}{n'}\frac{f_2}{f_1} \tag{14.59}$$

The objective/tube lens arrangement in Fig. 14.12(b) provides a way of altering the magnification M of the system by changing the focal lengths of either the objective or the tube lens. Changing the focal length f_2 of the tube lens simply makes the image bigger or smaller with no change in resolution. Changing the focal length f_1 of the objective amounts to zooming into or out of the image with lower or higher resolution. In many microscopes, one cycles through different objectives with different focal lengths f_1, effectively changing the magnification (10×, 100×, etc.).

To construct the field distribution at the image plane, it is necessary to know the properties of all the rays α that leave the exit pupil. The azimuthal angles ϕ, ϕ' at the entrance and exit pupils are identical. The polar exit angle θ' of the rays is given by the Abbe sine condition (14.56). The strength factor $\boldsymbol{E}'_s(\theta', \phi')$ of the ray at the exit pupil still needs to be found, per (14.15). Let us start with the polarization of $\boldsymbol{E}'_s(\theta', \phi')$. From the laws of geometrical optics, $\boldsymbol{E}'_s(\theta', \phi')$ lies in a plane perpendicular to the ray. A good approximation for the polarization can be obtained by making the same assumption as in Section 14.4.2, wherein it was argued (in reference to Fig. 14.6) that the angle ψ between the electric field vector and the meridional plane remains constant as the ray α traverses the system [1, 34]. This requires that the angles of incidence at each refracting surface be small. In highly corrected optical components with multiple lenses, the deviation of a ray at each surface is minimal, and therefore the above assumption is valid. With this assumption, the strength factors of the rays at the entrance and exit pupils make the same angle with the meridional plane, as shown in Fig. 14.11.

The magnitude of $\boldsymbol{E}'_s(\theta', \phi')$ follows from the intensity law of geometrical optics [117]. Let us track an infinitesimal tube of rays containing the ray α in the object and image spaces. These rays emanate from the object-side origin O and converge at the image-side origin O', per Fig. 14.11. If the aberrations of the collection refocusing system are small, the principal radii of curvature of the geometrical optics wavefront in the image space are both approximately equal to the distance r' from O' [33]. The light intensities on the ray α in the object and image spaces are:

$$I_1 = n|\boldsymbol{E}_s|^2/(\eta_0 r^2), \qquad I_2 = n'|\boldsymbol{E}'_s|^2/[\eta_0 (r')^2] \tag{14.60}$$

in which r and r' arc arbitrary distances from O and O', respectively. (Note that the light intensity is a direct measure of the signal collected by recording media such as photoresists and CCD cameras.) The infinitesimal areas on the spherical wavefronts intersected by the tubes of rays are:

$$dS_1 = r^2 \sin\theta \, d\theta \, d\phi, \qquad dS_2 = (r')^2 \sin\theta' \, d\theta' \, d\phi' \tag{14.61}$$

Assuming that the absorptive, reflective, and refractive losses in the collection refocusing system are negligible, conservation of energy dictates that the total powers crossing dS_1 and dS_2 be equal. Since the total power crossing an infinitesimal area dS equals intensity $\times\, dS$, this is equivalent to the intensity law of geometrical optics:

$$I_1 \, dS_1 = I_2 \, dS_2 \tag{14.62}$$

We therefore have

$$|\boldsymbol{E}'_s| = |\boldsymbol{E}_s| \sqrt{\frac{n \sin\theta \, d\theta \, d\phi}{n' \sin\theta' \, d\theta' \, d\phi'}} \tag{14.63}$$

From the Abbe sine condition, $n \sin\theta = M(n' \sin\theta')$. Using the chain rule, one can write:

$$\frac{d\theta}{d\theta'} = \frac{\cos\theta'}{\cos\theta} \frac{d(\sin\theta)}{d(\sin\theta')} = M \frac{n' \cos\theta'}{n \cos\theta} \tag{14.64}$$

Also noting that $d\phi' = d\phi$, (14.63) becomes:

$$|\boldsymbol{E}'_s| = M \sqrt{\frac{n' \cos\theta'}{n \cos\theta}} |\boldsymbol{E}_s| \tag{14.65}$$

If $\boldsymbol{E}_s(\theta, \phi)$ and $\boldsymbol{E}'_s(\theta', \phi')$ are expressed in spherical coordinates centered around O and O', respectively, a quick inspection of Fig. 14.11 reveals that the $\hat{\theta}'$ and $\hat{\phi}'$ components of $\boldsymbol{E}'_s(\theta', \phi')$ are given by:

$$\boldsymbol{E}'_s(\theta', \phi') \cdot \hat{\boldsymbol{\theta}}' = -M \sqrt{\frac{n' \cos\theta'}{n \cos\theta}} \boldsymbol{E}_s(\theta, \phi) \cdot \hat{\boldsymbol{\theta}} \tag{14.66a}$$

$$\boldsymbol{E}'_s(\theta', \phi') \cdot \hat{\boldsymbol{\phi}}' = -M \sqrt{\frac{n' \cos\theta'}{n \cos\theta}} \boldsymbol{E}_s(\theta, \phi) \cdot \hat{\boldsymbol{\phi}} \tag{14.66b}$$

Now, both the directions and the strength factors of the rays leaving the exit pupil are determined, and we are ready to construct the field at the image plane. The final step of the imaging process requires a connection between the geometrical optics field determined by the rays at the exit pupil and the electromagnetic field at the image plane $z' = 0$. This can be achieved by use of vectorial diffraction theory [33, 36, 94]. The vector field at the image plane is given by the *Debye–Wolf integral*:

$$\boldsymbol{E}_{\text{img}}(x', y') = \frac{jk'}{2\pi} \iint_{\Omega_{\text{img}}} \boldsymbol{E}'_s(s'_x, s'_y) \exp[-jk'(s'_x x' + s'_y y')] \, d\Omega \tag{14.67}$$

where $k' = n' k_0$ is the wavenumber in the image space. A change of variables is made from the angle variables (θ', ϕ') to the direction cosine variables (s'_x, s'_y):

$$(s'_x, s'_y) = (\cos\phi' \sin\theta', \sin\phi' \sin\theta') \tag{14.68}$$

In (14.67), Ω_{img} is the solid angle bounded by θ_{img}, and $d\Omega = ds'_x ds'_y / s'_z = ds'_x ds'_y / \cos\theta'$. It is straightforward to show that the Debye–Wolf integral in (14.67) is the "inverse" of the vectorial far-field expression (14.23) [33]. The Debye–Wolf integral in (14.67) can also be regarded as an infinite summation of plane waves incident from a spectrum of directions (θ', ϕ'). For this reason, it is also called the *angular spectrum* or the *plane-wave* representation of the image field.

The range of validity of the Debye–Wolf integral (14.67) warrants some discussion. If the exit pupil is at infinity, then the refocusing system is *image-side telecentric*, and (14.67) strictly applies [118]. However, if a certain geometrical condition is satisfied, (14.67) is also applicable for an exit pupil at a finite position.

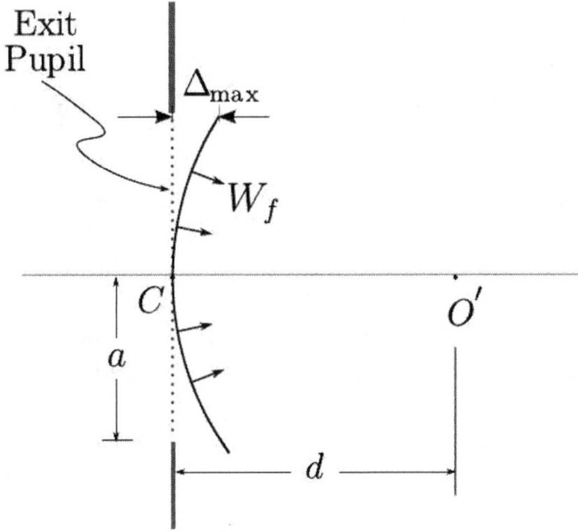

Fig. 14.13 Geometrical parameters used in the definition of the Fresnel number N_F.

Referring to Fig. 14.13, consider a spherical wavefront W_f passing through the center C of the exit pupil and converging toward the focal point O' at a distance d from the pupil. Let the radius of the exit pupil be a and the maximum distance between W_f and the exit pupil be denoted by Δ_{max}. To a good approximation, Δ_{max} is equal to $a^2/2d$. The *Fresnel number* N_F is a dimensionless quantity defined as Δ_{max} divided by one-half of the wavelength λ' in the image space:

$$N_F = \frac{\Delta_{max}}{\lambda'/2} \approx \frac{a^2}{\lambda' d} \tag{14.69}$$

The Fresnel number is approximately equal to the number of Fresnel zones that fill the aperture when viewed from the focal point O' [1]. It can be shown [119, 120] that the required condition for the validity of the Debye–Wolf integral (14.67) is that the Fresnel number be very large:

$$N_F >> 1 \tag{14.70}$$

For visible-range optical imaging systems employed in microscopy, photolithography, metrology, inspection, and alignment, the exit-pupil radius a is on the order of centimeters, so a/λ' is on the order of 10^4. The ratio a/d of the exit pupil radius to the pupil distance is equal to $\tan\theta_{img}$, which may range from 10^{-2} to infinity, depending on the magnification M. Therefore, it can safely be assumed that the Debye–Wolf integral (14.67) is a very accurate representation of the electromagnetic field in the image space for a wide range of optical systems.

If the image space is homogeneous, the Debye–Wolf integral in (14.67) gives the final result for the image field. If there is a nontrivial scattering topography in the image space such as a CCD or a photoresist, the integral (14.67) for the image field should be considered only as an incident field. Since plane-wave incidence is usually the easiest illumination scheme to consider, the angular-spectrum interpretation of the Debye–Wolf integral becomes quite useful in many cases. The incident field (14.67) is a coherent illumination beam (see Section 14.4.1) that can be written as the sum of plane-wave components:

$$d\boldsymbol{E}_{img}(x', y') = \frac{jk'}{2\pi}\boldsymbol{E}'_s(s'_x, s'_y)\exp[-jk'(s'_x x' + s'_y y')]\, d\Omega \tag{14.71}$$

If the image space is simply a stack of laterally infinite planar layers, each plane wave (14.71) can be propagated into this medium using standard Fresnel refraction formulas [117, 121–125]. For more complex topographies, the FDTD method can be used to obtain the field distribution in the image space [126].

The Debye–Wolf integral (14.67) can be generalized to include the aberrations of the collection refocusing optics by the inclusion of an additional phase factor in the exponential kernel of the integral:

$$\boldsymbol{E}_{img}(x', y') = \frac{jk'}{2\pi}\iint_{\Omega_{img}} \boldsymbol{E}'_s(s'_x, s'_y)\exp\left\{-jk'\left[s'_x x' + s'_y y' + \Phi(s'_x, s'_y)\right]\right\} d\Omega \tag{14.72}$$

where the *aberration function* $\Phi(s'_x, s'_y)$ is a measure of the deviation of the wavefront from a perfect spherical shape [33]. Regarding the aberration function as a small perturbation, the validity condition (14.70) can still be assumed to hold for (14.72). If the image space is homogeneous, the generalized Debye–Wolf integral (14.72) gives the final field distribution. We next discuss the numerical evaluation of (14.72) for a homogeneous image space, regarding the original equation (14.67) as a special case. The cases of periodic and nonperiodic scatterers will be considered separately.

14.7.2 Periodic Scatterers

We assume that, at the end of the collection step, the strength factor $\boldsymbol{E}_s(\theta, \phi)$ at the far zone has been found in the form of (14.31). Substituting the definitions of α, β in (14.22) and β_p, β_q in (14.28), the strength factor in (14.31) becomes:

$$\boldsymbol{E}_s(s_x, s_y) = jk2\pi \sum_p \sum_q c_{pq} \boldsymbol{R}_{pq}\, \delta(ks_x - ks_{x_i} + p2\pi/d_x)\, \delta(ks_y - ks_{y_i} + q2\pi/d_y) \tag{14.73}$$

where k is the wavenumber in the object space. A change of variables is made from the angle variables (θ, ϕ) to the direction cosines (s_x, s_y) at the entrance pupil:

$$(s_x, s_y) \;=\; (\alpha/k,\, \beta/k) \;=\; (\cos\phi \sin\theta,\, \sin\phi \sin\theta) \tag{14.74}$$

and the direction cosines of the incident plane wave in the object space are defined as:

$$(s_{x_i}, s_{y_i}) \;=\; (\cos\phi_i \sin\theta_i,\, \sin\phi_i \sin\theta_i) \tag{14.75}$$

The Abbe sine condition states that the direction cosines (s'_x, s'_y) of a ray at the exit pupil are $1/M'$ times the direction cosines (s_x, s_y) of the same ray at the entrance pupil, where M' is given by (14.57). Substituting (14.73) with $(s_x, s_y) = (M's'_x,\, M's'_y)$ into (14.66) and using the scaling property $\delta(ax) = \delta(x)/|a|$ of the Dirac delta function, we obtain the strength factor at the exit pupil:

$$\boldsymbol{E}'_s(s'_x, s'_y) \;=\; \frac{j2\pi}{M'k'} \sum_p \sum_q \left\{ \begin{matrix} (n'c_{pq}c'_{pq}/n)^{1/2} \boldsymbol{R}'_{pq}\, \delta[s'_x - s'_{x_i} + p2\pi/(M'kd_x)] \cdot \\ \delta[s'_y - s'_{y_i} + q2\pi/(M'kd_y)] \end{matrix} \right\} \tag{14.76}$$

where

$$(s'_{x_i}, s'_{y_i}) \;=\; (\cos\phi_i \sin\theta_i / M',\, \sin\phi_i \sin\theta_i / M') \tag{14.77}$$

and the cosine parameters c'_{pq} are defined the same way as in (14.32), with (β_p, β_q) replaced by $(\beta_p/M',\, \beta_q/M')$. The $\hat{\boldsymbol{\theta}}'$ and $\hat{\boldsymbol{\phi}}'$ components of the vector amplitude $\boldsymbol{R}'_{pq}$ are:

$$\boldsymbol{R}'_{pq_{\theta'}} = -\boldsymbol{R}_{pq_\theta}, \qquad \boldsymbol{R}'_{pq_{\phi'}} = -\boldsymbol{R}_{pq_\phi} \tag{14.78}$$

The image field is obtained by substituting (14.76) into the Debye–Wolf integral (14.72):

$$\boldsymbol{E}_{\text{img}}(x', y') \;=\; \exp[-jk'(s'_{x_i}x' + s'_{y_i}y')] \sum_p \sum_q \bar{\boldsymbol{R}}'_{pq} \exp\left[j\frac{2\pi}{M}\left(p\frac{x'}{d_x} + q\frac{y'}{d_y} \right) \right] \tag{14.79}$$

with the modified vector Floquet mode $\bar{\boldsymbol{R}}'_{pq}$ defined by:

$$\bar{\boldsymbol{R}}'_{pq} \;=\; -\sqrt{\frac{n'c_{pq}}{n\,c'_{pq}}}\, \frac{1}{M'}\, \boldsymbol{R}'_{pq} \exp\left[-jk'\Phi\left(s'_{x_i} - \frac{p2\pi}{M'kd_x},\; s'_{y_i} - \frac{q2\pi}{M'kd_y} \right) \right] \tag{14.80}$$

The phase factor in front of the summations in (14.79) is also present in the object-space field distribution (14.27). It is enforced by the plane wave incident on the periodic scatterer. An interesting consequence of (14.79) and (14.80) is that not only the image is inverted, but the *polarization* of the electromagnetic field in the image is inverted as well. This is seen more clearly if we assume $M' = 1$ and $\Phi = 0$ in (14.80), which gives $\bar{\boldsymbol{R}}'_{pq} = -\boldsymbol{R}'_{pq}$. This result is intuitively satisfying, since it implies a *vector inversion* of the electromagnetic field as a generalization of the classical image inversion of geometrical optics.

If the image field (14.79) is to be evaluated at a few (x', y') positions, it can be calculated directly by a brute-force method. If a whole region is of interest, then a DFT-based evaluation is more efficient. It will now be shown that, using the DFT (and its efficient computation by the fast Fourier transform, FFT), expression (14.79) for the image field can be evaluated at a discrete rectangular grid of (x', y') points with an arbitrary spacing in x' and y'. First, it is important to remember that the p and q indices in (14.79) belong to a finite set. They are the indices of the scattered Floquet modes $\boldsymbol{R}_{pq}$ that fall within the entrance pupil, and subsequently leave the exit pupil. The range of indices in (14.79) is thus defined by the following condition:

$$\left(s'_{x_i} - p\frac{2\pi}{M' k d_x} \right)^2 + \left(s'_{y_i} - q\frac{2\pi}{M' k d_y} \right)^2 < \sin^2 \theta_{\mathrm{img}} \tag{14.81}$$

This condition is shown geometrically in Fig. 14.14.

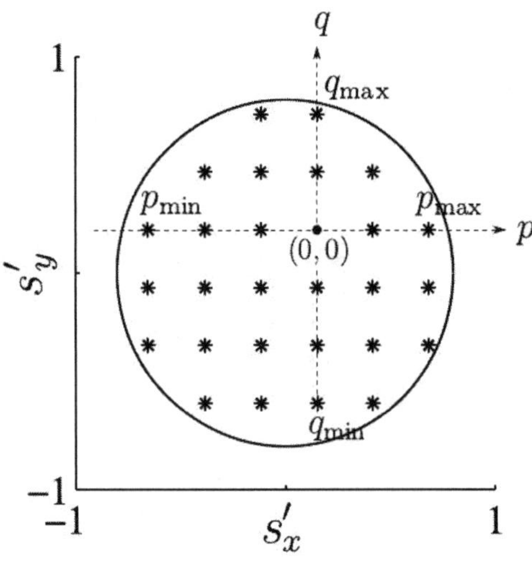

Fig. 14.14 Admissible Floquet modes for a periodic scatterer. Only those that fall within the exit pupil contribute to the image.

In Fig 14.14, a rectangular grid of points (s'_{x_p}, s'_{y_q}) is represented by the indices p and q. The origin $(p, q) = (0, 0)$ of the indices corresponds to (s'_{x_i}, s'_{y_i}). Only the direction cosines that fall within a circle of radius $\sin\theta_{\mathrm{img}}$ are counted in (14.79) because of the condition (14.81). Let $p_{\min}, p_{\max}$ denote the minimum and maximum permissible indices in p, and $q_{\min}, q_{\max}$ denote the corresponding indices for q. Let us write the summation term in (14.79) as a double summation over a rectangular region of indices limited by $p_{\min}$, $p_{\max}$, $q_{\min}$, and $q_{\max}$, with the implicit assumption that $\bar{\boldsymbol{R}}'_{pq}$ vanishes outside the range set by (14.81):

$$\sum_{p=p_{\min}}^{p_{\max}} \sum_{q=q_{\min}}^{q_{\max}} \bar{\boldsymbol{R}}'_{pq} \exp\left[j\frac{2\pi}{M}\left(p\frac{x'}{d_x} + q\frac{y'}{d_y} \right) \right] \tag{14.82}$$

which becomes, after shifting the indices by $p_{\min}$ and $q_{\min}$:

$$\exp\left[j\frac{2\pi}{M}\left(p_{\min}\frac{x'}{d_x} + q_{\min}\frac{y'}{d_y} \right) \right] \sum_{p=0}^{p_{\max}-p_{\min}} \sum_{q=0}^{q_{\max}-q_{\min}} \bar{\boldsymbol{R}}'_{p+p_{\min},\, q+q_{\min}} \exp\left[j\frac{2\pi}{M}\left(p\frac{x'}{d_x} + q\frac{y'}{d_y} \right) \right] \tag{14.83}$$

The summation term above will take a true DFT form if it is sampled at a discrete set of points. Consider the sampling:

$$(x', y') = (m\Delta x, n\Delta y) \qquad m = 0 \ldots P-1, \qquad n = 0 \ldots Q-1 \tag{14.84}$$

in which Δx and Δy are chosen such that the sampling covers the entire magnified periods in both x' and y':

$$P\Delta x = Md_x, \qquad Q\Delta y = Md_y \tag{14.85}$$

Substituting (14.84) in (14.83), the summation term becomes:

$$\sum_{p=0}^{p_{\max}-p_{\min}} \sum_{q=0}^{q_{\max}-q_{\min}} \bar{\boldsymbol{R}}'_{p+p_{\min},\, q+q_{\min}} \exp\left[j2\pi\left(\frac{pm}{P} + \frac{qn}{Q} \right) \right] \tag{14.86}$$

This expression is almost in the same form as a DFT, and can be evaluated using the same efficient FFT algorithms [50]. In fact, when divided by PQ, it is called the $(P \times Q)$-point *inverse DFT* of $\bar{\boldsymbol{R}}'_{p+p_{\min},\, q+q_{\min}}$. For an efficient FFT operation, P and Q should be chosen to have small multiple prime factors. Once the inverse-DFT expression (14.86) is computed, the image field (14.79) is directly obtained at the sampled image positions $(x', y') = (m\Delta x, n\Delta y)$. This sampling can be made arbitrarily fine by increasing P and Q.

14.7.3 Nonperiodic Scatterers

For a nonperiodic scatterer, the far-zone field is evaluated at a discrete set of observation directions. Two different arrangements for this discrete set were shown in Fig. 14.10. If the rectangular arrangement of Fig. 14.10(a) is chosen, then the direction cosines (s'_x, s'_y) are also distributed in a rectangular grid inside a circle of radius $\sin\theta_{\text{img}}$. The relationship between the object-side strength factor $\boldsymbol{E}_s(s_x, s_y)$ and the image-side strength factor $\boldsymbol{E}'_s(s'_x, s'_y)$ is given by (14.66). Once $\boldsymbol{E}'_s(s'_x, s'_y)$ is determined, the Debye–Wolf integral (14.72) for the image field can then be evaluated numerically using a DFT.

A quick comparison shows that (14.72) is in the same form as the inverse-Fourier-transform relation (14.17), except for a trivial change of variables $(s'_x, s'_y) \rightarrow (k's'_x, k's'_y)$. Since the double integral only covers the region $[(s'_x)^2 + (s'_y)^2]^{1/2} \leq \sin\theta_{\text{img}}$, we can extend the limits of the integral from $-\infty$ to ∞ and assign $\boldsymbol{E}'_s(s'_x, s'_y) = 0$ for $[(s'_x)^2 + (s'_y)^2]^{1/2} > \sin\theta_{\text{img}}$. Using the 2-D Fourier relation (14.16), we can invert the relation (14.72) to obtain:

$$\boldsymbol{E}'_s(s'_x, s'_y) = \frac{-jk'\cos\theta'}{2\pi}\exp[jk'\Phi(s'_x, s'_y)]\iint_{x,y}\boldsymbol{E}_{\text{img}}(x', y')\exp[jk'(s'_x x' + s'_y y')]\,dx'dy' \qquad (14.87)$$

Let the direction cosines inside the imaging cone be discretized by dividing the rectangular region defined by $-\sin\theta_{\text{img}} < s'_x < \sin\theta_{\text{img}}$ and $-\sin\theta_{\text{img}} < s'_y < \sin\theta_{\text{img}}$ into $P \times Q$ rectangular patches, and choosing (s'_{x_p}, s'_{y_q}) to be at the center of each patch:

$$\begin{aligned} s'_{x_p} &= s'_{x_0} + p\Delta s'_x \qquad & p = 0\ \ldots,\ P-1 \\ s'_{y_q} &= s'_{y_0} + q\Delta s'_y \qquad & q = 0\ \ldots,\ Q-1 \end{aligned} \qquad (14.88)$$

where

$$s'_{x_0} = -\sin\theta_{\text{img}}(1 - 1/P), \qquad s'_{y_0} = -\sin\theta_{\text{img}}(1 - 1/Q) \qquad (14.89)$$

and the sampling periods $\Delta s'_x$ and $\Delta s'_y$ are:

$$\Delta s'_x = 2\sin\theta_{\text{img}}/P, \qquad \Delta s'_y = 2\sin\theta_{\text{img}}/Q \qquad (14.90)$$

Now, we define an auxiliary variable $\boldsymbol{G}'(s'_x, s'_y)$ that combines the strength factor $\boldsymbol{E}'_s(s'_x, s'_y)$ with the $\cos\theta'$ factor and the aberration phase factor $\exp[-jk'\Phi(s'_x, s'_y)]$ in (14.87) as follows:

$$\boldsymbol{G}'(s'_x, s'_y) \triangleq \frac{\boldsymbol{E}'_s(s'_x, s'_y)}{\cos\theta'}\exp[-jk'\Phi(s'_x, s'_y)] \qquad (14.91)$$

This definition facilitates the direct application of a 2-D DFT to the numerical evaluation of (14.72). The 2-D array resulting from the sampling of the auxiliary variable $\boldsymbol{G}'(s'_x, s'_y)$ according to (14.88) is defined as:

$$\boldsymbol{G}'[p, q] = \boldsymbol{G}'(s'_{x_p}, s'_{y_q}) \qquad (14.92)$$

Now, it will be shown how the 2-D DFT of the array $\boldsymbol{G}'[p, q]$ can be related to the continuous image field $\boldsymbol{E}_{\text{img}}(x', y')$ in (14.72). Adopting the same convention as in Section 14.6, we define the 2-D DFT of $\boldsymbol{G}'[p, q]$ as:

$$\boldsymbol{E}[m,n] = \sum_{p=0}^{N_p-1}\sum_{q=0}^{N_q-1}\boldsymbol{G}'[p, q]\exp\left[-j2\pi\left(\frac{pm}{N_p} + \frac{qn}{N_q}\right)\right] \qquad (14.93)$$

The DFT lengths N_p and N_q are greater than or equal to P and Q, respectively. If they are greater, then $\boldsymbol{G}'[p, q]$ is zero-padded up to the required length. It is shown in Appendix 14C that $\boldsymbol{E}[m, n]$ is a sampled and periodically replicated (aliased) version of the continuous image field $\boldsymbol{E}_{\text{img}}(x', y')$:

$$\boldsymbol{E}[m,n] = \frac{2\pi}{jk'\Delta s'_x \Delta s'_y} \sum_{r=-\infty}^{\infty} \sum_{s=-\infty}^{\infty} \exp\left\{ j2\pi \left[\frac{s'_{x_0}}{\Delta s'_x N_p}(m+rN_p) + \frac{s'_{y_0}}{\Delta s'_y N_q}(n+sN_q) \right] \right\} \times \boldsymbol{E}_{\text{img}}(m\Delta_x + rD'_x,\ n\Delta_y + sD'_x) \tag{14.94}$$

in which the spatial sampling periods Δ_x, Δ_y and the aliasing periods D'_x, D'_y are defined as:

$$\Delta_x = \frac{2\pi}{k'\Delta s'_x N_p}, \qquad \Delta_y = \frac{2\pi}{k'\Delta s'_y N_q} \tag{14.95}$$

$$D'_x = N_p \Delta_x = \frac{2\pi}{k'\Delta s'_x}, \qquad D'_y = N_q \Delta_y = \frac{2\pi}{k'\Delta s'_y} \tag{14.96}$$

If the shifted replicas of $\boldsymbol{E}_{\text{img}}(m\Delta_x, n\Delta_y)$ in (14.94) do not overlap, then $\boldsymbol{E}_{\text{img}}(m\Delta_x, n\Delta_y)$ can be retrieved from (14.94) as follows:

$$\boldsymbol{E}_{\text{img}}(m\Delta_x, n\Delta_y) \approx \frac{jk'\Delta s'_x \Delta s'_y}{2\pi} \exp\left[-j2\pi \left(\frac{s'_{x_0}}{\Delta s'_x N_p} m + \frac{s'_{y_0}}{\Delta s'_y N_q} n \right) \right] \boldsymbol{E}[m,n] \tag{14.97}$$

for a range of m and n values centered around $m = n = 0$ over which $\boldsymbol{E}_{\text{img}}(m\Delta_x, n\Delta_y)$ is nonzero. The condition for this retrieval is that the aliasing periods D'_x, D'_y are greater than the x' and y' dimensions W'_x, W'_y of the "nonzero-field area" over which the amplitude of the electromagnetic field $\boldsymbol{E}_{\text{img}}(x', y')$ is non-negligible:

$$D'_x > W'_x, \qquad D'_y > W'_y \tag{14.98}$$

Barring diffraction effects on the order of $\sim\lambda$, the dimensions W'_x, W'_y are equal to the corresponding dimensions at the object-side multiplied by the magnification M of the system. At the end of Section 14.6, these object-side dimensions were defined as W_x and W_y. It follows that the dimensions W'_x, W'_y of the "nonzero-field area" at the image space are given by MW_x and MW_y. Using the definitions of the aliasing periods D'_x, D'_y in (14.96), the non-aliasing condition (14.98) becomes:

$$\Delta s'_x < \frac{2\pi}{k'MW_x}, \qquad \Delta s'_y < \frac{2\pi}{k'MW_y} \tag{14.99}$$

The Abbe sine condition (14.56) also relates the sampling periods $(\Delta s_x, \Delta s_y)$, $(\Delta s'_x, \Delta s'_y)$ at the entrance and exit pupils linearly through $M' = n'M/n$. Using this relationship and $k'/k = n'/n$ in (14.99), we obtain the sampling relations (14.55) given at the end of Section 14.6, reproduced here for convenience:

$$\Delta s_x < \frac{2\pi}{kW_x}, \qquad \Delta s_y < \frac{2\pi}{kW_y} \tag{14.100}$$

This condition places an upper limit on the distances Δs_x and Δs_y between the direction cosines of the angles at which the far-zone field is collected [see Fig. 14.10(a)].

Assuming that the retrieval (14.97) is accurate, the electric field $\boldsymbol{E}_{\text{img}}(x', y')$ is now known at discrete spatial positions $(m\Delta_x, n\Delta_y)$. We know that the vector field $\boldsymbol{E}_{\text{img}}(x', y')$ is spatially bandlimited, since $\boldsymbol{E}_s'(s_x', s_y')$ is only nonzero inside $[(s_x')^2 + (s_y')^2]^{1/2} \le \sin\theta_{\text{img}}$. The spatial bandwidth of the field is therefore $2k'\sin\theta_{\text{img}}$. From the Nyquist sampling theorem, this bandlimited field is completely determined by its sampled version $\boldsymbol{E}_{\text{img}}(m\Delta, n\Delta)$ if the sampling period is smaller than or equal to $\Delta = 2\pi/(2k'\sin\theta_{\text{img}}) = \lambda'/(2\sin\theta_{\text{img}})$, which corresponds to the traditional definition of the "diffraction limit." For the minimum allowable DFT lengths $N_p = P$ and $N_q = Q$ in (14.93), it follows from (14.90) that both sampling periods in (14.95) are equal to this limit, and the continuous field $\boldsymbol{E}_{\text{img}}(x', y')$ is represented by the least possible number of sampling points. To evaluate the continuous field $\boldsymbol{E}_{\text{img}}(x', y')$ at higher spatial precision, one can simply increase the DFT lengths N_p and N_q.

A second way to numerically evaluate (14.72) follows from a generalization of the Nijboer–Zernike aberration theory [1], called the extended Nijboer–Zernike (ENZ) theory by its developers [127–130]. The important features of the ENZ method are the evaluation of the series expansions in the original Nijboer–Zernike method with more terms, and the redefinition of the aberration function $\Phi(s_x', s_y')$ as a complex quantity, thereby accounting for the variations of the vector amplitude $\boldsymbol{E}_s'(s_x', s_y')$ on the wavefront. In the ENZ method, one starts by expanding the aberration function $\Phi(s_x', s_y')$ into a series in the form:

$$\Phi(s_x', s_y') = \sum\nolimits_{n,m} \alpha_{nm} R_n^m(s')\cos(m\phi') \tag{14.101}$$

in which $s' = [(s_x')^2 + (s_y')^2]^{1/2}$ and ϕ' is the azimuthal angle in the image space. In (14.101), $R_n^m(s')$ are the Zernike circle polynomials [1], and different terms in (14.101) are orthogonal to each other inside the unit circle. The coefficients α_{nm} can therefore be obtained by integrating $\Phi(s_x', s_y')R_n^m(s')\cos(m\phi')$ over the unit circle and making use of the orthogonality property. Because polar coordinates (s', ϕ') are used in (14.101), the polar arrangement of (s_x, s_y) in Fig. 14.10(b) at the collection step is more suitable for this integration.

Next, the generalized Debye–Wolf integral (14.72) is expanded into an infinite series in $\Phi(s_x', s_y')$, resulting in definite integrals involving the products of exponential and Bessel functions. These integrals can be evaluated off-line, and lookup tables can be generated for repeated use in the future. Using these lookup tables and the coefficients α_{nm} of the aberration function $\Phi(s_x', s_y')$, the generalized Debye–Wolf integral (14.72) can be calculated to a desired degree of accuracy. Any additional phase term in the aberration function $\Phi(s_x', s_y')$ requires only the recalculation of the coefficients α_{nm}. At this point, the use of Zernike circle polynomials in (14.101) becomes a real advantage. If the additional phase term for the aberration is expressed in terms of Zernike polynomials, only the α_{nm} terms that have the same indices as those polynomials will be affected. As a result, the effects of primary aberrations (spherical aberration, coma, astigmatism, etc.) on the imaging performance can be investigated extremely efficiently, without any need to carry out a 2-D numerical integration or numerical quadrature for each different aberration. The ENZ method has also been generalized to planar multilayered structures [131].

The ENZ formulation does not suffer from the aliasing artifacts encountered in the DFT-based formulation. It is inherently geared toward synthesizing the images of nonperiodic structures. In [53, 84, 85], the image-space field distributions (also called *aerial images* in photolithography) of nonperiodic masks were computed using the ENZ method. On the other hand, convergence and range-of-validity issues are of greater importance in the ENZ method because of the heavy use of series expansions. The relative theoretical complexity and the difficulty of constructing the lookup tables is another disadvantage.

Remember that the integral expression (14.72) for the image field is based upon a coherent illumination beam. If Köhler illumination is employed (see Section 14.4.2), this coherent illumination beam is one of the plane-wave components in Fig. 14.5. To obtain the total intensity at the image space, the image intensities corresponding to every incidence direction and polarization in Fig. 14.5 should be added. We recall that the image intensity due to a single plane wave is proportional to the absolute square of $\boldsymbol{E}_{\text{img}}(x', y')$ in (14.72).

One subtle point that needs to be addressed with regard to non-periodic scatterers is the presence of planar material layers in the object space. This issue was touched upon at the end of Section 14.6. The NFFFT surface S in Fig. 14.9(b) only collects the *scattered field* and not the incident or reflected beam. Therefore, the contribution to the image by the latter needs to be calculated separately. Let the incident or reflected beam be a plane wave with direction cosines (s_{x_i}, s_{y_i}) in the object space. Other coherent beams can be expressed as a sum of plane waves (see Section 14.4.1). Notationally, this plane wave can be regarded as the zeroth-order Floquet mode in (14.27):

$$\boldsymbol{E}_i(x,y) = \boldsymbol{R}_{00} \exp[-jk(s_{x_i}x + s_{y_i}y)] \tag{14.102}$$

With this notation, the results for the image field of periodic scatterers in Section 14.7.2 are immediately applicable. Only considering the $(p,q) = (0,0)$ mode in (14.79), the image field due to the incident or reflected plane wave is found as:

$$\boldsymbol{E}_{\text{img}}(x', y') = -\sqrt{\frac{n'\cos\theta_i}{n\cos\theta_i'}}\frac{1}{M'}\boldsymbol{R}'_{00} \exp[-jk'\Phi(s'_{x_i}, s'_{y_i})] \exp[-jk'(s'_{x_i}x' + s'_{y_i}y')] \tag{14.103}$$

in which $\boldsymbol{R}'_{00}$ is given by (14.78). Using (14.103), each incident or reflected plane wave can be propagated to the image space and added coherently to the image field (14.97) scattered from the object.

We conclude this section with a brief discussion of broadband imaging. Almost all of the formulation presented for the illumination, collection, and refocusing steps has been for harmonic time dependence $\exp(j\omega t)$, which corresponds to full temporal coherence and an infinitesimally narrow bandwidth. All the results so far can be immediately generalized to a broadband source. It was mentioned in Section 14.2 that a large class of optical sources can be modeled as statistically stationary in time. The total intensity at any position (x', y') at the image plane is the integral of the power spectral density $S_{\text{img}}(x', y'; \omega)$ across all frequencies. From the power spectral density relation (14.3), $S_{\text{img}}(x', y'; \omega)$ is the product of the power spectral density $S_i(\omega)$ of the excitation and the frequency response $H(\omega)$ of the system. As mentioned in Section 14.2, the frequency response $H(\omega)$ can be evaluated at multiple frequencies in a single simulation run if a broadband method such as FDTD is used to calculate the time-domain scattering response.

Another technical point related to broadband simulation is the effect of different excitation wavelengths on the numerical calculation of the image. If the discrete arrangement of the $\Delta s'_x$, $\Delta s'_y$ defined by (14.88) is kept fixed at every frequency, the spatial sampling periods Δ_x, Δ_y as well as the aliasing periods D'_x, D'_y in (14.95), (14.96) for the sampled image field (14.94) scale linearly with the wavelength λ. This complicates the direct summation of the image field in the image plane, since the field is evaluated at different spatial points at different wavelengths. To avoid this complication, it is advisable to *scale* the direction cosines Δs_x, Δs_y (therefore the $\Delta s'_x$, $\Delta s'_y$) linearly by the wavelength in the collection step. This implies that a different set of observation directions is recorded for each wavelength in the near-field-to-far-field transform step (see Section 14.6). An additional advantage of scaling the direction cosines with the wavelength is that the no-aliasing condition (14.99) becomes only dependent on the spatial extent of the object-side field distribution, and independent of the wavelength.

14.8 IMPLEMENTATION EXAMPLES: NUMERICAL MICROSCOPE IMAGES

In this section, we present examples of numerical microscope images synthesized using a software package (which we've named *Angora*) that implements techniques reviewed in the previous sections. The Maxwell's equations solver that generates the near-fields for each object to be computationally imaged is our in-house implementation of the 3-D FDTD method. Angora has been used to generate original results for previous publications [11, 37, 46, 112] and has been thoroughly tested and verified. It is freely available under the GNU Public License [132].

In each of the following examples, the object and image spaces are both free space, i.e., $n' = n = 1$. The object and image-space wavenumbers are equal ($k' = k = k_0$), and from (14.57), the angular de-magnification M' is equal to the magnification M.

14.8.1 Letters "N" and "U" Embossed on a Thin Dielectric Substrate

We consider first the numerical microscope image of a thin structure forming the letters "N" and "U" embossed on an infinite dielectric substrate. Figure 14.15(a) illustrates 2-D cross-sections of the simulation geometry across the x–y and x–z planes. The embossed structure was assumed to be a dielectric of refractive index $n = 1.38$ and dimensions $12 \times 6 \times 0.27$ μm, situated on a glass half-plane of refractive index $n = 1.5$. This represented an idealized situation with the embossed structure resting on a glass slide, with the illumination focused on the top surface of the slide and the thickness of the slide much larger than the focal depth of the illumination. Since even the smallest-NA illumination has a finite focal depth in practice, this condition was very easy to fulfill. Consequently, the scattered beam due to the bottom surface of the slide was far out of focus and spread over a large area, and could be neglected.

The structure was illuminated normally by a y-polarized plane wave having the modulated-Gaussian time waveform $\sin(2\pi f_0 t)\exp(-t^2/2\tau^2)$ with $f_0 = 5.89 \times 10^{14}$ Hz and $\tau = 2.13 \times 10^{-15}$ sec. The –20-dB free-space wavelengths of this beam were 400 and 700 nm. This approximated a polarized illumination beam with very low illumination NA (see Section 14.4.1). Incoherent and unpolarized illumination could be achieved by repeating this simulation for multiple incidence directions and polarizations (see Section 14.4.2, especially the discussion involving Fig. 14.5).

The scattered near fields were computed in a 12.635- × 6.65- × 0.5985-μm FDTD grid with 13.33-nm cubic unit cells and a time-step set at 0.98 times the Courant limit. Ten-cell-thick convolution perfectly matched layer (CPML) [133] absorbing boundaries simulated the grid extending to infinity. Plane-wave sourcing was via the total-field/scattered-field formulation [29] with multilayer capabilities [46], where the TF/SF interface was located 8 grid cells from the CPML inner surface. The scattered field was collected on a virtual surface located 4 grid cells from the CPML, and transformed to the far field using a frequency-domain vector-potential NFFFT algorithm [29] for a two-layered medium [110].

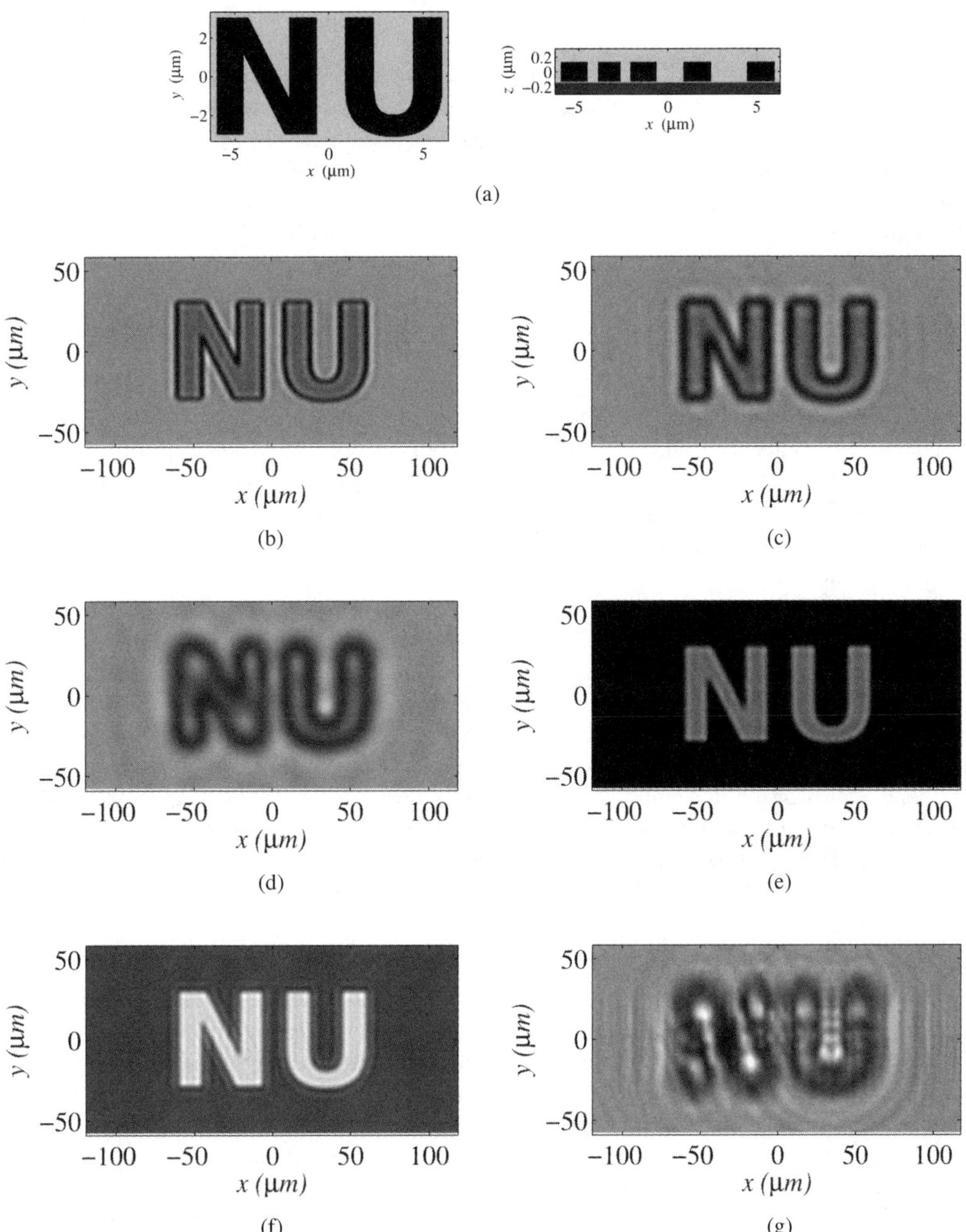

Fig. 14.15 Example of a 3-D FDTD computational imaging simulation, where the synthesized images are all in grayscale. Here, the letters "N" and "U" are assumed to be embossed on a glass substrate and imaged at magnification $M = 10$: (a) x–y and x–z cross-sections of the FDTD grid; (b) bright-field image for $NA_{obj} = 0.9$; (c) bright-field image for $NA_{obj} = 0.4$; (d) bright-field image for $NA_{obj} = 0.2$; (e) dark-field image; (f) phase-contrast image; (g) image of the off-focus plane $z = 6$ μm.

The far field was calculated at a set of observation directions (θ, ϕ) arranged as in Fig. 14.10(a), with equally spaced direction cosines (s_x, s_y) within a collection numerical aperture $\mathrm{NA}_{\mathrm{obj}} = 0.9$. The spacings of the direction cosines were $\Delta s_x = 0.0167$ and $\Delta s_y = 0.0333$. A smaller spacing was necessary in s_x because the structure was wider in the x-direction [see (14.100) in Section 14.7.3]. At each observation direction, the far field was calculated at 7 wavelengths between 400 and 700 nm with equal spacing in $k = 2\pi/\lambda$. It was assumed that the microscope had a magnification $M = 10$, and was free of aberrations $(\Phi = 0)$. The sampled field distribution $\boldsymbol{E}_{\mathrm{img}}(m\Delta_x, n\Delta_y)$ at the image plane was calculated using the DFT-based refocusing algorithm of Section 14.7.3, described by (14.92)–(14.97) with $\boldsymbol{E}_s'(s_{x_p}', s_{y_q}')$ given by (14.66). The continuous field $\boldsymbol{E}_{\mathrm{img}}(x', y')$ was oversampled with $N_p = N_q = 256$ in (14.93) for a smooth intensity image. Since the scattering geometry was two layered, the plane wave reflected from the air–glass interface had to be propagated to the image space using (14.102) and (14.103).

We now review sample results of the computational image synthesis. Unless otherwise noted, all intensity spectra in Figs. 14.15(b)–(g) are normalized by the intensity spectrum at a pixel corresponding to the glass. Figure 14.15(b) depicts in grayscale the mean of the normalized image intensity across all wavelengths (called the *bright-field image*) for a collection NA of 0.9. Here, the grayscale limits are black (0) and white (1.85). Figures 14.15(c) and 14.15(d) show how this image is blurred by reducing the collection NA to 0.4 and 0.2, respectively, with the same grayscale limits. In fact, this blurring is due to the degraded diffraction limit.

In Fig. 14.15(e), the plane wave reflected from the glass slide is subtracted from the image, resulting in a modality similar to *dark-field microscopy*. If the reflected plane wave is phase-shifted by 90° instead of being subtracted from the total image, the image in Fig. 14.15(f) is obtained. This is very similar to the procedure followed in *phase-contrast microscopy*. In both Figs. 14.15(e) and 14.15(f), the collection NA is 0.9 and the spectra are normalized by the same glass spectrum used to normalize the previous figures. However, the grayscale limits are 0 and 3.5 because of the higher intensity contrast. We can see that the phase-contrast image yields better contrast than the dark-field image. Finally, in Fig. 14.15(g), the normalized bright-field image of the off-focus plane $z = 6$ μm at the object space is shown for a collection NA of 0.9. The distortion of the image due to the lack of sharp focus is clearly visible.

14.8.2 Polystyrene Latex Beads

In the second example [11], we compare the numerical "microscope in a computer" images and spectra of polystyrene latex beads to physical laboratory measurements using a microscope coupled to a spectrometer and a CCD camera. In the laboratory measurements, two different sizes of polystyrene latex beads (2.1 and 4.3 μm in diameter, Thermo Scientific) were placed on a glass slide of refractive index $n = 1.5$. The refractive index of the latex beads was specified to be between 1.59 and 1.61 in the wavelength range of 486 to 589 nm. White-light illumination was provided by a xenon lamp. This illumination was passed through a diffuser to smoothen inhomogeneities, and then projected on the sample using a Köhler setup with illumination numerical aperture $\mathrm{NA}_{\mathrm{ill}} = 0.2$. The magnification of the microscope was $M = 40$, and the collection numerical aperture was $\mathrm{NA}_{\mathrm{obj}} = 0.6$. The image of the sample was projected on a spectrograph with a 10-μm-wide slit coupled to a CCD camera. This recorded the spectra along a column of pixels, resulting in a 2-D data array. In turn, a 3-D spectroscopic image was acquired by scanning the slit of the spectrograph over the image with 10-μm steps. More details of the optical measurements can be found in [134].

As in the previous example, the bottom surface of the glass slide was far out of focus, and the reflection from that interface was spread over a large area with much reduced intensity at the top interface. Therefore, the glass slide was modeled as a two-layered space. Consequently, the multilayer techniques used for the previous example were also used here.

The parameters for the FDTD-synthesized images shown in Fig. 14.16 were as follows: grid dimensions $5 \times 5 \times 5$ μm with 31-nm cubic unit cells, and a time-step equal to 0.98 times the Courant limit. The FDTD grid was terminated with a 10-cell-thick CPML absorbing boundary.

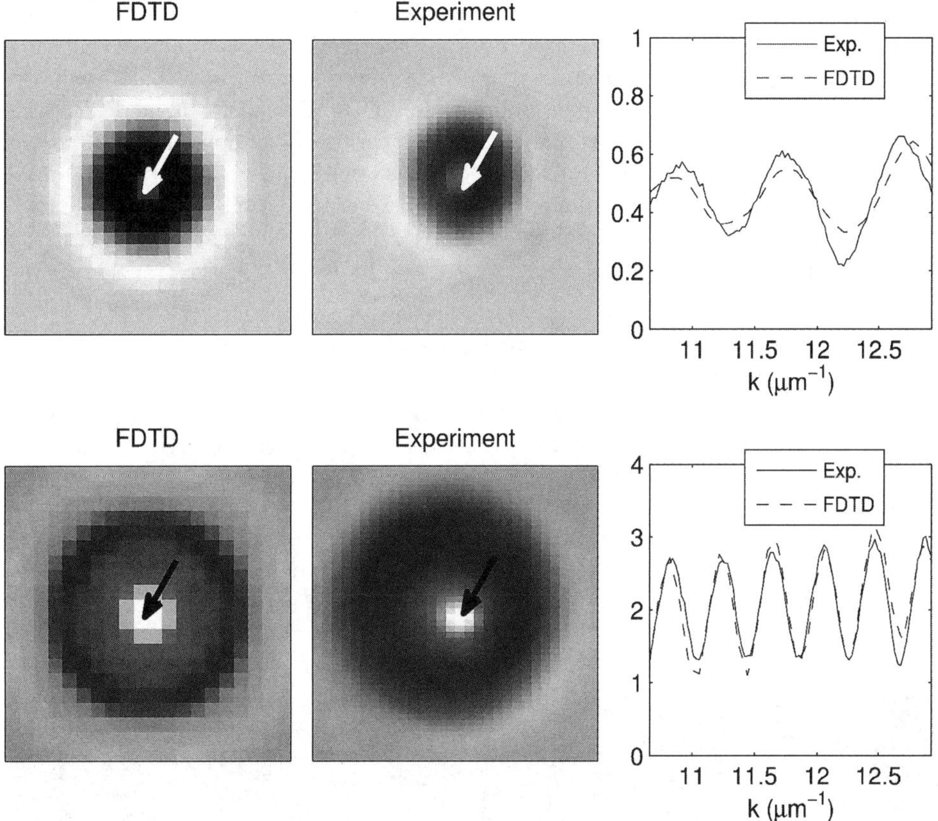

Fig. 14.16 Comparison of FDTD-synthesized and measured bright-field grayscale microscope images and pixel spectra for an isolated polystyrene latex bead. Top: 2.1-μm bead; bottom: 4.3-μm bead. *Source:* Capoglu et al., *Optics Lett.*, vol. 36, no. 9, 2011, pp. 1596–1598, ©2011 The Optical Society of America.

Regarding the synthesis procedure used to generate the images shown in Fig. 14.16, a fixed refractive index value of $n = 1.61$ was chosen for the polystyrene bead as a first approximation. Exploiting the rotational symmetry of the bead and the microscope setup, 204 incident plane waves (102×2 polarizations) were distributed in the Cartesian arrangement of Fig. 14.5 only within the first quadrant of the circle of illumination. The final image was synthesized by rotating the resulting image intensity by 0°, 90°, 180°, and 270°, and adding the intensities.

Each incident plane wave had a sine-modulated Gaussian waveform with –20-dB wavelengths at 486 and 589 nm. The scattered light was collected at a set of observation directions arranged with equally spaced 50 × 50 direction cosines (s_x, s_y) inside a collection numerical aperture $\mathrm{NA}_{\mathrm{obj}} = 0.6$. Here, the spacing in s_x and s_y was uniform and equal to $\Delta s_x = \Delta s_y = 0.024$. The far field at each direction was calculated at 30 wavelengths between 486 and 589 nm spaced linearly in $k = 2\pi/\lambda$. The sampled field distribution $\boldsymbol{E}_{\mathrm{img}}(m\Delta_x, n\Delta_y)$ at the image plane was calculated with no oversampling (at 50 × 50 points) using the refocusing algorithm (14.92)–(14.97) with $\boldsymbol{E}'_s(s'_{x_p}, s'_{y_q})$ given by (14.66). Both the measured/simulated spectra were normalized by the measured/simulated spectrum at a glass pixel.

The grayscale plots in the left and center columns of Fig. 14.16 show the FDTD-synthesized and measured bright-field images, respectively. The plots in the right column show the measured and simulated spectra between 486 and 589 nm at the pixels denoted by the arrows. Because the precise focal plane position in the measurement was unknown, the FDTD image at the optimum focusing depth was chosen for each comparison. The optimum focal positions at the object space were $z = 1.58$ μm for the 2.1-μm bead and $z = 1.44$ μm for the 4.3-μm bead. At these optimum focal positions, the root-mean-square errors in the spectra were 11.4% and 8.2%, respectively. This error was primarily caused by the dispersive nature of the beads and the variations in their sizes.

14.8.3 Pair of Contacting Polystyrene Microspheres in Air

Figure 14.17 illustrates an example of the excellent accuracy of the "microscope in a computer" in computationally synthesizing images of penetrable micron-scale objects, especially when using a high-resolution 3-D FDTD model having a uniform grid cell size of 10 nm. This figure compares the measured and computationally synthesized bright-field grayscale images of two adjacent contacting 4.3-μm-diameter polystyrene microspheres in air. A numerical aperture of 0.6 was used, assuming illumination wavelengths between 400 and 700 nm. Each image is normalized by its maximum intensity, and the same grayscale is used for both.

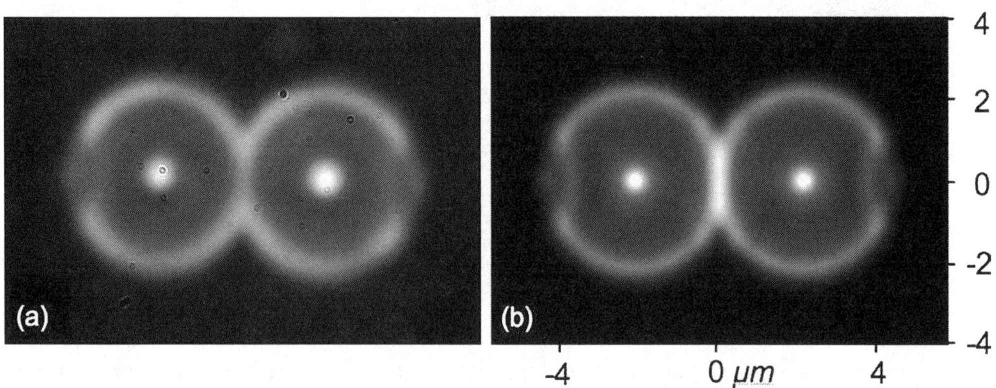

Fig. 14.17 Comparison of (a) measured and (b) FDTD-synthesized bright-field grayscale microscope images of a pair of adjacent contacting 4.3-μm-diameter polystyrene microspheres in air for NA = 0.6 and wavelengths between 400 and 700 nm.

14.8.4 Human Cheek (Buccal) Cell

This section demonstrates the feasibility of using the "microscope in a computer" to synthesize the bright-field true-color image of a biological cell. For this study, it was decided to computationally image a readily available human cell, a cheek (buccal) cell, swabbed from the interior cheek surface. The approximate overall dimensions of this cell were $80 \times 80 \times 1$ μm.

The surface profile of the cheek cell was measured with nanometer resolution using atomic force microscopy, and read into the FDTD grid. For the initial model, the cell was assumed to be filled with a homogeneous lossless dielectric of refractive index $n = 1.38$ (a value guided by previous experimental studies), resting on a glass slide ($n = 1.5$).

The 3-D FDTD grid was configured with $3400 \times 3400 \times 58$ cubic cells of size 25 nm, spanning an overall volume of $85 \times 85 \times 1.45$ μm. The grid was terminated with a 5-cell CPML, and the time-step was 0.04715 fs (2% below the Courant stability limit). Two illuminations having orthogonal polarizations were modeled separately, and the resulting intensities were added to obtain the final image. Each illumination was a normally incident plane wave having a sine-modulated Gaussian time-waveform with –20-dB wavelengths of 400 and 700 nm. FDTD computations were parallelized using the OpenMPI implementation of the message-passing interface (MPI) standard, and executed on 256 processors ($16 \times 16 \times 1$ Cartesian division) of the Quest supercomputer at Northwestern University. The total simulation time was ~30 hours.

The far field was computed at a range of directions arranged in a 2-D Cartesian pattern in the direction-cosine space. Direction cosines were spaced $\Delta s_x = \Delta s_y = 0.0048$ apart from each other inside a collection numerical aperture of $\mathrm{NA_{obj}} = 0.6$. For each observation direction, 10 wavelengths spaced linearly in wavenumber k were recorded from 400 to 700 nm. The final intensity spectrum was normalized by the spectrum of a glass pixel, resulting in a normalized spectroscopic reflectance image.

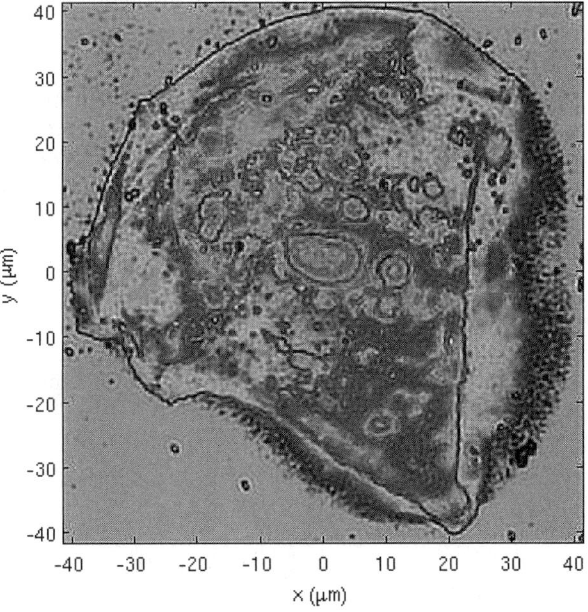

Fig. 14.18 FDTD-synthesized bright-field true-color microscope image of a human cheek (buccal) cell modeled in a 3-D grid having uniform 25-nm cubic cells.

Using the CIE color-matching functions documented for average human vision [135], a true-color image of the cheek cell was generated from the wavelength spectrum of the FDTD-synthesized image. This was done by reducing the wavelength spectrum into three numbers that represent the response of human photoreceptor cones sensitive to short, middle, and long wavelengths. The resulting bright-field true-color image of the human cheek-cell model is shown in Fig. 14.18.

14.9 SUMMARY

This chapter presented a comprehensive account of the theoretical principles and literature references that comprise the foundation for emerging numerical electromagnetic simulations of optical imaging systems based on three-dimensional full-vector solutions of Maxwell's equations. These techniques permit the computational synthesis of images formed by every current form of optical microscopy (bright-field, dark-field, phase-contrast, etc.), as well as optical metrology and photolithography. Focusing, variation of the numerical aperture, and so forth can be adjusted simply by varying a few input parameters – literally a "microscope in a computer." This permits a rigorous simulation of both existing and proposed novel optical imaging techniques over a 10^7:1 dynamic range of distance scales, i.e., from a few nanometers (the voxel size within the microstructure of interest over which Maxwell's equations are enforced) to a few centimeters (the location of the image plane where the amplitude and phase spectra of individual pixels are calculated).

We showed how a general optical imaging system can be segmented into four self-contained sub-components (illumination, scattering, collection and refocusing), and how each of these sub-components is mathematically analyzed. Approximate numerical methods used in the modeling of each sub-component were explained in appropriate detail. Relevant practical applications were cited whenever applicable. Finally, the theoretical and numerical results were illustrated via several implementation examples involving the computational synthesis of microscope images of microscale structures.

This chapter will hopefully constitute a useful starting point for those interested in modeling optical imaging systems from a rigorous electromagnetics point of view. A distinct feature of our approach is the extra attention paid to the issues of discretization and signal processing. This is a key issue in finite methods, where the electromagnetic field is only given at a finite set of spatial and temporal points.

ACKNOWLEDGMENT

The numerical simulations in Section 14.8 were made possible by a supercomputing grant on the Quest high-performance computing system at Northwestern University.

APPENDIX 14A: DERIVATION OF EQUATION (14.9)

The mutual coherence function $J^*(x_d; y_d)$ resulting from the finite collection of plane waves in Fig. 14.5 is:

$$J^*(x_d; y_d) = \Delta s_x \Delta s_y \sum_{m,n} P(s_{x_m}, s_{y_n}) \exp[-jk(s_{x_m} x_d + s_{y_n} y_d)] \qquad (14A.1)$$

where both m and n range from $-\infty$ to ∞. Expression (14.6) for the original mutual coherence function $J(x_d; y_d)$ is in the form of a Fourier transform, and can be inverted to yield the following for $P(s_x, s_y)$:

$$P(s_x, s_y) = \frac{k^2}{(2\pi)^2} \int_{-\infty}^{\infty} \int_{-\infty}^{\infty} J(x'_d; y'_d) \exp[jk(s_x x'_d + s_y y'_d)]\, dx'_d\, dy'_d \tag{14A.2}$$

Primed coordinates are used to avoid confusion in what follows. Using (14A.2), (14A.1) becomes:

$$\begin{aligned} J^*(x_d; y_d) &= \int_{-\infty}^{\infty} \int_{-\infty}^{\infty} dx'_d\, dy'_d\, J(x'_d; y'_d) \\ &\times \frac{k^2 \Delta s_x \Delta s_y}{(2\pi)^2} \sum_{m,n} \exp\left\{jk[s_{x_m}(x'_d - x_d) + s_{y_n}(y'_d - y_d)]\right\} \end{aligned} \tag{14A.3}$$

Substituting $s_{x_m} = m\Delta s_x$ and $s_{y_n} = n\Delta s_y$, the expression on the second line of (14A.3) becomes [136]:

$$\sum_{r=-\infty}^{\infty} \sum_{s=-\infty}^{\infty} \delta\left(x'_d - x_d - r\frac{2\pi}{k\Delta s_x}\right) \delta\left(y'_d - y_d - s\frac{2\pi}{k\Delta s_y}\right) \tag{14A.4}$$

which, when substituted into (14A.3), yields the desired relation (14.9).

APPENDIX 14B: DERIVATION OF EQUATION (14.38)

The original, periodic, continuous vector field **E**(x, y) can be written using the Floquet expansion (14.27) as follows:

$$\boldsymbol{E}(x, y) = \sum_{a=-\infty}^{\infty} \sum_{b=-\infty}^{\infty} \boldsymbol{R}_{ab} \exp[-j(\beta_a x + \beta_b y)] \tag{14B.1}$$

where the integral indices a, b are used to avoid later confusion with the Fourier indices p, q. Sampling this function at $m\Delta x$ and $n\Delta y$, where Δx and Δy are given by (14.35), and applying the phase shift as defined in (14.36), one obtains:

$$\begin{aligned} \bar{\boldsymbol{E}}[m,n] &= \sum_{a=-\infty}^{\infty} \sum_{b=-\infty}^{\infty} \boldsymbol{R}_{ab} \exp[-j(\beta_a m\Delta x + \beta_b n\Delta y)] \\ &\qquad \times \exp\left[jk\sin\theta_i\left(d_x \cos\phi_i \frac{m}{M} + d_y \sin\phi_i \frac{n}{N}\right)\right] \end{aligned} \tag{14B.2}$$

Substituting the expressions for the Floquet wavenumbers $\beta_a = k\cos\phi_i \sin\theta_i - a(2\pi/d_x)$ and $\beta_b = k\sin\phi_i \sin\theta_i - b(2\pi/d_y)$, and the sampling relations (14.35), the above expression simplifies to:

$$\bar{\boldsymbol{E}}[m,n] = \sum_{a=-\infty}^{\infty} \sum_{b=-\infty}^{\infty} \boldsymbol{R}_{ab} \exp\left[j2\pi\left(\frac{am}{M} + \frac{bn}{N}\right)\right] \tag{14B.3}$$

Now, the DFT of this array is given by:

$$\begin{aligned}\tilde{\bar{E}}[p,q] &= \sum_{m=0}^{M-1}\sum_{n=0}^{N-1} \bar{E}[m,n]\exp\left[-j2\pi\left(\frac{pm}{M}+\frac{qn}{N}\right)\right] \\ &= \sum_a\sum_b \boldsymbol{R}_{ab}\left\{\sum_{m=0}^{M-1}\sum_{n=0}^{N-1}\exp\left[-j2\pi\left(\frac{(p-a)m}{M}+\frac{(q-b)n}{N}\right)\right]\right\}\end{aligned} \tag{14B.4}$$

The double-summation expression in brackets in the second line of (14B.4) is equal to:

$$MN\sum_{r=-\infty}^{\infty}\sum_{s=-\infty}^{\infty}\delta(a,\, p+rM)\,\delta(b,\, q+sN) \tag{14B.5}$$

where $\delta(.\,,\,.)$ is the Kronecker delta symbol. Substituting (14B.5) in (14B.4), we obtain the desired relation (14.38).

APPENDIX 14C: DERIVATION OF EQUATION (14.94)

The sampled auxiliary variable $\boldsymbol{G}'[p,q]$ of (14.92) is, from (14.87)–(14.90):

$$\begin{aligned}\boldsymbol{G}'[p,q] &= \frac{-jk'}{2\pi}\int_{-\infty}^{\infty}\int_{-\infty}^{\infty} dx'\,dy'\,\boldsymbol{E}_{\text{img}}(x',y') \\ &\quad\times \exp[jk'(s'_{x_0}x'+s'_{y_0}y')]\exp[jk'(p\Delta s'_x x'+q\Delta s'_y y')]\end{aligned} \tag{14C.1}$$

The DFT of $\boldsymbol{G}'[p,q]$ is, from (14.93):

$$\begin{aligned}\boldsymbol{E}[m,n] &= \sum_{p=0}^{N_p-1}\sum_{q=0}^{N_q-1}\boldsymbol{G}'[p,q]\exp\left[-j2\pi\left(\frac{pm}{N_p}+\frac{qn}{N_q}\right)\right] \\ &= \frac{-jk'}{2\pi}\int_{-\infty}^{\infty}\int_{-\infty}^{\infty} dx'\,dy'\,\boldsymbol{E}_{\text{img}}(x',y')\exp[jk'(s'_{x_0}x'+s'_{y_0}y')] \\ &\quad\times\sum_{p=0}^{N_p-1}\sum_{q=0}^{N_q-1}\exp\left\{j\left[\left(k'\Delta s'_x x'-\frac{2\pi m}{N_p}\right)p+\left(k'\Delta s'_y y'-\frac{2\pi n}{N_q}\right)q\right]\right\}\end{aligned} \tag{14C.2}$$

The summations in p and q can be extended to infinity, since $\boldsymbol{G}'[p,q]$ is only nonzero for a finite number of p and q values. The resulting infinite summation is equal to an infinite series of Dirac delta functions [136]:

$$\sum_{p=-\infty}^{\infty}\sum_{q=-\infty}^{\infty}\exp\left\{j\left[\left(k'\Delta s_x' x' - \frac{2\pi m}{N_p}\right)p + \left(k'\Delta s_y' y' - \frac{2\pi n}{N_q}\right)q\right]\right\} =$$
$$(2\pi)^2\sum_{r=-\infty}^{\infty}\sum_{s=-\infty}^{\infty}\delta\left(k'\Delta s_x' x' - \frac{2\pi m}{N_p} - 2\pi r\right)\delta\left(k'\Delta s_y' y' - \frac{2\pi n}{N_q} - 2\pi s\right) \tag{14C.3}$$

Using the scaling property of the Dirac delta function, this becomes:

$$\frac{(2\pi)^2}{(k')^2\,\Delta s_x'\,\Delta s_y'}\sum_{r=-\infty}^{\infty}\sum_{s=-\infty}^{\infty}\delta\left(x' - m\Delta_x - rD_x'\right)\delta\left(y' - n\Delta_x - sD_y'\right) \tag{14C.4}$$

in which the spatial sampling periods Δ_x, Δ_y and the aliasing periods D_x', D_y' are given by (14.95) and (14.96). Substituting this expression into (14C.2) and using the sifting property of the delta function to evaluate the integral, the desired relation (14.94) is obtained.

APPENDIX 14D: COHERENT FOCUSED BEAM SYNTHESIS USING PLANE WAVES

Extending the discussion in Section 14.4.1, this appendix describes an efficient procedure for synthesizing a coherent focused beam in the context of the FDTD method [37]. Our procedure has its origins in the work of Richards and Wolf [34], who described the electromagnetic theory of the coherent focused beam in terms of integrals involving special functions. In fact, we take one of their intermediate steps as our starting point by interpreting the coherent focused beam as a sum of plane waves. However, we work directly in the time domain, which is immediately suited for implementation using FDTD. Figure 14D.1 illustrates the problem geometry.

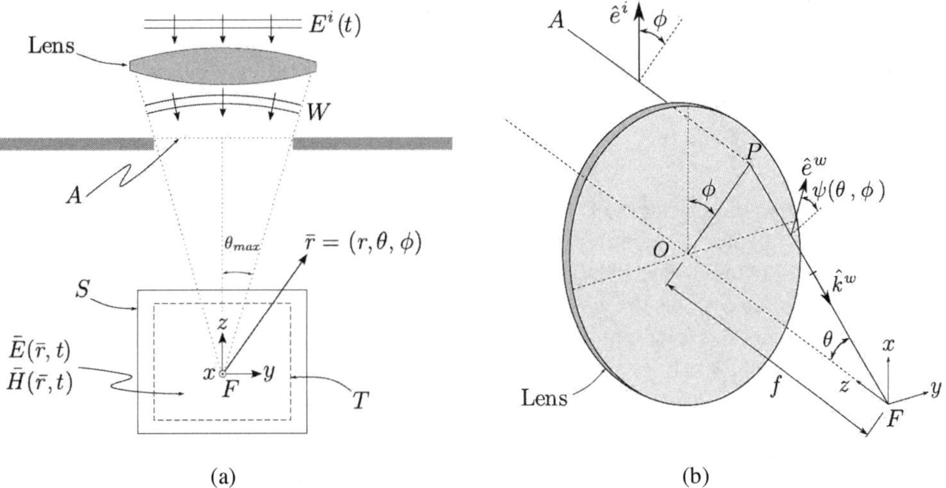

Fig. 14D.1 Geometry of the focused beam: (a) The converging lens, located far from the scatterer, creates a beam focused at *F* that is calculated only inside the FDTD grid, *S*, surrounding the scatterer. (b) The electric field on ray *APF* makes the same angle with the meridional plane before and after refraction through the lens. *Source:* Capoglu, Taflove, and Backman, *Optics Express*, Vol. 16, 2008, pp. 19208–19220. ©2011 The Optical Society of America.

In Fig. 14D.1(a), the FDTD simulation space, S, is a compact region surrounding the scatterer that contains the focal point F. Spherical coordinates (r, θ, ϕ) are centered at F, with the z-axis parallel to the optical axis. Refractive indices everywhere are assumed to be unity. An incident x-polarized plane wave with the electric field $\boldsymbol{E}^{\mathrm{i}}$ impinges on the converging lens, which then creates the spherical wavefront W converging at F. This wavefront is diffracted by the circular screen at A, which is assumed to be in the far field of S. The vector electric and magnetic fields in the vicinity of F are denoted by $\boldsymbol{E}(\boldsymbol{r}, t)$ and $\boldsymbol{H}(\boldsymbol{r}, t)$, respectively.

In Fig. 14D.1(b), the passage of a typical ray AP through the aplanatic lens is shown in greater detail. Using geometrical optics principles, we will now derive an expression for the electric field on refracted ray PF in terms of the electric field on ray section AP. The electric field on AP is assumed to be oriented toward the x-axis; therefore $\hat{\boldsymbol{e}}^{\mathrm{i}} = \hat{\boldsymbol{x}}$. The electric field $\boldsymbol{E}^{\mathrm{W}}(\boldsymbol{r},t)$ on the spherical converging wavefront W can be written as:

$$\boldsymbol{E}^{\mathrm{W}}(\boldsymbol{r},t) = \hat{\boldsymbol{e}}^{\mathrm{W}}(\theta, \phi) \cdot \frac{a(\theta, \phi, t + r/c)}{r} \tag{14D.1}$$

The unit vector $\hat{\boldsymbol{e}}^{\mathrm{W}}(\theta, \phi)$ determines the polarization of the electric field on PF, and $a(\theta, \phi, t)$ is the *strength factor* of the same ray [34]. Geometrical optics principles dictate that $\hat{\boldsymbol{e}}^{\mathrm{W}}(\theta, \phi)$ and $a(\theta, \phi, t)$ are *invariant* along a ray, except for a time advance by r/c applied to the latter. The relationship between $a(\theta, \phi, t)$ and $E^{\mathrm{i}}(t)$ follows from the intensity law of geometrical optics [34]:

$$a(\theta, \phi, t) = f \cdot \sqrt{\cos\theta}\, E^{\mathrm{i}}(t) \tag{14D.2}$$

where f is the focal length of the lens. If the incidence angles at each surface in the lens system are small, each ray stays on the same side of the meridional plane (the plane containing the ray and the optical axis) during its passage from the lens system. Furthermore, the angle that the electric field vector on a ray makes with the meridional plane is unchanged by refraction [34]. This greatly simplifies the calculation of $\hat{\boldsymbol{e}}^{\mathrm{W}}(\theta, \phi)$. If we define the *polarization angle* $\psi(\theta, \phi)$ of a ray as the angle that the electric vector on this ray makes with the meridional plane, per Fig. 14D.1(b), it follows that $\psi(\theta, \phi)$ is given by:

$$\psi(\theta, \phi) = \phi \tag{14D.3}$$

This relation greatly simplifies the FDTD implementation. In FDTD, the polarization of a plane wave is commonly expressed with respect to a reference plane, which can be made to coincide with the meridional plane in Fig. 14D.1(b) [37].

Combining the above results and applying the Kirchhoff boundary condition on A, the electric field at and around focal point F can be expressed as [34, 37]:

$$\boldsymbol{E}(\boldsymbol{r}, t) = \frac{f}{2\pi c} \int_{\phi=0}^{2\pi} \int_{\theta=0}^{\theta_{\max}} \hat{\boldsymbol{e}}^{\mathrm{W}}(\theta, \phi)\, \dot{E}^{\mathrm{i}}(t') \sqrt{\cos\theta}\, \sin\theta\, d\theta\, d\phi \tag{14D.4}$$

where $\dot{E}^{\mathrm{i}}$ denotes the time derivative of E^{i}, $\theta_{\max}$ is the half angular range of aperture A, $t' = t + \hat{\boldsymbol{r}} \cdot \boldsymbol{r}/c$ is the advanced or retarded time, and $\hat{\boldsymbol{r}}$ is the unit vector in the direction of $\boldsymbol{r}$. Expression (14D.4) is a decomposition of the electric field in the image region of the lens in terms of *plane waves* incident from a range of directions (θ, ϕ). The plane wave from (θ, ϕ) has the polarization $\hat{e}^{W}(\theta, \phi)$, the incidence vector $\hat{k}^{W}(\theta, \phi) = -\hat{r}$ [per Fig. 14D.1(b)], and the differential amplitude waveform (referred to the origin $r = 0$) given by:

$$dE(0,t) = \frac{\sin\theta \, d\theta \, d\phi}{2\pi c}\dot{a}(\theta,\phi,t) = \frac{f \cdot \sqrt{\cos\theta}\sin\theta \, d\theta \, d\phi}{2\pi c}\dot{E}^{i}(t) \tag{14D.5}$$

In FDTD, the double integral in (14D.4) is approximated by the following finite sum:

$$\mathbf{E}(\mathbf{r},t) \cong \frac{f}{2\pi c}\sum_{n,m}\alpha_{nm}\hat{e}^{W}(\theta_n,\phi_m)\,\dot{E}^{i}(t'_{nm})\sqrt{\cos\theta_n}\sin\theta_n \qquad n=1,\ldots N,\ m=1,\ldots M \tag{14D.6}$$

in which $0 < \theta_n < \theta_{max}$ and $0 < \phi_m < 2\pi$ are the incidence angles of the plane waves, and $t'_{nm} = t + \hat{r}_{nm} \cdot \mathbf{r}/c$ are the retarded times for these incidence angles. The amplitude factors α_{nm} are the numerical integration (or *quadrature*) factors for the double integral in (14D.4) in θ and ϕ. In [37], a quadrature method based on the Gauss–Legendre rule was used for the finite sum in (14D.6). Here, we present an alternate method, based on discrete sampling theory and a Cartesian placement of plane waves in direction-cosine space [137]. The direction cosines that correspond to each incidence direction (θ, ϕ) are defined as:

$$s_x = \sin\theta\cos\phi\,, \qquad s_y = \sin\theta\sin\phi \tag{14D.7}$$

shown in a Cartesian distribution in Fig. 14D.2.

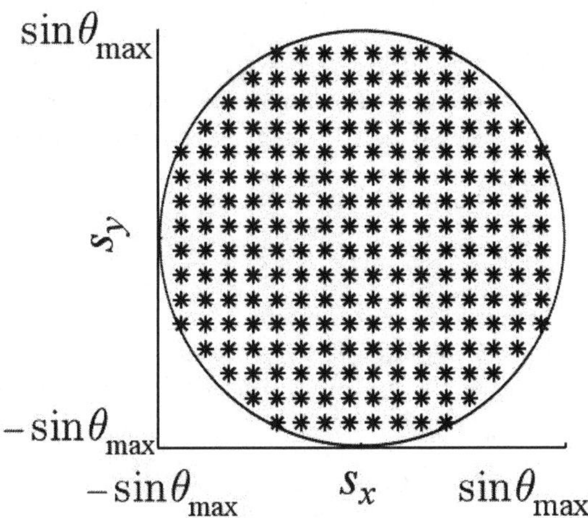

Fig. 14D.2 Cartesian distribution of plane-wave direction cosines within the illumination numerical aperture $\sin\theta_{max}$ for synthesis of the focused beam. *Source:* I. R. Capoglu, J. D. Rogers, A. Taflove, and V. Backman, *Progress in Optics*, Vol. 57, 2012, pp. 1–91, ©2012 Elsevier.

In Fig. 14D.2, the direction cosines are confined within a circle of radius $\sin\theta_{max}$ (also called the *numerical aperture* of the illumination), where θ_{max} is the illumination half-angle shown in Fig. 14D.1(a). The spacing between the direction cosines is $\Delta s = \Delta s_x = \Delta s_y$, which should be chosen sufficiently fine to faithfully represent the focused beam around the scatterer. This choice is based on considerations of discrete sampling theory [50]. It is easier to grasp the main idea behind this theory if we consider individual monochromatic components of the time-domain expression (14D.4) or (14D.6). For one of these monochromatic components with wavenumber $k = 2\pi/\lambda$, it can be easily shown that the discrete arrangement in Fig. 14D.2 results in a *spatially periodic* illumination in the focal plane with the period:

$$L_{\mathrm{p}} = \frac{2\pi}{k\Delta s} \tag{14D.8}$$

This period should be large enough such that the successive periods do not overlap or contaminate the edges of the simulation grid. We know that the focused beam is confined to within a diffraction spot that is several times the *spread parameter*:

$$W_{\mathrm{beam}} = \frac{\lambda}{\sin\theta_{\mathrm{max}}} \tag{14D.9}$$

To avoid overlap of the periodic replicas of the beam, we should therefore have:

$$L_{\mathrm{p}} >> W_{\mathrm{beam}} \tag{14D.10}$$

or equivalently:

$$\Delta s << \frac{2\pi}{kW_{\mathrm{beam}}} \tag{14D.11}$$

If the maximum lateral extent of the FDTD grid is much larger than W_{beam}, then it should replace W_{beam} in sampling condition (14D.11). For wideband illumination as in (14D.4), the spread parameter in (14D.9) should be calculated using the largest wavelength in the illumination waveform.

REFERENCES

[1] Born, M., and E. Wolf, *Principles of Optics: Electromagnetic Theory of Propagation, Interference and Diffraction of Light*, 7th ed., Cambridge: Cambridge University Press, 1999.

[2] Goodman, J. W., *Introduction to Fourier Optics*, 2nd ed., New York: McGraw-Hill, 1996.

[3] Cole, D. C., E. Barouch, E. W. Conrad, and M. Yeung, "Using advanced simulation to aid microlithography development," *Proc. IEEE*, Vol. 89, 2001, pp. 1194–1215.

[4] Neureuther, A., "If it moves, simulate it!," in *Optical Microlithography XXI*, Pts 1-3, *Proc. SPIE*, Vol. 6924, 2008, pp. 692402-1–15.

[5] Neureuther, A. R., "Simulation of optical lithography and inspection," *Microelectronics Engineering*, Vol. 17, 1992, pp. 377–384.

[6] Nikolaev, N. I., and A. Erdmann, "Rigorous simulation of alignment for microlithography," *J. Microlithographic Microfabrication of Microsystems*, Vol. 2, 2003, pp. 220–226.

[7] Hollmann, J. L., A. K. Dunn, and C. A. DiMarzio, "Computational microscopy in embryo imaging," *Optics Letters*, Vol. 29, 2004, pp. 2267–2269.

[8] Simon, B., and C. A. DiMarzio, "Simulation of a theta line-scanning confocal microscope," *J. Biomedical Optics*, Vol. 12, No. 6, 2007, pp. 064020-1–9.

[9] Tanev, S., J. Pond, P. Paddon, and V. V. Tuchin, "Optical phase contrast microscope imaging: An FDTD modeling approach," in *Optical Technologies in Biophysics and Medicine IX, Proc. SPIE*, 2008, pp. 7910E-1–11.

[10] Tanev, S., W. B. Sun, J. Pond, V. V. Tuchin, and V. P. Zharov, "Flow cytometry with gold nanoparticles and their clusters as scattering contrast agents: FDTD simulation of light-cell interaction," *J. Biophotonics*, Vol. 2, 2009, pp. 505–520.

[11] Capoglu, I. R., C. A. White, J. D. Rogers, H. Subramanian, A. Taflove, and V. Backman, "Numerical simulation of partially coherent broadband optical imaging using the FDTD method," *Optics Lett.*, Vol. 36, no. 9, 2011, pp. 1596–1598.

[12] Subramanian, H., P. Pradhan, Y. Liu, I. R. Capoglu, X. Li, J. D. Rogers, A. Heifetz, D. Kunte, H. K. Roy, A. Taflove, and V. Backman, "Optical methodology for detecting histologically unapparent nanoscale consequences of genetic alterations in biological cells," *Proc. National Academy of Sciences USA*, Vol. 105, 2008, pp. 20118–20123.

[13] Subramanian, H., P. Pradhan, Y. Liu, I. R. Capoglu, J. D. Rogers, H. K. Roy, R. E. Brand, and V. Backman, "Partial-wave microscopic spectroscopy detects sub-wavelength refractive index fluctuations: An application to cancer diagnosis," *Optics Letters*, Vol. 34, 2009, pp. 518–520.

[14] Hansen, R., "Focal region characteristics of focused array antennas," *IEEE Trans. Antennas and Propagation*, Vol. 33, 1985, pp. 1328–1337.

[15] Goodman, J. W., *Statistical Optics*, New York: Wiley, 2000.

[16] Christov, I. P., "Propagation of partially coherent light pulses," *Optics Acta*, Vol. 33, 1986, pp. 63–72.

[17] Wang, L. G., Q. Lin, H. Chen, and S. Y. Zhu, "Propagation of partially coherent pulsed beams in the spatiotemporal domain," *Physical Review E*, Vol. 67, No. 5, Part 2, 2003, pp. 056613-1–7.

[18] Lajunen, H., P. Vahimaa, and J. Tervo, "Theory of spatially and spectrally partially coherent pulses," *J. Optical Society of America A*, Vol. 22, 2005, pp. 1536–1545.

[19] Papoulis, A., *Probability, Random Variables, and Stochastic Processes*, New York: McGraw-Hill, 1991.

[20] Haykin, S., *Communication Systems*, 4th ed., New York: Wiley, 2001.

[21] Tanev, S., V. V. Tuchin, and P. Paddon, "Cell membrane and gold nanoparticles effects on optical immersion experiments with noncancerous and cancerous cells: Finite-difference time-domain modeling," *J. Biomedical Optics*, Vol. 11, 2006, pp. 064037-1–6.

[22] Salski, B., and W. Gwarek, "Hybrid finite-difference time-domain Fresnel modeling of microscopy imaging," *Applied Optics*, Vol. 48, 2009, pp. 2133–2138.

[23] Smith, G. S., *An Introduction to Classical Electromagnetic Radiation*, New York: Cambridge University Press, 1997.

[24] Salski, B., and W. Gwarek, "Hybrid FDTD-Fresnel modeling of microscope imaging," in *Proc. International Conf. on Recent Advances in Microwave Theory and Applications*, 2008, Jaipur, India, pp. 398–399.

[25] Salski, B., and W. Gwarek, "Hybrid FDTD-Fresnel modeling of the scanning confocal microscopy," in *Scanning Microscopy, Proc. SPIE*, 2009, pp. 737826-1–6.

[26] Salski, B., M. Celuch, and W. Gwarek, "FDTD for nanoscale and optical problems," *IEEE Microwave Magazine*, Vol. 11, No. 2, 2010, pp. 50–59.

[27] Yeh, C., S. Colak, and P. Barber, "Scattering of sharply focused beams by arbitrarily shaped dielectric particles: An exact solution," *Applied Optics*, Vol. 21, 1982, pp. 4426–4433.

[28] Petersson, L., and G. S. Smith, "On the use of a Gaussian beam to isolate the edge scattering from a plate of finite size," *IEEE Trans. Antennas and Propagation*, Vol. 52, 2004, pp. 505–512.

[29] Taflove, A., and S. C. Hagness, *Computational Electrodynamics: The Finite-Difference Time-Domain Method*, 3rd ed., Norwood, MA: Artech House, 2005.

[30] Judkins, J. B., and R. W. Ziolkowski, "Finite-difference time-domain modeling of nonperfectly conducting metallic thin-film gratings," *J. Optical Society of America A*, Vol. 12, 1995, pp. 1974–1983.

[31] Judkins, J. B., C. W. Haggans, and R. W. Ziolkowski, "Two-dimensional finite-difference time-domain simulation for rewritable optical disk surface structure design," *Applied Optics*, vol. 35, 1996, pp. 2477–2487.

[32] Choi, K., J. Chon, M. Gu, and B. Lee, "Characterization of a sub-wavelength-scale 3D void structure using the FDTD-based confocal laser scanning microscopic image mapping technique," *Optics Express*, Vol. 15, 2007, pp. 10767–10781.

[33] Wolf, E., "Electromagnetic diffraction in optical systems: I. An integral representation of the image field," *Proc. Royal Society of London A*, Vol. 253, 1959, pp. 349–357.

[34] Richards, B., and E. Wolf, "Electromagnetic diffraction in optical systems: II. Structure of the image field in an aplanatic system," *Proc. Royal Society of London A*, Vol. 253, 1959, pp. 358–379.

[35] Munro, P., and P. Török, "Vectorial, high numerical aperture study of Nomarski's differential interference contrast microscope," *Optics Express*, Vol. 13, 2005, pp. 6833–6847.

[36] Török, P., P. Munro, and E. E. Kriezis, "High numerical aperture vectorial imaging in coherent optical microscopes," *Optics Express*, Vol. 16, 2008, pp. 507–523.

[37] Capoglu, I. R., A. Taflove, and V. Backman, "Generation of an incident focused light pulse in FDTD," *Optics Express*, Vol. 16, 2008, pp. 19208–19220.

[38] Kempe, M., U. Stamm, B. Wilhelmi, and W. Rudolph, "Spatial and temporal transformation of femtosecond laser pulses by lenses and lens systems," *J. Optical Society of America B*, Vol. 9, 1992, pp. 1158–1165.

[39] Davidson, D. B., and R. W. Ziolkowski, "Body-of-revolution finite-difference time-domain modeling of space-time focusing by a three-dimensional lens," *J. Optical Society of America A*, Vol. 11, 1994, pp. 1471–1490.

[40] Ibragimov, E., "Focusing of ultrashort laser pulses by the combination of diffractive and refractive elements," *Applied Optics*, Vol. 34, 1995, pp. 7280–7285.

[41] Gu, M., and C. Sheppard, "Three-dimensional image formation in confocal microscopy under ultra-short-laser-pulse illumination," *J. Modern Optics*, Vol. 42, 1995, pp. 747–762.

[42] Veetil, S. P., H. Schimmel, F. Wyrowski, and C. Vijayan, "Wave optical modeling of focusing of an ultra short pulse," *J. Modern Optics*, Vol. 53, 2006, pp. 2187–2194.

[43] Coen, S., A. Chau, R. Leonhardt, J. D. Harvey, J. C. Knight, W. J. Wadsworth, and P. Russell, "Supercontinuum generation by stimulated Raman scattering and parametric four-wave mixing in photonic crystal fibers," *J. Optical Society of America B*, Vol. 19, 2002, pp. 753–764.

[44] Booth, M. J., R. Juskaitis, and T. Wilson, "Spectral confocal reflection microscopy using a white light source," *J. European Optical Society Rapid Publication*, Vol. 3, 2008, pp. 08026-1–6.

[45] Winton, S. C., P. Kosmas, and C. M. Rappaport, "FDTD simulation of TE and TM plane waves at nonzero incidence in arbitrary layered media," *IEEE Trans. Antennas and Propagation*, Vol. 53, 2005, pp. 1721–1728.

[46] Capoglu, I. R., and G. S. Smith, "A Total-field / scattered-field plane-wave source for the FDTD analysis of layered media," *IEEE Trans. Antennas and Propagation*, Vol. 56, 2008, pp. 158–169.

[47] Zhang, L., and T. Seideman, "Rigorous formulation of oblique incidence scattering from dispersive media," *Physical Review B*, Vol. 82, 2010, pp. 155117-1–15.

[48] Nolte, A., J. B. Pawley, and L. Höring, "Non-Laser Light Sources for Three-Dimensional Microscopy," in *Handbook of Biological Confocal Microscopy*, 3rd ed., J. B. Pawley, ed., Berlin: Springer, 2006.

[49] Press, W. H., B. P. Flannery, S. A. Teukolsky, and W. T. Vetterling, *Numerical Recipes in C: The Art of Scientific Computing*, 2nd ed., Cambridge: Cambridge University Press, 1992.

[50] Oppenheim, A. V., R. W. Schafer, and J. R. Buck, *Discrete-Time Signal Processing*, 2nd ed., Upper Saddle River, NJ: Prentice Hall, 1999.

[51] Wojcik, G. L., J. Mould, R. J. Monteverde, J. J. Prochazka, and J. Frank, "Numerical simulation of thick line width measurements by reflected light," in *Integrated Circuit Metrology, Inspection and Process Control V, Proc. SPIE*, Vol. 1464, 1991, pp. 187–203.

[52] Erdmann, A., and P. Evanschitzky, "Rigorous electromagnetic field mask modeling and related lithographic effects in the low k1 and ultrahigh numerical aperture regime," in *J. Micro/ Nanolithography, MEMS and MOEMS, Proc. SPIE*, Vol. 6, 2007, pp. 031002-1–16.

[53] Haver, S. v., O. T. Janssen, J. J. Braat, A. J. E. M. Janssen, H. P. Urbach, and S. F. Pereira, "General imaging of advanced 3D mask objects based on the fully-vectorial Extended Nijboer-Zernike (ENZ) theory," in *Optical Microlithography XXI*, Pts. 1–3, *Proc. SPIE*, Vol. 6924, 2008, pp. 69240U–1–8.

[54] Mandel, L., "Intensity fluctuations of partially polarized light," *Proc. Physical Society of London*, Vol. 81, 1963, pp. 1104–1114.

[55] Tervo, J., T. Setälä, and A. T. Friberg, "Theory of partially coherent electromagnetic fields in the space-frequency domain," *J. Optical Society of America A*, Vol. 21, 2004, pp. 2205–2215.

[56] Totzeck, M., "Numerical simulation of high-NA quantitative polarization microscopy and corresponding near-fields," *Optik*, Vol. 112, 2001, pp. 399–406.

[57] Yang, S.-H., T. Milster, J. R. Park, and J. Zhang, "High-numerical-aperture image simulation using Babinet's principle," *J. Optical Society of America A*, Vol. 27, 2010, pp. 1012–1023.

[58] Tanev, S., V. V. Tuchin, and J. Pond, "Simulation and modeling of optical phase contrast microscope cellular nanobioimaging," in *15th International School on Quantum Electronics; Laser Physics and Applications, Proc. SPIE*, Vol. 7027, 2008, pp. 702716-1–8.

[59] Yee, K. S., "Numerical solution of initial boundary value problems involving Maxwell's equations in isotropic media," *IEEE Trans. Antennas and Propagation*, Vol. AP-14, 1966, pp. 302–307.

[60] Taflove, A., "Application of the finite-difference time-domain method to sinusoidal steady-state electromagnetic penetration problems," *IEEE Trans. Electromagnetic Compatibility*, Vol. 22, 1980, pp. 191–202.

[61] Jin, J., *The Finite Element Method in Electromagnetics*, New York: Wiley, 2002.

[62] Teixeira, F. L., and W. C. Chew, "Lattice electromagnetic theory from a topological viewpoint," *J. Mathematical Physics*, Vol. 40, 1999, pp. 169–187.

[63] Berenger, J.-P., "A perfectly matched layer for the absorption of electromagnetic waves," *J. Computational Physics*, Vol. 114, 1994, pp. 185–200.

[64] Guerrieri, R., K. H. Tadros, J. Gamelin, and A. R. Neureuther, "Massively parallel algorithms for scattering in optical lithography," *IEEE Trans. Computer-Aided Design of Integrated Circuits and Systems*, Vol. 10, 1991, pp. 1091–1100.

[65] Wong, A. K., *Rigorous Three-dimensional Time-domain Finite-difference Electromagnetic Simulation*, Ph.D. thesis, University of California, Berkeley, 1994.

[66] Wong, A. K., and A. R. Neureuther, "Mask topography effects in projection printing of phase-shifting masks," *IEEE Trans. Electron Devices*, Vol. 41, 1994, pp. 895–902.

[67] Wong, A. K., R. Guerrieri, and A. R. Neureuther, "Massively-parallel electromagnetic simulation for photolithographic applications," *IEEE Trans. Computer-Aided Design of Integrated Circuits and Systems*, Vol. 14, 1995, pp. 1231–1240.

[68] Pistor, T. V., K. Adam, and A. Neureuther, "Rigorous simulation of mask corner effects in extreme ultraviolet lithography," *J. Vacuum Science Technology B*, Vol. 16, 1998, pp. 3449–3455.

[69] Pistor, T. V., and A. Neureuther, "Extreme ultraviolet mask defect simulation," *J. Vacuum Science Technology B*, Vol. 17, 1999, pp. 3019–3023.

[70] Pistor, T. V., and A. R. Neureuther, "Calculating aerial images from EUV masks," in *Emerging Lithographic Technologies III*, Pts 1 and 2, *Proc. SPIE*, Vol. 3676, 1999, pp. 679–696.

[71] Brukman, M., Y. F. Deng, and A. Neureuther, "Simulation of EUV multilayer mirror buried defects," in *Emerging Lithographic Technologies IV*, *Proc. SPIE*, Vol. 3997, 2000, pp. 799–806.

[72] Pistor, T. V., *Electromagnetic Simulation and Modeling with Applications in Lithography*, Ph.D. thesis, University of California, Berkeley, 2001.

[73] Azpiroz, J. T., *Analysis and Modeling of Photomask Near-fields in Sub-wavelength Deep Ultraviolet Lithography*, Ph.D. thesis, University of California, Los Angeles, 2004.

[74] Adam, K., and A. R. Neureuther, "Methodology for accurate and rapid simulation of large arbitrary 2D layouts of advanced photomasks," in *21st Annual Bacus Symposium on Photomask Technology*, Pts 1 and 2, *Proc. SPIE*, Vol. 4562, 2002, pp. 1051–1067.

[75] Tadros, K., A. R. Neureuther, and R. Guerrieri, "Understanding metrology of polysilicon gates through reflectance measurements and simulation," in *Integrated Circuit Metrology, Inspection, and Process Control V*, *Proc. SPIE*, Vol. 1464, 1991, pp. 177–186.

[76] Yin, X. M., A. Wong, D. Wheeler, G. Williams, E. Lehner, F. Zach, B. Kim, Y. Fukuzaki, Z. G. Lu, S. Credendino, and T. Wiltshire, "Sub-wavelength alignment mark signal analysis of advanced memory products," in *Metrology, Inspection, and Process Control for Microlithography XIV*, *Proc. SPIE*, Vol. 3998, 2000, pp. 449–459.

[77] Deng, Y. F., T. Pistor, and A. R. Neureuther, "Rigorous electromagnetic simulation applied to alignment systems," in *Optical Microlithography XIV*, Pts 1 and 2, *Proc. SPIE*, Vol. 4346, 2001, pp. 1533–1540.

[78] Furukawa, H., and S. Kawata, "Analysis of image formation in a near-field scanning optical microscope: Effects of multiple scattering," *Optics Commun.*, Vol. 132, 1996, pp. 170–178.

[79] Vasilyeva, E., and A. Taflove, "Three-dimensional modeling of amplitude-object imaging in scanning near-field optical microscopy," *Optics Lett.*, Vol. 23, 1998, pp. 1155–1157.

[80] Vasilyeva, E., and A. Taflove, "3-D FDTD image analysis in transmission illumination mode of scanning near-field optical microscopy," in *Proc. IEEE Antennas and Propagation Society International Symposium*, Atlanta, GA, Vol. 4, 1998, pp. 1800–1803.

[81] Simpson, S. H., and S. Hanna, "Analysis of the effects arising from the near-field optical microscopy of homogeneous dielectric slabs," *Optics Commun.*, Vol. 196, 2001, pp. 17–31.

[82] Krug, J. T., E. J. Sanchez, and X. S. Xie, "Design of near-field optical probes with optimal field enhancement by finite difference time domain electromagnetic simulation," *J. Chemical Physics*, Vol. 116, 2002, pp. 10895–10901.

[83] Symons, W. C., K. W. Whites, and R. A. Lodder, "Theoretical and experimental characterization of a near-field scanning microwave (NSMM)," *IEEE Trans. Microwave Theory and Techniques*, Vol. 51, 2003, pp. 91–99.

[84] Janssen, O. T., S. v. Haver, A. J. E. M. Janssen, J. J. Braat, P. Urbach, and S. F. Pereira, "Extended Nijboer-Zernike (ENZ) based mask imaging: Efficient coupling of electromagnetic field solvers and the ENZ imaging algorithm," in *Optical Microlithography XXI*, Parts 1–3, *Proc. SPIE*, Vol. 6924, 2008, pp. 692410-1–9.

[85] Haver, S. v., J. J. Braat, A. J. E. M. Janssen, O. T. Janssen, and S. F. Pereira, "Vectorial aerial-image computations of three-dimensional objects based on the extended Nijboer-Zernike theory," *J. Optical Society of America A*, Vol. 26, 2009, pp. 1221–1234.

[86] Liu, Q. H., "The pseudospectral time-domain (PSTD) method: A new algorithm for solutions of Maxwell's equations," in *IEEE Antennas and Propagation Society International Symposium Digest*, Montreal, Canada, Vol. 1, 1997, pp. 122–125.

[87] Tseng, S. H., "Virtual optical experiment: Characterizing the coherent effects of light scattering through macroscopic random media," *Japan J. Applied Physics, Part 1*, Vol. 46, 2007, pp. 7966–7969.

[88] Wojcik, G. L., J. Mould, E. Marx, and M. P. Davidson, "Numerical reference models for optical metrology simulation," in *Integrated Circuit Metrology, Inspection, and Process Control VI, Proc. SPIE*, Vol. 1673, 1992, pp. 70–82.

[89] Vallius, T., and J. Turunen, "Electromagnetic models for the analysis and design of complex diffractive microstructures," in *ICO20: Optical Information Processing, Parts 1 and 2, Proc. SPIE*, Vol. 6027, 2006, pp. 602704-1–4.

[90] Besbes, M., J. P. Hugonin, P. Lalanne, S. v. Haver, O. Janssen, A. M. Nugrowati, M. Xu, S. F. Pereira, H. P. Urbach, A. Nes, P. Bienstman, G. Granet, A. Moreau, S. Helfert, M. Sukharev, T. Seideman, F. I. Baida, B. Guizal, and D. V. Labeke, "Numerical analysis of a slit-groove diffraction problem," *J. European Optics Society Rapid Publication*, Vol. 2, 2007, pp. 07022-1–17.

[91] Erdmann, A., T. Fuhner, F. Shao, and P. Evanschitzky, "Lithography simulation: Modeling techniques and selected applications," in *Modeling Aspects in Optical Metrology II, Proc. SPIE*, Vol. 7390, 2009, pp. 739002-1–17.

[92] Harrington, R. F., *Time-Harmonic Electromagnetic Fields*, New York: Wiley-IEEE Press, 2001.

[93] Stratton, J. A., *Electromagnetic Theory*, New York: McGraw-Hill, 1941.

[94] Kline, M., and I. W. Kay, *Electromagnetic Theory and Geometrical Optics*, Huntington, NY: Krieger Pub. Co., 1979.

[95] Yeung, M. S., "Modeling high numerical aperture optical lithography," in *Optical/Laser Microlithography, Proc. SPIE*, Vol. 922, 1988, pp. 149–167.

[96] Yeung, M. S., D. Lee, R. Lee, and A. R. Neureuther, "Extension of the Hopkins theory of partially coherent imaging to include thin-film interference effects," in *Optical/Laser Microlithography VI, Proc. SPIE*, Vol. 1927, pt. 1, 1993, pp. 452–463.

[97] Totzeck, M., "Some similarities and dissimilarities of imaging simulation for optical microscopy and lithography," in *Proc. 5th International Workshop on Automatic Processing of Fringe Patterns*, Berlin, Germany, 2006, pp. 267–274.

[98] Lee, R. T., and G. S. Smith, "An alternative approach for implementing periodic boundary conditions in the FDTD method using multiple unit cells," *IEEE Trans. Antennas and Propagation*, Vol. 54, 2006, pp. 698–705.

[99] Peterson, A. F., S. L. Ray, and R. Mittra, *Computational Methods for Electromagnetics*, New York: IEEE Press, 1998.

[100] Maystre, D., and M. Neviere, "Electromagnetic theory of crossed gratings," *J. Optics*, Vol. 9, 1978, pp. 301–306.

[101] Bracewell, R. N., *The Fourier Transform and Its Applications*, 2nd ed., New York: McGraw-Hill, 1986.

[102] Chen, K.-M., "A mathematical formulation of the equivalence principle," *IEEE Trans. Microwave Theory and Techniques*, Vol. 37, 1989, pp. 1576–1581.

[103] Balanis, C. A., *Advanced Engineering Electromagnetics*, New York: Wiley, 1989.

[104] Umashankar, K., and A. Taflove, "A novel method to analyze electromagnetic scattering of complex objects," *IEEE Trans. Electromagnetic Compatibility*, Vol. EMC-24, 1982, pp. 397–405.

[105] Taflove, A., and K. Umashankar, "Radar cross section of general three-dimensional scatterers," *IEEE Trans. Electromagnetic Compatibility*, Vol. EMC-25, 1983, pp. 433–440.

[106] Yee, K. S., D. Ingham, and K. Shlager, "Time-domain extrapolation to the far field based on FDTD calculations," *IEEE Trans. Antennas and Propagation*, Vol. 39, 1991, pp. 410–413.

[107] Luebbers, R. J., K. S. Kunz, M. Schneider, and F. Hunsberger, "A finite-difference time-domain near zone to far zone transformation," *IEEE Trans. Antennas and Propagation*, Vol. 39, 1991, pp. 429–433.

[108] Felsen, L. B., and N. Marcuvitz, *Radiation and Scattering of Waves*, Piscataway, NJ: IEEE Press, 1994.

[109] Michalski, K. A., and J. R. Mosig, "Multilayered media Green's functions in integral equation formulations," *IEEE Trans. Antennas and Propagation*, Vol. 45, 1997, pp. 508–519.

[110] Demarest, K., Z. Huang, and R. Plumb, "An FDTD near-to-far-zone transformation for scatterers buried in stratified grounds," *IEEE Trans. Antennas and Propagation*, Vol. 44, 1996, pp. 1150–1157.

[111] Martin, T., and L. Pettersson, "FDTD time domain near-to-far-zone transformation above a lossy dielectric half-space," *Applied Computational Electromagnetics Society J.*, Vol. 16, 2001, pp. 45–52.

[112] Capoglu, I. R., and G. S. Smith, "A direct time-domain FDTD near-field-to-far-field transform in the presence of an infinite grounded dielectric slab," *IEEE Trans. Antennas and Propagation*, Vol. 54, 2006, pp. 3805–3814.

[113] Capoglu, I. R., *Techniques for Handling Multilayered Media in the FDTD Method*, Ph.D. thesis, Georgia Institute of Technology, 2007.

[114] Bochkanov, S., and V. Bystritsky, "Computation of Gauss-Legendre quadrature rule nodes and weights," August 2008. Online: http://www.alglib.net/integral/gq/glegendre.php

[115] Kingslake, R., *Lens Design Fundamentals*, San Diego, CA: Academic Press, 1978.

[116] Barrett, H. H., and K. J. Myers, *Foundations of Image Science*, New York: Wiley-Interscience, 2004.

[117] Flagello, D. G., T. Milster, and A. E. Rosenbluth, "Theory of high-NA imaging in homogeneous thin films," *J. Optical Society of America A*, Vol. 13, 1996, pp. 53–64.

[118] Sheppard, C., "The optics of microscopy," *J. Optics A: Pure and Applied Optics*, Vol. 9, 2007, pp. S1–S6.

[119] Wolf, E., and Y. Li, "Conditions for the validity of the Debye integral-representation of focused fields," *Optics Communications*, Vol. 39, 1981, pp. 205–210.

[120] Li, Y., and E. Wolf, "Focal shift in focused truncated Gaussian beams," *Optics Commun.*, Vol. 42, 1982, pp. 151–156.

[121] Bernard, D. A., and H. P. Urbach, "Thin-film interference effects in photolithography for finite numerical apertures," *J. Optical Society of America A*, Vol. 8, 1991, pp. 123–133.

[122] Török, P., P. Varga, Z. Laczik, and G. R. Booker, "Electromagnetic diffraction of light focused through a planar interface between materials of mismatched refractive indices: an integral representation," *J. Optical Society of America A*, Vol. 12, 1995, pp. 325–332.

[123] Török, P., and P. Varga, "Electromagnetic diffraction of light focused through a stratified medium," *Applied Optics*, Vol. 36, 1997, pp. 2305–2312.

[124] Van de Nes, A. S., L. Billy, S. F. Pereira, and J. J. Braat, "Calculation of the vectorial field distribution in a stratified focal region of a high numerical aperture imaging system," *Optics Express*, Vol. 12, 2004, pp. 1281–1293.

[125] Tang, X. G., F. H. Gao, Y. K. Guo, J. L. Du, S. J. Liu, and F. Gao, "Analysis and simulation of diffractive imaging field in thick film photoresist by using angular spectrum theory," *Optical Commun.*, Vol. 244, 2005, pp. 123–130.

[126] Gamelin, J., R. Guerrieri, and A. R. Neureuther, "Exploration of scattering from topography with massively parallel computers," *J. Vacuum Science Technology B*, Vol. 7, 1989, pp. 1984–1990.

[127] Janssen, A. J. E. M., "Extended Nijboer-Zernike approach for the computation of optical point-spread functions," *J. Optical Society of America A*, Vol. 19, 2002, pp. 849–857.

[128] Braat, J., P. Dirksen, and A. J. E. M. Janssen, "Assessment of an extended Nijboer-Zernike approach for the computation of optical point-spread functions," *J. Optical Society of America A*, Vol. 19, 2002, pp. 858–870.

[129] Braat, J. J., P. Dirksen, A. J. E. M. Janssen, and A. S. Nes, "Extended Nijboer-Zernike representation of the vector field in the focal region of an aberrated high-aperture optical system," *J. Optical Society of America A*, Vol. 20, 2003, pp. 2281–2292.

[130] Braat, J. J., P. Dirksen, A. J. E. M. Janssen, S. v. Haver, and A. S. Nes, "Extended Nijboer-Zernike approach to aberration and birefringence retrieval in a high-numerical-aperture optical system," *J. Optical Society of America A*, Vol. 22, 2005, pp. 2635–2650.

[131] Braat, J. J., S. v. Haver, A. J. E. M. Janssen, and S. F. Pereira, "Image formation in a multilayer using the extended Nijboer-Zernike theory," *J. European Optical Society Rapid Publ.*, Vol. 4, 2009, pp. 09048-1–12.

[132] Capoglu, I. R., "Angora: A free software package for finite-difference time-domain (FDTD) electromagnetic simulation," Online: http://www.angorafdtd.org

[133] Roden, J. A., and S. D. Gedney, "Convolution PML (CPML): An efficient FDTD implementation of the CFD-PML for arbitrary media," *Microwave & Optical Technology Lett.*, Vol. 27, 2000, pp. 334–339.

[134] Liu, Y., X. Li, Y. L. Kim and V. Backman, "Elastic backscattering spectroscopic microscopy," *Optics Lett.*, Vol. 30, 2005, pp. 2445–2447.

[135] Fairchild, M. D., *Color Appearance Models*, New York: Wiley, 2005.

[136] Oppenheim, A. V., A. S. Willsky and S. H. Nawab, *Signals and Systems*, Upper Saddle River, NJ: Prentice Hall, 1997.

[137] Capoglu, I. R., J. D. Rogers, A. Taflove, and V. Backman, "The microscope in a computer: Image synthesis from three-dimensional full-vector solutions of Maxwell's equations at the nanometer scale," *Progress in Optics*, Vol. 57, 2012, pp. 1–91.

Chapter 15

Computational Lithography Using the Finite-Difference Time-Domain Method

Geoffrey W. Burr and Jaione Tirapu Azpiroz

15.1 INTRODUCTION

Over the past four decades, Moore's law has driven the pace of development in the semiconductor industry. This technology trend, first published by Gordon Moore in 1965 [1, 2], calls for the number of transistors per unit-area (and thus the number per unit-cost) to double roughly every 18 months, providing lower-cost-per-function with the introduction of every new technology node. The term *technology node* refers to a distinct generation of integrated circuit scaling, historically quantified by the smallest achievable DRAM metal half-pitch. This label can be used to refer to all the technology (designs, equipment, and lithography) enabling that level of device integration. To establish industry standards, the International Technology Roadmap for Semiconductors (ITRS) [3] is updated roughly every 2 years as consensual guidance for the ever-increasing research and development investments of the semiconductor industry.

This relentless progress through technology nodes has been driven by better designs and better materials, but most importantly, by better optical lithography. Despite mounting challenges, optical lithography continues to enable the steady miniaturization of the billions of transistors (and the meters of microscopic metal wiring) that need to be patterned in order to fabricate modern very large-scale-integration (VLSI) chips. At the time of writing (late 2011), lithographers have already entered the development cycle of the 20-nm technology node (as measured by the DRAM M1 half-pitch in the 2011 ITRS [3]), in preparation for high-volume manufacturing currently planned for 2016.

To fabricate a VLSI chip using optical lithography, the schematic representations of the finalized circuit design must be translated into a set of geometrical shapes to be fabricated on the silicon substrate. These shapes are distributed among a number of discrete layers, which define the transistors on the surface of the silicon wafer and all the metal connections above (or through) the wafer. A reticle or photomask (with patterns inscribed by a high-resolution electron- or laser-beam writer) is repeatedly illuminated so that the desired geometrical shapes can be rapidly transferred into light-sensitive photoresist on the surface of wafer after wafer.

Projection optical lithography, the current workhorse of the semiconductor industry, involves diffraction from a photomask followed by imaging using a highly corrected optical system with a typical demagnification factor of 4×. Initially, patterning of the photoresist was done using *direct-contact lithography* with a 1× magnification. Here a hard mask placed in contact with the photoresist on the wafer surface is exposed to ultraviolet light, transferring the patterns to the resist. Patterning soon moved to use *proximity lithography*, similar to contact lithography except that a small separation is maintained between the mask and the resist. This separation reduces defects and damage to the mask, but at the cost of near-field scattering, degraded resolution, and increased positioning difficulty. In contrast, the sophisticated lens sitting between the photomask and the wafer in a projection lithography system allows high resolution and reduces defects, as well as significantly relaxed mask feature sizes due to the 4× demagnification factor. Combined with a reticle-scanning technique that mitigates problems associated with exposing the full mask [4], these advantages have allowed projection lithography to dominate the history of optical lithography.

15.1.1 Resolution

In a projection optical lithography system (Fig. 15.1), laser light illuminates a mask containing the template of the design to be transferred to the silicon wafer. The mask diffracts the light into a series of spatial frequencies (angles) representing the mask patterns; the projection lens then collects only a portion of these spatial frequencies. Only light propagating with angles θ falling within a cone of radius given by the *numerical aperture* of the lens, $\text{NA} = n_i \sin\theta$, can pass through the lens. Here n_i is the index of refraction through which this light propagates, which on the *image space* side near the wafer may be greater than one. This act of filtering out the highest spatial frequencies (largest angles) constitutes a fundamental limit on the achievable resolution in projection lithography.

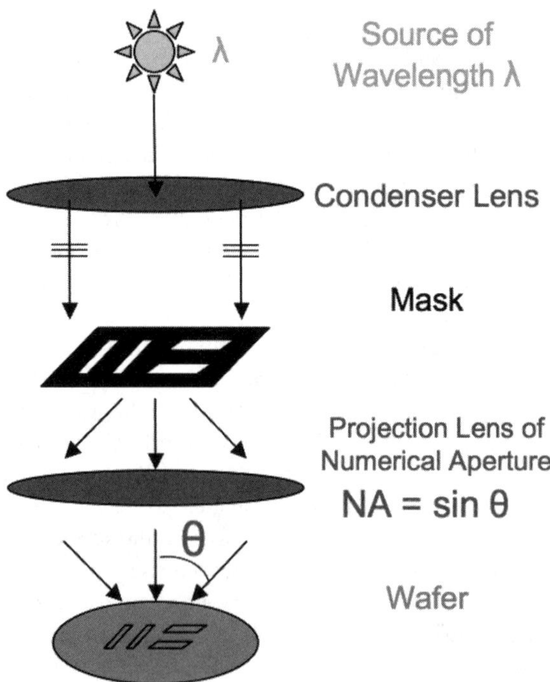

Fig. 15.1 A projection optical lithography system includes a light source, the source optics, a mask, the lithography lens, and a wafer coated with photoresist.

Resolution in optical lithography is defined in terms of the size of the minimum resolvable pitch on the wafer, as limited by the finite numerical aperture. (Quantitatively, resolution follows the inverse of this pitch, so that "high" resolution means many tight-pitch lines per unit distance.) Since an isolated feature produces a continuous spectrum of spatial frequencies, some portion of this broad spray of diffracted light can always pass through the entrance pupil to the image space. This makes it possible to print such isolated features even when they are very small.

In contrast, periodic patterns such as gratings only diffract light into a finite and discrete set of spatial frequencies at intervals $\Delta k = 2\pi / P$, where k denotes the spatial wavevector and P the grating period, also known as the grating pitch. These spatial frequencies, commonly known as the reticle *diffraction orders*, diverge from the object plane at discrete angles given by $\sin\theta_m = m\lambda / P$, with $m = 0, \pm 1, \pm 2, \ldots$. It is the interference between at least two of these diffraction orders that produces the spatial variation of the grating image at the wafer plane. Thus, the grating period P must be large enough to allow at least two diffraction orders to be collected by the lens pupil. As a result, the smallest resolvable grating pitch, $P_{\min}$, is determined by the ratio of the illumination wavelength to the numerical aperture of the projection optics, as:

$$P_{\min} = 2k_1 \frac{\lambda}{\mathrm{NA}} \tag{15.1}$$

This expression is frequently rewritten in terms of the *half-pitch* $F = P_{\min}/2$, since that dictates the smallest feature that can be patterned, be it a narrow metal line or the space between such lines. As a result of this limitation, the minimum area of any device requiring independent electrodes – such as a flash memory cell or any other device fabricated in large arrays at the highest possible device density – is $2F \times 2F = 4F^2$ [5].

Equation (15.1), known as the *Rayleigh resolution criterion* after Lord Rayleigh's original explorations on resolution in the 1880s [6–8], is probably the most important equation in lithography. It represents the minimum half-pitch printable with a system wavelength λ and numerical aperture NA, as scaled by the k_1 factor quantifying the complexity of the process (in Rayleigh's original derivations, $k_1 = 0.61$). This expression also describes how the minimum printable feature can be decreased by either moving to a shorter wavelength λ of light, increasing the numerical aperture NA of the optical system, or reducing the k_1 factor of the lithographic process.

The trend of higher density through gradually shorter wavelength and increased NA started early, with the light source being first the g-line (436 nm, 1970) and then the i-line (365 nm, 1984) of the mercury lamp. After the advent of the KrF excimer laser (248 nm, 1989) and then the ArF excimer laser (193 nm, ~2003), this highly successful trend was expected to continue on to the F_2 excimer laser (157 nm), followed by extreme ultraviolet (EUV at 13.5 nm). However, the development of 157-nm lithography (vacuum ultraviolet) was abandoned in 2003 due to challenges with the quality of the lens, pellicle, and resist materials [9].

Instead, immersion lithography was introduced at the same 193-nm wavelength (*deep ultraviolet*, DUV), with the space between the last lens element and the wafer filled by water — a fluid with an index of refraction significantly higher than air. This development allowed optical lithography to continue shrinking feature sizes, not by shrinking illumination wavelength, but instead by increasing the maximum angle that could reach the wafer. However, 193-nm immersion lithography cannot move to another immersion fluid better than water, due to difficulties in synthesizing suitable higher-index materials for both the lens and the immersion fluid [10, 11]. This limits the maximum available numerical aperture to 1.35.

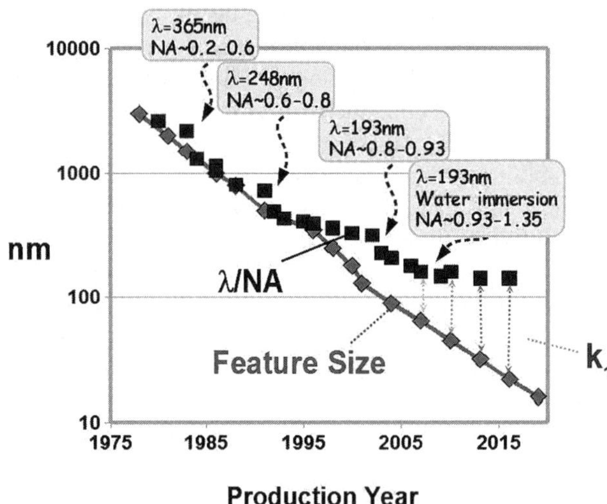

Fig. 15.2 Semiconductor technology scaling as measured by the technology node minimum feature size, compared to scaling of the fundamental optical resolution of lithographic scanners [3, 12].

As a result, the long-running and steady decline of the λ/NA ratio has come to a halt, at least temporarily, as shown in Fig. 15.2 [3, 12]. Moreover, while EUV technology may someday provide a much-needed wavelength leap down to 13.5 nm, repeated delays in its development are forcing lithographers to push hard to maximize the capabilities of 193-nm immersion lithography to print deeply sub-wavelength features. This gap between the improved resolution required by Moore's law and the stagnated optical resolution of 193-nm immersion has posed high demands on the lithography process. Yet, the gap is being bridged by steadily pushing the k_1 factor down to historical minima, principally through the development and implementation of innovative techniques of resolution enhancement. These accomplishments mark the advent of *computational lithography* — a new scientific discipline aimed at modeling and enhancing the resolution and overall performance of the lithography process.

15.1.2 Resolution Enhancement

Smaller values of the k_1 factor, demanded by the aggressive use of 193-nm lithography at 1.35 NA to print wafer features as small as 40-nm half-pitch, translate into higher complexity in the lithographic patterning process. As a result, the smallest geometrical shapes that can be patterned are actually significantly smaller than the wavelength of the light used. This achievement has been engineered through the development of a number of ingenious techniques referred to collectively as *resolution enhancement techniques* (RETs) [13–15].

In most resolution enhancement techniques, the photomask is not necessarily inscribed with the targeted geometrical shapes themselves, as seen in Fig. 15.3(a). Instead, the patterns that are inscribed are slightly distorted versions of these shapes. These distortions have been carefully optimized in order to robustly produce the desired shapes at the wafer surface, within the optical image of the reticle delivered by the sophisticated, but ultimately imperfect optical system. RETs work by precompensating for known and systematic distortions induced by the lithography process.

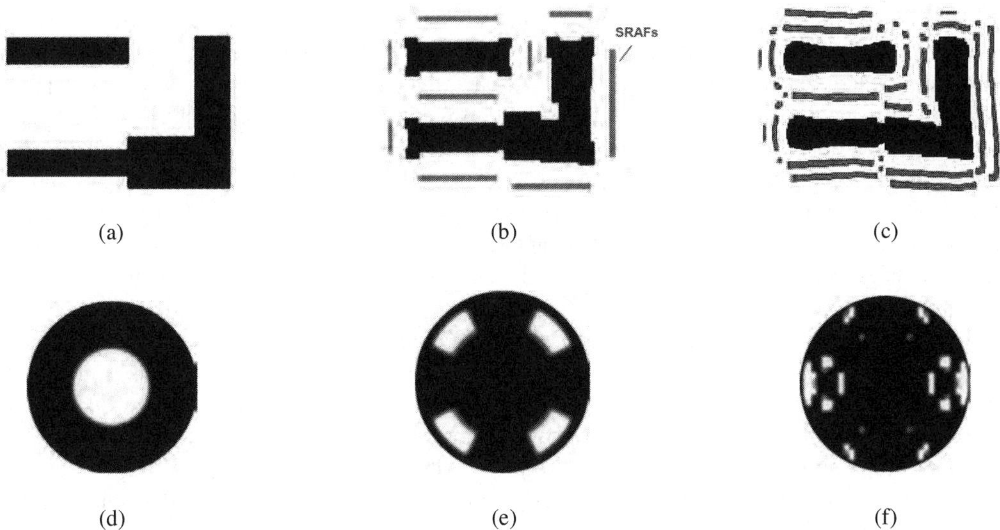

(a) (b) (c)

(d) (e) (f)

Fig. 15.3 Evolution of lithography masks and sources: (a) Mask layout without OPC; (b) mask layout with OPC and sub-resolution assist features (SRAFs); (c) mask layout after mask optimization; (d) conventional illumination source; (e) parametric "Quasar" source; (f) free-form programmable source.

For instance, optical proximity correction (OPC) uses iterative simulation of the aerial image to precompensate the mask shapes for local deviations from the desired image pattern induced by the filtering of the higher spatial frequencies by the diffraction-limited optics [16–18]. For instance, line ends and outsides of bends may be systematically emphasized, while insides of bends are de-emphasized to avoid the "rounding" that would otherwise occur at these spots.

Another resolution enhancement technique involves incorporating sub-resolution assist features (SRAFs), as shown in Fig. 15.3(b). These are added to the mask to ensure that nearby features print correctly even when the wafer is slightly out of focus. Because sparse isolated features are created by the interference of numerous diffraction orders, such features are typically more sensitive to focus variations than more densely populated regions. An artificially dense diffractive environment can be created by adding SRAFs to the mask in the vicinity of such isolated features. The location and size of these features must be carefully optimized to enhance performance of the main feature, yet ensure that any SRAF itself does not print into the resist.

In addition to this precompensation of the mask shapes, the shape of the illuminating source can be optimized to improve aerial image quality and process robustness across the wide variety of patterns to be printed by the same reticle. Historically, this optimization was performed across a small discrete set of parametrically defined shapes, as illustrated in Figs. 15.3(d) and 15.3(e). More recently, however, algorithms for the global optimization of both mask shapes and illumination source configurations have been developed [15, 19–22]. The implementation of these techniques of *source mask optimization* (SMO) have been enabled both by advances in mask fabrication technology and by the development of free-form programmable illuminators [23], as illustrated in Figs. 15.3(c) and 15.3(f), respectively. Higher contrast can also be achieved by engineering the mask to introduce phase shifts in the transmitted fields. Shifts of 180° between adjacent shapes induce interference and can decrease the minimum printable pitch by as much as $2\times$.

Finally, while the theoretical lower limit with a single exposure is $k_1 = 0.25$, the use of multiple exposures can reduce the effective k_1 factor below this value. Lithographically printing "isolated" small features at a large pitch is easier than printing "densely" at a small pitch. Thus, double-patterning—in which staggered small features are patterned using two successive 193-nm immersion exposures—is now in volume manufacturing, with triple- and perhaps even quadruple-patterning to follow soon afterwards. Despite the delayed arrival of EUVL, its shorter wavelength should eliminate the need for multiple exposures, at least in its first generation. However, EUVL will need to rely on OPC and accurate simulation in order to correct for its own idiosyncrasies (Section 15.6), and to deliver feature sizes that are much smaller than what was originally planned for its debut.

Optimization of such resolution enhancement techniques inherently involves the use of intensive and increasingly more complex simulations of the imaging process. These simulations prominently feature the computation of electromagnetic fields at various stages: illumination and diffraction from the photomask, propagation through the imaging system, and exposure of the photoresist and wafer topography. In the case of immersion lithography, deep ultraviolet light at a wavelength of 193 nm propagates through a transmissive mask and a sophisticated lithography lens system. In the case of extreme ultraviolet (EUV) lithography, both the flat photomask and the curved imaging mirrors rely on precision multi-layer dielectric stacks in order to efficiently reflect the 13.5-nm wavelength light. In either case, precompensation of entire chip layouts requires iterative simulation of the lithography process on millions of patterns. Fortunately, many of the computations needed to simulate these optical processes can be greatly simplified for propagation angles close to normal incidence. Thus reticle patterns containing billions of features can actually be completely evaluated in a reasonable time period. This new model paradigm blends both accuracy and speed, with simple but efficient models that are developed using rigorous simulations on small sets of patterns, followed by calibration to actual wafer line-width measurements.

As dimensions shrink and technology complexity rises, however, geometrical approximations of the underlying electromagnetic field interactions can become increasingly inaccurate. Accurate simulation of both immersion and EUV systems can be expected to be an increasingly critical component of semiconductor manufacturing for the foreseeable future in order to develop, calibrate, and benchmark better (yet still fast) approximation methods. While these rigorous computations of the exact three-dimensional (3-D) electromagnetic fields can be much slower than the approximate methods, rapid turnaround is still crucial. The finite-difference time-domain (FDTD) method offers advantages such as flexibility, speed, accuracy, parallelized computation, and the ability to simulate a wide variety of materials, making it a popular choice for numerical electromagnetic computations in optical lithography simulation software. In this chapter, we review the fundamental physical concepts and numerical considerations required to perform such electromagnetic computations for optical lithography in the context of semiconductor manufacturing.

15.2 PROJECTION LITHOGRAPHY

In a projection optical lithography system (Fig. 15.1), laser light illuminates a mask containing the template of the design to be transferred to the silicon wafer. The mask diffracts the light into a series of spatial frequencies (angles) representing the mask patterns. The projection lens then collects these spatial frequencies, but only over a finite range of angles close to the optical axis.

A 4× demagnified image is focused by the lens onto the wafer. (In an immersion lithography system, the gap between the last lens element and the wafer is filled with water rather than air.) The wafer has been precoated with a layer of *photoresist*, typically an organic polymer, which undergoes irreversible photochemical reactions when exposed to light. Chemical amplification during the exposure process causes the optical response of the photoresist to be highly nonlinear around a particular exposure threshold, converting the continuously varying incoming optical image into a sharp, binary pattern of exposed and unexposed material.

Afterwards, development of the positive (negative) resist removes the material that was sufficiently exposed (was not sufficiently exposed) to the incoming optical image. If all has gone well, the desired circuit layout has been replicated in the developed resist film, and the patterned wafer can now undergo an etch or ion-implantation operation. This patterning cycle is repeated as many as 30 or 40 times during fabrication of a chip, in order to define all the different layers that comprise an integrated circuit. These layers range from the ultra-small features of the transistors formed in the front end of the line (FEOL) on the surface of the silicon wafer, to the much larger wiring many layers above the silicon, patterned with older, lower-resolution tools at the far back end of the line (BEOL). Above the transistors, the fabrication process tends to alternate between wiring layers containing long trenches and *contact layers* containing many small holes. These *contact* holes are needed to form the metal vias that vertically connect each pair of wiring layers. For trenches, either positive or negative photoresist can be used; however, for contact masks, bright holes on the reticle and positive-tone photoresist are preferred. A modern lithography scanner exposes several thousand 300-mm-diameter wafers per day, every day.

In the following sub-sections, we discuss the key modules in a projection optical lithography system: the light source, the photomask, the lithography lens, the wafer, and the photoresist. We then discuss the effects of partial coherence and interference and polarization.

15.2.1 Light Source

The illumination system supplies a highly monochromatic beam of light of high and uniform intensity. Monochromatic light is important because refractive lenses can be designed to be nearly free of aberrations, but only over a very narrow bandwidth of wavelengths. Moreover, intense illumination delivers the threshold exposure dose each wafer must receive rapidly, leading to high throughput in terms of wafers exposed per hour. Thus, light sources for lithography must be very intense and have a very narrow bandwidth, which reduces the number of possibilities to just a handful of wavelengths, as described in the introduction.

The light from a lithographic illumination source is delivered to the mask by condenser optics, which must provide uniform illumination to all portions of the mask, perform various forms of spectral filtering, and control the spatial coherence properties of the source. In Köhler illumination [24], illustrated in Fig. 15.4, the mask sits at the exit pupil of the condenser optics, with the illumination source imaged into the entrance pupil of the main lithography lens located downstream from the mask. This configuration predominates in lithography because it can provide uniform illumination from a non-uniform source, provided that well-corrected condenser lenses are employed [25]. Since any given point within the extended source illuminates every portion of the mask, Köhler illumination delivers uniform intensity across the mask [26]. Although the light from this particular source point is slightly converging as it passes through the mask on the way to its image plane downstream, the radius of curvature is sufficiently large that the local wavefront at the mask can effectively be considered as a spatially coherent plane wave.

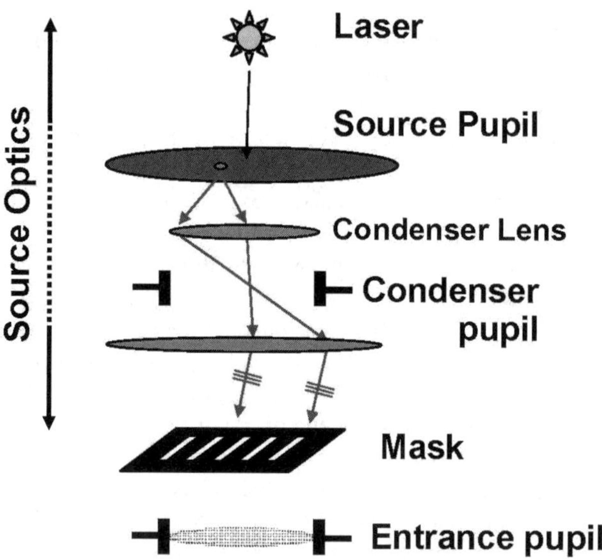

Fig. 15.4 In Köhler illumination, the source is imaged into the entrance pupil of the projection optics, with the mask located at the exit pupil of the condenser optics. Each source point generates a converging spherical wavefront of large radius of curvature at the mask plane, hence providing a uniform and nearly plane wavefront impinging on the mask at an angle determined by the position of the source point on the source pupil.

Since this plane wave is diffracted by the mask into multiple diffraction orders, which then propagate through the lithography lens system and arrive at the wafer, the original spatial coherence remains unaltered and interference occurs. However, other plane waves arriving at the mask from other portions of the source are spatially incoherent with respect to our plane wave and to each other (mutual incoherence). When these plane waves overlap on the mask or on the wafer, no stable interference pattern is produced. Perfectly coherent illumination of the photomask would not only be difficult to achieve in practice, but would also limit the resolution of features in the design with high spatial frequency content and lead to undesirable long-range interference effects (speckle).

In practice, the illumination system comprises an extended but finite area of spatially incoherent source points. Each point in this extended source is associated with a coherent, linearly polarized plane wave impinging on the mask at an oblique angle. We can represent this angle at the mask in terms of a source-point location relative to the optical axis on a *source pupil* [24], as if the source sat at the input Fourier transform plane of a perfect thin lens located one focal length upstream from the mask. For a lithographer, *coherent illumination* means that the source pupil contains only a single, infinitely small source point. When this source point is located at the center of the source pupil as illustrated in Fig. 15.5(a), the associated plane wave illuminating the mask arrives at normal incidence. (Since the illumination is really a gently converging spherical wave, the precise local incidence angle of this particular "plane wave" will vary slightly away from normal incidence for features located away from the center of the mask. However, this subtlety is typically ignored, even in full-bore computational lithography.)

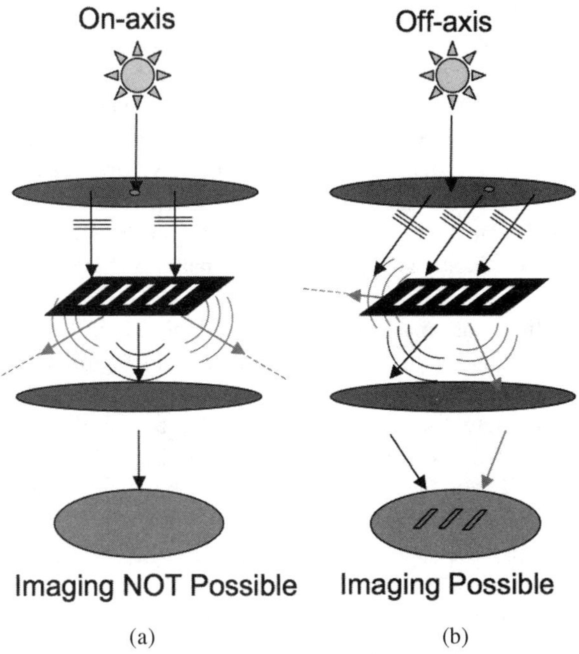

Fig. 15.5 Examples of (a) on-axis illumination and (b) off-axis illumination.

Since source points at off-center or "off-axis" positions illuminate the mask at an oblique angle as illustrated in Fig. 15.5(b), control over the "spatial coherence" of the source pupil provides an opportunity to transfer more information from the mask to the wafer than would otherwise be possible. For instance, under normally incident illumination, as the pitch of a simple grating decreases, then the diffraction angles of the first-order diffraction increase steadily and symmetrically until both the +1 and −1 orders fall outside the entrance pupil of the lithography lens, as in Fig. 15.5(a). At this point, only the zero-order light arrives at the wafer, so no interference pattern is formed and no information from the grating arrives at the wafer. Hence, the patterning of this grating pattern fails.

In contrast, the diffraction pattern from the mask created by an off-axis source point also propagates at an oblique angle toward the entrance pupil of the projection optics. We can choose an incidence angle that is about half as large as the diffraction angle of the tight-pitch grating, which we just failed to print at normal incidence. Now both the 0 (undiffracted) and −1 diffraction orders can pass through the lithography lens, allowing the same grating to print, simply by using an appropriate off-axis illumination, as in Fig. 15.5(b). This increase in the number of high-order diffracted beams that can be collected and imaged onto the wafer plane by the projection optics can be used to directly improve resolution of fine features on the mask.

15.2.2 Photomask

In projection lithography, the photomask, also known as the *reticle*, modulates the amplitude and phase of the incident illumination. In its simplest form, the mask contains an exact (but scaled 4× larger in size) template of the layout to be printed on the wafer, and the lithography lens simply images this onto the wafer surface. However, in order to push performance toward the physical limits of the system, the imaging process is more appropriately considered to be the diffraction by the mask of the set of illuminating plane waves, followed by the imaging of the limited range of spatial frequencies that reach the entrance pupil of the lithography lens into the thick photoresist stack sitting above the wafer. In this context, the most robust printing of a layout may necessitate mask patterns that differ noticeably (or sometimes, differ completely) from the intended layout. In almost all cases, lithography mask patterns tend to be *Manhattanized layouts*, in which even very complex polygons have all their edges oriented along one of two orthogonal directions. The simplest mask patterns therefore resemble collections of similarly oriented rectangles.

Photomasks are written onto mask blanks built on high-quality quartz templates. The most common size for reticles today is the so-called 6" (0.25"-thick) format, having the dimensions $152 \times 152 \times 6.35$ mm. The upstream face of the photomask has an anti-reflection coating to minimize reflection back into the source and stray light (flare) throughout the lithography system. The downstream face of a not-yet-patterned mask blank contains a film stack that is typically 50 to 100 nm thick, with shading layers to provide (any) absorbing properties, on top of an anti-reflection coating layer to reduce reflection of light back into the substrate. Careful tuning of the thicknesses and composition of the film stack provides tight control over the possible amplitude and phase changes of the transmitted fields.

Because the reticle carries a pattern that will be transferred to thousand of wafers, its quality and durability are of critical importance. The surface of a working reticle is typically protected by an ultra-thin pellicle (a suspended membrane that protects each reticle from damage), sitting far enough downstream from the mask surface that it is sufficiently out of focus. Mask patterns are etched onto the reticle either by a laser writer or by electron-beam lithography, both capable of better resolution than optical lithography albeit with orders of magnitude lower throughput. Since mask patterns will be de-magnified when projected onto the wafer, they are written at an enlarged scale and then inspected carefully. Continuity of mask inspection and repair capabilities is an important consideration when a change to a new and improved mask blank is being considered.

Several advanced mask types are shown in Fig. 15.6. If the film stack has <0.1% transmission (e.g., an optical density > 3.0), then the mask is called *binary*. An ideal binary mask, as in Fig. 15.6(a), offers a highly opaque absorber of nearly 0% transmission and 100% transmission through the mask etched openings, so that the incident illumination is modulated with a scalar binary function. Conventionally, such masks are formed by a thick (60–100 nm) layer of chrome. Although alternative materials are also available today [27, 28], the basic mask materials are shown in Table 15.1 [29]. Generally the goal with a binary mask is to end up with an absorber layer that is both very thin—to minimize "thick-mask" interactions at the edges of etched openings—while also being very opaque. This combination requires an extremely high absorption coefficient α (absorption per unit thickness).

Extensions of binary masks with more complex film stacks, materials, and mask profiles have been progressively explored and occasionally adopted to enhance resolution over the last two decades [30–34]. Phase-shifting masks (PSM) like those shown in Figs. 15.6(b) – 15.6(f) modulate the phase of the incident electromagnetic field and thus can improve image contrast by inducing destructive interference of the fields if the phase shifts are close to 180°. In many cases, both amplitude and phase are modulated. The most common versions of phase-shifting masks are the alternating PSM (Alt.–PSM) shown in Fig. 15.6(b) and the attenuated PSM (Atten.–PSM) shown in Fig. 15.6(c).

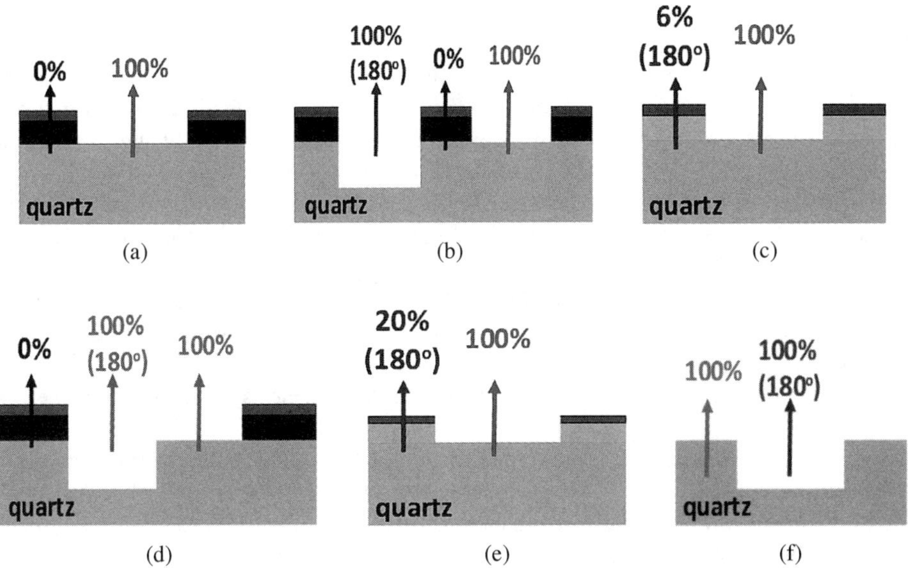

Fig. 15.6 Photolithography masks are built onto a transparent quartz substrate and may comprise a few layers of varying thickness and composition to produce the desired transmission and phase. Commonly used mask types in optical lithography include the (a) binary mask; (b) alternating phase-shifting mask; (c) attenuated phase-shifting mask; (d) Tritone mask; (e) high-transmission phase-shifting mask; and (f) chromeless mask.

TABLE 15.1

Common Material Constants at 193-nm Wavelength for Typical DUV Mask Types [27, 29, 35]

Mask	Material	n	α	Thickness
Binary	Chrome	0.842	1.647	70–110 nm
6% Atten. PSM	MoSi	2.26	0.575	70 nm
Quartz	SiO2	1.563	0.000	–

Mark Levenson's original alternating PSM concept [31], illustrated in Fig. 15.6(b), introduces a 180° phase difference between different successive apertures on the mask by etching the glass behind a defined subset of the openings to a depth equivalent to 180°:

$$d_{180^\circ} = \frac{\lambda}{2\left[n_{\text{glass}}(\lambda) - n_{\text{air}}(\lambda)\right]} \tag{15.2}$$

Since the index of refraction of quartz glass at DUV wavelengths is about $n_{\text{glass}} = 1.56$ as shown in Table 15.1, and that of air is $n_{\text{air}} = 1$, the required etch depths are on the order of the wavelength. On an Alt.–PSM mask, etched and unetched mask apertures must be separated by sections of opaque absorber. In contrast, in a Tritone mask, illustrated in Fig. 15.6(d), areas with different phase shifts need not be separated by sections of absorber material.

Another popular phase-shifting mask is the Attenuated PSM (Atten.–PSM) [30, 36], as illustrated in Fig. 15.6(c). Common Atten.–PSM masks are comprised of a layer of molybdenum silicon (MoSi) material approximately 70 nm thick with the material optical constants of Table 15.1. Here the absorber layer is designed to offer both partial transmission and a phase difference of 180° with respect to the transmission through the clear openings. This simplifies fabrication since there is no need to carefully control the etch depth across the reticle, yet destructive interference between fields still enhances the image contrast at contour boundaries. The original Atten.–PSM mask of Fig. 15.6(c) offered a transmission of only 6% to 7% with a phase shift of 180°. However, newer high-transmission variants, illustrated in Fig. 15.6(e), can offer the same phase shift with transmission ranging between 15% and 40%, thereby increasing the amount of light reaching the wafer and helping to maintain wafer throughput [35, 37].

Finally, the chromeless phase-shift lithography (CPL) mask is represented in Fig. 15.6(f). This mask does not employ any kind of absorbing material on the quartz substrate. Instead, diffraction is solely due to phase modulation of the incident illumination, achieved with selective etching into the quartz. A particular issue in CPL masks involves the vast regions of the reticle where the wafer should remain unexposed. For instance, sometimes a mask may contain a relatively small number of sparsely separated target features; yet locations far away from these spots must still remain dark at the wafer to avoid being inadvertently exposed. In a CPL mask, such regions of the reticle must be filled with small-pitch gratings in order to deflect almost all the incident light outside the entrance pupil of the lithography lens. Such gratings require careful attention to both etch depth and grating duty cycle (ratio of the etched and unetched portions of each small-pitch grating cycle), in order that the "dark" regions of the wafer do in fact stay dark. In this context, the absorber available in a Tritone mask becomes a highly attractive feature, because it provides many of the benefits of CPL while using absorption rather than high-spatial-frequency gratings for the dark regions.

Masks capable of modulating both the amplitude and the phase of the incident illumination such as a 6.5% Atten.–PSM [30] can provide enhanced contrast in the focused image relative to purely binary masks. Hard phase-shift masks such as Alt.–PSM [31] or chromeless masks are capable of providing much higher contrast and even frequency doubling of the image pitch. Electromagnetic scattering on the mask topography, however, can undermine some of the benefits provided by phase-shifting masks by inducing amplitude losses and shifts on the phase of the transmitted fields. In general, as topography gets thicker (as in masks with deep etches and thick absorbing layers), accurate first-principles computation of the electromagnetic fields transmitted through the reticle (as offered by FDTD) becomes more important for accurate results. This becomes especially true as the performance of the lithography system is pushed toward the fundamental physical limits. In contrast, advanced binary masks with thin absorbing layers and no etched glass have less mask topography and can still rely on approximations of the field transmission. Thus, the use of thin masks allows designers to obtain a significant fraction of the available performance while still avoiding computation of the exact electromagnetic field effects [28, 38].

However, since the maximum system performance offered by phase-and-amplitude masks exceeds that available from amplitude-only masks, one can expect that the careful evaluation of electromagnetic field effects (followed by appropriate approximate prediction) will only become more critical as the semiconductor industry continues to rely on the same 193-nm immersion lithography.

15.2.3 Lithography Lens

Both the large size and enormous cost of a modern lithography *stepper* are dominated by the lithographic imaging system. Stepper lenses can be up to 1 meter long and weigh up to 500 kg, and are firmly held in a concentric shaft inside the tool. The entire system consists of a complex arrangement of 25 to 40 glass elements which must provide a 4× reduction factor with nearly diffraction-limited imaging characteristics at the illumination wavelength. (Reduction factors of more than 4× would offer less sensitivity to small mask defects, but would also shrink the illuminated field size on the wafer and the maximum possible chip size, typically about 25 × 25 mm.) Very high-quality fused silica and strict lens design, fabrication, and assembly are required, thereby greatly increasing the cost of both the lens and the encompassing lithography tool. However, one of the advantages of such a near-perfect lens system is that we can model the propagation of light through the lenses using simple geometrical optics.

The term *stepper* refers to the action of exposing the reticle onto a small portion of the wafer, then stepping laterally by slightly more than the width of the reticle to expose the next portion of the wafer. This process is repeated until the entire wafer is patterned. A large number of semiconductor *die* are patterned on a typical 300-mm wafer. In general, a reticle can pattern more than one die (or even more than one kind of die) simultaneously, but a die generally cannot be larger than a single reticle.

To further reduce the maximum ray angle within the lens, the most recent lithography tools are *scanners* rather than steppers, wherein the illumination pattern spans only a one-dimensional slice across the width of the reticle. The reticle is then rapidly scanned along a direction orthogonal to this slice so that all portions of the reticle are illuminated. Simultaneously, on the other side of the lithography lens, the wafer is scanned in the opposite direction (because the lens flips the reticle image on the wafer) at precisely 4× the velocity. The advantage here is that aberrations now need only be corrected for the edge rather than the corner of the reticle. This decrease in ray distance from the optical axis by a factor of $\sqrt{2}$ can be well worth the additional engineering to coordinate the stages moving the reticle and wafer. As with steppers, once a single copy of the reticle is scanned onto a small portion of the wafer, the stages move the wafer laterally in preparation for the scanned exposure of the next die.

An important measure of the light-gathering power of a lens is its numerical aperture (NA), defined as:

$$\text{NA} = n \sin\theta \tag{15.3}$$

Here, as illustrated in Fig. 15.1, θ is the maximum ray angle leaving (entering) the lens, while n is the refractive index of the medium behind (in front of) the lens. The entrance side of the lens is referred to as object space, and the exit side is called image space. The numerical apertures in these two spaces are related through the magnification of the system, M, as:

$$\text{M} = \frac{\text{NA}_\text{o}}{\text{NA}_\text{i}} \tag{15.4}$$

A system with a typical demagnification factor of 4× thus has M = 1/4 .

In the image space, optomechanical design limits the maximum incidence angle relative to the optical axis to an upper limit of approximately 70°. This restricts the maximum NA available to roughly 0.93× the refractive index of the coupling medium. Despite this constraint, values of NA_i as high as 1.35 are now common in immersion lithography scanners, because water fills the space between the last lens surface and the resist surface. However, since fluids with a higher index than water (for which $n = 1.436$ at 193 nm) tend to be difficult to handle and purify, or are highly toxic to humans, it appears that immersion lithography will not evolve much farther than the present NA of 1.35.

15.2.4 Wafer

The lithographic lens delivers a focused aerial image of the reticle to a 2-D plane orthogonal to the optical axis, downstream from the last lens element. To print a focused rather than blurred image, the photoresist layer at the surface of the wafer must be at or near this plane. The range around the plane of best focus over which the image remains adequately sharp defines the *depth of focus* (DoF). Rayleigh's derivation of the DoF criterion at the resolution limit of (15.1) is given by [24]:

$$\mathrm{DoF} = \pm k_2 \frac{\lambda}{\mathrm{NA}^2} \tag{15.5}$$

where k_2 represents a second characteristic constant of the complexity of the lithography process. As the numerical aperture has steadily increased to shrink the size of features we can print, (15.5) shows that this has greatly reduced the depth of focus. Thus advanced lithography has succeeded to date only by maintaining tight control of the position of the wafer within the scanner. This requirement poses a critical challenge to the future ability to print resist features within the required tolerance constraints and hence yield functional chips.

The derivation of (15.5) involves a number of approximations that make it only a rough measure of the depth of focus available to print some of today's most critical chip levels. Accurate evaluation requires thorough characterization of the process through simulation and empirical wafer data. Values of DoF encountered in 193-nm immersion lithography are typically ±0.2 μm for the more critical levels in the 32-nm technology node [39]. However, the maximum tolerable DoF continues to shrink as performance is pushed from the 32-nm to 22-nm technology node and beyond, even though exactly the same imaging optics are being used. Since the photoresist itself has a finite thickness larger than the DoF (0.2–0.8 μm), the position of the plane of best focus must be controlled accurately to provide good image quality throughout the resist thickness.

15.2.5 Photoresist

Photoresists are available in two main categories: positive or negative. Positive resists become soluble in developer solution upon exposure to light, while negative resists lose their solubility in those areas exposed to light. This means that in order to produce contact holes on positive resists, the mask feature is simply an aperture in the absorbing layer (such as chrome). After chemical and thermal *development* of the exposed resist, a small hole in the photoresist is surrounded by cross-linked photoresist, allowing an etch, deposition, or implant to affect only that small portion of the wafer. On the other hand, to produce the same contact hole with a negative resist, one would need a mask consisting of an opaque chrome feature surrounded by glass. Since the latter mask is affected by diffraction much more strongly than the first, contact layers are almost always patterned with positive resist.

By the time it arrives at the photoresist sitting above the wafer, the incoming optical image has lost all of its high-frequency components, which were blocked by the limited numerical aperture of the lens. For example, given a series of densely packed parallel lines (a tight-pitch grating) on the reticle, the arriving optical pattern resembles nothing more than a 1-D sinusoid at this tight pitch. One of the unsung heroes of Moore's law is the way that photoresists restore the original sharp ON-and-OFF pattern of the reticle by *thresholding* the smoothly varying image into distinct topography.

Thus photoresists with a nearly stepwise solubility response to the intensity of light are essential. Due to their high sensitivity, chemically amplified resists have dominated the semiconductor industry for more than 20 years [40, 41]. Any energy deposited by the optical image generates photoacid, a catalyst. Since the presence of photoacid drives the generation of more photoacid, this process is highly nonlinear around a particular exposure threshold. In the portions of our sinusoid where the total optical exposure (the product of local optical irradiance, or energy per unit time per unit area, and the time duration of the exposure) exceeds this threshold, a large amount of photoacid is generated. In nearby regions where this threshold is not reached, the amount of photoacid remains small. Post-exposure baking drives de-blocking reactions in the polymer region with this catalyst, allowing the exposed resist to be dissolved away by developer (*developed*). Diffusion of photoacid during this baking process is one factor that can reduce resolution at advanced technology nodes, and thus is an important consideration within ongoing photoresist research.

Features to be printed by a lithography mask are often quantified by one or more *critical dimensions* (CDs), or feature widths that are particularly important for the subsequent electrical characteristics of the resulting semiconductor devices. For each of these CDs, there is an acceptable tolerance range around the nominal or target width. *Process window* is a term that refers to the variations that can be allowed in focus position and exposure dose while still printing within these acceptable CD tolerance ranges. In general, the goal of lithographers is to make the process window as large as possible, so that printing of the right-size features can still occur even when the wafer is slightly out of focus and the exposure dose is slightly imperfect.

Other important metrics for computational lithography include *line-edge roughness* and *mask error enhancement factor*, also known as *MEEF* [42, 43]. While nontrivial to quantify precisely [44, 45], line-edge roughness means exactly what it says: the extent to which the edges of lines are rough instead of straight. These small but rapid fluctuations of the edge position along the length of a resist line are caused by random variations in the optical and chemical processes and are often modeled stochastically.

MEEF refers to the impact that variations of the mask linewidth $\partial(LW_{\mathrm{m}})$ have on errors in the wafer-resist linewidth $\partial(LW_{\mathrm{w}})$, scaled by the magnification M (usually M = 1/4):

$$\mathrm{MEEF} = \frac{1}{\mathrm{M}}\frac{\partial(LW_{\mathrm{w}})}{\partial(LW_{\mathrm{m}})} \tag{15.6}$$

MEEF is a dimensionless number representing the multiplicative effect that errors on the mask have on the wafer due to both the optics and the photoresist. For instance, if the mask linewidth has a error of 12 nm and the MEEF is 2.5×, then the error at the wafer linewidth is 7.5 nm. A perfect imaging system, if available, would deliver an exact, 4× demagnified image of this slightly flawed mask to the wafer, yielding only a 3-nm error and hence showing an ideal value of MEEF equal to unity. Instead, the imperfect imaging system is tuned for performance in such a way that the slight mask error induces an actual wafer error that is 2.5× larger. Large MEEF values are a consequence of lithography processes that are tuned for low k_1 factor.

15.2.6 Partial Coherence

Following Section 15.2.3, the amount of partial coherence in a lithographic system is controlled by the maximum extent of the illuminating source, as quantified by the *partial coherence factor* σ. Lithographic convention is to normalize the value of σ to the object space NA_o, as in:

$$\sigma = \frac{NA_c}{NA_o} \tag{15.7}$$

where NA_c is the numerical aperture of the condenser optics. Typically, σ ranges between zero and one. Absent diffraction by the mask, light from the extreme periphery of the source pupil ($\sigma \sim 1$) just barely passes into (the opposite side of) the entrance pupil of the main lithography lens. Any larger values of σ would lead to a significant drop in the amount of optical power that actually makes it into the main lens.

Partial coherence is also an avenue for decreasing k_1 and thus improving resolution [13, 24, 46], with the minimum resolvable grating half-pitch potentially reduced by a factor of $1/(1+\sigma)$. For instance, when $\sigma \sim 0$, the only illumination is on-axis as shown in Fig. 15.5(a), and the smallest resolvable pitch results in ±1 diffraction orders that just enter the outer edges of the entrance pupil of the lens. In contrast, the maximum value of $\sigma = 1$ includes strongly off-axis illumination from source points at the extreme edge of the source pupil. This particular illumination produces a 0 diffraction order that enters the entrance pupil at its very outer edge as shown in Fig. 15.5(b). However, for the same grating just discussed, both the associated −1 and −2 diffraction orders would now be entering the lens as well. For instance, the −1 order could be entering along a direction near the optical axis and the −2 order could enter the far side of the entrance pupil. In fact, we can further decrease the pitch by a factor of 2× and still have both 0 and −1 diffraction orders entering into the entrance pupil, passing through the lens, and interfering at the wafer. Thus, by allowing maximal use of the entire source aperture, partial coherence leads to the minimum theoretically possible resolution with a single exposure, $k_1 = 0.25$. In practice, when k_1 is smaller than about 0.3, it is difficult to maintain a manufacturable single-exposure process with sufficient process window across all patterns comprising the layout.

Values of k_1 lower than 0.25 can be achieved through the use of multiple-patterning or multiple-exposure methods [47]. In a double-patterning approach, the wafer is first coated with a resist layer and exposed with the contents of one mask. Then, the resist is developed to reveal the printed patterns, which are next coated with a new layer of resist. Subsequently, this new resist layer is exposed with the contents of a second mask using a layout complementary to the first exposure. As a result, the density of the final printed layout on the wafer is increased. In a double-exposure approach, one single resist layer is successively exposed with the contents of two masks with no intermediate development step.

Such techniques add complexity to the design process, increase the number of costly masks and process steps, and mandate even tighter controls in the manufacturing process. In particular, tight *overlay* — the ability to laterally align the reticle in an exposure step to the pattern printed during a previous exposure — is critical for all multiple-patterning and multiple-exposure techniques. Despite these difficulties, these techniques have become essential tools for overcoming the scaling limitations inherent in conventional single-exposure imaging, enabling 193-nm immersion lithography to deliver effective k_1 factors as low as 0.19 [48].

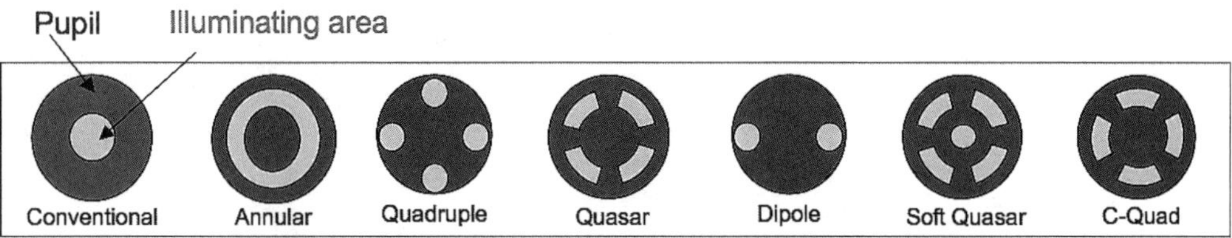

Fig. 15.7 Common illumination source configurations used in advanced lithography.

Of course, even the simplest lithographic reticles have significantly more complexity than a simple grating. As briefly mentioned in Section 15.1.2, the choice of the most appropriate source pupil to use for a given mask is an important design decision involving a significant amount of computation. Figure 15.7 shows a few of the illumination source configurations commonly used in advanced lithography. Dense periodic patterns benefit from large values of σ, while small values of σ provide better image quality with sparse or nearly isolated features [13, 49]. Large partial-coherent factors also help reduce proximity effects by collecting more orders through the lens, although at the expense of image contrast (because the source points add incoherently). However, this incoherent addition helps reduce long-range coherence effects known as speckle.

Recently, completely programmable source pupils [23] have been developed to allow full *source-mask optimization* (SMO) [15, 19, 21]. Here, the patterns on the reticle and the pattern of source pixels are optimized together to extend the achievable performance of existing 193-nm immersion tools [50]. While this involves solving the computationally intensive inverse problem of the lithography process in the frequency domain, it allows performance to be pushed close to the theoretical limits. While the resulting mask and source shapes may no longer resemble the original design, they produce an optimized wavefront at the wafer that is capable of resolving the exact target design patterns inside the resist layer. When only the mask shapes are optimized for a fixed source configuration, such techniques are known as *inverse lithography* [20].

15.2.7 Interference and Polarization

The *paraxial regime* covers the range where illumination and diffraction angles are small. In this regime, light propagates in nearly the same direction as the optical axis (which runs perpendicularly from the source through the reticle, lens, and wafer). Each of these electromagnetic waves has associated transverse electric and magnetic field vector components. In the paraxial regime, the electric field vectors for all the light diffracted from the mask are very similar: in a plane approximately parallel to the mask surface. This is because even though the image space NA is large (1.35), the object space NA is smaller by the 4× reduction factor, resulting in an NA of ~0.34 with maximum ray angles away from the optical axis of ~20°. However, after light passes through the lithography lens, the incidence angles are 4× larger and coupling can occur between the electromagnetic field components at the wafer. This results in a loss of contrast and a broadening of the image along the polarization direction [51].

As in many areas of optics, it is common to analyze this for two orthogonal polarization cases, and treat everything else as a linear combination of these fundamental components. In optical lithography, these two orthogonal field components are often aligned with the directions of Manhattan layouts.

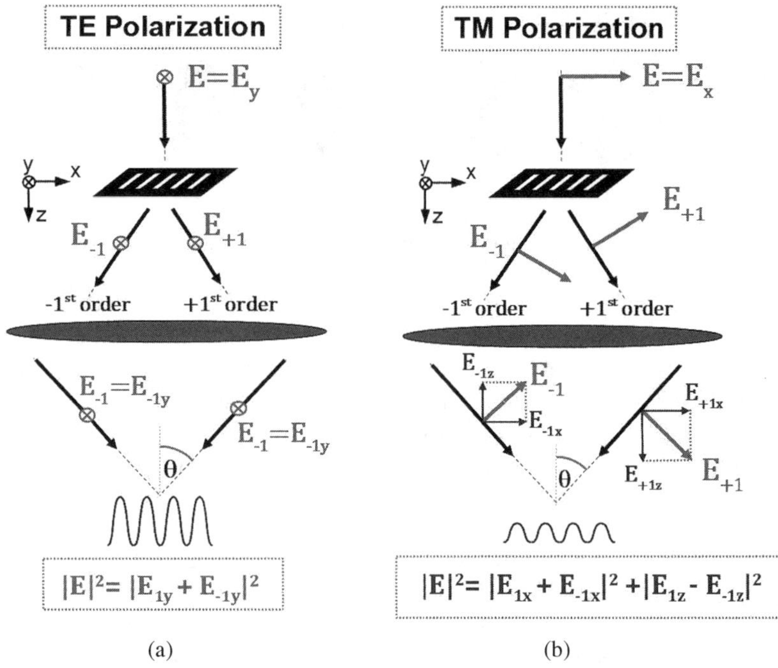

Fig. 15.8 Influence of the polarization of the incident field on imaging of a one-dimensional grating. (a) With TE polarization, the electric fields remain parallel to the mask surface for all beams and therefore interfere at 100%, providing maximal contrast for all grating pitches. (b) With TM polarization, the electric field of the incident wave is parallel to the mask surface. However, the diffracted and imaging waves include electric-field components parallel to the optical axis, reducing the contrast of the resulting interference pattern.

Figure 15.8(a) illustrates the example of a simple one-dimensional grating illuminated by a normal-incidence TE-polarized plane wave having the electric field vector parallel to the long dimension of the grating grooves, and thus normal to the plane of incidence. With this illumination, the diffraction by the mask does not change the polarization. Therefore, the electric fields of all diffracted orders remain normal to the plane of incidence. Similarly, the electric fields at the exit pupil of the projection system and those arriving at the wafer are parallel to each other and remain normal to the plane of incidence. The interference pattern between any two of these electromagnetic waves is weighted by the scalar product of their field vectors; since the field vectors are parallel, the scalar product is maximal. Thus, for TE polarization, interference contrast is always maximal, independent of the grating pitch, as illustrated in Fig. 15.9.

On the other hand, Fig. 15.8(b) illustrates the example of a one-dimensional grating illuminated at normal incidence with TM polarization. Here, the incident electric field lies within the plane of incidence and is parallel to the mask surface. The fields diffracted by the mask remain parallel to the plane of incidence, but the electric field does not necessarily remain parallel to the mask surface. In addition, the electric field vectors of different diffracted beams are decidedly not parallel, both just after diffraction and even more noticeably at the wafer. However, interference between the focusing beams is required to produce a high-contrast image intensity at the focus plane. (This combination process is coherent because all this light originated from the same source point within the source pupil. When completely combined, this superimposed light then sums incoherently with the contributions from other source points.)

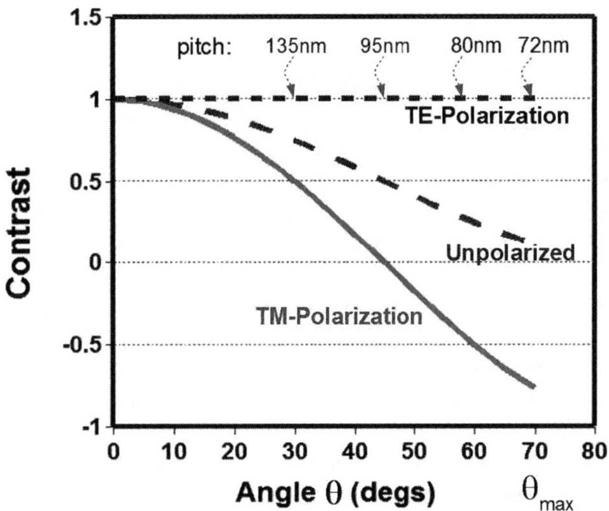

Fig. 15.9 Interference contrast as a function of angle and grating pitch for TE and TM polarizations. TE polarization guarantees 100% interference of the fields at the image plane, and thus provides maximal image contrast regardless of the grating pitch or incident angle. Image contrast for TM polarization, however, greatly depends on the exact direction of the electric fields of the imaging beams, which is a function of the grating pitch. This results in certain mask pitches where total field cancellation or even contrast reversal can occur (contrast < 0).

In the vectorial sum of the electric field of each interfering wave, some field components add while others subtract. This results in a loss of contrast in the final image intensity distribution. Worse yet, the magnitude of this contrast loss depends on the direction of propagation and thus the pitch of the pattern being imaged. This effect is illustrated in Fig. 15.9 for the case of TM polarization. In fact, complete cancellation or even contrast reversal of the fields can occur for certain grating pitches and incidence angle. Also shown in Fig. 15.9 is the average behavior when the light is unpolarized, since this implies that all polarizations are evenly represented. Since all incident polarizations can be decomposed into either TE or TM modes, unpolarized light is 50% TE-polarized and 50% TM-polarized. Clearly, careful control over polarization is another key aspect for ensuring high-contrast imaging. Source points at the ±X periphery designed to print tight-pitch horizontal gratings can be polarized along Y for TE-polarized illumination, while source points at the ±Y periphery print high-contrast vertical gratings with X-polarized illumination (also TE illumination). However, this configuration must be used with care since TM-polarized light (for instance, diffraction of the light from ±Y source-points by the horizontal gratings, or of the ±X source-points by the vertical gratings) can also readily pass through the entrance pupil if the spatial frequencies are sufficiently low. This is another aspect where computation of all the various implications is vital.

Numerous other effects can occur in the optical system that are beyond the scope of this chapter. These include polarization-dependent lens *aberrations* and the effects of the full polarization *Jones Pupils*, the role of *laser bandwidth* on focus blur, stress-induced *mask birefringence* (where the index of refraction depends on the incident polarization), and *flare* (where mask reflectivity and all the opportunities for light to scatter can generate an unwanted yet noticeable background illumination).

15.3 COMPUTATIONAL LITHOGRAPHY

The focusing of light with a lens has long been the subject of quantitative study in the framework of imaging systems such as microscopes, telescopes, cameras, and other optical systems [24, 52]. At least within the paraxial regime, scalar diffraction theory [24] can provide an accurate description of focused images. In the context of lithography, scalar diffraction refers both to the representation of the propagating fields as scalar quantities as well as to the representation of the boundary conditions on the mask surface as a scalar binary function of the mask shapes. Thus, the effects of the mask topography on the impinging illumination are completely ignored. This scalar formulation can provide adequate theoretical predictions for image-space numerical apertures as large as 0.7 [49, 53–55]. However, as numerical apertures increase and mask dimensions shrink, the aspects ignored by scalar diffraction theory — scattering from the mask topography as well as interactions between the various electromagnetic field vector components propagating at oblique angles through the optical system — become important. For lithography systems with numerical apertures larger than 0.7 or for mask features that approach a few wavelengths in dimension, vector diffraction derived from rigorous electromagnetic theory must be applied in order to accurately compute the fields at the wafer due to the mask and source.

15.3.1 Image Formation

In the Köhler illumination configuration, the illumination produced by two source points is mutually incoherent. *Source integration* or *Abbe's method* [56, 57] calls for the images generated by each source point to be incoherently added together to compute the final image. Moreover, since the absorbed energy within the nonmagnetic photoresist is simply proportional to the squared electric field [58], and since the actual exposure dose is chosen empirically, we need only consider the normalized electric field components.

For now, we compute the image in the empty output focal plane where the photoresist will eventually be located. The *aerial image intensity* at a position $\boldsymbol{r}_w$ in the vicinity of the focal point can be expressed as:

$$\mathbf{I}^{\text{image}}(\boldsymbol{r}_w) = \iint\limits_{s_x^2+s_y^2 \le \text{NA}_c^2} I_S(s_x, s_y) \left| \boldsymbol{E}^{\text{image}}(\boldsymbol{r}_w;\, s_x,\, s_y) \right|^2 ds_x\, ds_y \tag{15.8}$$

Here, $\boldsymbol{E}^{\text{image}}(\boldsymbol{r}_w; s_x, s_y)$ represents the normalized aerial image electric field distribution produced by the incident plane wave on the photomask with time-averaged intensity $I_S(s_x, s_y)$ and direction cosines $[s_x, s_y, (1 - s_x^2 - s_y^2)^{1/2}]$. The aerial image is then the incoherent summation across all the source-points (s_x, s_y) within the numerical aperture of the condenser optics, NA_c. In the following five sub-sections, we derive the expression for computing $\boldsymbol{E}^{\text{image}}(\boldsymbol{r}_w)$ from the mask diffraction, propagation through the projection optics, and focusing into the resist layer.

Thin-Mask Approximation

Many approaches for computing scattered electromagnetic fields trace their origins back to the *Huygens-Fresnel principle*, the idea that every point on a wavefront is a source of further propagating spherical waves, which then combine through interference. However, in many such problems, imperfect knowledge of the fields at boundaries complicates the mathematical analysis of this otherwise straightforward principle.

In 1882, Kirchhoff provided the first mathematical foundations to the Huygens–Fresnel principle using a problem still very relevant to modern optical lithography: diffraction of light from a transmitting aperture in an otherwise opaque screen. Kirchhoff used *physical optics* to approximate the fields on the plane of the screen. These so-called *Kirchhoff's boundary conditions* replace the fields within the aperture by those that would exist in the absence of the opaque screen, with the fields everywhere else on the screen assumed to be zero. In lithographic terms, this approximation assumes that the fields at the exit surface of a mask are simply the product of the incident field and the scalar transmission function of the mask patterns. This model, known in lithography as the *thin-mask approximation* (TMA), blithely assumes that the field passing through each aperture is unperturbed by the boundaries of these apertures. As a result, scattering induced by the mask topography as well as significant polarization effects are ignored. Figure 15.10 shows the differences in the electromagnetic near-field amplitudes at the exit surface of a mask between the thin-mask approximation, per Fig. 15.10(a), and the full solution of Maxwell's equations according to FDTD, per Fig. 15.10(b).

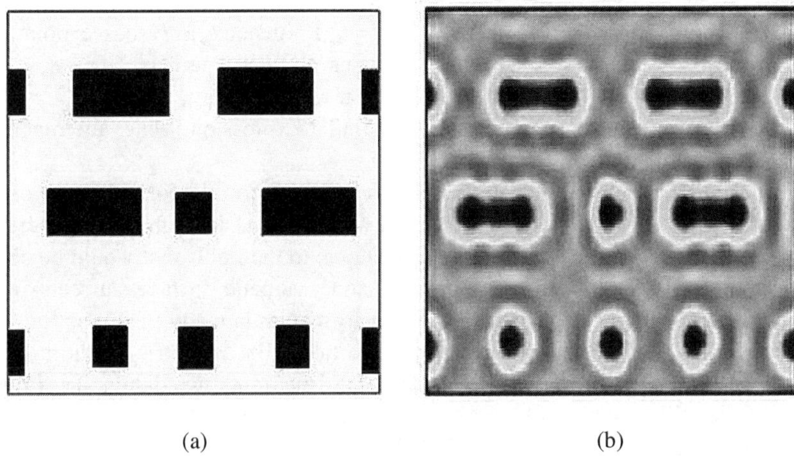

(a) (b)

Fig. 15.10 (a) Thin-mask approximation of the electromagnetic near-field amplitudes at the exit surface of a mask; (b) FDTD-computed solution of Maxwell's equations for the same field amplitudes.

Kirchhoff's boundary conditions introduce mathematical inconsistencies between the field and its derivative [52], and discontinuities in the values of the fields on the mask surface. Despite these problems, the thin-mask approximation is exceedingly simple to implement. Furthermore, it provides sufficiently accurate results when diffraction from apertures of diameter a much larger than the wavelength ($ka >> 1$) is evaluated at distances z several wavelengths from the aperture ($z > 2a^2/\lambda$). Thus, the thin-mask approximation is the default computational model used in commercial lithography simulators today, even for advanced computation of critical mask levels for the 32-nm technology node and beyond.

Mask Diffraction

To go beyond the thin-mask approximation, we must turn to vector diffraction theory. Rigorous vector diffraction theory was first applied to the lithographic imaging and exposure process by

Yeung [59], based on work by E. Wolf [60], and was subsequently expanded on by several other authors [51, 53, 58, 61]. Yeung's formulation of the vectorial Huygens' principle is due to Franz [62–64], which is also used here since it satisfies Maxwell's equations for both continuous and discontinuous electromagnetic fields.

The electromagnetic fields at some distant point $\boldsymbol{r}$ due to radiation from sources within a volume V can be derived in terms of the associated electric and magnetic vector potentials, $\boldsymbol{A}$ and $\boldsymbol{F}$ [65, 66], as follows:

$$\boldsymbol{E}(\boldsymbol{r}) = \frac{1}{j\omega\varepsilon}\nabla\times\nabla\times\boldsymbol{A} - \nabla\times\boldsymbol{F} \tag{15.9}$$

Here, the electric vector potential $\boldsymbol{A}$ is related to the electric current density $\boldsymbol{J}$ via the solution to the wave equation $\nabla^2\boldsymbol{A} + k^2\boldsymbol{A} = -\boldsymbol{J}$. This solution can be expressed as:

$$\boldsymbol{A}(\boldsymbol{r}) = \iiint_V \boldsymbol{J}(\boldsymbol{r}_m)G(\boldsymbol{r}-\boldsymbol{r}_m)\,d\boldsymbol{r}_m = \iiint_V \boldsymbol{J}(\boldsymbol{r}_m)\frac{e^{-jk|\boldsymbol{r}-\boldsymbol{r}_m|}}{4\pi|\boldsymbol{r}-\boldsymbol{r}_m|}\,d\boldsymbol{r}_m \tag{15.10}$$

where $\boldsymbol{r} = r\hat{\boldsymbol{r}}$ denotes the observation point, $\boldsymbol{r}_m = r_m\hat{\boldsymbol{r}}_m$ denotes a $\boldsymbol{J}$ source point within the volume V (bounded by the surface S) which contains all such sources, $G(\boldsymbol{r}-\boldsymbol{r}_m)$ is the free-space scalar Green's function [the solution to the scalar wave equation $\nabla^2 G + k^2 G = -\delta(\boldsymbol{r}-\boldsymbol{r}_m)$], and $k = 2\pi/\lambda$ is the wavevector in free space. A similar expression relates the magnetic vector potential $\boldsymbol{F}$ to the magnetic current density $\boldsymbol{K}$.

The *surface equivalence theorem*, one of the most rigorous formulations of Huygens' principle, states that the radiated fields observed at $\boldsymbol{r}$ external to V that are generated by the original $\boldsymbol{J}$ and $\boldsymbol{K}$ sources contained within V are identical to the fields that would be observed at $\boldsymbol{r}$ radiated by properly defined equivalent electric and magnetic surface currents on S. This theorem is implemented in the near-to-far-field transformation in many FDTD applications [65].

In the context of lithography, we need only the normalized electric fields in the far field. To this end, upon applying the surface equivalence theorem, we obtain the Franz surface integrals of vector diffraction at observation point $\boldsymbol{r}$ external to V due to equivalent current sources defined on the surface S of V [62, 63, 65]:

$$\boldsymbol{E}(\boldsymbol{r}) = \frac{1}{j\omega\varepsilon}\nabla\times\nabla\times\iint_S[\hat{\boldsymbol{n}}\times\boldsymbol{H}(\boldsymbol{r}_m)]G(\boldsymbol{r}-\boldsymbol{r}_m)\,d\boldsymbol{r}_m - \nabla\times\iint_S[\hat{\boldsymbol{n}}\times\boldsymbol{E}(\boldsymbol{r}_m)]G(\boldsymbol{r}-\boldsymbol{r}_m)\,d\boldsymbol{r}_m \tag{15.11}$$

where $\boldsymbol{r}_m$ denotes a source point for $\boldsymbol{H}$ or $\boldsymbol{E}$ on S, and $\hat{\boldsymbol{n}}$ is the unit outward normal vector to S at $\boldsymbol{r}_m$. Equation (15.11) represents the vectorial formulation of Huygens' principle, according to which the tangential fields on S can be considered to be the sources of the fields external to this surface. This formulation requires computation of only the tangential components of the fields on S, and is therefore used in many electromagnetic scattering problems.

Equation (15.11) also provides the most rigorous formulation for computing the fields diffracted by the lithography photomask. Consider dividing space in two regions separated by a surface S parallel to the mask surface, as illustrated in Fig. 15.11(a). We can replace the mask and the illumination contained within volume V in Region I with their equivalent sources, $\boldsymbol{J}$ and $\boldsymbol{K}$, on S as illustrated in Fig. 15.11(b). These sources then produce the equivalent diffracted fields in Region II outside volume V. This formulation is also valid for open surfaces and in the presence of discontinuous fields on S, such as those produced when Kirchhoff boundary conditions are applied on the reticle surface.

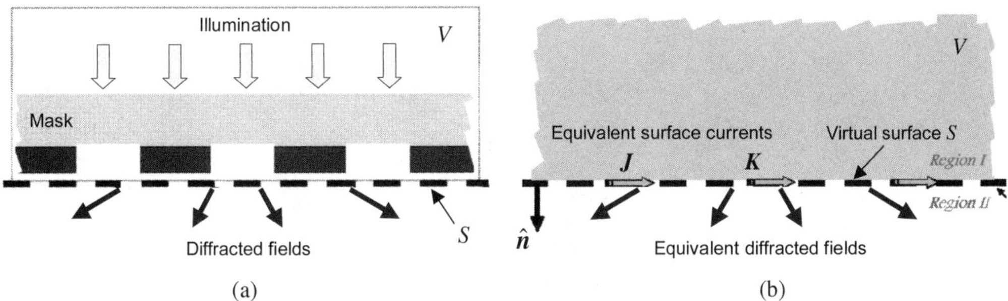

Fig. 15.11 Application of Huygens' principle to the computation of the fields diffracted from the photomask. The original problem (a) can be replaced by an equivalent problem (b) where the fields generated by the equivalent surface currents are identical to the original diffracted fields outside volume V, given that the boundary conditions are satisfied on the virtual (mathematical) surface S.

To exactly determine the fields in the source-free region, we need only use a numerical Maxwell's equations solver such as FDTD to accurately and efficiently compute the tangential electric and magnetic fields at a given surface just outside the mask. This can be the exit interface of the lithographic photomask or any plane parallel to this surface where the tangential components of the fields are known. However, if we were to consider the tangential fields only over the mask openings, as in the thin-mask approximation, then the solutions would not be exact.

The fields diffracted by the reticle propagate toward the entrance pupil (EP) of the projection optics of the lithography scanner, which is located many wavelengths away from the mask surface. We show in Appendix 15A that at an observation point $\boldsymbol{r}$ in the far-field region of the mask, the radiated electric field at the entrance pupil due to these surface currents is given by:

$$\boldsymbol{E}^{\mathrm{EP}}(\boldsymbol{r}) = -\nabla\times\boldsymbol{F} = -\nabla\times\int_{x_m}\int_{y_m}\left[-2\hat{z}\times\boldsymbol{E}(\boldsymbol{r}_m)\right]\frac{e^{-jk|\boldsymbol{r}-\boldsymbol{r}_m|}}{4\pi|\boldsymbol{r}-\boldsymbol{r}_m|}d\boldsymbol{r}_m \tag{15.12}$$

This can be represented in terms of the Fourier transform of the surface fields over (x_m, y_m) on S:

$$\boldsymbol{E}^{\mathrm{EP}}(\boldsymbol{r}) \approx \frac{j}{\lambda}\frac{e^{-jkr}}{r}\cos\theta\,\tilde{\boldsymbol{E}}^{\mathrm{M}}(p_x, p_y) = \frac{j}{\lambda}\frac{e^{-jkr}}{r}\cos\theta\int_{x_m}\int_{y_m}\boldsymbol{E}^{\mathrm{M}}(\boldsymbol{r}_m)\,e^{jk(p_x x_m + p_y y_m)}dx_m\,dy_m \tag{15.13}$$

Each term $\tilde{\boldsymbol{E}}^{\mathrm{M}}(p_x, p_y)$ represents a plane wave with a direction of propagation given by the wavevector $\boldsymbol{k} = k\hat{\boldsymbol{p}}$, which satisfies the transverse electromagnetic condition for plane waves; that is, $\boldsymbol{k}\cdot\tilde{\boldsymbol{E}}^{\mathrm{M}} = 0$. The spatial frequency $k[p_x, p_y, (1-p_x^2-p_y^2)^{1/2}]$ is associated with direction cosines (p_x, p_y, p_z) according to:

$$\hat{\boldsymbol{p}} = \frac{(\boldsymbol{r}-\boldsymbol{r}_m)}{|\boldsymbol{r}-\boldsymbol{r}_m|} = p_x\hat{\boldsymbol{x}} + p_y\hat{\boldsymbol{y}} + p_z\hat{\boldsymbol{z}} = \sin\theta\cos\varphi\,\hat{\boldsymbol{x}} + \sin\theta\sin\varphi\,\hat{\boldsymbol{y}} + \cos\theta\,\hat{\boldsymbol{z}} \tag{15.14}$$

where θ is the angle formed by $\hat{\boldsymbol{p}}$ relative to the optical axis (z-axis), and φ is the angle of the projection of $\hat{\boldsymbol{p}}$ on S relative to the x-axis, as illustrated in Fig. 15.12.

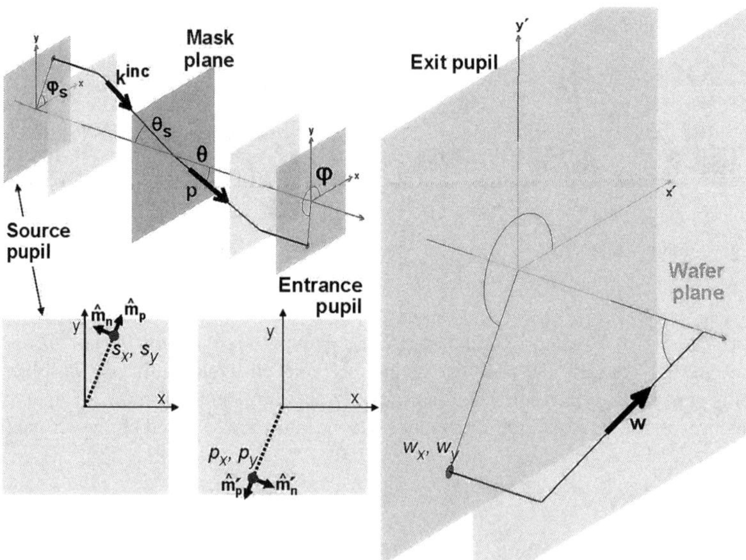

Fig. 15.12 Coordinates associated with an optical lithography system. Object-space coordinates are denoted (x, y, z) while image-space coordinates are denoted (x', y', z'). Each direction $\hat{p}$ of propagation of the waves diffracted by the mask, together with the optical axis, $\hat{z}$, forms the meridional plane. The corresponding propagation direction $\hat{w}$ of the wave converging toward the wafer plane from the exit pupil lies in the same meridional plane.

Equation (15.13) demonstrates without approximation that the diffracted fields can be completely formulated in terms of the tangential components of the electric fields on the mask surface. For a purely periodic mask layout, these diffracted far fields can be expressed as a spherical wavefront, multiplied by a series of delta functions that are non-zero only at the spatial frequencies corresponding to the diffracted orders. Locally, each diffracted order behaves as a plane wave, with a reduced amplitude factor (by $\cos\theta$) due to the increase in solid angle covered by the diffracted order.

Focusing of Fields

Lithographic scanners employ "diffraction-limited" lenses made of pure fused silica together with sophisticated systems to mitigate any source of aberrations that can occur during operation [67, 68]. Hence, we can assume that the diverging spherical wavefront at the entrance pupil in the object space produces a perfect converging spherical wavefront at the exit pupil in the image space, propagating toward its geometrical focal point F_1. Application of geometrical ray tracing to track wave propagation from the entrance to the exit pupils is both computationally efficient and sufficiently accurate. The amplitude and phase of the fields emerging at the exit pupil are thus closely related to those arriving at the entrance pupil due to mask diffraction. The diffraction formulation is then applied to the fields at the exit pupil in order to compute the fields near the focal point. Following the same steps as in the previous sub-section, the fields radiated from the exit pupil and arriving near the geometrical focal point at some distance from the aperture can be deduced by applying Huygens' principle at the exit pupil aperture, as illustrated in Fig. 15.13.

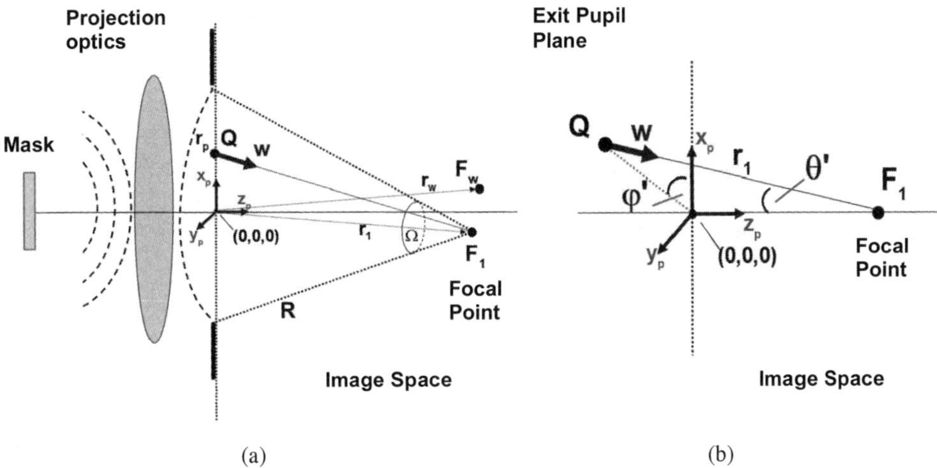

(a) (b)

Fig. 15.13 Schematic of image focusing of a converging spherical wave emanating from the exit pupil aperture toward the geometrical focus F_1, where the image fields are being measured at a point F_w; Ω is the solid angle subtended by the aperture at the focal point.

With lens diameters of hundreds of millimeters [49], numerical integrations over the exit pupil aperture area are both computationally intensive and inefficient, given that the largest contributions to the integral arise from a small discrete set of stationary points inside the aperture. Thus, an asymptotic evaluation of the focusing fields integral using the method of stationary phase is commonly applied to the modeling of imaging by lithographic systems, known as Debye's representation [52]. A number of approximations are used in the derivation of Debye's formulation of the focusing fields, which are justified under the conditions of large lens diameter and relatively long focal distances, as well as high-quality aberration-free optics.

Following Stamnes [52], Kirchhoff's boundary conditions are applied on the exit pupil surface, where it is assumed that the fields emanating from the exit pupil aperture are in the form of a spherical wavefront converging toward its geometrical focal point at $\boldsymbol{r}_1$, with the amplitude at each point of the aperture equal to $\boldsymbol{A}(x_p, y_p)$. Any phase deviation from the spherical wavefront induced by aberrations in the propagation through the projection optics system are given by the function $k'\phi(x_p, y_p)$. The aperture radius is proportional to the image space numerical aperture at the exit pupil (that is, $r_a = |\boldsymbol{r}_1| \mathrm{NA_i} = r_1 \mathrm{NA_i}$), and the wavevector in the image space is given by $k' = kn_i = (2\pi/\lambda)n_i$, with n_i being the local index of refraction. The fields at the exit aperture thus obey the expression:

$$\boldsymbol{E}^{\mathrm{XP}}(x_p, y_p) = \begin{Bmatrix} \boldsymbol{A}(x_p, y_p)\, e^{jk'\phi(x_p, y_p)}\, e^{jk'|\boldsymbol{r}_1 - \boldsymbol{r}_p|} \Big/ |\boldsymbol{r}_1 - \boldsymbol{r}_p| & \text{for } x_p^2 + y_p^2 \le r_a^2 \\ 0 & \text{for } x_p^2 + y_p^2 > r_a^2 \end{Bmatrix} \qquad (15.15)$$

The resulting integral representation is evaluated retaining only the lower-order asymptotic solution corresponding to stationary points in the interior of the pupil aperture. This approximation results in a formulation of the fields in the focal region as a superposition of all plane waves of wavevectors given by $\boldsymbol{k}' = k'\hat{\boldsymbol{w}} = k'[w_x, w_y, (1 - w_x^2 - w_y^2)^{1/2}]$, propagating within the cone subtended by the edge of the aperture as seen from the geometrical focal point, with a discontinuity in the angular spectrum of the fields on the exit pupil aperture edge as:

$$E^{\text{image}}(\boldsymbol{r}_w) = \frac{j}{\lambda} \iint_{w_x w_y \leq NA_i} \frac{e^{-jk'\cdot(\boldsymbol{r}_w - \boldsymbol{r}_1)}}{w_z} A(w_x, w_y)\, dw_x\, dw_y \tag{15.16}$$

Here, the observation point $\boldsymbol{r}_w$ and the geometrical focal point $\boldsymbol{r}_1$ are assumed to be close to each other compared to the distance to the exit pupil plane. The geometrical focal point can also be assumed to be on the optical axis, the z-axis, such that the vector cosines of each plane wave propagating toward the geometrical focal point are given by:

$$\hat{\boldsymbol{w}} \equiv \frac{(\boldsymbol{r}_1 - \boldsymbol{r}_p)}{|\boldsymbol{r}_1 - \boldsymbol{r}_p|} = w_x\hat{\boldsymbol{x}} + w_y\hat{\boldsymbol{y}} + w_z\hat{\boldsymbol{z}} = -\sin\theta'\cos\varphi'\hat{\boldsymbol{x}} - \sin\theta'\sin\varphi'\hat{\boldsymbol{y}} + \cos\theta'\hat{\boldsymbol{z}} \tag{15.17}$$

where the angles θ' and φ' are defined with respect to the focal point F_1 as in Fig. 15.13(b). The vector $(\boldsymbol{r}_w - \boldsymbol{r}_1)$ represents the directed distance between the observation point F_w and the geometrical focus F_1. The term $\boldsymbol{A}(w_x, w_y)$ represents the exit-pupil aperture fields at the position given by the stationary-point solution, $(x_s, y_s) \equiv r_1(w_x, w_y)$, where r_1 is the distance between the exit pupil and the geometrical focus. In other words, $\boldsymbol{A}(w_x, w_y)$ represents the fields within the exit-pupil aperture propagating with the direction cosines (w_x, w_y).

This approximation further simplifies the Kirchhoff's boundary conditions by omitting any influence due to the rim of the aperture, and retains only the geometrical optics contribution from the fields inside the aperture, assumed unperturbed by its boundary. First derived by P. Debye in 1909 and further developed and extended to a vectorial formulation by several authors [60, 69–71], this formulation has constituted the foundations of the imaging approach [51, 53, 58, 59, 61] widely used in computational lithography software.

Studies have been carried out to understand the impact of the fields diffracted at the boundary of the aperture [72–74], and the validity of the Debye formulation of the focusing fields [24, 52, 75]. These studies show that the Debye approximation is accurate at distances from the pupil plane that satisfy the condition that the Fresnel number $N = a^2/\lambda z_1 >> 1$, where a is the aperture radius and z_1 is the geometrical focal distance. Equivalently, $\text{NA} >> (\lambda/z_1)^{1/2}$ [75]. For high numerical imaging systems where the angular aperture is larger than 45°, a more general form of the validity condition, as stated by Wolf and Li [75], must be employed:

$$kz_1 >> \frac{\pi}{\sin^2(\theta'/2)} \tag{15.18}$$

With focusing distances of the order of 500 μm to a few millimeters [49], lens diameters about 250–300 mm, and numerical apertures higher than 0.85 in 193-nm wavelength lithography, this condition guarantees that the Debye integral formulation yields sufficiently accurate results for the field distribution in the neighborhood of the focal point.

High-NA and Vector Imaging

In the regime of image-space numerical apertures bounded by $\text{NA}_\text{i} < 0.6$, the maximum angle θ' formed by the diffracted light ray $\hat{\boldsymbol{w}}$ and the optical axis $\hat{\boldsymbol{z}}$ is also relatively small $[\theta_{\max} = \sin^{-1}(\text{NA}_\text{i})]$, and the paraxial or small-angle approximation $\cos\theta \approx 1$ yields accurate results. For higher values of $\text{NA}_\text{i} \approx 0.6 - 0.7$, the nonparaxial scalar diffraction theory can still provide adequate predictions of the diffracted fields, however with decreasing accuracy as NA_i

increases. For $NA_i > 0.7$, polarization effects and the inherent coupling between the vector components of the electromagnetic fields become noticeable. In this high-NA regime, vector diffraction theory is necessary [53–55, 61] to track the propagation of light from the mask surface to the image plane.

The nearly diffraction-limited imaging characteristics of the refractive lenses allows us to assume the imaging system is isoplanatic (phase invariant), wherein the image of an object point changes only in location but not in form as the source point moves through the object plane [76]. Under these circumstances and according to the Fresnel refraction formula [65], each ray traces a path that lies on its meridional plane, the plane formed by the ray wavevector direction, $\hat{\boldsymbol{p}}$, and the optical axis, $\hat{z}$, as illustrated in Fig. 15.12. To a good approximation, the associated polarization vector maintains a constant angle with the meridional plane along the entire path traced by the ray if the angles of incidence at the various surfaces in the system are small [59, 69]. For each source point, the fields at the entrance pupil in the object space are linearly polarized on the plane normal to the propagation direction of each diffracted ray. Each of these diffracted rays is incident on the entrance pupil at a nearly normal incident angle with the lens surface. Thus, the electric field vector forms a very small angle with the glass surface of the first lens. According to Fresnel theory, the fields obtained after refraction at the first lens surface are also linearly polarized and the polarization direction is effectively unchanged. Repeating this argument for each lens surface encountered throughout the imaging system, the polarization direction of the electric field vector maintains an approximately constant angle with the meridional plane.

Specifically, the polarization state of each ray exiting the last lens surface can be traced back to the polarization state at the entrance pupil by decomposing first the electric field into its projection on the meridional plane, that is, along the direction $\hat{\boldsymbol{m}}_p$ parallel to the meridional plane and normal to the ray direction, and its component along the direction $\hat{\boldsymbol{m}}_n$ normal to both the meridional plane and to the ray direction. The field amplitudes along these normal and parallel directions to the meridional plane, E_{m_n} and E_{m_p}, remain unchanged by refraction during propagation through the optical system, except for the geometrical factor arising from the conservation of energy to be discussed in the next sub-section.

The ray direction in the object space, $\hat{\boldsymbol{p}}$, however, rotates onto the ray direction in the image space, $\hat{\boldsymbol{w}}$, as it propagates through the system. Thus, the projection of the electric field polarization onto the global Cartesian axes $(\hat{x}, \hat{y}, \hat{z})$ changes and some coupling between the field Cartesian components can take place. This polarization rotation is accounted for by the polarization rotation tensor **T** derived in Appendix 15C. This matrix **T** can be used to determine the Cartesian components of each plane wave at the exit pupil from those at the entrance pupil. To deduce this tensor **T**, we decompose the fields at the entrance pupil into their local projections along the directions normal and parallel to its meridional plane, that is, E_n^{EP} and E_p^{EP}, and apply the condition that the polarization angle with respect to this plane remains approximately constant through the projection system. The unit vectors along the directions normal and parallel to the meridional plane can be calculated as $\hat{\boldsymbol{m}}_n = \hat{\boldsymbol{p}} \times \hat{z}$ and $\hat{\boldsymbol{m}}_p = \hat{\boldsymbol{m}}_n \times \hat{\boldsymbol{p}}$ in the object space, where $\hat{\boldsymbol{p}}$ is given by (15.14). By analogy, in the image space we have $\hat{\boldsymbol{m}}'_n = \hat{\boldsymbol{w}} \times \hat{z}$ and $\hat{\boldsymbol{m}}'_p = \hat{\boldsymbol{m}}'_n \times \hat{\boldsymbol{w}}$, where $\hat{\boldsymbol{w}}$ is given by (15.17). According to the meridional plane approximation, the field components along the directions normal, $\hat{\boldsymbol{m}}_n$, and parallel, $\hat{\boldsymbol{m}}_p$, to the meridional plane remain approximately the same as they propagate through the optical projection system. Thus, at the exit pupil, we can assume that $E_{m'_p}^{\mathrm{XP}} = E_{m_p}^{\mathrm{EP}}$ and $E_{m'_n}^{\mathrm{XP}} = E_{m_n}^{\mathrm{EP}}$.

In terms of the Cartesian axes shown in Fig. 15.12, using (15C.8a, b, c, d) from Appendix 15C, we can write:

$$\begin{bmatrix} E_x^{\mathrm{XP}} & E_y^{\mathrm{XP}} & E_z^{\mathrm{XP}} \end{bmatrix} = \begin{bmatrix} E_x^{\mathrm{EP}} & E_y^{\mathrm{EP}} & E_z^{\mathrm{EP}} \end{bmatrix} \cdot \mathbf{T} \qquad \text{where } \mathbf{T} = \begin{bmatrix} T_{xx} & T_{yx} & T_{zx} \\ T_{yx} & T_{yy} & T_{zy} \\ T_{zx} & T_{zy} & T_{zz} \end{bmatrix} \tag{15.19}$$

Matrix **T** includes the changes in the direction of the electric field from the entrance pupil to the wafer plane, as well as from the mask plane to the entrance pupil, without the common approximation that the polarization on the object space remains parallel to the polarization of the incident illumination [46, 51, 53]. The derivation of the components of **T** can be found in Appendix 15C.

Conservation of Energy and Obliquity Factor

The amplitude function $\boldsymbol{A}(w_x, w_y)$ in the exit pupil from (15.16) can be related to each diffracted ray at the entrance pupil, $\boldsymbol{E}^{\mathrm{EP}}(p_x, p_y)$ of (15.13) by tracing the geometrical ray propagation and polarization state through the optical system as seen in the previous sub-section and Appendix 15C.

On the other hand, the ray direction $\hat{\boldsymbol{p}}$, per (15.14), is rotated onto $\hat{\boldsymbol{w}}$, given by (15.17), as it reaches the last lens surface of the projection system. Assuming negligible losses due to reflection and absorption, the magnitude of the field vectors at the image region must satisfy the following conservation of energy relation between the fields on a small area element da on the diverging spherical wavefront of radius r on the entrance pupil, and the fields on the corresponding area element da' on the converging spherical wavefront of radius r_1 on the exit pupil:

$$n_{\mathrm{i}} \left| \boldsymbol{E}^{\mathrm{XP}}(x', y') \right|^2 da' = n_{\mathrm{o}} \left| \boldsymbol{E}^{\mathrm{EP}}(x, y) \right|^2 da \tag{15.20}$$

where $da = r^2 \sin\theta \, d\theta \, d\varphi$ and $da' = r_1^2 \sin\theta' d\theta' d\varphi'$. The angle θ between the incident ray to the entrance pupil and the optical axis at the object space, and the angle θ' between the outgoing ray at the exit pupil and the optical axis at the image space, are related through the Abbe's sine condition for the optical system [24]:

$$n_{\mathrm{o}} \sin\theta = \mathrm{M} \, n_{\mathrm{i}} \sin\theta' \tag{15.21}$$

where M denotes the demagnification of the lens (usually 1/4). As a consequence, the direction cosines of each plane wave on the entrance pupil relate to each plane wave on the exit pupil according to the relations:

$$p_x = -\mathrm{M} w_x \qquad p_y = -\mathrm{M} w_y \qquad p_z = \sqrt{1 - \left(\frac{\mathrm{M} n_{\mathrm{i}}}{n_{\mathrm{o}}} \sin\theta' \right)^2} \qquad w_z = \sqrt{1 - (\sin\theta')^2} \tag{15.22}$$

According to the system's isoplanatic condition, both rays at the entrance and exit pupils remain on the meridional plane and, therefore, $d\varphi = d\varphi'$. These two conditions result in the obliquity factor for the field magnitude at the exit pupil that guarantees conservation of energy as follows. Given that both the entrance and exit pupils are at large distances from the spherical wavefront geometrical focus, that is, $r \approx r_1 >> \lambda$, we can approximate $r \approx r_1$. Then, using the relation $n_{\mathrm{o}} \cos\theta \, d\theta = \mathrm{M} \, n_{\mathrm{i}} \cos\theta' d\theta'$, we obtain the following inclination factor:

$$|\boldsymbol{E}^{\mathrm{XP}}| = |\boldsymbol{E}^{\mathrm{EP}}|\,\mathrm{M}\sqrt{\frac{n_{\mathrm{i}}\cos\theta'}{n_{\mathrm{o}}\cos\theta}} \tag{15.23}$$

where the constants n_{o} and n_{i} denote the refractive indexes at the object and image spaces, respectively. Typically, $n_{\mathrm{o}} = 1$ for air in the object space and $n_{\mathrm{i}} = 1.436$ when employing water immersion in the image space. This inclination factor results from the condition of conservation of energy and is necessary to account for the angle magnification between the rays at the entrance pupil and the exit pupil.

Substituting (15.23) into (15.15) and using the entrance-pupil fields of (15.13), we arrive at the final expression for the image fields as:

$$\boldsymbol{E}^{\mathrm{image}}(\boldsymbol{r}_w) = \frac{1}{\lambda^2}\iint_{w_x w_y \le \mathrm{NA_i}} O(w_x, w_y)\,\mathbf{T}(w_x, w_y)\,\tilde{\boldsymbol{E}}^{\mathrm{M}}(\mathrm{M}w_x, \mathrm{M}w_y)\,e^{-jk'\cdot(r_w - r_1)}\,dw_x\,dw_y \tag{15.24}$$

for an observation point at $\boldsymbol{r}_w$ and the geometrical focal point at $\boldsymbol{r}_1$. The final obliquity factor $O(w_x, w_y)$ is given by:

$$O(w_x, w_y) = \mathrm{M}\sqrt{\frac{n_{\mathrm{i}}\cos\theta'}{n_{\mathrm{o}}\cos\theta}}\cdot n_{\mathrm{i}}n_{\mathrm{o}}\frac{\cos\theta}{\cos\theta'} = \mathrm{M}n_{\mathrm{i}}^2\sqrt{\frac{n_{\mathrm{o}}\cos\theta}{n_{\mathrm{i}}\cos\theta'}} = \mathrm{M}n_{\mathrm{i}}^2\sqrt{\frac{n_{\mathrm{o}}p_z}{n_{\mathrm{i}}w_z}} \tag{15.25}$$

This final obliquity factor in (15.25) results when a second inclination factor is applied to the propagating diffraction orders in the form of the ratio $n_{\mathrm{i}}n_{\mathrm{o}}\cos\theta/\cos\theta'$. This second inclination factor arises in the expressions for the diverging fields at the entrance pupil, (15.13), and the converging fields at the exit pupil, (15.16), after correcting for the refractive index in the object and image space, respectively. This factor must be applied to the fields passing from the exit pupil of the lens to the wafer plane to account for the extended solid angle covered by those waves propagating at oblique angles on both image and object spaces. The final oblique factor in (15.25) guarantees a brightness-invariant model of the lens system. More detailed analysis of the origin of these inclination factors can be found in [77].

15.3.2 Mask Illumination

The Köhler illumination configuration predominates in lithography because it provides uniform illumination from a source that in general is nonuniform, provided well-corrected condenser lenses are employed [25]. In Köhler illumination, the source can be considered to be located at the focal plane of the condenser and the object at the condenser exit pupil. As shown in Fig. 15.4, a telecentric condenser is used in refractive optical lithography, where the source cross-section is imaged on the entrance pupil of the projection lens system.

In the Köhler illumination configuration, both unpolarized and polarized sources can be numerically modeled through the superposition of the aerial images produced by two mutually incoherent polarized illuminations with orthogonal polarization states. Each source point is modeled as originating a coherent, linearly polarized plane wave emerging from the lens with an angle determined by the source-point location relative to the optical axis. While polarized illumination has become a common choice to print the most critical levels with k_1 factors below 0.3, unpolarized light is nonetheless often preferred to print less critical levels in lithography.

Unpolarized illumination reduces image sensitivity to residual polarization aberrations in the optics, reduces the long-term induction of birefringence in the illuminator optics, and helps mitigate long-range effects associated with coherent illumination. Modeling the effects of illuminating a topographic photomask with either a mixed-polarization state or with unpolarized light simply requires two separate rigorous electromagnetic simulations, one for each of the two orthogonal polarizations. For the mixed-polarization state, these results are combined coherently as fields; for unpolarized illumination, they are combined incoherently as intensities.

Pupil Coordinates

Following the conventions stated by the Zeiss-IBM joint formulation work [78], we can convert from spatial coordinates to pupil coordinates by decomposing the electric fields into their projections on the meridional plane (formed by the vectors $\hat{\boldsymbol{p}}$ and $\hat{z}$); that is, along the direction $\hat{\boldsymbol{m}}_p$ parallel to the meridional plane and normal to the propagation direction, and along $\hat{\boldsymbol{m}}_n$, perpendicular to both the meridional plane and to the ray direction.

Referring to Fig. 15.14, we now elaborate on the earlier discussion of obtaining the field amplitudes E_{m_n} and E_{m_p} along the normal and parallel directions to the meridional plane.

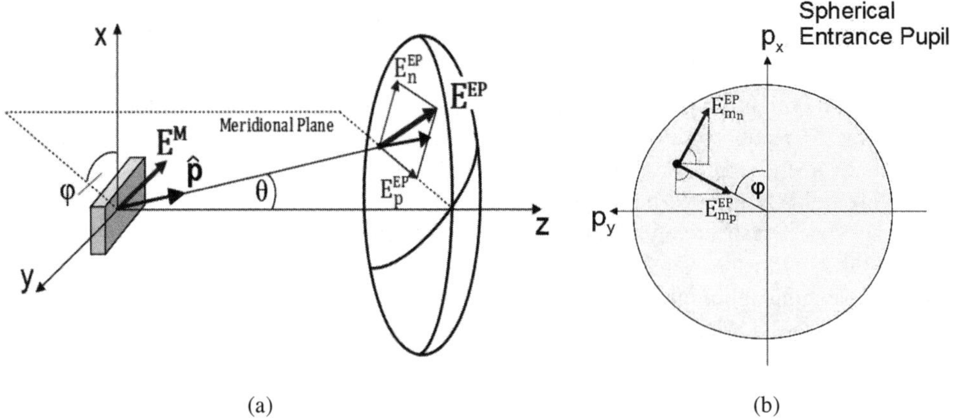

Fig. 15.14 (a) Geometry of the decomposition of the fields diffracted by the photomask in terms of the components parallel and normal to the meridional plane, which is formed by the direction of propagation $\boldsymbol{k} = k\hat{\boldsymbol{p}}$ and the optical axis $\hat{z}$. (b) Coordinates in the entrance pupil.

The following expression specifies E_{m_n} and E_{m_p}:

$$\begin{bmatrix} E_{m_p} \\ E_{m_n} \end{bmatrix} = \frac{1}{\sqrt{p_x^2 + p_y^2}} \begin{bmatrix} -p_z p_x & -p_z p_y & (p_x^2 + p_y^2) \\ p_y & -p_x & 0 \end{bmatrix} \cdot \begin{bmatrix} E_x \\ E_y \\ E_z \end{bmatrix} \tag{15.26}$$

where the angles that the ray forms with the Cartesian coordinate system relate to the direction-cosines $(p_x, p_y, p_z) = (\cos\varphi \sin\theta, \sin\varphi \sin\theta, \cos\theta)$ as:

$$\theta = \sin^{-1}\left(\sqrt{p_x^2 + p_y^2}\right) \qquad \varphi = \tan^{-1}(p_y / p_x) \tag{15.27}$$

Field components E_{m_n} and E_{m_p} propagate tangential to a spherical wavefront toward the entrance pupil with direction cosines as illustrated in Fig. 15.14(b). These field components can then be expressed in terms of their projections along the pupil coordinates or direction cosines (p_x, p_y) as follows:

$$\begin{bmatrix} E_{p_x} \\ E_{p_y} \end{bmatrix} = \begin{bmatrix} -\cos\varphi & \sin\varphi \\ -\sin\varphi & -\cos\varphi \end{bmatrix} \cdot \begin{bmatrix} E_{m_p} \\ E_{m_n} \end{bmatrix} \tag{15.28}$$

Replacing the vector $[E_{m_p}, E_{m_n}]$ by its expression in Cartesian coordinates as in (15.26), then performing the matrix multiplication, and finally noting that $p_x^2 + p_y^2 = (1-p_z)(1+p_z)$, we obtain the expression for the fields in pupil coordinates based on its Cartesian components:

$$\begin{bmatrix} E_{p_x} \\ E_{p_y} \end{bmatrix} = \Gamma^T \begin{bmatrix} E_x \\ E_y \\ E_z \end{bmatrix} = \frac{1}{1+p_z} \begin{bmatrix} 1+p_z-p_x^2 & -p_x p_y & -p_x(1+p_z) \\ -p_x p_y & 1+p_z-p_y^2 & -p_y(1+p_z) \end{bmatrix} \cdot \begin{bmatrix} E_x \\ E_y \\ E_z \end{bmatrix} \tag{15.29}$$

where the superscript T denotes a transpose matrix. This matrix represents the transformation needed at the entrance pupil to express the fields in pupil or direction-cosine coordinates. To convert from pupil coordinates to spatial Cartesian coordinates, we have:

$$\begin{bmatrix} E_x \\ E_y \\ E_z \end{bmatrix} = \Gamma \begin{bmatrix} E_{p_x} \\ E_{p_y} \end{bmatrix} \tag{15.30}$$

Source Pupil Map and Oblique Incidence

Arbitrarily shaped sources are becoming widespread in optical lithography and are often specified by a source map. Source maps are discretizations of the intensity distribution on the condenser optics exit pupil at uniformly spaced pupil coordinates, (σ_x, σ_y). Source-map coordinates vary in the range $(\sigma_x, \sigma_y)_{\max} = (\pm 1, \pm 1)$ as they have been normalized to the object-space numerical aperture, $\mathrm{NA_o} = \mathrm{M\,NA_i}$ with M = 1/4. Thus, these source-map coordinates are proportional to the direction cosines (s_x, s_y) of each incident plane wave onto the mask surface, $(\sigma_x, \sigma_y) = (\mathrm{M\,NA_i})^{-1}(s_x, s_y)$. One of the outermost points in any source map is always $(\sigma_x = 1,\ \sigma_y = 0)$. However, the particular angle that this introduces into the lithography system depends on the NA.

Source Irradiance

The discretization of the source is often based on a uniform grid of finite-sized pixels in direction-cosine space $\Delta s_x \Delta s_y$ and centered at the source point (s_x, s_y). As shown in Fig. 15.15, each source pixel subtends a different solid angle $\Delta\Omega_s$ at the mask plane as given by:

$$\Delta\Omega_s\big|_{s_x, s_y} = \frac{\Delta s_x \Delta s_y}{\sqrt{1 - s_x^2 - s_y^2}} \tag{15.31}$$

Since the value of the direction cosine $(1-s_x^2-s_y^2)^{1/2}$ changes across the source-pupil position (s_x, s_y), the mapping between direction-cosine space and direction solid angle is nonuniform.

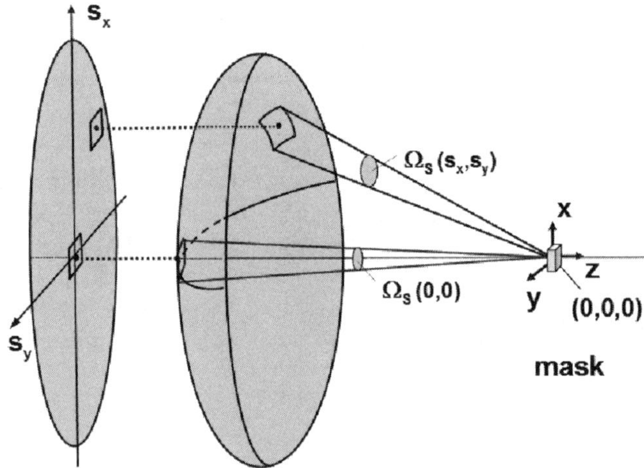

Fig. 15.15 Source maps are discretizations of the intensity in the illumination pupil into a uniform grid in direction-cosine space (s_x, s_y). Each source pixel centered at (s_x, s_y) of finite and constant size in direction-cosine space $\Delta s_x \Delta s_y$ subtends a different solid angle Ω_S at the mask plane.

Lithography simulators, on the other hand, commonly treat the source map as a tabulation of the illuminating intensity (as $I_{\text{source}} \equiv |\boldsymbol{E}|^2$) of the plane wave associated to the center direction cosine coordinates of each pixel [77]. The sampling of the source pupil usually employs steps as small as $\Delta\sigma \leq 0.01$ so that the image produced by any source point within that pixel is essentially identical to the image associated with the center source point. Each source pixel is then associated to an incident plane wave of electric field amplitude given by the pixel tabulation value, often normalized to the maximum intensity, hence representing the weighting of the corresponding source pixel contribution.

Following Rosenbluth [15, 77], the conventional radiometric value assigned to each pixel of a source map as supplied by the manufacturer or extracted from calibrating the scanner is interpreted as a measured integral of source radiance (brightness) over the pixel solid angle. The pixel intensity $I_s(s_x, s_y) \equiv |\boldsymbol{E}|^2$ relates to the brightness B according to $|\boldsymbol{E}|^2 = (8\pi / cn_g)\Delta\Omega_s \text{B}$, with c denoting the free-space speed of light. Since the source pixels subtend equal pixel sizes $(\Delta s_x, \Delta s_y)$ in direction-cosine space, they subtend different solid angles at the mask plane. To satisfy the brightness theorem and to avoid distortions of the simulated aerial image arising from an inaccurate weighting of the integrated effect of each source pixel in the image, the appropriate apodization effect of the solid-angle factor needs to be accounted for when extracting the value of $|\boldsymbol{E}|^2$ from the source map according to:

$$|\boldsymbol{E}|^2 = \frac{S}{\sqrt{1-s_x^2-s_y^2}} \tag{15.32}$$

Fortunately, even with 193-nm lithography scanners with NA as high as 1.35, the 4× reduction factor ensures that the obliquity factor $(1-s_x^2-s_y^2)^{1/2} \geq 0.94$ remains relatively close to 1.

In the above discussion, (15.32) assumes a wave incident on the mask surface propagating in a half-plane quartz substrate, as assumed by numerical electromagnetic simulations. Transmission mask absorbers, however, rest on a quartz substrate of finite thickness, illuminated by the real source wavefront propagating in air before impinging on the mask substrate backside interface. When the mask film-stack configuration is known, we can apply Fresnel transmission coefficients for TE and TM polarizations to calculate the apodization caused by the backside air–glass interface on the plane wave electric field amplitude that we use as an input to the Maxwell's equations solver [63]:

$$\tau_{\mathrm{TE}} = \frac{|E_g|}{|E_a|} = \frac{2n_a\cos\theta_a}{n_a\cos\theta_a + n_g\cos\theta_g}, \qquad \tau_{\mathrm{TM}} = \frac{|E_g|}{|E_a|} = \frac{2n_a\cos\theta_g}{n_a\cos\theta_g + n_g\cos\theta_a} \tag{15.33}$$

where θ_a and θ_g denote the angle formed by the incident plane-wave wavevector $\boldsymbol{k}_{\mathrm{inc}}$ and the optical axis $\hat{z}$ in air and glass, respectively; and n_a and n_g denote the refractive index of air and glass, respectively.

Source Polarization

Each incident plane wave impinging on the mask, the entrance pupil or wafer surface, has a polarization state given by the direction of the electric field with respect to the plane of incidence formed by the wavevector $\boldsymbol{k}$ and the vector z normal to the mask surface. In this scenario, we can define TE polarization when the electric field is normal to the plane of incidence as shown in Fig. 15.16(a), and TM polarization when the electric field is parallel to the plane of incidence as shown in Fig. 15.16(b).

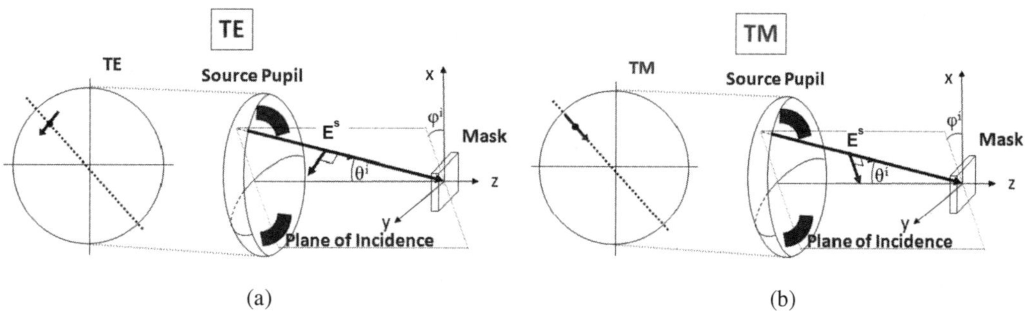

Fig. 15.16 Definition of source polarization in terms of the source pupil diagram and orientation of the corresponding electric field impinging on the mask surface. (a) TE or azimuthal polarization — electric field is normal to the plane of incidence; (b) TM or radial polarization — electric field is parallel to the plane of incidence.

With the usual so-called *Manhattan* layouts, other common orthogonal polarization states are the result of dividing the illumination into X-oriented and Y-oriented source polarizations, where the x-axis and y-axis are parallel to the edges of the mask polygons. The polarizations described above and other common polarization options enabled by the condenser optics for the illuminating source are illustrated in Fig. 15.17.

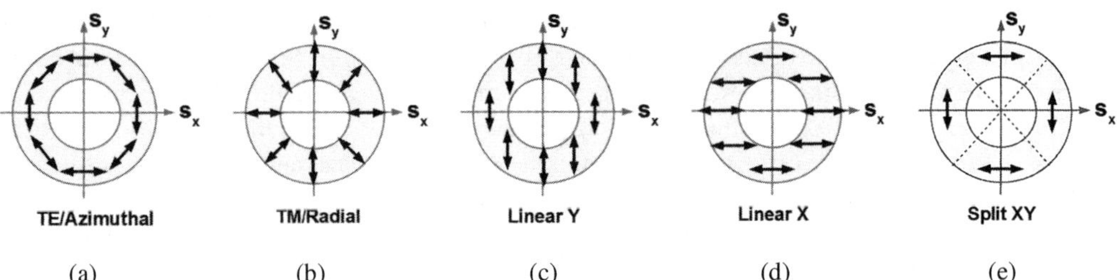

(a) (b) (c) (d) (e)

Fig. 15.17 Polarization options supported by the condenser optics to illuminate the photomask: (a) TE or azimuthal polarization; (b) TM or radial polarization; (c) linear Y polarization; (d) linear X polarization; (e) split X-Y polarization.

Unpolarized light can be mathematically modeled with the incoherent superposition of the separate results using two orthogonal polarizations, which can be TE-TM, X-Y, or any other combination of mutually orthogonal polarization states.

Finally, the polarization of the incident wave is determined from (15.30) applied to the source-pupil coordinates (s_x, s_y), by substituting the value of the pupil components of the electric field following the diagram of Fig. 15.14(b), where for the most common polarization optics we have:

X-polarization:
$$\begin{bmatrix} E^{\mathrm{s}}_{s_x} \\ E^{\mathrm{s}}_{s_y} \end{bmatrix} = \begin{bmatrix} 1 \\ 0 \end{bmatrix} \tag{15.34}$$

Y-polarization:
$$\begin{bmatrix} E^{\mathrm{s}}_{s_x} \\ E^{\mathrm{s}}_{s_y} \end{bmatrix} = \begin{bmatrix} 0 \\ 1 \end{bmatrix} \tag{15.35}$$

TE or azimuthal polarization:
$$\begin{bmatrix} E^{\mathrm{s}}_{s_x} \\ E^{\mathrm{s}}_{s_y} \end{bmatrix} = \begin{bmatrix} -\sin\varphi_{\mathrm{s}} \\ \cos\varphi_{\mathrm{s}} \end{bmatrix} \tag{15.36}$$

TM or radial polarization:
$$\begin{bmatrix} E^{\mathrm{s}}_{s_x} \\ E^{\mathrm{s}}_{s_y} \end{bmatrix} = \begin{bmatrix} \cos\varphi_{\mathrm{s}} \\ \sin\varphi_{\mathrm{s}} \end{bmatrix} \tag{15.37}$$

Unpolarized: Two separate simulations are necessary with orthogonal polarizations to be incoherently superposed in order to simulate unpolarized light. These two polarizations can be TE/TM or X/Y, or any other pair of orthogonal polarizations.

15.3.3 Partially Coherent Illumination: The Hopkins Method

Abbe's method as described in Section 15.3.1 is relatively straightforward to understand. Each source point leads to plane waves diffracted from the mask that pass through the lens to the aerial image plane. In the Köhler illumination configuration, the aerial images produced by two source points are mutually incoherent. The final aerial image due to an extended lithography source can then be calculated using (15.8) as the incoherent superposition of all the aerial image intensity distributions generated by each source point. When using a source map, this source integration is typically implemented as a sum over a finite grid of discrete (s_x, s_y) points, each representing a small square pixel of the source pupil. However, whenever the mask changes, optical proximity corrections are applied, or we move to a new location on the mask, one is forced to recompute the entire complicated interaction between the mask response and the relevant transfer function(s) of the source and lens system.

In contrast, in the *transfer cross coefficient* or *Hopkins method* [24, 79], the order of integrations is swapped so that the integration over the source can be carried out first. This allows the source and lens system to be succinctly described by a set of *transmission cross-coefficients* (TCC), which turn out to be independent of the mask shapes. As a result, we can calculate and store these coefficients once and then re-use them again and again with every mask shape across the entire reticle.

In (15.8) for the aerial image intensity distribution, $\boldsymbol{E}^{\text{image}}(\boldsymbol{r}_w; s_x, s_y)$, we are fully prepared if the set of diffracted plane waves, $\tilde{\boldsymbol{E}}^{\text{M}}_{s_x, s_y}$, should be completely different for each unique source point (s_x, s_y). However, since angles in the object space remain in or near the paraxial regime, it is reasonable to assume that the fields diffracted from the reticle vary only slightly as the angle of incidence varies. This allows us to use the Hopkins approximation, where mask diffraction is assumed to be independent of the incident angle, as in:

$$\tilde{\boldsymbol{E}}^{\text{M}}_{s_x, s_y}(\text{M}w_x, \text{M}w_y) \approx \tilde{\boldsymbol{E}}^{\text{M}}_{0,0}(\text{M}w_x - s_x, \text{M}w_y - s_y) \equiv \tilde{\boldsymbol{E}}^{\text{M}}_{0,0}(f, g) \tag{15.38}$$

The Hopkins approximation states that the diffracted pattern produced by an incident wave propagating in a direction defined by the vector cosines (s_x, s_y) is just a "tilted" version of the diffraction pattern produced by a wave, $\tilde{\boldsymbol{E}}^{\text{M}}_{0,0}$, at normal incidence, shifted in spatial frequency by exactly (s_x, s_y). Using the change of variables $(\text{M}w_x - s_x, \text{M}w_y - s_y) = (f, g)$, the Hopkins approximation allows the rearrangement of the integrals in the aerial image calculation (15.8) as:

$$I^{\text{image}}(\boldsymbol{r}) = \iiiint\limits_{f, g;\, f', g'} \textbf{TCC}(f, g; f', g')\, \tilde{\boldsymbol{E}}^{\text{M}}_{0,0}(f, g)\, \tilde{\boldsymbol{E}}^{\text{M}^*}_{0,0}(f', g')\, df\, dg\, df' dg' \tag{15.39}$$

where the integration over the source pupil, having been swapped to become the inner integral, provides the TCC coefficients:

$$\textbf{TCC}(f, g; f', g') = \iint\limits_{s_x^2 + s_y^2 \le \text{NA}_c^2} \begin{bmatrix} I_{\text{source}}(s_x, s_y)\, \boldsymbol{B}(f + s_x/\text{M},\ g + s_y/\text{M}) \cdot \\ \boldsymbol{B}(f' + s_x/\text{M},\ g' + s_y/\text{M})^*\, ds_x\, ds_y \end{bmatrix} \tag{15.40}$$

Here, the function $\boldsymbol{B}(w_x, w_y)$ represents the complex vectorial constant of proportionality between the transmitted fields at the mask and the fields at the wafer plane, as given by:

$$\boldsymbol{B}(w_x, w_y) = \frac{1}{\lambda^2} e^{-jk\hat{\boldsymbol{w}} \cdot \boldsymbol{r}_w}\, O(w_x, w_y)\, \boldsymbol{T}(w_x, w_y) \tag{15.41}$$

The TCC coefficients depend on the source contribution $I_{\text{source}}(s_x, s_y)$ and on the effects of propagating through the projection optics for each source point (s_x, s_y), but are independent of the mask diffraction contributions, $\tilde{\boldsymbol{E}}^{\text{M}}_{0,0}$, and thus from the mask layout. As a result, the TCC coefficients can be separately precalculated and stored in tables for later simulations of any arbitrary mask layout, as least until the illumination scheme, NA, or focus plane is changed. This formulation of the image intensity as the bilinear sum of all interfering pairs of waves exiting the projection lens, with each interfering pair contribution being weighted by the corresponding TCC coefficients, results in a much more computationally efficient formulation than the Abbe's method. As a result, it is the primary workhorse of most commercially available lithography simulators [17, 80].

15.3.4 Image in Resist Interference

The latent image refers to the image intensity distribution inside the photoresist resulting from the interference of the diffraction orders propagating from the mask and focused on the wafer surface by the projection optics, as illustrated in Fig. 15.18(a). Common calculations of the fields inside the resist follow thin-film formulations where each plane wave exiting the lens is used as the input to a matrix routine [51, 58, 59, 61, 81].

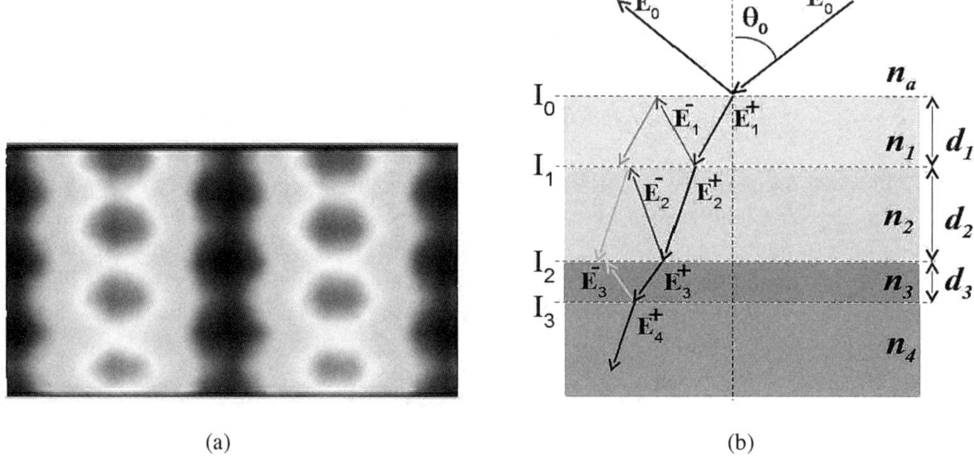

Fig. 15.18 (a) The photoresist layer is often enclosed within a film-stack configuration that is optimized to reduce reflection from the substrate. Despite this, interference between the imaging beams can still occur inside a complex thin-film stack, with reflection and refraction from film interfaces leading to standing waves. (b) Computation of the focused image intensity distribution within the resist requires consideration of all reflections throughout the stack.

Optical models usually do not include the photochemical reaction occurring in the resist during exposure, also known as bleaching [51, 53, 61, 82]. As a consequence of exposure, the resist absorption coefficient decreases with time modifying its refractive index and, thus, the field distribution. Yeung [59] included the effect of bleaching by calculating the refractive index at every time-step until the total exposure dose had been delivered. In his method, he employed Dill's model of positive photoresist under exposure [83]. Incorporation of Dill's model of the bleaching process into full 3-D electromagnetic simulation of the exposure can accurately predict the formation of the latent image profile within the resist. Unfortunately, the computational burden is very high.

The most common practice is to evaluate the latent image inside the resist with optical models, and then use this as input to models of the resist chemical phenomena occurring during resist development. These models of the resist development often lump various chemical steps into one single empirically calibrated model. Calibration of empirical models is a time- and resource-intensive effort, relying on extensive metrology of resist linewidth using scanning electron microscopy on hundreds to thousands of patterns at various process conditions. The resist model is then used to predict the resist contours produced by any latent image intensity distribution within the resist. More physical models would be desirable but they usually lack the computational speed necessary for full chip applications.

The layer of photoresist is often enclosed within a film stack configuration that is carefully optimized to reduce reflection at the expected incident angles. Nonetheless, interference between the imaging beams does occur inside the complex thin-film stack as in Fig. 15.18(b), with reflection and refraction from film interfaces leading to standing waves. Computation of the focused image intensity distribution inside the resist requires accurate computation of all reflections through the stack. When the material films coating the wafer have planar interfaces, an analytical thin-film model of the reflection and refraction at each interface of the film stack can be used to compute the final interference pattern. The computation of the propagation of the imaging orders through the resist film stack can then be included in (15.16) as described in Ref. [80], without any need for numerical electromagnetic simulations. Assuming that layer 2 in Fig. 15.18(b) corresponds to the resist layer, then the thin-film model needs only compute the forward and backward fields inside the resist layer, that is, $\boldsymbol{E}_2^+$ and $\boldsymbol{E}_2^-$ as a function of the fields incident on the top surface of the stack (interface I_0), $\boldsymbol{E}_0^+$. In the context of lithographic imaging on the resist, $\boldsymbol{E}_0^+$ corresponds to the plane waves leaving the exit pupil of the imaging system.

15.4 FDTD MODELING FOR PROJECTION LITHOGRAPHY

In Section 15.2, we showed that the natural evolution of modern lithography has brought it to an interesting inflection point. Even as improvements in NA and λ have slowed down, the use of optical proximity correction and other powerful resolution-enhancement techniques that depend on extensive lithography simulation have allowed feature sizes to continue to shrink by significant reductions in the k_1 factor. At the same time, chip sizes have not changed, meaning that the number of features in a VLSI design has increased exponentially. Despite this, reticle patterns containing billions of features can still be completely evaluated for on-wafer performance in a reasonable time frame. While some of this is due to the improved computational power provided by these VLSI products themselves, a critical aspect is that many of the computations needed to simulate these optical processes can be greatly simplified, especially for propagation angles close to normal incidence.

However, as dimensions continue to shrink, the underlying electromagnetic field interactions begin to have a significant impact on the actual patterns printed on the wafer. In particular, mask topography effects are a critical source of simulation errors, especially for alternating phase-shift and other masks [43, 84–86] that can offer higher performance through modulation of both amplitude and phase. In this regime, the electromagnetic field response of the reticle becomes a complicated function involving incident polarization, incident angle, mask topography, and mask composition [32, 87, 88]. These same effects were always present for amplitude-only masks and in previous lithographic generations, but at a level that could either be ignored or finessed.

The development, calibration, and benchmarking of fast approximation methods that can accurately accommodate these mask-topography issues require many rigorous computations of the exact 3-D electromagnetic fields. While these computations need not be as fast as the full-chip simulations, fast turn-around time is critical. The FDTD method based on Yee's algorithm [89, 90] provides various advantages over other algorithms when used in computational lithography applications. FDTD has long been used to simulate the electromagnetic response of the reticle in optical lithography [91] and of the distribution of the fields at the wafer plane when imaging into a surface with topography [92].

FDTD is highly flexible and does not require prior mesh generation, an important advantage when many different simulation objects (mask layouts) will be run. It can handle almost all relevant mask materials and all layout geometries, and is relatively easy to parallelize across multiple computational nodes, yet remains highly scalable as the number of these nodes grows. FDTD can also simulate 3-D geometries with better turn-around time than the next-most frequently used algorithm in optical lithography, rigorous coupled-wave analysis (RCWA). Other algorithms have also been proposed to calculate the field transmitted by the photomask [93, 94], but have not proved to be as straightforward to apply to large-area simulations as FDTD.

15.4.1 Basic FDTD Framework

Large-scale FDTD simulations inherently involve considerable memory and run-time constraints. For example, a good rule of thumb in FDTD calls for the simulation space to be discretized into at least 20 grid cells per wavelength, independent of the particular geometry. For a 3-D simulation of a 2-D test pattern spanning $2\ \mu m \times 2\ \mu m$ @wafer (and thus $8\ \mu m \times 8\ \mu m$ @mask), this results in a simulation with roughly $800 \times 800 \times 100$ grid cells, or a minimum of 2 GB of memory (single-precision computation, ~30 bytes/grid cell).

On the other hand, FDTD offers enormous flexibility. Once a simulation framework is in place, the same algorithm is capable of handling almost all relevant mask materials and layout geometries that can be suitably discretized into that framework. Furthermore, as a nearest-neighbor algorithm, FDTD is quite straightforward to parallelize across multiple computational nodes and remains highly scalable as the number of these nodes grows large. Considering the increasing availability of massively parallel supercomputers such as IBM's BlueGene/L (BG/L) and BlueGene/P with thousands of processors [95], parallelization offers a highly practical path toward significant increases in the size (and number) of electromagnetics simulations that can be performed in a reasonable turn-around time. Such capability allows the precise computation of the electric fields for small test patterns similar to those that are used for calibrating OPC and other RETs. The demonstrations shown in this chapter use our in-house FDTD implementation, parallelized for the BlueGene/L.

While many mask patterns in circuit designs are inherently periodic, most lithography simulators employ periodic boundary conditions even when simulating nonperiodic patterns. In such cases, the structure of interest is padded with enough unpatterned area that the periodicity is not problematic. Hence, FDTD simulations for lithography are performed over a very thin but very wide domain, as illustrated in Fig. 15.19, which sandwiches the mask in the z-dimension with quartz on the top and air on the bottom. This allows us to compute the near fields of the mask at a plane very close to the mask surface, as described in (15.13) and Fig. 15.11.

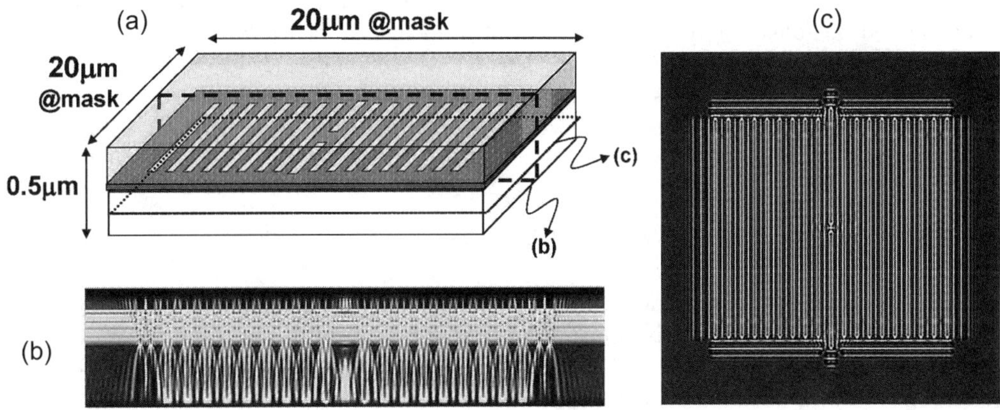

Fig. 15.19 (a) Perspective view of the FDTD simulation domain; (b) side view; (c) top view.

The incident wave is introduced inside the quartz substrate and propagates along the optical z-axis, and the diffracted orders of the mask emerge into the air downstream. Perfectly matched layer (PML) absorbing boundary conditions (ABCs) are used at the top and bottom of the FDTD grid to simulate an infinite extension of the computational domain along the z-axis. This virtually eliminates nonphysical back-reflections from the outermost z-boundaries of the grid [90, 96]. In our implementation, we use the convolutional PML [97]. The FDTD model includes at least 2–3λ of both the glass upstream and the air downstream of the mask before the PML ABCs. This helps ensure that light is coupled cleanly into the simulation space, and the diffracted light can be accurately monitored, as discussed below.

The FDTD simulation domain has a lateral size of $(L_x, L_y) = (L'_x/\text{M}, L'_y/\text{M})$, where (L'_x, L'_y) is the intended size of the pattern when printed on the wafer, and $1/\text{M} = 4$. If the pattern to be simulated is aperiodic, L_x and L_y should be large enough to avoid contaminating the results with the proximity effects from the "next" copies of the layout. As a rule of thumb, a minimum of 5λ of space at wafer scale, or (5/M)λ at mask scale, should be added around the outside of an aperiodic pattern.

Lateral periodic boundary conditions inherently constrain the allowed values of diffracted wavevectors to a discrete set of diffraction orders. The simulation domain (L_x, L_y) over which periodic boundary conditions are applied is significantly smaller than the size of the isoplanatic patch, the region over which the outgoing spherical waves can be assumed to be plane waves, as described previously. Thus, we can model the far-field mask diffraction with a Fourier transform of the near fields. The important point here is that we use FDTD to compute the exact fields downstream from the mask, and then Fourier transform from there to the far field. This allows us to correctly incorporate the effects of topography and polarization, unlike the direct Fourier transform of the mask transmission that is invoked in the thin-mask approximation.

Assuming an $L_x \times L_y$ FDTD computational simulation domain, the wavevector $\boldsymbol{k} = (k_x, k_y, k_z) = k[p_x, p_y, (1 - p_x^2 - p_y^2)^{1/2}]$, with $k = 2\pi/\lambda$, can only take the following values:

$$k_x = kp_x = q(2\pi/L_x), \quad q = 0, \pm 1, \pm 2, \ldots$$
$$k_y = kp_y = r(2\pi/L_y), \quad r = 0, \pm 1, \pm 2, \ldots \tag{15.42}$$

where (q, r) represents the integer indices of each diffracted order. This criterion holds for both incoming and outgoing waves. The discrete set of beams diffracted by the mask propagates toward the entrance pupil of the lithography lens with direction cosines ($p_x = \sin\varphi \sin\theta = q\lambda/L_x$, $p_y = \cos\varphi \sin\theta = r\lambda/L_y$), where θ represents the angle formed by the wavevector $\boldsymbol{k}$ and the optical axis $\hat{z}$, and φ is the angle between the projection of wavevector $\boldsymbol{k}$ onto the x-y plane and the positive x-axis.

15.4.2 Introducing the Plane-Wave Input

In FDTD, we establish such periodic boundary conditions along $\hat{\boldsymbol{x}}$ and $\hat{\boldsymbol{y}}$ by satisfying the Bloch condition. For instance, given our simulation of lateral extent L_x and L_y, then we need to enforce $\boldsymbol{E}(\boldsymbol{r} + L_x\hat{\boldsymbol{x}}) = \boldsymbol{E}(\boldsymbol{r} + L_y\hat{\boldsymbol{y}}) = \boldsymbol{E}(\boldsymbol{r})$ at the simulation boundaries. For the incidence angles corresponding to the grating orders of the simulation, then:

$$\exp\left(j\boldsymbol{k}\cdot L_x\hat{\boldsymbol{x}}\right) = \exp\left(j\boldsymbol{k}\cdot L_y\hat{\boldsymbol{y}}\right) = 1 \tag{15.43}$$

and the FDTD simulation requires only real fields. For all other incidence angles, complex field components must be simulated to provide both the real and imaginary components for correctly "wrapping" the periodic simulation boundaries [90, 98]. This allows the simulation to implement $\boldsymbol{E}(\boldsymbol{r} + L_x\hat{\boldsymbol{x}}) = \boldsymbol{E}(\boldsymbol{r})\exp(j\boldsymbol{k}\cdot L_x\hat{\boldsymbol{x}})$ and the analogous equation along the y-dimension. In essence, we are running two FDTD simulations at this point—one for the real or cosine component of this equation, and one for the imaginary (sine) component. Since this approximately doubles both the memory and run-time requirements, from the point of view of FDTD efficiency it is preferable to use only those angles of incidence corresponding to allowed orders.

However, there are two competing issues to consider. The first is that this constraint on input angles may not map well onto the areas of importance within the source pupil. The second is the need to rely on the Hopkins approximation, where a tilt in the illumination is treated simply as a tilt of the output orders (corresponding to a translation in the Fourier domain). While this is still a common assumption in lithography simulations, in the hyper-NA imaging regime, this approximation has been shown to potentially introduce non-negligible errors in the aerial image [32], as we shall discuss in Section 15.5.3. However, while the reticle electromagnetic response should be fully re-simulated for large changes in the incidence angle, negligible variations can be expected in the lithographic response for small changes, and thus interpolation over a small range using the Hopkins approximation should still be able to provide accurate results. For maximal flexibility in selecting a small set of incidence angles to cover the relevant portions of the source pupil, it may be useful to introduce arbitrary angles of incidence. This is certainly true when masks of different sizes (and thus a different set of allowed orders) need to be simulated for the same source pupil.

Since the 3-D FDTD simulation used to simulate a 2-D mask pattern has periodic boundary conditions along the x- and y-axes, we can accurately introduce an "infinitely" wide plane wave without any boundary effects. Furthermore, by using the total-field/scattered-field (TF/SF) sourcing method [90], the input wave driven by electric current (J) sources at the input plane can be subtracted at the H-field positions located one-half Yee-cell upstream. In this manner, the input light propagates only downstream toward the mask, further eliminating stray radiation from the simulation.

Typically, in a lithography simulation, the monochromatic (sinusoidal) source is initially ramped up in amplitude over 8–10 optical cycles, and the FDTD computation is continued until the fields throughout the simulation stabilize. This can either be monitored by the cycle-to-cycle change in the *E*-fields, or by the monitors used downstream of the mask to measure the diffracted orders (see Section 15.4.3 that follows). There is little point in running the simulation any longer than is necessary to obtain this convergence. Since lithography simulations tend to be fairly thin and the incidence angle will rarely exceed 20° from the normal, convergence typically takes between 25 and 50 optical cycles, depending on whether the materials are pure dielectrics or metals.

Before starting to introduce masks into a lithography simulation, it is crucial to ensure that the sourcing method is cleanly generating a single plane wave. For instance, if the simulation space is all glass, then after the fields have stabilized, the resultant *E*-fields should be perfect sinusoidal ribbons across the simulation, and no reflected power should be present in the glass upstream from the one-way TF/SF source. In the presence of just a simple air–glass interface, the ratio of the power reflected by this interface (as measured by a monitor upstream of the driving source) to the power transmitted past this interface can be readily compared to the Fresnel reflection coefficients [24, 63].

As discussed earlier, polarization properties can be assumed to be unaffected by the field propagation throughout the imaging system. Diffraction by the photomask, however, is greatly dependent on the polarization of the incident light and introduces coupling between orthogonal polarizations. To cover all possible input polarizations, we need only complete two FDTD simulations with orthogonal polarizations. For a simple grating, these might correspond to TE polarization, with the electric field polarized parallel to the long lines comprising the grating, and TM polarization with the electric field polarized normal to the grating lines (and thus parallel to the grating vector). We can also model unpolarized illumination simply by combining the intensity rather than field results of these two simulations at the wafer.

Thus for each input source point, the electromagnetic modeling procedure consists of two FDTD simulations (one for each orthogonal input polarization) sandwiched between two simple analytical transformations. The first transformation maps the input plane wave from the source pupil to the reticle input, the second transforms the reticle output to the entrance pupil. We discuss the first transformation here, and the second transformation in the next section (Section 15.4.3).

Following the conventions described in Section 15.3.2, the source map is specified in normalized pupil coordinates $(\sigma_x, \sigma_y) = (M\mathrm{NA_i})^{-1}(s_x, s_y)$. Given propagation along z and an $\exp(j\omega t)$ temporal dependence, the source point (s_x, s_y) maps into a wavevector $\boldsymbol{k}_\mathrm{s}$ upstream of the reticle propagating in air as:

$$\boldsymbol{k}_\mathrm{s} = \frac{2\pi}{\lambda}\left[\cos\varphi_\mathrm{s}\sin\theta_\mathrm{s}\hat{x} + \sin\varphi_\mathrm{s}\sin\theta_\mathrm{s}\hat{y} + \cos\theta_\mathrm{s}\hat{z}\right] \tag{15.44}$$

represented in terms of angles as

$$\varphi_\mathrm{s} = \tan^{-1}\left(\frac{\sigma_y}{\sigma_x}\right), \qquad \theta_\mathrm{s} = \sin^{-1}\left[\frac{\mathrm{NA_i}}{4}\sqrt{(\sigma_x)^2 + (\sigma_y)^2}\right] \tag{15.45}$$

Here, φ_s is the angle between the projection of wavevector $\boldsymbol{k}_\mathrm{s}$ onto the x-y plane and the positive x-axis, θ_s is the angle between $\boldsymbol{k}_\mathrm{s}$ and the positive z-axis [90] as shown in Fig. 15.12, and $\mathrm{NA_i}$ is the numerical aperture of the imaging system. To map the normalized input polarization in the source pupil, $\hat{\boldsymbol{e}}_{\mathrm{s}} = E^{\mathrm{s}}_{s_x}\hat{\boldsymbol{s}}_x + E^{\mathrm{s}}_{s_y}\hat{\boldsymbol{s}}_y$, onto the polarization vector, $\hat{\boldsymbol{e}}_{\mathrm{m}}$, at the mask entrance plane just upstream of the reticle, we can decompose into S and P components as:

$$\hat{\boldsymbol{e}}_{\mathrm{m}} = f_{\mathrm{S}}\,\hat{\boldsymbol{n}} + f_{\mathrm{P}}\left(\hat{\boldsymbol{n}} \times \hat{\boldsymbol{k}}_{\mathrm{s}}\right) \tag{15.46}$$

where $\hat{\boldsymbol{k}}_{\mathrm{s}}$ is the normalized k-vector in the incidence plane, and $\hat{\boldsymbol{n}} = \hat{\boldsymbol{k}}_{\mathrm{s}} \times \hat{\boldsymbol{z}}$ is the normal to the incidence plane. These S and P components relate back to the polarization vector in the source pupil, $\hat{\boldsymbol{e}}_{\mathrm{s}}$, as:

$$f_{\mathrm{S}} = \hat{\boldsymbol{n}} \cdot \hat{\boldsymbol{e}}_{\mathrm{s}}\,, \qquad f_{\mathrm{P}} = \sqrt{1 - \left(f_{\mathrm{S}}\right)^2} \tag{15.47}$$

Then, angle ψ_{s} between the electric field incident on the mask, $\hat{\boldsymbol{e}}_{\mathrm{m}}$, and the normal to the incidence plane, $\hat{\boldsymbol{n}}$ [90], can be written as:

$$\psi_{\mathrm{s}} = \sin^{-1}\left(\hat{\boldsymbol{e}}_{\mathrm{m}} \cdot \hat{\boldsymbol{n}}\right) \tag{15.48}$$

At this point, we have the two orthogonal polarizations for our two FDTD simulations—for a grating lying in the plane of incidence, the S component corresponds to TE polarization and the P component to TM polarization. Of course, for all other grating vectors, this incident angle is a mixture of TE and TM. However, FDTD correctly models the relevant polarization mixing induced by the mask.

Note that since the FDTD simulation introduces the plane wave into an infinite half-space of glass with index of refraction n_{g} representing the reticle substrate, (15.44) and (15.45) should be modified slightly as:

$$\boldsymbol{k}^{\mathrm{g}}_{\mathrm{s}} = \frac{2\pi n_{\mathrm{g}}}{\lambda}\left(\cos\varphi_{\mathrm{s}}\sin\theta^{\mathrm{g}}_{\mathrm{s}}\,\hat{\boldsymbol{x}} + \sin\varphi_{\mathrm{s}}\sin\theta^{\mathrm{g}}_{\mathrm{s}}\,\hat{\boldsymbol{y}} + \cos\theta^{\mathrm{g}}_{\mathrm{s}}\,\hat{\boldsymbol{z}}\right) \tag{15.49}$$

since

$$\theta^{\mathrm{g}}_{\mathrm{s}} = \sin^{-1}\left[\frac{\mathrm{NA_i}}{4n_{\mathrm{g}}}\sqrt{(\sigma_x)^2 + (\sigma_y)^2}\right] \tag{15.50}$$

according to Snell's law. Note, however, that only the z-component of the wavevector, the actual direction of the P-polarized component, and the angle $\theta^{\mathrm{g}}_{\mathrm{s}}$ are actually different. The x- and y-components of the wavevector are identical. Thus, the same Bloch conditions established above apply in both the infinitely wide glass-filled portion at the upstream side of the simulation and the infinitely wide air-filled portion at the downstream side.

15.4.3 Monitoring the Diffraction Orders

In the context of the lithographic optical train, the FDTD simulation needs only to compute the transfer function(s) between the input plane waves coming from source points in the source pupil, and the output plane waves associated with points in the entrance pupil of the lithographic imaging system. A unique FDTD simulation is needed for each of the two polarizations from each source point (s_x, s_y) to be simulated. We will discuss the choice of how many source points to simulate in Section 15.5.3.

We can compactly measure the diffracted orders during the FDTD simulation by taking a running Fourier transform. As the simulation runs, we simply update the real and imaginary coefficients of grating monitors placed downstream from the mask, as in the following for the E_x electric field vector component at location (x, y, z) within the FDTD grid:

$$\mathrm{Re}\left(\tilde{E}_x\right) = E_x \frac{\Delta_x \Delta_y}{L_x L_y} \cos\left(-xk_x - yk_y - zk_z + \omega t\right) \tag{15.51}$$

$$\mathrm{Im}\left(\tilde{E}_x\right) = E_x \frac{\Delta_x \Delta_y}{L_x L_y} \sin\left(-xk_x - yk_y - zk_z + \omega t\right) \tag{15.52}$$

where $\Delta_x \Delta_y$, is the local grid cell area, t is the instantaneous time in seconds within the FDTD simulation, and the particular diffracted order (q, r) to be obtained by correlation is:

$$k_x = \frac{2\pi}{\lambda} \sin\theta_s \cos\varphi_s + \frac{2\pi q}{L_x} \tag{15.53}$$

$$k_y = \frac{2\pi}{\lambda} \sin\theta_s \sin\varphi_s + \frac{2\pi r}{L_y} \tag{15.54}$$

$$k_z = \sqrt{\left(\frac{2\pi}{\lambda}\right)^2 - k_x^2 - k_y^2} \tag{15.55}$$

Similar equations hold for grating monitors $\tilde{E}_y$ and $\tilde{E}_z$ obtained for FDTD fields E_y and E_z.

Here, we are directly taking the Fourier transform into the far field for this particular diffraction order. If the only light in the simulation is a plane wave at angle (θ_s, φ_s) having refracted out of the glass from incident wave (θ_s^g, φ_s), then only the $(q, r) = 0, 0$ monitor will be nonzero. Since the only possible way for light to propagate downstream in such a periodic unit cell is in a diffraction order, if we monitor a large enough set of (q, r) values, we can account for all transmitted light. By monitoring in the glass behind the source (with a factor of n_g included wherever 2π appears), we can similarly monitor all the diffracted light. Since the acceptance angle of a lithography lens is fairly small due to the 4× demagnification factor, only those (q, r) values that fall within an angle range of ±20° need be monitored. Of course, as the simulation extent (L_x, L_y) increases, the number of such relevant diffraction orders increases as well since the angular spacing between each drops as $1/L_x$ and $1/L_y$.

Each monitor is summed over a single z-slice of the simulation without double-counting the periodic boundary cells. Each field component is evaluated at its own position, reflecting the slight (one-half grid cell) offsets within the Yee lattice. The k_x, k_y, (zk_z), and ω variables can be precomputed externally to reduce the number of internal computations during FDTD computation. Since the input light is monochromatic with the single frequency $\omega = 2\pi c/\lambda$, we need only update these monitors in quadrature, e.g., twice per optical cycle, at any two time-points separated by a quarter period of the sinusoidal wave:

$$T/4 = 2\pi/4\omega \tag{15.56}$$

For this reason, it is extremely convenient to pick the size of the FDTD time-step such that the number of time-steps in an optical cycle is exactly a multiple of four. This can be trivially done without violating the Courant stability limit.

One or more of these monitors can be observed to detect the convergence of the FDTD simulation and signal completion. For instance, the intensity (the spatial integral of the squared E-field) in the 0 diffracted order is almost always a good choice. The easiest way to detect convergence is to run until the standard deviation of this intensity over the previous 5–10 optical cycles has fallen below some suitable accuracy target (1% or 0.1%). It is often the case that, toward the end of such FDTD simulations, the monitors all show decaying oscillatory behavior, converging to a final saturated value but varying both above and below this value. Thus, even with fairly short FDTD runtimes, higher accuracy can potentially be obtained by not just averaging each monitor over the last 5–10 optical cycles before convergence, but by fitting this oscillatory behavior to extract the saturated real and imaginary coefficients of each diffraction monitor.

15.4.4 Mapping onto the Entrance Pupil

After the FDTD run completes, six real values per diffraction order will be available, corresponding to three complex-valued electric fields. Typically, the role of such an FDTD simulation is to replace the thin-mask approximation (TMA) code in a commercial lithography simulator with results that more accurately reflect the electromagnetic field effects. In such cases, the FDTD results must be reinserted into the commercial simulator for inclusion of the same TCC, obliquity, and Jones matrix factors that it uses, even for TMA, to compute the at-wafer or in-resist intensity. Some simulators can simply import the complex electric field amplitudes directly in the x, y, z coordinate system at the mask.

Other simulators, following the conventions stated by the Zeiss-IBM joint formulation work [78], ask for the transfer function between the x- and y-polarized light in the pupil plane and the x- and y-polarized light in the entrance (and thus exit) pupil, as was described in Section 15.3.2. Here, each FDTD simulation must correspond to either x- or y-input polarization at the source pupil plane. For either $s_x = 0$ or $s_y = 0$, these input waves retain their polarization at the mask; for all other source points, however, the resulting linear polarization to be introduced into the simulation must be carefully obtained through conversion into components normal and parallel to the incidence plane, as discussed in Section 15.4.2.

The benefit of this approach is that each of the two FDTD simulations then directly produces half of the required transfer coefficients. As for (15.29), for an x-polarized input, the complex amplitudes:

$$XX(q,r) = \tilde{E}_x \Gamma_{xx} + \tilde{E}_y \Gamma_{yx} + \tilde{E}_z \Gamma_{zx} \tag{15.57}$$

$$XY(q,r) = \tilde{E}_x \Gamma_{xy} + \tilde{E}_y \Gamma_{yy} + \tilde{E}_z \Gamma_{zy} \tag{15.58}$$

represent the coupling into x- and y-polarized output at the entrance pupil. Similarly for y-polarized input:

$$YX(q,r) = \tilde{E}_x \Gamma_{xx} + \tilde{E}_y \Gamma_{yx} + \tilde{E}_z \Gamma_{zx} \tag{15.59}$$

$$YY(q,r) = \tilde{E}_x \Gamma_{xy} + \tilde{E}_y \Gamma_{yy} + \tilde{E}_z \Gamma_{zy} \tag{15.60}$$

Since we performed the conversion between the source pupil and the mask plane in converting the x- (or y-) polarized input into an incident wave for driving the FDTD simulation, now we must convert the diffracted beams back into the entrance pupil. The appropriate conversion coefficients are:

$$\Gamma_{xx} = 1 - \frac{p_x^2}{1 + \sqrt{1 - p_x^2 - p_y^2}} \tag{15.61}$$

$$\Gamma_{xy} = \Gamma_{yx} = -\frac{p_x p_y}{1 + \sqrt{1 - p_x^2 - p_y^2}} \tag{15.62}$$

$$\Gamma_{yy} = 1 - \frac{p_y^2}{1 + \sqrt{1 - p_x^2 - p_y^2}} \tag{15.63}$$

$$\Gamma_{zx} = -p_x \tag{15.64}$$

$$\Gamma_{zy} = -p_y \tag{15.65}$$

where

$$p_x = \sin\theta_s \cos\varphi_s + q\lambda / L_x \tag{15.66}$$

$$p_y = \sin\theta_s \sin\varphi_s + q\lambda / L_y \tag{15.67}$$

Once these conversions are in place, then simulations can be performed through the commercial lithography simulator, with the FDTD step performed offline in the user's code and re-inserted.

Before simulating critical test features of interest, it is important to verify that FDTD simulations match reliable benchmarks as established by exercising either the TMA or electromagnetic field features of a standard commercial simulator. For example, Fig. 15.20 compares FDTD results with the outputs of the RCWA mode of a commercial simulator for the diffraction efficiency of a 1-D equal-line and space grating on a 6.5% attenuated PSM mask.

For the results shown in Fig. 15.20, the long dimension of the grating grooves is parallel to the y-axis, with normally incident, linearly polarized plane waves with TE or y-polarization, and TM or x-polarization. The scalar or thin-mask approximation of the 0th and 1st diffraction efficiency would predict constant values of value 0.14 and 0.16 for TE and TM polarization, respectively, independent of pitch. Attenuated PSM masks are known to induce larger transmission losses on the polarization component parallel to the topography edges (TE in 1-D layouts) than on the component normal to the topography edges (TM). This is argued to be a consequence of the continuity of the parallel field on the MoSi surface, which induces deeper penetration of the TE field in the lossy MoSi layer.

As the grating pitch decreases toward the illuminating wavelength, the diffraction intensity falls and rapidly oscillates. These oscillations result from Wood's anomaly [99, 100]—the interaction (at the right combination of incident angle and pitch) between diffracted orders that skim along the mask surface and the propagating orders. These pitch-dependent oscillations of the diffracted pattern can introduce significant variation from the aerial image intensity as predicted by the scalar models, causing problems both as a function of pitch (as seen in Fig. 15.20) and across the source pupil (Section 15.5.3).

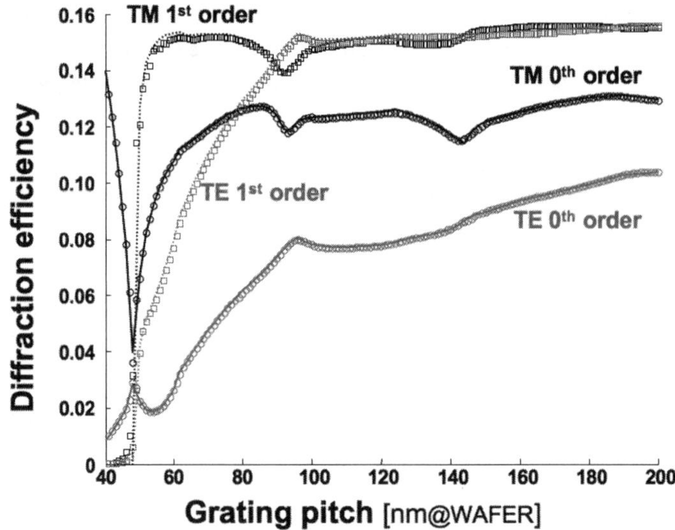

Fig. 15.20 Comparison of FDTD (symbols) and commercial electromagnetic solver (lines) results for the diffraction efficiency of the 0th and 1st TE and TM orders of a 1-D equal-line and space grating on a 6.5% attenuated PSM mask consisting of 70 nm of molybdenum silicide (MoSi) on quartz.

The mask near fields have to be captured at a plane parallel to, and downstream from, the mask surface, but not too close to it. If these fields are captured too close to the mask surface, then the measurement can potentially be influenced by evanescent surface waves, especially for mask pitches at or smaller than the wavelength.

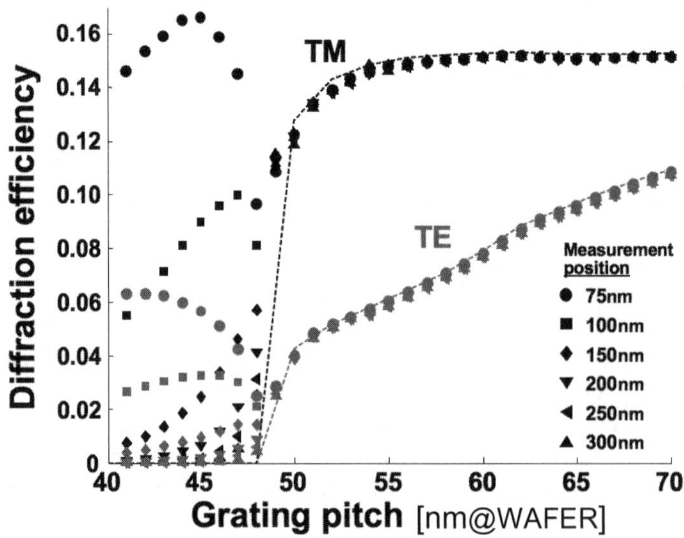

Fig. 15.21 Diffraction efficiencies of the TE and TM 1st orders of the grating of Fig. 15.20 are plotted for measurement planes at various distances from the quartz surface. Note that the first exit plane (at 75 nm) is only 5 nm beyond the surface of the 70-nm-thick MoSi stack.

Figure 15.21 displays an example of the TE and TM 1st diffraction orders of the same grating as Fig. 15.20, with a mask thickness of 70 nm. The first exit plane (at 75 nm) is located only 5 nm beyond the surface of the MoSi stack, and thus shows significant diffraction efficiency even when the pitch is below the cut-off value (48 nm) for evanescent waves. Since the location of the mask exit plane can clearly influence the monitored amplitudes, the mask exit plane should be located at least 50 – 100 nm beyond the last surface of the mask for any simulation, and even further when high accuracy is desired. Obviously, such increases in the overall thickness of the FDTD simulation space must be balanced against the increase in memory requirements and the impact on run time, since more field values must be computed at each FDTD time-step, and the convergence of each FDTD simulation may require several more optical cycles.

15.4.5 FDTD Gridding

The FDTD algorithm is theoretically capable of arbitrarily low error in simulating Maxwell's equations as the grid-cell size approaches zero. However, for reasonable choices of grid-cell size in conventional FDTD, material boundaries are "staircased" to the nearest grid cell. For instance, two patterns in which the absorber–air interfaces are offset by distances smaller than one grid cell could be represented by the same simulation geometry. Figure 15.22 shows the errors introduced by using 8-nm grid cells relative to using 1-nm grid cells, for the 0th-order diffraction efficiency of 40- to 100-nm pitch gratings at wafer scale on a 6.5% absorber-transmission Atten.-PSM mask blank, as the grating is moved laterally with respect to the grid by distances of 0 to 6 nm.

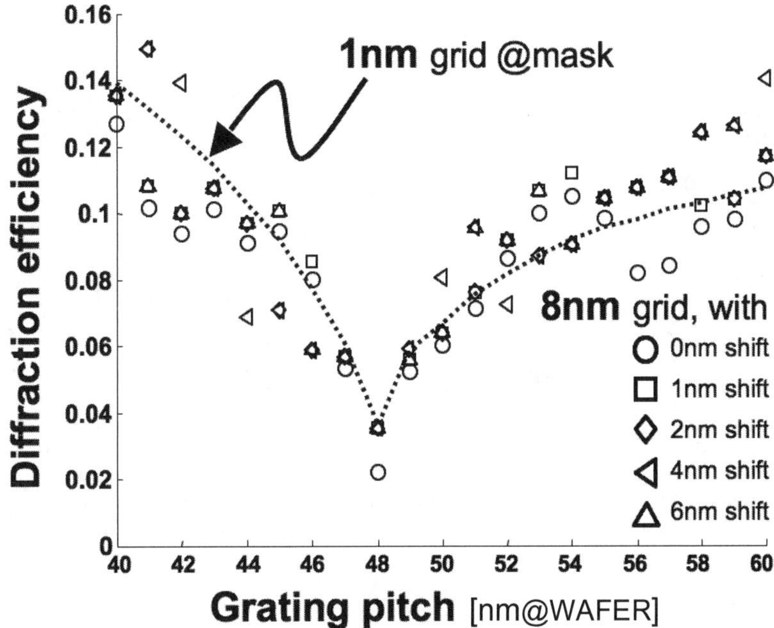

Fig. 15.22 Error introduced by standard FDTD using 8-nm grid cells relative to 1-nm grid cells for the diffraction efficiency of the 0th order of small-pitch gratings on a 6.5% absorber-transmission Atten.-PSM mask blank, as the grating is moved laterally with respect to the grid by distances of 0 to 6 nm.

To improve on the trade-off between computational complexity and simulation accuracy inherent within FDTD, we use a simple subpixel-smoothing technique [101, 102] (discussed in Chapter 6) that permits accurate simulation of dielectric interfaces not aligned with the grid. With this technique, in a preprocessing step, an effective dielectric constant is computed for every E-field component adjacent to a dielectric interface. The resulting FDTD update coefficients are stored for each E-component and used instead of the default values during time-stepping. Hence, there is no impact on the running time of the FDTD simulation. (However, a more complex algorithm is needed for arbitrarily oriented metallic interfaces [103].)

The correction of FDTD staircasing errors using subpixel smoothing allows precise modeling of the exact material interfaces at the mask and can offer noticeable accuracy improvement. For example, Fig. 15.23 shows the diffraction efficiency of the 0th order of the same 40- to 100-nm pitch gratings at wafer scale of Fig. 15.22 discretized to the same 8-nm grid, but simulated using subpixel smoothing. With this method, an accuracy level that is comparable to that obtained using fine 1-nm grid cells is achieved using coarse 8-nm grid cells, but with a much faster simulation time.

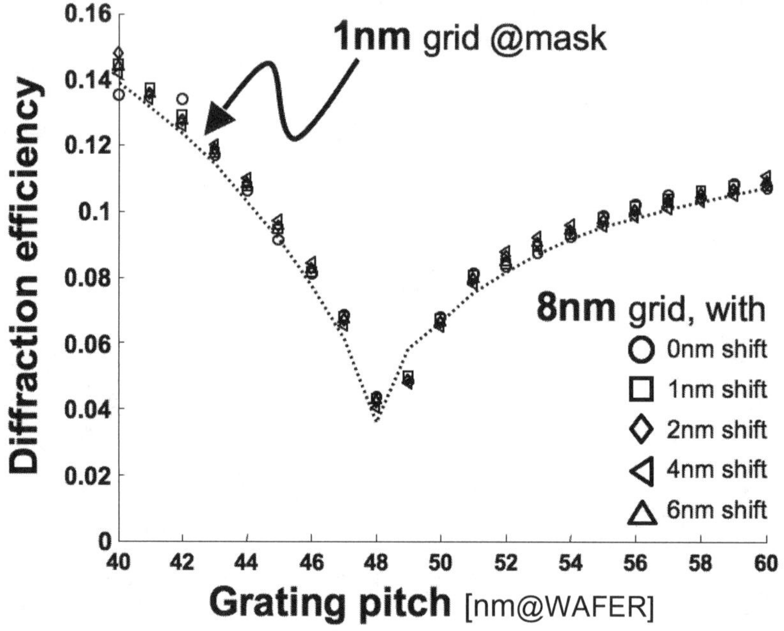

Fig. 15.23 The diffraction-efficiency example of Fig. 15.22 is repeated, but using a subpixel-smoothing FDTD technique (discussed in Chapter 6) to correct for the staircasing errors in the coarse grid comprised of 8-nm unit cells. Now, the coarse grid provides accuracy that is comparable to the fine grid comprised of 1-nm unit cells, but with a much faster simulation time.

15.4.6 Parallelization

High-performance computing can be used for highly accurate, fully 3-D electromagnetic simulation of very large mask layouts, conducted in parallel with a reasonable turn-around time. A 3-D simulation of a large 2-D layout spanning $5 \times 5\,\mu\text{m}$ at the wafer plane (and thus

20 × 20 × 0.5 µm at the mask) results in a simulation with roughly 12.5 GB of memory for a grid cell size of 10 nm at the mask, using single-precision arithmetic with ~30 bytes/grid cell. Such a large layout can be computed in approximately 1 hour using one IBM BlueGene/L "midplane" containing 512 dual-processor nodes with 256 MB of memory per processor. This allows numerous moderate-size simulations of practical interest in lithography development to be farmed out to multiple nodes and executed with reasonable runtimes.

Our scaling studies on BlueGene/L demonstrate that simulations larger than 10 × 10 µm at the mask can be computed in a few hours, provided only that enough compute nodes are available to allow the problem to fit in the available memory per node. Based on the scaling behavior shown in Fig. 15.24, one can estimate the resources necessary for a given simulation job to be completed with a predetermined running time.

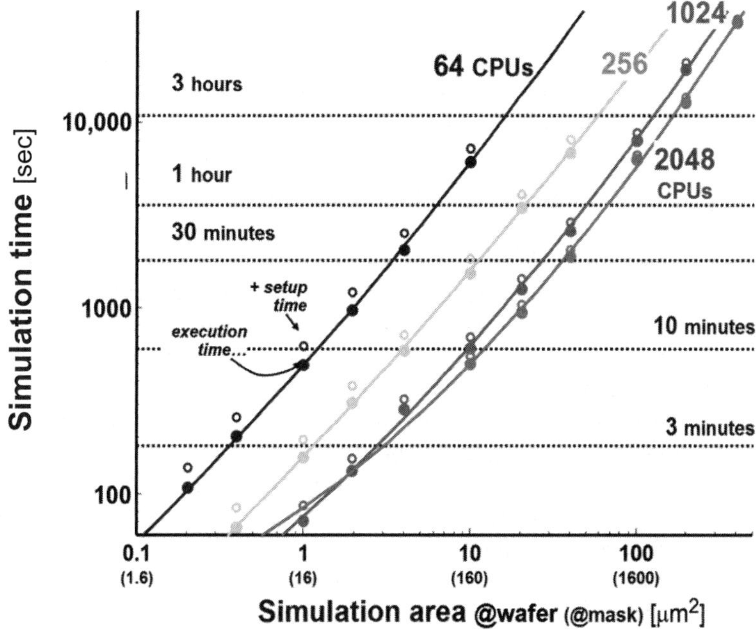

Fig. 15.24 FDTD simulation time using IBM BlueGene/L vs. simulation area and number of processors. Due to low BlueGene/L communications overhead, the total computation time scales with the computational load of each processor, not the total problem size. Thus, larger problem sizes can be simulated in the same time by simply increasing the number of processors.

We have found that FDTD computational lithography is both feasible and practical when executing the FDTD simulations (after subpixel-smoothing preprocessing) on the IBM BlueGene/L. With subpixel smoothing, refining the grid cell size from 8 nm to 2 nm at the mask produces only ~3.6% change in the diffraction pattern, but increases the running time to over 1 hour on 1024 CPUs. Further refining the grid cell size to 1 nm at the mask produces negligible (~0.25%) change in the diffraction pattern, but greatly expands the running time to 5 hours on 2048 CPUs. Therefore, in our experience, using 8-nm grid cells at the mask combined with subpixel-smoothing preprocessing offers a reasonable trade-off between the accuracy of the FDTD simulations, their running times, and the required computer resources.

15.5 APPLICATIONS OF FDTD

In previous sections, we showed how FDTD can be used to rigorously obtain the electromagnetic fields produced by the mask and then the resulting intensity image at the wafer. Unfortunately, the computational cost of applying FDTD on large areas prohibits its use at full-chip scale. Despite this limitation, FDTD has several important roles in computational lithography. The first is in gauging the impact of additional electromagnetic field effects when weighing the benefits and drawbacks of different kinds of mask blanks. These effects include the polarization-dependent interaction of the finite-thickness mask topography with the incident illumination. Benefits and drawbacks are gauged in terms of lithographic metrics such as pattern linewidth at best focus, shifts of the plane of best focus, and changes in contrast or depth of focus. Another role of FDTD is in designing augmentation schemes that insert awareness of field effects into the basic thin-mask approximation without significantly increasing its computational complexity. Typically, FDTD is essential in the design, calibration, and verification phases of these algorithms. Finally, FDTD is essential in verifying and maintaining the validity of the Hopkins approximation discussed in Section 15.3.3, which has become important as the source pupil has become an essential design variable in modern lithography.

15.5.1 Electromagnetic Field Impact of Mask Topography

Overall, while electromagnetic field effects show a complex dependence on the mask and illumination characteristics, they can be understood in terms of variations introduced to the amplitude and phase (or real and imaginary) components of the fields diffracted by the thin mask approximation. Transmission losses on the mask are responsible for amplitude errors on the aerial image intensity, and can often be approximated with a simple bias applied to the mask edge; phase errors tend to shift the position of best focus. The impact of the topography in modifying the diffracted fields from the ideal TMA predictions is known to be substantially more degrading for masks that modulate both amplitude and phase [28, 104]. Even within the same mask type group, these electromagnetic field effects can vary widely with mask composition, thickness, profile, and incident polarization [87, 105–107].

Alternating phase-shifting masks introduce a 180° phase difference between the transmission through deeply etched and unetched mask apertures separated by sections of opaque absorber, per Fig. 15.25(a). The electromagnetic scattering from this abrupt topography results in substantial losses and a clear transmission imbalance between etched and unetched apertures, as illustrated qualitatively in Fig. 15.25(b). This important case of mask topography impact on the lithographic aerial image, first recognized in 1992 by Wong and Neureuther [91], was one of the first cases where electromagnetic effects became noticeable enough that corrections were required on mask layouts.

Fortunately, a simple mask bias to increase the size of the phase-shifting openings is often enough to compensate for the loss in transmission. Studies demonstrated that the optimum bias to compensate for scattering losses is approximately constant for mask sizes down to the size of the wavelength [64, 108]. While phase-depth biasing and undercutting of the etched quartz can also enhance transmission and correct for phase errors, these techniques are much less practical. Thus, a constant mask bias has prolonged the applicability of computationally efficient TMA. However, as minimum pitch and critical dimension continue to shrink, effects such as the additional scattering losses at the edges of Alt.-PSM or Atten.-PSM mask film stack eventually lead to large CD errors that can no longer be compensated by dose or mask size adjustments.

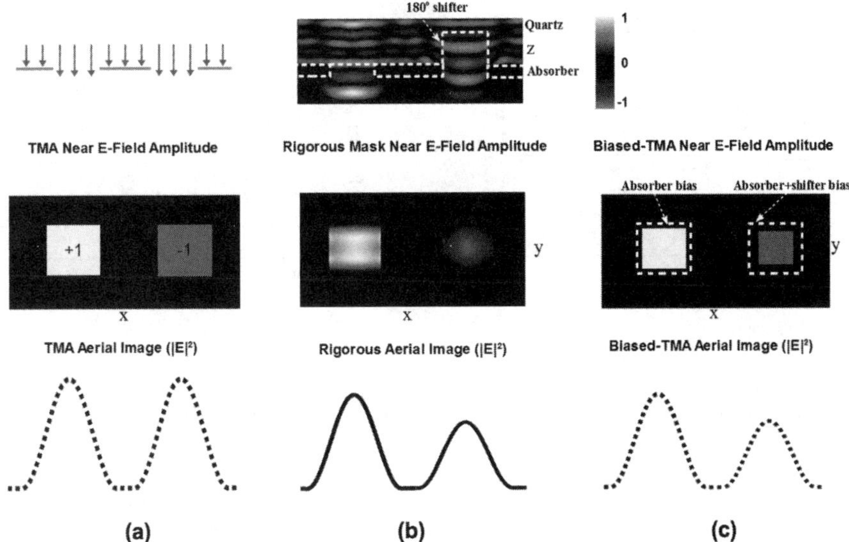

Fig. 15.25 Schematic illustration, for an Alt.-PSM mask, of the mask field and the aerial image produced by (a) the thin mask approximation (TMA); (b) the real electromagnetic fields on the mask; and (c) the thin mask approximation with absorber and shifter CD biases. The shifter bias is applied to the 180° shifter opening to model amplitude losses, and the absorber bias is applied to every edge (all polarities) to model scattering losses.

The phase component of the electromagnetic field-induced perturbation is particularly problematic because it shifts the plane of best focus at the wafer in a pitch-dependent manner. Unfortunately, a simple bias on the mask cannot compensate for this effect [109]. This pitch-dependent focus shift has been experimentally verified on 6.5% absorber-transmission Atten.-PSM gratings using a 193-nm aerial image microscope [104]. Best focus was measured for grating pitches between 90 and 150 nm, mask linewidths between 40 and 60 nm as described in wafer-plane dimensions, and partially coherent $0.7\sigma_{\text{in}} - 0.93\,\sigma_{\text{out}}$ dipole illumination with 1.2 NA. Figures 15.26(a) and 15.26(b) show the best focus obtained for each pitch averaged across all mask linewidths, measured separately for TE (parallel to line edges) and TM (perpendicular to line edges) polarizations. Also shown is verification by full electromagnetic field simulations, and the erroneous predictions of TMA.

We show in Appendix 15D that for a simple case of two-beam interference (i.e., 0th and 1st orders), an electromagnetic field-induced phase error of δ between interfering beams leads directly to a shift of best focus that is linearly proportional to this error:

$$z_{\text{BF}} = \frac{\delta}{k'(\cos\theta'_{\text{inc}} - \cos\theta'_d)} \tag{15.68}$$

where θ_{inc} is the propagation angle of the incident light, $\theta_d = \sin^{-1}(\sin\theta_1 - \sin\theta_{\text{inc}})$ is the propagation angle of the resulting 1st diffracted order in the object space, and θ'_{inc} and θ'_d represent the corresponding angles in the image space according to (15.21). The difference between these two angles is effectively the z-component of the grating vector coupling the incident wave into the diffracted wave.

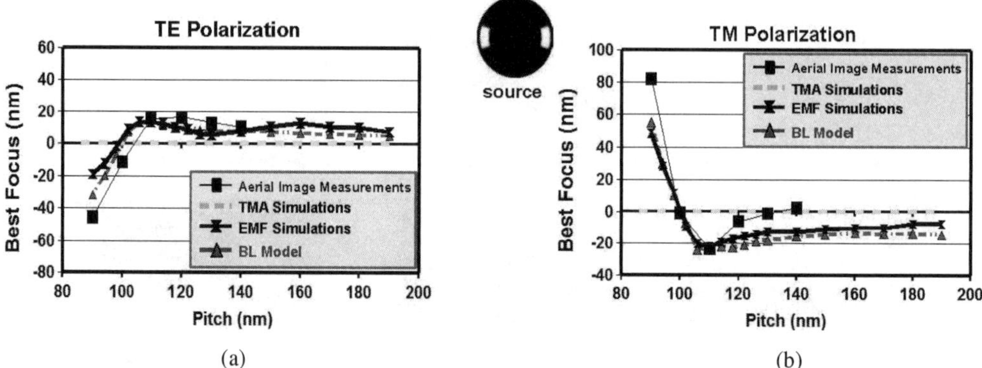

Fig. 15.26 Experimental measurements and full electromagnetic field (EMF) simulations show the pitch-dependent focus shift introduced for (a) TE and (b) TM polarization by the imaginary component of transmission errors associated with field effects on the reticle. Measurements were made for 6.5% absorber-transmission Atten.-PSM gratings with pitches between 90 and 150 nm using a Zeiss 193-nm polarization-capable AIMS microscope. For both AIMS measurements and full electromagnetic field simulations, the best focus per pitch was averaged across mask linewidths between 40 and 60 nm at wafer-scale dimensions.

According to (15.68), depending on the value of θ_d (and equivalently θ_d'), small variations of the transmitted phase through the mask can translate into a significant shift of the plane of best focus. While at normal incidence these focus shifts are manageably small, phase deviations are amplified by oblique illumination, especially for angles and pitch values close to the resolution limit. This effectively shrinks the common process window within which reliable printing can be obtained across a range of grating pitches. When $\theta_d = \theta_{\text{inc}}$ for a specific grating pitch and incident angle, the plane of best focus is effectively located at "∞," resulting in a grating image that is *isofocal*—effectively independent of focus.

15.5.2 Making TMA More Electromagnetic-Field Aware

Mask Bias for Amplitude Errors

As seen above, we can interpret the distortions of the diffracted electromagnetic fields induced by the mask topography as decomposed into an amplitude and a phase component. We refer to these distortions as errors in the sense that these are effects induced by interactions between the electromagnetic fields and the real 3-D mask topography that the simple TMA model fails to model correctly.

To compensate for transmission losses on the mask topography, rather than adding a physical bias when making the mask, one can instead add a bias in TMA simulations to mimic the electromagnetic field effects. Obviously, this can be tuned and adjusted more easily than the physical mask. A single value of mask bias can approximate the field-induced transmission losses with reasonable accuracy, even for mask spaces as small as the wavelength. However, this bias value is a strong function of the mask topography. As illustrated in Fig. 15.25(c), the bias necessary on the phase-shifting apertures is a combination of the shifter bias due to the etched glass and the absorber bias due to scattering on the absorber edge. The exact value of the bias per edge also depends on the polarization of the incident illumination.

An example of such a single mask bias is shown in Fig. 15.27. Here, the effective electromagnetic field (EMF) bias is defined and calculated as the difference in mask-space size necessary to print a 1-D line exactly with the target critical dimension (CD), as estimated using either rigorous electromagnetic simulations or the thin-mask approximation (TMA), while maintaining all other exposure settings such as dose and focus constant. In other words, the (positive) effective field bias is calculated as EMF bias = Mask space_{EMF} – Mask space_{TMA}. This is the size of the simple opaque bias needed, in simulation, to mimic how the field effects physically narrow the printed size of this space.

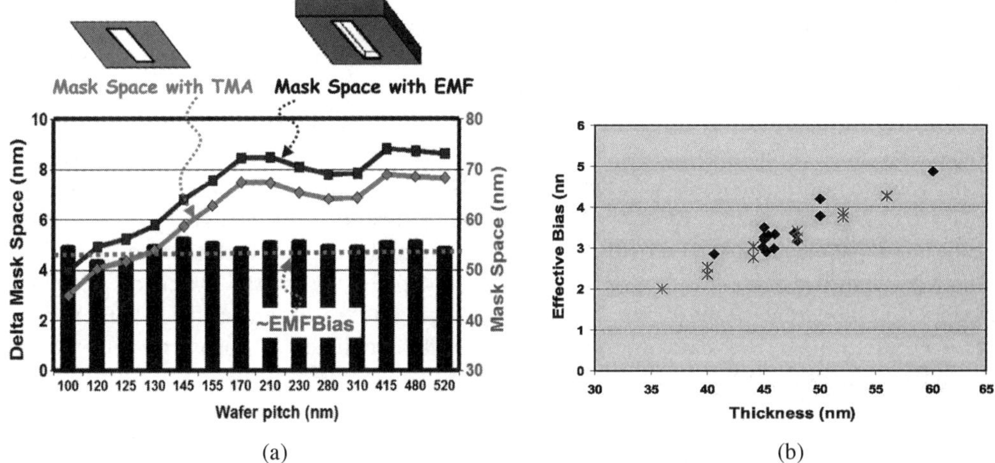

Fig. 15.27 (a) EMF bias calculated across a set of wafer pitches exposed with a 1.35 NA unpolarized C-quad illumination to print a 50-nm linewidth on wafer, showing the average across pitch providing the final bias value. (b) Relation between overall mask absorber film thickness and effective EMF bias for a set of binary masks.

Specifically, Fig. 15.27 shows the EMF bias calculated across a set of pitches exposed with a 1.35-NA unpolarized C-quad illumination and a mask-space size optimized to accurately print a 50-nm linewidth on wafer with a binary absorber mask [38]. Since the offset between the TMA and EMF curves in this figure is roughly constant across different pitches (and thus different mask spaces), a single average bias value can provide a first-order model of EMF-induced mask topography losses without any modification to the TMA algorithm. Although this result corresponds to unpolarized illumination, polarization-dependent electromagnetic field effects can be readily accommodated through slightly different bias values for each polarization. The unpolarized bias is then roughly the average of the two polarization-specific biases.

Due to the more severe impact of topography on phase-shifting masks, the tendency in recent years has been to move toward binary blanks. However, conventional chrome-on-glass binary masks can behave as TE polarizers for wafer pitches of 150 nm and below, with TM polarization suppressed for certain line-to-space ratios [110] including equal line-and-space gratings [87]. This behavior has been attributed to the excitation of surface plasmons on the chrome aperture sidewalls by the TM field [111]. Thinner binary masks, such as the recently developed opaque MoSi on glass (OMOG) mask blank, allow for effectively smaller EMF bias values, indicating reduced topography impact. In fact, across a number of different binary mask blanks, effective EMF bias scales nearly linearly with total thickness of the absorber film, regardless of mask transmission, per Fig. 15.27(b) [38]. Thinner masks also improve mask manufacturability, allowing smaller assist features and corner-to-corner gaps. Hence, binary blanks of thin topography are becoming more common among the most aggressive optical lithography technology [112].

Advanced Photomask Electromagnetic Field Models

As we just saw, a simple shift in mask-edge position (e.g., an "opaque" bias) can compensate for the amplitude component of these electromagnetic field errors with nearly no runtime penalty relative to the unbiased TMA model. In contrast, as seen in Fig. 15.26, a simple biased TMA model fails to predict the shifts in the plane of best focus at the wafer that are caused by the phase component of the field-induced perturbation.

Asymptotic solutions to the diffraction by an aperture show a geometrical optics component propagating through the aperture unaffected by the boundary, and a diffracted wave emanating from every point of the boundary of the aperture [24, 72–74]. Most models used in lithography to approximate the impact of the mask topography rest on this analytical and empirical observation that electromagnetic field effects act as a small local perturbation to the TMA model associated with the fields in the vicinity of the mask topography edges. One example of such models of lithographic mask field effects, known as the boundary-layer (BL) model [64, 109, 113, 114], offers a way to incorporate both the amplitude and phase parts of the error due to field effects more accurately than a simple TMA model.

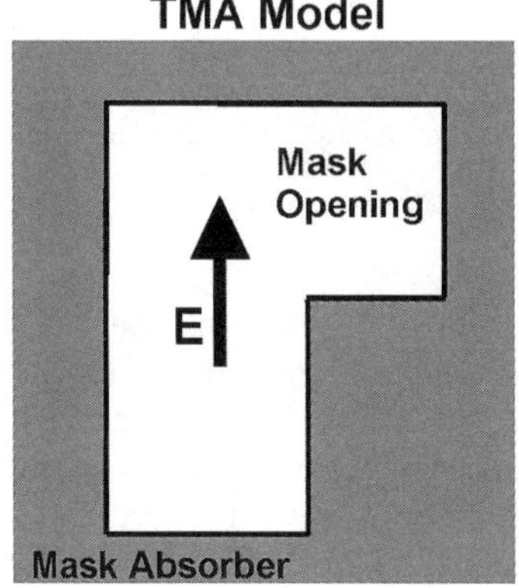

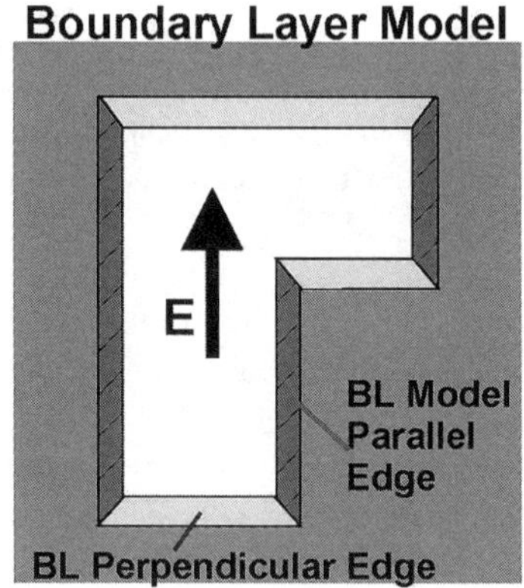

Fig. 15.28 Comparison of a boundary-layer (BL) model vs. a TMA model. In a BL model, a complex transmission strip is placed at each edge, designed to correct both electromagnetic field-induced amplitude and phase errors. BL parameters depend on edge orientation relative to polarization, and are calibrated through simulation of the real and imaginary (amplitude and phase) components of the aerial image fields and their deviation from the thin-mask approximation.

As shown in Fig. 15.28, the BL model consists of placing a strip in the vicinity of each edge with fixed width and complex transmission. The width of the BL (BL_{width}) controls the variations in peak amplitude; its imaginary transmission (BL_{imag}) corrects for quadrature-component electromagnetic field effects due to the finite-thickness topography. Two different sets of BL edge-correction parameters are needed. One set is needed for edges parallel to the polarization of the electric field incident on the mask, e.g., horizontal edges under x-polarized illumination and vertical edges under y-polarization. A different set is needed for edges perpendicular to the polarization of incident electric field, e.g., horizontal edges under y-polarization and vertical edges under x-polarization. Since this model requires only the incorporation of strip-like features of fixed width and complex transmission at the edge of each mask feature, the computational complexity is quite similar to the thin-mask approximation.

While $\mathrm{BL_{width}}$ and $\mathrm{BL_{imag}}$ can be obtained through calibration to resist linewidth measurements, a simpler method uses a rigorous electromagnetic simulation such as FDTD. To calibrate these parameters, we need only the difference in real and imaginary components between the aerial image electric field computed using the rigorous simulation and the image fields computed with TMA, preferably for a range of mask patterns of varying linewidth. Referring to (15.69), we see that the perturbation ΔE^{EMF} due to electromagnetic field effects (added to the TMA component E^{TMA}) scales inversely with the linewidth, LW. This allows $\mathrm{BL_{width}}$ and $\mathrm{BL_{imag}}$ to be extracted as:

$$E^{\mathrm{EMF}} = E^{\mathrm{TMA}} + \Delta E^{\mathrm{EMF}} = E^{\mathrm{TMA}}\left[1 + \frac{(-2\mathrm{BL_{width}})}{\mathrm{LW}} + j\left(-\mathrm{BL_{imag}}\frac{2\mathrm{BL_{width}}}{\mathrm{LW}}\right)\right] \tag{15.69}$$

In other words, the relative error in the real and imaginary components of the electric field due to the topography is inversely proportional to the aperture size, which means that it can be modeled as a simple boundary layer of fixed width and transmission. This confirms that electromagnetic field effects are largely associated with the edges of the aperture.

The $\mathrm{BL_{width}}$ factor is typically scaled to the local mask blank so that the overall real part of the transmission through the boundary layer is simply equal to the mask transmission. For purely opaque masks, the real part of the transmission through the boundary layer is zero and the transmission coefficient is purely imaginary. Then, $\mathrm{BL_{width}}$ completely controls the variations in peak amplitude, while $\mathrm{BL_{imag}}$ corrects for phase deviations due to the 3-D topography. By repeating this calibration for both TE and TM polarizations, different model parameters for each orthogonal polarization can be obtained. The result is an accurate model of the effect of mask topography as shown in Figs. 15.26(a) and 15.26(b), where a well-calibrated boundary-layer model accurately (and rapidly) predicts the electromagnetic field-induced pitch-dependent focus shift associated with the imaginary transmission component.

The exact values of the $\mathrm{BL_{width}}$ and $\mathrm{BL_{imag}}$ parameters are ultimately a function of the mask film stack and exact topography profile shape. Separate FDTD simulations are necessary to calibrate the boundary-layer model for different mask profiles and materials. Alternative methods have been proposed for this calibration using either simulations or linewidth measurements at the wafer plane [115–118]. Similarly, extensions to the original boundary-layer model have been proposed to improve accuracy for off-axis illumination [119], to add the effect of corners [120], and to improve computational efficiency [116, 121].

Another mask electromagnetic field model available for rapid computational lithography is the domain-decomposition method (DDM) [122], which replaces the simplistic step-like response at each edge of the mask with the rigorously obtained electromagnetic scattering response of that edge. This is usually described in terms of the difference between the TMA step and the FDTD simulation of a single isolated edge illuminated with a normally incident plane wave. This calculation is performed for the case of an edge parallel to the input polarization and is repeated for the orthogonal polarization. These difference fields are then stored and added to each mask edge of the TMA representation of a denser layout, relying on the approximation that the topography effects of each edge in a dense layout are quite similar to those of a similarly oriented but isolated edge. Since the correction factors provided by DDM contain both real and imaginary parts, they are capable of accurately modeling finite-thickness mask topography through focus. Extensions to the classical DDM model have been added to account for the changes in diffraction response at oblique incidence [123].

15.5.3 Hopkins Approximation

The effectiveness of many commercial lithography simulators rests on the accuracy of the Hopkins approximation, which states that a tilt in illumination merely tilts the output diffracted orders. While this assumption is inherent to the thin-mask approximation, it also allows electromagnetic modeling to restrict itself to simulating only the diffraction due to two normally incident plane waves of orthogonal polarization.

However, in the most advanced (NA = 1.35) immersion lithography systems, illumination can arrive at the mask surface with oblique angles of incidence of up to 20°. Under these conditions, the accuracy of the Hopkins approximation becomes questionable. Figure 15.29 displays an example of the dependency of the 0th diffracted order of an Atten.-PSM grating on the incident angle. Here, the change in diffraction efficiency with respect to normal incidence is shown for a 1-D attenuated PSM grating of 100-nm pitch and 50-nm linewidth at wafer scale, illuminated by a plane wave of TE polarization within the range of angles expected for NA = 1.35 and 4× magnification.

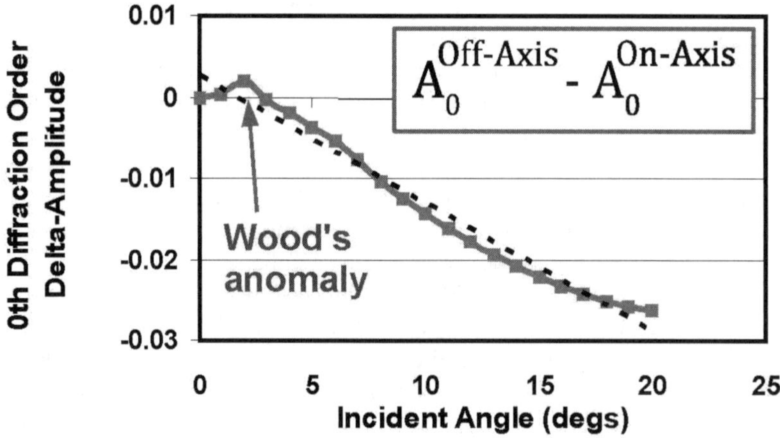

Fig. 15.29 Example of the dependence of the 0th diffracted order on incident angle as compared to normal incidence on a 1-D attenuated-PSM grating of 100-nm pitch and 50-nm linewidth at wafer scale illuminated by a plane wave with TE polarization.

Since the dependence of the diffraction efficiency on incident angle is a complex function of both pattern and incident polarization, both the linewidths of printed patterns as well as the position of the focal plane can change. Moreover, as mask pitches approach 80 to 100 nm at wafer scale, periodic arrays of repeated features on the mask can excite surface plasmons along the reticle plane. This effect, first observed as a function of wavelength by Wood in 1902 [99], gives rise to rapid variations in diffraction efficiencies with incident angle.

For instance, in Fig. 15.20, as the grating pitch decreases toward the illuminating wavelength, the diffraction intensity falls and rapidly oscillates as a function of the grating pitch. At the right combination of incident angle and pitch, there can be one or more diffracted orders that skim along the mask surface. Light that would have departed in one particular propagating diffraction order for a slightly different incidence angle can couple into this surface mode instead, and then couple back out into a completely different diffraction order. The resulting diffracted pattern can vary significantly from the aerial image intensity as predicted by the scalar models, both as a function of pitch [27, 124], as seen in Fig. 15.20, and across the source pupil.

The aerial image of small-pitch gratings (of the order of 80 to 100 nm for 193-nm, 1.35-NA immersion lithography and 4× magnification) is formed primarily by the interference of the 0th and 1st orders. The product of the 0th and 1st diffracted amplitudes directly modulates the image intensity at the wafer plane. Figure 15.30(a) is a plot of this $I_0 I_1$ product for an 80-nm pitch grating as a function of incident angle. The inflection point at the Rayleigh angle of ~12° corresponds to the transition of the second diffracted order from evanescent to propagating. Fortunately, other transitions that occur at larger incidence angles fall beyond the range of interest for a 1.35-NA lithography system.

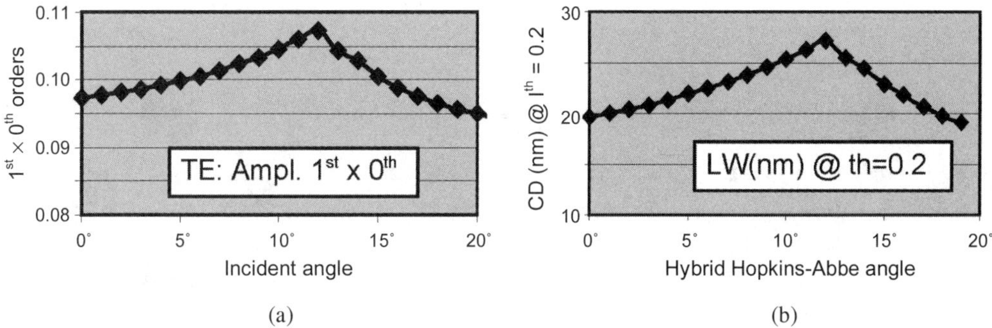

(a) (b)

Fig. 15.30 Example of the effect of Wood's anomaly on the choice of incidence angle on a 1-D Atten.-PSM grating of 80-nm pitch and 40-nm linewidth at wafer scale, when illuminated with a 0.2σ dipole with *y*-polarization, and imaged with 1.35-NA immersion lithography. (a) Effect on the product of 0th and 1st diffracted orders, proportional to the modulation of the image intensity; (b) change in CD value measured at constant threshold of 0.2 on the plane of best focus.

When the slower but more-accurate Abbe's formulation is used to calculate the aerial image linewidth at a constant intensity threshold for this same grating, we see, per Fig. 15.30(b), that this change in the interference term directly affects the printed linewidth. Here, the angle of incidence for the electromagnetic simulation of the mask transmission varies between 0° and 20°. Even when averaged over the particular source pupil to be used, it is clear that this effect results in significant differences between the predicted and printed linewidths, not just for TMA, but even for the predictions of full electromagnetic field simulations if performed only for normal incidence.

One option for adapting to this issue would be to revert to using Abbe's formulation for the computation of the image intensity distribution. Unfortunately, this would require rigorous electromagnetic simulations for all incident directions within the condenser numerical aperture, greatly increasing the computation time per mask pattern.

However, the Hopkins formulation can be extended to include the diffraction effects of obliquely incident angles through a hybrid Hopkins–Abbe formulation. First, the source pupil is divided into several subsections, with the Hopkins formulation applied across each subsection. Then, at most two rigorous electromagnetic diffraction computations are performed per subsection for the two orthogonal polarizations, at a single representative incidence angle strategically located within each subsection. Finally, separate aerial image simulations for each of the source-pupil subregions (assuming shift invariance over each subregion) form the composite aerial image through incoherent superposition [32, 123].

Such hybrid Hopkins–Abbe approaches are necessary to enhance accuracy of hyper-NA simulations where the effect of having incident illumination at very oblique angles is non-negligible. Since the runtime is increased by the number of subsections, the sampling of the source pupil plays an important role in the accuracy and runtime penalty of the hybrid Hopkins–Abbe method [32, 123]. If no prior knowledge of the illumination scheme is available, then the entire source pupil must be sampled on a uniform grid, leading to a potentially very large number of rigorous electromagnetic simulations, per Fig. 15.31(a). However, this method allows for computation of the diffraction orders for arbitrary sources through interpolation between the stored results for incident waves at the grid points.

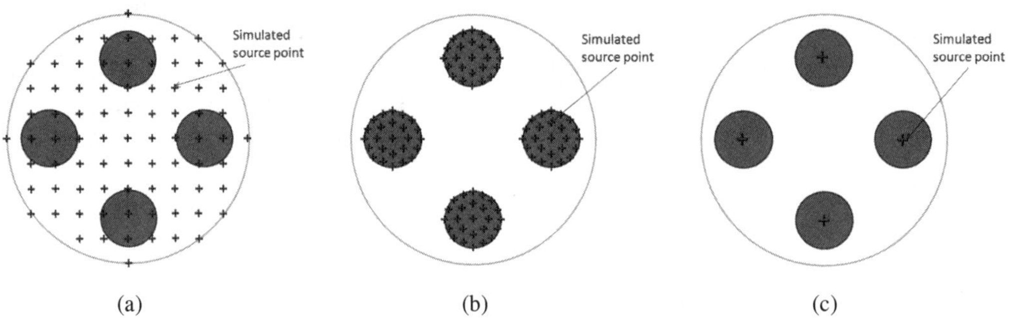

Fig. 15.31 Source sampling: (a) uniform grid sampling; (b) nonuniform sampling; (c) hybrid Hopkins-Abbe with nonuniform sampling.

In contrast, if knowledge of the illumination scheme is available, computation can be applied more efficiently. Here, we can customize the pupil sampling to the source shape to be used, computing the rigorous electromagnetic diffraction only for the incidence angles that fall within the source shape, per Fig. 15.31(b). The most efficient nonuniform sampling calls for just one incidence angle per subsection, as shown in Fig. 15.31(c). In addition to knowledge of the source, this method does not allow easy extension to arbitrary sources through interpolation. However, nonuniform pupil sampling can increase the accuracy while greatly reducing the computation runtime.

15.6 FDTD MODELING FOR EXTREME ULTRAVIOLET LITHOGRAPHY

Immersion lithography and double-patterning techniques have extended the use of 193-nm projection lithography through many more semiconductor nodes than was originally planned, or even believed to be possible. However, as double-patterning turns into triple- or even quadruple-patterning for linewidths smaller than 40-nm half-pitch, the additional masks and exposures will start to become costly, to say nothing of the extreme challenges in overlay and CD uniformity.

Extreme ultraviolet lithography (EUVL) is seen as the most promising patterning technique for replacing current deep ultraviolet lithography in high-volume manufacturing [125]. In fact, for all intents and purposes, it is the *only* promising replacement technique. Thus, even though significant challenges in source power, mask defects, and resist sensitivity continue to delay its long-awaited adoption, EUVL continues to receive significant investment by semiconductor industry and research consortia. Originally called "soft x-ray" lithography, EUVL operates around the 13.5-nm wavelength, and in many ways is an optical projection lithography technique sharing the principles of prior generations of lithography systems.

At the 13.5-nm wavelength range, however, most materials and gases are highly absorbing, including all refractive optical lenses and reticles. Thus, EUVL requires high-vacuum operation, with all refractive optics replaced by reflective mirrors including the reticle. This change has significant implications on the performance of the lithography process that must be well understood, calling again for highly accurate computations of the electromagnetic response.

15.6.1 EUVL Exposure System

The EUVL exposure system introduces a set of new optical elements and configurations, including multilayer mirrors, reflecting reticles, and non-telecentric illumination. These require accurate electromagnetic computations for full understanding and optimization.

Multilayer Mirror

In the EUV regime, index contrast between different materials tends to be low and absorption tends to be high. Thus, the highly reflective optical mirrors for EUVL illumination optics, projection optics, and reticles are implemented with precision multilayer thin-film coatings. In these thin-film stacks of alternating materials, the thicknesses are chosen to satisfy the Bragg condition, $2d \sim \lambda/\sin\theta$, while the materials are chosen for maximal index contrast and transmission. Here, θ denotes the angle of incidence on the multilayer, and d represents the thickness of each repeating bilayer unit as illustrated in Fig. 15.32(a). The index of refraction does not appear here because, as shown in Table 15.2, most materials have indices quite close to unity in the EUV regime.

When the Bragg condition is satisfied, weak reflections from all interfaces interfere constructively, which can produce a very high composite reflectivity when the stack contains enough layers. The large absorption of all materials in the EUV spectral range limits the maximum reflectivity achievable to about 70% at near-normal incidence, using at least 40 periods of molybdenum and silicon (Mo/Si) bilayers at the 13.5-nm wavelength [125, 126]. The high efficiency possible with this mirror stack enabled the development of reflective optical elements in this regime, and ultimately determined the wavelength of the illuminating source required for EUV lithography [125].

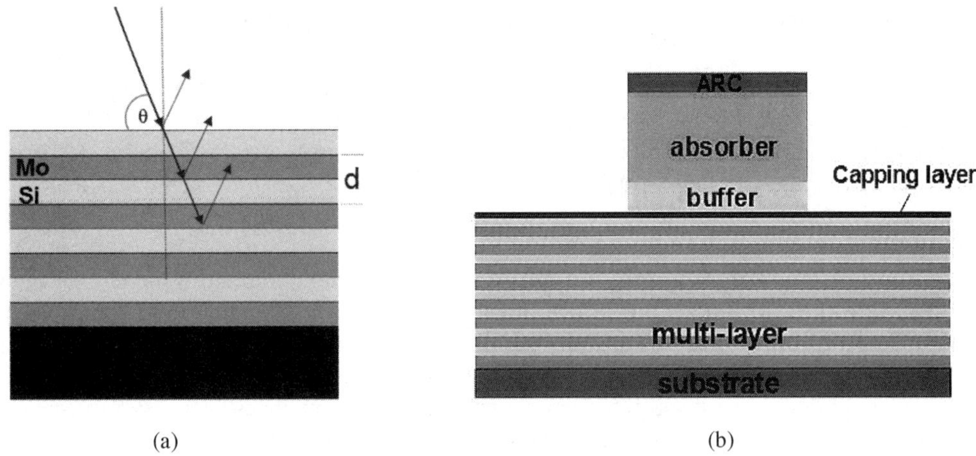

Fig. 15.32 (a) Cross-section of a multilayer thin-film reflector, where individual reflections from each interface interfere constructively to produce a high composite reflectivity; (b) cross-section of an EUVL mask stack.

TABLE 15.2

Typical Materials Used in EUVL Reflective Optics and Reticles, with Their Optical Constants at the EUV Wavelength

Layer	Material	n	α	Thickness
ARC	TaON	0.9083	0.05106	12 nm
Absorber	TaN	0.9272	0.0429	55 nm
Buffer	SiO2	0.9785	0.0106	10 nm
Capping	Ru	0.8888	0.0161	3.1 nm
ML-Mo	Mo	0.9253	0.0062	2.8 nm (40 layers)
ML-Si	Si	0.9995	0.0018	4.2 nm (40 layers)
Substrate	SiO2	0.9785	0.0106	50 nm

Multilayer film reflectors of constant bilayer unit period display a narrow bandwidth around the resonant high reflectivity wavelength and incident angle. Graded multilayer reflectors, where the bilayer unit period is gradually varied across the height of the stack, can provide broadband reflectivity. This increases the range of available incident angles on the various mirror surfaces in an EUVL optical design, often at the expense of peak reflectivity [127]. Finally, a 2- to 3-nm-thick ruthenium capping layer is often deposited on the top multilayer surface to protect against radiation-induced carbon growth and surface oxidation [125].

Reflective Optics and Projection System

Since each mirror surface in an EUVL optical system has a peak reflectivity of $R = 70\%$, the overall transmission of the system decreases rapidly with the number of reflections. Thus, projection-system designs with a small number of optical elements are highly preferred.

However, parallel scanning of both mask and wafer stages [128] without mechanical interference is possible only with an even number of reflections. A reflective projection system with as few as four mirrors is a viable solution for numerical apertures of 0.1 and 4× reduction imaging while also satisfying practical lithographic requirements [129]. A simplified representation of such a reflective optical system is illustrated in Fig. 15.33(a) [130]. The first EUVL systems for high-volume manufacturing are designed for a numerical aperture of 0.25 [131, 132] using a six-mirror design. Aspherical mirror surfaces provide enough degrees of freedom to correct for wavefront distortions.

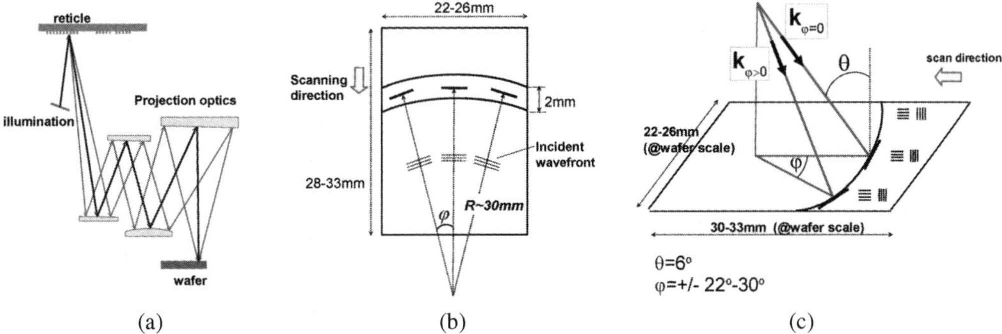

Fig. 15.33 (a) Schematic of a four-mirror EUVL optical system, designed to ensure telecentricity in the image space at the wafer at the expense of non-telecentric illumination of the reflective reticle. (b) Segmented ring-shaped slit field illuminating the mask in a scanning configuration. (c) Illumination wavefront on the mask is rotationally symmetric and the angle formed by the plane of incidence and the scanning direction at the mask surface changes with the position along the ring-shaped slit. This generates a position-dependent shadow effect on mask shapes.

An EUVL reflective projection system is designed to remain telecentric at the image space or wafer plane. Thus, the chief-ray of the image propagates parallel to the optical axis and impinges perpendicular to the wafer surface. Since EUVL reticles are also reflective with the patterned absorber material lying over the multilayer coating, illuminating the photomask at normal incidence would create an obscuration in the pupil. This obscuration is avoided by illuminating the mask surface at a small angle with respect to the mask's normal direction. Several projection optics designs have been proposed and patented over the years [128, 133] with values of chief-ray angle (CRA) at the object space ranging from 4° to 10°. The 0.25-NA, 4× reduction ratio, six-mirror design by Zeiss and ASML for their first high-volume manufacturing tool employs an incident illumination angle on the reticle of 6° [131]. This non-telecentricity in the object space leads to a shadowing of the mask topography, which is both pattern and orientation dependent, translating into an unintended deformation of the image pattern.

Source and Illuminator

Requirements for reflective optics with a minimum number of mirror surfaces in EUVL lead to systems where a high degree of aberration and distortion correction can only be achieved within a narrow, rotationally symmetric, ring-shaped area of the mirror surface. Thus, EUVL systems operate in a scanning mode with the scanned field in the shape of an arc having an axis of symmetry along the scan direction. The ring field in the image space has the same shape, but is 4× smaller in dimension. As illustrated in Fig. 15.33(b), in wafer-scale dimensions, the slit is 2 mm in height along the scan direction, 22 to 26 mm wide, and is scanned over a length of 28 to 33 mm to cover the reticle [129, 133, 134].

Illumination on the mask thus also forms a rotationally symmetric wavefront with the same radius of curvature as the arc-shaped slit. At each point along the arc-field perimeter, the illuminating wavefront can be represented locally as a plane wave in a rotated plane of incidence that remains perpendicular to the slit perimeter. This illumination configuration differs from the rectangular-shaped field used in DUVL, where each illuminating plane wave varies only slightly across the field. As a consequence, the exact plane of incidence of each illuminating plane wave impinging on the mask — defined by the chief-ray angle θ relative to the optical axis normal to the mask surface and the azimuthal angle φ relative to the scan direction as displayed in Fig. 15.33(c) — depends on the location within the mask relative to the center of the ring-field slit. The arc-shaped slit radius of curvature, R, defined relative to the optical axis (vertex of the aspheric mirrors), is usually about 30 mm at the wafer plane, resulting in an azimuthal angle varying in the range $\pm 22^{\circ}$ to $\pm 26^{\circ}$, depending on the specific optical design. As a consequence, the shadow cast by the mask topography due to the oblique mask illumination varies with the pattern orientation relative to the scan direction, and with the pattern location relative to the center of the arc-shaped slit.

The illumination system is responsible for converting the isotropic plasma source radiation into a uniform arc-shaped wavefront on the mask plane. This is achieved with an essential device in EUVL systems, the fly's-eye integrator [131, 134, 135]. Here, two segmented mirrors divide the collimated radiation from the plasma source into hundreds of discrete illumination channels that reflect arc-shaped wavefronts toward the mask plane [135, 136]. Each of these secondary images of the light source is focused onto the entrance pupil of the projection optics, thus implementing Köhler illumination. Partial coherence adjustment and more complex source pupils can be achieved with pupil apertures containing flexible mirror elements [132].

15.6.2 EUV Reticle

EUVL reticles are reflective, comprised of a multilayer mirror (as discussed in Section 15.6.1) deposited over a fused silica blank substrate, a protective capping layer, and an absorber stack. This absorber stack can, by itself, contain several layers, as illustrated in Fig. 15.32(b). EUVL reticles contain many more layers than transmissive DUVL reticles, adding to the complexity in simulating and manufacturing these structures. Very stringent specifications for flatness and defect-free surfaces are imposed on the EUVL blank substrates, since any roughness or particles on the substrate surface can lead to topography within the multilayer films, inducing phase errors in the reflected wavefront.

Since the fabrication of defect-free masks continues to be a challenge, rigorous simulations of defective mask cross-sections are often needed to understand the resulting impact on imaging. The most common problems are phase defects that result from substrate bumps, pits, and particles near the bottom of the stack, although imperfections during the multilayer deposition can introduce local changes in the films. Amplitude defects are bumps or particles located near the top of the multilayer stack that prevent the incident light from reflecting [125].

Above the mirror stack, a highly absorbing material is patterned with the design layout shapes to modulate the reflected wavefront with the design content. The EUVL absorber stack is often comprised of an absorber material with a high extinction coefficient, together with a buffer layer to protect the capping layer and multilayer mirror during mask fabrication. To reduce the reflectivity of the absorber, an anti-reflective coating (ARC) layer is often added to the top of the absorber stack. Chromium-based absorbers, common during the early stages of EUVL development, have since been replaced by tantalum-based absorbers with improved optical and thermal properties.

A prototypical mask stack, including the indices of refraction at the EUV wavelength of 13.5 nm, is described in Table 15.2 [137, 138]. Multilayer stacks for EUVL typically use between 40 and 60 Mo/Si bilayers, resulting in a minimum multilayer thickness of 280 nm. Common EUVL absorber stacks are of the order of 65 to 90 nm thick to guarantee enough opacity, although thinner absorber options are being explored as well [139]. A minimum of 40 bilayers must be modeled for the accurate simulation of a EUV reticle electromagnetic response. Thus, the active portion of an EUVL reflective reticle is significantly thicker (in wavelength units) than in a conventional transmissive mask.

Binary masks (where the absorber stack completely attenuates the incident illumination) are most common at this stage of EUVL technology development. Layout patterns are etched into the absorber stack down to the capping layer protecting the multilayer, as illustrated in Fig. 15.34(a). Alternative mask configurations to guarantee that EUVL can be extended to higher resolution are also being explored [140, 141]. One variant removes the need for an absorber stack by patterning deep into the multilayer stack, as shown in Fig. 15.34(b). Alternatively, similar to Atten.- and Alt.-PSM masks in DUVL, the multilayer stack can be etched so as to obtain 180° phase shifting of the reflected fields, as shown in Figs. 15.34(c) and 15.34(d).

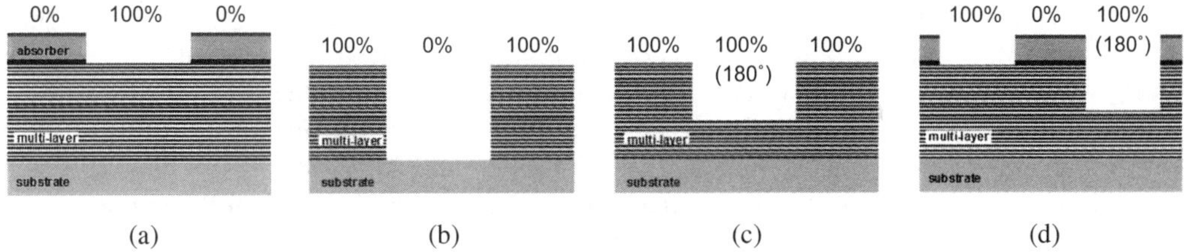

Fig. 15.34 Several EUVL mask types: (a) binary type 1; (b) binary type 2; (c) absorber-less phase-shifting mask; (d) alternating phase-shifting mask.

15.6.3 EUVL Mask Modeling

EUV lithography retains many characteristics common to DUVL such as uniform, partially coherent Köhler illumination with an ideally aberration-free optical system. The shorter wavelength also enables operating with much more relaxed k_1 factors, which reduces the need for aggressive corrections on the mask. For instance, at a 13.5-nm wavelength and 0.25 NA, features smaller than 20-nm half-pitch can be patterned with k_1 factors still above 0.5. Reflective masks with abrupt topography relative to the wavelength and the non-telecentric illumination, however, introduce new effects with high potential for distorting the image in new and position-dependent manners. Thus, in addition to standard diffraction-induced optical proximity corrections, EUV masks require precompensation, not only for the geometric shadowing, but also for the complex effects that arise from the interplay between the off-axis illumination, the multilayer mirror, and the abrupt absorber topography.

Specific EUVL effects that can severely impact the aerial image if left uncorrected include:

Image asymmetry, where the image intensity distribution is asymmetrical with respect to the position of best focus;

Focus shift, where the actual plane of best focus is shifted relative to the focal plane identified by the thin-mask approximation;

Pattern shifts, where the image cross-section contours shift relative to the desired position;

HV bias, where the printed linewidth is different for horizontal and vertical lines of otherwise identical mask shapes. HV bias can also refer to the difference in mask correction required by horizontal and vertical lines in order to print with the same linewidth at the wafer with identical process conditions.

Understanding the interactions between the increased complexity of the mask topography and the illumination conditions requires rigorous electromagnetic simulations. These simulations can be used to optimize the mask materials and profiles to minimize these effects, while also helping to develop simpler approximate models that can be compatible with full-chip simulations.

FDTD Modeling Considerations

Interactions between the illumination and the abrupt mask topography result in the same electromagnetic field issues discussed in Section 15.5. Changes in both the amplitude and phase of the reflected wavefront, which can occur even at normal incidence, can only be studied by rigorously solving Maxwell's equations. Moreover, non-telecentric illumination of the mask introduces asymmetry in the diffracted orders arising from the interplay between oblique illumination, the multilayer mirror, and the high absorber topography. This asymmetry in the mask near fields introduces small amounts of aerial image asymmetry and pattern shift through focus even when the mask-output fields are properly backpropagated to the object focus plane.

EUVL mask simulations with FDTD are quite similar to DUV simulations, except for some key differences in memory and computational run-time needs. The grid cell size must still be a fraction of the wavelength, even when using subpixel smoothing, and the multilayer mirror film must be accurately represented. Without the use of subpixel-smoothing techniques, some authors recommend a grid cell size of $\lambda/20$ [142], which for a 13.5-nm wavelength results in a grid cell size of ~0.5 nm. Many layouts of interest in EUVL are 1-D patterns of pitches not larger than about 400 nm at mask scale. Unfortunately, the simulation domain for EUVL reticles including the entire multilayer mirror and absorber thickness explodes in size very quickly, requiring the aggressive use of subpixel-smoothing techniques, parallelization, and a new kind of Fourier boundary condition [143] to make runtimes tractable.

EUVL simulations require oblique incident illumination with a 6° angle and, depending on the position on the slit, an azimuthal angle that can vary up to $\varphi = \pm 30°$. The excitation wave is generated from a plane located a few nanometers above the mask absorber, producing an obliquely incident plane wave as in the DUV simulations. After the computation reaches convergence, the output fields needed to compute the diffraction spectra are extracted at a plane in a uniform-medium region a few nanometers above the top absorber surface, in order to avoid influence by the topography, per Fig. 15.35(a). Thus, a total-field/scattered-field source configuration that introduces only downward-propagating light is essential here. Due to the non-telecentricity of the mask illumination and the high topography, the choice of the measurement plane introduces artificial asymmetry and nonphysical shifts in the image, requiring correction as will be described in the next subsection.

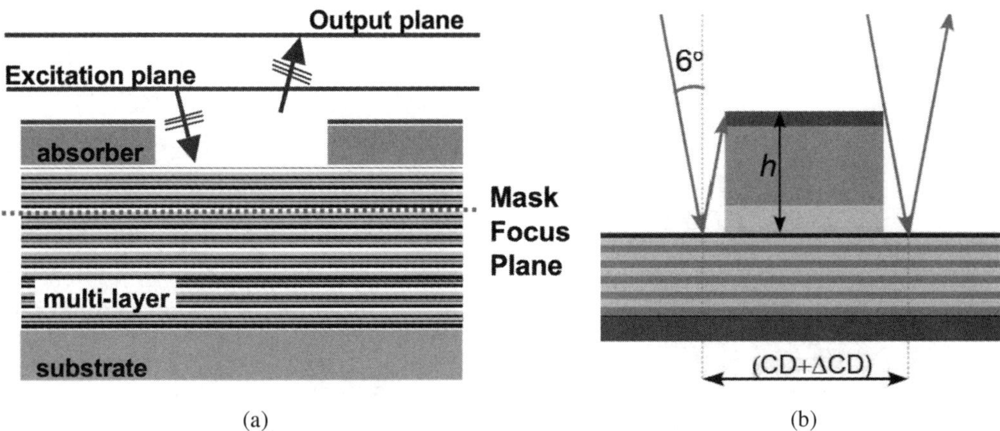

(a) (b)

Fig. 15.35 (a) Mask focal plane back-propagation; (b) geometrical shadow bias.

The abrupt topography of reflective EUV masks, combined with the oblique incident angle, results in different reflected near-fields depending on the orientation of the mask pattern relative to the incident illumination. This difference can be qualitatively understood in terms of a geometrical shadow as illustrated in Fig. 15.35(b), which causes changes in the printed linewidth of horizontal and vertical lines of otherwise identical mask lines. Because the underlying electromagnetic interactions are complex, rigorous field computations are necessary for accurate quantitative analysis.

Mask Non-Telecentricity and Mask Focus

Both the incoming illumination and the reflected fields are non-telecentric in EUVL, propagating at an oblique angle of about 6° with respect to the mask normal. Assuming a 1-D line and space pattern located at the center of the slit of the illuminating arc-field, we can assume for simplicity an azimuthal angle of 0° according to Fig. 15.33(b). Manhattan layouts at this location are always oriented either parallel or perpendicular to the plane of incidence of the incoming illumination. Line and space patterns oriented parallel to the plane of incidence (i.e., parallel to the scan direction) are referred to as vertical lines or V-lines, as illustrated in Fig. 15.36(a). Conversely, line and space patterns oriented perpendicular to the plane of incidence (i.e., perpendicular to the scan direction) are referred to as horizontal or H-lines, as shown in Fig. 15.36(b).

In the case of the V-lines of Fig. 15.36(a), both the –1 and +1 diffracted orders are symmetrical at the output plane depicted in the figure. Thus, vertical lines on the mask produce aerial images that are symmetrical for positive and negative defocus, and show no focus-dependent pattern shift. The image is centered at $y = 0$, just as in a conventional on-axis three-beam interference pattern generated by the fully telecentric illumination in a DUVL system.

However, in the case of the H-lines of Fig. 15.36(b), the –1 and +1 diffracted orders are asymmetrical, even when ignoring mask topography. This asymmetry depends on the separation of the measurement plane from the effective mask focus plane, as depicted in Fig. 15.35(a). Thus, the aerial image intensity is asymmetrical for positive and negative defocus conditions, and is not centered at $x = 0$. Instead, the aerial image pattern is shifted in x by an amount that depends on the focal position z at the wafer.

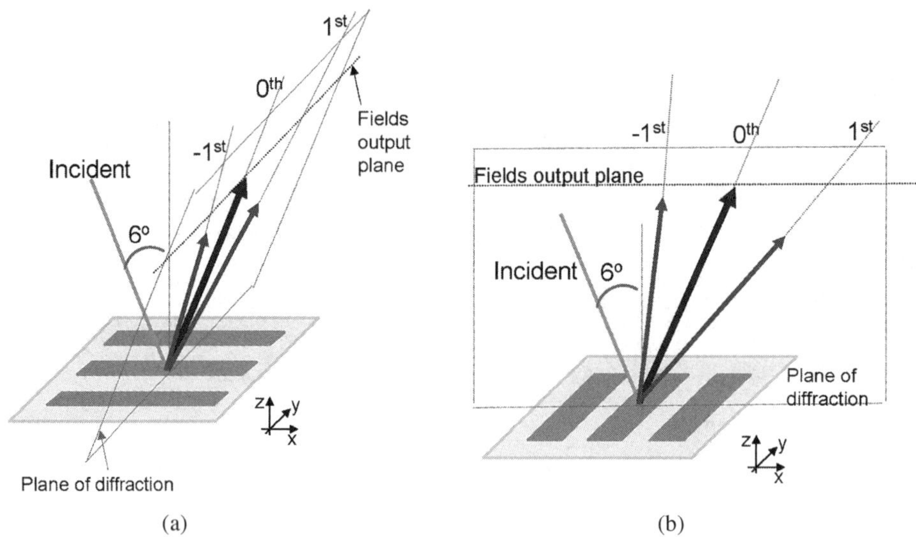

Fig. 15.36 Non-telecentric illumination of a reflective mask with a simple line and space pattern of pitch such that only three diffraction orders are collected by the optical projection system: (a) vertical lines (V-lines) parallel to the scan direction; (b) horizontal lines (H-lines) normal to the scan direction.

This asymmetrical aerial image and focus-dependent pattern shift is a consequence of the non-telecentricity on the mask plane when the output fields are not back-propagated to the mask focus plane. For reflective reticles, the effective mask focus plane is located at some distance inside the multilayer, as illustrated in Fig. 15.35(a). Failure to accurately account for the mask focus plane induces artificial pattern shifts at the wafer plane, shifts in the plane of best focus, and linewidth differences between horizontal and vertical lines.

It is quite straightforward to back-propagate the measured plane-wave diffraction orders to any given z-plane by $\boldsymbol{E}_{\text{final}} = \boldsymbol{E}_{\text{out}} \exp(jk_0 \Delta z_{\text{BF}_{\text{mask}}})$. However, to determine the back-propagation distance, $\Delta z_{\text{BF}_{\text{mask}}}$, the aerial image of a few mask patterns must be simulated using varying amounts of back-propagating distance until the aerial image pattern shift at the wafer plane of best focus vanishes [144, 145]. Several authors have shown that the effective mask focal plane for a 40-bilayer multilayer stack and a 70-nm-thick absorber stack is within the multilayer film, approximately 150 nm below the top of the absorber stack [144–146]. As a rule of thumb, the pattern shift and mask focus plane are related according to $\Delta x = \text{M}\Delta z \tan(\theta_{\text{inc}})$.

Mask Topography and Shadowing Effects

Mask shadowing is a characteristic feature of non-telecentric illumination of EUVL reflective mask topography that cannot be ignored by lithography models. It introduces significant printed linewidth errors that are both position and orientation dependent, unless precorrection is applied to the mask shapes. The most noticeable difference is between horizontal and vertical lines at the center of the slit where the azimuthal angle is $\varphi = 0^\circ$ as shown in Fig. 15.37. At this location the horizontal lines are perpendicular to the plane of incidence and experience the maximum shadowing effect, while vertical lines are parallel to the plane of incidence and are unaffected by the $\theta = 6^\circ$ oblique incidence.

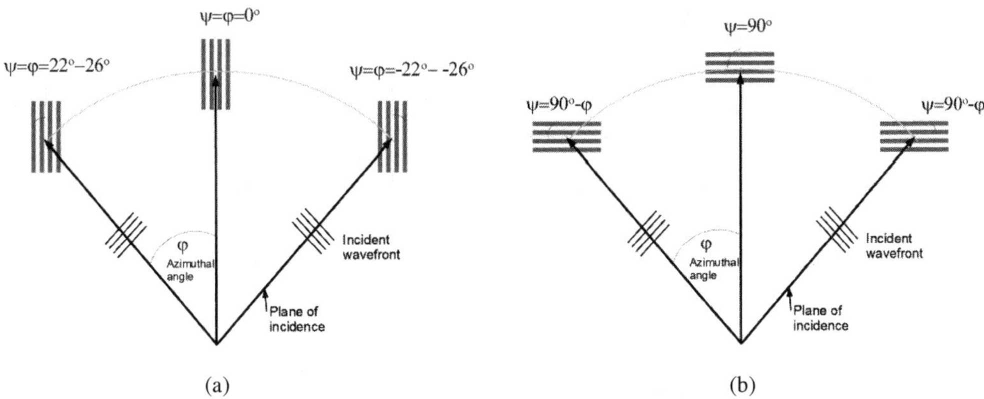

Fig. 15.37 Variation of the azimuthal angle of the plane of incidence along the arc-shaped scanning field relative to the orientation of simple line and space patterns in a Manhattan layout for (a) vertical lines (V-lines); (b) horizontal lines (H-lines).

This difference in the mask near fields translates into a difference in the printed linewidth of horizontal and vertical lines on the wafer. The measured HV linewidth difference in resist is also influenced by optical and resist factors (NA, source, MEEF) as well as the mask near fields. Hence, the HV linewidth difference at the wafer is not necessarily the same as the correction required for H-lines and V-lines on the mask to remove this difference. McIntyre et al. have introduced the following three definitions for quantifying EUVL mask effects [145, 147]:

1. *HV printed bias (HVPB)* — a measurement of the printed linewidth difference on the wafer between horizontal and vertical lines of equal size and proximity on the mask. This represents the interplay between the mask shadowing effects and the overall lithography process.

2. *HV bias (HVB)* — the difference in mask absorber width required for both horizontal and vertical lines to print lines of the same linewidth on the wafer under identical process conditions. This difference is greatest at the center of the slit, where vertical lines are parallel to the plane of incidence, and smallest at the slit ends.

At the center of the slit, the HV bias could be interpreted as a geometrical shadow per Fig. 15.35(b) that enlarges linewidths and shrinks spacewidths. However, the exact amount of correction cannot be extracted with simple geometry, and requires rigorous computation of the electromagnetic response of the entire mask topography, including the multilayer mirror. Several authors have demonstrated that a single HV mask bias of approximately 2.5 nm in absorber linewidth for the standard mask topography of Fig. 15.32(b) can account for the differences in printed linewidth between horizontal and vertical lines at the center of the slit [142, 145]. Moreover, this single bias value remains approximately constant for a range of line and space patterns, down to pitches as small as 90 nm at wafer scale. Scattering from topography edges in closer proximity, such as denser line and space mask patterns or 2-D layouts, begins to interact. The resulting HV bias deviates from the single constant value in manners that require more complex electromagnetic field models [146, 148].

Similarly, a simple geometrical interpretation of the change in HV bias as the azimuthal angle varies along the slit as per Fig. 15.37 is inaccurate when compared to rigorous electromagnetic simulations [142, 148]. In particular, some authors have observed that HV bias varies as $\sin^2\psi$, where ψ is the angle between the plane of incidence and the line orientation, per Figs. 15.37(a) and 15.37(b) [142, 147, 148]. The correction required for H or V mask lines as a function of the slit position can be described as $\Delta MaskCD(\psi) = HVbias|_{\text{center}} \sin^2 \psi$ for dimensions as small as 60-nm pitch. However, improved understanding and modeling of these effects is an area of active research.

3. *Delta CD (ΔCD)* — refers to the overall difference in printed linewidth in resist when calculated with rigorous electromagnetic simulations as compared to the printed linewidth computed using a thin-mask approximation.

15.6.4 Hybrid Technique Using Fourier Boundary Conditions

Full numerical computations for EUVL would require a simulation domain enclosing the substrate and multilayer amounting to at least 280 nm or about 20 EUV wavelengths. The absorber stack of approximately 70 nm adds another 5 to 6 wavelengths. With FDTD simulations that require 10 to 20 grid cells per wavelength, this domain would require at least 40×250 grid cells for even a simple 1-D grating layout of small pitch. Thus, full simulation of EUVL reticles requires much larger domains than simulation of DUV transmissive masks, which might have only 10×15 grid cells for a similar small-pitch, 1-D grating layout. Memory and runtime requirements are therefore significantly larger.

A hybrid technique using Fourier boundary conditions was developed by Pistor [143] to circumvent the need to simulate the reflection from the multilayer in every simulation. Referring to Fig. 15.38, the multilayer stack is replaced by its Fourier transform response, hence reducing the FDTD simulation domain size to something of the order of DUV transmissive masks. In this manner, the multilayer mirror is removed so that the incident fields propagate through the patterned absorber stack topography and down to the lower PML boundary, where ideally nothing is reflected. These fields are captured at a plane below the absorber stack denoted by h_o and decomposed into their plane-wave components through a Fourier transform.

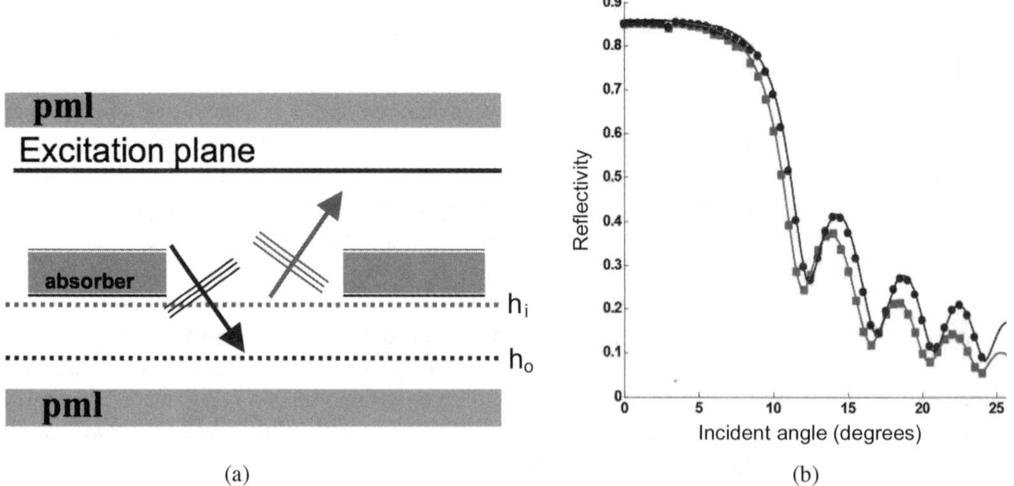

Fig. 15.38 (a) Diagram of the simulation domain of an EUVL reticle employing Fourier boundary conditions. (b) Reflectivity vs. incident angle of a multilayer mirror for TE and TM polarization where the solid lines correspond to the analytical solution and the dots represent the FDTD result.

A corresponding reflected plane wave is excited at the plane denoted h_i, weighted by the appropriate reflection coefficient at the angle corresponding to that plane wave. These re-introduced plane waves propagate upward into the simulation domain, passing through the patterned absorber topography toward the lens. The reflection coefficient A_r applied to each Fourier component is determined according to the reflectivity vs. incident-angle plot of Fig. 15.38(b). The phase difference in the waves due to the separation between planes h_i and h_o also is incorporated through a phase term $\exp[-jk_{m,n}(h_i - h_o)]$. These weighted and re-introduced plane waves represent the same reflected fields that would be generated by the entire multilayer mirror structure.

Since the multilayer mirror is a planar thin-film stack, the reflection coefficients can be computed analytically, greatly simplifying the computation of these Fourier boundary conditions. Figure 15.38(b) displays the reflectivity response of the 40-bilayer multilayer mirror as computed by the FDTD code (dots) compared to the reflectivity calculated analytically using thin-film theory for both TE and TM polarizations. As with DUVL, arbitrarily polarized or unpolarized illumination configurations require only the appropriate combination of the results of two simulations.

Insertion of EUV lithography for high-volume semiconductor manufacturing will help relax the k_1 factor, allowing larger values than the aggressively low values currently being used with DUV. This will also help reduce the amount of proximity correction needed for conventional correction of diffraction effects. However, accurate location- and orientation-dependent pre-correction of mask shapes — to compensate for EUVL-specific mask effects — will be mandatory. Moreover, as mask linewidth dimensions shrink and higher numerical apertures and more oblique incident angles are introduced, the optimization of the mask topography will only increase the demand for rigorous electromagnetic simulations.

15.7 SUMMARY AND CONCLUSIONS

As the characteristic dimensions of VLSI chip technology shrink and complexity rises, the usual geometrical approximations of the electromagnetic field interactions underlying optical lithographic technology can become increasingly inaccurate. Accurate simulation of both immersion and EUV lithographic systems is expected to be an increasingly critical component of semiconductor manufacturing for the foreseeable future. While rigorous computation of the required 3-D electromagnetic fields can be much slower than the approximate methods, rapid turnaround is still crucial. The FDTD method offers advantages such as flexibility, speed, accuracy, parallelized computation, and the ability to simulate a wide variety of materials, making it a popular choice for electromagnetic computations in optical lithography simulation software. This chapter reviewed the fundamental physical concepts and numerical considerations required to perform such electromagnetic computations for VLSI optical lithography in the context of semiconductor manufacturing.

We note that FDTD computation of the electromagnetic fields underlying VLSI optical lithography offers the best combination of accuracy and turn-around time to understand and model field effects involved with relatively small patterns. However, the computational resources required by FDTD are not compatible with full-chip simulations on currently available computers. Especially for EUVL mask technology, there remains a need for field solutions offering both sufficient accuracy and acceptable turn-around time for full-chip corrections. This remains an area of continuing research.

APPENDIX 15A: FAR-FIELD MASK DIFFRACTION

Huygens' surface equivalence principle enables replacing any material within V provided that the original boundary conditions remain satisfied by the equivalent sources distributed over the boundary. For instance, the medium inside V in Fig. 15.11(b) can be assumed to be a perfect electric conductor (PEC). The equivalent surface electric current, $\boldsymbol{J} = 0$, then vanishes and only a magnetic current density $\boldsymbol{K}$ is present, radiating in the presence of the electric conductor to produce the original diffracted fields in Region II.

Since S is assumed to extend to infinity through periodic boundary conditions applied on both lateral dimensions, it radiates equally in both directions. Thus, from image theory, in order to obtain the correct diffracted fields in the region of interest (Region II), the equivalent magnetic surface current should be doubled, $\boldsymbol{K} = -2\hat{\boldsymbol{z}} \times \boldsymbol{E}^{\mathrm{M}}$. Here, the fields on surface S are equivalent to the mask fields $\boldsymbol{E}^{\mathrm{S}} \equiv \boldsymbol{E}^{\mathrm{M}}$, and the vector normal to the surface is replaced by the unit vector along the z-axis, that is, $\hat{\boldsymbol{n}} \equiv \hat{\boldsymbol{z}}$. The fields in Region I are incorrect, but they are of no relevance to the image formation.

We can calculate the gradient of the Green's function as:

$$\nabla\left(\frac{e^{-jk|\boldsymbol{r}-\boldsymbol{r}_m|}}{|\boldsymbol{r}-\boldsymbol{r}_m|}\right) = -\hat{\boldsymbol{p}}\left(\frac{1}{|\boldsymbol{r}-\boldsymbol{r}_m|} + jk\right)\cdot\left(\frac{e^{-jk|\boldsymbol{r}-\boldsymbol{r}_m|}}{|\boldsymbol{r}-\boldsymbol{r}_m|}\right) \approx -jk\hat{\boldsymbol{p}}\frac{e^{-jk|\boldsymbol{r}-\boldsymbol{r}_m|}}{|\boldsymbol{r}-\boldsymbol{r}_m|} \tag{15A.1}$$

with the unit vector $\hat{\boldsymbol{p}}$ pointing in the direction of $(\boldsymbol{r} - \boldsymbol{r}_m)$. In optical lithography problems, the field location $\boldsymbol{r}$ is usually taken at the entrance pupil of the first lens of the projection imaging system, which is located many wavelengths away from the mask surface. This allows the approximation $k|\boldsymbol{r} - \boldsymbol{r}_m| >> 1$ in the expression $(1 + jk|\boldsymbol{r} - \boldsymbol{r}_m|)/|\boldsymbol{r} - \boldsymbol{r}_m| \approx jk$, leading to the final expression for the gradient of the spherical wavefront function. This means that, according to the method of stationary phase [24, 52], only points about the optical axis contribute significantly to the diffraction integral. Therefore, the origin of the unit vectors $\hat{\boldsymbol{p}}$ pointing toward the observation point $\boldsymbol{r}$ on the entrance pupil can be taken as the center of coordinates on the object plane, expressed as:

$$\hat{\boldsymbol{p}} = \frac{\boldsymbol{r}}{r} = p_x\hat{\boldsymbol{x}} + p_y\hat{\boldsymbol{y}} + p_z\hat{\boldsymbol{z}} = \sin\theta\cos\varphi\hat{\boldsymbol{x}} + \sin\theta\sin\varphi\hat{\boldsymbol{y}} + \cos\theta\hat{\boldsymbol{z}} \tag{15A.2}$$

This condition represents the Fraunhofer region of diffraction, where the binomial expression can be approximated by $|\boldsymbol{r} - \boldsymbol{r}_m| \approx r - \boldsymbol{r}\cdot\boldsymbol{r}_m$, reducing the Greens' function to its far-field expression:

$$\frac{e^{-jk|\boldsymbol{r}-\boldsymbol{r}_m|}}{4\pi|\boldsymbol{r}-\boldsymbol{r}_m|} \approx \frac{e^{-jkr}}{4\pi r}e^{jk\hat{\boldsymbol{p}}\cdot\boldsymbol{r}_m} \tag{15A.3}$$

Substituting (15A.3) into (15.12), and making use of relation (15A.1), we obtain the result for the fields at the entrance pupil (EP) as:

$$\boldsymbol{E}^{\mathrm{EP}}(\boldsymbol{r}) \approx \frac{e^{-jkr}}{4\pi r}(-j\boldsymbol{k})\times\int_{x_m}\int_{y_m}\left[2\hat{\boldsymbol{z}}\times\boldsymbol{E}^{\mathrm{M}}(\boldsymbol{r}_m)\right]e^{jk\hat{\boldsymbol{p}}\cdot\boldsymbol{r}_m}\,d\boldsymbol{r}_m \tag{15A.4}$$

where the coordinates (x_m, y_m) refer to points on surface S (equivalent to points on the mask plane).

The integral in (15A.4) can be identified as the spatial Fourier transform of the fields on surface S for the spatial frequencies $\boldsymbol{k} = k\hat{\boldsymbol{p}} = k[p_x, p_y, (1-p_x^2-p_y^2)^{1/2}]$ as defined in Section 15.3.1. This is expressed as:

$$\tilde{\boldsymbol{E}}^{\mathrm{M}}(p_x, p_y) = \int_{x_m}\int_{y_m} \boldsymbol{E}^{\mathrm{M}}(\boldsymbol{r}_m) e^{jk(p_x x_m + p_y y_m)}\, dx_m dy_m \tag{15A.5}$$

Applying this plane-wave property to (15A.4) via $-j\boldsymbol{k} \times [2\hat{\boldsymbol{z}} \times \tilde{\boldsymbol{E}}^{\mathrm{M}}(p_x, p_y)] = j2kp_z\tilde{\boldsymbol{E}}^{\mathrm{M}}(p_x, p_y)$, we obtain the far-field expression (15.13) of (15.12) used in lithography to represent the field diffracted by the photomask as collected at the entrance pupil of the projection optics:

$$\boldsymbol{E}^{\mathrm{EP}}(\boldsymbol{r}) \approx j2k\frac{e^{-jkr}}{4\pi r}e^{jkp_z z_m}p_z\tilde{\boldsymbol{E}}^{\mathrm{M}}(p_x, p_y) = \frac{j}{\lambda}\frac{e^{-jkr}}{r}\cos\theta\, \tilde{\boldsymbol{E}}^{\mathrm{M}}(p_x, p_y) \tag{15A.6}$$

As in the first Rayleigh-Sommerfeld scalar diffraction formulation [24, 52], an inclination factor equal to $\cos\theta$ is included in the integral. This factor, which in classical scalar diffraction integrals results from approximations regarding the boundary conditions, is seen here to derive directly from Maxwell's equations. The inclination factor accounts for the decrease in field density at oblique propagating angles [77].

APPENDIX 15B: DEBYE'S REPRESENTATION OF THE FOCUSING FIELDS

Following the same procedure as in Section 15.3.1, the fields radiated from the exit pupil at a distance from the aperture and nearby the geometrical focal point can be deduced by applying Huygens' principle to the fields, as illustrated in Fig. 15.13. At the plane of the exit pupil aperture, we can express the fields as:

$$\boldsymbol{E}^{\mathrm{image}}(\boldsymbol{r}_w) = -\nabla \times \int_{x_p}\int_{y_p}\left[-2\hat{\boldsymbol{n}} \times \boldsymbol{E}^{\mathrm{XP}}(\boldsymbol{r}_p)\right]\frac{e^{-jk'|\boldsymbol{r}_w-\boldsymbol{r}_p|}}{4\pi|\boldsymbol{r}_w-\boldsymbol{r}_p|}\, dx_p dy_p \tag{15B.1}$$

where $\boldsymbol{r}_w$ is the location of F_w, an observation point near $\boldsymbol{r}_1$, the location of the focal point F_1; and $\boldsymbol{r}_p = x_p\hat{\boldsymbol{x}} + y_p\hat{\boldsymbol{y}} + z_p\hat{\boldsymbol{z}}$ is the location of any point Q on the pupil aperture. According to the coordinate system illustrated in Fig. 15.13, the exit pupil is assumed for simplicity to be located exactly at the position $z_p = 0$. Similar to the object-space case, the wavevector in the image space is given by $k' = kn_{\mathrm{i}} = 2\pi n_{\mathrm{i}}/\lambda$, where n_{i} is the index of refraction of the image-space medium.

Adjusting the steps taken in deriving (15A.1) to the case of diffraction from the exit pupil, we have $\nabla = (\hat{\boldsymbol{x}}\,\partial/\partial x_w + \hat{\boldsymbol{y}}\,\partial/\partial y_w + \hat{\boldsymbol{z}}\,\partial/\partial z_w)$, which is applied to the expression for the spherical wavefront. We can write:

$$\nabla\left(\frac{e^{-jk'|\boldsymbol{r}_w-\boldsymbol{r}_p|}}{|\boldsymbol{r}_w-\boldsymbol{r}_p|}\right) = -\frac{1}{|\boldsymbol{r}_w-\boldsymbol{r}_p|}\begin{bmatrix}\hat{\boldsymbol{x}}(x_w-x_p)+\\ \hat{\boldsymbol{y}}(y_w-y_p)+\\ \hat{\boldsymbol{z}}(z_w-z_p)\end{bmatrix}\cdot\left(\frac{1}{|\boldsymbol{r}_w-\boldsymbol{r}_p|}+jk'\right)\cdot\left(\frac{e^{-jk'|\boldsymbol{r}_w-\boldsymbol{r}_p|}}{|\boldsymbol{r}_w-\boldsymbol{r}_p|}\right) \tag{15B.2}$$

Since exit-pupil point Q is assumed to be many wavelengths away from the focal point F_1 and the observation point F_w, while the observation point F_w is assumed to be in the proximity of F_1, we can approximate $(\boldsymbol{r}_w-\boldsymbol{r}_p) \approx (\boldsymbol{r}_1-\boldsymbol{r}_p) \equiv r_1\hat{\boldsymbol{w}}$, where $\hat{\boldsymbol{w}}$ points in the direction of $(\boldsymbol{r}_1-\boldsymbol{r}_p)$.

For simplicity, we can further assume that the geometrical focus lies on the optical axis, i.e., the z-axis in Fig. 15.13(b), and therefore $\boldsymbol{r}_1 = (x_1, y_1, z_1) = (0, 0, z_1)$. The final expression for $\hat{\boldsymbol{w}}$ is then defined as:

$$\begin{aligned} \hat{\boldsymbol{w}} &\equiv \frac{\boldsymbol{r}_1 - \boldsymbol{r}_p}{|\boldsymbol{r}_1 - \boldsymbol{r}_p|} = \frac{-x_p\hat{\boldsymbol{x}} - y_p\hat{\boldsymbol{y}} + z_1\hat{\boldsymbol{z}}}{|\boldsymbol{r}_1 - \boldsymbol{r}_p|} = w_x\hat{\boldsymbol{x}} + w_y\hat{\boldsymbol{y}} + w_z\hat{\boldsymbol{z}} \\ &= -\sin\theta'\cos\varphi'\,\hat{\boldsymbol{x}} - \sin\theta'\sin\varphi'\,\hat{\boldsymbol{y}} + \cos\theta'\,\hat{\boldsymbol{z}} \end{aligned} \tag{15B.3}$$

where the angles θ' and φ' are defined with respect to the focal point F_1 as in Fig. 15.13(b). Finally, using $(\boldsymbol{r}_w - \boldsymbol{r}_p) \approx r_1\hat{\boldsymbol{w}}$ with $k'r_1 >> 1$ in (15B.2), we obtain:

$$\nabla\left(\frac{e^{-jk'|\boldsymbol{r}_w-\boldsymbol{r}_p|}}{|\boldsymbol{r}_w - \boldsymbol{r}_p|}\right) = -jk'\hat{\boldsymbol{w}}\frac{e^{-jk'|\boldsymbol{r}_w-\boldsymbol{r}_p|}}{|\boldsymbol{r}_w - \boldsymbol{r}_p|} = -j\boldsymbol{k}'\frac{e^{-jk'|\boldsymbol{r}_w-\boldsymbol{r}_p|}}{|\boldsymbol{r}_w - \boldsymbol{r}_p|} \tag{15B.4}$$

with $\boldsymbol{k}' = k'\hat{\boldsymbol{w}}$ in the direction defined by the vector $(\boldsymbol{r}_1 - \boldsymbol{r}_p)$. Substituting in (15B.1) and noting that $\hat{\boldsymbol{n}} = \hat{\boldsymbol{z}}$, we have:

$$\boldsymbol{E}^{\text{image}}(\boldsymbol{r}_w) = -j\boldsymbol{k}'\times\iint_{x_p\,y_p}\left[2\hat{\boldsymbol{z}}\times\boldsymbol{E}^{\text{XP}}(\boldsymbol{r}_p)\right]\frac{e^{-jk'|\boldsymbol{r}_w-\boldsymbol{r}_p|}}{4\pi|\boldsymbol{r}_w - \boldsymbol{r}_p|}dx_p\,dy_p \tag{15B.5}$$

Deviating from the approach followed in Section 15.3.1 we apply now the angular-spectrum representation of the free-space Green's function to (15B.5):

$$\frac{e^{-jk'|\boldsymbol{r}_w-\boldsymbol{r}_p|}}{|\boldsymbol{r}_w - \boldsymbol{r}_p|} = \frac{j}{2\pi}\iint_{k_x\,k_y}\frac{1}{k_z}e^{-j\left[k_x(x_w-x_p)+k_y(y_w-y_p)+k_z|z_w-z_p|\right]}dk_x\,dk_y \tag{15B.6}$$

where $k_z = [(k')^2 - k_x^2 - k_y^2]^{1/2}$ and $(z_w - z_p) = z_w > 0$ always, obtaining:

$$\boldsymbol{E}^{\text{image}}(\boldsymbol{r}_w) = -j\boldsymbol{k}'\times\iint_{x_p\,y_p}\left[2\hat{\boldsymbol{z}}\times\boldsymbol{E}^{\text{XP}}(\boldsymbol{r}_p)\right]\frac{j}{8\pi^2}\iint_{k_x\,k_y}\frac{1}{k_z}e^{-j\left[k_x(x_w-x_p)+k_y(y_w-y_p)+k_z z_w\right]}dk_x\,dk_y\,dx_p\,dy_p \tag{15B.7}$$

Exchanging the order of the integration, we can express the image fields in terms of the spatial-frequency Fourier transform representation of the fields at the exit-pupil surface as follows:

$$\boldsymbol{E}^{\text{image}}(\boldsymbol{r}_w) = \frac{\boldsymbol{k}'}{8\pi^2}\times\left(2\hat{\boldsymbol{z}}\times\iint_{k_x\,k_y}\frac{1}{k_z}e^{-j(k_x x_w + k_y y_w + k_z z_w)}\,\tilde{\boldsymbol{E}}^{\text{XP}}(k_x, k_y)\,dk_x\,dk_y\right) \tag{15B.8}$$

where $\tilde{\boldsymbol{E}}^{\text{XP}}(k_x, k_y)$ is given by:

$$\tilde{\boldsymbol{E}}^{\text{XP}}(k_x, k_y) = \iint_{x_p\,y_p}\boldsymbol{E}^{\text{XP}}(\boldsymbol{r}_p)e^{j(k_x x_p + k_y y_p)}\,dx_p\,dy_p \tag{15B.9}$$

This gives the final expression for the fields diffracted from the exit pupil as a sum of an infinite number of plane waves emanating from the exit pupil as:

$$\boldsymbol{E}^{\text{image}}(\boldsymbol{r}_w) = \frac{k'}{4\pi^2} \times \left(\hat{z} \times \iint_{k_x\, k_y} \frac{1}{k_z} e^{-jk'\cdot r_w}\, \tilde{\boldsymbol{E}}^{\text{XP}}(k_x, k_y)\, dk_x dk_y \right) \tag{15B.10}$$

We have assumed that the fields emanating from the exit pupil aperture are in the form of a spherical wavefront converging toward its geometrical focal point at $\boldsymbol{r}_1$. The amplitude at each point of the aperture is equal to $\mathbf{A}(x_p, y_p)$, and any phase deviation from the spherical wavefront induced by aberrations in the propagation through the projection optics system is given by the function $k'\phi(x_p, y_p)$. The fields at the exit aperture thus obey the expression:

$$\boldsymbol{E}^{\text{XP}}(x_p, y_p, 0) = \begin{cases} \mathbf{A}(x_p, y_p)\, e^{jk'\phi(x_p,\, y_p)}\, e^{jk'|\boldsymbol{r}_1 - \boldsymbol{r}_p|} / |\boldsymbol{r}_1 - \boldsymbol{r}_p| & \text{for } x_p^2 + y_p^2 \le a^2 \\ 0 & \text{for } x_p^2 + y_p^2 > a^2 \end{cases} \tag{15B.11}$$

where a is the radius of the exit-pupil aperture. The fields are assumed to equal those that would exist in the absence of the aperture, unperturbed by the boundary rim, and vanishing outside the aperture on the rest of the plane. This corresponds to Kirchhoff's boundary condition approximation in a manner similar to the way in which the thin-mask approximation is applied to the fields on the mask exit plane. Substituting (15B.11) into (15B.9), the expression for the angular spectrum of the exit pupil fields, we obtain:

$$\tilde{\boldsymbol{E}}^{\text{XP}}(k_x, k_y) = \iint_{x_p^2 + y_p^2 \le a^2} \mathbf{A}(x_p, y_p)\, e^{j(k_x x_p + k_y y_p)} e^{jk'\phi(x_p,\, y_p)} e^{jk'|\boldsymbol{r}_1 - \boldsymbol{r}_p|} / |\boldsymbol{r}_1 - \boldsymbol{r}_p|\, dx_p\, dy_p \tag{15B.12}$$

A further approximation can be applied in order to derive an expression for the focusing fields that form the image in the neighborhood of the geometrical focal point. In particular, the method of stationary phase is applied to (15B.12) in order to derive its asymptotic solution.

The magnitude of the integrand of (15B.12), $Mag(x_p, y_p) = \mathbf{A}(x_p, y_p) / |\boldsymbol{r}_1 - \boldsymbol{r}_p|$, is assumed to vary slowly compared to the phase, $Phase(x_p, y_p) = (k_x x_p + k_y y_p) + k'\phi(x_p, y_p) + k'|\boldsymbol{r}_1 - \boldsymbol{r}_p|$, so that the lower-order asymptotic results are suitably accurate. This condition is usually satisfied if $k' = 2\pi n_i/\lambda$ is sufficiently large. Under these conditions, the only points contributing significantly to the integral are those around stationary points where the phase derivative is zero, and those at the end points of the integration limits [24, 52]. While other points satisfying certain conditions at the boundaries can also contribute to the integral, these are assumed negligible and ignored in this derivation. For more details, refer to the work by Stamnes [52].

In the case of 2-D integrals, the stationary points retained in this derivation are those inside the aperture where the phase satisfies $\partial Phase(x_p, y_p, 0)/\partial x_p = \partial Phase(x_p, y_p, 0)/\partial y_p = 0$. Furthermore, assuming, for simplicity, that the impact of aberrations is negligible such that $\phi(x_p, y_p, 0) = 0$, we get the final expression for the stationary points:

$$\begin{aligned} x_s &= x_1 - (z_1 - z_p)\cdot(k_x / k_z) \qquad & x_s &= r_1 w_x \ \text{ if } x_1 \sim 0 \\ y_s &= y_1 - (z_1 - z_p)\cdot(k_y / k_z) \qquad & y_s &= r_1 w_y \ \text{ if } y_1 \sim 0 \end{aligned} \tag{15B.13}$$

The contribution of this stationary point is zero if $x_s^2 + y_s^2 \ge a^2$. If $x_s^2 + y_s^2 \le a^2$, the final lower-order contribution of this stationary point is [52]:

$$\tilde{\boldsymbol{E}}_S^{\mathrm{XP}}(k_x, k_y) = \begin{cases} (j2\pi / k_z)\mathbf{A}(x_s, y_s)e^{j\boldsymbol{k}'\cdot\boldsymbol{r}_1} & \text{inside } \Omega \\ 0 & \text{outside } \Omega \end{cases} \tag{15B.14}$$

where Ω is the solid angle subtended by the aperture at the focal point, and the stationary points are given by (15B.13). The term $\mathbf{A}(x_s, y_s)$ represents the aperture fields at position (x_s, y_s) on the exit-pupil plane propagating with the direction cosines (w_x, w_y).

In this derivation of (15B.14), known as the Debye approximation [24, 52], the contribution of stationary points on the boundary to the integral is ignored. Hence, this approximation retains only the effect of the exit-pupil aperture spectrum due to the interior stationary points, and ignores the impact on the angular spectrum (and image fields) of edge diffraction at the aperture rim. Additional, but often weaker, contributions due to other stationary points are also ignored, as well as the consequences of having either non-negligible aberrations or stationary points that approach the aperture edges or fall close to each other [52].

As a final step in order to deduce the Debye expression for the image fields, the result of (15B.14) is inserted into (15B.10) to obtain:

$$\boldsymbol{E}_D^{\mathrm{image}}(\boldsymbol{r}_w) = \frac{k'}{4\pi^2} \times \left(\hat{z} \times \iint_{k_x k_y \in \Omega} \frac{j2\pi}{k_z^2} \mathbf{A}(w_x, w_y) e^{-j\boldsymbol{k}'\cdot(\boldsymbol{r}_w - \boldsymbol{r}_1)} dk_x \, dk_y \right) \tag{15B.15}$$

We can use the geometrical optics result that the amplitude of the fields at every point of the exit-pupil aperture derives from a plane-wave component traveling through the projection lens system and satisfying $\boldsymbol{k}'\cdot\mathbf{A}(w_x, w_y)$. Substituting the propagation wavevector in the image space given by $\boldsymbol{k}' = k'[w_x, w_y, (1 - w_x^2 - w_y^2)^{1/2}]$, we obtain:

$$\boldsymbol{E}^{\mathrm{image}}(\boldsymbol{r}_w) = \frac{j}{\lambda} \iint_{w_x w_y \leq \mathrm{NA_i}} \frac{1}{w_z} \mathbf{A}(w_x, w_y) e^{-j\boldsymbol{k}'\cdot(\boldsymbol{r}_w - \boldsymbol{r}_1)} dw_x \, dw_y \tag{15B.16}$$

which was used in Section 15.3.1 to extract the fields focused by the scanner optics onto the wafer plane.

APPENDIX 15C: POLARIZATION TENSOR

The Cartesian components of each plane wave at the exit pupil can be determined from those at the entrance pupil. For each ray direction (w_x, w_y, w_z), polarization rotation can be expressed as a tensor $\mathbf{T}$, obtained after decomposing the fields into their local projections along the directions normal and parallel to its meridional plane, E_{m_n} and E_{m_p}. We apply the condition that the polarization angle with respect to this plane remains approximately constant through the projection system. The directions $\hat{\boldsymbol{m}}_n$ and $\hat{\boldsymbol{m}}_p$ are given by:

$$\hat{\boldsymbol{m}}_n = \hat{\boldsymbol{p}} \times \hat{z} = \frac{p_y \hat{\boldsymbol{x}} - p_x \hat{\boldsymbol{y}}}{\sqrt{p_x^2 + p_y^2}} \tag{15C.1a}$$

$$\hat{\boldsymbol{m}}_p = \hat{\boldsymbol{m}}_n \times \hat{\boldsymbol{p}} = -\frac{\sqrt{1 - p_x^2 - p_y^2}\,(p_x \hat{\boldsymbol{x}} + p_y \hat{\boldsymbol{y}}) - (p_x^2 + p_y^2)\hat{z}}{\sqrt{p_x^2 + p_y^2}} \tag{15C.1b}$$

where $\hat{\boldsymbol{p}}$ is given by (15.14). The transformation matrix between both coordinate systems can be expressed as:

$$\begin{bmatrix} \hat{\boldsymbol{m}}_p \\ \hat{\boldsymbol{m}}_n \\ \hat{\boldsymbol{p}} \end{bmatrix} = \frac{1}{\sqrt{p_x^2 + p_y^2}} \begin{bmatrix} -p_z p_x & -p_z p_y & p_x^2 + p_y^2 \\ p_y & -p_x & 0 \\ p_x\sqrt{p_x^2 + p_y^2} & p_y\sqrt{p_x^2 + p_y^2} & p_z\sqrt{p_x^2 + p_y^2} \end{bmatrix} \cdot \begin{bmatrix} \hat{\boldsymbol{x}} \\ \hat{\boldsymbol{y}} \\ \hat{\boldsymbol{z}} \end{bmatrix} = \mathbf{T}_\text{o} \cdot \begin{bmatrix} \hat{\boldsymbol{x}} \\ \hat{\boldsymbol{y}} \\ \hat{\boldsymbol{z}} \end{bmatrix} \tag{15C.2}$$

such that on the entrance pupil:

$$\begin{aligned} \boldsymbol{E}^\text{EP} &= \begin{bmatrix} E_x^\text{EP} & E_y^\text{EP} & E_z^\text{EP} \end{bmatrix} \cdot \begin{bmatrix} \hat{\boldsymbol{x}} \\ \hat{\boldsymbol{y}} \\ \hat{\boldsymbol{z}} \end{bmatrix} = \begin{bmatrix} E_x^\text{EP} & E_y^\text{EP} & E_z^\text{EP} \end{bmatrix} \cdot \mathbf{T}_\text{o}^{-1} \begin{bmatrix} \hat{\boldsymbol{m}}_p \\ \hat{\boldsymbol{m}}_n \\ \hat{\boldsymbol{p}} \end{bmatrix} \\ &= \begin{bmatrix} E_{m_p}^\text{EP} & E_{m_n}^\text{EP} & 0 \end{bmatrix} \cdot \begin{bmatrix} \hat{\boldsymbol{m}}_p \\ \hat{\boldsymbol{m}}_n \\ \hat{\boldsymbol{p}} \end{bmatrix} \end{aligned} \tag{15C.3}$$

given that the fields are orthogonal to the ray direction of propagation in the far-field region of the reticle as observed in (15.13).

Repeating the steps on the image side for the vector $\hat{\boldsymbol{w}}$ given by (15.17), we have:

$$\hat{\boldsymbol{m}}_n' = \hat{\boldsymbol{m}}_n = \frac{-w_y \hat{\boldsymbol{x}} + w_x \hat{\boldsymbol{y}}}{\sqrt{w_x^2 + w_y^2}} \tag{15C.4a}$$

$$\hat{\boldsymbol{m}}_p' = \hat{\boldsymbol{m}}_n' \times \hat{\boldsymbol{w}} = \frac{\sqrt{1 - w_x^2 - w_y^2}\,(w_x \hat{\boldsymbol{x}} + w_y \hat{\boldsymbol{y}}) - (w_x^2 + w_y^2)\hat{\boldsymbol{z}}}{\sqrt{w_x^2 + w_y^2}} \tag{15C.4b}$$

resulting in the following transformation matrix:

$$\begin{bmatrix} \hat{\boldsymbol{m}}_p' \\ \hat{\boldsymbol{m}}_n' \\ \hat{\boldsymbol{w}} \end{bmatrix} = \frac{1}{\sqrt{w_x^2 + w_y^2}} \begin{bmatrix} w_z w_x & w_z w_y & -(w_x^2 + w_y^2) \\ -w_y & w_x & 0 \\ w_x\sqrt{w_x^2 + w_y^2} & w_y\sqrt{w_x^2 + w_y^2} & w_z\sqrt{w_x^2 + w_y^2} \end{bmatrix} \cdot \begin{bmatrix} \hat{\boldsymbol{x}} \\ \hat{\boldsymbol{y}} \\ \hat{\boldsymbol{z}} \end{bmatrix} = \mathbf{T}_\text{i} \cdot \begin{bmatrix} \hat{\boldsymbol{x}} \\ \hat{\boldsymbol{y}} \\ \hat{\boldsymbol{z}} \end{bmatrix} \tag{15C.5}$$

According to the meridional plane approximation, the field components along the directions normal, $\hat{\boldsymbol{m}}_n$, and parallel, $\hat{\boldsymbol{m}}_p$, to the meridional plane remain approximately the same as they propagate through the optical projection system, such that at the exit pupil we can assume $E_{m_p'}^\text{XP} = E_{m_p}^\text{EP}$ and $E_{m_n'}^\text{XP} = E_{m_n}^\text{EP}$. Thus, on the image side:

$$\begin{bmatrix} E_x^{\mathrm{XP}} & E_y^{\mathrm{XP}} & E_z^{\mathrm{XP}} \end{bmatrix} \cdot \begin{bmatrix} \hat{\boldsymbol{x}} \\ \hat{\boldsymbol{y}} \\ \hat{\boldsymbol{z}} \end{bmatrix} = \begin{bmatrix} E_{m_p'}^{\mathrm{XP}} & E_{m_n'}^{\mathrm{XP}} & 0 \end{bmatrix} \cdot \begin{bmatrix} \hat{\boldsymbol{m}}_p' \\ \hat{\boldsymbol{m}}_n' \\ \hat{\boldsymbol{s}} \end{bmatrix} = \begin{bmatrix} E_{m_p}^{\mathrm{EP}} & E_{m_n}^{\mathrm{EP}} & 0 \end{bmatrix} \cdot \begin{bmatrix} \hat{\boldsymbol{m}}_p' \\ \hat{\boldsymbol{m}}_n' \\ \hat{\boldsymbol{w}} \end{bmatrix}$$

$$= \begin{bmatrix} E_{m_p}^{\mathrm{EP}} & E_{m_n}^{\mathrm{EP}} & 0 \end{bmatrix} \cdot \mathbf{T}_{\mathrm{i}} \cdot \begin{bmatrix} \hat{\boldsymbol{x}} \\ \hat{\boldsymbol{y}} \\ \hat{\boldsymbol{z}} \end{bmatrix} = \begin{bmatrix} E_x^{\mathrm{EP}} & E_y^{\mathrm{EP}} & E_z^{\mathrm{EP}} \end{bmatrix} \cdot \mathbf{T}_{\mathrm{o}}^{-1} \cdot \mathbf{T}_{\mathrm{i}} \cdot \begin{bmatrix} \hat{\boldsymbol{x}} \\ \hat{\boldsymbol{y}} \\ \hat{\boldsymbol{z}} \end{bmatrix} \tag{15C.6}$$

with

$$\begin{bmatrix} E_x^{\mathrm{XP}} & E_y^{\mathrm{XP}} & E_z^{\mathrm{XP}} \end{bmatrix} = \begin{bmatrix} E_x^{\mathrm{EP}} & E_y^{\mathrm{EP}} & E_z^{\mathrm{EP}} \end{bmatrix} \cdot \mathbf{T} \tag{15C.7}$$

and

$$\mathbf{T}_{\mathrm{o}}^{-1} \cdot \mathbf{T}_{\mathrm{i}} = \mathbf{T} = \begin{bmatrix} T_{xx} & T_{yx} & T_{zx} \\ T_{yx} & T_{yy} & T_{zy} \\ T_{zx} & T_{zy} & T_{zz} \end{bmatrix} \tag{15C.8a}$$

where

$$T_{xx} = \frac{w_y^2 + w_x^2 \left[w_z p_z - \mathrm{M}(w_x^2 + w_y^2) \right]}{w_x^2 + w_y^2} \qquad T_{yy} = \frac{w_x^2 + w_y^2 \left[w_z p_z - \mathrm{M}(w_x^2 + w_y^2) \right]}{w_x^2 + w_y^2} \tag{15C.8b}$$

$$T_{yx} = \frac{-w_x w_y \left[1 - w_z p_z + \mathrm{M}(w_x^2 + w_y^2) \right]}{w_x^2 + w_y^2} \qquad T_{zz} = w_z p_z - \mathrm{M}(w_x^2 + w_y^2) \tag{15C.8c}$$

$$T_{zx} = -w_x (p_z + \mathrm{M} w_z) \qquad T_{zy} = -w_y (p_z + \mathrm{M} w_z) \tag{15C.8d}$$

These results are consistent with the so-called "polarization state matrix" based in the trigonometrically determined polarization transformation by Mansuripur [70], as long as the scalar approximation is applied in the object space, that is, $\mathrm{M}(w_x^2 + w_y^2)/w_z$, $\mathrm{M}w_z/w_z << p_z \sim 1$.

APPENDIX 15D: BEST FOCUS

To understand the relationship between the mask phase or imaginary errors and the shift in the plane of best focus, we assume a 1-D grating. The grating is illuminated by an oblique incident plane wave propagating parallel to the *x-z* plane with electric field polarized along the *y*-axis (TE polarization). The wavevectors of the incident wave and 0th and 1st diffracted orders (Fig. 15D.1), assuming the substrate has an index of refraction equal to air $n_s = 1$, are given by:

$$\boldsymbol{k}_{\mathrm{inc}} = -\hat{\boldsymbol{x}} k \sin\theta_{\mathrm{inc}} + \hat{\boldsymbol{z}} k \cos\theta_{\mathrm{inc}} \tag{15D.1a}$$

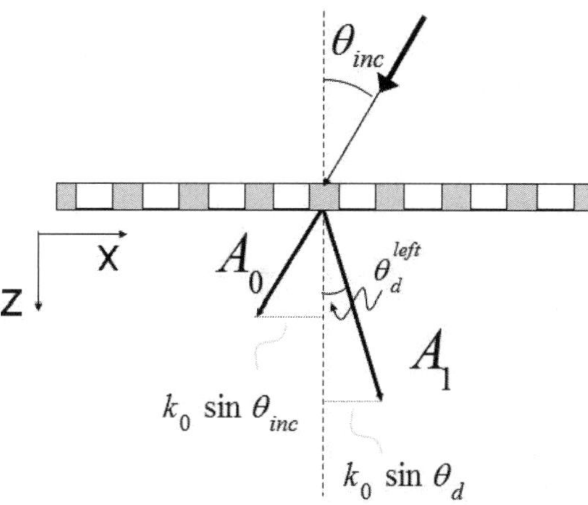

Fig. 15D.1 Incident and diffracted angles defined for a grating illuminated at oblique incidence.

$$\boldsymbol{k}_1 = \hat{x}k\sin\theta_1 + \hat{z}k\cos\theta_1 = \hat{x}k\frac{\lambda}{L_x} + \hat{z}k\sqrt{1-\left(\frac{\lambda}{L_x}\right)^2} \tag{15D.1b}$$

Here, the relation between the diffraction angle and the mask pitch, L_x, $p_x = \sin\theta_1\cos\varphi = \lambda/L_x$, has been used for the 1-D case where $\varphi = 0$. With oblique incidence, the diffracted pattern is shifted in spatial frequency by the value of the incident wavevector, that is, by $\boldsymbol{k}_{\text{inc}}$. At the small pitches at which the focus shift occurs, the aerial image fringes are chiefly formed by the interaction of these two lower diffracted orders, since these are the only orders collected by the entrance-pupil numerical aperture. For simplicity, we ignore high-NA vector effects and losses that might occur in the propagation through the imaging system. Finally the electric field at the image plane is given by the vector sum of the two propagating orders as follows:

$$\begin{aligned} \boldsymbol{E}^l_{\text{image}} &= \mathbf{A}_0 e^{-j\boldsymbol{k}'_{\text{inc}}\cdot\boldsymbol{r}} + \mathbf{A}_1 e^{-j(\boldsymbol{k}'_1+\boldsymbol{k}'_{\text{inc}})\cdot\boldsymbol{r}} \\ &= \mathbf{A}_0 e^{-jk'(x\sin\theta'_{\text{inc}}+z\cos\theta'_{\text{inc}})} + \mathbf{A}_1 e^{-jk'(-x\sin\theta'_d+z\cos\theta'_d)} \end{aligned} \tag{15D.2}$$

where the angle θ_d is defined in Fig. 15D.1 and given by:

$$\sin\theta_d = \sin\theta_1 - \sin\theta_{\text{inc}}, \qquad \cos\theta_d = \cos\theta_1 + \cos\theta_{\text{inc}} \tag{15D.3}$$

and the prime variables are the corresponding propagating angles in the image space. The aerial image is given by the intensity distribution:

$$I^l_{\text{image}} = |A_0|^2 + |A_1|^2 + 2\,\text{Re}\left[A_0 A_1^* e^{-jk'x(\sin\theta'_{\text{inc}}+\sin\theta'_d)} e^{-jk'z(\cos\theta'_{\text{inc}}-\cos\theta'_d)}\right] \tag{15D.4}$$

where due to EMF effects we have:

$$A_0 A_1 = \alpha e^{j\delta} \tag{15D.5}$$

and thus:

$$I^{l}_{\text{image}} = |A_0|^2 + |A_1|^2 + 2\alpha\cos\begin{bmatrix} k'x(\sin\theta'_{\text{inc}} + \sin\theta'_d) + \\ k'z(\cos\theta'_{\text{inc}} - \cos\theta'_d) - \delta \end{bmatrix} \tag{15D.6}$$

The plane z_{BF} of best focus is given by setting the z-derivative to zero ($\partial/\partial z = 0$), as:

$$k' z_{\text{BF}}(\cos\theta'_{\text{inc}} - \cos\theta'_d) - \delta = 0 \quad \rightarrow \quad z_{\text{BF}} = \frac{\delta}{k'(\cos\theta'_{\text{inc}} - \cos\theta'_d)} \tag{15D.7}$$

We see that when $\sin\theta_{\text{inc}} = 0.5\sin\theta_1$, then $\sin\theta_d = \sin\theta_{\text{inc}}$ (and therefore $\cos\theta_d = \cos\theta_{\text{inc}}$), and the best focus is at positive infinity (∞). At a fixed incident angle, this translates into a singularity in the plot of best focus vs. pitch at the exact pitch at which this condition is satisfied. The value of the best focus reaches toward $\sim\pm\infty$ for pitches just above the singularity, and reaches toward $\sim\mp\infty$ for pitches just below, depending on the sign of the phase term δ. Since angle θ_{inc} depends on the numerical aperture of the imaging system, the exact pitch at which the focus starts shifting depends on the numerical aperture and the exact shape of the illuminating source.

This result corresponds to nearly "coherent" illumination, where the source pupil contains only two source points. When the mask is instead illuminated with an extended source, each source pixel generates a plane wave with a slight range of incidence angles, $\theta_{\text{inc}} \pm \Delta\theta$. The precise pitch at which the relationship $\cos\theta_d = \cos\theta_{\text{inc}} \pm \Delta\theta$ varies across each source pixel and across the active portion of the source pupil. Thus, for partially coherent illumination, the empirical dependence of the plane of best focus on pitch does not exhibit a singularity. Instead, there is a substantial change in focus around the pitch value satisfying the condition $\cos\theta_d = \cos\theta_{\text{inc}}$, as seen in Figs. 15.26(a) and 15.26(b). The sign of this change in focus depends on the sign of the phase term δ, a direct consequence of the impact of mask topography on the diffracted orders.

Thus, shifts in the plane of best focus, caused by variations in the transmitted phase through the mask, are amplified by oblique illumination and pitch values close to the resolution limit. When the illumination is at normal incidence, then $z_{\text{BF}} = \delta/(k' - k_{1z})$, and the denominator can never reach 0. At normal incidence, the focus shifts are a function only of the phase term δ, resulting in very small deviations.

REFERENCES

[1] Moore, G. E., "Cramming more components onto integrated circuits," *Electronics*, Vol. 38, 1965.

[2] Moore, G. E., "Lithography and the future of Moore's law," *Proc. SPIE*, Vol. 2437, 1995.

[3] *International Technology Road Map for Semiconductors*, 2011, Online: http://www.itrs.net/Links/2011ITRS/Home2011.htm

[4] Van Zant, P., *Microchip Fabrication: A Practical Guide to Semiconductor Processing*, 4th ed., New York: McGraw-Hill, 2000.

[5] Burr, G. W., B. N. Kurdi, J. C. Scott, C. H. Lam, K. Gopalakrishnan, and R. S. Shenoy, "Overview of candidate device technologies for storage-class memory," *IBM J. Research & Development*, Vol. 52, 2008, pp. 449–464.

[6] Rayleigh, L., "Investigations in optics, with special reference to the spectroscope," *Philosophical Magazine*, Vol. 8, 1879, pp. 216–274, 403–411, 477–486.

[7] Rayleigh, L., "Investigations in optics, with special reference to the spectroscope," *Philosophical Magazine*, Vol. 9, 1880, pp. 40–55.

[8] Rayleigh, L., "On the theory of optical instruments, with special reference to the microscope," *Philosophical Magazine*, Vol. 42, 1886, pp. 167–195.

[9] Lammers, D., "Intel drops 157 litho from roadmap," May 23, 2003. Online: http://www.eetimes.com/electronicsnews/4092440/Intel-drops-157-litho-from-roadmap

[10] LaPedus, M., "Nikon scraps high-index tool project," May 15, 2008. Online: http://www.eetimes.com/electronicsnews/4077027/Nikon-scraps-high-index-tool-project

[11] Merritt, R., "ASML describes road map to 22 nm litho," Oct. 15, 2008. Online: http://www.eetimes.com/electronicsnews/4079388/Video-ASML-describes-road-map-to-22-nm-litho

[12] Liebmann, L., L. Pileggi, J. Hibbeler, V. Rovner, T. Jhaveri, and G. Northrop, "Simplify to survive, prescriptive layouts ensure profitable scaling to 32 nm and beyond," *Proc. SPIE*, Vol. 7275:72750A, 2009.

[13] Wong, A. K., *Resolution Enhancement Techniques in Optical Lithography*, SPIE Press, 2001.

[14] Schellenberg, F., ed., *Selected Papers in Resolution Enhancement Techniques*, SPIE Press, 2004.

[15] Rosenbluth, A. E., S. Bukofsky, C. Fonseca, M. Hibbs, K. Lai, A. Molless, R. N. Singh, and A. K. K. Wong, "Optimum mask and source patterns to print a given shape," *J. Microlithographic Microfabrication, Microsystems*, Vol. 1, 2002, pp. 13–30.

[16] Otto, O. W., J. G. Garofalo, K. K. Low, C.-M. Yuan, R. C. Henderson, C. Pierrat, R. L. Kostelak, S. Vaidya, and P. K. Vasudev, "Automated optical proximity correction: A rules-based approach," *Proc. SPIE*, Vol. 2197, 1994, pp. 278–293.

[17] Cobb, N., A. Zakhor, and E. Miloslavsky, "Mathematical and CAD framework for proximity correction," *Proc. SPIE*, Vol. 2726, 1996, pp. 208–222.

[18] Pati, Y. C., and T. Kailath, "Phase-shifting mask for microlithography automated design and mask requirements," *J. Optical Society of America A*, Vol. 11, 1994, pp. 2438–2452.

[19] Socha, R., X. Shi, and D. LeHoty, "Simultaneous source mask optimization (SMO)," *Proc. SPIE*, Vol. 5853, 2005, p. 180.

[20] Abrams, D. S., and L. Pang. "Fast inverse lithography technology," *Proc. SPIE*, Vol. 6154:61541J, 2006.

[21] Rosenbluth, A. E., D. O Melville, K. Tian, S. Bagheri, J. Tirapu Azpiroz, K. Lai, A. Waechter, T. Inoue, L. Ladanyi, F. Barahona, K. Scheinberg, M. Sakamoto, H. Muta, E. Gallagher, T. Faure, and M. Hibbs, "Intensive optimization of masks and sources for 22-nm lithography," *Proc. SPIE*, Vol. 7274:727409, 2009.

[22] Krasnoperova, A., J. A. Culp, I. Graur, and S. Mansfield, "Process window OPC for reduced process variability and enhanced yield," *Proc. SPIE*, Vol. 6154, 2006, p. 61543L–1.

[23] Mulder, M., A. Engelen, O. Noordman, G. Streutker, B. van Drieenhuizen, C. van Nuenen, W. Endendijk, J. Verbeeck, W. Bouman, A. Bouma, R. Kazinczi, R. Socha, D. Jurgens, J. Zimmermann, B. Trauter, J. Bekaert, B. Laenens, D. Corliss, and G. McIntyre, "Performance of FlexRay: A fully programmable illumination system for generation of freeform sources on high NA immersion systems," *Proc. SPIE*, Vol. 7640, 2010, 764005.

[24] Born, M., and E. Wolf, *Principles of Optics*, 6th ed., New York: Pergamon Press, 1987.

[25] Goodman, D. S., and A. E. Rosenbluth, "Condenser aberrations in Köhler illumination," *Proc. SPIE*, Vol. 922, 1988, pp. 108–134.

[26] Mack, C. A., *Fundamental Principles of Optical Lithography: The Science of Microfabrication*, West Sussex, England: Wiley, 2011.

[27] Erdmann, A., and P. Evanschitzky, "Rigorous electromagnetic field mask modeling and related lithographic effects in the low k_1 and ultrahigh numerical aperture regime," *J. Micro/Nanolithography, MEMS, and MOEMS*, Vol. 6, 2007, 031002.

[28] McIntyre, G., M. Hibbs, J. Tirapu-Azpiroz, G. Han, S. Halle, T. Faure, R. Deschner, B. Morgenfeld, S. Ramaswamy, A. Wagner, T. Brunner, and Y. Kikuchi, "Lithographic qualification of new opaque mosi binary mask blank for the 32-nm node and beyond," *J. Micro/Nanolithography, MEMS, MOEMS*, Vol. 9, 2010, 013010–1.

[29] Sato, T., A. Endo, A. Mimotogi, S. Mimotogi, K. Sato, and S. Tanaka, "Impact of polarization for an attenuated phase shift mask with arf hyper-NA lithography," *Proc. SPIE*, Vol. 5754, 2005, p. 1063.

[30] Flanders, D. C., and H. I. Smith, "Spatial period division exposing," U.S. Patent 4,360,586, 1982.

[31] Levenson, M. D., N. S. Viswanathan, and R. A. Simpson, "Improving resolution in photolithography with phase shifting mask," *IEEE Trans. Electron Devices*, Vol. ED-29, 1982, pp. 1828–1836.

[32] Erdmann, A., G. Citarella, P. Evanschitzky, H. Schermer, V. Philipsen, and P. De Bisschop, "Validity of the Hopkins approximation in simulations of hyper-NA (NA>1) line-space structures for an attenuated PSM mask," *Proc. SPIE*, Vol. 6154, 2006, pp. 167–178.

[33] Yoshizawa, M., V. Philipsen, L. H. A. Leunissen, E. Hendrickx, R. Jonckheere, G. Vandenberghe, U. Buttgereit, H. Becker, C. Koepernik, and M. Irmscher," Comparative study of bi-layer attenuating phase-shifting masks for hyper-NA lithography," *Proc. SPIE*, Vol. 6283, 2006, 1G.

[34] Cheng, W.-H., and J. Farnsworth, "Control of polarization and apodization with film materials on photomasks and pellicles for high NA imaging performance," *Proc. SPIE*, Vol. 6520, 2007, 65200O.

[35] Ham, Y. M., N. Morgana, P. Sixt, M. Cangemi, B. Kasprowicz, and C. Progler, "Optical properties and process impacts of high transmission EAPSM," *Proc. SPIE*, Vol. 6154, 2006, 615413.

[36] Lin, B. J., "The attenuated phase-shifting mask," *Solid State Technology*, Vol. 35, 1992, pp. 43–47.

[37] Conley, W., N. Morgana, B. S. Kasprowicz, M. Cangemi, M. Lassiter, L. C. Litt, M. Cangemi, R. Cottle, W. Wu, J. Cobb, Y. M. Ham, K. Lucas, B. Roman, and C. Progler, "High transmission mask technology for 45-nm node imaging," *Proc. SPIE*, Vol. 6154, 2006, 615411.

[38] Tirapu-Azpiroz, J., G. McIntyre, T. Faure, S. Halle, M. Hibbs, A. Wagner, K. Lai, E. Gallagher, and T. Brunner, "Understanding the trade-offs of thinner binary mask absorbers," *Proc. SPIE*, Vol. 7823, 2010, 78230M–1.

[39] Lai, K., S. Burns, S. Halle, L. Zhuang, M. Colburn, S. Allen, C. Babcock, Z. Baum, M. Burkhardt, V. Dai, D. Dunn, E. Geiss, H. Haffner, G. Han, P. Lawson, S. Mansfield, J. Meiring, B. Morgenfeld, C. Tabery, Y. Zou, C. Sarma, L. Tsou, W. Yan, H. Zhuang, D. Gil, and D. Medeiros, "32-nm logic patterning options with immersion lithography," *Proc. SPIE*, Vol. 6924, 2008, 69243C.

[40] Ito, H., C. G. Willson, and J. M. M. Frechet, "New UV resists with negative or positive tone," *Proc. IEEE Symposium on VLSI Technology*, 1982, p. 86.

[41] Ito, H., "Chemical amplification resist: History and development within IBM," *IBM J. Research & Development*, Vol. 41, 1997, p. 119.

[42] Wong, A. K., R. A. Ferguson, and L. W. Liebmann, "Lithographic effects of mask critical dimension error," *Proc. SPIE*, Vol. 3334, 1998, pp. 106–116.

[43] Pierrat, C., and A. Wong, "The MEF revisited: Low k_1 effects versus mask topography effects," *Proc. SPIE*, Vol. 5040, 2003, pp. 193–202.

[44] Gallatin, G. M., "Resist blur and line edge roughness," *Proc. SPIE*, Vol. 5754, 2005, pp. 38–52.

[45] Mack, C. A., "Stochastic modeling of photoresist development in two and three dimensions," *J. Micro/Nanolithography, MEMS, MOEMS*, Vol. 9, 2010, 041202.

[46] Wong, A. K., *Optical Imaging in Projection Microlithography*, SPIE Press, 2005.

[47] Maenhoudt, M., J. Versluijs, H. Struyf, J. Van Olmen, and M. Van Hove, "Double patterning scheme for sub-0.25 k_1 single damascene structures at NA = 0.75, λ= 193 nm," *Proc. SPIE*, Vol. 5754, 2005, pp. 1508–1518.

[48] Holmes, S. J., C. Tang, S. Burns, Y. Yin, R. Chen, C.-S. Koay, S. Kini, H. Tomizawa, S.-T. Chen, N. Fender, B. Osborn, L. Singh, K. Petrillo, G. Landie, S. Halle, S. Liu, J. C. Arnold, T. Spooner, R. Varanasi, M. Slezak, and M. Colburn, "Optimization of pitch-split double patterning photoresist for applications at the 16 nm node," *Proc. SPIE*, Vol. 7972, 2011, 79720G.

[49] Levinson, H. J., *Principles of Lithography*, SPIE Press, 2001.

[50] McGrath, D., "ASML describes road map to 22-nm litho," Feb 27, 2009. Online: http://www.eetimes.com/ electronics-news/4081494/What-is-source-mask-optimization

[51] Flagello, D. G., T. Milster, and A. E. Rosenbluth, "Theory of high-NA imaging in homogeneous thin films," *J. Optical Society of America A*, Vol. 13, 1996, pp. 53–64.

[52] Stamnes, J. J., *Waves in Focal Regions: Propagation, Diffraction and Focusing of Light, Sound and Water Waves*, Institute of Physics Publishing, Adam Hilger Series on Optics and Optoelectronics, 1986.

[53] Adam, K., Y. Granik, A. Torres, and N. Cobb, "Improved modeling performance with an adapted vectorial formulation of the Hopkins imaging equation," *Proc. SPIE*, Vol. 5040, 2003, pp. 78–90.

[54] Cole, D., E. Barouch, U. Hollerbach, and S. Orszag, "Derivation and simulation of higher numerical aperture scalar aerial images," *J. Applied Physics*, Vol. 31, 1992, pp. 4110–4119.

[55] Cole, D., E. Barouch, U. Hollerbach, and S. Orszag, "Extending scalar aerial image calculations to higher numerical apertures," *J. Vacuum Science Technology B*, Vol. 10, 1992, pp. 3037–3041.

[56] Abbe, E., "Beitrage zur theorie des mikroskops und der mikroskopischen wahrnehmung — Contributions to the theory of the microscope and the nature of microscopic vision," *Archiv. f. Mikroskopische Anat.*, Vol. 9, 1873, pp. 413–468.

[57] Goodman, J. W., *Statistical Optics*, New York: Wiley, 1985.

[58] Flagello, D. G., and A. E. Rosenbluth, "Lithographic tolerances based on vector diffraction theory," *J. Vacuum Science Technology B*, Vol. 10, 1992, pp. 2997–3003.

[59] Yeung, M. S., "Modeling high numerical aperture optical lithography," *Proc. SPIE*, Vol. 922, 1988, pp. 149–167.

[60] Wolf, E., "Electromagnetic diffraction in optical systems I: An integral representation of the image field," *Proc. Royal Society A*, Vol. 253, 1959, pp. 349–357.

[61] Yeung, M. S., D. Lee, R. Lee, and A. R. Neureuther, "Extension of the Hopkins theory of partially coherent imaging to include thin-film interference effects," *Proc. SPIE*, Vol. 1927, 1993, pp. 452–463.

[62] Tai, C. T., "Direct integration of field equations," *Progress in Electromagnetics Research*, Vol. 28, 2000, pp. 339–359.

[63] Orfanidis, S. J., "Electromagnetic waves and antennas," 2008. Online: http://www.ece.rutgers.edu/orfanidi/ewa

[64] Tirapu-Azpiroz, J., *Analysis and Modeling of Photomask Near Fields in Sub-Wavelength Deep Ultraviolet Lithography*, Ph.D. dissertation, Dept. of Electrical Engineering, University of California, Los Angeles (UCLA), 2004.

[65] Balanis, C., *Advanced Engineering Electromagnetics*, New York: Wiley, 1989.

[66] Peterson, A. F., S. L. Ray, and R. Mittra, eds., *Computational Methods for Electromagnetics*, New York: Wiley-IEEE Press, 1997.

[67] Matsuyama, T., Y. Ohmura, and D. M. Williamson, "The lithographic lens: Its history and evolution," *Proc. SPIE*, Vol. 6154, 2006, 615403.

[68] Bruning, J. H., "Optical lithography... 40 years and holding," *Proc. SPIE*, Vol. 6520, 2007, 652004.

[69] Richards, B., and E. Wolf, "Electromagnetic diffraction in optical systems II: Structure of the image in an aplanatic system," *Proc. Royal Society A*, Vol. 253, 1959, pp. 358–379.

[70] Mansuripur, M., "Distribution of light at and near the focus of high-numerical aperture objectives," *J. Optical Society of America A*, Vol. 3, 1986, pp. 2086–2093.

[71] Li, Y., and E. Wolf, "Three-dimensional intensity distribution near the focus in systems of different Fresnel numbers," *J. Optical Society of America A*, Vol. 1, 1984, pp. 801–808.

[72] Marchand, E. W., and E. Wolf. "Boundary diffraction wave in the domain of the Rayleigh-Kirchhoff diffraction theory," *J. Optical Society of America*, Vol. 52, 1962, pp. 761–767.

[73] Miyamoto, K., and E. Wolf. "Generalization of the Maggi-Rubinowicz theory of the boundary diffraction wave – Part I," *J. Optical Society of America*, Vol. 52, 1962, pp. 615–625.

[74] Miyamoto, K., and E. Wolf, "Generalization of the Maggi-Rubinowicz theory of the boundary diffraction wave – Part II," *J. Optical Society of America*, Vol. 52, 1962, pp. 626–637.

[75] Wolf, E., and Y. Li, "Conditions for the validity of the Debye integral representation of focused fields," *Optics Communications*, Vol. 39, 1981, pp. 205–210.

[76] Goodman, J. W., *Introduction to Fourier Optics*, New York: McGraw-Hill, 1996.

[77] Rosenbluth, A. E., J. Tirapu-Azpiroz, K. Lai, K. Tian, D. O. S. Melville, M. Totzeck, V. Blahnik, A. Koolen, and D. Flagello, "Radiometric consistency in source specifications for lithography," *Proc. SPIE*, Vol. 6924, 2008, 69240V.

[78] Totzeck, M., O. Dittmann, A. Gohnermeier, V. Kamenov, B. Geh, D. Krahmer, A. Rosenbluth, and K. Lai, "Suggestion for a file-format for Jones pupil data," Memorandum, Carl Zeiss AG, 2006.

[79] Hopkins, H. H., "The concept of partial coherence in optics," *Proc. Royal Society of London*, Vol. A208, 1951, pp. 263–277.

[80] Rosenbluth, A. E., G. Gallatin, R. L. Gordon, W. Hinsberg, J. Hoffnagle, F. Houle, K. Lai, A. Lvov, M. Sanchez, and N. Seong, "Fast calculation of images for high numerical aperture lithography," *Proc. SPIE*, Vol. 5377, 2004, 615.

[81] Qian, Q.-D., and F. A. Leon, "Fast algorithms for 3D high NA lithography simulation," *Proc. SPIE*, Vol. 2440, 1995, pp. 372–380.

[82] Mack, C. A., "A comprehensive optical lithography model," *Proc. SPIE*, Vol. 538, 1985, pp. 207–220.

[83] Dill, F. H., A. R. Neureuther, J. A. Tuttle, and E. J. Walker, "Modeling projection printing of positive photoresist," *IEEE Trans. Electron Devices*, Vol. ED-22, 1975, pp. 456–464.

[84] Borodovsky, Y., R. E. Schenker, G. A. Allen, E. Tejnil, D. H. Hwang, F.-C. Lo, V. K. Singh, R. E. Gleason, J. E. Brandenburg, and R. M. Bigwood, "Lithography strategy for 65-nm node," *Proc. SPIE*, Vol. 4754, 2002, pp. 1–14.

[85] Wong, A. K., and A. R. Neureuther, "Mask topography effects in projection printing of phase-shifting masks," *IEEE Trans. Electron Devices*, Vol. 41, 1994, pp. 895–902.

[86] Yeung, M. S., and E. Barouch, "Limitation of the Kirchhoff boundary conditions for aerial image simulation in 157 nm optical lithography," *IEEE Electron Device Lett.*, Vol. 21, 2000, pp. 433–435.

[87] Erdmann, A., "Mask modeling in the low k_1 and ultrahigh NA regime: Phase and polarization effects," *Proc. SPIE*, Vol. 5835, 2005, pp. 69–81.

[88] Hibbs, M. S., and T. A. Brunner, "Phase calibration for attenuating phase-shift masks," *Proc. SPIE*, Vol. 6152, 2006, 1L.

[89] Yee, K. S., "Numerical solution of initial boundary value problems involving Maxwell's equations in isotropic media," *IEEE Trans. Antennas and Propagation*, Vol. 14, 1966, pp. 302–307.

[90] Taflove, A., and S. C. Hagness, *Computational Electrodynamics: The Finite-Difference Time-Domain Method*, 3rd ed., Norwood, MA: Artech, 2005.

[91] Wong, A. K., and A. R. Neureuther, "Polarization effects in mask transmission," *Proc. SPIE*, Vol. 1674, 1992, pp. 193–200.

[92] Guerrieri, R., K. H. Tadros, J. Gamelin, and A. R. Neureuther, "Massively parallel algorithms for scattering in optical lithography," *IEEE Trans. Computer-Aided Design of Integrated Circuits and Systems*, Vol. 10, 1991, pp. 1091–1100.

[93] Yeung, M. S., "Three-dimensional mask transmission simulation using a single integral equation method," *Proc. SPIE*, Vol. 3334, 1998, pp. 704–713.

[94] Yeung, M. S., "Fast and rigorous three-dimensional mask diffraction simulation using Battle-Lemarie wavelet based multiresolution time-domain method," *Proc. SPIE*, Vol. 5040, 2003, pp. 69–77.

[95] Gara, A., M. A. Blumrich, D. Chen, G. L. T. Chiu, P. Coteus, M. E. Giampapa, R. A. Haring, P. Heidelberger, D. Hoenicke, G. V. Kopcsay, T. A. Liebsch, M. Ohmacht, B. D. Steinmacher-Burow, T. Takken, and P. Vranas, "Overview of the Blue Gene/L system architecture," *IBM J. Research & Development*, Vol. 49, 2005, pp. 195–212.

[96] Berenger, J.-P., "A perfectly matched layer for the absorption of electromagnetic waves," *J. Computational Physics*, Vol. 114, 1994, pp. 185–200.

[97] Roden, J. A., and S. D. Gedney, "Convolution PML (CPML): An efficient FDTD implementation of the CFS-PML for arbitrary media," *Microwave & Optical Technology Lett.*, Vol. 27, 2000, pp. 334–339.

[98] Burr, G. W., "Photonics," Chapter 16 in *Computational Electrodynamics: The Finite-Difference Time-Domain Method*, 3rd ed., Taflove, A., and S. C. Hagness, eds., Norwood, MA: Artech, 2005.

[99] Wood, R. W., "On a remarkable case of uneven distribution of light in a diffraction grating spectrum," *Phil. Magazine*, Vol. 4, 1902, pp. 396–408.

[100] Rayleigh, L., "On the dynamical theory of gratings," *Proc. Royal Society*, Vol. A79, 1907, pp. 399–416.

[101] Farjadpour, A., D. Roundy, A. Rodriguez, M. Ibanescu, P. Bermel, J. D. Joannopoulos, S. G. Johnson, and G. W. Burr, "Improving accuracy by subpixel smoothing in the finite-difference time domain," *Optics Lett.*, Vol. 31, 2006, pp. 2972–2974.

[102] Oskooi, A., C. Kottke, and S. G. Johnson, "Accurate finite-difference time-domain simulation of anisotropic media by subpixel smoothing," *Optics Lett.*, Vol. 34, 2009, pp. 2778–2780.

[103] Deinega, A., and I. Valuev. "Subpixel smoothing for conductive and dispersive media in the finite-difference time-domain method," *Optics Lett.*, Vol. 32, 2007, pp. 3429–3431.

[104] Tirapu-Azpiroz, J., G. W. Burr, A. E. Rosenbluth, and M. Hibbs, "Massively parallel FDTD simulations to address mask electromagnetic effects in hyper-NA immersion lithography," *Proc. SPIE*, Vol. 6924, 2008, 69240Y.

[105] Estroff, A., Y. Fan, A. Bourov, F. Cropanese, N. Lafferty, L. Zavyalova, and B. Smith, "Mask induced polarization," *Proc. SPIE*, Vol. 5377, 2004, pp. 1069–1080.

[106] Brok, J. M., and H. P. Urbach, "Rigorous model of the scattering of a focused spot by a grating and its application in optical recording," *J. Optical Society of America A*, Vol. 20, 2003, pp. 256–272.

[107] Cheng, W. H., J. Farnsworth, and T. Bloomstein, "Vectorial imaging of deep sub-wavelength mask features in high NA arf lithography," *J. Micro/Nanolithography, MEMS, & MOEMS*, Vol. 7, 2008, 013001.

[108] Burkhardt, M., R. L. Gordon, M. S. Hibbs, and T. A. Brunner, "Through-pitch correction of scattering effects in 193-nm alternating phase-shifting masks," *Proc. SPIE*, Vol. 4691, 2002, 348.

[109] Tirapu-Azpiroz, J., A. E. Rosenbluth, K. Lai, C. Fonseca, and D. Yang, "Critical impact of EMF effects on RET and OPC performance for 45 nm and beyond," *J. Vacuum Science & Technology B*, Vol. 25, 2007, pp. 164–168.

[110] Yamamoto, N., J. Kye, and H. J. Levinson, "Mask topography effect with polarization at hyper NA," *Proc. SPIE*, Vol. 6154 2006, 4F.

[111] Aksenov, Y., P. Zandbergen, and M. Yoshizawa, "Compensation of high-NA mask topography effects by using object modified Kirchhoff model for 65- and 45-nm nodes," *Proc. SPIE*, Vol. 6154, 2006, 1H.

[112] Faure, T., K. Badger, L. Kindt, Y. Kodera, T. Komizo, T. Mizuguchi, S. Nemoto, T. Senna, R. Wistrom, A. Zweber, K. Nishikawa, Y. Inuzuki, and H. Yoshikawa, "Development and characterization of a thinner binary mask absorber for 22-nm node and beyond," *Proc. SPIE*, Vol. 7823, 2010.

[113] Tirapu-Azpiroz, J., P. Burchard, and E. Yablonovitch, "Boundary layer model to account for thick mask effects in photolithography," *Proc. SPIE*, Vol. 5040, 2003, pp. 1611–1619.

[114] Tirapu-Azpiroz, J., and E. Yablonovitch, "Fast evaluation of photomask near-fields in sub-wavelength 193-nm lithography," *Proc. SPIE*, Vol. 5377, 2004, pp. 1528–1535.

[115] Bai, M., L. S. Melvin III, Q. Yan, J. P. Shiely, B. J. Falch, C.-C. Fu, and R. Wang, "Approximation of three-dimensional mask effects with two-dimensional features," *Proc. SPIE*, Vol. 5751, 2005, 446.

[116] Kim, S., Y. C. Kim, S. Suh, S. Lee, S. Lee, H. Cho, and J. Moon, "OPC to account for thick mask effects using simplified boundary layer model," *Proc. SPIE*, Vol. 6349, 2006, 63493I.

[117] Miller, M. A., A. R. Neureuther, D. P. Ceperley, J. Rubinstein, and K. Kikuchi, "Characterization and monitoring of photomask edge effects," *Proc. SPIE*, Vol. 6730, 2007, 67301U.

[118] Tirapu-Azpiroz, J., A. E. Rosenbluth, M. Hibbs, and T. Brunner, "EMF correction model calibration using asymmetry factor data obtained from aerial images or a patterned layer," Patent application FIS920080333US1, 2010.

[119] Tawfik, T. M., A. H. Morshed, and D. Khalil, "Modeling mask scattered field at oblique incidence," *Proc. SPIE*, Vol. 7274, 2009, 727433.

[120] Liu, P., Y. Cao, L. Chen, G. Chen, M. Feng, J. Jiang, H.-Y. Liu, S. Suh, S.-W. Lee, and S. Lee, "Fast and accurate 3-D mask model for full-chip OPC and verification," *Proc. SPIE*, Vol. 6520, 2007, 65200R.

[121] Tirapu-Azpiroz, J., A. E. Rosenbluth, I. Graur, G. W. Burr, and G. Villares, "Isotropic treatment of EMF effects in advanced photomasks," *Proc. SPIE*, Vol. 7488, 2009, 74882D.

[122] Adam, K., and A. Neureuther, "Methodology for accurate and rapid simulation of large arbitrary 2-D layouts of advanced photomasks," *Proc. SPIE*, Vol. 4562, 2002, pp. 1051–1067.

[123] Adam, K., and M. C. Lam, "Hybrid Hopkins-Abbe method for modeling oblique angle mask effects in OPC," *Proc. SPIE*, Vol. 6924, 2008, 69241E.

[124] Maystre, D., and M. Neviere, "Quantitative theoretical study on the plasmon anomalies of diffraction gratings," *J. Optics (Paris)*, Vol. 8, 1977, pp. 165–174.

[125] Wu, B., and A. Kumar, *Extreme UltraViolet Lithography*, New York: McGraw-Hill, 2009.

[126] Barbee, T. W., S. Mrowka, and M. C. Hettrick, "Molybdenum-silicon multilayers for the extreme ultraviolet," *Applied Optics*, Vol. 24, 1985, pp. 883–886.

[127] Kuhlmann, T., S. Yulin, T. Feighl, N. Kaiser, H. Bernitzki, and H. Lauth, "Design and fabrication of broadband EUV multilayer mirrors," *Proc. SPIE*, Vol. 4688, 2002, pp. 509–515.

[128] Hudyma, R., "An overview of optical systems for 30-nm-resolution lithography at EUV wavelengths," *Proc. SPIE*, Vol. 4832, 2002, pp. 137–148.

[129] Jewell, T. E., K. P. Thompson, and J. M. Rodgers, "Reflective optical designs for soft x-ray projection lithography," *Proc. SPIE*, Vol. 1527, 1991, pp. 163–173.

[130] Smith, D. G., "Modeling EUVL illumination systems," *Proc. SPIE*, Vol. 7103, 2008, 7103B.

[131] Lowisch, M., P. Kuerz, H.-J. Mann, O. Natt, and B. Thuering, "Optics for EUV production," *Proc. SPIE*, Vol. 7636, 2010, 763603.

[132] Murakami, K., T. Oshino, H. Kondo, M. Shiraishi, H. Chiba, H. Komatsuda, K. Nomura, and J. Nishikawa, "Fabrication of a fly-eye mirror for extreme ultraviolet lithography illumination system," *J. Micro/Nanolithography, MEMS, & MOEMS*, Vol. 8, 2009, 041507.

[133] Hudyma, R., H.-J. Mann, and U. Dinger, *Projection System for EUV Lithography*, U.S. Patent 6,985,210 B2, 2006.

[134] Antoni, M., W. Singer, J. Schultz, J. Wangler, I. Escudero-Sanz, and B. Kruizinga, "Illumination optics design for EUV lithography," *Proc. SPIE*, Vol. 4146, 2000, pp. 25–34.

[135] Komatsuda, H., "Novel illumination system for EUVL," *Proc. SPIE*, Vol. 3997, 2000, pp. 765–776.

[136] Takino, H., T. Kobayashi, N. Shibata, M. Kuki, A. Itoh, and H. Komatsuda, "Fabrication of a fly-eye mirror for extreme ultraviolet lithography illumination system," *Proc. SPIE*, Vol. 4343, 2001, pp. 576–584.

[137] Van Setten, E., S. Lok, J. van Dijk, C. Kaya, K. van Ingen Schenau, K. Feenstra, H. Meiling, and C. Wagner, "EUV imaging performance — moving towards production," *Proc. SPIE*, Vol. 7470, 2009, 74700G.

[138] Online: http://henke.lbl.gov/optical-constants/getdb2.html

[139] McIntyre, G., C. Zuniga, E. Gallagher, J. Whang, and L. Kindt, "The trade-offs between thin and thick absorbers for EUV photomasks," *Proc. SPIE*, Vol. 8166, 2011, 81663U.

[140] LaFontaine, B., A. R. Pawloski, Y. Deng, C. Chovino, L. Dieu, O. R. Wood II, and H. J. Levinson, "Architectural choices for EUV lithography masks: Patterned absorbers and patterned reflectors," *Proc. SPIE*, Vol. 5374, 2004, pp. 300–310.

[141] Deng, Y., B. La Fontaine, H. J. Levinson, and A. R. Neureuther, "Rigorous EM simulation of the influence of the structure of mask patterns on EUVL imaging," *Proc. SPIE*, Vol. 5037, 2003, pp. 302–313.

[142] Kang, H., S. Hansen, J. van Schoot, and K. van Ingen Schenau, "EUV simulation extension study for mask shadowing effect and its correction," *Proc. SPIE*, Vol. 6921, 2008, 69213I.

[143] Pistor, T. V., *Electromagnetic Simulation and Modeling with Applications in Lithography*, Ph.D. thesis, University of California, Berkeley, 2001.

[144] Schmoeller, T., and T. Klimpel, "EUV pattern shift compensation strategies," *Proc. SPIE*, Vol. 6921, 2008, 69211B.

[145] McIntyre, G., C.-S. Koay, M. Burkhard, H. Mizuno, and O. Wood, "Modeling and experiments of non-telecentric thick mask effects for EUV lithography," *Proc. SPIE*, Vol. 7271, 2009, 72711C–1.

[146] Word, J., C. Zuniga, M. Lam, M. Habib, K. Adam, and M. Oliver, "OPC modeling and correction solutions for EUV lithography," *Proc. SPIE*, Vol. 8166, 2011, 81660Q.

[147] Ruoff, J., "Impact of mask topography and multilayer stack on high NA imaging of EUV masks," *Proc. SPIE*, Vol. 7823, 2010, 78231N.

[148] Song, H., L. Zavyalova, I. Su, J. Shiely, and T. Schmoeller, "Shadowing effect modeling and compensation for EUV lithography," *Proc. SPIE*, Vol. 7969, 2011, 79691O.

Chapter 16

FDTD and PSTD Applications in Biophotonics[1]

Ilker R. Capoglu, Jeremy D. Rogers, César Méndez Ruiz, Jamesina J. Simpson, Snow H. Tseng, Kun Chen, Ming Ding, Allen Taflove, and Vadim Backman

16.1 INTRODUCTION

This chapter reviews recent research involving computational modeling of linear ("elastic") light interactions with biological tissues via the direct solution of Maxwell's equations. A primary potential application of this *biophotonics* modeling is the diagnosis of human disease, especially cancer. We focus on two specific computational techniques for Maxwell's equations implemented on Cartesian space grids: the finite-difference time-domain (FDTD) method [1, 2] and the pseudospectral time-domain (PSTD) method [3, 4]. To-date, these two techniques have shown the best promise for rigorously modeling full-vector three-dimensional (3-D) optical interactions with individual biological cells and clusters of cells.

Why are FDTD and PSTD computational solutions of Maxwell's equations important in biophotonics? First, FDTD and PSTD models account for how the shape and internal inhomogeneities of each biological cell interact with the impinging light to generate the local full-vector optical electromagnetic field, which can involve both propagating and evanescent components. Second, FDTD and PSTD account for the simultaneous interaction of the local optical electromagnetic fields of many closely spaced biological cells to obtain the composite macroscopic optical properties of the tissue. The comprehensive nature of such rigorous Maxwell's equations modeling (potentially over distance scales from ~10 nm to ~100 μm) yields a rich set of electromagnetic wave phenomena that cannot be calculated by previous approximate techniques employing heuristic simplifications based on radiative transfer theory [5]. Such approximate techniques omit key electromagnetic wave characteristics of light, and treat light propagation as an energy-transport problem. This omission can yield research findings of potentially questionable accuracy and validity [6, 7].

In the following sections, we shall review recent work in FDTD and PSTD computational Maxwell's equations modeling of light interactions with biological tissues. Overall, our goal is to demonstrate that FDTD and PSTD solution techniques for Maxwell's equations are providing means to strengthen the science base for cellular-level and tissue-level biophotonics, and to accelerate the development of corresponding novel clinical technologies.

---

[1]This chapter is updated and expanded with permission from S. H. Tseng, I. R. Capoglu, A. Taflove, and V. Backman, "Modeling of Light Scattering by Biological Tissues via Computational Solution of Maxwell's Equations," Chapter 3 in *Biomedical Applications of Light Scattering*, A. Wax and V. Backman, eds., New York: McGraw-Hill, 2010, ©2010 McGraw-Hill.

16.2 FDTD MODELING APPLICATIONS

16.2.1 Vertebrate Retinal Rod

Arguably the first application of FDTD to cellular-level biophotonics was reported in [8], wherein visible light interactions with a retinal photoreceptor were modeled for the two-dimensional (2-D) TM_z and TE_z polarization cases. The working hypothesis was that the detailed physical structure of a photoreceptor impacts the physics of its optical absorption and thereby, vision. One such photoreceptor was studied: the vertebrate retinal rod. The bulk structure of the retinal rod exhibits the physics of an optical waveguide, while the periodic internal disk-stack structure adds the physics of an optical interferometer. These effects combine to generate a complex optical standing wave within the rod, thereby creating a pattern of local intensifications of the optical field.

The FDTD model of the rod reported in [8] had the cross-section dimensions of 2×20 μm, corresponding to $(3.8\lambda_d - 5.7\lambda_d) \times (38\lambda_d - 57\lambda_d)$ over the range of impinging wavelengths considered, where λ_d denotes the optical wavelength within the rod's dielectric media. A uniform Cartesian space grid having 5.0-nm-square unit cells was utilized. This permitted resolution of the rod's 15-nm-thick outer wall and 15-nm-thick internal disk membranes. A total of 799 disks were distributed uniformly along the length of the rod, separated from each other by 10 nm of fluid, and separated from the outer wall membrane by 5 nm of fluid. The index of refraction of the membrane was chosen to be 1.43, and the index of refraction of the fluid was chosen to be 1.36, in accordance with generally accepted physiological data. These parameters implied a resolution within the dielectric media of $\lambda_d/70$ to $\lambda_d/105$, depending on the incident wavelength.

As reported in [8], Fig. 16.1 provides visualizations of the FDTD-computed magnitude of the normalized electric field values of the optical standing wave within the retinal rod model for TM_z illumination at the free-space wavelengths λ_0 = 714, 505, and 475 nm. Similar visualizations were obtained for the TE_z illumination case.

To assist in understanding the physics of the retinal rod as an optical structure, the standing-wave magnitude data at each λ_0 were reduced as follows. First, at each transverse plane located at a given y_0 in the rod, the electric field values, $E(x, y_0)$, of the optical standing wave were integrated over the x-coordinate to obtain a single number, $E_{int}(y_0)$. Second, a discrete spatial Fourier transform of the set of $E_{int}(y_0)$ values was performed over the y-coordinate. With the exception of isolated peaks unique to each λ_0, the spatial-frequency spectra for each polarization were found to be essentially independent of the illumination wavelength. This indicated that the retinal rod exhibits a type of frequency-independent electrodynamic behavior.

The agreement of the spatial-frequency spectra for the three incident wavelengths for each polarization was so remarkable that the overall procedure was tested for computational artifacts. The test involved perturbing the indices of refraction of the membrane and fluid from those of the vertebrate rod to those of glass and air, while leaving the geometry unchanged. It was found that the glass–air spectrum exhibited little correlation (i.e., numerous sharp high-amplitude oscillations) over the entire spatial-frequency range considered. On the other hand, the normalized membrane–fluid spectrum varied in a tight range near unity through spatial frequencies of 3.6 μm^{-1}. Reference [8] concluded that the vertebrate retinal rod indeed exhibits a type of wavelength-independent behavior based on its internal electromagnetic field structure.

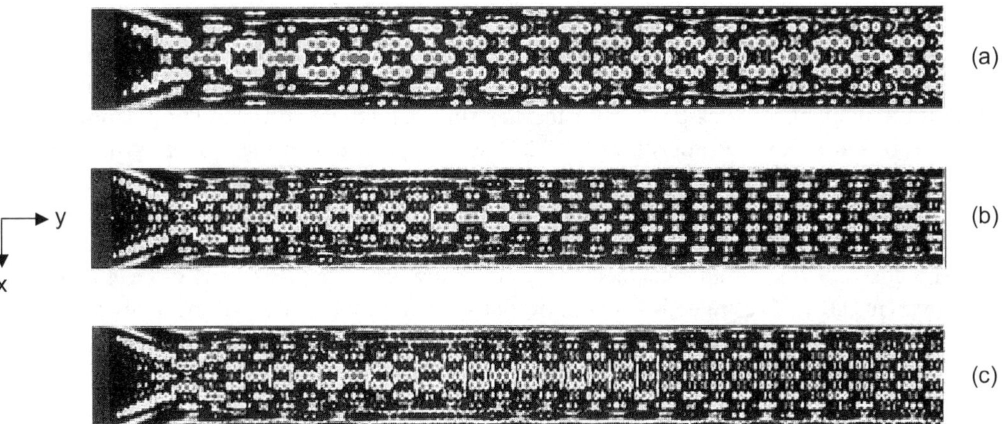

Fig. 16.1 Visualizations of the FDTD-computed optical E-field standing wave within the retinal rod model for TM_z illumination at free-space wavelengths: (a) $\lambda_0 = 714$ nm; (b) 505 nm; and (c) 475 nm. *Source:* Piket-May, Taflove, and Troy, *Optics Letters*, 1993, pp. 568–570, ©1993 The Optical Society of America.

16.2.2 Angular Scattering Responses of Single Cells

FDTD modeling of 3-D optical scattering by a single biological cell has been possible since the mid-1990s. Dunn and Richards-Kortum [9] reported one of the earliest such studies wherein they investigated the angular scattering patterns of biological cell models under near-infrared illumination. This regime was chosen because the interaction of light with the cell at these wavelengths is dominated by scattering rather than absorption. Figure 16.2(a) depicts a cross-section of their 3-D model to illustrate the geometry of the scattering problem.

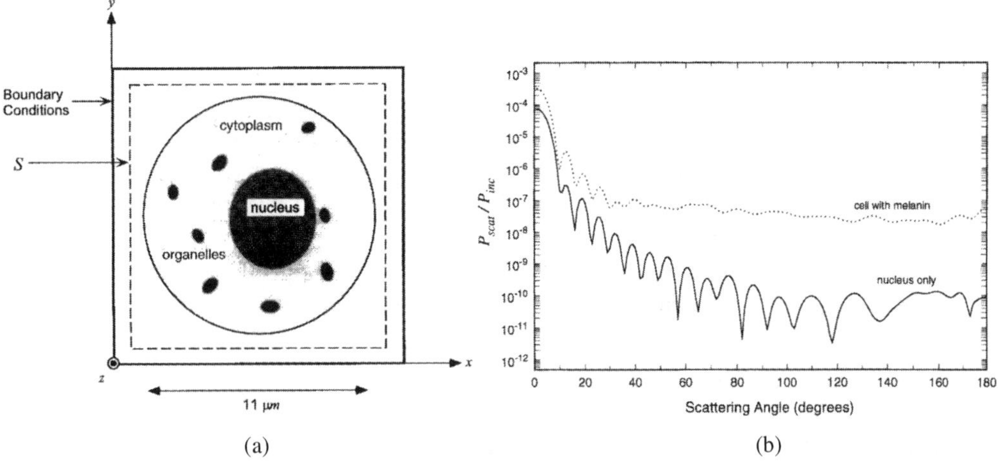

Fig. 16.2 (a) Geometry of an early 3-D FDTD model of optical scattering by a biological cell. (b) Comparison of the scattering responses of the model cell of (a) with only a nucleus (solid line) and with a nucleus and melanin (dotted line). *Source:* Dunn and Richards-Kortum, *IEEE J. Selected Topics in Quantum Electronics*, 1996, pp. 898–905, ©1996 IEEE.

The cell model of Fig. 16.2 was discretized in a 3-D FDTD grid terminated with absorbing boundary conditions. Refractive indices varying from 1.35 to 1.7 represented the extracellular fluid and the intracellular elements (cytoplasm, nucleus, mitochondria, and melanin). Plane-wave illumination was provided using the total-field/scattered-field (TF/SF) formulation, and a near-field-to-far-field transform (NFFFT) was used to obtain the scattering patterns [2]. The FDTD simulation was repeated for two different cases. In the first case, only the nucleus was modeled; in the other, both the nucleus and the melanin were modeled. The results, shown in Fig. 16.2(b), reveal the importance of melanin in the far-field scattering of the model cell.

Subsequently, the simulation methods of [9] were tested experimentally via goniometric measurements [10]. The angular scattering response of cultured OVCA-420 ovarian cancer cells was measured using the goniometry setup shown schematically in Fig. 16.3(a).

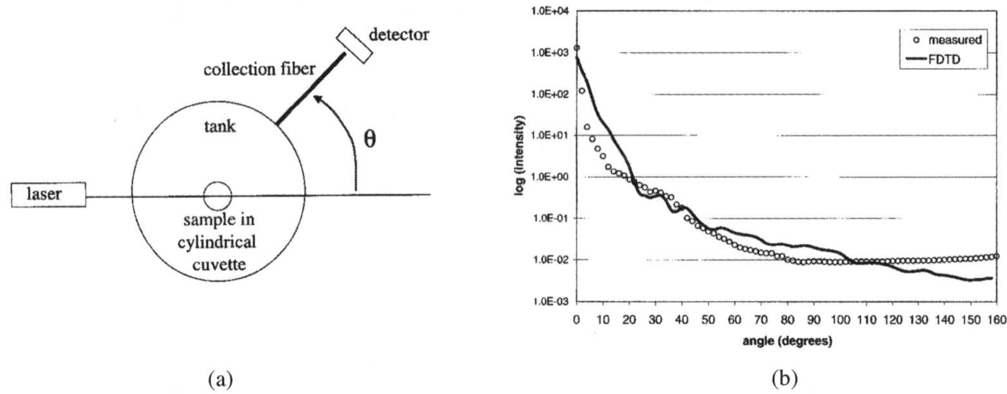

Fig. 16.3 Comparison of the cell-culture goniometry results to the FDTD-computed far-field patterns. (a) Schematic illustration of the goniometer setup; (b) comparison of the measured and FDTD data for OVCA-420 cells. *Source:* Drezek, Dunn, and Richards-Kortum, *Applied Optics*, 1999, pp. 3651–3661, ©1999 The Optical Society of America.

The cell suspension in the cylindrical cuvette was illuminated by an unpolarized 5-mW He-Ne laser. Using a stepper motor coupled to a collection fiber and a detector, the scattered light was measured over an angular range of 160° with 2° angular resolution. The cell solution was diluted progressively until the results were unaffected by further dilution. Next, the cell model parameters for the FDTD simulation were extracted from phase-contrast microscope observations of the OVCA-420 cells. This resulted in a cell model of diameter 10 μm with a 6-μm-diameter nucleus. Other organelles (mitochondria and melanin) were also added to the model, comprising 10% of the cell volume. Figure 16.3(b) shows good agreement between the goniometry measurements and the FDTD-calculated far-field patterns.

16.2.3 Precancerous Cervical Cells

The Richards-Kortum group pioneered FDTD modeling of light scattering from cervical cells during the earliest stages of cancer development [11, 12]. Investigations centered on how such light scattering is affected by changes in nuclear morphology, DNA content, and chromatin texture that occur during neoplastic progression. FDTD was applied to calculate the magnitude and angular distribution of scattered light as a function of pathologic grade.

We now consider work by the Richards-Kortum group on 2-D FDTD models of cellular scattering, as illustrated in Fig. 16.4 [11]. In this example, the cell cytoplasm, when present, had a diameter of 8 μm, and the nucleus had a diameter of 4 μm. Refractive index values for the cytoplasm and the nucleus were 1.37 and 1.40, respectively. Organelle refractive indices were in the 1.38 to 1.42 range, and organelle sizes ranged from 0.1 to 1 μm. Approximately 25% of the available space within the cell (i.e., space not already occupied by the nucleus) was filled with organelles. Wavelengths spanned from 600 to 1000 nm in 5-nm increments.

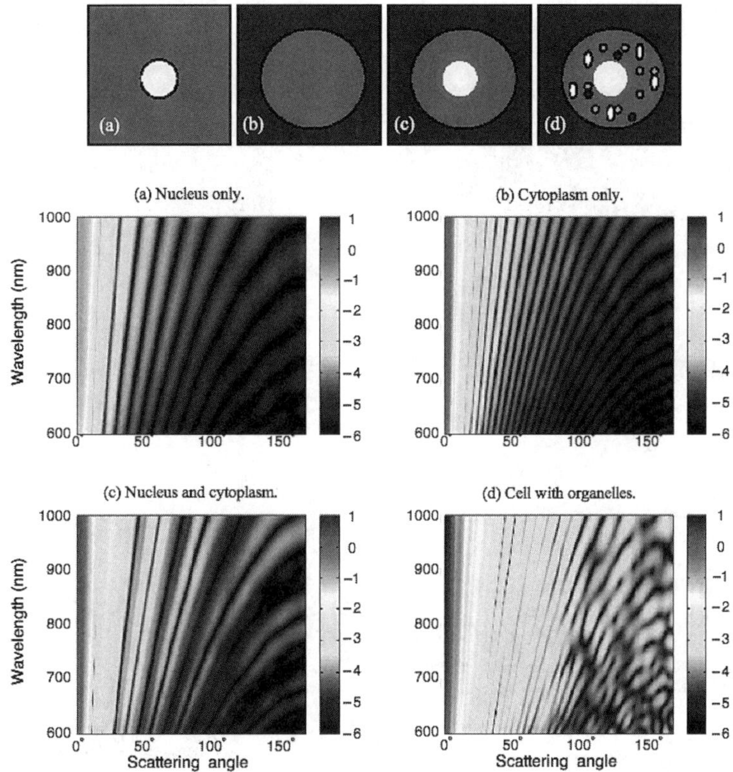

Fig. 16.4 Visualizations of the FDTD-computed optical scattering of four models of a cell: (a) nucleus only; (b) cytoplasm only; (c) nucleus and cytoplasm; and (d) nucleus and cytoplasm containing organelles. The color scale corresponds to the log of the scattered intensity. *Source:* Drezek, Dunn, and Richards-Kortum, *Optics Express*, 2000, pp. 147–157, ©2000 The Optical Society of America.

From Fig. 16.4, we note how the introduction of heterogeneities in the form of small organelles impacts scattering. Closely following the discussion of [11], the addition of cytoplasmic organelles begins to obscure the interference peaks visible in the simulations using homogeneous geometries. The effects of the heterogeneities are most noticeable at angles over 90°, partially because the scattered intensity values in this region are five to six orders of magnitude smaller than the scattered intensity values at low angles.

Reference [11] then proceeded to consider more complicated 2-D descriptions of cellular morphology. In the example illustrated in Fig. 16.5, two cells containing multiple sizes and shapes of organelles and heterogeneous nuclei were considered. In the first cell, the morphology was defined using histological features of normal cervical cells. In the second cell, the morphology was defined based on the features of cervical cells staged as high-grade dysplasia. To emphasize differences due to the internal contents, both cells were assumed to be circular with 9-μm diameters. The most significant differences between the dysplastic cell relative to the normal cell included increased nuclear size and nuclear-to-cytoplasmic ratio (normal 0.2, dysplastic 0.67), asymmetric nuclear shape, increased DNA content, and a hyperchromatic nucleus with areas of coarse chromatin clumping and clearing.

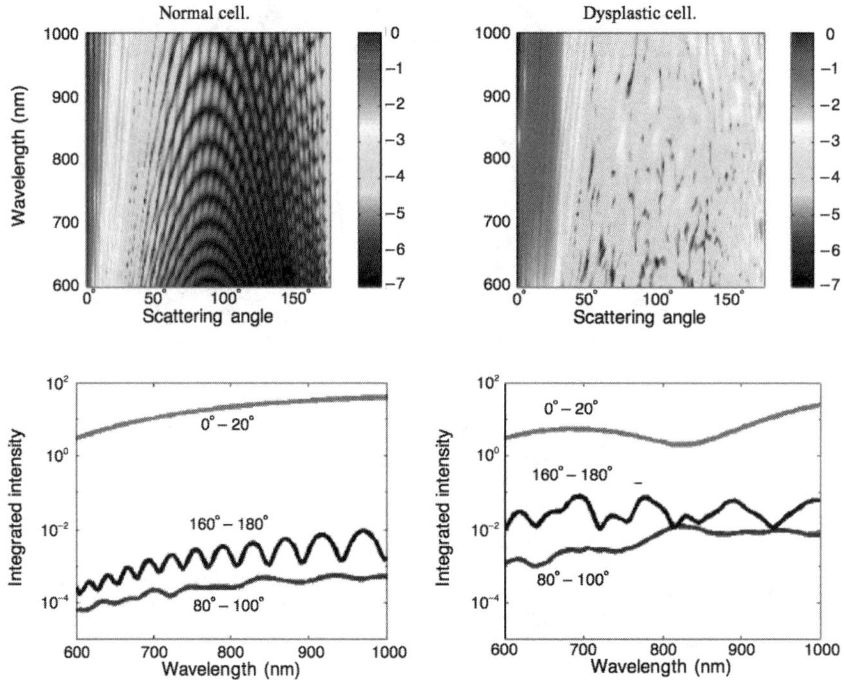

Fig. 16.5 Top: Visualizations of the FDTD-computed optical scattering from models of normal (left) and dysplastic (right) cervical cells. The color scale corresponds to the log of the scattered intensity. Bottom: Integrated scattered intensities over three angular ranges for normal (left) and dysplastic (right) cervical cells. *Source:* Drezek, Dunn, and Richards-Kortum, *Optics Express*, 2000, pp. 147–157, ©2000 The Optical Society of America.

For the normal cell considered in Fig. 16.5, nuclear refractive-index variations were assumed to be uniformly distributed in the range $n = 1.40 \pm 0.02$ at spatial frequencies from 10 to 30 μm^{-1}, thereby simulating a fine, heterogeneous chromatin structure. In the dysplastic cell, nuclear refractive-index variations were distributed in the range $n = 1.42 \pm 0.04$ at spatial frequencies from 3 to 30 μm^{-1}, thereby simulating a coarser, more heterogeneous chromatin structure. Both normal and dysplastic cells contained several hundred organelles (radii from 50 to 500 nm; n = 1.38 to 1.40) randomly distributed throughout the cytoplasm.

From Fig. 16.5, the FDTD results indicate that the dysplastic cell exhibits elevated scattering at small angles due to the larger nucleus, and also at large angles due to alterations in the chromatin structure, which results in increased heterogeneity of the refractive index [11]. Since the dysplastic cell contains a large heterogeneous nucleus that is comprised of an assortment of scatterer sizes and refractive indexes, distinct interference peaks are not present. Although heterogeneities are present in the structure of a normal cell, these are not significant enough to disrupt the peaks resulting from the cytoplasm and nuclear boundaries [11].

The bottom half of Fig. 16.5 displays the integrated scattered intensity as a function of wavelength for three angular ranges: 0° to 20°, 80° to 100°, and 160° to 180°. It is seen that changes in the wavelength dependence of the scattering between normal and dysplastic cells are most pronounced at large angles. Although obtained using a simple model, this observation can guide further studies on optimizing scattering-based optical techniques for determining dysplasia.

Arifler et al. [13] constructed normal and dysplastic cervical cell models via quantitative histopathology, and calculated angular scattering using FDTD. A distinct feature was the use of perfectly matched layer (PML) absorbing boundary conditions (ABCs) to improve the accuracy at scattering angles close to 180°, which had been previously subject to substantial numerical reflections from the outer grid boundaries. Cervical cells collected from human biopsies were inspected by pathologists and classified into normal, human papilloma virus (HPV), CIN 1 (mild dysplasia), CIN 2 (moderate dysplasia), or CIN 3 (severe dysplasia or squamous carcinoma in situ) categories. The cells were also categorized with respect to their positions in the epithelium as either basal, parabasal, intermediate, or superficial. A quantitative histopathological analysis was then performed on five normal and sixteen CIN 3 biopsies, and the nuclear and cellular shape statistics of the cells were extracted. Random cell samples were then constructed for normal and CIN 3 cells, and these samples were used in the FDTD simulations. Nine samples were generated for each of the normal and CIN 3 categories, with three samples belonging to each of the basal/parabasal, intermediate, and superficial epithelial depths.

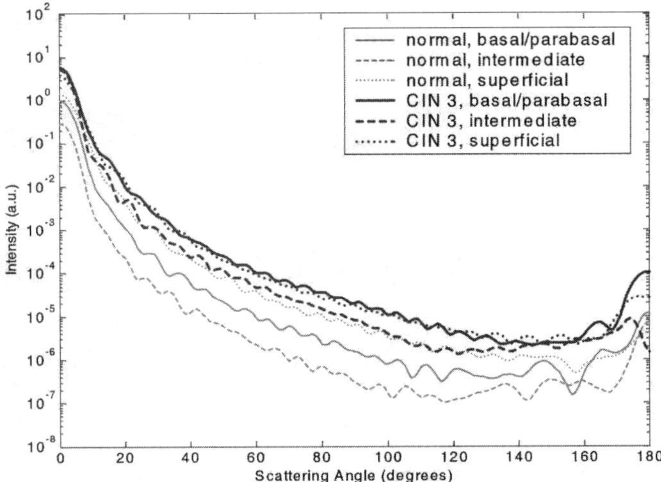

Fig. 16.6 Angular scattering responses of cervical-cell FDTD models constructed with the help of quantitative histopathology. Two different diagnostic categories (normal, CIN 3) and three different epithelial depths (basal/parabasal, intermediate, superficial) were considered. *Source:* Arifler et al., *J. Biomedical Optics*, 2003, pp. 484–494, ©2003 SPIE.

Figure 16.6 shows the azimuth-averaged scattering response of these cell models with respect to the θ scattering angle [13]. These curves were averaged over the three samples belonging to the same diagnostic (normal/CIN 3) and epithelial depth category. It is evident that the cells at the intermediate layer display the most marked difference in angular response between normal and CIN 3, whereas the superficial cells display the least difference. Attention is also drawn to the fact that the statistical variation of the scattering from normal cells is larger than that of the CIN 3 cells, despite the larger structural variety manifested by the latter category.

16.2.4 Sensitivity of Backscattering Signatures to Nanometer-Scale Cellular Changes

During the past decade, significant experimental results indicate that light-scattering signals can provide means for early-stage detection of colon cancer before any other biomarker [14]. In combination with the findings reported by the Richards-Kortum group, it is clear that light scattering is sensitive to small differences in tissue and cellular structures. An important question then arises: Which light-scattering properties provide the best sensitivity to detect cellular changes (likely at the nanometer scale) that indicate progression to cancer?

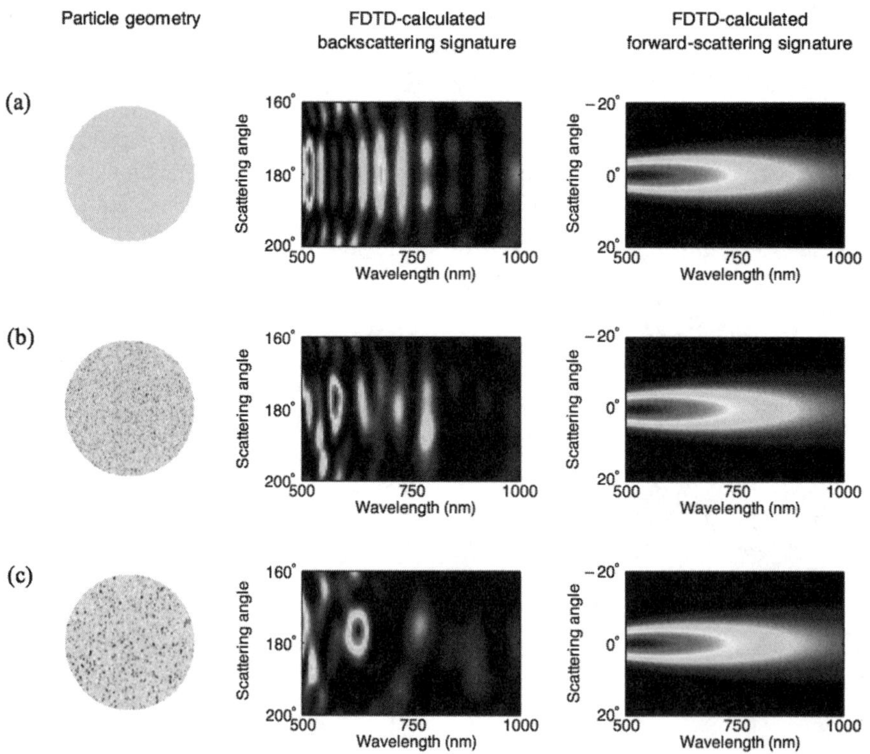

Fig. 16.7 Visualizations of the FDTD-computed optical scattering signatures of a spherical 4-μm-diameter particle with a volume-averaged refractive index $n_{avg} = 1.1$. (a) Homogeneous particle; (b) inhomogeneous particle with refractive index fluctuations $\Delta n = \pm 0.03$ spanning distance scales of ~50 nm; (c) inhomogeneous particle with refractive index fluctuations $\Delta n = \pm 0.03$ spanning distance scales of ~100 nm. *Source:* Li, Taflove, and Backman, *IEEE J. Selected Topics in Quantum Electronics*, 2005, pp. 759–765, ©2005 IEEE.

Figure 16.7 illustrates the application of FDTD to evaluate the sensitivity of optical backscattering and forward-scattering signatures to refractive index fluctuations spanning nanometer-length scales [15]. Here, the spectral and angular distributions of scattered light from inhomogeneous spherical dielectric particles with identical sizes and volume-averaged refractive indices are compared with corresponding data calculated for their homogeneous counterparts.

The optical backscattering signatures (shown in the center panels of Fig. 16.7) are of particular interest. These are visualizations of the FDTD-calculated backscattering intensity distributions as functions of wavelength and scattering angle within a ±20° range of direct backscatter. Relative to the homogeneous case of Fig. 16.7(a), we observe distinctive features of the backscattering signatures for the randomly inhomogeneous cases of Figs. 16.7(b) and 16.7(c). This is despite the fact that the inhomogeneities for these cases have characteristic sizes of only 50 and 100 nm, respectively, which are much smaller than the illumination wavelength of 750 nm. In contrast, the forward-scattering signatures shown in the right-hand panels exhibit no distinctive features.

These FDTD calculations strongly support the hypothesis that there exist signatures in backscattered light that are sufficiently sensitive to detect alterations in the cellular architecture at the nanometer scale. Importantly, this sensitivity is *not bound by the diffraction limit.* Potentially, backscattering signatures can serve as biomarkers to detect and characterize slight alterations in tissue structure that may be precursors of cancer [14].

16.2.5 Modeling Mitochondrial Aggregation in Single Cells

Observations of healthy cells compared to cancerous cells indicate that the mitochondria are typically aggregated near the nucleus in healthy cells, whereas in cancer cells, the mitochondria are more randomly distributed. This suggests that differences in mitochondrial distributions could be used as a diagnostic tool by characterizing light scattering from cells using, for example, flow cytometry.

Su et al. [16] reported 3-D FDTD models of light scattering from single cells having varying distributions of mitochondria. Each cell was modeled as a 10-μm-diameter sphere containing a central 6-μm-diameter spherical nucleus and 180 randomly positioned 0.8-μm-diameter spheres distributed throughout the cytoplasm representing mitochondria. Weighting laws were applied to simulate either a uniform distribution of the mitochondria within the cytoplasm, aggregation of the mitochondria around the nucleus, or aggregation of the mitochondria around the cell membrane. A refractive index $n = 1.38$ was assumed for the cytoplasm, $n = 1.39$ for the nucleus, $n = 1.42$ for the mitochondria, and $n = 1.334$ for the surrounding medium. Each FDTD simulation employed a Liao ABC, 40-nm cubic grid cells, and an illuminating wavelength of 632.8 nm.

Figure 16.8 displays example realizations of the cell models and the resulting FDTD-computed scattering distributions. By analyzing the simulated scattered light from eight realizations of each aggregation state, Su et al. determined that differences in the small-angle forward-scattered intensity may be most sensitive to differences in mitochondrial distributions, as shown in Fig. 16.9 [16]. Although this study used cell models that are much simpler than real cells, this work represents an interesting step toward simulating differences in light scattering due to variations in mitochondrial aggregation.

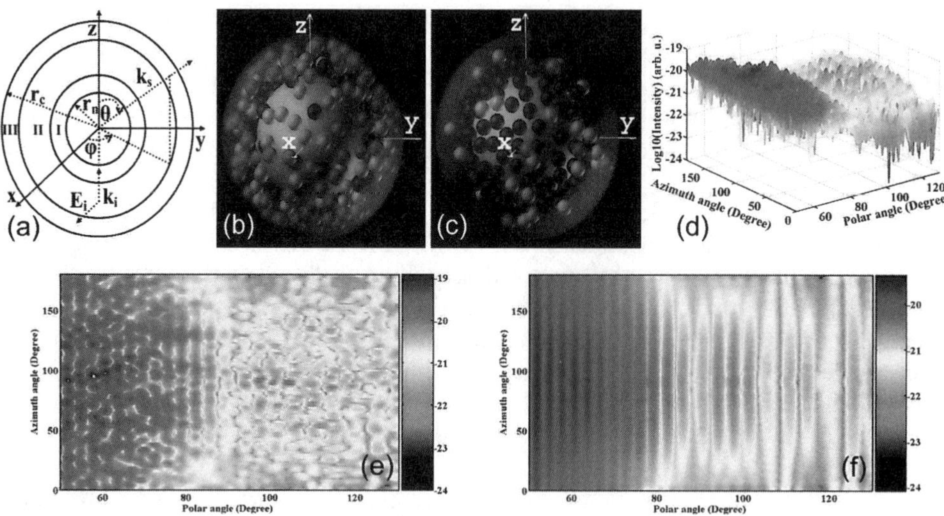

Fig. 16.8 (a) Schematic of the zones where mitochondria are concentrated for each condition; (b) example of randomly distributed mitochondria; (c) example of mitochondria aggregated near the nucleus; (d), (e) FDTD-computed light scattering distribution for the cell shown in (b); (f) FDTD-computed light scattering distribution for the cell model with no mitochondria. *Source:* Su et al., *Optics Express*, 2009, pp. 13381–13388, ©2009 The Optical Society of America.

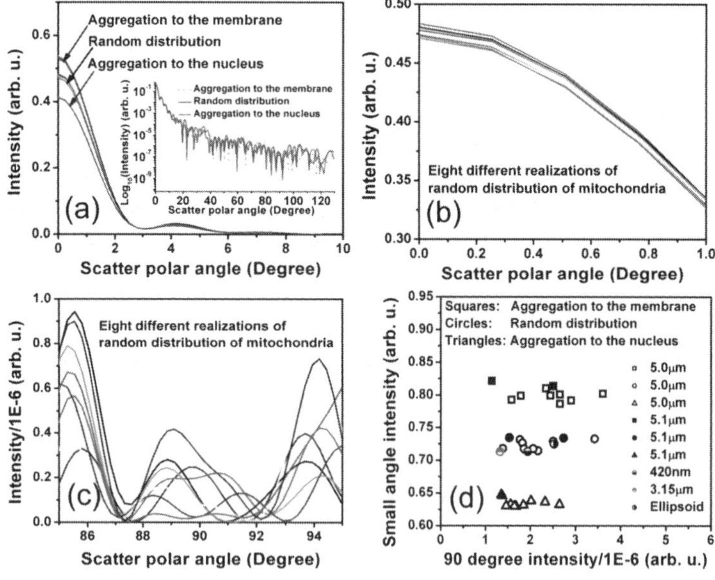

Fig. 16.9 (a) Differences in the FDTD-computed scattered intensity at small angles for the three aggregation scenarios; (b) zoomed-in view of (a) for the unaggregated case showing each of eight realizations; (c) scattered intensity around 90° scattering for un-aggregated case; (d) plot of small-angle vs. large-angle scattered intensity. *Source:* Su et al., *Optics Express*, 2009, pp. 13381–13388, ©2009 The Optical Society of America.

16.2.6 Focused Beam Propagation through Multiple Cells

In many cases, microscopy (in its various modalities) is the primary tool for studying cellular systems. However, microscopic imaging of tissue, especially in vivo, is challenging because scattering reduces contrast and degrades image quality for focal planes below the surface. Fortunately, optical sectioning methods like confocal microscopy and two-photon microscopy can be used to advantage. In confocal microscopy, scattered light from outside the focal plane is rejected by a pinhole. In multi-photon microscopy, tissue nonlinearity provides a fluorescence response that occurs primarily where the optical electric field is highest in the focal volume, thereby eliminating the out-of-focus signal. However, for all microscopy modalities, scattering plays a major role in the propagation of the focused beam through tissue. While Monte Carlo ray tracing can model statistical propagation and scattering in tissue, the FDTD method provides an excellent means to quantitatively investigate the effect of specific scattering bodies such as the nucleus and other organelles on focused beam propagation.

Starosta and Dunn [17] reported using the 3-D FDTD method to study the propagation of a focused beam of free-space wavelength $\lambda_0 = 800$ nm through a collection of many cells. Cell nuclei were modeled as 6.0- × 5.0-μm ellipsoids having either a uniform refractive index, $n = 1.40$, or internal inhomogeneities represented by randomly placed 1-μm-diameter overlapping spheres of index $n = 1.40 \pm 0.03$. Clusters of bodies of refractive index $n = 1.38$, representing mitochondria or lysosomes, were embedded in the cytoplasm ($n = 1.36$). Mitochondria were modeled as 1.5- × 0.5-μm ellipsoids, and lysosomes were modeled as 0.5-μm-diameter spheres. The background medium surrounding the cells was index matched to the cytoplasm. Figure 16.10 illustrates a typical three-cell cluster used in the simulations.

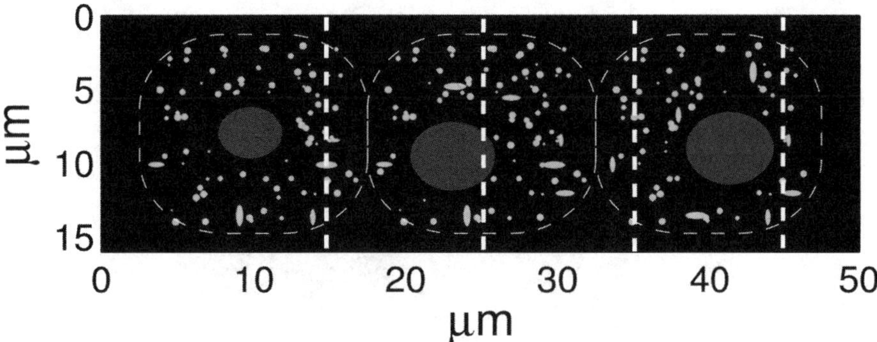

Fig. 16.10 Sample refractive-index configuration used in the 3-D FDTD simulations of focused beam propagation through multiple cells. The background medium was index matched to the cytoplasm at $n = 1.36$. Organelles (mitochondria) were modeled as small spheres or ellipsoids with $n = 1.38$, and nuclei were modeled as ellipsoids with $n = 1.40$. Vertical dashed lines show the location of the focal planes that were investigated. *Source:* Starosta and Dunn, *Optics Express*, 2009, pp. 12455–12469, ©2009 The Optical Society of America.

To track the field of a Gaussian focused beam with 0.5 numerical aperture (NA), the FDTD simulations of [17] used a 3-D grid having 29.4-nm cubic unit cells (~1/20 dielectric wavelength, λ_d) terminated with a Berenger PML ABC. Simulations were run until the steady state was reached, and the fields were recorded across planes of interest to show the lateral and axial field distributions.

Parametric studies in [17] were performed by varying the organelle density and focal depth. The results indicated that nuclei are the dominant source of scattering, and for shallow focal depths, the focal volume is not significantly degraded. The intensity at the focal volume was found to decrease in accordance with an exponential Beer-Lambert attenuation law when the nuclei were removed. When nuclei were included in the simulation, the intensity at the focal volume was shown to vary, and in some cases even increase due to focusing effects by the nuclei, but these effects were highly dependent on the relative positions of the nuclei and the focus spots. For greater focal depths, it was found necessary to include a larger number of cells in the lateral direction in order to accurately account for scattering from all parts of the beam, as shown in Fig. 16.11.

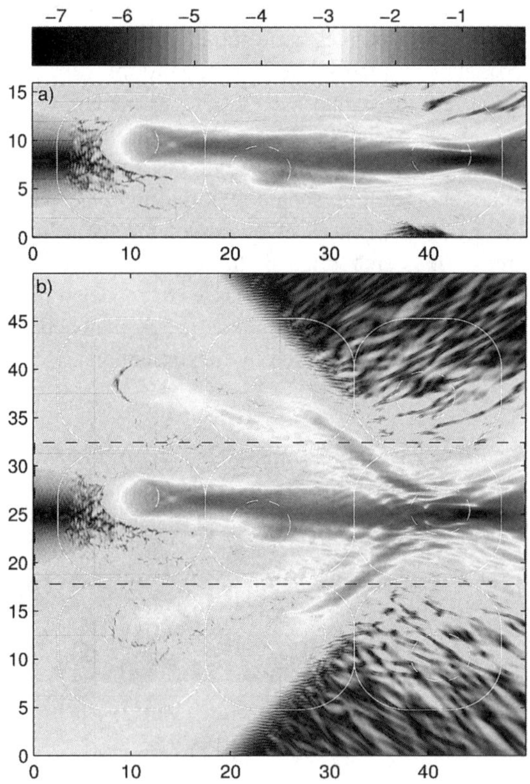

Fig. 16.11 (a) Visualization of the FDTD-computed intensity distribution for a simulation of three cells used in the parametric study of focused beam propagation through multiple cells. (b) Results of a much larger simulation of $3 \times 3 \times 3$ cells. *Source:* Starosta and Dunn, *Optics Express*, 2009, pp. 12455–12469, ©2009 The Optical Society of America.

The study of [17] demonstrated rigorously that while the point-spread function (PSF) of a focused beam does not significantly degrade at shallow penetration depths in biological tissues, the intensity of the beam attenuates exponentially with depth due to scattering by organelles. However, at greater focal depths, the PSF does begin to degrade. This study also highlighted the importance of including large volumes of cells when simulating optical propagation through tissues, and incorporating accurate knowledge of the intracellular refractive-index distributions.

16.2.7 Computational Imaging and Microscopy

Embryo Imaging System

Hollmann, Dunn, and DiMarzio [18] constructed a computational model of an optical system for imaging embryos to determine their viability. In this system, a third dimension is added to the imaging process using a procedure called Z-stacking, wherein the objective is focused at different depths in the cell and a 3-D image cube is recorded. Inverting the 3-D data and reconstructing the original 3-D refractive index distribution is generally a nontrivial task since the image at each focal depth is affected to some degree by the images at other focal depths. However, if a small amount of information is needed, such as the aggregation characteristics of mitochondria or the size of the nuclei, Z-stacking can yield adequate quantitative information.

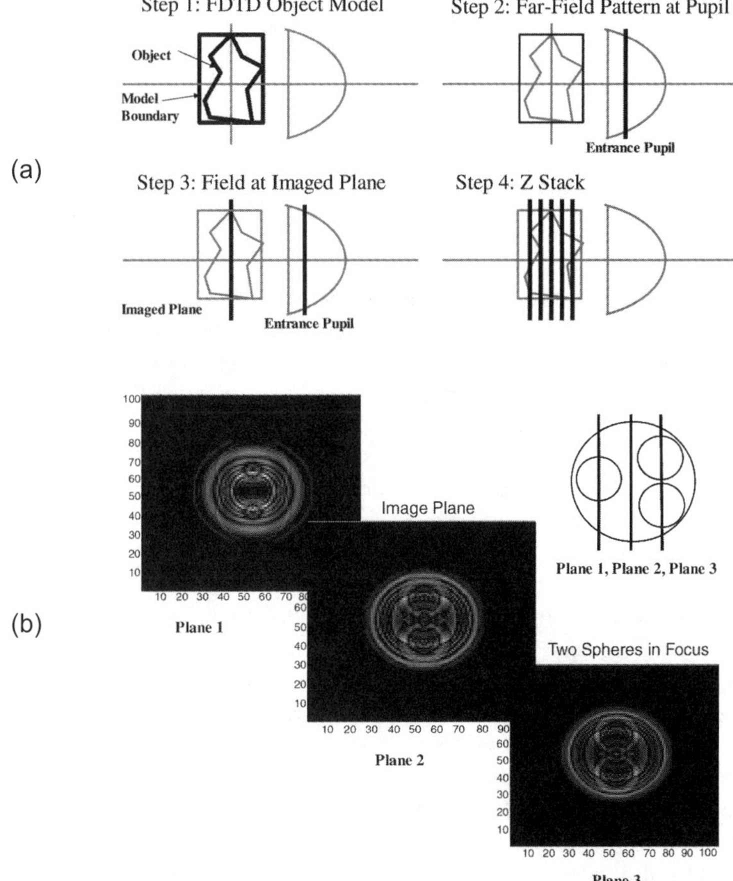

Fig. 16.12 Computational simulation of an embryo imaging system: (a) Schematic description of the Z-stacking procedure; (b) simulated intensity images of a test structure consisting of a large sphere enclosing three smaller spheres, where the images are computed at three different focal depths. *Source:* Hollmann, Dunn, and DiMarzio, *Optics Letters*, 2004, pp. 2267–2269, ©2004 The Optical Society of America.

Figure 16.12(a) illustrates the four-step Z-stacking procedure [18]. Step 1 involves FDTD modeling of plane-wave illumination of the object of interest. Step 2 propagates the FDTD-computed near fields obtained in Step 1 to the entrance pupil of the objective (which is assumed to be in the far-field of the object) via the near-field-to-far-field transform. In Steps 3 and 4, the wavefront at the entrance pupil is first propagated to the exit pupil and then refocused at varying focal depths. The result is a 3-D data array consisting of image intensities at different lateral positions and focusing depths. Refocusing of the field between Step 3 and Step 4 in Fig. 16.12(a) is calculated using the Fresnel-Kirchhoff integral from scalar diffraction theory.

Figure 16.12(b) depicts the computational imaging results of a low-dielectric-contrast test geometry constructed to assess the capabilities of the Z-stacking procedure [18]. The test structure, shown on the top right of Fig. 16.12(b), consisted of a large sphere of refractive index $n = 1.35$ that contained three smaller spheres of index $n = 1.37$. The Z-stacking algorithm was applied at three different depths indicated by planes 1, 2, and 3 in this figure, and the corresponding computed image intensities at these depths are shown in the adjacent gray-scale views. The outlines of the small spheres inside the large sphere are clearly visible. It was concluded that an FDTD scattering model linked to a simple Z-stacking procedure can provide realistic quantitative information concerning the internal 3-D structure of biological cells [18].

Confocal Imaging of Human Skin

Confocal microscopy is an imaging technique that uses point illumination combined with a pinhole at the detector to increase the axial and longitudinal image resolution [19]. Simon and DiMarzio [20] described a 2-D FDTD model of a confocal reflectance theta microscope, which is a variation on the usual confocal setup. Instead of a point source and a pinhole, a confocal theta microscope uses a line source and a 1-D array detector, thereby reducing the confocal response to a single dimension. The advantage of this scheme is smaller size and cost, and therefore a more clinically usable microscope.

Figure 16.13(a) is a schematic diagram of the computationally modeled theta microscope of Simon and DiMarzio [20]. This was assumed to have a bistatic configuration involving separate illumination and collection paths. Unlike the common confocal setup, the illumination and detection pinholes were not located at optically conjugate positions.

Figure 16.13(b) shows the computational model of [20] for human skin, along with the illumination and collection geometry of the confocal theta microscope system. The background of the computational area was assumed to be water (refractive index $n = 1.33$). The test target, positioned at the tip of the white dashed triangles in this figure, was a single bead ($n = 1.47$) of radius 0.8 μm. A dermis layer ($n = 1.4$) was positioned beneath the sinusoidal dermis–epidermis (DE) junction. Above the DE junction, an epidermis layer consisted of cells with cytoplasm ($n = 1.37$), nuclei ($n = 1.39$), mitochondria ($n = 1.42$), and melanin ($n = 1.7$). Intercellular fluid ($n = 1.34$) was assumed to fill the rest of the volume. The sizes and shapes of the epidermal cells and cell organelles in the skin model were obtained from physical cytometric measurements. Cells relatively close to the surface were assigned more elliptical shapes, while cells near the DE junction were assigned shapes closer to circles. Nuclei occupied 10% of the cells' volumes, and the 0.4-μm melanin clumps occupied 8.5%. Mitochondria, also modeled as ellipses, occupied 10% of the cells' volumes.

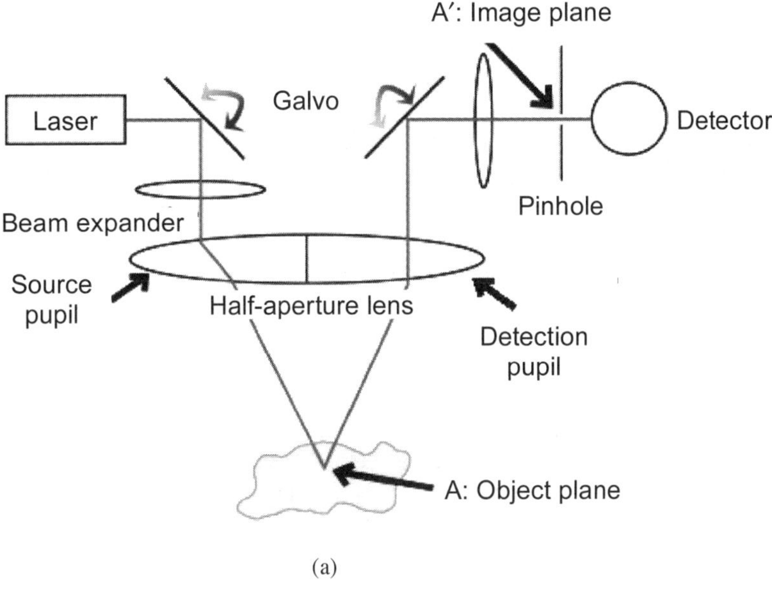

(a)

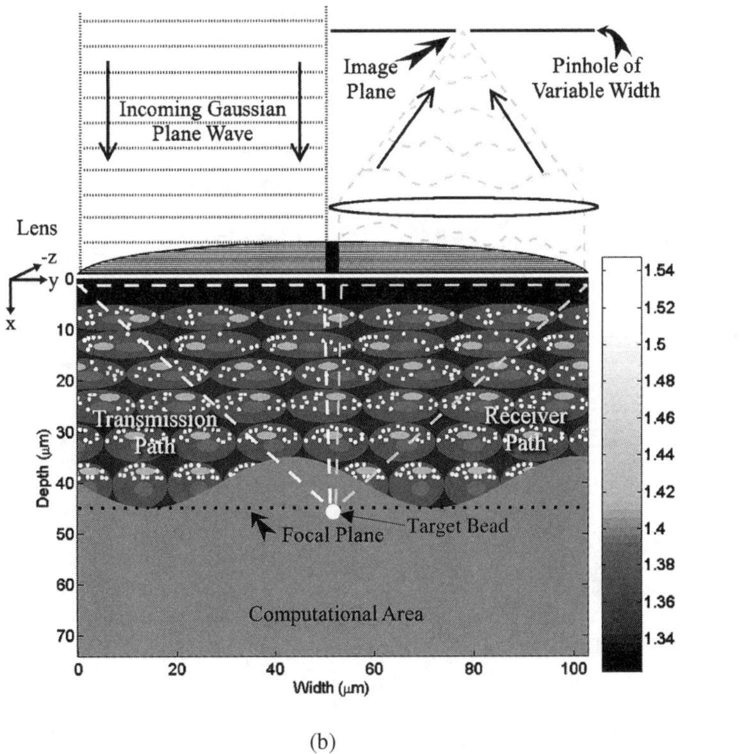

(b)

Fig. 16.13 (a) Schematic illustration of the confocal theta line-scanning microscope; (b) FDTD computational model for the human skin section and the confocal theta microscope. *Source:* Simon and DiMarzio, *J. Biomedical Optics*, 2007, 064020–1–9, ©2007 SPIE.

In Fig. 16.13(b), the upper surface of the computational area, marked by the top edges of the two white triangles, was assumed to coincide with the pupil of the microscope. Simulated illumination was sent through the left part of the pupil, while scattered light was collected from the right part. The illumination was modeled as a converging Gaussian beam with a width of 34.31 μm, or 33% of the width of the computational area. This beam propagated downward and focused toward the tip of the white triangle on the left. Scattered near fields were computed using a 2-D TM_z FDTD formulation, with the grid terminated by first-order Mur ABCs. The collected field at the right-half of the pupil plane at the top of the computational area was then propagated to the image plane using the 2-D Fresnel–Kirchhoff integral.

In one of the numerical experiments reported in [20], an examination was conducted of the signal-to-clutter ratio (SCR) of the theta microscope in imaging the 0.8-μm-radius test bead. This examination started with modeling the highest SCR case — the test bead suspended in water — and then progressively added various skin layers and components to the model to observe how the SCR would degrade. Figure 16.14 graphs the computed drop in the SCR for a range of pinhole diameters at the detector. Here, it is worth noting that the flat DE junction resulted in a lower (poorer) SCR than the sinusoidal DE junction. This occurred because the flat DE junction generated more clutter due to its specular reflection than did the sinusoidal DE junction. Figure 16.14 also shows that including melanin in the model significantly reduced the SCR performance.

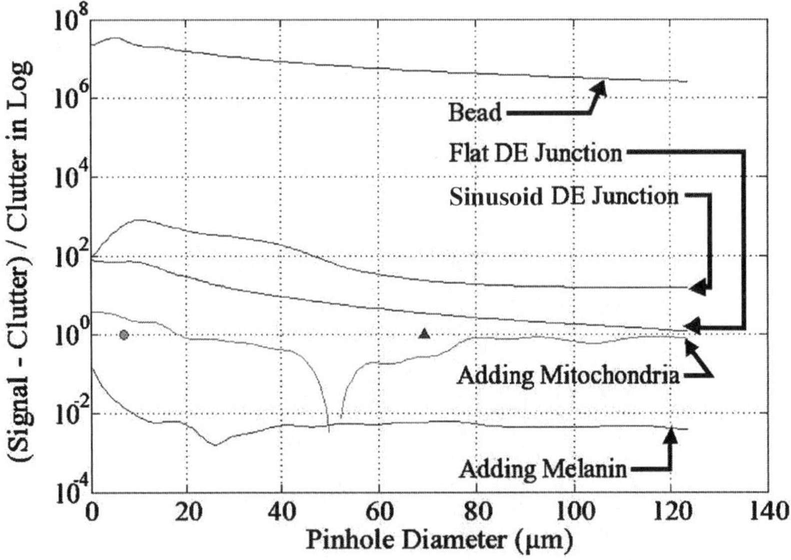

Fig. 16.14 Signal-to-clutter ratio (SCR) for varying levels of detail in the skin model, plotted with respect to the pinhole diameter. *Source:* Simon and DiMarzio, *J. Biomedical Optics*, 2007, 064020–1–9, ©2007 SPIE.

Simon and DiMarzio concluded that computational simulations could be very helpful in the functional optimization of the theta line-scanning microscope [20]. Furthermore, they concluded that computational simulations could also benefit studies of adaptive optics in skin imaging, help to analyze the speckle problem in reflectance confocal microscopes, and also quantify the signatures of the various components in normal and diseased skin tissues.

Optical Phase-Contrast Microscopy

Optical phase-contrast microscopy (OPCM) is an imaging modality that improves the contrast of weakly scattering objects viewed in transmission. This technique advances or retards the phase of an optical electromagnetic wave that is not scattered by a cell relative to the wave that is scattered or diffracted by the cell. The additional phase accumulated by a wave as it passes through the cell causes it to either constructively or destructively interfere with the wave that is not scattered by the cell, resulting in improved contrast.

This imaging modality was computationally simulated in [21] by separately calculating the optical electric field with and without the model cell in place. The 3-D FDTD simulation employed a uniaxial PML (UPML) ABC with Bloch-periodic boundary conditions in the lateral dimension, along with a modified total-field/scattered-field formulation and high-resolution ($\lambda_0/20$) grid cells. The simulated near field was transformed to the far field where phase shift was added to the field not scattered by the cell. This field was then added to the scattered far field and squared to calculate the image intensity distribution.

Figure 16.15 depicts simulated OPCM images of a model cell with refractive index matched to the cytoplasm and a phase offset of either –90° or +90°. This figure illustrates how OPCM in combination with the introduction of contrast enhancers such as gold nanoparticles can be used to label specific intracellular structures, in this case the nuclear membrane.

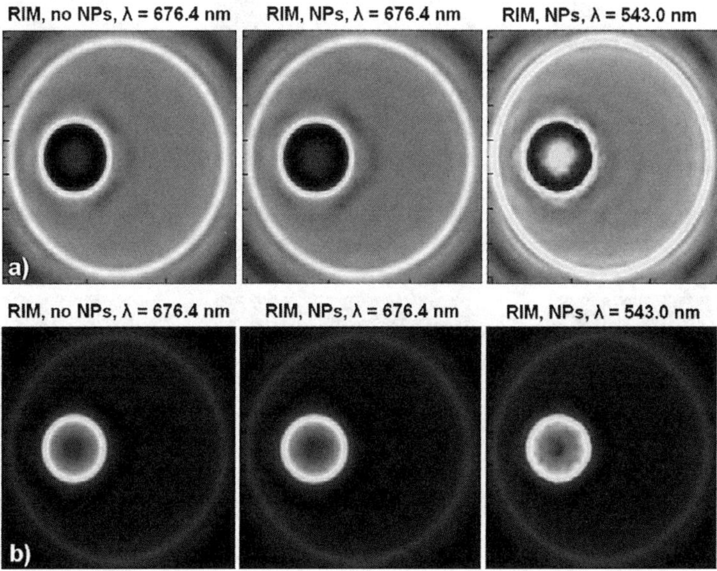

Fig. 16.15 FDTD-generated OPCM images of model cell with refractive index matched to the cytoplasm and a phase offset of either (a) –90° or (b) +90°. Left column: model cell with no internal gold nanoparticles. Middle column: model cell with gold nanoparticles randomly distributed on the nuclear membrane, but the illumination wavelength is off the resonance of the nanoparticles. Right column: same as the middle column, but the illumination wavelength is exactly at the resonance of the gold nanoparticles. *Source:* S. Tanev, W. B. Sun, J. Pond, V. V. Tuchin, and V. P. Zharov, "Flow cytometry with gold nanoparticles and their clusters as scattering contrast agents: FDTD simulation of light-cell interaction," *J. Biophotonics*, Vol. 2, 2009, pp. 505–520, ©2009 Wiley-VCH Verlag GmbH & Co. KGaA. Reproduced with permission.

Bright-Field, Dark-Field, Phase-Contrast, and Spectral Microscopy of a Human Cheek Cell

In [22] and in Chapter 14 of this book, Capoglu et al. showed in detail how 3-D FDTD modeling can be used in conjunction with appropriate field transformations to construct a comprehensive simulation of essentially all forms of optical microscopy, including at the minimum bright-field, dark-field, phase-contrast, and spectral microscopy. This subsection summarizes the application of the technique of Capoglu et al. to computationally synthesize microscope images of a human cheek (buccal) cell of approximate overall dimensions $80 \times 80 \times 1$ μm.

The first step involved generating the geometry of the cheek cell within a 3-D FDTD grid. The surface profile of the cell was measured with nanometer resolution using atomic force microscopy (AFM), and read into the FDTD grid. This profile is shown in grayscale in Fig. 16.16(a). The maximum height value, represented by the brightest shade of gray, is 990 nm. For the initial model, the cheek cell was assumed to be filled with a homogeneous lossless dielectric of refractive index $n = 1.38$ (a value guided by previous experimental studies), resting on a glass slide ($n = 1.5$).

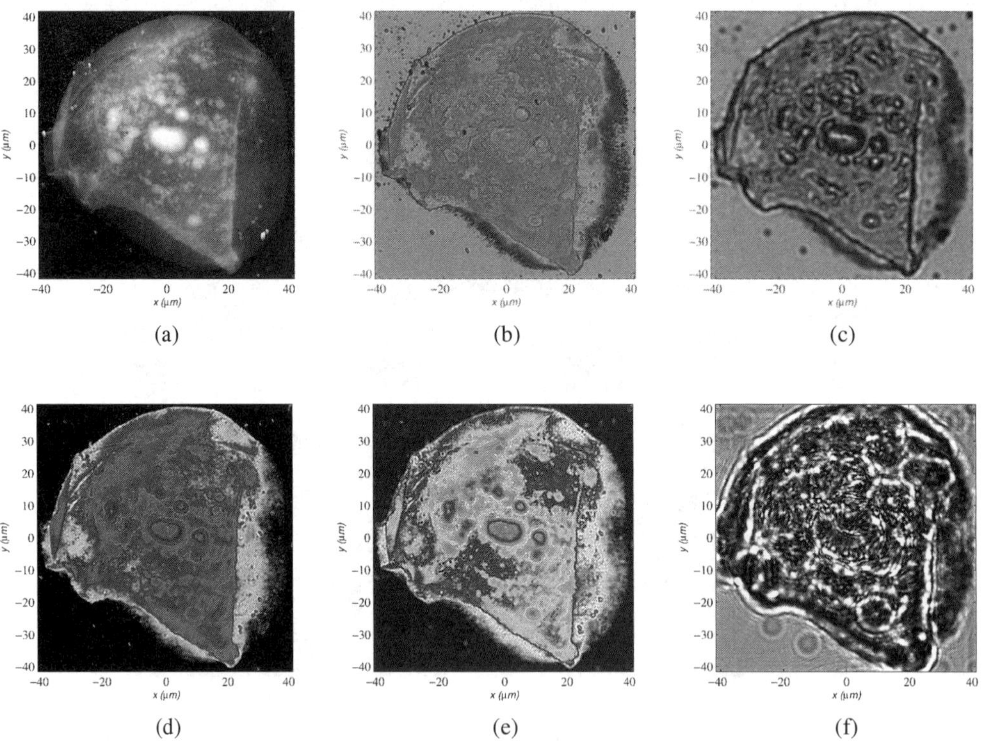

Fig. 16.16 FDTD-synthesized microscope images of a human cheek (buccal) cell model at magnification M = 1: (a) AFM profile of the cell surface; maximum height = 990 nm; (b) bright-field image for $NA_{obj} = 0.6$; (c) bright-field image for $NA_{obj} = 0.2$; (d) dark-field image; (e) phase-contrast image; (f) off-focus image at plane $z = 20$ μm. Grayscale values are −137 to +990 nm for (a); 0 to 2 for (b), (c), and (f); and 0 to 6 for (d) and (e). *Source:* I. R. Capoglu, J. D. Rogers, A. Taflove, and V. Backman, *Progress in Optics*, Vol. 57, 2012, pp. 1–91, ©2012 Elsevier.

The FDTD grid was configured with 3400 × 3400 × 58 cubic cells of size 25 nm, spanning a total volume of 85 × 85 × 1.45 μm. The grid was terminated with a 5-cell convolutional PML ABC, and the time-step was 0.04715 fs (2% below the Courant stability limit). Two illuminations having orthogonal polarizations were modeled separately, and the resulting intensities were added to obtain the final image. Each illumination was a normally incident plane wave having a sine-modulated Gaussian time waveform whose –20-dB wavelengths were 400 and 700 nm.

The far field was computed at a range of directions arranged in a 2-D Cartesian pattern in the direction-cosine space, as explained in detail in [22] and Chapter 14 of this book. Direction cosines were spaced $\Delta s_x = \Delta s_y = 0.0048$ apart from each other inside a collection numerical aperture of $\text{NA}_{\text{obj}} = 0.6$. For each observation direction, 10 wavelengths spaced linearly in wavenumber k were recorded from 400 to 700 nm. The final intensity spectrum was normalized by the spectrum of a glass pixel, resulting in a normalized spectroscopic reflectance image.

FDTD computations were parallelized using the OpenMPI implementation of the message-passing-interface (MPI) standard, and executed on 256 processors (16 × 16 × 1 Cartesian division) of the Quest supercomputer at Northwestern University. Each run took ~30 hours.

Figures 16.16(b) and 16.16(c) depict in grayscale the simulated bright-field reflectance images at $\text{NA}_{\text{obj}} = 0.6$ and 0.2, respectively. Here, the minimum and maximum brightness values correspond to 0 and 2. Figures 16.16(d) and 16.16(e) depict the simulated dark-field and phase-contrast images, respectively, with grayscale brightness values between 0 and 6. The dark-field microscope modality was modeled by coherently removing the reflection of the plane wave from the glass slide in post-processing. The phase-contrast microscopy modality was obtained by shifting the reflection of the plane wave from the glass slide by 90°, instead of removing it. Figure 16.16(f) demonstrates the effect of altering the plane of focus. Here, the bright-field image of the off-focus plane $z = 20$ μm is shown at the same grayscale level as in Fig. 16.16(b).

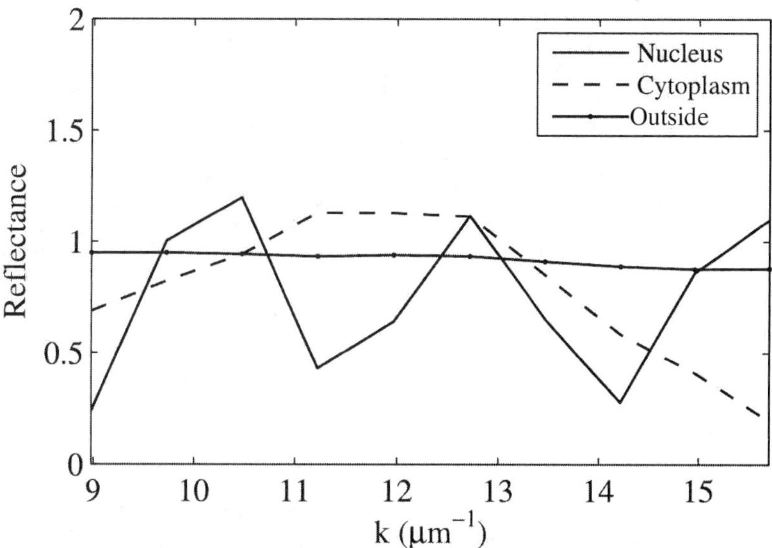

Fig. 16.17 Spectra of the reflected light between 400 and 700 nm at three pixels in Fig. 16.16(b): within the nucleus, within the cytoplasm, and outside the cheek cell. *Source:* I. R. Capoglu, J. D. Rogers, A. Taflove, and V. Backman, *Progress in Optics*, Vol. 57, 2012, pp. 1–91, ©2012 Elsevier.

Figure 16.17 graphs the normalized reflectance spectra at three pixels in Fig 16.16(b): within the nucleus, within the cytoplasm, and outside the cheek cell. The spectrum of the pixel located outside of the cheek cell is close to unity since all of these spectra are normalized by the spectrum of the reflection of the incident plane wave from the glass interface. The slight deviation from unity for this spectrum results from height variations and noise in the AFM measurements.

Using the CIE color-matching functions documented for average human vision [23], a true-color image of the cheek cell was generated from the wavelength spectrum of the FDTD-synthesized image. This was done by reducing the wavelength spectrum into three numbers that represent the response of human photoreceptor cones sensitive to short, middle, and long wavelengths. The resulting true-color image of the cheek-cell model is shown in Fig. 16.18.

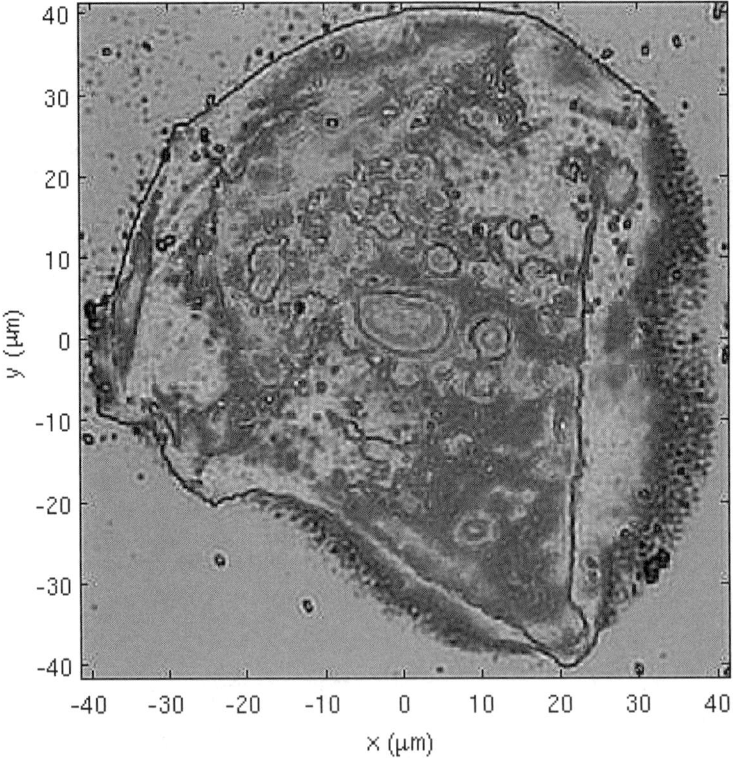

Fig. 16.18 FDTD-synthesized true-color image of the cheek cell model obtained by reducing the reflected spectrum to three numbers representing the responses of human retinal photoreceptors. The CIE color-matching functions [23] were used for this purpose.

This simulation study clearly demonstrates the power of FDTD numerical computations for simulating important optical microscopy modalities. The ever-increasing efficiency of computer processors and the widespread availability of high-performance computing resources will enable and encourage modeling of progressively more complicated biological media. Although modeling a large scatterer such as the cheek cell reported in this study is currently beyond the capability of an average personal workstation, this limitation is likely to be overcome soon.

16.2.8 Detection of Nanometer-Scale *z*-Axis Features within HT-29 Colon Cancer Cells Using Photonic Nanojets

According to a 2008 article in the *Proceedings of the National Academy of Sciences USA* [24], nanometer-scale refractive-index features along the *z*-axis (depth direction) within biological cells could undergo alterations during the initial stages of cancer development, well in advance of cellular changes that are currently considered to be histologically detectable. This section reviews two recent applications of 3-D FDTD modeling [25, 26] that indicate that such nanoscale *z*-axis refractive-index features can be detected within dielectric objects, including biological cells, by analyzing the backscattered light from impinging photonic nanojets [27–30].

The Photonic Nanojet

The photonic nanojet is a narrow, high-intensity beam of light that emerges from the shadow-side surface of a plane-wave-illuminated dielectric microsphere of diameter larger than the wavelength, λ. The microsphere can be homogeneous or graded in the radial direction to yield particular desirable nanojet characteristics.

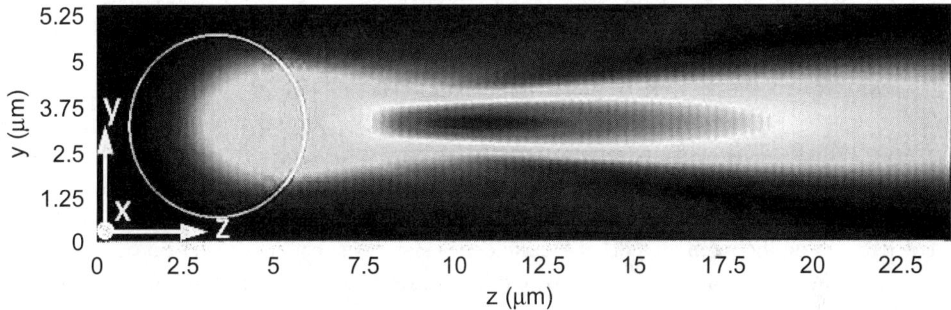

Fig. 16.19 Visualization of the electric field of a photonic nanojet generated by a plane-wave-illuminated, radially graded, 5-μm-diameter dielectric sphere. The incident wave is polarized along the *x*-axis and propagates along the *z*-axis. *Source:* Mendez-Ruiz and Simpson, *Optics Express*, Vol. 18, 2010, pp. 16805–16812, ©2010 The Optical Society of America.

Figure 16.19 illustrates an example photonic nanojet that results from an *x*-polarized, plane-wave-illuminated ($\lambda = 500\,\text{nm}$), 5-μm-diameter microsphere in free space, wherein six equally thick spherical refractive index shells are assumed [25]. In order, from the outer shell to the inner core, the refractive indices here are n = 1.02, 1.04, 1.06, 1.08, 1.10, and 1.12. (These values of n are physically realizable using recent technology [31].)

From the standpoint of this section, the key property of the photonic nanojet is its extreme sensitivity to the presence of a nanoparticle. For example, using an exact eigenfunction solution of Maxwell's equations, it was determined [28] that a 20-nm gold nanoparticle passing through the nanojet emitted by a 3.5-μm-diameter dielectric microsphere of refractive index $n = 1.59$ (illuminated in vacuum at $\lambda = 400$ nm) would cause a peak 40% increase of the microsphere's backscattered power. This perturbation, only 4 dB below the full backscattered power of the isolated microsphere, would be caused by a nanoparticle having 1/30,000th the microsphere's cross-section area. In effect, the nanojet would project the presence of the nanoparticle to the far field in a manner reminding one of the action of 1-D wave reflection.

FDTD Models of the Sensitivity of Nanojet Backscattering to a Nanometer-Thickness Film

Given the extreme sensitivity of nanojet backscattering to the presence of a nanoparticle, as shown in [28], it was desired to computationally investigate whether a similar sensitivity exists relative to the presence of a nanometer-thickness film located within the nanojet [25]. Here, pure scattered-field 3-D FDTD modeling was used to simulate the photonic nanojet of Fig. 16.19 impinging on the center of a 2-μm dielectric cube of refractive index $n = 1.3$. The dielectric cube was assumed to be either homogeneous or have embedded at its midpoint depth a 25-nm-thick ($\lambda/20$) film of index $n = 1.4$. The grid resolution was 25 nm, and a convolutional PML ABC was used to eliminate numerical wave reflections from the outer grid boundaries. For both simulation cases, the backscattered time-waveform was recorded at a point located 1.25 μm on the incident side of the microsphere for subsequent post-processing.

In addition, it was desired to test the hypothesis that the extreme sensitivity of nanojet backscattering is indeed a consequence of the nanojet approximating a true 1-D electromagnetic wave propagation/scattering system [25]. In a true 1-D universe, there would be no near field or far field. In such a universe, the spatial dependence of the Green's function would be entirely contained in a time-retarded Dirac delta function [32]. Hence, illuminated sub-wavelength-thickness dielectric films or other features would locally generate high-spatial-frequency electromagnetic field distributions that would propagate to the end of the 1-D universe with zero attenuation or spreading. Information regarding the existence and characteristics of such films could, in principle, be obtained at any remote location.

To test this hypothesis, pure scattered-field 1-D FDTD modeling with the same 25-nm grid resolution employed for the 3-D case was used to simulate the analogous planar layered geometry [25]: a plane wave normally incident on a 2-μm-thick dielectric slab of refractive index $n = 1.3$, either homogeneous or having embedded at its midpoint a 25-nm-thick film of refractive index $n = 1.4$. In this case, the backscattered time-waveform was recorded at a point located 1.25 μm on the incident side of the dielectric slab for subsequent post-processing.

Note that the 3-D and 1-D FDTD models described above and reported in [25] had *fundamentally* different levels of complexity. For the 1-D FDTD model, the dielectric slab was normally and directly illuminated by a simple plane wave arriving from an infinite free-space region. Both the slab (with or without the embedded film) and the impinging wave extended infinitely without change in the transverse directions. On the other hand, for the 3-D FDTD model, a graded dielectric microsphere was interposed between the illuminating plane wave and the dielectric cube target. Electromagnetic wave propagation and scattering were computed in all possible directions within and around the dielectric microsphere, and within and around the dielectric cube. Any level of agreement between the results of these two widely disparate models would support the 1-D interaction hypothesis for the photonic nanojet.

Figure 16.20 displays the results of these 3-D and 1-D FDTD models in terms of the cepstra of the backscattered time-waveforms [25]. The cepstrum is defined for this application as the result of taking the discrete Fourier transform (DFT) of the magnitude of the backscattered spectrum [33]. (The domain of the cepstrum is termed "quefrency" and integer multiples of the fundamental quefrency are termed "rahmonics" [33]. Although the units of the independent variable for the cepstrum are seconds, the cepstrum exists neither in the frequency domain nor the time domain.) In the backscattering analyses of [25, 26] reviewed here, the cepstrum was favored over the spectrum because the characteristics of the inhomogeneous media under investigation were more clearly extracted from the cepstral waveforms.

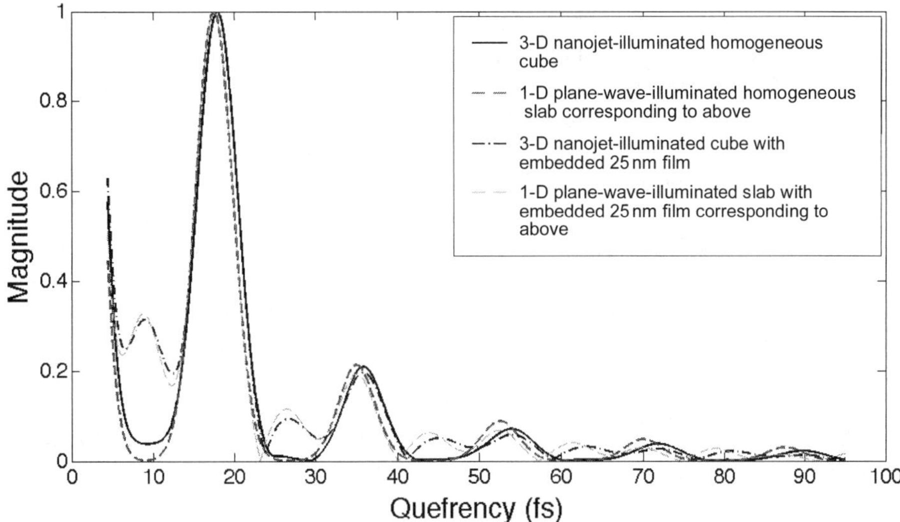

Fig. 16.20 FDTD-computed backscattered cepstra for a 3-D nanojet-illuminated $n = 1.3$ dielectric cube, with and without a 25-nm-thick ($\lambda/20$) $n = 1.4$ dielectric film embedded at its midpoint depth; and a 1-D plane-wave-illuminated $n = 1.3$ dielectric slab, with and without a 25-nm-thick ($\lambda/20$) $n = 1.4$ dielectric film embedded at its midpoint depth. *Source:* Mendez-Ruiz and Simpson, *Optics Express*, Vol. 18, 2010, pp. 16805–16812, ©2010 The Optical Society of America.

The FDTD-computed results displayed in Fig. 16.20 showed a strong correlation between the cepstrum of each 3-D nanojet-illuminated dielectric cube model and the cepstrum of its corresponding 1-D plane-wave-illuminated dielectric slab model. This provided significant support for the 1-D interaction hypothesis for the photonic nanojet proposed in [25]. That is, the photonic nanojet modeled in [25] indeed exhibited a nearly 1-D interaction with the 3-D dielectric cube, whether the cube was homogeneous or had an embedded thin film [25].

Furthermore, the results in Fig. 16.20 showed that embedding the thin 25-nm dielectric film within the dielectric cube or the dielectric slab caused significant modifications of the cepstra obtained for the homogeneous cube or slab case. Specifically, a second set of cepstral peaks arose, located between the original cepstral peaks. This provided significant support for the conjecture that photonic nanojet backscattering exhibits high sensitivity to the presence of a nanometer-thickness film located within the nanojet in a similar manner as it exhibits to the presence of a nanoparticle.

FDTD Models of Photonic Nanojet Backscattering by HT-29 Cells

Reference [26] extended the photonic nanojet backscattering studies of [25] to the more complex scenario of HT-29 human colon cancer cells. The goal was to investigate the potential of nanojet backscattering to characterize nanometer-scale z-axis features of these cancer cells for possible application to early-stage cancer detection.

Specifically, 3-D FDTD simulations were conducted of a nanojet impinging on an HT-29 cell, accounting for the cell's surface topography and internal refractive index fluctuations corresponding to varying levels of cancer aggressiveness. The backscattered time-waveforms were processed to obtain the corresponding cepstra, which were then analyzed in an attempt to distinguish between cells having greater internal refractive index fluctuations (presumed to be more aggressive) and smaller internal refractive index fluctuations (less aggressive).

First, the surface of an HT-29 cell was modeled in a 3-D FDTD grid by importing AFM measurement data [34]. The spacing between AFM measuring points was 34.8 nm in the transverse (x- and y-directions) and fractions of a nanometer in the longitudinal (depth, or z-direction). The HT-29 cell was modeled as being stationary and submerged in water, which permits studying living cells instead of dehydrated cells.

The intracellular solids of the HT-29 cell were modeled as a stationary process in the second-order cumulant approximation [35]. For this approximation, refractive index n varied randomly with position but was held constant (homogeneous) within each block of dimension equal to a parameter termed the correlation length, l_c. The mean refractive index (n_0) was 1.38. The variation of n (Δn) for typical biological tissue corresponding to the local concentration of intracellular solids can range from 0.02 to 0.1 [36]; here, the maximum Δn was set to 0.1. Note that Δn does not correlate with the carcinogenesis stage — the value of l_c determines the aggressiveness of a cancer cell line and distinguishes it from other lines [24].

Two cases of cancer cell lines were considered in [26]. One was the more aggressive C-terminal Src kinase (CsK) knockdown line that was represented with 600- × 600- × 100-nm rectangular homogeneous n-blocks. The other case was the less aggressive epidermal growth factor receptor (EGFR) knockdown line that was represented with 60- × 60- × 60-nm n-blocks [24]. Figure 16.21 shows an example 2-D geometry slice of a 3-D FDTD grid for the latter case, along with the microsphere for generating the nanojet, all immersed in water [26]. The microsphere was a homogeneous 1.5-μm-diameter silica sphere of $n = 1.42$ located 950 nm from the incident-side surface of the HT-29 cell at its geometric center in the transverse directions.

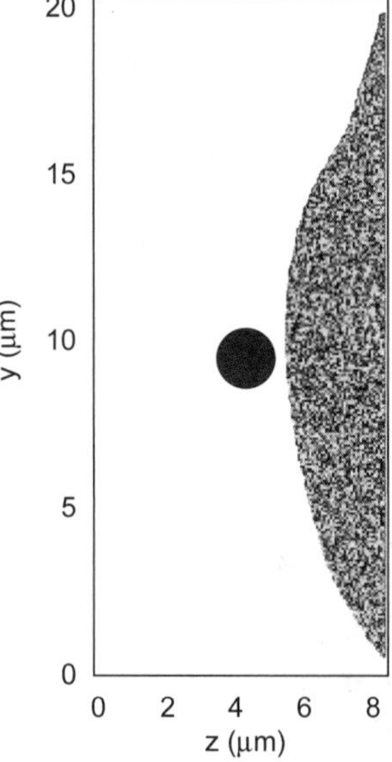

Fig. 16.21 Simulation geometry for the nanojet emitted by a 1.5-μm-diameter silica sphere impinging on an adjacent HT-29 cell model having a 60- × 60- × 60-nm pseudorandom refractive-index pattern. *Adapted from:* Mendez-Ruiz and Simpson, *J. Applied Computational Electromagnetics Society*, Vol. 27, 2012, pp. 215–222, ©2012 Applied Computational Electromagnetics Society.

For each simulation case, the backscattered time-waveform was recorded 3.8 μm on the incident side of the microsphere for subsequent post-processing. As part of the post-processing, the time-domain backscattered signal from the microsphere alone was subtracted from the results of the simulation of the microsphere and HT-29 cell. This permitted improved extraction of the backscattered signal from the HT-29 cell. An FDTD grid resolution of 10 nm in each Cartesian direction was chosen after performing iterative tests for numerical convergence.

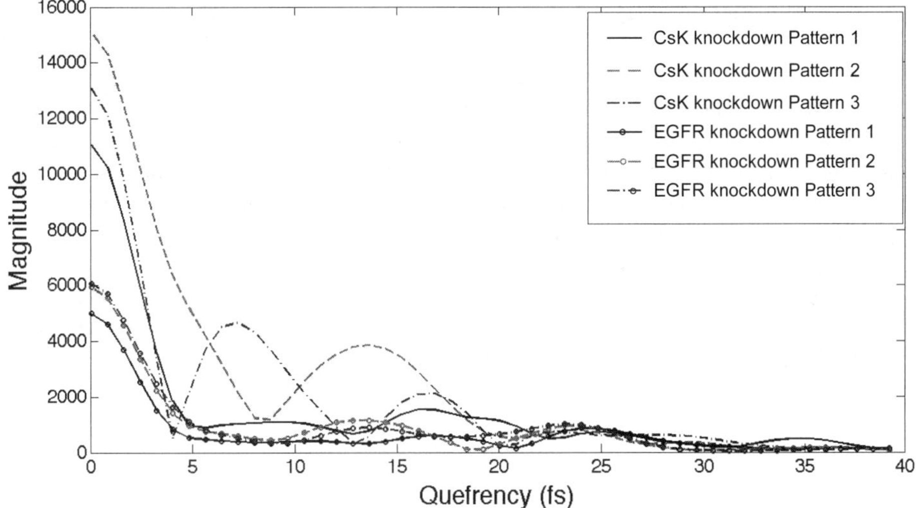

Fig. 16.22 Comparison of FDTD-computed photonic nanojet backscattering cepstral results of CsK and EGFR cell lines. Results for three pseudorandom internal fluctuation patterns of the index of refraction are shown for each cell line. *Source:* Mendez-Ruiz and Simpson, *J. Applied Computational Electromagnetics Society*, Vol. 27, 2012, pp. 215–222, ©2012 Applied Computational Electromagnetics Society.

Figure 16.22 shows the comparison of the FDTD-computed photonic nanojet backscattering cepstral results reported in [26] for three different sample pseudorandom internal fluctuation patterns of the index of refraction for each cell line. Below a quefrency of approximately 4 fs, the more aggressive CsK cell line exhibited a greater magnitude than the less aggressive EGFR line, but above 4 fs, it was difficult to unambiguously distinguish between the two cell lines. On the other hand, for all six cases, it was easily possible to distinguish one cell line from the other based on their magnitudes approaching 0 quefrency. Here, all three CsK sample cases had significantly higher magnitudes than all three sample EGFR cases.

The FDTD computational results of [26] indicate the possibility of using a cepstral analysis of the optical backscatter from nanojet-illuminated cancer cells to distinguish between the aggressiveness of these cells. This is based on presumed differences in these cells' internal refractive index fluctuations at sub-wavelength scales that are not accessible using conventional optical microscopy techniques [24]. Following the concept expressed in [24], it may thus be possible to use nanojet probes to screen for early-stage cancers by obtaining the backscattered cepstra of seemingly normal cells in the field of a developing malignancy; for example, cells swabbed from the rectum to screen for colon cancer, and cells swabbed from the interior surface of the cheek to screen for lung cancer.

16.2.9 Assessment of the Born Approximation for Biological Media

The Born approximation [37] is a common analytical solution technique used in scattering problems where the refractive index contrast is very small and the scattering is very weak. This approximation represents the first term in the Born series, which is a mathematical formulation of the multiple scattering processes inside the scatterer. Higher-order terms in this series are very seldom employed due to the exponentially increasing orders of the integrals involved. As a result, it is necessary to invoke other numerical solution methods to assess the accuracy of the Born approximation.

Capoglu et al. [38] used the FDTD method to quantify the accuracy of the Born approximation in describing the scattering properties of continuous biological random media. Specifically, they considered the scattering coefficient μ_s, defined as the average total scattered power per unit volume of scatterer under unit incident plane-wave intensity. For computational simplicity, they assumed a 2-D geometry; however, the theoretical and numerical methods are entirely applicable to 3-D media. In fact, they extrapolated their results to 3-D, and showed that the condition for the accuracy of the Born approximation is the same in 3-D.

In [38], Capoglu et al. defined the normalized refractive-index fluctuation of a medium by $\Delta n(\boldsymbol{\rho}) = [n(\boldsymbol{\rho}) - n_0] / n_0$ where n_0 is the average refractive index of the medium. This fluctuation was modeled by a 2-D homogeneous, isotropic Gaussian-distributed random field, described by the following two-point correlation function:

$$B_n(\Delta\rho) = \sigma_n^2 \frac{\Delta\rho}{l_c} K_1\left(\frac{\Delta\rho}{l_c}\right) \tag{16.1}$$

where $\Delta\rho$ is the lateral displacement between the two points, σ_n is the fluctuation strength, l_c is the correlation length, and $K_1(\cdot)$ is the modified Bessel function of the first order and second kind. The scattering coefficient μ_s for this medium was then derived in closed form under the Born approximation [39]. For TE-polarized monochromatic plane-wave illumination at frequency ω, μ_s is given by:

$$\mu_{s_{TE}} = \frac{\pi\sigma_n^2}{kl_c^2\left[1+4(kl_c)^2\right]^{3/2}} \begin{bmatrix} -1 - 4(kl_c)^4 + 8(kl_c)^6 + \sqrt{1+4(kl_c)^2} \\ + 2(kl_c)^2 \cdot \left(-3 + 2\sqrt{1+4(kl_c)^2}\right) \end{bmatrix} \tag{16.2}$$

where $k = \omega / c$ is the wavenumber in the surrounding medium.

Next, an accuracy condition was derived for (16.2). This condition was based on the following argument. Namely, the Born approximation becomes inaccurate in two extremes: (1) For a small sample size, there is insufficient spatial averaging for the correlation in (16.1) to take effect and influence the result; (2) for a large sample size, multiple scattering dominates, which is manifested by a decreased mean-free path, defined customarily as $l_s = 1/\mu_s$. Denoting the sample size by L, we should therefore have $l_c << L << l_s$, consequently, $l_c << l_s$. Using (16.2), the accuracy condition $\sigma_n^2 (kl_c)^2 << 1$ is obtained.

To test the proposed accuracy conditions, Capoglu et al. compared the results of FDTD computational models of scattering by simulated continuous biological random media with the Born approximation [38]. Random 2-D refractive-index samples from the statistical model of (16.1) were generated using an inverse Fourier transform approach. The spatial 2-D Fourier transforms of the random samples were generated first, and then transformed back into the spatial domain. [It was easier to generate a Fourier-transformed array first since the points in that array are statistically independent for a homogeneous (or stationary) random field.] Guided by previous refractometry studies, the average refractive index n_0 was chosen as 1.38. The ranges chosen for σ_n and l_c were also determined by the values reported in these studies. In the comparative FDTD simulations, a grid-cell size of 13.3 nm was chosen, and the grid was terminated by a 20-cell CPML ABC. A plane wave was sourced into the grid using the total-field/scattered-field technique, and the far field was computed using the near-field-to-far-field transform [2]. The electric field time-waveform of the incident plane wave was a sine-modulated Gaussian pulse with appreciable spectral content between 400 and 700 nm, which corresponds to the visible range. The scattered field was normalized by the incident wave amplitude and averaged over 200 samples.

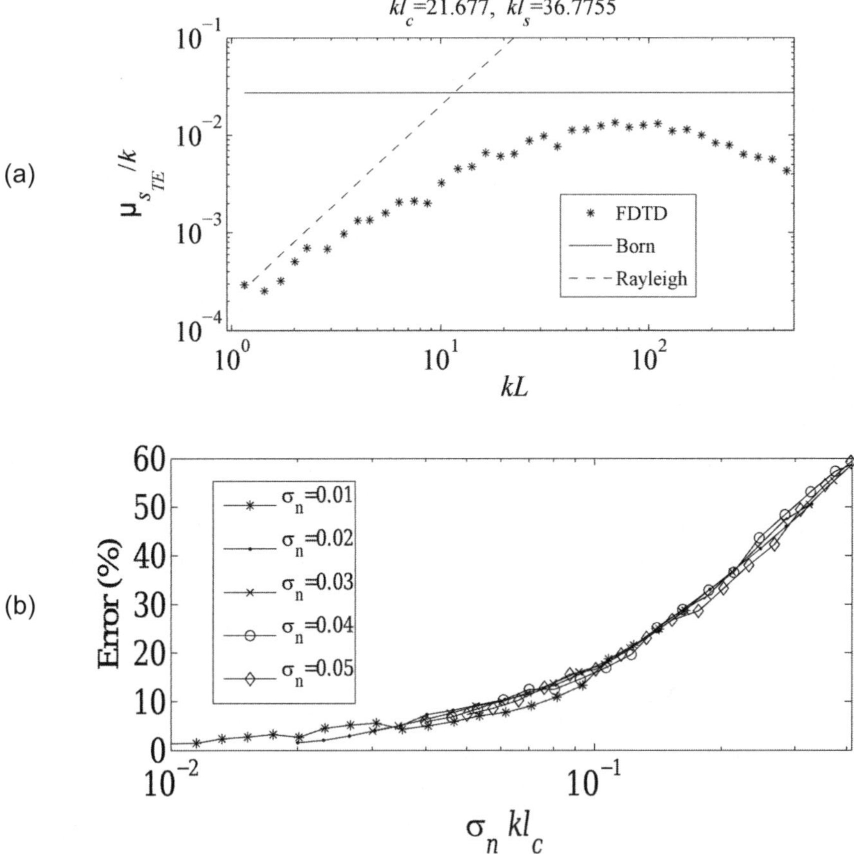

Fig. 16.23 Assessment of the accuracy of the Born approximation in 2-D continuous random media via comparison with FDTD results. (a) Normalized scattering coefficient vs. sample dimension L for $l_c \sim l_s$. In this regime, there exists no range of L in which the scattered power follows the Born approximation. (b) Normalized error in Born scattering coefficient vs. $\sigma_n k l_c$ for a range of σ_n values in a regime where the Born approximation transitions from accuracy. *Source:* I. R. Capoglu, J. D. Rogers, A. Taflove, and V. Backman, *Optics Letters*, Vol. 34, 2009, pp. 2679–2681 ©2009 The Optical Society of America.

Figure 16.23(a) compares the Born, Rayleigh, and FDTD-calculated μ_s values for a range of sample sizes L [38]. For this figure, the l_c and l_s values were deliberately chosen to be close to each other so that the condition $l_c << L << l_s$ for the Born approximation does *not* hold. For very small kL, the results tend asymptotically to Rayleigh scattering, which corresponds to the small-particle approximation. The linear dependence of the scattered power on the volume obviously breaks down at this extreme. This linear dependence also breaks down at the other extreme of kL, where L exceeds l_s. It is seen that there is no intermediate range of L for which the scattered power is linear with respect to the volume and follows the Born approximation result (16.2).

Figure 16.23(b) deals with a regime where the Born approximation transitions from accuracy to inaccuracy [38]. This figure graphs the normalized error in the scattering coefficient for TE illumination vs. $\sigma_n k l_c$ for a range of σ_n values. Here, the percentage error between the Born approximation and FDTD is defined as the minimum difference between (16.2) and the FDTD-calculated μ_s. The curves for different σ_n are seen to coincide, which lends further credence to the validity condition $\sigma_n^2 (k l_c)^2 << 1$.

16.3 OVERVIEW OF FOURIER-BASIS PSTD TECHNIQUES FOR MAXWELL'S EQUATIONS

In principle, FDTD can be used to model arbitrarily large collections of biological cells and model tissue-optics problems directly from Maxwell's equations. However, the size of problems amenable to FDTD modeling is limited, especially in three dimensions. Here, the database of electromagnetic field vector components used by Yee-algorithm FDTD rapidly exceeds available computer resources because of the fine-grained spatial resolution (10 to 20 or more grid cells per dielectric wavelength λ_d) that is required to achieve an acceptable predictive dynamic range.

To model large-scale electromagnetic wave interaction problems while retaining the advantages of FDTD, researchers have proposed replacing the second-order-accurate spatial derivative approximations with ones of higher accuracy. A promising class of such techniques for hyperbolic partial differential equations is spectral collocation [40–43], of which the pseudospectral time-domain (PSTD) method is specifically aimed to solve Maxwell's equations. Originally proposed by Liu [3], the global Fourier-basis PSTD method permits, in principle, relaxation of the spatial-resolution requirement to the Nyquist limit of two grid cells per wavelength. (Also see [2], especially its Section 4.9.4 and Chapter 17, for a comprehensive discussion of Liu's PSTD technique.) Liu has shown that, for large models in D dimensions that do not have geometrical details smaller than one-half wavelength, the global Fourier-basis PSTD method reduces computer-resource requirements by approximately 8^D:1 relative to the Yee-algorithm FDTD while achieving comparable accuracy [3]. This is the key to modeling large-scale biophotonics problems directly from Maxwell's equations.

Subsequently, Ding and Chen [4] introduced the SL-PSTD technique, which has improved computational efficiency and robustness relative to Liu's PSTD formulation. SL-PSTD is based on a multi-domain local Fourier transform that combines overlapping domain decomposition within the computation region and carefully constructed mathematical procedures to preserve numerical accuracy. Here, a reformatted derivative operator on the field-vector quantities enables retention of the familiar staggered FDTD space lattice, and furthermore eliminates the Gibbs phenomenon inherent in Liu's previous collocated-grid, global Fourier-basis PSTD formulation. See Chapter 1 of this book for a comprehensive discussion of the SL-PSTD technique.

16.4 PSTD AND SL-PSTD MODELING APPLICATIONS

In the following examples of applications of PSTD and SL-PSTD to biophotonics problems, we note that PML ABCs were used to terminate the computational grids. In addition, the surfaces of all of the modeled scattering shapes were approximated by simple staircasing at the coarse grid resolution permitted by these two techniques.

16.4.1 Enhanced Backscattering of Light by a Large Cluster of 2-D Dielectric Cylinders

We first consider PSTD modeling of optical enhanced backscattering (EBS), a phenomenon that has recently elicited attention as a potential means for the clinical diagnosis of disease [44–46]. Employing PSTD, Tseng et al. [47] reported the first simulation of EBS by numerically solving Maxwell's equations without heuristic approximations.

Figure 16.24 illustrates the geometry of the PSTD model of [47]: an 800- $\times$ 400-μm rectangular cluster of 10,000 or 20,000 randomly positioned, non-contacting, infinitely long, dielectric cylinders. Each cylinder was 1.2 μm in diameter with refractive index $n = 1.25$. An average surface-to-surface spacing of 2.8 μm was provided between adjacent cylinders. The rectangular cluster was illuminated by a coherent plane wave at $f_0 = 300$ THz ($\lambda_0 = 1$ μm) that was incident at 15° relative to the normal. Both the incident light and the backscattered light were polarized perpendicular to the plane of incidence, equivalent to collinear detection in EBS experiments. The PSTD grid had a uniform spatial resolution of 0.33 μm, equivalent to $0.42\lambda_d$ at the illumination wavelength. At this wavelength, the transport mean free path, l_s', was 5.59 μm.

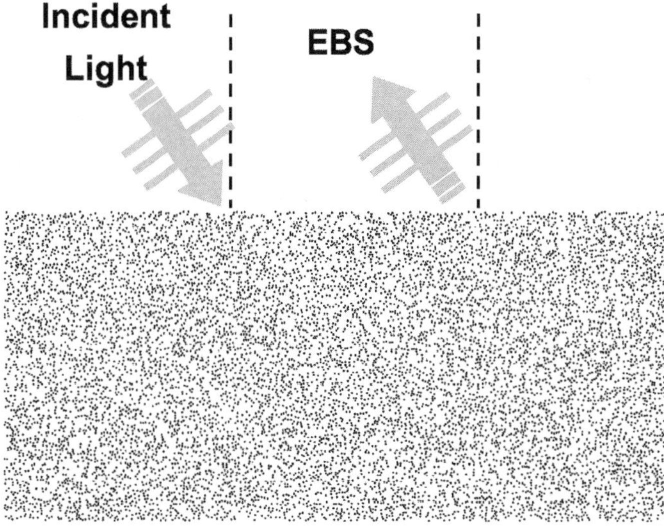

Fig. 16.24 Simulation geometry of PSTD-computed enhanced backscattering (EBS) in two dimensions: 800- $\times$ 400-μm rectangular cluster of 10,000 or 20,000 randomly positioned, non-contacting, 1.2-μm-diameter, $n = 1.25$ dielectric cylinders illuminated by a coherent plane wave at 15° from the normal. Both incident light and backscattered light were polarized perpendicular to the plane of incidence, equivalent to collinear detection in EBS experiments. *Source:* S. H. Tseng, A. Taflove, D. Maitland, V. Backman, and J. T. Walsh, *Optics Express*, Vol. 13, 2005, pp. 3666–3672, ©2005 The Optical Society of America.

In the modeling procedure reported in [47], it was desired to suppress speckle due to coherent interference effects of the random medium. To this end, the PSTD-computed scattered light intensity was ensemble-averaged over 40 simulations, each corresponding to a different random arrangement of cylinders within the rectangular cluster. Speckle was further suppressed by averaging over 50 different incident frequencies evenly spaced between $0.95f_0$ and $1.05f_0$. This was similar to experimental observations of EBS using non-monochromatic illumination with a temporal coherence length of 10 μm. An estimated 16 modes were averaged in the process.

Tseng et al. noted that their PSTD model could include only a finite-size random-medium region. However, they devised a means to efficiently account for this finite size by implementing a convolution of comparative benchmark analytical results for an infinite random region with an appropriate windowing function that represented the effective aperture of the finite random region in their model [47].

Figure 16.25 compares the angular distribution of the PSTD-computed EBS peak of [47] with that obtained using standard EBS theory based on the diffusion approximation [48]. The PSTD calculations were in very good agreement with the benchmark theory.

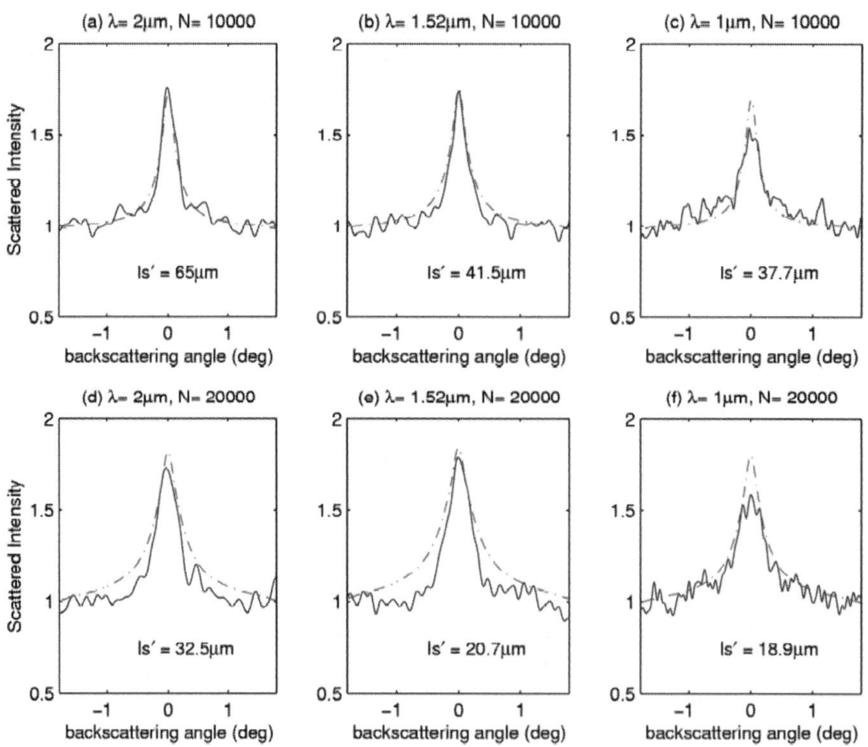

Fig. 16.25 Comparison of PSTD-computed EBS peaks (solid lines) for three wavelengths with theoretical benchmark results (dash-dotted lines) for rectangular clusters consisting of N cylinders. (a)–(c) correspond to $N = 10{,}000$ cylinders with $l_s' = 65.0$, 41.5, and 37.7 μm, respectively; (d)–(e) correspond to $N = 20{,}000$ cylinders with $l_s' = 32.5$, 20.7, and 18.9 μm, respectively. *Source:* S. H. Tseng, A. Taflove, D. Maitland, V. Backman, and J. T. Walsh, *Optics Express*, Vol. 13, 2005, pp. 3666–3672, ©2005 The Optical Society of America.

Recently, the phenomenon of low-coherence EBS was demonstrated to be promising for clinical diagnosis [46]. To simulate low-coherence EBS using PSTD, the frequency-averaging technique mentioned above can be modified.

16.4.2 Depth-Resolved Polarization Anisotropy in 3-D Enhanced Backscattering

In studies of enhanced backscattering, linear polarization gating and low-coherence enhanced backscattering are two powerful tools capable of selectively detecting low-order backscattered photons. Combining these two techniques can reveal even more properties of the nondiffusive layer and interrogate the underlying structure through polarization-induced coherence variations. For example, the linear polarization plane of the source beam bisects the space and breaks the cylindrical symmetry. Any spatial/angular variation exposed in the EBS signal indicates polarization-dependent scattering features of the structure. Experiments have shown that the size of constituent scatterers in discrete random media affects the cone shape of polarization-gated EBS. In linear polarization measurements on semi-infinite media, the co-polarized EBS cone of Rayleigh clusters has different widths on two orthogonal planes [49, 50], while they are equal for Mie clusters [50]. Thus, it is generally believed that azimuthal asymmetry is characteristic of the scattering of small particles, but absent in the case of large particles.

Numerical analyses of the azimuthal anisotropy effect in the polarized EBS cone must resort to direct methods that explicitly model vector-wave optical field interactions. To date, investigations have centered on Rayleigh scattering systems. In the electric-field Monte Carlo (EMC) approach [51], the electric field (rather than the Stokes vector) of light is traced, and the complete phase delay is accumulated along the random-walk path. For a turbid medium comprised of 100-nm-diameter polystyrene spheres illuminated by linearly polarized light at the 514.5-nm wavelength, EMC studies showed different angular azimuthal patterns in co- and cross-polarized EBS cones. In an approach based on superposition T-matrix solutions [52], simulations of Rayleigh clusters also reproduced the azimuthal asymmetry in co-polarized EBS cones.

However, the polarization evolution of the EBS signal in random Mie scattering systems can differ significantly from Rayleigh scattering systems. Ding and Chen applied the SL-PSTD technique to predict the polarization-dependent EBS response of a 3-D disordered medium consisting of Mie scatterers, especially to isolate effects localized in the superficial layer from those involving deeper regions [53]. Unlike previous work that simulated only bulk media [51, 52], SL-PSTD allows depth-resolved study of the polarization progression layer by layer.

Figure 16.26 illustrates the SL-PSTD modeling geometry of [53]. Here, 2-μm-diameter polystyrene beads were assumed to be randomly suspended in a 100- × 100-μm rectangular volume of water at a 2.9% volume concentration. The thickness of this water-particulate volume was varied to control the order of the scattering events, and the entire water-particulate volume was surrounded by an infinite homogeneous water background to avoid any interfaces with air. To deflect specular reflection away from the field of view, a pulsed linearly polarized incident plane wave was directed upon the top surface of the water-particulate volume at 5° from the normal. The observation coordinate system, $x'yz'$, was rotated by 5° about the y-axis of the medium coordinate system, xyz, as shown in Fig. 16.26. In this manner, the z'-axis was the exact backward direction, and hence the center of the EBS cone. The backscattering angle θ and the azimuth angle φ were then defined, respectively, as the inclination angle and the azimuth angle in the $x'yz'$ coordinate system. The incident polarization lay in the $x'z'$ plane, which was also the incidence plane. Backscattered light was decomposed into a component parallel to the $x'z'$ plane (the co-polarized $I_{\parallel}$) and its orthogonal component (the cross-polarized $I_{\perp}$).

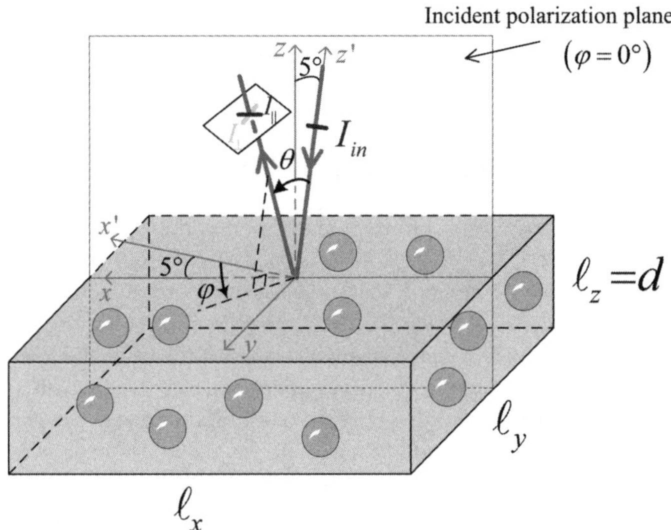

Fig. 16.26 Schematic diagram of the SL-PSTD model of depth-resolved polarization anisotropy in 3-D enhanced backscattering. *Source:* Ding and Chen, *Optics Express*, Vol. 18, 2010, pp. 27639–27649, ©2010 The Optical Society of America.

The central frequency f_0 of the illuminating pulse was set to 3.82×10^{14} Hz, corresponding to a 785-nm vacuum wavelength. Signals at 60 frequencies evenly distributed in $[0.95f_0, 1.05f_0]$ were extracted from the temporal response of the medium. To alleviate speckle, a two-stage average over 60 frequencies and 5 medium realizations was conducted. Random media of six different physical depths L were simulated, corresponding to the dimensionless optical depth parameter $\tau \equiv L/l_s$ ranging from 1 to 6, where l_s is the scattering mean-free-path length [53].

Figures 16.27(a) and 16.27(b) are polar-plot visualizations of the SL-PSTD-computed co-polarized and cross-polarized EBS intensity about the exact backscattered direction [53]. In these figures, the radial coordinate is the backscattering angle θ, ranging from 0° at the center of each plot to 8° at the outer border. The polar coordinate of each plot is the azimuth angle φ of the backscattering direction, with $\varphi = 0°$ defined as the incident polarization plane. Azimuthal anisotropy is evident in both the co-polarized and cross-polarized EBS plots for the optical thickness $\tau \leq 3$, but gradually disappears as τ increases.

As a consequence of breaking a high-order symmetry into a low-order one, the dissection of cylindrical symmetry by the polarization directions of incidence and detection discrimination reduces the cylindrical symmetry in the EBS cone into discrete symmetrical patterns. For the co-polarization case, the two directions coincide — the φ space is split into two and exhibits two-fold rotational symmetry. Therefore, the co-polarized EBS exhibits a two-fold symmetrical pattern. On the other hand, for the cross-polarization case, the two directions cross each other and split the angular space into a four-fold rotational symmetry. Therefore, the cross-polarized EBS shows a four-fold symmetrical pattern. Furthermore, the observed "×" pattern instead of a "+" pattern indicates that the influences of the initial polarization projection at incidence and the final polarization projection at discrimination are symmetrical, a manifestation of time-reversal symmetry. The "×" pattern means that all photons see the two directions equally, while the "+" pattern means that one-half of the photons favor the initial polarization direction, and vice versa. The second scenario is unlikely because all photons are identical.

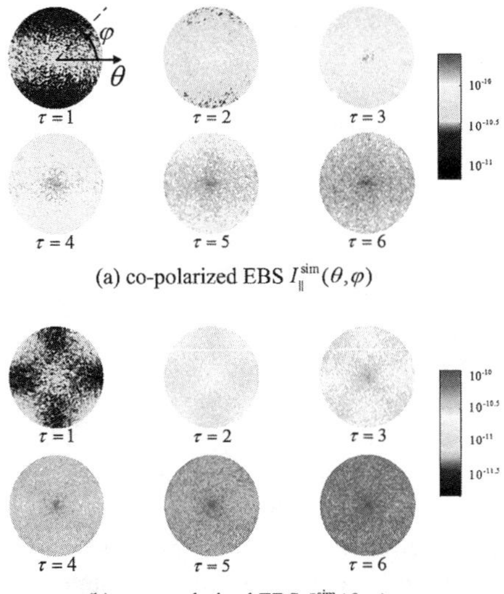

(a) co-polarized EBS $I_{\parallel}^{\mathrm{sim}}(\theta,\varphi)$

(b) cross-polarized EBS $I_{\perp}^{\mathrm{sim}}(\theta,\varphi)$

Fig. 16.27 Visualizations of the SL-PSTD-computed (a) co-polarized, and (b) cross-polarized EBS intensity about the exact backscattering direction from a 3-D random medium consisting of Mie scatterers under linearly polarized incidence. Azimuthal anisotropy appears as intensity fluctuations within each map. The patterns are most prominent for an optical depth $\tau = 1$ in both (a) and (b), but progressively fade away as τ goes to 6. *Source:* Ding and Chen, *Optics Express*, Vol. 18, 2010, pp. 27639–27649, ©2010 The Optical Society of America.

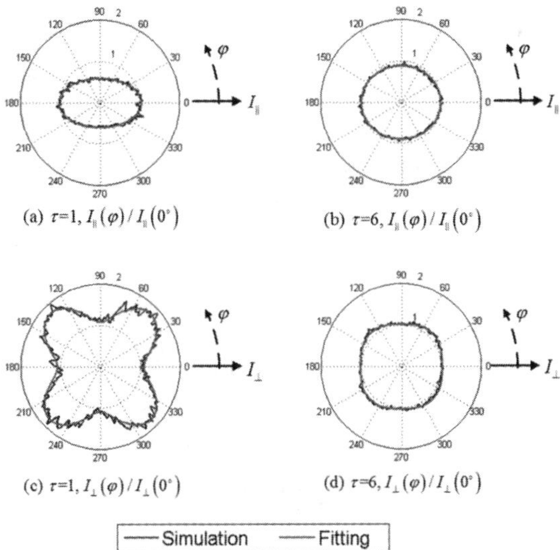

(a) τ=1, $I_{\parallel}(\varphi)/I_{\parallel}(0^{\circ})$

(b) τ=6, $I_{\parallel}(\varphi)/I_{\parallel}(0^{\circ})$

(c) τ=1, $I_{\perp}(\varphi)/I_{\perp}(0^{\circ})$

(d) τ=6, $I_{\perp}(\varphi)/I_{\perp}(0^{\circ})$

Fig. 16.28 Polar diagrams of the integrated SL-PSTD-computed EBS intensities and their empirical fittings: (a) optical depth $\tau = 1$, co-polarized; (b) $\tau = 6$, co-polarized; (c) $\tau = 1$, cross-polarized; (d) $\tau = 6$, cross-polarized. Because only the curve shape is relevant to the degree of anisotropy, each curve is individually normalized to its intensity at $\varphi = 0$. *Source:* Ding and Chen, *Optics Express*, Vol. 18, 2010, pp. 27639–27649, ©2010 The Optical Society of America.

To characterize the degree of anisotropy, the SL-PSTD-computed backscattering intensity can be integrated over θ at each azimuth angle φ as:

$$I_{\parallel}^{\text{sim}}(\varphi) \equiv \int_{0^\circ}^{8^\circ} I_{\parallel}^{\text{sim}}(\theta,\varphi)\sin\theta\, d\theta \tag{16.3}$$

$$I_{\perp}^{\text{sim}}(\varphi) \equiv \int_{0^\circ}^{8^\circ} I_{\perp}^{\text{sim}}(\theta,\varphi)\sin\theta\, d\theta \tag{16.4}$$

where the superscript "sim" denotes simulation. The results [normalized such that $I_{\parallel,\perp}^{\text{sim}}(0^\circ) = 1$] are presented as the noisy curves in the polar diagrams of Fig. 16.28, along with their empirical curve-fittings for the extreme cases of optical depth, $\tau = 1$ and $\tau = 6$ [53]. Here, the radial coordinate of each plot is the normalized integrated intensity, and the polar coordinate is the azimuth angle φ. The noise on each SL-PSTD-computed curve is due to residual speckle effects. We see that the co-polarized curve has an elliptical shape, whereas the cross-polarized curve is close to the superposition of a cosine function on top of a circle. Both curves converge quickly to circles as the optical depth τ increases. The deviation from the circular shape provides the most representative information about the angular anisotropy of the EBS cone.

In Fig. 16.28, two empirical formulas are introduced to fit the SL-PSTD-computed results:

$$I_{\parallel}^{\text{fit}}(\varphi) \equiv 1\Big/\sqrt{\frac{\cos^2\varphi}{a^2} + \frac{\sin^2\varphi}{a^2\varepsilon^2}} \tag{16.5}$$

$$I_{\perp}^{\text{fit}}(\varphi) \equiv b\left[1 - \gamma\cos(4\varphi)\right] \tag{16.6}$$

where the superscript "fit" denotes empirical fit. Here, the parameters ε and γ characterize the degree of anisotropy. (The factors a and b are irrelevant to the curve shapes.) For the co-polarized case, ε equals the ratio of the minor radius to the major radius of the ellipse. When ε reaches 1, the co-polarized EBS becomes azimuthally isotropic. For the cross-polarized case, γ describes the amplitude of the angular oscillation in the anisotropy. When γ reaches 0, the cross-polarized EBS becomes azimuthally isotropic. The parameter sets (a, ε) and (b, γ) are determined through minimization of the following two expressions:

$$\chi_{\parallel}^2 \equiv \frac{1}{M}\sum_{i=1}^{M}\left[I_{\parallel}^{\text{fit}}(\varphi_i) - I_{\parallel}^{\text{sim}}(\varphi_i)\right]^2 \tag{16.7}$$

$$\chi_{\perp}^2 \equiv \frac{1}{M}\sum_{i=1}^{M}\left[I_{\perp}^{\text{fit}}(\varphi_i) - I_{\perp}^{\text{sim}}(\varphi_i)\right]^2 \tag{16.8}$$

where M is the number of discrete azimuth angles in the SL-PSTD calculations.

The good agreement between the normalized integrated SL-PSTD-computed EBS intensities and the empirically fitted curves in Fig. 16.28 suggests that simple physical principles may govern the polarization evolution in low-order EBS effects. The underlying spherical symmetry of the constituent scattering structures, i.e., the Mie spheres, may play a role. The low-order enhanced backscattering of irregularly shaped scatterers may have angular patterns different from the elliptical or symmetrical-butterfly shapes.

Figure 16.29 illustrates how the parameters ε and γ, which characterize the degree of the anisotropy of the enhanced backscattering computed by SL-PSTD, vary with the optical depth τ [53]. As shown, ε and γ exhibit monotonic, but opposite trends as τ increases from 1 to 6. Namely, ε increases from 0.58 to 0.9, whereas γ decreases from 0.25 to 0.04. This is evidence that the anisotropy of the Mie scatterers is only associated with very low-order scattering for both co-polarized and cross-polarized enhanced backscattering.

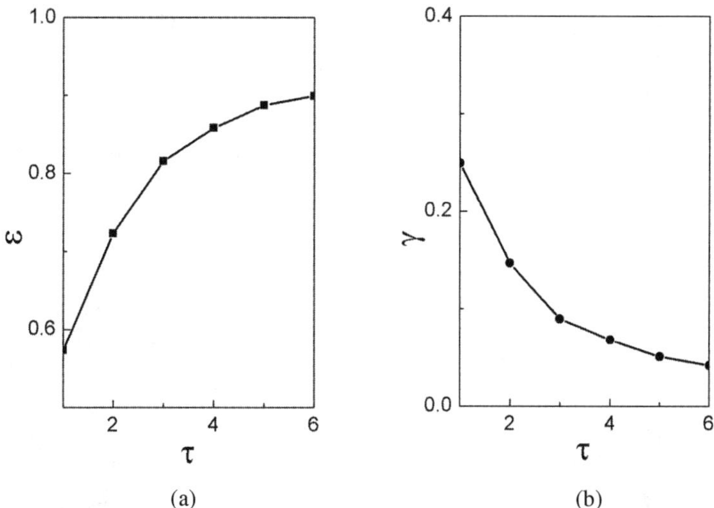

Fig. 16.29 Parameters characterizing the degree of the enhanced backscattering (EBS) anisotropy computed by SL-PSTD: (a) ε for co-polarized EBS; (b) γ for cross-polarized EBS. *Source:* Ding and Chen, *Optics Express*, Vol. 18, 2010, pp. 27639–27649, ©2010 The Optical Society of America.

Literature reports of laboratory measurements indicate the existence of an anisotropy in the co-polarized enhanced backscattering from bulk random media consisting of Rayleigh scatterers [49, 50], but not Mie scatterers [50]. Based on the SL-PSTD results summarized above, a potential explanation for this phenomenon involves competition between the enhanced backscattering signal from different depths. The total enhanced backscattering signal can be decomposed into low-order components ($\tau < 6$) and high-order components ($\tau > 6$). For both the Rayleigh- and Mie-scatterer random-media cases, the low-order components are anisotropic, but the anisotropy vanishes as the scattering order increases. Small Rayleigh scatterers carry small g factors and steer incoming photons nearly equally to all directions. Thus, the low-order components have large contributions and prevail in the total signal, showing an anisotropy in angle φ.

On the other hand, each individual Mie scattering event is highly forward oriented, and multiple scattering events are required to turn the incident photons into the backward direction. Consequently, the Mie enhanced backscattering signal is dominated by depolarized high-order components from the deep regions. Low-order components are buried in the noise, displaying no preference for any φ. New detection schemes employing low spatial coherence illumination can remove the high-order components and preserve the anisotropic low-order ones.

16.4.3 Sizing Spherical Dielectric Particles in a 3-D Random Cluster

Tseng et al. [54] reported the application of PSTD to model full-vector 3-D scattering of light by random clusters of spherical dielectric particles. A primary finding of this work was that the spectrum of scattered light from such a cluster contains information indicative of the size of the particles comprising the cluster, even for closely packed particles.

In their paper [54], Tseng et al. first described two validations of their 3-D PSTD modeling tool. The first validation, illustrated in Fig. 16.30(a), involves the monochromatic ($\lambda_0 = 0.75$ μm) plane-wave illumination of a single 8-μm-diameter dielectric sphere ($n = 1.2$) in free space. Here, the logarithm of the PSTD-computed differential scattering cross-section (DSCS) is graphed vs. the scattering angle and compared with the results of the corresponding analytical Mie expansion. The sphere had been modeled using a uniform 3-D PSTD lattice comprised of cubic 0.0833-μm ($\lambda_d/7.5$) space cells. Hence, the sphere's surface had been approximated in a staircased manner to this resolution. Despite the coarse resolution of the wavelength and the sphere-surface staircasing, the PSTD and Mie results are seen to agree very well over approximately 5 orders of magnitude for the complete range of scattering angles.

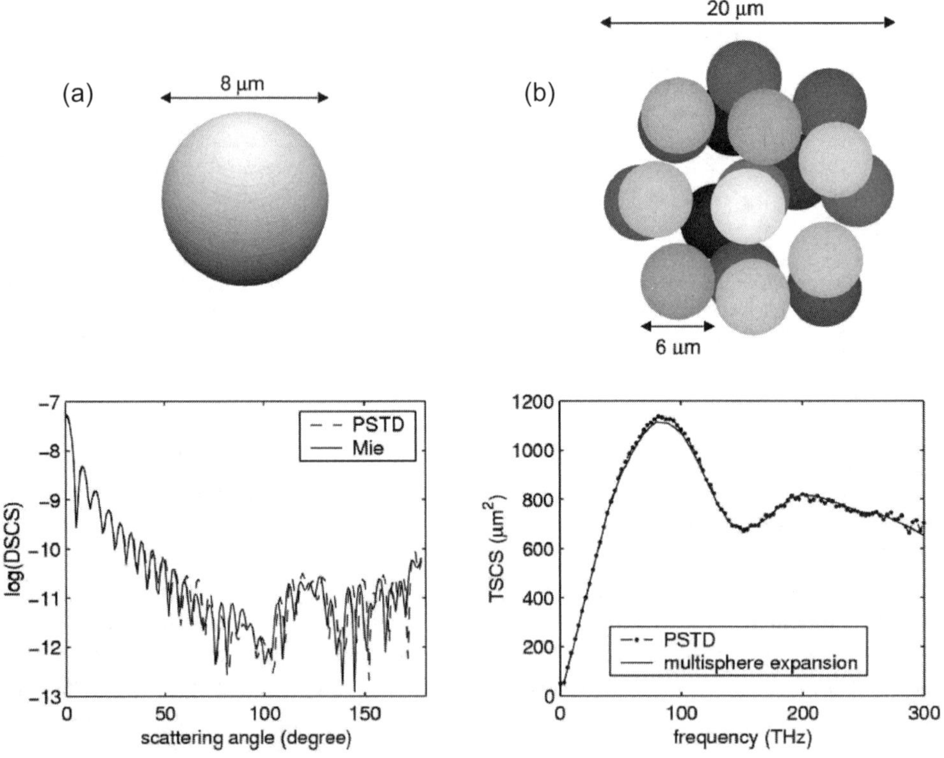

Fig. 16.30 Validations of the 3-D PSTD modeling tool. (a) Differential scattering cross-section of a single 8-μm-diameter dielectric sphere ($n = 1.2$) in free space for an incident wavelength $\lambda_0 = 0.75$ μm; (b) total scattering cross-section of a 20-μm-diameter cluster of 19 randomly positioned, non-contacting, 6-μm-diameter dielectric spheres ($n = 1.2$) in free space. *Source:* S. H. Tseng, A. Taflove, D. Maitland, and V. Backman, *Radio Science*, Vol. 41, 2006, RS4009, doi:10.1029/2005RS003408, ©2006 American Geophysical Union.

Figure 16.30(b) illustrates the second validation of the 3-D PSTD model [54]. Here, the PSTD-computed total scattering cross-section (TSCS) vs. frequency of a 20-μm-diameter cluster of 19 randomly positioned, non-contacting, d_0 = 6-μm-diameter dielectric spheres (n = 1.2) in free space is compared with a generalized multi-sphere Mie expansion. Here, each cubic PSTD space cell (and hence, each staircasing step of each sphere's surface) had been assigned a uniform size of 0.167 μm. From Fig. 16.30(b), we see that the PSTD and the multi-sphere Mie results agree very well for the complete range of frequencies investigated.

Having validated their 3-D PSTD modeling tool, Tseng et al. [54] then conducted a PSTD computational study of the TSCS spectra of three different clusters of spherical dielectric particles illuminated in free space. Each cluster had an overall spherical shape with a diameter of 25 μm. Within each cluster were closely packed N randomly positioned, non-contacting spherical particles (n = 1.2) of diameter d_0. A uniform PSTD space lattice resolution of 0.167 μm was employed for each model. Figure 16.31 illustrates the three cluster geometries: (a) N = 192 particles of diameter d_0 = 3 μm, optical thickness ~26; (b) N = 56 particles of diameter d_0 = 5 μm, optical thickness ~21; and (c) N = 14 particles of diameter d_0 = 7 μm, optical thickness ~10. Here, the optical thickness of each cluster is considered as being equivalent to the cluster diameter divided by the scattering mean free path.

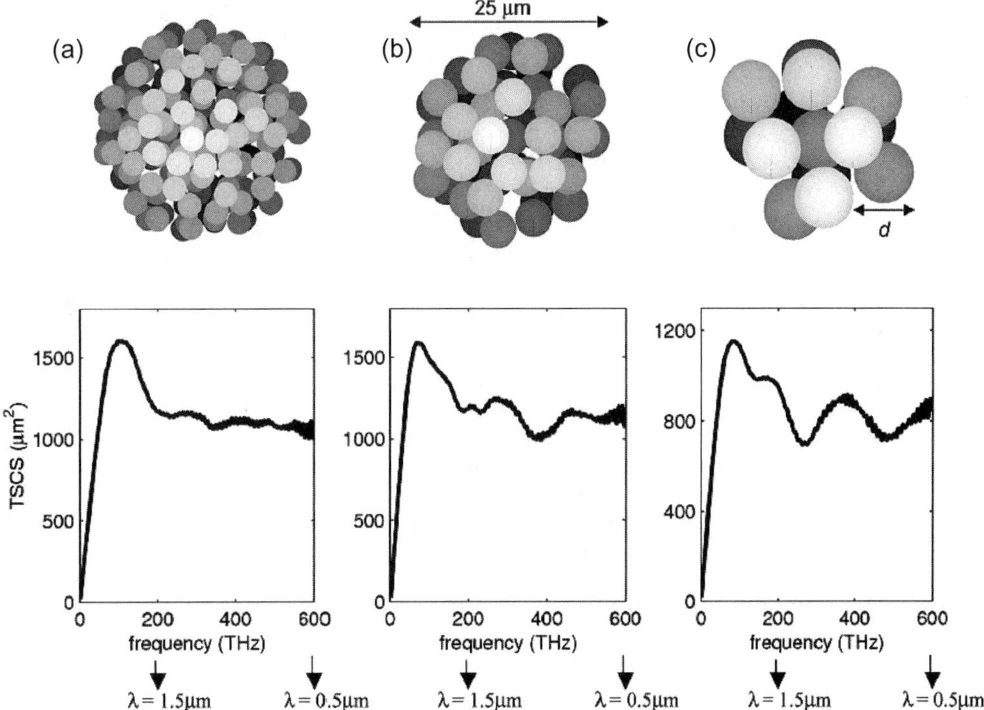

Fig. 16.31 PSTD-computed TSCS spectra of three different 25-μm-diameter clusters of N randomly positioned, closely packed, non-contacting, dielectric spheres (n = 1.2) of diameter d_0 in free space. (a) N = 192, d_0 = 3 μm, optical thickness ~26; (b) N = 56, d_0 = 5 μm, optical thickness ~21; (c) N = 14, d_0 = 7 μm, optical thickness ~10. *Source:* S. H. Tseng, A. Taflove, D. Maitland, and V. Backman, *Radio Science*, Vol. 41, 2006, RS4009, doi:10.1029/2005RS003408, ©2006 American Geophysical Union.

From Fig. 16.31, we observe that all three TSCS spectra are similar at wavelengths longer than ~3 μm. This suggests that for long wavelengths, incident light cannot discern microscopic structural differences among the three cluster geometries. However, for wavelengths shorter than ~1.5 μm, the TSCS spectra exhibit distinctive oscillatory features. Tseng et al. [54] offered the hypothesis that these oscillatory features could yield information regarding the diameter d_0 of the individual dielectric particles comprising each cluster, despite the close packing and mutual interaction of these particles.

To test this hypothesis, Tseng et al. [54] conducted a cross-correlation study of each TSCS spectrum of Fig. 16.31. Specifically, each TSCS spectrum of Fig. 16.31 was subjected to a cross-correlation with the TSCS spectrum of a single, isolated, dielectric spherical particle of trial diameter d. This cross-correlation was performed for several hundred trial values of d in the range of 2 to 10 μm, and the set of results was plotted as a function of d. The hypothesis would be strongly supported if the cross-correlation data were to peak exactly at $d = d_p = d_0$, the actual diameter of the constituent particles of the cluster. Figure 16.32 illustrates the results of this study [54].

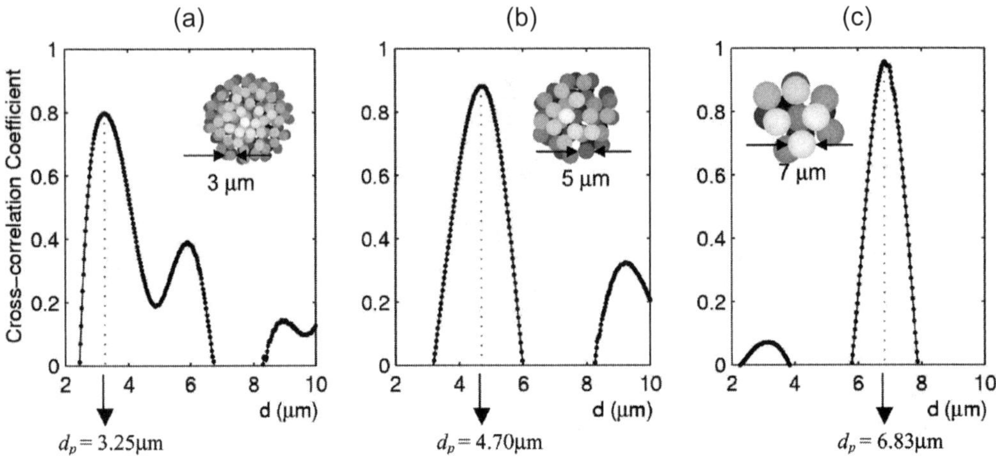

Fig. 16.32 Results of cross-correlating each TSCS spectrum shown in Fig. 16.31 with the TSCS spectrum of a single, isolated, dielectric spherical particle of trial diameter d. This cross-correlation was performed for several hundred trial values of d in the range of 2 to 10 μm, and the results were plotted as a function of d. In each case, the peak of the cross-correlation occurred at a trial diameter d_p approximately equal to the actual diameter d_0 of the particles comprising the cluster. *Source:* S. H. Tseng, A. Taflove, D. Maitland, and V. Backman, *Radio Science*, Vol. 41, 2006, RS4009, doi:10.1029/2005RS003408, ©2006 American Geophysical Union.

From Fig. 16.32, we observe that, despite the close packing of each cluster, the peak of its cross-correlation occurred at approximately the actual diameter of its constituent spherical particles: $d_p = 3.25$ μm vs. $d_0 = 3$ μm for cluster (a); $d_p = 4.70$ μm vs. $d_0 = 5$ μm for cluster (b); and $d_p = 6.83$ μm vs. $d_0 = 7$ μm for cluster (c). This supports the hypothesis that significant information regarding a cluster's constituent particles is embedded within the oscillatory features of its TSCS spectrum — even for optically thick clusters where the surface-to-surface spacing between adjacent constituent particles is less than the illuminating wavelength.

16.4.4 Optical Phase Conjugation for Turbidity Suppression

Tissue turbidity has been a formidable obstacle for optical tissue imaging and related applications. Multiply scattered photons are conventionally regarded as being random and stochastic in their trajectories. However, recent research indicates that such scattering is actually a causal and time-reversible process.

Recently, optical phase conjugation (OPC) has been experimentally demonstrated as a means to suppress tissue turbidity [55]. Because several aspects of the physical basis and application of OPC are not yet well understood, rigorous simulations are required to reveal information that cannot be easily obtained via laboratory experiments. PSTD is well suited for such simulations since it is based on the fundamental Maxwell's equations and furthermore can accommodate the macroscopic light-interaction regions which are involved in OPC.

Tseng and Yang reported the initial application of PSTD to simulate OPC [56]. Figure 16.33 illustrates the geometry of their 2-D PSTD simulation, which employed a uniform spatial resolution of 0.3 μm. Here, the incident light was a pulsed Gaussian beam with a cross-sectional width of 13.4 μm and a temporal duration of 4.472 fs. This beam underwent multiple scattering within a rectangular (560- × 260-μm) cluster of 2500 randomly positioned dielectric cylinders before reaching a phase-conjugate mirror. Each cylinder within the cluster had a 2.5-μm diameter and the refractive index n = 1.2.

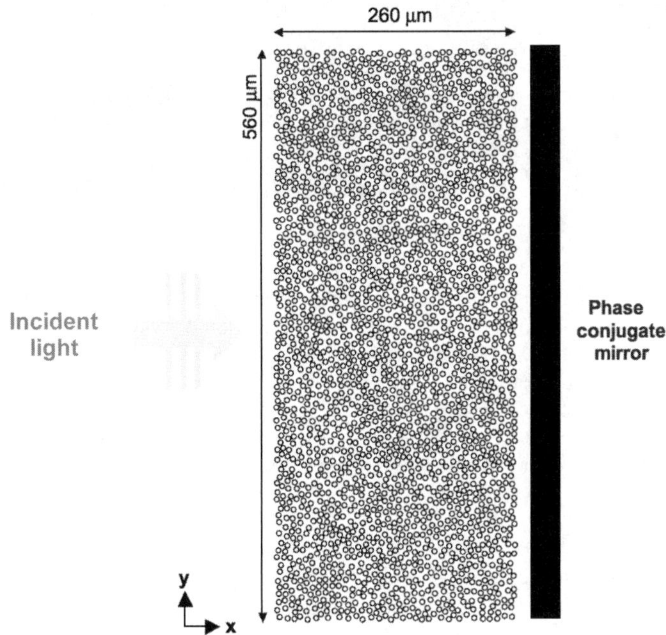

Fig. 16.33 PSTD OPC simulation geometry: rectangular (560- × 260-μm) cluster of 2500 randomly positioned 2.5-μm-diameter dielectric cylinders (n = 1.2) illuminated by a pulsed light beam. The light was multiply scattered within the cluster before reaching a phase-conjugate mirror. Then, the phase and propagation direction of the light were inverted, causing the light to propagate in the reverse direction and trace back to its origination point. *Source:* S. H. Tseng and C. Yang, *Optics Express*, Vol. 15, 2007, pp. 16005–16016, ©2007 The Optical Society of America.

Figure 16.34 visualizes the PSTD-computed evolution of the optical electric field for this model [56]. From this figure, we see that the wavefront of the incident light spread out due to diffraction as it propagated through the random cluster of dielectric cylinders. After reaching the phase-conjugate mirror, the phase and propagation direction of the impinging light were inverted, causing the light to propagate in the reverse direction and trace back to its origination point, where refocusing occurred. However, this refocusing effect was imperfect since some light was lost due to scattering out of the computation grid. This resulted in a wider and more reverberant refocused wavefront profile than the original. Nevertheless, the basic principle of turbidity compensation by employing a phase-conjugate mirror was well demonstrated.

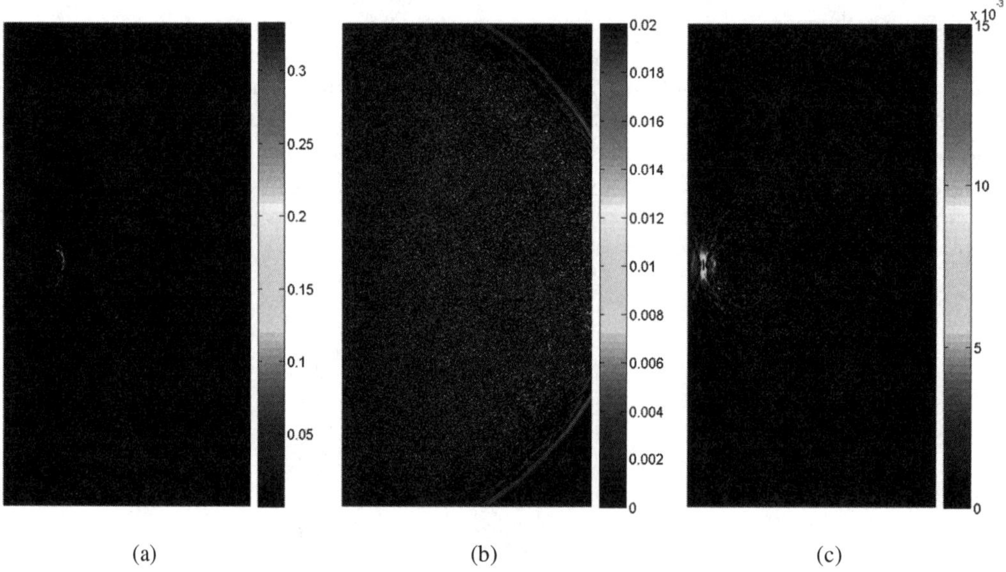

Fig. 16.34 PSTD simulation of optical phase conjugation for the scattering geometry of Fig. 16.33. The *E*-field at various time-steps throughout the evolution is shown: (a) 200 fs, (b) 1000 fs, and (c) 2400 fs. As light scatters through the cluster of dielectric cylinders, the wavefront spreads out due to diffraction [panel (b)]. After reflection by the phase-conjugate mirror at the far right side, the light back-traces and refocuses back to the original location where it first emerged [panel (c)]. *Source:* S. H. Tseng and C. Yang, *Optics Express*, Vol. 15, 2007, pp. 16005–16016, ©2007 The Optical Society of America.

In subsequent work, Tseng applied PSTD to investigate how the amplitude and cross-sectional width of the OPC-refocused light pulse vary with the optical thickness of the scattering medium [57]. In this study, he simulated light scattering by a cluster of lossless, randomly positioned dielectric cylinders of refractive index $n = 1.2$, number density $0.0186\ \mu\text{m}^{-2}$, and anisotropy factor $g = 0.85$. For a wavelength $\lambda = 1\ \mu\text{m}$, the reduced scattering coefficient was $\mu_s' = 0.0194\ \mu\text{m}^{-1}$. While maintaining a fixed number density, the thickness L of the scattering medium was varied from 40 to 320 μm — six times the transport mean-free-path at $\lambda = 1\ \mu\text{m}$.

Figure 16.35 depicts the PSTD-computed transverse profile of the OPC-refocused light pulse corresponding to this scattering medium for $L = 40$, 80, 120, 160, 200, 240, 280, and 320 μm [57]. We see that the amplitude and width of the transverse profile of the refocused light pulse exhibited only a slight sensitivity to the eight-fold increase of thickness L.

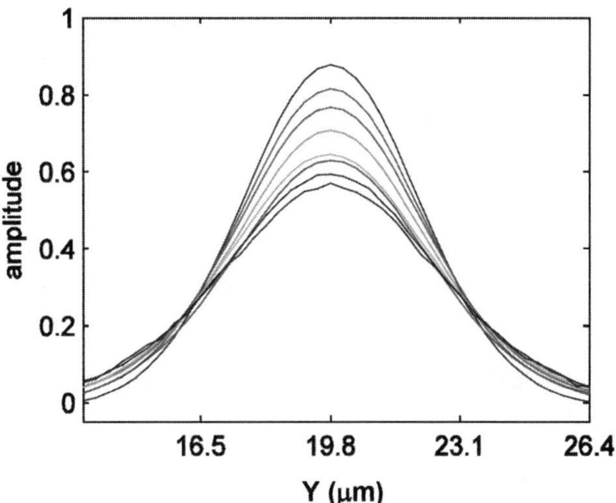

Fig. 16.35 PSTD-computed transverse profiles of the OPC-refocused light pulse for various thicknesses of the scattering medium. The top to bottom curves correspond to a scattering medium with a thickness of 40, 80, 120, 160, 200, 240, 280, and 320 μm, respectively. *Source:* S. H. Tseng, *IEEE J. Lightwave Technology*, Vol. 27, 2009, pp. 3919–3922, ©2009 IEEE.

Tseng also applied PSTD to simulate incident light pulses of various initial cross-sectional widths. Figure 16.36 plots the transverse width of the OPC-refocused light pulse vs. $\mu_s' \times L$, the optical thickness of the scattering medium measured in units of multiples of a single transport mean-free-path [57].

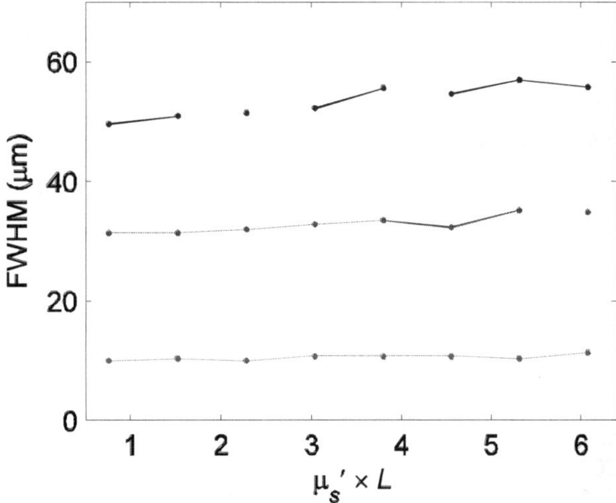

Fig. 16.36 PSTD-computed cross-sectional widths (FWHM) of OPC-refocused light pulses vs. the optical thickness $\mu_s' \times L$ of the scattering medium. Results are shown for incident light pulses of three initial cross-sectional widths: 9.32, 29.97, and 47.3 μm (bottom to top). *Source:* S. H. Tseng, *IEEE J. Lightwave Technology*, Vol. 27, 2009, pp. 3919–3922, ©2009 IEEE.

Figure 16.36 shows that the PSTD-computed full-width at half-maximum (FWHM) of the OPC-refocused light pulse displayed little sensitivity to the optical thickness of the scattering medium up to six times the transport mean-free-path [57]. The slowly decreasing intensity of the refocused pulse is a consequence of scattering rather than absorption effect of the turbid medium.

16.5 SUMMARY

This chapter reviewed qualitatively the technical basis and representative applications of FDTD and PSTD computational solutions of Maxwell's equations to biophotonics. The applications of FDTD highlighted in this chapter reveal its utility to provide ultrahigh-resolution models of optical interactions with individual biological cells, and furthermore to provide the physics kernel of advanced computational microscopy techniques.

The applications of PSTD highlighted in this chapter reveal its utility to model optical interactions with clusters of many biological cells and even macroscopic sections of biological tissues. This can help to achieve a better understanding of the physics of enhanced backscattering and turbidity suppression.

In all this, a key goal has been to alert and inform readers about how FDTD and PSTD can put Maxwell's equations to work in the analysis and design of a wide range of biophotonics technologies. The range of applications of these computational techniques to biophotonics will certainly grow as more researchers become aware of these powerful tools, and furthermore as computer capabilities continue to improve.

REFERENCES

[1] Yee, K. S., "Numerical solution of initial boundary value problems involving Maxwell's equations in isotropic media," *IEEE Trans. Antennas & Propagation*, Vol. 14, 1966, pp. 302–307.

[2] Taflove, A., and S. C. Hagness, *Computational Electrodynamics: The Finite-Difference Time-Domain Method*, 3rd ed., Norwood, MA: Artech, 2005.

[3] Liu, Q. H., "The PSTD algorithm: A time-domain method requiring only two cells per wavelength," *Microwave & Optical Technology Lett.*, Vol. 15, 1997, pp. 158–165.

[4] Ding, M., and K. Chen, "Staggered-grid PSTD on local Fourier basis and its applications to surface tissue modeling," *Optics Express*, Vol. 18, 2010, pp. 9236–9250.

[5] Chandrasekhar, S., *Radiative Transfer*, New York: Dover, 1960.

[6] Tseng, S. H., and B. Huang, "Comparing Monte Carlo simulation and pseudospectral time-domain numerical solutions of Maxwell's equations of light scattering by a macroscopic random medium," *Applied Physics Lett.*, Vol. 91, 2007, 051114.

[7] Marti-Lopez, L., J. Bouza-Dominguez, J. C. Hebden, S. R. Arridge, and R. A. Martinez-Celorio, "Validity conditions for the radiative transfer equation," *J. Optical Society of America A*, Vol. 20, 2003, pp. 2046–2056.

[8] Piket-May, M. J., A. Taflove, and J. B. Troy, "Electrodynamics of visible light interactions with the vertebrate retinal rod," *Optics Lett.*, Vol. 18, 1993, pp. 568–570.

[9] Dunn, A., and R. Richards-Kortum, "Three-dimensional computation of light scattering from cells," *IEEE J. Selected Topics in Quantum Electronics*, Vol. 2, 1996, pp. 898–905.

[10] Drezek, R., A. Dunn, and R. Richards-Kortum, "Light scattering from cells: Finite-difference time-domain simulations and goniometric measurements," *Applied Optics*, Vol. 38, 1999, pp. 3651–3661.

[11] Drezek, R., A. Dunn, and R. Richards-Kortum, "A pulsed finite-difference time-domain (FDTD) method for calculating light scattering from biological cells over broad wavelength ranges," *Optics Express*, Vol. 6, 2000, pp. 147–157.

[12] Drezek, R., M. Guillaud, T. Collier, I. Boiko, A. Malpica, C. Macaulay, M. Follen, and R. Richards-Kortum, "Light scattering from cervical cells throughout neoplastic progression: Influence of nuclear morphology, DNA content, and chromatin texture," *J. Biomedical Optics*, Vol. 8, 2003, pp. 7–16.

[13] Arifler, D., M. Guillaud, A. Carraro, A. Malpica, M. Follen, and R. Richards-Kortum, "Light scattering from normal and dysplastic cervical cells at different epithelial depths: Finite-difference time-domain modeling with a perfectly matched layer boundary condition," *J. Biomedical Optics*, Vol. 8, 2003, pp. 484–494.

[14] Roy, H. K., Y. Liu, R. K. Wali, Y. L. Kim, A. K. Kromin, M. J. Goldberg, and V. Backman, "Four-dimensional elastic light-scattering fingerprints as preneoplastic markers in the rat model of colon carcinogenesis," *Gastroenterology*, Vol. 126, 2004, pp. 1071–1081.

[15] Li, X., A. Taflove, and V. Backman, "Recent progress in exact and reduced-order modeling of light-scattering properties of complex structures," *IEEE J. Selected Topics in Quantum Electronics*, Vol. 11, 2005, pp. 759–765.

[16] Su, X.-T., K. Singh, W. Rozmus, C. Backhouse, and C. Capjack, "Light scattering characterization of mitochondrial aggregation in single cells," *Optics Express*, Vol. 17, 2009, pp. 13381–13388.

[17] Starosta, M. S., and A. K. Dunn, "Three-dimensional computation of focused beam propagation through multiple biological cells," *Optics Express*, Vol. 17, 2009, pp. 12455–12469.

[18] Hollmann, J. L., A. K. Dunn, and C. A. DiMarzio, "Computational microscopy in embryo imaging," *Optics Letters*, Vol. 29, 2004, pp. 2267–2269.

[19] Novotny, L., and B. Hecht, *Principles of Nano-Optics*, Cambridge, UK: Cambridge University Press, 2006.

[20] Simon, B., and C. A. DiMarzio, "Simulation of a theta line-scanning confocal microscope," *J. Biomedical Optics*, Vol. 12, 2007, pp. 064 020–1–9.

[21] Tanev, S., W. B. Sun, J. Pond, V. V. Tuchin, and V. P. Zharov, "Flow cytometry with gold nanoparticles and their clusters as scattering contrast agents: FDTD simulation of light-cell interaction," *J. Biophotonics*, Vol. 2, 2009, pp. 505–520.

[22] Capoglu, I. R., J. D. Rogers, A. Taflove, and V. Backman, "The microscope in a computer: Image synthesis from three-dimensional full-vector solutions of Maxwell's equations at the nanometer scale," *Progress in Optics*, Vol. 57, 2012, pp. 1–91.

[23] Fairchild, M. D., *Color Appearance Models*, New York: Wiley, 2005.

[24] Subramanian, H., P. Pradhan, Y. Liu, I. Capoglu, X. Li, J. Rogers, A. Heifetz, D. Kunte, H. Roy, A. Taflove, and V. Backman, "Optical methodology for detecting histologically unapparent nanoscale consequences of genetic alterations in biological cells," *Proc. National Academy of Sciences USA*, Vol. 105, 2008, pp. 20124–20129.

[25] Méndez-Ruiz, C., and J. J. Simpson, "Detection of embedded ultra-sub-wavelength-thin dielectric features using elongated photonic nanojets," *Optics Express*, Vol. 18, 2010, pp. 16805–16812.

[26] Méndez-Ruiz, C., and J. J. Simpson, "Cepstrum analysis of photonic nanojet-illuminated biological cells," *J. Applied Computational Electromagnetics Society*, Vol. 27, 2012, pp. 215–222.

[27] Chen, Z., A. Taflove, and V. Backman, "Photonic nanojet enhancement of backscattering of light by nanoparticles: A potential novel visible-light ultramicroscopy technique," *Optics Express*, Vol. 12, 2004, pp. 1214–1220.

[28] Li, X., Z. Chen, A. Taflove, and V. Backman, "Optical analysis of nanoparticles via enhanced backscattering facilitated by 3-D photonic nanojets," *Optics Express*, Vol. 13, 2005, pp. 526–533.

[29] Heifetz, A., S.-C. Kong, A. V. Sahakian, A. Taflove, and V. Backman, "Photonic nanojets," *J. Computational & Theoretical Nanoscience*, Vol. 6, 2009, pp. 1979–1992.

[30] Yang, S., A. Taflove, and V. Backman, "Experimental confirmation at visible light wavelengths of the backscattering enhancement phenomenon of the photonic nanojet," *Optics Express*, Vol. 19, 2011, pp. 7084–7093.

[31] Poco, J. F., and L. W. Hrubesh, "Method of producing optical quality glass having a selected refractive index," U.S. Patent 6,158,244, 2008.

[32] Nevels, R., and J. Jeong, "The time-domain Green's function and propagator for Maxwell's equations," *IEEE Trans. Antennas & Propagation*, Vol. 52, 2004, pp. 3012–3018.

[33] Bogert, B. P., M. J. R. Healy, J. W. Tukey, and M. Rosenblatt, "The quefrency analysis of time series for echoes: Cepstrum, pseudo-autocovariance, cross-cepstrum, and saphe cracking," Chapter 15 (pp. 209–243) in *Time Series Analysis*, M. Rosenblatt, ed., New York: Wiley, 1963.

[34] Richter, M., Personal communication to J. J. Simpson.

[35] Haley, S. B., and P. Erdos, "Wave propagation in one-dimensional disordered structures," *Physical Review B*, Vol. 45, 1992, pp. 8572–8584.

[36] Schmitt, J. M., and G. Kumar, "Optical scattering properties of soft tissue: A discrete particle model," *Applied Optics*, Vol. 37, 1998, pp. 2788–2797.

[37] Born, M., and E. Wolf, *Principles of Optics: Electromagnetic Theory of Propagation, Interference and Diffraction of Light*, 7th ed., Cambridge, UK: Cambridge University Press, 1999.

[38] Capoglu, I. R., J. D. Rogers, A. Taflove, and V. Backman, "Accuracy of the Born approximation in calculating the scattering coefficient of biological continuous random media," *Optics Letters*, Vol. 34, 2009, pp. 2679–2681.

[39] Ishimaru, A., *Wave Propagation and Scattering in Random Media*, New York: Wiley-IEEE Press, 1999.

[40] Kreiss, H. O., and J. Oliger, "Comparison of accurate methods for integration of hyperbolic equations," *Tellus*, Vol. 24, 1972, pp. 199–215.

[41] Orszag, S. A., "Comparison of pseudospectral and spectral approximation," *Studies in Applied Mathematics*, Vol. 51, 1972, pp. 253–259.

[42] Witte, D. C., and P. G. Richards, "The pseudospectral method for simulating wave propagation," pp. 1–18 in *Computational Acoustics, Vol. 3*, D. Lee, A. Cakmak, and R. Vichnevetsky, eds., New York: North-Holland, 1990.

[43] Fornberg, B., "Introduction to pseudospectral method via finite differences," Chap. 3 in *A Practical Guide to Pseudospectral Methods*, Cambridge, UK: Cambridge University Press, 1996.

[44] Roy, H. K., V. Turzhitsky, A. Gomes, M. J. Goldberg, J. D. Rogers, Y. L. Kim, T. K. Tsang, D. Shah, M. S. Borkar, M. Jameel, N. Hasabou, R. Brand, Z. Bogojevic, and V. Backman, "Prediction of colonic neoplasia through spectral marker analysis from the endoscopically normal rectum: An ex vivo and in vivo study," *Gastroenterology*, Vol. 134, No. 4, April 2008, P-1-P-332-Supplement 1, Article 751, p. A109.

[45] Subramanian, H., P. Pradhan, Y. L. Kim, and V. Backman, "Penetration depth of low-coherence enhanced backscattering of light in sub-diffusion regime," *Physical Review E*, Vol. 75, 2007, 041914.

[46] Kim, Y. L., V. M. Turzhitsky, Y. Liu, H. K. Roy, R. K. Wali, H. Subramanian, P. Pradhan, and V. Backman, "Low-coherence enhanced backscattering: Review of principles and applications for colon cancer screening," *J. Biomedical Optics*, Vol. 11, 2006, 041125-1-10.

[47] Tseng, S. H., A. Taflove, D. Maitland, V. Backman, and J. T. Walsh, "Simulation of enhanced backscattering of light by numerically solving Maxwell's equations without heuristic approximations," *Optics Express*, Vol. 13, 2005, pp. 3666–3672.

[48] Akkermans, E., P. E. Wolf, and R. Maynard, "Coherent backscattering of light by disordered media: Analysis of the peak line-shape," *Physical Review Lett.*, Vol. 56, 1986, pp. 1471–1474.

[49] van Albada, M. P., M. B. van der Mark, and A. Lagendijk, "Observation of weak localization of light in a finite slab: Anisotropy effects and light-path classification," *Physical Review Lett.*, Vol. 58, 1987, pp. 361–364.

[50] van Albada, M. P., M. B. van der Mark, and A. Lagendijk, "Polarisation effects in weak localisation of light," *J. Physics D: Applied Physics*, Vol. 21(10S), 1988, pp. S28–S31.

[51] Sawicki, J., N. Kastor, and M. Xu, "Electric field Monte Carlo simulation of coherent backscattering of polarized light by a turbid medium containing Mie scatterers," *Optics Express*, Vol. 16, 2008, pp. 5728–5738.

[52] Mishchenko, M. I., J. M. Dlugach, and L. Liu, "Azimuthal asymmetry of the coherent backscattering cone: Theoretical results," *Physical Review A*, Vol. 80, 2009, 053824.

[53] Ding, M., and K. Chen, "Numerical investigation on polarization characteristics of coherent enhanced backscattering using SL-PSTD," *Optics Express*, Vol. 18, 2010, pp. 27639–27649.

[54] Tseng, S. H., A. Taflove, D. Maitland, and V. Backman, "Pseudospectral time domain simulations of multiple light scattering in three-dimensional macroscopic random media," *Radio Science*, Vol. 41, 2006, RS4009, doi:10.1029/2005RS003408.

[55] Yaqoob, Z., D. Psaltis, M. S. Feld, and C. Yang, "Optical phase conjugation for turbidity suppression in biological samples," *Nature Photonics*, Vol. 2, 2008, pp. 110–115.

[56] Tseng, S. H., and C. Yang, "2-D PSTD simulation of optical phase conjugation for turbidity suppression," *Optics Express*, Vol. 15, 2007, pp. 16005–16016.

[57] Tseng, S. H., "Analysis of the cross-sectional width of the optical phase conjugation refocusing of light multiply scattered through macroscopic random media," *IEEE J. Lightwave Technology*, Vol. 27, 2009, pp. 3919–3922.

Chapter 17

GVADE FDTD Modeling of Spatial Solitons

Zachary Lubin, Jethro H. Greene, and Allen Taflove

17.1 INTRODUCTION

Designing photonic devices for emerging optical communications and computing applications requires understanding how light propagates and scatters in linear and nonlinear materials having wavelength-dependent and intensity-dependent properties. Due to the complexity of the underlying physics, inhomogeneous material compositions, and nanoscale feature sizes of future photonic devices, engineering design will require determining the optical electromagnetic field within these devices from first principles — the full-vector Maxwell's equations.

To this end, a recent extension of FDTD modeling, the *general vector auxiliary differential equation* (GVADE) method, has been reported [1]. Based on the time-domain Maxwell's curl equations, the GVADE FDTD method permits modeling the propagation of electromagnetic waves having multiple electric-field vector components in arbitrary inhomogeneous materials characterized by multiple wavelength-dependent and intensity-dependent polarizations.

This chapter derives the GVADE FDTD method and highlights recent applications to model spatial soliton propagation and scattering. First, previous analytical and computational approaches for nonlinear optics are reviewed, especially regarding spatial solitons. This is followed by a review of a classical mathematical treatment of nonlinear optics involving three distinct physical mechanisms contributing to the polarization term in the Maxwell–Ampere law. With the fundamentals established, the mathematical model and numerical algorithm of the GVADE FDTD method are then derived. The chapter concludes with illustrative applications of GVADE FDTD modeling including studies of the propagation of single narrow spatial solitons having both transverse and longitudinal electric field components; soliton–soliton interactions for various mutual phases of the optical carrier; soliton scattering by air gaps in the background medium; and soliton reflections from plasmonic media.

17.2 ANALYTICAL AND COMPUTATIONAL BACKGROUND

In recent years, there has been increased interest in the study of self-guided optical beams that propagate without confining waveguide structures. Such beams are commonly referred to as spatial optical solitons. Stable spatial optical solitons have been observed in planar material slabs where the slab provides confinement in one transverse dimension, and a self-focusing phenomenon due to the material nonlinearity provides confinement in the other transverse direction. An intriguing feature of spatial solitons is their particle-like behavior wherein two parallel in-phase (out-of-phase) solitons can attract (repel) each other. This behavior could potentially lead to the realization of all-optical logic circuitry having sub-picosecond switching times. The ability to understand and control the propagation and scattering of light in complex nonlinear materials is essential to realize such advanced technology.

To fully understand the electromagnetic wave phenomena responsible for spatial optical soliton propagation and interactions, it is necessary to model the vector effects inherent in the full set of Maxwell's equations. However, due to the complexity of solving these fundamental coupled partial differential equations, simplifying assumptions are often made. Specifically, the behavior of spatial solitons in cubic Kerr nonlinear materials is frequently described by the *nonlinear Schrödinger equation* (NLSE), a much simpler scalar wave equation [2]. Primarily, the paraxial approximation is made, which assumes that beams change their transverse profiles slowly during propagation relative to the carrier modulation, allowing separation of the dominant oscillating term from the remaining amplitude profile [2]. Furthermore, a scalar approximation is made wherein the electric field is assumed to be linearly polarized along a transverse coordinate, and the longitudinal electric-field component is neglected [3]. The NLSE is typically solved numerically for the field envelope using schemes such as the *split-step Fourier method* [4]. A rigorous derivation of the NLSE including higher-order terms is given in [5].

The simplifying assumptions regarding beam behavior in the NLSE draw into question its applicability to increasingly complex soliton interaction geometries. Furthermore, the NLSE is not directly applicable to realistic materials characterized by a multiplicity of linear and nonlinear polarizations.

The FDTD method provides an alternative approach to model the behavior of spatial optical solitons. FDTD is a direct time-domain solution of the full-vector Maxwell's curl equations with *no assumptions* other than its space-time sampling of the continuous optical electromagnetic field at the nanometer-attosecond scale. Furthermore, FDTD permits multiple linear and nonlinear dispersions to be incorporated in the model via the electric-field polarization for arbitrary inhomogeneous material geometries.

Early work by Goorjian et al. applied FDTD to model nonlinear optics and temporal solitons by incorporating material dispersion via the electric flux density, and by introducing auxiliary variables [6, 7]. This was subsequently extended by Nakamura et al. to include multiple poles in the chromatic linear Lorentz dispersion [8]. Kashiwa and Fukai proposed the auxiliary differential equation (ADE) method as a means to extend FDTD to incorporate polarization by time-stepping auxiliary differential equations concurrently with Maxwell's curl equations [9]. Using an efficient reformulation of the ADE method, which avoids the need to solve a system of equations at each time-step, Fujii et al. modeled propagation in a Kerr medium with Debye and Lorentz linear dispersions and the Raman nonlinear dispersion [10].

The GVADE FDTD method extends the technique of Fujii et al. to model nonlinear optics problems where the electric field has two or three orthogonal vector components. GVADE emphasizes the polarization current density $\boldsymbol{J}$ rather than the polarization $\boldsymbol{P}$, thereby bypassing the need for storing and updating the electric flux density $\boldsymbol{D}$.

17.3 MAXWELL–AMPERE LAW TREATMENT OF NONLINEAR OPTICS

This section reviews the mathematical foundation of the GVADE FDTD method: the Maxwell–Ampere law with associated polarization current density terms arising from three distinct linear and nonlinear dielectric processes occurring at optical wavelengths. Additional theory on nonlinear optics and solitons is described by various references, for example, [2, 4, 5, 11].

Maxwell's equations in differential form are:

$$\nabla \times \boldsymbol{H} = \frac{\partial \boldsymbol{D}}{\partial t} \tag{17.1}$$

$$\nabla \times \boldsymbol{E} = -\mu \frac{\partial \boldsymbol{H}}{\partial t} \tag{17.2}$$

where $\boldsymbol{H}$ and $\boldsymbol{E}$ are the instantaneous magnetic and electric vector fields, respectively, μ is the magnetic permeability (assumed to be a constant in this chapter), and $\boldsymbol{D}$ is the electric flux density. The right-hand side of (17.1) can be rewritten in terms of the polarization current density, $\boldsymbol{J}_{\mathrm{p}}$, which is the time-derivative of the electric polarization vector $\boldsymbol{P}$:

$$\frac{\partial \boldsymbol{D}}{\partial t} = \frac{\partial(\varepsilon_0 \varepsilon_\infty \boldsymbol{E} + \boldsymbol{P})}{\partial t} = \varepsilon_0 \varepsilon_\infty \frac{\partial \boldsymbol{E}}{\partial t} + \frac{\partial \boldsymbol{P}}{\partial t} = \varepsilon_0 \varepsilon_\infty \frac{\partial \boldsymbol{E}}{\partial t} + \boldsymbol{J}_{\mathrm{p}} \tag{17.3}$$

where we allow for a non-unity high-frequency relative permittivity, ε_∞. This yields:

$$\nabla \times \boldsymbol{H} = \varepsilon_0 \varepsilon_\infty \frac{\partial \boldsymbol{E}}{\partial t} + \boldsymbol{J}_{\mathrm{p}} \tag{17.4}$$

Let the polarization current density be decomposed into three components:

$$\boldsymbol{J}_{\mathrm{p}} = \boldsymbol{J}_{\mathrm{Lorentz}} + \boldsymbol{J}_{\mathrm{Raman}} + \boldsymbol{J}_{\mathrm{Kerr}} \tag{17.5}$$

where $\boldsymbol{J}_{\mathrm{Lorentz}}$ is the polarization current density due to the linear Lorentz dispersion, $\boldsymbol{J}_{\mathrm{Raman}}$ is due to the nonlinear Raman dispersion, and $\boldsymbol{J}_{\mathrm{Kerr}}$ is due to the instantaneous third-order electronic nonlinearity.

In (17.5), $\boldsymbol{J}_{\mathrm{Lorentz}}$ arises from the frequency dependence of a material's linear susceptibility, $\tilde{\chi}_0^{(1)}(\omega)$. This dependence can be expressed as a summation of resonance terms, where each term corresponds to the polarization due to a single pole of the Sellmeier expansion. Following this construction, the sinusoidal steady-state electric polarization $\tilde{\boldsymbol{P}}_{\mathrm{Lorentz}}$ can be expressed as a function of frequency ω in terms of the three-pole Sellmeier expansion [11]:

$$\tilde{\boldsymbol{P}}_{\mathrm{Lorentz}} = \varepsilon_0 \, \tilde{\chi}_0^{(1)}(\omega) \tilde{\boldsymbol{E}} = \varepsilon_0 \tilde{\boldsymbol{E}} \sum_{l=1}^{3} \frac{\beta_l \omega_l^2}{\omega_l^2 - \omega^2} \tag{17.6}$$

where β_l is the strength of the lth resonance, ω_l is the frequency of the lth resonance, and each tilde denotes a complex-valued, frequency-domain quantity. Differentiating (17.6) with respect to time results in the following frequency-domain expression for the polarization current density corresponding to the lth term of the Sellmeier expansion of the linear Lorentz dispersion:

$$\tilde{\boldsymbol{J}}_{\mathrm{Lorentz}_l} = \varepsilon_0 \, \beta_l \omega_l^2 \left(\frac{j\omega}{\omega_l^2 - \omega^2} \right) \tilde{\boldsymbol{E}} \tag{17.7}$$

In (17.5), $\boldsymbol{J}_{\text{Raman}}$ arises from Raman scattering caused by the conversion to vibrational energy of the energy of photons involved in inelastic collisions with molecules. The scattered photons have a lower energy, and therefore a lower frequency, than the incident photons. Raman dispersion has a significant effect on soliton integrity for ultra-short pulses on the order of 1 ps or less [4]. Wavelength shifts from Raman scattering can contribute new components in certain regions of the frequency band, modifying the pulse shape.

In (17.5), $\boldsymbol{J}_{\text{Kerr}}$ arises from the Kerr nonlinearity introduced by the third-order susceptibility. This response is assumed to be instantaneous and proportional to the cube of the electric field.

To formulate a mathematical model for $\boldsymbol{J}_{\text{Raman}}$ and $\boldsymbol{J}_{\text{Kerr}}$, consider the general definition of the time-domain electric polarization $\boldsymbol{P}$ for an isotropic medium, where the response is localized in space:

$$\begin{aligned} \boldsymbol{P} &= \varepsilon_0 \int_{-\infty}^{\infty} \chi_0^{(1)}(t-t')\boldsymbol{E}(\boldsymbol{r},t')\,dt' \\ &+ \varepsilon_0 \int_{-\infty}^{\infty}\int_{-\infty}^{\infty}\int_{-\infty}^{\infty} \chi_0^{(3)}(t-t_1, t-t_2, t-t_3) \vdots \boldsymbol{E}(\boldsymbol{r},t_1)\boldsymbol{E}(\boldsymbol{r},t_2)\boldsymbol{E}(\boldsymbol{r},t_3)\,dt_1\,dt_2\,dt_3 \end{aligned} \tag{17.8}$$

The second term in (17.8) is the nonlinear contribution to the polarization. It involves a tensor product and is denoted as $\boldsymbol{P}_{\text{NL}}$. For a simple model of the electron response accounting for nonresonant and incoherent nonlinear effects, $\boldsymbol{P}_{\text{NL}}$ can be described by the Born–Oppenheimer approximation [12]:

$$\boldsymbol{P}_{\text{NL}}(\boldsymbol{r},t) = \varepsilon_0 \chi_0^{(3)} \boldsymbol{E}(\boldsymbol{r},t) \int_{-\infty}^{\infty} g(t-\tau)\left|\boldsymbol{E}(\boldsymbol{r},\tau)\right|^2 d\tau \tag{17.9}$$

where $\chi_0^{(3)}$ is the strength of the third-order nonlinearity; the polarization is assumed to lie in the direction of the electric field; and the causal response function $g(t)$ is normalized so that:

$$\int_{-\infty}^{\infty} g(t)\,dt = 1 \tag{17.10}$$

To model the Kerr and Raman polarizations, $g(t)$ is given by [13]:

$$g(t) = \alpha\delta(t) + (1-\alpha)g_{\text{Raman}}(t) \tag{17.11}$$

where α is a real-valued constant in the range $0 \le \alpha \le 1$ that parameterizes the relative strengths of the Kerr and Raman contributions; $\delta(t)$ is a Dirac delta function that models Kerr nonresonant virtual electronic transitions on the order of 1 fs or less; and $g_{\text{Raman}}(t)$ is given by:

$$g_{\text{Raman}}(t) = \left(\frac{\tau_1^2 + \tau_2^2}{\tau_1\tau_2^2}\right) e^{-t/\tau_2} \sin(t/\tau_1)\, U(t) \tag{17.12}$$

where $U(t)$ is the Heaviside unit step function. Effectively, $g_{\text{Raman}}(t)$ models the impulse response of a single Lorentzian relaxation centered on the optical phonon frequency $1/\tau_1$ and having a bandwidth of $1/\tau_2$, the reciprocal phonon lifetime.

From (17.9) and (17.11), the Kerr polarization current density is given by:

$$\begin{aligned} \boldsymbol{J}_{\text{Kerr}}(t) &= \frac{\partial}{\partial t}\boldsymbol{P}_{\text{Kerr}}(t) = \frac{\partial}{\partial t}\left[\varepsilon_0\,\chi_0^{(3)}\alpha\,\boldsymbol{E}(t)\int_{-\infty}^{\infty}\delta(t-\tau)\left|\boldsymbol{E}(\tau)\right|^2 d\tau\right] \\ &= \frac{\partial}{\partial t}\left[\varepsilon_0\,\chi_0^{(3)}\alpha\left|\boldsymbol{E}(t)\right|^2\boldsymbol{E}(t)\right] \end{aligned} \tag{17.13}$$

and the Raman polarization current density is given by:

$$\boldsymbol{J}_{\text{Raman}}(t) = \frac{\partial}{\partial t}\boldsymbol{P}_{\text{Raman}}(t) = \frac{\partial}{\partial t}\left[\varepsilon_0\,\chi_0^{(3)}(1-\alpha)\boldsymbol{E}(t)\int_{-\infty}^{\infty}g_{\text{Raman}}(t-\tau)\left|\boldsymbol{E}(\tau)\right|^2 d\tau\right] \tag{17.14}$$

where the spatial dependence has been suppressed for ease of notation. This can be rewritten as the convolution:

$$\boldsymbol{J}_{\text{Raman}}(t) = \frac{\partial}{\partial t}\boldsymbol{P}_{\text{Raman}}(t) = \frac{\partial}{\partial t}\left\{\varepsilon_0\,\boldsymbol{E}(t)\left[\chi_{\text{Raman}}^{(3)}(t) * \left|\boldsymbol{E}(t)\right|^2\right]\right\} \tag{17.15a}$$

where

$$\chi_{\text{Raman}}^{(3)}(t) = \chi_0^{(3)}(1-\alpha)\,g_{\text{Raman}}(t) \tag{17.15b}$$

17.4 GENERAL VECTOR AUXILIARY DIFFERENTIAL EQUATION METHOD

Following [1], this section develops the auxiliary differential equations and corresponding time-stepping expressions that model all of the linear and nonlinear material dispersions described in Section 17.3. This time-stepping process is linked to the basic FDTD Maxwell's curl equations algorithm to yield updates for all electromagnetic fields and auxiliary variables.

17.4.1 Lorentz Linear Dispersion

Starting with (17.7), we first multiply both sides by $(\omega_l^2 - \omega^2)$, yielding:

$$\omega_l^2\tilde{\boldsymbol{J}}_{\text{Lorentz}_l} - \omega^2\tilde{\boldsymbol{J}}_{\text{Lorentz}_l} = \varepsilon_0\,\beta_l\,\omega_l^2\,j\omega\tilde{\boldsymbol{E}} \tag{17.16}$$

Transforming to the time domain, we obtain:

$$\omega_l^2\boldsymbol{J}_{\text{Lorentz}_l} + \frac{\partial^2\boldsymbol{J}_{\text{Lorentz}_l}}{\partial t^2} = \varepsilon_0\,\beta_l\,\omega_l^2\frac{\partial\boldsymbol{E}}{\partial t} \tag{17.17}$$

Applying finite-difference approximations for the time-derivatives centered at time-step n, (17.17) becomes:

$$\omega_l^2 \boldsymbol{J}^n_{\text{Lorentz}_l} + \frac{\boldsymbol{J}^{n+1}_{\text{Lorentz}_l} - 2\boldsymbol{J}^n_{\text{Lorentz}_l} + \boldsymbol{J}^{n-1}_{\text{Lorentz}_l}}{(\Delta t)^2} = \varepsilon_0 \beta_l \omega_l^2 \frac{(\boldsymbol{E}^{n+1} - \boldsymbol{E}^{n-1})}{2\Delta t} \tag{17.18}$$

where a semi-implicit approximation is invoked for the time-derivative of the electric field [14]. Solving for $\boldsymbol{J}^{n+1}_{\text{Lorentz}_l}$ yields:

$$\boldsymbol{J}^{n+1}_{\text{Lorentz}_l} = \alpha_l \boldsymbol{J}^n_{\text{Lorentz}_l} - \boldsymbol{J}^{n-1}_{\text{Lorentz}_l} + \gamma_l \frac{(\boldsymbol{E}^{n+1} - \boldsymbol{E}^{n-1})}{2\Delta t} \tag{17.19a}$$

where

$$\alpha_l = 2 - \omega_l^2 (\Delta t)^2 , \qquad \gamma_l = \varepsilon_0 \beta_l \omega_l^2 (\Delta t)^2 \tag{17.19b}$$

To obtain a semi-implicit updating relation for this polarization current density at time-step $n+1/2$, its values at $n+1$ and n are averaged:

$$\boldsymbol{J}^{n+1/2}_{\text{Lorentz}_l} = \frac{\boldsymbol{J}^{n+1}_{\text{Lorentz}_l} + \boldsymbol{J}^n_{\text{Lorentz}_l}}{2} \tag{17.20}$$

Substituting this into (17.19a) yields:

$$\boldsymbol{J}^{n+1/2}_{\text{Lorentz}_l} = \frac{1}{2}\left[(1+\alpha_l) \boldsymbol{J}^n_{\text{Lorentz}_l} - \boldsymbol{J}^{n-1}_{\text{Lorentz}_l} + \gamma_l \frac{(\boldsymbol{E}^{n+1} - \boldsymbol{E}^{n-1})}{2\Delta t} \right] \tag{17.21}$$

17.4.2 Kerr Nonlinearity

The update equation for the polarization current density due to the Kerr nonlinearity is obtained by approximating (17.13) with a temporal finite-difference centered around time-step $n+1/2$. This results in the semi-implicit updating relation:

$$\boldsymbol{J}^{n+1/2}_{\text{Kerr}} = \frac{\alpha \varepsilon_0 \chi_0^{(3)}}{\Delta t} \left(\left|\boldsymbol{E}^{n+1}\right|^2 \boldsymbol{E}^{n+1} - \left|\boldsymbol{E}^n\right|^2 \boldsymbol{E}^n \right) \tag{17.22}$$

17.4.3 Raman Nonlinear Dispersion

Equation (17.15a) contains a convolution. For convenience a new scalar variable $S(t)$ is introduced:

$$S(t) = \chi^{(3)}_{\text{Raman}}(t) * \left|\boldsymbol{E}(t)\right|^2 \tag{17.23}$$

This can be written in the frequency domain as:

$$\tilde{S}(\omega) = \tilde{\chi}^{(3)}_{\text{Raman}}(\omega) \cdot \mathcal{F}\left[\left|\boldsymbol{E}(t)\right|^2 \right] \tag{17.24}$$

where $\mathcal{F}$ denotes the Fourier transform. The third-order susceptibility due to the Raman nonlinearity can be written in the frequency domain as:

$$\tilde{\chi}^{(3)}_{\text{Raman}}(\omega) = \frac{(1-\alpha)\,\chi_0^{(3)}\,\omega_{\text{Raman}}^2}{\omega_{\text{Raman}}^2 + j2\omega\,\delta_{\text{Raman}} - \omega^2} \tag{17.25}$$

where

$$\omega_{\text{Raman}} = \sqrt{\frac{\tau_1^2 + \tau_2^2}{\tau_1^2\,\tau_2^2}} \tag{17.26a}$$

$$\delta_{\text{Raman}} = \frac{1}{\tau_2} \tag{17.26b}$$

Substituting (17.25) into (17.24) results in:

$$\tilde{S}(\omega) = \frac{(1-\alpha)\,\chi_0^{(3)}\,\omega_{\text{Raman}}^2}{\omega_{\text{Raman}}^2 + j2\omega\,\delta_{\text{Raman}} - \omega^2} \cdot \mathcal{F}\left[\left|\boldsymbol{E}(t)\right|^2\right] \tag{17.27}$$

Multiplying through by the denominator and expanding yields:

$$\omega_{\text{Raman}}^2\,\tilde{S}(\omega) + j2\omega\,\delta_{\text{Raman}}\,\tilde{S}(\omega) - \omega^2\tilde{S}(\omega) = (1-\alpha)\,\chi_0^{(3)}\,\omega_{\text{Raman}}^2 \cdot \mathcal{F}\left[\left|\boldsymbol{E}(t)\right|^2\right] \tag{17.28}$$

The corresponding auxiliary differential equation is obtained by transforming (17.28) to the time domain. This yields:

$$\omega_{\text{Raman}}^2\,S(t) + 2\delta_{\text{Raman}}\frac{\partial S(t)}{\partial t} + \frac{\partial^2 S(t)}{\partial t^2} = (1-\alpha)\,\chi_0^{(3)}\,\omega_{\text{Raman}}^2\left|\boldsymbol{E}(t)\right|^2 \tag{17.29}$$

Upon substituting $S(t)$ into (17.15a), we approximate the resulting expression with a temporal finite difference centered about time-step $n+1/2$. This yields the semi-implicit updating relation:

$$\boldsymbol{J}_{\text{Raman}}^{n+1/2} = \frac{\varepsilon_0}{\Delta t}\left(\boldsymbol{E}^{n+1}S^{n+1} - \boldsymbol{E}^{n}S^{n}\right) \tag{17.30}$$

To obtain an explicit time-stepping relation for S, we first approximate (17.29) with temporal finite differences centered about time-step n:

$$\omega_{\text{Raman}}^2\,S^n + 2\delta_{\text{Raman}}\left(\frac{S^{n+1} - S^{n-1}}{2\Delta t}\right) + \left[\frac{S^{n+1} - 2S^n + S^{n-1}}{(\Delta t)^2}\right] = (1-\alpha)\,\chi_0^{(3)}\omega_{\text{Raman}}^2\left|\boldsymbol{E}^n\right|^2 \tag{17.31}$$

Multiplying (17.31) by $2(\Delta t)^2$ and solving for S^{n+1} yields the desired explicit time-stepping relation for S:

$$S^{n+1} = \left[\frac{2 - \omega_{\text{Raman}}^2(\Delta t)^2}{\delta_{\text{Raman}}\Delta t + 1}\right]S^n + \left[\frac{\delta_{\text{Raman}}\Delta t - 1}{\delta_{\text{Raman}}\Delta t + 1}\right]S^{n-1} + \left[\frac{(1-\alpha)\,\chi_0^{(3)}\,\omega_{\text{Raman}}^2(\Delta t)^2}{\delta_{\text{Raman}}\Delta t + 1}\right]\left|\boldsymbol{E}^n\right|^2 \tag{17.32}$$

17.4.4 Solution for the Electric Field

Now, all of the polarization current density terms are known at time-step $n+1/2$. Upon applying a finite-difference approximation for the time-derivative in (17.4) centered at this time-step, and assuming that the high-frequency relative permittivity, ε_∞, is unity, we obtain:

$$\nabla \times \boldsymbol{H}^{n+1/2} = \frac{\varepsilon_0}{\Delta t}\left(\boldsymbol{E}^{n+1} - \boldsymbol{E}^{n}\right) + \sum_{l=1}^{3} \boldsymbol{J}^{n+1/2}_{\text{Lorentz}_l} + \boldsymbol{J}^{n+1/2}_{\text{Kerr}} + \boldsymbol{J}^{n+1/2}_{\text{Raman}} \tag{17.33}$$

Because (17.33) involves vector quantities having dependencies between the vector components, solving for the electric field requires solving a system of coupled nonlinear equations. This system is solved using a multidimensional numerical root-finding algorithm by introducing a vector objective function, $[X\ Y\ Z]^{\text{T}}$, and then minimizing this function via an iterative process.

Substituting the expressions for $\boldsymbol{J}^{n+1/2}_{\text{Lorentz}_l}$, $\boldsymbol{J}^{n+1/2}_{\text{Kerr}}$, and $\boldsymbol{J}^{n+1/2}_{\text{Raman}}$ derived in (17.21), (17.22), and (17.30), respectively, we obtain the following vector objective function for (17.33) [1]:

$$\begin{aligned}
[X\ Y\ Z]^{\text{T}} &= -\nabla \times \boldsymbol{H}^{n+1/2} + \frac{\varepsilon_0}{\Delta t}\left(\boldsymbol{E}^{n+1} - \boldsymbol{E}^{n}\right) \\
&\quad + \frac{1}{2}\sum_{l=1}^{3}\left[(1+\alpha_l)\boldsymbol{J}^{n}_{\text{Lorentz}_l} - \boldsymbol{J}^{n-1}_{\text{Lorentz}_l} + \gamma_l\left(\frac{\boldsymbol{E}^{n+1} - \boldsymbol{E}^{n-1}}{2\Delta t}\right)\right] \\
&\quad + \frac{\alpha\varepsilon_0\chi_0^{(3)}}{\Delta t}\left(\left|\boldsymbol{E}^{n+1}\right|^2\boldsymbol{E}^{n+1} - \left|\boldsymbol{E}^{n}\right|^2\boldsymbol{E}^{n}\right) + \frac{\varepsilon_0}{\Delta t}\left(\boldsymbol{E}^{n+1}S^{n+1} - \boldsymbol{E}^{n}S^{n}\right)
\end{aligned} \tag{17.34}$$

A multidimensional Newton's method can be used to solve this system of equations [1, 15]. Although the method reported in [1] and reviewed here is general and can be applied in three dimensions, only the two-dimensional (2-D) transverse-magnetic (TM) case with the fields (E_x, E_y, H_z) has been considered to date. For this case, the objective function is $[X\ Y]^{\text{T}}$. To iterate and refine guesses for Newton's method, we define $G_x^{[g+1]}$ and $G_y^{[g+1]}$ to be, respectively, the guesses for E_x^{n+1} and E_y^{n+1} at iteration $[g+1]$. These are derived from the previous guesses at iteration $[g]$ as [1]:

$$\begin{bmatrix} G_x^{[g+1]} \\ G_y^{[g+1]} \end{bmatrix} = \begin{bmatrix} G_x^{[g]} \\ G_y^{[g]} \end{bmatrix} - \left(\mathbf{M}^{-1}\cdot\begin{bmatrix} X \\ Y \end{bmatrix}\right)\Bigg|^{[g]} \tag{17.35}$$

where each term on the right-hand side is evaluated using values from the guess at iteration $[g]$, and $\mathbf{M}$ is the 2×2 Jacobian matrix with elements:

$$\text{M}_{11} = \frac{1}{4\Delta t}\left(\gamma_1+\gamma_2+\gamma_3\right) + \frac{\varepsilon_0}{\Delta t}\left\{1 + \alpha\chi_0^{(3)}\left[3\left(G_x^{[g]}\right)^2 + \left(G_y^{[g]}\right)^2\right] + S^{n+1}\right\} \tag{17.36a}$$

$$\text{M}_{22} = \frac{1}{4\Delta t}\left(\gamma_1+\gamma_2+\gamma_3\right) + \frac{\varepsilon_0}{\Delta t}\left\{1 + \alpha\chi_0^{(3)}\left[\left(G_x^{[g]}\right)^2 + 3\left(G_y^{[g]}\right)^2\right] + S^{n+1}\right\} \tag{17.36b}$$

$$\text{M}_{21} = \text{M}_{12} = \frac{2\varepsilon_0}{\Delta t}\alpha\chi_0^{(3)}G_x^{[g]}G_y^{[g]} \tag{17.36c}$$

where the γ_l are constants defined in (17.19b), and S^{n+1} is obtained via the explicit time-stepping relation (17.32). Iterations are repeated until the objective function is sufficiently close to zero. For efficient convergence, a numerical friction or dispersion can be added to reduce the distance between successive guesses if a solution is not found after a small number of iterations.

The complete iteration sequence is as follows. Starting from zero initial conditions, $\boldsymbol{H}^{n+1/2}$ is obtained via a standard explicit temporal central-difference realization of (17.2). Then, S^{n+1} is obtained explicitly via (17.32). Then, $\boldsymbol{E}^{n+1}$ is obtained via the iterative procedure of (17.35). Then, $\boldsymbol{J}^{n+1}_{\text{Lorentz}_l}$ is obtained explicitly via (17.19). The marching-in-time cycle then begins again.

17.4.5 Drude Linear Dispersion for Metals at Optical Wavelengths

Metals can exhibit dielectric-like effects at optical wavelengths and possess linear chromatic dispersion. This dispersion is most simply approximated using a single-pole Drude model. As shown in [14], the Drude dispersion can be readily incorporated into the traditional FDTD time-stepping of Maxwell's curl equations using an ADE scheme. This scheme, summarized below, results in a time-stepping algorithm that can also be conducted in parallel with the GVADE field updates of Section 17.4.4.

The complex-valued permittivity as a function of frequency for the single-pole Drude model is given by:

$$\tilde{\varepsilon}(\omega) = \varepsilon_0\left(\varepsilon_\infty - \frac{\omega_\text{p}^2}{\omega^2 - j\omega\gamma}\right) \tag{17.37}$$

where ω_p is the plasma frequency, ε_∞ is the high-frequency relative permittivity, and γ is the inverse of the free-electron-gas relaxation time. These parameters have been reported for multiple metals and wavelengths [16, 17]. The frequency-domain polarization current density $\tilde{\boldsymbol{J}}_\text{D}$ corresponding to (17.37) is:

$$\tilde{\boldsymbol{J}}_\text{D} = -j\omega\varepsilon_0\left(\frac{\omega_\text{p}^2}{\omega^2 - j\omega\gamma}\right)\tilde{\boldsymbol{E}} \tag{17.38}$$

Upon multiplying both sides of (17.38) by $\omega^2 - j\omega\gamma$, then dividing through by $-j\omega$, and finally transforming to the time domain, we obtain the desired auxiliary differential equation for $\boldsymbol{J}_\text{D}$:

$$\frac{\partial \boldsymbol{J}_\text{D}}{\partial t} + \gamma \boldsymbol{J}_\text{D} = \varepsilon_0\omega_\text{p}^2\boldsymbol{E} \tag{17.39}$$

Equation (17.39) is approximated with a temporal finite difference centered about time-step $n+1/2$, yielding the semi-implicit expression:

$$\left(\frac{\boldsymbol{J}_\text{D}^{n+1} - \boldsymbol{J}_\text{D}^{n}}{\Delta t}\right) + \gamma\left(\frac{\boldsymbol{J}_\text{D}^{n+1} + \boldsymbol{J}_\text{D}^{n}}{2}\right) = \varepsilon_0\omega_\text{p}^2\left(\frac{\boldsymbol{E}^{n+1} + \boldsymbol{E}^{n}}{2}\right) \tag{17.40}$$

Solving for the polarization current density at time-step $n+1$, we obtain:

$$\boldsymbol{J}_\text{D}^{n+1} = k\boldsymbol{J}_\text{D}^{n} + \beta\left(\boldsymbol{E}^{n+1} + \boldsymbol{E}^{n}\right) \tag{17.41}$$

where

$$k = \frac{1 - \gamma\Delta t/2}{1 + \gamma\Delta t/2}, \qquad \beta = \frac{\varepsilon_0 \omega_{\mathrm{p}}^2 \Delta t/2}{1 + \gamma\Delta t/2} \tag{17.42a, b}$$

We next use the results of (17.41) to provide the polarization current density term in (17.4). This yields the following expression for $\nabla \times \boldsymbol{H}$ evaluated at time-step $n+1/2$:

$$\begin{aligned} \nabla \times \boldsymbol{H}^{n+1/2} &= \varepsilon_0 \varepsilon_\infty \left(\frac{\boldsymbol{E}^{n+1} - \boldsymbol{E}^n}{\Delta t} \right) + \left(\frac{\boldsymbol{J}_{\mathrm{D}}^{n+1} + \boldsymbol{J}_{\mathrm{D}}^n}{2} \right) \\ &= \varepsilon_0 \varepsilon_\infty \left(\frac{\boldsymbol{E}^{n+1} - \boldsymbol{E}^n}{\Delta t} \right) + \left[\frac{(1+k)\boldsymbol{J}_{\mathrm{D}}^n + \beta\left(\boldsymbol{E}^{n+1} + \boldsymbol{E}^n\right)}{2} \right] \end{aligned} \tag{17.43}$$

Algebraic rearrangement of (17.43) allows solution for $\boldsymbol{E}^{n+1}$, and hence the required explicit time-stepping expression for $\boldsymbol{E}$ for the single-pole Drude dispersion:

$$\boldsymbol{E}^{n+1} = \left(\frac{2\varepsilon_0 \varepsilon_\infty - \beta\Delta t}{2\varepsilon_0 \varepsilon_\infty + \beta\Delta t} \right) \boldsymbol{E}^n + \left(\frac{2\Delta t}{2\varepsilon_0 \varepsilon_\infty + \beta\Delta t} \right) \cdot \left[\nabla \times \boldsymbol{H}^{n+1/2} - \frac{(1+k)\boldsymbol{J}_{\mathrm{D}}^n}{2} \right] \tag{17.44}$$

In regions of the FDTD grid where the linear Drude dispersion is used to model metal, the electric field is updated from previously computed values using simply (17.44). [Here, the iterative procedure specified in (17.35) is *not* needed.] Then $\boldsymbol{J}_{\mathrm{D}}$ is updated with (17.41), and finally $\boldsymbol{H}$ is updated with a traditional FDTD temporal finite-difference approximation of (17.2).

17.5 APPLICATIONS OF GVADE FDTD TO TM SPATIAL SOLITON PROPAGATION

This section reviews published applications of the GVADE FDTD method to model the propagation of bright spatial solitons in an infinite, homogeneous, nonlinear dispersive medium for the 2-D TM polarization case [18, 19]. Included in this review are GVADE FDTD models of a single, narrow, fundamental spatial soliton, an overpowered spatial soliton, and the interaction of pairs of copropagating spatial solitons having either $0°$, $90°$, or $180°$ relative carrier phasing.

In all of these examples, we consider $+x$-propagating spatial solitons having the field components (E_x, E_y, H_z). Each single spatial soliton or pair of co-propagating spatial solitons was excited by a *hard source* applied to the H_z field components located along the left outer boundary $(x = 0)$ of the FDTD grid. These components were assigned values oscillating sinusoidally as a function of the time-step number, and having amplitudes following a hyperbolic secant profile in space as a function of each component's y-position [20]:

$$H_z(0, y, t) = A_0 \left[\operatorname{sech}\left(\frac{y - y_1}{w_0} \right) \sin(\omega t) + C \operatorname{sech}\left(\frac{y - y_2}{w_0} \right) \sin(\omega t + \theta) \right] \tag{17.45}$$

Here, the parameter C was set to zero for the single-soliton case, and unity when a pair of co-propagating solitons was modeled. The terms y_1 and y_2 specified the locations of the transverse profile peaks, with the separation distance given by $d = |y_2 - y_1|$. The relative phase between the solitons was specified by θ. In this chapter, the transverse width of a soliton is described by the ratio of the full-width at half-maximum (FWHM) of the beam field amplitude to the carrier wavelength of the beam in the dielectric material, λ_d.

The propagation medium in [18, 19] was assumed to be the realistic model of fused silica reported in [8]. This model included a three-pole Sellmeier linear dispersion, an instantaneous Kerr nonlinearity, and a dispersive Raman nonlinearity having the following parameters:

Sellmeier: $\beta_1 = 0.69617,\ \omega_1 = 2.7537 \times 10^{16}$ rad/s
$\beta_2 = 0.40794,\ \omega_2 = 1.6205 \times 10^{16}$ rad/s
$\beta_3 = 0.89748,\ \omega_3 = 1.9034 \times 10^{14}$ rad/s

Kerr: $n_2 = 2.48 \times 10^{-20}$ m^2/W

Raman: $\tau_1 = 12.2$ fs, $\tau_2 = 32.0$ fs, $\alpha = 0.7$

17.5.1 Single Narrow Fundamental TM Spatial Soliton

We first review the GVADE FDTD simulation, reported in [18], of the propagation of a single, narrow, fundamental, bright TM spatial soliton. Here, a 50- × 20-μm ($x \times y$) modified FDTD grid [14] containing 5000 × 2000 square grid cells of size $\Delta = 10$ nm was used. In this grid, the E_x and E_y components were collocated in space so that $\boldsymbol{E}$ and $|\boldsymbol{E}|$ [required in (17.22), (17.30), and (17.34)] could be obtained without interpolating spatially staggered E-field components. Spatial staggering of $\boldsymbol{E}$ relative to H_z was retained, however. No outer absorbing boundary was used because the grid was sufficiently extended to avoid spurious reflections over the simulated time of interest. A vacuum carrier wavelength, $\lambda_0 = 800$ nm, was used in the simulations, corresponding to $\lambda_d = 552$ nm (resolved 55:1 by the 10-nm grid cells). The grid resolution was chosen empirically based on studies of computational convergence and efficiency. The same considerations were involved in choosing the time-step, $\Delta t = 4.17 \times 10^{-18}$ s, a value that was 5.65 times below the nominal Courant stability limit for a conventional FDTD grid with square 10-nm grid cells and standard linear field updates [14].

The spatial soliton was excited at $x = 0$ with a characteristic width $w_0 = 261$ nm per (17.45), equivalent to a FWHM beamwidth of 690 nm. This resulted in a FWHM/λ_d ratio of 690/552 = 1.25. An amplitude $A_0 = 2.0 \times 10^8$ A/m was found empirically to balance diffraction and self-focusing, and hence lead to a spatial soliton. Figure 17.1 is a visualization of the magnitude of the GVADE-FDTD-computed electric field of the soliton at 60,000 time-steps [18]. Here, the transverse profile of the simulated spatial soliton remained intact over a propagation distance exceeding 120 diffraction lengths L_{diff}, where $L_{\text{diff}} = \pi w_0^2 / \lambda_d$.

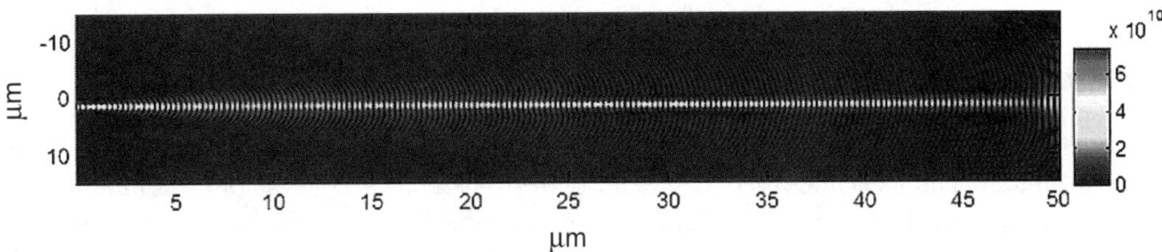

Fig. 17.1 Visualization of the GVADE-FDTD-computed $|\boldsymbol{E}|$ of a narrow (FWHM/λ_d = 1.25) bright, spatial soliton [field components (E_x, E_y, H_z)] propagating from left to right over a distance exceeding 120 diffraction lengths. *Source:* Z. Lubin, J. H. Greene, and A. Taflove, *IEEE Microwave and Wireless Components Lett.*, Vol. 21, 2011, pp. 228–230, ©2011 IEEE.

17.5.2 Single Wide Overpowered TM Spatial Soliton

We next review the GVADE FDTD simulation, reported in [19], of the propagation of single, wide, bright, TM spatial solitons of free-space carrier wavelength $\lambda_0 = 433$ nm. Each soliton was excited with an amplitude $A_0 = 9.54 \times 10^7$ A/m, twice the level indicated by the NLSE for a fundamental soliton solution. Based on numerical convergence tests for this case, square grid cells of size $\Delta = 8$ nm were used, with a time-step of $\Delta t = 3.34 \times 10^{-18}$ s.

Figure 17.2 provides visualizations of the magnitude of the GVADE-FDTD-computed electric field for three different initial excitation beamwidths at $x = 0$: $8.95\lambda_d$, $5.97\lambda_d$, and $4.88\lambda_d$. We see that the computed beams exhibited periodic focusing and defocusing ("breathing"), and the period of this expansion and contraction decreased as the beamwidth was reduced. This non-stationary non-paraxial behavior is not predicted by the NLSE [19].

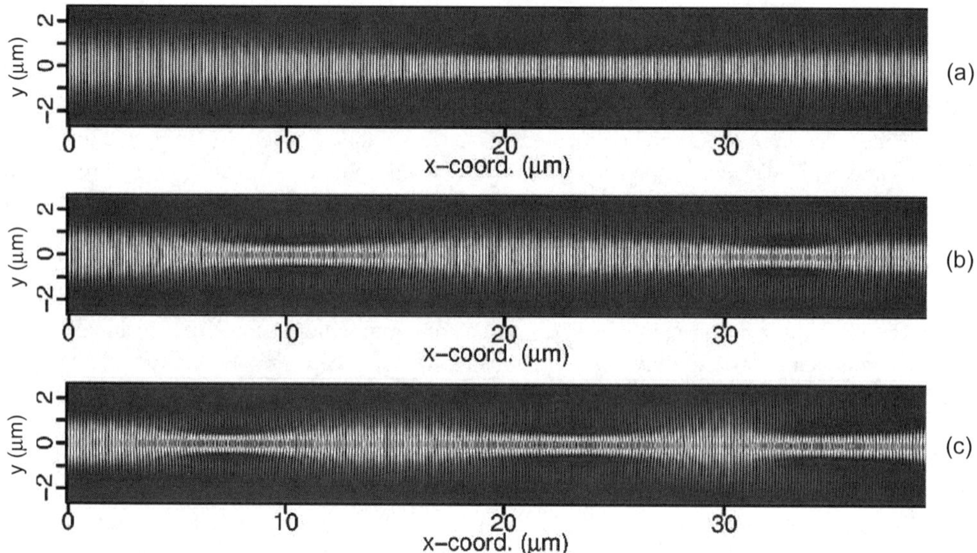

Fig. 17.2 Visualizations of the GVADE-FDTD-computed $|\boldsymbol{E}|$ of a single, overpowered, wide, bright spatial soliton [field components (E_x, E_y, H_z)] propagating from left to right for initial beamwidths of (a) $8.95\lambda_d$; (b) $5.97\lambda_d$; and (c) $4.88\lambda_d$. *Source:* J. H. Greene and A. Taflove, *IEEE Microwave and Wireless Components Lett.*, Vol. 17, 2007, pp. 760–762, ©2007 IEEE.

17.5.3 Interactions of Copropagating Narrow TM Spatial Solitons

In principle, the interactions of copropagating spatial solitons can be exploited to implement routing, guiding, and switching functions in all-optical interconnects and photonic circuits [21]. This has been studied experimentally [22, 23] and via computational simulations [20] for pairs of closely spaced spatial solitons. However, until the publication of the GVADE FDTD simulations of copropagating TM solitons in [18], first-principles simulations based on Maxwell's equations had modeled only transverse electric (TE) polarization solitons with only a single electric field component [20]. This section reviews interactions of pairs of ultranarrow TM spatial solitons with beamwidths on the order of one wavelength, as reported in [18].

Equal-Amplitude In-Phase Solitons

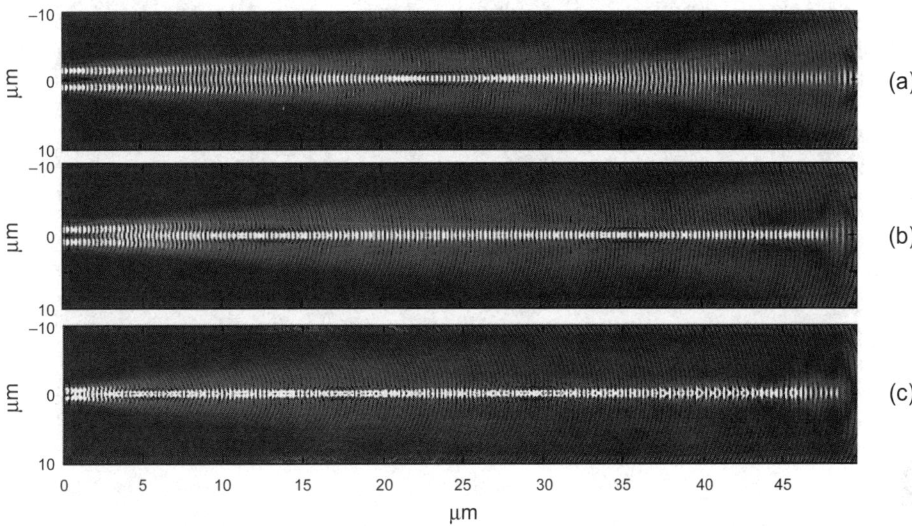

Fig. 17.3 Visualizations of the GVADE-FDTD-computed $|\boldsymbol{E}|$ of a pair of copropagating, narrow (FWHM / λ_d = 1.25), bright, in-phase spatial solitons [field components (E_x, E_y, H_z)], each having beam parameters as in Fig. 17.1. Normalized separation distance $d_n = d$/FWHM: (a) 3.5, (b) 2.5, (c) 1.5. The scale is the same as in Fig. 17.1. *Source:* Z. Lubin, J. H. Greene, and A. Taflove, *IEEE Microwave and Wireless Components Lett.*, Vol. 21, 2011, pp. 228–230, ©2011 IEEE.

Reference [18] considered pairs of spatial solitons copropagating in the nonlinear dispersive silica medium discussed in Section 17.4.1. Each soliton had the same beam parameters as that visualized in Fig. 17.1, and the simulations were conducted with the same space and time steps.

Figure 17.3 visualizes the GVADE-FDTD-computed soliton–soliton interactions for the equal-amplitude, in-phase case [$C = 1$ and $\theta = 0$ in (17.45)] for a normalized separation distance $d_n = d$/FWHM of 3.5, 2.5, and 1.5 [18]. Because the solitons interfered constructively, the electric field and refractive index were higher between the beams. The solitons bent toward one another, and what started as an independent propagation of the soliton pair transformed via mutual attraction to a single soliton-like beam. This beam persisted, extending more than 120 diffraction lengths. The distance to the soliton merging depended on the initial separation: ~15 μm for $d_n = 3.5$, but only ~3 μm for $d_n = 1.5$. While there were slight periodic expansions and focusing of the fused transverse profile, the beams never reemerged as separate solitons.

Equal-Amplitude Solitons with Relative Phases of π and π/2

Figures 17.4 and 17.5 visualize the GVADE-FDTD-computed equal-amplitude soliton–soliton interactions for $\theta = \pi$ and $\theta = \pi/2$, respectively [18]. Both cases exhibited a mutual repulsion phenomenon such that the angle of repulsion increased as d_n decreased. However, for $\theta = \pi/2$, as d_n decreased, the lower soliton gained energy at the expense of the upper, which eventually dispersed in Fig. 17.5(c). This soliton interaction behavior is consistent with the experimental findings of [22, 23]. The work of [18] was the first reported computational study of this problem for the TM polarization case from first principles, i.e., the full-vector Maxwell's equations.

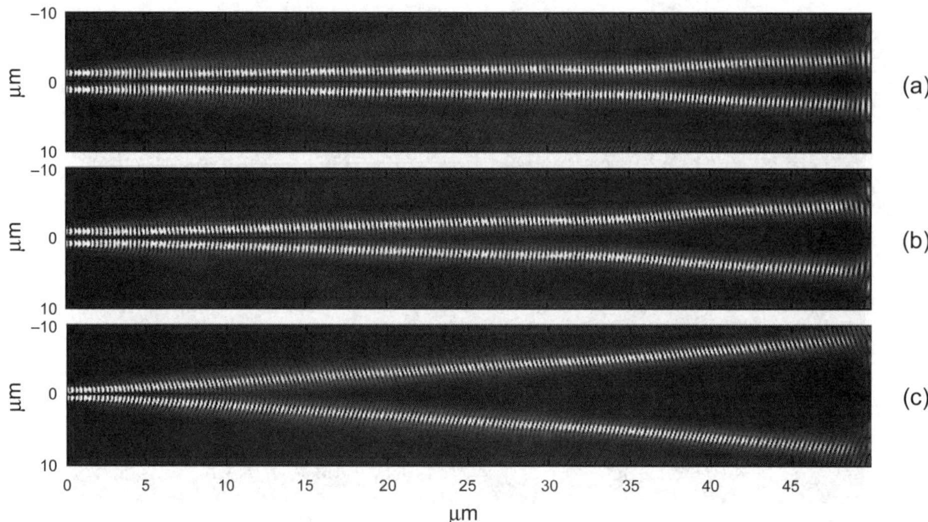

Fig. 17.4 Visualizations of the GVADE-FDTD-computed $|\boldsymbol{E}|$ of a pair of copropagating, narrow (FWHM/λ_d = 1.25), bright, anti-phased ($\theta = \pi$) spatial solitons [field components (E_x, E_y, H_z)], each having the same initial beam parameters as the soliton of Fig. 17.1. Normalized separation distance $d_n = d/\text{FWHM}$: (a) 3.5, (b) 2.5, and (c) 1.5. The scale is the same as shown in Fig. 17.1. *Source:* Z. Lubin, J. H. Greene, and A. Taflove, *IEEE Microwave and Wireless Components Lett.*, Vol. 21, 2011, pp. 228–230, ©2011 IEEE.

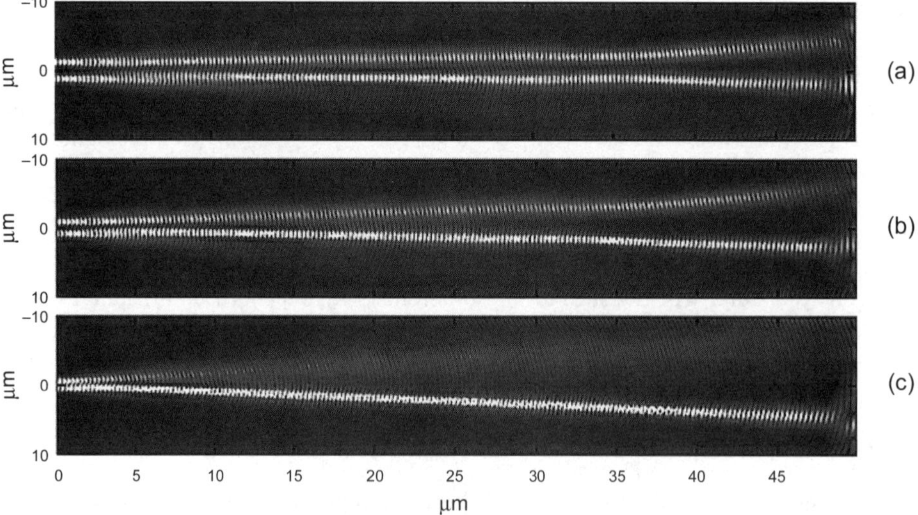

Fig. 17.5 Visualizations of the GVADE-FDTD-computed $|\boldsymbol{E}|$ of a pair of copropagating, narrow (FWHM/λ_d = 1.25), bright spatial solitons in phase-quadrature ($\theta = \pi/2$) [field components (E_x, E_y, H_z)], each having the same initial beam parameters as the soliton of Fig. 17.1. Normalized separation distance $d_n = d/\text{FWHM}$: (a) 3.5, (b) 2.5, and (c) 1.5. The scale is the same as shown in Fig. 17.1. *Source:* Z. Lubin, J. H. Greene, and A. Taflove, *IEEE Microwave and Wireless Components Lett.*, Vol. 21, 2011, pp. 228–230, ©2011 IEEE.

17.6 APPLICATIONS OF GVADE FDTD TO TM SPATIAL SOLITON SCATTERING

The GVADE FDTD modeling applications reviewed in Section 17.5 involved beam propagation in an infinite homogeneous material space. However, the GVADE technique can be applied to any material geometry. Hence, it permits the simulation of spatial soliton scattering interactions with arbitrary material inhomogeneities, as derived from the full-vector Maxwell's equations.

This section reviews applications of the GVADE FDTD method to model the scattering of TM bright spatial solitons by inhomogeneities embedded within a nonlinear dispersive medium. Two types of inhomogeneities are discussed: sub-wavelength square air holes [19], and plasmonic metal films. It appears that neither of these scatterers, which are broadly representative of the micro- and nanostructures in future photonic circuits utilizing spatial solitons, can be modeled using existing NLSE approaches.

17.6.1 Scattering by a Square Sub-Wavelength Air Hole

We first review the GVADE FDTD simulation, reported in [19], of the scattering of the wide, overpowered soliton of Fig. 17.2(b) by a square, sub-wavelength air hole. Incorporating the air hole into the model involved simply setting to zero the polarization current density within the air hole, and time-stepping the field components located there with the basic FDTD algorithm.

In [19], the overpowered soliton of Fig. 17.2(b), with FWHM/λ_d = 5.97, was launched from the far-left side of the computational grid in the realistic fused silica material described in Section 17.5. The grid cell size was Δ = 10 nm and the time-step was $\Delta t = 4.17 \times 10^{-18}$ s.

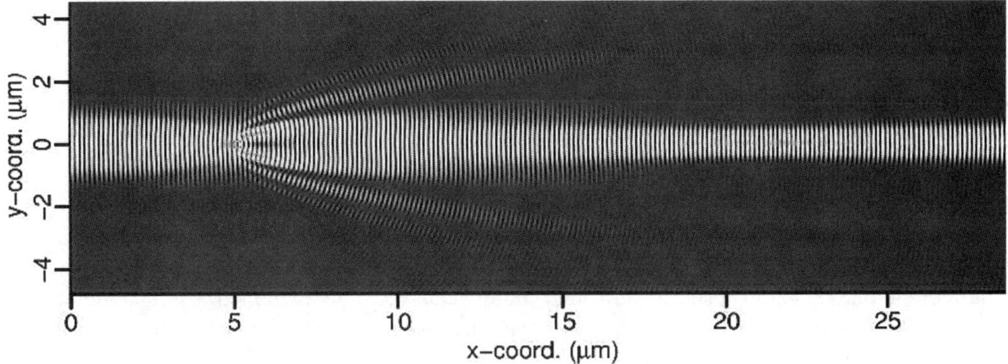

Fig. 17.6 Visualization of the GVADE-FDTD-computed $|\boldsymbol{E}|$ of the scattering of the bright, overpowered spatial soliton of Fig. 17.2(b) [field components (E_x, E_y, H_z)] by a square 250-nm air hole located at $x = 5$ μm. *Source:* J. H. Greene and A. Taflove, *IEEE Microwave and Wireless Components Lett.*, Vol. 17, 2007, pp. 760–762, ©2007 IEEE.

Figure 17.6 shows GVADE FDTD simulation results [19] for a square 250- × 250-nm air hole placed 5 μm along the propagation axis of the breathing, overpowered soliton of Fig. 17.2(b), at which point the soliton was at its average width. After scattering by the air hole, energy coalesced into a relatively lower-power soliton beyond a distance of ~30 λ_d. Sufficient energy was retained for this soliton to remain overpowered relative to a solution of the NLSE, and the soliton continued to exhibit periodic envelope modulation, qualitatively similar to Fig. 17.2.

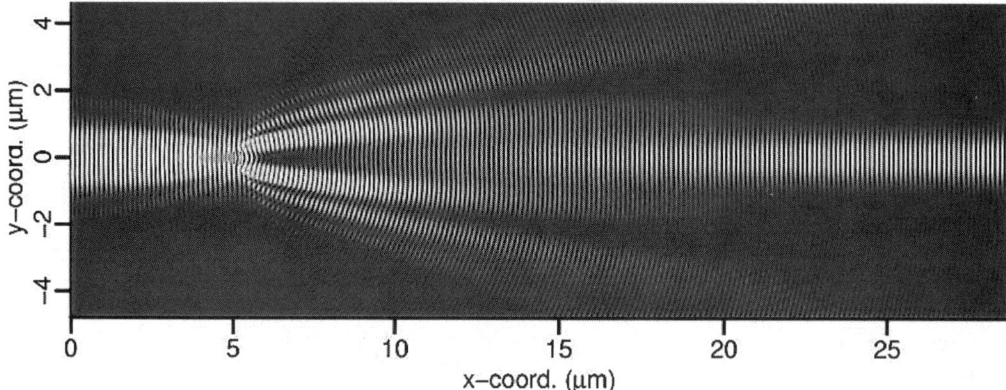

Fig. 17.7 Visualization of the GVADE-FDTD-computed $|\boldsymbol{E}|$ of the scattering of the bright, overpowered spatial soliton of Fig. 17.2(b) [field components (E_x, E_y, H_z)] by a square 350-nm air hole located at $x = 5$ μm. *Source:* J. H. Greene and A. Taflove, *IEEE Microwave and Wireless Components Lett.*, Vol. 17, 2007, pp. 760–762, ©2007 IEEE.

Figure 17.7 shows GVADE FDTD simulation results [19] for a square 350- × 350-nm air hole placed 5 μm along the propagation axis of the soliton of Fig. 17.2(b). Here, more energy was lost due to scattering than for the smaller 250- × 250-nm air hole of Fig. 17.6. While a soliton did eventually reform at ~60 λ_d beyond the air hole, it was no longer overpowered relative to an NLSE solution, and its transverse profile did not change as it propagated.

17.6.2 Interactions with Thin Plasmonic Gold Films

Recent studies have explored soliton interactions with metals, sometimes referred to as soliton plasmonics. For the optical regime, a metal can be modeled as a plasma of free electrons [24]. Incident electromagnetic fields on metal surfaces trigger oscillations of these electrons and can excite surface waves known as *surface plasmon polaritons* (SPPs). When the incident frequency is close to the natural plasma resonance frequency, enhanced sub-wavelength evanescent fields can occur at the surface, which are useful for applications such as molecular biosensing and nanofabrication [25, 26]. These enhanced fields can also be exploited for spatial soliton applications. For example, metal waveguides can increase power efficiency in soliton generation [27]. Also, a type of sub-wavelength soliton known as a discrete soliton can be formed in periodic metal-dielectric slabs from SPP tunneling [28]. Other studies have examined SPP focusing by nonlinearities at interfaces [29], and also control of SPPs by adjacent solitons [30]. Because future all-optical photonic circuits using spatial solitons are likely to integrate metal structures in order to exploit these surface interactions for generation and guiding, soliton interactions with metal particles and films represent an important problem.

This section presents recent applications of the GVADE FDTD method to model TM bright spatial solitons in fused silica that impinge obliquely on embedded thin plasmonic gold films. Specifically, we discuss simulation of a soliton beam-splitting phenomenon that occurs for oblique incidence on a 20-nm-thick gold film. We also quantify the Goos–Hänchen lateral displacement of the reflected soliton in the plane of incidence at the surface of a 2-μm-thick gold film.

Launching an Obliquely Propagating Spatial Soliton

To launch a single spatial soliton of amplitude A_0 and carrier frequency ω_0 at an oblique downward angle ϕ relative to the x-axis within the GVADE FDTD grid, the following hard source was applied at the left outer boundary, $i = 0$, of the grid:

$$H_z(i=0, j, l) = A_0 \operatorname{sech}\left[\frac{(j-j_1)\Delta}{w_0}\right] \sin\left[\omega_0(l\Delta t - j\tau)\right] U(l\Delta t - j\tau) \tag{17.46}$$

Here, j is the grid index in the y-dimension, j_1 is the index in the y-dimension corresponding to the peak of the soliton transverse profile envelope, Δ is the grid-cell size, Δt is the time-step, l is the time-step counter, $U(t)$ is the unit step-function, and $j\tau$ is a linear time-delay taper across the H_z sources wherein $\tau = n_0 \Delta \sin\phi / c$, for n_0 the refractive index of the medium at ω_0. Expression (17.46) is a modification of (17.45) that operates similarly to a linear phased-array antenna wherein the radiating elements are assigned appropriate time delays (equivalently, phase shifts).

Soliton and Material Model

Except for the oblique propagation direction, the soliton model discussed in this section assumed the same electromagnetic field components (E_x, E_y, H_z), beam parameters ($\lambda_0 = 800$ nm, $w_0 = 261$ nm, FWHM/$\lambda_d = 1.25$), and nonlinear dispersive fused silica background medium, as for the $+x$-directed TM spatial soliton of Section 17.5.1. The gold film was modeled by the linear Drude dispersion described in Section 17.4.5 with $\omega_p = 1.3544 \times 10^{16}$ rad/s, $\gamma = 1.1536 \times 10^{14}$ rad/s, and $\varepsilon_\infty = 9.0685$ [17]. The grid resolution was again $\Delta = 10$ nm with $\Delta t = 4.17 \times 10^{-18}$ s.

Soliton Beam-Splitting and Reflection Phenomena for Optical Routing

We first discuss a GVADE FDTD simulation of a spatial soliton beam-splitting phenomenon that was observed for the soliton and material model summarized above. Here, the soliton was directed to impinge obliquely on a 20-nm-thick gold film at 33° relative to the film's surface normal. For this case, the gold film thickness was approximately one skin depth.

Figure 17.8 visualizes the GVADE-FDTD-computed electric field magnitude after the spatial soliton had both penetrated through, and reflected off, the gold film [31]. Although some diffraction and beam spreading resulted from this interaction, both the reflected and transmitted beams remained intact after propagation beyond 10 μm from the film, equivalent to ~30 diffraction lengths. The electric field amplitude of the transmitted beam was observed to be approximately one-half that of the incident beam, in general agreement with plane-wave Fresnel theory [32].

Figure 17.9 illustrates how the beam-splitting phenomenon depicted in Fig. 17.8 could serve as a means for routing spatial solitons [31]. In this GVADE-FDTD simulation, the incident soliton was again directed to impinge obliquely on a 20-nm-thick gold film, where it was split into approximately equal-strength transmitted and reflected beams. Here, however, the reflected beam was intercepted by a second gold film, 2 μm thick and vertically displaced from the first film. The second gold film caused an essentially complete re-reflection of the beam reflected from the first gold film. This generated a pair of copropagating spatial solitons having an adjustable separation depending on the lateral displacement of the second gold film.

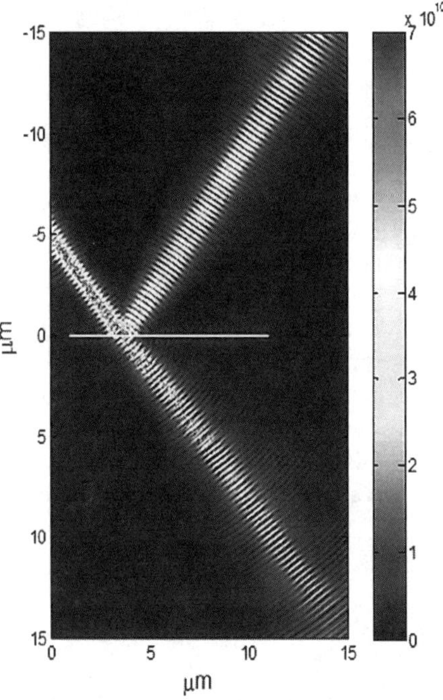

Fig. 17.8 Visualization of the GVADE-FDTD-computed $|E|$ of a bright, narrow, TM spatial soliton [field components (E_x, E_y, H_z)] impinging at 33° on a 20-nm-thick gold film embedded in fused silica. *Source:* Z. Lubin, J. H. Greene, and A. Taflove, *Microwave and Optical Technology Lett.*, Vol. 54, 2012, pp. 2679–2684, ©2012 Wiley Periodicals, Inc.

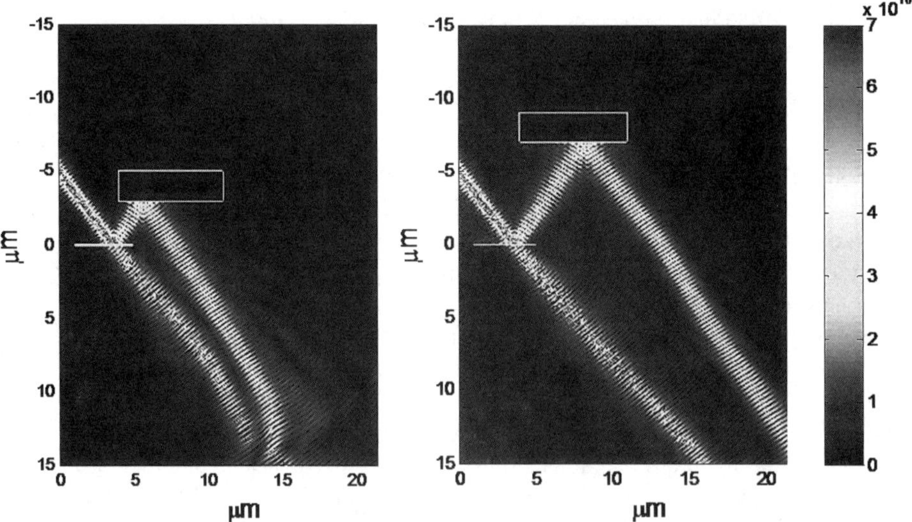

Fig. 17.9 Visualization of the GVADE-FDTD-computed $|E|$ of a bright, narrow, TM spatial soliton [field components (E_x, E_y, H_z)] interacting with two gold films embedded in fused silica. Lower gold film: 20 nm thick; upper gold film: 2 µm thick and vertically displaced by either 3 or 7 µm. *Source:* Z. Lubin, J. H. Greene, and A. Taflove, *Microwave and Optical Technology Lett.*, Vol. 54, 2012, pp. 2679–2684, ©2012 Wiley Periodicals, Inc.

Goos–Hänchen Lateral Displacement of the Reflected Soliton

We last consider the Goos–Hänchen shift, a lateral displacement of a reflected beam in the plane of incidence from an interface. This displacement is due to the combination of the multiple plane waves that effectively comprise the incident beam, each having a different phase shift after reflection. For beams in linear media, it is dependent on the wavelength, angle of incidence, polarization, and media dielectric properties.

Motivated by recent findings that the Goos–Hänchen shift can occur for linear TM optical beams incident on gold surfaces [33], we explored a possible similar effect for narrow, nonparaxial, TM spatial solitons. Figure 17.10 shows a lateral displacement of ~275 nm in the reflected beam for a soliton obliquely incident at 70° to the surface normal of a 2-μm-thick gold film [31]. This is in good agreement with a reported shift of 300 nm for a linear beam at a wavelength of 826 nm [33].

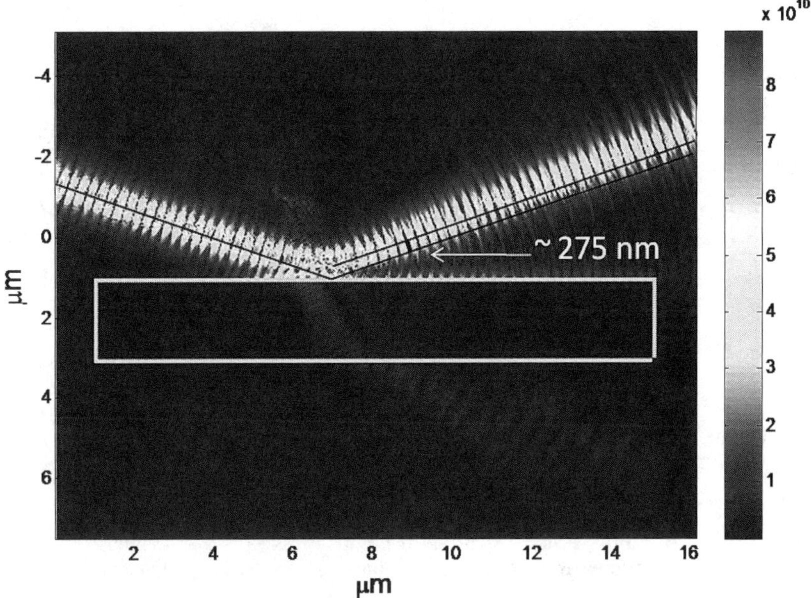

Fig. 17.10 Visualization of the GVADE-FDTD-computed $|E|$ of a bright, narrow, TM spatial soliton [field components (E_x, E_y, H_z)] reflecting from a 2-μm-thick gold film, demonstrating an approximate 275-nm Goos–Hänchen lateral shift of the reflected beam. *Source:* Z. Lubin, J. H. Greene, and A. Taflove, *Microwave and Optical Technology Lett.*, Vol. 54, 2012, pp. 2679–2684, ©2012 Wiley Periodicals, Inc.

17.7 SUMMARY

The GVADE FDTD method is a powerful tool for full-vector solutions of electromagnetic wave propagation in materials having linear and nonlinear dispersions. This technique provides a direct time-domain solution of Maxwell's equations without simplifying assumptions such as paraxial or scalar approximations. Furthermore, it can be applied to arbitrary inhomogeneous material geometries, and both linear and nonlinear polarizations can be incorporated through the Maxwell–Ampere law.

This chapter derived the GVADE FDTD time-stepping equations for the TM electromagnetic field in a realistic model of fused silica characterized by a three-pole Sellmeier linear dispersion, an instantaneous Kerr nonlinearity, and a dispersive Raman nonlinearity. Next, the technique was extended to model a plasmonic metal characterized by a linear Drude dispersion. The GVADE FDTD method was then applied to model single nonparaxial and overpowered spatial solitons; the interaction of closely spaced, copropagating, nonparaxial spatial solitons; spatial soliton scattering by sub-wavelength air holes; and interactions between nonparaxial spatial solitons and thin gold films.

We conclude that emerging applications in optical communications and computing utilizing microscale and nanoscale photonic circuits require controlling complex electromagnetic wave phenomena in linear and nonlinear materials with important sub-wavelength features. To this end, the GVADE FDTD method extends the existing toolset of approximate photonics simulation techniques by providing means to construct rigorous first-principles models directly from Maxwell's equations.

REFERENCES

[1] Greene, J. H., and A. Taflove, "General vector auxiliary differential equation finite-difference time-domain method for nonlinear optics," *Optics Express*, Vol. 14, 2006, pp. 8305–8310.

[2] Kivshar, Y., and G. Agrawal, *Optical Solitons: From Fibers to Photonic Crystals*, New York: Academic Press, 2003.

[3] Matuszewski, M., "Self-consistent treatment of the full vectorial nonlinear optical pulse propagation equation in an isotropic medium," *Optics Communications*, Vol. 221, 2003, pp. 337–351.

[4] Agrawal, G., *Nonlinear Fiber Optics*, 4th ed., New York: Academic Press, 2006.

[5] Boyd, R. W., *Nonlinear Optics*, 3rd ed., New York: Academic Press, 2008.

[6] Goorjian, P. M., and A. Taflove, "Direct time integration of Maxwell's equations in nonlinear dispersive media for propagation and scattering of femtosecond electromagnetic solitons," *Optics Lett.,* Vol. 17, 1992, pp. 180–182.

[7] Goorjian, P. M., A. Taflove, R. M. Joseph, and S. C. Hagness, "Computational modeling of femtosecond optical solitons from Maxwell's equations," *IEEE J. Quantum Electronics*, Vol. 28, 1992, pp. 2416–2422.

[8] Nakamura, S., N. Takasawa, and Y. Koyamada, "Comparison between finite-difference time-domain calculation with all parameters of Sellmeier's fitting equation and experimental results for slightly chirped 12fs laser pulse propagation in a silica fiber," *IEEE J. Lightwave Technology*, Vol. 23, 2005, pp. 855–863.

[9] Kashiwa, T., and I. Fukai, "A treatment by FDTD method of dispersive characteristics associated with electronic polarization," *Microwave and Optical Technology Lett.,* Vol. 3, 1990, pp. 203–205.

[10] Fujii, M., M. Tahara, I. Sakagami, W. Freude, and P. Russer, "High-order FDTD and auxiliary differential equation formulation of optical pulse propagation in 2-D Kerr and Raman nonlinear dispersive media," *IEEE J. Quantum Electronics*, Vol. 40, 2004, pp. 175–182.

[11] Saleh, B., and M. Teich, *Fundamentals of Photonics, 2nd Ed.*, New York: Wiley, 2007.

[12] Hellworth, R. W., "Third-order optical susceptibility of liquids and solids," *J. Progress in Quantum Electronics*, Vol. 5, 1977, pp. 1–68.

[13] Blow, K., and D. Wood, "Theoretical description of transient stimulated Raman scattering in optical fibers," *IEEE J. Quantum Electronics*, Vol. 25, 1989, pp. 2665–2673.

[14] Taflove, A., and S. C. Hagness, *Computational Electrodynamics: The Finite-Difference Time-Domain Method*, 3rd ed., Norwood, MA: Artech, 2005.

[15] Heath, M., *Scientific Computing: An Introductory Survey*, 2nd ed., New York: McGraw-Hill, 2002.

[16] Johnson, P. B., and R. W. Christy, "Optical constants of the noble metals," *Physical Review B*, Vol. 6, 1972, pp. 4370–4379.

[17] Vial, A., A. Grimault, D. Macias, D. Barchiesi, and M. Chapelle, "Improved analytical fit of gold dispersion: Application to the modeling of extinction spectra with a finite-difference time-domain method," *Physical Review B*, Vol. 71, 2005, 085416-1-7.

[18] Lubin, Z., J. H. Greene, and A. Taflove, "FDTD computational study of ultra-narrow TM nonparaxial spatial soliton interactions," *IEEE Microwave and Wireless Components Lett.*, Vol. 21, 2011, pp. 228–230.

[19] Greene, J. H., and A. Taflove, "Scattering of spatial optical solitons by sub-wavelength air holes," *IEEE Microwave and Wireless Components Lett.*, Vol. 17, 2007, pp. 760–762.

[20] Joseph, R. M., and A. Taflove, "Spatial soliton deflection mechanism indicated by FD-TD Maxwell's equations modeling," *IEEE Photonics Technology Lett.*, Vol. 6, 1994, pp. 1251–1254.

[21] Alcantara, L. D. S., M. A. C. Lima, A. C. Cesar, B. V. Borges, and F. L. Teixeira, "Design of a multifunctional integrated optical isolator switch based on nonlinear and nonreciprocal effects," *Optical Engineering*, Vol. 44, 2005, 124002-1-9.

[22] Aitchison, J. S., A. M. Weiner, Y. Silberberg, D. E. Leaird, M. K. Oliver, J. L. Jackel, and P. W. E. Smith, "Experimental observation of spatial soliton interactions," *Optics Lett.*, Vol. 16, 1991, pp. 15–17.

[23] Shalaby, M., F. Reynaud, and A. Barthelemy, "Experimental observation of spatial soliton interactions with a $\pi/2$ relative phase difference," *Optics Lett.*, Vol. 17, 1992, pp. 778–780.

[24] Maier, S., *Plasmonics: Fundamentals and Applications*, Berlin: Springer, 2007.

[25] Genet, C., and T. W. Ebbesen, "Light in tiny holes," *Nature*, Vol. 445, 2007, pp. 39–46.

[26] Ozbay, E., "Plasmonics: Merging photonics and electronics at nanoscale dimensions," *Science*, Vol. 311, 2006, pp. 189–193.

[27] Feigenbaum, E., and M. Orenstein, "Plasmon-soliton," *Optics Lett.*, Vol. 32, 2007, pp. 674–676.

[28] Marini, A., A. V. Gorbach, and D. V. Skryabin, "Coupled-mode approach to surface plasmon polaritons in nonlinear periodic structures," *Optics Lett.*, Vol. 35, 2010, pp. 3532–3534.

[29] Davoyan, A. R., I. V. Shadrivov, and Y. S. Kivshar, "Self-focusing and spatial plasmon-polariton solitons," *Optics Express*, Vol. 17, 2009, pp. 21732–21737.

[30] Bliokh, K. Y., Y. P. Bliokh, and A. Ferrando, "Resonant plasmon-soliton interaction," *Physical Review A*, Vol. 79, 2009, 41803-1-3.

[31] Lubin, Z., J. H. Greene, and A. Taflove, "FDTD computational study of nanoplasmonic guiding structures for nonparaxial spatial solitons," *Microwave and Optical Technology Lett.*, Vol. 54, 2012, pp. 2679–2684.

[32] Heavens, O. S., *Optical Properties of Thin Solid Films*, Dover, 1991.

[33] Merano, M., A. Aiello, G. W. 't Hooft, M. P. van Exter, E. R. Eliel, and J. P. Woerdman, "Observation of Goos-Hänchen shifts in metallic reflection," *Optics Express*, Vol. 15, 2007, pp. 15928–15934.

Chapter 18

FDTD Modeling of Blackbody Radiation and Electromagnetic Fluctuations in Dissipative Open Systems[1]

Jonathan Andreasen

18.1 INTRODUCTION

A proper treatment of blackbody radiation constitutes an essential part of the exact quantum-mechanical theory of the interaction of light with matter. For an electromagnetic (EM) field inside an open resonant cavity, dissipation may occur due to output coupling (leakage) through the boundary. The EM field couples to an external reservoir by exiting the cavity. This external reservoir may be a blackbody, i.e., a body that absorbs all incident radiation. To remain at thermal equilibrium, however, this blackbody must emit radiation in addition to absorbing it. The emitted temperature- and frequency-dependent radiation is known as blackbody radiation, which couples back into the open cavity. Therefore, in addition to dissipation, the light field in an open cavity experiences fluctuation because of its coupling to the external environment.

In fact, an empty cavity that reaches thermal equilibrium has the same intensity and frequency distribution as the blackbody [1]. The dependence on frequency implies that temporal correlations exist in the blackbody radiation.

Due to its intrinsic dependence on time and well-established absorbing boundary conditions, the finite-difference time-domain (FDTD) method is well suited to study systems with open boundary conditions and their interaction with external reservoirs. This chapter reviews a promising FDTD modeling technique for simulation of the EM fluctuations inherent in such dissipative open systems, following the work originally reported in [2].

18.2 STUDYING FLUCTUATION AND DISSIPATION WITH FDTD

The relation between fluctuation and dissipation in systems coupled to external reservoirs is so fundamental that it spans many areas of science, with roots in Brownian motion [3, 4]. Quantum fluctuations due to the spontaneous emission of atoms were first introduced as electric field noise

---

[1] This chapter is adapted from Ref. [2], J. Andreasen, H. Cao, A. Taflove, P. Kumar, and C. qi Cao, "Finite-difference time-domain simulation of thermal noise in open cavities," *Physical Review A*, Vol. 77, 2008, 023810, ©2008 American Physical Society.

in FDTD simulations of microcavity lasers [5]. Fluctuations of the atomic properties themselves were later incorporated into FDTD [6] based on the quantum Langevin equation [7] and applied to random laser studies [8]. The light field in an open cavity can experience separate quantum fluctuations because of its coupling to radiative reservoirs. The influence of these reservoirs is typically negligible because the EM energy at visible frequencies is much larger than the thermal energy at room temperature. At higher temperatures or longer wavelengths (e.g., mid-infrared wavelengths), however, this noise becomes significant.

The fluctuation-dissipation theorem states that the damping of a system is determined by the fluctuating forces of the reservoir, which also introduce fluctuations into the system [9]. There are two dissipation mechanisms for the cavity field: (1) intracavity absorption and (2) output coupling or leakage. Early laser theory introduced thermal noise via a heat bath comprised of lossy oscillators or absorbing atoms [10, 11], accounting for light absorption inside the cavity. FDTD simulations introduced noise related to intracavity absorption [12, 13]. A fluctuating electric field was added as a soft source at every grid point inside the cavity with its rms amplitude proportional to the local absorption coefficient. Results showed that a material with a specific geometry can sufficiently modify the thermal noise that it approaches blackbody radiation.

Output coupling is not a local loss and thus, fluctuations cannot be added to the local EM field (at every grid point). For an open cavity whose loss only comes from the output coupling, thermal noise within the system is attributed to the thermal radiation that penetrates the cavity through the coupling [14, 15]. In the modal picture, widely used in quantum optical studies, thermal noise is introduced so that the quantum operator of a leaky cavity mode satisfies the commutation relation [16]. Without such noise, the operator becomes a classical quantity and contradicts quantum mechanics.

However, the use of modes in quantum optical studies has several key shortcomings. First, the amount of thermal noise depends on the mode decay rate, which must be known in order to solve the Langevin equation for the field operator. For complex open cavities (such as ones comprised of random structures [8]), this information is *a priori* unknown. Second, if the cavity is very leaky, the significant overlap of modes in frequency makes it difficult to distinguish one mode from another. Third, in the presence of nonlinearity, modes do not exist in the strict sense. Overall, it would be desirable to study the noise of a cavity field without prior knowledge of the cavity modes.

The FDTD method affords exactly this possibility. With FDTD, modal behavior is an emergent property of the EM field that results from its temporal evolution. Consequently, noise can be introduced to the EM field without invoking modes. This opens a promising, novel approach to study the quantum-mechanical aspects of radiation in macroscopic systems via classical electrodynamics simulations. Such an approach has the potential to permit rigorous theoretical investigations of noise in the context of quantum optics and open systems such as chaotic open cavities. In particular, the dynamics of such systems are very difficult to study using the standard frequency-domain methods.

18.3 INTRODUCING BLACKBODY RADIATION INTO THE FDTD GRID

In FDTD simulations, light escaping from an open system effectively exits the computation space by the action of an absorbing boundary condition (ABC) located at the outer grid boundary. In addition to allowing the simulation of open systems, e.g., leaky optical cavities, in any dimension, the ABC also acts as an external reservoir with which the system interacts. The perfectly matched layer (PML) is a well-known example of such an ABC. Since essentially all impinging fields are absorbed, the ABC can be modeled as a blackbody. To remain in thermal equilibrium, the blackbody must radiate into the system. The blackbody radiation from the ABC propagates into the cavity and acts as noise to the cavity field, wherein the amount of noise penetrating the cavity depends on the output coupling. Thus, the strength of the EM fluctuations within the cavity intrinsically incorporates the cavity leakage rate.

To simulate the blackbody radiation, the FDTD grid is surrounded with a series of random noise sources adjacent to the ABC. These soft sources radiate EM waves into the grid having spectral properties consistent with blackbody radiation. Although we focus on one-dimensional (1-D) systems in this discussion, extension to two-dimensional (2-D) and three-dimensional (3-D) systems is straightforward. The 1-D grid is discretized with grid-cells of size Δx, and the EM fields are updated each time-step, Δt. As shown in Fig. 18.1, two point sources are placed at the extremities of the grid. Each source generates a random electric field value E_s at every time-point, $t_k = k\Delta t$.

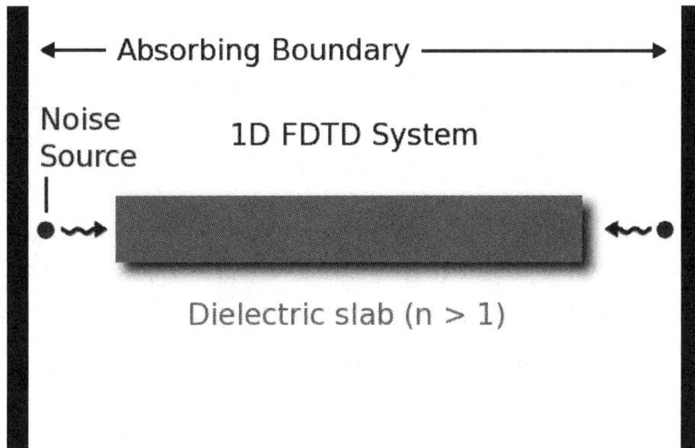

Fig. 18.1 The FDTD absorbing boundary condition (ABC) is modeled as a blackbody that radiates into the system. Noise sources are placed next to the ABC to simulate blackbody radiation. No restrictions exist on the material structure in the FDTD grid. A dielectric slab is shown here.

A Fourier transform of the temporal correlation function of the electric field, $\langle E_s(t_1)E_s(t_2)\rangle$, gives the noise spectrum $D(\omega, T)$. If $E_s(t_k)$ is uncorrelated in time, i.e., $\langle E_s(t_1)E_s(t_2)\rangle \propto \delta(t_2 - t_1)$, then $D(\omega, T)$ is the white-noise spectrum. This is incorrect because $D(\omega, T)$ should be equal to the energy density of blackbody radiation [1].

Planck first derived the frequency distribution of blackbody radiation by considering a large number of identical resonators with discrete energies [17]. The energy density of blackbody radiation in one dimension is found here by considering the grand potential Λ of a Bose–Einstein ideal gas with an affinity of zero (due to the fact that photons are not conserved) [1]. Consider photons of frequency ω in a 1-D space of length L. The density of states is $(L/\pi c)\,d\omega$, where c is the speed of light in vacuum. The grand potential is expressed as:

$$\Lambda = -\frac{L}{\pi c}\int \log(1 - e^{-\beta\hbar\omega})\,d\omega \tag{18.1}$$

where $\beta = 1/k_B T$ and k_B is the Boltzmann constant. The radiation energy can be found as $U_e = -\partial\Lambda/\partial\beta$, giving the energy density of radiation:

$$\frac{U_e}{L} = \frac{\hbar}{\pi c}\int \frac{\omega e^{-\beta\hbar\omega}}{1-e^{-\beta\hbar\omega}}\,d\omega = \frac{\hbar}{\pi c}\int \frac{\omega}{e^{\beta\hbar\omega}-1}\,d\omega \tag{18.2}$$

Thus, the energy density of radiation per unit frequency is:

$$D(\omega, T) = \frac{\hbar}{\pi c}\left(\frac{\omega}{e^{\hbar\omega/k_B T}-1}\right) \tag{18.3}$$

For computational convenience, the range of ω is extended from $(0, \infty)$ to $(-\infty, \infty)$. Since the electric field in the FDTD simulation is a real number, $D(-\omega, T)$ must be equal to $D(\omega, T)$ for $\omega > 0$. Therefore, $D(\omega, T) = D(|\omega|, T)$ and is normalized as:

$$D_{\text{norm}}(|\omega|, T) = \frac{6\hbar^2}{\pi k_B^2 T^2}\left(\frac{|\omega|}{e^{\hbar|\omega|/k_B T}-1}\right) \tag{18.4}$$

so that $\int_{-\infty}^{\infty} D_{\text{norm}}(|\omega|, T)d\omega = 2\pi$. Consequently, the temporal correlation function for the source electric field is given by:

$$\langle E_s(t_1)E_s(t_2)\rangle = \frac{\delta^2}{2\pi}\int_{-\infty}^{\infty} d\omega\, D_{\text{norm}}(|\omega|, T)e^{j\omega(t_2-t_1)} \tag{18.5}$$

where δ is the rms amplitude of the noise field whose value is to be determined later. For blackbody radiation, the field correlation function is given specifically by:

$$\langle E_s(t_1)E_s(t_2)\rangle = \frac{3\delta^2}{\pi^2}\left\{\zeta[2, 1-j(t_2-t_1)k_B T/\hbar] + \zeta[2, 1+j(t_2-t_1)k_B T/\hbar]\right\} \tag{18.6}$$

where the ζ function is given by:

$$\zeta(s, a) = \sum_{l=0}^{\infty}(l+a)^{-s} \tag{18.7}$$

The method of Freilikher et al. [18] provides an efficient and straightforward way of generating random numbers for $E_s(t_k)$ so that (18.5) is satisfied. This technique was originally developed in the context of creating random surfaces with specific height correlations. The end result takes advantage of the fast Fourier transform (FFT) that is used to generate the source electric field:

$$E_s(t_k) = \frac{\delta}{\sqrt{\tau_{\text{sim}}}}\sum_{l=-N}^{N-1}(P_l + jQ_l)D_{\text{norm}}^{1/2}(|\omega_l|, T)\, e^{j\omega_l t_k} \tag{18.8}$$

where $2N$ is the total number of time-steps; $\tau_{\text{sim}} = 2N\Delta t$ is the total simulation time; $\omega_l = 2\pi l/\tau_{\text{sim}}$; and P_l and Q_l are independent Gaussian random numbers of zero mean and unity variance having the symmetries $P_l = P_{-l}$ and $Q_l = -Q_{-l}$. These Gaussian random numbers can be generated by the Marsaglia and Bray modification of the Box–Müller transformation [19], a very fast and reliable method, assuming that the uniformly distributed random-number generator being used is also fast and reliable.

When setting the grid-cell size Δx and time-step Δt to be used in the FDTD simulations, the characteristics of blackbody radiation must be considered. The temporal correlation time or coherence time τ_c of the thermal noise is defined as the full-width at half-maximum (FWHM) of the temporal field correlation function. If Δt is set too close to τ_c, E_s exhibits a sudden jump at each time-step. This presents a problem, since the 1-D FDTD algorithm cannot accurately propagate such jumps if the Courant factor $S \equiv c\Delta t/\Delta x$ is other than unity [20]. For $S > 1$, the FDTD algorithm is unstable and yields exponentially increasing values. For $S < 1$, pulses with step discontinuities become distorted with fringes corresponding to both retarded and superluminal propagation. This distortion occurs because of numerical dispersion artifacts arising from the approximate spatial derivatives applied to the sampled electric and magnetic fields. Fortunately, setting $S = 1$ eliminates these artifacts and yields a numerically stable algorithm [20]. In addition, when Δx is selected to properly sample the highest useful spatial frequencies in the grid, $S = 1$ implies that $\Delta t << \tau_c$, which provides a dense temporal sampling relative to the correlation/coherence time of the thermal noise.

18.4 SIMULATIONS IN VACUUM

We now examine blackbody radiation in a 1-D FDTD vacuum grid. The electric field sources, implemented via (18.8), generate both an electric field $E(x, t)$ and a magnetic field $H(x, t)$, which propagate into the grid. $E(x,\omega)$ and $H(x,\omega)$ are obtained via a discrete Fourier transform (DFT) of $E(x,t)$ and $H(x,t)$. Since both $E(x,t)$ and $H(x,t)$ are real valued, $E(x,\omega) = E(x,-\omega)$ and $H(x,\omega) = H(x,-\omega)$. The EM energy density at frequency ω includes $E(x,\omega)$, $E(x,-\omega)$, $H(x,\omega)$, and $H(x,-\omega)$. For a vacuum grid, the steady-state energy density at every position x should equal the blackbody radiation density. The rms amplitude δ of the source field E_s is determined by:

$$\frac{1}{2}\left[\varepsilon_0 \left|E\big(x,|\omega|\big)\right|^2 + \mu_0 \left|H\big(x,|\omega|\big)\right|^2\right] = \frac{\hbar}{\pi c}\cdot\frac{|\omega|}{e^{\hbar|\omega|/k_B T} - 1} \tag{18.9}$$

To obtain an accurate noise spectrum with the DFT, both the frequency and temporal resolutions must be chosen carefully. The two problems affecting the reliability of the DFT are aliasing and leakage due to the use of a finite simulation time [21]. The solution to these problems is to increase the number of time-steps $2N$ and decrease the time-step Δt. This takes the DFT closer to a perfect analytical Fourier transform, but run-time and memory limitations must be considered as well. Taking advantage of the FFT algorithm significantly reduces the time required by both noise generation and spectral analysis.

Further limitations can be enforced for more efficient computation. Although the thermal noise spectrum can be very broad, only noise within a certain frequency range is relevant to a specific problem. Let $\omega_{\min}$ and $\omega_{\max}$ denote the lower and upper limits of the frequency range of interest, and $\Delta\omega$ the frequency resolution needed within this range. To guarantee the accuracy of the noise simulation in $\omega_{\min} < \omega < \omega_{\max}$, the total running time τ_{sim} must exceed $2\pi/\omega_{\min}$ and $2\pi/\Delta\omega$. Time-step Δt has an additional requirement, $\Delta t < \pi/\omega_{\max}$.

In one dimension, two independent noise signals $E_s(t_k)$ are generated via (18.8). One is added as a soft source adjacent to the left ABC; the other is added as a soft source adjacent to the right ABC. Both have equal rms amplitude δ, adjusted so that (18.9) is satisfied. Thus, the average EM flux to the left equals the flux to the right at any position x in the grid. Since the system is one dimensional, the EM flux at any distance away from the source has the same magnitude. For the EM energy density radiated by one source to equal $D\big(|\omega|, T\big)$, we have:

$$\delta = \sqrt{\frac{2}{\varepsilon_0}\frac{1}{6\hbar c}k_B T} \tag{18.10}$$

Figure 18.2 depicts examples of the electric-field noise source $E_s(t_k)$ at $T = 30{,}000\,\text{K}$ and $T = 50{,}000\,\text{K}$. The two point noise sources at the ABC interface radiate into both the grid and the ABC. Since the sources on either side of the grid are uncorrelated, their energy densities, not their field amplitudes, add in the FDTD grid. Thus, no modification of δ from that given by (18.10) is needed to satisfy (18.9). After the noise fields in the grid reach a steady state, the noise spectrum at any grid point is obtained by a DFT. The frequency range of interest is set as $\omega_{\text{min}} = 2 \times 10^{15}$ Hz, $\omega_{\text{max}} = 2.5 \times 10^{16}$ Hz, and the frequency resolution is $\Delta\omega = 1 \times 10^{12}$ Hz. From the condition $\Delta t < \pi/\omega_{\text{max}}$, Δx is less than 37 nm.

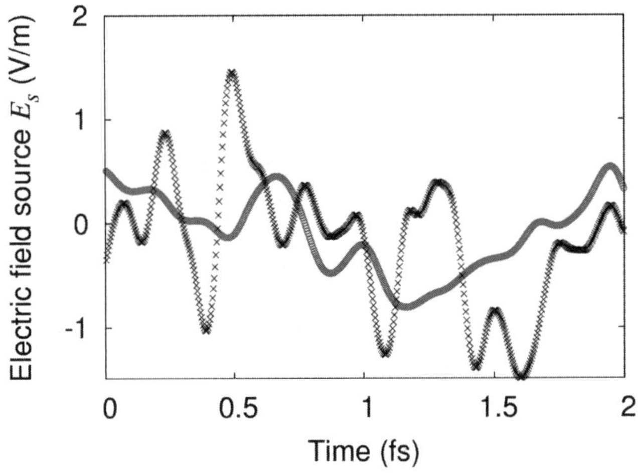

Fig. 18.2 Noise source electric field $E_s(t_k)$ generated for $T = 30{,}000\,\text{K}$ (∘ symbols) and $T = 50{,}000\,\text{K}$ (× symbols). The noise amplitude increases, but the correlation time τ_c decreases, for higher temperatures: $\tau_c \approx 0.337$ fs for $T = 30{,}000\,\text{K}$, $\tau_c \approx 0.203$ fs for $T = 50{,}000\,\text{K}$. The FDTD grid cell size is $\Delta x = 1$ nm, and the total FDTD simulation time is $\tau_{\text{sim}} = 7$ ps. *Source:* J. Andreasen, H. Cao, A. Taflove, P. Kumar, and C. qi Cao, *Physical Review A*, Vol. 77, 2008, 023810, ©2008 American Physical Society.

Figure 18.3 compares the FDTD-calculated energy density to that of thermal radiation density $D(\omega, T)$ for $T = 30{,}000\,\text{K}$ and $T = 50{,}000\,\text{K}$. From Fig. 18.3, we see that using $\Delta x = 10\,\text{nm}$ creates a slight discrepancy at high frequencies: namely, at $\omega > 1 \times 10^{16}$ Hz the mean error exceeds ~2.5%. To reduce the error to below 2.5% at $\omega_{\text{max}} = 2.5 \times 10^{16}$ Hz, the resolution is refined. For example, using $\Delta x = 4\,\text{nm}$ reduces the error at ω_{max} to 1.6%; using $\Delta x = 1$ nm further reduces the error to less than 0.1%.

Of course, refining Δx requires a corresponding reduction of Δt to maintain numerical stability of the FDTD algorithm. If the total number of time-steps (here, $2N = 2^{21}$) is fixed, this decrease of Δt leads to a reduced total simulation time of $\tau_{\text{sim}} = 2N\Delta t$. However, the conditions $2\pi/\tau_{\text{sim}} < \omega_{\text{min}}$ and $2\pi/\tau_{\text{sim}} < \Delta\omega$ must still be satisfied. With $\Delta x = 1$ nm, $2\pi/\tau_{\text{sim}}$ increases to 9×10^{11} Hz, which is still below the set values of ω_{min} and $\Delta\omega$. Therefore, using the value of δ in (18.10) and carefully choosing Δx and Δt, the blackbody spectrum at every point in the FDTD grid can be properly computed within the frequency range of interest.

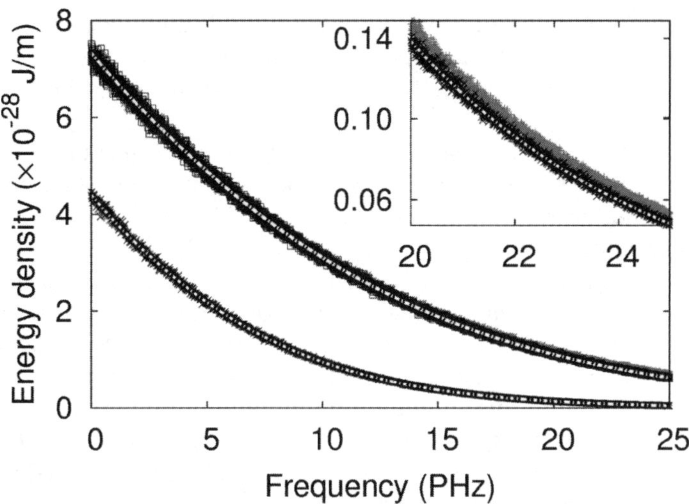

Fig. 18.3 FDTD-calculated energy density of blackbody radiation propagating in a 1-D vacuum versus frequency ω for temperatures $T = 30{,}000\,\text{K}$ (lower) and $T = 50{,}000\,\text{K}$ (upper). The inset shows the energy density for temperature $T = 30{,}000\,\text{K}$ at higher frequencies. Data are obtained by averaging over 2000 simulations with grid resolution $\Delta x = 10$ nm (+ symbols) and $\Delta x = 1$ nm (• symbols). The source spectra $D(\omega, T)$ are plotted as solid lines ( ) on top of the numerical spectra. *Source:* J. Andreasen, H. Cao, A. Taflove, P. Kumar, and C. qi Cao, *Physical Review A*, Vol. 77, 2008, 023810, ©2008 American Physical Society.

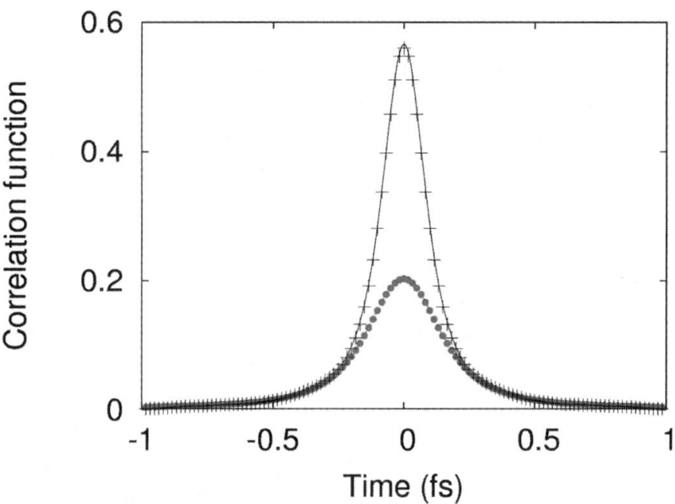

Fig. 18.4 FDTD-computed temporal correlation function $\langle E_s(t_1)E_s(t_2)\rangle$ vs. $(t_2 - t_1)$ for the noise-source electric field at $T = 30{,}000\,\text{K}$ (• symbols) and $T = 50{,}000\,\text{K}$ (+ symbols). Noise correlation times are $\tau_c \approx 0.337$ fs for $T = 30{,}000\,\text{K}$, $\tau_c \approx 0.203$ fs for $T = 50{,}000\,\text{K}$. Solid curve (——) represents $\langle E_s(t_1)E_s(t_2)\rangle$ given by (18.6) for $T = 50{,}000\,\text{K}$. Every 5th data point is taken from the FDTD results to more clearly show the agreement with the analytical solution. The FDTD grid cell size is $\Delta x = 1$ nm, and the total FDTD simulation time is $\tau_{\text{sim}} = 7$ ps. *Source:* J. Andreasen, H. Cao, A. Taflove, P. Kumar, and C. qi Cao, *Physical Review A*, Vol. 77, 2008, 023810, ©2008 American Physical Society.

Since the spectrum of EM energy density at any point in a 1-D grid is identical to that at the source, no distortion of the noise spectrum can arise because of the propagation of the noise fields in a vacuum. This has been numerically confirmed: δ does not depend on the length of the 1-D system.

Figure 18.4 displays the FDTD-calculated temporal correlation function of the electric field at $T = 30{,}000\,\mathrm{K}$ and $T = 50{,}000\,\mathrm{K}$. With increasing temperature, the correlation time τ_c drops quickly; the quantitative dependence is $\tau_c \approx 1.32\hbar / k_B T$. This $1/T$ dependence does not change for a dimensionality higher than one; only the prefactor changes [22]. As τ_c decreases, the time-step Δt is reduced to maintain the temporal resolution of the correlation function. The subsequent reduction of total running time τ_{sim} does not affect the numerical accuracy, as long as the total number of time-steps $2N$ is fixed. A decrease of $2N$ would result in an increased mean-square error in the correlation function due to reduced sampling.

In Fig. 18.4, the excellent agreement of the FDTD-computed temporal correlation function with the analytical expression for blackbody radiation given by (18.6) confirms that introducing noise sources with the characteristics of blackbody radiation adjacent to the FDTD absorbing boundary effectively simulates blackbody radiation in vacuum.

18.5 SIMULATIONS OF AN OPEN CAVITY

We now consider the interaction of blackbody radiation with the 1-D dielectric slab of length L and refractive index $n > 1$ shown in Fig. 18.1. This slab constitutes an open cavity in that EM field leakage occurs from both surfaces of the slab into an exterior region. Due to this leakage, cavity modes within the slab have a decay time (photon lifetime) given by τ.

We begin by examining FDTD simulations of blackbody radiation for a good cavity where $\tau >> \tau_c$, the coherence time of the thermal radiation. In this case, the average amount of thermal noise in one cavity mode is found to agree with the solution of the quantum Langevin equation under the Markovian approximation. Subsequently, we consider the transition from the Markovian regime to the non-Markovian regime, and demonstrate that the buildup of the intracavity noise field depends on the ratio of τ_c to τ. This result is explained qualitatively by the interference effect.

18.5.1 Markovian Regime ($\tau >> \tau_c$)

For a lossless 1-D dielectric cavity, the cavity mode frequency is $\omega_m = m(\pi c/nL)$, where m is an integer and c is the speed of light in vacuum. The frequency spacing of adjacent modes is $d\omega = \pi c/nL$, which is independent of m. Decay of the cavity field is caused only by its escape from the cavity. All of the cavity modes have roughly the same decay time $\tau = -2nL/[c\ln(r^2)]$, where $r = (1-n)/(1+n)$ is the reflection coefficient at the boundary of the dielectric slab. The mode linewidth is $\delta\omega = 2/\tau$. Only good cavities whose modes are well separated in frequency, namely, $\delta\omega < d\omega$, are considered here. Since $\delta\omega \propto 1/L$, the ratio $\delta\omega/d\omega$ is independent of L, and is only a function of n.

The Langevin equation for the annihilation operator $\hat{a}_m(t)$ of photons in the mth cavity mode is given by:

$$\frac{d\hat{a}_m(t)}{dt} = -\frac{1}{\tau}\hat{a}_m(t) + \hat{F}_m(t) \tag{18.11}$$

where $\hat{F}_m(t)$ is the Langevin force. If $\tau >> \tau_c$, $\hat{F}_m(t)$ can be considered δ-correlated in time. The Markovian approximation gives $\langle \hat{F}_m^\dagger(t)\hat{F}_m(t')\rangle = D_F\,\delta(t-t')$. According to the fluctuation-dissipation theorem, $D_F = (1/\tau)n_T(\omega_m)$, where $n_T(\omega_m) = 1/[\exp(\hbar\omega_m/k_B T) - 1]$ is the number of thermal photons in a vacuum mode of frequency ω_m at temperature T.

From (18.11), the average photon number in one cavity mode $\langle \hat{n}_m(t)\rangle \equiv \langle \hat{a}_m^\dagger(t)\hat{a}_m(t)\rangle$ satisfies:

$$\frac{d}{dt}\langle \hat{n}_m(t)\rangle = -\frac{2}{\tau}\langle \hat{n}_m(t)\rangle + \frac{2}{\tau}n_T(\omega_m) \tag{18.12}$$

At steady state, $\langle \hat{n}_m\rangle = n_T(\omega_m)$ in each cavity mode. Here, the number of thermal photons is determined by the Bose–Einstein distribution $n_T(\omega_m)$. Because the amount of thermal fluctuation entering the cavity increases at the same rate as the intracavity energy decays, $\langle \hat{n}_m\rangle$ is independent of the cavity mode decay rate.

Since there is neither a driving field (e.g., a pumping field) nor excited atoms in the cavity, the EM energy stored in one cavity mode comes entirely from the blackbody radiation of the ABC, which is coupled into that particular mode. Assuming well-separated modes ($\delta\omega < d\omega$), the following expression allows calculation of the steady-state number of photons in the mth cavity mode from the FDTD-computed intracavity EM energy within the frequency range $\omega_{m-1/2} < \omega < \omega_{m+1/2}$, where $\omega_{m\pm 1/2} = (m \pm 1/2)\pi c/nL$:

$$n_m \equiv \langle \hat{n}_m\rangle = \frac{1}{\hbar\omega_m}\int_{\omega_{m-1/2}}^{\omega_{m+1/2}} d\omega \int_0^L dx \left[\frac{1}{2}\varepsilon\left|E(x,\omega)\right|^2 + \frac{1}{2}\mu_0\left|H(x,\omega)\right|^2\right] \tag{18.13}$$

Similarly, after the intracavity EM field reaches the steady state, the average thermal energy density U inside the cavity is calculated as:

$$U(\omega) = \frac{1}{L}\int_0^L dx \left[\frac{1}{2}\varepsilon\left|E(x,\omega)\right|^2 + \frac{1}{2}\mu_0\left|H(x,\omega)\right|^2\right] \tag{18.14}$$

In the FDTD simulations discussed below, the temperature of the thermal sources adjacent to the ABC is $T = 30{,}000$ K, generating radiation with a coherence time of $\tau_c = 0.337$ fs. The dielectric slab has the length $L = 2400$ nm and a refractive index $n = 6$. This choice of slab parameters results in a cavity lifetime, $\tau = 143$ fs, that is much longer than τ_c, and furthermore provides sufficiently separated cavity modes in frequency. The grid resolution Δx is 1 nm to ensure that $\Delta x << \lambda/n$. Finally, each simulation is run for $2N = 2^{21}$ time-steps.

Figure 18.5 shows the FDTD-computed intracavity noise spectrum $U(\omega)$ for the dielectric slab model. This spectrum exhibits peaks at the cavity resonant frequencies ω_m. Because $n > 1$, EM energy is also stored in the dielectric slab at frequencies away from the cavity resonances. For example, $U(\omega_{m\pm 1/2})$ is higher than that in vacuum by a factor of $2n^2/(n^2+1)$. Thus, the entire intracavity noise spectrum lies above the vacuum blackbody radiation spectrum.

Upon applying (18.13) to the FDTD-computed fields in this example, the modal photon number n_m is determined to equal $n_T(\omega_m)$ with a mean error less than 0.1%. This result confirms that the FDTD computational model of thermal noise in this open dielectric cavity is consistent with the prediction of quantum-mechanical theory, that is, n_m coincides with the Bose–Einstein distribution n_T for $\tau >> \tau_c$.

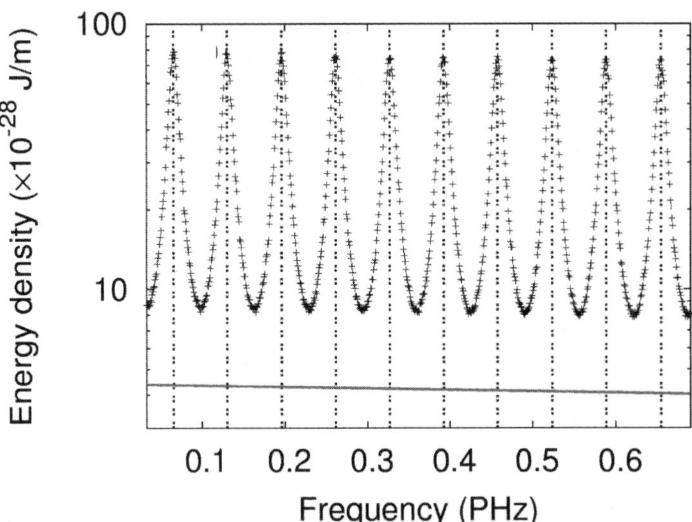

Fig. 18.5 FDTD-computed spatially averaged EM energy density $U(\omega)$ vs. frequency ω in a dielectric slab cavity of length $L = 2400\,\text{nm}$, refractive index $n = 6$, and cavity decay time $\tau = 143\,\text{fs} >> \tau_c$. Vertical dashed lines mark the frequencies ω_m of the cavity modes; the spectrum of impinging blackbody radiation $D(\omega, T)$ is the lower sloping (——) line. *Source:* J. Andreasen, H. Cao, A. Taflove, P. Kumar, and C. qi Cao, *Physical Review A*, Vol. 77, 2008, 023810, ©2008 American Physical Society.

18.5.2 Non-Markovian Regime ($\tau \sim \tau_c$)

The non-Markovian regime is approached by changing the cavity configuration so that the field lifetime τ is reduced, eventually becoming comparable to the coherence time, τ_c, of the thermal radiation impinging onto the cavity. For the 1-D dielectric slab cavity example, this can be readily accomplished by keeping the slab's refractive index at $n = 6$ while decreasing the cavity length L. This is a simple way of increasing the mode linewidth $\delta\omega$ while keeping the modes separated in frequency, i.e., keeping $\delta\omega / d\omega$ constant. Meanwhile, the increased mode linewidth and mode spacing allows for coarser frequency resolution, namely, an increase of $\Delta\omega$.

Figure 18.6 shows the number of thermal photons, n_m, in a cavity mode for the dielectric slab cavity model. This is calculated by applying (18.13) to the FDTD-computed fields as L decreases from 2400 nm to reduce τ. As τ approaches τ_c, n_m is no longer independent of τ, but starts increasing from $n_T(\omega_m)$. This means that the number of thermal photons captured by a cavity mode increases with the decrease of τ. In effect, as the cavity field lifetime approaches the coherence time of thermal radiation impinging on the cavity, the constructive interference of the thermal field increases within the cavity, leading to a larger buildup of intracavity energy.

We note that the modal photon numbers in Fig. 18.6 are time-averaged values. From the standpoint of quantum mechanics, the fact that these values are much less than unity implies that, most of the time, there is no photon in the cavity mode.

Figure 18.7 shows the FDTD-computed intracavity noise spectrum $U(\omega)$ for the case of the cavity length $L = 20$ nm ($\tau = 1.19$ fs). Here, the FDTD modeling parameters are $\Delta x = 0.1$ nm, $\Delta t = 3.33 \times 10^{-19}$ s, and $2N = 2^{21}$, with $\Delta\omega = 9 \times 10^{12}$ Hz. (Hence, the accuracy requirement $\Delta t << \tau$ is maintained.) Similar to Fig. 18.5, peaks occur at the cavity resonant frequencies ω_m, and the entire spectrum lies above the vacuum blackbody radiation spectrum.

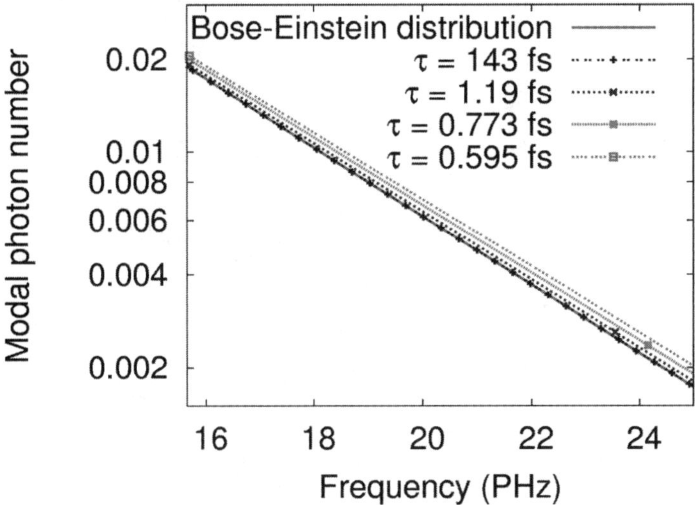

Fig. 18.6 FDTD-computed number of thermal photons, n_m, in individual cavity modes for the dielectric slab cavity model. The refractive index of the dielectric is held constant at $n = 6$, but the cavity length L decreases from 2400 nm to reduce τ. The impinging blackbody radiation has $T = 30{,}000$ K and $\tau_c = 0.337$ fs, so that the values of τ/τ_c are 424, 3.53, 2.29, and 1.77. Lines are drawn to connect the data points at the mode frequencies $\omega_m = m(\pi c/nL)$ to illustrate frequency dependence. For $\tau >> \tau_c$, the photon number n_m coincides with the Bose–Einstein distribution n_T. However, when $\tau \sim \tau_c$, n_m deviates from n_T. *Source:* J. Andreasen, H. Cao, A. Taflove, P. Kumar, and C. qi Cao, *Physical Review A*, Vol. 77, 2008, 023810, ©2008 American Physical Society.

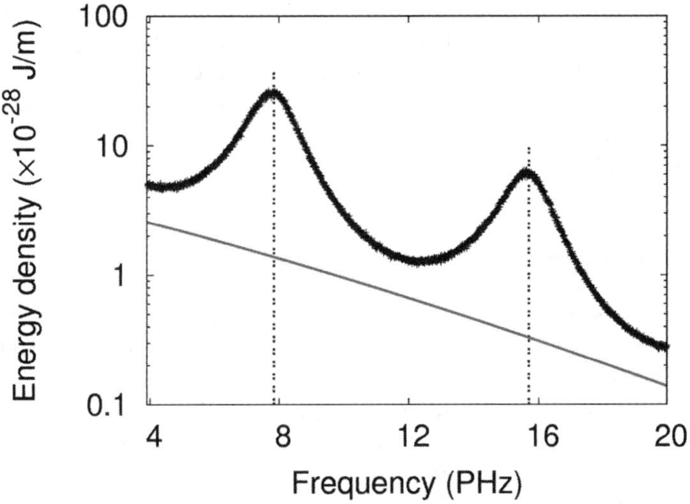

Fig. 18.7 FDTD-computed spatially averaged EM energy density $U(\omega)$ vs. frequency ω in a dielectric slab cavity of length $L = 20$ nm, refractive index $n = 6$, and $\tau = 1.19$ fs, comparable to τ_c. Vertical dashed lines mark the frequencies ω_m of the cavity modes. The spectrum of the impinging blackbody radiation $D(\omega, T)$ is the lower sloping (——) line. *Source:* J. Andreasen, H. Cao, A. Taflove, P. Kumar, and C. qi Cao, *Physical Review A*, Vol. 77, 2008, 023810, ©2008 American Physical Society.

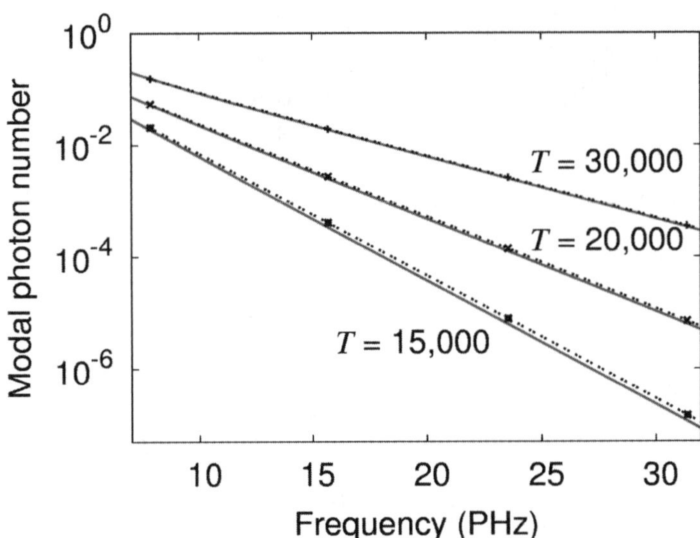

Fig. 18.8 FDTD-computed number of thermal photons, n_m, in individual cavity modes for a dielectric slab cavity with $n = 6$ and $L = 20$ nm (cavity decay time $\tau = 1.19$ fs). The temperature of blackbody radiation is varied to change τ_c, yielding τ/τ_c ratios of 3.45 at $T = 30{,}000$ K, 2.33 at $T = 20{,}000$ K, and 1.79 at $T = 15{,}000$ K. Dotted lines connect data points at the mode frequencies $\omega_m = m(\pi c/nL)$ to illustrate frequency dependence. For comparison, the Bose–Einstein distribution $n_T(\omega)$ (——) is shown for each temperature. *Source:* J. Andreasen, H. Cao, A. Taflove, P. Kumar, and C. qi Cao, *Physical Review A*, Vol. 77, 2008, 023810, ©2008 American Physical Society.

Figure 18.8 shows that the non-Markovian regime can also be approached by reducing the temperature T so that τ_c increases. In this scenario, there is decreased energy density of the thermal radiation, i.e., fewer thermal photons impinging on the cavity, and hence a smaller number of thermal photons, n_m, in a cavity mode. Nevertheless, $n_m > n_T(\omega_m)$ at the same temperature because the longer coherence time of the thermal field results in enhanced constructive interference within the cavity.

18.5.3 Analytical Examination and Comparison

The FDTD simulations summarized above demonstrate that, in the non-Markovian regime, the buildup of the intracavity noise field depends on the ratio of the cavity field lifetime, τ, to the coherence time of thermal radiation, τ_c. The advantage of using FDTD is that the thermal noise is introduced in the time domain without requiring prior knowledge of the cavity modes.

To gain a better understanding of the FDTD simulation results in the non-Markovian regime, the effect of τ_c on the amount of thermal noise within an open cavity is now examined analytically. The ratio of the intracavity EM energy at frequency ω to the energy density of the thermal source outside the cavity is given by:

$$W(\omega) = \frac{1}{D(\omega,T)} \int_0^L dx \left[\frac{1}{2}\varepsilon \left| E(x,\omega) \right|^2 + \frac{1}{2}\mu_0 \left| H(x,\omega) \right|^2 \right] \tag{18.15}$$

For a dielectric slab of refractive index n and length L, the expression for $W(\omega)$ is obtained using the transfer-matrix method [23]:

$$W(\omega) = \frac{2nc}{\omega}\left[\frac{2\omega nL(1+n^2)/c + (n^2-1)\sin(2\omega nL/c)}{1+6n^2+n^4 - (n^2-1)^2\cos(2\omega nL/c)}\right] \tag{18.16}$$

This can be used to calculate the ratio $B_m(\tau, \tau_c) \equiv n_m / n_T(\omega_m)$ as:

$$n_m = \frac{1}{\hbar\omega_m}\int_{\omega_{m-1/2}}^{\omega_{m+1/2}} d\omega\, W(\omega)\, D(\omega, T) \tag{18.17}$$

In the Markovian regime, $\tau \gg \tau_c$. Here, $D(\omega, T)$ is nearly constant over the frequency interval of one cavity mode, and can be removed from the integral in (18.17). This yields:

$$B_m(\tau, \tau_c) = \frac{D(\omega_m, T)}{\hbar\omega_m n_T(\omega_m)}\int_{\omega_{m-1/2}}^{\omega_{m+1/2}} d\omega\, W(\omega) = \frac{1}{\pi c}\int_{\omega_{m-1/2}}^{\omega_{m+1/2}} d\omega\, W(\omega) \tag{18.18}$$

Using the same parameters as in the FDTD simulations ($n = 6$, $L = 2400$ nm, and $\tau = 143$ fs), the integration of $W(\omega)$ in (18.18) gives a value close to πc. Hence, $B_m(\tau, \tau_c) \approx 1$ in this regime, with the deviation from unity being greater for smaller m. One possible reason for the latter is that the condition $\delta\omega \ll \omega_m$ no longer holds for small m, and there is a large uncertainty in defining the frequency of a cavity mode whose linewidth is comparable to its center frequency. In other words, the calculation of n_m using (18.17) becomes questionable.

As the ratio τ/τ_c falls to enter the non-Markovian regime, $D(\omega, T)$ acquires sufficient variation over the frequency range of a cavity mode to prevent its removal from the integral in (18.17). Consequently, $B_m(\tau, \tau_c)$ increases in a manner that is found to be consistent with the FDTD simulations [2]. The progression to the non-Markovian regime can be studied in two ways. First, if τ is fixed and τ_c is increased by lowering the temperature T, the absolute number of thermal photons n_m in a cavity mode drops, but its ratio to the number of thermal photons in a vacuum mode $n_T(\omega_m)$ increases. Conversely, if τ_c is fixed and τ is decreased by shortening the cavity length L, both n_m and $n_m / n_T(\omega_m)$ increase. The departure of n_m from $n_T(\omega_m)$ is a direct consequence of the breakdown of the Markovian approximation. When the coherence time of thermal radiation is comparable to the cavity decay time, the Langevin force $\hat{F}_m(t)$ in (18.11) is no longer δ-correlated in time, and (18.12) is invalid.

Overall, in both the Markovian and non-Markovian regimes, $B_m(\tau, \tau_c)$ depends only on the ratio τ/τ_c and not on τ or τ_c individually. Cavity modes of higher m have a larger value of $B_m(\tau, \tau_c)$ in both regimes.

18.6 SUMMARY AND OUTLOOK

In this chapter, fluctuations of electromagnetic fields in open cavities due to output coupling were simulated using the FDTD method. The foundation of this discussion was the fluctuation-dissipation theorem, which dictates that cavity field dissipation by leakage is accompanied by thermal noise, simulated here by classical electrodynamics. The absorbing boundary of the

FDTD grid was treated as a blackbody, which radiates into the grid. Noise sources were synthesized with spectra equivalent to that of blackbody radiation at various temperatures. It was found possible to select parameters in the 1-D FDTD simulations to avoid distortion of the noise spectra caused by the numerical wave propagation. The FDTD-computed noise fields propagating in vacuum retained the blackbody spectra and temporal correlation functions.

When an open dielectric cavity was placed in the FDTD grid, the thermal radiation was coupled into the cavity and contributed to the thermal noise for the cavity field. In the Markovian regime where the cavity photon lifetime τ is much longer than the coherence time of thermal radiation τ_c, the FDTD-calculated amount of thermal noise in a cavity mode agreed with that given by the quantum Langevin equation. This validated the numerical model of thermal noise that originates from cavity openness or output coupling. FDTD simulations also demonstrated that, in the non-Markovian regime, the steady-state number of thermal photons in a cavity mode exceed that in a vacuum mode. This was attributed to the constructive interference of the thermal field inside the cavity.

The advantage of the FDTD numerical model is that the thermal noise is added in the time domain without prior knowledge of cavity modes. It can be applied to simulate complex open systems whose modes are not known prior to the FDTD calculations. This approach is especially useful for very leaky cavities whose modes overlap strongly in frequency, as the thermal noise related to the cavity leakage is introduced naturally without distinguishing the modes. Therefore, the method developed here can be applied to a whole range of quantum optics problems.

Although the FDTD calculation of thermal noise was performed on 1-D systems here, extension to 2-D and 3-D systems is straightforward. Note that this implementation of blackbody radiation is *not* equivalent to the simulation of zero-point fluctuations that have a different physical origin than thermal noise. However, the numerical method reviewed here *can* be used to study the dynamics of EM fields excited by arbitrarily correlated noise sources. One potential application is noise radar [24, 25]. The propagation, reflection, and scattering of ultrawideband signals utilized by noise radar can be readily simulated using the present technique.

REFERENCES

[1] Garrod, C., *Statistical Mechanics and Thermodynamics*, New York: Oxford University Press, 1995.

[2] Andreasen, J., H. Cao, A. Taflove, P. Kumar, and C. qi Cao, "Finite-difference time-domain simulation of thermal noise in open cavities," *Physical Review A*, Vol. 77, 2008, 023810.

[3] Brown, R., "A brief account of microscopical observations made in the months of June, July, and August, 1827, on the particles contained in the pollen of plants; and on the general existence of active molecules in organic and inorganic bodies," *Philosophical Magazine*, Vol. 4, 1828, pp. 161–173.

[4] Langevin, M. P., "On the theory of Brownian motion," *C. R. Academic Science* (Paris), Vol. 146, 1908, pp. 530–533.

[5] Slavcheva, G. M., J. M. Arnold, and R. W. Ziolkowski, "FDTD simulation of the nonlinear gain dynamics in active optical waveguides and semiconductor microcavities," *IEEE J. Selected Topics in Quantum Electronics*, Vol. 10, 2004, pp. 1052–1062.

[6] Andreasen, J., and H. Cao, "Finite-difference time-domain formulation of stochastic noise in macroscopic atomic systems," *J. Lightwave Technology*, Vol. 27, 2009, pp. 4530–4535.

[7] Drummond, P. D., and M. G. Raymer, "Quantum theory of propagation of nonclassical radiation in a near-resonant medium," *Physical Review A*, Vol. 44, 1991, pp. 2072–2085.

[8] Andreasen, J., and H. Cao, "Numerical study of amplified spontaneous emission and lasing in random media," *Physical Review A*, Vol. 82, 2010, 063835.

[9] Louisell, W. H., *Quantum Statistical Properties of Radiation*, New York: Wiley, 1973.

[10] Lax, M., "Quantum noise IV: Quantum theory of noise sources," *Physical Review*, Vol. 145, 1966, 110.

[11] Haken, H., *Laser Theory*, Berlin: Springer-Verlag, 1983.

[12] Luo, C., A. Narayanaswamy, G. Chen, and J. Joannopoulos, "Thermal radiation from photonic crystals: A direct calculation," *Physical Review Lett.*, Vol. 93, 2004, 213905.

[13] Chan, D., M. Soljacic, and J. Joannopoulos, "Direct calculation of thermal emission for three-dimensionally periodic photonic crystal slabs," *Physical Review E*, Vol. 74, 2006, 036615.

[14] Lang, R., and M. Scully, "Fluctuations in mode-locked 'single-mode' laser oscillation," *Optics Communications*, Vol. 9, 1973, pp. 331–337.

[15] Ujihara, K., "Quantum theory of a one-dimensional laser with output coupling: Linear theory," *Physical Review A*, Vol. 16, 1977, pp. 652–658.

[16] Haken, H., *Light: Waves, Photons, Atoms*, New York: North-Holland Physics Publishing, 1981.

[17] Planck, M., "On the law of distribution of energy in the normal spectrum," *Annalen der Physik*, Vol. 4, 1901, p. 553 ff.

[18] Freilikher, V., E. Kanzieper, and A. Maradudin, "Coherent scattering enhancement in systems bounded by rough surfaces," *Physics Rep.*, Vol. 288, 1997, pp. 127–204.

[19] Brysbaert, M., "Algorithms for randomness in the behavioral sciences: A tutorial," *Behavioral Research Methods Ins. C.*, Vol. 23, 1991, pp. 45–60.

[20] Taflove, A., and S. C. Hagness, *Computational Electrodynamics: The Finite-Difference Time-Domain Method*, 3rd ed., Norwood, MA: Artech, 2005.

[21] Hamming, R., *Numerical Methods for Scientists and Engineers*, New York: Dover, 1986.

[22] Kano, Y., and E. Wolf, "Temporal coherence of black body radiation," *Proc. Physical Society*, Vol. 80, 1962, pp. 1273–1276.

[23] Born, M., and E. Wolf, *Principles of Optics*, New York: Pergamon Press, 1975.

[24] Horton, B. M., "Noise-modulated distance measuring systems," *Proc. IRE*, Vol. 47, 1959, pp. 821–828.

[25] Theron, I., E. Walton, S. Gunawan, and L. Cai, "Ultrawideband noise radar in the VHF/UHF band," *IEEE Trans. Antennas and Propagation*, Vol. 47, 1999, pp. 1080–1084.

Chapter 19

Casimir Forces in Arbitrary Material Geometries[1]

Ardavan Oskooi and Steven G. Johnson

19.1 INTRODUCTION

In recent years, Casimir forces arising from quantum vacuum fluctuations of the electromagnetic field [1–3] have become the focus of intense theoretical and experimental effort [4–21]. This effect has been verified via many experiments [22–25], most commonly in simple, one-dimensional (1-D) geometries involving parallel plates or approximations thereof, with some exceptions [26]. A particular topic of interest is the geometry and material dependence of the force, a subject that has only recently begun to be addressed in experiments [26] and by promising new theoretical methods [27–38]. For example, recent work has shown that it is possible to find unusual effects arising from many-body interactions or from systems exhibiting strongly coupled material and geometric dispersion [39–43]. These numerical studies have been mainly focused on two-dimensional (2-D) [13, 44–46] or simple three-dimensional (3-D) constant-cross-section geometries [33, 40, 47] for which numerical calculations are tractable.

This chapter reviews a simple and general method, originally reported in [48–50], to compute Casimir forces in arbitrary geometries and for arbitrary materials. The new method is based on the finite-difference time-domain (FDTD) solution of Maxwell's equations [51]. A time-domain approach offers a number of advantages over previous methods. First, it enables the use of powerful free and commercial FDTD software with no modification. This permits exploration of the material and geometry dependence of Casimir forces, especially for inhomogeneous/anisotropic dielectrics [52] specified in 3-D. Second, a time-domain formulation offers a fundamentally different viewpoint on Casimir phenomena, and thus new opportunities for theoretical understanding of these forces in complex geometries.

---

[1]This chapter is a synthesis derived from Refs. [48–50]: (1) A. W. Rodriguez, A. P. McCauley, J. D. Joannopoulos, and S. G. Johnson, "Casimir forces in the time domain: Theory," *Physical Review A*, Vol. 80, 2009, 012115, ©2009 The American Physical Society; (2) A. P. McCauley, A. W. Rodriguez, J. D. Joannopoulos, and S. G. Johnson, "Casimir forces in the time domain: Applications," *Physical Review A*, Vol. 81, 2010, 012119, ©2010 The American Physical Society; and (3) K. Pan, A. P. McCauley, A. W. Rodriguez, M. T. H. Reid, J. K. White, and S. G. Johnson, "Calculation of nonzero-temperature Casimir forces in the time domain," *Physical Review A*, Vol. 83, 2011, 040503(R), ©2011 The American Physical Society.

For consistency of notation relative to these three source papers, in this chapter the symbol i is used to designate $\sqrt{-1}$, rather than the symbol j; and a phasor is denoted as $e^{-i\omega t}$.

This chapter first reviews the theory presented in [48] to compute Casimir forces via FDTD modeling. This is followed by a review of the harmonic-expansion technique reported in [49] that substantially increases the speed of the computation for many systems. This technique allows Casimir forces to be efficiently computed, even on single computers, and further expands the range of accessible problems when using commonly available parallel-processing FDTD software. The FDTD-Casimir force models reported in [49] for representative 2-D and 3-D geometries are then reviewed, culminating in a fully 3-D model of the stable levitation of a silica microsphere in a high-permittivity dielectric fluid above a spherically indented metal surface. The chapter concludes with a review of the work reported in [50], which extended the techniques of [48, 49] to nonzero temperatures.

The work reported in [48–50] demonstrated both the validity of the basic FDTD-Casimir approach and the desirable properties of the harmonic expansion technique. Simulations were conducted using a versatile and freely available FDTD code [53] that is scriptable or programmable to automatically run the sequence of simulations required to determine the Casimir force.

19.2 THEORETICAL FOUNDATION

Reference [48] reported the derivation of a new numerical technique to compute the Casimir force on a body using the FDTD method. The approach involved a modification of the well-known stress-tensor method [2]. By this method, the force on an object can be found by integrating the Minkowski stress tensor around a surface S surrounding the object, and over all frequencies. Reference [48] abandoned the frequency domain altogether in favor of a purely time-domain scheme in which the force on an object is computed via a series of independent FDTD calculations in which sources are placed at each point on S. The electromagnetic response to these sources is then integrated in time against a predetermined function $g(-t)$.

19.2.1 Stress-Tensor Formulation

Following the discussion in [48], the Casimir force on a body can be expressed [2] as an integral over any closed surface S (enclosing the body) of the mean electromagnetic stress tensor $\langle T_{ij}(\omega,\boldsymbol{x})\rangle$, where ω denotes frequency and $\boldsymbol{x}$ denotes spatial position. In particular, the Casimir force in the ith direction is given by:

$$F_i = \int_0^\infty d\omega \oiint_S \sum_j \langle T_{ij}(\omega,\boldsymbol{x})\rangle \, dS_j \tag{19.1}$$

The stress tensor is expressed in terms of correlation functions of the field operators $\langle E_i(\omega,\boldsymbol{x})\, E_j(\omega,\boldsymbol{x}')\rangle$ and $\langle H_i(\omega,\boldsymbol{x})\, H_j(\omega,\boldsymbol{x}')\rangle$:

$$\begin{aligned}\langle T_{ij}(\omega,\boldsymbol{x})\rangle &= \mu(\omega,\boldsymbol{x})\left[\langle H_i(\boldsymbol{x})\, H_j(\boldsymbol{x})\rangle_\omega - 0.5\delta_{ij}\sum_k \langle H_k(\boldsymbol{x})\, H_k(\boldsymbol{x})\rangle_\omega\right] \\ &+ \varepsilon(\omega,\boldsymbol{x})\left[\langle E_i(\boldsymbol{x})\, E_j(\boldsymbol{x})\rangle_\omega - 0.5\delta_{ij}\sum_k \langle E_k(\boldsymbol{x})\, E_k(\boldsymbol{x})\rangle_\omega\right]\end{aligned} \tag{19.2}$$

where both the electric-field and magnetic-field correlation functions can be written as derivatives of a vector potential operator $\boldsymbol{A}^E(\omega,\boldsymbol{x})$:

$$E_i(\omega,\boldsymbol{x}) = -i\omega A_i^E(\omega,\boldsymbol{x}) \tag{19.3}$$

$$\mu H_i(\omega,\boldsymbol{x}) = (\nabla\times)_{ij} A_j^E(\omega,\boldsymbol{x}) \tag{19.4}$$

A superscript is explicitly placed on the vector potential in order to refer to the choice of gauge [(19.3) and (19.4)], in which $\boldsymbol{E}$ is obtained as a time-derivative of $\boldsymbol{A}$.

The fluctuation-dissipation theorem relates the correlation function of $\boldsymbol{A}^E$ to the photon Green's function $G_{ij}^E(\omega;\boldsymbol{x},\boldsymbol{x}')$:

$$\left\langle A_i^E(\omega,\boldsymbol{x})\, A_j^E(\omega,\boldsymbol{x}')\right\rangle = -\frac{\hbar}{\pi}\operatorname{Im}\left[G_{ij}^E(\omega;\boldsymbol{x},\boldsymbol{x}')\right] \tag{19.5}$$

where G_{ij}^E is the vector potential A_i^E in response to an electric-dipole current $\boldsymbol{J}$ along the $\hat{\boldsymbol{e}}_j$ direction:

$$\left[\nabla\times\frac{1}{\mu(\omega,\boldsymbol{x})}\nabla\times -\omega^2\varepsilon(\omega,\boldsymbol{x})\right]\boldsymbol{G}_j^E(\omega;\boldsymbol{x},\boldsymbol{x}') = \delta(\boldsymbol{x}-\boldsymbol{x}')\hat{\boldsymbol{e}}_j \tag{19.6}$$

Given G_{ij}^E, one can use (19.3) and (19.4) in conjunction with (19.5) to express the field correlation functions at points $\boldsymbol{x}$ and $\boldsymbol{x}'$ in terms of the photon Green's function:

$$\left\langle E_i(\omega,\boldsymbol{x})\, E_j(\omega,\boldsymbol{x}')\right\rangle = \frac{\hbar}{\pi}\omega^2\operatorname{Im}\left[G_{ij}^E(\omega;\boldsymbol{x},\boldsymbol{x}')\right] \tag{19.7}$$

$$\left\langle H_i(\omega,\boldsymbol{x})\, H_j(\omega,\boldsymbol{x}')\right\rangle = -\frac{\hbar}{\pi}(\nabla\times)_{il}(\nabla'\times)_{jm}\operatorname{Im}\left[G_{lm}^E(\omega;\boldsymbol{x},\boldsymbol{x}')\right] \tag{19.8}$$

To find the force via (19.1), $G_{ij}^E(\omega;\boldsymbol{x},\boldsymbol{x}'=\boldsymbol{x})$ must first be computed at every $\boldsymbol{x}$ on the surface of integration S, and for every ω [2]. Equation (19.6) can be solved numerically in a number of ways, such as by a finite-difference discretization [30]. This involves discretizing space and solving the resulting matrix eigenvalue equation using standard numerical linear algebra techniques [54, 55]. We note that finite spatial discretization automatically regularizes the singularity in G_{ij}^E at $\boldsymbol{x}=\boldsymbol{x}'$, making G_{ij}^E finite everywhere [30].

19.2.2 Complex Frequency Domain

As shown, (19.6) is of limited computational utility because it gives rise to an oscillatory integrand with non-negligible contributions at all frequencies, making numerical integration difficult [30]. However, the integral over ω can be re-expressed as the imaginary part of a contour integral of an analytic function by commuting the ω integration with the Im operator in (19.7) and (19.8). The primitive causality constraint implies that there can be no poles in the integrand in the upper-half of the complex plane. The integral, considered as a complex contour integral, is then invariant if the contour of integration is deformed above the real-frequency axis and into the first quadrant of the complex-frequency plane, via some mapping $\omega\to\omega(\xi)$.

Now, a positive imaginary component can be added to the frequency, causing the force integrand to decay rapidly with increasing ξ. In particular, upon deformation, (19.6) is mapped to:

$$\left[\nabla \times \frac{1}{\mu(\xi,\boldsymbol{x})} \nabla \times -\omega^2(\xi)\,\varepsilon(\xi,\boldsymbol{x})\right] \boldsymbol{G}_j^E(\xi;\boldsymbol{x},\boldsymbol{x}') = \delta(\boldsymbol{x}-\boldsymbol{x}')\hat{\boldsymbol{e}}_j \tag{19.9}$$

and (19.7) and (19.8) are mapped to:

$$\left\langle E_i(\xi,\boldsymbol{x})\, E_j(\xi,\boldsymbol{x}')\right\rangle = \frac{\hbar}{\pi}\omega^2(\xi) G_{ij}^E(\xi;\boldsymbol{x},\boldsymbol{x}') \tag{19.10}$$

$$\left\langle H_i(\xi,\boldsymbol{x})\, H_j(\xi,\boldsymbol{x}')\right\rangle = -\frac{\hbar}{\pi}(\nabla\times)_{il}\,(\nabla'\times)_{jm}\, G_{lm}^E(\xi;\boldsymbol{x},\boldsymbol{x}') \tag{19.11}$$

Equation (19.1) becomes:

$$F_i = \text{Im}\left[\int_0^\infty d\xi \frac{d\omega}{d\xi} \oint\!\!\!\oint_S \sum_j \left\langle T_{ij}(\xi,\boldsymbol{x})\right\rangle dS_j\right] \tag{19.12}$$

A finite spatial grid (as used in this approach) requires no further regularization of the integrand, and the finite value of all quantities means that there is no difficulty in commuting the Im operator with the integration.

It is possible to choose from a general class of contours, provided that they satisfy $\omega(0) = 0$ and remain above the real ξ-axis. The standard contour $\omega(\xi) = i\xi$ is a Wick rotation, which is known to yield a force integrand that is smooth and exponentially decaying in ξ [2]. In general, the most suitable contour depends on the numerical method being employed. A Wick rotation guarantees a strictly positive-definite and real-symmetric Green's function, making (19.6) solvable by the most efficient numerical techniques (e.g., the conjugate-gradient method) [55]. One can also solve (19.6) for arbitrary $\omega(\xi)$, but this generally involves the use of direct solvers or more complicated iterative techniques [54]. However, the class of contours amenable to an efficient time-domain solution is more restricted. For instance, a Wick rotation turns out to be unstable in the time domain because it implies the presence of gain [48].

19.2.3 Time-Domain Approach

It is possible to solve (19.6) in the time domain by evolving Maxwell's equations in response to a delta-function current impulse $\boldsymbol{J}(t,\boldsymbol{x}) = \delta(t-t')\delta(\boldsymbol{x}-\boldsymbol{x}')\hat{\boldsymbol{e}}_j$ in the direction of $\hat{\boldsymbol{e}}_j$. The term G_{ij}^E can then be directly computed from the Fourier transform of the resulting $\boldsymbol{E}$ field. However, obtaining a smooth and decaying force integrand requires expressing the mapping $\omega \rightarrow \omega(\xi)$ in the time-domain equations of motion.

A simple way to see the effect of this mapping is to notice that (19.9) can be viewed as the Green's function at real "frequency" ξ and complex dielectric permittivity [48]:

$$\varepsilon_c(\xi,\boldsymbol{x}) = \frac{\omega^2(\xi)}{\xi^2}\,\varepsilon(\boldsymbol{x}) \tag{19.13}$$

where, for simplicity, μ and ε are taken to be frequency independent. At this point, it is important to emphasize that the original physical system ε at a frequency ω is the one in which Casimir forces and fluctuations appear. The dissipative system ε_c at a frequency ξ is merely an artificial technique introduced to compute the Green's function.

Integrating along a frequency contour $\omega(\xi)$ is therefore equivalent to making the medium dispersive in the form of (19.13). Consequently, the time-domain equations of motion under this mapping correspond to evolution of the fields in an effective dispersive medium given by $\varepsilon_c(\xi,\boldsymbol{x})$.

To be suitable for FDTD, this medium should have three properties: (1) It must respect causality; (2) it cannot support gain, which leads to exponential blowup in the time domain; and (3) it should be easy to implement. A Wick rotation is very easy to implement in the time domain, corresponding to setting $\varepsilon_c = -\varepsilon$. However, a negative epsilon represents gain (the refractive index is $\pm\sqrt{\varepsilon}$, where one of the signs corresponds to an exponentially growing solution). Therefore, a more general frequency-dependent ε_c must be considered.

Implementing arbitrary dispersion in FDTD generally requires the introduction of auxiliary fields or higher-order time-derivative terms into Maxwell's equations [51]. In general, this becomes computationally expensive. The precise implementation depends strongly on the choice of contour $\omega(\xi)$. However, almost any dispersion is suitable, as long as it is causal and dissipative (excluding gain). A simple choice is an $\varepsilon_c(\xi,\boldsymbol{x})$ corresponding to a medium with frequency-independent conductivity σ:

$$\varepsilon_c(\xi,\boldsymbol{x}) = \varepsilon(\boldsymbol{x})\left(1 + \frac{i\sigma}{\xi}\right) \tag{19.14}$$

This has three main advantages. First, it is implemented in many FDTD solvers currently in use. Second, it is numerically stable. Third, it can be efficiently implemented without an auxiliary differential equation [51]. In this case, the equations of motion in the time domain are given by:

$$\frac{\partial \mu \boldsymbol{H}}{\partial t} = -\nabla \times \boldsymbol{E} \tag{19.15}$$

$$\frac{\partial \varepsilon \boldsymbol{E}}{\partial t} = \nabla \times \boldsymbol{H} - (\sigma\varepsilon)\boldsymbol{E} - \boldsymbol{J} \tag{19.16}$$

Writing the conductivity term as $\sigma\varepsilon$ is nonstandard, but is convenient here for numerical reasons. In conjunction with (19.3) and (19.4) and a Fourier transform in ξ, this yields a photon Green's function given by:

$$\left[\nabla \times \frac{1}{\mu(\boldsymbol{x})} \nabla \times -\xi^2 \varepsilon(\boldsymbol{x})\left(1 + \frac{i\sigma}{\xi}\right)\right] \boldsymbol{G}_j(\xi;\boldsymbol{x},\boldsymbol{x}') = \delta(\boldsymbol{x}-\boldsymbol{x}')\hat{\boldsymbol{e}}_j \tag{19.17}$$

This corresponds to picking a frequency contour of the form:

$$\omega(\xi) \equiv \xi\sqrt{1 + \frac{i\sigma}{\xi}} \tag{19.18}$$

Note that, in the time domain, the frequency of the fields is ξ, and not ω, i.e., their time dependence is $e^{-i\xi t}$. The only role of the conductivity σ here is to introduce an imaginary component to (19.17) in correspondence with a complex-frequency mapping. It also explicitly appears in the final expression for the force, (19.12), as a multiplicative (Jacobian) factor.

The standard FDTD method involves a discretized form of (19.15) and (19.16), from which one obtains $\boldsymbol{E}$ and $\boldsymbol{B}$, not G_{ij}^{E}. However, in the frequency domain, the photon Green's function, being the solution to (19.6), solves exactly the same equations as those satisfied by the electric field $\boldsymbol{E}$, except for a simple multiplicative factor in (19.3). Specifically, G_{ij}^{E} is given in terms of $\boldsymbol{E}$ by:

$$G_{ij}^{E}(\xi;\boldsymbol{x},\boldsymbol{x}') = -\frac{E_{i,j}(\xi,\boldsymbol{x})}{i\xi\mathcal{J}(\xi)} \tag{19.19}$$

where $E_{i,j}(\xi,\boldsymbol{x})$ is the electric field in the ith direction due to the impulsive electric current source, $\boldsymbol{J}(t,\boldsymbol{x}) = \delta(t)\,\delta(\boldsymbol{x}-\boldsymbol{x}')\,\hat{\boldsymbol{e}}_j$.

In principle, the electric-field and magnetic-field correlation functions can now be computed using (19.10) and (19.11) with $\omega(\xi)$ given by (19.18), and by setting $\boldsymbol{x}=\boldsymbol{x}'$ in (19.11). Since a discrete spatial grid is assumed, no singularities arise for $\boldsymbol{x}=\boldsymbol{x}'$, and in fact any $\boldsymbol{x}$-independent contribution is canceled after integration over S. This is straightforward for (19.7) since the $\boldsymbol{E}$-field correlation function only involves a simple multiplication by $\omega^2(\xi)$. However, the $\boldsymbol{H}$-field correlation function, (19.8), involves derivatives in space. Although it is possible to compute these derivatives numerically as finite differences, it is conceptually much simpler to pick a different vector potential, analogous to (19.3) and (19.4), in which $\boldsymbol{H}$ is the time-derivative of a vector potential $\boldsymbol{A}^{H}$. As discussed in the appendix of [48], this choice of vector potential implies a frequency-independent magnetic conductivity, $\sigma\mu$, and a magnetic current, $\boldsymbol{J}$. The resulting time-domain equations of motion are:

$$\frac{\partial \mu \boldsymbol{H}}{\partial t} = -\nabla\times\boldsymbol{E} + (\sigma\mu)\boldsymbol{H} - \boldsymbol{J} \tag{19.20}$$

$$\frac{\partial \varepsilon \boldsymbol{E}}{\partial t} = \nabla\times\boldsymbol{H} \tag{19.21}$$

In this gauge, the new photon Green's function $G_{ij}^{H} = \left\langle A_i^{H}(\xi,\boldsymbol{x})\,A_j^{H}(\xi,\boldsymbol{x}')\right\rangle$ is given by:

$$G_{ij}^{H}(\xi;\boldsymbol{x},\boldsymbol{x}') = -\frac{H_{i,j}(\xi,\boldsymbol{x})}{i\xi\mathcal{J}(\xi)} \tag{19.22}$$

where $H_{i,j}(\xi,\boldsymbol{x})$ is the magnetic field in the ith direction due to the impulsive magnetic current source, $\boldsymbol{J}(t,\boldsymbol{x}) = \delta(t)\,\delta(\boldsymbol{x}-\boldsymbol{x}')\,\hat{\boldsymbol{e}}_j$. The magnetic-field correlation function, given by:

$$\left\langle H_i(\xi,\boldsymbol{x})\,H_j(\xi,\boldsymbol{x}')\right\rangle = \frac{\hbar}{\pi}\omega^2(\xi)\,G_{ij}^{H}(\xi;\boldsymbol{x},\boldsymbol{x}') \tag{19.23}$$

is now defined as a frequency multiple of G_{ij}^{H} rather than by a spatial derivative of G_{ij}^{E}.

This approach to computing the magnetic-field correlation function has the advantage of treating the electric and magnetic fields on the same footing, and also allows examination of only the field response at the location of the current source. The removal of spatial derivatives also greatly simplifies the incorporation of discretization into the equations, as discussed in the appendix of [48]. Although unphysical, magnetic currents and conductivities are easily implemented numerically. Alternatively, one could simply interchange ε and μ, and $\boldsymbol{E}$ and $\boldsymbol{H}$, and run the simulation entirely as in (19.15) and (19.16).

The full Casimir force integral is then expressed in the symmetric form:

$$F_i = \mathrm{Im}\left\{ \frac{\hbar}{\pi} \int_{-\infty}^{\infty} g(\xi)\left[\Gamma_i^E(\xi) + \Gamma_i^H(\xi)\right] d\xi \right\} \tag{19.24}$$

where

$$\Gamma_i^E(\xi) \equiv \oint\!\!\!\oint_S \sum_j \varepsilon(\boldsymbol{x})\left[E_{i,j}(\boldsymbol{x}) - 0.5\,\delta_{ij}\sum_k E_{k,k}(\boldsymbol{x})\right] dS_j \tag{19.25}$$

$$\Gamma_i^H(\xi) \equiv \oint\!\!\!\oint_S \sum_j \frac{1}{\mu(\boldsymbol{x})}\left[H_{i,j}(\boldsymbol{x}) - 0.5\,\delta_{ij}\sum_k H_{k,k}(\boldsymbol{x})\right] dS_j \tag{19.26}$$

represent the surface-integrated E-field and H-field responses in the frequency domain. For notational simplicity in these expressions, $E_{i,j}(\boldsymbol{x}) \equiv E_{i,j}(\xi;\boldsymbol{x})$, $H_{i,j}(\boldsymbol{x}) \equiv H_{i,j}(\xi;\boldsymbol{x})$, and:

$$g(\xi) \equiv \frac{\omega^2}{i\xi\mathcal{J}(\xi)}\frac{d\omega}{d\xi}\Theta(\xi) \tag{19.27}$$

Here, the path of integration has been extended to the entire real ξ-axis with the use of the unit-step function $\Theta(\xi)$ for later convenience.

The product of the fields with $g(\xi)$ naturally decomposes the problem into two parts: computation of the surface integral of the field correlations Γ and of the function $g(\xi)$. The Γ_i's contain all the structural information and are straightforward to compute from the output of any available FDTD solver with no modification to the code. This output is then combined with $g(\xi)$, which is easily computed analytically, and integrated in (19.24) to obtain the Casimir force. As discussed in [48], the effect of spatial and temporal discretization enters explicitly only as a slight modification to $g(\xi)$ in (19.24), leaving the basic conclusions unchanged.

19.2.4 Expression for the Casimir Force as a Time-Domain Integration

The convolution theorem can be used to re-express the integral of the product of $g(\xi)$ and $[\Gamma^E(\xi) + \Gamma^H(\xi)]$ in (19.24) as an integral over time t of their Fourier transforms $\mathrm{g}(-t)$ and $[\Gamma^E(t) + \Gamma^H(t)]$. Taking advantage of the causality conditions $[\Gamma^E(t), \Gamma^H(t) = 0 \text{ for } t < 0]$ yields the following expression for the Casimir force as a time-domain integration:

$$F_i = \mathrm{Im}\left\{ \frac{\hbar}{\pi} \int_0^{\infty} g(-t)\left[\Gamma_i^E(t) + \Gamma_i^H(t)\right] dt \right\} \tag{19.28}$$

In (19.28), $\Gamma^E(t)$ and $\Gamma^H(t)$ rapidly decay with time due to the assumed finite conductivity and lack of sources for $t > 0$. The following is a compact expression given in [49] for $\Gamma_i^E(t)$ in terms of the electric field response $E_{ij}(t; \boldsymbol{x}, \boldsymbol{x}')$ in direction i at $(t, \boldsymbol{x})$ to source current $J(t, \boldsymbol{x}) = \delta(t)\delta(\boldsymbol{x} - \boldsymbol{x}')$ in direction j:

$$\begin{aligned} \Gamma_i^E(t) &\equiv \oiint_S \left[E_{ij}(t; \boldsymbol{x}, \boldsymbol{x}) - 0.5\,\delta_{ij} \sum_k E_{kk}(t; \boldsymbol{x}, \boldsymbol{x}) \right] dS_j(\boldsymbol{x}) \\ &\equiv \oiint_S \Gamma_{ij}^E(t; \boldsymbol{x}, \boldsymbol{x})\, dS_j(\boldsymbol{x}) \end{aligned} \tag{19.29}$$

where differential area element $dS_j(\boldsymbol{x}) \equiv n_j(\boldsymbol{x})\, dS(\boldsymbol{x})$, and $\boldsymbol{n}(\boldsymbol{x})$ is the unit normal vector to S at $\boldsymbol{x}$. An analogous definition holds for $\Gamma_i^H(t)$ involving the H-field Green's function H_{ij}.

The computation of the Casimir force described in [48] required finding both the $\Gamma^E(t; \boldsymbol{x}, \boldsymbol{x})$ and $\Gamma^H(t; \boldsymbol{x}, \boldsymbol{x})$ field responses with a separate time-domain simulation for every point $\boldsymbol{x} \in S$. This required a large number of simulations, with the precise number being dependent on the resolution and shape of S, making the computation potentially very costly in practice. Reference [49] reported a means to dramatically reduce the number of required simulations to compute the Casimir force by reformulating this force in terms of harmonic expansions in $\Gamma^E(t; \boldsymbol{x}, \boldsymbol{x})$ and $\Gamma^H(t; \boldsymbol{x}, \boldsymbol{x})$ involving distributed field responses to distributed currents. This will be reviewed in Section 19.3.

19.2.5 Evaluation of $g(-t)$ in (19.28)

Using the frequency-independent conductivity contour (19.18), corresponding to (19.15) and (19.16), the following explicit form for $g(\xi)$ is found [48]:

$$g(\xi) = -i\xi \sqrt{1 + i\sigma/\xi}\, \left(1 + i\sigma/2\xi\right) \Theta(\xi) \tag{19.30}$$

Fourier transformation of $g(\xi)$ yields the geometry-independent function, $g(-t)$, which is time-integrated against the FDTD-computed fields in (19.28) to obtain the correct Casimir force. Key aspects of the evaluation of $g(-t)$ are now reviewed, per the discussion in [49].

Treatment of the High-Frequency Divergence

According to [49], $g(\xi)$ has a behavior such that it diverges in the high-frequency limit. Namely, for large ξ, $g(\xi)$ has the form:

$$g(\xi) \to g_1(\xi) \equiv \frac{\xi}{i}\Theta(\xi) + \sigma\Theta(\xi) \qquad \text{as } \xi \to \infty \tag{19.31}$$

Viewing $g_1(\xi)$ as a function, its Fourier transform $g_1(t)$ could only be computed by introducing a cutoff in the frequency integral at the Nyquist frequency. This is because the time signal is only defined up to a finite sampling rate, and the integral of a divergent function may appear to be undefined in the limit of no cutoff [49].

Applying this procedure to compute $g(-t)$ yields a function having strong oscillations at the Nyquist frequency. The amplitude of these oscillations can be quite high, increasing the time needed to obtain convergence and also making any physical interpretation more difficult [49].

These oscillations are entirely from the high-frequency behavior of $g(\xi)$, where $g(\xi) \sim g_1(\xi)$. However, $g(t)$ and $g(\xi)$ only appear when they are being integrated against smooth, rapidly decaying field functions $\Gamma(\boldsymbol{x}, t)$ or $\Gamma(\boldsymbol{x}, \xi)$. In this case, g can be viewed as a tempered distribution (such as the δ function) [56]. Although $g(\xi)$ diverges for large ξ, this divergence is only a power law, so it is a tempered distribution and its Fourier transform is well defined without any truncation. In particular, the Fourier transform of $g_1(\xi)$ is given by [49]:

$$g_1(-t) = \frac{i}{2\pi}\left(\frac{1}{t^2} + \frac{\sigma}{t}\right) \tag{19.32}$$

Adding and subtracting the term $g_1(\xi)$ from $g(\xi)$, the remaining term decays to zero for large ξ and can be Fourier transformed numerically without the use of a high-frequency cutoff. This allows $g(-t)$ to be computed as the sum of $g_1(t)$ plus the Fourier transform of a well-behaved function, yielding a well-behaved time dependence [49].

Treatment of the Low-Frequency Singularity and Further Simplification

In addition to the treatment of the high-frequency divergence discussed above, Ref. [49] found it convenient to Fourier-transform the low-frequency singularity of $g(\xi)$ analytically. As discussed in [48], the low-frequency limit of $g(\xi)$ is given by:

$$g(\xi) \to g_2(\xi) \equiv \frac{\sqrt{i}}{2}\frac{\sigma^{3/2}}{\xi^{1/2}}\Theta(\xi) \qquad \text{as } \xi \to 0 \tag{19.33}$$

The Fourier transform of $g_2(\xi)$, viewed as a distribution, is:

$$g_2(-t) = \frac{i}{4\sqrt{\pi}}\frac{\sigma^{3/2}}{t^{1/2}} \tag{19.34}$$

After removing both the high-frequency and low-frequency divergences of $g(\xi)$, a numerical Fourier transform is performed on the function $\delta g(\xi) \equiv g(\xi) - g_1(\xi) - g_2(\xi)$, which is well-behaved in both the high- and low-frequency limits [49].

In the present discussion, the concern is only with real sources, in which case all fields $\Gamma(\boldsymbol{x}, t)$ are real. Here, the only contributor to the force is the imaginary part of $g(-t)$, given by:

$$\mathrm{Im}\left[g(-t)\right] = \mathrm{Im}\left[\delta g(-t)\right] + \frac{1}{2\pi}\left(\frac{1}{t^2} + \frac{\sigma}{t}\right) + \frac{1}{4\sqrt{\pi}}\frac{\sigma^{3/2}}{t^{1/2}} \tag{19.35}$$

Special Case: 3-D z-Invariant Systems Involving Only Perfect Conductors

As discussed in [30], the stress-tensor frequency integral for a 3-D z-invariant system involving only vacuum and perfect conductors is identical in value to the integral of the stress tensor for the associated 2-D system (corresponding to taking a $z = 0$ cross-section), with an extra factor of

$i\omega/2$ in the frequency integrand. In the time domain, this corresponds to solving the 2-D system with a new $g(-t)$. For this case, the Fourier transform can be performed analytically, yielding [49]:

$$\mathrm{Im}\left[g(-t)\right] = \frac{1}{2\pi}\left(\frac{2}{t^3} + \frac{3\sigma}{2t^2} + \frac{\sigma^2}{2t}\right) \tag{19.36}$$

19.3 REFORMULATION IN TERMS OF A HARMONIC EXPANSION

This section reviews the efficient reformulation of the Casimir force in terms of a harmonic expansion in $\Gamma^E(t;\boldsymbol{x},\boldsymbol{x})$, as reported in [49]. An analogous development holds for $\Gamma^H(t;\boldsymbol{x},\boldsymbol{x})$.

Because S is assumed to be a compact surface, $\Gamma_{ij}(t;\boldsymbol{x},\boldsymbol{x})$ can be rewritten as an integral over S:

$$\Gamma^E_{ij}(t;\boldsymbol{x},\boldsymbol{x}) = \oiint_S \Gamma^E_{ij}(t;\boldsymbol{x},\boldsymbol{x}')\,\delta_S(\boldsymbol{x}-\boldsymbol{x}')\,dS(\boldsymbol{x}') \tag{19.37}$$

where dS is a scalar unit of area, and δ_S denotes a δ function with respect to integrals over the surface S. Given a set of basis functions $\{f_n(\boldsymbol{x})\}$ defined on S, the following expansion of the δ function in (19.3) is valid for all points $\boldsymbol{x}, \boldsymbol{x}' \in S$:

$$\delta_S(\boldsymbol{x}-\boldsymbol{x}') = \sum_n \bar{f}_n(\boldsymbol{x})\,f_n(\boldsymbol{x}') \tag{19.38}$$

where, for simplicity, $\{f_n(\boldsymbol{x})\}$ is assumed to be complete and orthonormal on S, but is otherwise arbitrary. Inserting this expansion into (19.37) and rearranging terms yields:

$$\Gamma^E_{ij}(t;\boldsymbol{x},\boldsymbol{x}) = \sum_n \bar{f}_n(\boldsymbol{x})\left[\oiint_S \Gamma^E_{ij}(t;\boldsymbol{x},\boldsymbol{x}')\,f_n(\boldsymbol{x}')\,dS(\boldsymbol{x}')\right] \tag{19.39}$$

The term in square brackets can be understood in a physical context: it is the electric-field response at position $\boldsymbol{x}$ and time t to a current source on S of the form $J(\boldsymbol{x},t) = f_n(\boldsymbol{x})\,\delta(t)$. Reference [49] denoted this quantity by :

$$\Gamma^E_{ij;n}(t;\boldsymbol{x}) \equiv \oiint_S \Gamma^E_{ij}(t;\boldsymbol{x},\boldsymbol{x}')\,f_n(\boldsymbol{x}')\,dS(\boldsymbol{x}') \tag{19.40}$$

where the n subscript indicates that this is a field in response to a current source determined by $f_n(\boldsymbol{x})$. In fact, $\Gamma^E_{ij;n}(t;\boldsymbol{x})$ is exactly what can be calculated in an FDTD simulation using a current $J(\boldsymbol{x},t) = f_n(\boldsymbol{x})\,\delta(t)$ for each n. This equivalence is illustrated in Fig. 19.1.

The procedure specified in [49] to calculate the Casimir force was only slightly modified from the one outlined in [48]: after first defining the problem geometry and an integration surface S, one additionally needs to specify a set of harmonic basis functions $\{f_n(\boldsymbol{x})\}$ on S. For each harmonic moment n, one inserts a current function $J(\boldsymbol{x},t) = f_n(\boldsymbol{x})\,\delta(t)$ on S, and uses FDTD to compute the field response $\Gamma_{ij;n}(\boldsymbol{x},t)$. Summing over all harmonic moments yields the total Casimir force.

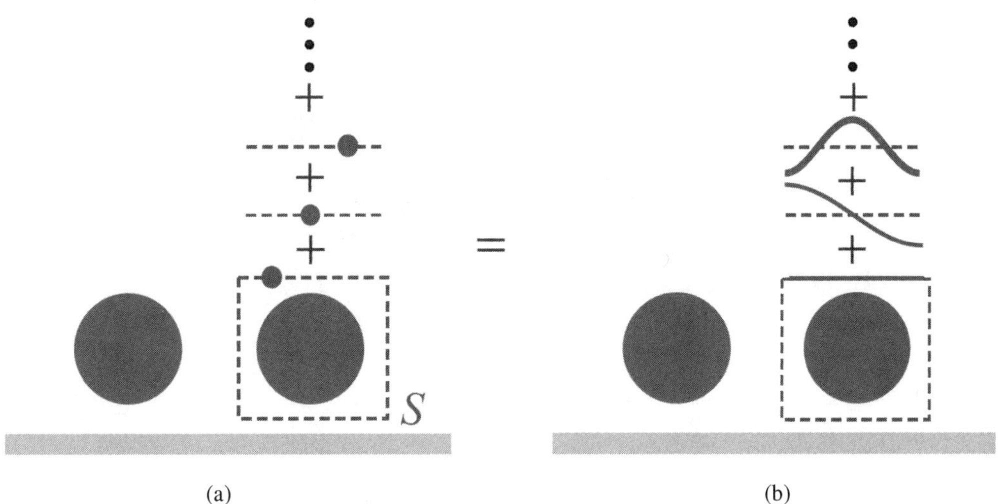

Fig. 19.1 Differing harmonic expansions of the source currents on surface S: (a) expansion using point sources, where each dot represents a different simulation; (b) use of $f_n(\boldsymbol{x}) \sim \cos(x)$ for each side of S. Either forms a complete basis for all functions in S. *Source:* A. P. McCauley, A. W. Rodriguez, J. D. Joannopoulos, and S. G. Johnson, *Physical Review A*, Vol. 81, 2010, 012119, ©2010 The American Physical Society.

In principle, any surface S and any harmonic source basis can be used. Point sources, as discussed in [48], are a simple, although highly inefficient, example. However, many common FDTD algorithms involve Cartesian grids where S is naturally rectangular. For such grids, as reported in [49], the field integration can be performed with high accuracy using a Fourier cosine series separately defined on each face of S. The Fourier cosine series on a discrete grid is essentially a discrete cosine transform, a well-known discrete orthogonal basis with rapid convergence properties [57]. This is in contrast to discretizing some basis such as spherical harmonics that are only approximately orthogonal when discretized on a Cartesian grid.

19.4 NUMERICAL STUDY 1: A 2-D EQUIVALENT TO A 3-D CONFIGURATION

This section reviews the FDTD simulation reported in [49] of the Casimir force between two perfect metallic cylinders ($\varepsilon = -\infty$) sandwiched between two sidewalls, per the geometry illustrated in Fig. 19.2. Previously, a high-precision scattering calculation of this problem had been published for the case of perfect metallic sidewalls [32], employing a specialized exponentially convergent basis suitable for cylinder or plane geometries. Results of the study of [32] had indicated an interesting non-monotonic dependence of the cylinder-to-cylinder Casimir force on the normalized sidewall-to-cylinder distance, h/a.

The precision of the study of [32] and its interesting results made it an excellent comparative benchmark in [49] for testing the accuracy of its proposed harmonic-expansion FDTD-Casimir technique. In addition, Ref. [49] reported simulating the same geometry for the case of perfect magnetic conductor sidewalls ($\mu = -\infty$) to demonstrate that the FDTD-Casimir technique can be used to model configurations containing magnetic materials.

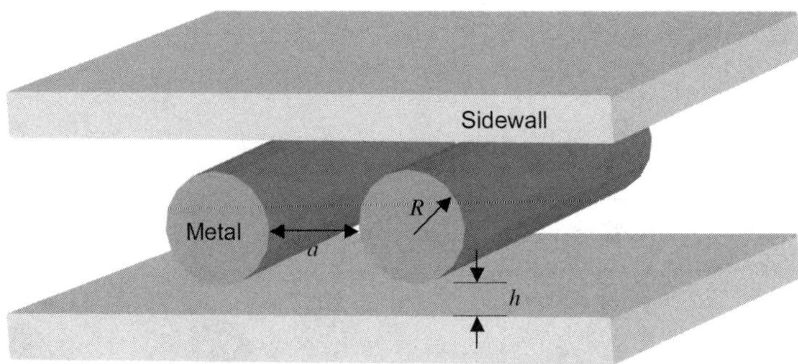

Fig. 19.2 3-D configuration modeled for the Casimir force as an equivalent 2-D geometry. *Adapted from:* A. P. McCauley, A. W. Rodriguez, J. D. Joannopoulos, and S. G. Johnson, *Physical Review A*, Vol. 81, 2010, 012119, ©2010 The American Physical Society.

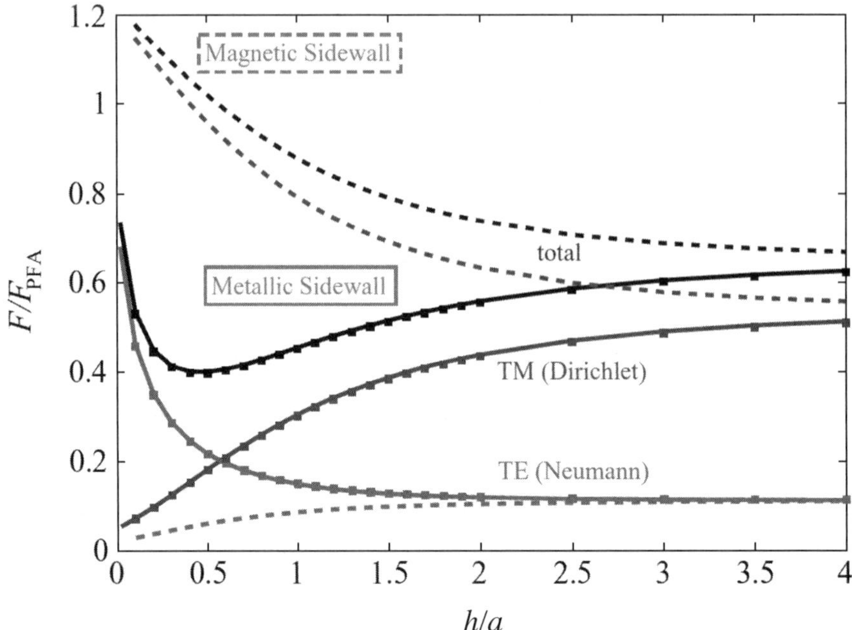

Fig. 19.3 Cylinder-to-cylinder Casimir force F for the geometry of Fig. 19.2 as a function of sidewall separation h/a, normalized by the proximity force approximation (PFA) $F_{PFA} = \hbar c\zeta(3)d/8\pi a^3$. Square data points show FDTD-computed results for the TE, TM, and total force in the presence of perfect electric conductor sidewalls. Solid lines show the results from the high-precision scattering calculations of [32], which are in excellent agreement with the FDTD results. Dashed lines indicate FDTD results for the same force components, but in the presence of perfect magnetic conductor sidewalls. Note that the total force is non-monotonic for perfect electric conductor sidewalls, but is monotonic for perfect magnetic conductor sidewalls. *Source:* A. P. McCauley, A. W. Rodriguez, J. D. Joannopoulos, and S. G. Johnson, *Physical Review A*, Vol. 81, 2010, 012119, ©2010 The American Physical Society.

Although 3-D in nature, the system of Fig. 19.2 was translation invariant in the z-direction and involved only perfect metallic or magnetic conductors. Hence, this configuration was treated as a 2-D problem using the form of $g(-t)$ given in (19.36) [49]. Here, surface S consisted of four faces, each of which was a line segment of length L parameterized by a single variable, x. A cosine basis was employed for the harmonic expansion on each side of S. The basis functions for each side were:

$$f_n(x) = \sqrt{\frac{2}{L}} \cos\left(\frac{n\pi x}{L}\right), \qquad n = 0, 1, ..., \tag{19.41}$$

where $f_n(x) = 0$ for all points x not on that side of S. These functions, and their equivalence to a computation using δ-function sources as basis functions, are shown schematically in Fig. 19.1.

In the case of the FDTD algorithm of [49], space was discretized on a Yee grid [51], and in most cases x was situated between two grid points. It was found sufficient to place suitably averaged currents on neighboring grid points, as several available FDTD implementations provide features to accurately interpolate currents from any location onto the grid.

As reported in [49], Fig. 19.3 shows the FDTD-computed cylinder-to-cylinder Casimir force for the structure of Fig. 19.2 vs. the normalized vertical sidewall separation, h/a. Results for the transverse electric (TE) and transverse magnetic (TM) field components are shown for two cases: perfect electric conductor sidewalls, and perfect magnetic conductor sidewalls.

From Fig. 19.3, we see that the FDTD results reported in [49] were in excellent agreement with previous high-precision scattering calculations for the case of perfect electric conductor sidewalls [32], with both techniques indicating a non-monotonic behavior of the Casimir force in h/a. Invoking the method of images for conducting walls, Ref. [32] had explained that this behavior results from a competition between the TM force (which dominates for large h/a, but is suppressed for small h/a) and the TE force (which has the opposite behavior).

The FDTD results in Fig. 19.3 for the case of perfect magnetic conductor sidewalls were reported in [49] without a benchmark comparison, because no comparative data were available at the time of publication. Reference [49] explained the monotonic behavior of the cylinder-to-cylinder Casimir force for this case as a result of the image currents flipping sign for perfect magnetic conductor sidewalls compared to perfect electric conductor sidewalls. That is, for small h/a, the assumed magnetic conductor sidewalls enhanced the TM force and suppressed the TE force.

As reported in [49], Fig. 19.4 shows the convergence of the harmonic expansion for the configuration of Figs. 19.2 and 19.3 as a function of n. Asymptotically for large n, an n^{-4} power law is clearly discernible. The explanation for this convergence follows readily from the geometry of S. Here, the electric field $\boldsymbol{E}(\boldsymbol{x})$, when viewed as a function along S, has nonzero first derivatives at the corners. However, the cosine series used here always has a vanishing derivative. This implies that its cosine-transform components decay asymptotically as n^{-2} [55]. As Γ^E is related to the correlation function $\langle \boldsymbol{E}(\boldsymbol{x})\boldsymbol{E}(\boldsymbol{x})\rangle$, their contributions decay as n^{-4}.

One could instead consider a Fourier series defined around S, but the convergence rate would be the same because the derivatives of the fields would be discontinuous at the corners of S. While a circular surface would have no corners in the continuous case, on a Cartesian grid it would effectively have many corners, and hence poor convergence with resolution.

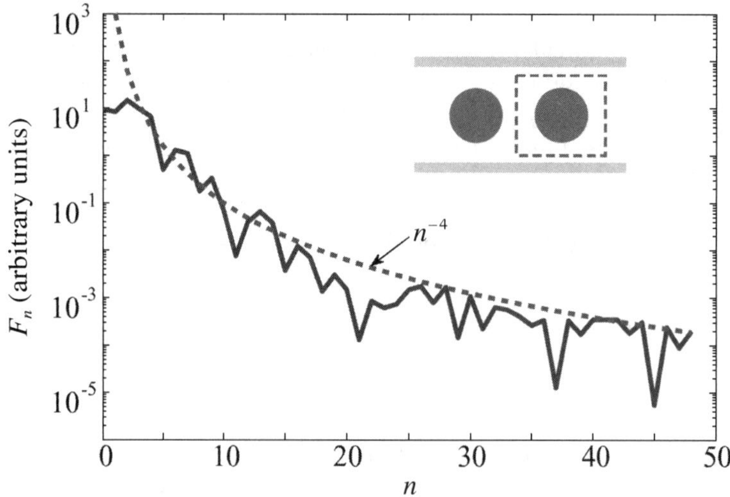

Fig. 19.4 Relative contribution of harmonic moment n in the cosine basis to the total Casimir force for the configuration of Figs. 19.2 and 19.3 (shown in the inset). *Source:* A. P. McCauley, A. W. Rodriguez, J. D. Joannopoulos, and S. G. Johnson, *Physical Review A*, Vol. 81, 2010, 012119, ©2010 The American Physical Society.

19.5 NUMERICAL STUDY 2: DISPERSIVE DIELECTRIC MATERIALS

Dispersion in FDTD in general requires fitting an actual dispersion to a simple model (e.g., a series of Lorentzian or Drude contributions). Assuming this has been done, these models can then be analytically continued onto the complex conductivity contour.

As an example of an FDTD-Casimir calculation involving dispersive materials, Ref. [49] considered the geometry of Fig. 19.5: two identical silicon waveguides suspended in empty space. Previously, this geometry had been investigated to determine the classical optical force between the waveguides [58, 59].

Reference [49] modeled silicon as a dispersive dielectric having the relative permittivity:

$$\varepsilon(\omega) = \varepsilon_f + \frac{\varepsilon_f - \varepsilon_0}{1 - (\omega/\omega_0)^2} \qquad (19.42)$$

where $\omega_0 = 6.6 \times 10^{15}$ rad/s, $\varepsilon_0 = 1.035$, and $\varepsilon_f = 11.87$. This dispersion was implemented in FDTD by the standard technique of auxiliary differential equations [51] mapped into the complex-ω plane, as explained in [48]. A perfectly matched layer (PML) absorbing boundary condition [51] was used to simulate the unbounded vacuum region surrounding the waveguides.

The system of Fig. 19.5 was translation-invariant in the z-direction. The technique of the previous section could have been applied if this system had contained only perfect conductors (i.e., only one 2-D simulation required to compute the Casimir force). However, the presence of dielectrics hybridized the two polarizations and required an explicit k_z integral [30]. Each value of k_z corresponded to a separate 2-D simulation with Bloch-periodic boundary conditions. Here, the force for each k_z was smooth and rapidly decaying, so only a few k_z points were needed.

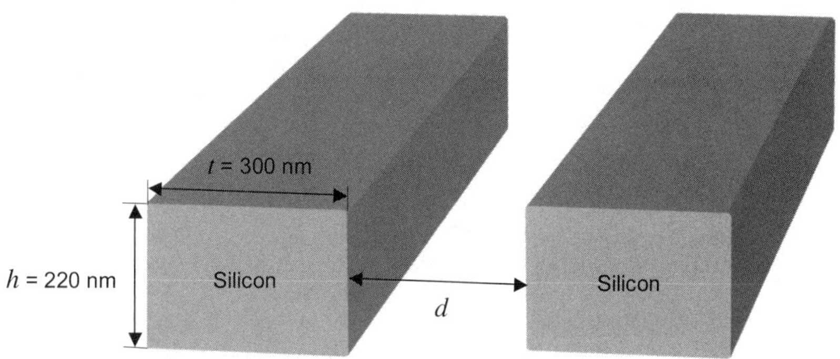

Fig. 19.5 Geometry of a pair of suspended waveguides modeled for the Casimir force. *Adapted from:* A. P. McCauley, A. W. Rodriguez, J. D. Joannopoulos, and S. G. Johnson, *Physical Review A*, Vol. 81, 2010, 012119, ©2010 The American Physical Society.

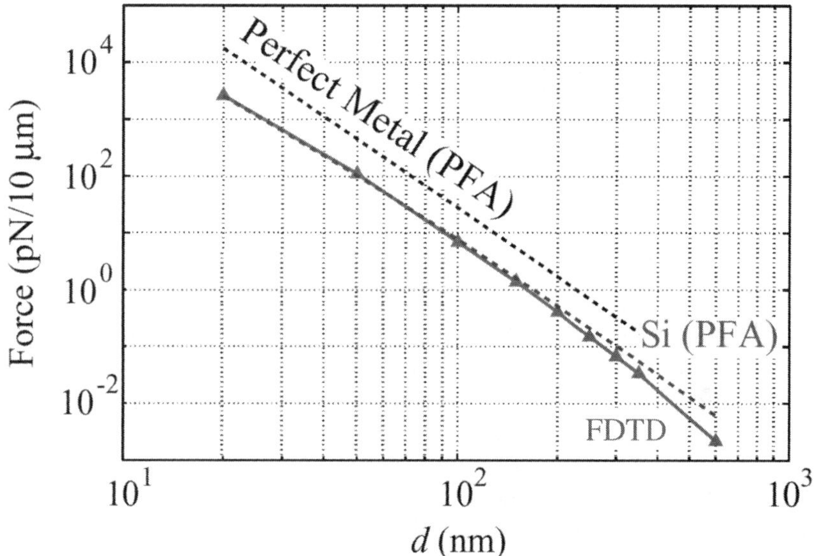

Fig. 19.6 FDTD-computed Casimir force per unit length between the suspended silicon waveguides of Fig. 19.5 (triangles) compared to 1-D proximity force approximations for two parallel silicon plates of finite thickness, and two parallel perfect metal plates. *Source:* A. P. McCauley, A. W. Rodriguez, J. D. Joannopoulos, and S. G. Johnson, *Physical Review A*, Vol. 81, 2010, 012119, ©2010 The American Physical Society.

As reported in [49], Fig. 19.6 compares the FDTD-computed Casimir force per unit length between the suspended silicon waveguides of Fig. 19.5 to simplified 1-D proximity force approximations (PFAs) [60, 61] for two parallel silicon plates of finite thickness, and two parallel perfect metal plates. In each PFA, the silicon or metal plates were assumed to be separated by d and extend infinitely in both directions perpendicular to the Casimir force. The approximate Casimir force per unit length was then integrated only over h, i.e., only over a plate area equal to the interaction area of the waveguides. In addition, the silicon plates were assumed to have the same thickness t as the waveguides in the direction parallel to the force.

As seen in Fig. 19.6, the FDTD waveguide computations and the silicon-plate PFA values reported in [49] agreed well at separation distances $d < h$, as expected. However, the PFA results became more inaccurate as d increased, being off by 50% at $d = 300$ nm. The metal-plate PFA values exhibited an exponential decay with d similar to that of the silicon plates, but were displaced upward by one-half to one order of magnitude over the range of d considered in [49].

19.6 NUMERICAL STUDY 3: CYLINDRICAL SYMMETRY IN THREE DIMENSIONS

For a 3-D geometry with cylindrical symmetry, Ref. [49] employed a cylindrical surface S and a complex exponential basis $e^{im\phi}$ in the ϕ-direction. Assuming a separable source with $e^{im\phi}$ dependence, the resulting fields could also be separable with the same ϕ dependence, and the unknowns could reduce to a 2-D (r, z) problem for each m. This resulted in a substantial reduction in computational costs compared to a full 3-D computation. Here, the operative expression for the Casimir force (as derived in Appendix 19A) was:

$$F_i = \sum_n \int_0^\infty \mathrm{Im}\left[g(-t)\right] dt \oint\!\!\oint_S \Gamma_{ij;n}(\boldsymbol{x}, t)\, ds_j(\boldsymbol{x}) \tag{19.43}$$

where the m dependence was absorbed into the definition of Γ:

$$\Gamma_{ij;n}(\boldsymbol{x}, t) \equiv \Gamma_{ij;n,m=0}(\boldsymbol{x}, t) + 2\sum_{m>0} \mathrm{Re}\left[\Gamma_{ij;n,m}(\boldsymbol{x}, t)\right] \tag{19.44}$$

and $ds_j = n_j(\boldsymbol{x})ds$ for ds being a 1-D Cartesian line element. As derived in Appendix 19A, the Jacobian factor r obtained from converting to cylindrical coordinates canceled out, so that the 1-D (r-independent) measure ds was appropriate to use in the surface integration. Also, the $2\mathrm{Re}[\ldots]$ arose from the fact that the $+m$ and $-m$ terms were complex conjugates. Although the exponentials $e^{im\phi}$ were complex, only the real part of the field response appeared in (19.44), allowing use of $\mathrm{Im}[g(-t)]$ alone in (19.43).

Given the assumed $e^{im\phi}$ dependence of the fields, Ref. [49] observed that one can write Maxwell's equations in cylindrical coordinates to obtain a 2-D equation involving only the fields in the (r, z) plane. This simplification was incorporated into the FDTD solver [53] used to obtain the results of [49], where the computational grid was restricted to the (r, z) plane and m appeared as a parameter. Here, implementing cylindrical symmetry was almost identical to the 2-D situation, differing only in that there was an additional index m over which the Casimir force was summed.

Reference [49] illustrated the use of the cylindrical-symmetry FDTD-Casimir algorithm with an examination of the configuration shown in Fig. 19.7. Here, rotational (ϕ) invariance was imposed instead of translational (z) invariance. In this case, the sidewalls were joined to form a cylindrical tube, and PML was used to terminate the tube ends. The force between the two pistons was computed as a function of h/a (the $h = 0$ case having been solved analytically [62]).

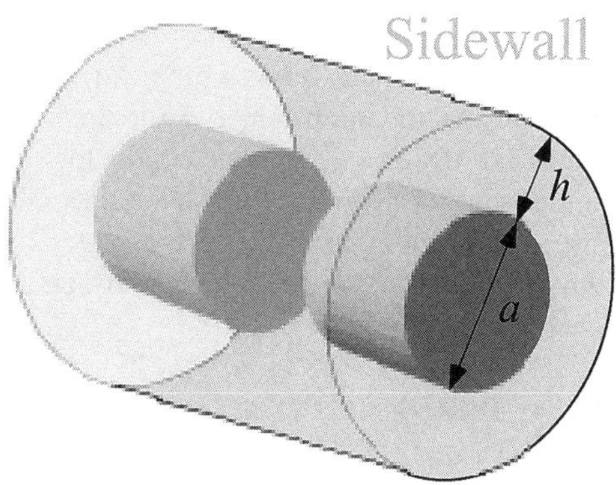

Fig. 19.7 Geometry of the cylindrically symmetric configuration modeled for the piston-to-piston Casimir force. Both pistons were perfect metallic conductors, and the sidewalls were either perfect metallic or perfect magnetic conductors. *Adapted from:* A. P. McCauley, A. W. Rodriguez, J. D. Joannopoulos, and S. G. Johnson, *Physical Review A*, Vol. 81, 2010, 012119, ©2010 The American Physical Society.

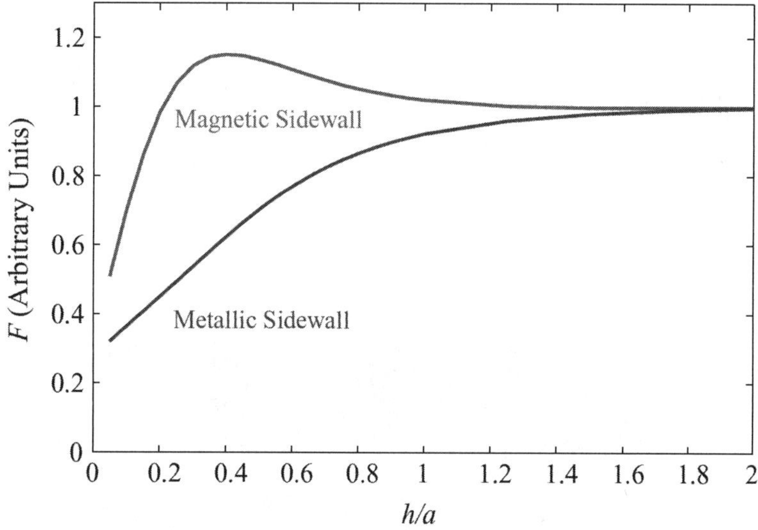

Fig. 19.8 FDTD-computed piston-to-piston Casimir force as a function of the normalized sidewall spacing h/a for the configuration of Fig. 19.7. *Source:* A. P. McCauley, A. W. Rodriguez, J. D. Joannopoulos, and S. G. Johnson, *Physical Review A*, Vol. 81, 2010, 012119, ©2010 The American Physical Society.

As reported in [49], Fig. 19.8 shows the FDTD-computed piston-to-piston Casimir force as a function of h/a for a relatively low-resolution model, accurate to within a few percent. Because of the 2-D nature of this problem, the computation time was comparable to that of the 2-D geometry of the previous section: about 5 min running on eight processors for each value of h/a. Only indices $n, m \in \{0, 1, 2\}$ were needed for each result to converge to within 1%, after which the error was dominated by the spatial discretization.

In direct contrast to the case of Section 19.4 of two adjacent metal cylinders with translational symmetry, Fig. 19.8 shows that the Casimir force between two adjacent metal cylinders with rotational symmetry is monotonic in h/a for perfect metal sidewalls, and non-monotonic for sidewalls comprised of perfect magnetic conductors. Apart from this interesting result, reporting the modeling of perfect magnetic conductor sidewalls in [49] demonstrated the capability of the FDTD-Casimir algorithm to examine the material dependence of this force. In fact, no additional code would be required to model dispersive and/or anisotropic materials.

19.7 NUMERICAL STUDY 4: PERIODIC BOUNDARY CONDITIONS

This section reviews computations reported in [49] involving a periodic array of dispersive silicon dielectric waveguides above a silica substrate, shown in Fig. 19.9. As discussed in [30], the Casimir force for periodic systems can be computed as an integral over all Bloch wavevectors in the directions of periodicity. Here, two directions, x and z, were periodic, the latter being the limit of the period approaching zero. The Casimir force was then given by:

$$\int_0^\infty \int_0^\infty F_{k_z, k_x} \, dk_z \, dk_x \tag{19.45}$$

where F_{k_z, k_x} denoted the force computed from one simulation of the unit cell using Bloch-periodic boundary conditions with wavevector $\boldsymbol{k} = (k_x, 0, k_z)$. In this example, the unit cell was of period 1 μm in the x-direction and of zero length in the z-direction, so the computations were effectively two dimensional (although they had to be integrated over k_z).

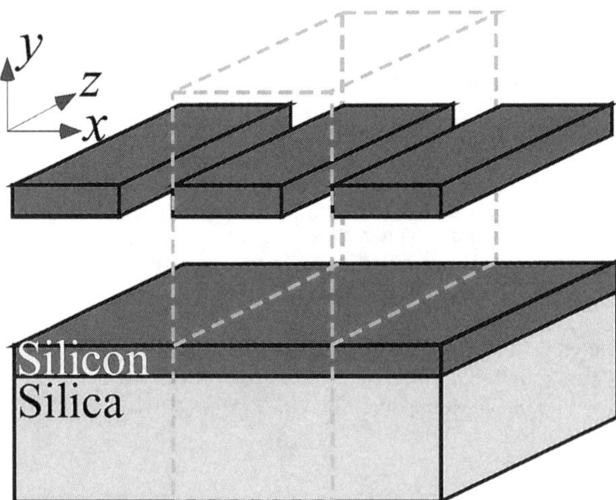

Fig. 19.9 Geometry of the model of the Casimir force between a periodic array of silicon waveguides and a silicon-silica substrate. The system is periodic in the x-direction and translation-invariant in the z-direction, so the computation involves a set of 2-D simulations. *Adapted from:* A. P. McCauley, A. W. Rodriguez, J. D. Joannopoulos, and S. G. Johnson, *Physical Review A*, Vol. 81, 2010, 012119, ©2010 The American Physical Society.

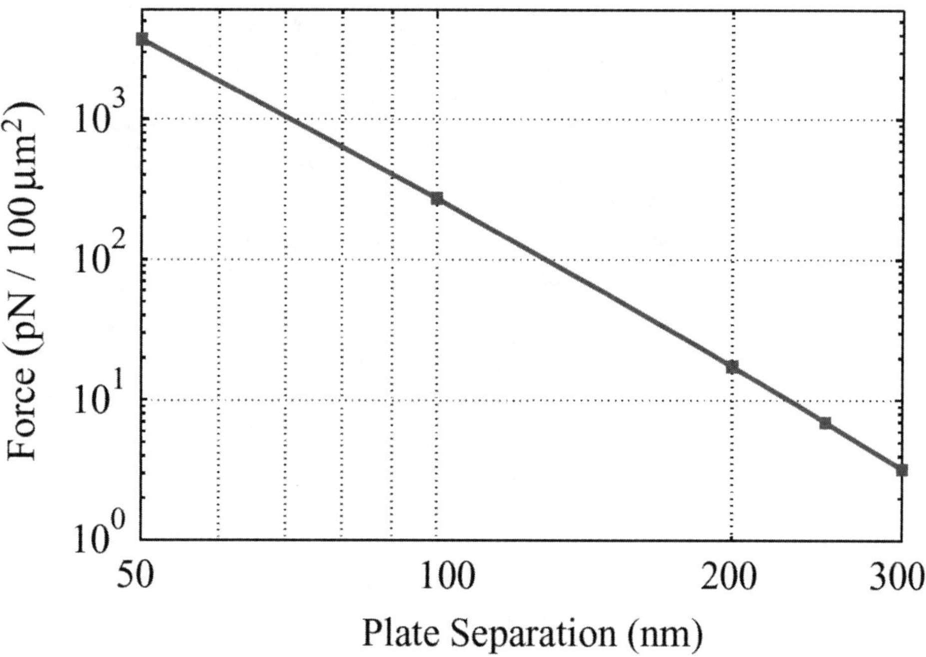

Fig. 19.10 FDTD-computed Casimir force for the system of Fig. 19.9 as a function of the separation between the periodic array of silicon waveguides and the silicon-silica substrate. *Source:* A. P. McCauley, A. W. Rodriguez, J. D. Joannopoulos, and S. G. Johnson, *Physical Review A*, Vol. 81, 2010, 012119, ©2010 The American Physical Society.

Reference [49] used the dispersive model of (19.42) for silicon. For silica, it used [40]:

$$\varepsilon(\omega) = 1 + \sum_{j=1}^{3} \frac{C_j\,\omega_j^2}{\omega_j^2 - \omega^2} \tag{19.46}$$

where $(C_1, C_2, C_3) = (0.829, 0.095, 1.098)$ and $(\omega_1, \omega_2, \omega_3) = (0.867, 1.508, 203.4) \times 10^{14}$ (rad/s). As reported in [49], Fig. 19.10 shows the FDTD-Casimir results for the variation of the force as the separation between the periodic waveguide array and the substrate was varied.

19.8 NUMERICAL STUDY 5: FULLY 3-D FDTD-CASIMIR COMPUTATION

Reference [49] reported the FDTD computation of the Casimir force for the system shown in Fig. 19.11: a silica microsphere immersed in bromobenzene located directly above a spherical indentation in a perfect metal plane. Because this system satisfied $\varepsilon_{\mu\text{sphere}} < \varepsilon_{\text{fluid}} < \varepsilon_{\text{plane}}$, the Casimir force was expected to be repulsive [19]. This afforded the possibility of calculating an equilibrium height of the microsphere balancing the upward Casimir repulsion and the downward force of gravity, with lateral confinement provided by Casimir repulsion from the surface of the spherical indentation. Such a stable levitation phenomenon had been explored previously in [42], and remains of significant interest.

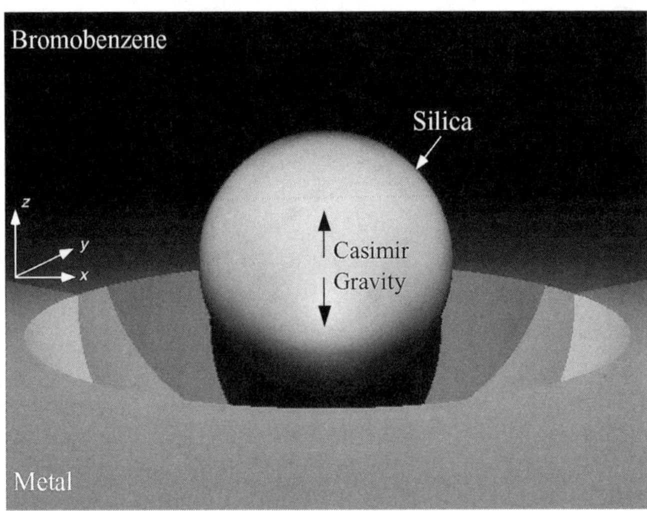

Fig. 19.11 Model of the Casimir force between a silica microsphere (immersed in bromobenzene) and a perfect metal plane with a spherical indentation. *Source:* A. P. McCauley, A. W. Rodriguez, J. D. Joannopoulos, and S. G. Johnson, *Physical Review A*, Vol. 81, 2010, 012119, ©2010 The American Physical Society.

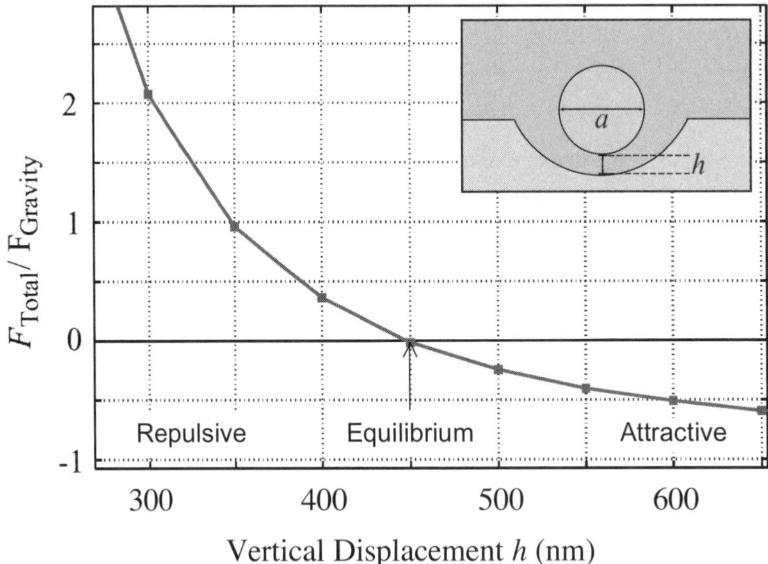

Fig. 19.12 FDTD-computed total vertical force (Casimir + gravity) acting on the silica microsphere of Fig. 19.11 (also see the inset) of radius 500 nm above a spherical metal indentation of radius 1 μm. As the assumed height h of the microsphere's surface above the bottom of the indentation was increased from zero, the initially computed net repulsive force diminished; passed through 0 at the vertical equilibrium point, $h = 450$ nm; and thereafter became attractive. *Adapted from:* A. P. McCauley, A. W. Rodriguez, J. D. Joannopoulos, and S. G. Johnson, *Physical Review A*, Vol. 81, 2010, 012119, ©2010 The American Physical Society.

Reference [49] neglected dielectric dispersion in this model and used the zero-frequency permittivity values for silica ($\varepsilon = 2.02$) and bromobenzene ($\varepsilon = 4.30$). Densities of 1.49 g/cm^3 and 1.96 g/cm^3 were used for bromobenzene and silica, respectively.

The first step in determining the stable levitation point was to calculate the total vertical force (Casimir + gravity) acting on the silica microsphere, assuming alignment of its axis with the symmetry axis of the metal indentation. Because this configuration was cylindrically symmetric, it could be efficiently computed as in the previous section.

As reported in [49], Fig. 19.12 shows the FDTD-computed results for the total vertical force assuming a microsphere radius of 500 nm and a metal-indentation radius of 1 μm. For this specific case, the gravity force was found to balance against the Casimir force at a microsphere height of $h = 450$ nm above the bottom of the indentation.

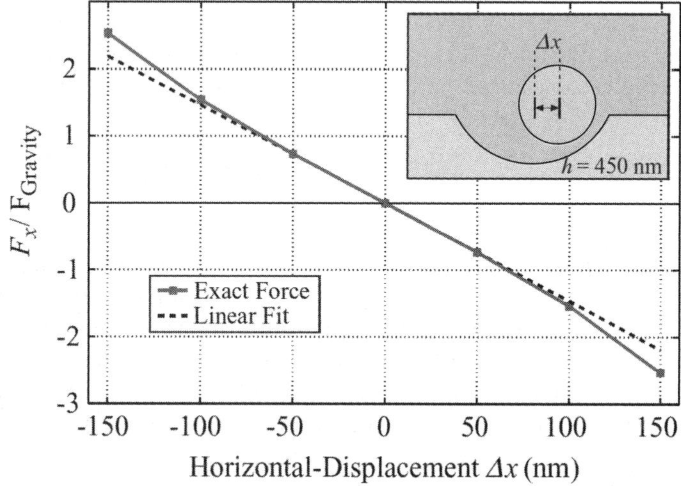

Fig. 19.13 FDTD-computed Casimir restoring force on the silica microsphere of Fig. 19.11 as a function of its lateral displacement Δx with h fixed at 450 nm, the vertical equilibrium position for $\Delta x = 0$. *Source:* A. P. McCauley, A. W. Rodriguez, J. D. Joannopoulos, and S. G. Johnson, *Physical Review A*, Vol. 81, 2010, 012119, ©2010 The American Physical Society.

To determine the strength of the lateral confinement, Ref. [49] reported a fully 3-D FDTD-Casimir computation in which the center of the microsphere was displaced laterally from equilibrium by a distance Δx while the vertical position was held fixed at the equilibrium value, $h = 450$ nm. As reported in [49], Fig. 19.13 shows that, over the range $|\Delta x| < 100$ nm, the Casimir restoring force varied approximately linearly with Δx, increasing more rapidly for larger displacements. At these larger lateral displacements, the vertical force was no longer zero because of the curvature of the metal indentation, and hence needed to be recomputed if the total force at these positions was required.

The fully 3-D FDTD-Casimir computations performed for Fig. 19.13 parallelized very easily. Every source term, polarization, and k-point could be computed in parallel, and the individual FDTD calculations could be parallelized in the available software [53]. In this specific example, each force point required less than one hour on a 2009-vintage computer with 1000+ processors.

19.9 GENERALIZATION TO NONZERO TEMPERATURES

This section reviews the technique reported in [50] to compute Casimir forces at nonzero temperatures ($T > 0$) with time-domain electromagnetic simulations. This technique generalizes the computational approach based on the FDTD method previously demonstrated for $T = 0$ in [48, 49], and reviewed in Sections 19.2 through 19.8. Compared to the previous $T = 0$ method, only a small modification is required, but some care is needed to properly capture the zero-frequency contribution. The validations reported in [50] against analytical and numerical benchmarks are reviewed. The latter study showed a surprising high-temperature disappearance of the non-monotonic behavior computed for a similar geometry, reviewed in Section 19.4.

19.9.1 Theoretical Foundation

The Casimir force arises from fluctuations at all frequencies ω, and the $T = 0$ force can be expressed as an integral $F(0) = \int_0^\infty f(\xi)\,d\xi$ over Wick-rotated imaginary frequencies $\omega = i\xi$ [2]. At $T > 0$, this integral becomes the following sum over "Matsubara frequencies" $\xi_n = n\pi\omega_T$ for integers n:

$$F(T) = \pi\omega_T\left[0.5f(0^+) + \sum_{n=1}^{\infty} f(n\pi\omega_T)\right] \tag{19.47}$$

where $\omega_T = 2k_B T/\hbar$ and k_B is Boltzmann's constant [2]. At room temperature, $\xi = \pi\omega_T$ corresponds to a "wavelength" $2\pi/\xi = 7\,\mu\text{m}$, much larger than most experimental separations, so usually $T > 0$ corrections are negligible [2]. However, experiments are pushing toward >1-μm separations in attempts to measure this phenomenon [63, 64], recently culminating in an experiment at several micrometers that appears to clearly observe the $T > 0$ corrections [65]. In addition, much larger T corrections have been predicted with certain materials and geometries [66].

As reviewed in Section 19.2.1, the fluctuation-dissipation theorem can be used to compute Casimir forces. Here, the mean-square electric and magnetic fields $\langle E^2\rangle$ and $\langle H^2\rangle$ can be computed from classical Green's functions [2], and the mean stress tensor can be computed and integrated to obtain the Casimir force. Using this technique, Equation (19.7) is modified to account for the $T > 0$ temperature dependence of the correlation function $\langle E^2\rangle$ at each ω [50]:

$$\left\langle E_i(\omega,\boldsymbol{x})\,E_j(\omega,\boldsymbol{x}')\right\rangle = -\frac{\hbar}{\pi}\,\mathrm{Im}\left[\omega^2 G_{ij}^E(\omega;\boldsymbol{x},\boldsymbol{x}')\right]\coth\left(\omega/\omega_T\right) \tag{19.48}$$

where the temperature dependence appears as a coth factor from a Bose–Einstein distribution. (The magnetic-field correlation $\langle H^2\rangle$ has a similar form.) If this is Wick rotated to imaginary frequency $\omega = i\xi$, the poles in the coth function give the sum (19.47) over Matsubara frequencies [24]. In the present electromagnetic simulations, what is actually computed is the electric or magnetic field in response to an electric or magnetic dipole current, respectively. This is related to G_{ij} by:

$$E_{ij}(\omega;\boldsymbol{x},\boldsymbol{x}') = -i\omega G_{ij}^E(\omega;\boldsymbol{x},\boldsymbol{x}') \tag{19.49}$$

where $E_{ij}(\omega;\boldsymbol{x},\boldsymbol{x}')$ denotes the electric field response in the ith direction due to a dipole current source $\boldsymbol{J}(\omega;\boldsymbol{x},\boldsymbol{x}') = \delta(\boldsymbol{x}-\boldsymbol{x}')\,\hat{\boldsymbol{e}}_j$.

19.9.2 Incorporating $T > 0$ in the Time Domain

Because the standard $T > 0$ analysis of (19.47) is expressed in the frequency domain, one can, in principle, derive the time-domain approach via Fourier transformation. A starting point is the development of $g(\xi)$ in Section 19.2.5. At real ω, the effect of $T > 0$ is to include an additional factor $\coth[\omega(\xi)/\omega_T]$ in the $\omega(\xi)$ integral from (19.48). Thus, a naïve approach is to replace $g(\xi)$ of Section 19.2.5 with:

$$\begin{aligned} g(\xi) &\rightarrow g(\xi)\coth\left[\,\omega(\xi)/\omega_T\right] \\ &= -i\xi\sqrt{1 + i\sigma/\xi}\,\left(1 + i\sigma/2\xi\right)\coth\left[\,\omega(\xi)/\omega_T\right] \end{aligned} \tag{19.50}$$

and then Fourier-transform this $g(\xi)$ to obtain $g(t)$.

However, there is a problem with this approach. The $1/\omega$ singularity in $\coth[\omega(\xi)/\omega_T]$ means that (19.50) is not locally integrable around $\xi = 0$, and therefore its Fourier transform is not well defined. If this problem is ignored, and a discrete Fourier transform is implemented by simply assigning an arbitrary finite value for the $\xi = 0$ term, then the computed Casimir force for $T > 0$ is incorrect. Reference [50] quantified the resulting error for the case of parallel perfect-metal plates in one dimension by direct comparison to the analytical Lifshitz formula [67].

Instead, a natural solution is to treat $\omega \neq 0$ by using the coth factor as in (19.50), but in addition, subtracting the $\omega = 0$ pole and handling it separately [50]. The correct $\omega = 0$ contribution can be extracted from (19.47), converted to the time domain, and added back in manually as a correction to $g(t)$. In particular, the $\coth[\omega(\xi)/\omega_T]$ function has poles at $\omega = in\pi\omega_T$ for integers n. When the ω integral is Wick rotated, the residues of these poles give the Matsubara sum (19.47) via contour integration [24]. Subtracting the $n = 0$ pole from the coth function yields [50]:

$$g_{n>0}(\xi) = g(\xi)\left\{\coth\left[\frac{\omega(\xi)}{\omega_T}\right] - \frac{\omega_T}{\omega(\xi)}\right\} \tag{19.51}$$

Then, the result of the time-domain integration of $g_{n>0}(t)\Gamma(t)$ corresponds to all of the $n > 0$ terms in (19.47), and there is no problem with the Fourier transformation to $g_{n>0}(t)$ [50].

To handle the $\omega = 0$ contribution, one begins with the real-ω $T = 0$ force expression, following the notation of Section 19.2:

$$F_i = \operatorname{Im}\left[\frac{\hbar}{\pi}\int_0^\infty g_R(\omega)\,\Gamma_i(\omega)\,d\omega\right] \tag{19.52}$$

where $g_R(\omega) = -i\omega$ is the weighting factor for the $\sigma = 0$ real-ω contour, and $\Gamma_i(\omega) = \Gamma_i^E(\omega) + \Gamma_i^H(\omega)$ is the surface-integrated stress tensor (electric- and magnetic-field contributions). From (19.47), the $\omega = 0$ contribution for $T > 0$ is then [50]:

$$F_{i,(n=0)} = \lim_{\omega\to 0^+}\operatorname{Im}\left[\frac{\hbar}{\pi}\frac{1}{2}(-i\omega)\Gamma_i(\omega)\frac{2\pi k_B T}{\hbar}\right] = \lim_{\omega\to 0^+}\operatorname{Re}\left[-\omega\Gamma_i(\omega)\,k_B T\right] \tag{19.53}$$

In (19.53), $\hbar$ cancels in the $\omega = 0$ contribution; this term dominates in the limit of large T where the fluctuations can be thought of as purely classical thermal phenomena [50]. To relate (19.53) to what is actually computed in the FDTD method requires some care because of the transformation to the $\omega(\xi)$ contour. The quantity $\Gamma_i^E(\omega)$ is proportional to an integral of $E_{ij}(\omega) = -i\omega G_{ij}(\omega)$, from (19.49). However, the $\omega(\xi)$ transformed system computes $\tilde{\Gamma}_i^E(\xi) \sim \tilde{E}_{ij}(\xi) = -i\xi\tilde{G}_{ij}(\xi)$, where $\tilde{G}(\xi)$ solves (19.6) with $\omega^2\varepsilon(\boldsymbol{x}) \rightarrow \xi^2(1 + i\sigma/\xi)\varepsilon(\boldsymbol{x})$. What is actually wanted is $-i\omega G_{ij}(\omega)\big|_{\omega=\omega(\xi)} = -i\omega(\xi)\tilde{G}_{ij}(\xi)$. Therefore, the correct $\omega = 0$ contribution is given by [50]:

$$\lim_{\omega\to 0^+} \Gamma_i^E(\omega) = \lim_{\xi\to 0^+} \frac{\omega(\xi)}{\xi} \tilde{\Gamma}_i^E(\xi) \tag{19.54}$$

Combined with the $\omega(\xi)k_B T$ factor from (19.53), this gives an $n = 0$ contribution of $\tilde{\Gamma}\big|_{\xi=0^+}$ multiplied by $-[\omega(\xi)]^2 k_B T/\xi\big|_{\xi=0^+} = \sigma k_B T$. This $\omega = 0$ term corresponds to a simple expression in the time domain, since $\tilde{\Gamma}\big|_{\xi=0^+}$ is simply the time integral of $\tilde{\Gamma}(t)$, and the coefficient $\sigma k_B T$ is merely a constant. Therefore, while $g_{n>0}(t)\tilde{\Gamma}(t)$ was originally integrated to obtain then $n > 0$ contributions, the $n = 0$ contribution is included if, instead, the following expression is integrated:

$$\left[g_{n>0}(t) + \sigma k_B T\right]\tilde{\Gamma}(t) \tag{19.55}$$

The term $[g_{n>0}(t) + \sigma k_B T]$ generalizes the original $g(t)$ from Section 19.2 to any $T \geq 0$ [50].

19.9.3 Validations

Parallel Perfect Metallic Plates

Reference [50] reported checking (19.55) for the 1-D case of the Casimir force between two infinite, parallel, perfect metallic plates. The benchmark calculation was provided by the analytical Lifshitz formula [68]. It was verified that the $g_{n>0}$ term of (19.51) correctly provides the $n > 0$ terms, and the $\sigma k_B T$ term provides the correct $n = 0$ contribution. Hence, the total Casimir force was correct.

Parallel Square Perfect Metallic Rods between Parallel Perfect Metallic Sidewalls

Reference [50] also reported the more-complicated geometry shown schematically in the inset of Fig. 19.14. Here, two perfect metallic rods of square cross-section were assumed to be located between two parallel perfect metallic sidewalls. This geometry could be solved for the 2-D case of z-invariant fluctuations in a manner similar to the configuration of Fig. 19.2, which involved metallic rods of circular cross-section located between parallel metallic sidewalls. Benchmark calculations were provided by a frequency-domain boundary-element method (BEM) [37].

In Fig. 19.14, the solid lines were computed using the BEM assuming $T = 0$, whereas the open circles were computed by the $T = 0$ FDTD method, as in Section 19.4 and Fig. 19.3. Both methods agreed. The Casimir force at $T = 1 \times \pi c\hbar/k_B a$ was also computed, where the $\xi = 0^+$ term dominated. Here, the FDTD method with the $T > 0$ modification of (19.55) (diamonds) agreed with the BEM results (dashed lines), where the latter were computed using the Matsubara sum of (19.47) to model $T > 0$.

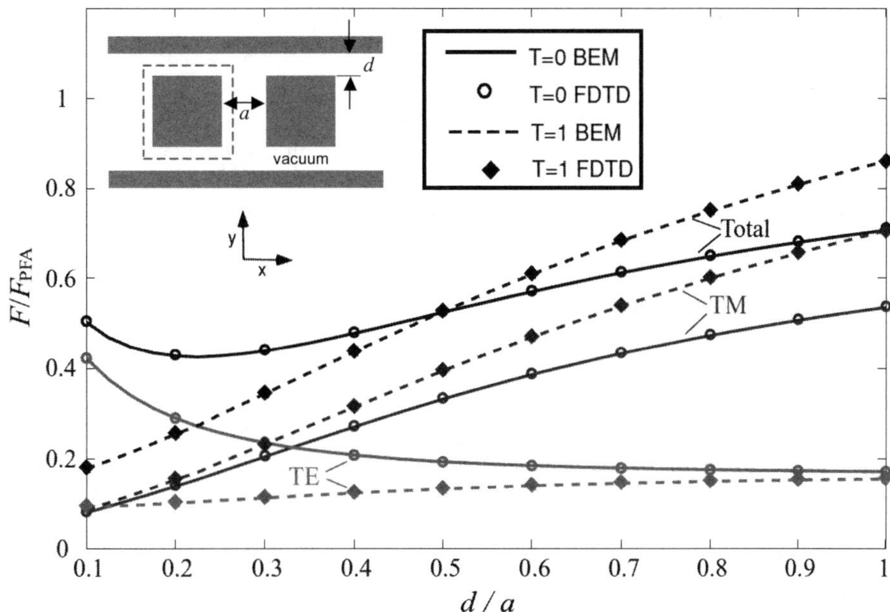

Fig. 19.14 Comparison between FDTD (open circles and filled diamonds) and frequency-domain BEM (solid and dashed lines) calculations of the 2-D Casimir force (z-invariant fluctuations) between two perfect-metal square rods located between two perfect-metal sidewalls, normalized by the proximity force approximation, F_{PFA}. At $T = 0$ (open circles and solid lines), the total force varied non-monotonically with d/a due to competition between the TE and TM polarizations. At $T = 1\times\pi c\hbar / k_{\mathrm{B}}a$ (dashed lines and filled diamonds), the BEM and FDTD results again matched, but the non-monotonicity disappeared. *Source:* K. Pan, A. P. McCauley, A. W. Rodriguez, M. T. Homer Reid, J. K. White, and S. G. Johnson, *Physical Review A*, Vol. 83, 2011, 040503(R), ©2011 The American Physical Society.

At $T = 0$, such geometries exhibit an interesting non-monotonic variation of the Casimir force between the two rods as a function of the sidewall separation, as shown by the solid lines and open circles in Fig. 19.14, as well as in Fig. 19.3. This does not arise in the simple pairwise-interaction heuristic picture of the Casimir force. Non-monotonicity arises from a competition between forces from transverse-electric (TE, $\boldsymbol{E}$ in-plane) and transverse-magnetic (TM, $\boldsymbol{E}$ out-of-plane) polarizations [40], which can be explained by a method-of-images argument [32].

19.9.4 Implications

This section has shown that a relatively simple modification to the time-domain method reviewed in Section 19.2 allows off-the-shelf FDTD software to easily calculate Casimir forces at nonzero temperatures. Recent predictions of realistic geometry and material effects [66], combined with the fact that temperature effects in complex geometries are almost unexplored at present, provide hope that future work will reveal further surprising temperature effects that are observable in emerging micro-mechanical systems.

Various authors have debated which model of $\varepsilon(\omega)$ most accurately reflects experiments [63–65]. However, the FDTD-Casimir technique stands apart from such debates, since standard FDTD techniques are available to account for complex material dispersions [51]. For example, although the perfect metal assumed in this section is obviously an artificial material, it has been argued that if the perfect metal is viewed as the limiting case of a Drude model, then the $n = 0$ contribution should be omitted [23] for the TM mode. With the technique reviewed in this section, this is accomplished simply by dropping the second term from (19.55) during the TM calculation. Even so, the non-monotonicity still disappears in a similar manner [50].

19.10 SUMMARY AND CONCLUSIONS

This chapter first reviewed the theoretical foundation established in [48] for employing the FDTD method to compute Casimir forces. This was followed by a review of the efficient implementation of the FDTD-Casimir technique reported in [49] that employed a harmonic expansion in source currents. Finally, this chapter reviewed the extension of the FDTD-Casimir technique reported in [50] to account for nonzero temperatures. In addition to providing simulations of fundamental physical phenomena, these developments may permit the design of novel micro- and nano-mechanical systems comprised of complex materials.

In practice, the harmonic expansion of [49] converges rapidly with higher harmonic moments, making the overall computational complexity of the FDTD method $O(N^{1+1/d})$ for N grid points and d spatial dimensions. This arises from the $O(N)$ number of computations needed for one FDTD time-step, while the time increment used varies inversely with the spatial resolution [51], leading to $O(N^{1/d})$ time-steps per simulation. In addition, there is a constant factor proportional to the number of terms retained in the harmonic expansion, as an independent simulation is required for each term. For comparison, without a harmonic expansion, one would have to run a separate simulation for each point on S. In that case, there would be $O(N^{(d-1)/d})$ points, leading to an overall computational cost of $O(N^2)$ [48].

Reference [49] did not claim that its harmonic-expansion FDTD technique is the most efficient means to compute Casimir forces. In fact, in other recent work, the frequency-domain boundary-element method [37] has also demonstrated excellent capabilities to model arbitrary 3-D geometries. However, such methods and their implementations are structure-specific. That is, they must be substantially revised when new types of materials or boundary conditions are desired that change the underlying Green's function (e.g., going from metals to dielectrics, periodic boundary conditions, or isotropic to anisotropic materials). On the other hand, very general FDTD codes such as [53] require no such modifications and are readily available. Furthermore, simulations using these codes can be executed with high efficiency on large-scale parallel-processing supercomputers, allowing problem sizes to continually expand as the number of processors typically available in such machines increases from year to year.

ACKNOWLEDGMENTS

The work reported in this chapter, compiled from [48–50], was supported in part by the Army Research Office through the ISN under Contract No. W911NF-07-D-0004, by the MIT Ferry Fund, by U.S. Department of Energy Grant No. DE-FG02-97ER25308, by the Defense Advanced Research Projects Agency under Contract No. N66001-09-1-2070-DOD, and by the Singapore-MIT Alliance Computational Engineering Flagship program.

APPENDIX 19A: HARMONIC EXPANSION IN CYLINDRICAL COORDINATES

For problem geometries that are rotationally invariant about the z-axis, working in cylindrical coordinates (r, ϕ, z) is appropriate. Here, S is a surface of revolution consisting of the rotation of a parameterized curve $[r(s),\ \phi = 0,\ z(s)]$ about the z-axis. The most practical harmonic expansion basis consists of functions of the form $f_n(x)e^{im\phi}$. Given a ϕ dependence, many FDTD codes solve a modified set of Maxwell's equations involving only the (r, z) coordinates. In this case, for each m, the model is reduced to a 2-D problem where both sources and fields are specified only in the (r, z) plane.

Once the fields are determined in the (r, z) plane, the Casimir force contribution for each m is given by [49]:

$$\begin{aligned}&\int_0^{2\pi} d\phi \int_S r(\boldsymbol{x}) e^{-im\phi} ds_j(\boldsymbol{x}) \int_0^{2\pi} d\phi' \\ &\quad \times \int_S r(\boldsymbol{x}') e^{im\phi'} \delta_S(\boldsymbol{x}-\boldsymbol{x}')\, \Gamma^E_{ij;m}(t;\boldsymbol{x},\boldsymbol{x}')\, ds(\boldsymbol{x}')\end{aligned} \tag{19A.1}$$

where $\boldsymbol{x}$ ranges over the full 3-D (r, ϕ, z) system. Here, the Cartesian line element ds along the 1-D surface S was introduced in anticipation of the cancellation of the Jacobian factor $r(\boldsymbol{x})$ from the integration over S. While this is only the contribution for Γ^E, the contribution for Γ^H is identical in form [49].

In cylindrical coordinates, the representation of the δ function is:

$$\delta(\boldsymbol{x}-\boldsymbol{x}') = \frac{1}{2\pi r(\boldsymbol{x})}\,\delta(\phi-\phi')\,\delta(r-r')\,\delta(z-z') \tag{19A.2}$$

For simplicity, assume that S consists entirely of $z = \text{const}$ and $r = \text{const}$ surfaces (the more general case follows by an analogous derivation). In these cases, the surface δ function δ_S is given by:

$$\delta_S(\boldsymbol{x}-\boldsymbol{x}') = \frac{1}{2\pi r(\boldsymbol{x})}\,\delta(\phi-\phi')\,\delta(r-r'), \qquad z = \text{const} \tag{19A.3a}$$

$$\delta_S(\boldsymbol{x}-\boldsymbol{x}') = \frac{1}{2\pi r(\boldsymbol{x})}\,\delta(\phi-\phi')\,\delta(z-z'), \qquad r = \text{const} \tag{19A.3b}$$

In either case, substituting either form of δ_S into (19A.1) results in a cancellation with the first $r(\boldsymbol{x})$ factor. Now, one picks an appropriate decomposition of δ_S into functions f_n (a choice of $r = \text{const}$ or $z = \text{const}$ merely implies that the f_n are either functions of z or r, respectively). Either case is denoted as $f_n(\boldsymbol{x})$, with the r and z dependence implicit [49].

The contribution for each value of n is now considered. The integral over $\boldsymbol{x}'$ is [49]:

$$\Gamma^E_{ij;nm}(t,\boldsymbol{x}) = \int_0^{2\pi} d\phi' \int_S r(\boldsymbol{x}')\, \Gamma^E_{ij;nm}(t,\boldsymbol{x},\boldsymbol{x}')\, f_n(\boldsymbol{x}')\, e^{im\phi'} ds(\boldsymbol{x}') \tag{19A.4}$$

As noted in the text, $\Gamma^E_{ij;nm}(t,\boldsymbol{x})$ is simply the field computed in the FDTD simulation due to a 3-D current source of the form $f_n(\boldsymbol{x})e^{im\phi}$. With the assumed cylindrical symmetry, this field must have a ϕ dependence of the form $e^{im\phi}$:

$$\Gamma^E_{ij;nm}(t,r,z,\phi) = \Gamma^E_{ij;nm}(t,r,z)e^{im\phi} \tag{19A.5}$$

This factor of $e^{im\phi}$ cancels with the remaining $e^{-im\phi}$. The integral over ϕ then produces a factor of 2π that cancels the one introduced by δ_S. After removing these factors, the problem is reduced to one of integrating the field responses entirely in the (r,z) plane. The contribution for each n and m is then [49]:

$$\int_S f_n(\boldsymbol{x})\,\Gamma^E_{ij;nm}(t,r,z)\,ds_j(\boldsymbol{x}) \tag{19A.6}$$

If one chooses the $f_n(\boldsymbol{x})$ to be real valued, the contributions for $+m$ and $-m$ are related by complex conjugation. The sum over m can then be rewritten as the real part of a sum over only nonnegative values of m. The final result for the Casimir force from the electric-field terms is then [49]:

$$F_i = \int_0^\infty dt\,\mathrm{Im}\left[g(-t)\right]\sum_n\int_S f_n(r,z)\,\Gamma^E_{ij;n}(t,r,z)\,ds_j(r,z) \tag{19A.7}$$

where the m dependence has been absorbed into the definition of $\Gamma^E_{ij;n}$ as follows:

$$\Gamma^E_{ij;n}(t,r,z) \equiv \Gamma^E_{ij;n,m=0}(t,r,z) + 2\sum_{m>0}\mathrm{Re}\left[\Gamma^E_{ij;nm}(t,r,z)\right] \tag{19A.8}$$

The dependence on r and z is explicitly included to emphasize that the integrals are confined to the 2-D (r,z) plane. The Casimir force receives an analogous contribution from the magnetic-field terms.

REFERENCES

[1] Casimir, H. B. G., "On the attraction between two perfectly conducting plates," *Proc. K. Ned. Akad. Wet.*, Vol. 51, 1948, pp. 793–795.

[2] Landau, L. D., E. M. Lifshitz, and L. P. Pitaevskii, *Statistical Physics: Part 2*, 3rd ed., Oxford, UK: Pergamon, 1980.

[3] Milonni, P. W., *The Quantum Vacuum: An Introduction to Quantum Electrodynamics*, San Diego, CA: Academic Press, 1993.

[4] Boyer, T. H., "Van der Waals forces and zero-point energy for dielectric and permeable materials," *Physical Review A*, Vol. 9, 1974, pp. 2078–2084.

[5] Lamoreaux, S. K., "Demonstration of the Casimir force in the 0.6 to 6 μm range," *Physical Review Lett.*, Vol. 78, 1997, pp. 5–8.

[6] Mohideen, U., and A. Roy, "Precision measurement of the Casimir force from 0.1 to 0.9 μm," *Physical Review Lett.*, Vol. 81, 1998, pp. 4549–4552.

[7] Chen, F., U. Mohideen, G. L. Klimchitskaya, and V. M. Mostepanenko, "Demonstration of the lateral Casimir force," *Physical Review Lett.*, Vol. 88, 2002, 101801.

[8] Chan, H. B., V. A. Aksyuk, R. N. Kleinman, D. J. Bishop, and F. Capasso, "Quantum mechanical actuation of microelectromechanical systems by the Casimir force," *Science*, Vol. 291, 2001, pp. 1941–1944.

[9] Iannuzzi, D., M. Lisanti, and F. Capasso, "Effect of hydrogen-switchable mirrors on the Casimir force," *Proc. National Academy of Sciences USA*, Vol. 101, 2004, pp. 4019–4023.

[10] Iannuzzi, D., M. Lisanti, J. N. Munday, and F. Capasso, "The design of long range quantum electrodynamical forces and torques between macroscopic bodies," *Solid State Communications*, Vol. 135, 2005, pp. 618–626.

[11] Maia Neto, P. A., A. Lambrecht, and S. Reynaud, "Roughness correction to the Casimir force: Beyond the proximity force approximation," *Europhysics Lett.*, Vol. 69, 2005, pp. 924–930.

[12] Brown-Hayes, M., D. A. R. Dalvit, F. D. Mazzitelli, W. J. Kim, and R. Onofrio, "Towards a precision measurement of the Casimir force in a cylinder-plane geometry," *Physical Review A*, Vol. 72, 2005, 052102.

[13] Bordag, M., "Casimir effect for a sphere and a cylinder in front of a plane and corrections to the proximity force theorem," *Physical Review D*, Vol. 73, 2006, 125018.

[14] Onofrio, R., "Casimir forces and non-Newtonian gravitation," *New Journal of Physics*, Vol. 8, 2006, 237.

[15] Emig, T., "Casimir-force-driven ratchets," *Physical Review Lett.*, Vol. 98, 2007, 160801.

[16] Munday, J. N., and F. Capasso, "Precision measurement of the Casimir-Lifshitz force in a fluid," *Physical Review A*, Vol. 75, 2007, 060102(R).

[17] Miri, M., and R. Golestanian, "A frustrated nanomechanical device powered by the lateral Casimir force," *Applied Physics Lett.*, Vol. 92, 2008, 113103.

[18] Genet, C., A. Lambrecht, and S. Reynaud, "The Casimir effect in the nanoworld," *European Physics J. Special Topics,* Vol. 160, 2008, pp. 183–193.

[19] Munday, J., F. Capasso, and V. A. Parsegia, "Measured long-range repulsive Casimir-Lifshitz forces," *Nature*, Vol. 457, 2009, pp. 170–173.

[20] Klimchitskaya, G. L., U. Mohideen, and V. M. Mostapanenko, "The Casimir force between real materials: Experiment and theory," *Review of Modern Physics*, Vol. 81, 2009, pp. 1827–1885.

[21] Dobrich, B., M. DeKieviet, and H. Gies, "Scalar Casimir-Polder forces for uniaxial corrugations," *Physical Review D*, Vol. 78, 2008, 125022.

[22] Bordag, M., U. Mohideen, and V. M. Mostepanenko, "New developments in the Casimir effect," *Physics Rep.*, Vol. 353, 2001, pp. 1–205.

[23] Milton, K. A., "The Casimir effect: Recent controversies and progress," *J. Physics A*, Vol. 37, 2004, pp. R209–R277.

[24] Lamoreaux, S. K., "The Casimir force: Background, experiments, and applications," *Rep. Progress in Physics*, Vol. 68, 2005, pp. 201–236.

[25] Capasso, F., J. N. Munday, D. Iannuzzi, and H. B. Chan, "Casimir forces and quantum electrodynamical torques: Physics and nanomechanics," *IEEE J. Selected Topics in Quantum Electronics,* Vol. 13, 2007, pp. 400–415.

[26] Chan, H. B., Y. Bao, J. Zou, R. A. Cirelli, F. Klemens, W. M. Mansfield, and C. S. Pai, "Measurement of the Casimir force between a gold sphere and a silicon surface with a nanotrench array," *Physical Review Lett.,* Vol. 101, 2008, 030401.

[27] Emig, T., A. Hanke, R. Golestanian, and M. Kardar, "Probing the strong boundary shape dependence of the Casimir force," *Physical Review Lett.,* Vol. 87, 2001, 260402.

[28] Gies, H., K. Langfeld, and L. Moyaerts, "Casimir effect on the worldline," *J. High Energy Physics*, Vol. 2003, 018.

[29] Gies, H., and K. Klingmüller, "Worldline algorithms for Casimir configurations," *Physical Review D*, Vol. 74, 2006, 045002.

[30] Rodriguez, A., M. Ibanescu, D. Iannuzzi, J. D. Joannopoulos, and S. G. Johnson, "Virtual photons in imaginary time: Computing exact Casimir forces via standard numerical electromagnetism techniques," *Physical Review A*, Vol. 76, 2007, 032106.

[31] Rahi, S. J., T. Emig, R. L. Jaffe, and M. Kardar, "Casimir forces between cylinders and plates," *Physical Review A*, Vol. 78, 2008, 012104.

[32] Rahi, S. J., A. W. Rodriguez, T. Emig, R. L. Jaffe, S. G. Johnson, and M. Kardar, "Nonmonotonic effects of parallel sidewalls on Casimir forces between cylinders," *Physical Review A*, Vol. 77, 2008, 030101(R).

[33] Emig, T., N. Graham, R. L. Jaffe, and M. Kardar, "Casimir forces between arbitrary compact objects," *Physical Review Lett.,* Vol. 99, 2007, 170403.

[34] Dalvit, D. A. R., P. A. Maia Neto, A. Lambrecht, and S. Reynaud, "Probing quantum-vacuum geometrical effects with cold atoms," *Physical Review Lett.*, Vol. 100, 2008, 040405.

[35] Kenneth, O., and I. Klich, "Casimir forces in a T-operator approach," *Physical Review B*, Vol. 78, 2008, 014103.

[36] Reynaud, S., P. A. Maia Neto, and A. Lambrecht, "Casimir energy and geometry: Beyond the proximity force approximation," *J. of Physics A: Mathematical and Theoretical,* Vol. 41, 2008, 164004.

[37] Reid, M. T. H., A. W. Rodriguez, J. White, and S. G. Johnson, "Efficient computation of Casimir interactions between arbitrary 3D objects," *Physical Review Lett.,* Vol. 103, 2009, 040401.

[38] Pasquali, S., and A. C. Maggs, "Numerical studies of Lifshitz interactions between dielectrics," *Physical Review A*, Vol. 79, 2009, 020102(R).

[39] Antezza, M., L. P. Pitaevski, S. Stringari, and V. B. Svetovoy, "Casimir-Lifshitz force out of thermal equilibrium and asymptotic nonadditivity," *Physical Review Lett.,* Vol. 97, 2006, 223203.

[40] Rodriguez, A., M. Ibanescu, D. Iannuzzi, F. Capasso, J. D. Joannopoulos, and S. G. Johnson, "Computation and visualization of Casimir forces in arbitrary geometries: Nonmonotonic lateral-wall forces and the failure of proximity-force approximations," *Physical Review Lett.*, Vol. 99, 2007, 080401.

[41] Zaheer, S., A. W. Rodriguez, S. G. Johnson, and R. L. Jaffe, "Optical-approximation analysis of sidewall-spacing effects on the force between two squares with parallel sidewalls," *Physical Review A*, Vol. 76, 2007, 063816.

[42] Rodriguez, A. W., J. N. Munday, J. D. Joannopoulos, F. Capasso, D. A. R. Dalvit, and S. G. Johnson, "Stable suspension and dispersion-induced transitions from repulsive Casimir forces between fluid-separated eccentric cylinders," *Physical Review Lett.,* Vol. 101, 2008, 190404.

[43] Milton, K. A., P. Parashar, and J. Wagner, "Exact results for Casimir interactions between dielectric bodies: The weak-coupling or van der Waals limit," *Physical Review Lett.,* Vol. 101, 2008, 160402.

[44] Emig, T., A. Hanke, R. Golestanian, and M. Kardar, "Normal and lateral Casimir forces between deformed plates," *Physical Review A*, Vol. 67, 2003, 022114.

[45] Hertzberg, M. P., R. L. Jaffe, M. Kardar, and A. Scardicchio, "Attractive Casimir forces in a closed geometry," *Physical Review Lett.*, Vol. 95, 2005, 250402.

[46] Rodrigues, R. B., P. A. Maia Neto, A. Lambrecht, and S. Reynaud, "Lateral Casimir force beyond the proximity-force approximation," *Physical Review Lett.*, Vol. 96, 2006, 100402.

[47] Dalvit, D. A. R., F. C. Lombardo, F. D. Mazzitelli, and R. Onofrio, "Exact Casimir interaction between eccentric cylinders," *Physical Review A*, Vol. 74, 2006, 020101(R).

[48] Rodriguez, A. W., A. P. McCauley, J. D. Joannopoulos, and S. G. Johnson, "Casimir forces in the time domain: Theory," *Physical Review A*, Vol. 80, 2009, 012115.

[49] McCauley, A. P., A. W. Rodriguez, J. D. Joannopoulos, and S. G. Johnson, "Casimir forces in the time domain: Applications," *Physical Review A*, Vol. 81, 2010, 012119.

[50] Pan, K., A. P. McCauley, A. W. Rodriguez, M. T. Homer Reid, J. K. White, and S. G. Johnson, "Calculation of nonzero-temperature Casimir forces in the time domain," *Physical Review A*, Vol. 83, 2011, 040503(R).

[51] Taflove, A., and S. C. Hagness, *Computational Electrodynamics: The Finite-Difference Time-Domain Method*, 3rd ed., Norwood, MA: Artech, 2005.

[52] Rosa, F. S. S., D. A. R. Dalvit, and P. W. Milonni, "Casimir interactions for anisotropic magnetodielectric metamaterials," *Physical Review A*, Vol. 78, 2008, 032117.

[53] Meep (free FDTD simulation software package developed at MIT to model electromagnetic systems), Online: http://ab-initio.mit.edu/wiki/index.php/Meep

[54] Barrett, R., M. Berry, T. F. Chan, J. Demmel, J. Donato, J. Dongarra, V. Eijkhout, R. Pozo, C. Romine, and H. Van der Vorst, *Templates for the Solution of Linear Systems: Building Blocks for Iterative Methods*, 2nd ed., Philadelphia, PA: SIAM, 1994.

[55] Trefethen, L. N., and D. Bau, *Numerical Linear Algebra*, Philadelphia, PA: SIAM, 1997.

[56] Rudin, W., *Real and Complex Analysis*, New York: McGraw-Hill, 1966.

[57] Rao, R. K., and P. Yip, *Discrete Cosine Transform: Algorithms, Advantages, Applications*, Boston: Academic Press, 1990.

[58] Li, M., W. H. P. Pernice, and H. X. Tang, "Tunable bipolar optical interactions between guided lightwaves," *Nature Photonics*, Vol. 3, 2009, pp. 464–468.

[59] Povinelli, M. L., M. Loncar, M. Ibanescu, E. J. Smythe, S. G. Johnson, F. Capasso, and J. D. Joannopoulos, "Evanescent-wave bonding between optical waveguides," *Optics Lett.*, Vol. 30, 2005, pp. 3042–3044.

[60] Lifshitz, E. M., "Influence of temperature on molecular attraction forces between condensed bodies," *Dokl. Akad. Nauk SSSR*, Vol. 100, 1955, 879 (in Russian).

[61] Dzyaloshinskii, I. E., E. M. Lifshitz, and L. P. Pitaevskii, "The general theory of van der Waals forces," *Advances in Physics*, Vol. 10, 1961, pp. 165–209.

[62] Marachevsky, V. N., "Casimir interaction of two plates inside a cylinder," *Physical Review D*, Vol. 75, 2007, 085019.

[63] Decca, R. S., D. Lopez, E. Fischbach, G. L. Klimchitskaya, D. E. Krause, and V. M. Mostepanenko, "Tests of new physics from precise measurements of the Casimir pressure between two gold-coated plates," *Physical Review D*, Vol. 75, 2007, 077101.

[64] Masuda, M., and M. Sasaki, "Limits on nonstandard forces in the submicrometer range," *Physical Review Lett.,* Vol. 102, 2009, 171101.

[65] Sushkov, A. O., W. J. Kim, D. A. R. Dalvit, and S. K. Lamoreaux, "Observation of the thermal Casimir force," *Nature Physics*, Vol. 7, 2011, pp. 230–233.

[66] Rodriguez, A. W., D. Woolf, A. P. McCauley, F. Capasso, J. D. Joannopoulos, and S. G. Johnson, "Achieving a strongly temperature-dependent Casimir effect," *Physical Review Lett.*, Vol. 105, 2010, 060401.

[67] Landau, L. D., and E. M. Lifshitz, *Statistical Physics: Part 1*, 3rd ed., Oxford, UK: Butterworth-Heinemann, 1980.

[68] Milton, K. A., *The Casimir Effect: Physical Manifestations of Zero-Point Energy*, River Edge, NJ: World Scientific, 2001.

Chapter 20

Meep: A Flexible Free FDTD Software Package[1]

Ardavan Oskooi and Steven G. Johnson

20.1 INTRODUCTION

One of the most common computational tools in classical electromagnetism is the finite-difference time-domain (FDTD) algorithm, which divides space and time into a regular grid and simulates the time evolution of Maxwell's equations [1–4]. This chapter reviews the description in [5] of the free, open-source implementation of the FDTD algorithm developed at the Massachusetts Institute of Technology (MIT): *Meep* (an acronym for MIT Electromagnetic Equation Propagation), available online at http://ab-initio.mit.edu/meep.

Meep is a full-featured software package. It includes modeling capabilities for arbitrary anisotropic, nonlinear, and dispersive electric and magnetic media; a variety of boundary conditions including symmetries and perfectly matched layers (PMLs); distributed-memory parallelism; spatial grids in Cartesian coordinates in one, two, and three dimensions as well as in cylindrical coordinates; and flexible output and field computations. Meep also includes some unusual features, such as advanced signal processing to analyze resonant modes, accurate subpixel averaging, a frequency-domain solver that exploits the time-domain code, complete scriptability, and integrated optimization facilities.

Rather than review the well-known FDTD algorithm itself (which is thoroughly covered in [1–4]), this chapter focuses on the particular design decisions that went into the development of Meep whose motivation may not be apparent from textbook FDTD descriptions. This includes the tension between abstraction and performance in FDTD implementations, and the unique or unusual features of the Meep software.

While more than 25 commercial FDTD software packages are available for purchase [6], the needs of research often demand the flexibility provided by access to the source code, and relaxed licensing constraints to speed porting to new clusters and supercomputers. Many groups end up developing their own FDTD code to serve their needs, a duplication of effort that seems wasteful. Most of these are not released to the public, and the handful of other free-software FDTD programs that could be downloaded when Meep was first released in 2006 were not nearly full featured enough. Since then, Meep has been cited in over 600 journal publications and has been downloaded more than 54,000 times, reaffirming the demand for such a package.

---

[1]This chapter is adapted from Ref. [5], A. Oskooi, D. Roundy, M. Ibanescu, P. Bermel, J. D. Joannopoulos, and S. G. Johnson, "Meep: A flexible free-software package for electromagnetic simulations by the FDTD method," *Computer Physics Communications*, Vol. 181, 2010, pp. 687–702, ©2010 Elsevier.

20.1.1 Alternative Computational Tools

FDTD algorithms are, of course, only one of many computational tools that have been developed for modeling electromagnetic wave interactions. In fact, FDTD may perhaps seem primitive in light of relatively sophisticated techniques such as finite-element methods (FEMs) with high-order accuracy and/or adaptive unstructured meshes [7, 8], or boundary-element methods (BEMs) that discretize only interfaces between homogeneous materials rather than volumes [9–12]. However, each tool has its strengths and weaknesses, and no single one is a panacea. For example, the nonuniform unstructured grids of FEMs have compelling advantages for metallic structures where micrometer wavelengths may be paired with nanometer skin depths. On the other hand, this flexibility comes at a price of substantial software complexity, which may not be worthwhile for dielectric devices at infrared wavelengths (such as in integrated optics or fibers) where the refractive index (and hence the typical resolution required) varies by less than a factor of 4 between materials, while small features such as surface roughness can be accurately handled by perturbative techniques [13].

BEMs, based on frequency-domain integral-equation formulations of electromagnetism, are especially powerful for scattering problems involving small objects in a large volume, since the volume need not be discretized and no artificial "absorbing boundaries" are needed. On the other hand, BEMs have a number of limitations: (1) They may still require artificial absorbers for interfaces extending to infinity, such as input/output waveguides [14]; (2) any change to the Green's function, such as introduction of anisotropic materials, imposition of periodic or symmetry boundary conditions, or a switch from three to two dimensions, requires re-implementation of large portions of the software (e.g., singular panel integrations and fast solvers) rather than purely local changes as in FDTD or FEM; (3) continuously varying materials are inefficiently modeled; and (4) frequency-domain solutions are inadequate for nonlinear or active systems in which frequency is not conserved—using BEM for such problems requires an expensive solver that is nonlocal in time as well as in space [11].

And then, of course, there are specialized tools that solve only a particular type of electromagnetic problem, such as MIT's MPB software, which only computes eigenmodes (e.g., waveguide modes) [15]. Such tools are powerful and robust within their domain, but are not a substitute for a general-purpose Maxwell's equations simulation.

FDTD has the advantages of simplicity, generality, and robustness: (1) It is straightforward to implement the full time-dependent Maxwell equations for nearly arbitrary materials, including nonlinear, anisotropic, dispersive, and time-varying materials; (2) it allows a wide variety of boundary conditions; (3) it allows modeling the temporal dynamics of the system, including key physical processes coupled to Maxwell's equations such as the electron energy transitions involved in lasing [16–20]; and (4) the algorithm is easily parallelized to run on clusters. FDTD is very attractive to researchers whose primary concern is investigating new physics, and for whom programmer time and the training of new students is more expensive than computer time.

20.1.2 The Initial-Value Problem Solved by Meep

The starting point for any FDTD solver, including Meep, is the set of two time-dependent Maxwell's curl equations, which in their simplest form can be written:

$$\frac{\partial \boldsymbol{B}}{\partial t} = -\nabla \times \boldsymbol{E} - \boldsymbol{J}_B \tag{20.1}$$

$$\frac{\partial \boldsymbol{D}}{\partial t} = \nabla \times \boldsymbol{H} - \boldsymbol{J} \tag{20.2}$$

where (respectively) $\boldsymbol{E}$ and $\boldsymbol{H}$ are the macroscopic electric and magnetic fields, $\boldsymbol{D}$ and $\boldsymbol{B}$ are the electric displacement and magnetic induction fields [21], $\boldsymbol{J}$ is the electric-charge current density, and $\boldsymbol{J}_B$ is a fictitious magnetic-charge current density (sometimes convenient in calculations, e.g., for magnetic-dipole sources).

In Meep FDTD calculations, one typically solves the initial-value problem where the fields and currents are zero for $t < 0$, and then nonzero values evolve for $t > 0$ in response to currents $\boldsymbol{J}(\boldsymbol{x}, t)$ and/or $\boldsymbol{J}_B(\boldsymbol{x}, t)$ having arbitrary time variations. In contrast, a frequency-domain solver assumes a time dependence of $e^{-i\omega t}$ for all currents and fields, and solves the resulting linear equations for the sinusoidal steady-state response or eigenmodes [22, Appendix D].) Hence, frequency-domain techniques solve for only the forced response of (20.1) and (20.2) to a sinusoidal excitation, whereas Meep/FDTD computes the homogeneous solution to this system as well as any forced response, i.e., the complete time-dependent solution.

Meep employs dimensionless units $\varepsilon_0 = \mu_0 = c = 1$. This choice emphasizes both the scale invariance of Maxwell's equations [22, Chapter 2], and also the fact that the most meaningful quantities to calculate are almost always dimensionless ratios (such as scattered power over incident power, or wavelength over some characteristic length scale). A user can pick any desired unit of distance a, either an SI unit such as $a = 1$ μm or some typical length scale of a given problem. Then, all distances are given in units of a, all times in units of a/c, and all frequencies in units of c/a. Meep allows modeling linear dispersionless media having the constitutive relations $\boldsymbol{D} = \varepsilon \boldsymbol{E}$ and $\boldsymbol{B} = \mu \boldsymbol{H}$, where ε and μ are possibly tensor relative permittivity and permeability values. Meep also allows modeling nonlinear and/or dispersive media.

20.1.3 Organization of This Chapter

This chapter is organized as follows. Section 20.2 discusses the discretization and coordinate system used in Meep. In addition to the standard Yee discretization [1], this raises the question of how exactly the FDTD grid is described and divided into "chunks" for parallelization, PML, and other purposes. Section 20.3 describes a central principle of Meep's design: pervasive interpolation, which provides (as much as possible) the illusion of continuity in specifying sources, materials, and outputs. This led to the development of several techniques unique to Meep, such as a scheme for subpixel material averaging designed to eliminate the first-order error usually associated with averaging techniques or staircasing of interfaces. Section 20.4 motivates and describes the techniques used in Meep for modeling nonlinear and dispersive materials. This includes a slightly unusual method to implement nonlinear materials using a Padé approximant that eliminates the need to solve cubic equations for every pixel. Section 20.5 describes how typical computations are performed in Meep, such as memory-efficient transmission spectra or sophisticated analysis of resonant modes via harmonic inversion. This section also describes how the time-domain code has been adapted, almost without modification, to solve frequency-domain problems with much faster convergence to the steady-state response than merely time-stepping. The user interface of Meep is discussed in Section 20.6, explaining the considerations that led to a scripting interface (rather than a GUI or CAD interface). Section 20.7 describes some of the trade-offs between performance and generality in this type of code and the specific compromises chosen in Meep. Finally, Section 20.8 provides a summary and concluding remarks.

20.2 GRIDS AND BOUNDARY CONDITIONS

The starting point for the FDTD algorithm is the discretization of space and time into a grid. In particular, Meep uses the standard Yee grid discretization, which staggers the electric and magnetic fields in time and in space, with each field component sampled at different spatial locations offset by half a pixel, allowing the time and space derivatives to be formulated as central-difference approximations [23]. This much is common to nearly every FDTD implementation and is described in detail elsewhere [1].

To parallelize Meep, efficiently support simulations with symmetries, and efficiently store auxiliary fields only in certain regions (for PML absorbing layers), Meep further divides the FDTD grid into "chunks" that are joined together into an arbitrary topology via boundary conditions. (In the future, different chunks may have different resolutions to implement a nonuniform grid [24–27].) Furthermore, Meep distinguishes two coordinate systems: one consisting of integer coordinates on the Yee grid, and one of continuous coordinates in "physical" space that are interpolated as necessary onto the grid (see Section 20.3). This section describes these concepts as they are implemented in Meep, since they form the foundation for the remaining sections and the overall design of the Meep software.

20.2.1 Coordinates and Grids

The two spatial coordinate systems in Meep are described by the `vec`, a continuous vector in R^d (in d dimensions), and the `ivec`, an integer-valued vector in Z^d describing locations on the Yee grid. The origin of these systems is an arbitrary `ivec` that can be set by the user, but is typically the center of the computational space. If $\boldsymbol{n}$ is an `ivec`, the corresponding `vec` is given by $0.5\Delta x\boldsymbol{n}$, where Δx is the spatial resolution (the same along x, y, and z). That is, the integer coordinates in an `ivec` correspond to *half*-pixels, as shown in the right panel of Fig. 20.1. This is to represent locations on the spatial Yee grid, which offsets different field components in space by half a pixel as shown in two dimensions (2-D) in the right panel of Fig. 20.1. In three dimensions (3-D), the E_x and D_x components are sampled at `ivecs` $(2\ell+1, 2m, 2n)$, E_y and D_y are sampled at `ivecs` $(2\ell, 2m+1, 2n)$, and so on; H_x and B_x are sampled at `ivecs` $(2\ell, 2m+1, 2n+1)$, H_y and B_y are sampled at `ivecs` $(2\ell+1, 2m, 2n+1)$, and so on.

In addition to these field-component grids, an occasional reference is made to the *centered* grid at odd `ivecs` $(2\ell+1, 2m+1, 2n+1)$ corresponding to the "center" of each pixel. The philosophy of Meep, as described in Section 20.3, is that as much as possible the user should be concerned only with continuous physical coordinates (`vecs`), and the interpolation/discretization onto `ivecs` occurs internally as transparently as possible.

20.2.2 Grid Chunks and Owned Points

An FDTD simulation must occur within a finite volume of space terminated with some boundary conditions, possibly by absorbing PML regions as described below. This rectilinear computational region is further subdivided into convex rectilinear chunks. On a parallel computer, different chunks may be stored at different processors. To simplify the calculations for each chunk, Meep employs the common technique of padding each chunk with extra "boundary" pixels that store the boundary values [28] (shown as gray regions in Fig. 20.1). This means that the chunks overlap in the interior of the computational space, where the overlaps require communication to synchronize the values.

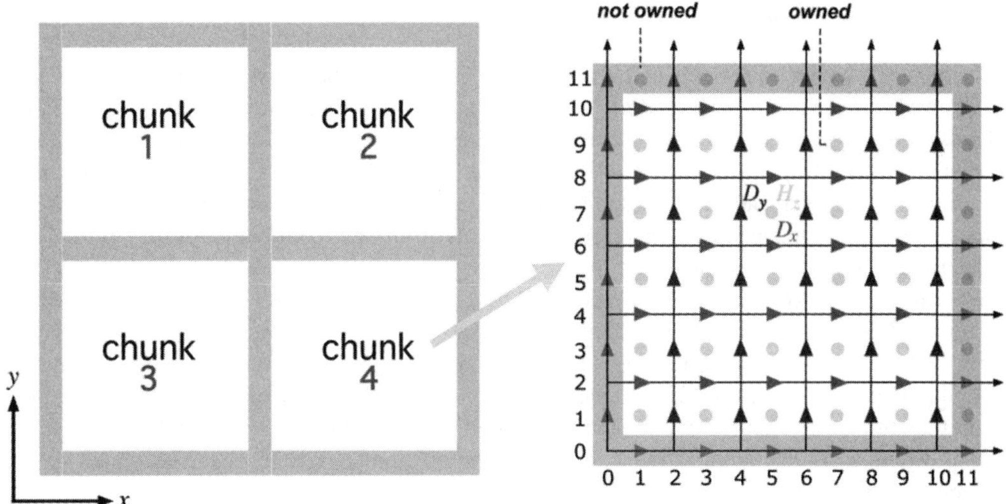

Fig. 20.1 In Meep, the FDTD computational space is divided into chunks (left) that have a one-pixel overlap (gray regions). Each chunk (right) represents a portion of the Yee grid, partitioned into owned points (chunk interior) and not-owned points (gray regions around the chunk edges) that are determined from other chunks and/or via boundary conditions. Every point in the interior of the computational cell is owned by exactly one chunk, the chunk responsible for time-stepping that point. *Source:* A. Oskooi, D. Roundy, M. Ibanescu, P. Bermel, J. D. Joannopoulos, and S. G. Johnson, *Computer Physics Communications*, Vol. 181, 2010, pp. 687–702, ©2010 Elsevier.

More precisely, the grid points in each chunk are partitioned into "owned" and "not-owned" points. The not-owned points are determined by communication with other chunks and/or by boundary conditions. The owned points are time-stepped within the chunk, independently of the other chunks (and possibly in parallel), and *every grid point inside the computational space is owned by exactly one chunk.*

The question then arises: how does one decide which points within the chunk are owned? In order for a grid point to be owned, the chunk must contain all the information necessary for time-stepping that point, once the not-owned points have been communicated. For example, for a D_y point $(2\ell, 2m+1, 2n)$ to be owned, the H_z points at $(2\ell \pm 1, 2m+1, 2n)$ must both be in the chunk in order to compute $\nabla \times \boldsymbol{H}$ for time-stepping $\boldsymbol{D}$ at that point. This means that the D_y points along the left (minimum-x) edge of the chunk (as shown in the right panel of Fig. 20.1) *cannot* be owned: there are no H_z points to the left of them. An additional dependency is imposed by the case of anisotropic media: if there is an ε_{xy} coupling E_x to D_y, then updating E_x at $(2\ell+1, 2m, 2n)$ requires the four D_y values at $(2\ell+1 \pm 1, 2m \pm 1, 2n)$ (these are the surrounding D_y values, as seen in the right panel of Fig. 20.1). This means that the E_x and D_x points along the *right* (maximum-x) edge of the chunk (as shown in the right panel of Fig. 20.1) cannot be owned either: there are no D_y points to the right of them. Similar considerations apply to $\nabla \times \boldsymbol{D}$ and anisotropic μ.

All of these considerations result in the shaded-gray region of Fig. 20.1 (right) being not-owned. That is, if the chunk intersects $k+1$ pixels along a given direction starting at an `ivec` coordinate of 0 (e.g. $k = 5$ in Fig. 20.1), the endpoint `ivec` coordinates 0 and $2k+1$ are not-owned, and the interior coordinates from 1 to $2k$ (inclusive) are owned.

20.2.3 Boundary Conditions and Symmetries

All of the not-owned points in a chunk must be determined by boundary conditions of some sort. The simplest boundary conditions are when the not-owned points are owned by some other chunk, in which case the values are simply copied from that chunk (possibly requiring communication on a multiprocessor system) each time they are updated. To minimize communications overhead, all communications between two chunks are batched into a single message (by copying the relevant not-owned points to/from a contiguous buffer) rather than sending one message per point to be copied.

At the outer edges of the computational space, some user-selected boundary condition must be imposed. For example, one can use perfect electric or magnetic conductors where the relevant electric/magnetic-field components are set to zero at the boundaries. One can also use Bloch-periodic boundary conditions. Here, the fields on one side of the computational space are copied from the other side of the computational space, optionally multiplied by a complex phase factor $e^{ik_i\Lambda_i}$, where k_i is the propagation constant in the ith direction and Λ_i is the length of the computational space in the same direction. To simulate the extension of the computational space to infinity, Meep employs PML absorbers placed adjacent to the outer grid boundaries [1].

Bloch-periodic boundary conditions are useful in periodic systems [22], but this is only one example of a useful symmetry that can be exploited via boundary conditions. One can also have mirror and rotational symmetries. For example, if the materials and the field sources have a mirror symmetry, the computational costs could be halved by storing chunks only in half the computational space and applying mirror boundary conditions to obtain the not-owned pixels adjacent to the mirror plane.

As a more unusual example, consider the S-shaped structure in Fig. 20.2. This has no mirror symmetry, but is symmetric under 180° rotation, called C_2 symmetry [29]. Meep can exploit this case as well (assuming the field sources have the same symmetry), storing only half of the computational space and inferring the not-owned values along the dashed line by a 180° rotation. In the simple case where the stored region is a single chunk, this means that the not-owned points are determined by owned points in the same chunk, requiring copies, possibly with sign flips.

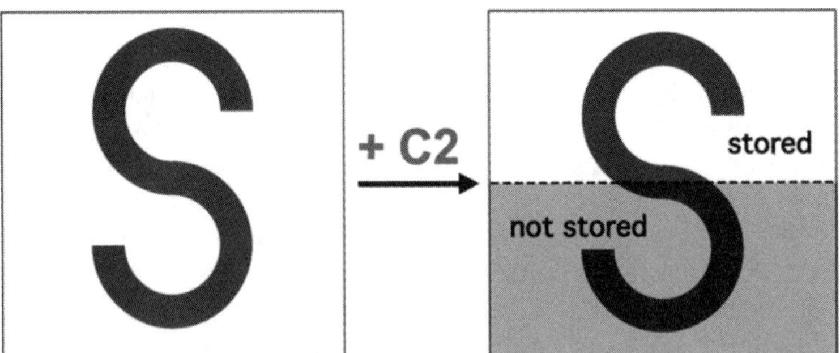

Fig. 20.2 Meep can exploit mirror and rotational symmetries such as the 180° (C_2) rotational symmetry of this S-shaped structure. Although Meep maintains the illusion that the entire structure is stored and simulated, internally only half of the structure is stored (as shown at right), and the other half is inferred by rotation. The rotation gives a boundary condition for the not-owned grid points along the dashed line. *Source:* A. Oskooi, D. Roundy, M. Ibanescu, P. Bermel, J. D. Joannopoulos, and S. G. Johnson, *Computer Physics Communications*, Vol. 181, 2010, pp. 687–702, ©2010 Elsevier.

Depending on the sources, the fields can be even or odd under mirror flips or C_2 rotations [22], so the user can specify an additional sign flip for the transformation of the vector fields (and pseudovector $\boldsymbol{H}$ and $\boldsymbol{B}$ fields, which incur an additional sign flip under mirror reflections [21, 22]). Meep also supports four-fold rotation symmetry (C_4) where the field can be multiplied by factors of 1, i, -1, or $-i$ under each 90° rotation [29]. Other rotations, such as three-fold or six-fold, are not supported because they do not preserve the Cartesian Yee grid. In 2-D, the x-y plane is itself a mirror plane, unless in the presence of anisotropic materials, and the symmetry decouples TE $\{E_x, E_y, H_z\}$ modes from TM $\{H_x, H_y, E_z\}$ modes [22]. In this case, Meep only allocates those fields for which the corresponding sources are present.

A central principle of Meep is that symmetry optimizations must be transparent to the user once the desired symmetries have been specified. Meep maintains the illusion that the entire computational space is computed. For example, the fields in the entire computational space can still be queried or exported to a file; flux planes and similar computations can still extend anywhere within the computational space; and so on. The fields in the non-stored regions are simply computed behind the scenes (without ever allocating memory for them) by transforming the stored chunks as needed. A key enabling factor for maintaining this illusion efficiently is the "loop-in-chunks" abstraction employed by the Meep code, described in Section 20.7.

Meep also supports continuous rotational symmetry around a given axis, where the structure is invariant under rotations and the fields transform as $e^{im\phi}$ for some m [22]. This is implemented separately by providing the option to simulate Maxwell's equations in the (r, z)-plane with cylindrical coordinates, for which operators like $\nabla\times$ change form.

20.3 APPROACHING THE GOAL OF CONTINUOUS SPACE-TIME MODELING

A core design philosophy of Meep is to approach the goal of electromagnetic field modeling in continuous space and time by mitigating the effects of the underlying FDTD discretization. There are two components to this approach: the input and the outputs. Continuously varying inputs, such as the geometry, materials, and source currents, must lead to continuously varying outputs. Furthermore, the effects of these inputs on the resulting outputs must converge as quickly as possible to the exact solution as the resolution increases. This section reviews how Meep approaches these goals for geometry/material inputs, current inputs, and field outputs.

20.3.1 Subpixel Smoothing

The second-order accuracy of FDTD generally degrades to first order if one directly discretizes a discontinuous material boundary [30, 31]. Moreover, directly discretizing a discontinuity in ε or μ leads to "staircased" interfaces that can only be varied in discrete pixel jumps. As discussed in Chapter 6, and as implemented in Meep, both of these problems can be solved by using an appropriate *subpixel smoothing* of ε and μ [31–33]. With this technique, before discretizing, dielectric discontinuities are smoothed into continuous transitions over a distance of one pixel, Δx, using a carefully designed averaging procedure. This achieves the goal of continuously varying results as the geometry is continuously varied. In the case of Meep, this is illustrated in Fig. 20.3 for a 2-D photonic crystal comprised of a square lattice of dielectric rods [5]. Here, the lowest TE-polarization (in-plane $\boldsymbol{E}$) eigenfrequency (computed as in Section 20.5) varies continuously with the eccentricity of the elliptical rods for subpixel smoothing, whereas the non-smoothed discontinuous discretization produces a stair-stepped discontinuous eigenfrequency.

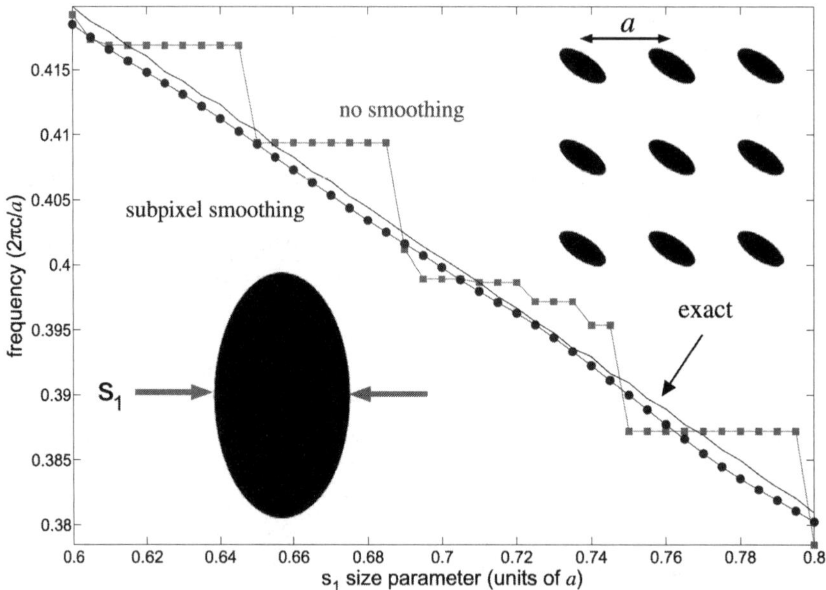

Fig. 20.3 A key principle of Meep is that continuously varying inputs yield continuously varying outputs. Here, the FDTD-computed eigenfrequency of a 2-D photonic crystal varies continuously with the assumed eccentricity of the elliptic cross-section dielectric rods comprising the crystal. This is accomplished by subpixel smoothing of the material parameters, whereas the non-smoothed result is "stair-stepped." (See Chapter 6 for a review of the theory and implementation of subpixel smoothing.) Specifically, this plot shows a TE (in-plane $\boldsymbol{E}$) eigenfrequency of a 2-D square lattice (period a) of dielectric ellipses ($\varepsilon = 12$) in air versus one semi-axis diameter of the ellipse (in gradations of $0.005a$) for no smoothing (squares, resolution = 20 pixels/a); subpixel smoothing (circles, resolution = 20 pixels/a); and "exact" results (line, no smoothing, resolution = 200 pixels/a). *Source:* A. Oskooi, D. Roundy, M. Ibanescu, P. Bermel, J. D. Joannopoulos, and S. G. Johnson, *Computer Physics Communications*, Vol. 181, 2010, pp. 687–702, ©2010 Elsevier.

However, it is not enough to realize continuously varying outputs from continuously varying geometrical/material inputs. In addition, the resulting outputs must converge as quickly as possible to the exact solution as the resolution increases. One must be alert to the possibility that a proposed subpixel smoothing technique could slow the convergence to the exact solution, or even degrade accuracy because the proposed smoothing would sufficiently alter the original structure to introduce additional error into the simulation [31].

To design an accurate smoothing technique, Meep exploits recent results in perturbation theory that show how a particular subpixel smoothing can be chosen to yield zero first-order error [13, 31–33]. Sample results are shown in Figs. 20.4 and 20.5 [5]. For both computation of the eigenfrequencies of an anisotropic photonic crystal in Fig. 20.4 and the scattering loss from a bump on a strip waveguide in Fig. 20.5, the errors in Meep's results decreased quadratically $[O(\Delta x)^2]$, whereas doing no averaging led to erratic linear convergence $[O(\Delta x)]$. Furthermore, Fig. 20.4 compares the subpixel smoothing of [31–33] to other potential subpixel-averaging schemes, including the obvious strategy of simply averaging ε within each pixel [34]. The other schemes led to first-order convergence, essentially little (if any) better than no averaging at all.

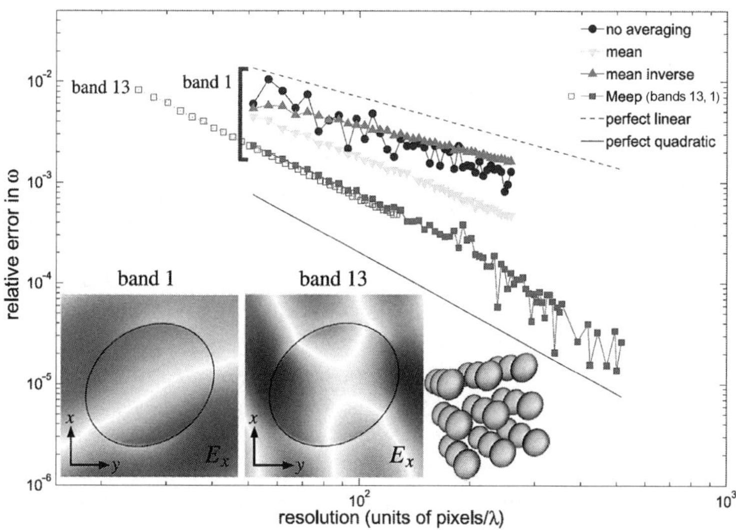

Fig. 20.4 Relative error $\Delta\omega/\omega$ for eigenmode calculation for a cubic lattice (period a) of 3-D anisotropic-ε ellipsoids (right inset) vs. grid resolution for several subpixel smoothing techniques. Dashed line shows linear convergence, solid line shows quadratic convergence. Most data are for eigenvalue band 1 (left inset shows x-y cross-section of unit cell) with vacuum wavelength $\lambda = 5.15a$. Hollow squares show Meep/FDTD results for band 13 (middle inset) with $\lambda = 2.52a$. *Source:* A. Oskooi, D. Roundy, M. Ibanescu, P. Bermel, J. D. Joannopoulos, and S. G. Johnson, *Computer Physics Communications*, Vol. 181, 2010, pp. 687–702, ©2010 Elsevier.

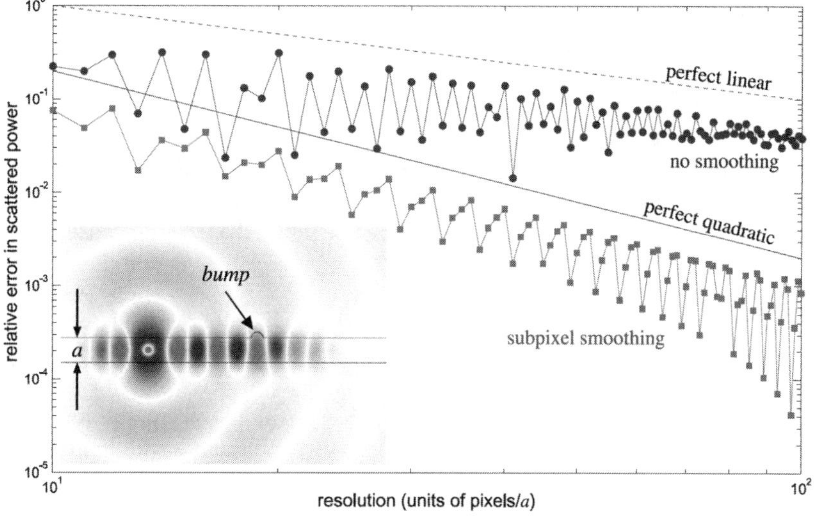

Fig. 20.5 Relative error in the scattered power from a small semicircular bump in a dielectric waveguide ($\varepsilon = 12$), excited by a point-dipole source in the waveguide, as a function of the grid resolution. Appropriate subpixel smoothing of the dielectric interfaces leads to roughly quadratic convergence (squares), whereas the unsmoothed structure has only first-order convergence (circles). *Source:* A. Oskooi, D. Roundy, M. Ibanescu, P. Bermel, J. D. Joannopoulos, and S. G. Johnson, *Computer Physics Communications*, Vol. 181, 2010, pp. 687–702, ©2010 Elsevier.

The theory and implementation of the subpixel smoothing used in Meep was introduced in [31–33] and is reviewed in Chapter 6; hence, it is only briefly summarized here. In order for a smoothing scheme to yield zero first-order perturbation, it must be anisotropic. Even if the initial interface is between isotropic materials, one obtains a tensor ε (or μ) which uses the mean ε for fields parallel to the interface, and the harmonic mean (inverse of the mean of ε^{-1}) for fields perpendicular to the interface. This was initially proposed heuristically [35] and later shown to be justified via perturbation theory [13, 31]. (If the initial materials are anisotropic, a more complicated formula is needed [32, 33].) The key point is that, even if the physical structure consists entirely of isotropic materials, the discretized structure uses anisotropic materials. Stable simulation of anisotropic media requires an FDTD variant recently proposed in [36].

There are several limitations to the subpixel smoothing used in Meep. First, perfect metal interfaces require a different approach [37, 38] not yet implemented in Meep; nor are interfaces with dispersive materials. However, there is numerical evidence that similar accuracy improvements could be obtained [39], possibly using the unconjugated form of perturbation theory for the complex-symmetric Maxwell equations in reciprocal media with losses [40]. Second, once the subpixel smoothing eliminates the first-order error, the presence of sharp corners (associated with field singularities) introduces an error intermediate between first and second order [31]. Third, even with subpixel smoothing, the fields directly on the material interface are still first-order accurate, at best. However, these localized errors are equivalent to currents that radiate zero power to first order [32, 41]. Therefore, the improved accuracy from subpixel smoothing is obtained when evaluating nonlocal properties such as resonant frequencies and eigenfrequencies (Fig. 20.4), the scattered flux integrated over a virtual surface away from the interface (Fig. 20.5), and overall integrals of fields and energies to which the interface contributes only $O(\Delta x)$ of the integration domain, and hence first-order errors on the interface have a second-order effect.

20.3.2 Interpolation of Field Sources

A field source such as a current density is specified at some point (for dipole sources) or in some region (for distributed current sources) in continuous space, and then must be restricted to a corresponding current source on the Yee grid. Meep performs this restriction using the loop-in-chunks abstraction of Section 20.7. Mathematically, Meep exploits a well-known concept, originating in multi-grid methods, that restriction can be defined as the transpose of interpolation [42]. This is illustrated by a 2-D example in Fig. 20.6.

Referring to Fig. 20.6(a), assume that the values f_1, f_2, f_3, and f_4 have been obtained at four grid points, and that $f(\boldsymbol{x}) = 0.32f_1 + 0.48f_2 + 0.08f_3 + 0.12f_4$ has been calculated using a bilinear interpolation of $\{f_1, f_2, f_3, f_4\}$. Then, as shown in Fig. 20.6(b), if a point-dipole current source J is placed at $\boldsymbol{x}$, it is restricted on the grid to values $J_1 = 0.32J$, $J_2 = 0.48J$, $J_3 = 0.08J$, and $J_4 = 0.12J$. In fact, this restriction is the transpose of the interpolation, and uses the same coefficients. Such a restriction preserves the sum (integral) of the currents, and typically leads to second-order convergence of the resulting fields as the grid resolution increases.

Figure 20.7 shows an example of the utility of this restriction process via the phenomenon of Cherenkov radiation [43]. Here, a point charge q moving at a constant velocity $v = 1.05c/n$, exceeding the phase velocity of light in a medium of index $n = 1.5$, emits a shockwave-like radiation pattern. This was directly modeled in Meep by a continuously moving current source $\boldsymbol{J} = -\boldsymbol{v}q\,\delta(\boldsymbol{x} - \boldsymbol{v}t)$ [44]. In contrast, discretizing the motion into jumps to the nearest grid point resulted in visible numerical artifacts in the radiation, as seen in the right panel of Fig. 20.7.

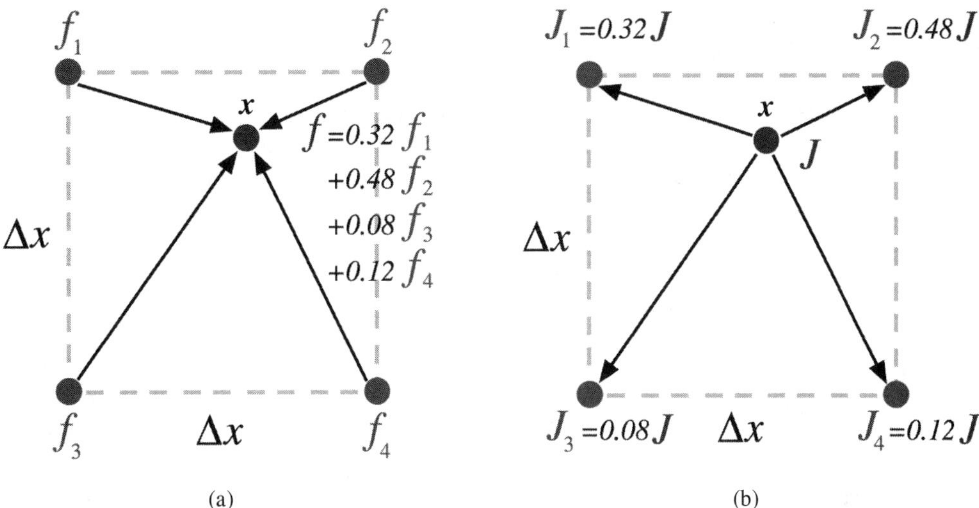

Fig. 20.6 (a) Bilinear interpolation of values $f_{1,2,3,4}$ on the grid to the value f at an arbitrary point $\boldsymbol{x}$; (b) reverse process (restriction) taking $J(\boldsymbol{x})$, e.g., a current source, and converting into values $J_{1,2,3,4}$ on the grid. Restriction can be viewed as the transpose of interpolation, and uses the same coefficients. *Source:* A. Oskooi, D. Roundy, M. Ibanescu, P. Bermel, J. D. Joannopoulos, and S. G. Johnson, *Computer Physics Communications*, Vol. 181, 2010, pp. 687–702, ©2010 Elsevier.

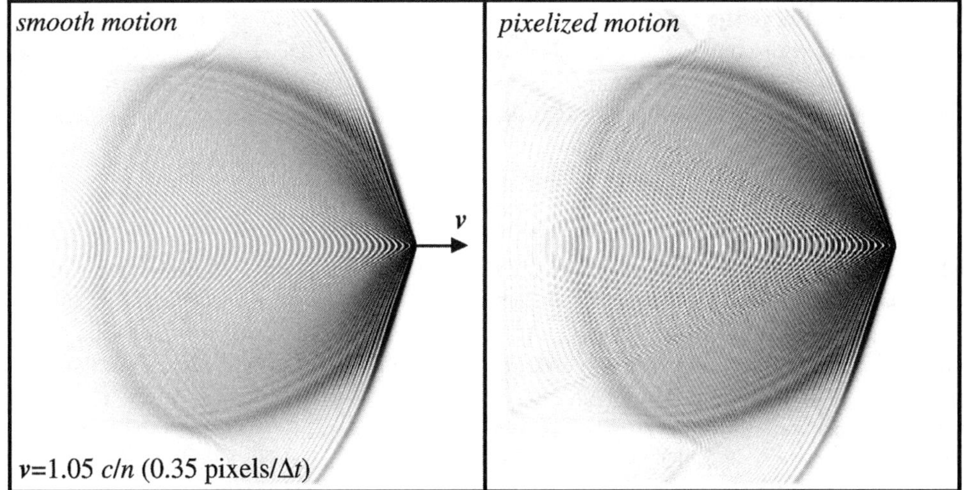

Fig. 20.7 Meep/FDTD-computed Cherenkov radiation emitted by a point charge moving at a speed $v = 1.05c/n$ exceeding the phase velocity of light in a homogeneous medium of index $n = 1.5$. Thanks to Meep's interpolation (or technically restriction), the smooth motion of the source current (left panel) can be expressed as continuously varying currents on the grid, whereas the non-smooth pixelized motion (no interpolation) (right panel) reveals high-frequency numerical artifacts of the discretization, counter-propagating wavefronts behind the moving charge. *Source:* A. Oskooi, D. Roundy, M. Ibanescu, P. Bermel, J. D. Joannopoulos, and S. G. Johnson, *Computer Physics Communications*, Vol. 181, 2010, pp. 687–702, ©2010 Elsevier.

20.3.3 Interpolation of Field Outputs

At any point in the problem space, Meep can evaluate any FDTD-computed electric and magnetic field component; combinations of these field components to evaluate the local power flux and energy density; integrations of field components and their combinations; and potential user-defined functions. In general, this requires spatial and/or temporal interpolations from the stored field values. Since the underlying FDTD central-difference algorithm has second-order accuracy in both space and time, Meep linearly interpolates fields as needed, a process that also has second-order accuracy for smooth functions.

As a specific example, Meep provides an interface to integrate any function of the computed fields over any convex rectilinear region (e.g., boxes, planes, or lines) by accumulating linearly interpolated values within the desired region. There are two subtleties in this process, summarized next.

First, an additional spatial interpolation is required to integrate quantities like $\boldsymbol{E} \times \boldsymbol{H}$ that mix fields that are evaluated and stored at different grid points. Here, the fields are first interpolated onto the centered grid (Section 20.2). Then, the integrand is computed. Finally, the linearly interpolated values of the integrand are accumulated over the specified region.

Second, a temporal interpolation is required to compute quantities like $\boldsymbol{E} \times \boldsymbol{H}$ that mix fields that are evaluated and stored at different times. In this case, $\boldsymbol{H}$ is stored at times $(n - 0.5)\Delta t$, while $\boldsymbol{E}$ is stored at times $n\Delta t$ [1]. Simply using these time-offset fields together provides only first-order accuracy. If second-order accuracy is desired, Meep provides the option to temporarily synchronize the electric and magnetic fields. Here, the magnetic fields are saved to a backup array, time-stepped by Δt, and then averaged with the backup array to obtain the magnetic fields at $n\Delta t$ with $O(\Delta t)^2$ accuracy. (The fields are restored from backup before resuming general time-stepping.) This yields second-order accuracy at the expense of an extra one-half time-step of computation, which is usually negligible because such outputs are rarely required at every time-step of a simulation. Section 20.5 reviews how Meep performs typical transmission simulations and other calculations efficiently.

20.4 MATERIALS

Time-dependent methods for electromagnetism, given their generality, allow for the simulation of a broad range of material systems. Certain classes of materials, particularly active and nonlinear materials that do not conserve frequency, are ideally suited for modeling by such methods. Materials are represented in Maxwell's equations (20.1) and (20.2) via the relative permittivity $\varepsilon(\boldsymbol{x})$ and permeability $\mu(\boldsymbol{x})$, which in general depend on position, frequency (material dispersion), and the fields themselves (nonlinearities). Meep currently supports arbitrary anisotropic material tensors, anisotropic dispersive materials (Lorentz–Drude models and conductivities, both magnetic and electric), and nonlinear materials (both second- and third-order nonlinearities). Taken together, these permit investigations of a wide range of physical phenomena.

The implementation of these materials in Meep is mostly based on standard techniques [1], so this discussion will focus on two places where Meep differs from the usual approach. For nonlinearities, Meep uses a Padé approximant to avoid solving cubic equations at each step. For PML absorbing media in cylindrical coordinates, Meep uses a "quasi-PML" [45] based on a Cartesian PML. This performs comparably to a true PML while requiring less computational effort.

20.4.1 Nonlinear Materials

Optical nonlinearities arise when large field intensities induce changes in the local ε or μ to produce a number of interesting effects: temporal and spatial soliton propagation, optical bistability, self-focusing of optical beams, second- and third-harmonic generation, and many other effects [46, 47]. Such materials are usually described by a power-series expansion of $\boldsymbol{D}$ in terms of $\boldsymbol{E}$ and various susceptibilities. In many common materials, or when considering phenomena in a sufficiently narrow bandwidth (such as the resonantly enhanced nonlinear effects [48] well suited to FDTD calculations), these nonlinear susceptibilities can be accurately approximated via nondispersive (instantaneous) effects [49]. Meep supports instantaneous isotropic (or diagonal anisotropic) nonlinear effects of the form:

$$D_i - P_i = \varepsilon^{(1)} E_i + \chi_i^{(2)} E_i^2 + \chi_i^{(3)} |\boldsymbol{E}|^2 E_i \tag{20.3}$$

where $\varepsilon^{(1)}$ represents all the linear nondispersive terms, and P_i is a dispersive polarization $\boldsymbol{P} = \chi^{(1)}_{\text{dispersive}}(\omega)\boldsymbol{E}$ from dispersive materials such as Lorentz media [1]. (A similar equation relates $\boldsymbol{B}$ and $\boldsymbol{H}$.) Implementing this equation directly, however, would require one to solve a cubic equation at each time step [1, Section 9.6], since $\boldsymbol{D}$ is updated from $\nabla \times \boldsymbol{H}$ before updating $\boldsymbol{E}$ from $\boldsymbol{D}$.

However, (20.3) is merely a power-series approximation for the true material response, valid for sufficiently small field intensities, so it is not necessary to insist that it be solved exactly. Instead, Meep approximates the solution of (20.3) by a Padé approximant [50], which matches the "exact" cubic solution to high-order accuracy by the rational function:

$$E_i = \left\{ \frac{1 + \dfrac{\chi^{(2)}}{[\varepsilon^{(1)}]^2}\tilde{D}_i + \dfrac{2\chi^{(3)}}{[\varepsilon^{(1)}]^3}\|\tilde{\boldsymbol{D}}\|^2}{1 + \dfrac{2\chi^{(2)}}{[\varepsilon^{(1)}]^2}\tilde{D}_i + \dfrac{3\chi^{(3)}}{[\varepsilon^{(1)}]^3}\|\tilde{\boldsymbol{D}}\|^2} \right\} [\varepsilon^{(1)}]^{-1} \tilde{D}_i \tag{20.4}$$

where $\tilde{D}_i = D_i - P_i$. For the case of isotropic $\varepsilon^{(1)}$ and $\chi^{(2)} = 0$, so that one has a purely Kerr ($\chi^{(3)}$) material, this matches the "exact" cubic E to $O(D^7)$ error. With $\chi^{(2)} \neq 0$, the error is $O(D^4)$.

For more complicated dispersive nonlinear media or for arbitrary anisotropy in $\chi^{(2)}$ or $\chi^{(3)}$, one approach that Meep may implement in the future is to incorporate the nonlinear terms in the auxiliary differential equations for a Lorentz medium [1].

20.4.2 Absorbing Boundary Layers: PML, Pseudo-PML, and Quasi-PML

A perfectly matched layer (PML) is an artificial absorbing medium that is commonly used to truncate computational grids for simulating wave equations (e.g., Maxwell's equations), and is designed to have the property that interfaces between the PML and adjacent media are reflectionless in the exact wave equation [1]. There are various interchangeable formulations of PML for FDTD methods [1], which are all equivalent to a coordinate stretching of Maxwell's equations into complex spatial coordinates. Meep implements a version of the uniaxial PML (UPML), expressing the PML as an effective dispersive anisotropic ε and μ [1].

For a given round-trip passage of light through PML, Meep provides support for arbitrary user-specified absorption profiles, which have an important influence on overall reflections due to discretization error and other effects. PML absorption is specified in terms of the amplitude of the light passing impinging on it, reflecting off the outer edge of the computational space, and propagating back [51, 52]. As discussed in Chapter 5, for the case of periodic media such as photonic crystals, the medium is not analytic, and the premise of PML's reflectionless property is violated. In this case, a "PML" material overlapped with the photonic crystal is actually a "pseudo-PML" (pPML) that is reflectionless only in the limit of a sufficiently thick and gradual absorber, and control over the absorption profile is important [51, 52].

For the radial direction in cylindrical coordinates, a true PML can be derived by coordinate stretching. However, this requires more storage and computational effort than the Cartesian UPML [53, 54], as well as increasing code complexity. Instead, in cylindrical coordinates, Meep implements a quasi-PML [45], which simply consists of using the Cartesian UPML materials as an approximation for the true radial PML. This approximation becomes more and more accurate as the outer radius of the computational space increases, because the implicit curvature of the PML region decreases with radius and approaches the Cartesian case. Furthermore, one recalls that every PML has reflections once space is discretized [1], which can be mitigated by gradually turning on the absorption over a finite thickness. Quasi-PML reflections are likewise mitigated by using a gradual absorption profile. The only question is this: how thick does the quasi-PML need to be to achieve low reflections, compared to a true PML?

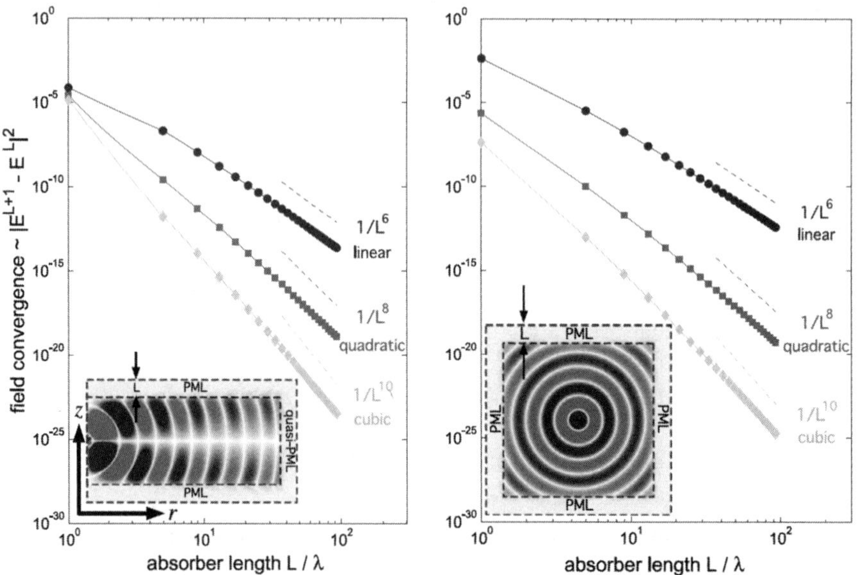

Fig. 20.8 The performance of a quasi-PML in the radial direction (cylindrical coordinates, left panel) at a resolution of 20 pixels/λ is nearly equivalent to that of a true PML (in Cartesian coordinates, right panel). The plot shows the difference in the electric field E_z (insets) from a point source between simulations with PML thickness L and L+1, which is a simple proxy for the PML reflections [51, 52]. The different curves are for PML conductivities that turn on as $(x/L)^d$ for $d = 1, 2, 3$ in the PML, leading to different rates of convergence of the reflection [51]. *Source:* A. Oskooi, D. Roundy, M. Ibanescu, P. Bermel, J. D. Joannopoulos, and S. G. Johnson, *Computer Physics Communications*, Vol. 181, 2010, pp. 687–702, ©2010 Elsevier.

Figure 20.8 shows that, for a typical calculation, the performance of the quasi-PML in cylindrical coordinates (left panel) is comparable to that of a true PML in Cartesian coordinates (right panel) [5]. This figure plots the "field convergence" factor of the PML as a function of its absorption length, L, for a fixed round-trip reflection, as postulated in [51, 52] and reviewed in Chapter 5. The field convergence factor is defined as the difference between the E-field at a given point in the computation space for simulations with PML (or quasi-PML) absorption lengths L and $L+1$. For the PML conductivity $\sigma(x)$ turned on gradually as $(x/L)^d$ for $d = 1, 2, 3$, it can be shown that the reflections decrease as $1/L^{2d+2}$, and the field-convergence factors decrease as $1/L^{2d+4}$ [51, 52]. Precisely these decay rates are observed in Fig. 20.8, with similar constant coefficients. Here, as the grid resolution was increased, approaching the exact wave equations, the constant coefficient in the Cartesian PML plot decreased, approaching zero reflection. On the other hand, the constant coefficient of the quasi-PML saturated at some minimum, corresponding to its finite reflectivity in the exact wave equation for a fixed L. This difference seems of little practical concern, however, because the reflection from a one-wavelength-thick quasi-PML at a moderate resolution (20 pixels/λ) is already so low.

20.5 ENABLING TYPICAL COMPUTATIONS

Simulating Maxwell's equations in the time domain enables the investigation of problems inherently involving multiple frequencies, such as nonlinearities and active media. However, it is also well adapted to solving frequency-domain problems since it can solve large bandwidths at once, for example analyzing resonant modes or computing transmission/reflection spectra. This section reviews techniques that Meep uses to efficiently compute scattering spectra and resonant modes in the time domain. Furthermore, this section reviews how the time-domain method can be adapted to a purely frequency-domain solver while sharing almost all of the underlying code.

20.5.1 Computing Flux Spectra

A principal task of a time-domain computational electrodynamics tool is the investigation of transmission or scattering spectra from arbitrary structures. Here, one wants to compute the transmitted or scattered power in a particular direction as a function of the frequency of the incident light. One can solve for the power at many frequencies in a single time-domain simulation by Fourier transforming the response to a short pulse.

Specifically, for a given surface S, one wishes to compute the integral of the Poynting flux:

$$P(\omega) = \mathrm{Re}\left[\iint_S \boldsymbol{E}(\boldsymbol{x},\omega)^* \times \boldsymbol{H}(\boldsymbol{x},\omega)\, dS\right] \tag{20.5}$$

where $\boldsymbol{E}(\boldsymbol{x},\omega)$ and $\boldsymbol{H}(\boldsymbol{x},\omega)$ are the complex-valued electric and magnetic field vector phasors of frequency ω calculated at $\boldsymbol{x}$, and Re denotes the real part of the expression. In the time domain, the basic idea is to use a short-pulse source covering a wide bandwidth including all frequencies of interest, and compute $\boldsymbol{E}(\boldsymbol{x},\omega)$ and $\boldsymbol{H}(\boldsymbol{x},\omega)$ from the Fourier transforms of $\boldsymbol{E}(\boldsymbol{x},t)$ and $\boldsymbol{H}(\boldsymbol{x},t)$, respectively. There are several ways to compute these Fourier transforms. For example, one could store the electric and magnetic fields throughout S over all times, and at the end of the simulation perform a discrete-time Fourier transform (DTFT) of the fields:

$$E(x,\omega) = \sum_{n} E(x,n\Delta t)\, e^{i\omega n\Delta t}\Delta t \tag{20.6}$$

for all frequencies ω of interest, possibly exploiting a fast Fourier transform (FFT) algorithm. Such an approach has the following computational cost. For a simulation having T time-steps, $F << T$ frequencies to compute, N_S fields in the flux region, and N pixels in the entire computational space, this approach requires $\Theta(N + N_S T)$ storage and $\Theta(NT + T\log T)$ time using an FFT-based chirp-z algorithm [55]. (Here, Θ has the usual meaning of an asymptotic tight bound [56].) The difficulty with this approach is that, if a long simulation (large T) is required to obtain a fine frequency resolution by the usual uncertainty relation [57], then the $\Theta(N_S T)$ storage requirements for the fields $E(x,t)$ and $H(x,t)$ at each point x in S become excessive.

Instead, Meep accumulates the DTFT summation of the fields at every point in S as the simulation progresses. Once time-stepping has terminated, (20.5) can be evaluated using these Fourier-transformed fields. The computational cost of this approach is $\Theta(N + N_S F)$ storage, which is much less than $\Theta(N_S T)$ if $F << T$; and $\Theta(NT + N_S FT)$ time. Although this approach works well, another possible approach under consideration for Meep is to use a Padé approximation. Here, one would store the fields at every time-step on S, but instead of using the DTFT, one would construct a Padé approximant to extrapolate the infinite-time DTFT from a short time series [58]. This would require $\Theta(N + N_S T)$ storage (but T would be potentially much smaller) and $O(NT + T\log^2 T)$ time [59].

20.5.2 Analyzing Resonant Modes

Another major goal of time-domain simulations is analysis of resonant phenomena, specifically by determining the resonant frequency ω_0 and the quality factors Q (i.e., the number of optical cycles, $2\pi/\omega_0$, for the field to decay by $e^{-2\pi}$) of one or more resonant modes. A straightforward and common approach to compute ω_0 and Q is to calculate the DTFT of the field at some point in the cavity in response to a short pulse [1]. Here, ω_0 is the center of a peak in the DTFT, and $1/Q$ is the fractional width of the peak at half-maximum. The problem with this approach is that the Fourier uncertainty relation (equivalently, spectral leakage from the finite time window [57]) means that resolving the peak in this way requires a simulation much longer than Q/ω_0 (problematic for structures that may have very high Q, even 10^9 or higher [60]). Alternatively, one can perform a least-squares fit of the field time-series within the cavity to an exponentially decaying sinusoid, but this leads to an ill-conditioned, non-convex, nonlinear fitting problem, and is especially difficult if more than one resonant mode may be present. If only a single resonant mode is present, one can perform a least-squares fit of the energy in the cavity to a decaying exponential in order to determine Q, but a long simulation is still required to accurately resolve a large Q, as shown in Fig. 20.9.

A more accurate and efficient approach, requiring only a short simulation even for very large Q values, is the technique of *filter diagonalization* originally developed for NMR spectroscopy. This transforms the time-series data into a small eigenproblem that is solved for all resonant frequencies and quality factors at once, even for multiple overlapping resonances [61]. Chapter 16 of [1] compared the DFT peak-finding method with filter diagonalization by attempting to resolve two near-degenerate modes in a microcavity, and demonstrated the latter's ability to accurately resolve closely spaced peaks with as much as a factor of 5 times fewer time-steps. Filter diagonalization has also been applied to compute quality factors of 10^8 or more using simulations only a few hundred optical cycles in length [60].

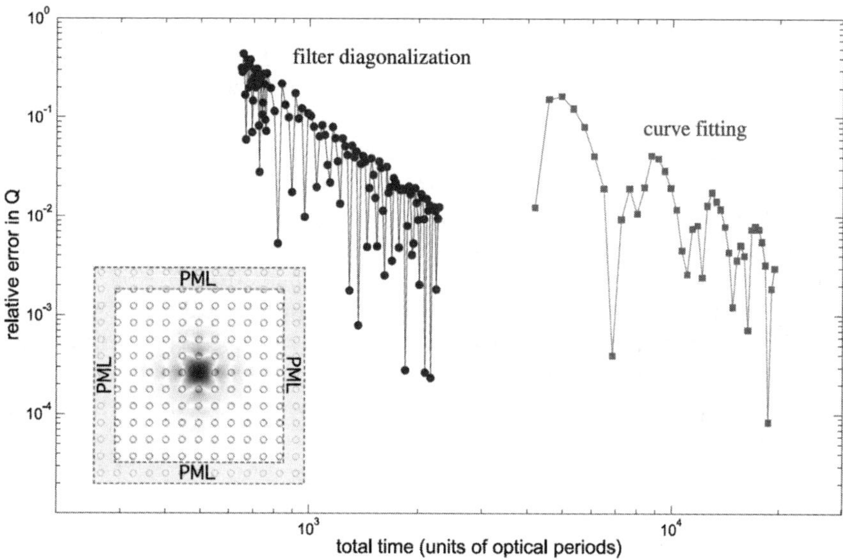

Fig. 20.9 Comparative relative error in the Meep/FDTD-computed quality factor Q of a defect-mode photonic-crystal resonant cavity (inset, period a) with $Q \sim 10^6$, vs. simulation time in units of optical periods of the resonance. Circles – filter-diagonalization method; squares – least-squares fit of energy in the cavity to a decaying exponential. Filter diagonalization requires many fewer optical periods than the decay time, Q, whereas curve fitting requires a simulation long enough for the fields to decay significantly. *Source:* A. Oskooi, D. Roundy, M. Ibanescu, P. Bermel, J. D. Joannopoulos, and S. G. Johnson, *Computer Physics Communications*, Vol. 181, 2010, pp. 687–702, ©2010 Elsevier.

Figure 20.9 quantifies the ability of filter diagonalization to resolve $Q \sim 10^6$ of a defect-mode cavity in a 2-D photonic crystal. Here, the cavity was formed by removing one rod in a square lattice (period a) of dielectric rods ($\varepsilon = 12$, radius $= 0.2a$) in air [22]. This figure compares the relative error in the calculated value of Q versus simulation time for filter diagonalization and the least-squares energy-fit method. From Fig. 20.9, we can see that filter diagonalization was able to identify Q using almost an order-of-magnitude fewer time-steps than the curve-fitting method. (Another possible technique to identify resonant modes uses Padé approximants, which can also achieve high accuracy from a short simulation [58, 62].)

20.5.3 Frequency-Domain Solver

A common computational electrodynamics problem involves determining the magnitude and phase of the sinusoidal steady-state fields that are developed within, or scattered by, a structure in response to a continuous sinusoidal source at a single frequency ω. In principle, the solution of such problems need not involve time at all, but instead require formulating and solving a system of linear algebraic equations for complex-valued vector-phasor currents and/or fields [22, Appendix D]. Such *frequency-domain* techniques include finite-element methods [7, 8], boundary-element methods [9–12], and finite-difference frequency-domain methods [63]. However, if one already has a full-featured parallel FDTD solver, it is attractive to exploit that solver for frequency-domain problems when they arise.

The most straightforward approach is to simply run the FDTD solver for a continuous sinusoidal source. When all transient effects from the source turn-on disappear, the result is the desired frequency-domain response. The difficulty with this approach is that a very long simulation may be required, especially if long-lived resonant modes are present at nearby frequencies (in which case a time much greater than Q/ω is required to reach steady state). Instead, Meep employs a means to use the FDTD time-step to directly plug a frequency-domain problem into an iterative linear solver, finding the frequency-domain response in the equivalent of many fewer time-steps while exploiting the FDTD code almost without modification.

The concepts underlying this approach are now summarized. Note that the central component of any FDTD algorithm is the time-step: an operation that advances the field by Δt in time. To extract a frequency-domain problem from this operation, the time-step can be expressed as an abstract linear operation. Namely, if $\mathbf{f}^n$ represents *all* of the fields (electric and magnetic) at time-step n, then (in a linear time-invariant structure) the time-step operation can be expressed in the form:

$$\mathbf{f}^{n+1} = \hat{T}_0 \mathbf{f}^n + \mathbf{s}^n \tag{20.7}$$

where $\hat{T}_0$ is the time-step operator with no sources and $\mathbf{s}^n$ are the source terms (currents) from that time-step. Now, suppose that one has a time-harmonic source $\mathbf{s}^n = \mathbf{s}e^{-i\omega n\Delta t}$ and wish to solve for the resulting time-harmonic (sinusoidal steady-state) fields $\mathbf{f}^n = \mathbf{f}e^{-i\omega n\Delta t}$. Substituting these into (20.7) yields the following linear equation for the field amplitudes $\mathbf{f}$:

$$\left(\hat{T}_0 - e^{-i\omega\Delta t}\right)\mathbf{f} = -\mathbf{s} \tag{20.8}$$

Equation (20.8) can be solved by an iterative method for expressions of the form $Ax = b$, where A is represented by $\hat{T}_0 - e^{-i\omega\Delta t}$, and $\mathbf{s}$ is obtained by executing a single time-step (20.7), with sources, starting from zero field, $\mathbf{f} = 0$. A key fact is that iterative techniques for $Ax = b$ only require one to supply a function that multiplies the linear operator A by a vector [64]. Hence, one can simply use a standard iterative method by calling the unmodified time-step function from FDTD to provide the linear operator. Since, in general, this linear operator is not Hermitian, especially in the presence of PML absorbing regions, Meep employs the BiCGSTAB-L algorithm. This is a generalization of the stabilized biconjugate gradient algorithm, where increasing the integer parameter L trades off increased storage for faster convergence [65, 66].

This technique allows all of the features implemented in Meep's FDTD solver (arbitrary materials, subpixel averaging, and other physical aspects; also parallelization, visualization, and the user interface) to be immediately available as a frequency-domain solver. Reference [5] reported two demonstrations of the advantages of applying this technique to obtain frequency-domain data relative to simply running FDTD simulations to the sinusoidal steady state. Here, the root-mean-square (rms) error in the field was computed as a function of the number of time-steps (or evaluations of $\hat{T}_0$ by BiCGSTAB-L). The first simulation, shown in Fig. 20.10, consisted of a point source in vacuum surrounded by PML (inset). The error in the frequency-domain results (squares) exhibited a rapid, near-exponential convergence, whereas the error in the time-domain method (circles) decreased far more gradually, only polynomially.

Figure 20.11 shows results for a much more challenging problem: obtaining the frequency-domain response of a ring cavity with multiple closely spaced, high-Q resonant modes, excited at one of the resonances [5]. In the time domain, these modes require a long simulation to reach the steady-state, whereas in the frequency domain the resonances correspond to poles (near-zero eigenvalues of A) that increase the condition number and hence slow convergence [64].

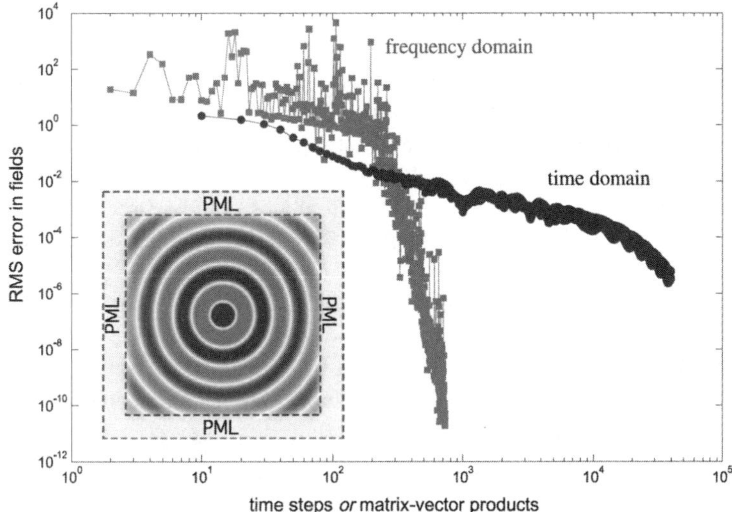

Fig. 20.10 Evolution of the rms error in the computed fields for a sinusoidal point source in vacuum (inset). Squares — frequency-domain solver adapted from the Meep FDTD code; circles — FDTD running as the transients decay. *Source:* A. Oskooi, D. Roundy, M. Ibanescu, P. Bermel, J. D. Joannopoulos, and S. G. Johnson, *Computer Physics Communications*, Vol. 181, 2010, pp. 687–702, ©2010 Elsevier.

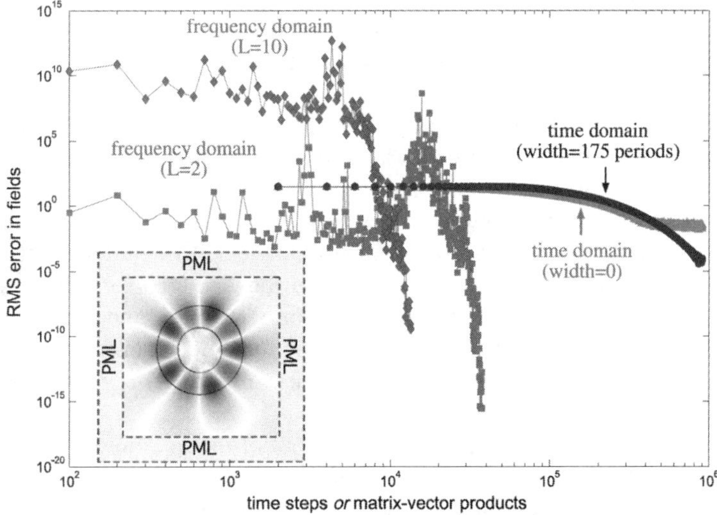

Fig. 20.11 Evolution of the rms error in the computed fields for a sinusoidal point source exciting one of several resonant modes of a dielectric ring resonator (inset, $\varepsilon = 11.56$). Squares — frequency-domain solver adapted from the Meep FDTD code; diamonds — frequency-domain solver for 5-times more storage, accelerating convergence; triangles — FDTD running as the transients decay; circles — FDTD running for a more gradual turn-on of the source, which avoids exciting long-lived resonances at other frequencies. *Source:* A. Oskooi, D. Roundy, M. Ibanescu, P. Bermel, J. D. Joannopoulos, and S. G. Johnson, *Computer Physics Communications*, Vol. 181, 2010, pp. 687–702, ©2010 Elsevier.

As expected, Fig. 20.11 shows that both the frequency-domain method derived from FDTD and the continuously running FDTD simulation took longer to converge to the sinusoidal steady state than for the nonresonant case of Fig. 20.10. However, the advantage of the exponential convergence of the former method is even more clear [6]. Furthermore, its convergence could be accelerated at the expense of using 5 times more storage ($L = 10$ vs. $L = 2$). When continuously running FDTD, convergence was limited by the decay of high-Q cavity modes at other frequencies. It was found that the impact of these modes could be reduced by turning on the sinusoidal source gradually, in this example over 175 optical periods using a hyperbolic tangent profile [6].

This is by no means the most sophisticated possible frequency-domain solver. For example, Meep currently does not use any preconditioner for the iterative scheme [64]. In 2-D, a sparse-direct solver may be far more efficient than an iterative scheme [64]. Nevertheless, the key point is that programmer time is much more expensive than computer time, and this technique allows Meep to obtain substantial improvements in solving frequency-domain problems with only minimal changes to the existing FDTD software.

20.6 USER INTERFACE AND SCRIPTING

The design of the style of user interaction in Meep was guided by two principles. First, in research or design, one rarely needs just one simulation. In fact, one usually performs a series of simulations for a class of related problems, exploring the parameter dependencies of the results, optimizing some output as a function of the input parameters, or looking at the same geometry under a sequence of different stimuli. Second, there is the Unix philosophy: "Write programs that do one thing and do it well" [67]. Meep should perform electromagnetic simulations, while for additional functionality, it should be combined with other programs and libraries via standard interfaces like files and scripts.

Both of these principles argue against incorporating in Meep a graphical CAD-style interface that is common in commercial FDTD software. First, while graphical interfaces provide a quick and attractive route to setting up a single simulation, they are not so convenient for a series of related simulations. One commonly encounters problems where (1) the size/position of certain objects is determined by the size/position of other objects; (2) where the number of objects is itself a parameter (such as a photonic-crystal cavity surrounded by a variable number of periods [22]); (3) where the length of the simulation is controlled by a complicated function of the fields; (4) where one output is optimized as a function of some parameter; and (5) many other situations that become increasingly cumbersome to express via a set of graphical tools and dialog boxes.

Second, writing a mediocre CAD program was out of the question. Had a CAD program been incorporated in Meep, it would have been of professional quality; the design would have been exported to a standard interchange format; and a conversion program would have been written to turn this format into what Meep expects.

Instead, it was decided that the most flexible and self-contained interface is to allow the user to control the simulation via an arbitrary program. Meep allows this style of interaction at two levels: via a low-level C++ interface, and via a standard high-level scripting language (Scheme) implemented by an external library (GNU Guile). The potential slowness of the scripting language is irrelevant because all of the expensive parts of the FDTD calculation are implemented in C/C++.

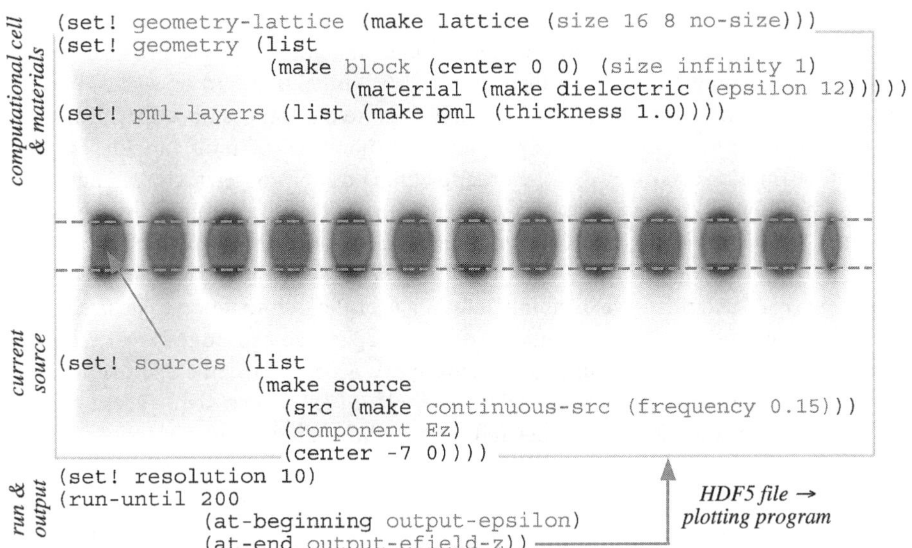

computational cell & materials
(set! geometry-lattice (make lattice (size 16 8 no-size)))
(set! geometry (list
(make block (center 0 0) (size infinity 1)
(material (make dielectric (epsilon 12))))))
(set! pml-layers (list (make pml (thickness 1.0))))
current source
(set! sources (list
(make source
(src (make continuous-src (frequency 0.15)))
(component Ez)
(center -7 0))))
run & output
(set! resolution 10)
(run-until 200
(at-beginning output-epsilon)
(at-end output-efield-z))
HDF5 file →
plotting program

Fig. 20.12 A simple Meep example showing the E_z field in a dielectric waveguide ($\varepsilon = 12$) from a point source at a given frequency. A visualization of the resulting field is in the background, and in the foreground is the input file in the high-level scripting interface (the Scheme language). *Source:* A. Oskooi, D. Roundy, M. Ibanescu, P. Bermel, J. D. Joannopoulos, and S. G. Johnson, *Computer Physics Communications*, Vol. 181, 2010, pp. 687–702, ©2010 Elsevier.

The high-level scripting interface to Meep is documented in detail, with several tutorials, on the Meep web page (http://ab-initio.mit.edu/meep), so this review is restricted to a single short example in order to convey the basic flavor. This example, shown in Fig. 20.12, computes the 2-D fields in response to a point source located within a dielectric waveguide [5]. The size of the computational space is first set to 16×8 (via `geometry-lattice`, so-called because it determines the lattice vectors in the periodic case). Here, the interpretation of the unit of distance is arbitrary and up to the user. For example, it could be 16×8 μm, in which case the frequency units are c/μm, or 16×8 mm with frequency units of c/mm, or any other convenient distance unit. Call this arbitrary unit of distance a.

Then, the geometry being modeled is specified as a list of objects like blocks, cylinders, etc.; in this case, by a single block defining the waveguide with $\varepsilon = 12$. Optionally, the geometry could be defined by arbitrary user-defined functions $\varepsilon(x, y)$ and $\mu(x, y)$. A PML is then specified at the outer boundary of the space with thickness 1. This lies *inside* the computational space, and overlaps the waveguide to absorb waveguide modes when they reach the edge of the space. A z-directed point electric-current source, $\boldsymbol{J}$, is added (sources of arbitrary spatial profile can also be specified). The time dependence of the source is specified as a sharp turn-on to a continuous sinusoid, $\cos(\omega t)$, at the beginning of the simulation; gradual turn-on profiles, Gaussian pulses, or arbitrary user-specified functions of time can also be specified. The frequency is 0.15 in units of c/a, corresponding to a vacuum wavelength $\lambda = a/0.15$ (e.g., $\lambda \approx 6.67$ μm if $a = 1$ μm).

Finally, the grid resolution is set to 10 pixels per unit distance (10 pixels/a), so that the entire computational space is 160×80 pixels. The simulation is set to run for 200 time units (units of a/c), corresponding to $200 \times 0.15 = 30$ optical periods, with the dielectric function output at the beginning, and the E_z field output at the end.

Although at a basic level, the input format can be considered to be just a file format with many parentheses, in fact one can use it to control the simulation. This is because Scheme is a full-fledged programming language. For example, it is straightforward to write loops and use arithmetic to define the geometry of objects and their relationships, or to perform parameter sweeps. It is also straightforward to expose external libraries for multivariable optimization, integration, root-finding, and other tasks in order that they can be coupled with simulations.

Another technique to achieve flexibility is the use of higher-order functions [68]. Namely, wherever it is practical, Meep's functions take functions as arguments instead of (or in addition to) numbers. Thus, for example, instead of specifying special input codes for all possible source distributions in space and time, Meep simply allows a user-defined function to be used.

More subtly, in Fig. 20.12, the arguments `output-epsilon` and `output-efield-z` of the `run-until` function are actually functions themselves. Meep allows the user to pass arbitrary "step functions" to `run-until` that are called after every FDTD time-step. These can perform arbitrary computations on the fields as desired, or halt the computation if a desired condition is reached. Such step-functions can be modified by transformation functions like `at-end`, which take step functions as arguments and return a new step function that only calls the original step functions at specified times (for example, at the beginning or end of the simulation, or at certain intervals). In this way, great flexibility in the output and computations is achieved. For example, by composing several of these transformations, one can output a particular field component at specific time intervals after a given time, and only within a certain sub-volume or slice of the computational space. One can even output an arbitrary user-defined function of the fields instead of predetermined components.

There is an additional subtlety when it comes to field output because the Yee lattice stores different field components at different points. Presented in this way to the user, it would be difficult to perform post-processing involving multiple field components, or even to compare plots of different field components. Therefore, as mentioned in Section 20.3 and again in Section 20.7.2, the field components are automatically interpolated from the Yee grid onto a fixed "centered" grid in each pixel when exported to a file.

Note that, in keeping with the Unix philosophy, Meep is not a plotting program. Instead, it outputs fields and related data to the standard HDF5 format [69], which can be read by many other programs and visualized in various ways. Meep also provides a way to effectively "pipe" the HDF5 output to an external program within Meep; for example, to output the HDF5 file, convert it immediately to an image with a plotting program, and then delete the HDF5 file. This is especially useful for producing animations consisting of hundreds of frames.

Finally, parallelism in Meep is completely transparent to the user. Exactly the same input script is fed to the serial version of Meep as to the parallel version of Meep, which is written with the MPI message-passing standard for distributed-memory parallelism [70]. Furthermore, the distribution of the data across processors and the collection of results are handled automatically.

20.7 ABSTRACTION VERSUS PERFORMANCE

In an FDTD simulation, essentially just one thing has to be fast: the inner loops over all the grid points, or some large fraction thereof. Everything else is negligible in terms of computation time. (For example, the use of a Scheme interpreter as the user interface has no performance consequences because the inner loops are not written in Scheme.) Importantly, for the inner loops, there is a distinct tension between abstraction and performance. This section reviews some of the trade-offs that result from this tension and the choices that have been made in Meep.

20.7.1 Primacy of the Inner Loops

The primacy of inner loops means that some popular principles of abstraction must be discarded. Consider a hypothetical, idealistic attempt to write an FDTD code in textbook object-oriented C++ style. Every pixel in the grid would be an object, and every type of material would be a subclass overriding the necessary time-stepping and field-access operations. Time-stepping would consist of looping over the grid, calling some "step" method of each object, so that objects of different materials (magnetic, dielectric, nonlinear, etc.) would dynamically apply the corresponding field-update procedures. In fact, the result of such a noble experiment would be a performance *failure*. The performance overhead of object de-referencing, virtual method dispatch, and function calls in the inner loop would overwhelm all other considerations.

In Meep, each field's components are stored as simple linear arrays of floating-point numbers in row-major (C) order (parallel-array data structures worthy of Fortran 66), and there are separate inner loops for each type of material. In a simple experiment on a 2.8-GHz Intel Core 2 CPU, merely moving the `if` statements for the different material types into these inner loops decreased Meep's performance by a factor of 2 in a typical 3-D calculation, and by a factor of 6 in 2-D (where the calculations are simpler and hence the overhead of the conditionals is more significant). The cost of the conditionals, including the cost of mispredicted branches and subsequent pipeline stalls [71] along with the frustration of compiler unrolling and vectorization, easily overwhelmed the small cost of computing, e.g., $\nabla \times \boldsymbol{H}$ at a single point. The bottom line: *avoid conditionals in any code inserted into Meep that can be considered to be of the inner-loop type*.

Meep provides intrinsic standard functions such as energy or flux that operate on the field-component arrays. Because these functions are routinely evaluated at many grid points over many time-steps during a simulation, they are tantamount to inner-loop operations, and hence are written in C/C++ to preserve performance. Users who insert their own functions of this type are strongly advised to write these functions in C/C++ to avoid a significant slowdown. The bottom line: *any code inserted into Meep that can be considered to be of the inner-loop type should be written in C/C++.*

20.7.2 Time-Stepping and Cache Trade-Offs

One of the dominant factors in performance on modern computer systems is not arithmetic, but memory: random memory access is far slower than arithmetic, and the organization of memory into a hierarchy of caches is designed to favor locality of access [71]. That is, one should organize the computation so that (1) as much work as possible is done with a given datum once it is fetched (temporal locality); and (2) subsequent data that are read or written are located nearby in memory (spatial locality). However, the optimal strategies to exploit both kinds of locality lead to sacrifices of abstraction and code simplicity so severe that Meep has been designed to sacrifice some potential performance in the name of simplicity.

As it is typically described, the FDTD algorithm has very little temporal locality: the field at each point is advanced in time by Δt, and then is not modified again until *all* the fields at every other point in the computational space have been advanced. To gain temporal locality, one must employ *asynchronous time-stepping*. Here, fields in compact regions of space are advanced several steps in time before advancing fields located outside of these regions, since over this time interval the effects of the remote fields cannot be felt. A detailed analysis of the characteristics of this problem, as well as a beautiful "cache-oblivious" algorithm that automatically exploits a cache of any size for grids of any dimensionality, is described in [72]. On the other hand, an important part of Meep's usability is the abstraction that the user can perform arbitrary computations or outputs using the fields in any spatial region at any time. This seems incompatible with the fields at different points in space being out-of-sync until a predetermined end of the computation. The bookkeeping difficulty of reconciling these two viewpoints is why Meep does not employ the asynchronous approach, despite its potential benefits.

However, there may appear to be at least a small amount of temporal locality in the synchronous FDTD algorithm: first $\boldsymbol{B}$ is advanced from $\nabla \times \boldsymbol{E}$, then $\boldsymbol{H}$ is computed from $\boldsymbol{B}$ and μ, then $\boldsymbol{D}$ is advanced from $\nabla \times \boldsymbol{H}$, then $\boldsymbol{E}$ is computed from $\boldsymbol{D}$ and ε. Since most fields are used at least once after they have been advanced, surely the updates of the different fields can be merged into a single loop, for example, advancing $\boldsymbol{D}$ at a point and then immediately computing $\boldsymbol{E}$ at the same point — the $\boldsymbol{D}$ field need not even be stored. Furthermore, since by merging the updates one is accessing several fields around the same point at the same time, perhaps one can gain *spatial* locality by interleaving the data, say, by storing an array of $(\boldsymbol{E}, \boldsymbol{H}, \varepsilon, \mu)$ tuples instead of separate arrays.

Meep does not do either of these things, however, for two reasons, the first of which is more fundamental. As is well known, one cannot easily merge the $\boldsymbol{B}$ and $\boldsymbol{H}$ updates with the $\boldsymbol{D}$ and $\boldsymbol{E}$ updates at the same point, because the discretized $\nabla\times$ operation is nonlocal (involves multiple grid points). This is why one normally updates $\boldsymbol{H}$ everywhere in space before updating $\boldsymbol{D}$ from $\nabla \times \boldsymbol{H}$, because in computing $\nabla \times \boldsymbol{H}$, one uses the values of $\boldsymbol{H}$ at different grid points, and all of them must be in sync. A similar reasoning applies to updating $\boldsymbol{E}$ from $\boldsymbol{D}$ and $\boldsymbol{H}$ from $\boldsymbol{B}$ once the possibility of anisotropic materials is included. Because the Yee grid stores different field components at different locations, any accurate modeling of off-diagonal susceptibilities must also inevitably involve fields at multiple points, as in [36]. To treat this, $\boldsymbol{D}$ must be stored explicitly, and the update of $\boldsymbol{E}$ from $\boldsymbol{D}$ must take place after $\boldsymbol{D}$ has been updated everywhere, in a separate loop. Hence, since each field is updated in a separate loop, the spatial-locality motivation to merge the field data structures rather than using parallel arrays is largely removed.

Of course, not all simulations involve anisotropic materials, although they do appear in many Meep models of nominally isotropic materials thanks to the subpixel smoothing discussed in Section 20.3.1. However, this leads to the second practical problem with merging the $\boldsymbol{E}$ and $\boldsymbol{D}$ (or $\boldsymbol{H}$ and $\boldsymbol{B}$) update loops: the combinatorial explosion of the possible material cases.

Specifically, with such merging, the update of $\boldsymbol{D}$ from $\nabla \times \boldsymbol{H}$ would require accommodating 16 possible cases, each of which would be a separate loop, as per the above discussion of the cost of putting conditionals inside the main loops. These would be as follows: (a) 4 cases — with or without PML, depending on the number (0, 1, 2, or 3) of PML conductivities and their orientations relative to the field; (b) 2 cases — with or without conductivity; and (c) 2 cases — with the derivative of either two $\boldsymbol{H}$ components (3-D or 2-D TM polarization), or only one $\boldsymbol{H}$ component (2-D TE polarization). Similarly, the update of $\boldsymbol{E}$ from $\boldsymbol{D}$ would require accommodating 12 possible cases: (a) 2 cases — with or without PML, distinct from those in the $\boldsymbol{D}$ update; (b) 3 cases — variable number (0, 1, or 2) of off-diagonal ε^{-1} components; and (c) 2 cases — with or without nonlinearity. A total of $16 \times 12 = 192$ cases would result if all of the above options were joined into a single loop, yielding a code-maintenance headache.

The performance penalty of separate $\boldsymbol{E}$ and $\boldsymbol{D}$ (or $\boldsymbol{H}$ and $\boldsymbol{B}$) updates appears to be modest. Even if, by somehow merging the loops, one assumes that the time to compute $\boldsymbol{E} = \varepsilon^{-1}\boldsymbol{D}$ could become *zero*, benchmarking the relative time spent in this operation indicates that a typical 3-D transmission calculation would be accelerated by only around 30%, and less in 2-D.

20.7.3 The Loop-in-Chunks Abstraction

This section reviews an abstraction that, while not directly visible to end-users of Meep, is key to the efficiency and maintainability of large portions of the software (field output, current sources, flux/energy computations, other field integrals, etc.). The purpose of this abstraction is to mask the complexity of the partitioning of the computational space into overlapping chunks connected by symmetries, communications, and other boundary conditions as described in Section 20.2.

Consider the output of the fields at a given time-step to an HDF5 data file. Meep provides a routine `get-field-pt` that, given a point in space, interpolates it onto the Yee grid and returns a desired field component at that point. In addition to interpolation, this routine must also transform the point onto a chunk that is actually stored (using rotations, periodicity, etc.) and communicate the data from another processor if necessary. If the point is on a boundary between two chunks, the interpolation process may involve multiple chunks, multiple rotations, etc., and communications from multiple processors. Because this process involves only a single point, it is not easily parallelizable.

Now, to output the fields everywhere in some region to a file, one approach is to simply call `get-field-pt` for every point in a grid for that region, and output the results. However, this is very slow because of the repeated transformations and communications for every single point. Nevertheless, one wants to interpolate fields for output rather than dumping the raw Yee grid, because it is much easier for post-processing if the different field components are interpolated onto the same grid. Also, to maintain transparency of features such as symmetry, one would like to be able to output the whole computational space (or an arbitrary subset), even if only a part of it is stored. Almost exactly the same problems arise for integrating quantities like flux ($\boldsymbol{E} \times \boldsymbol{H}$), energy, or user-defined functions of the fields (noting that functions combining multiple field components require interpolation); and also for implementing line, surface, or volume sources that must be projected onto the grid in some arbitrary volume.

One key to solving this difficulty is to realize that, when the field in some volume V is needed (for output, integration, and so on), the rotations, communications, etc. for points in V are identical for all the points in the intersection of V with some chunk (or one of its rotations/translations). The second key is to realize that, when interpolation is needed, there is a particular grid for which interpolation is easy. Namely, for *owned* points of the *centered* grid (Section 20.2) lying at the center of each pixel, it is always possible to interpolate from fields on any Yee grid without any inter-chunk communication, and by a simple equal-weight averaging of at most 2^d points in d dimensions.

The combination of these two observations leads to the *loop-in-chunks* abstraction. Given a convex, rectilinear volume V and a given grid (either centered or one of the Yee-field grids), it computes the intersection of all the chunks and their rotations/translations with V. For each intersection it invokes a caller-specified function, passing the portion of the chunk, the necessary rotations (etc.) of the fields, and interpolation weights (if needed, for the boundary of V). That function then processes the specified portion of the chunk, for example, outputting it to the corresponding portion of a file, or integrating the desired fields. All of this can proceed in parallel, with each processor considering only those chunks stored locally. This is relatively fast because the rotations, interpolations, and so on are computed only once per chunk intersection, while the inner loop over all grid points in each chunk can be as tight as necessary. Moreover, all of the complicated and error-prone logic involved in computing V's intersection with the chunks is localized to one place in the source code. (Special care is required to ensure that each conceptual grid point is processed exactly once despite chunk overlaps and symmetries.) Field output, integration, sources, and other functions of the fields are isolated from this complexity.

20.8 SUMMARY AND CONCLUSIONS

This chapter reviewed a number of the unusual implementation details of Meep that distinguish this software package from standard textbook FDTD methods, as first reported in [5]. The review began with a discussion of the fundamental structural unit of "chunks" that constitute the Yee grid and enable parallelization. Next, an overview was provided of Meep's core design philosophy of approaching the goal of continuous space-time modeling for inputs and outputs. The review continued with an explanation and motivation of the somewhat unusual design intricacies of nonlinear materials and PMLs, the important aspects of Meep's computational methods for flux spectra and resonant modes, and a demonstration of the formulation of a frequency-domain solver requiring only minimal modifications to the underlying FDTD algorithm. In addition to the inner workings of Meep's internal structure, the review included how such features are accessible to users via an external scripting interface.

A free/open-source, full-featured FDTD package like Meep could play a vital role in enabling new research in electromagnetic phenomena. Not only does it provide a low barrier to entry for standard FDTD simulations, but the simplicity of the FDTD algorithm combined with access to the source code offers an easy route to investigate new physical phenomena coupled with classical electrodynamics. For example, colleagues at MIT are working on coupling multi-level atoms to electromagnetism within Meep for modeling lasing and saturable absorption, adapting published techniques [16–20], but also including new physics such as the diffusion of excited gases. Other MIT colleagues have modified Meep for modeling gyromagnetic media in order to design new classes of "one-way" waveguides [73]. Meep has even been used to simulate the phenomena of Casimir forces arising from quantum vacuum fluctuations, which can be computed from classical Green's functions [74, 75] (see also Chapter 19). In fact, this was possible without any modifications of the Meep code due to the flexibility of Meep's scripting interface. It is hoped that other researchers, with the help of the understanding of Meep's architecture that Ref. [5] and this chapter provide, will be able to adapt Meep to future phenomena that have not yet been envisioned.

ACKNOWLEDGMENTS

The work reviewed in this chapter was supported in part by the Materials Research Science and Engineering Center program of the National Science Foundation under Grant Nos. DMR-9400334 and DMR-0819762; by the Army Research Office through the Institute for Soldier Nanotechnologies under contract DAAD-19-02-D0002; and by Dr. Dennis Healy of DARPA MTO under award N00014-05-1-0700 administered by the Office of Naval Research. The authors are also grateful to A. W. Rodriguez and A. P. McCauley for their efforts to generalize Meep for quantum-Casimir problems; to S. L. Chua for his work on dispersive and multi-level materials; and to Y. Chong for early support in Meep for gyrotropic media.

REFERENCES

[1] Taflove, A., and S. C. Hagness, *Computational Electrodynamics: The Finite-Difference Time-Domain Method*, 3rd ed., Norwood, MA: Artech, 2005.

[2] Kunz, K. S., and R. J. Luebbers, *The Finite-Difference Time-Domain Method for Electromagnetics*, Boca Raton, FL: CRC Press, 1993.

[3] Sullivan, D. M., *Electromagnetic Simulation Using the FDTD Method*, New York, NY: Wiley–IEEE Press, 2000.

[4] Yu, W., R. Mittra, T. Su, Y. Liu, and X. Yang, *Parallel Finite-Difference Time-Domain Method*, Norwood, MA: Artech, 2006.

[5] Oskooi, A., D. Roundy, M. Ibanescu, P. Bermel, J. D. Joannopoulos, and S. G. Johnson, "Meep: A flexible free-software package for electromagnetic simulations by the FDTD method," *Computer Physics Communications*, Vol. 181, 2010, pp. 687–702.

[6] Online: http://en.wikipedia.org/wiki/Finite-difference_time-domain_method

[7] Jin, J., *The Finite Element Method in Electromagnetics*, 2nd ed., New York: Wiley–IEEE Press, 2002.

[8] Yasumoto, K., ed., *Electromagnetic Theory and Applications for Photonic Crystals*, Boca Raton, FL: CRC Press, 2005.

[9] Rao, S., D. Wilton, and A. Glisson, "Electromagnetic scattering by surfaces of arbitrary shape," *IEEE Trans. Antennas and Propagation*, Vol. 30, 1982, pp. 409–418.

[10] Umashankar, K., A. Taflove, and S. Rao, "Electromagnetic scattering by arbitrary shaped three-dimensional homogeneous lossy dielectric objects," *IEEE Trans. Antennas and Propagation*, Vol. 34, 1986, pp. 758–766.

[11] Bonnet, M., *Boundary Integral Equation Methods for Solids and Fluids*, New York: Wiley, 1999.

[12] Chew, W. C., J.-M. Jin, E. Michielssen, and J. Song, eds., *Fast and Efficient Algorithms in Computational Electromagnetics*, Norwood, MA: Artech, 2000.

[13] Johnson, S. G., M. Ibanescu, M. A. Skorobogatiy, O. Weisberg, J. D. Joannopoulos, and Y. Fink, "Perturbation theory for Maxwell's equations with shifting material boundaries," *Physical Review E*, Vol. 65, 2002, 066611.

[14] Zhang, L., J. Lee, A. Farjadpour, J. White, and S. Johnson, "A novel boundary element method with surface conductive absorbers for 3-D analysis of nanophotonics," *Microwave Symposium Digest*, 2008 IEEE MTT-S International Symposium, 2008, pp. 523–526.

[15] Johnson, S. G., and J. D. Joannopoulos, "Block-iterative frequency-domain methods for Maxwell's equations in a plane wave basis," *Optics Express*, Vol. 8, 2001, pp. 173–190.

[16] Ziolkowski, R. W., J. M. Arnold, and D. M. Gogny, "Ultrafast pulse interactions with two-level atoms," *Physical Review A*, Vol. 52, 1995, pp. 3082–3094.

[17] Nagra, A. S., and R. A. York, "FDTD analysis of wave propagation in nonlinear absorbing and gain media," *IEEE Trans. Antennas and Propagation*, Vol. 46, 1998, pp. 334–340.

[18] Chang, S.-H., and A. Taflove, "Finite-difference time-domain model of lasing action in a four-level two-electron atomic system," *Optics Express*, Vol. 12, 2004, pp. 3827–3833.

[19] Huang, Y., and S.-T. Ho, "Computational model of solid-state, molecular, or atomic media for FDTD simulation based on a multi-level multi-electron system governed by Pauli exclusion and Fermi–Dirac thermalization with application to semiconductor photonics," *Optics Express*, Vol. 14, 2006, pp. 3569–3587.

[20] Bermel, P., E. Lidorikis, Y. Fink, and J. D. Joannopoulos, "Active materials embedded in photonic crystals and coupled to electromagnetic radiation," *Physical Review B*, Vol. 73, 2006, 165125.

[21] Jackson, J. D., *Classical Electrodynamics*, 3rd ed., New York: Wiley, 1998.

[22] Joannopoulos, J. D., S. G. Johnson, R. D. Meade, and J. N. Winn, *Photonic Crystals: Molding the Flow of Light*, 2nd ed., Princeton, NJ: Princeton Univ. Press, 2008.

[23] Yee, K. S., "Numerical solution of initial boundary value problems involving Maxwell's equations in isotropic media," *IEEE Trans. Antennas and Propagation*, Vol. 14, 1966, pp. 302–307.

[24] Berger, M. J., and J. Oliger, "Adaptive mesh refinement for hyperbolic partial differential equations," *J. Computational Physics*, Vol. 53, 1984, pp. 484–512.

[25] Kim, I. S., and W. J. R. Hoefer, "A local mesh refinement algorithm for the time domain finite difference method using Maxwell's curl equations," *IEEE Trans. Microwave Theory and Techniques*, Vol. 38, 1990, pp. 812–815.

[26] Zivanovic, S. S., K. S. Yee, and K. K. Mei, "A subgridding method for the time-domain finite-difference method to solve Maxwell's equations," *IEEE Trans. Microwave Theory and Techniques*, Vol. 39, 1991, pp. 471–479.

[27] Okoniewski, M., E. Okoniewska, and M. A. Stuchly, "Three-dimensional subgridding algorithm for FDTD," *IEEE Trans. Antennas and Propagation*, Vol. 45, 1997, pp. 422–429.

[28] Lin, C., and L. Snyder, *Principles of Parallel Programming*, New York: Addison-Wesley, 2008.

[29] Inui, T., Y. Tanabe, and Y. Onodera, *Group Theory and Its Applications in Physics*, Berlin: Springer-Verlag, 1996.

[30] Ditkowski, A., K. Dridi, and J. S. Hesthaven, "Convergent Cartesian grid methods for Maxwell's equations in complex geometries," *J. Computational Physics*, Vol. 170, 2001, pp. 39–80.

[31] Farjadpour, A., D. Roundy, A. Rodriguez, M. Ibanescu, P. Bermel, J. Joannopoulos, S. Johnson, and G. Burr, "Improving accuracy by sub-pixel smoothing in the finite difference time domain," *Optics Lett.*, Vol. 31, 2006, pp. 2972–2974.

[32] Kottke, C., A. Farjadpour, and S. G. Johnson, "Perturbation theory for anisotropic dielectric interfaces, and application to subpixel smoothing of discretized numerical methods," *Physical Review E*, Vol. 77, 2008, 036611.

[33] Oskooi, A., C. Kottke, and S. G. Johnson, "Accurate finite-difference time-domain simulation of anisotropic media by subpixel smoothing," *Optics Lett.*, Vol. 34, 2009, pp. 2778–2780.

[34] Dey, S., and R. Mittra, "A conformal finite-difference time-domain technique for modeling cylindrical dielectric resonators," *IEEE Trans. Microwave Theory and Techniques*, Vol. 47, 1999, pp. 1737–1739.

[35] Meade, R. D., A. M. Rappe, K. D. Brommer, J. D. Joannopoulos, and O. L. Alerhand, "Accurate theoretical analysis of photonic band-gap materials," *Physical Review B*, Vol. 48, 1993, pp. 8434–8437; Johnson, S. G., "Erratum," *Physical Review B*, Vol. 55, 1997, 15942.

[36] Werner, G., and J. Cary, "A stable FDTD algorithm for non-diagonal anisotropic dielectrics," *J. Computational Physics*, Vol. 226, 2007, pp. 1085–1101.

[37] Mezzanotte, P., L. Roselli, and R. Sorrentino, "A simple way to model curved metal boundaries in FDTD algorithm avoiding staircase approximation," *IEEE Microwave and Guided Wave Lett.*, Vol. 5, 1995, pp. 267–269.

[38] Anderson, J., M. Okoniewski, and S. S. Stuchly, "Practical 3-D contour/staircase treatment of metals in FDTD," *IEEE Microwave and Guided Wave Lett.*, Vol. 6, 1996, pp. 146–148.

[39] Deinega, A., and I. Valuev, "Subpixel smoothing for conductive and dispersive media in the finite-difference time-domain method," *Optics Lett.*, Vol. 32, 2007, pp. 3429–3431.

[40] Leung, P., S. Liu, and K. Young, "Completeness and time-independent perturbation of the quasinormal modes of an absorptive and leaky cavity," *Physical Review A*, Vol. 49, 1994, pp. 3982–3989.

[41] Johnson, S. G., M. L. Povinelli, M. Soljacic, A. Karalis, S. Jacobs, and J. D. Joannopoulos, "Roughness losses and volume-current methods in photonic-crystal waveguides," *Applied Physics B*, Vol. 81, 2005, pp. 283–293.

[42] Trottenberg, U., C. W. Oosterlee, and A. Schuller, *Multigrid*, San Diego, CA: Academic Press, 2000.

[43] Landau, L., L. Pitaevskii, and E. Lifshitz, *Electrodynamics of Continuous Media*, 2nd ed., Oxford: Butterworth-Heinemann, 1984.

[44] Luo, C., M. Ibanescu, S. G. Johnson, and J. D. Joannopoulos, "Cerenkov radiation in photonic crystals," *Science*, Vol. 299, 2003, pp. 368–371.

[45] Liu, Q. H., and J. Q. He, "Quasi-PML for waves in cylindrical coordinates," *Microwave and Optical Technology Lett.*, Vol. 19, 1998, pp. 107–111.

[46] Bloembergen, N., *Nonlinear Optics*, New York: W. A. Benjamin, 1965.

[47] Agrawal, G. P., *Nonlinear Fiber Optics*, 3rd ed., San Diego, CA: Academic Press, 2001.

[48] Rodriguez, A., M. Soljacic, J. D. Joannopoulos, and S. G. Johnson, "$\chi^{(2)}$ and $\chi^{(3)}$ harmonic generation at a critical power in inhomogeneous doubly resonant cavities," *Optics Express*, Vol. 15, 2007, pp. 7303–7318.

[49] Boyd, R. W., *Nonlinear Optics*, London: Academic Press, 1992.

[50] Baker, J., A. George, and P. Graves-Morris, *Padé Approximants*, 2nd ed., Cambridge, UK: Cambridge University Press, 1996.

[51] Oskooi, A., L. Zhang, Y. Avniel, and S. G. Johnson, "The failure of perfectly matched layers, and towards their redemption by adiabatic absorbers," *Optics Express*, Vol. 16, 2008, pp. 11376–11392.

[52] Oskooi, A., and S. G. Johnson, "Distinguishing correct from incorrect PML proposals and a corrected unsplit PML for anisotropic, dispersive media," *J. Computational Physics*, Vol. 230, 2011, pp. 2369–2377.

[53] Teixeira, F. L., and W. C. Chew, "Systematic derivation of anisotropic PML absorbing media in cylindrical and spherical coordinates," *IEEE Microwave and Guided Wave Lett.*, Vol. 7, 1997, pp. 371–373.

[54] He, J.-Q., and Q.-H. Liu, "A nonuniform cylindrical FDTD algorithm with improved PML and quasi-PML absorbing boundary conditions," *IEEE Trans. Geoscience and Remote Sensing*, Vol. 37, 1999, pp. 1066–1072.

[55] Bailey, D. H., and P. N. Swartztrauber, "A fast method for the numerical evaluation of continuous Fourier and Laplace transforms," *SIAM J. Scientific Computing*, Vol. 15, 1994, pp. 1105–1110.

[56] Cormen, T. H., C. E. Leiserson, R. L. Rivest, and C. Stein, *Introduction to Algorithms*, 3rd ed., Cambridge, MA: MIT Press, 2009.

[57] Oppenheim, A. V., and R.W. Schafer, *Discrete-Time Signal Processing*, 3rd ed., New York: Prentice-Hall, 2009.

[58] Guo, W.-H., W.-J. Li, and Y.-Z. Huang, "Computation of resonant frequencies and quality factors of cavities by FDTD technique and Padé approximation," *IEEE Microwave and Wireless Communications Lett.*, Vol. 11, 2001, pp. 223–225.

[59] Cabay, S., and D.-K. Choi, "Algebraic computations of scaled Padé fractions," *SIAM J. Computing*, Vol. 15, 1986, pp. 243–270.

[60] Rodriguez, A., M. Ibanescu, J. D. Joannopoulos, and S. G. Johnson, "Disorder-immune confinement of light in photonic-crystal cavities," *Optics Lett.*, Vol. 30, 2005, pp. 3192–3194.

[61] Mandelshtam, V. A., and H. S. Taylor, "Harmonic inversion of time signals and its applications," *J. Chemical Physics*, Vol. 107, 1997, pp. 6756–6769.

[62] Dey, S., and R. Mittra, "Efficient computation of resonant frequencies and quality factors of cavities via a combination of the finite-difference time-domain technique and the Padé approximation," *IEEE Microwave and Guided Wave Lett.*, Vol. 8, 1998, pp. 415–417.

[63] Christ, A., and H. L. Hartnagel, "Three-dimensional finite-difference method for the analysis of microwave-device embedding," *IEEE Trans. Microwave Theory and Techniques*, Vol. 35, 1987, pp. 688–696.

[64] Barrett, R., M. Berry, T. Chan, J. Demmel, J. Donato, J. Dongarra, V. Eijkhout, R. Pozo, C. Romine, and H. V. der Vorst, *Templates for the Solution of Linear Systems: Building Blocks for Iterative Methods*, Philadelphia, PA: SIAM, 1994.

[65] Sleijpen, G. L. G., and D. R. Fokkema, "BiCGSTAB(*L*) for linear equations involving unsymmetric matrices with complex spectrum," *Electronics Trans. Numerical Analysis*, Vol. 1, 1993, pp. 11–32.

[66] Sleijpen, G. L. G., H. A. van der Vorst, and D. R. Fokkema, "BiCGstab(*L*) and other hybrid Bi-CG methods," *Numerical Algorithms*, Vol. 7, 1994, pp. 75–109.

[67] Salus, P. H., *A Quarter Century of UNIX*, Reading, MA: Addison–Wesley, 1994.

[68] Abelson, H., and G. J. Sussman, *Structure and Interpretation of Computer Programs*, Cambridge, MA: MIT Press, 1985.

[69] Folk, M., R. E. McGrath, and N. Yeager, "HDF: An update and future directions," *Proc. 1999 Geoscience and Remote Sensing Symposium (IGARSS)*, Hamburg, Germany, Vol. 1, IEEE Press, 1999, pp. 273–275.

[70] Forum, T. M., "MPI: A message passing interface," *Proc. Supercomputing '93*, Portland, OR, 1993, pp. 878–883.

[71] Hennessy, J. L., and D. A. Patterson, *Computer Architecture: A Quantitative Approach*, 3rd ed., San Francisco: Elsevier, 2003.

[72] Frigo, M., and V. Strumpen, "The memory behavior of cache oblivious stencil computations," *J. Supercomputing*, Vol. 39, 2007, pp. 93–112.

[73] Wang, Z., Y. D. Chong, J. D. Joannopoulos, and M. Soljacic, "Reflection-free one-way edge modes in a gyromagnetic photonic crystal," *Physical Review Lett.*, Vol. 100, 2008, 013905.

[74] Rodriguez, A. W., A. P. McCauley, J. D. Joannopoulos, and S. G. Johnson, "Casimir forces in the time domain: Theory," *Physical Review A*, Vol. 80, 2009, 012115.

[75] A. P. McCauley, A. W. Rodriguez, J. D. Joannopoulos, and S. G. Johnson, "Casimir forces in the time domain: Applications," *Physical Review A*, Vol. 81, 2010, 012119.

Acronyms and Common Symbols

1-D	one-dimensional
2-D	two-dimensional
3-D	three-dimensional
A	ampere
ABC	absorbing boundary condition
ADE	auxiliary differential equation
ADI	alternating-direction implicit
AFM	atomic force microscopy
AFP	analytic field propagator
Alt.–PSM	alternating phase-shift mask
ARC	anti-reflective coating
Atten.–PSM	attenuated phase-shift mask
Au	gold
B	magnetic flux density
BEM	boundary-element method
BEOL	back end of the line
BG/L	BlueGene/L
BL	boundary layer
c	free-space (vacuum) speed of light
C	capacitance
CCD	charge-coupled device
CD	critical dimension
CFS-PML	complex frequency-shifted perfectly matched layer
CMOS	complementary metal oxide semiconductor
CPL	chromeless phase-shift lithography
CPML	convolutional perfectly matched layer
CRA	chief-ray angle
CsK	C-terminal Src kinase
D	electric flux density
dB	decibel(s)
DDM	domain-decomposition method
DE	dermis epidermis
DFT	density functional theory
DFT	discrete Fourier transform
DIC	differential interference contrast
DoF	depth of focus
DOS	density of states
DRAM	dynamic random-access memory
DSCS	differential scattering cross-section
DSSC	dye-sensitized solar cell
DTFT	discrete-time Fourier transform
DUV	deep ultraviolet
DUVL	deep-ultraviolet lithography

E	electric field intensity
EBS	enhanced backscattering
ED	electrodynamics
EGFR	epidermal growth factor receptor
EM	electromagnetic
EMC	electric-field Monte Carlo
EMF	electromagnetic field
ENZ	extended Nijboer-Zernike
EP	entrance pupil
ETRS	enforced time-reversible symmetry
EUV	extreme ultraviolet
EUVL	extreme-ultraviolet lithography
FDTD	finite-difference time-domain
FD-TDDFT	frequency-domain time-dependent density functional theory
FDTD-DPW	finite-difference time-domain discrete plane wave
FE	finite element
FEM	finite-element method
FEOL	front end of the line
FFT	fast Fourier transform
FIR	finite impulse-response
fs	femtosecond
FWHM	full-width at half-maximum
G	conductance
GaAs	gallium arsenide
GB	gigabyte
GHz	gigahertz
G-PSTD	global pseudospectral time domain
GTH	Goedecker-Teter-Hutter
GVADE	general vector auxiliary differential equation
H	magnetic field intensity
H-pol	horizontal polarization
HOMO	highest occupied molecular orbital
HPV	human papilloma virus
HVB	horizontal-vertical bias
HVPB	horizontal-vertical printed bias
I	current, amperes
IFA	incident field array
IIR	infinite impulse response
ITRS	International Technology Roadmap for Semiconductors
J	electric current density
L	inductance
L-C	inductor-capacitor
LCR	inductor-capacitor-resistor
LDOS	local density of states
LSPR	localized surface-plasmon resonance
LUMO	lowest unoccupied molecular orbital

M	magnetic current density
MB	megabyte
Meep	MIT Electromagnetic Equation Propagation
MHz	megahertz
MIT	Massachusetts Institute of Technology
MKL	Math Kernel Library
MoSi	molybdenum silicon
MPI	message-passing interface
NA	numerical aperture
NEGF	nonequilibrium Green's function
NFFT	near-to-far-field transform
NFFFT	near-field-to-far-field transform
NLSE	nonlinear Schrödinger equation
nm	nanometer
NRI	negative refractive index
NRI-TL	negative-refractive-index transmission line
ns	nanosecond
NSOM	near-field scanning optical microscope
NUIT	Northwestern University Information Technology
ODE	ordinary differential equation
OMOG	opaque molybdenum silicon on glass
OPC	optical phase conjugation
OPC	optical proximity correction
OPCM	optical phase-contrast microscopy
PBC	periodic boundary condition
PBE	Perdew-Burke-Ernzerhof
PBG	photonic bandgap
PC	photonic crystal
PDE	partial differential equation
PEC	perfect electric conductor
PFA	proximity force approximation
PMC	perfect-magnetic conductor
PML	perfectly matched layer
pPML	pseudo-PML
PRI	positive refractive index
ps	picosecond
PSF	point-spread function
PSM	phase-shifting mask
PSTD	pseudospectral time-domain
PVDZ	polarized valence double zeta
PWFD	plane-wave frequency-domain
Q	quality factor
QED	quantum electrodynamics
QM	quantum (mechanics, mechanical)
QM/ED	quantum mechanics/classical electrodynamics
R	resistance
rad	radian
RCS	radar cross-section

RCWA	rigorous coupled-wave analysis
RET	resolution enhancement technique
RF	radio-frequency
RLC	resistor inductor capacitor
rms	root-mean-square
RPA	random-phase approximation
RT-TDDFT	real-time time-dependent density functional theory
SAOP	statistical averaging of orbital potentials
SCR	signal-to-clutter ratio
SEM	scanning electron microscope
S-EP	staircased effective permittivity
SERS	surface-enhanced Raman (scattering, spectroscopy)
S-FDTD	stochastic finite difference time domain
SiC	silicon carbide
SL-PSTD	staggered-grid, local-Fourier-basis pseudospectral time domain
SMJ	single-molecule junction
SMO	source-mask optimization
SPP	surface plasmon polariton
SRAF	sub-resolution assist feature
SRF	scattering response function
SRR	split-ring resonator
SSC	Shanghai Supercomputer Center
TCC	transmission cross-coefficient
TCS	total cross-section
TDSE	time-dependent Schrödinger equation
TE	transverse electric
TEM	transverse electromagnetic
TF/SF	total-field/scattered-field
TM	transverse magnetic
TMA	thin-mask approximation
TSCS	total scattering cross-section
UPML	uniaxial perfectly matched layer
UWB	ultrawideband
V	voltage, volts
VLSI	very large-scale integration
Z	impedance
ZUML	unsplit-field PML proposal by Zhao
λ	wavelength
λ_0	free-space (vacuum) wavelength
λ_B	Bragg wavelength
λ_d	dielectric wavelength
μm	micrometer (micron)
μs	microsecond
ω	angular frequency

About the Authors

Allen Taflove (Editor and Chapters 14, 16, and 17) is a full professor of electrical engineering and computer science at the McCormick School of Engineering and Applied Science of Northwestern University, Evanston, Illinois. His academic degrees include the B.S., M.S., and Ph.D., all in electrical engineering from Northwestern University, in 1971, 1972, and 1975, respectively. Prior to joining Northwestern's faculty in 1984, he held three research-engineering positions at IIT Research Institute, Chicago, Illinois.

Since 1972, Prof. Taflove and his students and colleagues have developed fundamental theoretical approaches, algorithms, and applications of finite-difference time-domain (FDTD) computational solutions of the fundamental Maxwell's equations of classical electrodynamics. He coined the descriptors "finite difference time domain" and "FDTD" in a 1980 IEEE paper, and in 1990 was the first person to be named an IEEE Fellow in the FDTD technical area.

In 2002, the Institute of Scientific Information included Prof. Taflove in its original listing of the most-cited researchers worldwide. As of November 2012, according to Google Scholar®, his books, journal papers, and patents have received a total of more than 21,500 citations, and the exact phrase "finite difference time domain" which he coined has appeared in over 50,000 publications. Specifically, Google Scholar ranks the three editions of his book, *Computational Electrodynamics: The Finite-Difference Time-Domain Method*, 7th on its all-time list of the most-cited books in physics; and Microsoft Academic Search® ranks the first two editions of this book 9th on its all-time list of the most-cited publications in engineering. In May 2010, *Nature Milestones: Photons* named Prof. Taflove as one of the two principal pioneers of numerical methods for solving Maxwell's equations.

Currently, continuing a collaboration that began in 2003, Prof. Taflove is working with Prof. Vadim Backman of Northwestern's Biomedical Engineering Department in research aimed at reliable, minimally invasive, low-cost detection of early-stage cancers of the colon, lung, and pancreas. Here, FDTD modeling of the electrodynamics of visible-light backscattering by biological cells is providing essential insights into the physics of the novel cancer-screening techniques being pioneered by Prof. Backman—techniques that may one day save many lives.

Ardavan Oskooi (Co-editor and Chapters 4, 5, 6, 19, and 20) is currently a Japan Society for the Promotion of Science postdoctoral research fellow in the Quantum Optoelectronics Laboratory of Prof. Susumu Noda in the Department of Electronic Science and Engineering at Kyoto University. Dr. Oskooi's academic degrees include the B.S. in engineering science from the University of Toronto in 2004, the M.S. in computation for design and optimization from the Massachusetts Institute of Technology (MIT) in 2008, and the Ph.D. in materials science and engineering in 2010, also from MIT, where he worked with Profs. Steven G. Johnson of Applied Mathematics and John Joannopoulos of Physics on computational research in nanophotonic device design. Dr. Oskooi's research interests span the design, fabrication, and optimization of nanophotonic devices, particularly photonic-bandgap materials, for applications in energy and information. He is one of the core developers of MIT's suite of FDTD Maxwell's equations solvers: *Meep*, available online at http://ab-initio.mit.edu/meep. Free and open source, Meep has been cited in more than 600 journal papers and has been downloaded more than 54,000 times.

Steven G. Johnson (Co-editor and Chapters 4, 5, 6, 19, and 20) is an associate professor of applied mathematics at MIT, where he joined the faculty in 2004. His academic credentials include three B.S. degrees from MIT in 1995 (physics, mathematics, and computer science), and the Ph.D. in physics from MIT in 2001. He is the author or co-author of over 150 journal articles, a 2008 textbook, and 27 issued patents. His publications deal primarily with the design and understanding of nanophotonic systems and photonic crystals, in which wavelength-scale structures are used to modify the behavior of light. He is also known for several free numerical software packages, including the *MPB* and *Meep* electromagnetic modeling tools, and also the *FFTW* package for fast Fourier transforms (for which he received the 1999 Wilkinson Prize for Numerical Software).

Iftikhar Ahmed (Chapter 8) is a research scientist in the Department of Electronics and Photonics of the Institute of High-Performance Computing, A*STAR, Singapore, which he joined in 2006. His academic degrees include the B.S. (1995) from the University of Engineering and Technology, Taxila, and the Ph.D. (2006) from Dalhousie University, Canada, both in electrical engineering. His research has involved computational electromagnetics (CEM) modeling from RF/microwave to optical frequencies, and from macro-scale down to nanometer-scale devices, the latter via novel algorithms for coupling Maxwell's and Schrödinger's equations. In addition, he is also involved in multiphysics and multiscale CEM algorithms and their hardware acceleration. Dr. Ahmed has authored and coauthored over 60 journal and conference papers. His current research interests include computational electromagnetics and computational nanophotonics.

Jonathan Andreasen (Chapter 18) is a postdoctoral research associate in the College of Optical Sciences at the University of Arizona. He received a B.S. degree in computational physics from Illinois State University (2003), where he was awarded the Outstanding Graduating Physics Senior prize. He received a Ph.D. degree in physics from Northwestern University (2009) for his study of electromagnetic noise in open systems. In 2010, he received the Chateaubriand Postdoctoral Fellowship from the Embassy of France in the United States to study in the Laboratoire de Physique de la Matière Condensée, Université de Nice—Sophia Antipolis, France. Dr. Andreasen's research interests are computational electrodynamics, parallel computing, random lasers, electromagnetic noise, and extreme nonlinear optics.

Roberto B. Armenta (Chapter 11) received the B.Eng. degree (with distinction) in 2005 from the University of Victoria, Victoria, British Columbia, Canada; and the M.A.Sc. and Ph.D. degrees from the University of Toronto, Toronto, Ontario, Canada, in 2007 and 2012, respectively, all in electrical engineering. His research interests are in the areas of electromagnetic theory and numerical techniques for solving Maxwell's equations at microwave and optical frequencies. His work on finite-difference methods has sought to combine the tools of differential geometry with high-order finite-difference approximations to create more flexible and efficient algorithms.

Dr. Armenta received a Canada Graduate Scholarship for doctoral studies from the Natural Sciences and Engineering Research Council of Canada (NSERC) in 2008, the IEEE Microwave Theory and Techniques Society Graduate Fellowship in 2010, the IEEE Antennas and Propagation Society Doctoral Research Award in 2010, the Doctoral Completion Award from the School of Graduate Studies at the University of Toronto in 2011, and a Postdoctoral Fellowship from NSERC in 2012.

Jaione Tirapu Azpiroz (Chapter 15) received her Ph.D. in electrical engineering in 2004 at the University of California, Los Angeles (UCLA), where she investigated sub-wavelength optical lithography modeling in Prof. Eli Yablonovitch's photonics and optoelectronics research group. Before joining UCLA, Dr. Tirapu Azpiroz spent a year in the Electromagnetics Division of the European Space Agency in Noordwijk, The Netherlands, after having received her M.S. in telecommunications engineering at the Public University of Navarre, in Spain.

Dr. Tirapu Azpiroz joined IBM Microelectronics in 2004 as part of the Computational Lithography Group, working in the Semiconductor Research and Development Center in Hopewell Junction, New York. Here, her research involved modeling advanced optical lithography systems and engineering novel techniques for image-resolution enhancement. Her research achievements facilitated the development of advanced semiconductor manufacturing technology. She was part of the Source Mask Optimization project team that received an IBM Research Technical Accomplishment Award in 2009, and Best Paper Awards at the 2008 and 2011 SPIE Photomask Technology Conferences. In July 2012, Dr. Tirapu Azpiroz joined IBM Research-Brazil, where she is currently investigating "lab-on-a-chip" and "point-of-care" devices.

Vadim Backman (Chapters 14 and 16) is a full professor of Biomedical Engineering at the McCormick School of Engineering and Applied Science of Northwestern University, Evanston, Illinois. He is also Program Leader, Cancer and Physical Sciences, at the Lurie Comprehensive Cancer Center of Northwestern University, and the Director of a Bioengineering Research Partnership funded by the U.S. National Institutes of Health (NIH). His academic degrees include the B.S. and M.S. in physics from St. Petersburg Polytechnic Institute, St. Petersburg, Russia; the M.S. in physics from MIT; and the Ph.D. in medical engineering and medical physics from Harvard University and MIT.

Prof. Backman's research focuses on bridging advances in biophotonics into biomedical research and medicine. He develops novel biophotonics technologies for characterizing and imaging biological tissues with a focus on the nanoscale, microscale, and molecular levels. His research interests span from technology development to cancer biophysics to large-scale clinical trials.

Prof. Backman has received numerous awards, including being selected as one of the top 100 young innovators in the world by MIT's *Technology Review Magazine*, and is a Fellow of the American Institute for Medical and Biological Engineering. He has authored over 130 journal publications, 9 books and book chapters, and 17 patents. In the past three years, he has served as the principal investigator on 16 grants from NIH and the U.S. National Science Foundation (NSF).

Geoffrey W. Burr (Chapter 15) is a Research Staff Member at the IBM Almaden Research Center, San Jose, California. After receiving his Ph.D. degree in electrical engineering from the California Institute of Technology in 1996, he joined IBM Almaden as a postdoctoral researcher, assuming his present position in 1999. At IBM Almaden, he has worked at the forefront of a number of diverse fields. For the first part of his career, Dr. Burr worked extensively on holographic data storage, including systems studies, signal processing and coding, and the demonstration of cutting-edge experimental platforms. He then focused on computational electromagnetics to investigate aspects of nanophotonics, photonic crystals, scanning near-field optical microscopy, and computational lithography. Most recently, his research has turned to non-volatile or storage-class memory. In this technical area, he is involved in numerical modeling, materials science, and electrical testing for phase-change memory, resistance RAM, access devices for non-volatile memories, and reprogrammable logic.

Ilker R. Capoglu (Chapters 14 and 16) is a postdoctoral research fellow in the Biomedical Engineering Department of Northwestern University, Evanston, Illinois. He received his Ph.D. in electrical engineering from the School of Electrical and Computer Engineering of the Georgia Institute of Technology, Atlanta, Georgia, in 2007. His research interests include the development and optimization of numerical methods in electromagnetics, electromagnetic problems in multilayered and inhomogeneous media, bio-electromagnetics and bio-optics, and rigorous electromagnetic modeling of optical imaging systems. He is an associate member of URSI Commission K (Electromagnetics in Biology and Medicine.)

Dr. Capoglu is the sole developer of the free, open-source, FDTD software suite, *Angora*, available online at http://www.angorafdtd.org. Among other capabilities, *Angora* enables the rigorous simulation of all current forms of optical microscopy. Its elements implement computational electrodynamics models that span many orders-of-magnitude in characteristic physical dimensions from nanometer-size dielectric voxels comprising a biological cell (where Maxwell's equations are solved using FDTD) to individual pixels comprising the microscope image (where spectra are computed by exercising multiple integral transformations of the FDTD computed near fields).

Bin Chen (Chapter 2) is a professor at the Key Laboratory of Science and Technology on Electromagnetic Environmental Effects and Electro-optical Engineering, Nanjing, China. He received the B.S. and M.S. degrees in electrical engineering from the Beijing Institute of Technology, Beijing, China, in 1982 and 1987, respectively, and the Ph.D. degree in electrical engineering from the Nanjing University of Science and Technology, Nanjing, China, in 1997. Prof. Chen was the recipient of three National Science and Technique Awards of China in 1993, 1995, and 2000, respectively. His research interests include computational electromagnetics, electromagnetic wave propagation, and electromagnetic pulse.

Hailin Chen (Chapter 2) is a lecturer at the Key Laboratory of Science and Technology on Electromagnetic Environmental Effects and Electro-optical Engineering, Nanjing, China. He received the B.S. degree in electrical engineering from Qingdao University, Shandong, China, in 2001, and the M.S. and Ph.D. degrees in electrical engineering from Nanjing Engineering Institute, Nanjing, China, in 2004 and 2008, respectively. His research interests include computational electromagnetics, lightning protection, and electromagnetic materials.

Hanning Chen (Chapter 10) is an assistant professor of chemistry at George Washington University, Washington, D.C. He received the Ph.D. in theoretical and computational chemistry from the University of Utah in 2008 under the direction of Prof. Gregory Voth. From 2008–12, he was a postdoctoral fellow at the Argonne National Laboratory – Northwestern University Solar Energy Research Center under the supervision of Profs. George Schatz and Mark Ratner.

Prof. Chen's research interests include electronic structure calculations on functional materials under various nonequilibrium conditions, such as light irradiation and bias voltage. Previously, he won the Chemical Computing Group excellence award of the American Chemical Society, the Student Travel Grant Award of the American Biophysical Society, and the Cheves T. Walling outstanding graduate research award of the University of Utah. He is also designated by NSF as the XSEDE Campus Champion at the George Washington University.

Kun Chen (Chapters 1 and 16) is a professor at the Shanghai Institute of Optics and Fine Mechanics of the Chinese Academy of Sciences, Shanghai, China. He received the B.S. and M.S. degrees from the University of Science and Technology of China, in 1991 and 1993, respectively, and the Ph.D. in physics from MIT in 2000. In his early career, he majored in theoretical particle physics and worked on computational modeling of hadron structures. He shifted to the interdisciplinary research field of biomedical optics in 1995, and has since been active in near-infrared breast cancer imaging, light-scattering spectroscopy, bio-imaging using nanoparticles, laser-tweezers, Raman spectroscopy, single-cell cancer diagnosis, application of computational electromagnetics to modeling light-tissue interactions, and parallel computing algorithms. He joined the faculty of the Chinese Academy of Sciences in 2004 and was honored a "100-Talents" Scholar title.

James B. Cole (Chapter 12) is a professor at the University of Tsukuba, Japan. He received the Ph.D. in particle physics from the University of Maryland, and subsequently conducted postdoctoral research at the NASA Goddard Space Flight Center. There, he began his work with numerical simulations by developing models of cosmic ray antiproton flux. Later, he worked on stochastic simulations at the Army Research Laboratory, and visited the NTT Basic Research Laboratory, Japan, for one year. As a research physicist at the Naval Research Laboratory, working on the Connection Machine, he developed the earliest nonstandard finite difference (NS-FD) models for acoustic simulations. After joining the faculty of the University of Tsukuba he extended NS-FD models to computational electromagnetics and optics.

Ming Ding (Chapters 1 and 16) is a research and development engineer of the Tachyon model group in Brion Technologies, ASML. He received the B.S. degree in physics from Wuhan University, Wuhan, China, in 2006, and the Ph.D. degree in optics from the Institute of Optics and Fine Mechanics, Chinese Academy of Sciences, Shanghai, China, in 2011. Dr. Ding's research interests include computational electromagnetics and its applications to biomedical optics, parallel computing, and lithography.

Yantao Duan (Chapter 2) is a lecturer at the Key Laboratory of Science and Technology on Electromagnetic Environmental Effects and Electro-optical Engineering, Nanjing, China. He received the B.S. and M.S. degrees in electric systems and automation from the Nanjing Engineering Institute, Nanjing, China, in 2002 and 2006, respectively, and the Ph.D. degree in electrical engineering from PLA University of Science and Technology, Nanjing, China, in 2010. His research interests include computational electromagnetics and electromagnetic pulse.

Cynthia M. Furse (Chapter 7) is a full professor of electrical and computer engineering at the University of Utah, Salt Lake City, Utah, and serves as the Associate Vice President for Research at her university. Her academic degrees include the B.S., M.S., and Ph.D., all in electrical engineering from the University of Utah, in 1985, 1988, and 1994, respectively. Her expertise in electromagnetics is applied to sensing and communications in complex lossy scattering media with applications in bioelectromagnetics, geophysical prospecting, ionospheric plasmas, and aircraft wiring networks. All of these applications involve statistical variations, which spawned the research reviewed in Chapter 7. Prof. Furse is a Fellow of the IEEE and a founder of LiveWire Innovation, which creates equipment to locate electrical faults on live wires. She has instructed electromagnetics, wireless communication, computational electromagnetics, microwave engineering, and antenna design, and has received numerous teaching and research awards, including the 2009 IEEE Harriett B. Rigas award for educational excellence.

Stephen K. Gray (Chapter 9) is a senior scientist at Argonne National Laboratory, Argonne, Illinois, where he also serves as group leader of the Theory and Modeling Group in the Center for Nanoscale Materials. His academic degrees include the Ph.D. from the University of California-Berkeley. He has been active in research for over thirty years with more than 175 peer-reviewed publications, and is a Fellow of the American Physical Society. His current research interests include theoretical and computational modeling of dynamical processes in nanosystems. He places particular emphasis on modeling light interactions with metallic nanostructures via rigorous electrodynamics simulations, and modeling the quantum dynamics of molecular systems within nanoscale environments.

Jethro H. Greene (Chapter 17) is a member of the Quantitative Research Department at Citadel, LLC, Chicago, Illinois. He received the B.S. and Ph.D. degrees from the Department of Electrical Engineering at Northwestern University. He developed the general vector auxiliary differential equation (GVADE) FDTD method for full-vector Maxwell's equations modeling of electromagnetic wave propagation in nonlinear dispersive materials. His interests include computational electromagnetics, numerical algorithms, and distributed-memory parallel computing. Previously, he was a member of the Computational Methods Department at Impact Forecasting, LLC, building numerical models for hurricane and earthquake loss.

Eng Huat Khoo (Chapter 8) is a research scientist in the Department of Electronics and Photonics of the Institute of High Performance Computing (IHPC), A*STAR, Singapore. His academic degrees include the B.Eng. with first-class honors (2003) and the Ph.D. (2008), both in electrical and electronics engineering from Nanyang Technological University, Singapore. Prior to his present position, during 2009–10 he was a research fellow at Harvard University. He has been actively involved in the computational simulation of active and passive photonic-crystal devices and plasmonic polarizers, and has developed algorithms for simulating active photonic devices using the principles of solid-state physics. His work on active magnetic photonic laser extraction was selected for the cover page of *Applied Physics Letters* in 2009. Dr Khoo has authored and coauthored over 40 journal papers. His current research interests include plasmonics, photonic-crystal devices, passive and active photonics, magneto-optics, quantum mechanics, and solid-state physics.

Er Ping Li (Chapter 8) is a principal scientist and Director of the Electronics and Photonics Department of the Institute of High Performance Computing (IHPC), A*STAR, Singapore. He is also a chair professor in the Department of Information Science and Electronic Engineering at Zhejiang University, China. His academic degrees include the Ph.D. in electrical engineering (1992) from Sheffield Hallam University, Sheffield, U.K. His previous positions include senior research fellow, principal research engineer, associate professor, and Technical Director at the Singapore Research Institute, University and Industry.

Dr Li has authored/coauthored one book and more than 200 papers published in refereed international journals. He is a Fellow of IEEE and a Fellow of the Electromagnetics Academy, and has received numerous international recognitions including the Technical Achievement Award of the IEEE Electromagnetic Compatibility Society (EMC), the Institution of Engineers (Singapore) Prestigious Engineering Achievement Award, the IEEE Sustained Outstanding

Service Award, and the IEEE EMC Society Distinguished Lecturer. He is currently an associate editor of *IEEE Transactions on EMC* and *IEEE Transactions on Components, Packaging, and Manufacturing Technology*. He has served as general chair and technical program chair for a number of international conferences, and has presented invited and plenary talks at a number of international conferences and forums. His current research interests include computational electromagnetics, high-speed electronics, and nanoplasmonics.

Zachary Lubin (Chapter 17) is a senior systems engineer with Northrop Grumman Corporation, Electronic Systems, in Rolling Meadows, Illinois. His degrees (all in electrical engineering) include the B.S. from the University of Illinois at Chicago (2000); the M.S. from Illinois Institute of Technology, Chicago (2004); and the Ph.D. from Northwestern University, Evanston, Illinois (2012). Dr. Lubin has over ten years' experience in the development of radio-frequency (RF) systems for both commercial and defense applications. His technical activities at Northrop Grumman have included statistical signal processing and machine-learning algorithms; modeling and simulation for radar, electronic warfare, and communication systems; microwave circuit and antenna modeling, design, and testing; leading software-development teams and research-and-development projects; and architecting RF electronic warfare systems. Dr. Lubin has served as chairman of the IEEE Microwave Theory & Techniques / Antennas & Propagation Chicago joint chapters, and vice-chairman of the IEEE Signal Processing Chicago chapter. He has received numerous recognition awards from Northrop Grumman, and has presented at company symposia.

Jeffrey M. McMahon (Chapters 9 and 10) is a postdoctoral research associate at the Institute for Condensed Matter Theory at the University of Illinois at Urbana-Champaign. Previously, in 2005, he received a B.S. in chemistry, with minors in physics and mathematics, from Western Washington University; and in 2010, a Ph.D. in chemistry from Northwestern University. His current research interests include theoretical and computational electrodynamics, and the development and application of *ab initio* methods to understand basic properties of materials of physical and chemical importance.

Naoki Okada (Chapter 12) is a Ph.D. student in the Department of Computer Science, Graduate School of Systems and Information Engineering, University of Tsukuba, Japan. He received the B.S. and M.S. degrees in engineering from the University of Tsukuba in 2009 and 2011, respectively. Since 2011, he has been a Fellow of the Japan Society for the Promotion of Science (JSPS) in the field of computational electromagnetics. His current research interests include the modeling, design, and development of numerical electromagnetics algorithms; metamaterials; plasmonic materials; and optical waveguides. In particular, he focuses on the use of the FDTD techniques. He is also actively involved in graphics processing unit (GPU) parallel computation, and photorealistic rendering of structural colors such as the Morpho butterfly. His work can be seen on his website, http://nsfdtd.org/.

Mike Potter (Chapter 3) is an associate professor in the Department of Electrical and Computer Engineering, University of Calgary. Dr. Potter received the B.Eng. degree in engineering physics from the Royal Military College of Canada in 1992, served with Royal Canadian Navy as a Combat Systems Engineer for several years, and received the Ph.D. from the University of Victoria, Canada, in 2001. He also held a postdoctoral fellowship at the University of Arizona from 2001–02. He is a Member of the IEEE, and is a member of the IEEE Antennas and Propagation Society Education Committee. In 2011, along with Tengmeng Tan, he received the

IEEE Antennas and Propagation Society Schelkunoff Transactions Prize Paper Award. His current research interests include FDTD methods on alternative grids for electromagnetic and acoustic/elastic modeling, and computational electromagnetics in general.

Mark A. Ratner (Chapter 10) is the Dumas University Professor at Northwestern University, Evanston, Illinois, and Co-Director of the Initiative for Sustainability and Energy at Northwestern (ISEN). His academic degrees include a B.A. (1964) at Harvard and a Ph.D. (1969) at Northwestern. Prior to joining Northwestern's faculty, he held postdoctoral positions at Aarhus University and the University of Munich, and a faculty position at New York University.

Prof. Ratner is a member of the U.S. National Academy of Sciences, the American Academy of Arts and Sciences, the International Academy of Quantum Molecular Sciences, and the Royal Danish Academy of Sciences. He is the recipient of the Feynman Prize in Nanotechnology, the Langmuir Award of the American Chemical Society, and honorary Sc.D. degrees from Hebrew University and the University of Copenhagen.

Prof. Ratner is a materials chemist whose work focuses on the interplay between molecular structure and molecular properties. His current research is in seven primary areas: nonlinear optical response properties of molecules; electron transfer and molecular electronics; quantum dynamics and relaxation in the condensed phase; mean-field models for extended systems (including proteins and molecular assemblies); photonics in nanoscale systems; excitons in molecule-based photovoltaics; and hybrid classical/quantum representations.

Jeremy D. Rogers (Chapters 14 and 16) is a research assistant professor in the Biomedical Engineering Department of Northwestern University, Evanston, Illinois, where he studies light scattering in tissue as a means of cancer screening. He earned the B.S. (1999) in physics from Michigan Technological University, Houghton, Michigan, where he worked on laser trapping of aerosol particles in hollow fibers and stellar coronographs for extra-solar planet detection. He earned the M.S. (2003) and the Ph.D. (2006) in optical sciences from the University of Arizona, Tucson, Arizona. His graduate research combined asymmetric optical design for stray light reduction, grayscale lithography, optical metrology, and microfabrication to construct an endo-microscope with structured illumination for *in vivo* imaging of tissue. His research interests have allowed him to work on projects ranging from the study of bleaching in reef-building corals to designing and building a telescope to field detection of extremophile microbial life in the Canadian high arctic 740 miles from the North Pole.

César Méndez Ruiz (Chapter 16) is an electrical engineer at Intel Corporation. He received the B.S. in electromechanical engineering from the Universidad Panamericana, Zapopan, Jalisco, México, in 2003; the M.S. in electronics engineering from the Universidad de Guadalajara, Guadalajara, Jalisco, México, in 2007; and the Ph.D. in electrical engineering from the University of New Mexico-Albuquerque in 2011. In 2010, he received the IEEE Antennas and Propagation Society Doctoral Research Award to support his work on photonic nanojets. He was also named the Outstanding Graduate Student in Electrical Engineering at the University of New Mexico for the spring 2011 semester. During his Ph.D. program he held a scholarship from the Mexican National Council on Science and Technology. His current research interests include electromagnetic modeling of high-speed digital interconnects and associated signal-integrity problems.

Costas D. Sarris (Chapters 11 and 13) is an associate professor and the Eugene V. Polistuk Chair in Electromagnetic Design with the Edward S. Rogers Sr. Department of Electrical and Computer Engineering of the University of Toronto, Toronto, Ontario, Canada. He is also Associate Chair of the Division of Engineering Science of the University of Toronto. He received the M.Sc. degree in applied mathematics and the Ph.D. degree in electrical engineering from the University of Michigan, Ann Arbor, in 2002. His research interests are in the area of numerical electromagnetics, with emphasis on high-order, multiscale/multi-physics computational methods, modeling under stochastic uncertainty, as well as applications of time-domain analysis to wireless channel modeling, wave propagation in complex media and meta-materials, and electromagnetic compatibility/interference problems.

Prof. Sarris received the Ontario Government's Early Researcher Award. His students authored "Best Papers" at the 2008 Applied Computational Electromagnetics Society Conference and the 2009 IEEE International Microwave Symposium; and "Honorable Mentions" at the 2008 and 2009 IEEE Antennas and Propagation International Symposia. He was the Technical Program Committee co-/vice-chair for the 2010 IEEE Antennas and Propagation International Symposium and the 2012 IEEE International Microwave Symposium, respectively. Currently, he serves as an associate editor for *IEEE Transactions on Microwave Theory and Techniques*.

George C. Schatz (Chapters 9 and 10) is the Morrison Professor of Chemistry at Northwestern University, Evanston, Illinois. His academic degrees include a B.S. (1971) at Clarkson University and the Ph.D. (1976) at Caltech, both in chemistry. He was a postdoc at MIT, and has been at Northwestern since 1976. He has published three books and over 490 papers.

Prof. Schatz is a member of the U.S. National Academy of Sciences, the American Academy of Arts and Sciences, and the International Academy of Quantum Molecular Sciences. He is the recipient of the Bourke Medal of the Faraday Division of the Royal Society of Chemistry, the Feynman Prize of the Foresight Institute, and the Peter Debye Award of the American Chemical Society. He is Editor-in-Chief of the *Journal of Physical Chemistry*.

Prof. Schatz' group conducts theoretical and computational research applied to nanotechnology, materials physics and properties, macromolecular structures and dynamics, molecular self-assembly, optics, and biophysics. Interests include electronic structure methods, quantum and classical theories of dynamical processes, and the use of these methods to study the reactions of molecules at interfaces.

Jamesina J. Simpson (Chapter 16) is an associate professor in the Department of Electrical and Computer Engineering at the University of Utah, Salt Lake City, Utah. Her academic degrees include the B.S. (2003) and the Ph.D. (2007) in electrical engineering from Northwestern University, Evanston, Illinois. During her Ph.D. studies, she was awarded a National Science Foundation (NSF) Graduate Research Fellowship, research awards from both the IEEE Antennas and Propagation Society and the IEEE Microwave Theory and Techniques Society, and the Best Ph.D. Dissertation Award from Northwestern's Department of Electrical Engineering and Computer Science.

Prof. Simpson's research encompasses applications of FDTD computational solutions of Maxwell's equations to model electromagnetic phenomena at frequencies over 15 orders of magnitude (~1 Hz vs. ~600 THz), with applications ranging from space weather to early-stage cancer detection. In particular, she has pioneered FDTD modeling of electromagnetic wave propagation within the global Earth-ionosphere system to investigate the effects of solar coronal mass ejections upon terrestrial continental power systems.

Prof. Simpson has been the recipient of multiple research and teaching recognitions, including a 2010 NSF CAREER Award, a 2011 Air Force Summer Faculty Fellowship, and the 2012 IEEE Antennas and Propagation Society Donald G. Dudley, Jr. Undergraduate Teaching Award. She currently serves as an associate editor of *IEEE Transactions on Antennas and Propagation*.

Steven M. Smith (Chapter 7) is a senior staff engineer with L-3 Communications, Communications System West, and an adjunct assistant professor with the University of Utah. He received the B.S., M.S. and Ph.D. degrees in electrical and computer engineering from the University of Utah in 1985, 2007, and 2011, respectively. With expertise in RF and microwave circuit design, he has worked in the land-mobile radio industry, and currently works in the design of transceiver circuits for high-speed modems. His current research interests include computational electromagnetics, the development of faster stochastic methods such as that for FDTD, and low-noise circuit design and simulation.

Tengmeng Tan (Chapter 3) received the B.S. degree in electrical and computing engineering from University of Calgary, and subsequently the Ph.D. degree (in 2010) supervised by Prof. Mike Potter. The significance of Dr. Tan's Ph.D. research was recognized in 2011 with the IEEE Antennas and Propagation Society Schelkunoff Transactions Prize Paper Award. Subsequently, he was also awarded the NSERC (Natural Sciences and Engineering Research Council of Canada) Postdoctoral Fellowship. This fellowship supported him to collaborate with Profs. Allen Taflove and Vadim Backman of Northwestern University, Evanston, Illinois, in developing advanced numerical solutions for random-media electromagnetic wave phenomena to help identify novel biophotonics markers of early-stage human cancers. Dr. Tan is currently collaborating with Prof. Qing Liu at Duke University, North Carolina, investigating industrial applications of wave propagation enhanced by nanoparticles. His research interests include the theory and application of wave-propagation phenomena.

Snow H. Tseng (Chapter 16) is an associate professor in the Department of Electrical Engineering of National Taiwan University, Taipei, Taiwan. His academic degrees include the B.S. in physics (1994) from National Taiwan University, the M.S. in physics (2001) from the University of Chicago, Chicago, Illinois, and the Ph.D. in electrical engineering (2005) from Northwestern University, Evanston, Illinois. Subsequently, in 2006, he joined the faculty of the Graduate Institute of Photonics and Optoelectronics and the Department of Electrical Engineering of National Taiwan University.

Prof. Tseng's current research involves applying pseudospectral time-domain (PSTD) computational solutions of Maxwell's equations to rigorously investigate electromagnetic wave propagation and scattering in random media. He is particularly interested in how light interacts with biological tissues. A key goal in this regard is to enhance the penetration of light into biological tissues in support of emerging diagnostic and therapeutic techniques in biomedical optics.

Index

Recent Titles in the Artech House Antennas and Propagation Series

Jennifer T. Bernhard, Series Editor

Adaptive Array Measurements in Communications, M. A. Halim

Advances in Computational Electrodynamics: The Finite-Difference Time-Domain Method, Allen Taflove, editor

Advances in FDTD Computational Electrodynamics: Photonics and Nanotechnology, Allen Taflove, editor; Ardavan Oskooi, and Steven G. Johnson, coeditors

Analysis Methods for Electromagnetic Wave Problems, Volume 2, Eikichi Yamashita, editor

Antenna Design with Fiber Optics, A. Kumar

Antenna Engineering Using Physical Optics: Practical CAD Techniques and Software, Leo Diaz and Thomas Milligan

Antennas and Propagation for Body-Centric Wireless Communications, Second Edition, Peter S. Hall and Yang Hao, editors

Analysis of Radome-Enclosed Antennas, Second Edition, Dennis J. Kozakoff

Applications of Neural Networks in Electromagnetics, Christos Christodoulou and Michael Georgiopoulos

AWAS for Windows Version 2.0: Analysis of Wire Antennas and Scatterers, Antonije R. Djordjević, et al.

Broadband Microstrip Antennas, Girsh Kumar and K. P. Ray

Broadband Patch Antennas, Jean-François Zürcher and Fred E. Gardiol

CAD of Microstrip Antennas for Wireless Applications, Robert A. Sainati

The CG-FFT Method: Application of Signal Processing Techniques to Electromagnetics, Manuel F. Cátedra, et al.

Computational Electrodynamics: The Finite-Difference Time-Domain Method, Third Edition, Allen Taflove and Susan C. Hagness

Electromagnetic Modeling of Composite Metallic and Dielectric Structures, Branko M. Kolundzija and Antonije R. Djordjević

Electromagnetic Waves in Chiral and Bi-Isotropic Media, I. V. Lindell, et al.

Electromagnetics, Microwave Circuit and Antenna Design for Communications Engineering, Peter Russer

Engineering Applications of the Modulated Scatterer Technique, Jean-Charles Bolomey and Fred E. Gardiol

Fast and Efficient Algorithms in Computational Electromagnetics, Weng Cho Chew, et al., editors

Fresnel Zones in Wireless Links, Zone Plate Lenses and Antennas, Hristo D. Hristov

Handbook of Antennas for EMC, Thereza MacNamara

Introduction to Antenna Analysis Using EM Simulators, Hiroaki Kogure, Yoshie Kogure, and James C. Rautio

Iterative and Self-Adaptive Finite-Elements in Electromagnetic Modeling, Magdalena Salazar-Palma, et al.

LONRS: Low-Noise Receiving Systems Performance and Analysis Toolkit, Charles T. Stelzried, Macgregor S. Reid and Arthur J. Freiley

Measurement of Mobile Antenna Systems, Second Edition, Hiroyuki Arai

Microstrip Antenna Design Handbook, Ramesh Garg, et al.

Microwave and Millimeter-Wave Remote Sensing for Security Applications, Jeffrey A. Nanzer

Mobile Antenna Systems Handbook, Third Edition, Kyohei Fujimoto, editor

Multiband Integrated Antennas for 4G Terminals, David A. Sánchez-Hernández, editor

Noise Temperature Theory and Applications for Deep Space Communications Antenna Systems, Tom Y. Otoshi

Phased Array Antenna Handbook, Second Edition, Robert J. Mailloux

Phased Array Antennas with Optimized Element Patterns, Sergei P. Skobelev

Plasma Antennas, Theodore Anderson

Quick Finite Elements for Electromagnetic Waves, Giuseppe Pelosi, Roberto Coccioli, and Stefano Selleri

Radiowave Propagation and Antennas for Personal Communications, Second Edition, Kazimierz Siwiak

Solid Dielectric Horn Antennas, Carlos Salema, Carlos Fernandes, and Rama Kant Jha

Switched Parasitic Antennas for Cellular Communications, David V. Thiel and Stephanie Smith

Understanding Electromagnetic Scattering Using the Moment Method: A Practical Approach, Randy Bancroft

Wavelet Applications in Engineering Electromagnetics, Tapan Sarkar, Magdalena Salazar Palma, and Michael C. Wicks

For further information on these and other Artech House titles, including previously considered out-of-print books now available through our In-Print-Forever® (IPF®) program, contact:

Artech House
685 Canton Street
Norwood, MA 02062
Phone: 781-769-9750
Fax: 781-769-6334
e-mail: artech@artechhouse.com

Artech House
16 Sussex Street
London SW1V HRW UK
Phone: +44 (0)20 7596-8750
Fax: +44 (0)20 7630 0166
e-mail: artech-uk@artechhouse.com

Find us on the World Wide Web at: www.artechhouse.com